Organic Reactions and their Mechanisms

Organic Reactions and their Mechanisms

Third Edition

P S KALSI

Formerly
Dean of Colleges, Punjab Technical University
Jalandhar, India
and
Professor & Head, Department of Chemistry
College of Basic Sciences and Humanities
Punjab Agricultural University, Ludhiana, India

New Age Science Limited

The Control Centre, 11 A Little Mount Sion
Tunbridge Wells, Kent TN1 1YS, UK
www.newagescience.co.uk • e-mail: info@newagescience.co.uk

The Control Centre, 11 A Little Mount Sion, Tunbridge Wells, Kent TN1 1YS, UK
www.newagescience.co.uk • e-mail: info@newagescience.co.uk
Tel: +44(0) 1892 55 7767, Fax: +44(0) 1892 53 0358

ISBN : 978 1 906574 36 9

British Library Cataloguing in Publication Data
A Catalogue record for this book is available from the British Library

Printed and bound in India by Replika Press Pvt. Ltd.

NCL

National Chemical Laboratory
Division of Organic Chemistry (Synthesis)
Pune – 411 008, INDIA

Dr. Ganesh Pandey
FNA, FNASc, FASc

Foreword

The organic chemistry that serves the needs of society is becoming increasingly spohisticated. The students of organic chemistry want not only to creativity enrich the existing scientific knowledge for the betterment of mankind but also apply it for the sustainable development of the subject. However, they are hard pressed to find a general text to support their learning during the first year at University. The general organic chemistry texts have been written to accompany traditional curricular courses and with rather precisely defined requirements. Thus, the students are left with a limited scope to learn chemistry which encourages creativity.

One of the greatest challenges of organic chemistry is to make complex organic molecules. Effective synthetic strategy requires the development of novel selective reactions and reagents. The area has been playing an increasingly important role in serving as a source of understanding organic reactions and their mechanisms. The present book reveals author's belief that students benefit most of all from a book which leads from familiar concepts to unfamiliar ones, not just encouraging them to *know* but to understand and to understand *why*. A practitioner of organic chemistry must be aware of the fundamental reactions along with their mechanisms to have a thorough knowledge and understanding of this area. It is this understanding of organic reactions which provides an impetus into developing new tools for organic synthesis. For the same reason, an indispensable mechanistic insight is provided which is crucial to those who wish to apply these existing tools rationally and to contribute to the further developement of novel reagents and methodologies.

Compared with the earlier edition of the book, the present edition offers much more material to be learnt. Thus, this book on organic reactions is far from just a remake or update of a successful earlier version but, as the Publisher notes, is essentially a new book. Previous chapters have been extensively reworked and updated.

The "Organic Reactions and their Mechanisms" is a timely account of the current depth of this area of chemistry. The systematic presentation of the organic reactions will introduce students to the taste of ever growing organic chemistry. In the present work, Prof. Kalsi has set himself the goal of organizing the splendid array of organic reactions with their mechanisms. This provides a concise summary that should be of enormous assistance to those searching for a selective reaction to achieve a desired transformation. Coming from one of the leading authors, "Organic Reactions and their Mechanisms" is an authoritative source for a rapidly expanding field. One must admire Prof. Kalsi's courage in undertaking this monumental task.

Ganesh Pandey
Head, Division of Organic Chemistry

Preface to the Third Edition

A common approach to teach modern organic chemistry is through organic reactions and their mechanisms.

The present book is aimed to develop ideas to maintain interest of the today's student, his needs and the modern chemistry courses in different universities.

I have therefore, attempted to make this as a modern teaching and a learning text. Different topics and material are presented so as to develop ideas and in order to do this material in a particular chapter threads through the entire text. This book is therefore, a departure from some-long standing textbook traditions and does not present the material in an encyclopedic manner.

Several introductory chapters, *e.g.*, an aromatic character, chemical bonding, acids and bases provide a vehicle for discussing different organic reactions and the proposed mechanisms in the later parts of the book. These introductory chapters incorporate new ideas, new examples so as to develop maturity of thinking well before their use in a particular topic.

Some background and key information needed to understand the main text better is put in boxes. Interesting problems with their solutions have also been put in boxes throughout the text. You are advised to check your competance by solving these problems as soon as you finish reading a particular portion of a topic. You are advised not to look at the answers initially.

References and further reading is given at the end of the book.

I sincerely hope that the discussion in this text is useful and interesting and that you will let me know your experiences.

P. S. KALSI

Preface to the First Edition

Organic chemistry is a rapidly changing field and each year, new exciting advances are made. It may seem to the student that he needs ever more to learn year by year. However, fortunately, this is not so.

The purpose of the present book is to incorporate advances in the area of reaction mechanisms even as basic rules and concepts are emphasized in teaching organic reaction mechanisms. A deep understanding of the reaction mechanisms helps a student to appreciate as to how and why reactants go to products. The mechanistic principles are relatively few, and yet these account for the wide range of reactions of organic compounds. A conceptual understanding of the mechanisms of organic reactions, therefore, helps a student to interrelate and remember the various reactions.

I have been fortunate to have had the opportunity to teach a great variety of students at both the graduate and undergraduate level. Out of this teaching and my ongoing desire to teach the subject-matter in a lively and understandable manner, this, yet new book, on organic reactions and their mechanisms was born.

A fairly comprehensive review of organic structures and material which provides background to the study of mechanisms is presented in the first few chapters. The further presentation follows so that the student appreciates that despite a large number of organic reactions, a relatively few principles suffice to explain all of them.

The study of organic chemistry is much like learning a language where the reactons are the vocabulatary and their mechanisms the grammar. The design of the present textbook is, therefore, to ensure that the student has an intellectual grasp of the subject to prepare him not only for his qualifying examination but for various competitive examinations as well. In line with this objective, the references have been kept to a minimum. However, a student may like to pursue individual topics further, thus relevant reviews and books are noted, but the references to the original literature are limited to points of outstanding interest and some recent work. The objective has been to convey a deep understanding of reactions and their mechanisms rather than to bring out a reference text.

An attempt has been made to incorporate several important and recent developments in the subject. Every chapter has been brought up to date to include these. Some of these topics are the use of organo-transition-metal reagents, newer reagents and role of organosilicon compounds.

The text is extensively cross-referenced in order to call the students attention to the related material already presented in earlier chapters or to the material that is to come.

The best way to learn organic chemistry is by solving problems. In each chapter problems are presented not only to create thinking in students mind, but also also for the introduction of new material. Answers to these problems can be found at the end.

P. S. Kalsi

Important Reactions and Rearrangements

Some More Important Reagents

Contents

CHAPTER 1

Fundamental Principles and Special Topics

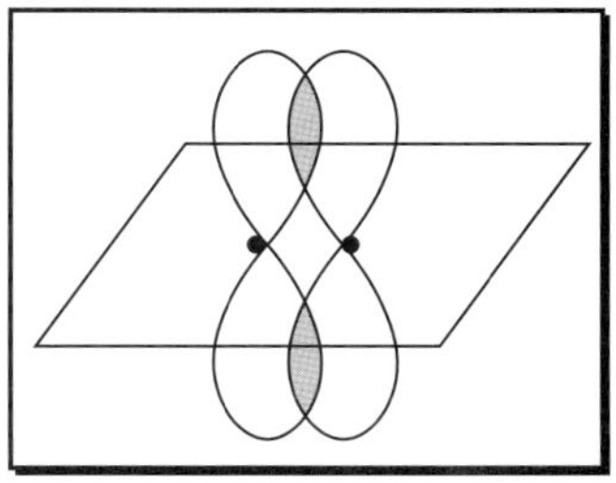

One begins a study of reaction mechanisms by examining some of the basic principles. A basic understanding of these concepts helps largely in understanding of reactions and their mechanisms. Thiols undergo an oxidative coupling when treated with mild oxidizing agents to give disulphides: $2RS—H + H_2O_2 \longrightarrow RS—RS + 2H_2O$. The understanding of this reaction requires a knowledge of bond dissociation energy. The bond dissociation energy of the S—H bond of thiols (~ 80 kcal/mol) is much lower than the O—H bond of alcohols (~ 100 kcal/mol). It is this weakness of the S—H bond which allows thiols to undergo an oxidative coupling, and the alcohols do not display this reaction. On treatment with oxidizing agents, oxidation at the weaker C—H bond (~ 85 kcal/mol) takes place rather than at the strong O—H bond. Thus a knowledge of the nature and strength of bonds is essential for the chemical investigation of organic molecules. Similarly the properties of molecules are influenced by their structure.

1.1 STRUCTURE AND BONDING IN ORGANIC COMPOUNDS

A. Atomic Orbitals

The motion of the electrons around the nucleus can be described by wave equations. The solutions to these equations are atomic orbitals, which roughly delineate regions in space where there is a high probability of finding the electron. An *s* orbital is spherical; a *p* orbital looks like two touching spheres, or a "spherical figure eight" (Scheme 1.1). The sign of the orbital can be positive, negative, or zero (node). These signs do not represent positive or negative charges, since both lobes of an electron cloud must be negatively charged. They refer to the signs of

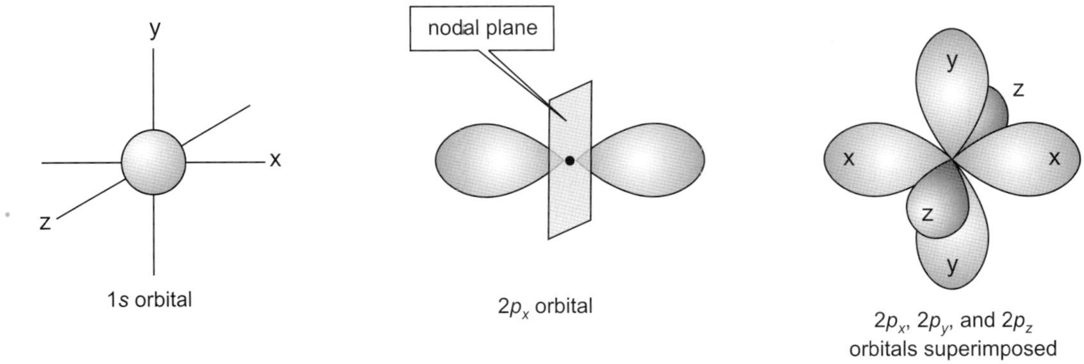

SCHEME 1.1

the wave function ψ. When two parts of an orbital are separated by a node, ψ always has opposite signs on the two sides of the node. With increasing energy, there is an increasing number of nodes. Each orbital can be occupied by a maximum of two electrons of opposite spin (Pauli exclusion principle, Hund's rule).

B. Molecular Orbitals and Bonding

In the molecular orbital method, a bond is formed when two atomic orbitals overlap. Atomic orbitals of the same sign overlap to give a bonding molecular orbital of lower energy. Atomic orbitals of opposite sign give rise to an antibonding molecular orbital of higher energy and containing a node. The number of molecular orbitals is equal to the number of atomic orbitals from which they derive. The overlap of atomic orbitals leads to the formation of sigma and *pi* bonds. Bonds made by overlap along the internuclear axis are called σ bonds (as in H_2, I; HF, II; F_2, III, Scheme 1.2) and those made by overlap of *p* orbitals perpendicular to the internuclear axis are called π bonds. (Scheme 1.2).

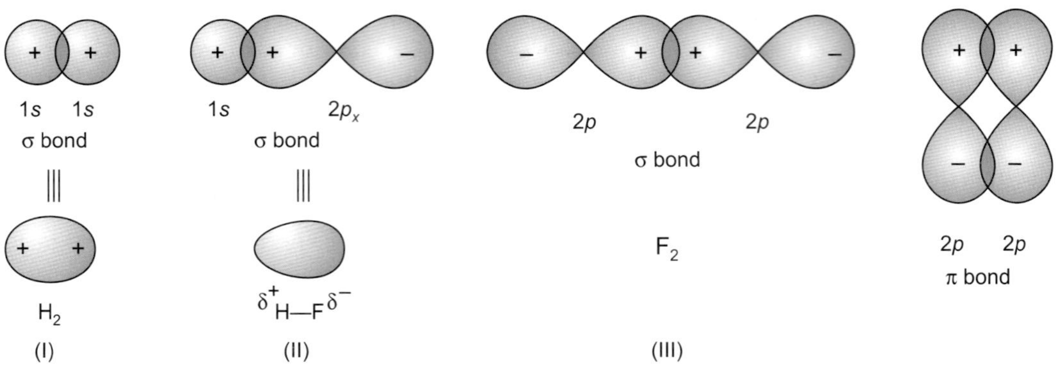

SCHEME 1.2

The hydrogen molecule is cylindrically symmetrical about a straight line drawn through the two nuclei and a cross-section of the molecular orbital when cut perpendicular to the bond axis is circular. This type of molecular orbital is termed σ orbital or σ bond. Similarly in the molecule of H—F, the σ bond is again cylindrically symmetrical about a line passing through the two nuclei. The electrons in the σ bond of H—F are however, not shared equally between the two dissimilar atoms of different electronegativities and unlike in H_2 or F_2 in H—F, the electron distribution is highly polarized (Scheme 1.2).

C. Hybrid Orbitals: Bonding in Complex Molecules

Hybridization of atomic orbitals accounts for observed bond angles and molecular geometries of the molecules.

(i) sp Hybrids give Linear Structures

In *e.g.*, beryllium hydride BeH_2, formation, consider the following points. In its ground state the beryllium atom has $1s^2\ 2s^2$ electronic configuration. Only a small amount of energy is needed to promote one electron from the 2*s* orbital to one of the 2*p* levels. In the $1s^2\ 2s^1\ 2p^1$ configuration (I, Scheme 1.3) beryllium could enter into bonding, as now two singly filled atomic orbitals are available for overlap. Energy lost in possible promotion of an electron from the 2*s* orbital to one of the 2*p* levels could be regained by bond formation. A bond formation could occur by overlap of the 2*s* orbital of Be with the 1*s* orbital of one H, on the one hand and the 2*p* orbital of Be with second H, on the other (Scheme 1.3). This possible arrangement would give two different bonds of unequal length and at an angle. Theory of electron repulsion, however predicts that compounds like BeH_2 to have linear structure with the bonds to Be of equal length (Scheme 1.3).

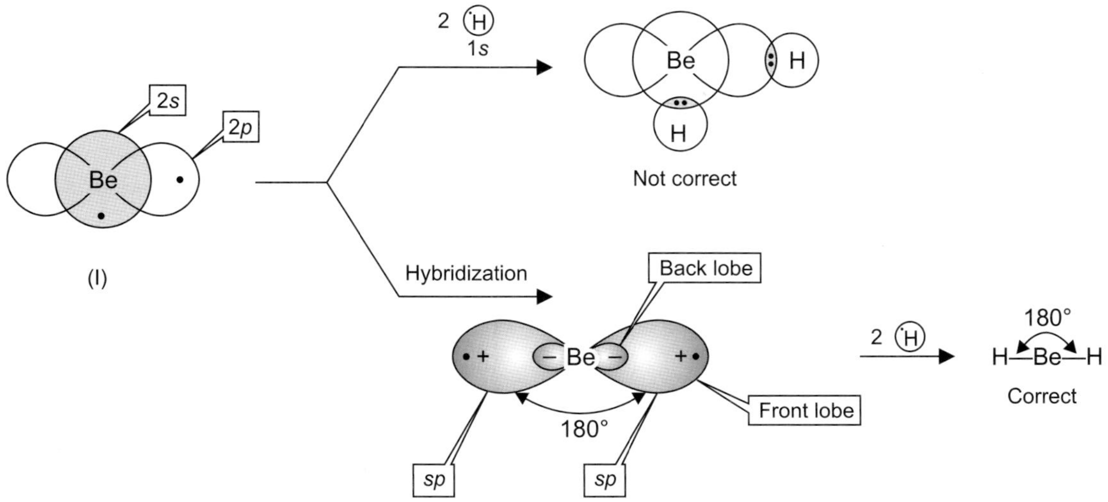

Hybridization in beryllium to create two *sp* hybrids

SCHEME 1.3

A way to explain the geometry of BeH_2 and other molecules is the approach called orbital hybridization. Like mixing of atomic orbitals on different atoms to give molecular orbitals, the mixing of atomic orbitals on the same atom gives new hybrid orbital. In beryllium, mixing the 2*s* and one of the 2*p* wave functions gives two new hybrids, called *sp* orbitals, made up of 50% *s* and 50% *p* character. This process rearranges the orbital lobes in space (Scheme 1.3). The major parts of the orbitals (front lobes) point away from each other at an angle of 180°. There are two additional minor back lobes (one for each *sp* hybrid) with opposite sign. The remaining two *p* orbitals are unchanged and overlap with the two hydrogen 1*s* orbitals gives linear BeH_2.

Thus a 2*s* and a 2*p* orbital mix in a *sp* hybridization to give two linear *sp* hybrids and the remaining two *p* orbitals remain unchanged. This bonding is found both in alkynes and nitriles. The nitrogen and carbon atom of a nitrile group (C≡N) are both *sp* hybridized.

(ii) sp^2 Hybrids give rise to Trigonal Structures

Structure of Borane (BH_3) has a triangular (trigonal planar) shape with the equivalent boron-hydrogen bonds. In its ground state boron has the electronic configuration $1s^2\,2s^2\,2p^1$. Promotion of a 2*s* electron to one of the 2*p* levels gives three singly filled atomic orbitals (one 2*s* two 2*p*) necessary for the formation of three bonds (Scheme 1.4). Mixing these three orbitals gives three equivalent hybrid orbitals which are sp^2. These have one part the character of an *s* orbital and two parts the character of *p* orbital. These orbitals are pointed toward the corners of an equilateral triangle with angles of 120° between their axes. The formation of borane is via the overlap of each of these three sp^2 orbitals with *s* orbitals of three hydrogen atoms. The sp^2 hybridization also offers a satisfactory model for carbon atoms which form double bonds.

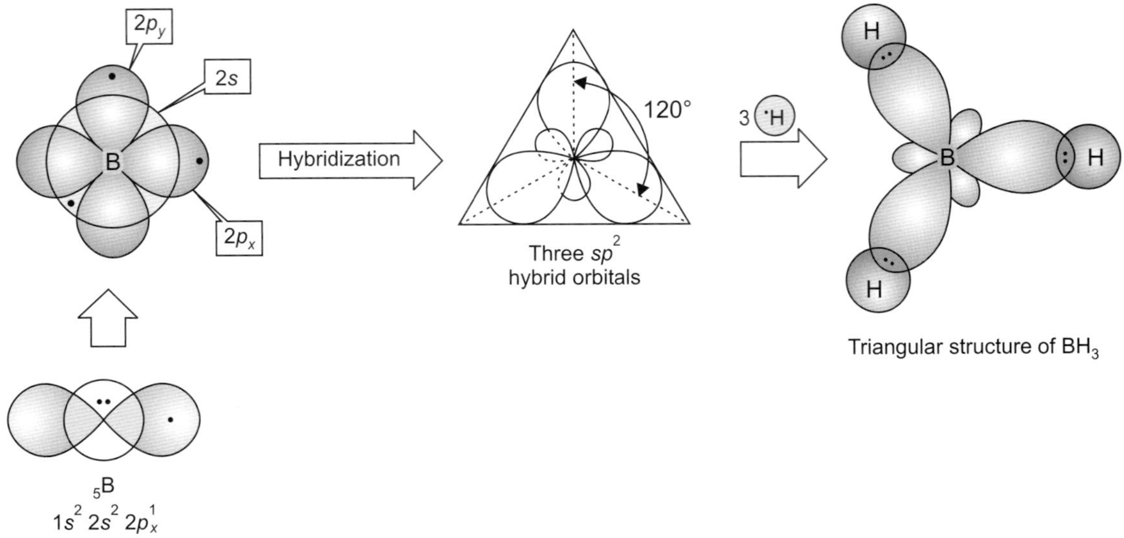

SCHEME 1.4

Trigonal boranes BH_3 and BF_3 have an empty $2p$ orbital. There is no positive charge on these compounds, but both are Lewis acids and react like cations (Scheme 1.4*a*).

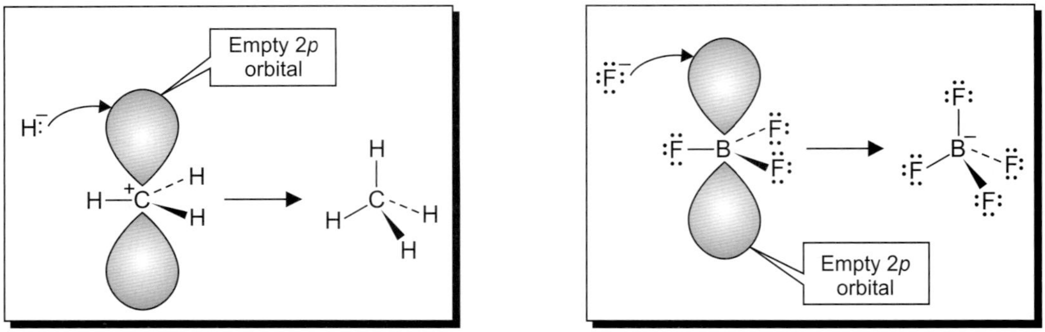

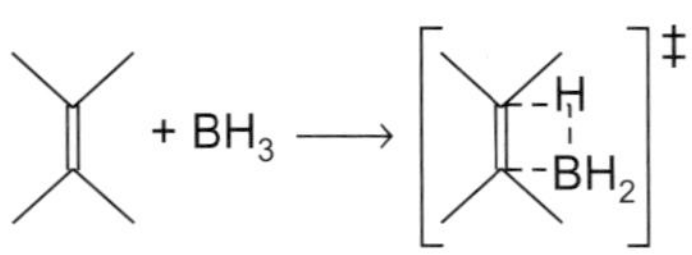

The π bond is electron rich and borane electron poor. In hydroboration the π bond of alkene acts as nucleophile and reacts with the strong Lewis acid BH_3.

SCHEME 1.4a

(iii) sp^3 Hybridization gives Tetrahedral Shape

In the case of carbon the promotion of one electron from $2s$ to $2p$ leads to four singly filled orbitals for bonding (Scheme 1.4*b*). The shape of the four C—H bonds of methane in space with minimum electron repulsion is tetrahedral. For this geometry the $2s$ orbital on carbon is hybridized with all three $2p$ orbitals to give four equivalent sp^3 orbitals with a tetrahedral arrangement (symmetrical arrangement) and each occupied by one electron. The overlap with *e.g.,* four hydrogen $1s$ orbitals gives methane.

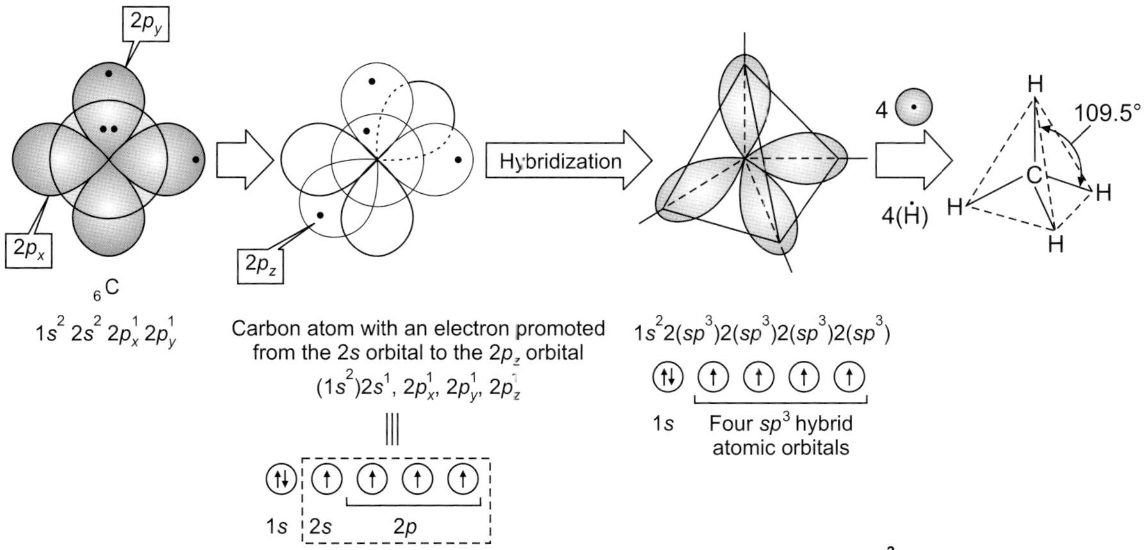

Structure of methane, CH_4 showing four equivalent σ bonds (s—sp^3 molecular bond orbitals) and tetrahedral structure

SCHEME 1.4b

A Double Bond

Consider the sp^2 (trigonal) hybridized carbon in e.g., a carbon-carbon double bond in ethene which is a flat molecule with bond angles close to 120°. Hybridization of the 2s orbital and two of the 2p orbitals leads to three equivalent sp^2 hybrid orbitals and one unhybridized 2p orbital (Scheme 1.4c) on each carbon. In ethene the C—C bond is an sp^2–sp^2 molecular σ bond while the C—H bonds are s–sp^2 molecular σ bonds. The π bond results from the parallel overlap through space of the p orbitals.

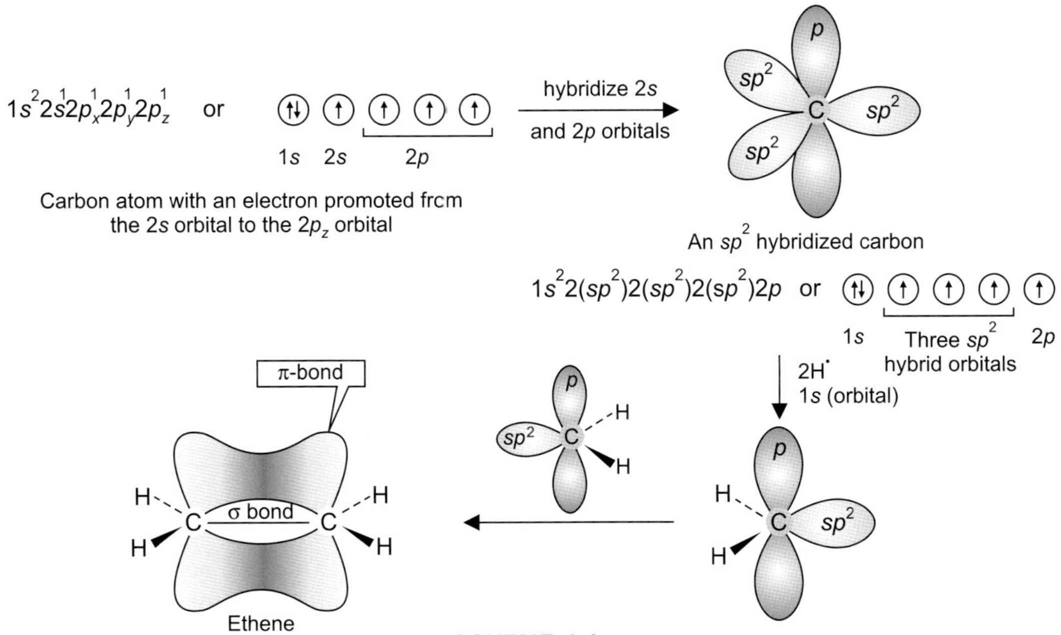

SCHEME 1.4c

D. Bond Angles, Shapes of Molecules (VSEPR model) and Reactivity

(i) Structure of Methane

Using the valence-shell—electron-pair repulsion (VSEPR) model, an atom is surrounded by an outer shell of valence electrons. These valence electrons may be involved in the formation of single, double, or triple bonds, or they may be unshared. Each of these combinations leads to a negatively charged region of space, and since like charges repel each other, the various regions of electrons density around an atom is spread out so that each is as far away from the others as possible. The Lewis structure for CH_4 shows a carbon atom surrounded by four separate regions of electron density, each of which consists of a pair of electrons forming a bond to a hydrogen atom. Using a VSEPR model, the four regions radiate from carbon in a way so that they are as far away from each other as possible. This is possible when the angle between any two pairs of electrons is 109.5°. Therefore, all H—C—H bond angles are predicted to be 109.5°, and the shape of the molecule is predicted to be tetrahedral. The H—C—H bond angles in methane as measured experimentally are 109.5°. Thus, the bond angles and shape of methane predicted by the VSEPR model are identical to those observed (Scheme 1.4*b*).

(ii) Structure of Ammonia and Water

In the molecule of NH_3 the N—H bonds are formed *via* the overlap of an sp^3 orbital of nitrogen with the *s* orbital of a hydrogen. The lone pair of electrons is on an sp^3 orbital. In nitrogen all the three 2*p* orbitals are available for bonding (Scheme 1.4*d*), however, direct overlap with 1*s* orbital of hydrogen would give, three N—H bonds perpendicular to each other with bond angles of 90°, the geometry of *p* orbitals. The bond angles in NH_3 are 107.3°, to show that nitrogen uses hybrid orbitals to form covalent bonds.

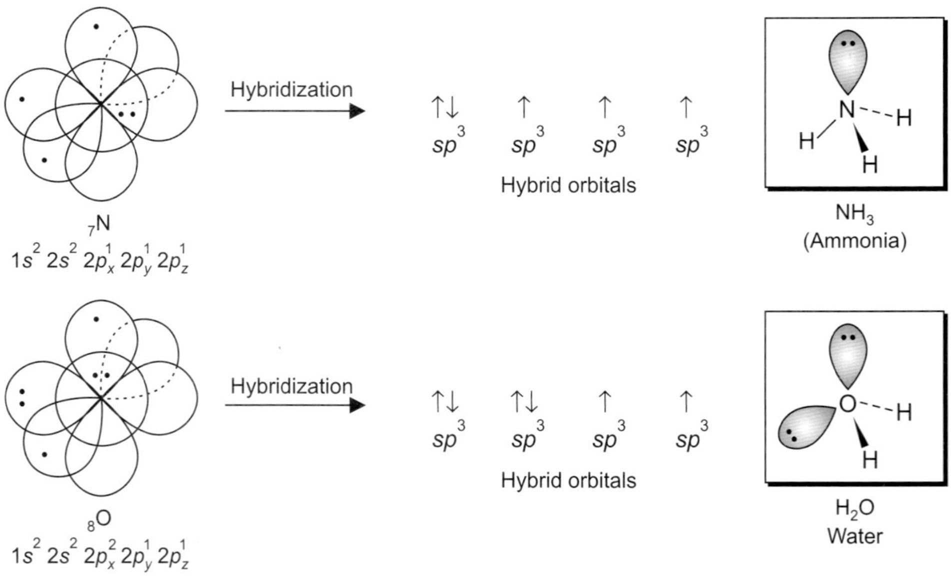

SCHEME 1.4d

Oxygen has two *p* orbitals which can be used for bonding. Again as seen in NH_3 an attempt to overlap 1*s* orbitals of two hydrogen atoms would lead to H—O—H angle of 90°. However, the bond angle in water is 104° close to that in NH_3. One can account for the bond angle of 104° in water provided oxygen uses the hybrid orbitals to form covalent bonds (Scheme 1.4*d*).

The Lewis structure of NH_3 shows nitrogen surrounded by four regions of electron density. Three regions contain single pairs of electrons forming covalent bonds with hydrogen atoms. The fourth region contains an unshared pair of electrons. These four regions of electron density are arranged in a tetrahedral fashion around the central nitrogen atom (Scheme 1.4*d*). The four regions of electron density (VSEPR model) around nitrogen are arranged in a tetrahedral manner with H—N—H bond angles 109.5°. The observed bond angles are however, 107.3°, this small difference between the predicted and observed angles can be explained due to repulsion between the unshared pair of electrons on nitrogen and the bonding pairs. This repulsion is greater than the electron repulsion between the two bonding pairs.

In water molecule, oxygen is surrounded by four separate regions of electron density. Two of these regions contain pair of electrons used to form covalent bonds with hydrogen; the remaining two contain unshared electron pairs. The four regions of electron density around oxygen (VSEPR model) are arranged in a tetrahedral manner, and the predicted H—O—H bond angle is 109.5°. The actual measured bond angle is 104.5°, a value smaller than predicted and even smaller than in NH_3. The explanation is the same as used to explain bond angles in NH_3, in the case of water the still smaller angle is due to the influence of now two lone pairs on oxygen instead of one on nitrogen.

(iii) Structure of a Double Bond

A double bond according to the VSEPR model, is treated as a single region of electron density. In formaldehyde (methanal), carbon is surrounded by three regions of electron density, two of which contain single pair of electrons forming single bonds to hydrogen atoms, while the third region of electron density has two pairs of electrons forming a double bond to oxygen. (Scheme 1.4*e*). In ethene (ethylene) each carbon atom is again surrounded by three regions of

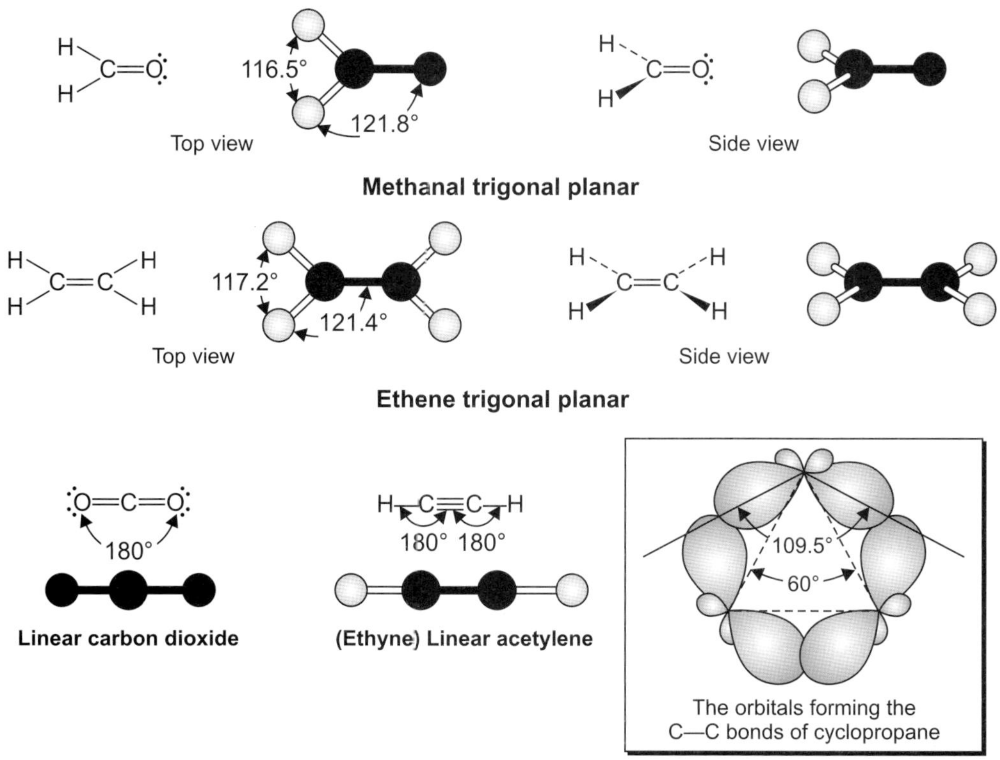

SCHEME 1.4e

electron density: two contain single pairs of electrons, and the other contains two pairs of electrons. Three regions of electron density about an atom are farthest apart provided these are in the same plane and make angles of 120° with each other. Thus, the predicted H—C—H and H—C—O bond angles in methanal are 120°; the predicted H—C—H and H—C—C bond angles in ethene are also 120°. Such an arrangement of an atom is called trigonal planar.

(*iv*) *Structure of Linear Molecules*

In other types of molecules, a central atom is surrounded by only two regions of electron density. In carbon dioxide *e.g.,* carbon is surrounded by two regions of electron density: each contains two pairs of electrons and forms a double bond to an oxygen atom. Same is the case with ethyne where each carbon is surrounded by two regions of electron density: one contains a single pair of electrons and forms a single bond to a hydrogen atom, and the other contains three pairs of electrons and forms a triple bond to a carbon atom. In each case, the two regions of electron density are farthest apart if they form a straight line through the central atom and generate an angle of 180°. Carbon dioxide and ethyne are therefore, linear molecules (Scheme 1.4*e*). The bond angle is thus, dependent on the orbital used by carbon in bond formation. The greater the amount of *s* character in the orbital the larger the bond angle. For example, sp^3 hybridized carbons have bond angles of 109.5°, sp^2 hybridized carbons have bond angles of 120°, and *sp* hybridized carbons have bond angles of 180°. The bond angles of sp^3 carbon are tetrahedral only when the four groups are identical as in CH_4 or CCl_4. In most of the cases the angles deviate a little from the tetrahedral value. For example, the C—C—Br angle in 2-bromopropane is 114.2°. Similarly, slight variations are generally found from the ideal values of 120° and 180° for sp^2 and *sp* carbon, respectively. These deviations are due to slightly different hybridizations, *i.e.*, a carbon bonded to four other atoms hybridizes one *s* and three *p* orbitals, but the four hybrid orbitals thus formed are generally not exactly equivalent, nor does each contain exactly 25% *s* and 75% *p* character. With the four atoms with different electronegativities, each makes its own demand for electrons from the carbon atom. In strained molecules the bond angles are largely distorted from the normal values.

VSEPR MODEL/SHAPE OF MOLECULES

One can predict bond angles of molecules and polyatomic ions using Lewis structures and the ***valence-shell electron-pair repulsion (VSEPR) model****. The atoms surrounded by four regions of electron density, predict bond angles of 109.5°; by three regions of electron density, predict bond angles of 120° while for two regions of electron density, predict bond angles of 180°. Thus benzene must be a flat hexagon since each carbon in benzene has three areas of electron density around it, thus the carbon atoms are trigonal planar, C—C—C bond angles 120° as well as H—C—C bond angles also 120° and all the carbon atoms in the ring are* sp^2 *hybridized. Similarly according to VSEPR theory a carbocation e.g., t-butyl cation is predicted to have a trigonal planar geometry (there are three areas of electron density around the central carbon atom).*

(*v*) *Strained Molecules*

In cyclopropane (Scheme 1.4*e*), for geometric reasons, the internuclear C—C—C angle is 60°. The carbon-carbon bonds in cyclopropane have more *p* character than normal sp^3. The orbitals thus, from bent bonds, which are weaker than those in normal alkanes. To compensate for the extra *p* character for C—C bonds extra *s* character is used for the C—H bonds, therefore, these bonds are shorter and stronger than alkyl C—H bonds (C—H bonds in ethane 1.10 Å).

(vi) Role of Bond Angles in Reaction Mechanism

Halocycloalkanes undergo S_N2 reactions, however, with significant rate differences depending on the size of the ring (Table 1.1). The strained cyclopropyl bromide does not undergo substitution, due to prohibitive strain in the transition state. In a S_N2 reaction the reacting carbon adopts an sp^2 configuration as the nucleophile replaces the leaving group (see, Scheme 5.7). The normal bond angle of sp^2 hybridization is 120° and the cyclopropane cannot be squeezed much from 60° (in cyclopropane which has a shape of regular triangle, the internal angle must be 60°).

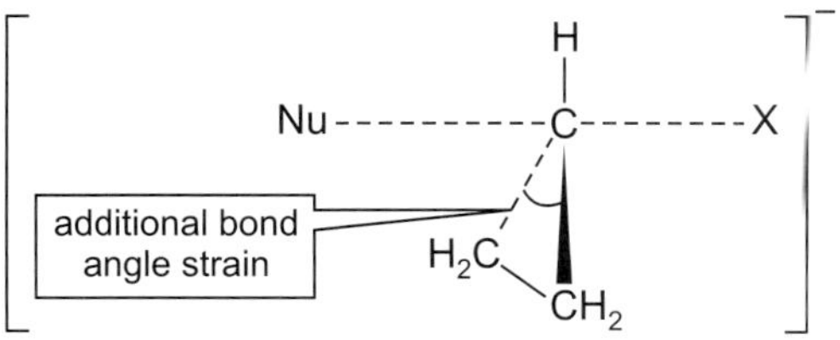

Transition state for S_N2 reaction on cyclopropyl halide

Table 1.1 Relative reactivities of cycloalkyl bromides in the S_N2 reaction

Alkyl Group	*Relative Rate*
Isopropyl	1.0
Cyclopropyl	negligible
Cyclobutyl	0.008
Cyclopentyl	1.6
Cyclohexyl	0.01
Cycloheptyl	1.0

SCHEME 1.4f

The ring provides a rigid framework which should be capable to tolerate this additional strain. Thus the low reactivity of cyclobutyl bromide is also due to these reasons. Cyclopentyl and cyclohexyl bromides display S_N2 reactivity which is close to their a cyclic counterparts, since these rings are more capable to attain sp^2 hybridization at the reacting carbon (see also Scheme 5.9).

E. Hybridization and Length and Strength of a Bond

Both the length and strength of a carbon-hydrogen bond is dependent on the hybridization of the carbon atom to which the hydrogen is attached. When there is more *s* character in the orbital used by carbon to form the bond, the shorter and stronger bond results. An *s* orbital is closer to the nucleus than a *p* orbital and thus the carbon-hydrogen bond formed by an *sp* (50% *s*) hybridized carbon is shorter and stronger than carbon-hydrogen bond formed by an sp^2 (33.3% *s*) hybridized carbon and this in turn in shorter and stronger than a carbon-hydrogen bond formed by an sp^3 (25% *s*) hybridized carbon.

The following points may be noted:

- In general shorter bonds are stronger bonds. Increasing *s* character shortens bonds, thus bonds strength increase with increasing *s* character.
- More the bonds holding two carbon atoms together, the shorter and stronger is the carbon-carbon bond. Triple bonds are shorter (C≡C, 1.20 Å) and stronger (C≡C, 200 kcal/mol) than double bonds (C=C, 1.33 Å, 152 kcal/mol), which are shorter and stronger than single bonds (C—C, 1.54 Å, 88 kcal/mol).

  Thus double bonds are both shorter and stronger than corresponding single bonds, however, not twice as strong, since π overlap is less than σ overlap. This means that a σ bond is stronger than a π bond. The difference in energy between a single bond, say C—C, and the corresponding double bond is the amount of energy necessary to cause rotation around the double bond.

F. Shapes of Some Reactive Intermediates

A carbocation contains a positively charged carbon atom bearing three substituents to suggest a trigonal planar arrangement as *e.g.,* in *tert*-butyl cation. In a carbocation the carbon is sp^2 hybridized and the unhybridized $2p$ orbital lies perpendicular to the sigma bond framework and has no electrons (is vacant). In BF_3 as well, one has the same situation, the boron atom is sp^2 hybridized and has a vacant p orbital perpendicular to BF_3 plane, and same is the case with BH_3.

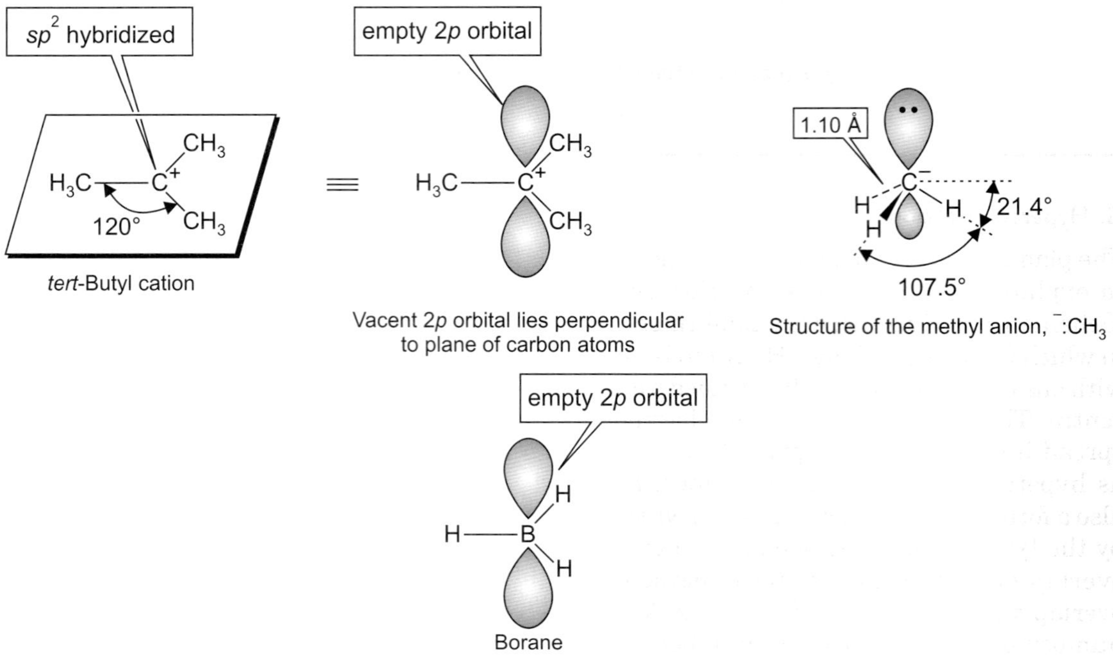

SCHEME 1.4g

A methyl anion has two more electrons than the cation. The orbitals have density at the positively charged nucleus, thus a negatively charged electron would be more stable (lower in energy) in an orbital with more s character. A pyramidal structure seems reasonable for the methyl anion however, the molecule is not a perfect tetrahedron. Hybridization and bond angles are closely related. With more s character in the orbital containing the nonbonding electrons, the pyramidal shape is likely to increase.

As per the calculations the structure of methyl anion is as shown (Scheme 1.4*g*). With the three sigma bonds and a lone pair, a carbanion is electronically similar to an amine. Consider the hybridization change on formation of methyl radical from methane (Scheme 1.4*h*). At present it is however not possible to choose between a planar or a pyramidal structure for a free radical *e.g.,* for neutral methyl radical which can be obtained by the removal of hydrogen atom from methane. Spectral measurement have however, shown that methyl radical has nearly planar configuration described by sp^2 hybridization with unpaired electron in the remaining p orbital perpendicular to the molecular plane (Scheme 1.4*h*). In the pyramidal arrangement, the carbon would be sp^3 hybridized and odd electron would occupy an sp^3 hybridized orbital in one corner of the tetrahedron.

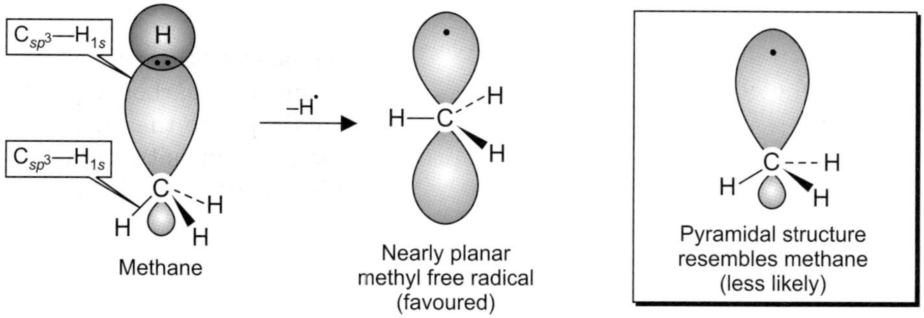

Hybridization change on forming methyl radical from methane

SCHEME 1.4h

G. Hyperconjugation

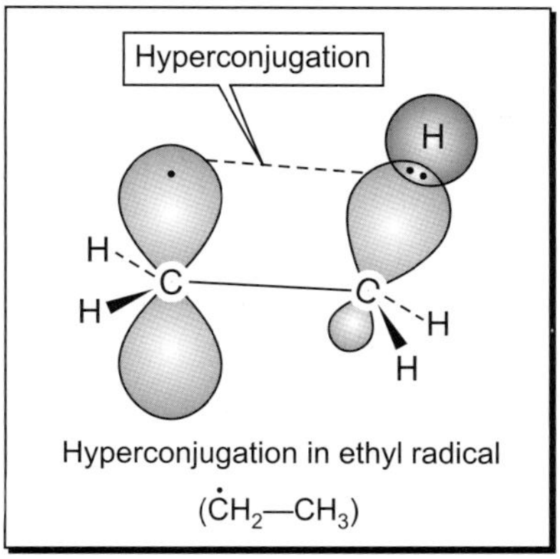

SCHEME 1.4i

The planar structure of methyl and other alkyl radicals helps to explain their relative stabilities by hyperconjugation (Scheme 1.4*i*). Thus there is a conformer in *e.g.,* ethyl radical in which C—H bond of the CH_3 group is aligned and overlaps with one of the lobes of singly occupied *p* orbital on the radical centre. Thus the bonding pair of electrons in the σ orbital spread into the partly empty *p* lobe a phenomenon known as hyperconjugation. Like resonance, hyperconjugation is also a form of electron delocalization which are distinguished by the type of orbital. Resonance generally refers to π type overlap of *p* orbitals, while hyperconjugation involves overlap with the orbital of σ bonds. With the increase in number of alkyl groups, the number of hyperconjugation interactions increases as in isopropyl group (more details are in sec. 2.14).

H. Sigma and *pi* Bonds

A C—C sigma bond is formed by the overlap of hybrid orbitals. In ethane *e.g.,* this bond consists of two sp^3 hybrids, when these roughly sp^3 hybridized carbons approach for the overlap of singly occupied orbitals (Scheme 1.4*j*).

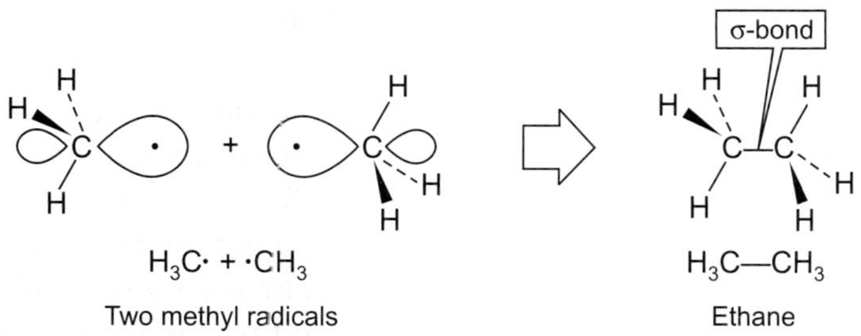

SCHEME 1.4j

Bonds made by overlap along the internuclear axis are called σ bonds, while those made by overlap of *p* orbitals perpendicular to the internuclear axis are called *pi* bonds (Scheme 1.4*k*).

In contrast to the sigma orbital, the *pi* molecular orbital has zero electron density along the molecular axis, but has the maximum electron density above and below the internuclear line. All the sigma bonds around the *pi* bonds are coplanar, the bond angles being 120°. The plane of the molecule is the nodal plane of the *pi* bond.

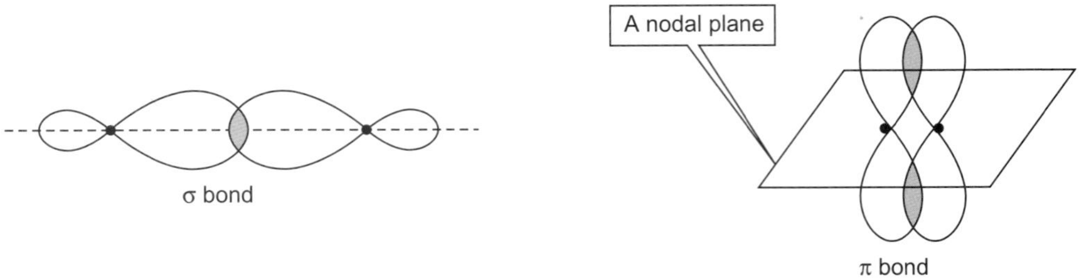

SCHEME 1.4k

1.2 ELECTRONEGATIVITY—DIPOLE MOMENT

Carbon is unique among the elements, since it is able to form a huge number of compounds by bonding to itself and to the atoms of other elements *e.g.,* hydrogen, oxygen, nitrogen, sulphur and the halogens. This bonding is almost always covalent.

The sharing of electrons in a covalent bond is not exactly equal when the linked elements are different. The relative attractive power exerted by an element on the electrons in a covalent bond can be expressed by its electronegativity. According to one quantitative definition of electronegativity, there is an increase in electronegativity along the series towards fluorine as shown (Scheme 1.5).

Hydrogen with electronegativity 2.0 is close in this respect with carbon (for further details on electronegativity see, Scheme 3.9). When two atoms with different electronegativities form a covalent bond, the atom with greater electronegativity draws the electron pair to it and a polar covalent bond results as in hydrogen chloride and can be represented by the usual symbol (I, Scheme 1.5) when necessary. In fact the hydrogen chloride molecule is a resonance hybrid of two resonating structures.

C	N	O	F
2.5	3.0	3.5	4.0

Relative electronegativities

$$\text{H—Cl} \longleftrightarrow \overset{+}{\text{H}}\,\overset{-}{\text{Cl}} \equiv \overset{\delta+}{\text{H}}\text{—}\overset{\delta-}{\text{Cl}}$$

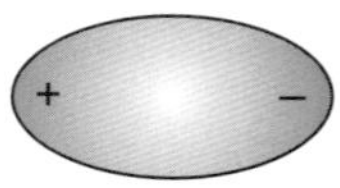

a dipole
(I)

SCHEME 1.5

As a consequence of a partially positive end (δ^+) and a partially negative end (δ^-) in HCl molecule represents a dipole (II, Scheme 1.5) and therefore, has a dipole moment (a physical property). Thus the dipole moment is a property of the molecule which is due to charge separations. It is defined as the product of the magnitude of the charge (e) in electrostatic units (esu) and the distance (d) which separates them in centimeters (cm):

$$\mu = e \times d$$

Dipole moments are typically of the order of 10^{-18} esu cm, since charges are typically of the order of 10^{-10} esu and the distance is of the order of 10^{-8} cm. For convenience this unit (1×10^{-18} esu cm) is defined as one Debye (abbreviated D). The direction of polarity of a polar bond is usually symbolized by a vector quantity (Scheme 1.6).

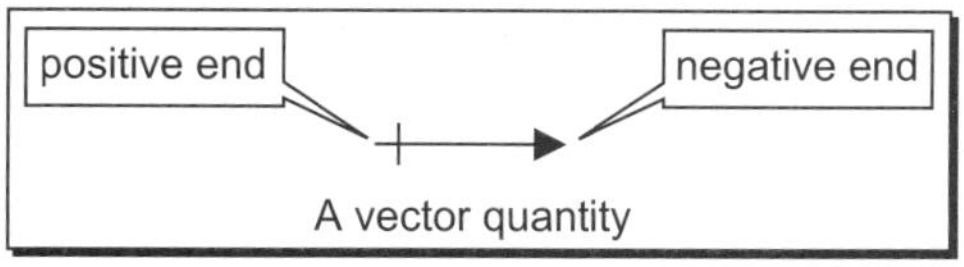

Scheme 1.6

The arrow head points to be the negative part of the molecule, while the crossed end is the positive end. A molecule with polar bonds, may, however, not possess a dipole moment *i.e.,* the molecule itself may be non-polar. This is so when a particular molecule has a shape (or symmetry) so that the dipoles of the individual bonds cancel each other. Thus one is concerned with the total moment of the molecule which is the vectorial sum of the individual bond moments as *e.g.,* in the case of 1, 2-dichloroethene isomers (Scheme 1.7).

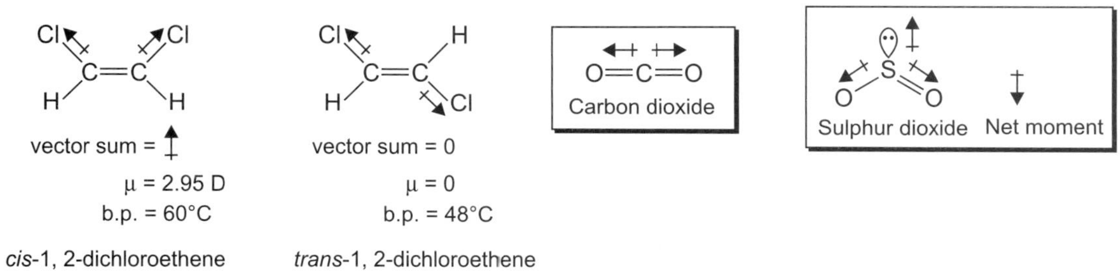

SCHEME 1.7

Similarly carbon in CO_2 is *sp* hybridized and the molecule is linear. The C—O bond moments oppose each other and cancel, in SO_2 however, S is sp^2 hybridized with two σ bonds to O and one with unshared electron pair. The O—S—O bond angle is around 120° and S—O moments do not cancel. Thus unlike CO_2, SO_2 has $\mu = 1.6$ D.

The carbonyl group is polar. The carbon atom is bonded to the more electronegative oxygen atom. The resulting imbalance in the electron density leads to a permanent dipole of 2–3 Debyes (D) in the case of simple carbonyl compounds (Scheme 1.8).

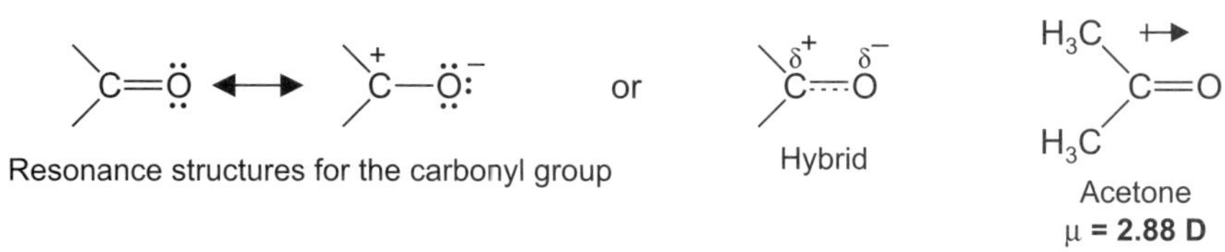

SCHEME 1.8

2-chloroethanol is much more acidic than ethanol. This can be explained due to electrostatic interaction of the C—Cl dipole with the negative charge of the alkoxide ion (Scheme 1.9). The negative charge on oxygen is nearer to the positive end of the dipole than it is to the negative end. Consequently, electrostatic attraction exceeds repulsion, leading to the stabilization of the anion. This stabilization of the anion increases its ease of formation and the conjugate acid, 2-chloroethanol, is more acidic than ethanol itself.

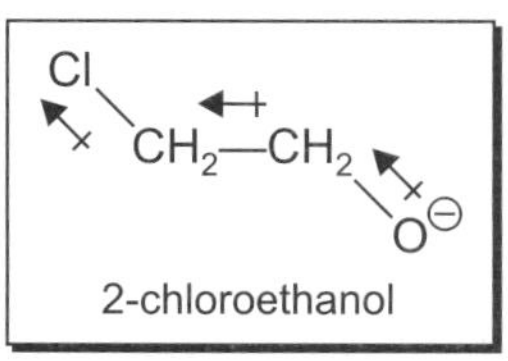

SCHEME 1.9

In the equilibrium for 2-halocyclohexanones (Scheme 1.10) there is an increase in the per cent of axial conformer on going from 1, 4-dioxane to heptane as solvent. The C—Cl and C=O dipoles reinforce each other in the equatorial form, however, these cancel to some extent in the axial form. Thus the equatorial form is more polar and should be favoured by the more polar solvents.

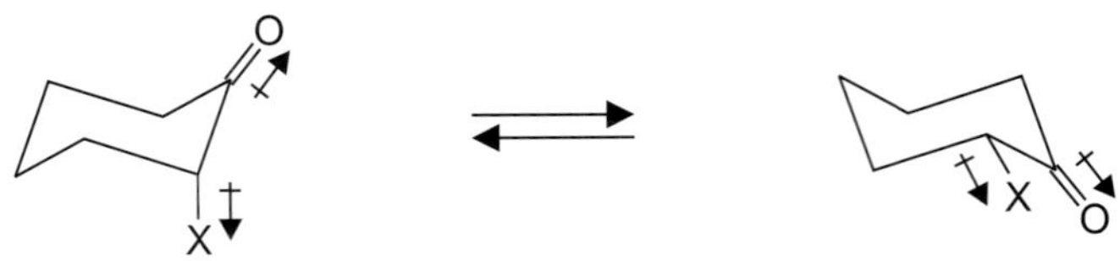

The equatorial form is more polar

Halogen	% Axial (heptane)	% Axial (1, 4-dioxane)
Br	85	62
Cl	76	37

2-halocyclohexanones

SCHEME 1.10

To cite one example of the involvement of carbon-halogen dipole is the electrophilic aromatic substitution in a halobenzene. A halogen is *o*, *p* — directing substituent. Substitution at the *meta* position of a halobenzene can lead to three resonance structures (Scheme 1.11). All the three structures are strongly destabilized by electrostatic interaction of the positive charge in the ring with the carbon-halogen dipole. As a consequence the *meta* position in a halobenzene is strongly deactivated. Though similar situation is also obtained during *o* and *p* attack, however, in these cases additional stable halonium ion structures make the *o*, *p* attack for more facile (Scheme 8.42).

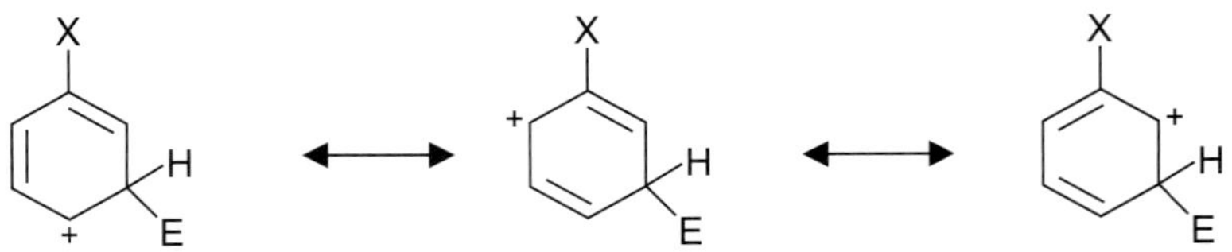

Reaction at the *meta* position

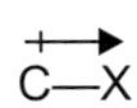

SCHEME 1.11

1.3 INDUCTIVE AND FIELD EFFECTS

One may observe a change in the rate constant or equilibrium constant of a reaction by replacing a hydrogen atom by another atom or group of atoms. These substituent effects may be the result of the size of the substituent (steric effect) and/or its influence on the availability of electrons (electronic effect) on the site of the reaction. The electronic effect which a substituent can exert may be either electron releasing or electron withdrawing. These electronic effects are further subdivided into an inductive and a resonance (mesomeric effect). The inductive effect (*I*) is a result of a substituents' intrinsic ability to supply electrons (electron donation, + *I* effect) or withdraw electrons (electron withdrawing) – *I* effect, *i.e.*, the inductive effect depends on the electronegativity of the substituent. The inductive effect is transmitted through σ bonds and weakens as the distance between the substituent and the reactive center increases. Thus the effect is greatest for the adjacent bond and may be felt weakly farther away (see, Scheme 3.18). The effect may be represented for ethyl chloride (Scheme 1.12). In this case chlorine atom has *–I* effect and thus C-1 atom looses some of its electron density and as a result C-1, Cl bond is polarized and a slight positive charge is generated on C-2. In this way the replacement of hydrogen atom by a more electronegative atom results in electron displacements throughout the molecule.

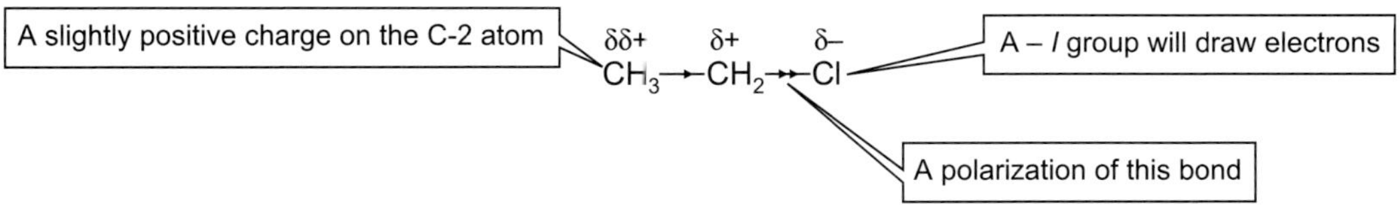

SCHEME 1.12

The other effect operates through space (and not through σ bonds) or through solvent molecules and is called the field effect. Normally the field effect depends on the geometry of the molecule whereas the inductive effect depends only on the nature of the bonds. As an example of the field effect (long range polar interactions) the two acids (I and II, Scheme 1.13) have different pK_a values. The inductive effect of the chlorine atoms on the position of the electrons in the COOH group must be same since the same bonds intervene. Consequently, the acidity must have been equal. However, this difference in pK_a value shows the operation of field effect, since the two chlorine atoms are placed closer in space to the COOH group in I than in II.

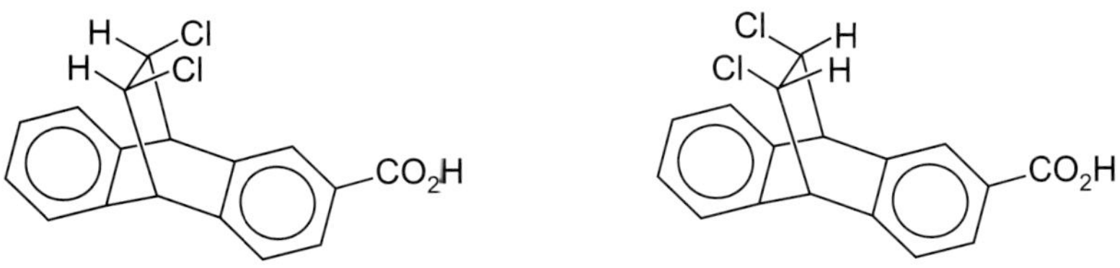

(I) (II)

SCHEME 1.13

The inductive effect (*+I*) of alkyl group has been invoked to explain the carbocation stability. This effect also helps in explaining the orientation and reactivity during electrophilic substitution on benzene derivatives (Sec. 8.8).

The resonance effect (see, Schemes 2.14 and 2.15) involves delocalization of electrons through resonance via the π system. Atoms and functional groups may be arranged according to their ability to donate or withdraw electrons. The inductive and resonance effects of many groups are in the same direction. Other groups display opposite effects in the two cases. Normally atoms which are more electronegative than carbon and which also have non-bonding electrons possess opposing characteristics. The halogens illustrate these opposing effects (Scheme 1.14). A good example is found during electrophilic substitution when the inductive effect of a halogen on the benzene ring slows the rate of further substitution (Scheme 1.14).

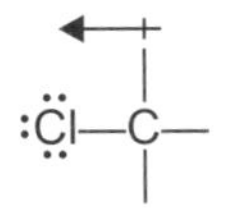

the inductive effect

resonance

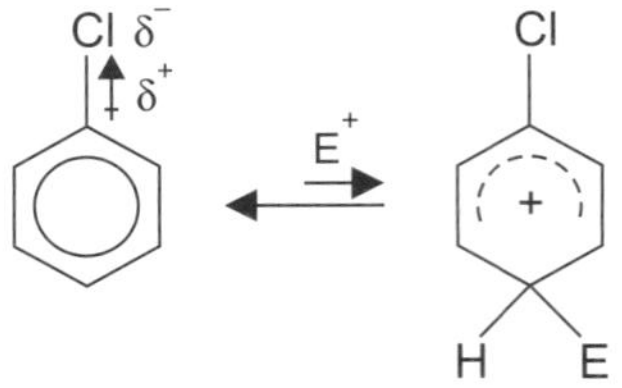

The C—Cl bond is strongly polarized. The δ^+ on the carbon slows a substitution reaction which places more positive charge on the ring. A halogen, however denotes electrons in the ring via resonance. Thus a halogen in o, p director.

SCHEME 1.14

EXERCISE 1.1

How one can explain that acetic acid (pK_a = 4.7) is stronger acid than 2, 2, 2-trifluoroethanol (pK_a = 12.8) and ethanol is the least acidic (pK_a = 15.9) from among these three compounds?

ANSWER. *Consider the species after the loss of a proton from each of these compounds. In the case of ethanol the negative charge resides on its single oxygen i.e., the charge is localized (Scheme 1.14a). In the carboxylate ion both inductive withdrawal of electrons and the ability of two atoms to share the negative charge via resonance renders the conjugate base of the carboxylic acid more stable than the conjugate base from ethanol. 2, 2, 2-Trifluoroethanol is much stronger acid than ethanol, since in the former the highly electronegative fluorines help in the stabilization of its alkoxide ion.*

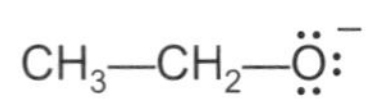

Charge is localized on the alkoxide ion

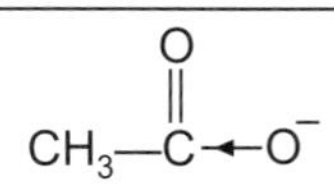

–*I* effect of carbonyl group in acetate anion also stabilizes it

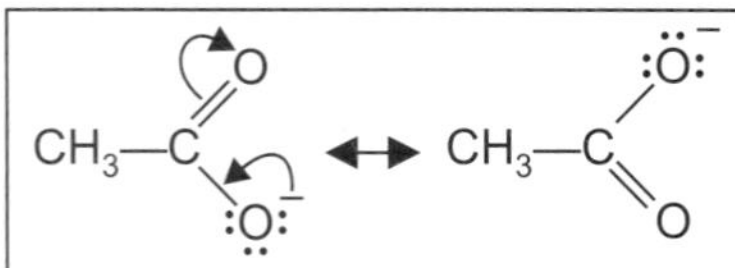

Charge is shared by oxygen atoms via resonance

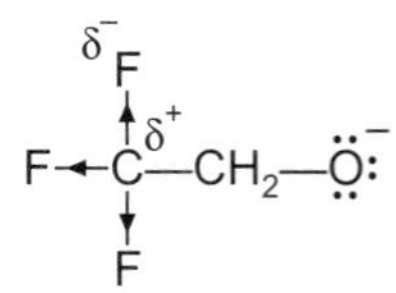

Strong–*I* effect of fluorines stabilizes the alkoxide ion

SCHEME 1.14a

EXERCISE 1.2

Acidity order of alcohols in aqueous solution is :

$CH_3OH > CH_3CH_2OH > (CH_3)_2CHOH > (CH_3)_3COH$. *Can inductive effect explain this order?*

ANSWER. *Electron donating inductive effect (+ I) of alkyl groups will retard the formation of an alkoxide to reduce the acidity of an alcohol. Thus t-Butanol is the weakest acid (Scheme 1.14b).*

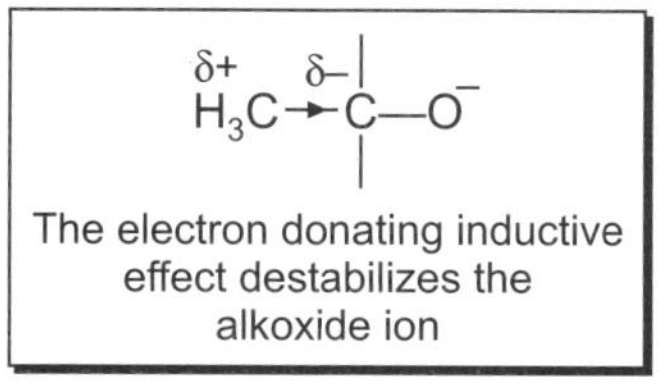

SCHEME 1.14b

However, in the gas phase it is found that the acidity order of alcohols is opposite to that found in solution. Thus it is probably not the + I effect of alkyl groups that is important but stabilizing effect of the solvent. A smaller alkoxide ion is approached more easily by the solvent to solvate it (see Scheme 3.18).

EXERCISE 1.3

Which of the alkenes (Scheme 1.14c) is expected to react faster with HX?

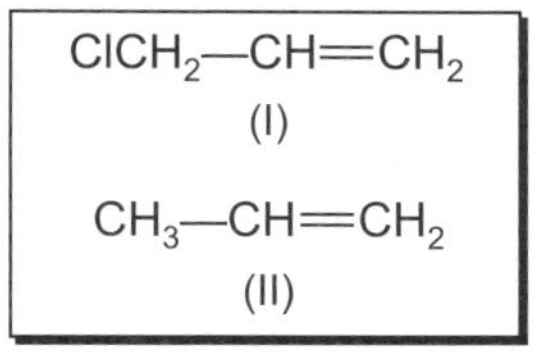

SCHEME 1.14c

ANSWER. *Consider the protonation of the double bond in (I, Scheme 1.14c) which is according to Markovnikov rule.*

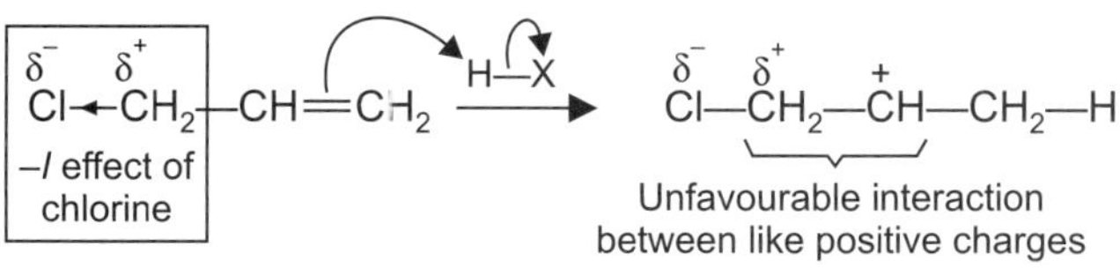

SCHEME 1.14d

Due to the –I effect of chlorine a positive charge on the methylene group would be in opposition to the expected carbocation formed during the addition of HX. No such effect is operative in (II).

EXERCISE 1.4

Discuss in terms of resonance and inductive effect the addition of HX to methyl vinyl ether.

ANSWER. *The –I effect of oxygen generates a partial positive charge on the adjacent carbon, which will get enhanced on protonation (Markovnikov rule) during the* *first step of addition of HX i.e., during the formation of intermediate carbocation which is an unfavourable situation. However, resonance stabilization dominates the inductive effect and the addition occurs smoothly (Scheme 1.14e).*

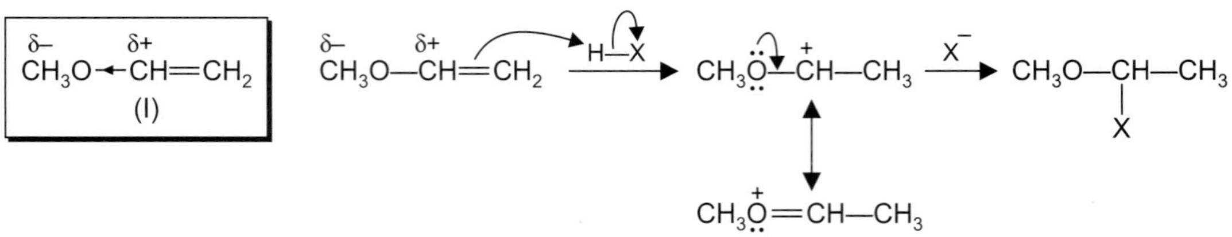

SCHEME 1.14e

1.4 HYDROGEN BOND

The hydrogen atom which is bonded to an electronegative atom can form a hydrogen bond to a second electronegative atom. The hydrogen bond, is thus a force of attraction between opposite partial charges, *e.g.,* δ^+ charge on H in the OH group and δ^- charge on the O of another group (Scheme 1.15).

No such partial charges exist in the molecules of alkanes since C and H have nearly same electronegativities. Only three elements, F, O and N, have atoms that are electronegative enough to participate significantly in hydrogen bonds. A hydrogen bond requires a hydrogen bond donor and a hydrogen bond acceptor as in the alcohol molecule (Scheme 1.15).

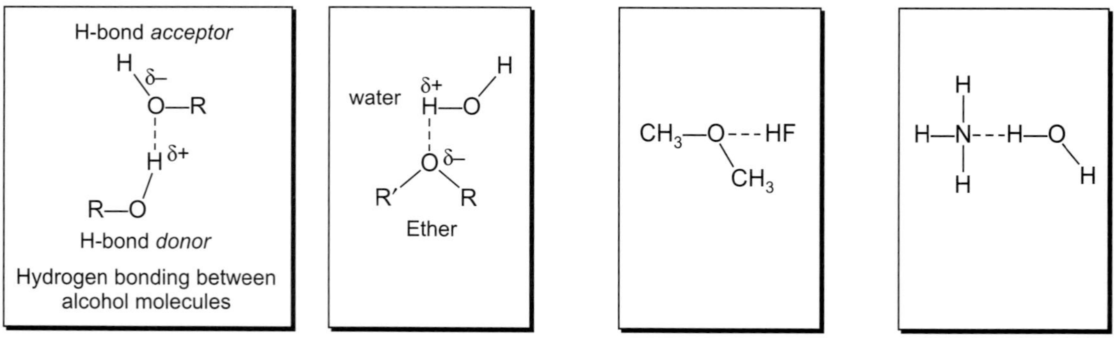

SCHEME 1.15

An ether has no O—H proton, therefore, the ether group cannot donate hydrogen bonds and thus cannot form a hydrogen bond with another ether molecule. Since, ether molecules are not held together by hydrogen bonds, they are more volatile than alcohols of the same molecular weight. The oxygen of the ether group can however, form hydrogen bonds with an alcohol or a other hydrogen bond donor *e.g.,* water (Scheme 1.15). So ethers are more soluble in water than in alkanes. The hydrogen bond is conventionally represented by a dotted line.

The hydrogen bond (bond dissociation energy about 1–9 kcal/mol) is weaker than an ordinary covalent bond. When there are many such bonds as in carbohydrates, the total strength is very great. The bond may be formed both between molecules of the same type as in alcohols (Scheme 1.15) and carboxylic acids (Scheme 1.16) and molecules of different type as in an ether and alcohol (Scheme 1.15) or as in the interaction between the proton of an alcohol and the oxygen of a carbonyl group. Two types of hydrogen bonding have been recognized: intramolecular (within the same molecule) and intermolecular (between two or more molecules).

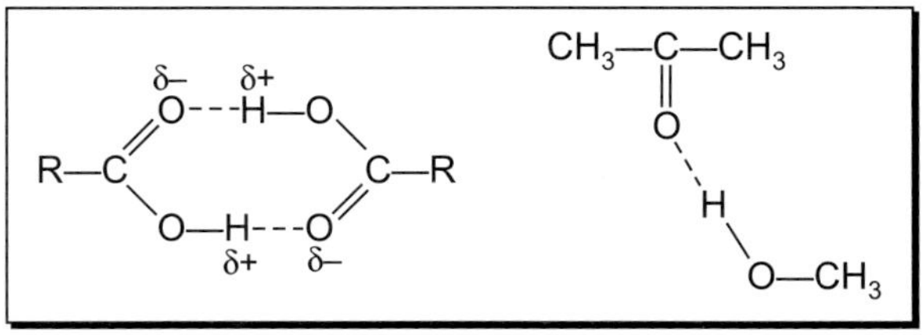

SCHEME 1.16

Due to hydrogen bonding, there is an increase in intermolecular 'aggregation' forces which is reflected in the boiling point and solubility of the organic compound. There is an increase in the boiling point since energy is required to separate the hydrogen bonded molecules in their translation to the gaseous state. Hydrogen bonds exist in the liquid and solid phases and in solution. Compounds which form strong hydrogen bonds may be associated even in the gas phase. Thus acetic acid exists as a dimer in the gas phase.

Intramolecular hydrogen bonds may also be formed and these have particular significance. When the resulting ring is five or six membered then the phenomenon is termed chelation. An example of chelation is for the enolic form of acetylacetone (Scheme 2.50). Since on chelation, intermolecular aggregation forces are not operative, chelated compounds have normal boiling points (Scheme 1.17). Thus, *o*-nitrophenol is much more volatile than its *p*-isomer, since only the latter can form intermolecular hydrogen bonds.

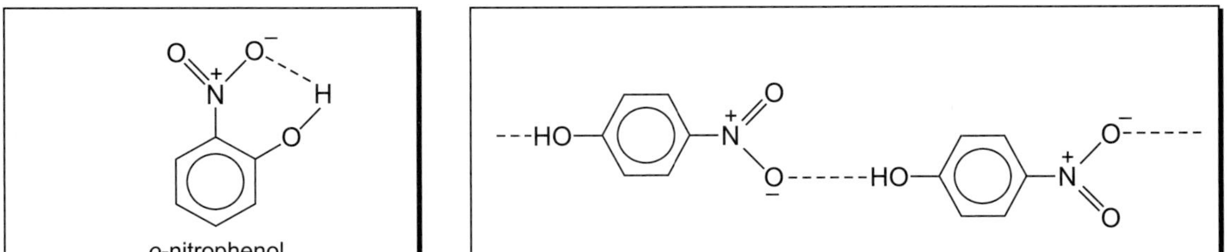

o-nitrophenol (more volatile because of intramolecular hydrogen bonding)

p-nitrophenol (less volatile because of intermolecular hydrogen bonding)

SCHEME 1.17

An important way to detect hydrogen bonding is *via* IR and NMR spectroscopy. A free OH group of an alcohol or a phenol shows a sharp infrared absorption around 3600 cm^{-1} (O—H stretching vibrations). On hydrogen bonding the band becomes broad and is shifted to lower frequencies (around 3400 cm^{-1}). In several cases in dilute solutions, there may be partial hydrogen bonding, *i.e.,* some hydroxyl groups are free and others bonded. In these cases one therefore, observes two bands, one sharp band at high frequency (around 3600 cm^{-1}) and another broad band at lower frequency (around 3400 cm^{-1}). A distinction can also be made between inter- and intramolecular hydrogen bonding on the basis of infrared spectroscopy. In very dilute solution, formation of intermolecular hydrogen bonds does not take place as the molecules are widely separated. Increasing the concentration of the alcohol or phenol causes the sharp band around 3600 cm^{-1} to be replaced by a broad and lower frequency band which is assigned to OH groups that are associated through intermolecular hydrogen bonding. Intramolecular hydrogen bonds remain unaffected and as a result the absorption band also remains unaffected. In the case of *o*-nitrophenol the OH band (intramolecular hydrogen bonding) is at 3200 cm^{-1} in

KBr pellet as well as in $CHCl_3$ solution, whereas in the *p*-isomer, the values are different in the two media KBr (pellet 3330 cm^{-1}; $CHCl_3$ solution 3520 cm^{-1}). In the ^1H NMR spectrum a hydrogen bonded hydroxyl group shows a downfield shift of its proton.

EXERCISE 1.5

(a) Why the O—H stretching frequency for t-butyl alcohol is a more sharper band compared with methanol?

(b) Which of the following norbornane systems can be detected by IR spectroscopy:

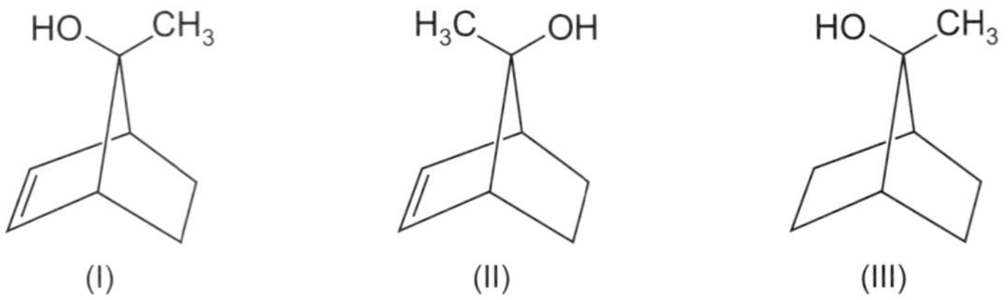

SCHEME 1.18

ANSWER. *(a) Due to steric effects it is far more difficult for t-butyl alcohol to involve in intermolecular hydrogen bond formation.*

(b) In (I, Scheme 1.18a), intramolecular hydrogen bonding would be detected.

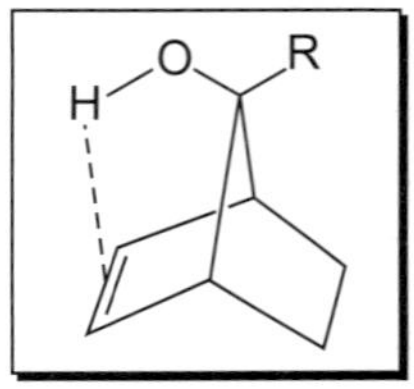

SCHEME 1.18a

Hydrogen bonding effects structure (chemical properties) and molecular shape of molecules. Thus *e.g.,* the role of intramolecular hydrogen bonding is reflected in the large amount of enol present in some tautomeric equilibria (see, Scheme 2.48). It also influences conformation of molecules. The six membered heterocycles of oxygen closely resemble the chair conformation of cyclohexane. In heteroxyclic rings the steric repulsions for axial substituents are reduced due to the replacement of a methylene groups of cyclohexane by oxygen or nitrogen. Since the divalent oxygen has no substituents, therefore, the 1, 3-diaxial interactions which are the main unfavourable interactions for axial substituents in cyclohexanes are absent (Scheme 1.18). With the presence of a polar substituent, interactions between the substituent and the ring heteroatom can become important. Thus, the preferred conformation of 5-hydroxy-1, 3-dioxane (Scheme 1.18*b*) has the hydroxyl group in the axial position. This conformation is favoured due to hydrogen bonding of the hydroxyl group with the ring oxygen which is possible only with the axial hydroxyl group to serve as a stabilizing force for this conformation.

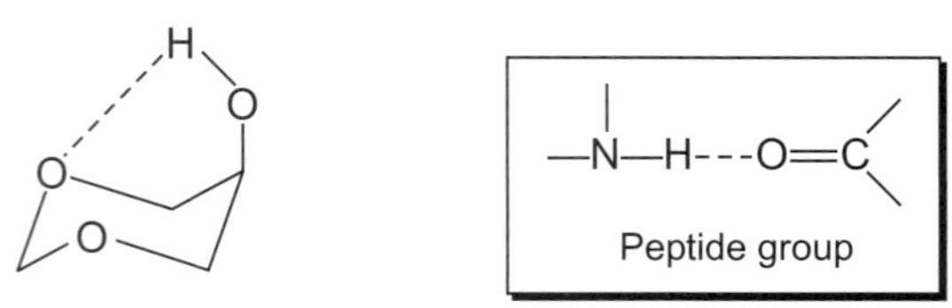

SCHEME 1.18b

The three dimensional structures of proteins and nucleic acid molecules is due to hydrogen bonding. In α-helices, hydrogen bonds extend from the H atoms of polar NH units in peptide groups to oxygen atoms of polar carbonyl units (Scheme 1.18b).

The nucleophiles may be solvated by hydrogen bonding to become less reactive in a nucleophilic substitution.

1.5 OTHER WEAKER BONDS

A. Charge Transfer and π Complexes

In several cases the molecules of the starting materials remain more or less intact and weak bonds hold the molecules together. Electron donor-acceptor (EDA) complexes provide an excellent example, where weaker bonds operate. In the case of EDA complexes, one always has a donor molecule and an acceptor. The donor may donate an unshared pair (an *n* donor) or a pair of electrons in a π orbital of a double bond or aromatic system (a π donor). Formation of an EDA complex is confirmed from the electronic spectrum. These complexes normally display a spectrum (a charge transfer spectrum) that is not the same as the sum of the spectra of the two individual molecules. This is due to the fact that the first excited state of the complex is relatively close in energy to the ground state, there is usually a peak in the visible or near uv-region and EDA complexes are often coloured. Generally in EDA complexes the donor and acceptor molecules are present in an integral ratio, most often 1 : 1. Several metal ions (acceptor) notably Ag^+ form stable complexes (which are often solids) with olefins, dienes or aromatic rings (donors) as shown (Scheme 1.19).

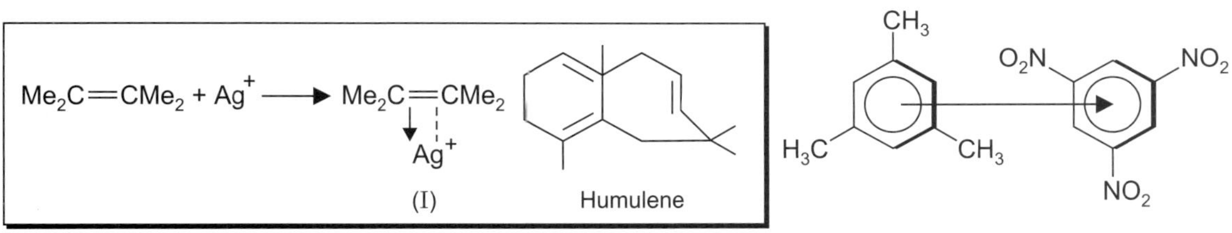

SCHEME 1.19

In the case of an olefin-silver ion complex of the type (I, Scheme 1.20) there are two bonds between the metal ion and the olefin. One of these is a σ bond which is formed by overlap of the filled π orbital of the olefin with the empty 5*s* orbital of the silver ion, while the other is a π bond formed by overlap of a filled 4*d* orbital of the silver ion and an empty antibonding π* orbital of the olefin. The bond is not from the silver ion to one atom but to the whole π center. Consequently, some electron density is transferred from the olefin to the metal ion. In several cases olefins are isolated and purified through their metal complexes. Thus the sesquiterpene humulene (Scheme 1.19) is purified through its solid adduct with $AgNO_3$, from which humulene is regenerated by steam distillation.

Silica gel, impregnated with $AgNO_3$ displays a highly selective adsorption properties regarding the geometry, degree of substitution and the number of double bonds in a compound. The impregnated adsorbent is highly useful for the sharp separation of such compounds.

1, 3, 5-Trinitrobenzene and other poly-nitro compounds form complexes with a variety of organic compounds including aromatic hydrocarbons, aromatic amines, olefins etc.

(Scheme 1.19). The complex with picric acid *e.g.*, a complex between picric acid (acceptor) and 1, 3, 5-trimethylbenzene (donor) is called a picrate. Picrates are solids with definite melting points and are used as derivatives of the compounds.

The bonding in these compounds is more difficult to explain. The difficulty is that although the donor has a pair of electrons to contribute (both *n* donors and π donors) the acceptor does not have a vacant orbital. One may explain the bonding in these complexes results from attractive forces between electron-rich and electron-poor substance (attraction of the dipole induced dipole type). The designation, charge transfer complex, orginates from a resonance description where the structure of the complex receives contribution from resonance forms involving transfer of an electron from the donor to the acceptor molecule. The name π complex is often used since normally atleast one component of the complex has a π electron system.

Metallocenes, *e.g.,* ferrocene (see, Scheme 2.41) and benzenechromium tricarbonyl (see Scheme 9.21) provide further examples of this class.

In complexes where the acceptor is iodine or bromine, the acceptor molecule accepts electrons from both *n* and π donors, probably by expansion of the outer shell to hold ten electrons. It is because of this complex formation that iodine does not have its normal purple colour in solvents like acetone or benzene. As an evidence for charge transfer complex formation, the iodine-benzene complex has a dipole moment, while iodine and benzene are individually non-polar.

B. Crown Ether Complexes and Cryptates

Crown ethers are large ring polyethers. The compounds are cyclic polymers of ethylene glycol, $(OCH_2CH_2)_n$, and are named in the form *x*-crown-*y*, where *x* is the total number of atoms in the ring and *y* is the number of oxygens. An example is [18]-crown-6, the cyclic hexamer of ethylene glycol (Scheme 1.20), Crown ethers are used as *cheleting agents* (from Greek *Chele*, "Crab's claw") and are important for their property of forming complexes with positive ions. Normally the cations are held tightly in the center of the cavity and each crown ether binds different ions depending on the size of the cavity. In each case the "cavity size" is a good match for the ionic diameter of the cation.

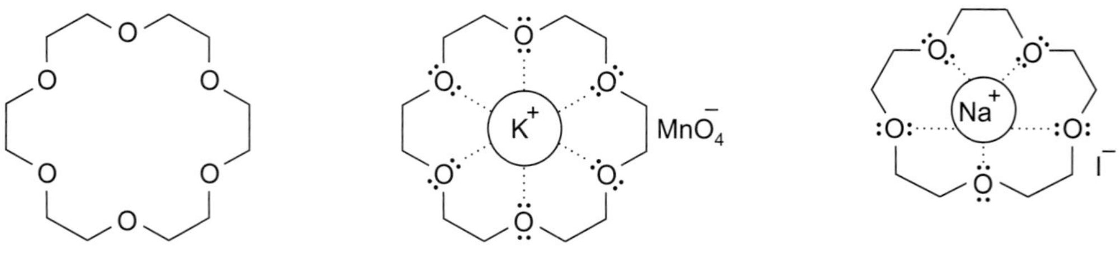

[18]-crown-6   [18]-crown-6 *chelates K*$^+$ [15]-crown-5 *chelates Na*$^+$

SCHEME 1.20

Thus 18-crown-6 binds K^+ (ionic diameter, 2.66 Å) but not Li^+ (ionic diameter 1.20 Å), similarly 12-crown-4, binds Li^+ but not K^+. Crown ethers have use in separating mixtures of cations and much use in organic synthesis. Potassium permanganate with deep violet colour is completely insoluble solid in benzene, it however, readily dissolves in benzene if 18-crown-6 is added to give a violet coloured solution. This solution is useful because it allows oxidations with $KMnO_4$ to be carried out in organic solvents. The solubility of $KMnO_4$ in benzene in the presence of [18]-crown-6 to give pink benzene is due to the fact that the six oxygens in 18-crown-6 are ideally situated to solvate a potassium cation. In the resulting complex the cation is solvated by the polar oxygens, but the exterior has hydrocarbon properties. As a result, the complexed ion is soluble in non-polar solvents (for their role in nucleophilic substitutions see Scheme, 5.17).

Crown ethers not only bind simple metal ions, but they can also bind ammonium and alkyl ammonium cations. This binding involves three N^+—H----O hydrogen bonds as well as electrostatic interactions (Scheme 1.20*a*). Thus crown ethers can act as receptor molecules for ammonium and alkyl ammonium cations as well. [18] Crown-6 forms a strong complex however, no such opportunity is made available by a smaller crown ether and therefore, in (II) there is a mismatch for the formation of three N^+—H----O hydrogen bonds.

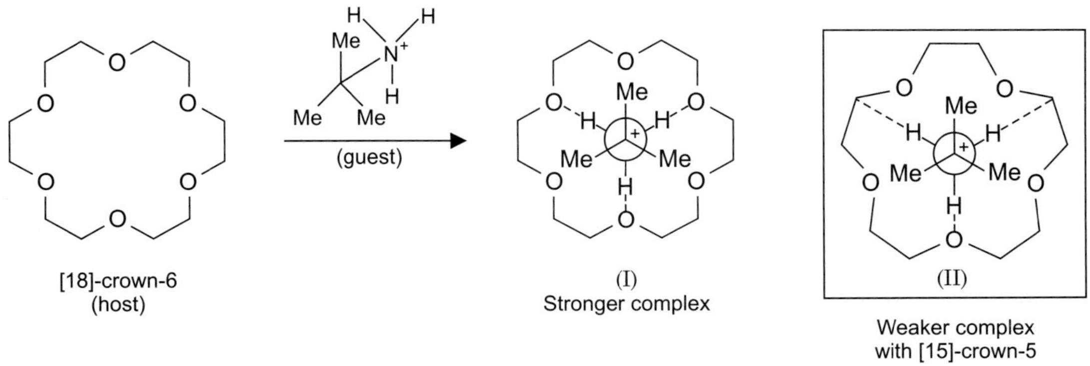

SCHEME 1.20a

The bonding of ammonium and alkyl ammonium cations by crown ethers gives a perching complex. Azacrowns are even better at binding ammonium ions because the nitrogen atoms are more basic than oxygen atoms. Another example shows a crown ether with a pyridine ring in the core structure (Scheme 1.20*b*). The complex (I, Scheme 1.20*a*) has to be viewed on the same lines as the perching complex (Scheme 1.20*b*). One calls this as "perching" on the crown ether, whereas K^+ can "nest" in it.

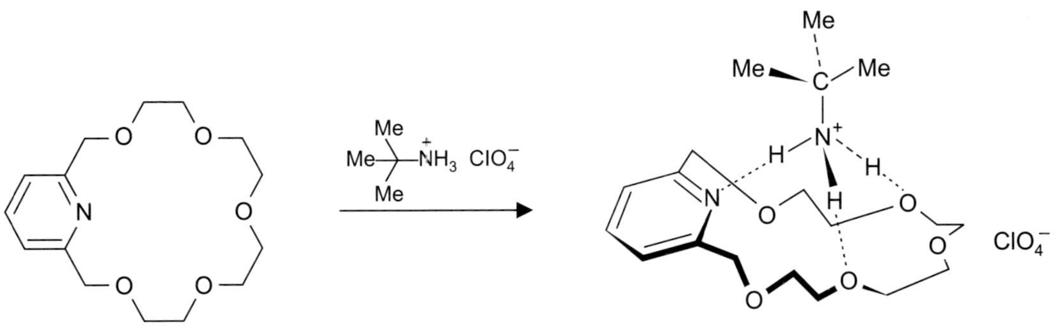

SCHEME 1.20b

The chiral crown ether (Scheme 1.20*c*) is capable of enantioselective recognition of chiral ammonium salts. This represents an excellent application of the separation (resolution) of chiral pharmaceuticals. A pharmaceutical industry has to often resolve two enantiomers of a chiral drug which display completely different activities.

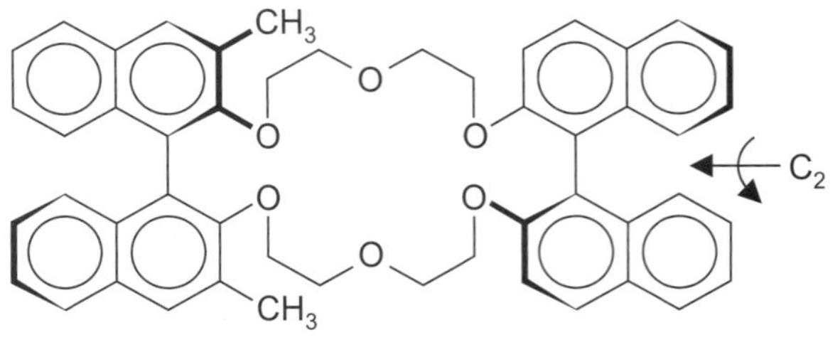

A chiral (*R*, *R*)-binaphthyl crown ether

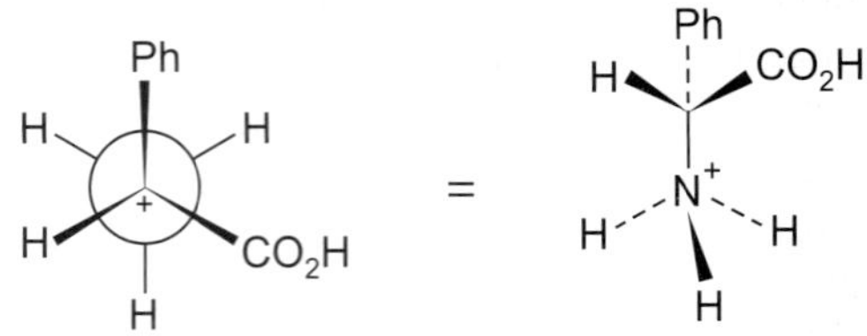

D-enantiomer of phenylglycine cation

SCHEME 1.20c

The *R*, *R* receptor preferentially binds salts of D-α amino acids and esters and the complexes can be extracted by $CDCl_3$ from an aqueous (D_2O) solution. The D-amino acid salts fit better in the chiral circular cavity than their enantiomers. In every case it was found that D-enantiomer was selectively complexed. The chiral recognition between the enantiomers of *e.g.*, PhCH(COOH) NH_3^+ is a consequence of steric interactions between the bulky substrate substituents with the methyl groups of the chiral crown ether (guest). Thus the selective binding is due to the fact that in the L-enantiomer of the phenylglycine cation the COOH group will come into close proximity to maximize steric interactions with CH_3 substituent of the host (Scheme 1.20*d*).

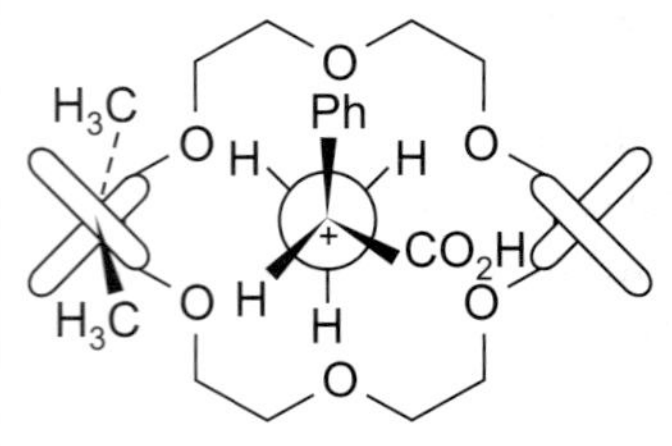

R, *R*-binaphthyl crown ether Scheme 1.20c with *D*-phenylglycinate cation.

SCHEME 1.20d

Cryptands are cage like bicyclic molecules which are like crown ethers except that these contain an additional "strap". Crown ethers are essentially two-dimensional molecules while cryptands are their three dimensional versions which like crown ethers can incorporate positive ions into the roughly spherical cavity within the cage. The crown ethers as well as cryptands

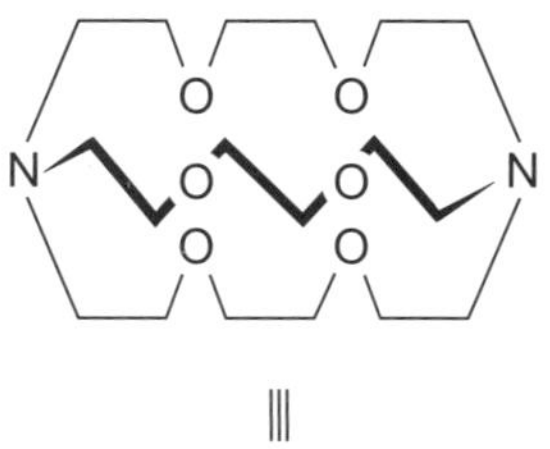

[18]-crown-6 is a floppy molecule which becomes rigid on binding a K^+ ion, thus some entropy is lost on binding. The three dimensional cryptand is more rigid than the crown ether leading to less entropy loss on binding.

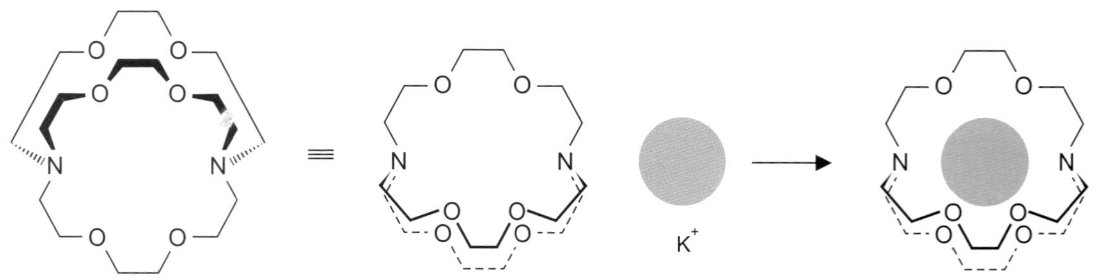

a [2.2.2] cryptand

a cryptate

SCHEME 1.20e

reflect on new techniques to mimic metalloprotein core chemistry. These are named depending on the number of oxygen atoms in each nitrogen-nitrogen linker (Scheme 1.20*e*) and [2.2.2] cryptand is the most important. The binding of K^+ by [2.2.2] cryptand in methanol *e.g.,* is some 10^4 times stronger than its crown analogue [18] crown-6. The [2.2.2] cryptand is based on a similar sized ring to [18]-crown-6 and thus this cryptand shows selectivity for K^+ over the other/alkali metal ions. This is largely due to the more flexible crown ether (entropy) than the rigid cryptand as well as the three dimensional nature of the cavity in the later.

A cyclodextrin consists of six, seven, eight, or more D-glucose units joined through 1,4-*alpha* linkages in a way so as to form a ring—a chain bracelet each link of which is a pyranose hexagon. These rings are doughnut-shaped, somewhat like crown ethers but with a number of important differences. The smallest of them, α-cyclodextrin, has a diameter about twice that of 18-crown-6, and its hole (4.5 Å across) is also about twice as broad. The molecule may be viewed as a pail without a bottom (Scheme 1.20*f*). The faces of a cyclodextrin are termed as primary and secondary faces, the primary face is the narrow end of the "pail like" structure and comprises the CH_2OH groups. The interior of the cavity of this arrangement is non-polar and can bind different guests. Hydrogen bonds or charge stabilization from a large number of OH groups are likely factors which contribute to the binding strength of the guest and may alter the selectivity. The following points may be noted:

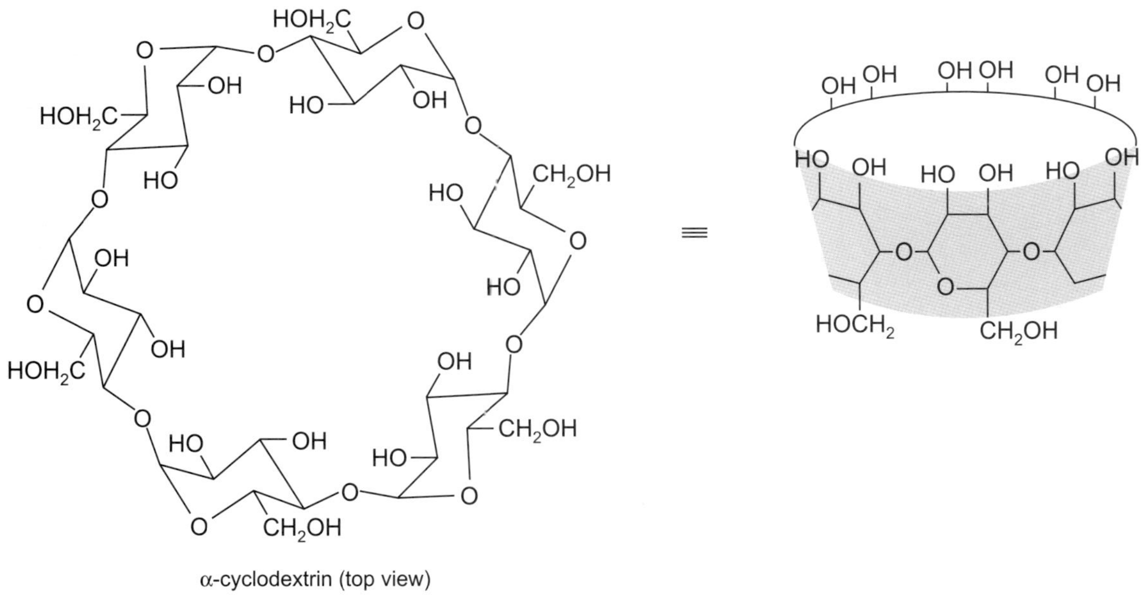

SCHEME 1.20f

- Just like a crown ether a cyclodextrin behaves as a host to a variety of guest molecules and it was with cyclodextrins, the host-guest relationship was initially recognized.
- A cyclodextrin has a polar hydrophobic outside and a relatively non-polar lipophilic inside.
- The lipophilic interior of a cyclodextrin holds a guest but not an ion and the guest can be a non-polar organic molecule or the non-polar part of an organic molecule.
- The hydrophilic exterior confers water solubility on the resulting complex. How well a guest molecule is accommodated depends upon its size and polarity, and the size of the particular cyclodextrin.

- A cyclodextrin is thus capable of hiding a part of the guest molecule in the cavity and expose other parts for reactivity.
- For all practical purposes a cyclodextrin can be drawn as in (Scheme 1.20*g*).

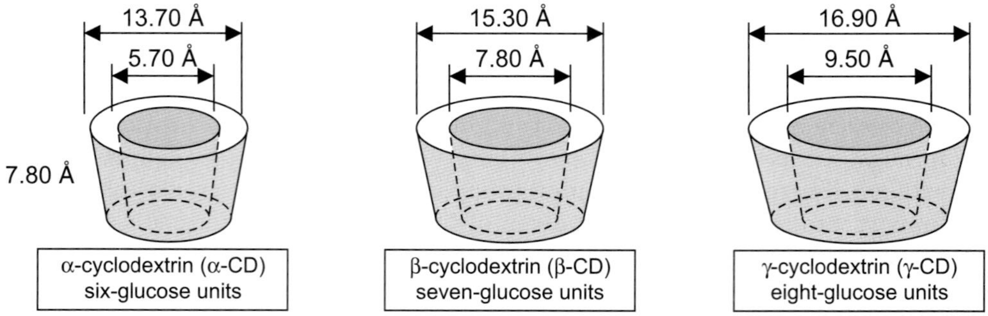

SCHEME 1.20g

p-Nitrophenol esters undergo hydrolysis at strikingly increased rates in the presence of β-CD.

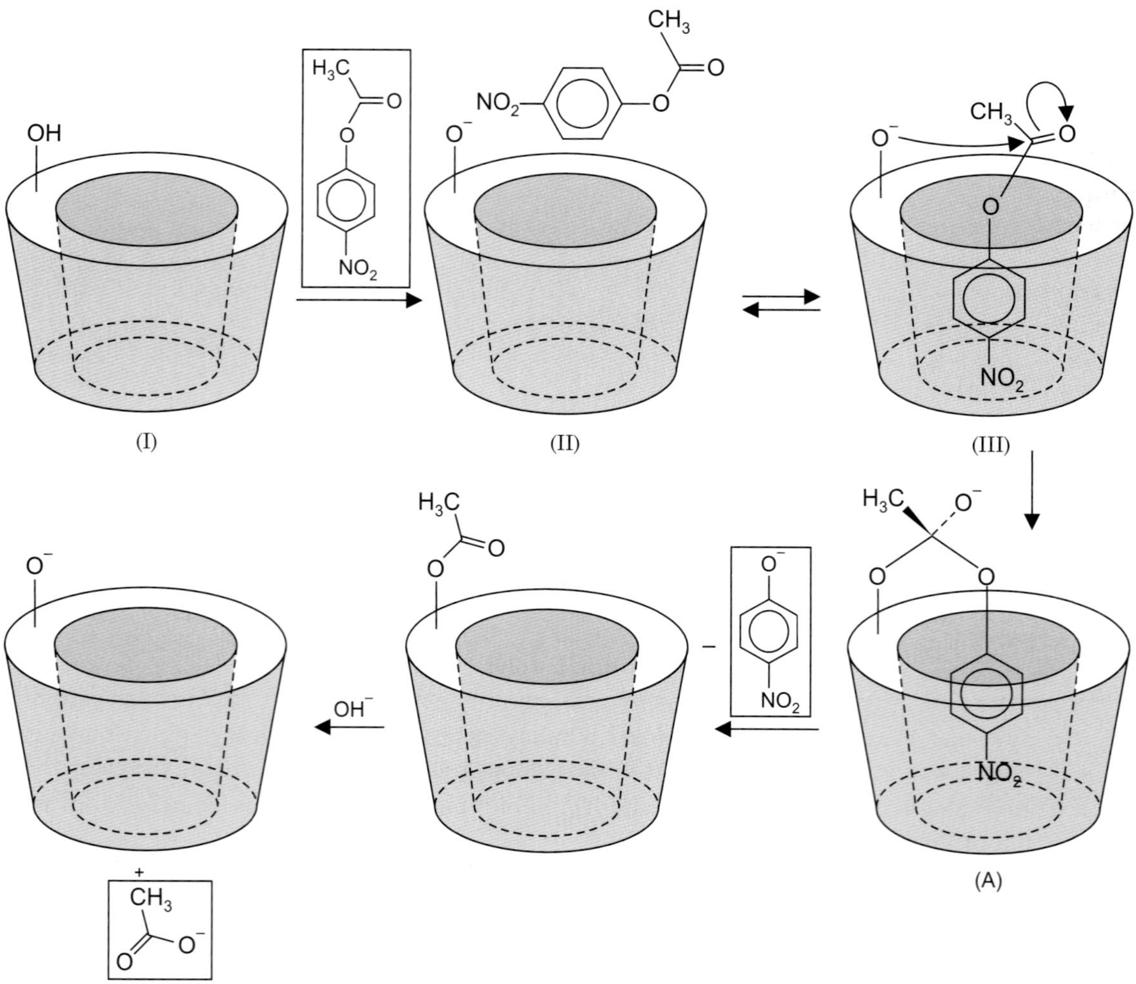

SCHEME 1.20h

When one of the OH groups of cyclodextrin is deprotonated it yields a nucleophilic functional group (II, Scheme 1.20*h*). This nucleophile is in close proximity to the hydrophobic cavity which binds a substrate. Thus the situation is similar to a substrate in the active site of the enzyme undergoing catalytic activity by the suitably placed amino acid residues. The mechanism of hydrolysis of *p*-nitrophenyl acetate is given (Scheme 1.20*h*) which follows the usual mechanistic pathways and involves the formation of a tetrahedral intermediate (A, Scheme 1.20*h*).

Selective aromatic substitution has been brought about by using α-cyclodextrin. When *e.g.,* anisole (10^{-4} M) in water is treated with HOCl (10^{-2} M) and an excess of α-cyclodextrin at room temperature, 96% chlorination was observed at the *p*-position of anisole ring. The hydroxyl group of α-cyclodextrin converted into hypochlorite brings about the chlorination of the exposed *para* position, while the other positions are protected by the cavity of cyclodextrin where the anisole molecule fits (Scheme 1.20*i*). This is an example of noncovalent catalyst, when the host provides the cavity for the reaction to take place without the involvement of a covalent intermediate.

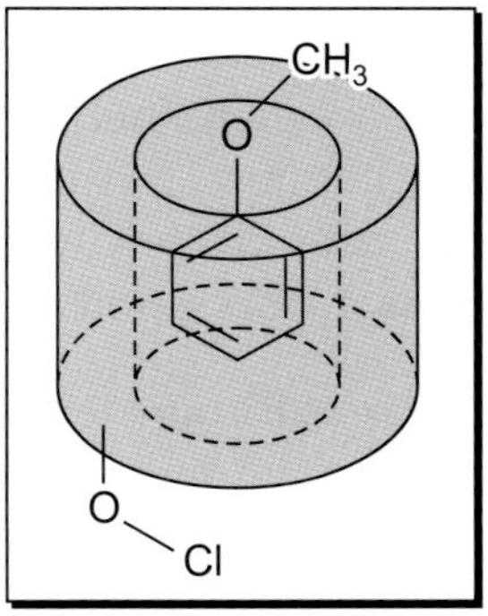

SCHEME 1.20i

C. Inclusion Compounds

Inclusion complexes are formed when the *host* compound is capable to form a crystal lattice in which there are spaces (of the shape of long tunnels or channels) which are of suitable size for the second compound (the *guest*) to fit. There is no bonding between the compounds acting as the host and guest but for the van der Waals forces. For the successful formation of an inclusion compound, the guest molecule must fit into the space properly. In case the guest molecule is either too large or too small, will not go into the lattice and the addition compound will not form.

Urea is an important host molecule for inclusion compound formation. Although ordinary crystalline urea is tetragonal, however when a suitable guest molecule is present it crystallizes in a hexagonal lattice with the guest trapped in long channels. The diameter of the channel is about 5 Å, and the shape and size of the molecule determines if or not it can be a guest (no chemical or any electronic effects are involved), *e.g.,* octane and 1-bromooctane form complexes with urea while 2-bromooctane, 2-methylheptane and 2-methyloctane are not suitable guests. The size and shape of the guest thus determines its entrance in the channels where it has to remain firmly in position.

Oleic acid (Scheme 1.21) gives *erythro* 9, 10-dihydroxy steric acid (m.p. 132°) on *syn* hydroxylation. The *threo* isomer has (m.p. 95°). The configurational assignment to these diols was made on the basis of formation of a urea inclusion complex. The low melting *threo* isomer forms a urea complex, while the *erythro* isomer does not. When one writes the staggered conformation of the two isomers it is the *erythro* diol in which the hydroxyl groups are on opposite side of the chain. The presence of these protruding groups prevents entrance of the *erythro* diol into the urea channel.

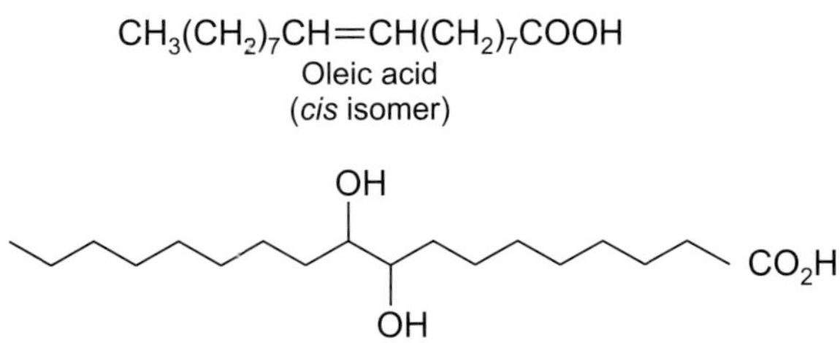

SCHEME 1.21

D. Clathrate Compounds

Inclusion compounds have spaces in the form of long tunnels or channels, whereas in clathrates (cage compounds) the spaces are completely enclosed. In sharp contrast to the inclusion compounds, the crystal lattices in clathrates can exist partially empty. Three molecules of hydroquinone (Scheme 1.22) *e.g.,* when held together by hydrogen bonding result in a cage structure where the guest molecule has to fit in. The guests in this case are methanol (not ethanol), SO_2, CO_2 and argon (not neon). Water is another host. Normally six molecules of water form a cage where guests *e.g.,* chlorine and methyl iodide can fit. These clathrates are solids at low temperatures, and decompose at room temperature.

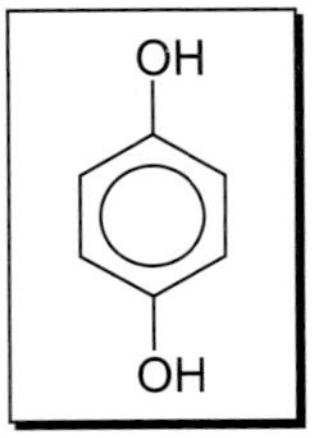

SCHEME 1.22

E. Catenanes and Rotaxanes

A catenane (I, Scheme 1.23, $n \geq 18$) is a compound consisting of two interlocking rings, arranged like two links of a chain. Its unusual feature being the absence of any covalent bonds holding the two rings together.

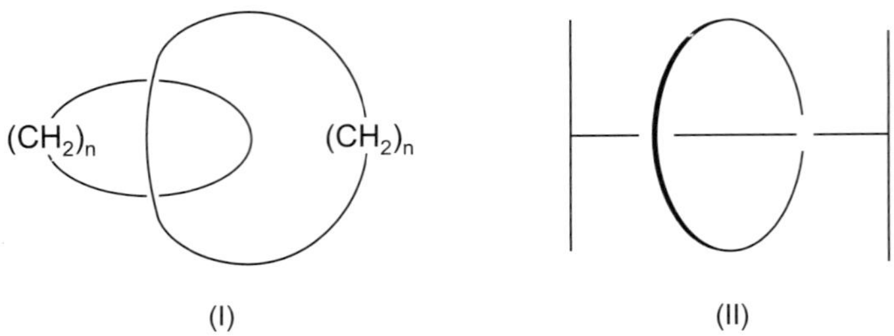

SCHEME 1.23

In a rotaxane (II, Scheme 1.23) a linear portion is threaded through a ring, and cannot get out due to bulky end groups.

1.6 BOND DISSOCIATION ENERGY

When atoms combine to form molecules, energy is released as covalent bonds are formed *e.g.,* hydrogen atoms combine to form hydrogen molecules, this reaction is exothermic and evolves 104 kcal of heat for every mole of hydrogen formed. For covalent bond cleavage energy must be supplied. Reactions in which only bond breaking occurs are always endothermic. The energy required to break the covalent bonds of hydrogen homolytically exactly equals, the energy evolved when the separate atoms combine to form molecules. The energies required to break covalent bonds homolytically are called homolytic bond dissociation energies which are generally designated by the symbol $DH°$. These energies (Table 1.2) vary widely, from weak (I—I 36 kcal/mol) to very strong bonds (H—F 136 kcal/mol).

Bonds made by overlapping orbitals which are closely matched both in energy and size are stronger than those not meeting these criteria. Consider the strength of bonds between hydrogen and halogens which decrease in the order F > Cl > Br > I, since the *p* orbital of the halogen which contributes to bonding becomes more larger and more diffuse with each element, the degree of overlap with the relatively smaller 1*s* orbital on hydrogen diminishes along the series (Table 1.2).

Table 1.2: Single-bond homolytic dissociation energies *DH*° kcal/mol at 25°C

$A : B \rightarrow A^{\cdot} + B^{\cdot}$		*DH*° = Homolytic bond dissociation energy or D(A – B)	
H—H	104	CH_3—H	104
D—D	106	C_2H_5—H	98
F—F	38	$CH_3CH_2CH_2$—H	98
Cl—Cl	58	$(CH_3)_2CHCH_2$—H	98
Br—Br	46	$(CH_3)_2CH$—H	95
I—I	36	$(CH_3)_3C$—H	93
H—F	136	CH_3—CH_3	90
H—Cl	103	C_2H_5—CH_3	86
H—Br	87.5		
H—I	71		

The C—H bond in an alkane has a strength of about 98 kcal/mol, while a C—C bond is relatively weaker as seen for the central bond in ethane, *DH*° = 90 kcal/mol (Table 1.1). In cyclopropane *e.g.*, the strain introduced by the reduction of the ideal tetrahedral angle of 109.5° to 60° is shown by unusual reactivity of the C—C bonds (hydrogenation to propane, bromination to give 1, 3 dibromopropane) and bond dissociation energy (Scheme 1.24) which is only 65 kcal/mol as compared to 90 kcal/mol for ethane.

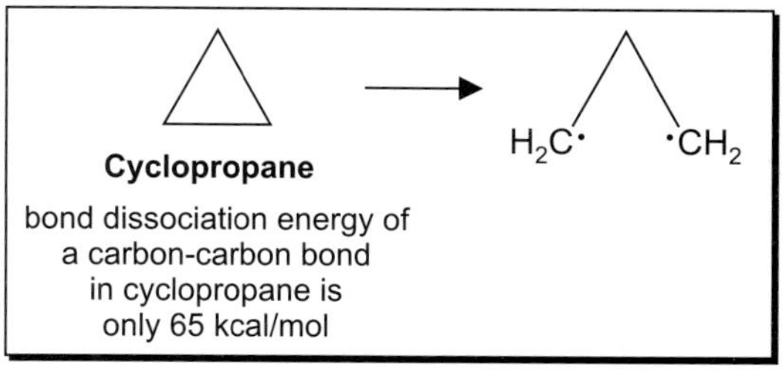

SCHEME 1.24

In alkanes *e.g.*, the *DH*° depends on the character of the radical products. The *DH*° for dissociation of a terminal C—H of an alkane is always around 98 kcal/mol. The product of such cleavage being a primary alkyl radical. On the other hand a C—H bond at a branch point is the weakest type of C—H bond (*DH*° 93 kcal/mol). This fission gives a tertiary free radical. The relative stability of alkyl radical is:

tertiary > secondary > primary > methyl.

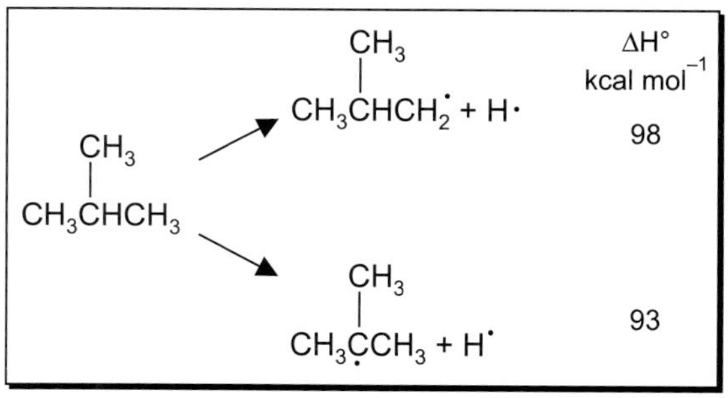

SCHEME 1.25

Bond dissociation energies reflect on the stability of alkyl free radicals. If one studies the fission of two types of C—H bonds in 2-methylpropane (Scheme 1.25), one of the products ($H^{\cdot}$) is the same in each case. The difference in ΔH° (ΔH° is the enthalpy of a bond dissociation reaction) of these reations provides a direct measure of the difference in stability of two alkyl radicals. Thus, *t*-butyl radical is more stable than the isobutyl radical by 5 kcal/mol.

The bond dissociation energy of an allylic and a benzylic C—H bond is indeed low (around 85 kcal/mol) and this information is helpful in the study of the mechanism of free radical allylic substitution (problem 2.3, Scheme 2.19).

Heterolytic bond dissociation energies (AB $\rightarrow$ A^+ + $B:^-$) are also known (H—H, 401 kcal/mol; H—F, 370 kcal/mol). On heterolysis of a neutral molecule a positive and a negative ions are formed. Separation of oppositely charged particles needs energy (~ 100 kcal/mol). In gas phase therefore, bond dissociation normally occurs via an easier route *i.e.,* homolysis. In an ionizing solvent, heterolysis is the preferred pathway for fission.

1.7 THE HAMMETT EQUATION—LINEAR FREE ENERGY RELATIONSHIP

For the study of reaction mechanism, one has to collect several evidences which point to the mechanism of a reaction. An important concept is that a given structural feature will effect related reactions in generally the same way. Thus if replacement of a hydrogen by chlorine makes acetic acid to become a stronger acid, then introduction of a chlorine at the α-position of propanoic acid will also result in an increase in its acidity. A linear free energy relationship is simply a quantitation of this concept. The first and most important linear free energy relationship is the Hammett σρ equation which is based on the acidities of aromatic carboxylic acids.

Acidities of benzoic and phenylacetic acids were measured by changing the substituent group on the aromatic ring. In these experiments the positions of acid-base equilibria were measured as functions of the substituent groups. In case the different acidity values for each series of compounds are only due to the influence of the substituents, then a relation between the sets of data should exist. When the pK_a values obtained from the two series of compounds were plotted against each other, a linear relation (expressed by the mathematical equation, I, Scheme 1.26) was obtained.

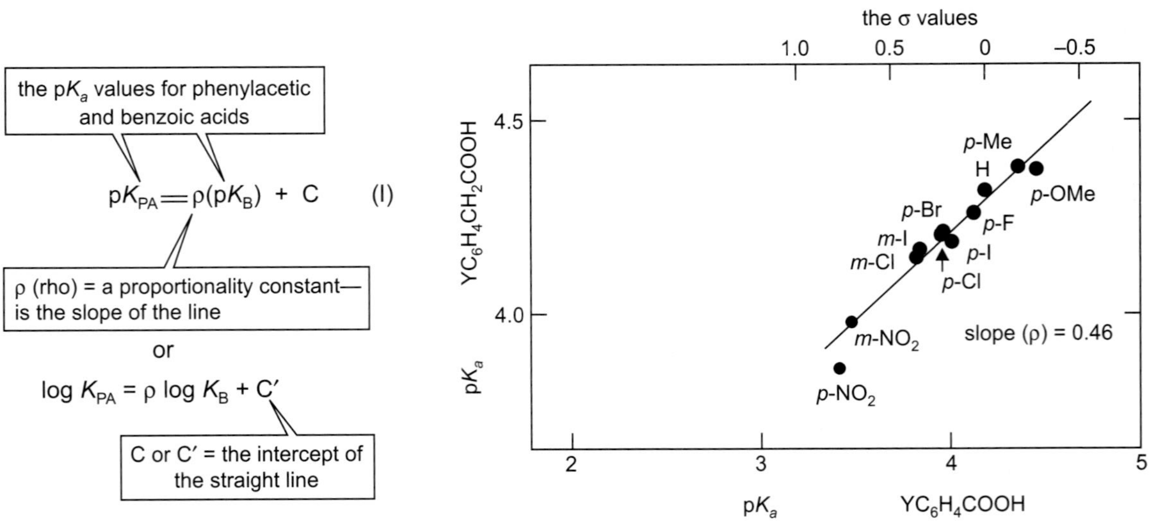

SCHEME 1.26

The acidity values of unsubstituted carboxylic acids (pK_0—substituent = H) are taken as standards with which the effect of a substituent is compared and the equation (II, Scheme 1.27) is for this standard. An expression (III, Scheme 1.27) is obtained on subtracting one equation from the other, and this expression relates the substituent effects on the two series of compounds (the K and K_0 represent the equilibrium constants for substituted and unsubstituted compounds.

$$\log K_{0PA} = \rho \log K_{0B} + C' \quad \text{(II)}$$

$$\boxed{\log \frac{K_{PA}}{K_{0PA}} = \rho \log \frac{K_B}{K_{0B}}} \quad \text{(III)}$$

SCHEME 1.27

sigma the substituent constant

$$\sigma = \log \frac{K_B}{K_{0B}} \quad \text{(IV)}$$

SCHEME 1.28

$$\log \frac{K_{PA}}{K_{0PA}} = \sigma\rho \quad \text{(V)}$$

Hammett equation

SCHEME 1.29

The acidities of different benzoic acids in aqueous media (25°C) are the standard measure of the effect of each substituent group and the substituent constant sigma (σ) for every substituent group is defined in the expression (IV, Scheme 1.28). The Hammett equation—the linear relation is thus expressed as in equation (V, Scheme 1.29). Sigma (σ), the substituent constant is a measure of the effect of a substituent on the acidity of benzoic acid. Those substituents which enhance the acidity relative to unsubstituted benzoic acid will show positive values ($\sigma > 0$), the hydrogen atom having a sigma value of zero. The Hammett equation can thus be used:

1. To account for the influence of substituents on molecular reactivity.
2. It explains the influence of polar *meta*- or *para*-substituents on the side chain reactions of benzene derivatives. Table 1.3 summarizes σ values for different *meta* and *para* substituents. The σ constant is generally independent of the nature of the reaction and is a quantitative measure of the polar effects in a given reaction by a *m*- or *p*-substituent relative to hydrogen. A negative σ value signifies an electron donating group whereas a positive value of σ signifies an electron attracting group. The larger the magnitude of σ, the greater is the effect of the substituent. The ρ constant (ρ is the reaction constant) is dependent on the nature of the reaction and on conditions. It is a measure of the sensitivity of a given reaction series to the polar effect of ring substituents *i.e.,* to the changes in the σ values of the substituent.

Table 1.3: Hammett substituent constant values of common groups

Substituent	σ_m	σ_r
NH_2	– 0.16	– 0.66
CH_3	– 0.07	– 0.17
OH	0.12	– 0.37
C_6H_5	0.06	– 0.01
OCH_3	0.12	– 0.27
SCH_3	0.15	0.0
F	0.34	0.06
I	0.35	0.18
Cl	0.37	0.23
Br	0.39	0.23
CF_3	0.43	0.54
CN	0.56	0.66
NO_2	0.71	0.78

Scheme 1.26 is a Hammett plot with ρ = 0.46. The value of ρ indicates the sensitivity of a reaction or an equilibrium to a particular substituents. A positive ρ-value shows that the reaction or equilibrium is aided by electron attracting substituents (withdrawal of electrons from the reaction site). In the case of phenyl acetic acids, for ionization, the ρ of + 0.46 points that a given electron attracting substituent facilitates ionization but has only 0.46 of the effect that the same substituent has in facilitating ionization of benzoic acid.

3. Reactions that are assisted by high electron density at the reaction site have negative ρ values.
4. The Hammett equation is however, not applicable to the influence of *ortho* substituents, since these exert steric effects.

When one considers the rates of hydrolysis of substituted benzoates with hydroxide ion (in aqueous acetone), one observes a straight line on plotting against the Hammett σ-constants with a slope (ρ) of 2.23. These data show that the substituent groups which facilitate ionization of benzoic acid, facilitate the hydrolysis of benzoate as well. For ester hydrolysis the transition state (Scheme 1.30, the reaction involves nucleophilic attack by the hydroxide ion on the carbon atom of the carbonyl group), has considerable negative charge, since positive ρ indicates stabilisation by electron attracting group. Indeed, in keeping with this observation the mechanism of ester hydrolysis which proceeds through an anionic tetrahedral intermediate gets support. Moreover, this hydrolysis will be further facilitated when electron withdrawing group (σ is positive, Table 1.3) is attached to the aromatic ring. With an electron releasing group (σ is negative) the reaction will be retarted.

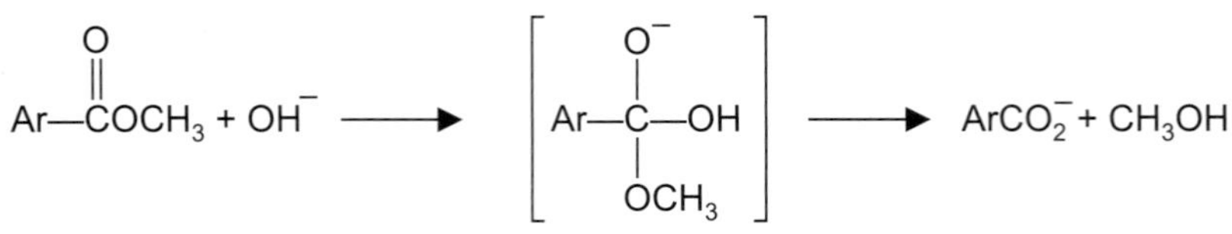

SCHEME 1.30

Hydration of styrenes ($HClO_4$, 25°C) shows a $-\rho$ value, to show that the transition state of the reaction is like a carbocation intermediate.

1.8 TAFT EQUATION

Taft equation is yet another linear free energy relationship which represents a structure-reactivity equation and correlates only field effects (inductive effects). The Hammett equation fails with aliphatic compounds (and *o*-substituted benzene derivatives) due to the different conformations the chains can adopt and because substituents may interact sterically with the reaction center.

A large number of aliphatic reaction rates can be correlated by this equation. Taft assumed that the hydrolysis of the esters (RCOOR′) will be subject to the same steric and resonance effects whether the hydrolysis is catalysed by acids or bases. Rate difference would thus be caused only by the field effects of R and R′ in RCOOR′. The transition state for acid-catalyzed hydrolysis (I, Scheme 1.31) has a greater positive charge (destabilized by $-I$ and stabilized by $+I$ substituents) than the starting ester, while the transition state for base-catalyzed hydrolysis (II, Scheme 1.31) has a greater negative charge than the starting ester. Field effects (inductive effects) of substituents X can thus be determined by measuring the rates of acid- and base-catalyzed hydrolysis from a series XCH_2COOR' where R′ is kept same. From these rate constants a new parameter σ^* (the polar substituent constant) which is believed to represent the I effect only, is introduced. This can be evaluated from the equation:

$$\sigma^* = \frac{1}{2.48}\,[\log (k/k_0)_B - \log (k/k_0)_A]$$

where k = rate constant for the hydrolysis of XCH_2COOR

k_0 = rate constant for the hydrolysis of the reference ester CH_3COOR'. The subscripts B and A denote the base and acid catalysed hydrolysis at the same temperature. The factor 2.48 is an arbitrary constant, which is introduced to bring σ^* values on the same approximate scale as the Hammett σ values.

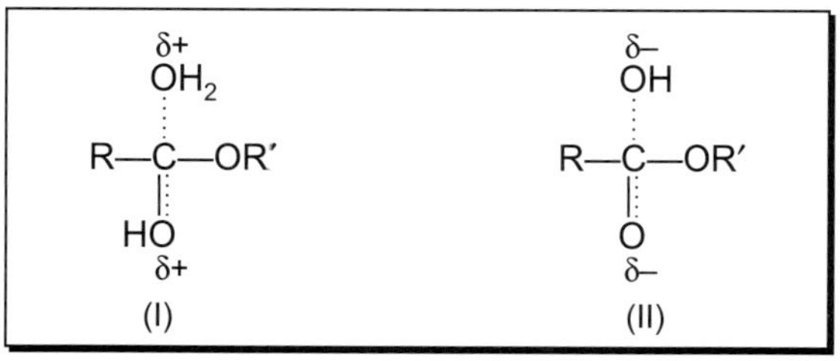

SCHEME 1.31

The data of the reaction can be correlated by the Taft equation:

$$\log \frac{k}{k_0} = \rho^* \sigma^*$$

when ρ^* is the reaction constant.

1.9 STERIC EFFECTS, STRAIN AND BREDT RULE

Strain causes a permanent deformation in the structure of a molecule and this raises its energy when compared to a structure which is not deformed. Steric strain is brought about in a molecule when two or more non-bonded atoms approach close to a position where their electron clouds start to repel each other. Example is of boat deformation of cyclohexane.

Steric effects in a molecule arise due to the presence of bulky groups near the reaction site. A decrease in the reaction rate due to this steric hindrance is a result of entirely a physical blockage to the attack of the reagent. However, in some cases the reaction may be much faster than expected on the basis of electrical effects alone. Lastly a reaction may take a different course due to steric effects.

When one considers nitration of toluene ($HNO_3 + Ac_2O$) at 0°C, *o*-, *m*- and *p*-nitrotoluene is formed in the ratios (expressed as percentages 61.5 : 1.5 : 37 (Scheme 1.32). The total reactivity of toluene in comparison with benzene is 27, and combination of this result with the isomer distribution of the nitrotoluenes gives the following data for the relative reactivities of each nuclear position in toluene compared with one position in benzene: *o*, 50; *m*, 1.3; *p*, 60. (This measure of the reactivity of a particular nuclear carbon in a given reaction is referred to as the *partial rate factor*). These data show the directive and activating effect of a substituent, CH_3, of +*I* type. Other alkyl-benzenes also give predominantly the *ortho* and *para* derivatives, as shown by the partial rate factors for the nitration in acetic anhydride (Scheme 1.33). A significant point is a sharp fall in the *ortho*; *para* ratio as the size of the alkyl group is increased and is a consequence of steric hindrance to *ortho*-substitution.

Substitution at carbonyl groups is very sensitive to steric hindrance. Thus tertiary acids such as $(CH_3)_3$ C—CO_2H, unlike primary and secondary acids, cannot be esterified by an alcohol in the presence of acid, nor can their esters be hydrolyzed by treatment with hydroxide ion or aqueous acid.

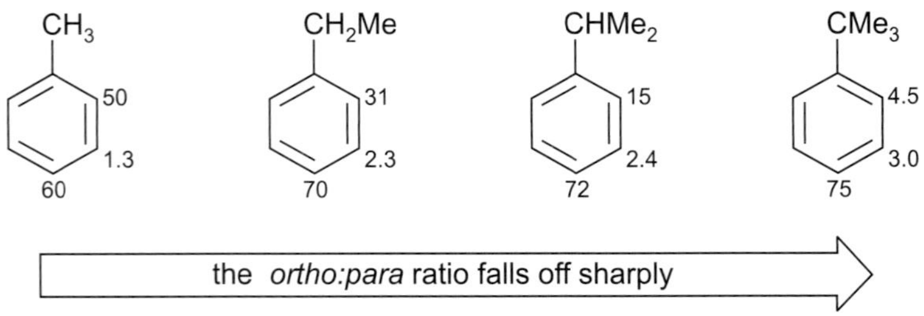

SCHEME 1.32

Alcohols are synthesized by reacting aldehydes and ketones with Grignards reagents. In the case of sterically congested cases when either the ketone or the Grignard reagent have bulky groups then enolization or reduction may be dominating reactions and the normal addition reaction is retarded the following points may be noted:

- Methyl magnesium bromide reacts with diisopropyl ketone to afford the corresponding tertiary alcohol in high yield (Scheme 1.33), however, with the bulkier reagents *e.g.*, isopropyl and *t*-butyl Grignards reagents the same ketone fails to react.

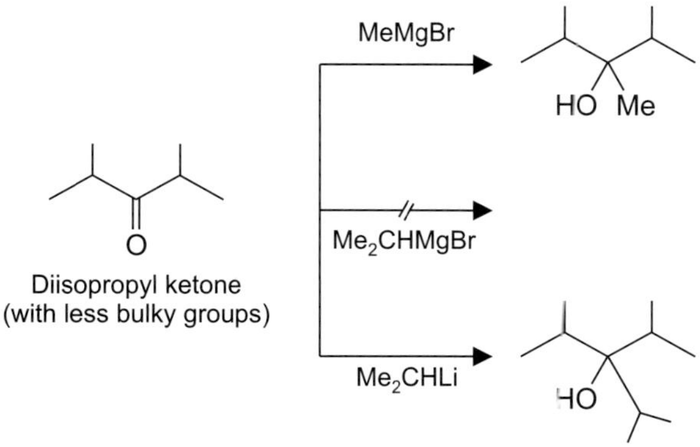

> *A Grignard reagent e.g., CH_3MgI does not react with ketones with very bulky groups, however, an organolithium undergoes a successful reaction.*

SCHEME 1.33

- Organolithium compounds, however, undergo a successful reaction and enolization and reduction are not observed.
- During enolization in a reaction between methylmagnesium bromide and a sterically hindered ketone which has an α-hydrogen (activated proton) the Grignard reagent acts as a base and not as a nucleophile. The Grignard reagent abstracts an α-proton from the ketone to yield an enolate which on reaction with acid gives the starting ketone (Scheme 1.33*a*).

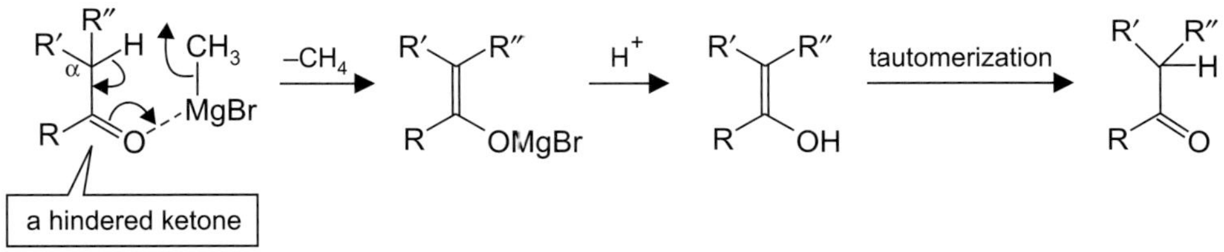

SCHEME 1.33a

- When the Grignards reagent is hindered and has a β-hydrogen, the carbonyl compound is reduced by hydride ion transfer. The reduction occurs with the Grignard reagent coordinated with the carbonyl compound and involves a six-membered cyclic transition state (Scheme 1.34).

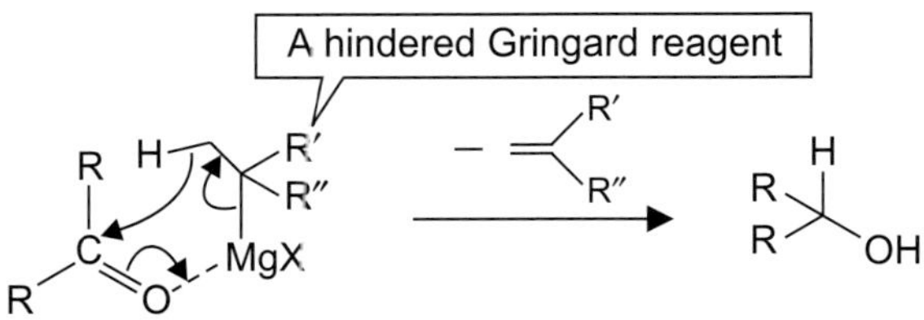

SCHEME 1.34

- Normally predominant 1, 2-addition of a Grignard reagent to a relatively unhindered aldehyde carbon is observed. With increased steric hindrance at the carbonyl carbon an α, β-unsaturated ketone gives both products of 1, 2- as well as 1, 4-addition (see, Scheme 2.14*b*).

When an intermediate carbocation is formed in a reaction, elimination of a proton competes with the addition of a nucleophile. Elimination is favoured when the addition is sterically hindered (Scheme 1.35).

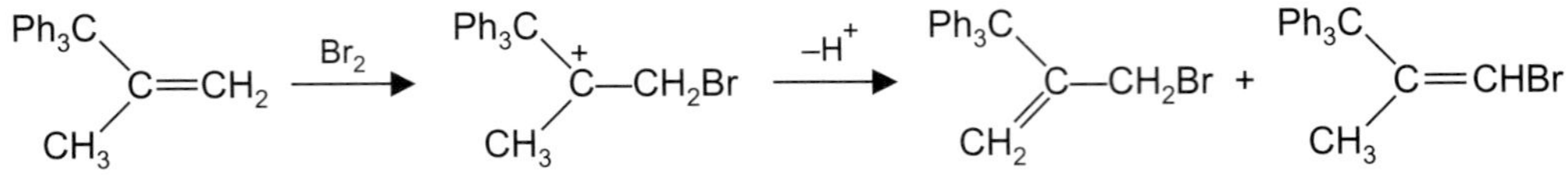

SCHEME 1.35

In the case of a tertiary alkyl halide (I, Scheme 1.36), when one or more alkyl groups are bulky (*e.g.*, *R* = *t*-butyl) these will be pushed together to generate steric hindrance among themselves and therefore, strain (*B* strain, for back strain). This is due to the fact that the central carbon being sp^3-hybridized, has angles of 109.5°. The rate of ionization in such molecules with *B* strain and consequently the solvolysis rate (S_N1 mechanism) is often accelerated. Thus in such compounds on solvolysis the central carbon is converted into a carbocation, the hybridization becomes sp^2 (angle 120°) and therefore, the strain is relieved.

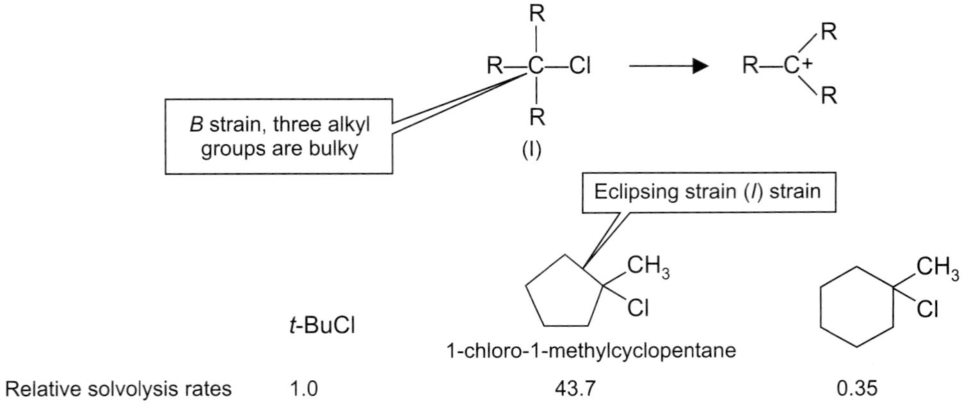

SCHEME 1.36

Several cyclic compounds may suffer from internal strain (*I* strain), a strain which arises from changes in ring strain which results from conversion of a tetrahedral to a trigonal carbon or vice versa. The S_N1 solvolysis of 1-chloro-1-methylcyclopentane (25°C, 80% EtOH) occurs about 44 times faster compared to the reference compound *t*-butyl chloride (Scheme 1.36). On solvolysis of 1-chloro-1-methyl-cyclopentane the eclipsing strain in the molecule is relieved. This enhancement in the rate of solvolysis is not observed in the corresponding cyclohexyl derivative, since it is not subject to eclipsing strain.

Small rings *e.g.*, three and four membered have internal bond angles which are smaller than the preferred tetrahedral angle of 109.5°. Cyclopropane (Scheme 1.36 a) *e.g.*, reflects this strain by high reactivity of its C—C bonds.

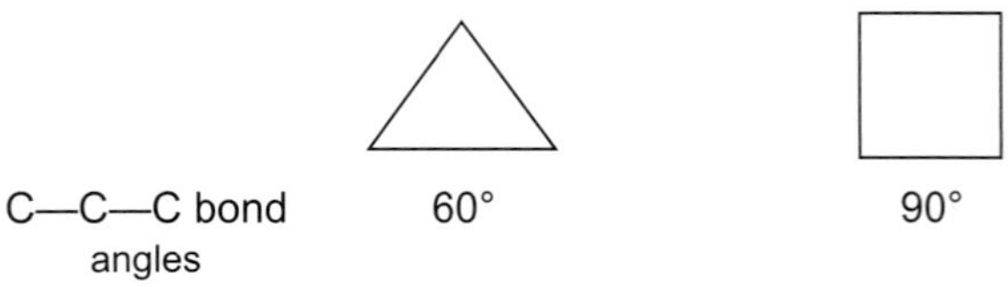

SCHEME 1.36a

The boat conformation of cyclohexane suffers from torsional strain from eclipsed bonds on the side of the molecule and van der Waals strain involving the flagpole hydrogens. This makes boat form less stable than the chair conformation (Scheme 1.36*b*). In a chair conformer a substituent in the axial position makes it less stable than when it is in equatorial position.

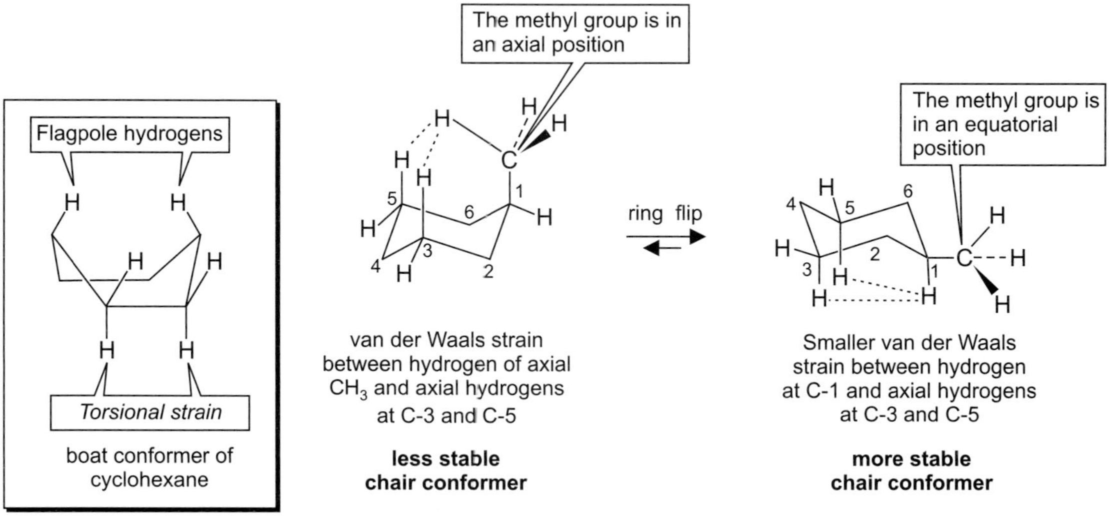

SCHEME 1.36b

Unlike (I, Scheme 1.37), II does not exist, this being a carbon-carbon bridgehead alkene (Bredt rule). Similarly, C—N bridgehead alkenes cannot exist. The system in II cannot exist due to excessive steric strain. This bridgehead carbon cannot flatten out since the rigid cage prevents this. As the bridge gets longer as in (III, Scheme 1.37), flexibility return to the system and such bridgehead alkenes (Anti Bredt alkenes) are capable to exist.

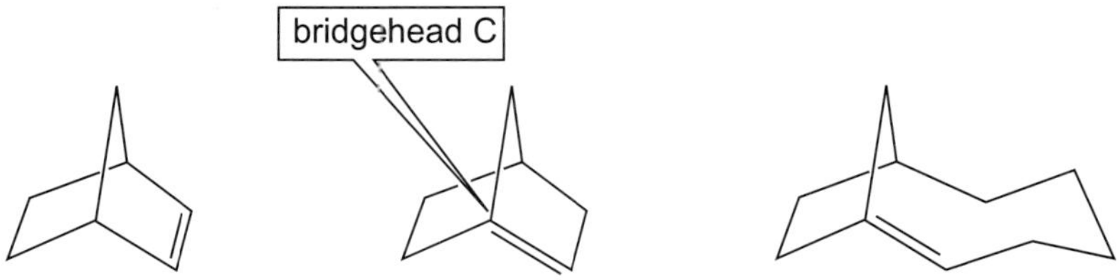

(I) Bicyclo[2.2.1]hept-2-ene (II) Bicyclo[2.2.1]hept-1-ene (III) Bicyclo[5.2.1]dec-1-ene

SCHEME 1.37

β-Keto acids undergo decarboxylation very easily, via the enol form of the ketone (see, Scheme 6.52). The bridgehead β-Keto acids are, however, resistant to decarboxylation, since the product would be a highly strained bridgehead olefin (see, Scheme 6.53).

PROBLEMS

1.1 Why alkanes have very small dipole moments?

1.2 Why cyclic 1, 2-diketones exist mainly in the enolic form?

1.3 What is the usual upper limit of dipole moment of organic molecules?

1.4 The reaction of substituted dimethylanilines with methyl iodide (in aqueous acetone) gives the trimethylanilinium iodide with $\rho = -3.30$. Explain the result.

1.5 What is the sign of ρ expected for the following reaction?

$$ArO^{-} + EtI \xrightarrow[EtOH]{OH^{-}} [Ar\overset{\delta-}{O}\cdots Et\cdots \overset{\delta-}{I}] \longrightarrow ArOEt + I^{-}$$

(S_N2) reaction

1.6 Predict the sign of ρ expected for the following reactions:

(*i*) ionization of benzoic acids (water), 25°C.

(*ii*) ionization of benzoic acids (40% aq. EtOH) 25°C.

(*iii*) ionization of phenols (water) 25°C.

ANSWERS TO THE PROBLEMS

1.1 Due to the small difference in electronegativities of carbon and hydrogen.

1.2 The main driving force for enolization is relief of the electrostatic repulsion that occurs when the two electrophilic carbonyl groups are adjacent to each other.

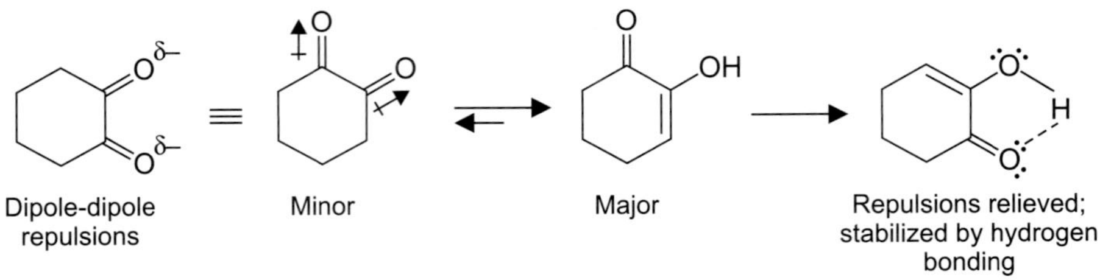

1.3 7D

1.4 The reaction proceeds through the usual S_N2 transition state. The electron donating substituents help in stabilizing the developing positive charge close to the ring and lead to a negative ρ value.

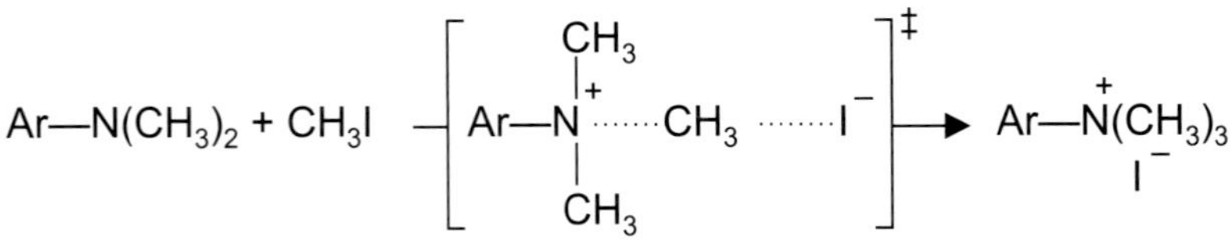

S_N2 Transition state

1.5 A negative ρ value. (actual value, $\rho - 0.99$). Thus electron-releasing groups will increase the rate and electron withdrawing group will decrease the rate. In this reaction ArO^- is the nucleophile, the more the charge on the oxygen atom, the faster will be the reaction. Thus electron release will increase the charge on the oxygen atom, whereas electron withdrawing group will decrease the charge on oxygen atom.

1.6 All are expected to show positive ρ.

CHAPTER 2

Delocalized Chemical Bonding

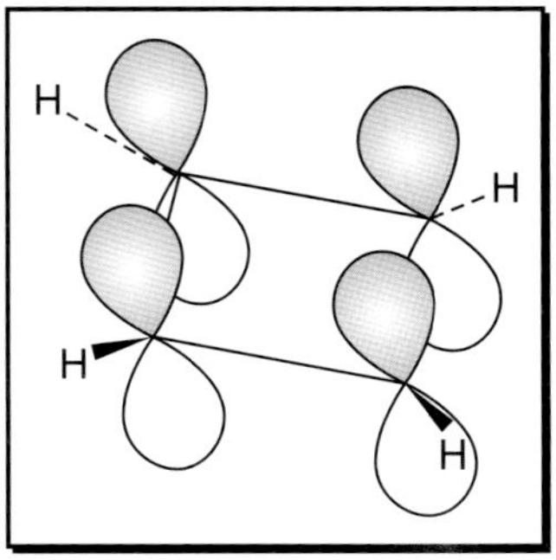

The bonding in some compounds can be adequately described by a single Lewis structure (a problem with Lewis structure is that these impose an artificial location on the electrons). However, in several other compounds a Lewis structure does not represent a correct representation for a molecule or an ion. These compounds contain one or more bonding orbitals which are not restricted to two atoms, but which are spread out over three or more. This type of bonding is said to be delocalized. Consider a conjugated molecule 1, 3-butadiene, *e.g.*,

2.1 1, 3-BUTADIENE A TYPICAL CONJUGATED SYSTEM

The heat of hydrogenation of a terminal alkene is about –30 kcal/mol as in the case of 1-butene. A compound with two non-interacting (*i.e.*, separated by one or more saturated carbon atoms) terminal double bonds should be about twice this value (– 60 kcal/mol). This is found to be so in the case of 1, 4-pentadiene $CH_2{=}CH{-}CH_2{-}CH{=}CH_2$ ($\Delta H°$ = –60.5 kcal/mol). However, on hydrogenation a conjugated diene *e.g.*, 1, 3-butadiene generates less energy. The difference, about 3.6 kcal/mol, is a result of stabilizing interaction between the two double bonds and is called resonance energy of 1, 3-butadiene, (Scheme 2.1).

$$\underset{\text{1, 3-butadiene}}{H_2C{=}CH{-}CH{=}CH_2} \xrightarrow{H_2,\ Pt} CH_3{-}CH_2{-}CH_2{-}CH_3 \qquad \Delta H° = -\ 56.6 \text{ kcal}$$

$$\underset{\text{1-butene}}{H_2C{=}CH{-}CH_2{-}CH_3} \xrightarrow{H_2,\ Pt} CH_3{-}CH_2{-}CH_2{-}CH_3 \qquad \Delta H° = -\ 30.1 \text{ kcal}$$

$$\times\ 2 = -\ 60.2 \text{ kcal}$$

Resonance energy of 1, 3-butadiene = 60.2 kcal – 56.6 kcal = 3.6 kcal

SCHEME 2.1

The stability of a conjugated diene depends on two factors. Firstly it is the hybridization of the orbitals forming the single bonds. The single bond in 1, 3-butadiene is formed due to overlap of an sp^2 orbital with another sp^2 orbital, while the single bonds in 1, 4-pentadiene result from the overlap of an sp^3 orbital with an sp^2 orbital (Scheme 2.1*a*). As on average a 2*s* electron is closer to the nucleus than a 2*p* electron, the electrons in an sp^2 orbital (with 33.3% *s* character) are closer to the nucleus than the electrons in an sp^3 orbital (with 25% *s* character). The length of a bond depends on the closeness of the electrons in the bonding orbital to the nucleus. Thus, the more the *s* character in the orbitals forming the σ bond, the shorter is the bond. This means that a σ bond formed by an sp^2-sp^2 overlap is shorter than a σ bond formed by an sp^3-sp^2 overlap. Shorter bonds are more stable, therefore, the molecule is more stable. A change in hybridization also affects the length of a carbon-hydrogen bond but not as much it affects the length of a carbon-carbon bond.

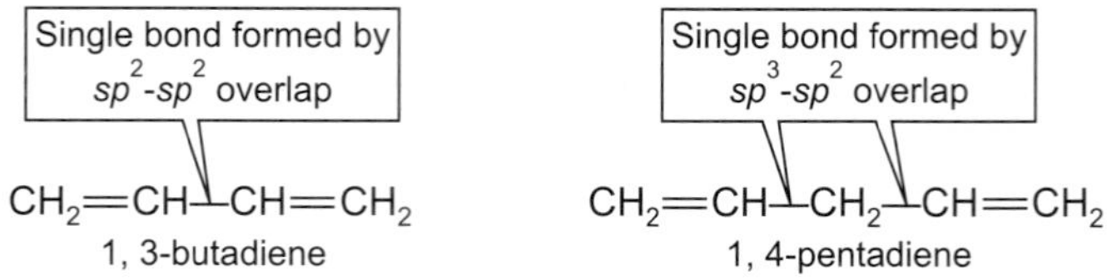

SCHEME 2.1a

Secondly a conjugated diene is more stable than an isolated diene due to resonance which means that the compound had delocalized electrons. The π electrons in a conjugated double bond are not localized between two carbons, but instead over four carbons. As a consequence of resonance the single bond in 1, 3-butadiene is not a pure single bond and has a partial double bond character. This feature contributes to the fact that an sp^2-sp^2 bond is shorter and more stable than an sp^2-sp^3 bond (Scheme 2.1*b*).

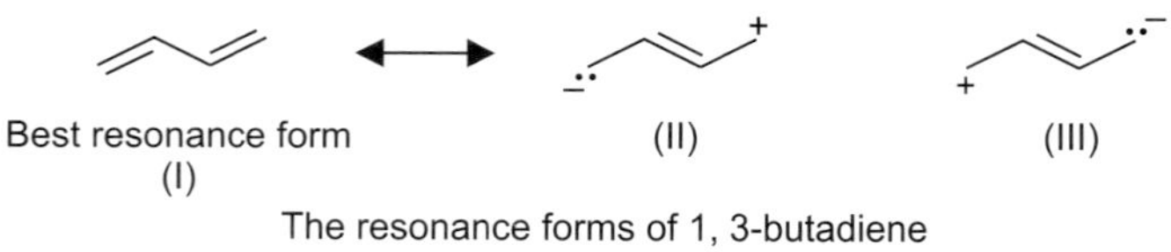

SCHEME 2.1b

The planar conformation of butadiene with the aligned *p* orbitals of the two double bonds allows overlap between the *pi* bonds. Precisely the electrons in the two double bonds are delocalized over the entire molecule to create some *pi* overlap and *pi* bonding in the C2—C3 bond. The length of this bond is intermediate between the normal length of a single bond and that of a double bond. (Scheme 2.1*c*) Lewis structures are not adequate to depict delocalized molecules like 1, 3-butadiene. To represent the bonding in conjugated systems such as 1, 3-butadiene accurately, one must consider molecular orbitals that represent the entire conjugated *pi* system and not just one bond at a time. Apart from adding stability to

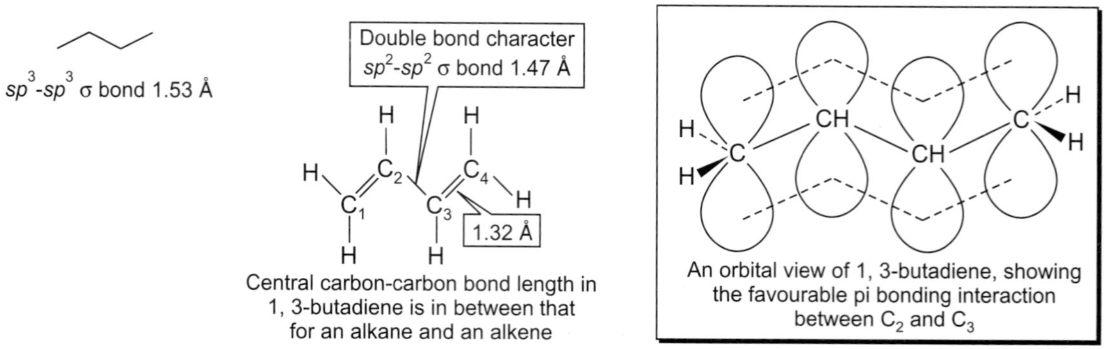

SCHEME 2.1c

1, 3-butadiene this π interactions raises the barrier to rotation about the single bond. Thus, of the two extreme coplanar conformations (Scheme 2.2), the *s-cis* form is about 3 kcal/mol less stable than the *s-trans* conformation.

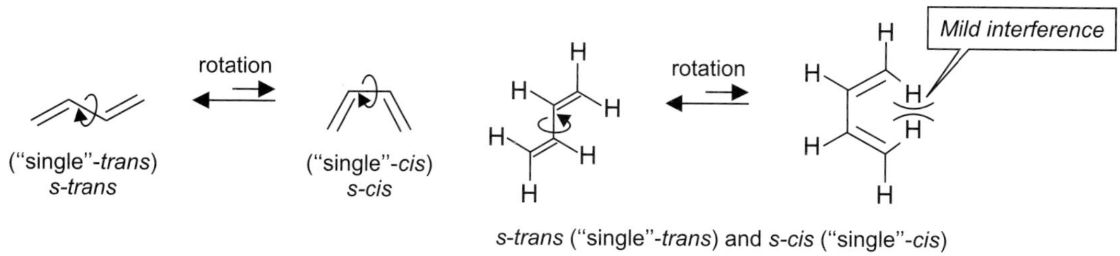

Interconversion of *s-trans* and *s-cis* conformations of 1, 3-butadiene

SCHEME 2.2

2.2 RESONANCE

(A) What is Resonance?

Two or more structures of the same molecule or an ion with same geometry, with the same number of paired electrons, but differing in the pairing arrangement of these electrons are termed resonance structures. These structures are conventionally shown as related to each other by a single double-headed arrow (↔) to emphasize that these structures have no physical reality or independent existence and do not represent different substances in equilibrium. Consider *e.g.*, the carbonate ion (CO_3^{2-}, Scheme 2.3), it has no unique Lewis structure, but one can draw three different but equivalent structures (I–III, Scheme 2.3). These three structures are resonance structures which have the characteristic property of being interconvertible by electron pair movement only and the nuclear positions in the molecule (or an ion) remain unchanged.

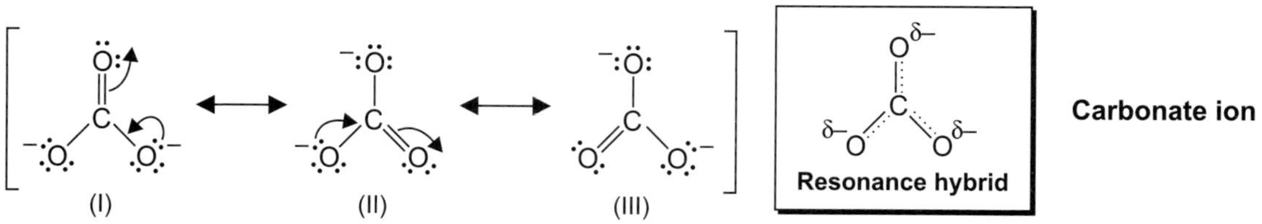

SCHEME 2.3

X-ray studies indicate that carbon-oxygen double bonds are shorter than single bonds and the carbonate ion shows that all of its carbon-oxygen bonds are of equal length. One is not shorter than the others as expected from either of the representations (I–III). Carbonate is perfectly symmetrical with a trigonal central carbon atom with all C—O bonds of equal length (between that of a double and a single bond). The negative charge is evenly distributed over all three oxygens *i.e.*, it is delocalized. Thus none of the Lewis structures written for this molecule is structurally correct. One arrives at a correct description if one creates conceptually an average structure out of all three (a resonance hybrid). Carbonate is thus a resonance hybrid of the three resonance structure I, II and III. Since all three structures are equivalent, they contribute equally to the true structure of the molecule, but none of them accurately represents it.

Resonance structures are thus not structures for an actual molecule or ion; these exist only in theory and thus can never be isolated. No single contributor adequately represents the molecule or ion. In resonance theory one views the carbonate ion, a real entity, as having a structure (Scheme 2.3) that is a hybrid of these three hypothetical resonance structures.

One may, therefore, mentally fashion a hybrid of all the resonance structures or may depict it on the paper (Scheme 2.3) as a non-Lewis structure which attempts to depict the hybrid. Thus *e.g.*, for carbonate ion the bonds are shown by a combination of a solid-dashed lines. This is to show that the bonds are something in between a single bond and a double bond. One also places a δ^- (partial minus) beside each oxygen to show that something less than a full negative charge is present on each oxygen atom.

(B) Resonance Structures of Benzene—Dewar Structures

Several possible resonance structures can be drawn for the molecule of benzene (Scheme 2.4). In fact the real structure is the resonance hybrid. Of these resonance structure. (I and II, Scheme 2.4) are called Kekulè structures *i.e.*, the structures of cyclic conjugated systems represented by conventional alternating single and double bonds. The rest of the resonance structures (III–V) of benzene (or other aromatics) with a single "long" bond between opposite atoms are called Dewar structures. The following points come to light when the wave equation is solved:

1. The energy value obtained by considering that (I and II) participate equally is lower than that for I or II individually.
2. On considering the Dewar structures, the value gets still lower.
3. Each of the Kekulé structures (I and II) contributes 39% to the actual molecule, while others contribute 7.3% each.

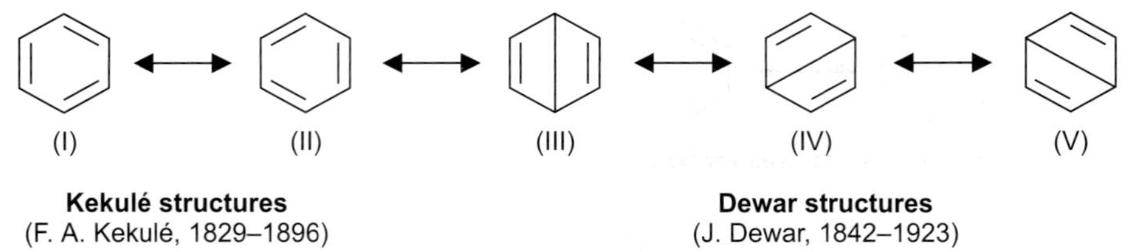

SCHEME 2.4

(C) Canonical Forms—Claus Benzene

In the valence bond method, various possible Lewis structures called canonical forms are drawn and the molecule is taken to be a weighted average of them *i.e.*, the resonance hybrid. Canonical form is the term which is almost, but not quite synonymous with resonance structures. The mathematical requirements (*e.g.*, orthogonality) for a canonical set of structures of a cyclic compound preclude structures with bonds that intersect. The rules for drawing proper resonance forms when applied to the π system of benzene permit not only the three Dewar structures, but also the unfamiliar crossed structure known as Claus benzene (Scheme 2.5). However, such a structure is not part of a canonical set.

Various investigators proposed many incorrect structures for benzene such as Dewar benzene, Claus benzene, Ledenburg prismane, and benzvalene (Scheme 2.5), Dewar benzene, prismane, and benzvalene (but not Claus benzene) have been synthesized. These compounds are unstable and isomerize to benzene via exothermic transformations.

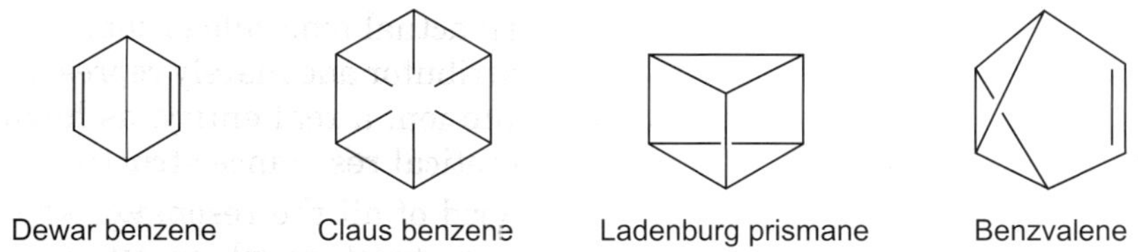

SCHEME 2.5

(D) Writing One Best Structure—Representation of Resonance Hybrid Structure

Benzene is the classic example used to illustrate resonance structures. Benzene is best described by two equivalent cyclohexatrienic resonance structures (I and II, Scheme 2.4). Similarly the three structures of carbonate are equivalent, *i.e.*, indistinguishable. The importance of resonance structures becomes apparent in explaining that each C—C bond in benzene is intermediate between a single and double bond, as is each C—O bond in CO_3^{2-}.

The problem of writing one best structure to position the $p\pi$ electrons in compounds or ions or radicals which may be represented by two or more equivalent structures is frequently solved by using broken lines connecting the atoms over which the $p\pi$ electrons are delocalized. Thus for benzene the single structure, (Scheme 2.6) and for CO_3^{2-} the single structure, (Scheme 2.3) are representations showing π electron distribution. In the case of a cyclic sextet of delocalized π electrons a solid circle inside the ring, as shown for benzene (Scheme 2.6) is used. Originally the circle notation referred to a sextet of electrons but now it has been extended to indicate any monocyclic aromatic system with $(4n + 2)$ π electrons as well as to show the electrons in the rings of polycyclic aromatic compounds.

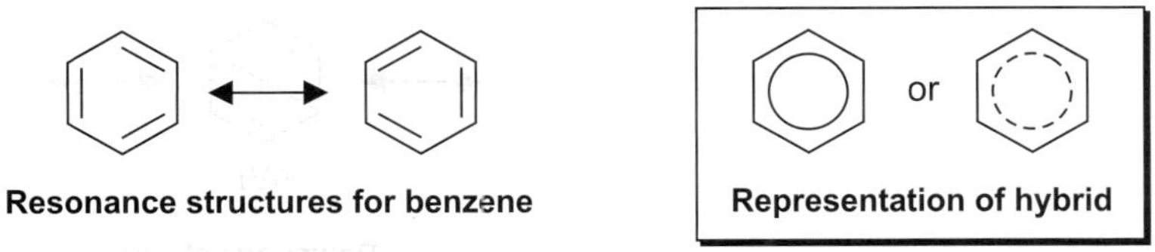

SCHEME 2.6

(E) Nonequivalent Resonance Structures

The resonance structures for the carbonate and the Kekulé structures of benzene are equivalent. However, many molecules can be described by a set of Lewis structures that are not equivalent; three of the resonance structures of butadiene (Scheme 2.1*b*) indicate this. The first structure (I) is a lower energy structure than the other two (II and III) which are charge-separated equivalent species. The structure (I) is said to make a larger contribution to the ground state structure than either of the charged structures. Thus the uncharged structure is the single best representation of butadiene. The fact that, experimentally, butadiene shows some double bond character between atoms 2 and 3 indicates that the charged structures (II and III) also make a contribution to the ground state.

For the enolate ion (Scheme 2.7, this is an oxygen analog of the allyl ion), the two resonance structures differ in the location of double bond and the charge. As in the case of butadiene, here as well, one of the structure is closer to the real one than the other and neither structure represents the actual molecule. One can use several guidelines to deal with such situations. In the two examples discussed above, two of the many guidelines help to pick up a preferred structure. According to the first, the more covalent bonds a structure has, the more stable it is (forming a covalent bond lowers the energy of atoms). Thus structure (I, Scheme 2.1*b*) for 1, 3-butadiene is by far the most stable and makes the largest contribution since it contains one more bond. Moreover, since separating opposite charges requires energy, thus structures in which opposite charges are separated have greater energy. For this reason as well, the structure (I) of butadiene is the most stable. Secondly, charges should be preferentially located on atoms with compatible electronegativity. For the enolate ion, structure (I, Scheme 2.7) is preferred where the negative charge is present on the more electronegative oxygen atom.

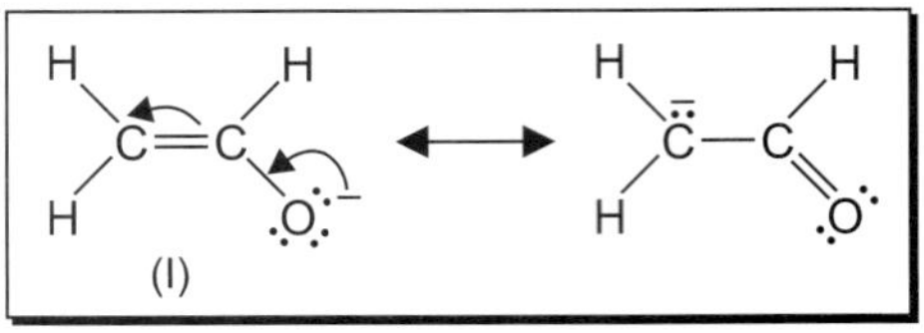

Enolate ion

SCHEME 2.7

Resonance contributors can be obtained by moving π electrons towards a π bond, moving π electrons towards a positive charge or moving a lone pair towards a π bond.

(F) Rules for Resonance-Contribution of a Resonance Structure to the Overall Hybrid

1. Resonance structures (canonical forms) exist only in one's imagination

One useful way to depict the actual structure of a molecule, a radical or an ion containing delocalized bonds is to draw several possible canonical forms. These structures are connected by double-headed arrows, and one assumes that the real molecule, radical or ion is like a hybrid of all of them. The resonance structures *i.e.*, canonical forms have no existence except on paper or in our imagination.

2. Only electron movement is allowed

The nuclei in each of the canonical structures must be in the same relative positions. Thus, structure (II, Scheme 2.8) does not contribute to the structure of isobutylene (I) which is instead the isomeric compound, *trans*-2-butene.

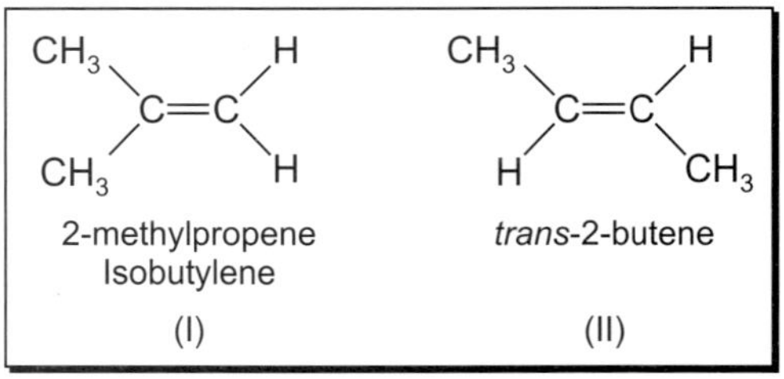

SCHEME 2.8

3. All resonance structures must have the same number of unpaired electrons

Thus, *e.g.*, $\dot{C}H_2$—CH=CH—$\dot{C}H_2$ is not a valid canonical form for butadiene, but is one representation of an electronically excited state of this molecule in which the complete pairing of the ground state is lost due to spin inversion.

4. All the atoms taking part in resonance i.e., the atoms that are a part of delocalized systems must lie in a plane or be nearly planar

Consequently resonance may be prevented or reduced in systems in which the involved atoms are sterically forced out of planarity (steric inhibition of resonance). In picryl iodide the bond lengths (1 and 2, Scheme 2.9) for the *ortho* and *para*-nitro groups are different. This shows that the oxygens of the *para*-nitro group are in resonance with the benzene ring, these being in the same plane. However, the oxygens of the *ortho*-nitro group are not, these being pushed out of the plane by the bulky iodine atom. Steric inhibition of resonance, explains the effect of *ortho*-substituents on the basic character of anilines (See, Scheme 3.35*b*).

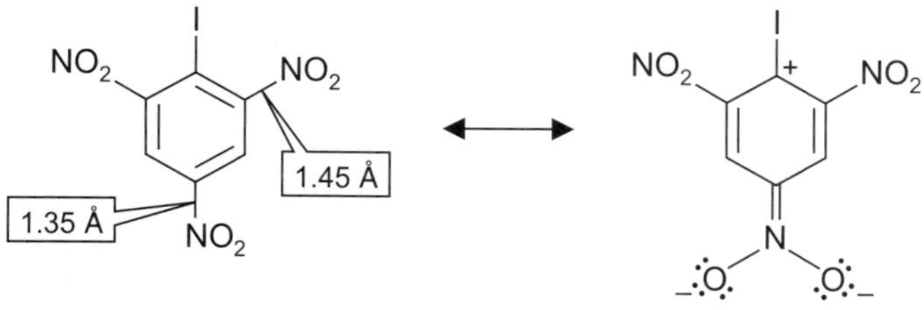

Steric inhibition of resonance

SCHEME 2.9

In several molecules benzene ring is forced out of planarity (I and II, Scheme 2.10). In I, a short bridge forces the benzene ring to become boat shaped. Although the compounds are still aromatic (NMR) but the properties are different from those of ordinary compounds with benzene rings.

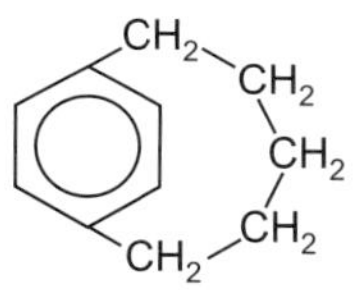

(I)
[5] Paracyclophane

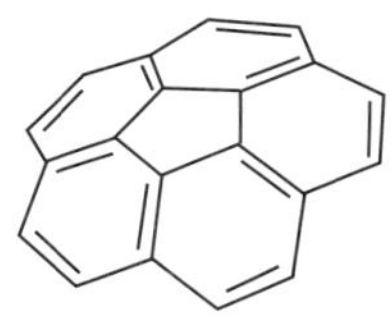

(II)
Corannulene

SCHEME 2.10

5. *The energy of the actual molecule (resonance hybrid) is lower than the energy that might be estimated from any contributing structure*

Allylic radicals, and cations are unusually stable and in Lewis terms, this stabilization is explained by electron delocalization (resonance stabilization, Scheme 2.11). This stability also reflects on the effect of a neighbouring double bond on the reactivity of a carbon center. Thus, even though the allyl cation is a primary carbocation it is more stable than a primary carbocation. In contrast with saturated primary haloalkanes ($CH_3—CH_2—CH_2Cl$), 3-chloropropene ($CH_2=CH—CH_2Cl$) dissociates fast under S_N1 (solvolysis) conditions to undergo rapid unimolecular substitution through a carbocation intermediate *e.g.*, heating with CH_3OH it gives 3-methoxypropene ($CH_2=CH—CH_2—OCH_3$, S_N1 product) through the stable allyl cation intermediate.

The primary C—H bond in propene is relatively weak (weaker than CH bond of ethane) and it is even weaker than a tertiary C—H bond (see, Scheme 1.25) to show the special stability associated with the allyl radical due to resonance.

The pK_a of propene is about 40 and it is more acidic than propane (pK_a = 50) to show that allyl anion formed by deprotonation is a facile process.

In terms of molecular orbitals each of three carbons in allyl system is sp^2 hybridized and bears a p orbital perpendicular to the molecular plane. The three p orbitals overlap to give a symmetric structure with delocalized electrons.

6. *All resonance structures do not contribute equally to the hybrid*

Resonance has significance when the canonical forms are of comparable energy. Thus the two Kekulé structures for benzene (Scheme 2.4) which are equivalent canonical forms contribute equally consequently stabilization energy is considerable. The Dewar structures (Scheme 2.4) which are more energetic are less important, however, their inclusion as the canonical forms results in still greater reduction in the total energy. The total stabilization energy being 150 kJ mol^{-1}. Thus each resonance structure contributes in proportion to its stability.

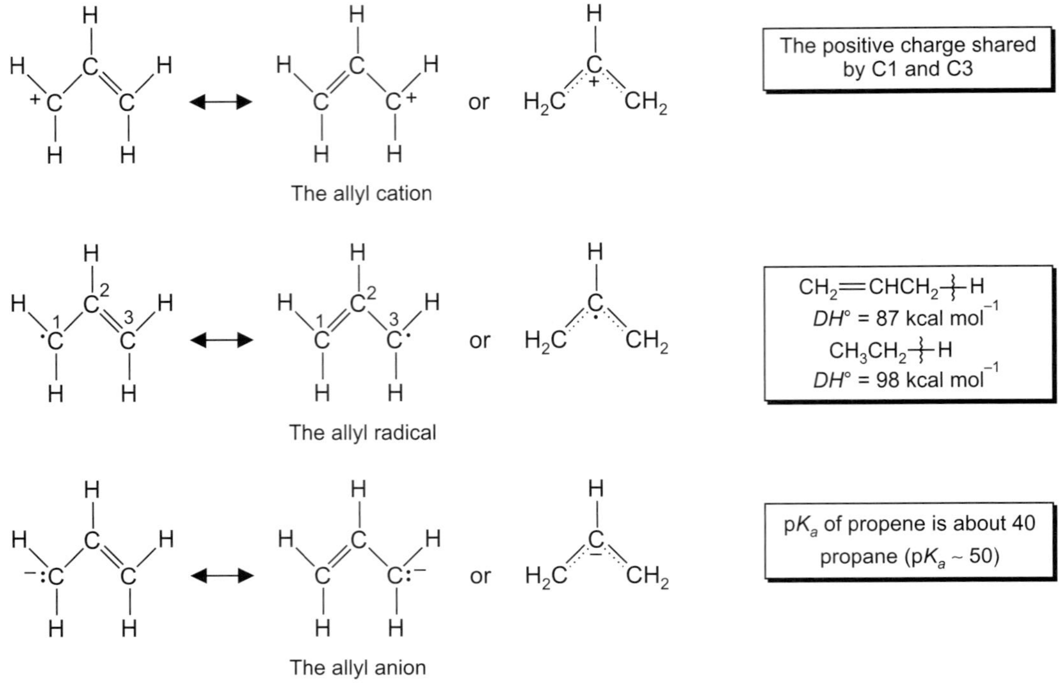

Delocalization in the 2-propenyl (allyl) system

SCHEME 2.11

(G) The Stabilities of Resonance Structures

All canonical forms do not contribute equally to the resonance hybrid, while the others may not contribute at all. For ethylene the high energy form, $\overset{-}{C}H_2$—$\overset{+}{C}H_2$ may be neglected when compared to CH_2=CH_2. Some of the following rules tend to help in deciding the relative stabilities of resonance (imaginary) structures. To find the most important resonance contributor to a given molecule, consider the octet rule, make sure that there is a minimum of charge separation, and place on the relatively more electronegative atoms as much negative and as little positive charge as possible.

1. Structures with a maximum of octets are preferred

This is so in the case of *e.g.*, nitrosyl cation NO^+ (Scheme 2.12). Additionally, the major resonance contributing structure has more covalent bonds than the minor contributor, *cf.* rule 2.

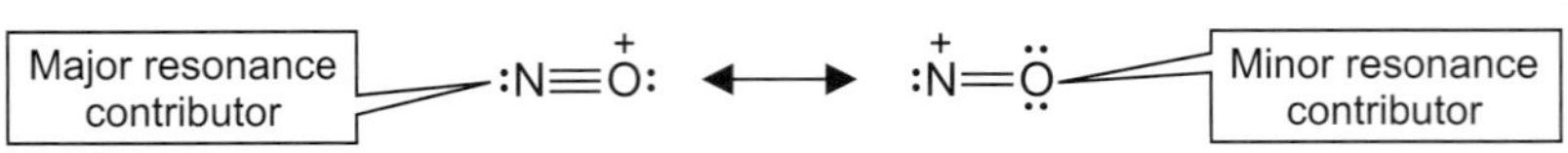

Nitrosyl cation

Structures with a maximum of octets are most important

SCHEME 2.12

2. The more covalent bonds a structure has, the more stable it is

From among the resonance structures of 1, 3-butadiene (Scheme 2.1*b*), the uncharged structure is the single best representation of butadiene.

3. Charge separation decreases stability

Separating opposite charges requires energy, thus resonance structure in which opposite charges are separated have lower stability. Of the three structures for 1, 3-butadiene, the two charge separated structures (II and III, Scheme 2.1*b*) make a smaller contribution to the hybrid.

4. Charge separation may be enforced by the octet rule

In some situations, however, charge separation is acceptable to ensure octet Lewis structures as in carbon monoxide (Scheme 2.13). When several charge-separated resonance structures can be drawn the most favorable is the one in which the charge distribution is compatible with the relative electronegativities of the component atoms. In diazomethane, nitrogen is more electronegative than carbon (Scheme 2.13).

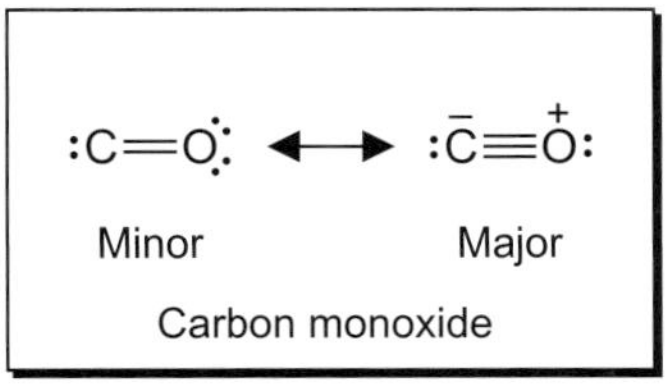

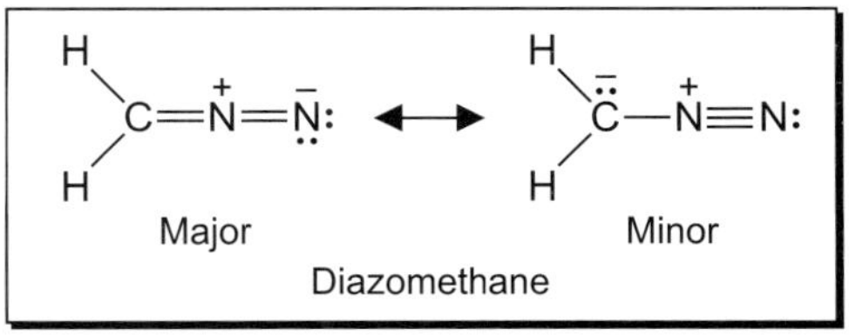

SCHEME 2.13

Thus in summary, the greater the predicted stability of a resonance structure, the more it will contribute to the resonance hybrid.

(H) The Resonance Effect—Mesomeric Effect

The resonance effect (mesomeric effect) involves delocalization of electrons through resonance *via* the π system. This effect involves decrease in electron density at one position with corresponding increase elsewhere. The resonance effect depends upon the overlap of certain orbitals. One may consider the operation of this effect in conjugated compounds with, π-bond-π-bond conjugation and π-bond-*p*-orbital conjugation. In the compounds of the first type, butadiene is the simplest example and has been described (Scheme 2.1*b*). It is a symmetrical molecule and therefore, conjugation does not lead to the appearance of a dipole. Thus contributions from the ionic structures $\overset{+}{C}H_2—CH=CH—\overset{-}{C}H_2$ and $\overset{-}{C}H_2—CH=CH—\overset{+}{C}H_2$ are necessarily equal and consequently their dipoles nullify each other. This however, is not so when π bonds of different types are in conjugation. Consider the α, β-unsaturated compound 3-buten-2-one (Scheme 2.14) in which structure (I) contributes the most to the resonance hybrid. This is due to the octet rule satisfied at every atom and there are no formal charges on this structure. However, structures (II and III, Scheme 2.14) are important contributors to the resonance hybrid, though their contribution is minor. In these structures C-2 and C-4 are electron deficient and this feature reflects on the chemical behaviour of these compounds. One thus finds that a β-carbon of an α, β-unsaturated compound is electrophilic like the carbonyl carbon itself and a nucleophile can bond with either of these two electrophilic carbons to give 1, 2- or a 1, 4-addition (Scheme 2.14, see also Scheme 7.83).

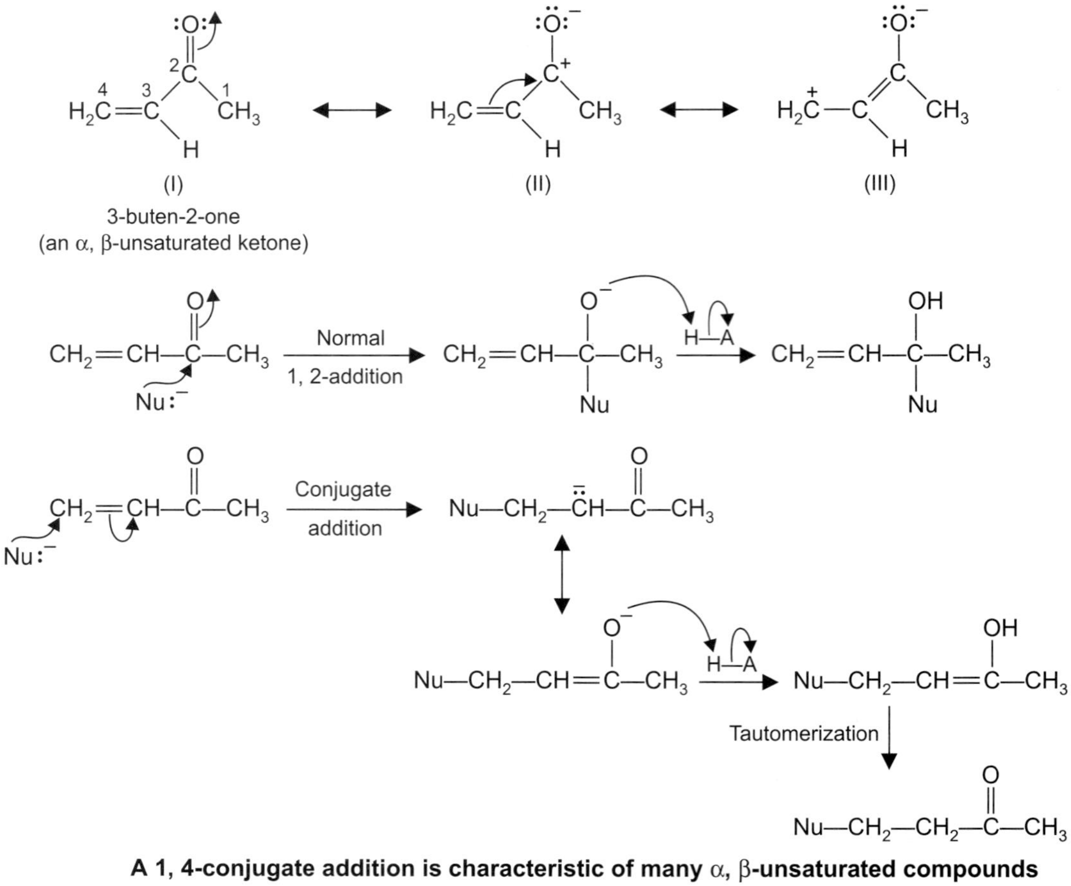

SCHEME 2.14

Excellent yields of 1, 4-addition products can be obtained by reacting an α, β-unsaturated compound with a lithium diorganocuparate reagent, a reagent which is called Gilman reagent with an organic group attached to a copper atom (Scheme 2.14*a*, also see Scheme 7.91).

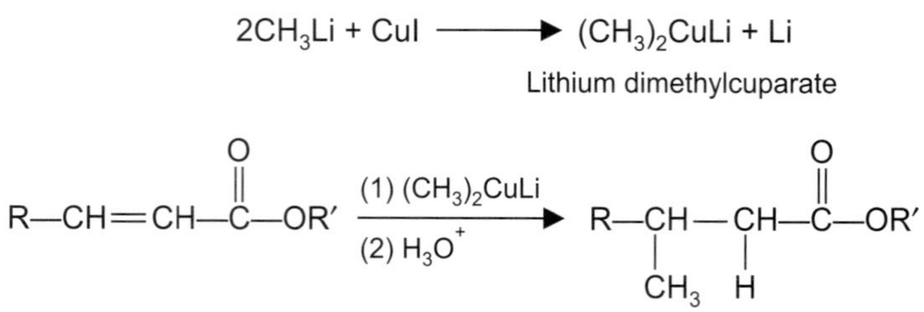

SCHEME 2.14a

When a Grignard reagent adds to an α, β-unsaturated aldehyde, 1, 2-addition generally predominates by the addition of the nucleophile to the unhindered aldehyde carbon (Scheme 2.14*b*), while with increased steric hindrance as shown in the reaction of a Grignard reaction with a ketone leads to both 1, 2- as well as 1, 4-addition.

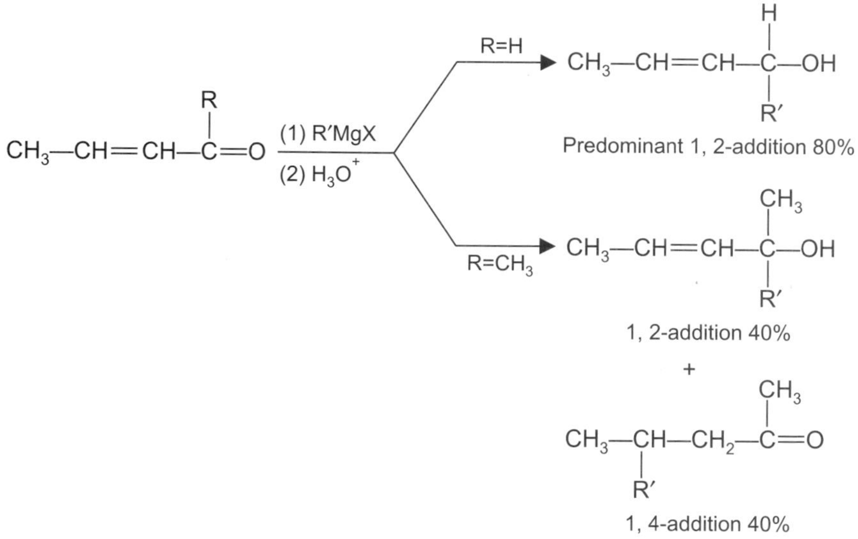

SCHEME 2.14b

Vinyl chloride (Scheme 2.15) provides a good example of π bond-*p* orbital conjugation. In this compound, the *p* orbital on the carbon which is linked to chlorine can overlap both with the *p* orbital on the second carbon atom and also with one of the *p* orbitals on chlorine (Scheme 2.15). As the *p* orbital of chlorine is initially filled, therefore, its participation in a delocalized π system requires a partial removal of electrons from chlorine leading to a dipole moment directed from chlorine towards carbon, which is in opposition to the dipole generated in the C—Cl σ bond because of the –*I* effect of chlorine. The net result is that the dipole moment of vinyl chlorine (1.44 D) is much smaller than that of ethyl chloride (2.0 D) where only the – *I* effect is operative. This donation of electrons by chlorine into a molecular π system is described as a +*M* effect. Using valence bond method, vinyl chloride may be represented as a hybrid of the structures (I and II). Both representations indicate that the C—Cl bond should be shorter than that in a saturated alkyl halide, and this is found to be so. The sp^2 carbon orbital involved is likely to produce a shorter and stronger bond than the ethyl sp^3 orbital. Moreover, an additional factor which leads to a still shorter and stronger bond is π overlap between the π orbital of the double bond and the lone-pair orbital of the halogen (Scheme 2.15).

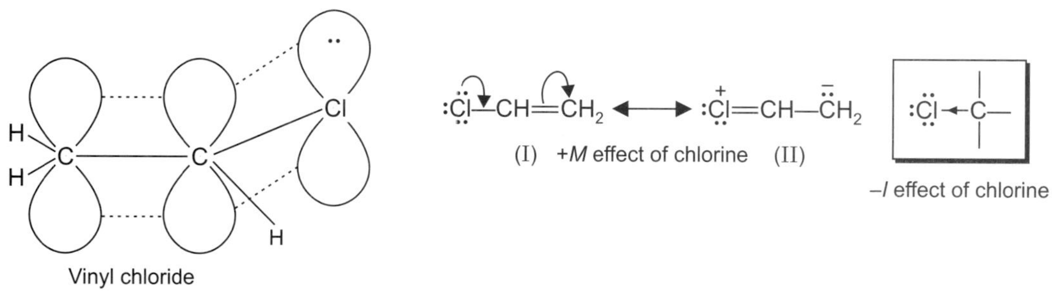

SCHEME 2.15

Resonance effects (*i.e.*, mesomeric effects +*M* or –*M*) of substituents help in explaining directive and rate controlling factors in electrophilic aromatic substitution.

(I) Resonance Energy

The difference in energy between the actual molecule and the Lewis structure (resonance structure) of lowest energy is called resonance energy (stabilization energy). The resonance stabilization arises as a result of the delocalization of the electrons over a conjugated system. Benzene *e.g.*, is found to be more stable than a system in which each of its three double bonds were individually similar to the one present in cyclohexene. This difference has been shown experimentally, *e.g.*, by measuring the heat of hydrogenation of benzene to cyclohexane (208.4 kJ mol^{-1} or 49.8 kcal mol^{-1}) and comparing it with the heat liberated by the hydrogenation of three moles of cyclohexane 361 kJ (3 mol)$^{-1}$ or 86.4 kcal (3 mol)$^{-1}$. The difference in these heats (36.6 kcal) is equivalent to the resonance energy of 1 mole of benzene. Usually the greater number of equivalent structures which can be written for a compound, the greater is its resonance energy.

(J) Resonance and Some Aspects of Chemical Behaviour and Other Properties

Theory of resonance helps greatly in explaining molecular structure and unusual chemical reactivity. Thus declocalization in the allyl system gives rise to the unusual reactivity of allylic bonds. Allylic halides undergo relatively rapid S_N1 and S_N2 reactions, as well as a new type of biomolecular substitution S_N2' (See Scheme 2.20).

Conjugation is reflected in the molecular structure of 1, 3-butadiene, revealing a relatively short central carbon-carbon bond with a small barrier to rotation of about 4 kcal/mole. Moreover, conjugated dienes are electron rich and are attacked by electrophiles to yield intermediate allylic cations on the way to 1, 2- or 1, 4-addition products (Scheme 4.18*a*). Acyclic extended conjugated systems display increasing reactivity since in these, many sites are open to attack by reagents and the ease of formation of delocalized intermediates. The conjugated system of benzene, is however, unusually unreactive because of its cyclic form.

Delocalization of electrons from F to B (Scheme 2.15*a*) helps in reducing the electron deficiency of electrons on B in BF_3 (increased stability) which is known to exist. On the other hand, BH_3 does not exist, since in this case no delocalization can occur. Thus, there is some double bond character in BF_3 and as expected it has a shorter B—F bond than in BF_4^-.

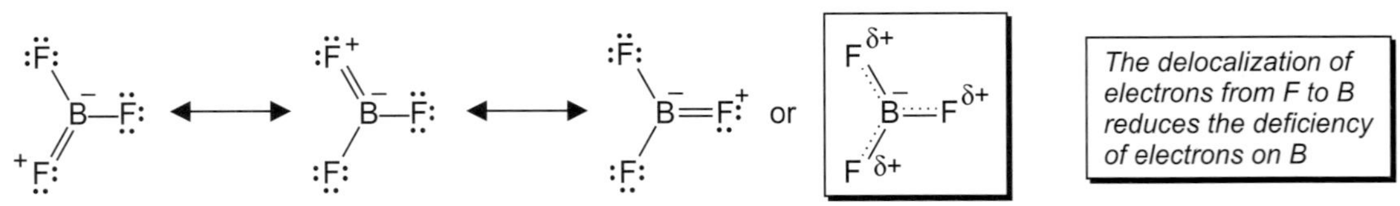

SCHEME 2.15a

The following selected examples are only presented to show the strength of the method of delocalization in explaining structure and reactivity:

1. Resonance and acid strength

The major reason that a carboxylic acid *e.g.*, acetic acid is acidic is due to the resonance stabilization of the carboxylate ion. The two resonance structures of this ion (Scheme 2.16) are equivalent and moreover the negative charge is shared between the two oxygens equally. The resonance stabilization of anion is greater compared to acetic acid. This resonance stabilization is not available to an alkoxide ion (Scheme 2.17). Phenols are also acidic (pK_a = 8–10), due to resonance, the negative charge in the conjugate base called the phenoxide ion is stabilized by delocalization into the ring (see, Scheme 3.20).

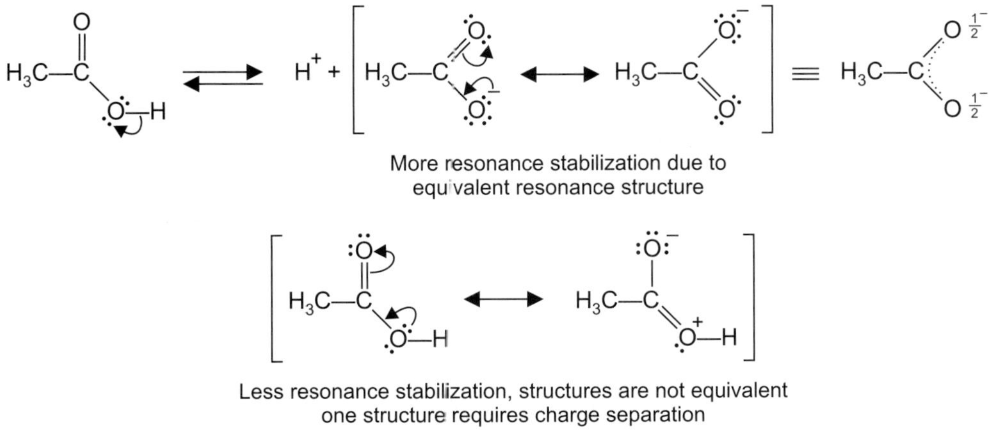

SCHEME 2.16

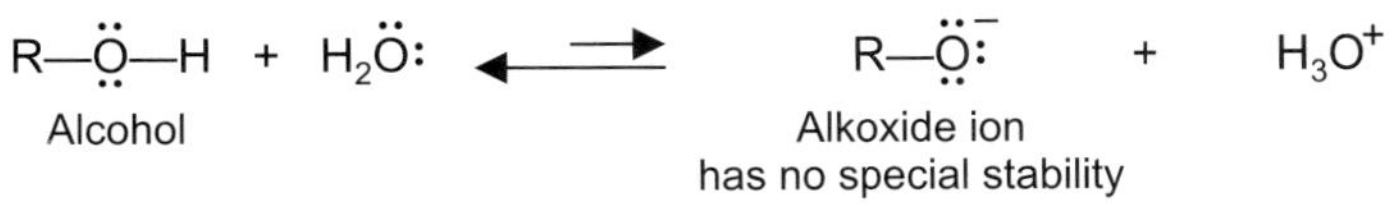

SCHEME 2.17

2. *Allylic rearrangement (allylic shift)*

Allylic reactants *e.g.*, allylic halides undergo nucleophilic substitution reactions readily and are usually accompanied by a rearrangement known as an allylic rearrangement (allylic shift). When allylic reactants are treated with nucleophiles under S_N1 conditions two products are usually obtained one of which is normal and the other is rearranged. This is seen in the case of either 1-chloro-2-butene or 3-chloro-1-butene (Scheme 2.18). Two products are formed since the intermediate carbocation is a resonance hybrid (C-1 and C-3 each carry a partial positive charge and both are attacked by the nucleophile). An allylic rearrangement cannot, however,

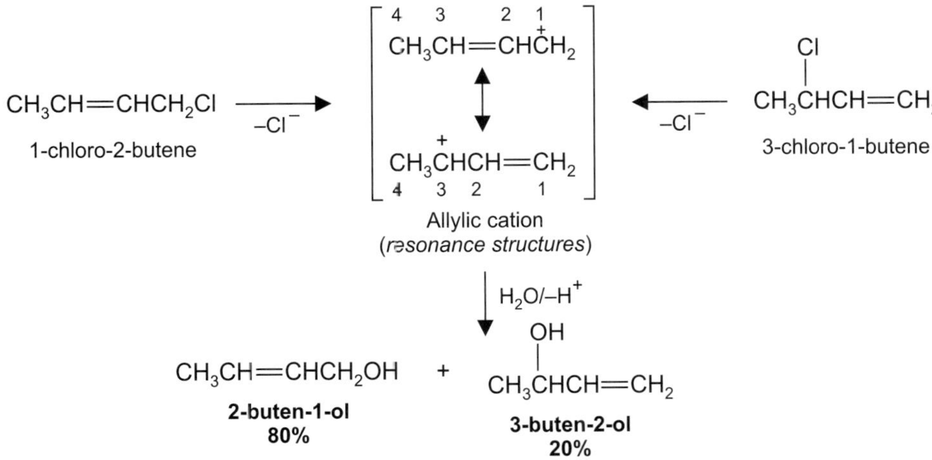

SCHEME 2.18

be detected in the case of symmetrical allylic cations. Moreover, different allylic halides may give identical products upon solvolysis provided they dissociate to the same allylic cation (Scheme 2.18). This mechanism has been called S_N1' mechanism (*i.e.*, substitution unimolecular with rearrangement). The regioselectivity when nucleophile attacks depends on steric hindrance, the attack at the less hindered end of the allylic system being faster.

An allylic shift is also observed during free-radical allylic bromination reactions on unsymmetrical alkenes *e.g.*, in 1-butene (Scheme 2.19). The reaction proceeds through a free radical mechanism involving abstraction of a hydrogen from comparatively weaker allylic C—H bond than those in saturated systems (see, Schemes 1.26, 2.11 and 16.26, cyclohexene being a symmetrical substrate, reaction at either position gives 3-bromocyclohexene). Free radical allylic bromination requires a very low concentration of bromine in the reaction mixture to enhance allylic substitution over ionic addition. Adding bromine would make the concentration too high, and ionic addition of bromine to the double bond would occur. A convenient bromine source for allylic bromination is *N*-bromosuccinimide (NBS, see, Scheme 16.8).

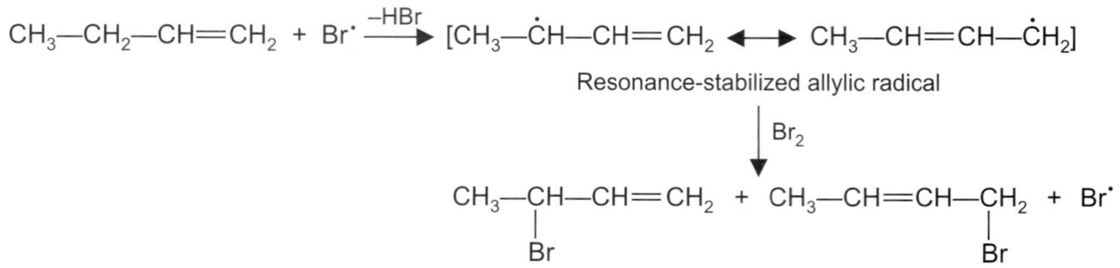

SCHEME 2.19

Allylic halides and tosylates also display enhanced reactivity toward nucleophilic displacement reactions by the S_N2 mechanism which occur without allylic shifts or other rearrangements. Thus allyl bromide reacts with nucleophiles by the S_N2 mechanism about 40 times faster than *n*-propyl bromide (see, Scheme 5.18). Nucleophilic substitution at an allylic carbon can take place by an S_N2 mechanism when no allylic rearrangement takes place. However, under S_N2 conditions an allylic rearrangement can take place when the nucleophile attacks the γ carbon (C-3) rather than the usual carbon (Scheme 2.20) and is then termed S_N2'. The S_N2' reactions may occur on substrates of the types C=C—CH_2Cl, while compounds of the type C=C—CR_2Cl retard S_N2 reaction because of steric hindrance at α position and then these give S_N2' rearrangement exclusively (Scheme 2.21, also see Scheme 5.20).

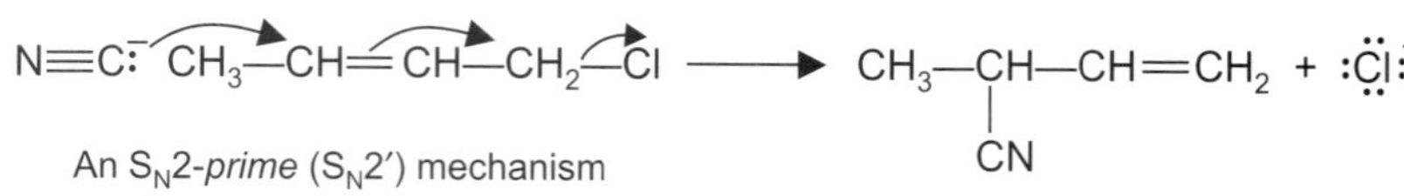

SCHEME 2.20

3. Resonance helps to provide a clear picture of species

In case *e.g.*, the structure of allyl radical was the fixed Lewis structure (classical structure), the ESR spectroscopy would be expected to display four signals for the non-equivalent hydrogens on it (Scheme 2.22). ESR spectrum of a free radical provides information about different "kinds" of hydrogen on it (see, Sec. 16.8, VIII). In the classical structure of the allyl radical, the two vinylic hydrogens (H_a and H_b) on the terminal carbon would be non-equivalent, since one is *cis*

An S_N2-*prime* (S_N2') mechanism

SCHEME 2.21

and the other *trans* to —CH_2. The two hydrogens (H_c) of —CH_2 would be equivalent because of rotation around the carbon carbon single bond. The vinylic hydrogen H_d on the middle carbon is different from all the others. The theory of resonance predicts a highly symmetrical structure for the allyl radical (Scheme 2.21a) and as expected displays only three ESR signals.

Allyl radical
Classical structure
expect 4 ESR signals

Allyl radical
Actual structure
gives 3 ESR signals

SCHEME 2.22

2.3 AROMATICITY

Aromatic compounds are those that resemble benzene. The aromatic properties of the benzene ring (it has physical and chemical properties entirely different than those expected of a conjugated triene) are related to the presence of a closed loop of electrons (the aromatic sextet). The special properties are reflected in equivalent C—C bond distances (1.39 Å, rather closer to a normal double bond 1.34 Å than to a normal single bond 1.54 Å), large resonance energy *i.e.*, delocalization energy (Sec. 2.1, I), the ultraviolet spectrum and in its relative chemical inertness. These properties arise because the π-electrons are delocalized over all carbon atoms of ring.

Thus for a compound to be aromatic like benzene it must obey the following conditions:

- It must be cyclic and planar
- It must have uninterrupted cloud of π electrons
- The π cloud must contain an odd number of pairs of π electrons

The molecular-orbital describes benzene to have six sp^2-hybridized carbon atoms, each of which forms σ-bonds with two carbon atoms and one hydrogen atom. The six remaining electrons are in p orbitals each of which overlaps with two neighbours. The planarity of the ring (strainless C—C—C angles 120°), allows maximum overlap of these$p$ obitals. Six delocalized π-orbitals are thus established of which the three of lowest energy (*i.e.*, bonding orbitals) are occupied. Their relative energies are shown (Scheme 2.22*a*).

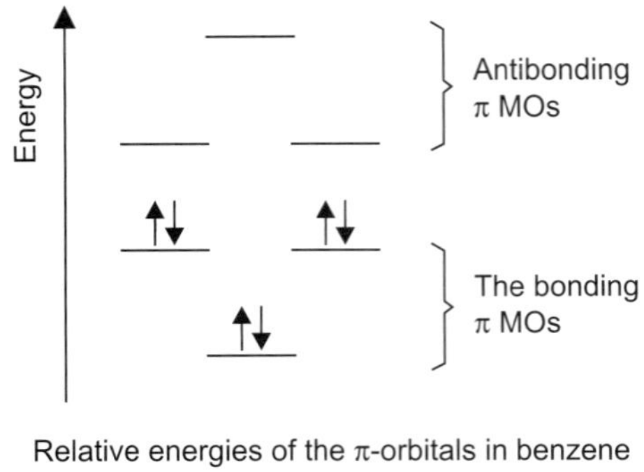

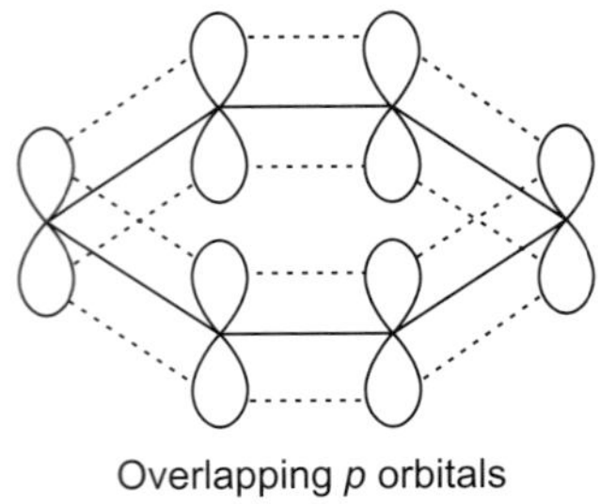

SCHEME 2.22a

Interestingly, unlike the situation in butadiene, where the freedom of movement of the π electrons in the delocalized MOs is opposed by electron repulsion, the π electrons in benzene can circulate round the ring in a synchronized manner without increasing their mutually repulsive forces. Consequently the delocalization energy in benzene is much more than in butadiene.

In addition to accounting for the stability of benzene, both the resonance theory and orbital theory explain the equivalence of the bonds and account for the inertness of benzene to addition.

(A) What are Aromatic Compounds?

An aromatic compound must meet the following criteria:

- The molecule must be cyclic, (aromaticity is the property of ring compounds) having conjugated *pi* bonds.
- Each atom in the ring must have an unhybridized *p* orbital. (The ring atoms are usually sp^2 hybridized, or occasionally *sp* hybridized).
- The unhybridized *p* orbitals must overlap to form a continuous ring of parallel orbitals. The structure must be planar (or nearly planar) for effective overlap.
- For a compound to be aromatic, it will be cyclic, fully conjugated, planar and contain $(4n + 2)\pi$ electrons where n is an integer, 0, 1, 2, 3. Common aromatic systems, thus have 2, 6 and 10π electrons for $n = 0$, 1 and 2. Benzene is cyclic with a continuous ring of overlapping *p* orbitals. It obeys Hückel's rule (6π electrons), is planar and is therefore, aromatic. 1, 3, 5-Cycloheptatriene is not aromatic. It has the correct number of π-electrons to be aromatic, however, the *p* orbitals at the end of the cycle do not overlap and the orbital connectivity is thus broken by the CH_2 group (Scheme 2.22*b*, see also Scheme 2.34), the molecule behaves as a normal triene.

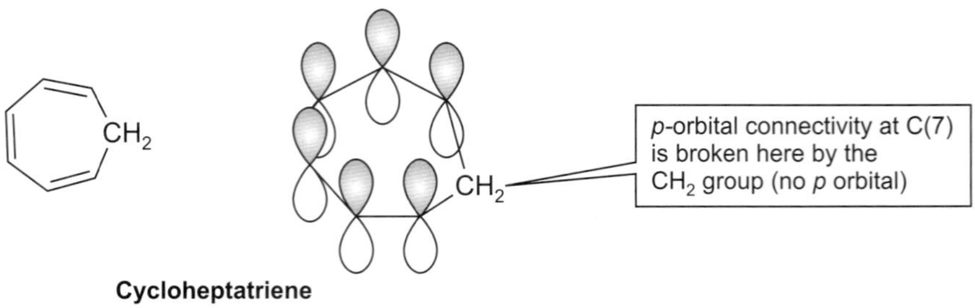

SCHEME 2.22b

Cyclooctatetraene is a conjugated eight carbon ring system (Scheme 2.23), the molecule is not planar but is tub-shaped. A regular octagon has angles of 135° while sp^2 angles are close to 120°. To avoid this angle strain the molecule adopts a non-planar conformation that avoids most of the overlap between adjacent π bonds *i.e.*, non-planarity uncouples the *p* orbitals and interrupts orbital connectivity.

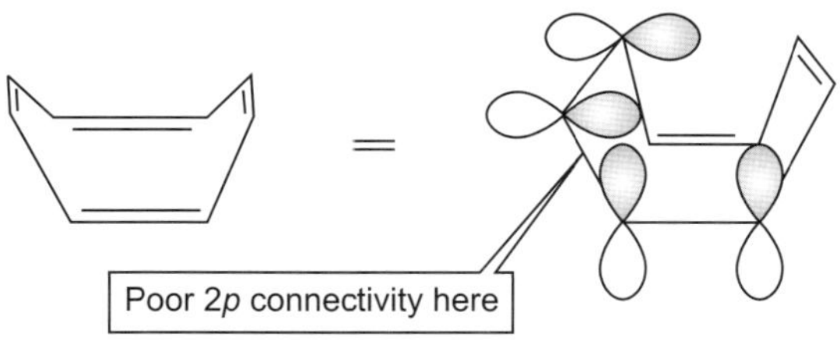

Cyclooctatetraene

SCHEME 2.23

The bond lengths in cyclooctatetraene are characteristic of localized single (1.46 Å) and double bond (1.33 Å). Cyclooctatetraene undergoes addition reactions that are typical of alkenes and is therefore, not an aromatic compound.

On the other hand cyclobutadiene (Scheme 2.24) is cyclic, planar and fully conjugated, however, Hückel's rule predicts no aromatic character since cyclobutadiene is a $4n$ molecule and not a $4n + 2$ molecule. Cyclobutadiene is very unstable (unlike an aromatic compound) and dimerizes by a Diels Alder reaction. It is however, stable in complexes with metals when electron density is withdrawn from the ring by the metal. These cyclobutadiene metal complexes are to be looked as systems with an aromatic duet, the ring is square planar and these undergo aromatic substitution.

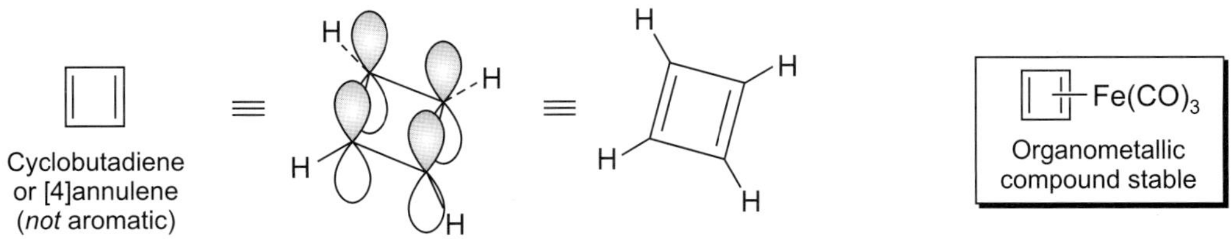

Cyclobutadiene

SCHEME 2.24

(B) NMR Spectroscopy and Aromaticity

The ring hydrogens of an aromatic system resonate at a very low field δ 7–8, significantly downfield from the already rather deshielded alkenyl hydrogens δ 4.5–5.5 (Scheme 2.24*a*). This downfield shift is due to a "ring current" that results from circulation of the *pi* electrons when the molecule is placed in the external applied magnetic field B_0 of the NMR instrument. The circulating electrons generate a magnetic field which is opposed to the external magnetic field in the center of the ring (shielding) but is parallel to the external magnetic field outside the ring in the region where the hydrogens are placed (deshielding). Since the induced field is parallel to the external field where the hydrogens of the benzene rings are located, less external field is needed to reach the total field required for the absorption of the electromagnetic radiation and the hydrogens appear at a downfield position. However, if a hydrogen is positioned near the center of the ring, then an upfield shift is observed. Thus one of the CH_2 groups in (A, Scheme 2.24*a*) is placed directly over the center of the aromatic ring. These protons appear upfield from TMS at the extremely high field position of δ –0.01. These unusual chemical shifts are associated with diamagnetic anisotropy.

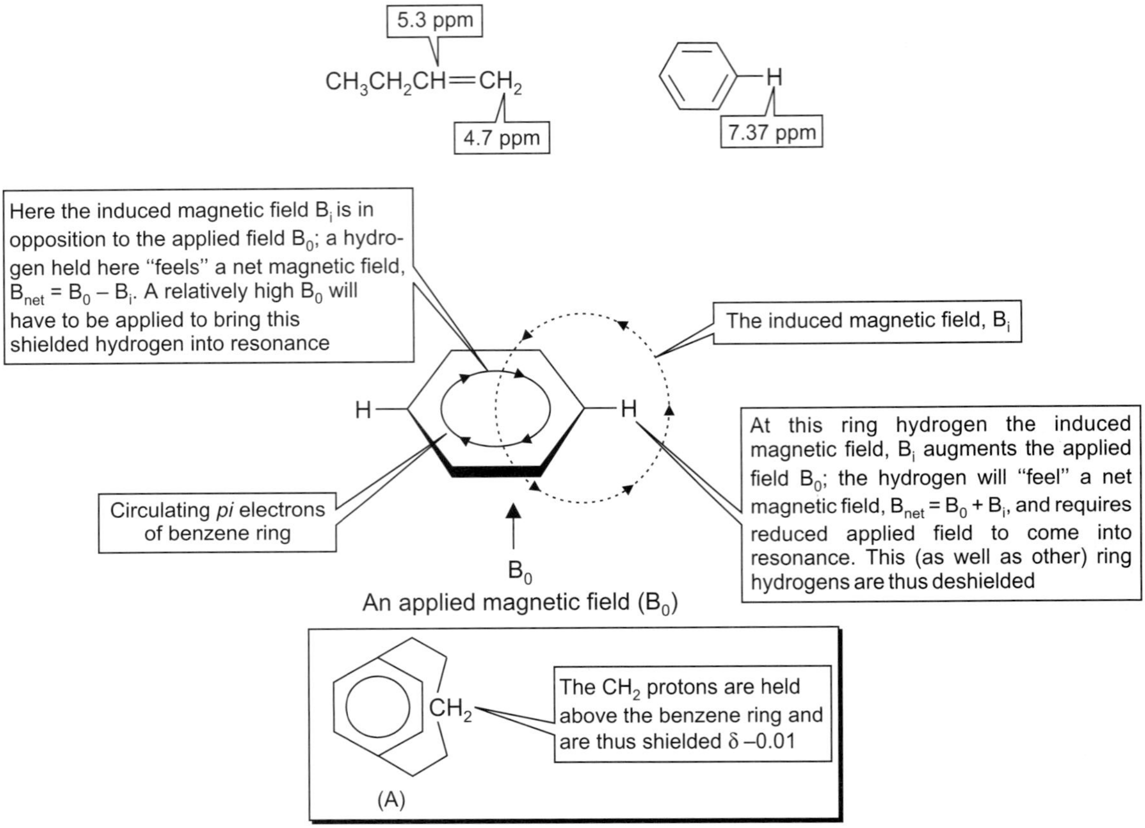

SCHEME 2.24a

All the six hydrogens of benzene therefore, show an expected unusual deshielding and appear around δ = 7.37. Cyclooctatetraene is tub shaped. ^{1}H NMR spectrum of cyclooctatetraene shows a sharp singlet at δ = 5.69 typical of an alkene. Its chemical reactivity is typical of a polyene, it can be catalytically hydrogenated to cyclooctane, and undergoes electrophilic additions and cycloaddition reactions.

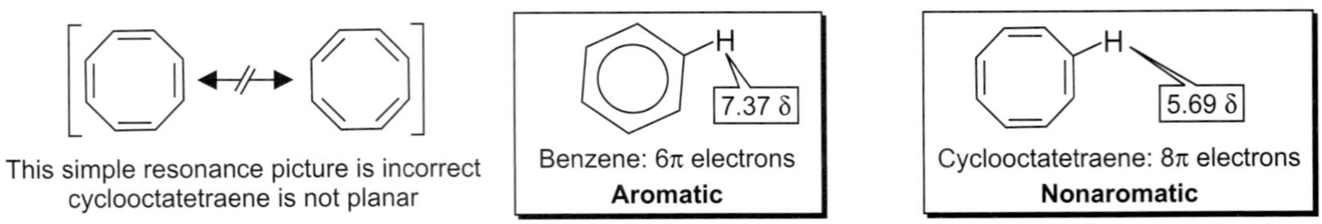

SCHEME 2.25

Diamagnetic Anisotropy

Diamagnetic anisotropy describes an environment in which different magnetic fields are found at different points in space. Anisotropy in Greek stands for "different in different directions". Thus the protons in the vicinity of an aromatic ring are deshielded while those which are placed above or within the ring would be highly shielded. Examples are benzene, paracyclophane (A, Scheme 2.24a) and [18] annulene (see, Scheme 2.25a).

Conditions for Aromaticity

In fact a unified theory has been developed that relates ring currents, resonance energies and aromatic character. For a compound to be aromatic one looks for, diamagnetic ring current, equal or approximately equal bond distances, planarity, chemical stability and the ability to undergo aromatic substitution.

2.4 THE TERMS AROMATIC, ANTIAROMATIC AND NONAROMATIC

(A) Aromatic Compounds

A planar cyclic system of unsaturated atoms containing $(4n + 2)$ π-electrons will be aromatic, where n is a positive integer or zero. An aromatic compound will have extra stabilization. Some slight deviation from planarity is allowed. Thus benzene a cyclic compound with a continuous ring of overlapping p orbitals obeying the Hückel's rule $(4n + 2\pi$ electron system) is aromatic.

The presence of diamagnetic ring current *i.e.*, deshielding of the hydrogens outside the ring while shielding of protons held over or within the ring further provides a powerful experimental criterion for the presence of aromaticity in a compound. Thus like benzene, [18] annulene with $(4n + 2)$ π electrons can achieve a planar conformation and has aromatic behaviour. It shows two signals in its ^{1}H NMR spectrum one at δ9 and other at very high field (beyond TMS signal) at δ3 (Scheme 2.25*a*).

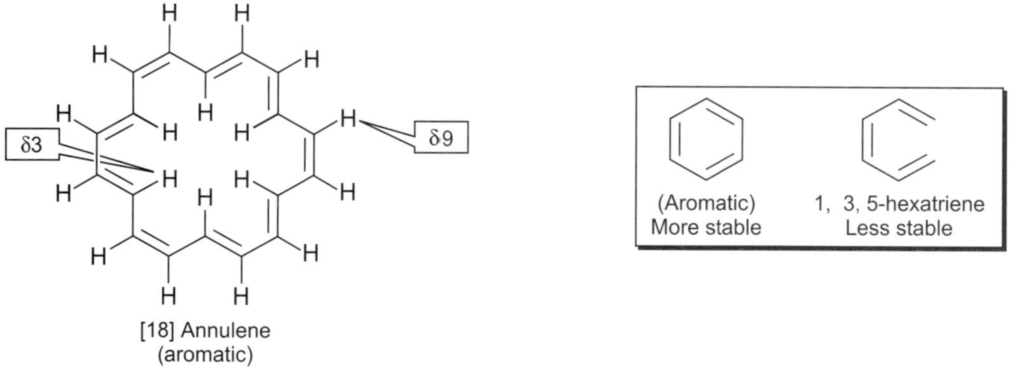

SCHEME 2.25a

Aromatic structures are more stable than their open chain counterparts, thus benzene is more stable than 1, 3, 5-hexatriene.

(B) Antiaromatic Compounds

(i) Cyclobutadiene

These are planar cyclic compounds with an uninterrupted ring of *p* orbital bearing atoms and the π cloud must contain an even number of pairs of π electrons *i.e.*, these are 4*n* π systems.

Antiaromatic compounds display a different ring current called paramagnetic, which induces a magnetic field that is parallel to the external magnetic field in the center of the ring while opposed to it outside the ring. Thus the hydrogens on the outside of the ring appear upfield from the position of a normal alkene hydrogen, a result exactly the opposite to the effect found for aromatic compounds (diamagnetic ring current).

Cyclobutadiene is an example of an antiaromatic compound (see Scheme 2.24), it is planar, cyclic and fully conjugated. It fails only the criterion, that it does not have a Hückel number of (4*n* + 2) π electrons, it being a 4*n* π system. Cyclobutadiene is less stable than its open-chain counterpart (1, 3-butadiene) and it is antiaromatic and expected to be very unstable. In fact its structure is rectangular and not square (Scheme 2.25*b*) with alternating short and long bonds. Thus the two diene forms (I and II, Scheme 2.25*b*) are isomers which equilibrate through a symmetrical transition state (III), rather than resonance forms.

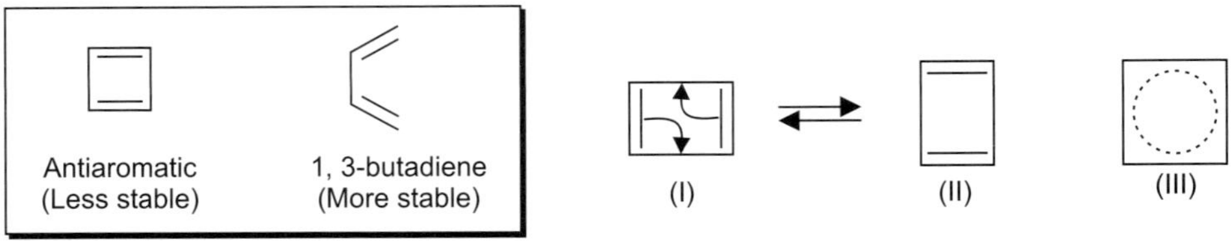

SCHEME 2.25b

Its highly reactive nature is evident from the fact that it is very difficult to isolate, it reacts with itself to give a dimer (Scheme 2.25*c*). One thus calls the molecule of cyclobutadiene as antiaromatic. It adopts the structure which minimizes the delocalization of its electrons, *i.e.*, in a rectangular cyclobutadiene, *p* orbital connectivity is minimized.

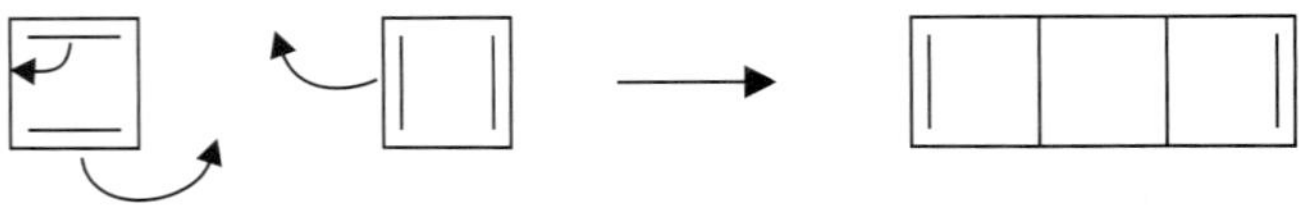

SCHEME 2.25c

(ii) Stable cyclobutadienes, push-pull effect (captodative effect)

Consider the case of tri-*tert*-butylcyclobutadiene, it is stable at room temperature for a short-time because the bulky *tert*-butyl groups hinder the dimerization reaction that destroys less hindered cyclobutadienes. The ring hydrogen of this compound displays its ^{1}H NMR signal at 5.38 δ, a position upfield from that of the hydrogens of a nonaromatic model compound such as cyclobutene (5.95 δ). This upfield shift is due to paramagnetic ring current (Scheme 2.25*d*).

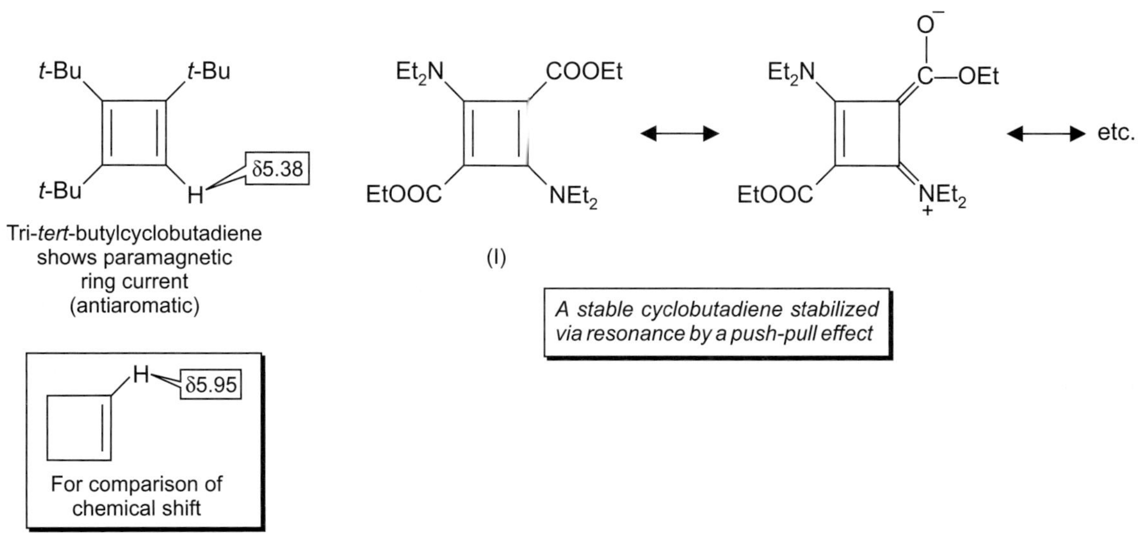

SCHEME 2.25d

The cyclobutadiene (I, Scheme 2.25*d*) is stable due to two electron-donating and two-electron withdrawing groups and due to resulting resonance (a push-pull or captodative effect). Indeed the ring is a distorted square with bond lengths of 1.46 Å and angles of 87° and 93° and represents another case of antiaromaticity.

(C) Nonaromatic Compounds

Recall that Hückel's rule is commonly used to depict a compound which is aromatic $(4n + 2)$ π electrons or antiaromatic $4n$ π electrons provided it has a continuous planar ring of overlapping *p* orbitals. If this is not the case then the compound is called nonaromatic. Cyclooctatetraene is a $4n$ cycle, it is however, non-planar and thus is not antiaromatic. Cyclooctatetraene by not being planar avoids being antiaromatic. Similarly 1, 3-cyclohexadiene does not have a continuous overlapping ring of *p* orbitals and is thus nonaromatic compound and is almost as stable as *cis*, *cis*-2, 4-hexadiene (Scheme 2.25*e*).

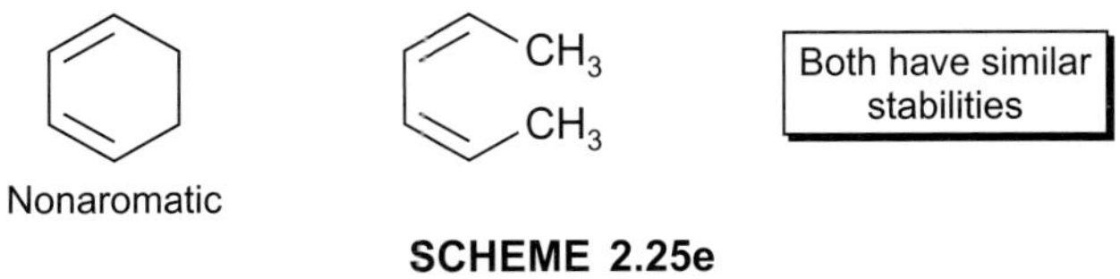

SCHEME 2.25e

2.5 ANNULENES

Annulene is a general name for completely conjugated monocyclic hydrocarbons. The ring size of an annulene is indicated by a number in brackets. Since the carbon atoms occur as doubly bonded pairs, an annulene must have an even number of carbon atoms. The aromaticity of cyclobutadiene a [4] annulene, Benzene a [6] annulene and cyclooctatetraene a [8] annulene has already been discussed. Large-ring annulenes display aromaticity or antiaromaticity depending on whether, these belong to $(4n + 2)$ or $4n$ systems respectively and whether the molecule can adopt the necessary planar conformation.

(A) [10] Annulenes—Systems of Ten Electrons

[10] Annulene can have three geometrical isomers, all *cis*, mono *trans* and di *trans*. In the all *cis* [10] annulene (I, Scheme 2.26), the planar conformation would have excessive amount of angle strain. For a regular decagon the angles would be 144°, much larger than 120° required of sp^2 angles. The [10] annulene isomer (III) with two *trans*-double bonds with all the angles 120° cannot adopt a planar conformation due to interaction (strain) between the two interior hydrogens. The ^{1}H NMR spectrum of the all *cis* form shows its protons in the olefinic region. When conflicting hydrogens of cyclodecapentaene (III) are replaced by a bridging methylene group, the resulting molecule is a bridged 10π electron system. This hydrocarbon which is a bridged cyclodecapentaene (IV) is not a completely coplanar π-system, even then cyclic overlap is substantial to give the compound significant aromatic character *i.e.*, it undergoes aromatic substitution, and displays diamagnetic ring current (Scheme 2.26).

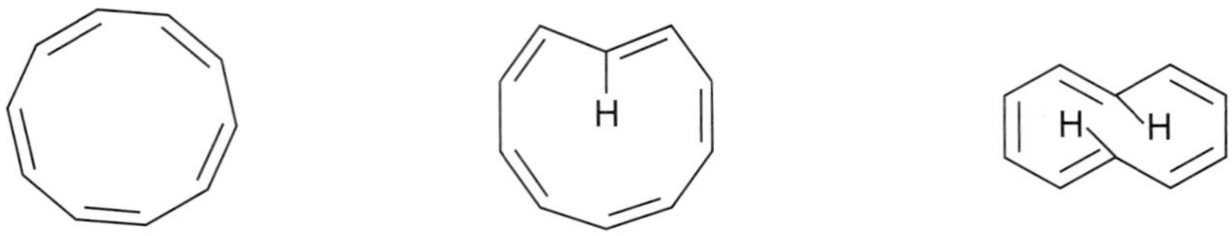

(I) *cis, cis, cis, cis, cis*-Cyclodecapentaene (nonaromatic)

(II) *trans, cis, cis, cis, cis*-Cyclodecapentaene (nonaromatic)

(III) *trans, cis, trans, cis, cis*-Cyclodecapentaene

[10] Annulenes-none is aromatic because none is planar

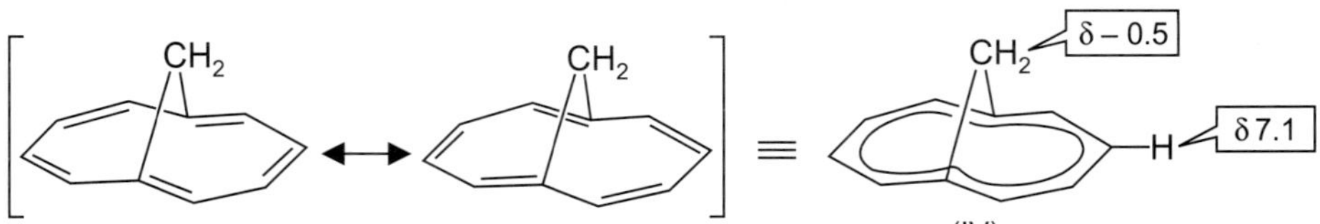

(IV)

Bridged cyclodecapentaene (aromatic)

SCHEME 2.26

Compounds (I and II, Scheme 2.26) have been isolated as crystalline solids at – 80°C. The mono *trans* form (II) suffers from the same problems as faced by the all *cis* form (I) though

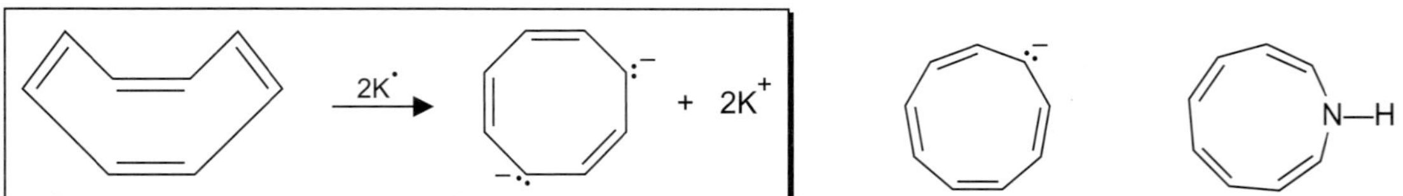

Cyclooctatetraene | Cyclooctatetraenyl dianion | Cyclononatetraenyl anion | Azonine

10π electron systems (aromatic)

SCHEME 2.26a

there is only one hydrogen inside the ring, there is severe angle strain and its planar form for aromaticity is again destabilized. Its ^{1}H NMR spectrum again shows that all its hydrogens lie in the olefinic region.

That aromaticity can compensate for strain effects is shown by the synthesis of a stable aromatic dianion (Scheme 2.26*a*) derived from cyclooctatetraene with ten π electrons (see, Scheme 2.36). Its planar structure is also favoured since the repulsive destabilization of two negative charges is avoided. Other 10π electron systems which are aromatic are in (Scheme 2.26*a*). Interesting case is of azonine which is planar and aromatic. The oxygen analog of azonine (oxonin) is non-planar and thus nonaromatic.

(B) Higher Annulenes

Both [12], and [16] annulenes, are not aromatic these being $4n$ compounds (Scheme 2.27) (n = 3 and 4 respectively) and not $4n + 2$ compounds. Both [12] annulene and [16] annulene do not display antiaromaticity either since these have the flexibility to adopt non-planar conformations *e.g.*, [16] annulene is non-planar and shows a pattern of alternating short and long bonds. As expected these show reactivity of conjugated polyenes.

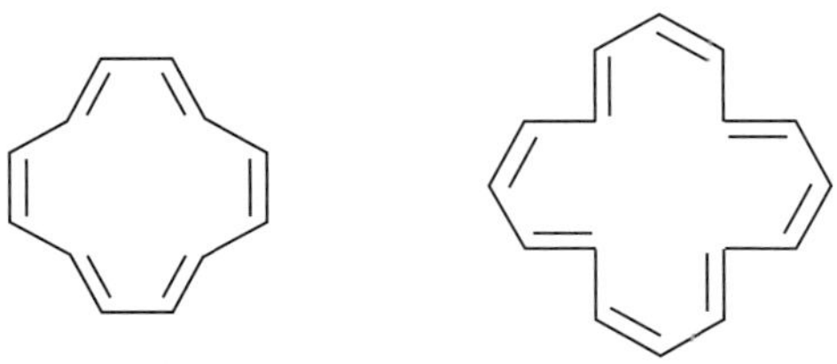

These 4n systems adopt non-planar conformations and are thus not antiaromatic. These behave as partially conjugated dienes

[12] Annulene [16] Annulene

Not aromatic and not even antiaromatic

SCHEME 2.27

On the other hand some of the larger annulenes *e.g.*, [14] and [18] annulenes ($4n + 2$ compounds) are aromatic compounds (Scheme 2.28) since these can achieve planar conformations (also see problems 2.7 and 2.8). Comparison of [14] annulene with [18] annulene is interesting. Both are aromatic since both can achieve planar conformations and as expected both display diamagnetic ring current. [14] annulene is somewhat destabilized by steric strain caused by hydrogens inside the ring while in [18] annulene repulsions among six interior hydrogens are almost removed since the ring is large enough. These differences are displayed by somewhat stronger diamagnetic ring current effects in [18] annulene compared to [14] annulene. Significantly comparable diamagnetic ring current is displayed in the bridged [14] annulene (I, scheme 2.28) which is now free from steric interactions among inside hydrogens, and is a stable aromatic compound with all bond distances close to 1.4 Å and it displays substitution rather than addition.

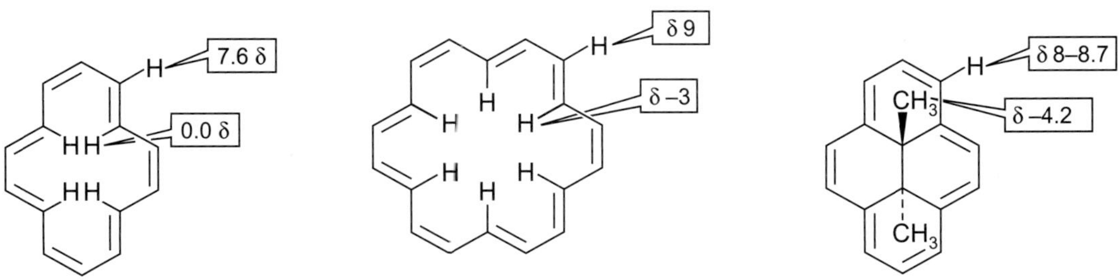

[14] annulene [18] annulene A bridged [14] annulene

SCHEME 2.28

2.6 THE FROST CIRCLE—MOLECULAR ORBITAL DESCRIPTION OF AROMATICITY AND ANTIAROMATICITY

Without the use of mathematics, the relative energies of the molecular orbitals of planar, fully conjugated molecules can be determined. One *simply inscribes a polygon corresponding to the ring of the compound being considered as a circle so that one corner of the polygon is at the bottom*. The points where the corners of the polygon touch the circle correspond to the energy levels of the π molecular orbitals of the system. The nonbonding line divides the hexagon exactly in the half, and also divides the bonding orbitals from antibonding orbitals. When an orbital falls on this line it is nonbonding orbital.

If one considers benzene, to locate molecular orbitals, one inscribes a hexagon in a circle and following the procedure described above one finds that there are no nonbonding molecular orbitals and a picture (Scheme 2.28*a*) exactly seen before (see, Scheme 2.22*a*) emerges. This means that an aromatic system like benzene has completely filled bonding orbitals with no electrons in either nonbonding or antibonding orbitals (stability of aromatic compounds).

For cyclobutadiene, a set of four molecular orbitals emerges of which one is bonding, one is antibonding while the other two are equienergetic (degenerate) nonbonding molecular orbitals (Scheme 2.28*a*). Thus cyclobutadiene has a pair of π electrons left after the bonding orbitals are filled.

The unfilled bonding orbitals or the unpaired electrons in nonbonding or antibonding orbitals lead to destabilization of antiaromatic molecules.

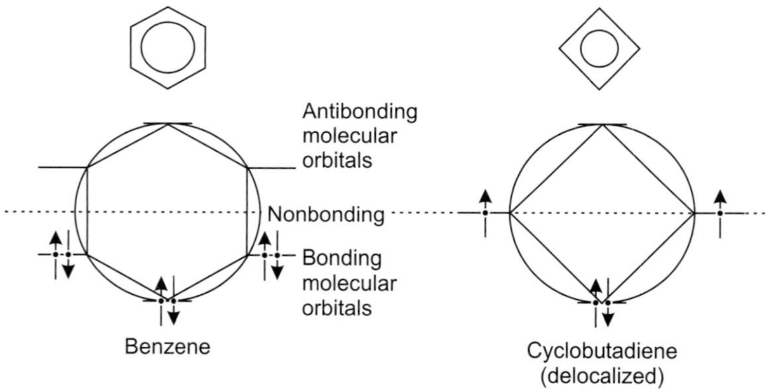

SCHEME 2.28a

EXERCISE 2.1

Using Frost circle method show why cyclooctatetraene is not aromatic ?

ANSWER. *Cyclooctatetraene has eight electrons (4n system). If the molecule was planar like cyclobutadiene it must have two electrons in nonbonding orbitals (Scheme 2.28b). Since stability is not attained by being planar, cyclooctatetraene, prefers to adopt instead a non-planar tub shape (In non-planar form most of the angle strain is relieved).*

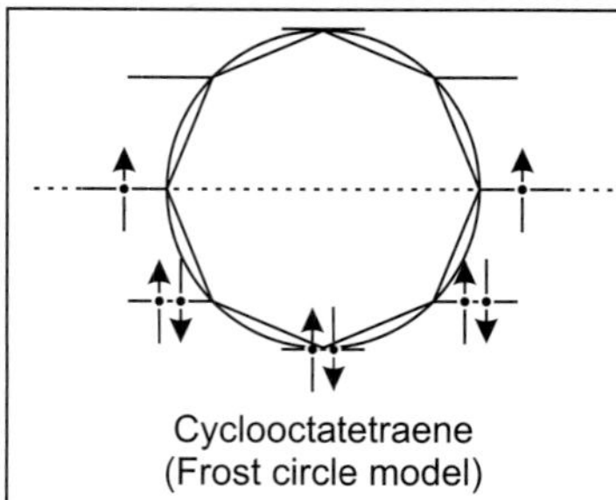

SCHEME 2.28b

2.7 AROMATIC AND ANTIAROMATIC IONS

(A) 2π Electron Systems

Cyclopropene (I, Scheme 2.28*c*) is not aromatic since it does not have an uninterrupted ring of *p* orbital bearing atoms, one of its ring atoms is sp^3 hybridized (Compare with cycloheptatriene Scheme 2.22*b*). A theorectical loss of a hydride ion from cyclopropene gives the cyclopropenyl cation (II). This cation (sp^2 hybridized) with a trigonal planar geometry has an empty *p* orbital and the ion has a cycle of three *p* orbitals. Thus cyclopropenyl carbocation has two π electrons distributed over three carbon atoms and fits Hückels rule and is aromatic.

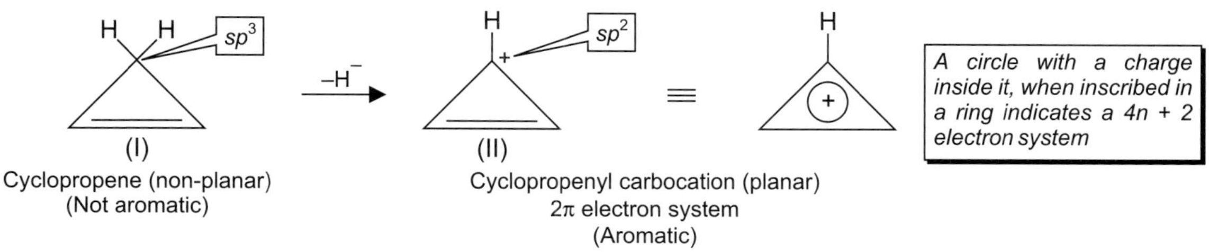

The cyclopropenyl cation is aromatic since it has an uninterrupted ring of *p* orbital-bearing atoms and the π cloud has one (odd number) pair of delocalized π electrons

SCHEME 2.28c

Significantly these cyclopropenyl carbocations are far more stable than other carbocations despite the strain associated with the internal bond angles of only 60°. As an example generally carbocations react rapidly with water which is a weak nucleophile. On the other hand tri-*tert* -butylcyclopropenyl perchlorate (I, Scheme 2.28*d*) which is a carbocation salt is sufficiently

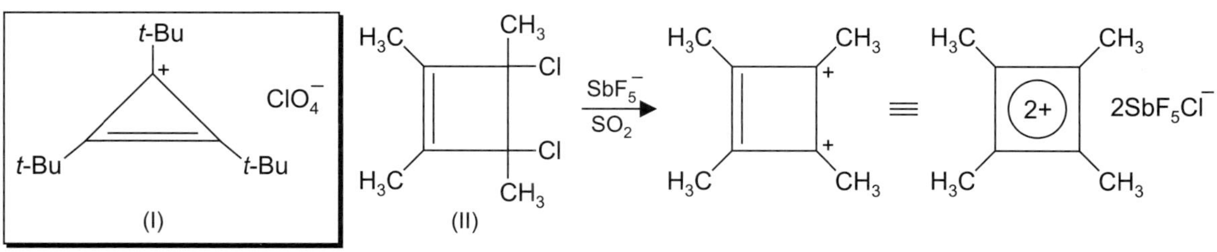

SCHEME 2.28d

stable (Scheme 2.28*d*) and is crystallized from water. One has already seen that cyclobutadiene is unstable antiaromatic 4*n* system. Loss of two chloride ions when (II, Scheme 2.28*d*) is dissolved in SbF_6/SO_2 generates a 2π electron aromatic system which is a square stable cyclobutenyl dication. As expected ^{13}C NMR shows that all methyl groups of the ring are magnetically equivalent as are the four ring carbons. The ring carbons are far down (δ 209). These are deshielded even more due to the low electron density on the ring.

Cyclopropenyl Cation and Anion

Cyclopropenyl cation (cyclopropenium ion) can be generated readily since it is an aromatic system. The Frost circle depicts the molecular orbital system. In the cyclopropenyl cation, there are only two π electrons. Therefore the bonding molecular orbital is full and there are no electrons in the antibonding orbitals. (Scheme 2.28e). Cyclopropenyl anion (Scheme 2.28e) is antiaromatic with antibonding molecular orbitals occupied. It is an unstable system.

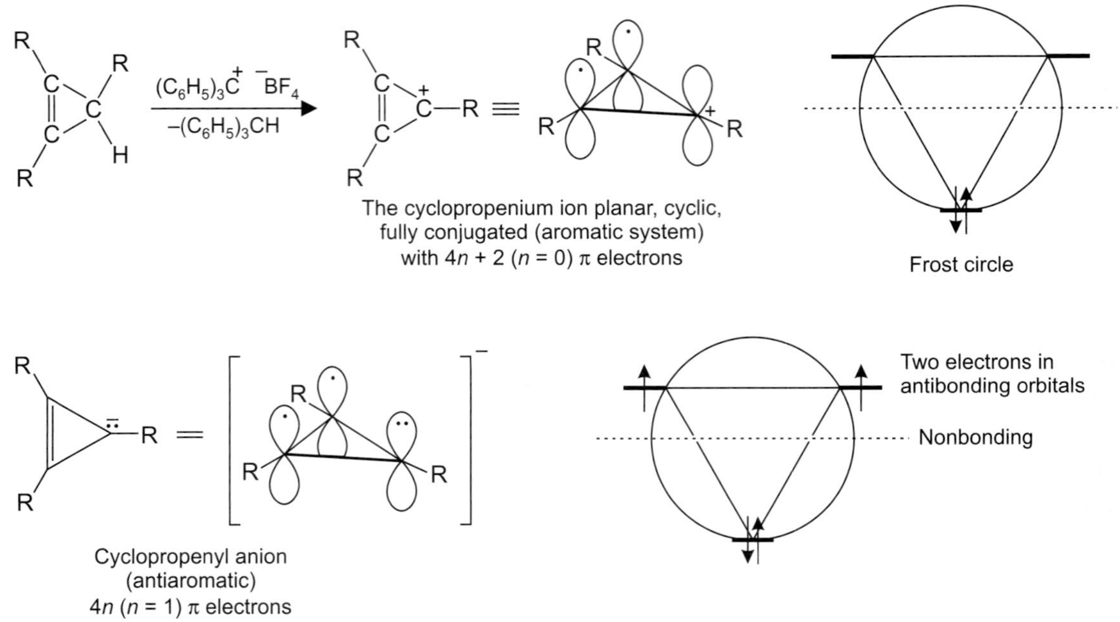

SCHEME 2.28e

(B) The Cyclopentadienyl Ions

Cyclopentadiene is not aromatic, but is unusually acidic for a hydrocarbon. It is found to have a pK_a of 16. Recall that a hydrogen bonded to an *sp* hybridized carbon is more acidic than a hydrogen bonded to an sp^3 hybridized carbon. Thus acetylene has a pK_a of 25 while that of ethane is 50. The pK_a of cyclopentadiene is 16 which is surprisingly acidic for a hydrogen bonded to an sp^3 hybridized carbon. The reason for this low pK_a of cyclopentadiene is that it can be converted to its anion, by deprotonation with moderately strong bases (Scheme 2.29). On deprotonation (loss of a proton), the resulting carbanion (cyclopentadienyl anion) is greatly stabilized by resonance, its resonance energy being 24–27 kcal/mol. ^{1}H NMR spectroscopy has shown that all five hydrogen atoms in the anion are equivalent. The Hückel's rule predicts that the cyclopentadienyl anion (4*n* + 2 system) is aromatic. Cyclopentadiene itself is not aromatic. Firstly it does not have the proper number of π electrons and secondly the π-electrons cannot be delocalized about the entire ring because of the intervening sp^3 hybridized $—CH_2—$ group with no available *p* orbital. One may recall the definition of an aromatic and antiaromatic compounds. To be aromatic or antiaromatic, a cyclic structure should have some number of conjugated p bonds, each atom in the ring must have an unhybridized p orbital. The unhybridized p orbitals must be able to overlap to give a continuous ring of parallel orbitals.

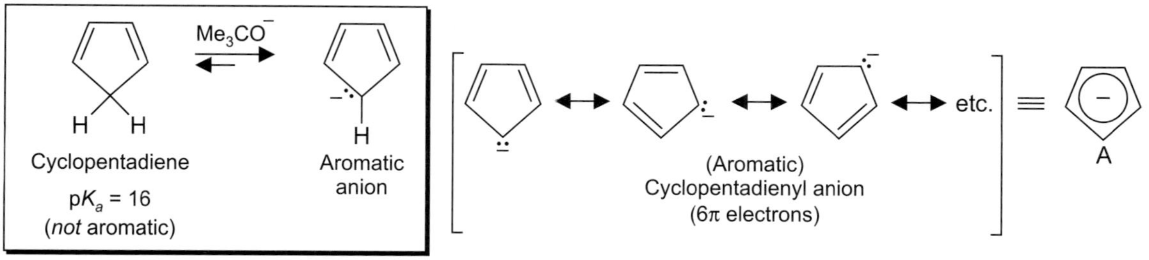

SCHEME 2.29

When the —CH_2— carbon of cyclopentadiene becomes sp^2 hybridized after the proton loss, the two electrons left behind can occupy the new p orbital which is formed. Now, this new p orbital can overlap with the p orbitals on either side of it to give a ring with six delocalized π electrons to make cyclopentadienyl anion aromatic. The six electrons of the anion can be put into molecular orbitals and like benzene these occupy, the set of three molecular orbitals (Scheme 2.30) without requiring antibonding or nonbonding molecular orbitals to be occupied. As a carbanion, cyclopentadienyl anion however, is quite reactive and it reacts readily with electrophiles. Thus, when one says that this ion is aromatic it does not mean that its stability is like benzene. Here the comparison is with the corresponding open chain ion (Scheme 2.30).

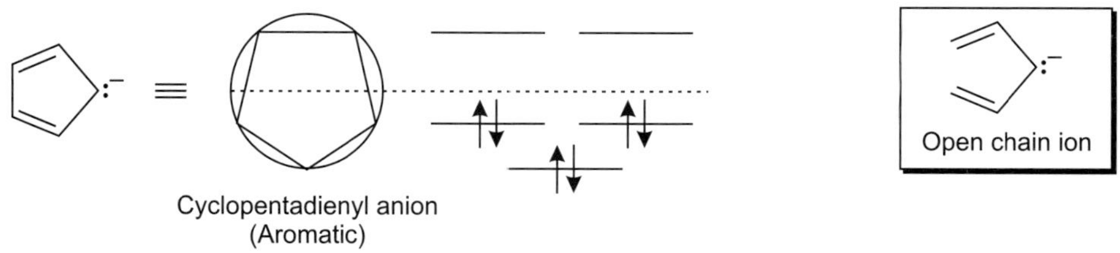

SCHEME 2.30

Hückel's rule predicts that cyclopentadienyl cation with four π electrons is antiaromatic and as a consequence, the cyclopentadienyl cation is not easily formed (Scheme 2.31). The fact that one can draw equivalent contributing structures for cyclopentadienyl cation like that for the anion does not reflect aromaticity. An aromatic system should meet the Hückel's criteria of aromaticity, ($4n + 2\pi$ electrons). Here one has a situation which is similar to planar cyclobutadiene which has two pairs of π electrons and is highly unstable antiaromatic compound (see, Scheme 2.24). The cyclopentadienyl cation has also two pairs of π electrons, thus this cation is antiaromatic and unstable.

Recall aromaticity reflects stability while antiaromaticity is characterized by instability and the relative stabilities are: aromatic compound > cyclic compound with localized electrons > antiaromatic compound.

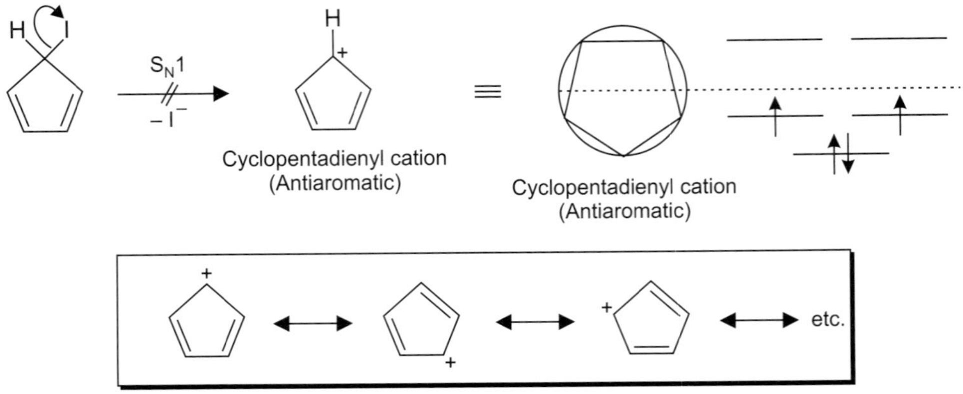

SCHEME 2.31

EXERCISE 2.2

On reaction with $AgBF_4$, 3-chlorocyclopropene precipitates AgCl. The crystalline organic material is soluble in polar solvents like nitromethane but has no solubility in hexane (non-polar solvent). The organic material dissolved in nitromethane containing KCl gives back the original reactant (3-chlorocyclopropene). Explain why 5-chloro-1, 3-cyclopentadiene does not react under these conditions?

ANSWER. *Due to the ready formation of an aromatic carbocation as a salt (Scheme 2.32). 5-chloro-1, 3-cyclopentadiene does not react under similar conditions since the corresponding salt if formed would be antiaromatic.*

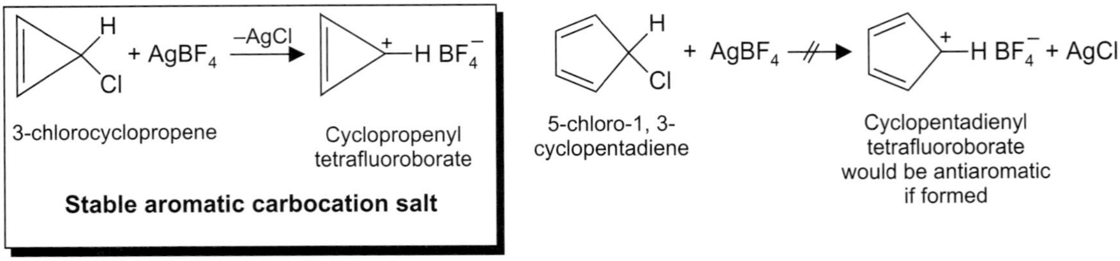

SCHEME 2.32

(C) The Cycloheptatrienyl Ions

Cycloheptatriene has six π electrons however, these cannot be fully delocalized due to the presence of the $-CH_2-$ group which does not have an available *p*-orbital (see the example of cyclopentadiene). The loss of a hydride ion from cycloheptatriene is facile and the resulting ion *i.e.*, cycloheptatrienyl cation (tropylium ion) is found to be highly stable. Tropylium bromide (Scheme 2.33) is actually an ionic compound (unlike aliphatic bromides) to show that this ion is aromatic and therefore, strongly resonance stabilized. On removal of a hydride ion from $—CH_2—$ group of cycloheptatriene, a vacant *p* orbital is generated, and the carbon becomes sp^2 hybridized. The resulting tropylium ion, *i.e.*, the cycloheptatrienyl cation has seven overlapping *p* orbitals with six delocalized π electrons. This situation therefore, makes tropylium cation, an aromatic species (Scheme 2.33). Its ^1H NMR spectrum shows that all the seven hydrogen atoms are equivalent.

The tropylium ion is an aromatic ion and as expected it is much less reactive than many carbocations and several tropylium salts can be isolated and stored and do not undergo decomposition. The Frost circle device shows that the three bonding molecular orbitals are fully occupied (Scheme 2.34) while the antibonding molecular orbitals are unoccupied. This molecular orbital picture of tropylium ion is in close resemblance to that of benzene (see, Scheme 2.28*a*). However, the tropylium ion is not as stable as benzene. The aromaticity of tropylium ion implies that the cyclic ion is more stable than the corresponding open chain ion.

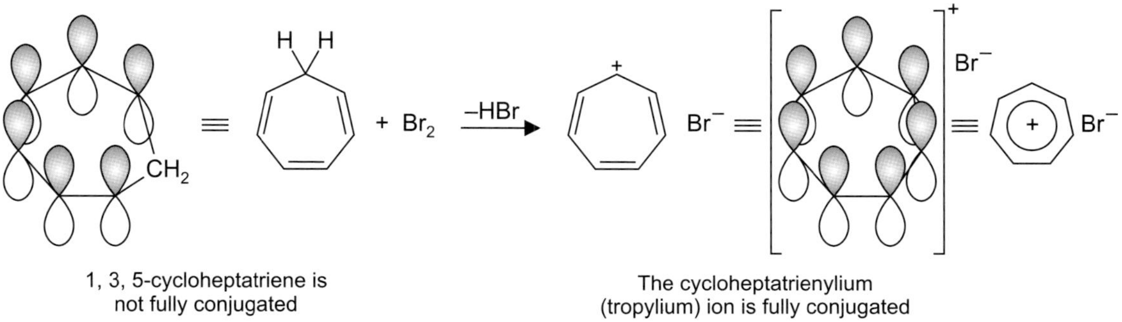

Cycloheptatrienyl bromide is an ionic compound, since its cation is aromatic

SCHEME 2.33

In sharp contrast to cyclopentadiene, one finds that cycloheptatriene has no unusual acidity. Thus unlike in cyclopentadiene, here the corresponding anion (Scheme 2.35) is difficult to make, since it is antiaromatic on the basis of Hückel's rule and as expected very reactive.

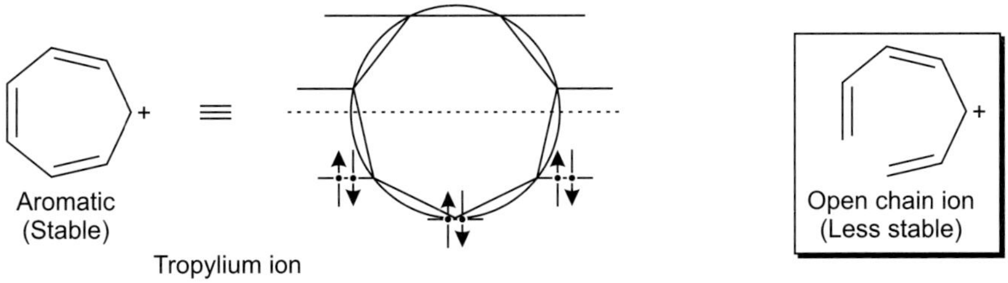

SCHEME 2.34

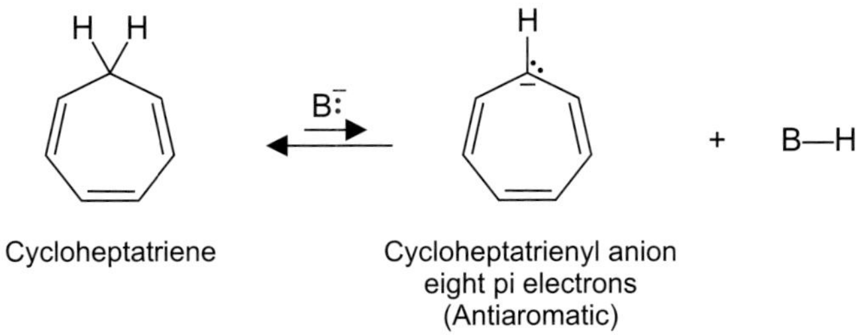

SCHEME 2.35

(D) The Cyclooctatetraene Dianion

Unlike cyclooctatetraene (Scheme 2.23), which is a tub shaped molecule, its dianion (Scheme 2.36, a $4n + 2$ species) is planar and is an aromatic dianion. Its successful preparation shows that the angle strain (135°) is not insurmountable. The dianion has a planar regular octagonal structure with all bond lengths equal. This example is another case where the cost in strain energy to achieve planarity is lower than the extra stability which would be gained from the aromatic ring.

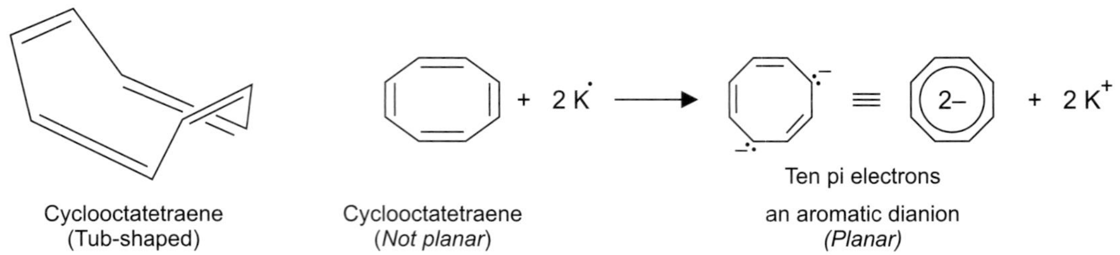

SCHEME 2.36

2.8 OTHER NON-BENZENOID AROMATIC COMPOUNDS

[6] Annulene (*i.e.*, benzene), naphthalene, phenanthrene and anthracene are examples of benzenoid aromatic compounds. On the other hand, the cyclopentadienyl anion, the cycloheptatrienyl cation and some aromatic annulenes (other than [6] annulene) are the examples of nonbenzenoid aromatic compounds.

The aromatic system of electrons ($4n + 2$) can be spread over two rings only provided 10 electrons (and not 8 or 12) are available for aromaticity. Attempts to prepare pentalene and heptalene (Scheme 2.37) have failed since these do not contain ($4n + 2$) electrons. Azulene is found to be a stable compound while the other two are not and this has been confirmed experimentally. Azulene has a resonance energy of 49 kcal/mol. It has considerable dipole moment (1.0D) while the dipole moment of the isomeric compound naphthalene is 0. The dipole moment of azulene suggests that charge separation exists in the molecule and that each ring approximates to a six π-electron system. Azulene (I, Scheme 2.37) may be regarded as a combination of aromatic cyclopentadienyl anion and aromatic cycloheptatrienyl cation. Thus in valence bond terms, the ionic structure of azulene (a non-benzenoid aromatic compound) is an important contributor to the resonance hybrid.

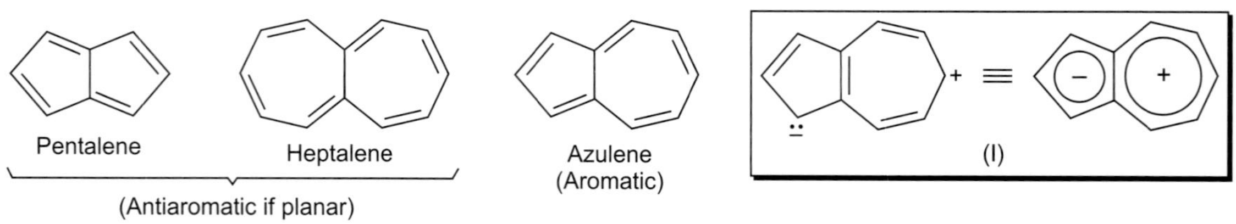

SCHEME 2.37

EXERCISE 2.3

Explain the following observations (Scheme 2.38)

(a) Why the triene (I) is readily deprotonated with butyllithium twice?

(b) The compound (II) has a high dipole moment.

(c) Why 2, 4, 6-cycloheptatrienone is very stable and can be isolated while 2, 4-cyclopentadienone is unstable and cannot be isolated?

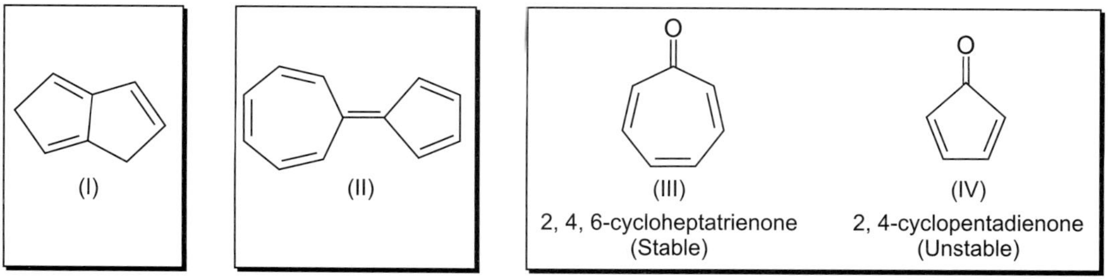

SCHEME 2.38

ANSWER. *(a) Due to the formation of an stable aromatic dianion with 10 π electrons (V, Scheme 2.39); (b) since its resonance structure (VI) has both of its rings aromatic; (c) the resonance structure of (III) shown in (VII, Scheme 2.39) is aromatic while from (IV) shown in (VIII) is antiaromatic.*

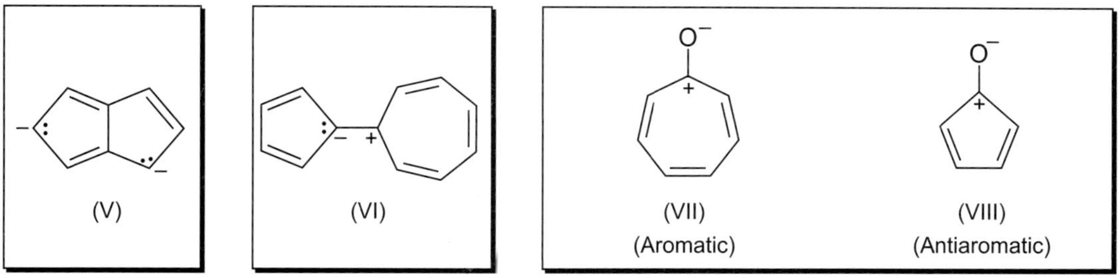

SCHEME 2.39

EXERCISE 2.4

Diphenylcyclopropenone (Scheme 2.39a) is a stable compound which displays aromatic character. Its reaction with hydrogen bromide gives a stable ionic salt. Explain.

ANSWER. *Diphenylcyclopropenone is a stable aromatic compound due to the resonance structure with two π electrons which is aromatic (compare Scheme 2.39). Of the different sites where protonation could occur, the Lewis basic carbonyl oxygen is the most unusual. Recall that only on this addition stable aromatic cyclopropenyl cation would be generated.*

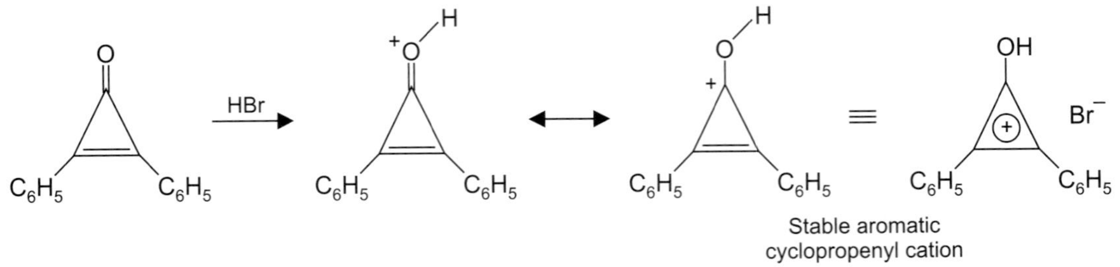

SCHEME 2.39a

Chemical behaviour of cyclic conjugated compounds can be largely understood by considering aromaticity or antiaromaticity of a reactant, intermediate or the final product by looking for a planar cycle of p orbitals.

2.9 HETEROCYCLIC AROMATIC COMPOUNDS

Heterocyclic compounds can also be aromatic since for the application of Hückel's rule what one needs is a ring of atoms, all with unhybridized *p* orbitals, in a planar arrangement in order that the *p* orbitals overlap in a continuous ring. Thus, the heterocyclic compounds (Scheme 2.39*b*) are all aromatic. Pyrrole, furan and thiophene infact represent 1-hetero 2, 4-cyclopentadienes and contain a butadiene unit bridged by a heteroatom bearing lone electron pairs. In electronic structure, these three compounds are similar to cyclopentadienyl anion (see, Scheme 2.29).

Both benzene and pyridine have a similar Kekulé structure. Pyridine with a resonance energy of 27 kcal (113 kJ) per mole shows typical characters of an aromatic compound. The nitrogen atom in pyridine is sp^2 hybridized. The sp^2 hybridized nitrogen donates one electron to the π system and this along with one each from the five carbon atoms provides pyridine a sextet of electrons similar to that in benzene. The nonbonding electrons of the nitrogen (in an sp^2 orbital which lies in the plane of the ring) do not interact with the π system of the ring. The unshared pair of nonbonding electrons confers on pyridine the properties of a weak base. Thus pyridine protonates to yield the pyridinium ion which retains its aromatic character since the process does not disturb the electrons of the aromatic sextet.

In pyrrole only four π electrons are contributed by the carbon atoms of the ring. To make an aromatic sextet the sp^2 hybridized nitrogen further contributes two electrons (Scheme 2.39*b*).

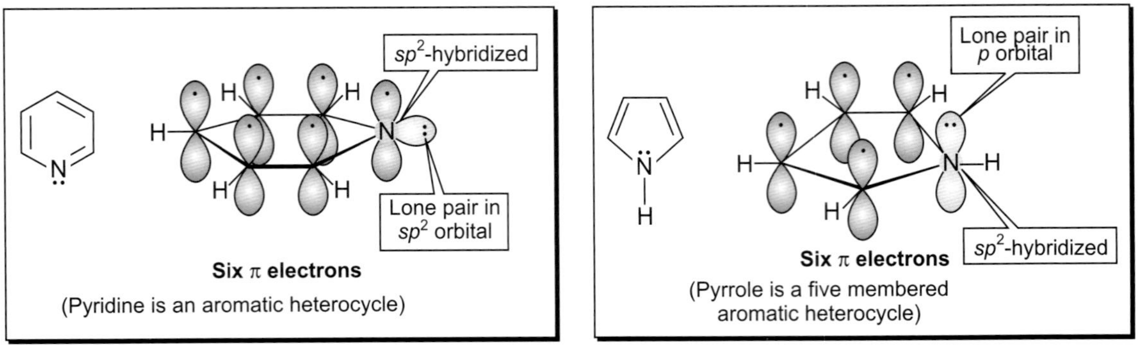

SCHEME 2.39b

Pyrrole is far less basic than pyridine (pK_b = 8.8). Because these apparently unshared electrons are in the aromatic π cloud and are not available for bonding with a proton.

Furan (Scheme 2.39*c*) is similar to pyrrole. In furan which is an aromatic five-membered heterocycle; the oxygen atom has two lone pairs of electrons. One of these pairs is made available to provide two electrons needed to satisfy the $4n + 2$ rule. The oxygen atom is sp^2 hybridized. One lone pair is placed in the unhybridized *p* orbital which combines with the four electrons in the double bonds to provide an aromatic sextet. The second electron lone pair occupies one of the sp^2 hybrid orbitals in the plane and thus with no opportunity to achieve overlap.

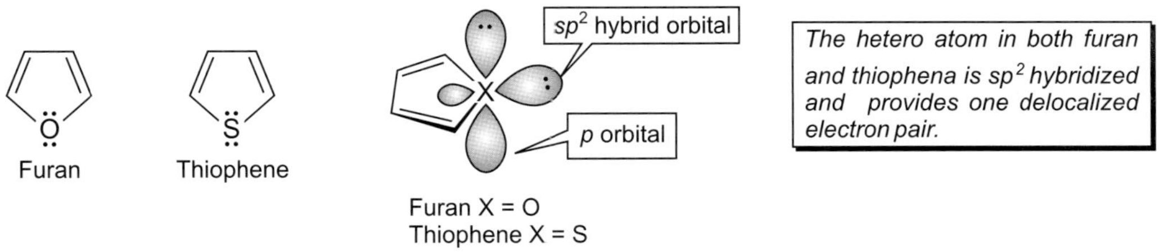

SCHEME 2.39c

In thiophene, (which is similar to furan) as well one of the lone electron pair occupies one of the sp^2 hybrid orbitals which is again in the plane and has no opportunity to achieve overlap. In thiophene however, the sulphur atom instead uses an unhybridized 3*p* orbital to contribute a pair of electrons to the π cloud to achieve aromaticity.

All the three heterocyclopentadienes (pyrrole, furan and thiophene) display unusual stability and undergo electrophilic aromatic substitution. All display ring currents and the consequent deshielding of the protons in their ^{1}H NMR spectra.

2.10 METALLOCENES AND RELATED COMPOUNDS

Ferrocene is prepared by reacting cyclopentadienyl anions with ferrous ion Fe^{2+} in the ratio of 2 : 1 respectively (Scheme 2.40). Thus the Grignard reagent of cyclopentadiene on reaction with ferrous chloride gives ferrocene as an orange solid. Ferrocene like other compounds with similar structure has two cyclopentadienide rings forming a sandwich around an iron atom. These compounds as a class are called metallocenes. Ferrocene is an aromatic compound (nonbenzenoid) which is highly stable and displays electrophillic aromatic substitutions. Thus ferrocene on heating with acetic anhydride and phosphoric acid gives acetyl ferrocene (Scheme 2.40, one may note that all C—H bonds on ferrocene are equivalent).

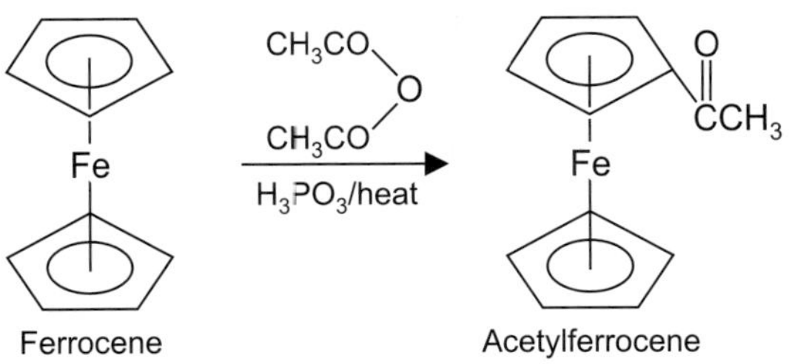

SCHEME 2.40

The carbon-iron bonding in ferrocene arises from overlap between the inner lobes of the p orbitals of the cyclopentadienyl anions and $3d$ orbitals of the iron atom. Moreover, this arrangement is such to allow the rings of ferrocene to rotate freely around the axis that passes through the iron atom and is perpendicular to the rings. A noteworthy feature of many organic derivatives of transition metals is that the organic group is bonded to the metal through the π system rather than by a σ bond as in (benzene tricarbonylchromium, which interestingly undergoes nucleophilic attack (see, chapter 9).

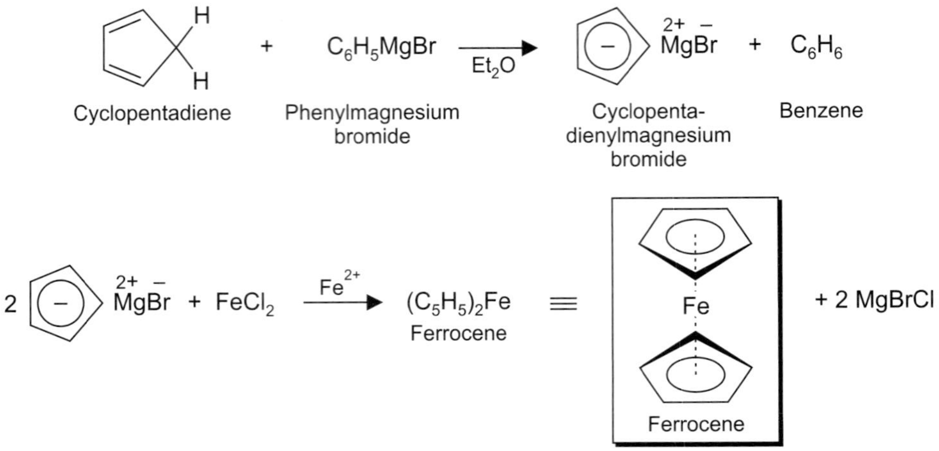

SCHEME 2.41

The iron of ferrocene with 18 valence electrons is coordinately saturated and one calculates this number like this.

Iron has 8 valence electrons in the elemental state. The oxidation state of iron in ferrocene is + 2, thus $d^n = 6$ ($d^n = 8 - 2 = 6$).

Each cyclopentadienyl (Cp) ligand of ferrocene donates 6 electrons to the iron and the valence electron count for iron is thus 18.

[total number of valence electrons = $d^n + 2(\text{Cp}) = 6 + 2(6) = 18$]

2.11 FUSED BENZENOIDS AND FULLERENES

(A) Naphthalene, Phenanthrene and Anthracene—Concept of Partial Bond Fixation

Naphthalene is a fused benzenoid hydrocarbon and is aromatic, ^{1}H NMR spectrum explains its aromaticity as it shows two symmetric multiplets at $\delta = 7.40$ and 7.77, characteristic of aromatic hydrogens deshielded by the ring current effect of the π-electron loop.

The UV spectrum displays peaks (as long as 320 nm) typical of an extended conjugated system.Thus it shows that the added four π electrons enter into efficient overlap with those of the attached benzene ring. Consequently, several resonance forms can be drawn (Scheme 2.41a). When an aromatic compound is ingested or inhaled, it gets converted into an arene oxide (enzymatically). Naphthalene gives only one arene oxide which also reflects on its aromatic character (Scheme 2.41a).

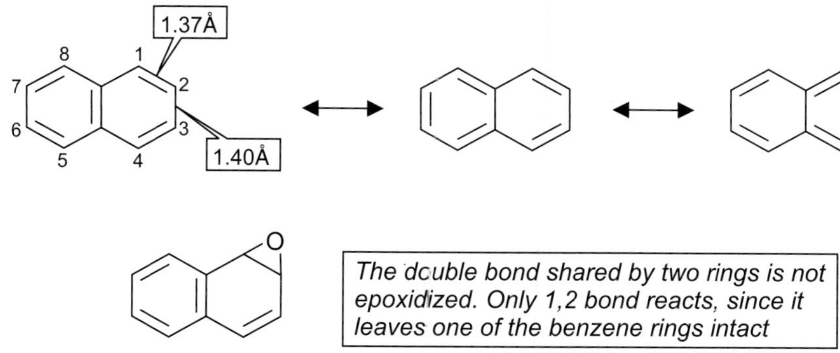

Naphthalene oxide

SCHEME 2.41a

If one assumes that the three resonance forms of naphthalene (not considering Dewar forms and the forms with charge separation) contribute equally, the 1, 2 bond has more double bond character than the 2, 3 bond and this is shown by the bond distances. Clearly these bond distances are different from pure single (1.54 Å) and double bonds (1.33 Å) and the 1, 2- and 1, 3-bond distances in naphthalene deviate in length from those in benzene (1.39 Å). Ozone attacks preferentially the 1, 2-bond in naphthalene. This nonequivalency of bonds is termed *partial bond fixation* which is displayed by almost all fused aromatic systems. When one considers phenanthrene (Scheme 2.41*b*) only in structure (V) 9, 10-bond is single bond, partial

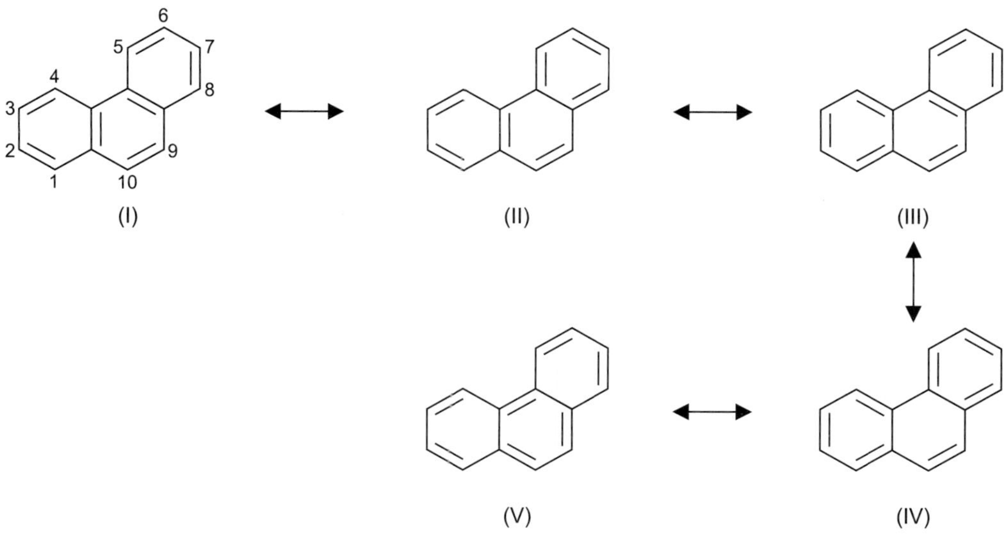

SCHEME 2.41b

bond fixation is thus at its maximum and as a consequence 9, 10-bond of phenanthrene is attacked easily by many reagents. Moreover, in four of the five resonance structures, the 9, 10 bond is double and its length is almost the same as of an alkenic C=C bond.

Phenanthrene is a hybrid of the five canonical forms and is more stable than anthracene which is best regarded as the resonance hybrid of four structures (Scheme 2.41*c*).

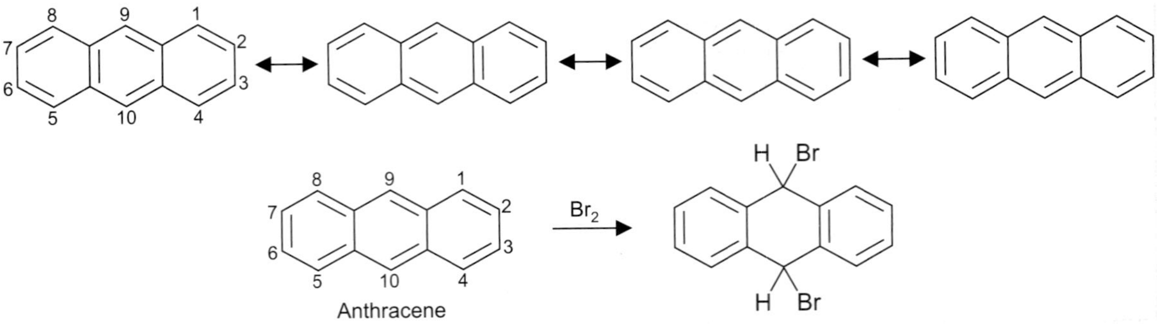

SCHEME 2.41c

An inspection of canonical forms shows that in anthracene all the three rings cannot be benzenoid at the same time. Moreover, the central ring contains a four-carbon fragment with a relatively high degree of double bond character.

> **Resonance Energies and the Number of Principal Canonical Forms**
>
> *Recall that the resonance energies of fused systems increase as the number of principal resonance structures increases. Thus, for benzene, naphthalene, anthracene, and phenanthrene, for which one can draw, respectively, two, three, four, and five main canonical forms, the resonance energies are: 36, 61, 84, and 92 kcal/mol respectively. Phenanthrene, which has a total resonance energy of 92 kcal/mol after reaction at 9, 10 bond e.g., by bromine two complete benzene rings remain, each with 36 kcal/mol that would be lost if other position is attacked.*

For the same reasons, in anthracene as well a reagent *e.g.*, bromine adds via 1, 4-addition in the central ring.

> **The Numbering System in Anthracene**
>
> *The numbering system shown in anthracene is somewhat unusual and was introduced during early chemical studies to indicate special behaviour associated with the 9, 10 bond.*

EXERCISE 2.5

Anthracene and phenanthrene represent aromatic hydrocarbons containing three fused benzene rings. If one continues this building process one can draw a fused system (I, Scheme 2.41d). Comment on its aromaticity?

ANSWER. *The system as drawn with circles is misleading. When one draws this system in Kekulé form then there is no way for each carbon to be sp^2 hybridized if each carbon has four valences i.e., double bonds cannot be distributed so that each carbon has one double and one single bond. Thus the molecule is not stable, being not fully aromatic. This molecule (phenalene) as expected is acidic and on treatment with base gives the aromatic anion (II, Scheme 2.41d).*

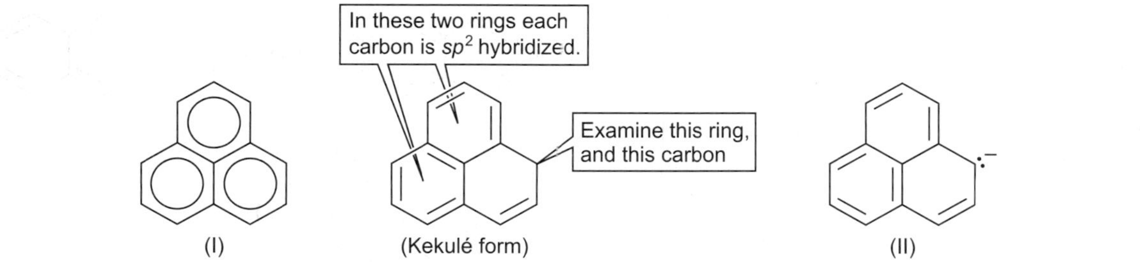

SCHEME 2.41d

(B) Fused Ring Systems and Annelation

When one draws all the fused polynuclear aromatic systems from four benzene rings, one faces several problems with some arrangements and only two of these are discussed (Scheme 2.41*e*). The situation in (I) is similar to the one discussed in Scheme 2.41*d*. A consideration of (II, Scheme 2.41*e*) shows that in a fused system each ring cannot get six electrons. One has

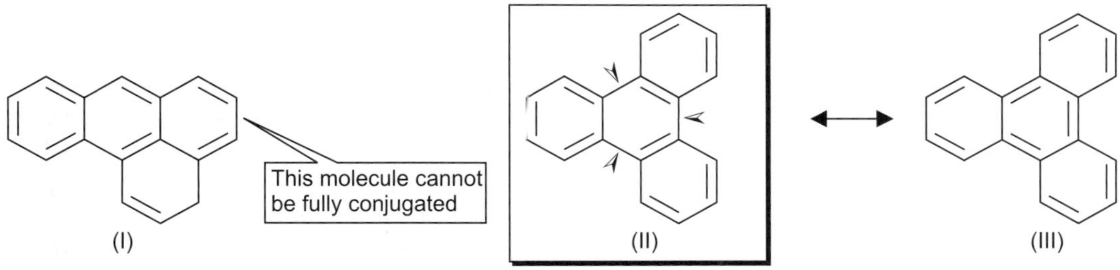

SCHEME 2.41e

already seen that in naphthalene only one of the rings has six electrons while the other has only four. Naphthalene is more reactive than benzene due to the fact that it contains one benzene ring and the other ring has only a butadiene system. This type of situation is extreme in the case of triphenylene (II, Scheme 2.41*e*) where one can draw eight canonical forms like (II) and (III). In all the canonical forms of the type II, the three bonds marked by arrows are only single bonds and none of these represents a double bond. In the resonance structure (III) only the middle ring is like benzene while the outer rings have butadiene type system. Thus considering (II, Scheme 2.41*e*) along with maximum number of canonical forms similar to it, the molecule represents a system in which 18 electrons are arranged so as to give only the outer rings a sextet each and the middle ring is "empty". The middle ring thus behaves as if it has given up a part of its aromaticity to adjacent rings and is called annelation, an effect which can be observed by UV spectroscopy (also see problem 2.7).

(C) Fullerene C_{60}

The two well known crystalline modifications of carbon are graphite and diamond. Graphite is a completely fused polycyclic benzenoid π system consisting of layers. These sheets are fully delocalized with all carbons sp^2 hybridized and thus graphite is conducting. In diamond all the carbons are sp^3 hybridized and form an insulating network of cross-linked cyclohexane chair conformers (Scheme 2.41*f*).

There is yet another spherical allotrope of carbon, the molecule C_{60} with a shape of a soccer ball and is called "footballene" or "soccerballene". Due to the similarity of the structure with a geodesic dome designed by the architect, Buckminster Fuller, the compound C_{60} has been named "buckminster fullerene". It is made by evaporating graphite electrodes into an atmosphere of helium followed by extraction of the soot like product with benzene and purified by chromatography. C_{60} is a closed shell polygon with 60 vertices and 32 faces of which 20 are hexagonal (aromatic) and 12 are pentagonal.

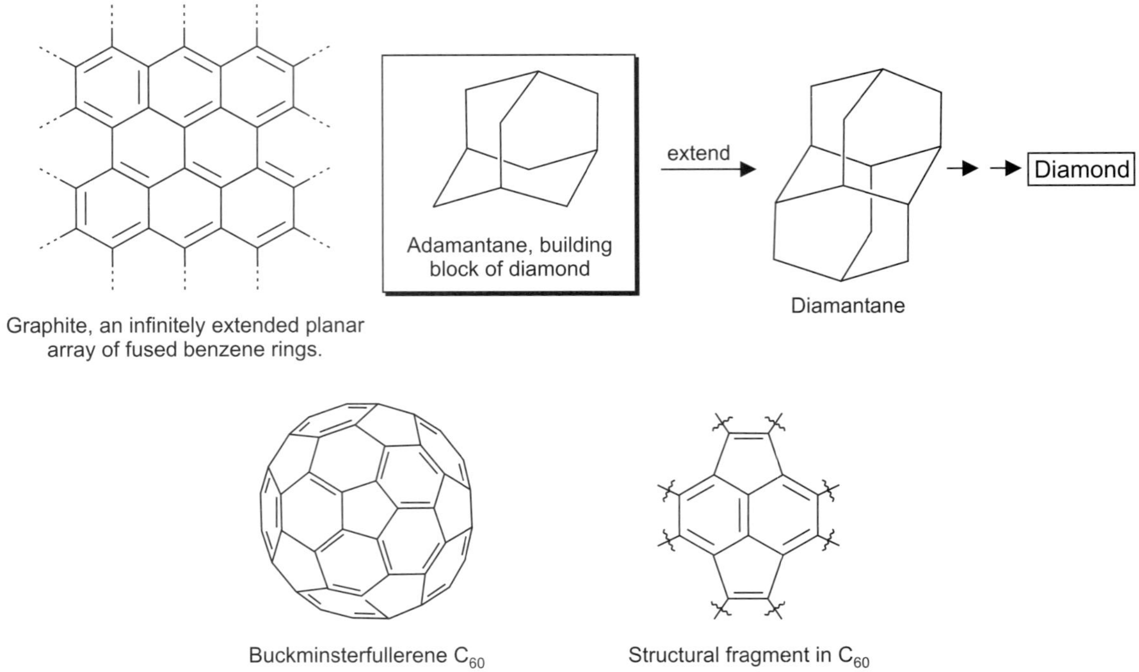

SCHEME 2.41f

C_{60} is a member of an interesting new group of aromatic compounds called fullerenes. Each carbon of fullerene is sp^2 hybridized and each six membered ring has three π-bonds through which delocalization occurs over the entire molecule (Scheme 2.41*f*). However C_{60} is not aromatic as benzene.

C_{60} does not react by electrophilic aromatic substitution as no hydrogen atoms are available to substitute, however, addition does take place. Interestingly OsO_4 adds to C_{60} in the presence of 4-*tert*-butylpyridine (4-*tert*-butylpyridine enhances the rate or osmylation). This reaction (Scheme 2.41*g*) is typical of alkenes. It may be noted that due to curvature, the constituent benzene rings in C_{60} are strained and consequently interesting addition reactions

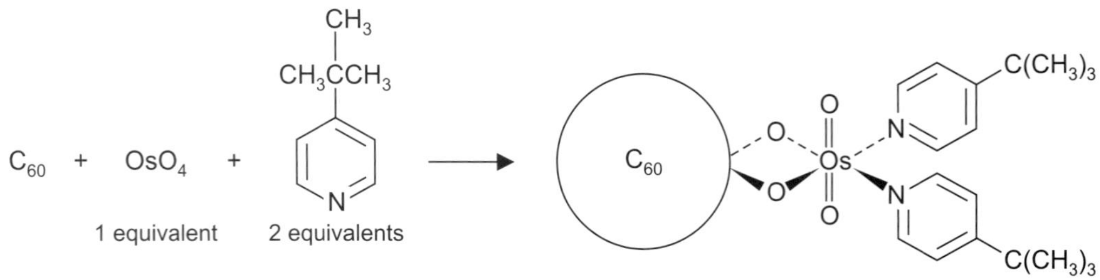

SCHEME 2.41g

are expected. For example, the ^{13}C-NMR spectrum of C_{60} shows a single peak to show that all of the carbon atoms in C_{60} are equivalent. The position of this peak at 142.6 ppm is more typical of a strained aromatic structure, than it is of a structure like naphthalene which gives a peak at 133.7 ppm.

As another example of enhanced reactivity, C_{60} undergoes a facile Birch reduction (with Li in liquid NH_3–*t*-BuOH) to give a hydrocarbon $C_{60}H_{36}$ as a major product. This reaction is a reduction of each cyclohexatriene fragment into a cyclohexene moiety, the latter being stable to Birch conditions. Treatment of (II, Scheme 2.41*h*) with 2, 3-dichloro-5, 6-dicyanobenzoquinone (DDQ) gives back the starting C_{60}.

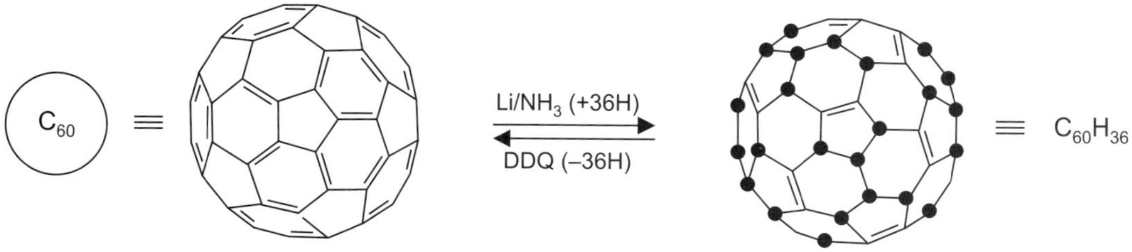

(I) (II)

SCHEME 2.41h

Electrophiles *e.g.*, bromine add to the double bonds in C_{60}, addition of chlorine however, occurs at higher temperatures (Scheme 2.41*i*). The chlorinated compound on treatment with potassium hydroxide in methanol gives a product which shows a broad peak at δ 3.7 in the ^{1}H NMR spectrum. This shows that nucleophilic substitution of chloride ion by methoxide ion. The molecular weight of 1526 amu of the compound points to the formula $C_{60}(OCH_3)_{26}$.

$$C_{60} + Br_2 \xrightarrow[25°C]{} C_{60}Br_2 + C_{60}Br_4$$

$$C_{60} + Cl_2 \xrightarrow[> 300°C]{} \underset{n = \sim 24 - 36}{C_{60}Cl_n}$$

$$C_{60}Cl_n + CH_3OH + KOH \xrightarrow{CH_3OH} C_{60}(OCH_3)_n + \overset{+}{K}\overset{-}{Cl}$$

SCHEME 2.41i

Metal ions can be captured inside the cage, *e.g.*, when graphite is soaked in a solution of the metal salt and dried and vaporized, some of the C_{60} cages that are generated have metal ions in them.

2.12 HOMOAROMATIC COMPOUNDS

In a homoaromatic compound there are one or more sp^3-hybridized carbon atoms in an otherwise conjugated cycle. An example is provided by tub shaped cyclooctatetraene which adds a proton to one of the double bonds to give a homotropylium ion (Scheme 2.41*j*) in which the aromatic sextet is spread over only seven carbon atoms as in tropylium ion. The eighth sp^3 carbon is thus forced to get placed vertically above the plane of aromatic system of seven carbons. One of these two protons is placed directly above the aromatic sextet and as expected is highly shielded (diamagnetic anisotropy).

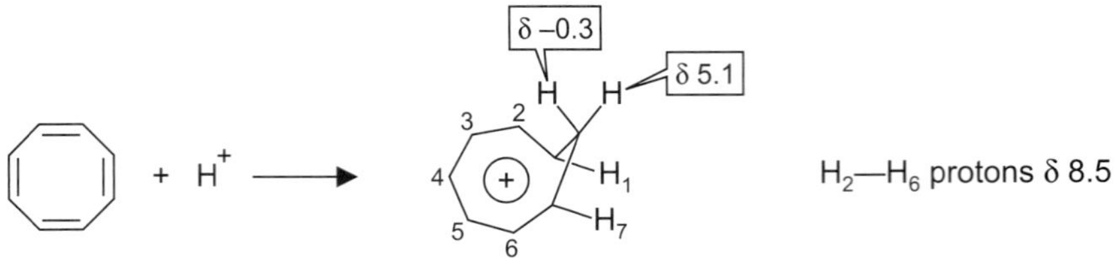

SCHEME 2.41j

2.13 HYPERCONJUGATION

The tertiary carbocations are more stable than secondary carbocations which in turn are more stable than primary carbocations (see, Scheme, 4.28). If one assumes that a methyl group can stabilize a carbocation by hyperconjugation (Scheme 2.42) then three methyl groups

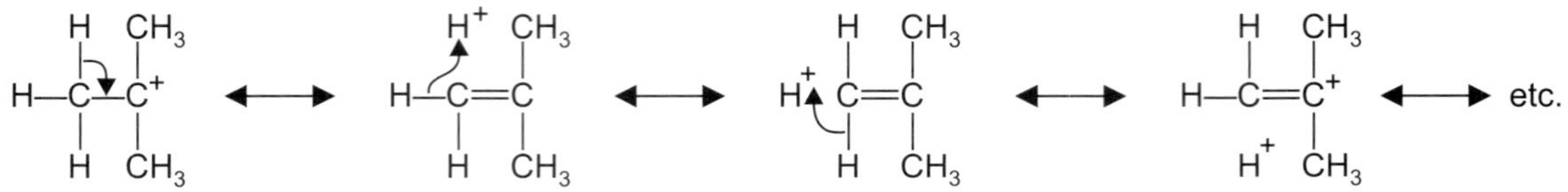

Isovalent hyperconjugation

SCHEME 2.42

will be the most effective (no change *i.e.,* "Sacrifice" of a bond, thus the name isovalent hyperconjugation). Hyperconjugation *i.e.,* the overlap between a *p* orbital and a sigma bond reflects on the carbocation stability. Alkyl groups have filled sp^3 orbitals which can overlap with the empty *p*-orbital on the positively charged carbon atom leading to stability of the carbocation (Scheme 2.43). Though the attached alkyl group rotates, one of its sigma bonds is always aligned with the *p*-orbital on the carbocation (II, Scheme 2.43). The electron pair of the

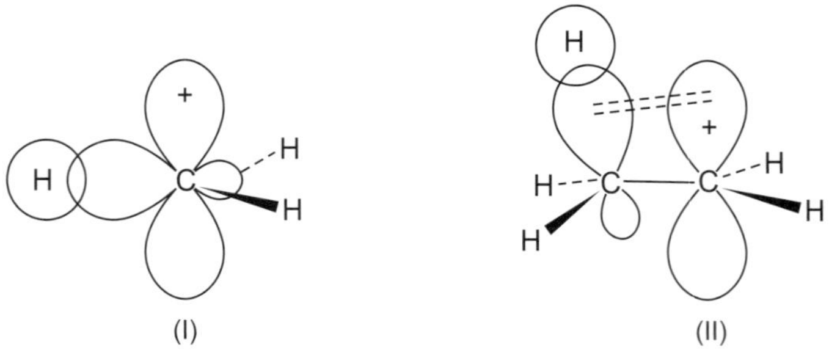

SCHEME 2.43

sigma bond slightly spreads out into the empty *p*-orbital to provide stabilization. In the methyl cation CH_3^+ (I, Scheme 2.43), however, the C—H bonds lie in the nodal plane of the vacant $2p_z$-orbital and prevent an overlap with it. The relative stabilities of tertiary, secondary and primary alkyl radicals are also explained similarly. In case of alkyl radicals there is overlap between the *p* orbital occupied by the odd electron and a σ orbital of the alkyl group (hyperconjugation). In terms of resonance theory, the ethyl radical for *e.g.,* is a hybrid of the four structures (Scheme 2.44, also see Scheme 1.4*i*).

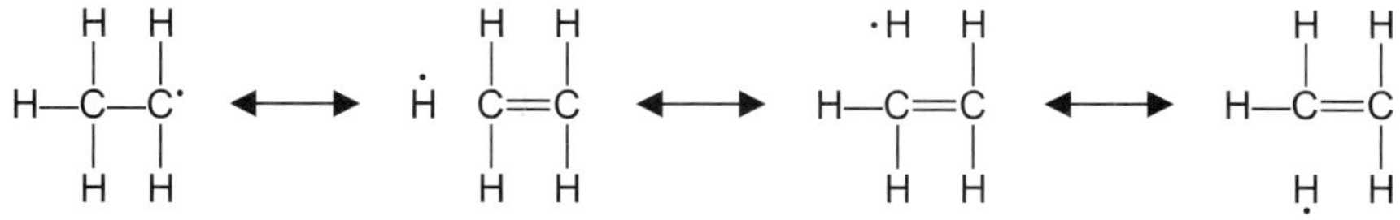

The ethyl radical (hyperconjugation in an alkyl free radical)

SCHEME 2.44

2.14 HEXAHELICENE

Two benzenes when fused together give naphthalene and this process when continued in a linear fashion gives anthracene and so on. All these molecules are achiral due to plane of symmetry. When the new benzene rings are added to naphthalene generating a curve one gets phenanthrene and ultimately hexahelicene which has three naphthalene units (Scheme 2.45) and is a chiral molecule without a stereocenter.

The terminal benzene rings in hexahelicene, are not able to occupy the same plane without coming in serious conflict with one another. The molecule of hexahelicene in thus forced to adopt a non-planar shape in which one side of the molecule must lie above the other because of crowding.

Hexahelicene is chiral due to its helical shape which could be either left-or-right-handed in orientation. The entire molecule is infact less than one full turn of the helix, but this is enough to generate chirality in hexahelicene. It has been resolved into remarkably stable enantiomers (Scheme 2.45) which display very high optical activity and correspond to 'right' and 'left-handed' spirals.

In hexahelicene the middle rings (3 and 4) lie in a plane, while the terminal rings (1 and 6) fall above and below this plane.

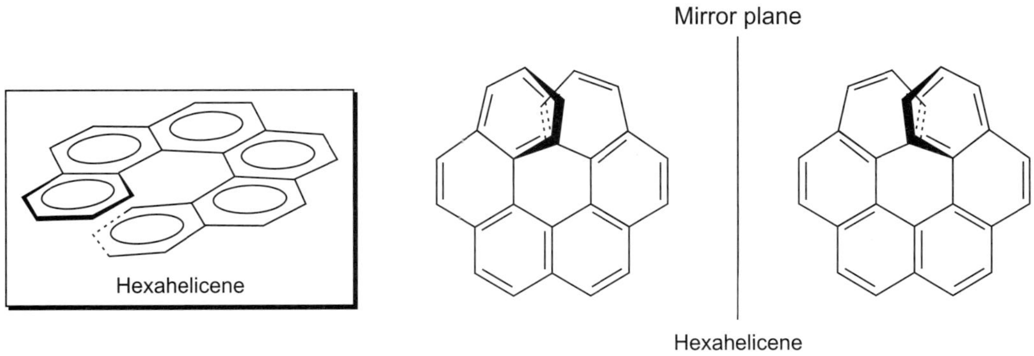

SCHEME 2.45

2.15 TAUTOMERISM

(A) Keto-enol Tautomerism

An example of this phenomenon is found in the hydration of acetylene (Scheme 2.46 H_2O, H_2SO_4, Hg SO_4). The initially formed vinyl alcohol spontaneously rearranges to the isomeric carbonyl compounds. However, the equilibrium greatly favours the keto structure. Another case is of acetone in aqueous solution.

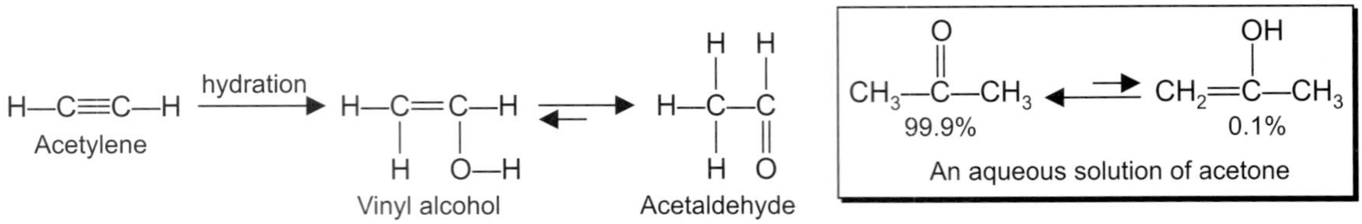

Keto-enol tautomerism

SCHEME 2.46

The greater stability of keto forms in the case of a monocarbonyl compound *e.g.,* acetone is assigned to the greater strength of the carbon-oxygen π bond (~ 364 kJ/mol) than the carbon-carbon π bond (~ 250 kJ/mol).

The interconversion of keto and enol forms is catalyzed both by acids and bases. In a basic solution, the base *e.g.,* hydroxide ion removes a proton from the α-carbon and the electrons are delocalized onto oxygen, to give an enolate. Its protonation by water gives the enol. In an acidic solution, the carbonyl oxygen is protonated and water removes a proton from the α-carbon, forming the enol (Scheme 2.47).

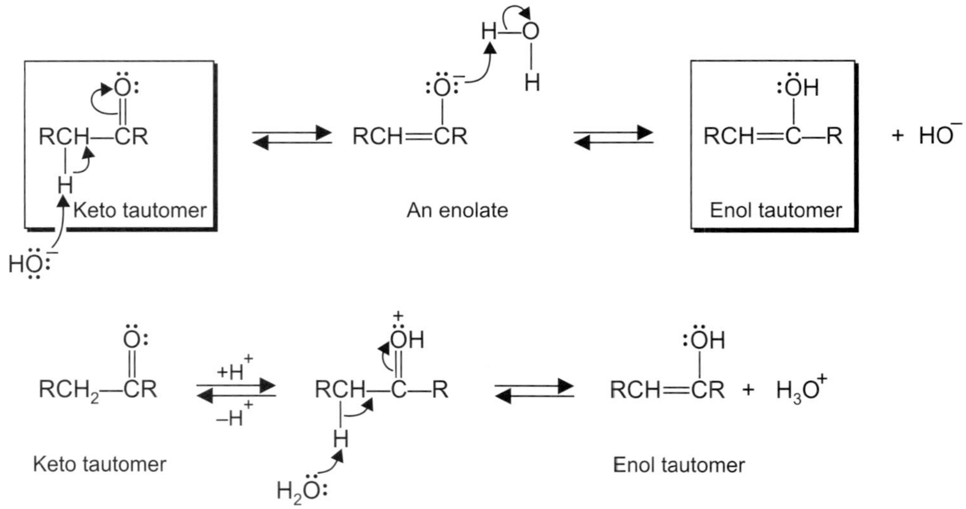

Keto-enol interconversion

SCHEME 2.47

The steps are reversed in the base- and acid-catalyzed mechanisms. In the base-catalyzed mechanism, the first step is removal of an α hydrogen and the second step is protonation of the oxygen. In the acid-catalyzed mechanism, the first step is protonation of the oxygen and the second step is removal of an α-hydrogen.

Phenol is an unusual compound since its enol tautomer is more stable than its keto tautomer. That is because the enol tautomer is aromatic but the keto tautomer is not (Scheme 2.48).

In the case of compounds having a second carbonyl group on the β-carbon *i.e.,* in 1,3-dicarbonyl compounds for *e.g.,* a β-diketone, the amount of enol present at equilibrium is far more. Thus acetylacetone exists in the enol form to the extent of about 76%. This stability of the enol form is explained by resonance stabilization of the conjugated double bonds and also via hydrogen bonding present in the cyclic form (Scheme 2.48).

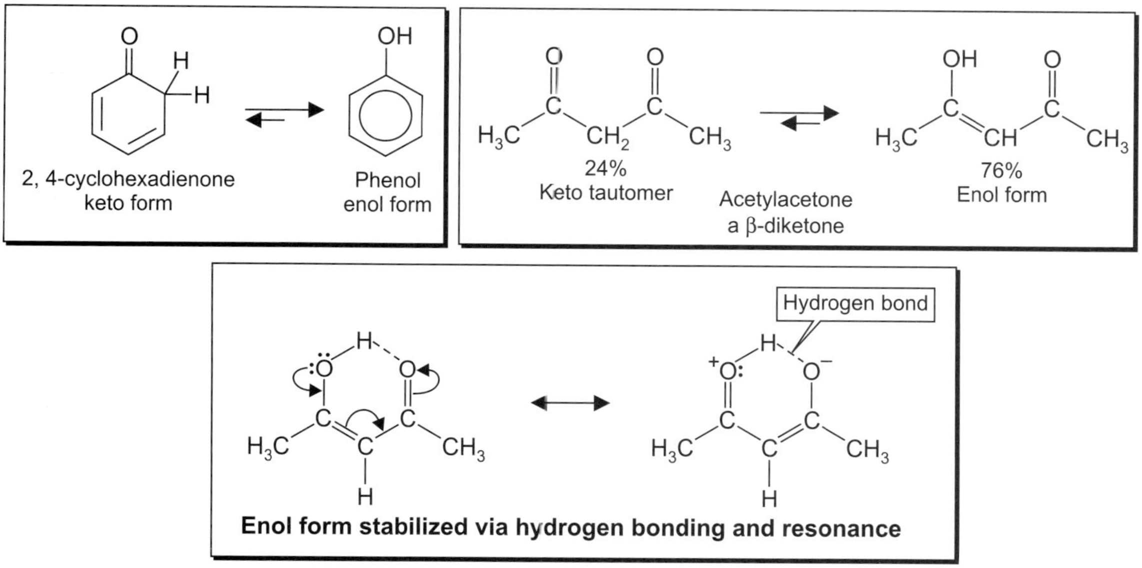

SCHEME 2.48

(B) Other Proton Shift Tautomerism

For simple phenols, equilibrium favours the enolic over keto form (Scheme 2.49). Phenol does not show any of the properties of a ketone. Compared with the situation for acetone, the enolic form of phenol has the stabilization energy of the aromatic ring with which oxygen is conjugated (*ca.* 150 kJ mol^{-1}). The ketonic form is conjugated but not aromatic, and thus has a very much smaller stabilization energy (*ca.* 20 kJ mol^{-1}). The enthalpy change on enolization is thus approximately 80 + 20 – 150 = – 50 kJ mol^{-1}, which favours the enol form. However, the keto form may predominate in some situations, *e.g.,* when a second OH group is present as in resorcinol (Scheme 2.49). In resorcinol two strongly bonded carbonyl groups are present in the keto form to overweigh the aromatic stabilization energy of the enol. Resorcinol has properties typical of both a phenol and a ketone. Therefore, this undergoes electrophilic substitutions and is reduced by sodium amalgam to 1, 3-cyclohexanedione a characteristic reaction of α, β-unsaturated carbonyl compounds. In contrast to phenol, β-naphthol displays certain ketonic

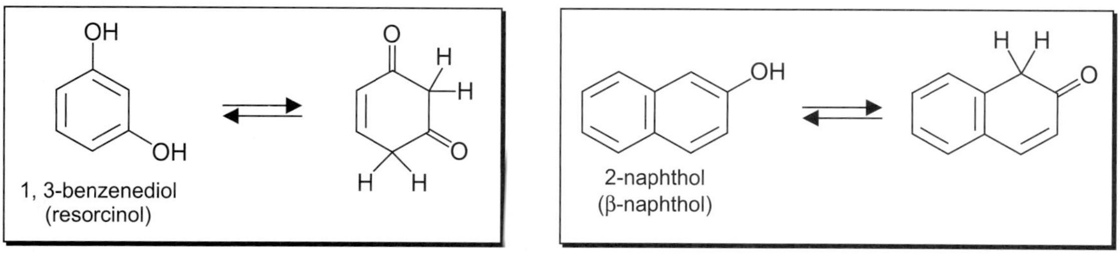

SCHEME 2.49

properties. In this case however, both the tautomers have aromatic stabilization energy (Scheme 2.49). The loss of this energy of ketonization is approximately the difference in stabilization energies of naphthalene and benzene (105 kJ mol^{-1}). Thus when compared with the ketonization of phenol for which the corresponding loss is 150 kJ mol^{-1}, the ketonization of β-naphthol is more favourable by about 45 kJ mol^{-1}.

In several heterocyclic compounds in the liquid phase or in solution the keto structure displays more stability as in the case of 4-pyridone (Scheme 2.50). In the ethanolic solution, only the keto form is detectable while the enolic form is the predominant product in the vapour phase.

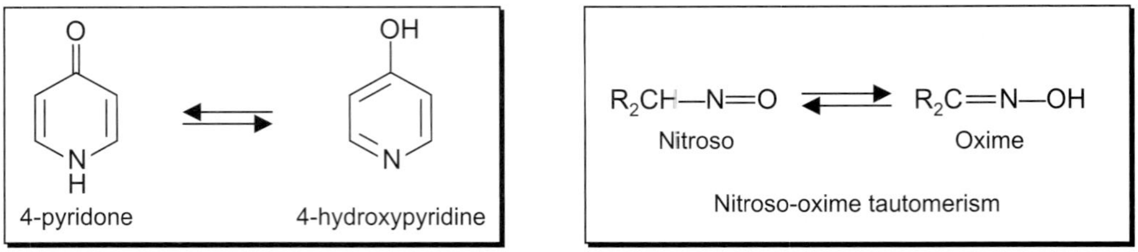

SCHEME 2.50

Nitroso compounds are stable only when these do not have α-hydrogen, otherwise the oxime form predominates (Scheme 2.50).

The aliphatic nitro compounds exist in equilibrium with *aci* form (Scheme 2.51). The *aci* form is much less stable than the nitro form. Unlike nitroso-oxime situation, in this case the enol structure of the *aci* form is less stable. This is due to resonance stabilization of the nitro form as against in nitroso case where such resonance is not possible.

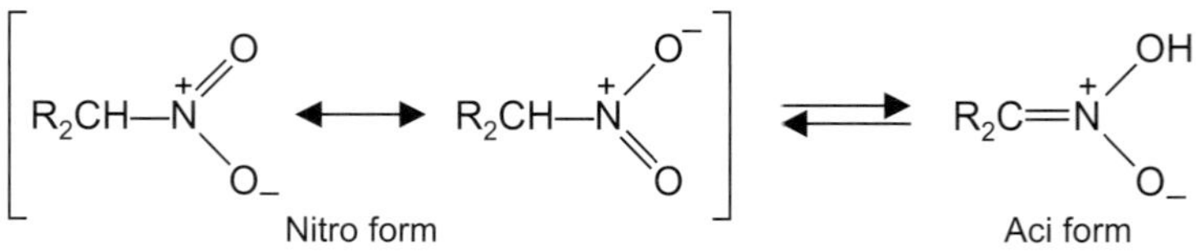

SCHEME 2.51

Primary amines react with aldehydes and ketones to give compounds with a carbon nitrogen double bond known as imines or Schiff bases (Scheme 2.52). Enamines are usually stable only when there is no hydrogen on the nitrogen.

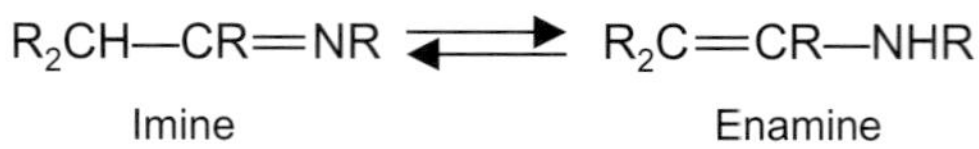

Imine-enamine tautomerism

SCHEME 2.52

Proton shift tautomerism is found in sugars where it is named ring chain tautomerism. In these cases equilibrium generally lies far to the right (Scheme 2.53). The compound that is formed as a result of an addition of an alcohol to the carbonyl group of an aldehyde is called a hemiacetal where the same carbon carries an ether and an alcohol group [R—CH(OH)(OR)].

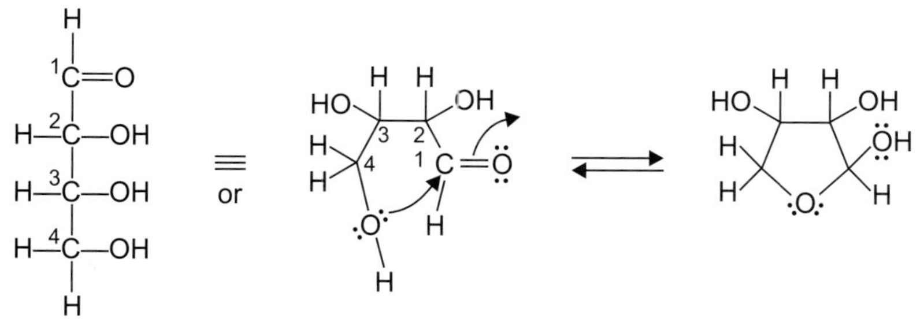

Aldose sugar Chain form Ring form

Pyran Furan

An aldose e.g., can form a five or a six membered ring which exists predominantly as a cyclic hemiacetal in solution.

SCHEME 2.53

PROBLEMS

2.1 After 1,3-butadiene accepts a proton, an allylic cation is formed. Which out of the structures I–III does not represent a resonance structure any why?

$$CH_3-\overset{+}{C}H-CH=CH_2 \quad CH_3-CH=CH-\overset{+}{C}H_2 \quad \overset{+}{C}H_2-CH_2-CH=CH_2$$

(I) (II) (III)

2.2 Structures (I–III) represents the resonance forms of a single compound. Can structures (IV and V) be considered as resonance structures of (I–III).

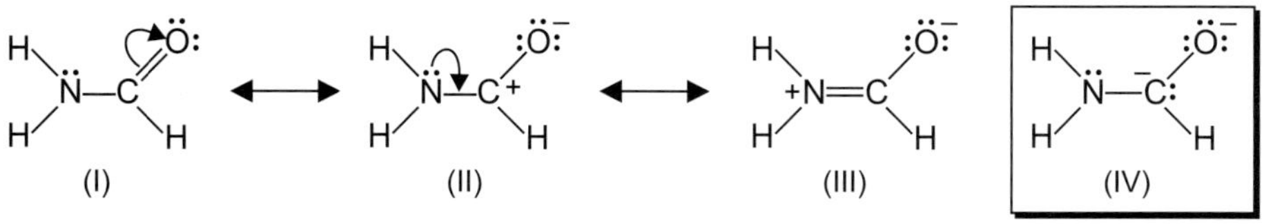

2.3 Under certain defined conditions, reaction with NBS under the influence of heat or light the following bromination of cyclohexene can be accomplished. Explain.

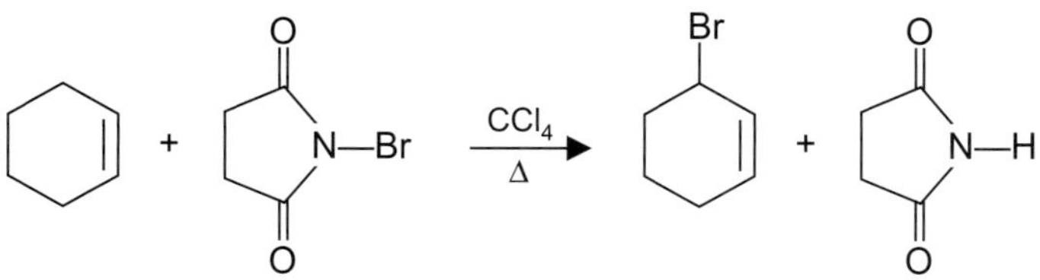

N-bromosuccinimide (NBS) Succinimide

2.4 Comment on the resonance structures (I–III of nitric acid) and (IV–V of cyanate ion).

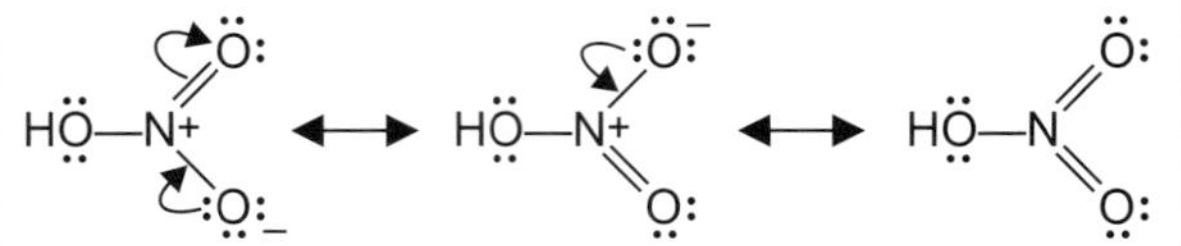

(I) (II) (III)

Nitric acid

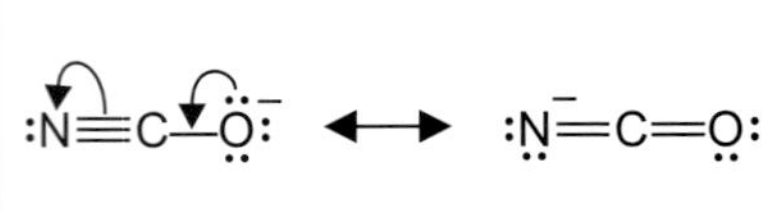

(IV) (V)

Cyanate ion

2.5 Comment on the aromaticity of the compounds (I and II), and tropone.

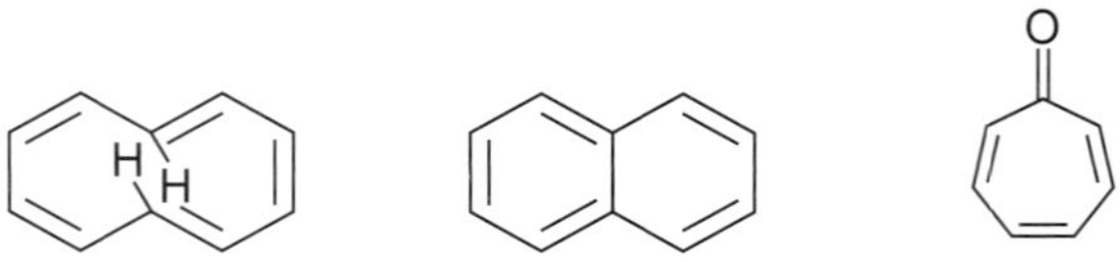

(I) (II) Cycloheptatrienone tropone

2.6 Why indene and fluorene are acidic (pK_a around 20 and 23 respectively) but less than cyclopentadiene (pK_a = 16). Why the acidity of (I) is more than of nitric acid?

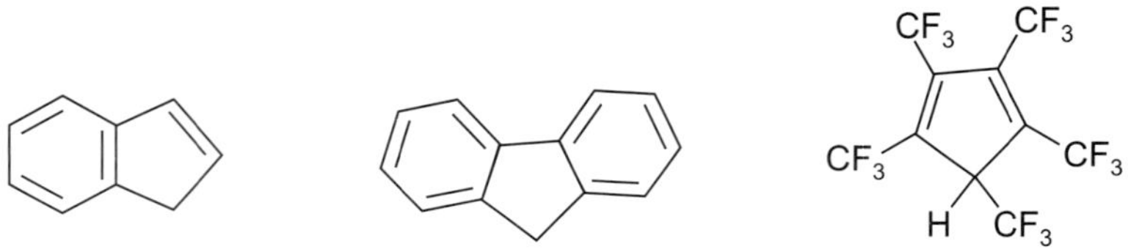

Indene Fluorene (I)

2.7 Why compared to [14] annulene (Scheme 2.28), [18] annulene is more stable?

2.8 Comment on the reactions (I and II).

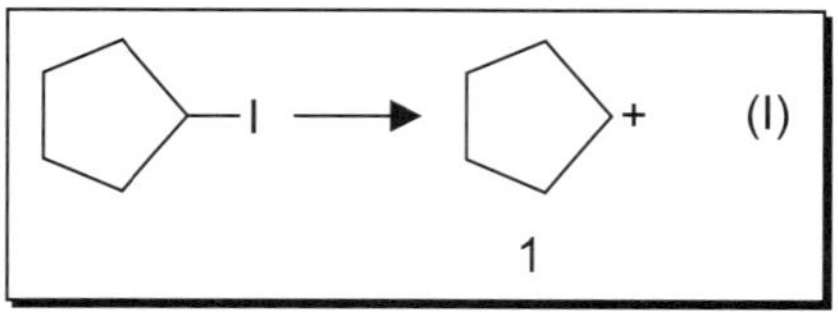

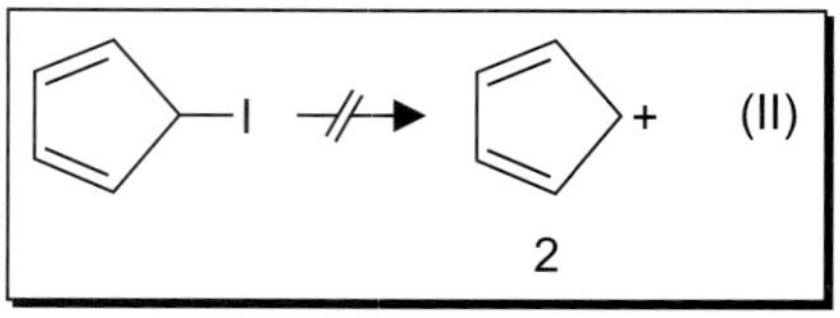

2.9 Why cyclopropenone is a stable compound while cyclopentadienone has not been prepared? Why dehydro [14] annulene is more stable aromatic compound compared to [14] annulene itself?

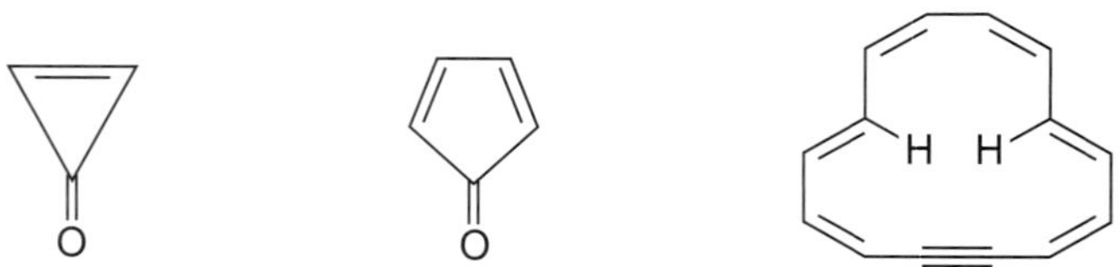

Cyclopropenone Cyclopentadienone Dehydro[14]annulene

2.10 Why the optically active ketone racemizes on treatment with a trace of acid or a base?

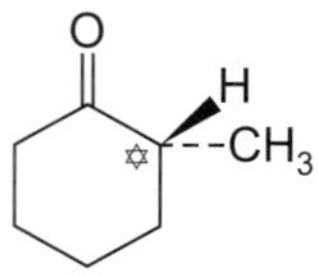

ANSWERS TO THE PROBLEMS

2.1 Structure (III) is not the proper structure since a hydrogen atom has been moved. In writing resonance structures, the positions of the nuclei of the atoms remain the same in all the resonance structures.

2.2 All structures (I–III) have 18 valence electrons and a net charge of 0, even though they differ in respect to formal charges on individual atoms. Structure IV has 20 valence electrons and a net charge of – 2. Thus it cannot be a resonance structure of (I–III). Arrangement in (V) though has the same number of electrons and the same atomic positions, but this structure has two unpaired electrons whereas all electrons in (I–III) are paired. Thus (V) also is not a resonance structure of (I–III).

2.3 The reaction is successful due to the low bond dissociation energy of allylic carbon hydrogen bonds in free radical halogenation. The reaction is initiated with a small amount of B$\dot{r}$ formed by the dissociation of the N–Br bond of NBS. The reaction steps are similar are shown for propene.

2.4 In structure (III) there are ten electrons around nitrogen thus it is not allowed resonance structure. Structures (I–II) are major contributors since the negative charge is on oxygen.

2.5 The [10] annulene (I $4n + 2$ compound) cannot adopt a planar conformation because two hydrogen atoms interfere with each other. The aromatic compound (II, naphthalene) is formed when these hydrogens are replaced by a bond.

Tropone would have an aromatic sextet if two of the C=O electrons keep away from the ring and are located near the electronegative oxygen. Thus the dipolar structures which provide a aromatic tropylium system provide a better picture of tropone. Such dipolar structures also explain the lack of ketonic properties in tropone.

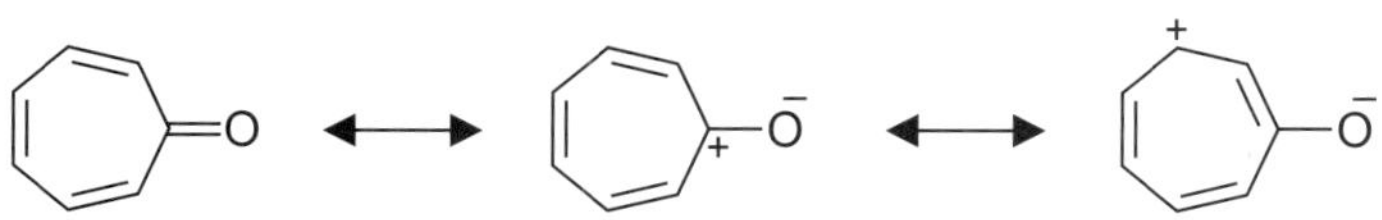

2.6 Annelation effects the electrons to be less available to the five membered ring both in indene and fluorene. For example, in fluorene as shown in the structure drawn under problem both the terminal rings are two benzene rings while the five membered ring looks like empty (annelation effect). On the removal of acidic H from the CH_2 group of the five membered ring gives an aromatic anion (now each of the outer six membered ring contains only a butadiene system). The strong acidity of (I) is due to the formation of the aromatic conjugate base which is also stabilized by the – *I* effect of four trifluoromethyl groups.

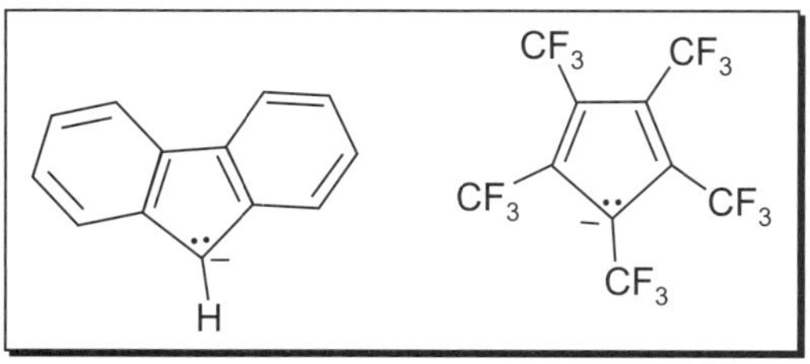

2.7 [18] Annulene is a larger annulene, therefore, the inner hydrogens do not interfere with each other. This makes [18] annulene almost planar and with its $4n + 2$ system of electrons is a stable aromatic compound.

2.8 The reaction (I) is a solvolysis reaction where the carbocation is the intermediate. The related reaction (II) does not occur under the conditions of solvolysis used for reaction (I, treatment with silver perchlorate in propionic acid), since unlike 1, species 2 are antiaromatic.

2.9 Electronegative oxygen atom pulls electrons towards it and in the process leaves only four electrons in cyclopentadienone which is thus unstable. Cyclopropenone for similar reasons should represent a potential aromatic system of two electrons. In $4n + 2$ systems there is decrease in aromaticity with decreasing planarity. [14] Annulene is not completely planar due to conflicting inner hydrogens. In dehydro [14] annulene the presence of the linear triple bond eliminates the hydrogen interferences present in the parent [14] annulene molecule. The two extra electrons of the triple bond donot play any role in the aromatic system and should not be counted.

2.10 Enolizable hydrogens are available on the ketone (I). Since the stereogenic carbon has also an enolizable hydrogen, its configuration will be inverted via the enol.

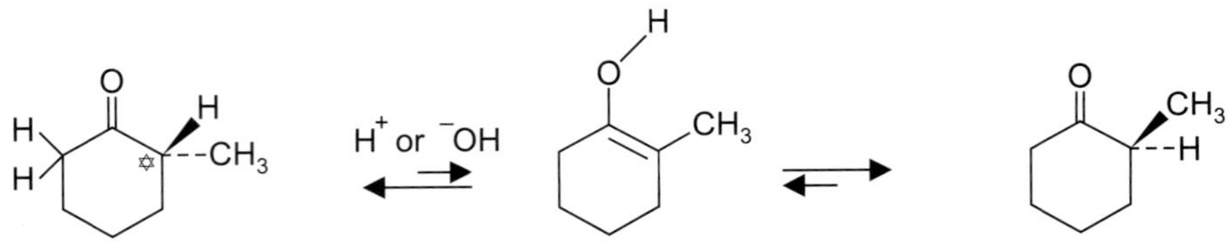

(*R*)-configuration Achiral (*S*)-configuration

CHAPTER 3

Organic Acids and Bases

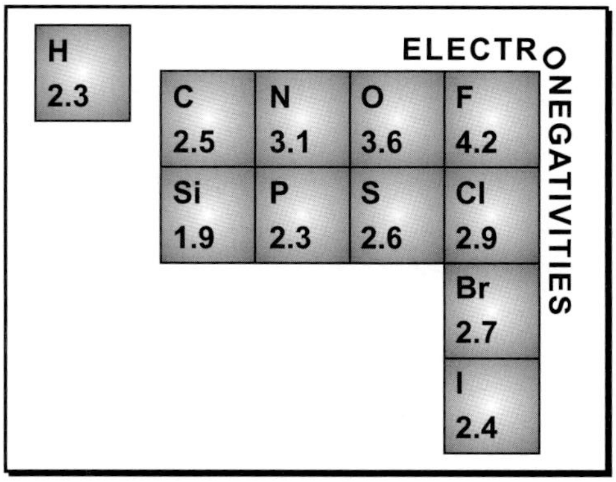

Most of the reactions in organic chemistry are either acid-base reactions outright, or they involve acid-base reaction at some stage. A knowledge of acid-base chemistry allows us to understand the relationship between structure of molecules and their reactivity, mechanisms of reactions, and throws light on several other aspects of organic chemistry.

3.1 THE BRONSTED-LOWRY CONCEPTS OF ACIDS AND BASES

According to this concept an acid is any species that can donate a proton and a base is any species that can accept a proton. Acids and bases are necessarily conjugate entities (Scheme 3.1). After accepting a proton a base becomes capable of returning that proton *i.e.,* it becomes an acid. When an acid donates its proton, it can accept that proton back to become a base. The molecule or ion which is formed when an acid loses its proton is called the conjugate base of that acid. The chloride ion is thus the conjugate base of HCl. The molecule or ion which is formed when a base accepts a proton is termed the conjugate acid of that base. The hydronium ion is thus called the conjugate acid of water.

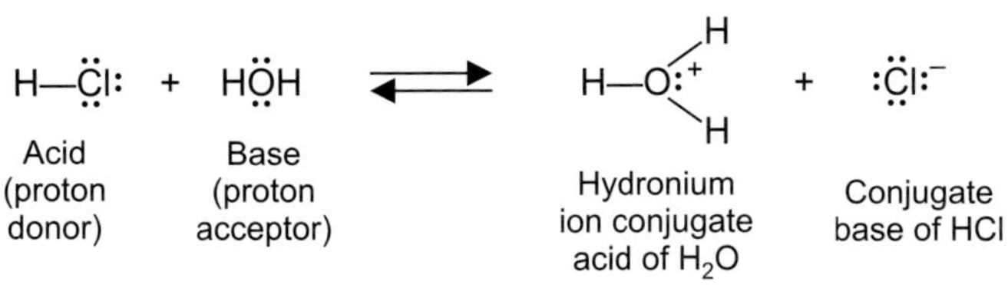

SCHEME 3.1

Like HCl, H_2SO_4 is also a strong acid and it completely transfers a proton when dissolved in water. Since H_2SO_4 has two protons which it can transfer to a base (Scheme 3.2), it is called a diprotic or dibasic acid.

$$H_2SO_4 + H_2O \longrightarrow H_3O^+ + HSO_4^-$$

$$HSO_4^- + H_2O \rightleftharpoons H_3O^+ + SO_4^{2-}$$

SCHEME 3.2

The following points may be noted:

- *The stability of the conjugate base formed on the loss of a proton determines the strength of an acid. The more stable the base, the stronger will be its conjugate acid.*
- *A stabler base is the one which can readily accomodate the electrons which it had shared with the proton.*
- *Weak bases are stable bases, as these cannot share their electrons well. Thus when the base is weaker, its conjugate acid will be strong.*

The proton transfer occurs in a stepwise fashion; the first proton transfer occurs completely, while the second transfer is only to the extent of ~ 10%. The strength of an acid is expressed by the extent of its ionization in water. The general reaction of a hypothetical acid (HA) with water is given (Scheme 3.3). The equilibrium constant for this reaction (Scheme 3.3) is written as the concentrations of the products divided by the concentrations of the reactants. The concentration of water is not considered as water is used as the solvent and its concentration is almost constant.

$$\underset{\text{Acid}}{HA} + H_2O \xrightleftharpoons{K_a} H_3O^+ + \underset{\text{Base}}{A^-}$$

$$K_a = \frac{[H_3O^+][A^-]}{[HA]}$$

$$pK_a = -\log K_a$$

$$pH = -\log [H_3O^+]$$

SCHEME 3.3

K_a is called the acidity constant and its size reflects on the relative strength of the acid. The stronger the acid the more it dissociates, giving a large value of K_a. Strong acids are almost completely ionized in water and their dissociation constants are larger than 1. Most organic acids are weak acids with values of K_a that are less than 10^{-4}. Many organic compounds are very weak acids; *e.g.*, methane and ethane are essentially nonacidic with K_a values less than 10^{-40}. Chemists usually express the acidity constant K_a as its negative logarithm pK_a. This is analogous to the hydronium ion concentration as pH. Strong acids generally have values of pK_a around 0, and weak acids, such as most organic acids, have values of pK_a that are greater than 4. Notice that the values of pK_a decrease as the values of K_a increase. The larger the value of pK_a, the weaker is the acid (Scheme 3.4).

The pH value indicates the concentration of hydrogen ions in a solution. The pH scale ranges from 0 to 14. The lower the pH, the more acidic the solution. The pH of a solution can be changed on adding acid or base to the solution (adding of base increases the pH by removing protons from the solution as a result of reacting with them to form water). The pK_a is characteristic of a particular compound, like its melting point or a boiling point, it tells how easily the compound gives up a proton. The pH scale is used to describe the acidity of a solution; pK_a describes a compound.

CH_3CO_2H	<	CF_3CO_2H	<	HCl
Weak acid				**Very strong acid**
pK_a = 4.76		pK_a = 0		pK_a = – 7

Increasing acid strength →

SCHEME 3.4

The acidity or basicity of a species depends on the structure of the species and on the nature of the solvent. Aniline is a weak base in water (which is a weak proton-donor) but a strong base in sulphuric acid; amide ion is a far stronger base in water than in liquid ammonia (since water is a stronger proton-donor than ammonia). There is a relationship between the strength of an acid and that of its conjugate base. For a given hypothetical acid (HA) to be strong, its conjugate base (A^-) must be stable in its anionic form; otherwise, HA would be reluctant to lose its proton. Thus, the stronger the acid, the weaker will be its conjugate base. One can therefore, relate the strength of a base to the pK_a of its conjugated acid. The larger the pK_a of the conjugated acid, the stronger is the base (Scheme 3.5).

$$HCl + H_2O \rightleftharpoons H_3O^+ + Cl^-$$

Strong acid — Weak base

$$CH_3\text{—}\ddot{O}H + H_2O \rightleftharpoons H_3O^+ + CH_3\ddot{O}{:}^-$$

Weak acid — Strong base

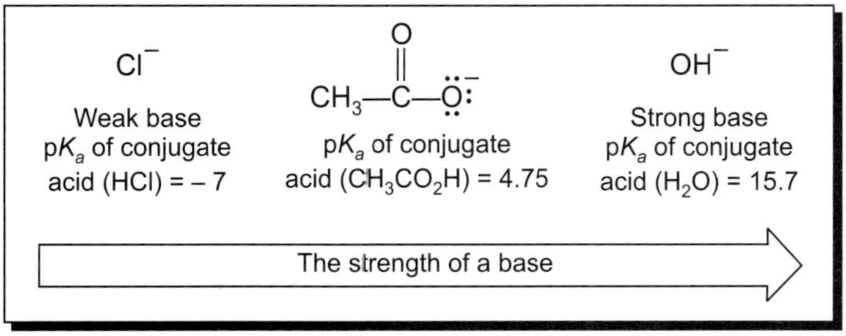

SCHEME 3.5

Thus acetic acid has greater acidity in water than in methanol, which is due to the solvents' abilities to solvate the product ions. The more polar water is better solvator of ions than is MeOH, thereby shifting the equilibrium in water more to the right. (Scheme 3.5*a*).

$$HOAc + H_2O \rightleftharpoons OAc^- + H_3O^+$$

Equilibrium lies more to the right

$$HOAc + MeOH \rightleftharpoons OAc^- + MeOH_2^+$$

SCHEME 3.5a

One finds that $HSbF_6$ is a very strong acid ($pK_a > -12$). It is so strong that it is called a "super acid", its conjugated base SbF_6^- is the weakest base; $CH_3—CH_3$ is the weakest acid (pK_a 50), while its conjugated base $CH_3CH_2^-$ is the strongest base.

3.2 THE LEWIS DEFINITION OF ACIDS AND BASES

Lewis acids are defined as electron-pair acceptors while Lewis bases are electron-pair donors (Scheme 3.6). The neutralization of HCl by NaOH in aqueous solution, involves the transfer of a proton from H_3O^+ to OH^-. The transfer of a proton can be written in an oversimplified way (Scheme 3.6). BF_3 has an empty orbital (2*p*) in the valence shell of boron and is a Lewis acid. NH_3 has an unshared pair of electrons on nitrogen and is the Lewis base. In this example, each atom takes on a formal charge; the resulting structure, however, has no net charge.

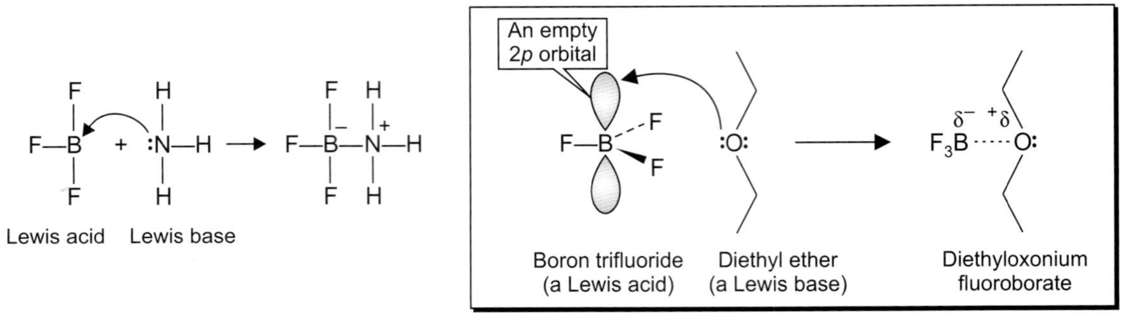

SCHEME 3.6

EXERCISE 3.1

Reduction of an imine with $NaBH_4$ in CH_3OH is not very effective. On adding BF_3 to the mixture, the reduction proceeds efficiently. Explain.

ANSWER. *Many reactions of amines and nitriles are analogous to those of aldehydes and ketones. Reductive aminations involve hydride addition (reduction) to carbon-nitrogen double bond of imine (Scheme 3.6a). BF_3 complexes with the lone pair in imine to make it more electron deficient and the complexed form behaves as a much better electrophile.*

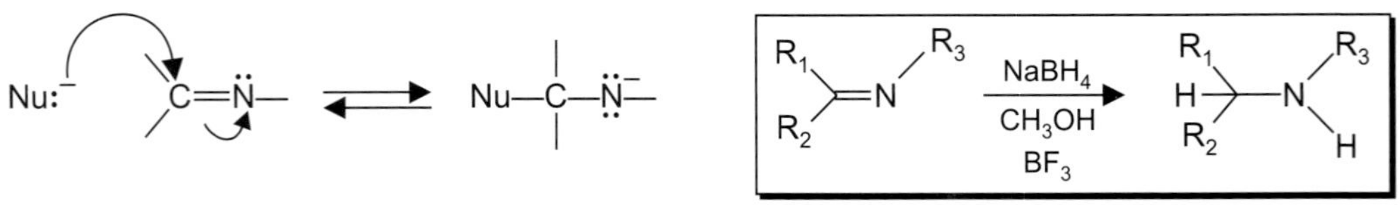

SCHEME 3.6a

Several compounds with Group IIIA elements *e.g.*, boron and aluminum are Lewis acids since Groups IIIA atoms have only a sextet of electrons in their outer shell. Several other compounds with atoms which have vacant orbitals also act as Lewis acids. Zinc and iron (III) halides (ferric halides) are often used as Lewis acids in organic reactions (Scheme 3.7). Bromine as such is not sufficiently electrophilic to attack benzene. Bromine in the presence of FeBr donates a pair of electrons to the Lewis acid $FeBr_3$ to give a reactive intermediate, which now has a weakened Br—Br bond with a partial positive charge on one of the bromine atoms (Scheme 3.7).

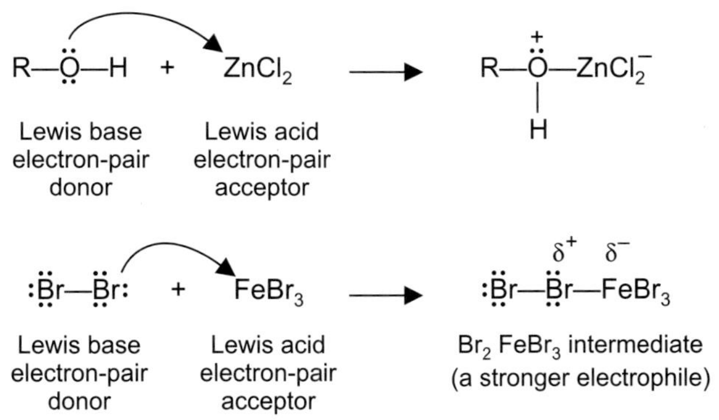

SCHEME 3.7

When the acid base reaction involves formation of a bond to some other element (especially carbon), one refers to the electron donor as a nucleophile (Lewis base) and the electron acceptor as an electrophile (Lewis acid). The nucleophile is said to "attack" the electrophile, and a curved arrow is used to indicate the flow of an electron pair from the electron donor to the electron acceptor. The movement of each pair of electrons involved in making or breaking bonds is indicated by its own separate arrow (Scheme 3.8).

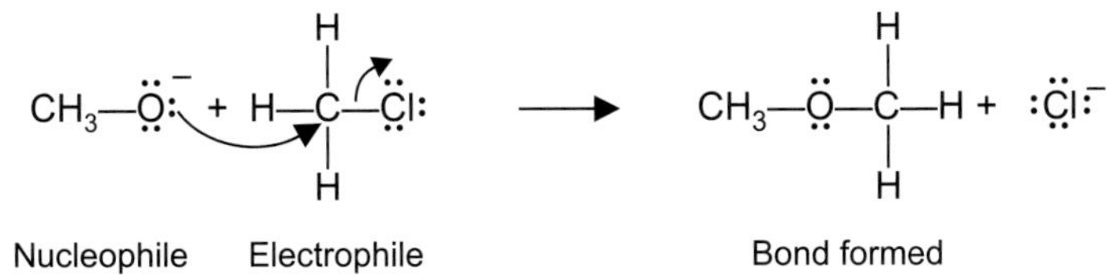

SCHEME 3.8

3.3 THE RELATION BETWEEN STRUCTURE AND ACIDITY

(A) The Strength of the Acid in Relation to Size of Atom to which Proton is Attached/Strength of the Bond to a Proton

When atoms of different sizes are involved, the stronger acid has its proton attached to the largest atom. Going down a column, the elements get larger with a decrease in their electronegativity. As a consequence the acidity of the hydrogen attached to it increases. For example, with an increase in the size of the halide ion its charge is spread over a larger volume of space and its stability increases (electron density decreases). One may also recall that the acid strength of a compound depends on the extent to which a proton can be released from it and transferred to a base. This involves the removal of a proton leading to cleavage of a bond to it to make the conjugate base more electrically negative. The strength of a bond to a proton gains significance in compounds in a vertical column of the periodic table as in the case of hydrogen halides (Scheme 3.9).

Thus among the halogen acids HF is the weakest and HI is the strongest acid; H—F bond being the strongest and H—I bond the weakest. As HI, HBr, and HCl are strong acids their conjugate bases (I^-, Br^-, Cl^-) are all weak bases (Scheme 3.9). One will notice a similar

	H—F	H—Cl	H—Br	H—I
pK_a	3.2	– 7	– 9	– 10

Increasing acid strength →

SCHEME 3.9

trend in acidities and basicities in other vertical columns of the periodic table. Thus considering the column headed by oxygen, the acidity increases in the order $H_2O < H_2S < H_2Se$. As O—H is the strongest bond, H_2O is the weakest acid, Se—H bond is the weakest, therefore H_2Se is the strongest acid, and the basicity increases in the order $SeH^- < SH^- < OH^-$.

Thiols are the sulphur analogues of alcohols. A thiol is less basic and more acidic (and more nucleophilic) than the corresponding alcohol. Unlike alcohols, thiols do not undergo substitution reactions since it is difficult to make SH group a good leaving group. Sulphur compounds being less basic are difficult to protonate and this is due to larger size of sulphur. Further S—H bonds are weaker than O—H bonds, thus thiols are stronger acids. Most alcohols have pK_a's of 16–18 while the corresponding value for a thiol is around 11. Consequently a thiol can be quantitatively converted into its conjugate base (RS^- a thiolate ion) by hydroxide (Scheme 3.9*a*).

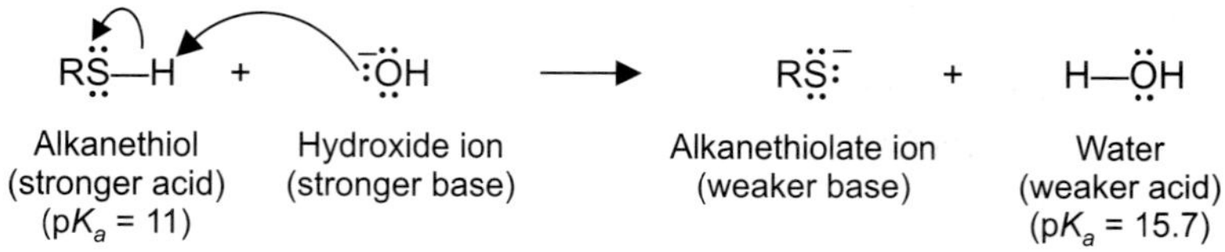

SCHEME 3.9a

EXERCISE 3.2

How do you explain the outcome of the substitution reactions (Scheme 3.9b)?

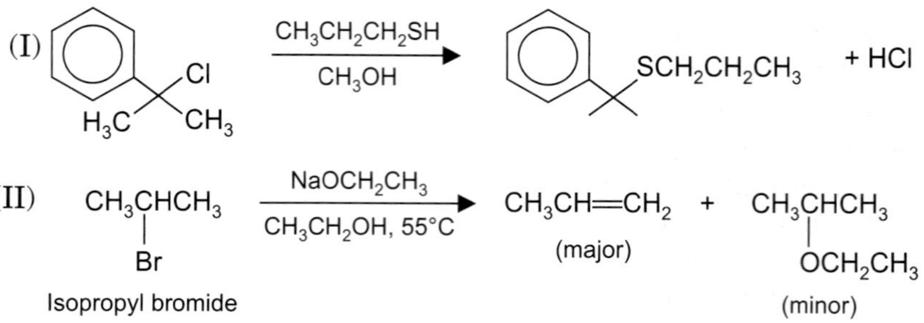

(III) 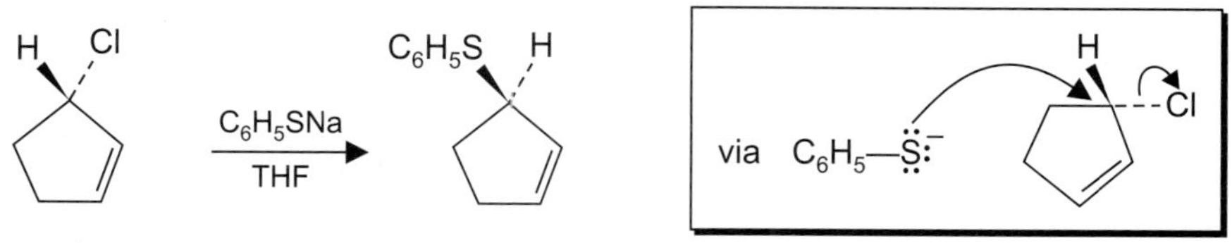

3-chlorocyclopentene

2-cyclopentenyl phenyl sulphide (75%)

SCHEME 3.9b

ANSWER. *(I) Because thiols are weaker bases, these act as nucleophiles in S_N1 reactions and an alcohol (CH_3OH) is used as a protic solvent without competition.*

(II) The major pathway for reaction of alkoxide ions with secondary alkyl halides is E2 elimination and not S_N2 reaction.

(III) Alkane thiolate ions RS^- are weaker bases than alkoxide ions RO^- and bring about S_N2 reactions even with secondary alkyl halides.

(B) Electronegativity of the Atom Bonded to the Hydrogen

When the atoms are almost similar in size, the more acidic compound will be that in which the hydrogen is attached to the more electronegative atom. Recall that elements in the second row of the periodic table have almost the same size, but these differ in electronegativities (C < N < O < F). The acidities of the hydrides increase in the order $CH_4 < NH_3 < H_2O < HF$ and therefore, the bases formed from these compounds have the relative stabilities ($CH_3^- < {}^-NH_2 < {}^-OH < {}^-F$), the more electronegative atom can bear the negative charge better. These data help to explain as to why an alcohol is more acidic compared to an amine since oxygen is more electronegative than amine. In the same way the protonated form of an alcohol is more acidic than a protonated amine (Scheme 3.9*c*).

	CH_3OH	CH_3NH_2	$CH_3\overset{+}{O}H_2$	$CH_3\overset{+}{N}H_3$
pK_a	15.5	40	– 2.5	10.7

SCHEME 3.9c

(C) The Effect of Hybridization

The hybridization of the carbon atom attached to a hydrogen greatly effects the acidity of that hydrogen. With more *p* character the acidity of a C—H bond decreases and thus acidity increases appreciably from alkanes through alkenes to alkynes. The pK_a of ethane (with an sp^3 hybridized C—H bond) is about 50, that of ethene (with an sp^2 hybridized C—H bond) is about 44, and that of ethyne (with an *sp* hybridized C—H bond) is about 25.

The *sp* hybridized atoms are more electronegative than sp^2 or sp^3 hybridized atoms. A lone pair in an sp^3 hybrid orbital (25% *s* character) is held farther from the nucleus than in an sp^2 hybrid orbital (33% *s* character) which is farther from the nucleus than the one in an *sp* hybrid orbital (50% *s* character). Since it is more favourable for the negative charge of an anion to be in an orbital closer to the positively charged nucleus, an *sp* hybridized anion is more stable than an sp^2 hybridized anion, which is more stable than an sp^3 hybridized anion (Scheme 3.10).

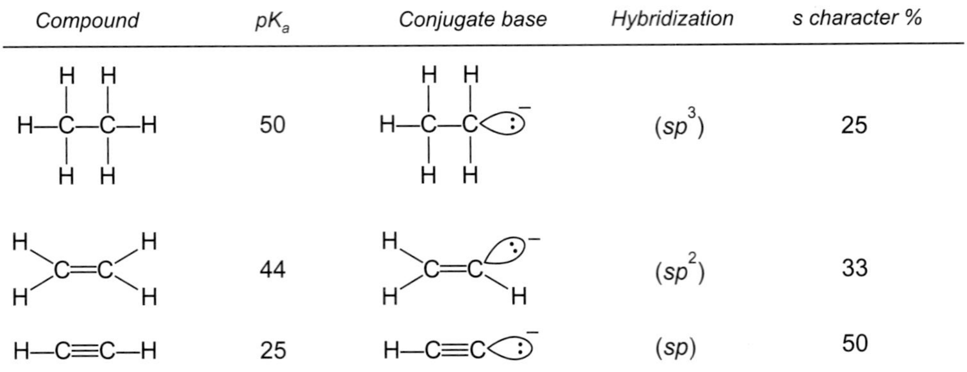

SCHEME 3.10

> *Generally, acid-base reactions always favour the formation of the weaker acid and the weaker base.*

As a result of the acidity of C—H bond in an alkyne, it is ionized on treatment with a strong base *e.g.*, amide ion in liquid ammonia to form its anion. This is a acid-base reaction which as expected favours the formation of the weaker acid and the weaker base. The position of the equilibrium for this reaction lies largely towards the right (Scheme 3.11).

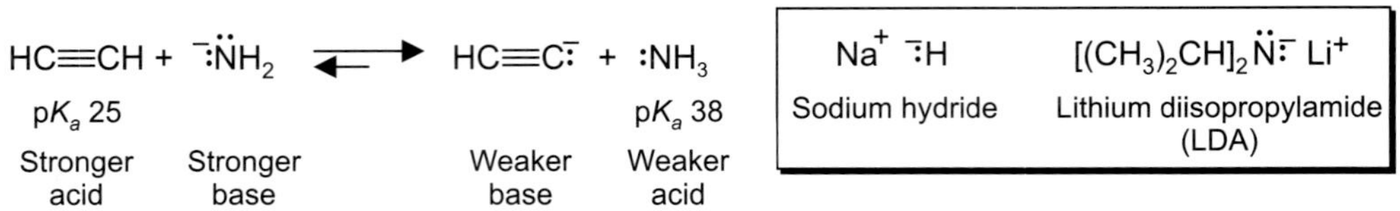

SCHEME 3.11

Other strong bases used to form acetylide anions are sodium hydride and lithium diisopropylamide (LDA). Water is a stronger acid than acetylene thus the hydroxide ion is not sufficiently strong base to convert acetylene to acetylide anion (Scheme 3.11*a*). Other examples of acid-base reactions to determine whether reactants or products are favoured at equilibrium are given (Scheme 3.11*a*).

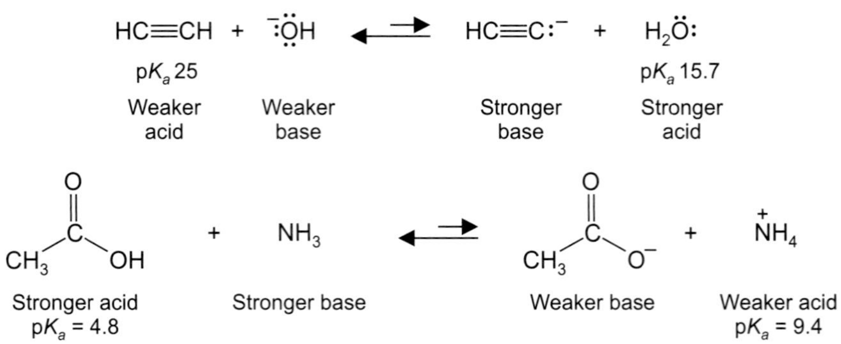

$$\underset{\substack{\text{Weaker acid}\\ pK_a = 15.9}}{CH_3CH_2OH} + \underset{\text{Weaker base}}{CH_3NH_2} \rightleftharpoons \underset{\text{Stronger base}}{CH_3CH_2O^-} + \underset{\substack{\text{Stronger acid}\\ pK_a = 10.7}}{CH_3\overset{+}{N}H_3}$$

SCHEME 3.11a

Synthesis Using Acetylide Ions

A useful reaction to form carbon-carbon bond is the reaction of an acetylide ion in an S_N2 pathway. Because acetylide anions are strong bases as well as good nucleophiles, alkylation of acetylide anions is successful only with methyl and primary halides (Scheme 3.11b). With secondary and tertiary halides, E2 elimination is the major reaction. By choosing an appropriate alkyl halide, terminal alkynes can be converted into alkynes of any desired chain length. This therefore, provides an example of alkylation reaction i.e., any reaction in which a new carbon-carbon bond to an alkyl group is formed.

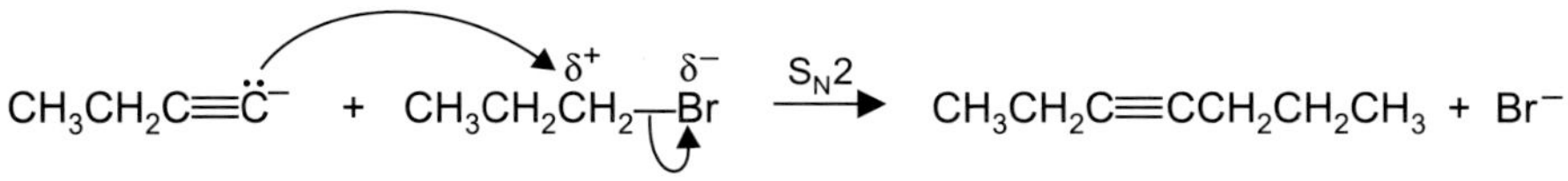

A useful synthetic reaction

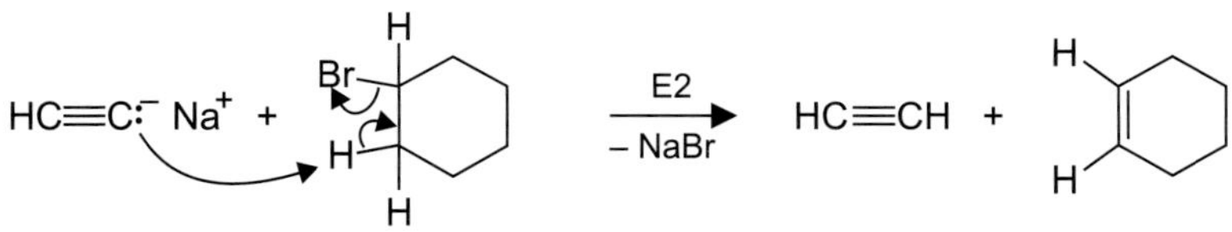

Sodium acetylide — Bromocyclohexane — Acetylene — Cyclohexene

Scheme 3.11b

(D) Inductive and Mesomeric Effects

The most important elements in organic systems which can donate protons are oxygen, sulphur, nitrogen and carbon. The acidities of the groups (Scheme 3.12) vary widely with the structure of the remainder of the molecule. One principle of significant importance in determining acidity is that any factor that stabilizes the anion of an acid relative to the acid itself increases the strength of the acid.

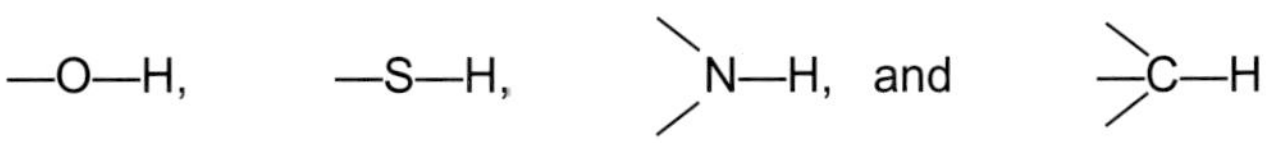

SCHEME 3.12

---

1. Acidity of O—H groups

An alcohol *e.g.*, ethanol (pK_a = 16) is a weak acid compared to a carboxylic acid *e.g.*, acetic acid (pK_a = 4.8). The resonance stabilization is not available to an alkoxide ion as is available to carboxylate anion (see Schemes 2.16 and 2.17).

One may account the acid-strengthening effect of the carbonyl group due to the difference in electronegativity between carbon and oxygen. There is a partial positive charge on the carbonyl carbon which induces a polarization of electrons in the O—H bond away from hydrogen. This electron-withdrawing inductive effect of the carbonyl group weakens the O—H bond to help the ionization of a carboxylic acid compared with an alcohol. (*i.e.*, CH_3COOH and CH_3CH_2OH).

Electron-withdrawing substituents near the carboxyl group increase the acidity of carboxylic acids. This is seen in the acidities of acetic acid and the halogen-substituted acetic acids. As the electronegativity of the halogen increases, its inductive effect increases (–*I* effect.: $I < Br \sim Cl < F$) and the strength of the halogen-substituted acid increases. Fluoroacetic acid as expected is the strongest of the monohalogenated acetic acids (Scheme 3.13). Thus a halogen atom in chloroacetic acid stabilizes the chloroacetate ion which is formed on proton loss by dispersing its negative charge. The negative charge is more spread out in the chloroacetate ion since it resides partially on the chlorine atom. The dispersal of charge makes a species more stable. An even stronger electron-withdrawing substituent such as nitro group (NO_2) enhances the acidity of a carboxylic acid even more. The pK_a of nitroacetic acid (NO_2CH_2COOH) is 1.68, while that of chloroacetic acid $ClCH_2COOH$ is 2.9.

	CH_3CO_2H	ICH_2CO_2H	$BrCH_2CO_2H$	$ClCH_2CO_2H$	FCH_2CO_2H
	Acetic acid	Iodoacetic acid	Bromoacetic acid	Chloroacetic acid	Fluoroacetic acid
pK_a	4.76	3.18	2.90	2.86	2.59

Increasing acid strength →

SCHEME 3.13

Multiple halogen substitution increases the acidity further (Scheme 3.14). Trichloroacetic acid, the strongest of the three acids, is a stronger acid than H_3PO_4.

	$ClCH_2CO_2H$	Cl_2CHCO_2H	Cl_3CCO_2H
	Chloroacetic acid	Dichloroacetic acid	Trichloroacetic acid
pK_a	2.86	1.48	0.70

Increasing acid strength →

SCHEME 3.14

The inductive effect of halogen substitution falls off rapidly with distance from the carboxyl group (Scheme 3.15). Electron releasing substituents intensify the negative charge to destabilize the anion and thus decreases acidity (Scheme 3.16).

	$CH_3CH_2CH_2CO_2H$	$CH_2(Cl)CH_2CH_2CO_2H$	$CH_3CH(Cl)CH_2CO_2H$	$CH_3CH_2CH(Cl)CO_2H$
	Butanoic acid	4-chlorobutanoic acid	3-chlorobutanoic acid	2-chlorobutanoic acid
pK_a	4.82	4.52	3.98	2.83

Increasing acid strength →

SCHEME 3.15

	H—CO_2H	CH_3—CO_2H	CH_3—CH_2—CO_2H	$(CH_3)_3C$—CO_2H
pK_a	3.77	4.75	4.88	5.05

Alkyl groups are acid-weakening to an extent depending on their + *I* effects:
$(CH_3)_3C > CH_3—CH_2 > CH_3 > H$

SCHEME 3.16

In case of alcohols (Scheme 3.17), the acidity decreases (pK_a increases) from methanol to primary, secondary and then finally tertiary systems. This has been assigned to steric

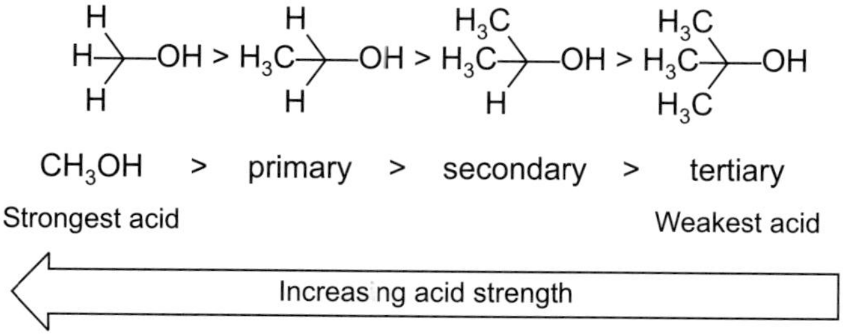

Relative pK_a values of alcohols (in solution)

SCHEME 3.17

disruption of solvation and to hydrogen bonding in the alkoxide. Both solvation and hydrogen bonding stabilize the negative charge on oxygen. Any interference with these processes leads to an increase in pK_a. The smaller the alkoxide ion the easier it is for the solvent molecules to approach and stabilize it. (Scheme 3.18, also see Scheme 3.33*a*).

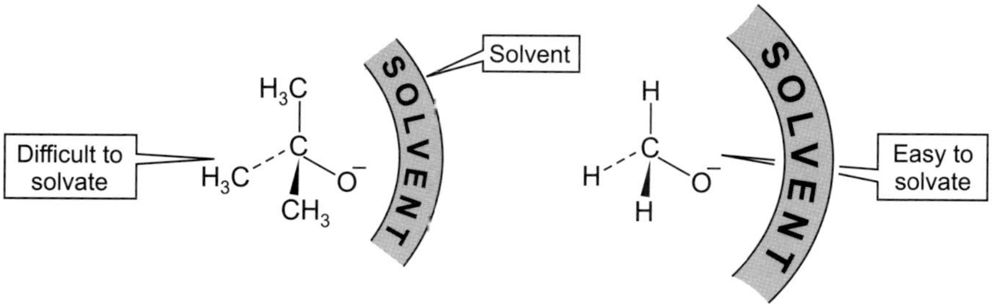

Solvent molecules can easily approach a smaller alkoxide ion to stabilize it.
SCHEME 3.18

Electron withdrawal (–*I* effect) by the halogen also stabilizes the negative charge on the alkoxide oxygen by electrostatic attraction. Thus, both 2-chloroethanol and 2, 2, 2-trifluoroethanol are more acidic (pK_a 14.3 and 12.4 respectively see Scheme 3.19) than ethanol (pK_a= 15.9).

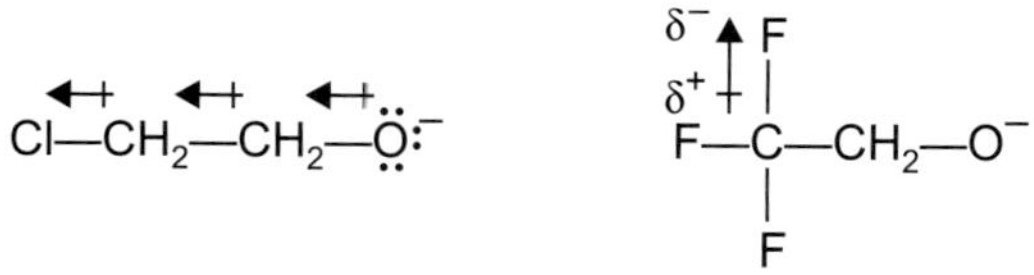

The alkoxide ion
(2, 2, 2-trifluoroethanol)

SCHEME 3.19

Compared to ethanol, phenol is more acidic (pK_a = 9.95). When phenol ionizes, the phenoxide ion is a resonance hybrid (Scheme 3.20).

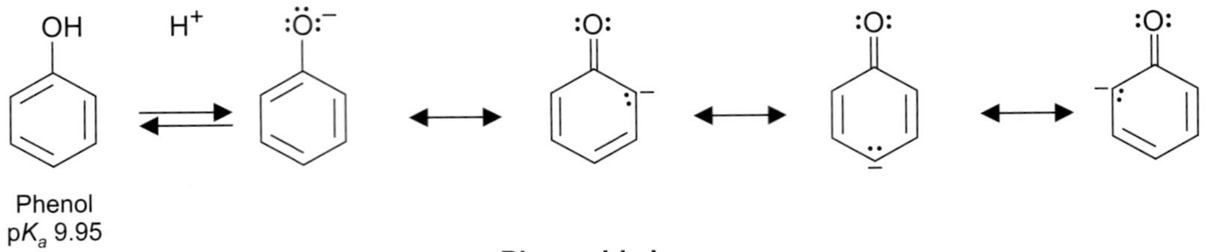

Phenoxide ion

SCHEME 3.20

Ring substituents have marked effects on the acidities of phenols by exerting both inductive and resonance effects. The inductive effect is due to electron polarization caused by differences in the relative electronegativities of bonded atoms and is relayed through sigma bonds. Consider the relative acidities of alkyl phenols and halophenols in terms of inductive effects. *p*-Cresol is a weaker acid than phenol, while *m*-chlorophenol is a stronger acid than phenol (Scheme 3.20*a*). The alkyl substituents are "electron-releasing" toward the aromatic ring. Because they are electron-releasing, they destabilize phenoxide ion-contributing resonance structure and reduce the acidity of alkyl substituted phenols (Scheme 3.20*a*). The inductive effect of the halogens operates in opposite direction than that of alkyl substituents. Because the halogens are more electronegative than carbon, they withdraw electron density from the aromatic ring and thereby stabilize the phenoxide ion and enhance the acidic-strengthening effect in halophenols. Both inductive and resonance effects are operative in nitrophenols. In *m*-nitrophenolate anion the negative charge is stabilized by the inductive effect of the electron-withdrawing nitro group therefore, *m*-nitrophenol is a stronger acid than phenol (Scheme 3.20*a*).

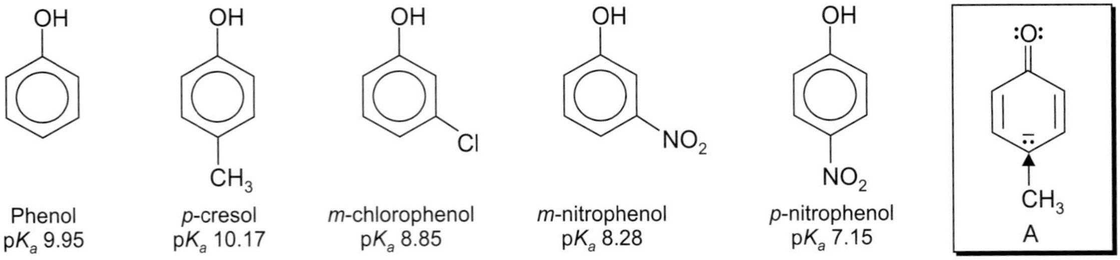

SCHEME 3.20a

The negative charge of the *p*-nitrophenolate anion is stabilized both by the inductive effect as well as a resonance effect. Additional delocalization of charge, beyond to that available in the phenolate anion and in the *m*-nitrophenolate anion is possible for the *p*-nitrophenolate anion. This extra stabilization of the conjugate base is reflected in the greater acidity of *p*-nitrophenol. (Scheme 3.20*b*). Increasing the number of nitro groups on phenol enhances the acidity, picric acid (2, 4, 6-trinitrophenol) with pK_a 0.38 is a strong acid, even stronger than phosphoric acid (H_3PO_4) and is comparable with trifluroacetic acid. (Scheme 3.20*b*).

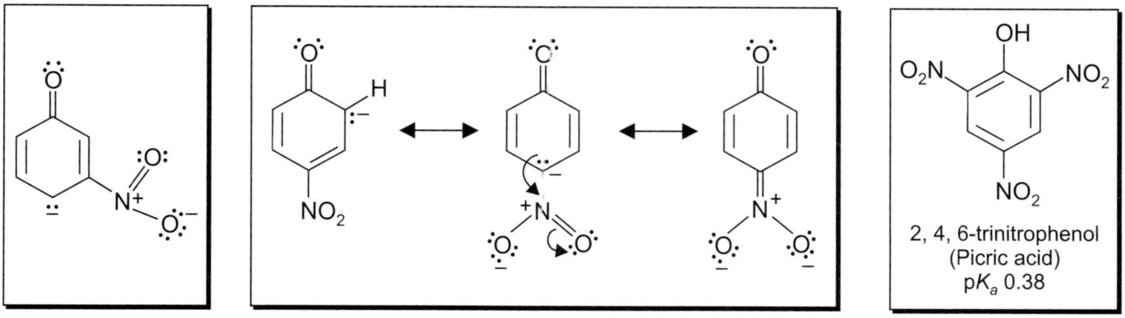

SCHEME 3.20b

The negative charge of the carboxylate ion is shared by the two carboxylate oxygen atoms and is not effectively delocalized by the aromatic ring (Scheme 3.21). In the phenoxide ion even though the negative charge is delocalized by the aromatic ring, benzoic acid is a stronger acid than phenol. The negative charge in benzoate ion, is equally shared by two electronegative oxygen atoms while in the phenoxide ion most of the negative charge resides only on the single oxygen atom.

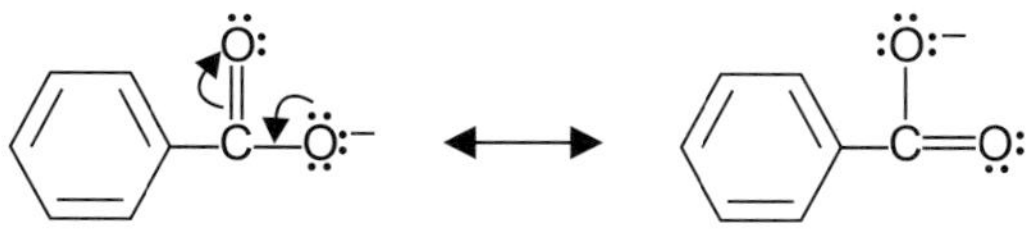

SCHEME 3.21

Since the benzene ring is not involved in resonance stabilization of the carboxylate group, substituents on a benzene ring influence acidity mainly by the inductive effect. An electron-withdrawing group *e.g.*, the —NO_2 group, that is substituted either in the *meta* or the *para* position increases the acidity of a benzoic acid while an electron-releasing group in the same positions decreases acid strength (Scheme 3.22).

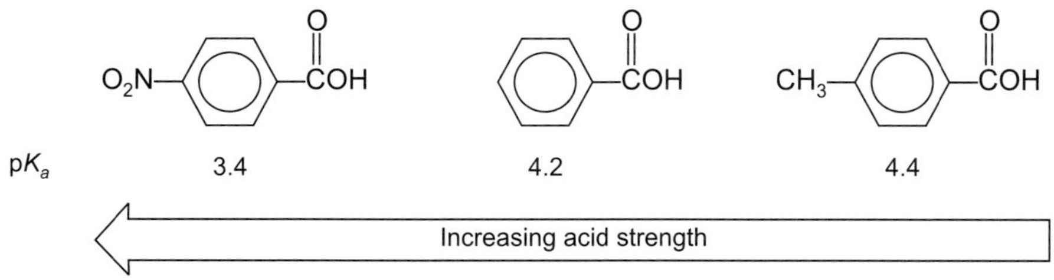

SCHEME 3.22

For nitrosubstituted benzoic acids, the *ortho* isomer is the most acidic pK_a = 2.17 (The pK_a of *p* isomer is 3.4, Scheme 3.22). A nitro group withdraws electrons inductively. It also withdraws electrons *via* resonance provided it is *ortho* or *para* to the COOH group. The *ortho* substituent is far stronger due to greater inductive electron withdrawal from a closer position.

EXERCISE 3.3

Which of the two benzoates (I or II, Scheme 3.22a) will show a faster rate of ionization.

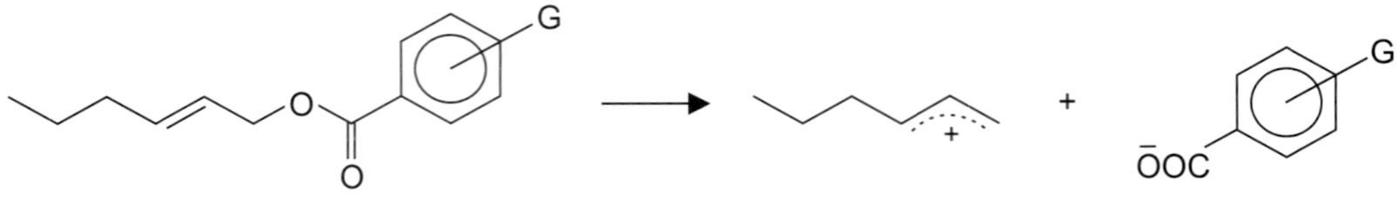

(I) G = NO_2, (II) G = CH_3

SCHEME 3.22a

ANSWER. *The substituent which stabilizes the benzoate ion will make the benzoic acid more acidic (decreased pK_a). A nitro group is an electron-withdrawing group and it increases the acidity of a benzoic acid. Thus since NO_2 group stabilizes the benzoate ion with its presence, the rate of ionization will go up, consequently (I) will display a faster rate of ionization. Methyl ion is an electron-releasing group, it will make the benzoate ion less stable to lower the rate of ionization.*

2. Acidity of sulphonic acids

Benzenesulphonic acid and its derivatives *e.g.*, *p*-toluene sulphonic acid are useful as strong acid catalysts in organic synthesis, as useful alternatives to sulphuric acid for, while being strong acids they do not bring about the side reactions, *e.g.*, oxidation and sulphonation which is characteristic of sulphuric acid. The strong –*I* effect of sulphone group (-SO_2-, sulphur is slightly more electronegative than carbon, and thus sulphonate anion is stabilized more than the carboxylate anion) and greater delocalization of the charge in the sulphonate anion compared to that in the carboxylate anion makes sulphonic acids very much stronger than carboxylic acids (Scheme 3.23).

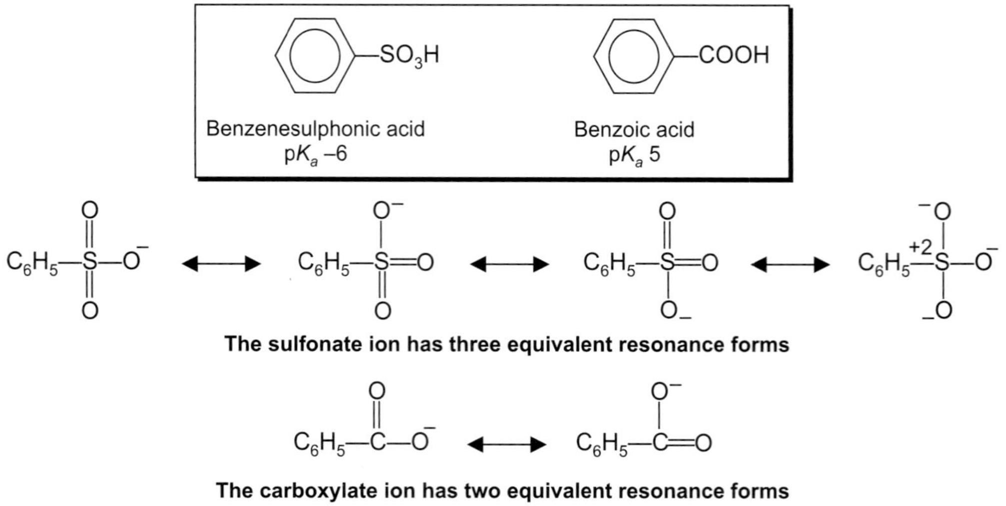

SCHEME 3.23

Super Acids

Pure sulphuric acid is a stronger acid than sulphuric acid in water since aqueous sulphuric acid is in fact H_3O^+. *Super acids are defined as compounds which are even stronger acids than 100%* H_2SO_4. *An example of a superacid is fluorosulphonic acid. The inductive effect of the electronegative fluorine makes this a stronger acid than sulphuric acid. This super acid is made further stronger on addition of a Lewis acid e.g.,* SbF_5. *Antimony pentafluoride complexes with the conjugate base of fluorosulphonic acid, decreasing its basicity. This mixture, termed magic acid is very strong and can protonate extremely weak bases e.g., electron pair of a carbon-carbon pi bond (Scheme 3.23a). The carbocation thus generated is stable enough in the solution of magic acid to be studied.*

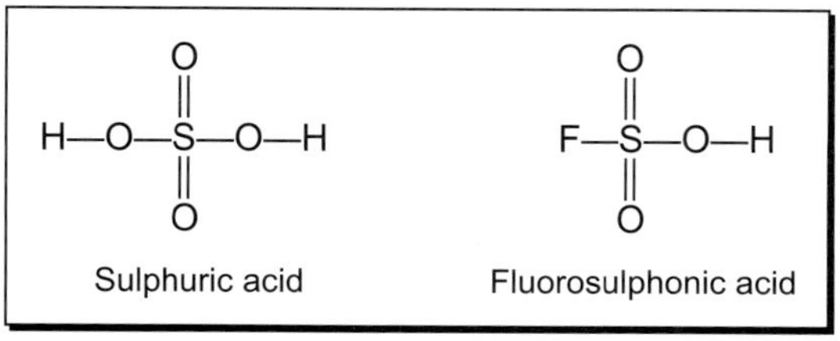

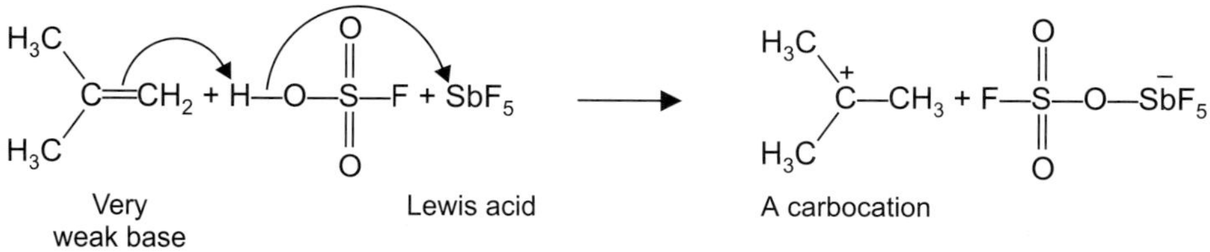

SCHEME 3.23a

G. Olah received Nobel Prize (1994) for this work on carbocations and superacids.

3. Acidity of N—H groups

Because nitrogen is much less electronegative than oxygen, the pK_a *e.g.,* of an amide is more than ten units larger than that of a carboxylic acid. Amides with one or two protons on the nitrogen are deprotonated at nitrogen to form an amidate ion which is resonance stabilized. The amidate ion is synthetically useful nucleophile, which is initially formed during Hoffmann rearrangement (see, Scheme 15.15). When N—H group is bonded to two carbonyl groups, its acidic properties are enhanced because the negative charge of the conjugate base is delocalized over both oxygens and the nitrogen. The pK_a of phthalimide is 8.3 and in aqueous

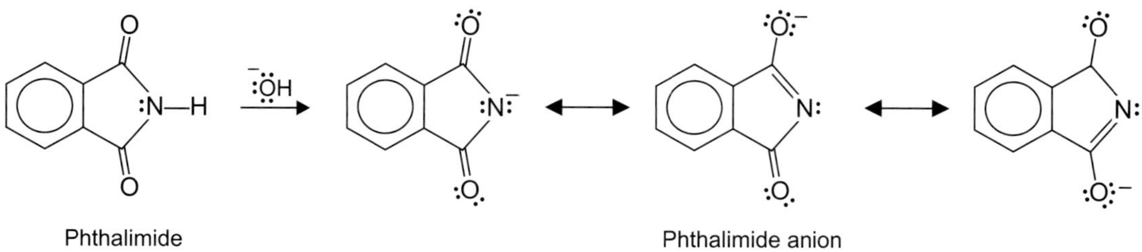

SCHEME 3.24

basic solution, the imide is converted almost completely into the anion (Scheme 3.24). Use is made of the acidity of imides in a method for forming C—N bonds *e.g.,* in Gabriel synthesis of amines. Phthalimide anion has nucleophilic properties and can enter into displacement reactions with alkyl halides.

4. Acidity of C—H groups

The sp^3 C—H bonds are less acidic than N—H bonds so that strong bases must be used to break them. Allyl cations and radicals are stabilized by resonance (see, Scheme 2.11). The sp^3 hybridized C—H bond in propene (pK_a = 40) is more acidic than in propane (pK_a = 50). The allyl anion formed by deprotonation of propene is stabilized by resonance (Scheme 3.25). When the conjugation of the allyl ion is extended further, the additional resonance contributors further stabilize the anion. The deprotonation of 1, 3-pentadiene (CH_2=CH—CH=CH—CH_3, pK_a = 33) is thus easier when compared to propene (Scheme 3.25).

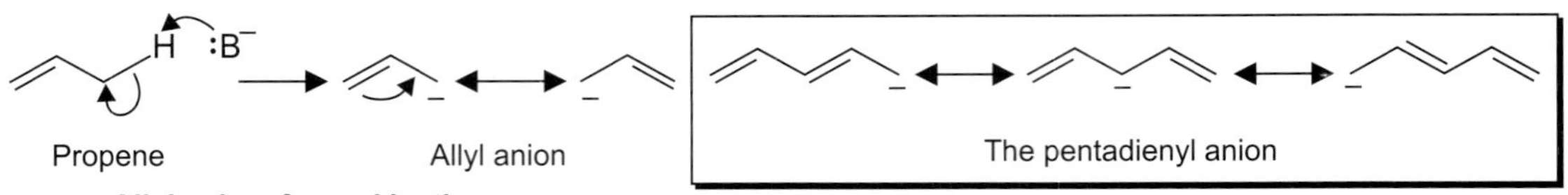

Allyl anion, formed by the deprotonation of propene

SCHEME 3.25

An anion formed from deprotonation of a C—H bond adjacent to a carbonyl group (α-proton) is also stabilized by resonance (Scheme 3.26).

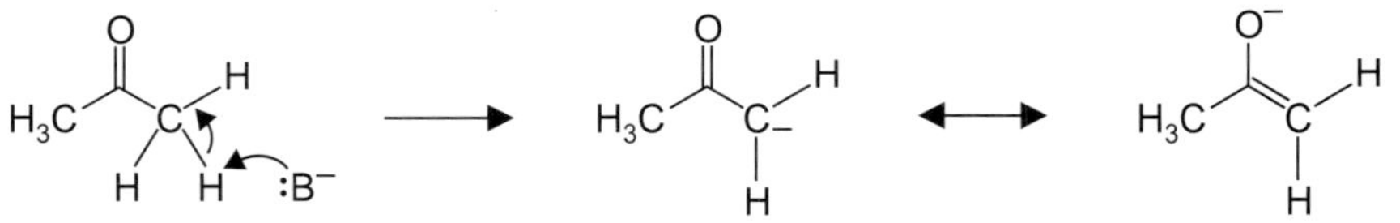

SCHEME 3.26

The resulting anion is stabilized not only because of delocalization as in the case of allyl anion, but also because one of the resonance contributors has negative charge on the more electronegative oxygen atom. Thus the pK_a values of aldehydes and ketones (19–21) are not only much lower than those in alkanes (Scheme 3.27) but much lower than the pK_a values of

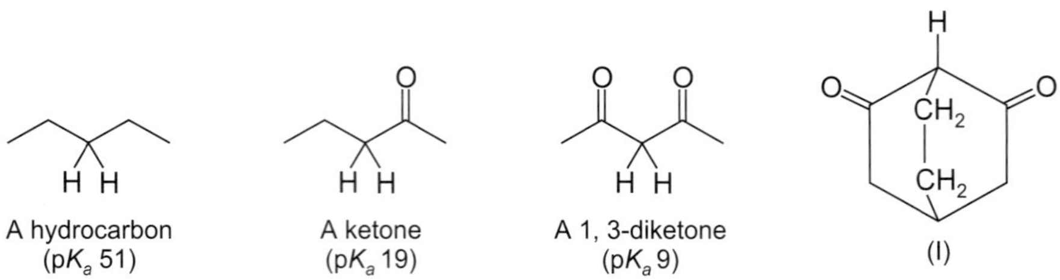

SCHEME 3.27

ethene (pK_a = 44) and ethyne (pK_a = 25), but higher than those of alcohols (pK_a = 15–18). When two carbonyl groups are adjacent to the same carbon atom as in 1, 3 relationship (in β-diketones deprotonation gives an anion with charge delocalization over three atoms (two oxygen atoms and one carbon atom) much the same way as in the pentadienyl anion. (Scheme 3.28). Other β-dicarbonyl compounds are β-keto esters ($RCOCH_2COOR$) and β-diesters ($ROCOCH_2COOR$). The anion resulting from deprotonation of a 1, 3-diketone is more stable than a simple enolate anion, and as a result, 1, 3-diketones (and other 1, 3-dicarbonyl compounds) are more acidic than simple ketones. However, the effect of the second carbonyl group on acidity is not as great as that of the first. The 1, 3-diketone (I, Scheme 3.27) lacks the normal acidic properties of a 1, 3-diketone, since the corresponding enolate would have a bridgehead double bond (Bredt rule violation, see Scheme 1.37).

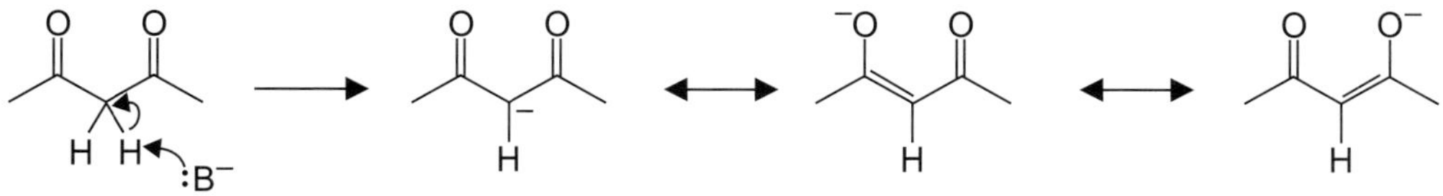

Resonance structures of the enolate anion of a 1, 3-diketone

SCHEME 3.28

Generation of aromaticity can significantly promote the loss of a proton. Deprotonation of cyclopentadiene generates the cyclopentadienyl anion, which contains six (that is, $4n + 2\pi$) electrons in a Hückel aromatic system. The pK_a of cyclopentadiene (16) is much lower than that of 1, 4 pentadiene (40) and is, in fact, very close to that of water (15.7), despite the cleavage of a C—H rather than an O—H bond (Scheme 3.29). However, 8π electron cycloheptatrienyl anion is very difficult to prepare. Thus cycloheptatriene is much less acidic than cyclopentadiene. The decreased acidity is due to the fact that the cycloheptatrienyl anion with 8π electrons is not a stable Hückel aromatic system.

When the C—H bond is adjacent to one group of $-M$ type; pK values lie in the range 10–20. Reactions involving anions derived from these substrates are the aldol condensation (>CH—CHO and >CH—COR), the Claisen condensation (>CH—CO_2R) and the Thorpe reaction (>CH—CN). When the C—H bond is adjacent to two groups of $-M$ type as in ethyl acetoacetate, $CH_3COCH_2CO_2Et$, and diethyl malonate, $CH_2(CO_2Et)_2$, the pK values of such compounds are in the range 4–12. When the C—H bond is part of cyclopentadiene or a derivative, the compounds are acidic because the derived anion, having six π-electrons and fulfilling the criteria for aromaticity, is strongly resonance stabilized (see, Scheme 3.29).

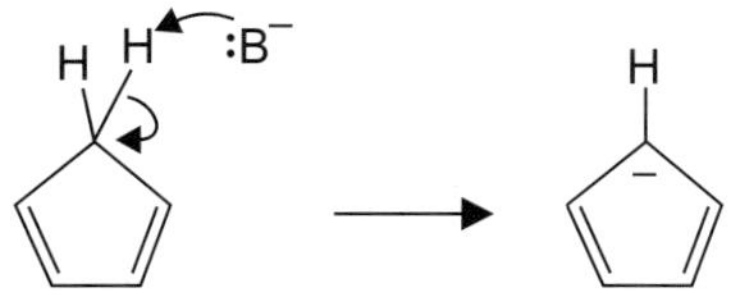

Cyclopentadiene has a pK_a of 16

Aromatic anion

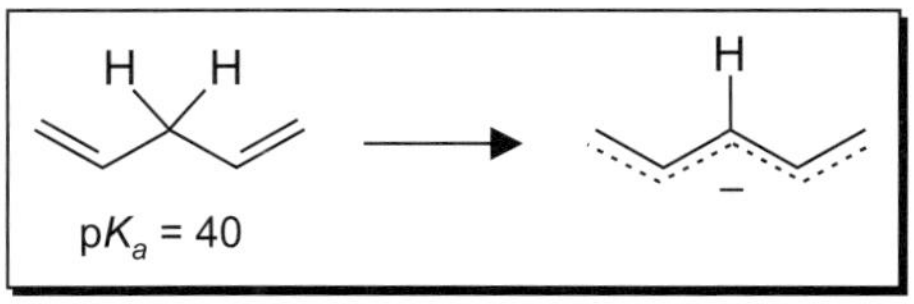

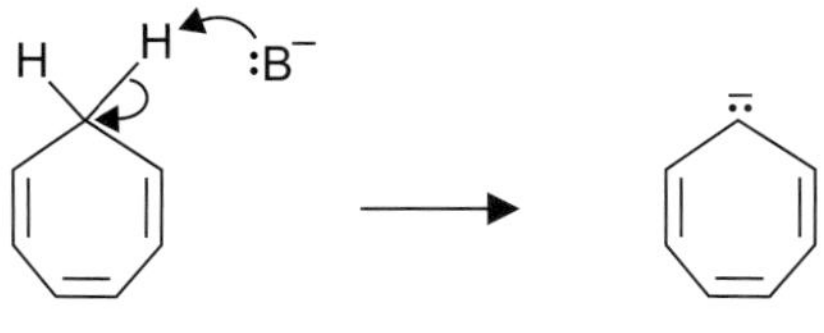

Cycloheptatriene a pK_a of about 40

Cycloheptatrienyl anion antiaromatic

SCHEME 3.29

Carbonyl is one of the number of groups of $-M$ type which have a marked acid promoting influence. The cyano group (–C≡N) and the nitro group ($-NO_2$) also withdraw electrons and cause α-hydrogens to be somewhat acidic. The resulting carbanions are stabilized by resonance (Scheme 3.30).

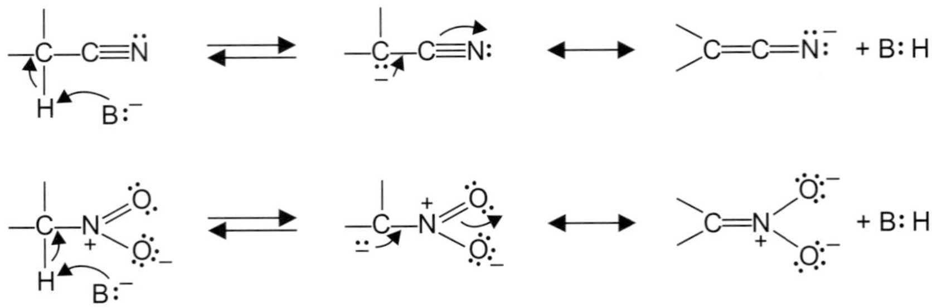

SCHEME 3.30

The α-hydrogen atoms to an alkyl side chain at the 2 or 4 (α or γ) position of a pyridine ring are comparable in acidity with methyl ketones and therefore, readily undergo base catalysed reactions. Thus, reaction of a carbonyl group with the alkyl anion formed by base provides a method for extending the pyridine side chain as in the case of γ-picoline. The acidity at this position is due to the formation of a resonance stabilized anion with negative charge on nitrogen (Scheme 3.30*a*).

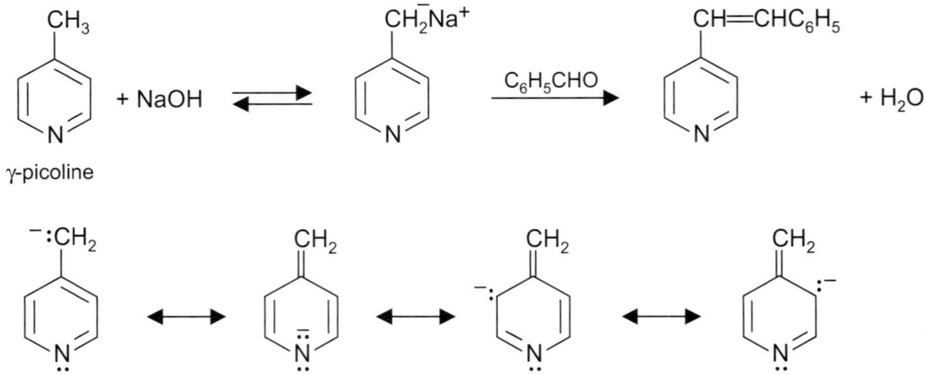

Anion from γ-picoline

SCHEME 3.30a

The enhanced acidity of the hydrogen atoms of an alkyl side chain at α positions is also due to the delocalization of negative charge in the intermediate anion onto the ring, particularly onto the nitrogen (Scheme 3.30*b*).

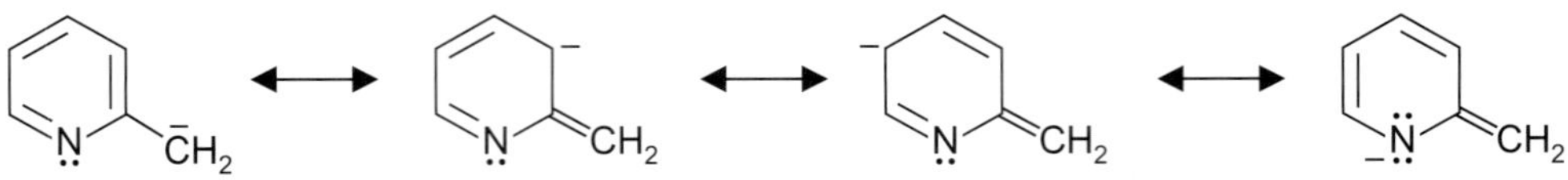

SCHEME 3.30b

However, the removal of a proton from β-picoline gives only a higher energy anion with negative charge only on carbons and none on nitrogen.

The conjugate base of 1, 3-dithiane is valuable in synthetic applications as a nucleophile. The anion is produced by deprotonation of 1, 3-dithiane using *n*-butyl lithium (Scheme 3.30*c*). 1, 3-dithiane is a weak proton acid (pK_a = 32). The hydrogens on the methylene group positioned between two sulphur atoms in 1, 3-dithiacyclohexane are relatively acidic, since the negative charge in the carbanion is stabilized by delocalization to each *S* by *p*-*d* π bonding. A 1, 3-dithiane can be prepared by the reaction of 1, 3-propanedithiol with an aldehyde (HCHO in this case). Among the many synthetic uses, aldehydes *e.g.*, may be prepared by alkylating 1, 3-dithiane at C-2 followed by hydrolysis of the resulting thioacetal (see Schemes 7.31–7.34).

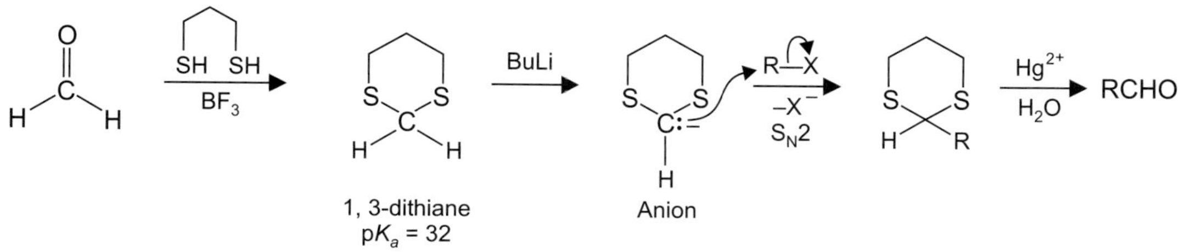

SCHEME 3.30c

(E) Effect of Hydrogen Bonding

When an acidic hydrogen is involved in a hydrogen bond with another atom in the same molecule then the strength of the acid is decreased. It is more difficult for a base to remove the proton since the hydrogen bond must also be broken along with sigma bond to the hydrogen. However, this effect is complicated by the inductive effect of the group involved in the hydrogen bond. The acetyl group is electron withdrawing and its location on the benzene ring opposite the carboxylic acid group increases the strength of this acid in comparison to benzoic acid (Scheme 3.31). However, when this group is present adjacent to COOH group it leads to an acid weakening effect. Infact its pK_a now becomes almost similar to that of benzoic acid. This is due to acid weakening effect of the hydrogen bonding.

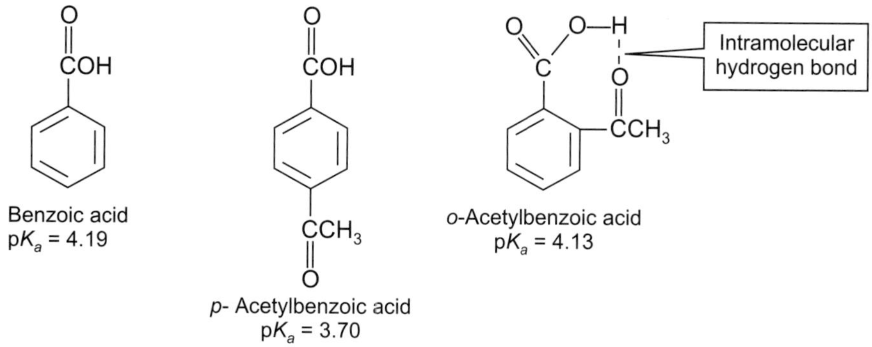

SCHEME 3.31

3.4 BASES

Nitrogen is the most important basic element in uncharged bases. An amine is nucleophile since its lone pair of nonbonding electrons can form a bond with an electrophile. An amine also behaves like a base by accepting a proton from a proton acid (I, Scheme 3.32). A convenient expression for relating basicities is a quantity called the basicity constant K_b or its negative logarithm pK_b. When an amine dissolves in water, the equilibrium (Scheme 3.32) is established. The larger the value of K_b (or the smaller the value of pK_b), the greater is the tendency of the amine to accept a proton from water and, thus, the greater will be the concentrations of RNH_3^+ and OH^- in the solution. Larger values of K_b, therefore, are associated with those amines that are stronger bases, and smaller values of K_b are associated with those amines that are weaker bases. (Just the opposite is true for values of pK_b). A structural feature which tends to stabilize the ammonium ion (relative to the free amine) makes the amine a stronger base. A structural feature which stabilizes the free amine (relative to the ammonium ion) makes the amine a weak base.

$$R\ddot{N}H_2 + HX \rightleftharpoons R\overset{+}{N}H_3 + X^- \quad \ldots (I)$$

$$R\ddot{N}H_2 + H_2O \overset{K_b}{\rightleftharpoons} R\overset{+}{N}H_3 + OH^-$$

$$K_b = \frac{[RNH_3^+][OH^-]}{[RNH_2]} \qquad pK_b = -\log K_b$$

SCHEME 3.32

Strength of Bases in Terms of their Conjugate Acids

Amines e.g., have very high pK_a values and therefore, these do not behave as acids. Ammonia also has a high pK_a(Scheme 3.32a). Thus amines behave as typical organic bases. Instead of specifying the strength of a base in terms of its pK_b value, it is convenient to express the strength of its conjugate acid and indicate it by its pK_a value. Thus the stronger the acid, the weaker is its conjugate base. Protonated methylamine is a stronger acid than protonated ethylamine and this reflects that methylamine is a weaker base compared to ethylamine.

CH_3NH_2 Methylamine $pK_a = 40$	NH_3 Ammonia $pK_a = 36$
$CH_3\overset{+}{N}H_3$ Protonated methylamine $pK_a = 10.7$	$CH_3CH_2\overset{+}{N}H_3$ Protonated ethylamine $pK_a = 11.0$

SCHEME 3.32a

3.5 RELATION BETWEEN STRUCTURE AND BASICITY

(A) General Discussion

The anions, *e.g.*, NH_2^- and EtO^- and neutral molecules with at least one unshared pair of electrons *e.g.*, NH_3 and EtOH represent the common bases. For the former types, the basicity is weakened by any factor which stabilizes the negative charge of the anion, while for the latter, the basicity is increased by any factor which stabilizes the positive charge on the conjugate acid of the base. Anionic bases are far more strong than their neutral analogues (*e.g.*, $NH_2^- >> NH_3$).

In the case of anionic bases the charge is generally associated with oxygen, nitrogen or carbon. The electronegativities of these elements follow the order O > N > C, the order of basicities is therefore, $R_3C^- > R_2N^- > RO^-$. Thus, amide ion is a far stronger base than hydroxide ion, and methide ion (CH_3^-) is so high in energy that it does not exist in organic media. However in Ph_3C^- where the negative charge is delocalized by the aromatic rings is more stable and is used (as sodium triphenylmethyl) in some reactions which need a particularly powerful base. Basicities of oxy-anions follow the order: $(CH_3)_3CO^- > CH_3O^- > PhO^- > CH_3CO_2^-$, and these are based on the principles which govern acidity: *t*-butoxide ion is a stronger base than methoxide ion since the three electron-releasing methyl groups in the former destabilize the negative charge; phenoxide ion is a weaker base than the alkoxide ion because the charge is delocalized over the aromatic ring; and acetate ion is still weaker because the charge delocalization by oxygen is much more effective.

(B) Substitution by Alkyl Groups

If one examines the pK_b values of the amines (Scheme 3.33) it is seen that most primary aliphatic amines (*e.g.*, methylamine and ethylamine) are somewhat stronger bases than ammonia (Recall that decreasing value of pK_b indicate increasing base strength, just as the higher the pK_a, the weaker the acid). One may account for this on the basis of the electron donating ability of an alkyl group. An alkyl group releases electrons, and it stabilizes the ammonium ion that results from the acid-base reaction by dispersing its positive charge. It stabilizes the ion to a greater extent than it stabilizes the amine. Thus in gas phase, the basicities of these amines increase on methyl substitution *i.e.*, $(CH_3)_3N > (CH_3)_2NH > CH_3NH_2 > NH_3$ and thus the trend is regular.

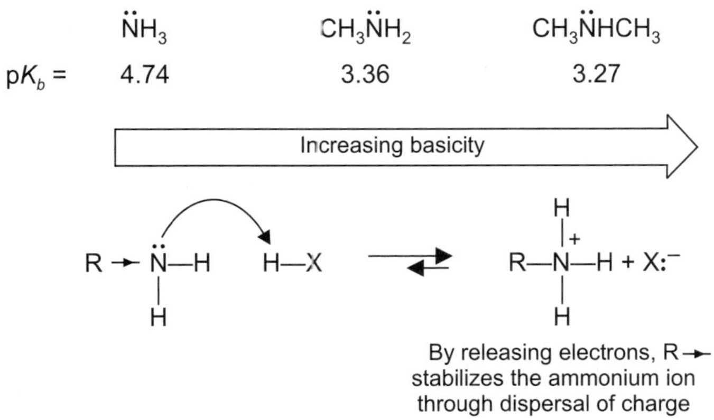

SCHEME 3.33

Interestingly, however, in aqueous solution the order of basicities is different and it is found that instead trimethylamine (pK_b 4.19) is a weaker base than dimethylamine (pK_b 3.27). In gas phase there is no solvation and only the stabilizing effect of methyl group(s) remains and each replacement of hydrogen with a methyl group has its stabilizing effect and therefore, the trend of basicities is regular. The ammonium ion is also stabilized by solvation in the presence of solvent (Scheme 3.33*a*). Trimethylamine is more hindered and its ammonium ion is less stabilized by solvation. Thus, these effects operate in different directions. In solution an alkyl group effects the stability of the ammonium ion in two ways it stabilizes by dispersing the charge and also it destabilizes the ammonium ion by interfering with solvation.

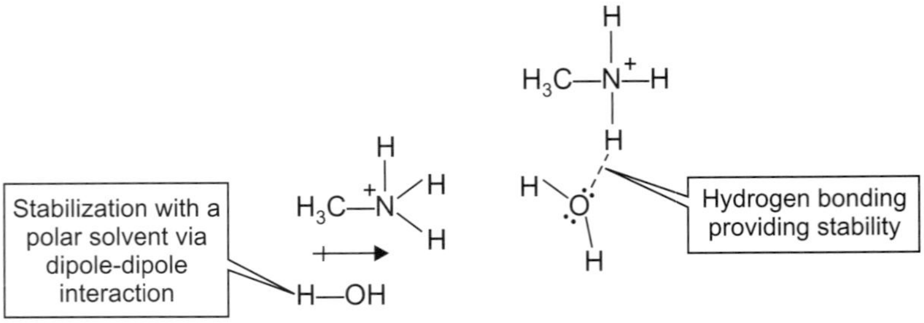

Stabilization of an ammonium ion by solvation

SCHEME 3.33a

(C) Resonance Effects

Aromatic amines (the anilines and their derivaties) are much weaker bases than the simple aliphatic amines. This reduced basicity is due to resonance stabilization of the nonbonding electrons in the free aromatic amine. Thus in aniline, the lone pair of nonbonding electrons on nitrogen is stabilized by overlap with the *p* orbitals of the ring (Scheme 3.34). On accepting a proton aniline is converted to an anilinium ion (Scheme 3.35) and now one can write only two resonance structures for the ion.

	Cyclo-$C_6H_{11}NH_2$	$C_6H_5NH_2$	*p*-$CH_3C_6H_4NH_2$
pK_b =	3.36	9.42	8.92
	Cyclohexylamine nonaromatic amine	Aromatic amines	

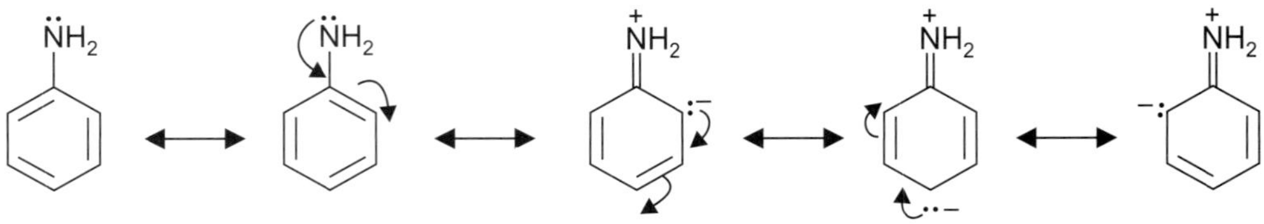

SCHEME 3.34

$$C_6H_5\ddot{N}H_2 + H_2O \rightleftharpoons C_6H_5\overset{+}{N}H_3 + \bar{O}H$$

Anilinium ion

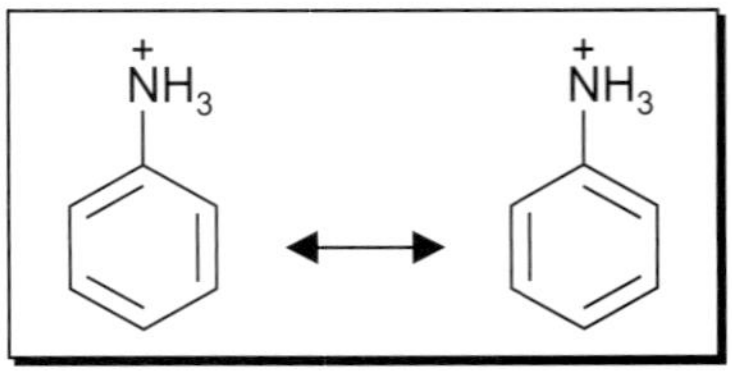

SCHEME 3.35

Thus aniline is stabilized in comparison to the ion, and aniline is not as basic as the aliphatic amines.

Electron-releasing groups with +*I* effect *e.g.,* alkyl groups increase the basicity of aromatic amines, while electron-withdrawing groups *e.g.,* halogen, nitro, carbonyl decrease their basicity. Decrease in basicity by halogen substitution is due to the electron-withdrawing inductive effect of the electronegative halogen. Decrease in basicity due to the presence of $-NO_2$ on the aromatic ring reflects a combination of inductive and resonance effects.

The basicity-decreasing effect of nitro substitution in the 3 position (Scheme 3.35*a*) is due to its inductive effect, whereas nitro substitution in the 4 position is due to both inductive as well as resonance effects. In the case of *para* substitution (and *ortho* substitution as well), delocalization of the lone pair on the amino nitrogen involves not only the carbons of the aromatic ring but also oxygen atoms of the nitro group.

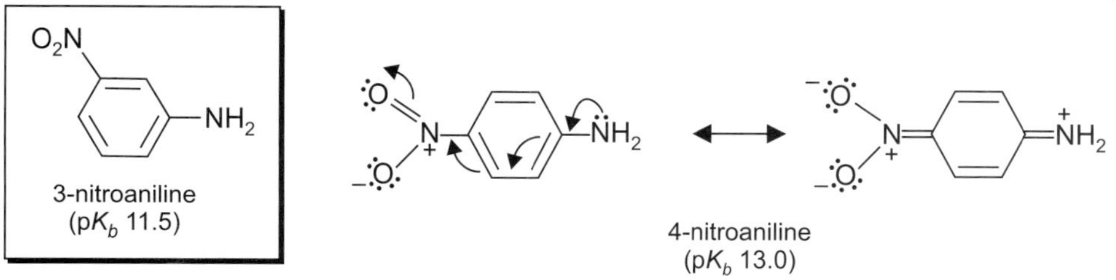

SCHEME 3.35a

One easily understands as to why compared with *N*, *N*-dimethylaniline (I, Scheme 3.35*b*) 2, 6-dimethyl *N*, *N*-dimethylaniline (II) is much more basic. The extended π bonding between the amino nitrogen and the ring can be attained (base weakening structural feature) only if the σ bonds on N attain coplanarity with the ring and its *ortho* bonds, as is so in the case of (I, Scheme, 3.35*b*). The presence of bulky substituents in the *ortho* position hinders the attainment of planar geometry. Due to this steric inhibition of resonance (II, Scheme 3.35*b*) is a stronger base than I (see also Scheme 2.9).

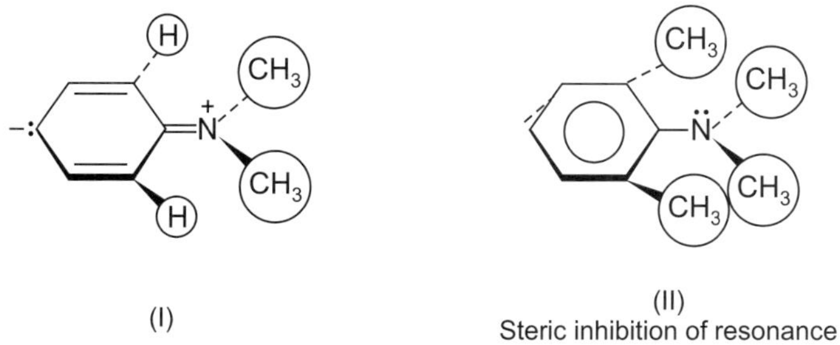

SCHEME 3.35b

Pyrrole is a very weak base (pK_b ~ 15). Pyrrole is aromatic because the lone pair of electrons on nitrogen is located in a *p* orbital, and these electrons contribute to the aromatic sextet. Pyrrole is thus extremely non-basic. Very strong acid is required to bring about protonation which does not occur on nitrogen but on C-2. Protonation at nitrogen gives an ammonium ion with no resonance stabilization, while protonation at the α-carbon (Scheme 3.36) gives a cation which can be described by three resonance forms.

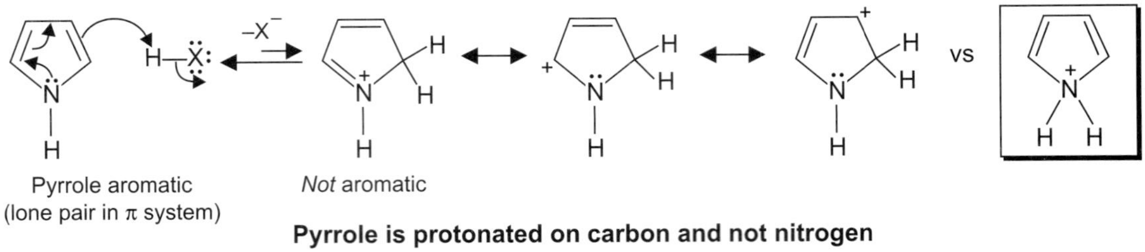

SCHEME 3.36

Amides (pK_b ~ 14) are far less basic and even less basic than aryl amines. The lower basicity of amides than amines is explained by resonance. An amide is stabilized by resonance involving the nonbonding pair of electrons on the nitrogen atom, and an amide protonated on its nitrogen atom does not display this type of resonance stabilization (Scheme 3.37). A more

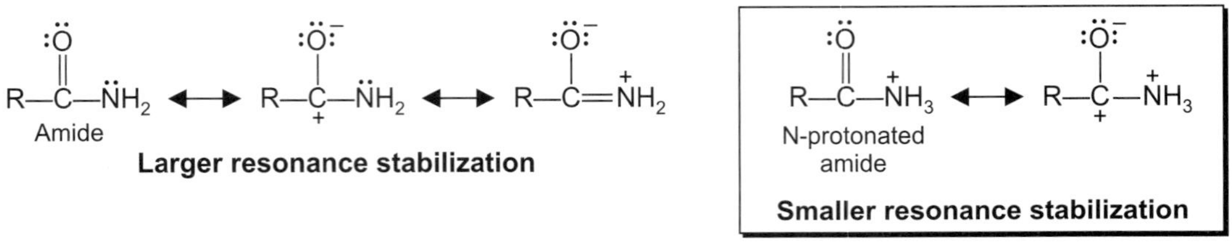

SCHEME 3.37

important factor to account for amides being weaker bases than amines is the powerful electron-withdrawing effect of the carbonyl group of the amide. This means that the equilibrium (II, Scheme 3.38) lies almost to the left as compared to reaction (I, Scheme 3.38). The nitrogen

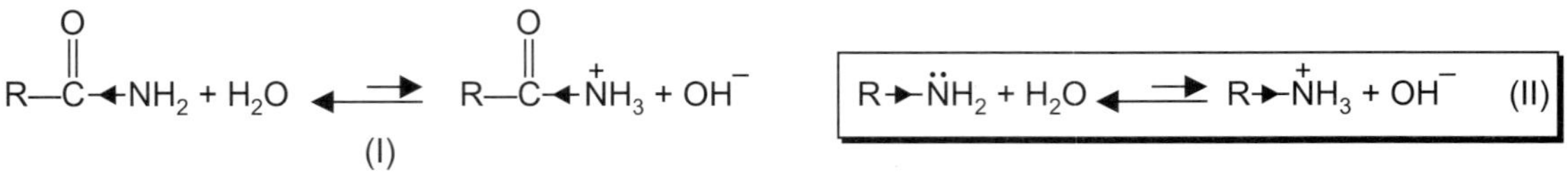

SCHEME 3.38

atoms of amides are so weakly basic that if an amide accepts a proton, it does so on its oxygen atom. Protonation on the oxygen atom occurs even though oxygen atom (because of their greater electronegativity) is typically less basic than nitrogen atom. If an amide accepts a proton on its oxygen atom, resonance stabilization involving the nonbonding electron pair of the nitrogen atoms is still operative (Scheme 3.39).

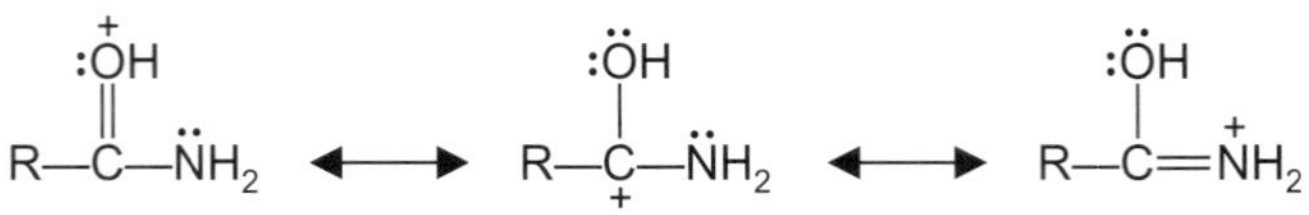

SCHEME 3.39

(D) Hybridization Effects

The study of terminal alkynes showed that electrons are held more tightly by orbitals with more *s* character. This principle is also used to explain the relative basicities of unsaturated amines. Pyridine is a weaker base than the simple aliphatic amines. In pyridine the nonbonding electrons occupy a sp^2 hybrid orbital with greater *s* character (Scheme 3.40) and has more tightly held electrons than those in the sp^3 orbital of an aliphatic amine, therefore, pyridine's nonbonding electrons are not that available for bonding to a proton. Since pyridine retains its aromaticity upon protonation, it is even more stronger base than pyrrole (Scheme 3.36). The nitrile's lone pair occupies an *sp*-hybridized orbital with 50 per cent *s* character. This orbital is close to the nucleus and these electrons are tightly bound and relatively unreactive to make it a very weak base.

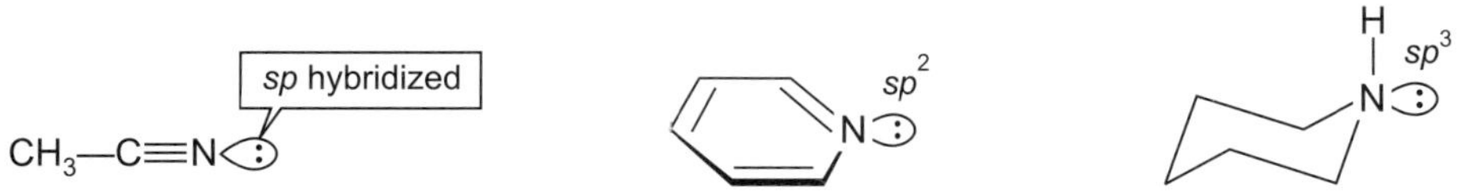

Acetonitrile $pK_b = 24$ | Pyridine $pK_b = 8.75$ | Piperidine $pK_b = 2.88$

SCHEME 3.40

Consider, for example guanidine (Scheme 3.41), which probably is the strongest, organic nitrogen containing base ($K_b = 1$), as strong as the alkali metal hydroxides. Considering the *s* character, the N (sp^3) of NH_2 has less *s* character compared with N (sp^2) of the imino group (= NH). However, actually it is the imino nitrogen which is protonated, since this would lead to the formation of a symmetrical resonance stabilized system with three equivalent contributing structures. The delocalization energy is therefore large and the conjugate acid formed by guanidine is unusally stable.

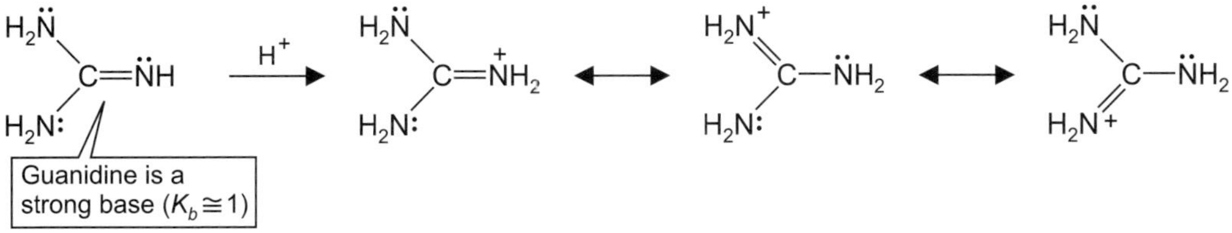

SCHEME 3.41

(E) Some Typical Bases and Their Reactions

The amount of enolate formed from a carbonyl compound depends on the pK_a of the carbonyl compound and the base used to remove the α-hydrogen. Thus, when hydroxide ion (the pK_a of its conjugate acid is 16) removes an α-hydrogen from cyclohexanone ($pK_a = 17$), only a small amount of the enolate is formed since hydroxide ion is a weaker base than the base being formed (I, Scheme 3.42). However, the use of strong non-nucleophilic bases like LDA ($pK_a = 35$) allows the conversion of carbonyl compounds (pK_a of α protons 20–25) completely to enolate anions (nucleophiles). This may be necessary in several situations *e.g.*, during aldol condensation (see Scheme 6.9) when the carbonyl compound has no chance to condense with itself. Moreover, generation of enolates with LDA provides a useful technique to alkylate ketones regioselectively (Scheme 3.42*a*, for details see Scheme 3.43*d*).

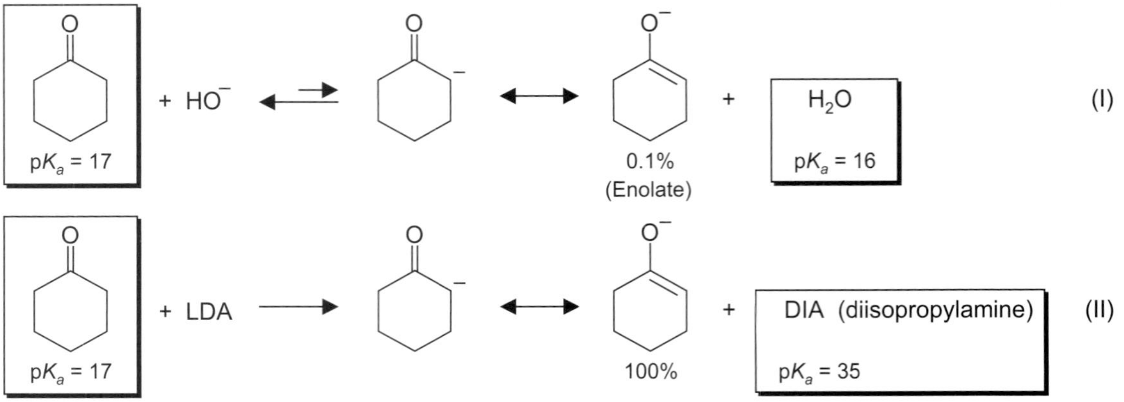

LDA is a base of choice for converting a carbonyl compound completely to an enolate, before it reacts with an electrophile

SCHEME 3.42

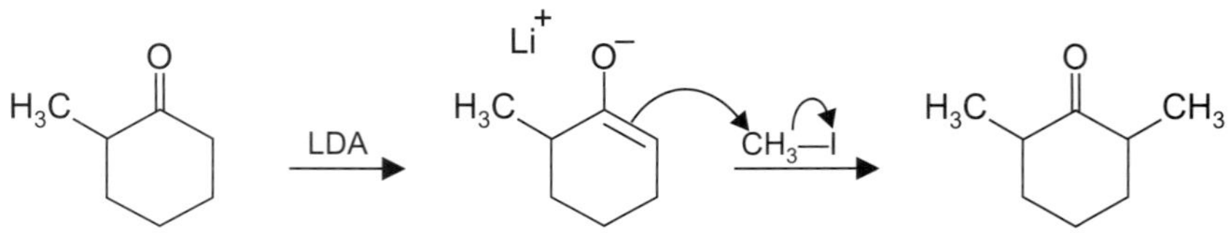

SCHEME 3.42a

Alkylation of the α-carbon of a carbonyl compound is an important reaction. Alkylation is carried out by first removing a proton from the α-carbon with LDA and then adding the appropriate alkyl halide. The reaction works best with primary alkyl halides (being an S_N2 reaction, Scheme 3.42*a*).

Dehydrohalogenation of an alkyl halide gives a mixture of alkenes in which either the more substituted (more stable) or less substituted alkene may be formed as the major product (Scheme 3.43). The steric bulk of the base is partly responsible for this behaviour. For the bulky *t*-butoxide ion it is easier to remove a more exposed (1°) hydrogen atom than the internal (2°) hydrogen atom.

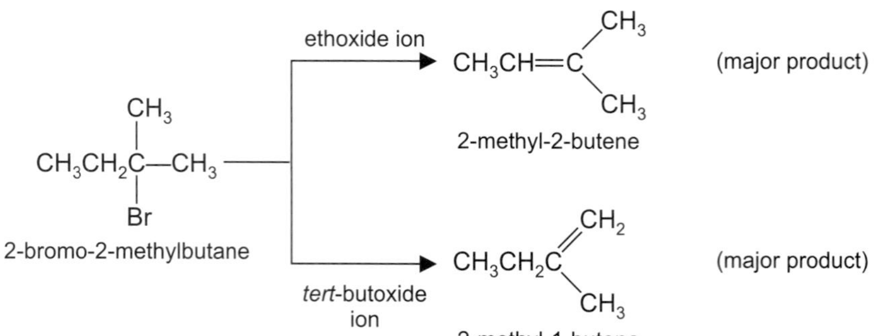

SCHEME 3.43

Formation of Carbanionic Nucleophiles (Role of Very Strong Bases)

Several structural features make a proton attached to a carbon acidic which can be removed by a base. Two such structural features are in phenylacetylene and acetophenone. For the proton removal one chooses a base with a higher pK_a *than the p*K_a *of the proton to be removed in order to achieve a complete conversion to carbanionic nucleophile. The p*K_a *values are reviewed (Scheme 3.43a).*

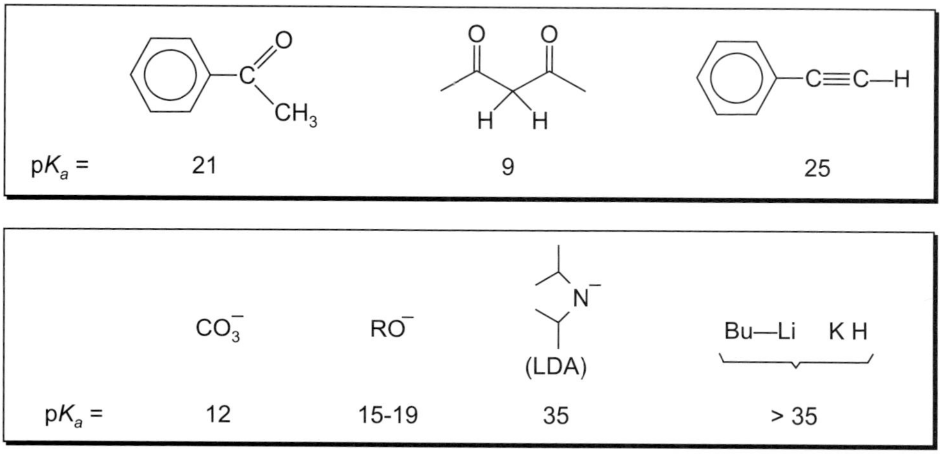

SCHEME 3.43a

Significantly several bases e.g., BuLi can act as nucleophiles. If the structural feature which makes the C—H proton acidic is itself an electrophile then a nucleophilic base cannot be ued. Butyl lithium (p$K_a > 45$*) converts phenylacetylene (p*$K_a \sim 25$*) to its conjugate base by proton removal but, it reacts as a nucleophile with the carbonyl group of acetophenone although the* α *protons of acetophenone have p*$K_a = 21$ *and are thus more acidic than the terminal proton in phenylacetylene (Scheme 3.43b).*

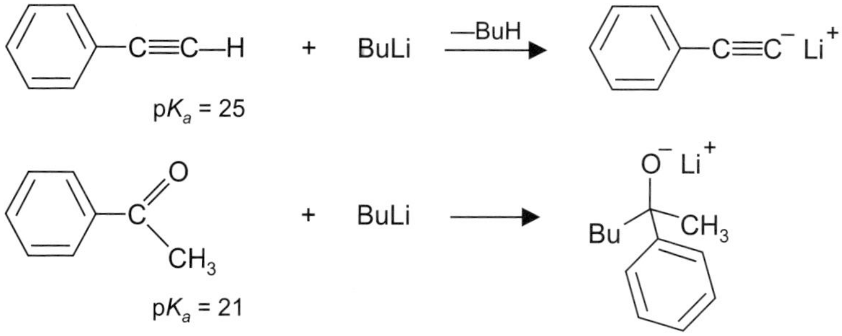

SCHEME 3.43b

It is to solve such nucleophilicity problems that hindered and very strong bases like LDA are used for the removal of acidic protons but themselves these are poor nucleophiles. Thus LDA can remove a proton from acidic C—H bond, but it does not attack the carbonyl group or other electrophilic centers.

3.6 SYNTHETIC APPLICATIONS OF LITHIUM DIISOPROPYLAMIDE (LDA)

(A) Preparation

It is prepared easily by adding butyl lithium to diisopropylamine (DIA) at –78°C in THF (Scheme 3.43*c*).

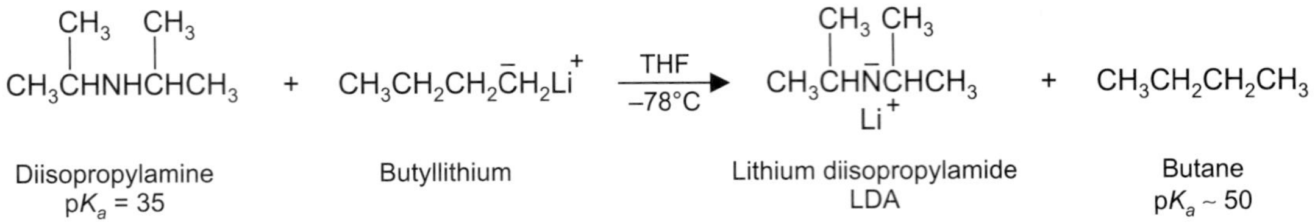

SCHEME 3.43c

(B) Enolate Regiomers

With an unsymmetric ketone, reaction with excess LDA at –78°C removes the proton from the less highly substituted α-carbon to give a kinetic enolate (Scheme 3.43*d*). The less substituted α-position has slightly more acidic protons, thus both steric and electronic factors lead to the formation of kinetic enolate. The enolate with more highly substituted double bond, the thermodynamic enolate is however, more stable. If a slight excess of ketone is used around room temperature 25°C (instead of excess LDA) or trace of protic impurities are present, an equilibrium is established to form thermodynamic enolate.

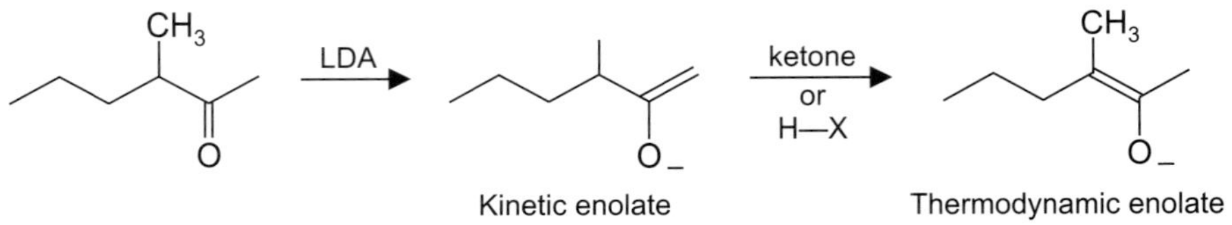

SCHEME 3.43d

The kinetic or thermodynamic enolates can be captured as enol trimethylsilyl ethers, purified and regenerated (see, Scheme 6.5) for use in aldol condensations.

(C) Synthesis of Alkynes

Successive E2 eliminations can be carried out on geminal or vicinal dihalides for their conversion into alkynes. The first elimination gives a vinyl halide and then to form an alkyne a powerful base is needed to carry out the second dehydrohalogenation reaction. Previously $NaNH_2$ in liquid ammonia was employed, now LDA is the common base used (Scheme 3.43*e*) and both the steps can be carried out with LDA itself.

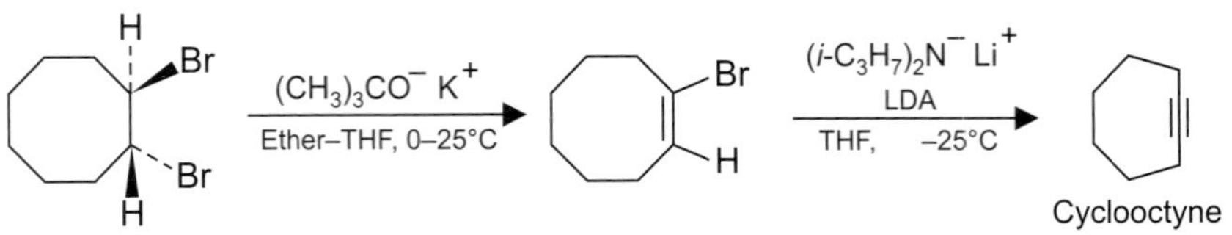

SCHEME 3.43e

(D) Enolates from Esters

As expected enolates from esters can be generated which can be then alkylated (Scheme 3.43*f*).

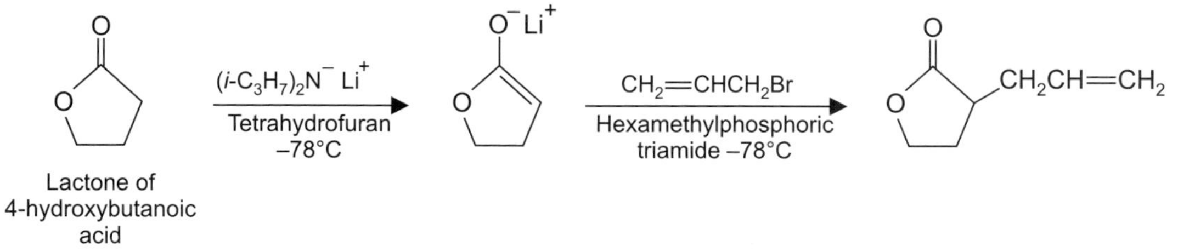

SCHEME 3.43f

(E) Synthesis of Fused Rings

Intramolecular alkylation reactions of enolate anions are used in the synthesis of fused rings. Thus *e.g.*, the cyclization of a disubstituted cyclopentane gives two different fused ring compounds depending on whether the reaction conditions favour the formation of kinetic enolate or the thermodynamic enolate (Scheme 3.43*g*).

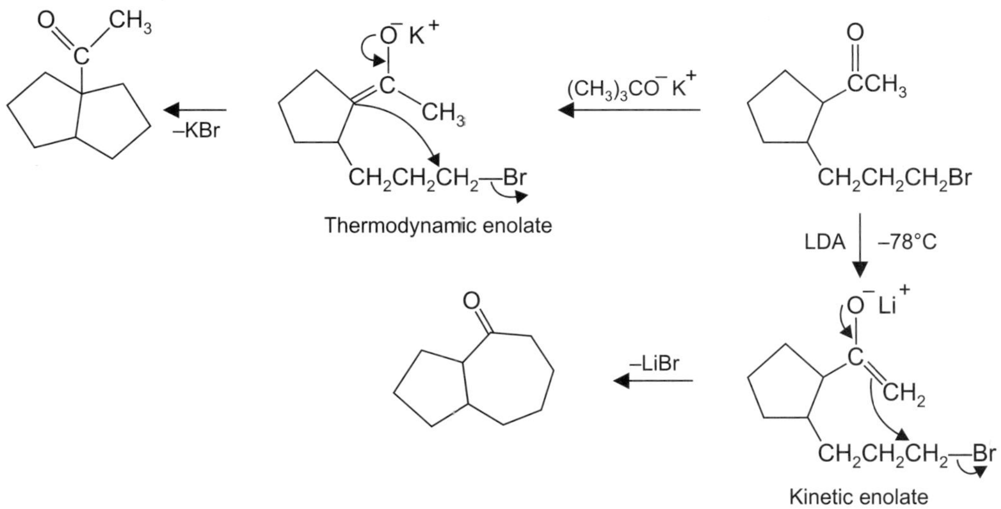

SCHEME 3.43g

EXERCISE 3.4

Predict the product from the reaction (Scheme 3.43h) at two different temperature conditions?

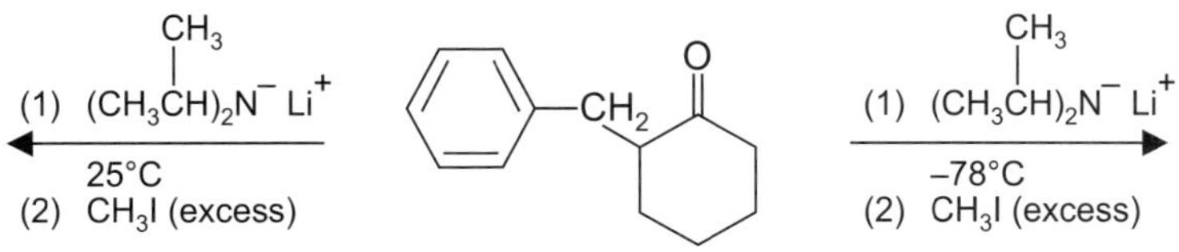

SCHEME 3.43h

ANSWER. *At lower temperature, kinetic enolate is exclusive which gives the corresponding alkylated product (Scheme 3.43i). At higher temperature (equilibrium conditions) the major product is derived from thermodynamic enolate.*

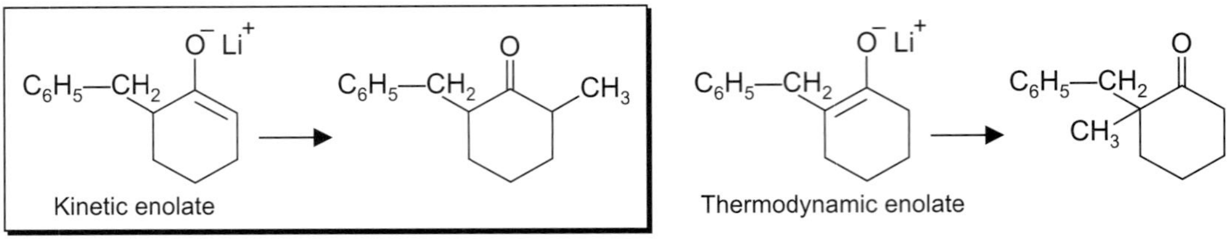

SCHEME 3.43i

The alkylations of enolates (see Scheme 3.42*a*) are typically S_N2 reactions which work well with primary alkyl halides/CH_3I. Recall that enolates are strong bases and use of secondary and tertiary alkyl halides leads to predominant elimination. One can work successfully with tertiary halides by using silyl enol ethers (see Scheme 6.43*b*).

3.7 ACID-BASE REACTIONS

In general, acid-base reactions always favour the formation of the weaker acid and the weaker base. The reason for this is that the outcome of an acid-base reaction is determined by the position of an equilibrium. Acid-base reactions are thus under equilibrium control to favour the formation of the most stable (lowest potential energy) species. The weaker acid and weaker base are more stable (lower in potential energy) than the stronger acid and stronger base. Thus phenol reacts with sodium hydroxide to form water soluble salts, while it does not react with a weaker base like sodium bicarbonate (Scheme 3.44).

This concept is used in the synthesis of deuterium and tritium labeled compounds often needed in the study mechanism of organic reactions (Scheme 3.44). Similar reactions involve the replacement of the α-hydrogen by deuterium via a carbanion formation (Scheme 3.45). The reaction occurs when an aldehyde or a ketone is dissolved in deuterium oxide containing sodium deuteroxide as the base. These isotopic exchange studies provide a strong evidence for α-carbanion formation.

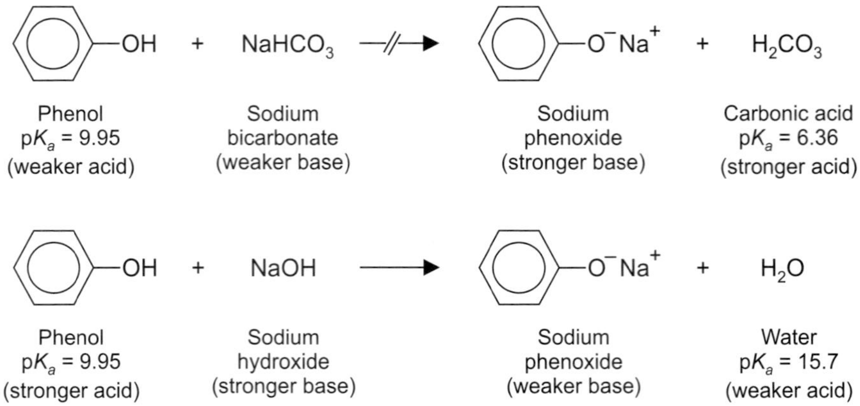

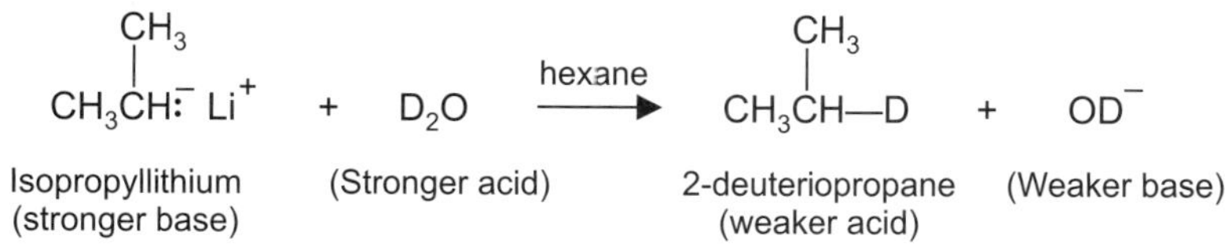

SCHEME 3.44

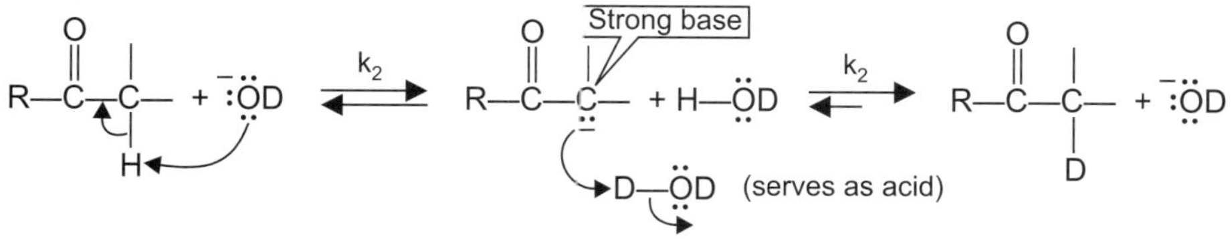

SCHEME 3.45

3.8 THE EFFECTS OF THE SOLVENT ON ACID AND BASE STRENGTH

One has already seen that inductive stabilization and steric hindrance of solvation act in opposite directions in the case of amines. These results are helpful in explaining the basicity of amines. A solvent can exert considerable influence on acid and base strengths by differential solvation. In the gas phase where the solvation effects are absent the basicity order of amines toward the proton are $R_3N > R_2NH > RNH_2 > NH_3$. This change in their basicities in the gas phase can now be explained by the electron donating effect of alkyl groups.

A more important aspect of the effect of solvent deals with its orientation (*i.e.*, entropy change) when an acid or a base is converted into its conjugate. If one considers the effect of a solvent on acidity, it is seen that in the absence of a solvent (*i.e.*, in the gas phase) most acids are far weaker than these are in solution. Acetic acid in gas phase has a pK_a of about 130. When one considers the conversion of CH_3COOH to CH_3COO^- in aqueous solution, one deals with solvation of both CH_3COOH and CH_3COO^- by water *via* hydrogen bonding. The solvent molecules arrange themselves around the CH_3COO^- group in a much more orderly fashion than they arrange around the CH_3COOH itself (the hydrogen bonding to CH_3COO^- is far more stronger than to CH_3COOH since the water molecules are more attracted to the negative charge). Thus on this ionization, there is decrease in entropy (loss of freedom, solvation of a species decreases the entropy of the solvent). Recall that it is only the positive entropy change (from order to disorder) which makes a negative contribution to $\Delta G°$ and is energetically favourable for the formation of products (see, Scheme 4.9). Thus the entropy change ($\Delta S°$) for the ionization of acetic acid is negative *i.e.*, the $-T\,\Delta S°$ term in equation $\Delta G° = \Delta H° - T\,\Delta S°$ leads to an acid-weakening positive contribution to $\Delta G°$. A fact which has been proved experimentally from the thermodynamic values for the dissociation of acetic acid and its comparison with chloroacetic acid. In the case of simple aliphatic and halogenated aliphatic acids in aqueous solutions at room temperature, it has indeed been shown, that entropy ($T\,\Delta\,S$) usually contributes much more to the total free energy change ΔG compared to enthalpy ΔH. Resonance and electron withdrawing effects of functional groups in RCOOH, effect its acidity in two distinct ways. These groups effect the enthalpy (electron withdrawal increases the acidity by stabilizing $RCOO^-$ by dispersing charge) and also the entropy.

Chloroacetic acid is a stronger acid compared to acetic acid and this increased acidity is due to the presence of electro-withdrawing chlorine atom. This electron withdrawal helps to spread the negative charge all over the chloroacetate ion. The chlorine atom therefore, makes the chloroacetate ion $ClCH_2COO^-$ less prone to bring about an ordering of the solvent compared to CH_3COO^-, since it now requires less stabilization through solvation (*i.e.*, during the ionization of $ClCH_2COOH$, the solvent molecules have much more freedom and consequently a higher entropy).

3.9 LEVELING EFFECT

The acid strengths of two strong acids, HY and HX cannot be compared in H_2O. Strong acids, especially in dilute solutions are practically completely ionized in H_2O, consequently the only acid present is H_3O^+. This is known as the leveling effect. It is necessary to compare them in a solvent which is much less basic than water so that an equilibrium is established with both the unionized acids and their conjugate bases present. The solvent of choice is 100% (glacial) acetic acid (Scheme 3.46).

$$HX + H_2O \longrightarrow H_3\overset{+}{O} + \overset{-}{X} \quad \text{In } H_2O$$

$$HX + HOAc \rightleftharpoons \overset{-}{X} + H_2\overset{+}{O}Ac \quad \text{In HOAc}$$

Leveling effect of solvent

SCHEME 3.46

3.10 HARD AND SOFT ACIDS AND BASES

A qualitative classification is used to characterize the acidity and basicity of Lewis acids and bases. Hard acids (electron pair acceptors) generally have a small electron acceptor site of high positive charge and do not possess unshared pairs of electrons in their valence shells. Hard acids are characterized by high electronegativity and have less polarizable sites examples are: H^+, BF_3, CO_2, SO_3, Li^+, Na^+. Hard bases (electron pair donors) are generally difficult to oxidize and have no empty low energy orbitals available. Hard bases are characterized by a highly electronegative donor atom of low polarizability. Example are: H_2O, OH^-, ROH, RNH_2, F^-. Soft acids are characterized by a large electron acceptor atom of high polarizability (*e.g.*, BH_3, Br_2, I_2, Cu^+, Ag^+, Pd^{2+}).

Soft bases (*e.g.*, CN^-, C_2H_4, C_6H_6, H^-, CO, R_2S) normally have electrons which are easily removed by oxidizing agents and have empty orbitals of low energy. Soft bases are characterized by a polarizable donor atom. Soft bases combine best with soft acids whereas hard acids combine best with hard acids. BH_3 (a soft acid) forms more stable complexes with soft bases like carbon monoxide and olefins than with hard bases. On the other hand, the hard acid-hard base complex $BF_3 . OR_2$ is more stable than $BF_3 . SR_2$, a hard acid-soft base complex. In BF_3, boron is largely B^{3+} because of the electronegative fluorines, hence BF_3 is hard; in BH_3, boron is largely neutral, hence BH_3 is soft. The softer bases react best in displacement reactions, thus the attacked carbon in an alkyl halide RX must also be soft like.

PROBLEMS

3.1 How one explains the acidity of nitromethane (CH_3NO_2, pK_a = 10.2)?

3.2 Comment on the acidity of C—H bond in a haloform.

3.3 Why amidines are far stronger base than amines?

3.4 Write the structure of the intermediate and the product formed by the based catalysed ($NaNH_2$) reaction of the following pyridine with methyl iodide.

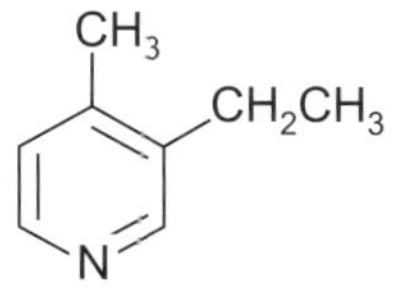

3.5 Why the following reaction proceeds under milder conditions compared to the one in Scheme 3.30*a*?

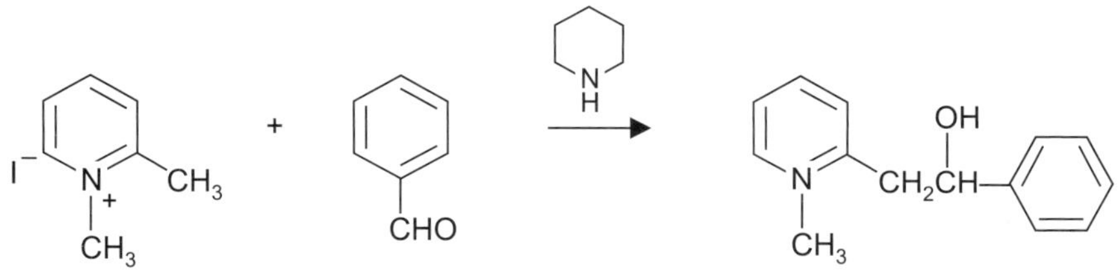

3.6 Why isocyanic acid, HN=C=O, and cyanic acid, N≡C–OH, have the same conjugate base?

3.7 Which out of the following pair is more basic?

(I) 4-cyanoaniline and 4-nitroaniline

(II) Diethylamine and pyrrolidine

(III) Amine and alcohol

(IV) *o*-chloroaniline and *p*-chloroaniline.

ANSWERS TO THE PROBLEMS

3.1 Loss of proton gives an anion whose charge is delocalized onto the oxygen atoms of the nitro group.

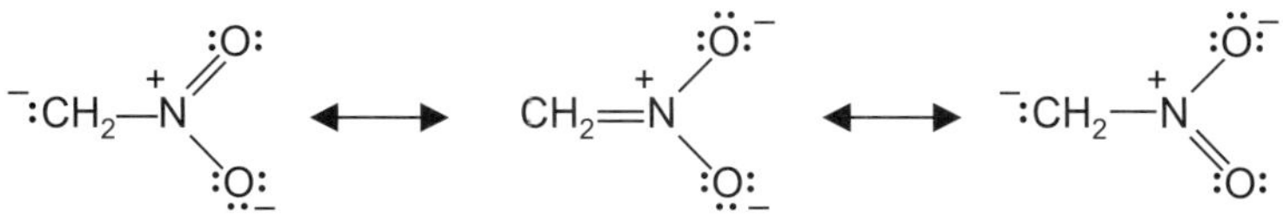

Resonance contributors for the carbanion from nitromethane

3.2 This bond is acidic, since the conjugate base is stabilized both by the inductive effects of three halogen atoms and by charge-delocalization, for the halogens (other than fluorine) have unfilled and relatively low-lying *d* orbitals. These anions, by loss of a halide ion, give carbenes which are reactive intermediates.

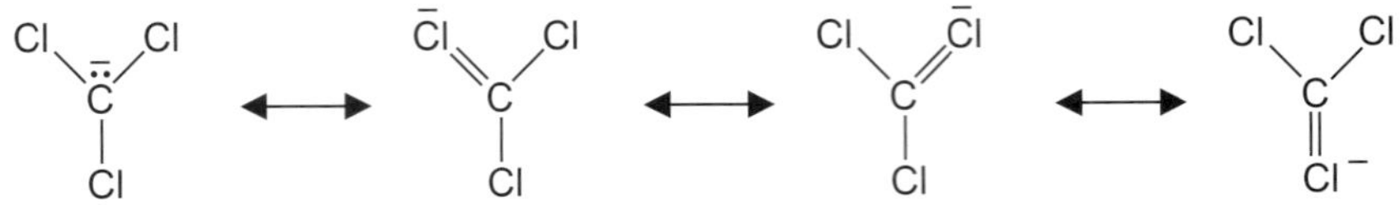

3.3 Because both the base and its conjugate acid are resonance-stabilized, the stabilization energy of the latter, whose principle canonical structures are equivalent, is greater than that of the former, in which one of the corresponding structure is dipolar and of high energy.

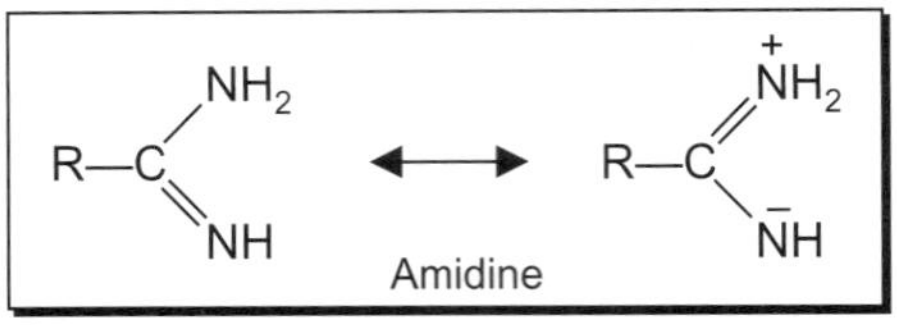

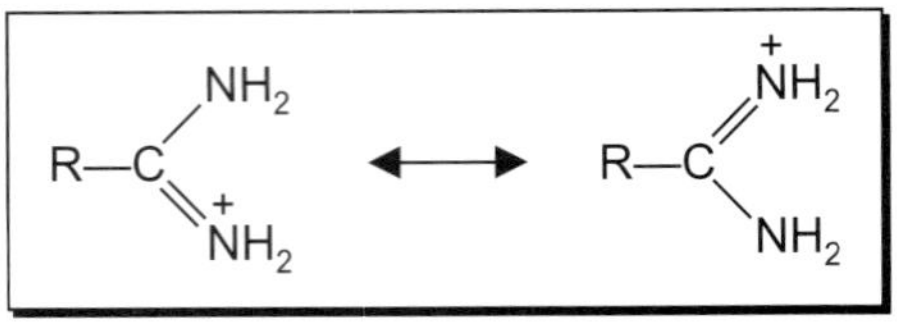

3.4 Only the hydrogens of the alkyl group either in 2- or 4-position are acidic.

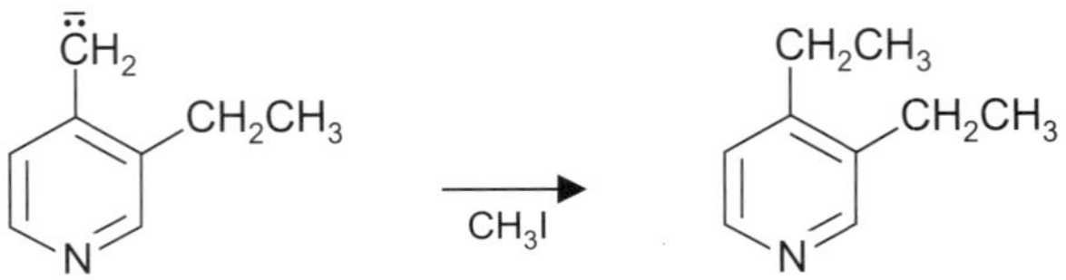

3, 4-diethylpyridine

3.5 This is due to enhancement of the side chain acidity in the N-alkylpyridinium compound.

3.6 The loss of proton from both the acids gives the conjugate bases which in fact are the contributing structures of the same resonance hybrid.

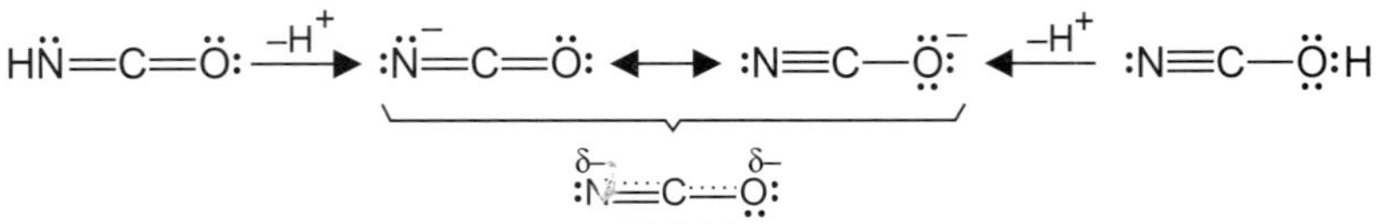

Hybrid

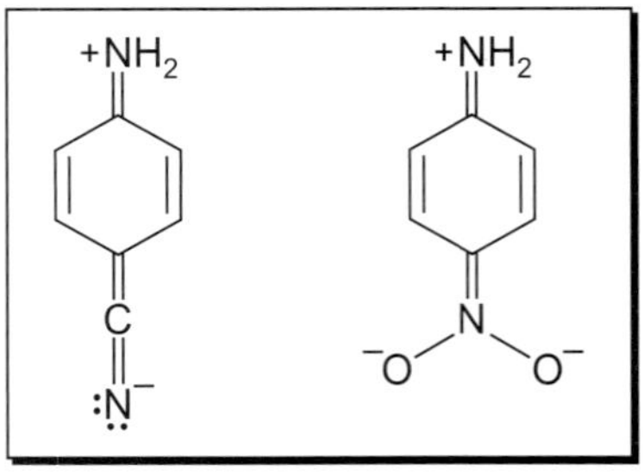

3.7 (I) 4-Cyanoaniline. In 4-nitroaniline, the base weakening electron delocalization is more effective, the negative charge here ends up with more electronegative oxygen than with the CN group where it ends up with less electronegative nitrogen.

(II) Diethylamine pK_b = 3.01 is a weaker base than nonaromatic heterocyclic amine pyrrolidine pK_b = 2.73. In the later the alkyl groups are "tied back" away from the unshared electrons of the nitrogen. This infact is the case with nonaromatic heterocyclic amines which are more basic compared with open chain secondary amines of same size.

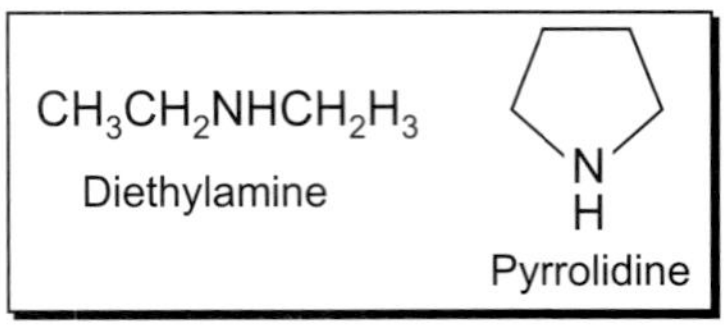

(III) Amines are much stronger bases compared with alcohols. One can focus not on the basicity of the amine, but on the bronsted acidity of its conjugate acid, the ammonium ion. In case the ammonium ion is a strong acid, the related amine must be a weak base. This is due to the fact that oxygen is more electronegative atom compared to nitrogen and thus can accomodate the negative charge much better.

$^{+}NH_4$ Ammonium ion	$^{+}OH_3$ Oxonium ion
pK_a = 9.2	pK_a = – 1.7
$^{+}N(CH_3)_4\ Cl^-$ Tetramethylammonium chloride, a stable solid	$^{+}O(CH_3)_3\ BF_4^-$ Trimethyloxonium fluoroborate, very reactive

(IV) The inductive effect of chlorine in the *ortho* position is more (being closer to the amino group) than when it is present in the *para* position. Thus the more effective inductive effect removes more electron density from nitrogen and therefore *ortho* isomer is the weaker base.

CHAPTER 4

Organic Reactions and the Determination of their Mechanisms

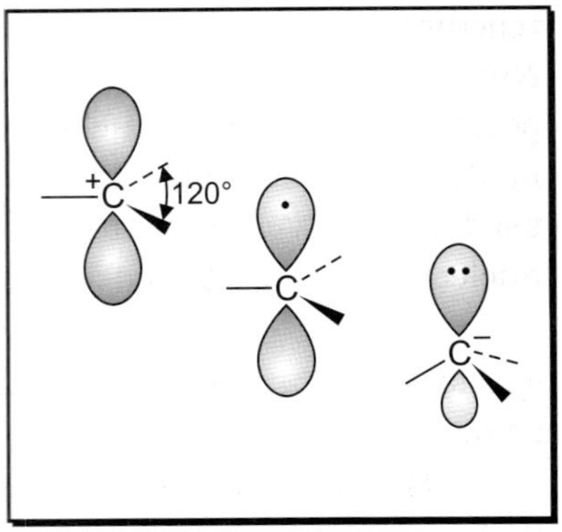

Most organic reactions can be placed in one of the six classes:

1. Substitution,
2. Addition,
3. Elimination,
4. Rearrangement,
5. Pericyclic reactions, and
6. Complex reactions.

Each of these terms describes an operation which occurs during a reaction on an organic compound. The organic compound undergoing structural or functional group changes is called a reactant or a substrate. A detailed and a stepwise description of the pathway by which reactants are converted to products is termed as the reaction mechanism. An acceptable mechanism has to account not only for structural changes and stereochemical outcome but the energy changes as well that take place at every stage of the reaction.

4.1 MECHANISTIC CLASSIFICATION

(A) Substitution Reactions

In a substitution reaction one atom, ion or a group is substituted in a reactant by another. The substituting species may be either a nucleophile, an electrophile or a free radical. Typical of the aliphatic substitution reactions are S_N2 type (Scheme 4.1). Aromatic electrophilic and nucleophilic substitutions represent a typical class of their own. The species that attacks a

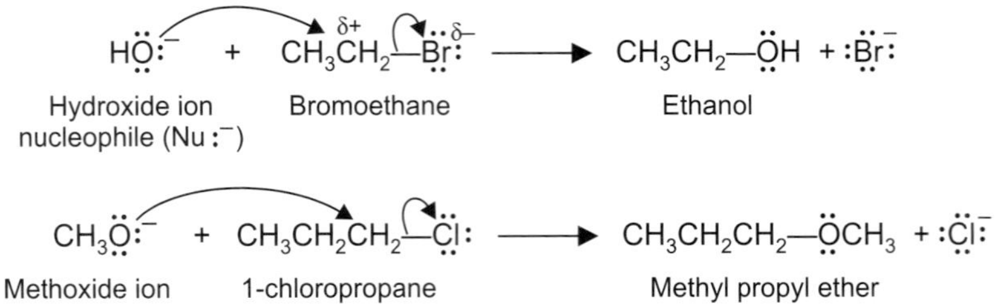

The sigma-bond electrons leave with the halogen.

Nucleophilic S_N2 substitution reactions

SCHEME 4.1

reactant *e.g.,* an alkyl halide (Scheme 4.1) in a substitution reaction is called a nucleophile Nu:$^-$ (literally, "nucleus lover"). A nucleophile is generally any species that is attracted to a positive center and a nucleophile is a Lewis base. Most nucleophiles are anions while some neutral polar molecules like H_2O, CH_3OH, and CH_3NH_2 which have unshared electrons can be used to form sigma bonds and act as nucleophiles. Substitutions by nucleophiles are called nucleophilic substitutions, or nucleophilic displacements.

The opposite of a nucleophile is an electrophile E^+ ("electron lover"). An electrophile is a species that is attracted toward a negative center and thus an electrophile represents a Lewis acid, such as H^+ or $ZnCl_2$.

Substitution on an aromatic ring is almost always the result of electrophilic reactions (Scheme 4.1*a*). Here the aromatic stabilization dictates the reaction mechanism which is substitution rather than addition. A large amount of resonance energy would have been lost if instead of substitution, addition had occurred. In this example, the electrophile substitutes for

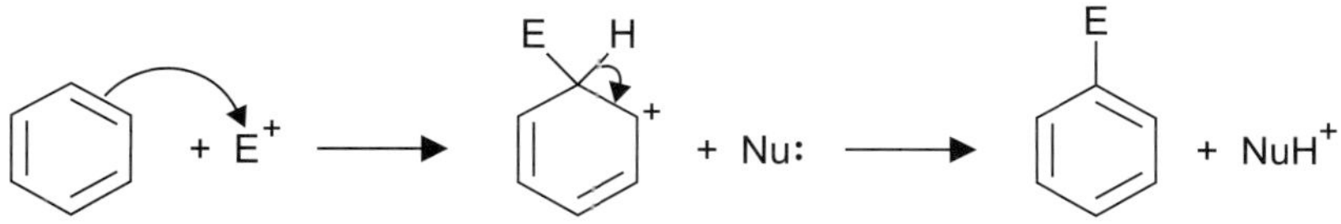

Electrophilic aromatic substitution

SCHEME 4.1a

a proton and an example is that of nitration of benzene with HNO_3/H_2SO_4. However, other groups can also leave during electrophilic aromatic substitution.

An electrophile can form a bond to an aromatic carbon atom already having a substituent other than hydrogen and the loss of that substituent results in Ipso substitution. Ipso substitution is limited to those reactants in which the group originally on the ring can be somewhat a good leaving group (Scheme 4.1*b*, also see Scheme 4.3*g*).

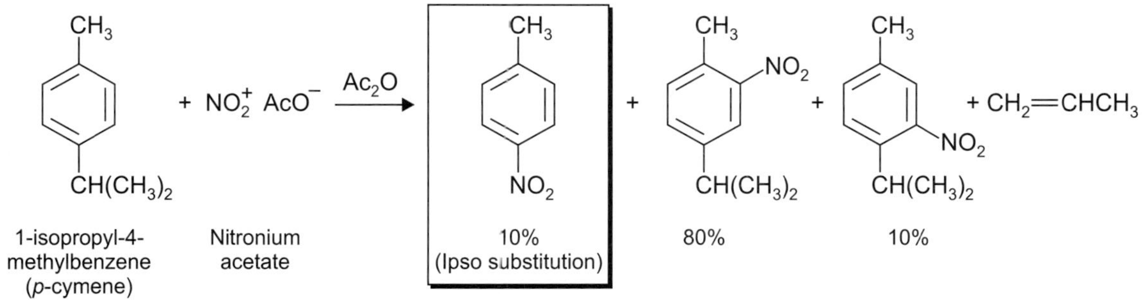

***Ipso* electrophilic aromatic substitution**

SCHEME 4.1b

Generally aromatic compounds undergo electrophilic substitution, however, nucleophilic substitution is also an important reaction. The early industrial processes for the formation of phenol and aniline (Scheme 4.1*c*) were nucleophilic substitution reactions.

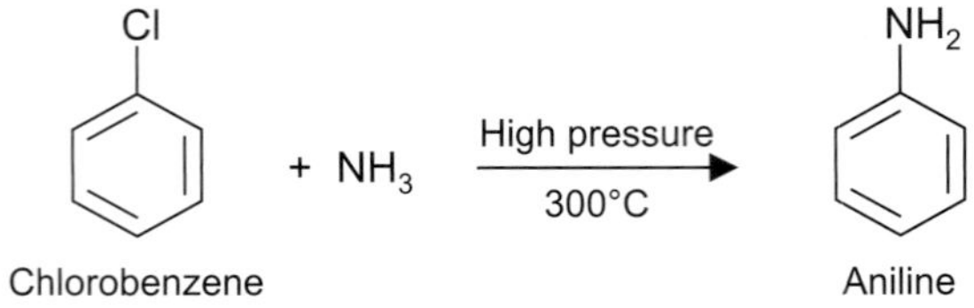

Nucleophilic aromatic substitution

SCHEME 4.1c

Nucleophilic aromatic substitution may involve addition of the nucleophile to the aromatic ring followed by loss of a leaving group. This sequence is reminiscent of nucleophilic substitution on carboxylic acid derivatives. Other reactions are believed to involve an aromatic cation or in some cases by initial 1, 2-elimination. The common to all the processes is that the aromaticity is retained in the product.

One mechanism of nucleophilic aromatic substitution is in (Scheme 4.1*d*). Here an amine acts as a nucleophile and the substrate is chlorobenzene substituted with an electron with drawing group like a nitro group. The reaction, however does not take place with chlorobenzene itself under similar conditions. The nitro group, evidently, lowers the activation energy of the reaction by stabilizing the negative charge generated by the addition of nucleophile to the aromatic ring. A series of resonance structures can be drawn to show this stabilization. Loss of chloride completes the substitution reaction. By analogy with the mechanism of electrophilic aromatic substitution, an addition-elimination mechanism looks reasonable for nucleophilic aromatic substitution (Scheme 4.1*d*).

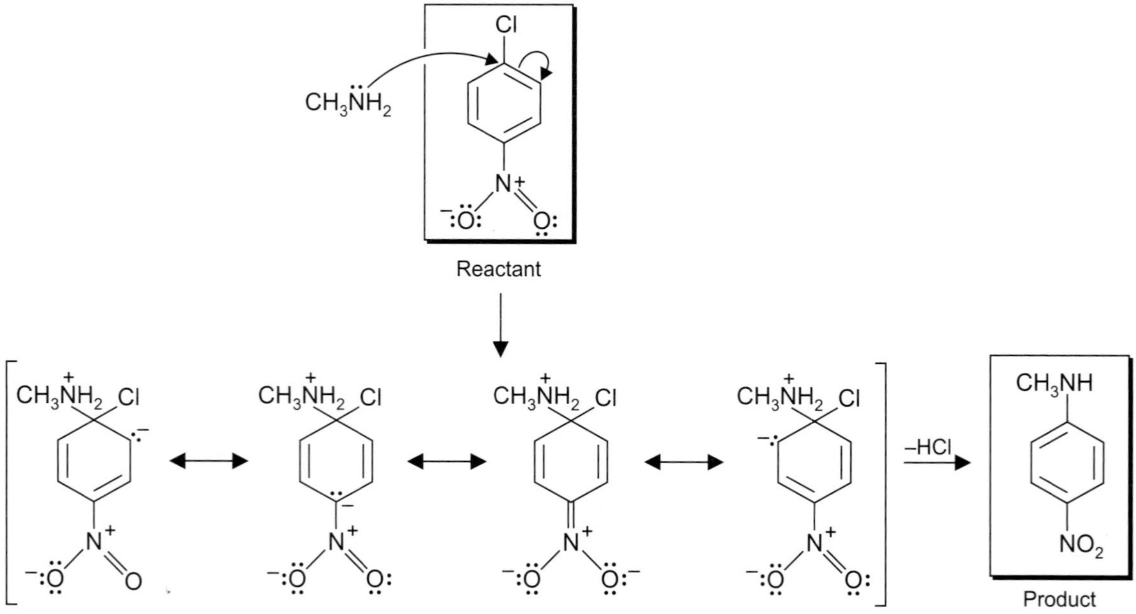

SCHEME 4.1d

(B) Addition Reactions

Addition reactions involve an increase in the number of groups attached to the substrate and thus a decrease in the degree of unsaturation of the substrate. Addition reactions are the reverse of elimination reactions. Generally an addition involves the gain of two groups or atoms (one electrophile and one nucleophile) at each end of a π bond (1, 2-addition) or ends of π system (*e.g.,* 1, 4 or 1, 6-addition). There are, however, examples of addition to certain highly reactive σ bonds (*e.g.,* cyclopropane additions).

Consider the electrophilic addition to an alkene where the π electrons (like that of a benzene ring) present an electron-rich region of potential reactivity. The first step is addition of an electrophile to an unsaturated carbon atom of an alkene (Scheme 4.2). The carbocation derived from the alkene usually adds a nucleophile to give the product of overall addition (the intermediate generated from the aromatic substrate on the other hand loses a cation and the product of substitution is formed see Scheme 4.1*a*).

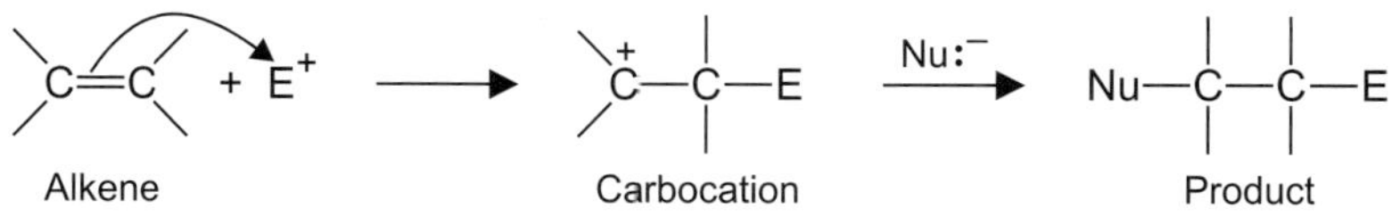

Electrophilic addition to an alkene

SCHEME 4.2

(C) Elimination Reactions

An elimination reaction proceeds *via* the removal of two atoms or group from the same molecule. In most of the cases (Scheme 4.3) the loss is from adjacent atoms so as to form a new double or triple bond. Some nucleophiles *e.g.,* ^-OH and ^-OR, are also strong bases. A tertiary alkyl halide is unable to undergo an S_N2 backside displacement because of steric hindrance; however, when heated with a strong base, usually $K^{+-}OH$ dissolved in ethanol, a tertiary alkyl halide undergoes an elimination reaction to yield an alkene. This elimination proceeds by a different path from that of the E1 mechanism and is termed E2 elimination. An E2 reaction is a one step reaction, like an S_N2 reaction. The strong base abstracts a proton from the alkyl halide, the electron pair forms a *pi* bond, and the halide ion leaves, all in one step.

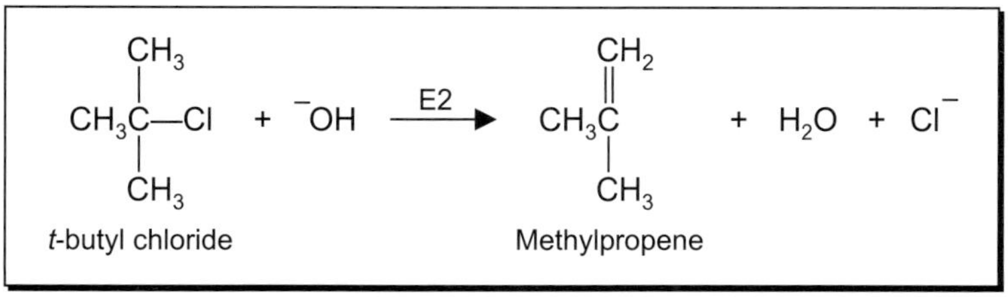

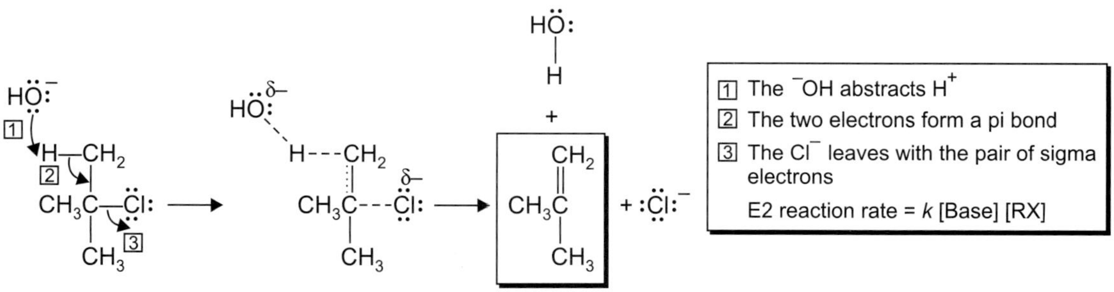

Mechanism of E2 elimination

SCHEME 4.3

A reaction proceeding by an E2 mechanism is a bimolecular elimination because two particles (^{-}OH and RX) are involved in the transition state of the only step (thus the rate-limiting step) of the reaction.

Nucleophilic Aromatic Substitution—An Example of Elimination-Addition Mechanism—Benzyne Formation

Reaction of chlorobenzene with sodamide in liquid ammonia at –33°C gives aniline. Chlorobenzene labelled with ^{14}C at the position bearing chlorine gives an equimolar mixture of unrearranged and rearranged products and this along with other evidence, shows the formation of a symmetrical species a benzyne. Benzyne is formed from chlorobenzene by an E2-type elimination and adds the nucleophile to give aniline (Scheme 4.3a).

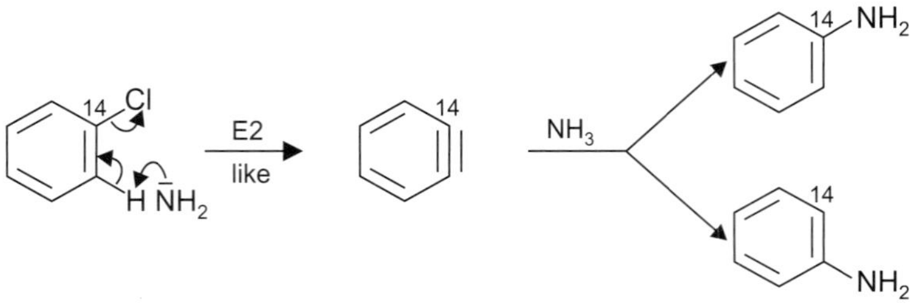

SCHEME 4.3a

This facile reaction is due to the strong basicity of amide ion while hydroxide ion, which is a much weaker base, reacts with chlorobenzene only at 340°C to give phenol via benzyne. Bromobenzene and iodobenzene react in a similar fashion while fluorobenzene does not yield benzyne directly with base due to greater strength of C—F bond.

EXERCISE 4.1

Why 2, 6-dimethylchlorobenzene does not undergo a nucleophilic aromatic substitution (Scheme 4.3b).

ANSWER. *The reactant does not have a H atom in the β-position to the halogen and therefore, benzyne cannot be formed via the elimination reaction.*

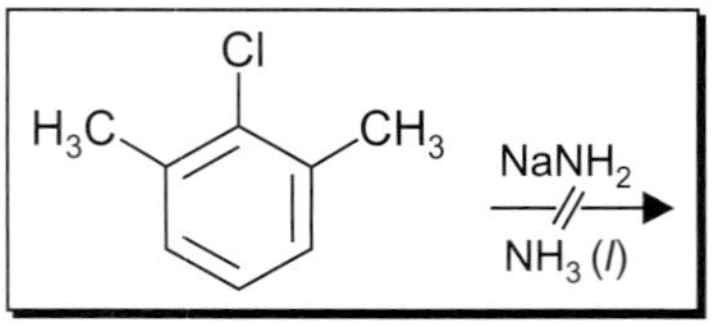

SCHEME 4.3b

(D) Molecular Rearrangements

Most of the molecular rearrangements involve the migration of an atom or a group from one atom to another. The type of migration depends on the number of electrons the migrating atom or group carries with it. The most common type are 1, 2 rearrangements in which the migrating group moves to the adjacent atom with its bonding pair of electrons. The electron-deficient carbocation attracts electron density from adjacent bonds which are rendered weak. The ready loss of a proton to a basic solvent molecule can lead to elimination. In some systems, another important side reaction, rearrangement can occur. The hydrogen attached by the weakened bond with *its bonding electrons* can move to the cationic center, to create a new carbocation (Scheme 4.3*c*).

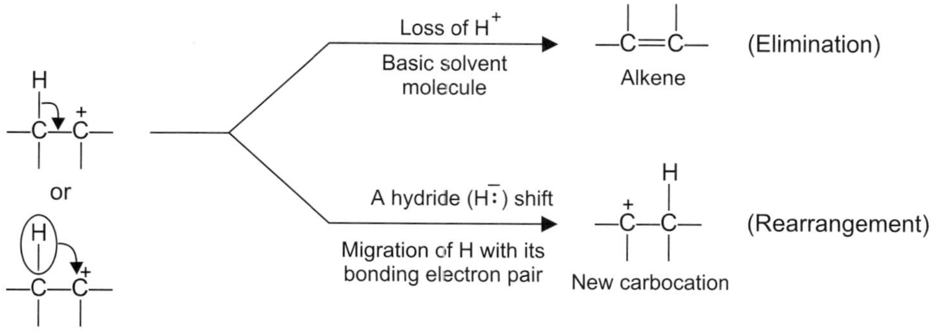

A rearrangement involves a change in the way the atoms are connected. A carbocation rearrangement may involve a 1, 2-hydride shift, a 1, 2-methyl shift or migration of a ring residue

SCHEME 4.3c

During a rearrangement (Scheme 4.3*c*) the positive charge moves to the carbon to which the hydrogen was originally attached. Such rearrangements are particularly important when the new carbocation is more stable than the initially formed carbocation, but the rearrangement still occurs even if the two carbocations have comparable stability. These reactions are commonly observed for secondary carbocations but almost never involve primary carbocations. In primary systems it is the S_N2 process which is generally favourable and the highly unstable primary carbocation is never formed. Rearrangements are less common in tertiary systems.

An example of a carbocation rearrangement is during the reaction of 3-methyl-2-butanol with HBr when the only product formed is 2-bromo-2-methylbutane. The intermediate *sec*-carbocation rearranges much faster compared to its reaction with bromide ion (Scheme 4.3*d*).

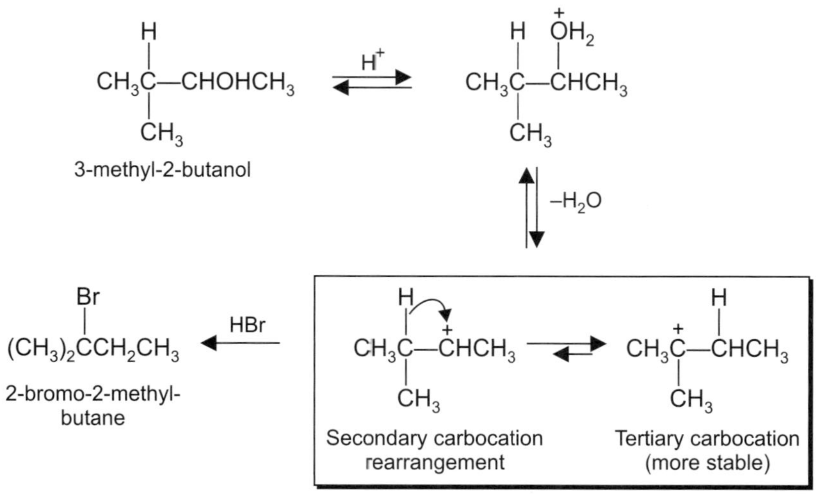

SCHEME 4.3d

EXERCISE 4.2

Explain as to why 2-pentanol and 3-pentanol on reaction with HBr give the mixture of 2-and 3-bromopentane (Scheme 4.3e)?

$$CH_3CH_2CH_2CHOHCH_3 \xrightarrow{HBr} \underset{\text{Major}}{CH_3CH_2CH_2CHBrCH_3} + \underset{\text{Minor}}{CH_3CH_2CHBrCH_2CH_3}$$

$$CH_3CH_2CHOHCH_2CH_3 \xrightarrow{HBr} \underset{\text{Minor}}{CH_3CH_2CH_2CHBrCH_3} + \underset{\text{Major}}{CH_3CH_2CHBrCH_2CH_3}$$

SHCEME 4.3e

ANSWER. *In both cases a secondary carbocation is formed and their rate of equilibration is not complete. The rate of reaction of the bromide ion with carbocation is comparable to the rate of rearrangement (Scheme 4.3f).*

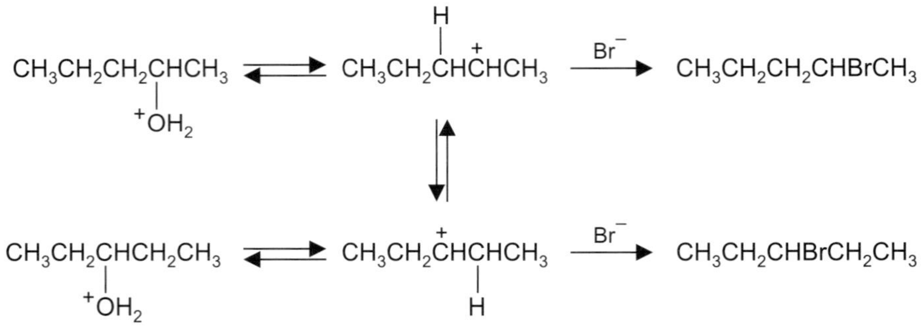

SCHEME 4.3f

Rearrangement During Ipso Attack

In a monosubstituted benzene the orientation is discussed in terms of attack at the ortho, meta or para position, however, attack at the position bearing the substituent (called ipso position) may also occur. The ipso attack has been generally studied in nitration reactions. The arenium ion can loose the substituent as a cationic species and the end result is aromatic substitution with a leaving group other than H. The electrophilic group (NO_2^+) may also undergo a 1, 2-migration followed by loss of proton. This migration occurs in those cases where the substituent already present on the ring is not a good leaving group (Scheme 4.3g). On rearrangement of the electrophile a proton becomes available which is lost to give normal substitution product.

Thus an electrophile can form a bond to an aromatic carbon atom to which a substituent other than hydrogen is already attached and the process is ipso addition which is the first step of ipso substitution. The loss of that substituent leads to ispo substitution.

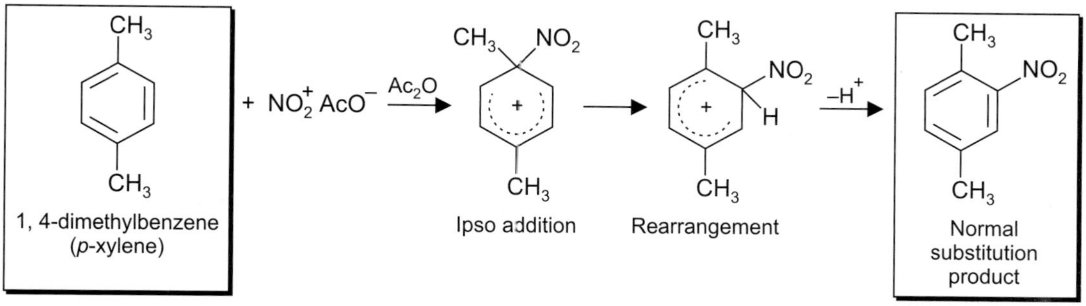

SCHEME 4.3g

(E) Pericyclic Reactions

Several reactions and rearrangements may however, follow pericyclic mechanisms (see Chapter 17).

(F) More Examples of Complex Reactions

Complex reactions can be combination of substitution, addition, elimination and rearrangement reactions, some of these had already been discussed. In all cases reactions may be broken down into a series of steps each involving one of these basic reactions. Consider for example, the Friedel-Crafts alkylation of benzene with 1-chloropropane (Scheme 4.4). It is an electrophilic substitution reaction, where the electrophilic carbocation or polarized haloalkane attacks the benzene ring followed by the loss of a proton. In reaction mixtures carbocations are capable of rearrangement *i.e.,* a change of structure, if these can yield more stable carbocations. Thus, the polarized primary haloalkane (a potential carbocation) shifts a hydride ion H:$^-$ to the adjacent positive carbon to yield the more stable secondary carbocation (Scheme 4.5).

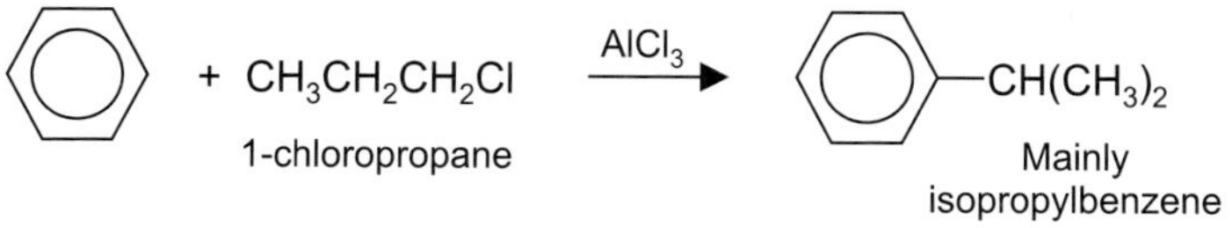

SCHEME 4.4

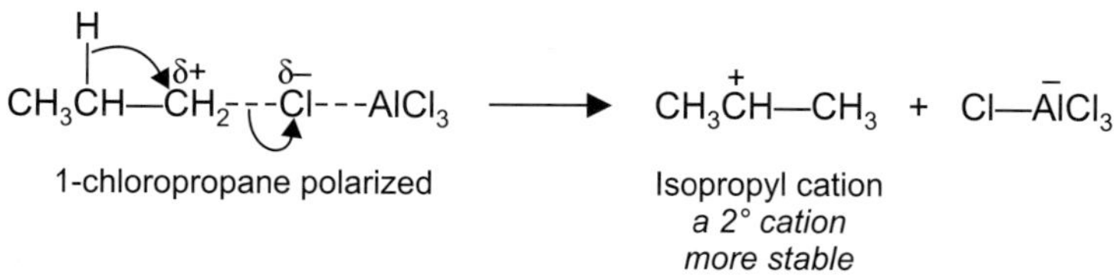

SCHEME 4.5

S_N1 substitution reactions of alkyl halides are accompanied by elimination reactions and these yield alkenes. These reactions are called dehydrohalogenation reactions. In these elimination reactions, the nucleophile acts as a base—a proton acceptor. This elimination reaction proceeds by an E1 mechanism (elimination, unimolecular). The first step in an E1 reaction is the ionization of the alkyl halide, (as in reaction by the S_N1 path). The first step in

E1 elimination is the spontaneous dissociation of the alkyl halide to give an intermediate carbocation in the slow rate limiting step. The second step in an E1 reaction is the loss of a proton (H^+) to the solvent. The electrons in the C—H bond are used to form a *pi* bond. The product is a stable uncharged alkene (Scheme 4.6).

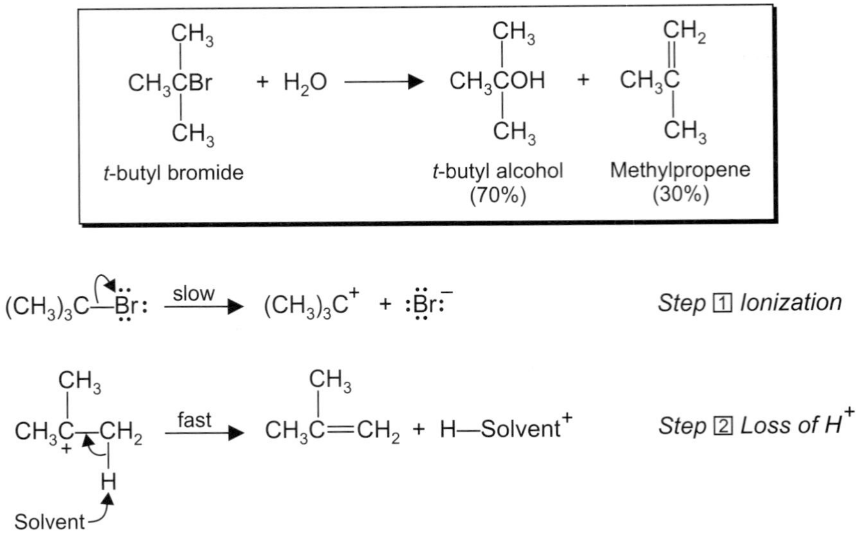

SCHEME 4.6

4.2 NUCLEOPHILES AND ELECTROPHILES

The nucleophile ("nucleus lover") is an electron rich ion or molecule that reacts at a positively charged site in a compound. $Nu:^-$ represents a negatively charged nucleophile *e.g.,* HO^-, RO^-, X^-, CN^- and Nu represents an uncharged nucleophile *e.g.,* H_2O, ROH. An electrophile (E, "electron lover") may be electrically neutral BF_3 or $AlCl_3$ or positively charged like H^+, Br^+, NO_2^+. Both BF_3 and BH_3 are Lewis acids, they do not have a positive charge but have empty 2*p* orbitals and react like cations (electrophiles, see Scheme 1.4*a*). To form the bromonium ion, a pair of nonbonding electrons on bromine (the nucleophile) attacks the empty *p* orbital on the positive carbon (the Lewis acid) to form a normal, two-electron bond, (Scheme 4.6*a*). In an ionic reaction a nucleophile (Nu:) shares an electron pair with an electrophile (E) in the process of bond formation (Nu: + E → Nu—E). Thus nucleophiles are Lewis bases; *i.e.,* electron pair donors and electrophiles are lewis acids, *i.e.,* electron pair acceptors.

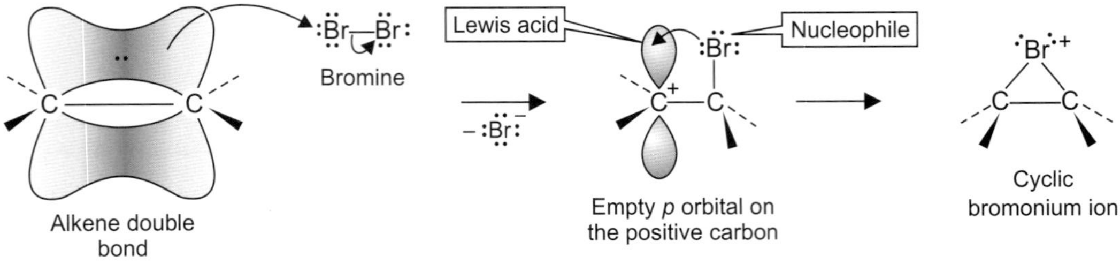

SCHEME 4.6a

4.3 ELECTRON MOVEMENT

When a bond breaks homolytically, each bonded atom retains one of the bond's two electrons. In contrast, when a bond breaks so that one of the two atoms retains both electrons, one says that heterolytic cleavage has occurred. Homolytic cleavage forms free radicals, while heterolytic cleavage forms ions. A heterolytic cleavage is sometimes called an ionic cleavage. A curved arrow is used to show the movement of the electron pair in an ionic cleavage. Half-arrows show the separation of the individual electrons in a homolytic cleavage (Scheme 4.7). Energy is released when bonds are formed, and energy is always consumed when the bonds cleave.

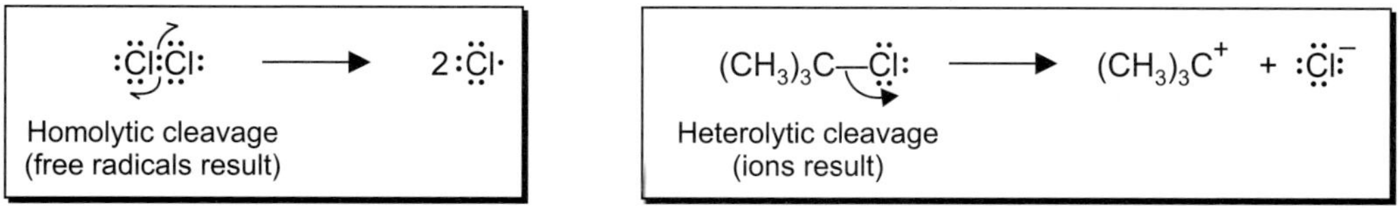

SCHEME 4.7

4.4 EQUILIBRIA AND FREE ENERGY

All chemical reactions are reversible and reactants and products interconvert to different degrees. When the concentrations of reactants and products do not undergo a change, the reaction is said to be in a state of equilibrium. In several cases, equilibrium lies largely (say, more than 99.9%) to the side of the products. When this happens the reaction is said to have gone to completion. (In such cases, the arrow indicating the reverse reaction is usually ommitted). Equilibria are described by equilibrium constant, K. To find an equilibrium constant one divides the product of the concentrations of the components on the right side of the reaction by the product of the concentrations of the components on the left, all given in units of mole litre^{-1}. A large value for K shows that a reaction will go to completion and has a large driving force.

Some chemical equilibria are:

$$A \underset{}{\overset{K}{\rightleftharpoons}} B \qquad K = \frac{[B]}{[A]}$$

$$A + B \overset{K}{\rightleftharpoons} C \qquad K = \frac{[C]}{[A]\,[B]}$$

$$A + B \overset{K}{\rightleftharpoons} C + D \qquad K = \frac{[C]\,[D]}{[A]\,[B]}$$

The value of K is determined by the change in free energy (sometimes called Gibbs free energy) that accompanies the reaction. Free energy is represented by G, and the change (Δ) in free energy of reactants and products in their standard states (25°C, 1 atm) is represented by $\Delta G°$. The relationship between $\Delta G°$ and K is in the expression (eq. I, Scheme 4.8).

$$\Delta G^\circ = -RT \ln K = -2.303\ RT \log K \text{ (in kcal mol}^{-1}\text{)} \qquad \text{(I)}$$

ΔG° = (Free energy of products) – (Free energy of reactants)

R = 1.986 cal/deg-mol (8.309 J/deg-mol), the gas constant

T = Temperature in degrees kelvin

$$\Delta G^\circ = -1.36 \log K \text{ (in kcal mol}^{-1}\text{)} \qquad \text{(II)}$$

SCHEME 4.8

When a reaction "goes to completion" or has "a large driving force", a certain amount of energy is released. A negative value of ΔG° shows a release of energy. It follows that a large value for K indicates a large favourable free energy change. At room temperature (25°C, 298 K), the equation (I, Scheme 4.8) becomes (eq. II, Scheme 4.8), and from this expression, an equilibrium constant of 10 would have a ΔG° of –1.36 kcal mol^{-1}, and conversely, a K of 0.1 would have a ΔG° = +1.36 kcal mole^{-1}. Because the relation is logarithmic, doubling the ΔG° value increases the K value exponentially. When K = 1, starting reactants and products are present in equal concentrations and ΔG° is zero.

4.5 FREE ENERGY CHANGE IN RELATION TO BOND STRENGTHS AND DEGREE OF ORDER IN A SYSTEM

The Gibbs standard free energy may be dissected into enthalpy and entropy components ΔH° and ΔS° respectively (Scheme 4.9).

$$\Delta G^\circ = \Delta H^\circ - T\Delta S^\circ$$

ΔG° = (Free energy of products) – (Free energy of reactants)

ΔH° = (Enthalpy of products) – (Enthalpy of reactants)

ΔS° = (Entropy of products) – (Entropy of reactants)

SCHEME 4.9

The enthalpy change ΔH° is defined as the heat of a reaction at constant pressure. Enthalpy changes in an organic chemical reaction relate mainly to changes in bond strengths during the course of the reaction. The value of ΔH° = ΔH° for bonds being broken $-\Delta H^\circ$ for bonds being formed.

The values of ΔH° can be calculated from bond dissociation energies (Table 1.2). Since values of ΔH° are easy to calculate organic chemists therefore, evaluate reactions from this quantity.

When weaker bonds are broken and stronger bonds are formed, heat is released and the reaction is then exothermic (negative value of ΔH°). In an exothermic reaction, the enthalpy term contributes to a favourable negative value of ΔG°. However, if stronger bonds are broken and weaker bonds are formed then energy is consumed in the reaction which becomes endothermic (positive value of ΔH°). In an endothermic reaction, the enthalpy term contributes to an unfavourable positive value of ΔG°.

The value of $\Delta H°$ for the chlorination of methane is about –25 kcal/mol (105 kJ/mol). Thus it is a highly exothermic reaction, with the decrease in enthalpy serving as the primary driving force. Another example of an exothermic reaction is the combustion of methane, (the main component of natural gas), to carbon dioxide and water. This process has a $\Delta H°$ value of –213 kcal mol^{-1} ($CH_4 + CO_2 \rightarrow CO_2 + H_2O$). The exothermic nature of this reaction is due to very strong bonds formed in the products. The entropy change $\Delta S°$ provides a measure of the changes of the order of a system or freedom of motion of a system. Reactions tend to favour products with the greatest entropy, since there is a negative sign in the entropy term of the free-energy expression. A positive value of $\Delta S°$ thus indicates that the products have more freedom of motion than the reactants, contributes to a favourable (negative) value of $\Delta G°$ (see, Sec. 3.6).

In most of the situations the enthalpy change is much larger than the entropy change, and the enthalpy term dominates the equation for $\Delta G°$. Therefore, a negative value of $\Delta S°$ does not for sure mean that the reaction has an unfavourable value $\Delta G°$. The formation of strong bonds (the change in enthalpy) is usually the most important component in the driving force for a reaction.

For example, consider the chlorination of ethane with chlorine to give chloroethane and hydrogen chloride. The reaction (Scheme 4.10) has a $\Delta H°$ of –28 kcal mol^{-1} and only a small $\Delta S°$ of +0.5 e.u. ("entropy units"). This shows that at room temperature (298 K) the contribution of $-T\,\Delta S°$ to $\Delta G°$ is only on the order of – 0.15 kcal mol^{-1} *i.e.*, almost negligible. The considerable driving force for the chlorination lies largely in the large negative value of $\Delta H°$.

$$CH_3CH_3 + Cl_2 \longrightarrow \underset{\text{Chloroethane}}{CH_3CH_2Cl} + HCl \qquad \Delta H° = -28 \text{ kcal mol}^{-1} \quad \Delta S° = +0.5 \text{ e.u.}$$

SCHEME 4.10

The activation energy of a reaction is composed of an enthalpy contribution, to which ring strain adds a positive (unfavourable) increment, and an entropy contribution, which is almost entirely influenced by the proximity of the reacting centers and relative rigidity of the transition state. These two factors display systematically variation however, in opposite directions from smaller to larger rings. At first sight, it may look surprising that usually, three membered rings are formed faster than four-membered rings. An example is of Williamson synthesis of cyclic ethers *via* intramolecular S_N2 reaction on halo alcohols (Scheme 4.10*a*).

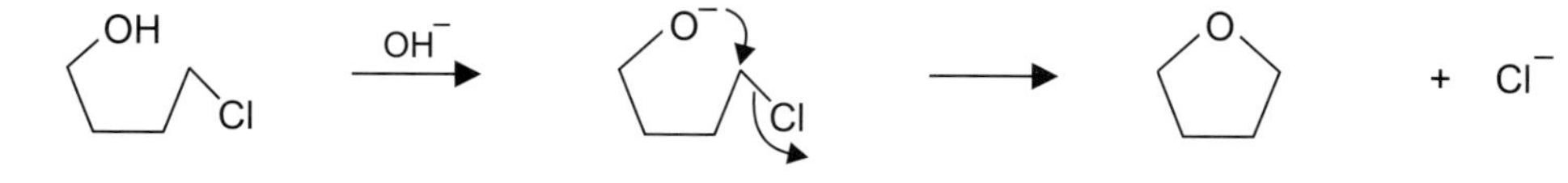

4-chlorobutanol An alkoxide intermediate tetrahydrofuran (THF)

Intramolecular S_N2 reaction—the Williamson synthesis

SCHEME 4.10a

A comparison of the relative rates of cyclic ether formation reveals that three membered rings (epoxides) form quickly. Thus the preparation of an epoxycyclopropane from a 2-bromo-alcohol is entropically highly favourable, since the nucleophile and the leaving group are as close to each other as possible. Although the ring strain is worst in this case. The transition state energy is relatively small, since a favourable entropy contribution allows relatively a rapid ring formation.

On increasing the chain of a halo alcohol, the reacting centers fall apart (unfavourable entropy factor) which contributes more flexibility (degrees of freedom) to the substrate. This flexibility has to be given up in the transition state of ring closure. Moreover, five-membered rings are formed easily. (The relative rates of cyclic ether formation with respect to ring size are $3 \geq 5 > 6 > 4 \geq 8$). In the formation of five-membered ring, compared to *e.g.*, a four-membered ring, though the reacting centers are far apart in the former, however, the unfavourable strain contribution to the enthalpy of activation is largely decreased. Five-membered rings form faster than six-membered rings, although the six-membered rings are almost strain free. In this case the relative greater distance between alkoxide and electrophilic carbon and consequently greater degree of freedom of the chain reflects on this.

The more negative the $\Delta H°$, the more positive the $\Delta S°$, the more negative the $\Delta G°$, the more exothermic (favourable) will be the reaction. While dealing with conformational changes one discusses the total changes in free energy that occur ($\Delta G°$). However, while discussing most chemical reactions involving bond breaking and forming, one normally discusses changes in enthalpy ($\Delta H°$). In many organic reactions, the entropy change is most often very small in relation to the change in enthalpy then the relationship holds: $\Delta G° \cong \Delta H°$.

In cyclohexane *e.g.*, because of its relatively increased rigidity, there are fewer degrees of vibrational and rotational freedom compared to a straight-chain hexane. Ring opening therefore, means a gain in entropy and ring closing a loss.

4.6 REACTION RATES

Reaction rates usually depend on the concentrations of the reactants. When the concentrations of the reactants is large they will collide more often and the greater will be the chance of reaction. The concentration of reactants influences the rate of a reaction. Consider the addition of reagent A to reagent B to give C ($A + B \rightarrow C$). In many transformations of this type, it is observed that increasing the concentration of either reactant increases the rate of the reaction. In some cases, the transition state is formed as the result of collision of molecules A and B. The rate is then expressed (eq. I, Scheme 4.11). In this equation the proportionality constant, k, is also called the rate constant of the reaction. The initial rate equals the rate constant when the two starting materials are at one molar concentration. A reaction for which the rate depends on the concentrations of two molecules in this way is termed second order.

There are processes whose rate depends on the concentration of only one of the reactant, such as in the reaction ($A \rightarrow B$, eq. II, Scheme 4.11). A reaction of this type is called first order. Rotation around a carbon-carbon bond follows such a rate law.

$$\text{Rate} = k\,[A][B] \text{ in units of moles l}^{-1}\text{ sec}^{-1} \qquad \text{(I)}$$

$$\text{Rate} = k\,[A] \text{ in units of moles l}^{-1}\text{ sec}^{-1} \qquad \text{(II)}$$

SCHEME 4.11

4.7 THE TRANSITION STATE—ACTIVATION ENERGY

A reaction which proceeds with a negative free energy change is called exergonic, and the one which proceeds with a positive free energy change is termed endergonic. The reaction between methyl chloride and hydroxide ion in aqueous solution is a very favourable exergonic process (at 60°C (333 K), $\Delta G° = -24$ kcal mol^{-1}. The equilibrium constant for the reaction is extremely large (near 10^{16}). Even then a dilute solution of methylchloride in aqueous base requires weeks to approach the position of equilibrium. Many factors must contribute together for the reaction to occur. These are:

- Reactants must have adequate energy to collide.
- The reactants must come together in a proper orientation favourable for reaction (*i.e.,* the specific orientation in the reaction of methylchloride with hydroxide ion in the S_N2 reaction).
- Sufficient energy must be available to break the bonds undergoing change. If *e.g.,* covalent bonds are broken in a reaction, the reactants must go up an energy hill first, before they can go downhill. This will be true even if the reaction is exergonic.

The configuration of reactants in which all the stringent requirements for an effective collision are met is commonly known as the transition state (activated complex), a hypothetical description of the atoms at the point of highest energy along the reaction pathway. By contrast, intermediates are capable of detection and in some cases even isolation. One cannot isolate transition states or even detect them. However, the reaction intermediates like a carbocation can not only be detected but can be even isolated.

Thus in short, just because a reaction has a negative $\Delta G°$ does not necessarily mean that it will take place during a reasonable period of time. A negative $\Delta G°$ is a necessary however, not a sufficient condition for a reaction to occur spontaneously. For a reaction to take place the transition state must be attained and energy must be supplied to the reactants for this purpose. This energy, the difference between the free energy of the reactants and of the transition state, is the free energy of activation $\Delta G^‡$. It is this activation energy that controls the rate of a chemical reaction. Typical organic reactions have activation energies of 10–50 kcal/mol (40–200 kJ/mol) and for these to occur free energy of activation $\Delta G^‡$ must be added.

4.8 TRANSITION STATE THEORY—MEASUREMENT OF ACTIVATION ENERGY

The transition state theory of rates of reactions makes as assumption in an elementary reaction the reactants must pass through a transition state. The reactants are in equilibrium with an activated complex, *e.g.,* for a bimolecular reaction (Scheme 4.12).

$$A + B \rightleftharpoons AB^‡ \longrightarrow \text{Products}$$

$AB^‡$ is an activated complex at the transition state.

SCHEME 4.12

One usually measures a somewhat different energetic barrier to reactions in the laboratory *i.e.,* the Arrhenius activation energy E_a. Values of Arrhenius activation energies are related to the enthalpies of activation $\Delta H^‡$, (eq. I, Scheme 4.13). Enthalpies of activation are related to free energies of activation $\Delta G^‡$ (eq. II, Scheme 4.13). In most of the situations the values of E_a is an acceptable approximation of $\Delta G^‡$.

The Swedish Chemist Arrhenius found the dependence of reaction rate on temperature at which the reaction is carried out. For chemical reactions the measurement of the variation of the rate of reaction at a number of different temperatures helps in calculating activation energy E_a, from the Arrhenius equation (Scheme 4.14), where k is the rate constant and A is a frequency factor for that reaction. The activation energy, E_a, is related to the activation enthalpy $\Delta H^\ddagger$ (Scheme 4.13).

$$E_a = \Delta H^\ddagger + RT \qquad \text{(I)}$$

$$\Delta G^\ddagger = \Delta H^\ddagger - T\Delta S^\ddagger \qquad \text{(II)}$$

SCHEME 4.13

Any reaction (with an activation energy) is faster at higher temperatures, as may be readily derived from the Arrhenius equation. When the temperature increases, a larger fraction of molecular collisions have enough kinetic energy for the reaction to occur and this increases the reaction rate. A typical distribution of energies for a collection of molecules at two temperatures where $T_2 > T_1$ is shown (Scheme 4.15). The activation energy for a hypothetical reaction is marked as E_a. The number of molecules with sufficient energy for reaction is given by the shaded areas, which is much greater at T_2 *i.e.,* the higher temperature. Since there are more molecules with energies sufficient for reaction to occur, at higher temperature, the reaction is faster. A rule of thumb is that with an increase of 10°C in temperature the reaction rate approximately doubles.

$$k = Ae^{-E_a/RT}$$

k is a rate constant
E_a is the energy of activation
T is the absolute temperature
R is the gas constant
A is a constant known as the pre-exponential factor

Arrhenius equation
The Arrhenius equation is generally used to determine energies of activation from rate data

SCHEME 4.14

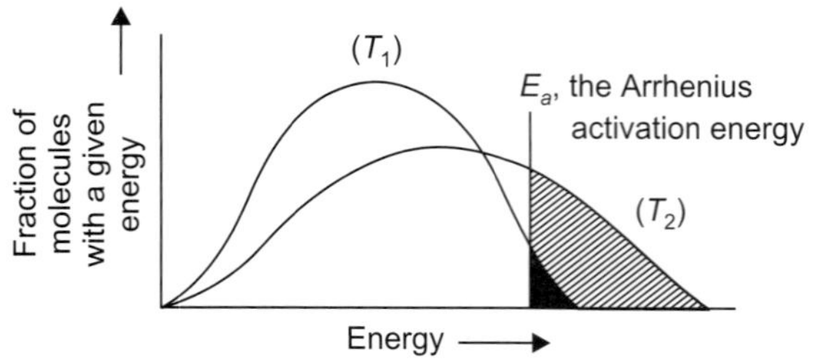

The distribution of energies at two temperatures, T_1 and T_2 ($T_2 > T_1$). The number of collisions with energies greater than the free energy of activation is shown by shading the area under each curve

$$E_a = \Delta H^\ddagger + RT$$

SCHEME 4.15

4.9 REACTION PROFILE DIAGRAMS

Organic chemists generally represent the course of a reaction through an energy profile diagram. The interpretation of a simple S_N2 reaction is shown in (Scheme 4.16) which throws light on the concepts of transition state and activation energy. This is the usual energy profile diagram for a one step reaction without an intermediate. For the reaction to occur, free energy of activation $\Delta G^{\ddagger}$ must be added.

The vertical axis of the energy profile depicts the total free energy of all the species involved in the reaction. The horizontal axis is called the reaction coordinate which reflects the progress of the reaction, going from the reactants on the left to the products on the right and usually considered to be related to the changes in molecular geometry *e.g.*, bond angle and bond length. The S_N2 reaction has only one transition state and no intermediate. The transition state has a definite geometry and charge distribution and has no finite existence. The system at this point is termed an activated complex. In the transition state theory the starting materials and the activated complex are in equilibrium and then the equilibrium constant is designated by $K^{\ddagger}$. Thus the rate constant of the reaction is dependent only on the position of the equilibrium between the starting materials and the activated complex, *i.e.*, on the value of $K^{\ddagger}$ ($\Delta G^{\ddagger} = -2.3\ RT \log K^{\ddagger}$). The interpretation of such figures can be explained by reference to (Scheme 4.16). The reaction represented will proceed spontaneously as the activation energy is relatively small and the free energy of the products is lower compared to that of the reactants.

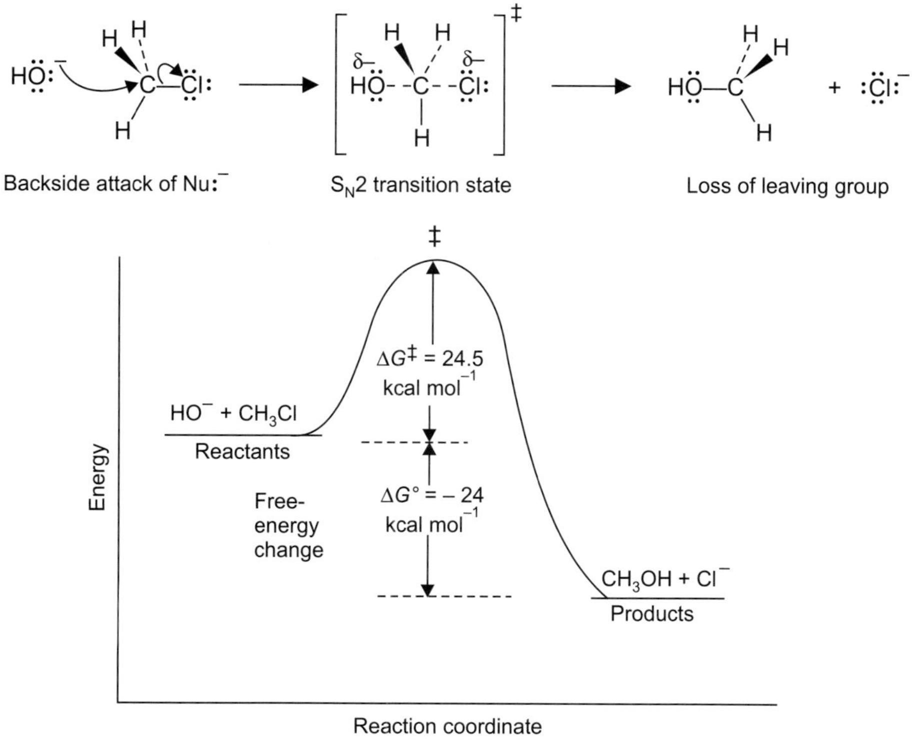

A potential energy diagram for the reaction of methyl chloride with hydroxide S_N2 mechanism, backside attack with inversion

SCHEME 4.16

By contrast, a reaction which has a high activation energy will take place very slowly at normal temperatures. Thus the S_N2 reaction of a sterically hindered secondary bromide will raise the energy of the transition state (compared with unhindered methyl bromide) and consequently the reaction will become slow.

Entropy of activation $\Delta S^{\ddagger}$ (difference in entropy between the starting compounds and the transition state) gains importance (see Scheme 4.13, eq. II) when the reacting molecules have to approach each other in a defined orientation for the reaction to occur. Consider the elimination reaction of an acyclic alkyl chloride with hydroxide ion to give an alkene. Under the usual E2 reaction conditions it is essential that OH^- be near the hydrogen, moreover the hydrogen must also be oriented *anti* to the chlorine atom (Scheme 4.17). For the success of this elimination, the reactants therefore, must surrender the freedom normally associated with them. This leads to a considerable loss of entropy *i.e.*, $\Delta S^{\ddagger}$ is negative.

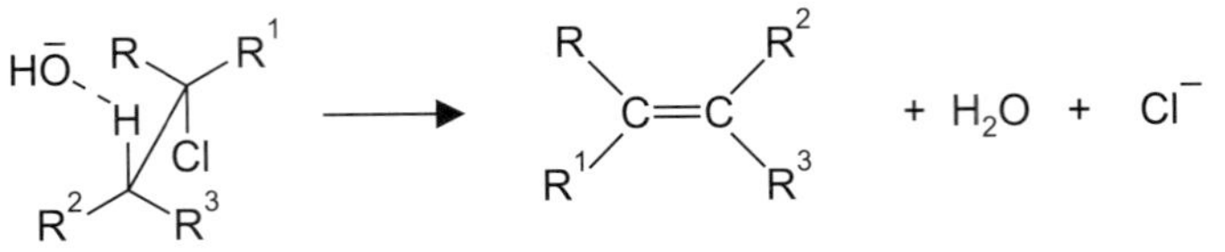

SCHEME 4.17

The reaction profile of a particular reaction can be altered by the use of a catalyst. A catalyst is a species that can change the reaction mechanism and thereby lower the activation energy by providing an alternative, lower energy pathway. A catalyst can therefore, increase the rate of a reaction but can have no effect on the equilibrium constant. At the end of a reaction a catalyst is recovered chemically unchanged.

The mechanistic description of some reactions may involve more than one steps. Reactions with intermediates are two step (or more) processes. In these reactions *e.g.*, in nucleophilic substitution of chloride by hydroxy in *tertiary*-butyl chlordie (S_N1 reaction), there are two transition states, each with an energy higher than the carbocation intermediate (Scheme 4.18). In such a reaction, one sees an energy "well" and deeper the well the more stable is the intermediate. The first step (Scheme 4.18) is the formation of an unstable carbocation intermediate. This high energy intermediate reacts with water *i.e.*, the nucleophile in a rapid second step to afford the protonated alcohol. The step 3 in the solvolysis of an alkyl halide, is the loss of a proton by a protonated alcohol. This reaction is an acid-base reaction and is not actually part of the S_N1 mechanism. The S_N1 path is a two-step sequence: (1) ionization of the alkyl halide to yield the intermediate carbocation and (2) combination of the carbocation with the nucleophile. The energy, profile diagram displays two energy maxima and the transition state associated with the formation of the carbocation and the transition state for the formation of the protonated alcohol *i.e.*, the formation of the new carbon-oxygen bond.

4.10 THE RATE DETERMINING STEP

In a multistep reaction, every step has its own characteristic rate. There can be only one overall reaction rate however, which is determined by the rate-determining step. In general, the highest-energy step of a multistep reaction is the "bottleneck", and it determines the overall rate. On an energy profile diagram the step involving formation of the highest energy transition state is designated as the rate determining step. In the reaction of methyl chloride with hydroxide the single step must be the rate controlling step (Scheme 4.16). In the two-step reaction (Scheme 4.18) the first step is rate-controlling.

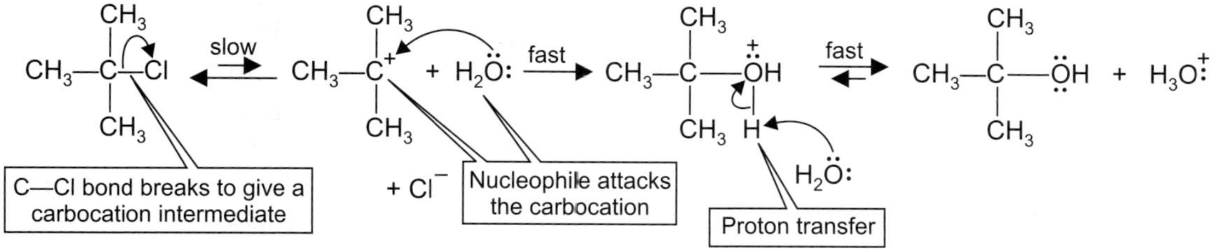

Mechanism of the S_N1 reaction

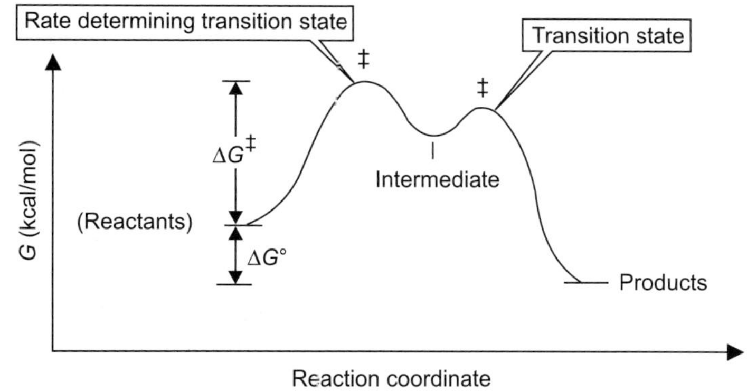

Reaction coordinate diagram for an S_N1 reaction

SCHEME 4.18

4.11 THERMODYNAMIC AND KINETIC CONTROL

When more than one product is formed during a reaction, the product which is formed most rapidly is called the kinetic product while the most stable product is called the thermodynamic product. Kinetically controlled reactions produce kinetic product as the major component, while thermodynamically controlled reactions produce thermodynamic product as the major component.

Under reversible reaction conditions the ratio of possible products is determined by the relative stability of each product which is measured by its standard free energy ($\Delta G°$). The composition of the equilibrium mixture does not depend on how fast ($\Delta G^{\ddagger}$) each product is formed in the reaction. This is termed thermodynamic (or equilibrium control). On the other hand when the quantity of each possible product is determined by how fast each product is formed and is not a function of the relative stability $\Delta G°$ of each product, one calls it kinetic control. Consider the addition of HBr to 1, 3-butadiene which gives two products both derived from a common intermediate (Scheme 4.18*a*).

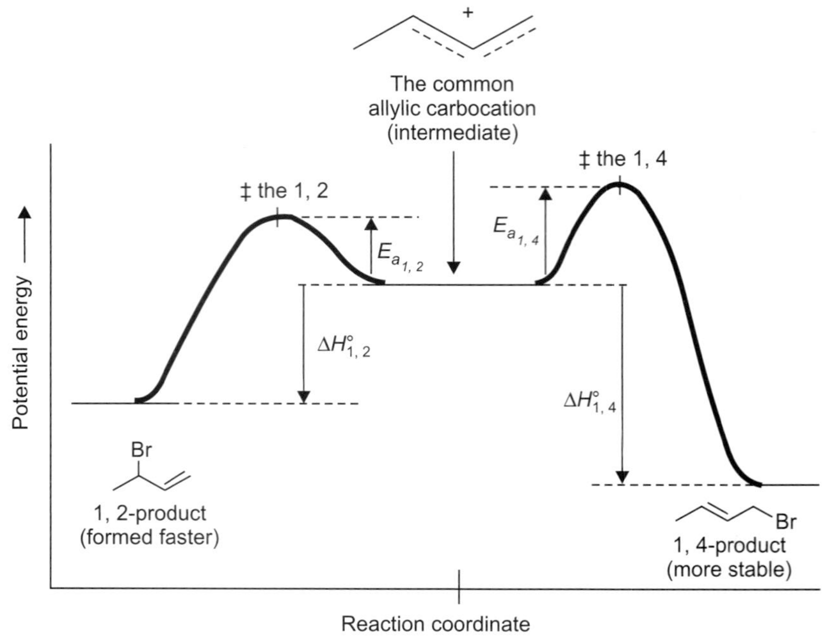

Reaction coordinate diagram for HBr addition to 1, 3-butadiene

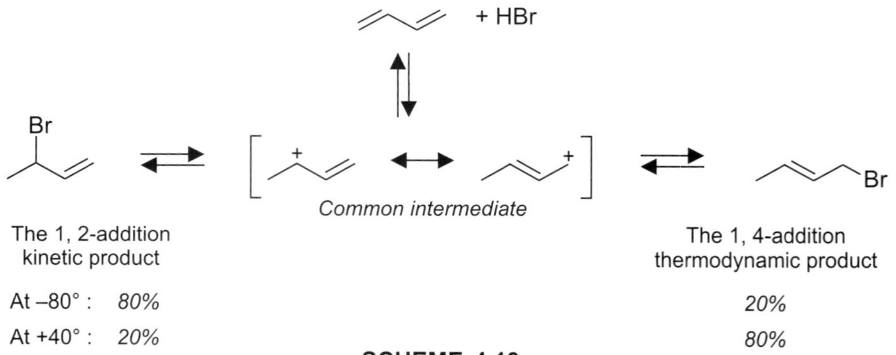

SCHEME 4.18a

The following points may be noted:

- The reaction proceeds through the same common intermediate which is a resonance-stabilized allylic cation that both products have in common. The formation of this intermediate is the first step of the reaction.
- Under sufficiently low temperature conditions, – 80°C (mild conditions) the reaction is irreversible and HBr adds to 1, 3-butadiene to give 1, 2-addition product as the major component. The reaction-energy diagram for the second step of this addition is in (Scheme 4.18*a*, recall that organic chemists analyze reactions by considering $\Delta H°$). The transition state for 1, 2-addition has lower energy than transition state for 1, 4-addition, consequently 1, 2-addition is associated with a lower activation energy (E_a). The 1, 2-addition occurs by the attack on the more highly substituted carbon (one may note here, that the less stable product has a more stable transition state).
- The attack by bromide on the allylic cation represents a highly exothermic process and therefore, the reverse reaction has a large activation energy. At low temperatures

(–80°C) enough energy is available and the reactants overcome the energy barrier for the first step of the reaction to form the intermediate, the resonance-stabilized allylic cation. Moreover, enough energy is also available for the intermediate to yield two addition products.

- At low temperature (– 80°C), however enough energy is not available for the reverse reaction and the product which is formed faster (the 1, 2-product) predominates.
- At higher temperature (40°C), however, enough energy is available for the products to go back to the intermediate *i.e.,* for the reverse reaction to occur. At 40°C an equilibrium is set up and the most stable species predominate which is the 1, 4-product.

4.12 APPLICATIONS OF KINETIC PRINCIPLES

(A) Hammond Postulate

The precise structure of an intermediate is far better understood than the transition state. Often it is helpful to regard an intermediate as a model for the transition state which reflects on the rate of a reaction. Thus the transition state may have some character of the intermediate formed in a reaction. The Hammond postulate states that the structure of the transition state for a reaction step is closer to that of the species (reactant or product of that step) to which it is closer in energy. Recall the application of Hammond postulate to explain regioselectivity of bromination versus chlorination.

Consider Markovnikov's rule (a regioselective reaction), it states that during ionic addition of an unsymmetrical reagent (*e.g.,* HX) to a double bond, the positive portion of the adding reagent attaches itself to that unsaturated carbon so as to give the more stable carbocation as an intermediate. As an example to explain Hammond's postulate, consider the addition of HBr to propylene (Scheme 4.18*b*). Of the two possible products, II predominates (Markovnikov's rule).

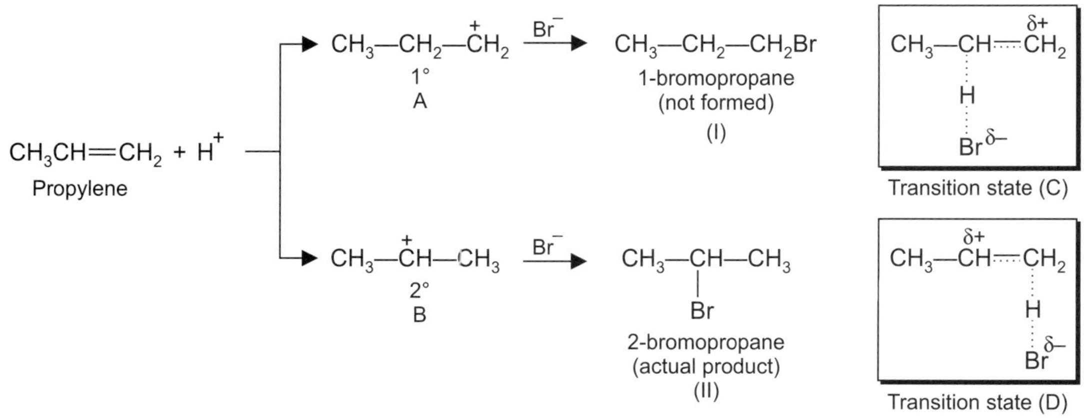

SCHEME 4.18b

The rate-determining step in the addition is the protonation of one of the carbon atoms of the double bond the give a carbocation which then reacts rapidly with the anion Br^- from HBr. One of two such carbocations A or B may be formed in competing ways. The secondary ion B is however, more stable. The rate of formation of these two ions is not immediately dependent on their stabilities, but rather on the relative free energies of the transition states for carbocation formation. The detailed structures of the transition states are uncertain; the new C—H bond is partly formed at the transition state, however, to an unknown extent. But one knows that the carbocations have relatively high free energies compared with the reactants (Scheme 4.18*b*) and there is thus a smaller change in energy when the transition states are transformed into the intermediate carbocations than when they revert to the reactants. Therefore, one may conclude that there is a smaller reorganization of the molecular geometry in passage from transition state to intermediate than in passage from transition state to reactants. Thus the transition states have resemblence to the carbocations which arise from them. The two transition states may be represented as (C and D, Scheme 4.18*c*). One therefore, concludes that the factors which determine the relative stabilities of the carbocations are also effective in determining the relative stabilities of the transition states and therefore, the more stable of the two ions is formed the faster and corresponds to a lower maximum on the energy profile.

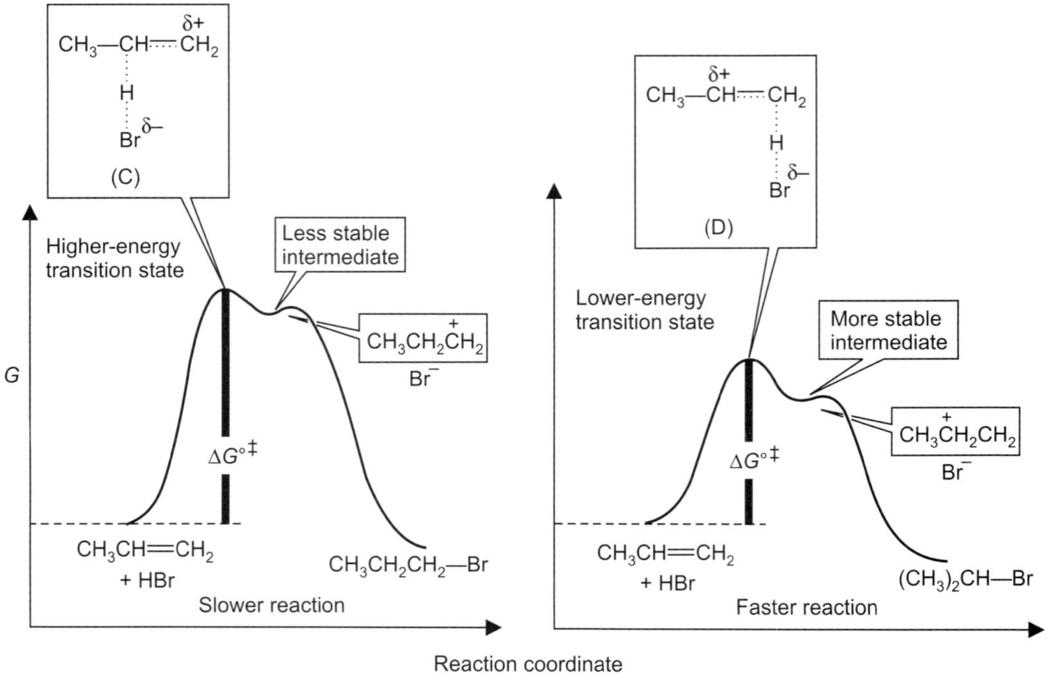

Energy profiles—the addition of HBr to propylene

SCHEME 4.18c

(B) The Effect of Solvent on Rate of a Reaction—Stabilization of Transition State

The nature of the solvent affects the rate of a reaction almost always. If one considers again the solvolysis of *t*-butyl chloride (see Scheme 4.18, the rate determining step is the formation of the carbocation intermediate), two characteristics of the solvent have their role in determining the relative free energies of reactant and the transition state and consequently the rate of reaction. Firstly, energy is needed to separate the unlike charges and therefore, the reaction rate increases with the dielectric constant.

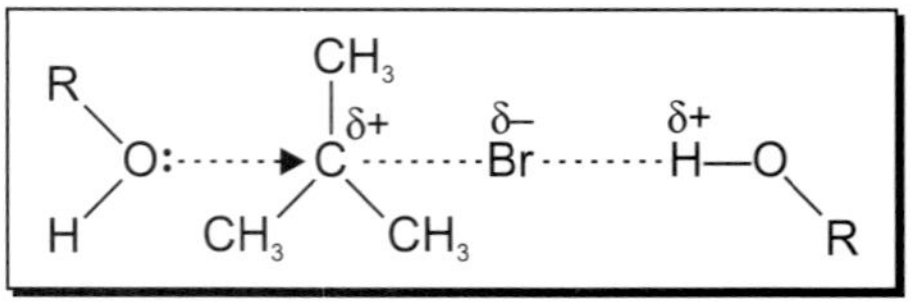

SCHEME 4.18d

Secondly the solvating power of the solvent plays an important role, since the transition state can be stabilized by solvation of the developing positive and negative ions. This solvation is significant with water and other hydroxylic compounds *e.g.,* alcoholic solvents. The electron-rich hydroxylic oxygen solvates the developing carbocation (Scheme 4.18*d*) while the developing chloride ion is solvated by hydrogen bonding to the hydroxylic hydrogen (also see Scheme 5.15). Polar aprotic solvents like DMSO and DMF are less effective solvents in this case, since they can only stabilize cations (see, Scheme 5.14). As a result the rate of solvolysis of *t*-butyl chloride increases in water and other alcoholic solvents, since the transition state is stabilized by solvation of both the developing positive and developing negative ions.

The following points may be noted.

- In the S_N1 reaction (Scheme 4.18*d*) the transition state is more polar (with more charge separation) than the reactant, therefore, a change to a more polar solvent stabilizes the transition state far more than it stabilizes the reactant. The rate of S_N1 reaction is very fast in a more polar solvent.
- In an S_N2 reaction, with a nucleophile with a negative charge, this charge is more dispersed in the transition state (the transition state is a much larger ion). In the nucleophile however, the charge is more concentrated and it is thus more effectively solvated and stabilized compared to the transition state. Thus the rate of an S_N2 reaction involving a negative nucleophile occurs more slowly when the anion solvating power of the solvent increases (Scheme 4.18*e*). In such cases the use of aprotic solvents *e.g.,* DMSO, DMF or acetone which cannot form a hydrogen bond with the nucleophile (no H bonded nitrogen or oxygen in these solvents is available) increase the reactivity of the nucleophile and thus enhance the rate of S_N2 reaction. For example, the reaction $Cl^- + CH_3I \rightarrow CH_3Cl + I^-$ is a million time faster in an aprotic solvent like DMSO compared to a protic solvent like methanol.

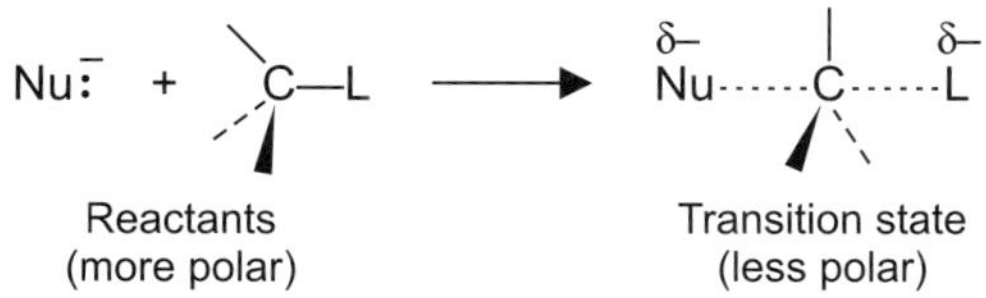

SCHEME 4.18e

- When in an S_N2 reaction, the nucleophile is neutral (Scheme 4.18*f*). The transition

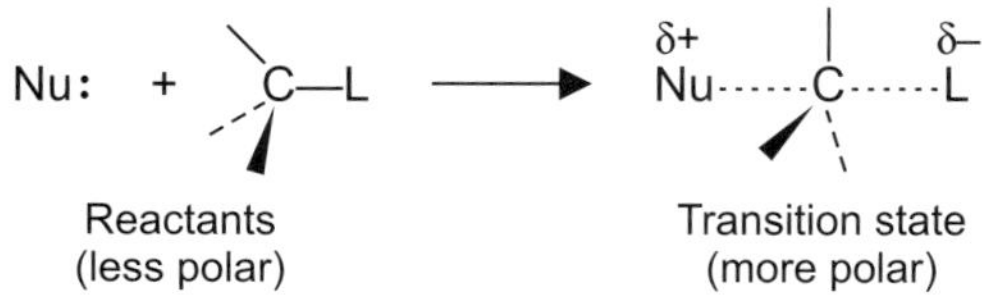

SCHEME 4.18f

state in such cases involves the generation of opposite charges (Scheme 4.18*f*). Employing a solvent like an alcohol solvates both positive and negative charges of the transition state, thus the reactions are faster in protic solvents than in aprotic solvents.

EXERCISE 4.3

How the rate of the reaction (Scheme 4.18g) with the shown transition state will be effected by using solvents with increasing solvating power?

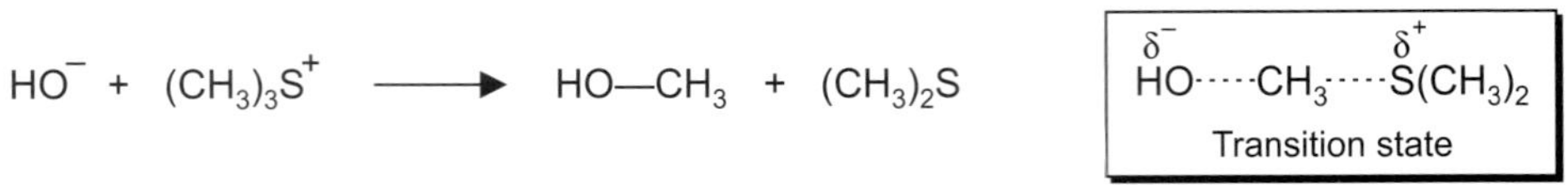

SCHEME 4.18g

ANSWER. *The unlike charges of the reactants are partially neutralized in the transition state. Thus such reaction occurs more slowly with the increasing solvating power of the solvent.*

EXERCISE 4.4

Consider the transition state involved during the reaction (ipso) of bromobenzene with an alkoxide ion (Scheme 4.18h). Why the reaction with t-butoxide ion in DMSO is very fast when compared to that in t-butanol where the rate is extremely slow?

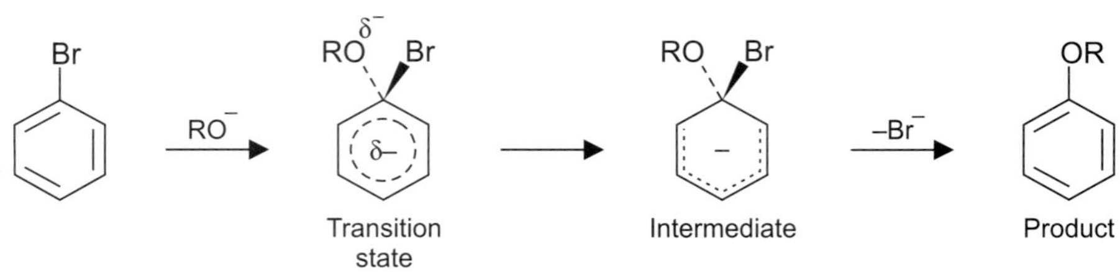

SCHEME 4.18h

ANSWER. *t-butanol solvates the anion strongly while DMSO cannot provide similar stabilization and in DMSO as solvent the nucleophile $(CH_3)_3CO^-$ is essentially free to take part in the reaction.*

4.13 THE CURTIN-HAMMETT PRINCIPLE—IMPORTANCE OF TRANSITION STATE

Consider examples of *anti* E2 elimination (see Schemes 12.13 and 12.14) which may be stereoselective or stereospecific. Consider the stereoselective elimination (Scheme 12.13). The Curtin-Hammett principle implies that in a chemical reaction that gives one product from one conformer and a different product from another conformer (provided the products do not interconvert while the two conformers are rapidly interconverting relative to the rate of product formation). The product composition is not determined by the relative populations of the ground state conformations but depend almost entirely on the relative energies of the representative transition states involved (Scheme 4.18*i*). There are three possible staggered conformations

(I–III, Scheme 4.18*i*). Only the conformations (I and II) have H and Br in *trans*-coplanar orientation for E2 elimination, the transition state from conformer (I) gives the minor product, since the transition state is of higher energy (compared to the transition state from conformer II) due to repulsive forces between the two methyl groups.

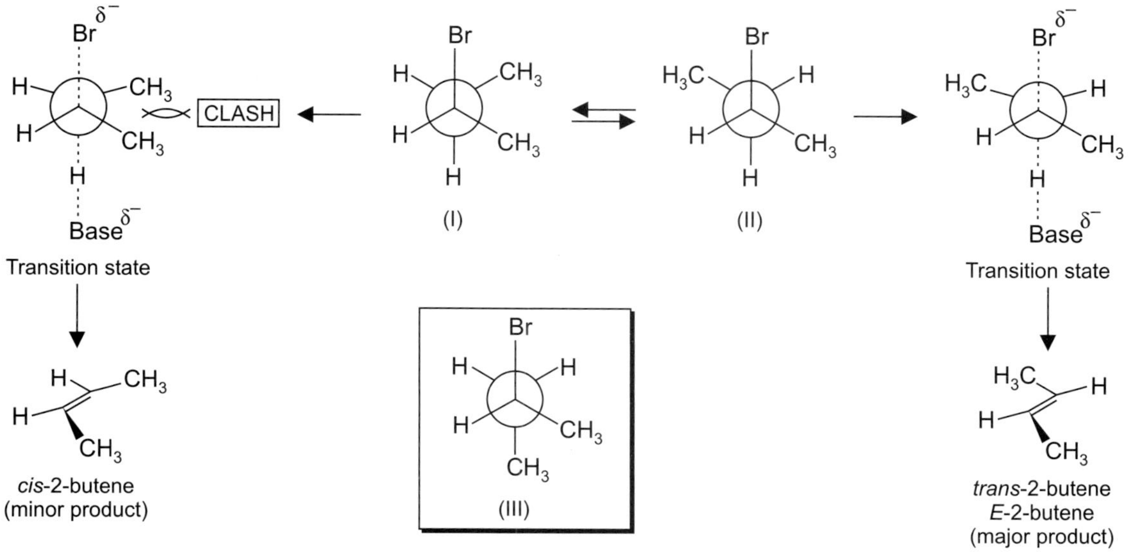

SCHEME 4.18i

4.14 MICROSCOPIC REVERSIBILITY

At equilibrium individual molecular processes and the exact reverse of these processes have an equal probability of occurring. If a certain number of molecules follow one path (of many possible paths) in the forward direction, the same number follow that path in the reverse direction. Microscopic reversibility is also known as the Principle of Detailed Balancing. In the reaction (Scheme 4.18*j*) if B is an intermediate in going from A to C, then under the same conditions B must be an intermediate in going from C to A. Consider *e.g.*, a step from the chain reaction during chlorination of CH_4. The attack by $Cl^{\bullet}$ on methane molecule gives a methyl radical and HCl (Scheme 4.18*j*) via a transition state in which the methyl group begins to flatten out from its original tetrahedral arrangement [subsequently methyl radical gives methyl chloride on reaction with chlorine]. The same mechanism is involved during the reverse process, $HCl + \dot{C}H_3 \longrightarrow CH_4 + Cl^{\bullet}$, during which the carbon of the methyl radical attacks the hydrogen of HCl from the side opposite the H—Cl bond and the C—H bond begins to form.

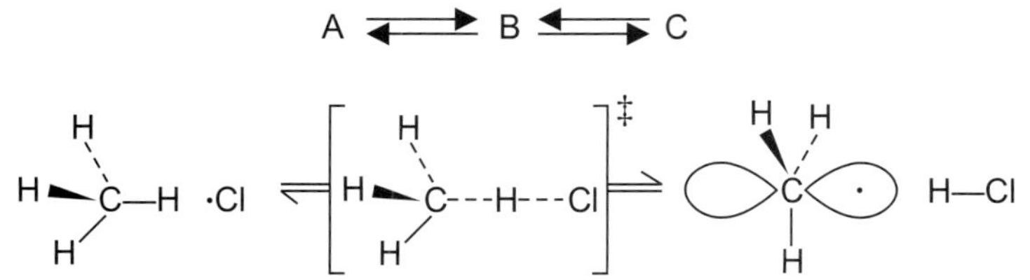

SCHEME 4.18j

4.15 METHODS OF DETERMINING MECHANISMS

(A) Identification of Products

A mechanism proposed for a reaction must account for all the products as well as for their relative proportions. Olefinic double bonds react with peracids to give epoxides and the reaction is an electrophilic attack on the olefin. Electron-releasing groups in the olefin and electron attracting groups in the peracid facilitate the reaction. The epoxidation reaction is first order with respect to the olefin and the peracid *i.e.,* it is a bimolecular process. One may, therefore,

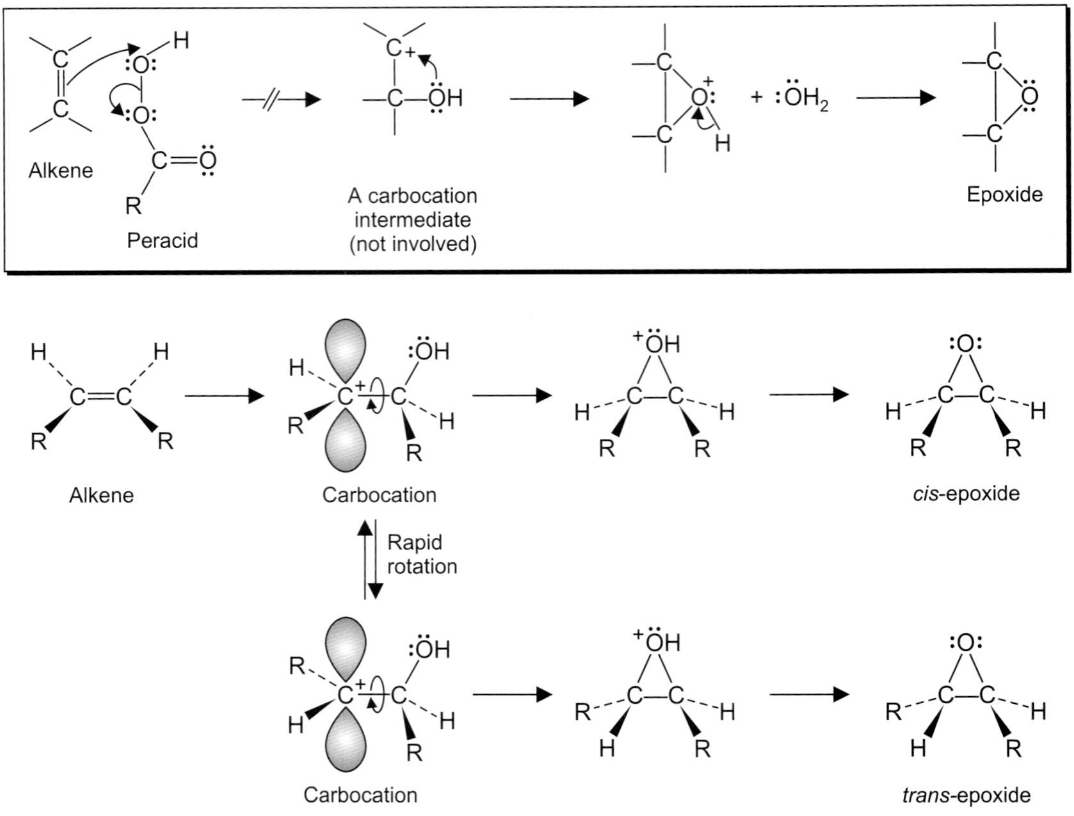

SCHEME 4.19

suggest an ionic mechanism (Scheme 4.19). A mechanism involving a carbocationic intermediate is however, not operative, since the initially formed carbocation would undergo a rapid rotation to yield a mixture of *cis* an well as *trans*-epoxides. The epoxide formation however, retains the stereochemical relationships present in the reacting alkene and the epoxidation reaction is found to be stereospecific. Thus *cis*-2-butene gives only *cis*-epoxide while *trans*-2-butene gives only *trans*-epoxide. A reaction in which the two new bonds are formed at the same time cannot change the stereochemical relationships of the groups in the original alkene. Thus the stereospecificity of the reaction indicates that ring formation occurs essentially in one step and does not involve a free carbocation. The currently accepted mechanism is a one step *syn* addition which involves transfer of the oxygen atom β to the carbonyl group from the peracid to the carbon-carbon double bond (Scheme 4.19*a*). Moreover, a large negative entropy of activation also supports the concerted mechanism.

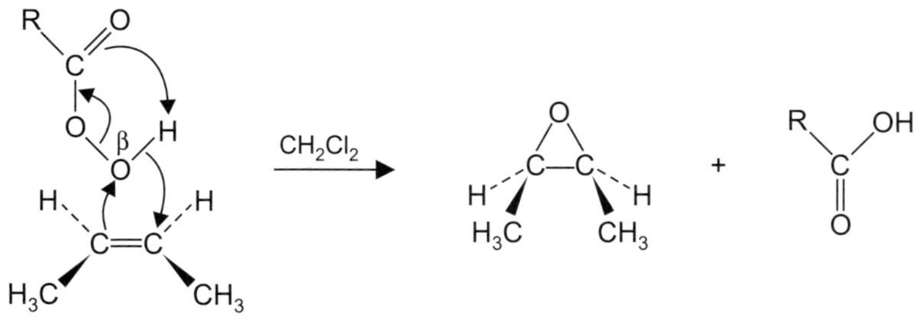

One step *syn*-addition of a peracid to an alkene

SCHEME 4.19a

(B) Detection of Intermediates

1. Direct Isolation

In several reactions an intermediate can be isolated either by using mild conditions or by stopping the reaction after a short time. In *e.g.,* Hoffmann rearrangement an acid amide is converted to an isocyanate and then to an amine on treatment of the amide with bromine in alkaline solution. The overall reaction formally consists of decarboxylation of an amide. In support of the mechanism (Scheme 4.19*b*) the involvement of an isocyanate intermediate has been suggested.

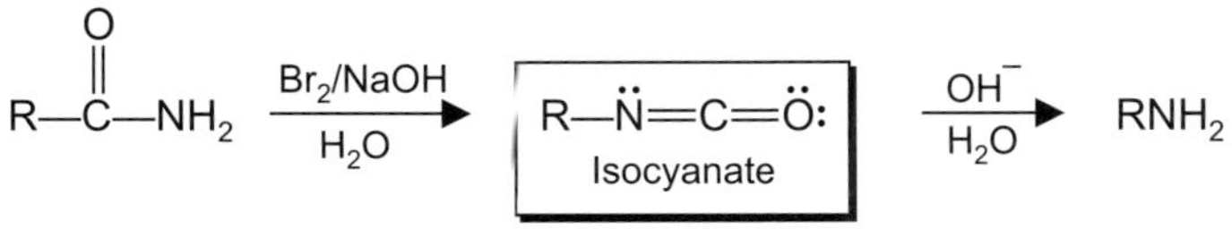

Hoffmann rearrangement

SCHEME 4.19b

The detailed mechanism is given (see Scheme 5.15). In a related curtius rearrangement an isocyanate has indeed been isolated (see Scheme 5.17).

EXERCISE 4.5

Why Hoffmann rearrangement is limited to amides of the type RCONH$_2$?

ANSWER. *Unless two H atoms are present on nitrogen of an amide, the isocyanate intermediate cannot be formed (Scheme 4.19c).*

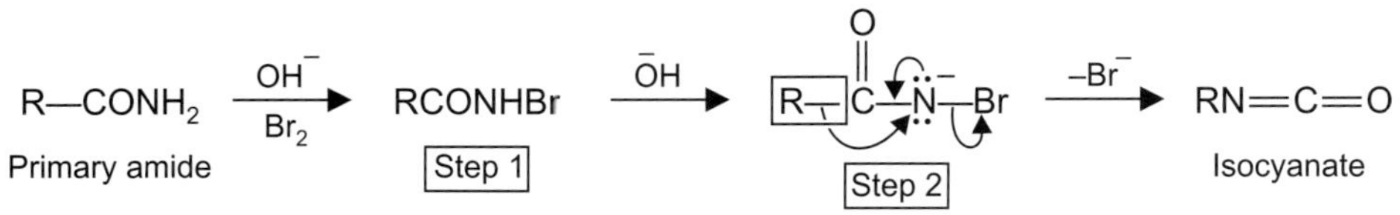

SCHEME 4.19c

Removal of first H gives N-bromoamide while removal of second H gives a bromo amide anion which undergoes a molecular rearrangement.

Application of Host-guest chemistry forms the basis of isolation of benzyne at low temperature inside the molecular container called a hemicarcerand. The incarcerated benzyne was sufficiently stable for NMR spectral measurements, before it ultimately underwent a Diels-Alder reaction with the host container molecule.

2. Spectroscopic Determination

The involvement of a nitronium ion (NO_2^+) in the electrophilic nitration of benzene was found on the basis of its detection by Raman spectroscopy (Sec. 8.2).

Benzyne is a reactive intermediate and as represented is strained acetylene in which the acetylene bond is badly bent. Some of its IR and ^{13}C NMR spectra (see Scheme. 4.50) confirm these observations.

3. Trapping

An intermediate can be identified by trapping it by the addition of a compound with which it is known to react. This is the way, benzynes are identified, since these react with dienes in the Diels-Alder reaction. Thus during the diazotization of anthranilic acid in the presence of furan a Diels-Alder adduct is isolated (Scheme 4.19*d*).

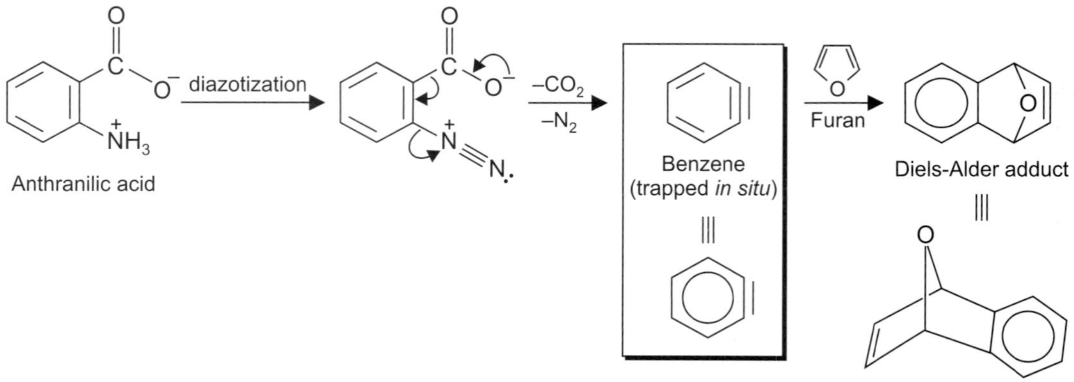

SCHEME 4.19d

4. Indirect Evidence

Intermediates may also be detected from indirect evidence. Thus when free radicals ($R^{\bullet}$) are involved, often a chain-reaction mechanism operates, which is easily recognizable (see Scheme 16.23). However, if carbocations (R^+) are involved, the result is often rearrangements in the carbon skeleton of the starting materials. Consider the general reaction of the formation of one alkyl halide from another alkyl halide. Reaction between bromoethane and sodium iodide (eq. I, Scheme 4.20), gives iodomethane and Br^- (as *e.g.,* NaBr) as the only two products. When this reaction is carried on 1-bromopropane; 1-iodopropane is the exclusive product (ignoring NaBr). There is no rearrangement to produce, for example, 2-iodopropane. Thus carbocations are not involved in the mechanism (Scheme 4.20) and 1-iodopropane is not formed *via* the 1° carbocation. In case 1° carbocation was formed, it would have rearranged to the more stable 2° carbocation from which some 2-iodopropane should have been formed. The reaction therefore, must be proceeding through a concerted (one step) mechanism (S_N2 mechanism).

$$CH_3{-}Br + :\ddot{\underset{..}{I}}:^- \longrightarrow CH_3{-}I + :\ddot{\underset{..}{Br}}:^- \qquad (I)$$

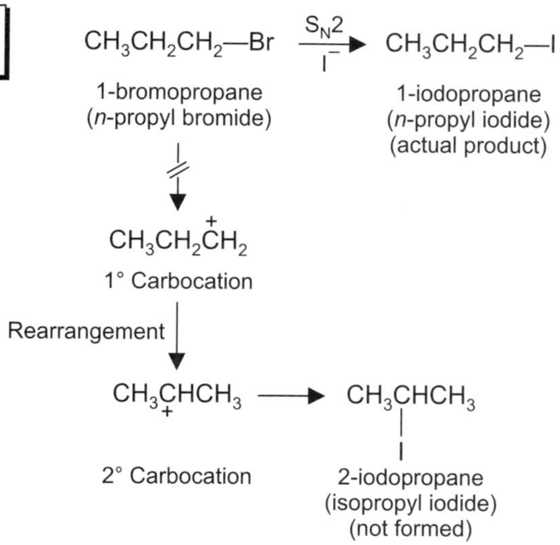

SCHEME 4.20

(C) Stereochemical Evidence

When a reaction gives products which can exist in more than one stereoisomeric form, their identification throws much light on the mechanism. The S_N2 reaction mechanism involves a one-step backside displacement of a leaving group by a nucleophile. It is a bimolecular reaction since two particles (molecules or ions) are involved in the transition state of the slowest (rate determining) reaction step, which is the only step in this reaction.

In the reaction of methyl chloride with hydroxide ion, one cannot detect inversion of configuration because both the starting material and the product are achiral. When a pure enantiomer of a chiral alkyl halide is used as the starting material, the inversion can be detected due to the change in configuration. These data show that S_N2 displacements to proceed by the one step backside attack by the nucleophile on the substrate (Scheme 4.21).

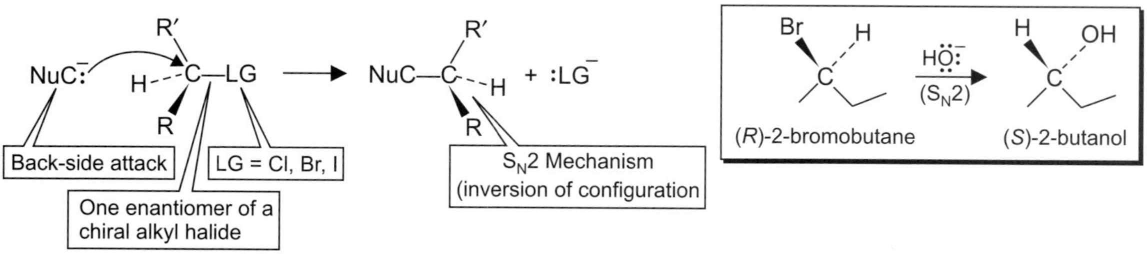

SCHEME 4.21

When an S_N2 reaction occurs at a stereogenic carbon, the configuration of that carbon is inverted in the product. S_N1 reaction, on the other hand, is unimolecular (kinetic evidence) and a two step sequence involving an intermediate formation of a planar carbocation. Thus such a reaction at the stereocenter of an optically active alkyl halide should lead to racemization (Scheme 4.22). Racemization is indeed observed by the expected attack of the nucleophile on either side of the planar achiral carbocation.

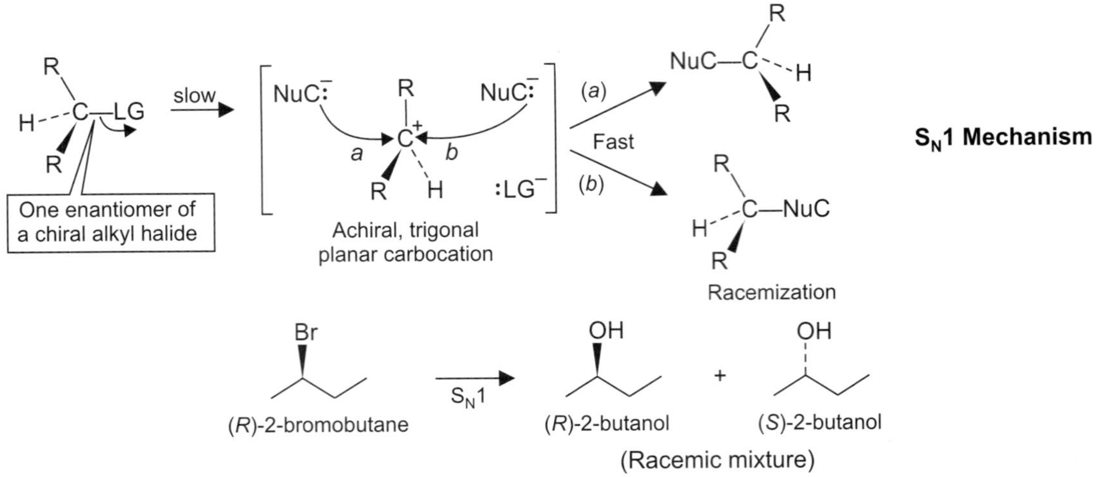

SCHEME 4.22

A cycloalkene adds bromine to yield only a *trans*-1, 2-dibromocycloalkane. This shows the *anti* mode of addition *via* a bridged ion intermediate. A carbocation intermediate should have given products of both *syn* as well as *anti* addition (Scheme 4.23).

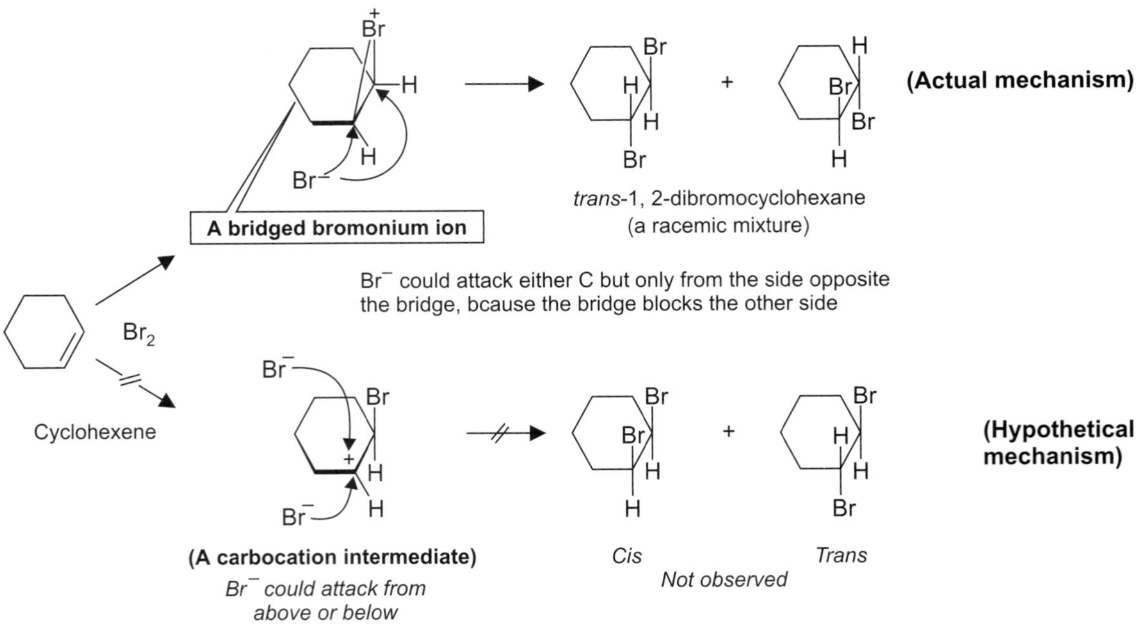

SCHEME 4.23

(D) Kinetic Isotope Effects

Chemists use different experimental techniques to reach at the most plausible mechanism for a reaction. The mechanisms of the S_N1, S_N2, E1, and E2 reactions are based on a knowledge of the rate law of the reaction, the relative reactivities of the reactants, and the structures of the products formed.

Another powerful experimental evidence to investigate the mechanism of a reaction is the deuterium kinetic isotope effect—the ratio of the rate constant observed for a compound having hydrogen to the rate constant observed for a related compound in which one or more of the hydrogens are replaced by deuterium, an isotope of hydrogen.

$$\text{Deuterium kinetic isotope effect} = \frac{k_H}{k_D} = \frac{\text{Rate constant of H-containing reactant}}{\text{Rate constant for D-containing reactant}}.$$

Because isotopic substitution has only a very small effect on the chemical behaviour of a compound, the isotopically modified compound undergoes the same reactions and follows the same mechanisms as its parent compound. Thus although the chemical properties of deuterium and hydrogen are similar; but, a C—D bond is about 1.2 kcal/mol (5 kJ/mol) stronger than a C—H bond. It is thus more difficult to break a C—D bond than a related C—H bond.

At a given temperature the rate of reaction for a compound containing a heavy isotope is slower than the compound with a lighter isotope. This holds *only* if breaking of that bond is involved at the transition state of the rate-determining step. If breaking of this bond occurs prior to or after the rate-determining step, isotopic substitution does not lead to a large change in the rate. This effect is most pronounced for hydrogen/deuterium, which has the largest mass difference of any isotopic pair. When a bond to hydrogen (or deuterium) cleaves in the rate-determining step, then k_H/k_D values of 2–8 are observed and are called *primary* kinetic deuterium isotope effects.

When the C—H(D) bond does not break in the rate-determining step, there are sometimes smaller effects on the rate resulting from isotopic substitution that are called *secondary* kinetic deuterium isotope effects. The k_H/k_D values are 1 – 1.3 for these effects. When a kinetic deuterium isotope effect is larger than about 1.5, it is a primary kinetic deuterium isotope effect to show that C—H(D) bond breaks in the rate-determining step. If a kinetic deuterium isotope effect is found to be between 1 and 1.5, it is a secondary kinetic deuterium isotope effect to indicate that C—H(D) bond cleavage is not involved in the rate-determining step.

This helps in determination of mechanism. In the bromination of acetone (Scheme 4.24), the rate-determining step is the tautomerization of acetone which involves cleavage of a C—H bond. In case this mechanistic assignment is correct, one should observe a substantial isotope effect on the bromination of deuterated acetone. Indeed k_H/k_D was found to be around 7. This is a primary kinetic deuterium isotope effect to indicate that removal of a proton is involved in the rate determining step. The base promoted bromination of ketones is a second order

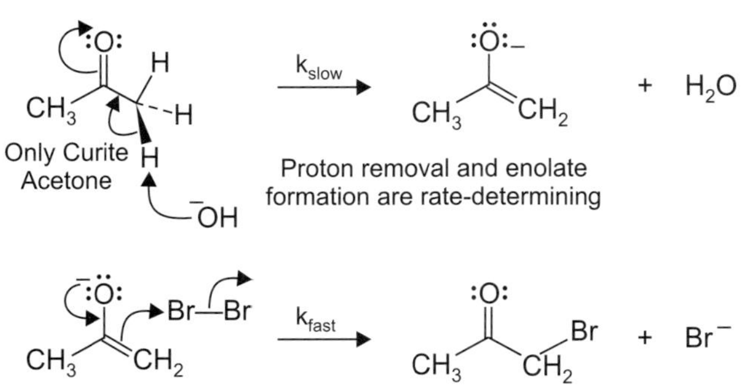

Base-promoted bromination of ketones

SCHEME 4.24

process, first order in ketone and first order in base (rate = k [ketone] [base]). The bromine concentration does not appear in the rate law. The lack of rate dependence on bromine requires that bromine is added to the molecule after the rate-determining step. A mechanism in keeping with these facts is proton removal and enolate formation as the rate determining steps.

Several mechanisms get support from kinetic isotope effect. Some of these are, oxidation of alcohols with chromic acid (see Scheme 13.1) and electrophilic aromatic substitution (see Scheme 8.6). An example of a secondary isotope effect is given (Scheme 4.25), where it is sure that the C—H bond does not break at all in the reaction. Secondary isotope effects for k_H/k_D are generally between 1.0–1.5.

$$(CZ_3)_2CHBr + H_2O \longrightarrow (CZ_3)_2CHOH + HBr$$

The solvolysis of isopropyl bromide where Z = H or D, k_H/k_D is 1.34. Secondary isotope effect

SCHEME 4.25

The substitution of tritium for hydrogen gives isotope effects which are numerically larger ($k_H/k_T = 16$).

E2 elimination like S_N2 process takes place in one step (without the formation of any intermediates). As the attacking base begins to abstract a proton from a carbon next to the leaving group, the C—H bond begins to break (see Scheme 4.3), a new carbon-carbon double bond begins to form and leaving group begins to depart. In keeping with this mechanism (Scheme 4.3), the base induced elimination of HBr from (I, Scheme 4.25*a*) proceeds 7.11 times faster than the elimination of DBr from (II). Thus C—H or C—D bond is broken in the rate limiting step. If it was not so, there would not have been any rate difference.

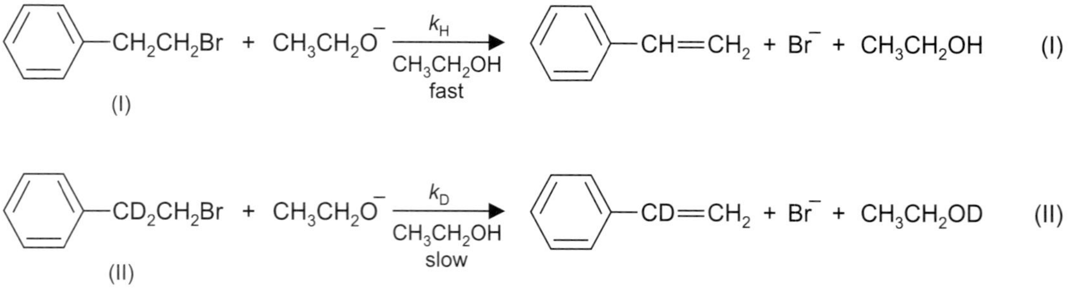

SCHEME 4.25a

No deuterium isotope effect is observed in E1 reactions since the rupture of C—H (or C—D) bond occurs only after the rate limiting step, rather than during it. Thus no rate difference can be measured between a deuterated and a non-deuterated substrate.

EXERCISE 4.6

The reactions (Scheme 4.25b) may follows the E2 mechanism, however, these occur at almost identical rates. What conclusion one can draw from these rate data?

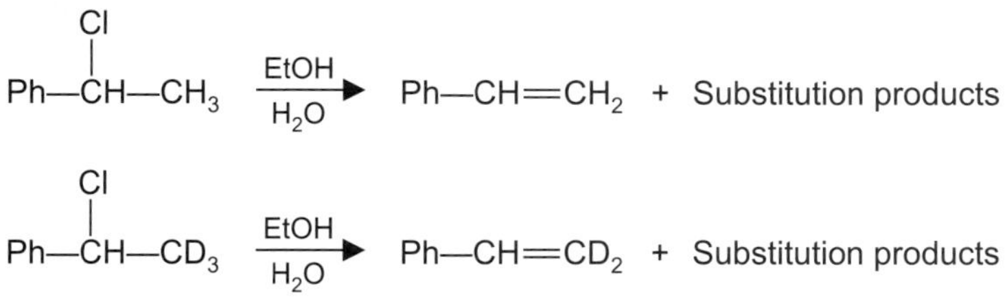

SCHEME 4.25b

ANSWER. *Since the rate of reaction of the undeuterated compound is almost similar to that of the deuterated analog, indicates that the C—D bond is not cleaved in the rate-determining step. Therefore, this reaction is not proceeding by an E2 mechanism.*

Consider the electrophilic nitration of benzene using acetyl nitrate whereby a hydrogen on benzene ring is replaced by a nitro group (Scheme 4.25c). The reaction is second order overall, first order in benzene, and first order in the nitrating agent, rate = $k[C_6H_6]$ [acetyl nitrate]. When fully deuterated benzene is used then $k_H/k_D = 1$ to prove that nitrating agent attacks the benzene ring in the rate-determining step, and the C—H bond cleavage does not occur in the rate-determining step. The formation of nitrobenzene does require the loss of hydrogen, and its loss does not take place in the rate-determining step of the reaction but after the rate-determining step.

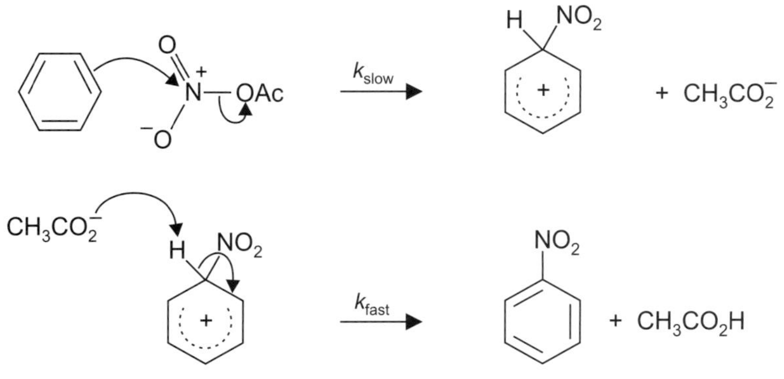

SCHEME 4.25c

(E) Kinetic Evidence

When one determines the kinetics of the reaction between *e.g.*, bromomethane and sodium iodide, the rate expression derived from the experimental data (Scheme 4.26) is obtained. The rate of the reaction depends on the concentration of both reactants therefore, both are involved in the rate-determining step.

$$CH_3{-}Br + :\ddot{\underset{..}{I}}:^- \longrightarrow CH_3{-}I + :\ddot{\underset{..}{Br}}:^-$$

$$\text{Rate} = k\,[CH_3Br]\,[:\ddot{\underset{..}{I}}:^-]$$

SCHEME 4.26

This is consistent with our earlier findings. It appears that the reaction is a one-step process with no detectable intermediates, it follows that both reactants must be involved in one step. The kinetic evidence, thus provides added evidence to the original assignment of mechanism.

(F) Isotopic Labelling

Working with molecules which are isotopically labelled followed by the tracing the path of reaction gives useful information on mechanism. Thus *e.g.*, the alkaline hydrolysis of esters labelled with ^{18}O helped to prove that the mechanism is bond cleavage between oxygen and the acyl group (Scheme 6.58). The benzyne mechanism for nucleophilic aromatic substitution is supported by isotopic labelling experiments (see Schemes 9.13 – 9.15).

(G) Crossover Experiments

Intra- or intermolecular nature of a rearrangement can often be demostrated by carrying out the reaction on a mixture and then analyzing the products. For example, when the Hoffmann reaction ($RCONH_2 \longrightarrow RNH_2$), was carried out on a mixture of $C_6H_4DCONH_2$ (3-deuteriobenzamide) and $C_6H_4CO\,{}^*NH_2$ (^{15}N-benzamide) mixed anilines could not be isolated. Thus the migrating group does not separate during the rearrangement (Scheme 15.16*a*).

4.16 REACTIVE INTERMEDIATES

Synthetic intermediates are stable products which are prepared, isolated and purified and subsequently used as starting materials in a synthetic sequence. Reactive intermediates, on the other hand, are short lived and their importance lies in the assignment of reaction mechanisms on the pathway from the starting substrate to stable products. These reactive intermediates are not isolated, but are detected by spectroscopic methods, or trapped chemically or their presence is confirmed by indirect evidence.

(A) Carbocations

Carbocations are the key intermediates in several reactions and particularly in nucleophilic substitution reactions.

1. Structure

Generally, in the carbocations the positively charged carbon atom is bonded to three other atoms and has no nonbonding electrons. A carbocation is an sp^2 hybridized carbon with a planar structure and bond angles of about 120°. There is a vacant unhybridized p orbital which *e.g.*, in the case of CH_3^+ lies perpendicular to the plane of C—H bonds (Scheme 4.27).

Triphenylmethanol reacts like any other alcohol with a strong aid fluoroboric acid to give a stable salt-triphenylmethyl fluoroborate containing the triphenylmethyl cation. This cation is propeller shaped through the central carbon and the three ring carbons connected to it lie in a plane (the three phenyl rings are at an angle of 54° to the plane of the trigonal carbon). The three benzene rings cannot be in the same plane as a result of van der Walls

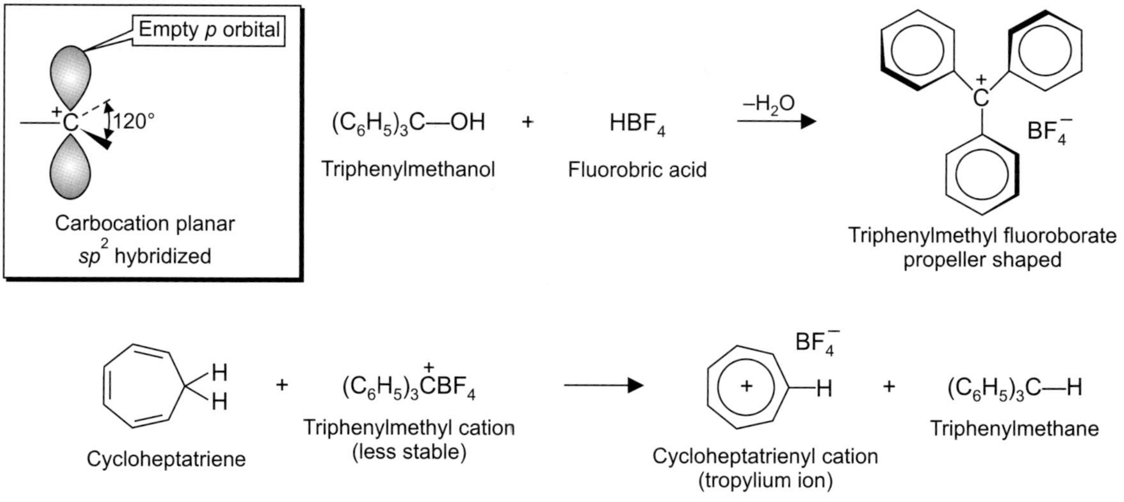

SCHEME 4.27

repulsions between the *ortho* hydrogens. The triarylmethyl cations are particularly stable due to the conjugation with the aryl groups which delocalize the positive charge. The triphenylmethyl cation reacts with cycloheptatriene to form the cycloheptatrienyl cation (tropylium ion) by abstraction of a hydride ion, the triphenylmethyl cation is converted to triphenylmethane in the process. The cycloheptatrienyl cation (an aromatic ion) is more stable than the triphenylmethyl cation, thus providing the driving force for its formation.

2. Stability

There is an increase in carbocation stability with additional alkyl substitution (Scheme 4.28). Thus one finds that addition of HX to three typical olefins decreases in the order $(CH_3)_2\,C{=}CH_2$ > $CH_3{—}CH{=}CH_2$ > $CH_2{=}CH_2$. This is due to the relative stabilities of the carbocations formed in the rate determining step (Scheme 4.28) which in turn follows from the fact that the stability is increased by the electron releasing methyl group (+*I*), three such groups being more effective than two, and two more effective than one.

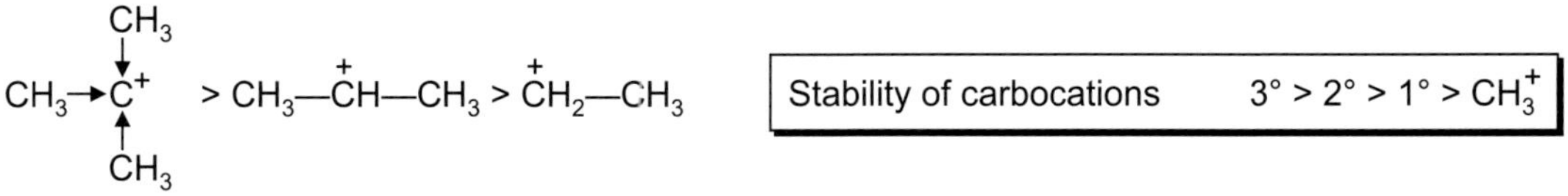

SCHEME 4.28

Further, a structural feature which reduces the electron deficiency at the tricoordinate carbon stabilizes the carbocation. When the positive carbon is in conjugation with a double bond, the stability is more. This is so, because due to resonance the positive charge is spread over two atoms instead of being concentrated only on one. This explains the stability associated with the allylic cation. The benzylic cations are stable, since one can draw canonical forms as for allylic cations (Scheme 4.29). Among the allylic and benzylic cations the relative stabilities are as expected (Scheme 4.29). An alkyl halide can be detected by reacting it with $AgNO_3$ in ethyl alcohol under S_N1 conditions when a precipitate of silver halide is formed (eq. A, Scheme 4.29, the mechanism is S_N1 since ethanol is polar and no other strong nucleophile is present). Keeping the carbocation stability in mind compound (I) as expected forms a precipitate immediately as compared to (II) since the carbocation from (I, Scheme 4.29) is a resonance stabilized benzylic cation.

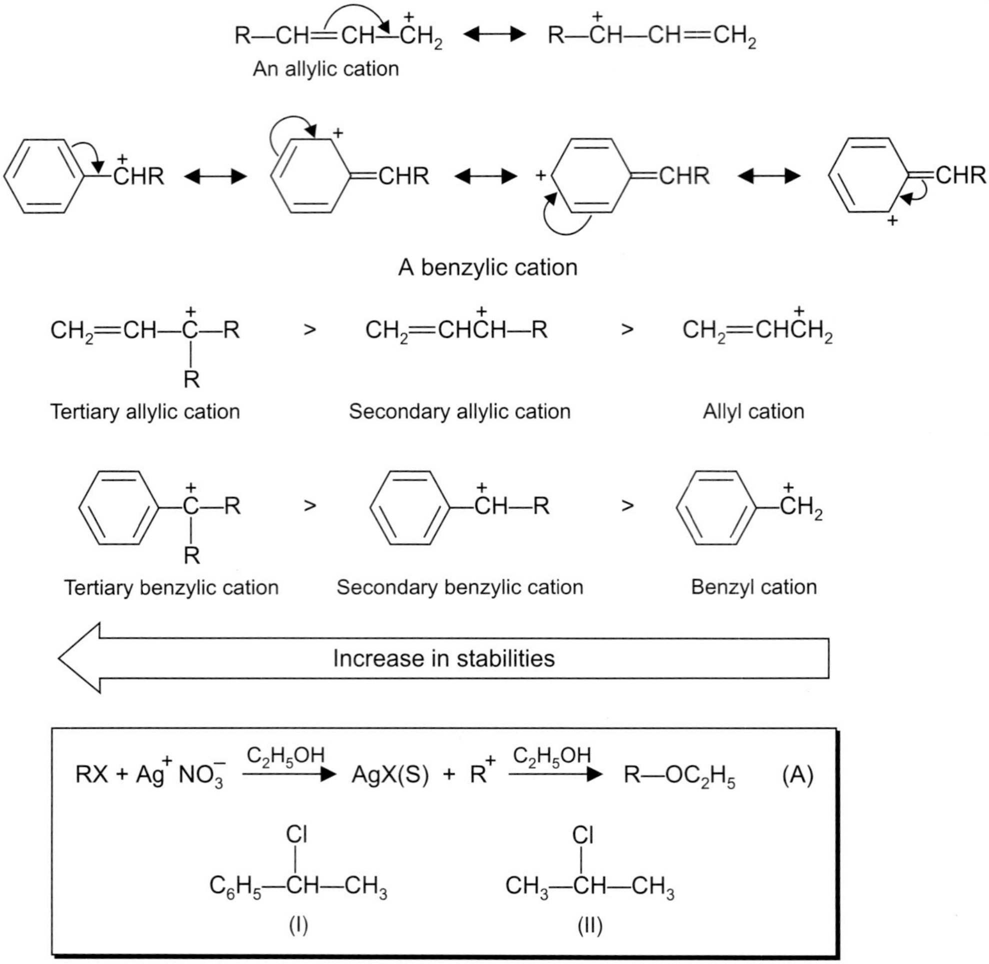

SCHEME 4.29

Aromaticity provides further stability to a carbocation when the vacant *p* orbital is part of the conjugated cyclic system. This aspect is completely discussed in Chapter 2. One may note that cyclopropenylium and cycloheptatrienylium ions are stable carbocations (Scheme 4.29*a*). The cyclic conjugated antiaromatic system *e.g.,* cyclopentadienylium ion is relatively more stable than a methyl cation, but it is far less stable than its aromatic analogs.

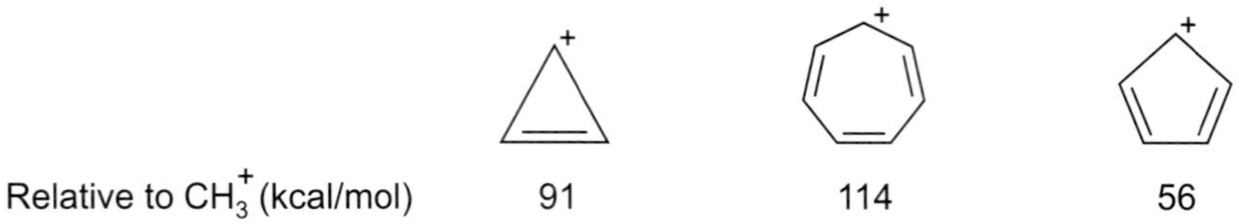

SCHEME 4.29a

The benzyl cation stability is affected by the presence of substituents on the ring. Electron donating *p*-methoxy (Scheme 4.29*b*) and *p*-amino groups stabilize the carbocation by 14 and 26 kcal/mole, respectively. The electron withdrawing groups like *e.g.*, *p*-nitro destabilize by 20 kcal/mol.

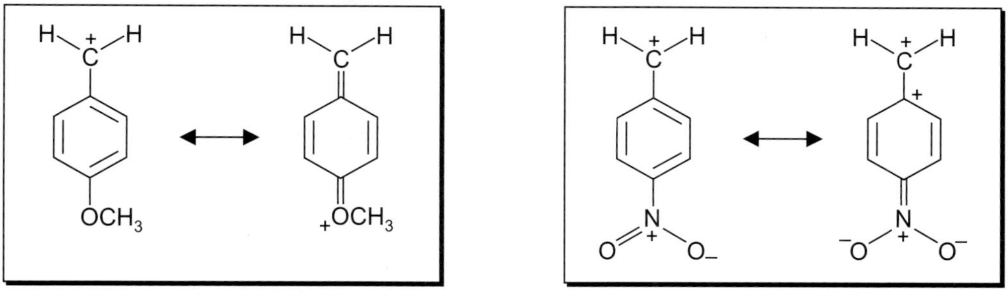

SCHEME 4.29b

A hetero atom with an unshared pair of electrons when present adjacent to the cationic center strongly stabilizes the carbocation (Scheme 4.30). The methoxy-methyl cation has been

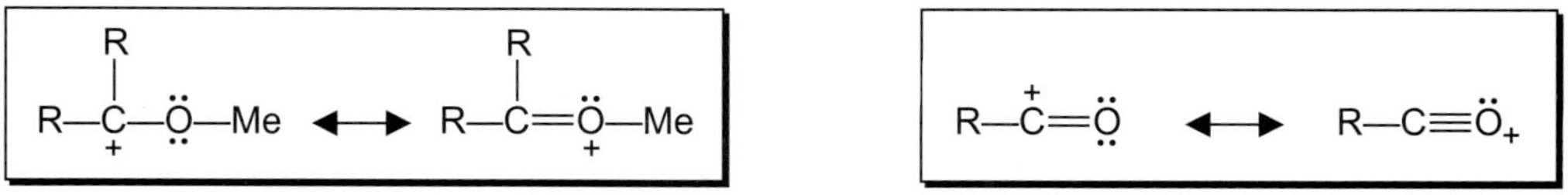

SCHEME 4.30

obtained as a stable solid $CH_3O^+CH_2$ SbF_6^-. Similarly acyl cations RCO^+ have been prepared and acetyl cation CH_3CO^+ is almost as stable as *t*-butyl cation. These ions are also stabilized by resonance, however the positive charge is largely located on carbon.

It is for this reason that a primary halide (I, Scheme 4.30*a*) reacts very fast under S_N1 conditions. This is due to the resonance stabilization of the carbocation formed after the loss of Cl. The carbocation has a resonance structure (III) where the octet rule is satisfied on all atoms and this structure provides very large resonance stabilization.

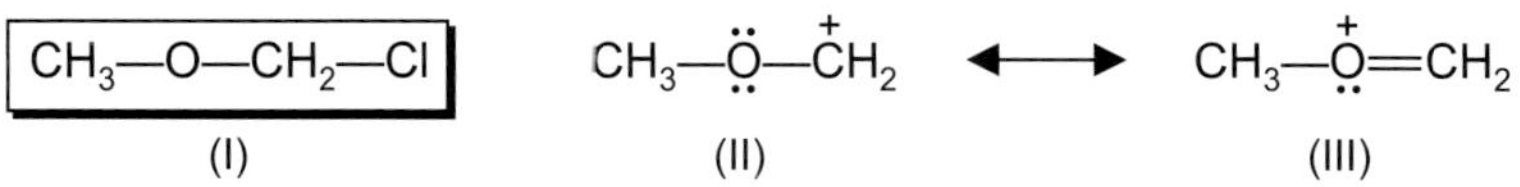

SCHEME 4.30a

Alcohols are protected to prevent their further reaction in synthesis and use of acetals *e.g.*, tetrahydropyranyl (THP) derivatives which are made from reaction with dihydropyran and an acid catalyst are useful for this purpose (Scheme 4.30*b*). Protonation of dihydropyran gives a carbocation which is highly stabilized since the positive charge is located on a carbon next to oxygen atom. The nucleophilic alcohol then attacks this carbocation and loss of proton gives the THP derivative.

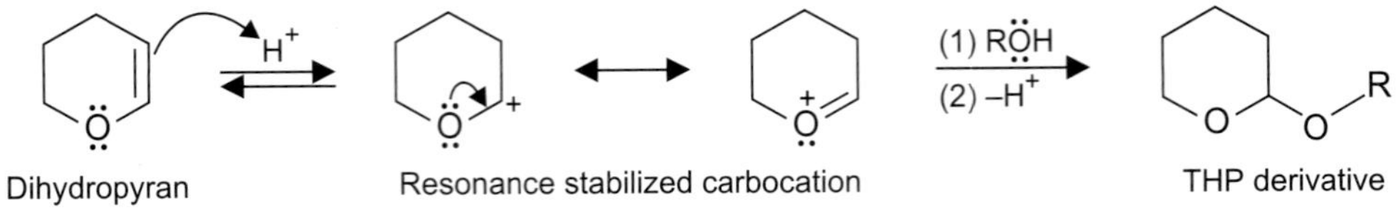

SCHEME 4.30b

EXERCISE 4.7

How the following relative rates of solvolysis (Scheme 4.30c) under S_N1 conditions can be explained?

	$CH_3OCH_2CH_2Cl$	$CH_3CH_2CH_2CH_2Cl$	$CH_3CH_2OCH_2Cl$
S_N1 solvolysis (relative rates)	0.2	1.0	10^9

SCHEME 4.30c

ANSWER. *The fastest rate of solvolysis in the case of chloromethylethyl ether is due to the specially stabilized carbocation which is formed after the loss of chlorine atom. 2-Chloroethylmethyl ether reacts the slowest since the derived carbocation in destabilized due to adjacent partially positive carbon atom (Scheme 4.30d).*

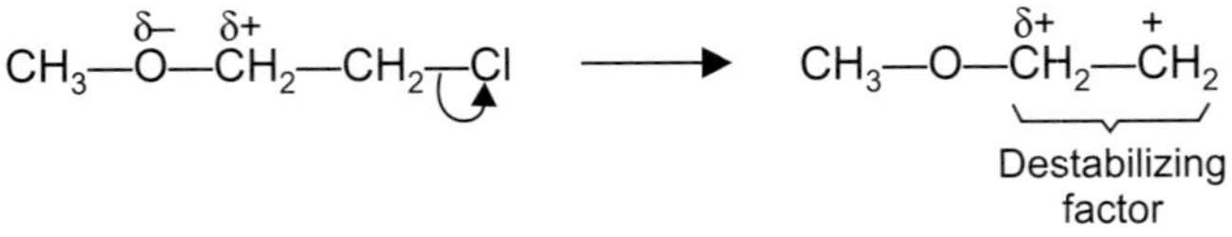

SCHEME 4.30d

EXERCISE 4.8

3-Butyne-1-ol can be made as shown (Scheme 4.30e), an acetylide ion is both a strong base and a good nucleophile. How 3-hexyn-1, 6-diol can be mode?

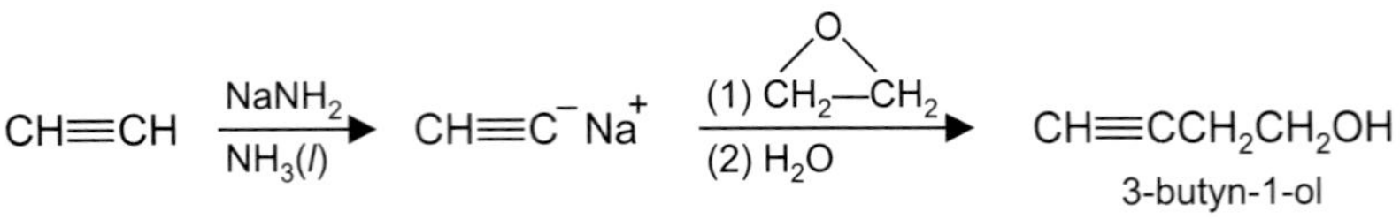

SCHEME 4.30e

ANSWER. *Due to the acidity of OH group which will interfere with the formation of acetylide ion in base, the hydroxyl group must first be protected (Scheme 4.30f).*

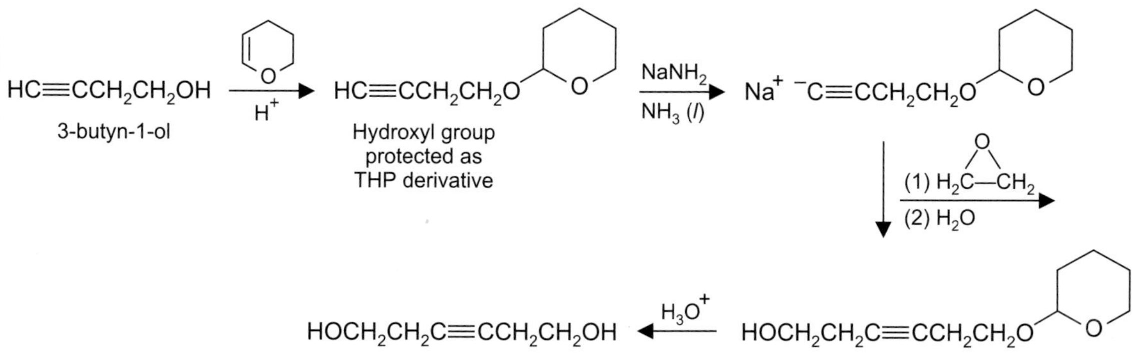

SCHEME 4.30f

Cyclopropylmethyl cations (I, Scheme 4.31) are even more stable than the benzyl cations. This stability increases with every cyclopropyl group and consequently tricyclopropylmethyl cation (II, Scheme 4.31) is even more stable than the triphenylmethyl cation. This special stability is a result of conjugation between the bent orbitals of the cyclopropyl rings (see Scheme 1.2*a*) and the vacant *p* orbital of the cationic carbon (III, Scheme 4.31).

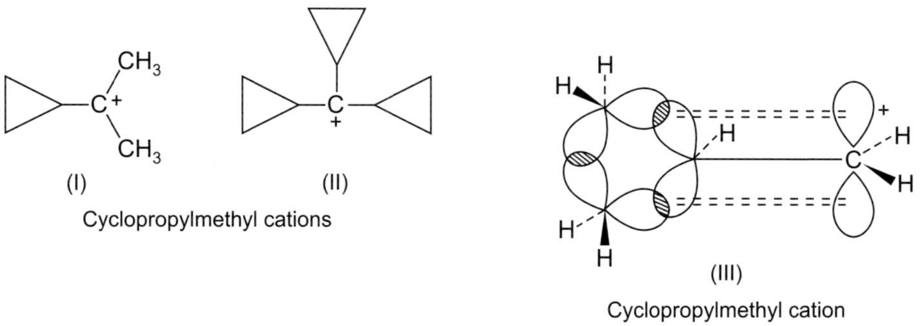

SCHEME 4.31

Participation of the cyclopropyl ring is reflected in the solvolysis of cyclopropylcarbinyl tosylate (I, Scheme 4.31*a*), which proceeds 10^6 times faster than solvolysis of the tosylate of 2-methyl-1-propanol (II). This is due to special stability which the cyclopropyl group imparts to the adjacent carbocationic center. However, cyclopropylcarbinyl cations undergo opening of the strained three membered ring to give a homoallylic cation (IV) and a cyclobutyl cation (V).

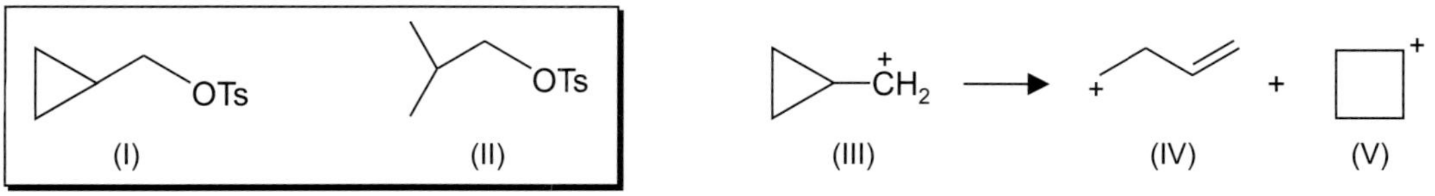

SCHEME 4.31a

Oxaspiropentanes (I, Scheme 4.31*b*) can be prepared by the reaction of ketones with sulphur stabilized cyclopropyl ylides (see Scheme 7.26*a*). These oxaspiropentanes on treatment with acid give a cationic carbon conjugated to the cyclopropyl ring (II) which undergoes rearrangement via ring opening.

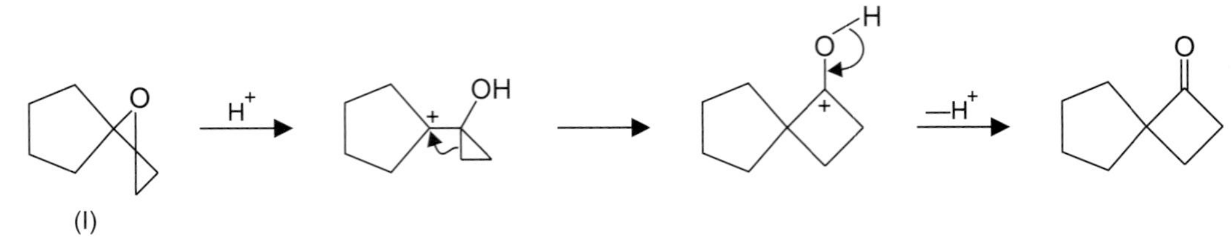

SCHEME 4.31b

That the carbocations are planar is shown by the fact that these are difficult or impossible to form at bridgeheads, where they cannot be planar. Thus, *t*-chloride, 1-chloroapocamphane is inert to S_N1 substitution. The cause of inertness is due to the presence of a rigid bridged system which prevents rehybridization to a planar sp^2 carbon. When the structure is flexible, the bridgehead carbocations can be prepared. Thus the bridgehead bromide, 1-bromoadamantane (II, Scheme 4.32) undergoes solvolysis. In fact (III) has been prepared as the SF_6^- salt. The unstability of bridgehead carbocation (IV, Scheme 4.32) which has also been prepared in super acid solution at – 78°C is due to stability gain from the conjugation with the three cyclopropyl groups.

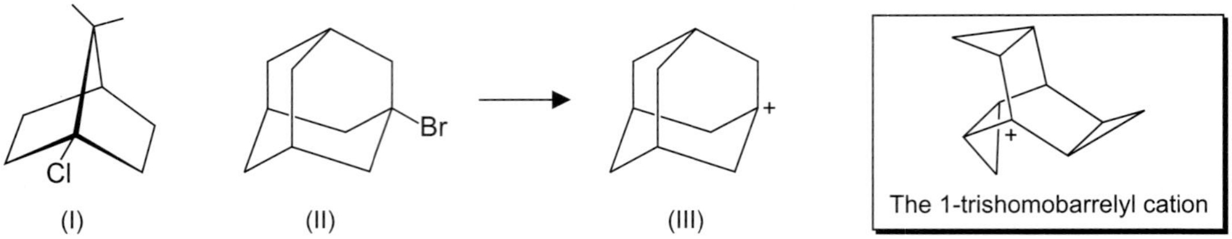

SCHEME 4.32

The stability order of a carbocation is explained by hyperconjugation (see Scheme 2.44). In vinyl cations ($CH_2{=}C^+H$), resonance stability lacks, completely and these therefore are very much less-stable. Such cations may be the intermediates in solvolysis reactions, however, the simple vinyl cations have not been directly observed (see Scheme 5.18*a*).

A carbocation is planar with an empty *p* orbital. A phenyl cation is very unstable and the *p* orbital is full (the part of the aromatic ring) and the empty orbital is an sp^2 orbital which is outside the ring. Thus normally *e.g.,* bromobenzene does not undergo S_N1 reaction since it would involve the formation of a highly unstable phenyl cation. However, significantly, the phenyl cation can be formed and captured by heating a diazonium salt which involves the loss a best leaving group—a molecule of nitrogen (Scheme 4.32*a*, see also Scheme 4.32*e*).

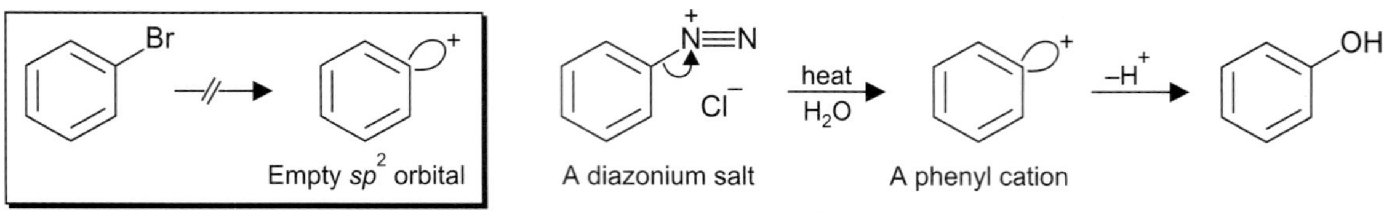

SCHEME 4.32a

3. Generation and Fate

The carbocations are formed as reactive intermediates in a variety of reactions. Thus during a direct ionization, a group attached to the carbon atom leaves with its pair of electrons to give a carbocation (see Scheme 4.6). After its formation, a carbocation may rearrange to give other more stable carbocation (see Scheme 4.5), may react with a charged or an uncharged nucleophile (see Scheme 4.22) or may lose a proton from the adjacent atom (see Scheme 4.6).

The following points may be noted regarding methods of preparation of carbocations and the reactions undergone by them.

- By the addition of an acidic reagent (HX) to an alkene (Scheme 4.32*b*), to form a more stable carbocation followed by its trapping with X^-.

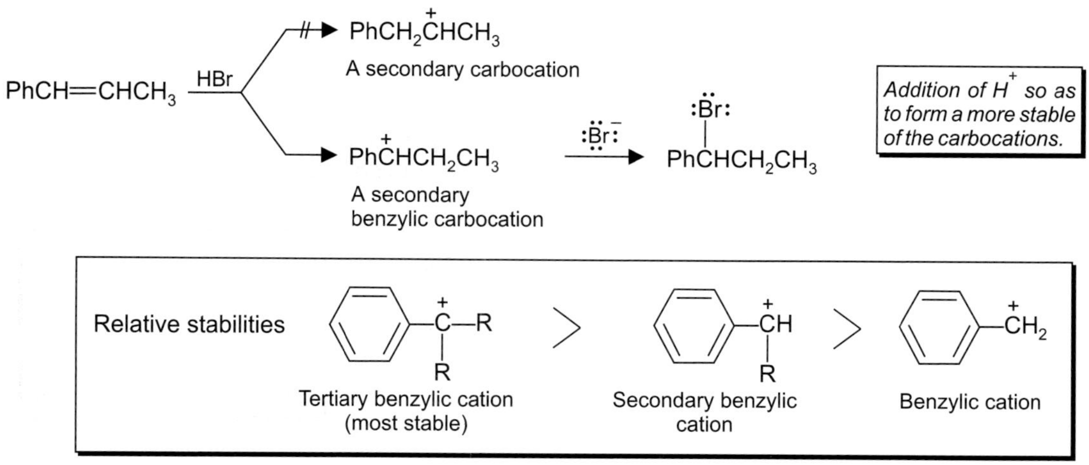

SCHEME 4.32b

- By treatment of an alcohol with an acidic reagent *e.g.*, HCl (Scheme 4.32*c*). The reactant (I) gives an onium salt which gives a 3° carbocation (A) by the loss of water. The relief

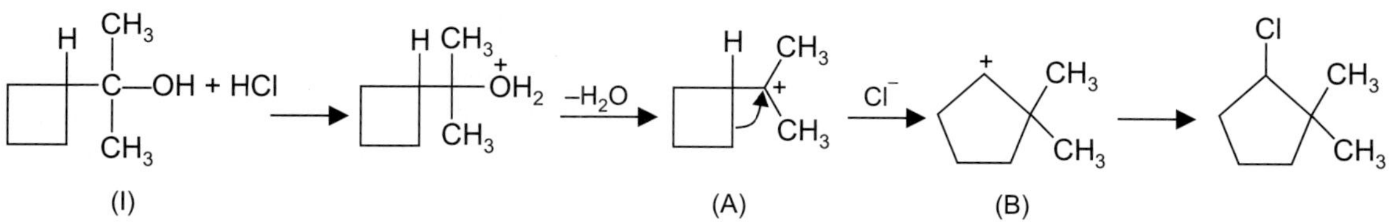

SCHEME 4.32c

of ring strain provides the driving force to give a less stable 2° carbocation (B) which is trapped by Cl^- to give the product. A similar situation obtains when the alcohol (I, Scheme 4.32*d*) is treated with HCl. The initially formed secondary carbocation (II) is trapped by chloride ion to directly give the S_N1 product (III). The carbocation (II) rearranges to the more stable benzylic cation (IV) by involving a 1, 2-hydride shift and finally the rearranged product (V).

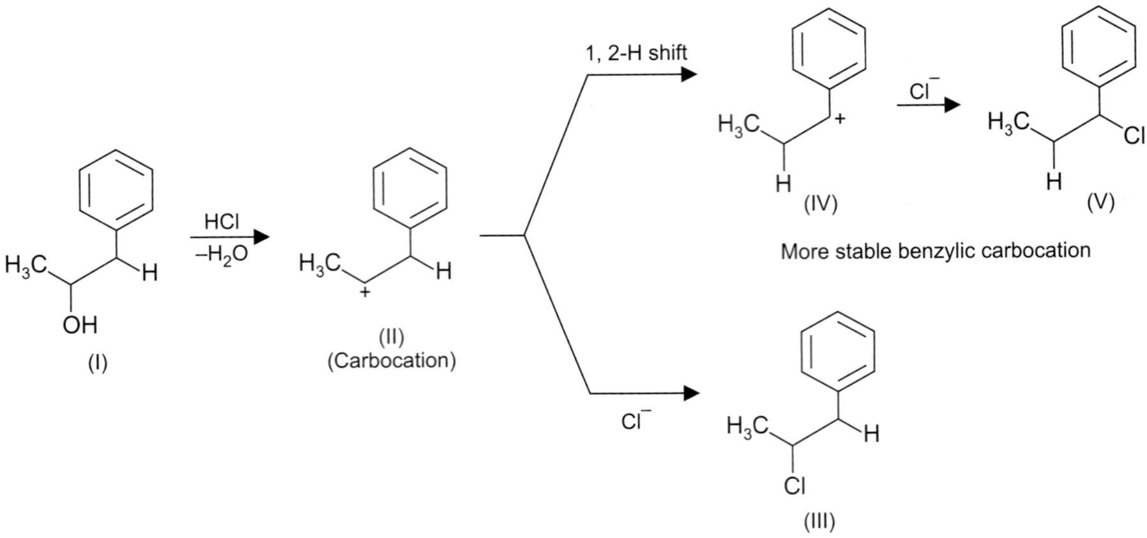

SCHEME 4.32d

- Amine (I, Scheme 4.32*e*) on reaction with nitrous acid (HNO_2 give a diazonium salt which readily decomposes to a cation which subsequently undergoes ring enlargement (Scheme 4.32*e*).

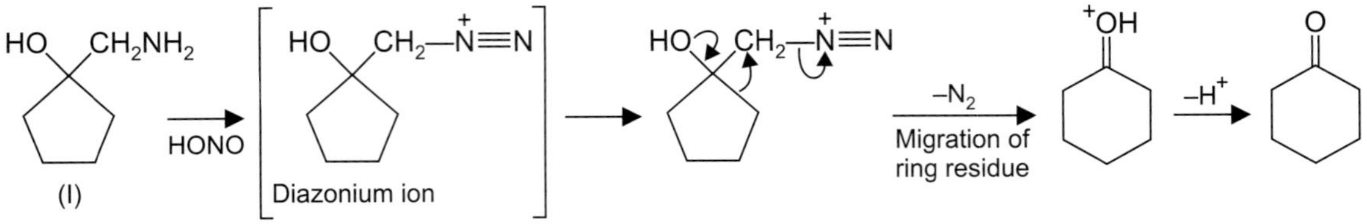

(A ring enlargement)

SCHEME 4.32e

- Epoxides on reaction with acid generate carbocations which may also undergo a rearrangement (Scheme 4.32*f*).

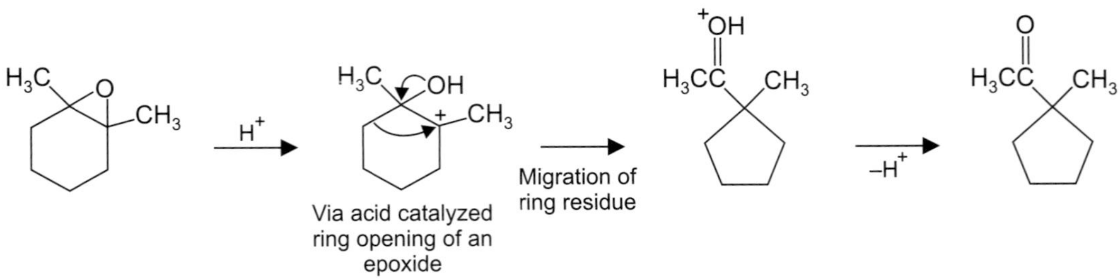

(A ring contraction)

SCHEME 4.32f

- Aldehydes and ketones on treatment with acid catalyst yield the oxygen stabilized cation (Scheme 4.32*g*). These cations can be stabilized on treatment of an aldehyde or a ketone with boron trifluoride.

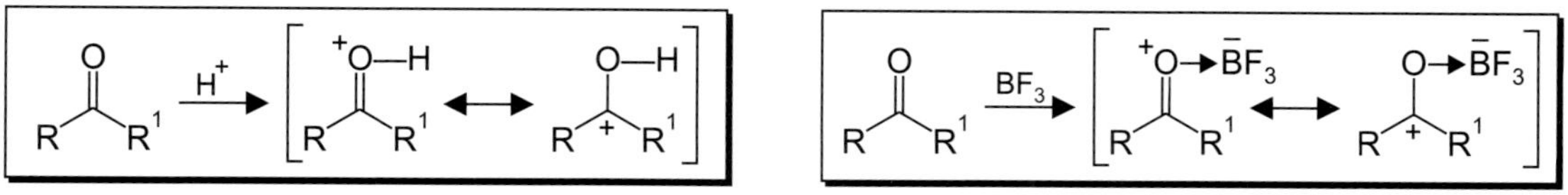

SCHEME 4.32g

- Resonance stabilized acylium ion can be generated by the removal of chlorine from an acid chloride with Lewis acids *e.g.*, aluminium chloride (Scheme 4.32*h*). The acylium ion *e.g.*, can attack the *pi* bond of benzene ring to yield an aryl ketone in Friedel Crafts acylation reaction.

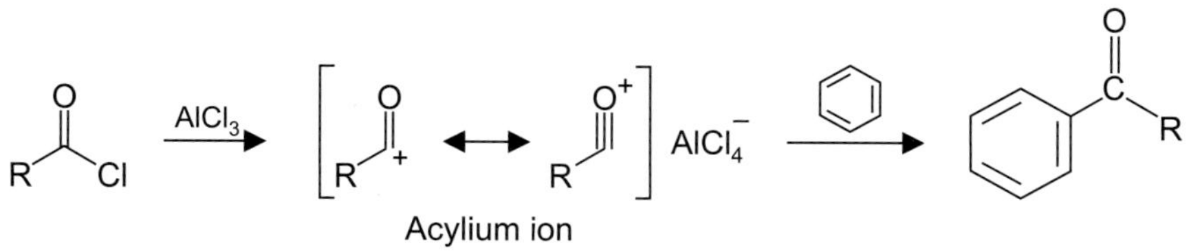

SCHEME 4.32h

- Primary reactants undergo S_N2 displacement reaction rather than generating carbocations and subsequent rearrangement (Scheme 4.32*i*).

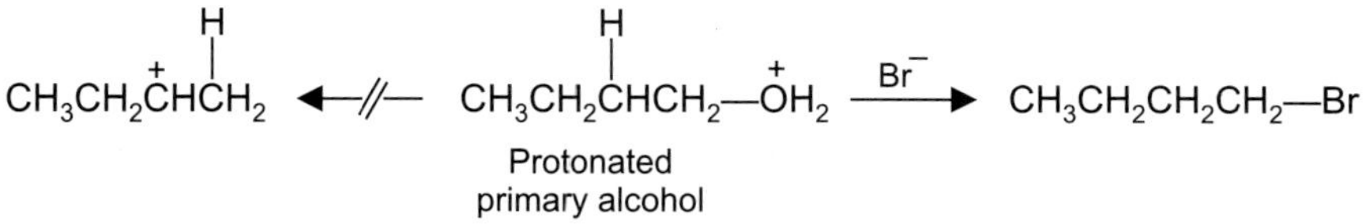

SCHEME 4.32i

Recall the steric requirements of S_N2 reactions. In the case of isobutyl alcohol a reaction with HBr/H_2SO_4 gives isobutyl bromide as the major product and *t*-butyl bromide is formed as a result of the formation of *t*-butyl cation formed via rearrangement (Scheme 4.32*j*). However, the neopentyl type of systems with a quaternary carbon next to the alcohol carbon display complete rearrangement. S_N2 reactions on such systems are completely hindered and thus the rearrangement reaction becomes the dominating reaction (Scheme 4.32*j*).

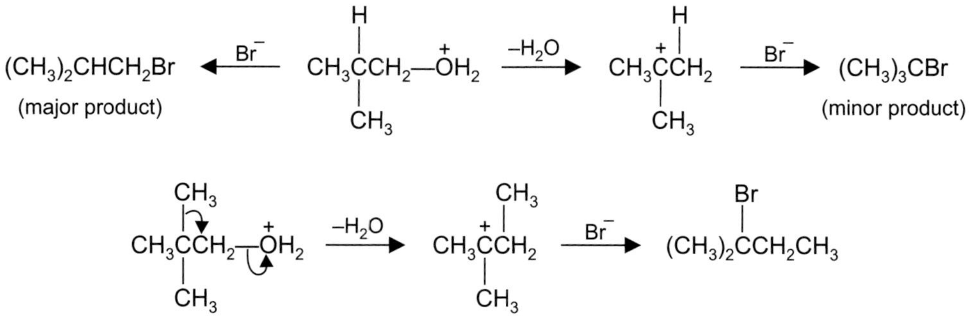

SCHEME 4.32j

Methods have been developed by Olah for preparing carbocations under conditions where they are stable enough to be studied by ^{1}H NMR spectroscopy. In liquid sulphur dioxide, alkyl fluorides react with the Lewis acid antimony pentafluoride to give solutions of carbocations (I, Scheme 4.33). The ^{1}H NMR spectrum of *tert*-butyl fluoride in liquid sulphur dioxide displays the nine protons as a doublet centered as δ 1.35. On adding antimony pentafluoride to the solution, the doublet at δ 1.35 is replaced by a singlet at δ 4.35. Both the change in the splitting pattern of the methyl protons and the downfield shift are in keeping with the formation of a *tert*-butyl cation (II, Scheme 4.33). On treating a solution of isopropyl fluoride in liquid sulfur dioxide with antimony pentafluoride a more remarkable downfield shift is seen (III, Scheme 4.33).

$$R{-}F + SbF_5 \xrightarrow{\text{liq. } SO_2} \overset{+}{R}\,SbF_6^- \qquad \text{(I)}$$

Solutions of carbocations

$$(CH_3)_3C{-}F + SbF_5 \xrightarrow{\text{liq. } SO_2} (CH_3)_3\overset{+}{C} \quad SbF_6^- \qquad \text{(II)}$$

δ 1.35 Doublet → δ 4.35 Singlet

$$(CH_3)_2{-}CHF + SbF_5 \longrightarrow (CH_3)_2{-}\overset{+}{C}H \quad SbF_6^- \qquad \text{(III)}$$

δ 1.23 (quartet), δ 4.64 (multiplet) → δ 5.03 (doublet), δ 13.50 (septet)

SCHEME 4.33

The tropylium ion and cyclopropenyl ions (see Schemes 2.34 and 2.37) are examples of cations stabilized by being part of a delocalized aromatic system. These cations are aromatic due to Hückel's rule.

(B) Carbanions

1. Structure

A carbanion possesses an unshared pair of electrons and thus represents a base. A best description is, that the central carbon atom is sp^3 hybridized with the unshared pair occupying one apex of the tetrahedron. Carbanions would thus have pyramidal structures similar to those of amines. It is believed that carbanions undergo a rapid interconversion between two pyramidal forms (Scheme 4.34).

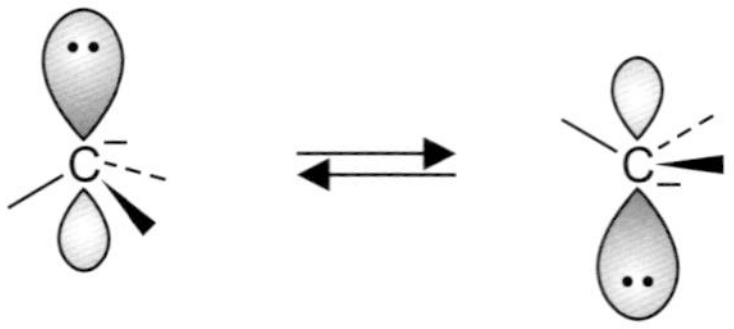

The two possible forms are rapidly inverting carbanion (pyramidal).

SCHEME 4.34

There is evidence for the sp^3 nature of the central carbon and for its tetrahedral structure. At bridgehead a carbon does not undergo reactions in which it must be converted to a carbocation. However, the reactions which involve carbanions at such centers take place with ease, and stable bridgehead carbanions are known. In case this structure is correct and if all three R groups on a carbanions are different, the carbanion should be chiral. All reactions therefore, which involve the formation of chiral carbanion should give retention of configuration. However, this never happens and has been explained due to an umbrella effect as in amines (Scheme 4.34). Thus the unshared pair and the central carbon rapidly oscillate from one side of the plane to the other.

2. Stability and Generation

The Grignard reagent is the best known member of a broad class of substances, called organometallic compounds (Chapter 7) where carbon is bonded to a metal: lithium, potassium, sodium, zinc, mercury, lead, thallium—almost any metal known. Whatever the metal, it is less electronegative than carbon, and the carbon-metal bond like the one in the Grignard reagent (Scheme 4.35) is highly polar. Although the organic group is not a full-fledged carbanion—an anion in which carbon carries negative charge, it however, has carbanion character. Organometallic compounds can serve as a source from which carbon is readily transferred with its electrons. On treatment with a metal, in RX the direction of the original dipole moment is reversed (reverse polarization). Thus, metallation turns an electrophilic carbon into a nucleophilic center (also see Sec. 7.5B, Scheme 7.31, *i.e.,* umpolung).

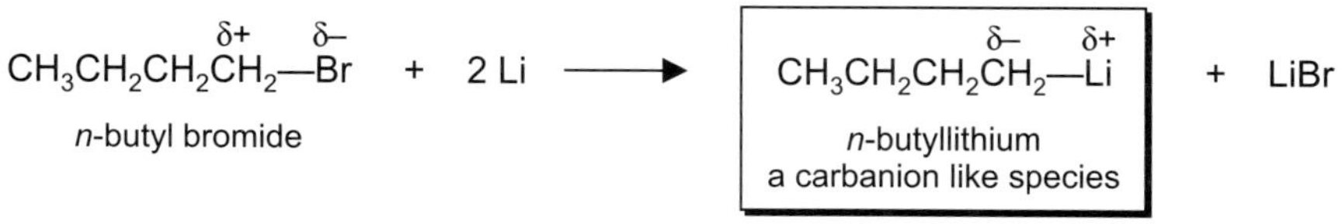

SCHEME 4.35

Acetylene is ionized on treatment with amide ion in liquid ammonia to form a sodium acetylide; this has a little covalent character and may be regarded as a true carbanion (see Scheme 3.11). This property is used in making substituted alkynes by reaction of an acetylide anion with an alkyl halide via an S_N2 reaction. The stability order of carbanions points to their high electron density. Alkyl groups and other electron-donating groups in fact destabilize a carbanion. The order of stability is the opposite of that for carbocations (Scheme 4.35*a*) and

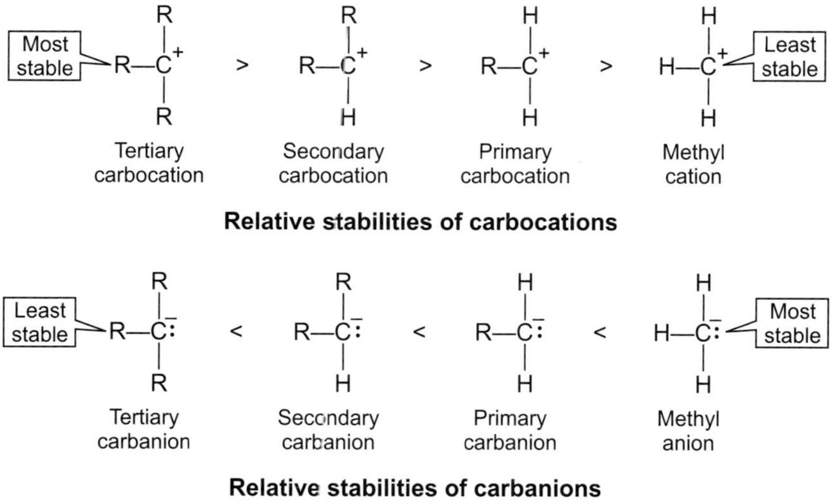

SCHEME 4.35a

free radicals, which are electron deficient and are stabilized by alkyl groups. Based on this stability order it is easy to understand that carbanions that occur as intermediates in organic reactions are almost always bonded to stabilizing groups. An important method of preparation thus involves a loss of proton from a haloform to afford a stabilized carbanion (Problem 3.2). Another factor which leads to stability is resonance *e.g.*, a carbonyl group stabilizes an adjacent carbanion via overlap of its *pi* bond with the nonbonding electrons of the carbanion. Carbanions generated from carbonyl compounds are often called enolate anions. Among other functional groups which exert a strong stabilizing effect on carbanions are nitro and cyano groups. Thus carbanions can be generated from substrates with a C—H bond adjacent to a carbonyl, C ≡ N or NO_2 groups (see Schemes 3.26 and 3.30).

The second row elements, particularly phosphorus and sulphur stabilize the adjacent carbanions. A very important nucleophilic carbon species constitute the phosphorus and sulphur ylides. The preparation of ylides is a two stage process, each stage of which belongs to a familiar reaction type: nucleophilic attack on an alkyl halide, and abstraction of a proton by a base (Scheme 4.36).

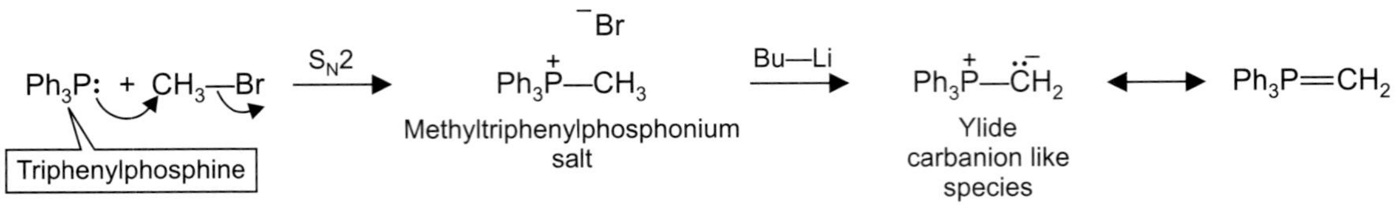

SCHEME 4.36

The phosphorus ylides (used in Wittig reaction) have hybrid structures, and it is the negative charge on carbon *i.e.,* the carbanion character of ylides which is responsible for their characteristic reactions (see Scheme 7.9). The sulfur atoms also stabilize carbanions and as a last example of formation and reactivity of an acyl anion equivalent is in (Scheme 4.36*a*).

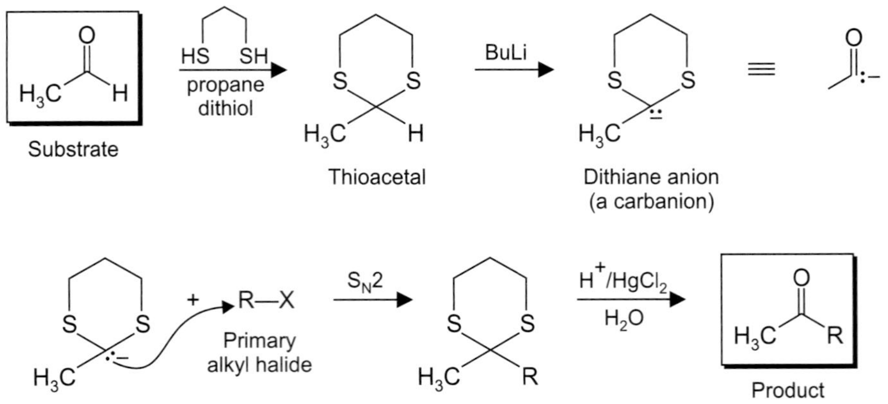

SCHEME 4.36a

The proton(s) on the carbon bearing two sulfur atoms is acidic (pK_a = 31) and can be removed with a strong base like *n*-butyl lithium to afford a stable carbanion. This carbanion is a strong nucleophile and displays S_N2 reactivity (also see Scheme 3.30*c*).

When a double or triple bond is located α to the carbanionic carbon, the ion is stabilized by resonance as in the case of benzylic type carbanions (Scheme 4.37). Thus toluene is more acidic (pK_a = 41) than alkanes (ethane, pK_a = 50).

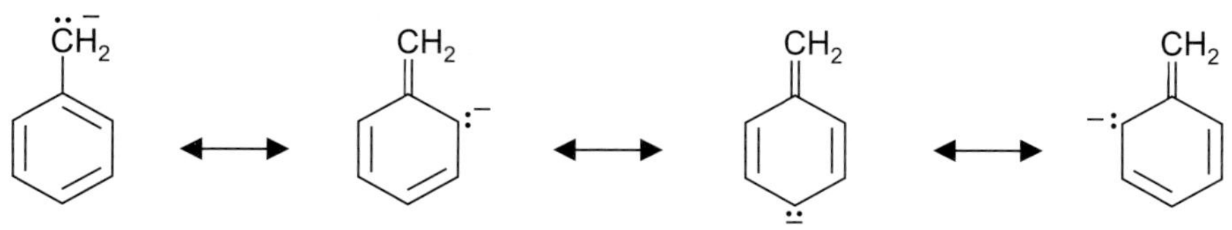

Resonance in benzylic carbanion

SCHEME 4.37

3. Properties

Carbanions are nucleophilic and basic and in this behaviour these are similar to amines, since the carbanion has a negative charge on its carbon, to make it a powerful base and a stronger nucleophile than an amine. Consequently, a carbanion is enough basic to remove a proton from ammonia (see Scheme 3.11).

EXERCISE 4.8a

Alkyl fluorides undergo E2 dehydrohalogenation to give less substituted alkene (Hoffmann regioselectivity) as the major product unlike E2 dehydrohalogenation of alkyl chlorides, alkyl bromides and alkyl iodides. Explain giving an example.

ANSWER. *E2 dehydrohalogenation of alkyl fluorides involves a carbanion-like transition state and thus carbanion stability has to be considered (Scheme 4.37a)*

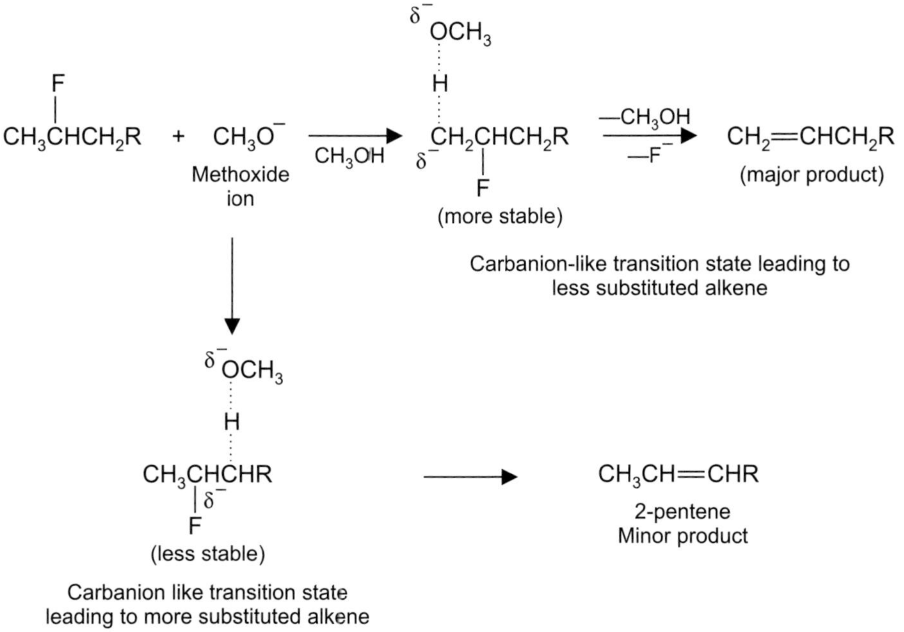

SCHEME 4.37a

Fluoride ion is the strongest base and therefore, the poorest leaving group. Thus when a base starts removing a proton from an alkyl fluoride there is less tendency for the fluoride ion to leave and negative charge develops on the carbon losing the proton. The transition state resembles a carbanion (Scheme 4.37a) rather than an alkene. To determine which of the carbanion-like transition states is more stable, one must determine which carbanion would be more stable. When a hydrogen and a chlorine, bromine, or iodine are eliminated from an alkyl halide, the halogen starts to leave as soon as the base begins to remove the proton. A negative charge therefore, does not build up on the carbon that is losing the proton. The resulting transition state then resembles an alkene rather than a carbanion.

(C) Free Radicals—An Introduction (For Details see chapter 16)

1. Structure and Geometry

A free radical is a species which has one or more unpaired electrons. In the species where all electrons are paired the total magnetic moment is zero. In radicals, however, since there are one or more unpaired electrons, there is a net magnetic moment and the radicals as a result are paramagnetic. Free radicals are usually detected by electron spin resonance (esr), which is also termed electron paramagnetic resonance (epr, see Scheme 2.22).

Simple alkyl radicals have a planar (trigonal) structure, *i.e.,* these have sp^2 bonding with the odd electron in a *p* orbital. The pyramidal structure is another possibility (Scheme 4.38),

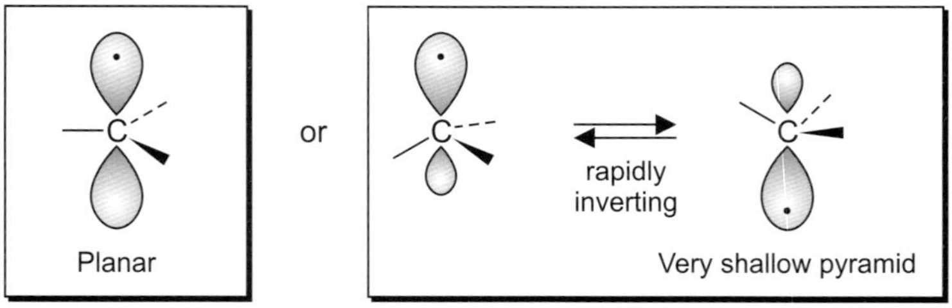

Methyl radical is planar while the trifluoromethyl radical is pyramidal. The oxygenated species $\dot{C}H_2OH$ and $\dot{C}Me_2OH$ lie somewhere in between.

SCHEME 4.38

when the bonding may be sp^3 and the odd electron is in an sp^3 orbital. The planar structure is in keeping with loss of activity when a free radical is generated at a stereocentre. On adding HBr to 2-methyl-l-butene in the presence of peroxide, the product has a single stereocenter. However, equal amounts of the *R* and *S* enantiomers are obtained. The product is a racemic mixture (Scheme 4.39). The carbon atom in the radical intermediate that bears the unpaired electron is planar and sp^2 hybridized. This means that the three substituents bonded to it are all in the same plane. Consequently, the *R* enantiomer and the *S* enantiomer will be obtained in equal amounts, because HBr has equal access to both sides of the achiral radical (Scheme 4.39). Unlike carbocations the free radicals can be generated at bridgeheads to show that pyramidal geometry for radicals is also possible and the free radicals need not be planar.

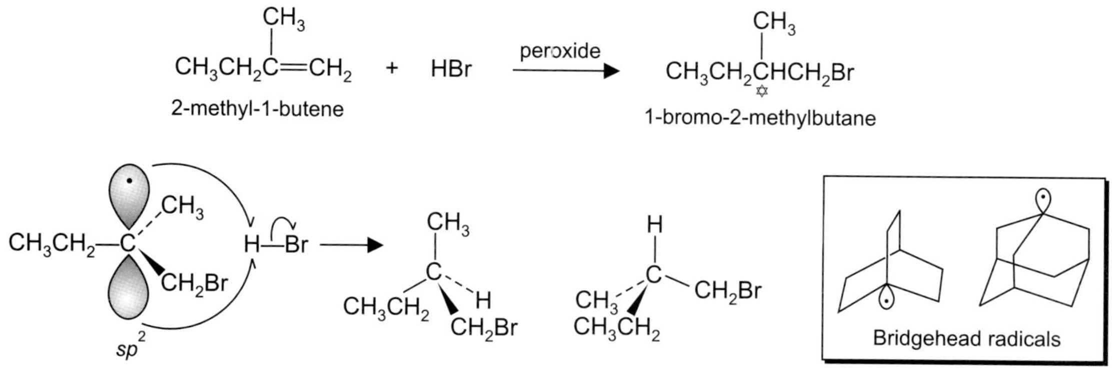

SCHEME 4.39

2. *Stability*

As in the case of carbocations, the stability of free radicals is tertiary > secondary > primary (see Scheme 4.35*a*) and is explained on the basis of hyperconjugation (see Scheme 1.4*i*). The stabilizing effects in allylic radicals (see Scheme 2.11) and benzyl radicals (Scheme 4.40), is due to resonance structures. The strength of the bonds between the same atoms can give an idea about the stability of the radicals formed by their homolysis. The following points may be noted:

- The strength of C—H bonds decrease in R—H when R changes from primary, secondary and tertiary. Thus tertiary alkyl radicals are the most stable.
- A C—H bond next to conjugating groups like allyl or benzyl are very weak, thus allyl and benzyl radicals are more stable than even alkyl radicals.
- The C—H bonds to alkenyl or phenyl groups are strong thus a vinyl or a phenyl radical is less stable than an alkyl radical.

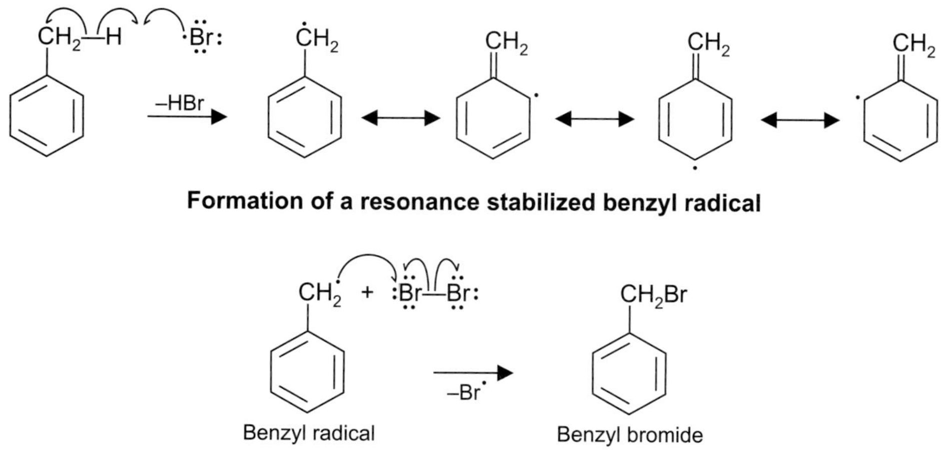

SCHEME 4.40

The triphenylmethyl type radicals are no doubt stabilized by resonance (Scheme 4.41), however, the major cause of their stability is the steric hindrance to dimerization (also see Schemes 16.3 and 16.3a). The dimeric product is a cyclohexadiene derivative (Scheme 4.42, also see, Scheme 16.2).

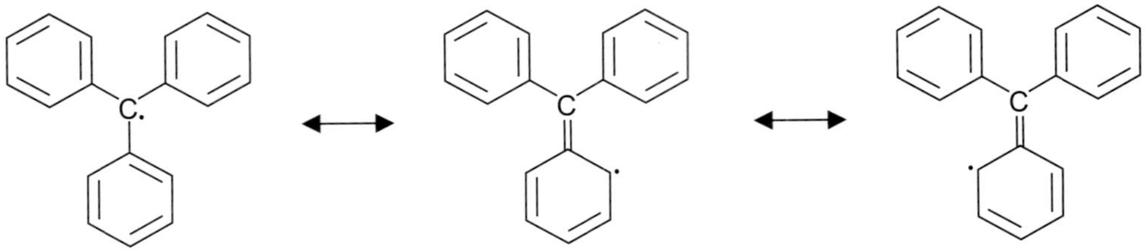

Free triphenylmethyl radical

SCHEME 4.41

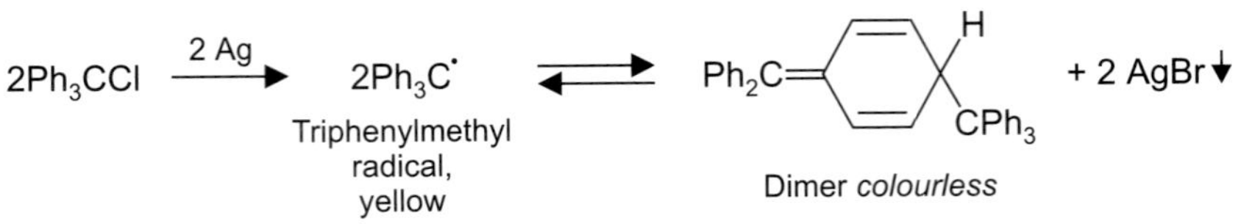

SCHEME 4.42

3. Generation and Properties

These are discussed in detail in Chapter 16. It may be sufficient here to mention that free radicals are formed on thermal and photochemical cleavage. Thus azo and peroxy compounds on heating give free radicals (Scheme 4.42*a*).

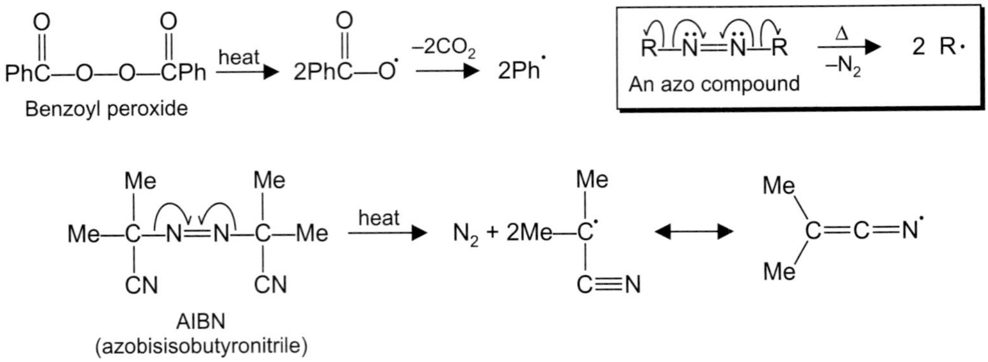

SCHEME 4.42a

The energy of light of 600–300 nm (48–96 kcal/mol) is around the order of magnitude of covalent bond energies. Thus on photochemical reaction, chlorine gives radicals ($Cl_2 \longrightarrow 2Cl^{\bullet}$). For further details on methods of preparation (see, Schemes 16.5–16.6).

The following points may be noted:

- AIBN is a source of relatively unreactive, nitrile stabilized radical. This radical is employed when weaker bonds (*e.g.*, Sn—H) are to be cleaved.
- Peroxides, however generate highly reactive RO• radicals and thus can abstract hydrogen from any position leading to loss of selectivily.

Some of the more general radical reactions are depicted (Scheme 4.42*b*), further details are in chapter 16.

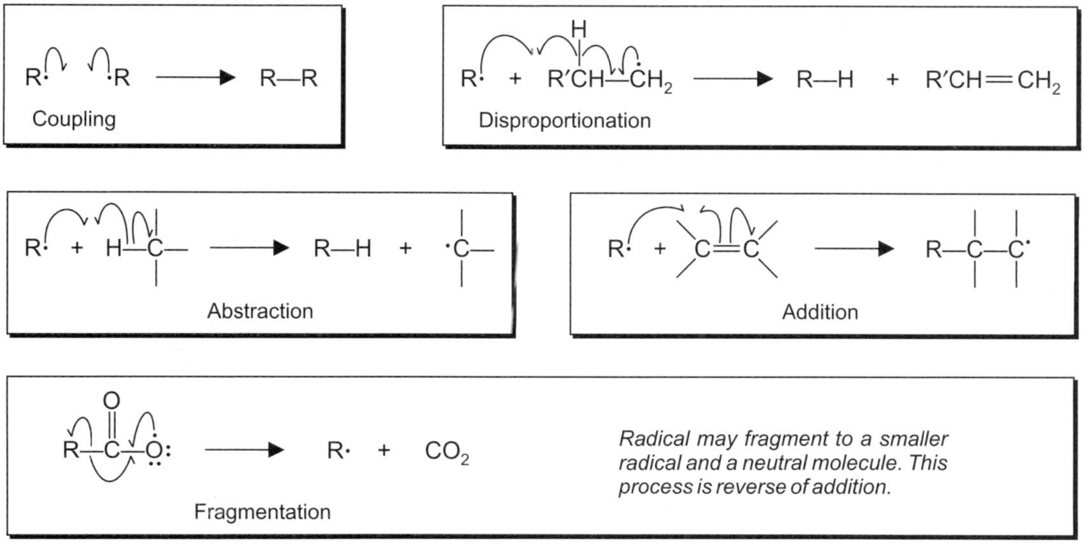

SCHEME 4.42b

(D) Carbenes

Carbenes are neutral intermediates having bivalent carbon, in which a carbon atom is covalently bonded to two other groups and has two valency electrons distributed between two nonbonding orbitals. When the two electrons are spin paired the carbene is a singlet, if the spins of the electrons are parallel it is a triplet (Scheme 4.43).

1. Structure

A singlet carbene is thought to possess a bent sp^2 hybrid structure in which the paired electrons occupy the vacant sp^2 orbital. A triplet carbene can be either bent sp^2 hybrid with an electron in each unoccupied orbital, or a linear *sp* hybrid with an electron in each of the unoccupied *p*-orbitals. It has however, been shown that several carbenes are in a non-linear triplet ground state. However, the dihalogenocarbenes and carbenes with oxygen, nitrogen and sulphur atoms attached to the bivalent carbon, exist probably as singlets. The singlet and triplet states of a carbene display different chemical behaviour. Thus addition of singlet carbenes to olefinic double bonds to form cyclopropane derivatives is much more stereoselective than addition of triplet carbenes (see Schemes 11.13*a* and 11.13*b*).

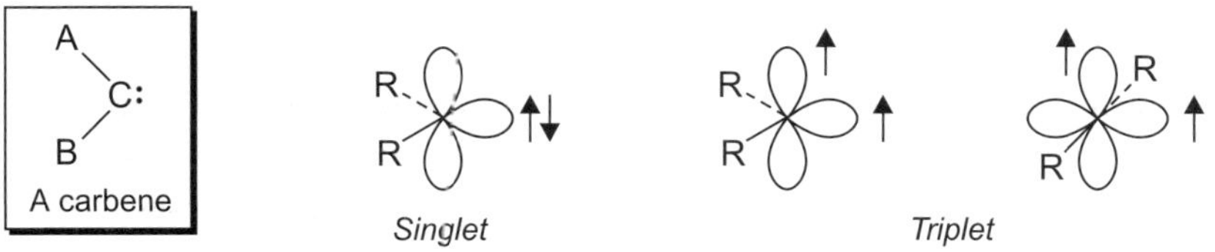

SCHEME 4.43

2. Generation

Carbenes are obtained by thermal or photochemical decomposition of diazoalkanes. These can also be obtained by α-elimination of a hydrogen halide from a haloform with base, or of a halogen from a *gem* dihalide with a metal (Scheme 4.44).

$$RCHN_2 \xrightarrow{h\nu} [RCH:] + N_2 \qquad N_2CHCO_2C_2H_5 \xrightarrow{\Delta} [:CHCO_2C_2H_5] + N_2$$

$$CHCl_3 \xrightarrow{B:^-} BH + :CCl_3^- \longrightarrow :CCl_2 + Cl^- + BH$$

SCHEME 4.44

3. Reactions

These add to carbon double bonds (see Schemes 11.13*a* and 11.13*b*) and also to aromatic systems and in the later case the initial product rearranges to give ring enlargement products (Scheme 4.45, see also exercise 8.8).

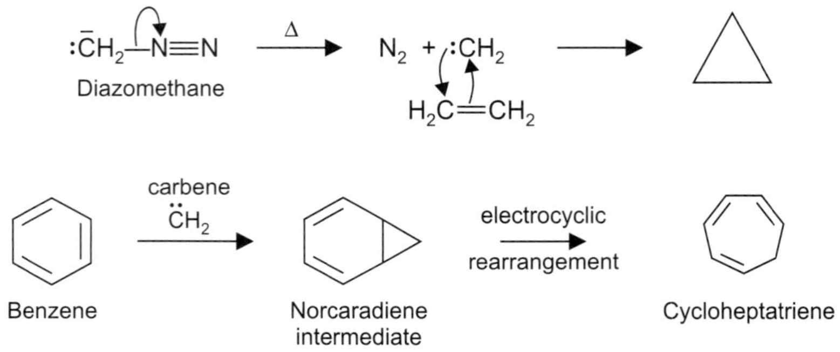

SCHEME 4.45

Benzvalene can be prepared by the reaction of lithium cyclopentadienide with a carbene prepared in *situ* from dichloromethane and methyl lithium (Scheme 4.45*a*). Carbenoids—organometallic or complexed intermediates which, while not free carbenes afford products expected from carbenes are usually called carbenoids (Scheme 4.45*b*).

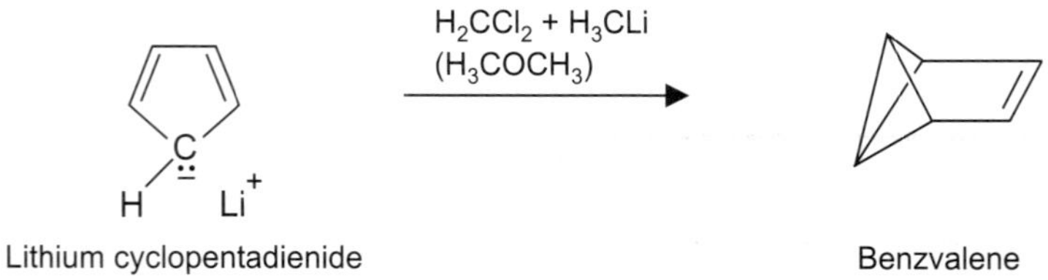

SCHEME 4.45a

$$CH_2I_2 + Zn \xrightarrow[Cu]{\text{diethyl ether}} ICH_2ZnI$$

Iodomethyl zinc iodide (a carbenoid)

SCHEME 4.45b

When a carbene is generated on a three membered ring, allenes are formed by rearrangement (Scheme 4.46). However, a similar formation at a cyclopropyl-methyl carbon gives ring expansion (Scheme 4.46, see also Scheme 10.22). Carbenes are also involved in Reimer-Tiemann reaction (see Schemes 8.52 and 8.53 and exercise 8.8).

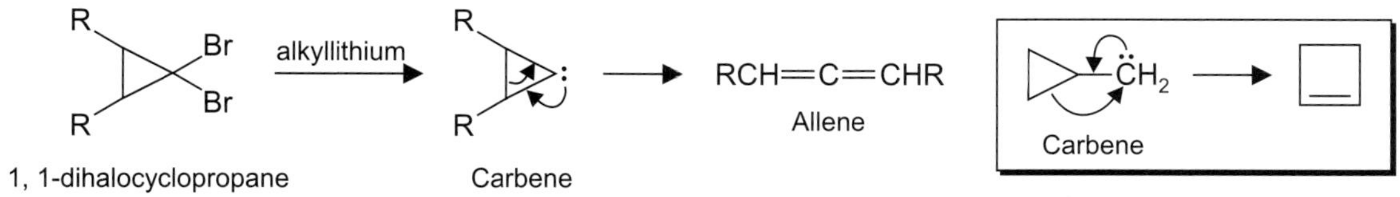

SCHEME 4.46

(E) Nitrenes

1. Structure and Generation

The nitrenes R—N represent the nitrogen analogs of carbenes and may be generated in the singlet (R—N̈:) or triplet state (R—N̈·). A nitrene can be generated *via* elimination (Scheme 4.47), or by the thermal decomposition of azides ($R—N{=}N^+{=}N^- \longrightarrow RN + N_2$).

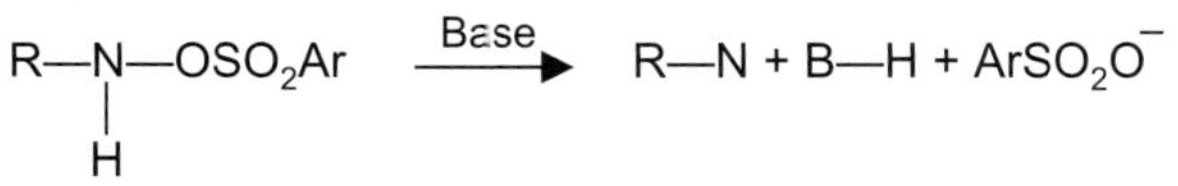

SCHEME 4.47

2. Reactions

In their chemical behaviour, nitrenes are similar to carbenes. Nitrenes, (in particular acyl nitrenes) get inserted into some bonds *e.g,* a C—H bond to give an amide (Scheme 4.48). Aziridines are formed when nitrenes add to C=C bonds (Scheme 4.48). Alkyl nitrenes do not display the typical reactions shown in Scheme 4.48, however, these undergo an instant rearrangement (Scheme 4.49).

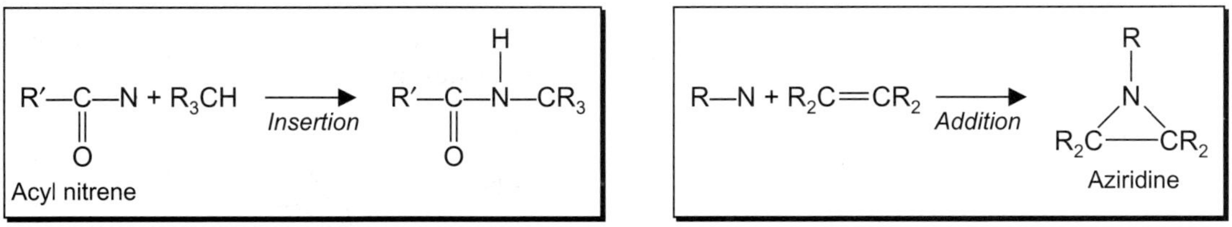

SCHEME 4.48

Alkyl nitrene *rearrangement*

SCHEME 4.49

(F) Arenium Ions

A considerable amount of experimental evidence indicates that electrophiles attack the π system of benzene to form a delocalized non-aromatic carbocation known as arenium ion or sometimes a σ complex (see Schemes 8.2 and 8.4). CMR spectroscopic evidence is available in favour of σ complex.

(G) Benzyne

It is a reactive intermediate in some nucleophilic aromatic substitutions. It is a benzene with two hydrogen atoms removed (see Sec. 9.3 and 9.4). It is usually drawn with a highly strained triple bond in the six membered ring. Benzyne intermediate has been observed spectroscopically and trapped (see Scheme 9.16*a*).

Benzyne is too reactive to be isolated and is observed only spectroscopically. Photolysis of benzocyclobutenedione gives benzyne. Although benzyne is usually represented as a cycloalkyne (Scheme 4.50), its triple bond exhibits an IR stretching frequency of 1846 cm^{-1} which is intermediate between the double (ethene, 1655 cm^{-1}) and triple (ethyne, 1974 cm^{-1}) bonds. The ^{13}C NMR values for these carbons (δ = 182.7 ppm) are also not of pure triple bonds. This evidence shows a contribution by a cumulated triene resonance structure. This bond is however, weak due to poor *p* orbital overlap in the plane of the ring (see Scheme 9.15).

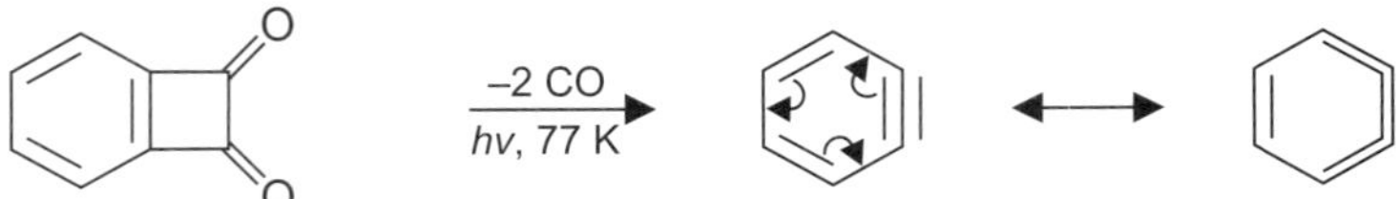

Benzocyclobutene-1, 2-dione

Benzyne

SCHEME 4.50

PROBLEMS

4.1 Why the addition of HCl to the following olefin takes place in the opposite manner as predicted by Markonikov's rule (*i.e.*, the hydrogen atom of HX adds to that carbon which has the greatest number of hydrogens?

$$CH_2{=}CH{-}CO_2Et + HCl \longrightarrow CH_2Cl{-}CH_2{-}CO_2Et$$

4.2 Considering *H* and *S* terms, how one can predict the favoured direction of the reaction $R:R \rightleftharpoons 2R^{\cdot}$?

4.3 Why compared to the slow esterification of acetic acid with methanol in the absence of an acid catalyst, the lactonization of 4-hydroxy-butyric acid is almost spontaneous?

4.4 Why the synthesis of larger rings is usually impeded?

4.5 Indicate if the following statements are true or false:

(*a*) More stable species have lower energy.

(*b*) In case products are favoured at equilibrium $\Delta G°$ is negative while K_{eq} is larger than 1.

(*c*) With a smaller rate constant, the reaction is fast.

(*d*) The products formed with stronger bonds and with larger freedom of motion causes $\Delta G°$ to be negative.

(*e*) Bond are only partially formed in a transition state while in intermediates the bonds are fully formed.

4.6 Write a mechanism for the following conversion:

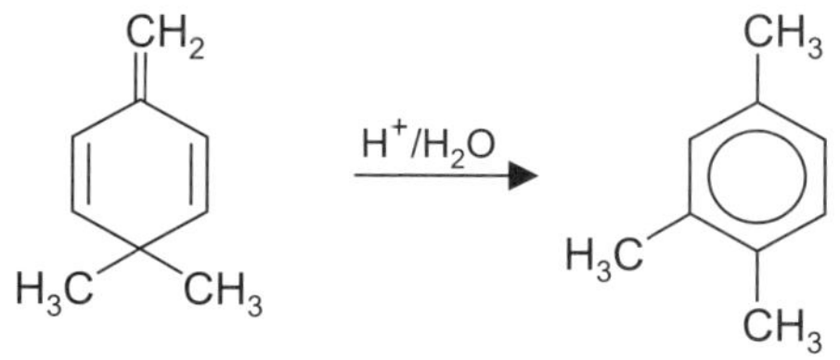

ANSWERS TO THE PROBLEMS

4.1 Of the two possible intermediate I and II, I is of lower energy than II. In, II the positive charge is adjacent to the strongly electron withdrawing carboethoxy group.

$$\overset{+}{C}H_2\text{—}CH_2\text{—}CO_2Et \qquad\qquad CH_3\text{—}\overset{+}{C}H\text{—}CO_2Et$$

(I) (II)

4.2 An molecule R—R gives two R$^\bullet$ radicals and their greater randomness makes ΔS positive and favourable for reaction. However, this term is much smaller compared to large unfavourable positive ΔH, this favors the reverse reaction when R—R is more stable than two radicals. Consequently ΔG is positive and the reverse reaction is favoured.

4.3 The rate determining step in each case is the formation of the new C—O bond, and in the transition state this bond is partially formed. Thus the $\Delta H^\ddagger$ terms are same in magnitude in both cases. In the esterification case, the passage to the transition state requires two molecules coming together to form one compound, thus $\Delta S^\ddagger$ is large and negative compared with the lactonization reaction. Therefore, $\Delta G^\ddagger$ is smaller for lactonization, consequently this process occurs faster.

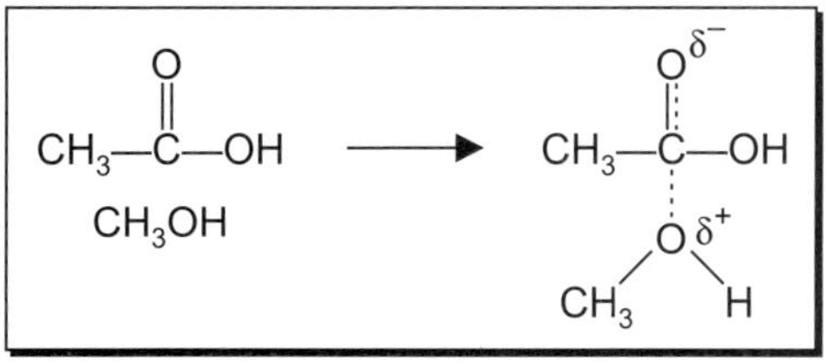

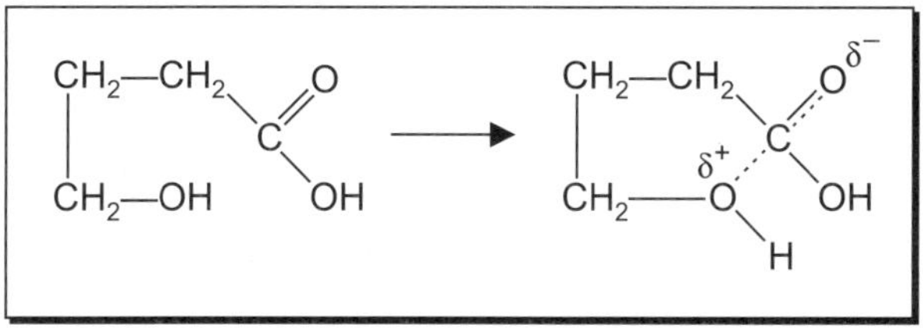

4.4 This is due to the involvement of adverse entropy effects, eclipsing and other strain factors during the formation of larger rings.

4.5 (*a*) true; (*b*) true; (*c*) false; (*d*) true; (*e*) true.

4.6

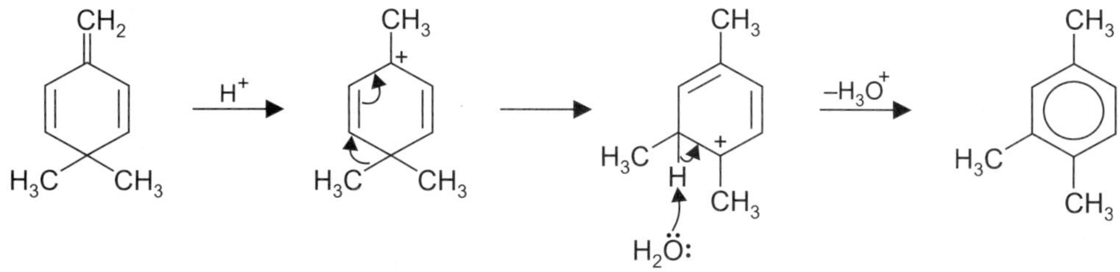

CHAPTER 5

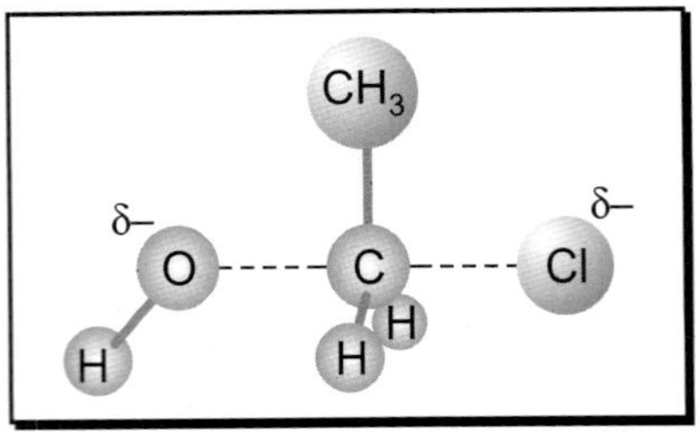

Aliphatic Nucleophilic Substitution and its Synthetic Applications

5.1 INTRODUCTION

In the nucleophilic substitution reactions a group attached to a carbon is replaced by another group which may involve several mechanisms. Bimolecular nucleophilic substitution S_N2 along with its unimolecular counterpart are the most important and involve reactions of alkyl halides, alcohols, and related compounds. Good nucleophiles (^-OH, CH_3O^-) encourage S_N2 reactions, while poor nucleophiles (H_2O, CH_3OH) encourage S_N1 reactions.

Competition between S_N1 and S_N2 Reactions

- *S_N2 reaction has high synthetic value since it leads to a single substitution product. An S_N1 reaction has a far less synthetic value since it leads to two substitution products if the leaving group is bonded to a stereogenic carbon.*
- *An S_N1 reaction is further complicated by carbocation rearrangements.*
- *An S_N1 reaction pathway (the initial species formed is a carbocation) is favoured when the carbocation is stabilized (tertiary or resonance stabilized carbocations), and the solvent is polar (to stabilize the transition state).*
- *The S_N2 pathway (a concerted pathway) is favoured when the electrophilic carbon is not sterically hindered and the solvent is aprotic (to make the nucleophile more reactive).*
- *The rate law for the reaction of a reactant e.g., an alkyl halide that can display both S_N2 and S_N1 reactions simultaneously is the sum of the individual rate laws.*

  Rate law (S_N2 reaction) = k_2 [alkyl halide] [nucleophile]

  Rate law (S_N1 reaction) = k_1 [alkyl halide]

Contribution to the rate *via* an S_N2 reaction

Contribution to the rate *via* an S_N1 reaction

Rate = k_2 [alkyl halide] [nucleophile] + k_1 [alkyl halide]

Thus on increasing the concentration of the nucleophile increases the rate of an S_N2 reaction but the rate of an S_N1 reaction is not effected.

- *The outcome of an S_N2 reaction on a chiral substrate is inversion of configuration relative to the configuration of the reactant. The S_N1 reaction on the other hand gives one product of retained configuration while the other with inverted configuration relative to the configuration of the reactant (Scheme 5.1).*

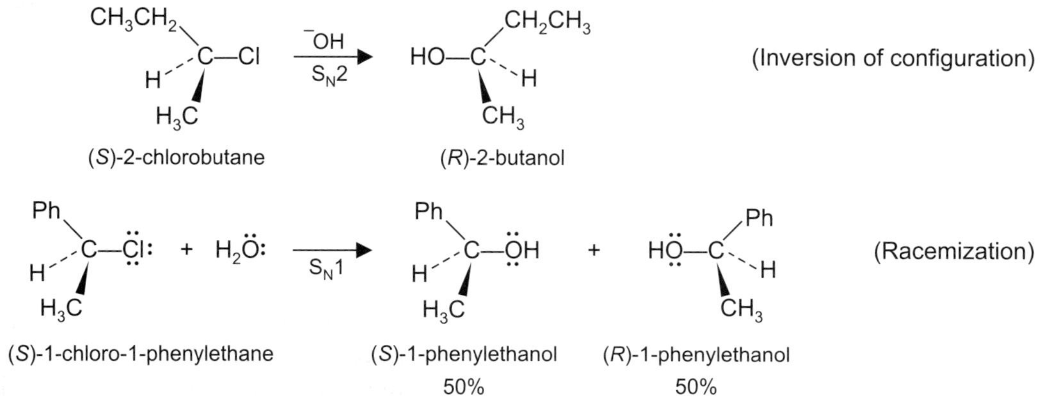

SCHEME 5.1

Substitution reactions face a competition from elimination reactions, since all nucleophiles are also bases. This is seen *e.g.*, with a secondary alkyl halide with cyanide ion (Scheme 5.2).

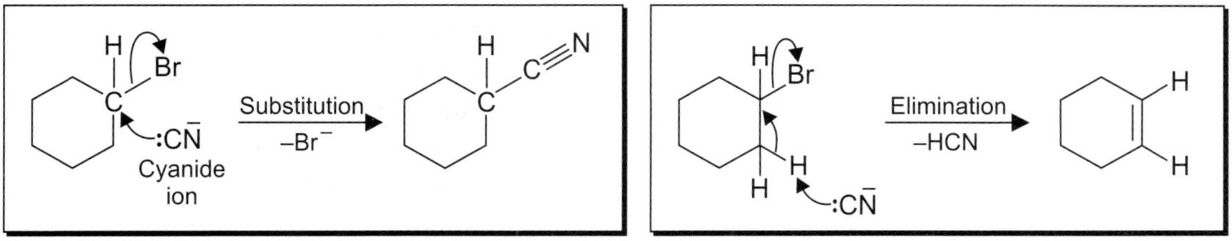

SCHEME 5.2

The nucleophilic substitution at the saturated carbon is the alkylation of the nucleophile. The nucleophile forms a bond to a saturated carbon atom from which a leaving group departs. These substitutions provide very useful synthetic pathways and an example is the preparation of methyl esters from carboxylic acids. Diazomethane in the presence of a proton source is converted into a very reactive methylating agent. The high reactivity is due to the presence of an excellent leaving group, gaseous nitrogen (Scheme 5.2*a*).

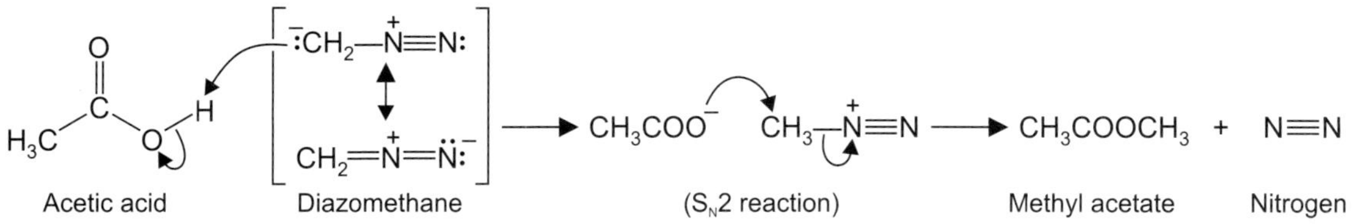

SCHEME 5.2a

The balance between the nucleophilic and basic characters of a reagent is greatly dependant on the size of the orbital containing the lone pair of electrons, and thus the polarizability of these electrons. The reagents with highly polarizable centers of electron density *e.g.*, halide ions are better nucleophiles than bases as these prefer to attack carbon (substitution) rather than the considerably smaller proton (elimination). Thus, I^- is the most nucleophilic of the halide ions and is the least basic, (and H—I is the strongest of the halogen acids). On the other hand, the high concentration of electron density in NH_2^-, OH^-, $C{\equiv}N^-$, and $^-C{\equiv}C{-}H$ makes these reagents good bases and only moderately active nucleophiles.

It is important to select the reagents which will give substitution products in high yield with less elimination, the nature of the nucleophile reflects on the desired outcome of the substitution reaction. To convert an alkyl halide to an alcohol, the nucleophile must be OH^- or H_2O. It is therefore, sometimes difficult to achieve only substitution, without a competing elimination reaction. The following points may be noted:

- Alcohols can be synthesized from alkyl halides by the use of either water or hydroxide ion as the nucleophile (Scheme 5.2*b*).

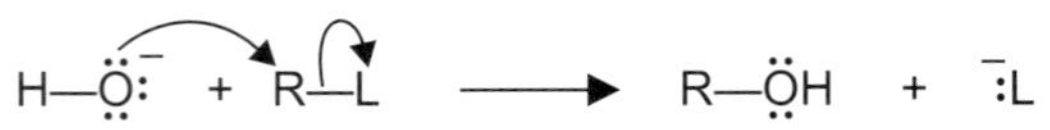

SCHEME 5.2b

- Good yields are obtained from S_N2 reactions of hydroxide ion with secondary alkyl halides that are allylic, benzylic, or adjacent to a carbonyl group (Scheme 5.2*c*).

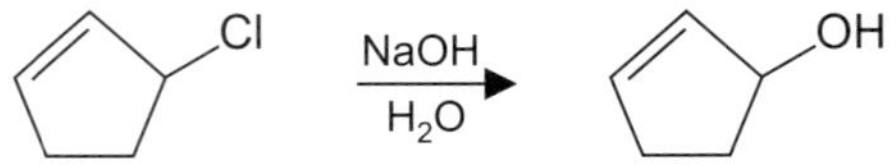

SCHEME 5.2c

- Hydroxide ion is not normally employed as a nucleophile with unactivated secondary halides and never with tertiary halides due to competing E2 elimination reactions. For these reactants replacement of the halide with OH can be accomplished by using water as a nucleophile and S_N1 conditions (Scheme 5.2*d*).

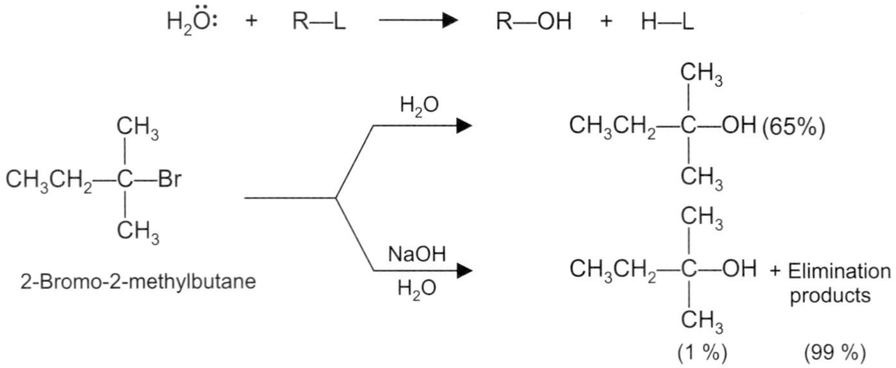

SCHEME 5.2d

EXERCISE 5.1

Which of the following routes (Scheme 5.2e) is suitable for the synthesis of ethyl isopropyl ether?

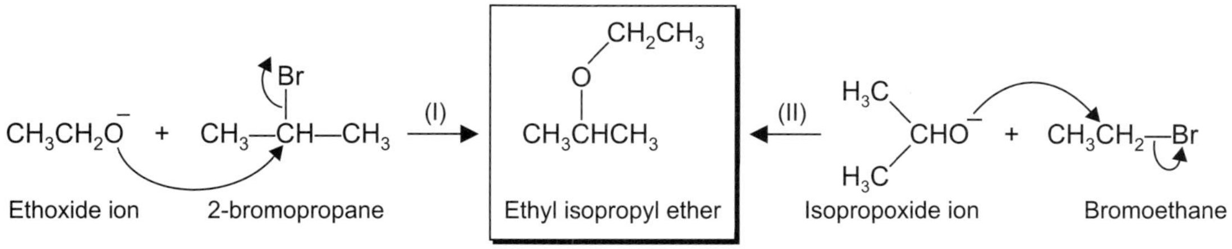

SCHEME 5.2e

ANSWER. *The route (II) using the primary halide. In either route, the base employed is an alkoxide which is a strong base. In route (I) therefore, using a secondary alkyl halide would mean that the major reaction would be E2 elimination rather than S_N2 substitution.*

The conversion of alcohols into alkyl halides is another important substitution reaction. As seen in E2 eliminations, a hydroxy group is not displaced as its anion, it being of high energy. Consequently alcohols are inert to nucleophiles, unless special conditions are employed. Thus a hydroxyl group can be displaced from an alcohol, when it is protonated using a strong acid and the process termed as electrophilic catalysis (Scheme 5.3).

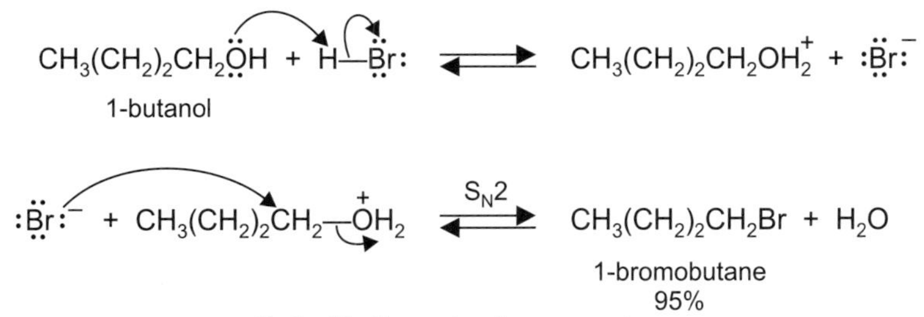

Substitution at primary carbon

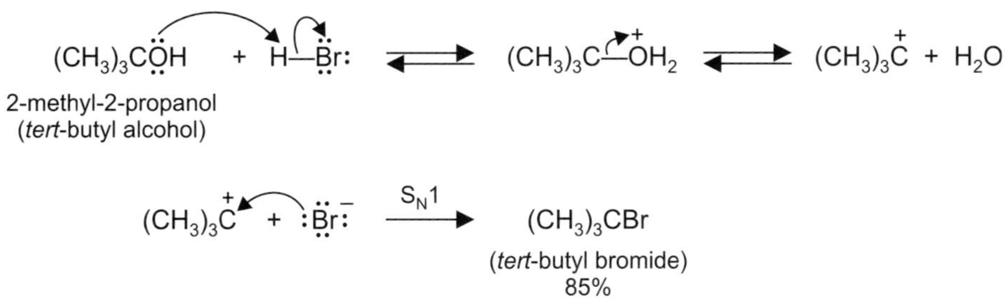

Substitution at tertiary carbon

SCHEME 5.3

Sulphonate esters, *e.g.*, *p*-toluenesulphonate esters known as tosylates (represented as –OTs in structures) are excellent leaving groups and react very much like alkyl halides in both S_N1 and S_N2 reactions (Scheme 5.4). In fact *p*-toluenesulphonate ion is about 100 times better than a chloride ion as a leaving group. Another related ester is methane sulphonate ester. During tosylate formation from an alcohol, the C—O bond of the alcohol remains intact and the alcohol, (as its tosylate ester) retains its stereochemical configuration. As expected during the S_N2 reaction on a tosylate the inversion of configuration is observed (Scheme 5.4).

The inversion of configuration was confirmed from the ester prepared from the parent alcohol by reaction with acetyl chloride (this ester has the same relative configuration as the alcohol or its tosylate ester) and comparing its [α] with that prepared by the acetolysis of the

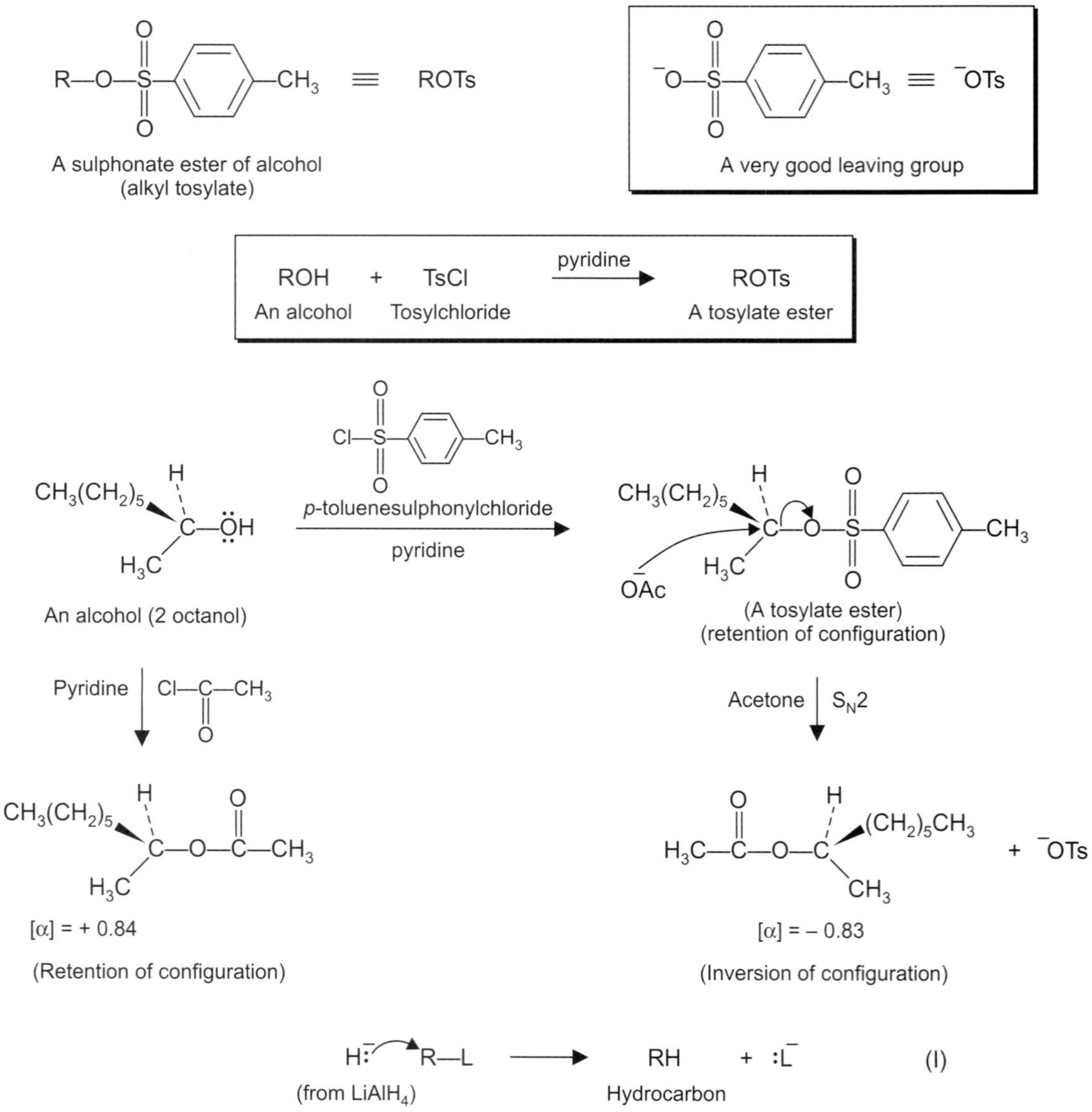

SCHEME 5.4

tosylate (S_N2 reaction). Conversion of an alcohol to its hydrocarbon via reduction of its tosylate occurs via S_N2 attack by a hydride ion, H^-: on the tosylate (eq. I, Scheme 5.4).

Complexation with the oxygen of hydroxyl group by adding Lewis acid, $ZnCl_2$ creates an even better leaving group than that formed by protonotion of the oxygen. The complex either reacts with the nucleophile by S_N1 or S_N2 mechanism (Scheme 5.5).

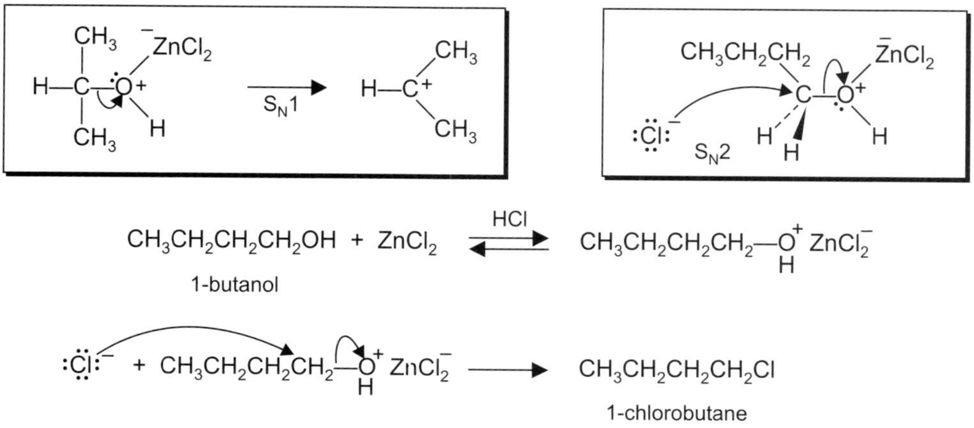

SCHEME 5.5

Chloroalkanes can be made the same way as bromoalkanes (see Scheme 5.3) using hydrochloric acid, but hydrochloric acid is not as reactive with alcohols as is hydrobromic acid. Hydrogen chloride is a weaker acid than hydrogen bromide and formation of the protonated alcohol (an **oxonium ion**) is less favourable with HCl as the reagent. Moreover, chloride is a poorer nucleophile than bromide and thus the displacement step is also less favourable. A Lewis acid *e.g.,* zinc chloride is often added to bring about an enhancement of the reaction of alcohols with HCl (Scheme 5.5).

S_N1 reactions involve the formation of carbocations, therefore rearrangements are often observed during these reactions. S_N2 reactions being one step processes give no possibility of rearrangement. Thus neopentyl bromide on boiling with ethanol (gives a rearrangement product (Scheme 5.6).

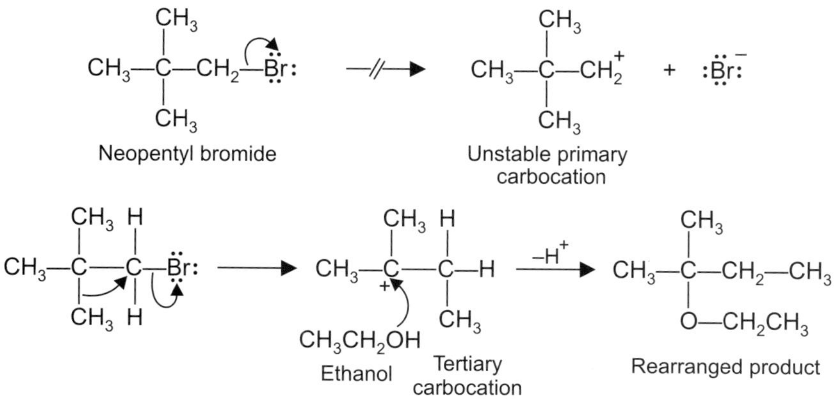

SCHEME 5.6

5.2 SYNCHRONOUS SUBSTITUTION—S_N2 PROCESS

(A) Mechanism—S_N2 Process

The mechanism of an S_N2 reaction is a concerted one step process without an intermediate and involves backside displacement of the leaving group by a nucleophile (see Scheme 5.4). The configuration at the carbon atom undergoing substitution inverts (Walden inversion) since the nucleophile approaches along a line diametrically opposite the bond to the leaving group. This stereochemical outcome of the reaction points to a transition state (Scheme 5.7) in which the carbon undergoing substitution involves a temporary rehybridization from sp^3 to sp^2 and finally back to sp^3, sp^2-hybridized carbon has a *p* orbital perpendicular to the plane of the bonds which partly overlaps with an orbital of the leaving group as well as with that of the incoming nucleophile.

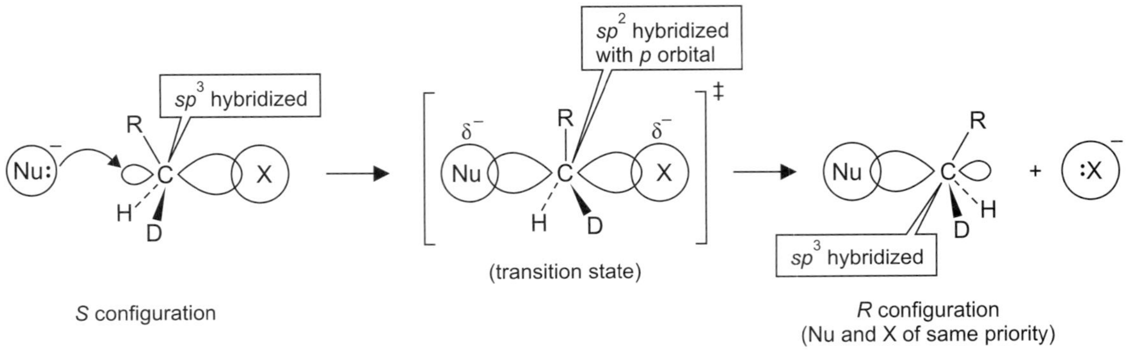

SCHEME 5.7

Thus the S_N2 reaction transition state with a given reactant may be represented properly (Scheme 5.7*a*).

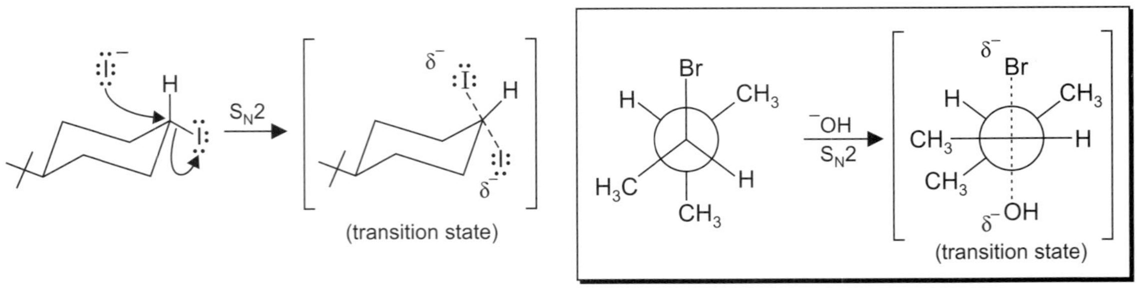

SCHEME 5.7a

(B) Structure of the Substrate

The rate of direct displacement *i.e.*, an S_N2 reaction is highly sensitive to the steric bulk of the substituents present on the carbon undergoing this reaction. As expected the degree of coordination increases at the reacting carbon atom and from the steric point of view, the optimum substrate would be CH_3—X. Each replacement of hydrogen by a more bulky alkyl group should decrease the rate of reaction. Consequently, the order of reactivity of alkyl groups is expected to be methyl > primary > secondary > tertiary and this is observed. Table 5.1 gives the relative rates of typical S_N2 reactions. Methyl halides react most rapidly and tertiary halides react so slowly as to be unreactive by the S_N2 mechanism. It may be noted that in E2 reactions the order of reactivities (see Scheme 12.11) is the opposite to this, therefore, the S_N2/E2 ratio is largest for a primary halide while it is least for a tertiary halide and this is seen during the reactions of alkyl bromides with ethoxide ion in ethanol at 55°C (Table 5.1). Thus tertiary halides do not give any significant yield in the S_N2 reactions. One fails to prepare *e.g.*, *t*-butyl cyanide from *t*-butyl chloride and cyanide ion as the product is derived only from elimination, $(CH_3)_2C{=}CH_2$.

Table 5.1: Relative rates of reactions of alkyl halides in S_N2 reactions

Substituent	*Compound*	*Relative rate*
Methyl	$CH_3{-}X$	30
1°	$CH_3{-}CH_2{-}X$	1
2°	$(CH_3)_2CH{-}X$	0.02
3°	$(CH_3)_3C{-}X$	0

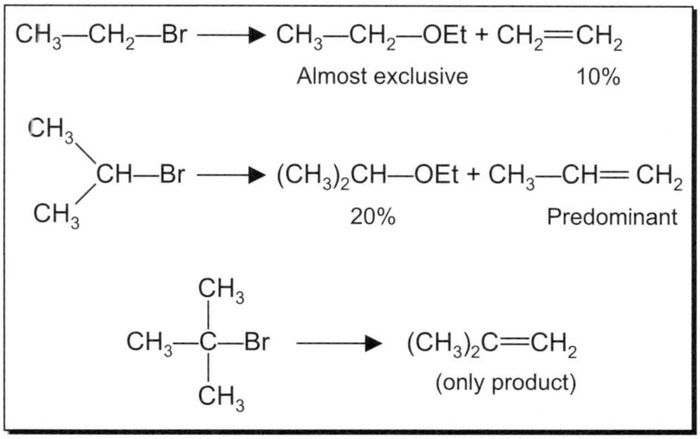

Neopentyl halides being primary halides are also unreactive in S_N2 reactions. This situation shows that steric hindrance effects operate even if the β-carbon is substituted by alkyl groups. A general statement is therefore, that S_N2 type displacements are retarded by increased steric repulsions at the transition state. In substrates of the type $R{-}CH_2{-}X$, where X is a leaving group, showed that steric effects of R are the dominant factor in determining rates (good yields of neopentyl iodide can be obtained via Grignard reagent, see Scheme 7.82). The bulky substituents on or near the carbon atom undergoing S_N2 reaction hinder the approach of the nucleophile to a distance within bonding range. In an extreme case within the series *i.e.*, in neopentyl system (compared to methyl), the approach of the nucleophile along the line of the C—X bond is hindered by a methyl group, whatever geometry is attained by rotation about the single bonds (Scheme 5.8).

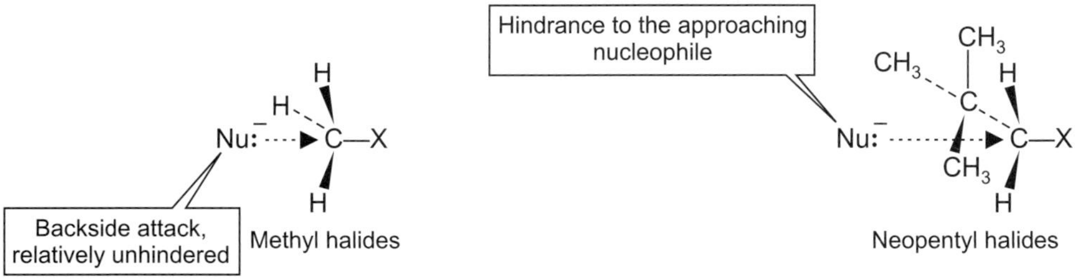

Steric effects in acyclic systems in the S_N2 reaction

SCHEME 5.8

The halocycloalkanes show considerable rate differences during S_N2 reactions depending on the size of the ring (for detailed discussion (see Table 1.1). Halocyclohexanes although seemingly more capable of attaining sp^2 hybridization at the reacting carbon are however, slower in S_N2 reaction (see Table 1.1). When a nucleophile approaches an equatorial halide, it faces an inhibiting effect *i.e.*, steric hindrance by the two axial hydrogens at C-3 and C-5 carbons (Scheme 5.9). In a conformation where the leaving group is axial, then its exit itself faces steric hindrance.

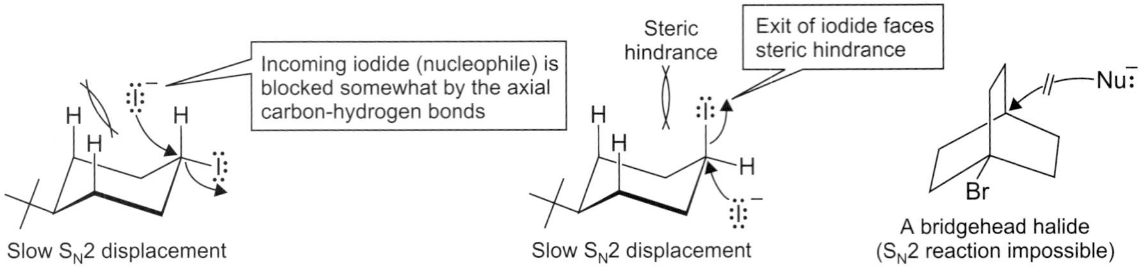

SCHEME 5.9

A bridgehead halide is inert since the three bridges prevent the backside attack necessary for S_N2 reaction. Moreover, inversion of such a system is impossible.

Epoxides are most important three-membered heterocycles which undergo reactions with nucleophiles such as hydroxide ion or alkoxides. The nucleophile attacks the less hindered carbon as expected from an S_N2 attack (1° > 2° > 3°) and the alkyl group is transferred with a heteroatom in the β-position (Scheme 5.9*a*).

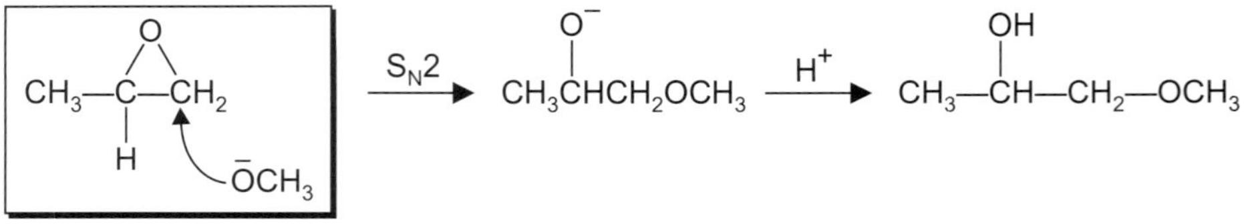

SCHEME 5.9a

The C—O bond of an epoxide is weaker than those of other ethers due to strained three membered ring. Organolithium and organomagnesium compounds, therefore, act as nucleophiles in S_N2 reactions with epoxides and otherwise these reagents are normally poor nucleophiles and being very strong bases promote elimination reactions (see Scheme 7.78).

(C) Nucleophiles and their Relative Nucleophilic Strength—Nucleophilicity—Polarizability of an Atom

The rate of an S_N2 reaction is directly related to the effectiveness of the nucleophile in displacing the leaving group. As a consequence, of a pair of nucleophiles containing the same reactive atom *e.g.*, methanol and methoxide ion, the species with a negative charge is the more powerful nucleophile. Put in other words, of a base and its conjugate acid, the base is always more nucleophilic. Therefore, OH^- is a stronger nucleophile than H_2O, SH^- a stronger nucleophile than $(CH_3)_2$ SH and NH_2^- is stronger nucleophile than NH_3. This finding is reasonable as a stronger bond between the nucleophilic atom and carbon would lead to a more stable transition state and thus a reduced activation energy. Because the S_N2 reaction is concerted the strength of the partially formed new bond will be reflected in the energy of the transition state; the more negative the attacking species, the faster will be the reaction.

There seems to be a good correlation between basicity and nucleophilicity. However, it would be a mistake to explain that methoxide is a much better nucleophile since it is much more basic. Basicity and nucleophilicity are two fundamentally different properties. Basicity (a measure of a thermodynamic phenomenon) is defined by the equilibrium constant for abstracting a proton (Scheme 5.10). On the other hand, nucleophilicity (a measure of a kinetic phenomenon) is defined by the rate of attack on an electrophilic carbon.

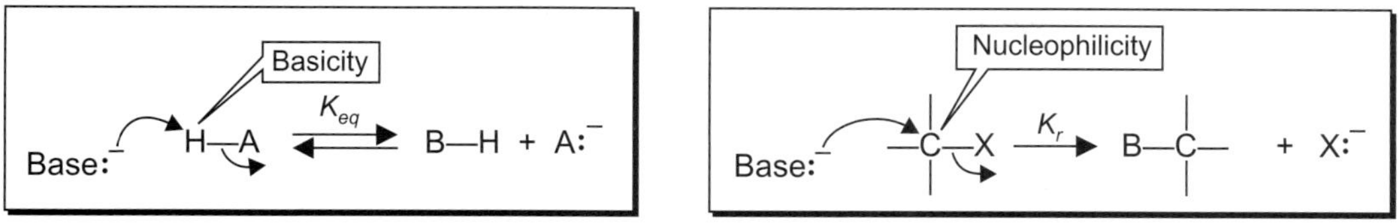

SCHEME 5.10

The hydroxide ion (OH^-) is a stronger base than a cyanide ion (CN^-); at equilibrium it has a greater affinity for a proton (the pK_a of H_2O is ~ 16, whereas the pK_a of HCN is ~ 10). However, cyanide ion is a stronger nucleophile, *i.e.*, it reacts more rapidly with a carbon having a leaving group than a hydroxide ion. In both cases (Scheme 5.10) however, a new bond is formed. When the new bond is to a proton, the species has reacted as a base; in case the bond is to a carbon it has reacted as a nucleophile. Moreover, since nucleophilicity is used to describe trends in the kinetic aspects of reactions, the relative nucleophilicity of a given species may differ from substrate to substrate.

A sterically bulky nucleophile is less reactive than a smaller one. An S_N2 reaction is not only sensitive to the bulky groups on the electrophile, but to the bulky groups on the nucleophile as well. This is due to nonbonded repulsions which would develop in the transition state. Thus although *t*-butoxide ion is a stronger base than ethoxide ion, but the bulky *t*-butoxide is a weaker nucleophile. Basicity is little effected by steric hindrance, since the attack is on an unhindered proton. The strength of a base depends only on how well the base shares its electrons with a proton (Scheme 5.11).

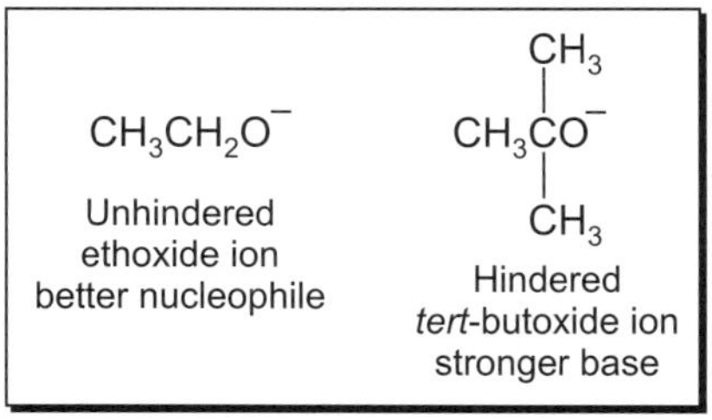

SCHEME 5.11

Nucleophilicity decreases on going from left to right in the periodic table. This follows the increase in electronegativity from left to right. Thus high electronegativity is unfavourable, due to tightly held electrons which are therefore, relatively less available for donation to the substrate for S_N2 process. Thus OH^- is more nucleophilic than F^-; NH_3 is more nucleophilic than H_2O.

Nucleophilicity increases going down the periodic table *i.e.*, Br^- is more nucleophilic than Cl^- and SH^- is more nucleophilic than OH^-. For the halide ions the nucleophilic reactivities follow the order:

$$I^- > Br^- > Cl^- > F^-$$

Thus I^- is the best nucleophile while F^- is poorest. This order is the reverse of the basicity order of the halide ions:

$$F^- > Cl^- > Br^- > I^-$$

Thus one finds that the correlation between basicity and nucleophilicity does not hold for the halides. Down a column in the periodic table the atoms become larger. Thus one has more electrons at a greater distance from the nucleus. These electrons are rather loosely held and the atom with this situation is more polarizable. The increase in polarizability leads to an ease of the distortion of the electron cloud of the attacking atom of the nucleophile to allow far more effective overlap in the transition state, with the back lobe of the slowly rehybridizing sp^3 hybrid used to maintain bonding to the leaving group. For these very reasons the basicity of the larger elements is relatively poor than smaller elements as the overlap with the hydrogen 1 *s* orbital is poor. Thus, one can use successfully the extent of polarizability to explain as to why I^- is a better nucleophile than F^- (Scheme 5.12). It is only with I^- that the partial bond formation begins at relatively large distances as the electron cloud of the nucleophile gets "pulled" to the carbon atom where the S_N2 reaction occurs. Thus the activation energy of the substitution is decreased. This however, is not the case with F^- which has tightly bound electrons

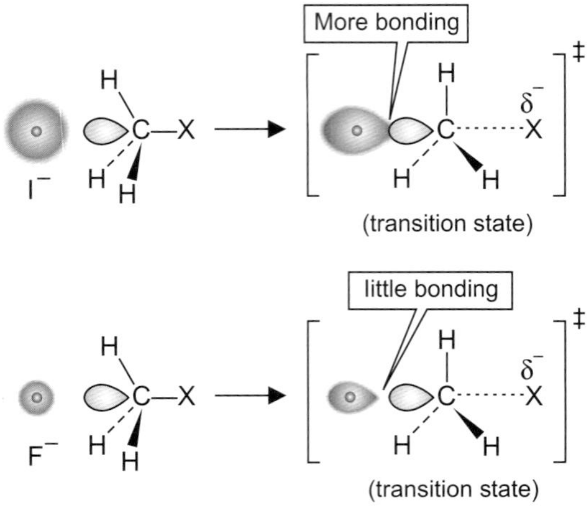

The larger I^- is more polarizable than a fluoride ion. Thus, the relatively loosely held electrons of the iodide ion can overlap even when farther away with the orbital of carbon undergoing nucleophilic attack. The tightly bound electrons of the fluoride ion cannot start to overlap until the atoms are closer together.

SCHEME 5.12

that cannot begin to form a C—F bond until the atoms are very close together. One may appreciate that the degree of nucleophilicity increases on going down in the periodic table even for uncharged nucleophiles *i.e.*, H_2 Se > H_2S > H_2O and PH_3 > NH_3. In these cases particularly, the fact that the increasing polarizability improves nucleophilic power explains the nucleophilicity trend. In these uncharged nucleophiles, the solvent effects must, however, be less pronounced.

Synthesis of amines involves the reaction of nucleophiles like NH_3 or an amine on alkyl halides or sulphonate esters in an S_N2 reaction (Scheme 5.12*a*). The product of the reaction with ammonia or an amine is an amine salt. The free amine is isolated on treatment with a base such as NaOH.

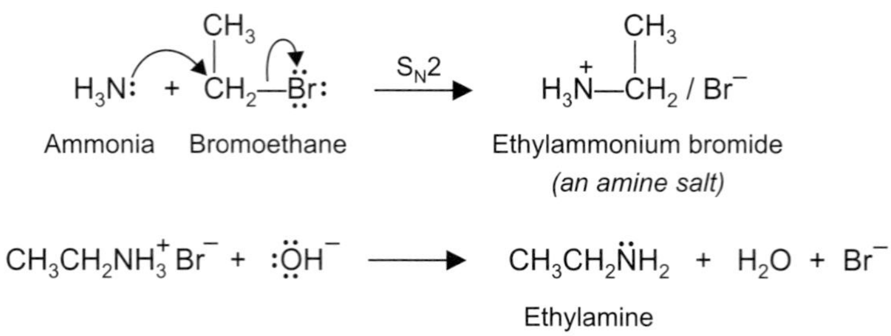

SCHEME 5.12a

The order of reactivity of alkyl halides is typical for S_N2 reactions: CH_3X > 1° > 2°. Tertiary alkyl halides react with ammonia or amines to give elimination products.

A disadvantage of this route to amines is that the product amine salt can exchange a proton with the starting ammonia or amine (Scheme 5.12*b*). This leads to two or more nucleophiles

$$CH_3CH_2NH_3^+\ Br^- + NH_3 \rightleftharpoons \underset{\textit{(also a nucleophile)}}{CH_3CH_2NH_2} + NH_4^+\ Br^-$$

SCHEME 5.12b

competing in the reaction with the reactant. Recall that a primary amine is a stronger base and a stronger nucleophile than is NH_3. For these reasons *e.g.*, methyl iodide and ethylamine in basic solution lead to quaternary ammonium iodide as the final product (Scheme 5.12*c*).

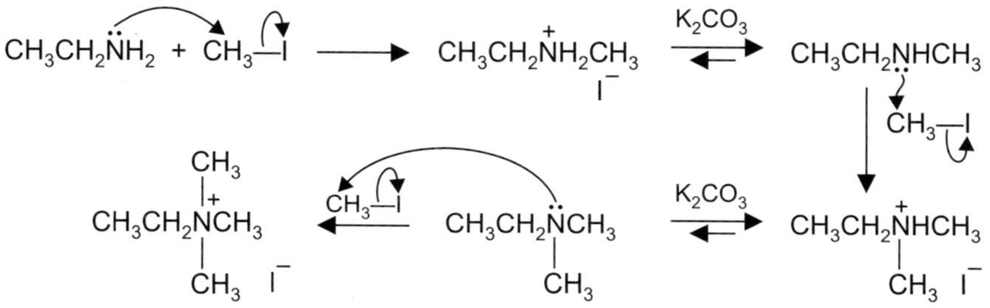

SCHEME 5.12c

EXERCISE 5.2

How an alkyl halide reacts with the azide ion? How this method can be used to synthesise primary amines?

ANSWER. *Azide ion is a very good nucleophile which reacts with an alkyl halide to give an alkyl azide. An alkyl azide is not however, nucleophilic and its catalytic hydrogenation gives the primary amine in good yield (Scheme 5.12d).*

$$CH_3CH_2CH_2Br + \underset{\text{Sodium azide}}{Na^+\ :\overset{-}{\ddot{N}}{=}\overset{+}{N}{=}\overset{-}{\ddot{N}}:} \longrightarrow CH_3CH_2CH_2N_3 \xrightarrow[Pt]{H_2} CH_3CH_2CH_2NH_2 + N_2$$

SCHEME 5.12d

EXERCISE 5.3

Explain the Gabriel synthesis of a primary amine.

ANSWER. *In phthalimide the electron pair on the nitrogen is neither basic nor nucleophilic (due to resonance). Recall that the hydrogen on nitrogen is very acidic and can be removed by a base to yield phthalimide anion which is a good nucleophile. Reaction with an alkyl halide gives an N-alkylphthalimide which does not alkylate further. Hydrolysis then produces the alkylamine (Scheme 5.12e).*

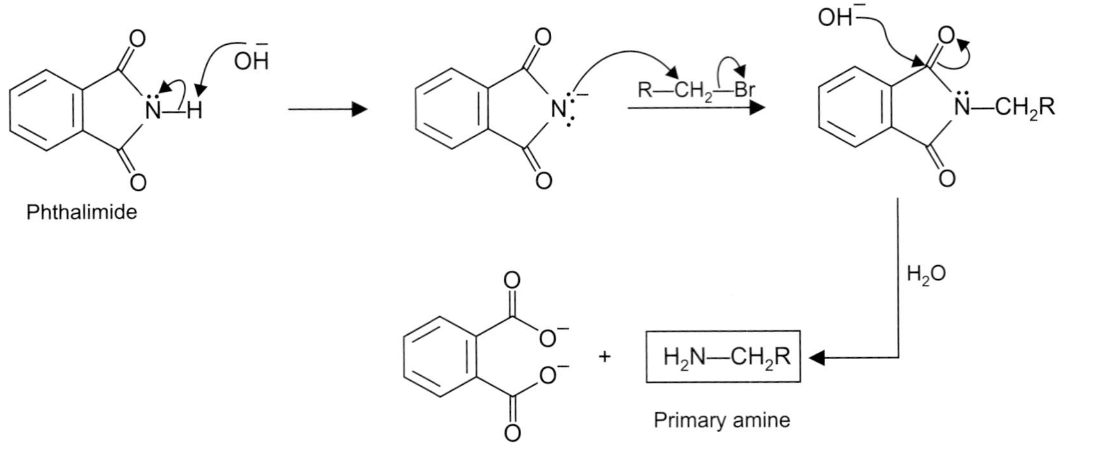

SCHEME 5.12e

(D) Nucleophilicity and Solvent Effects

Nucleophilicity is impeded by solvation. A molecule of a solvent *e.g.*, water or an alcohol (protic solvent) can form hydrogen bonds to an anionic nucleophile (Scheme 5.13). Smaller anions (concentrated charge) are more strongly solvated than larger ones. This phenomenon, therefore, creates a solvent induced barrier to attack at the substrate. For a solvated anion to act as a nucleophile in an S_N2 reaction energy is required to "strip off" some of the solvent molecules. This energy is much larger in the case of a small strongly solvated ion *e.g.*, F^- as compared to I^-. Consequently an iodide ion is a better nucleophile than fluoride ion in protic solvent in an S_N2 reaction. Table 5.2 lists some common nucleophiles which are listed in decreasing order of nucleophilicity in protic solvents *e.g.*, water or alcohol. It is found that when benzyl tosylate

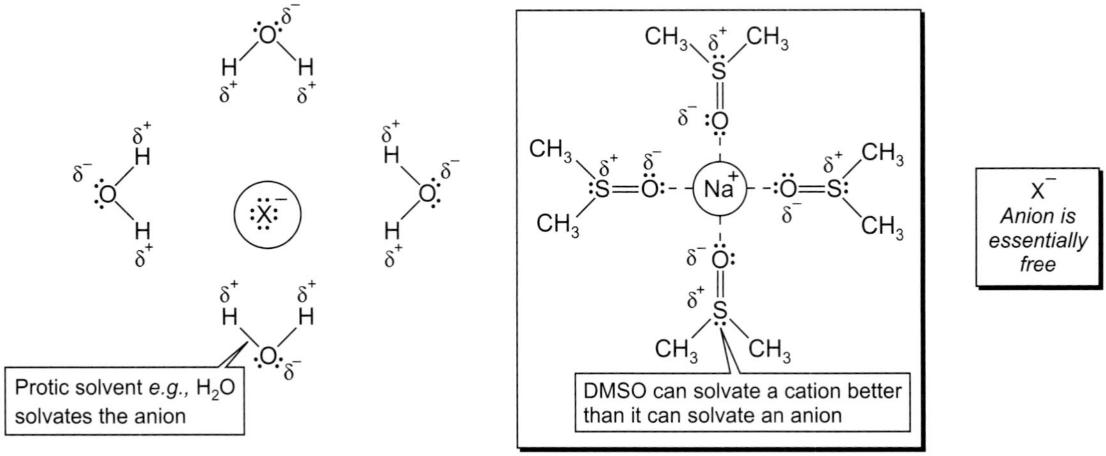

DMSO is a polar aprotic solvent and dissolves many organic compounds and salts. Nucleophiles are less solvated and are more free to react.

SCHEME 5.13

Table 5.2: Some common nucleophiles

Strong nucleophiles	*Weak nucleophiles*
SH^-	F^-
I^-	H_2O
$(CH_3CH_2)_2NH$	CH_3OH
$^-:C{\equiv}N$	
$(CH_3CH_2)_3N$	
OH^-	
CH_3O^-	
Moderate nucleophiles: Br^-, NH_3, Cl^-, OAc^-	

is heated in methanol, then benzyl methyl ether is formed (Scheme 5.13*a*). On adding bromine to the reaction, the reaction rate does not change but, now benzyl bromide is formed. It is an S_N1 reaction, therefore, the rate does not depend on nucleophile. Bromide ion is a better nucleophile

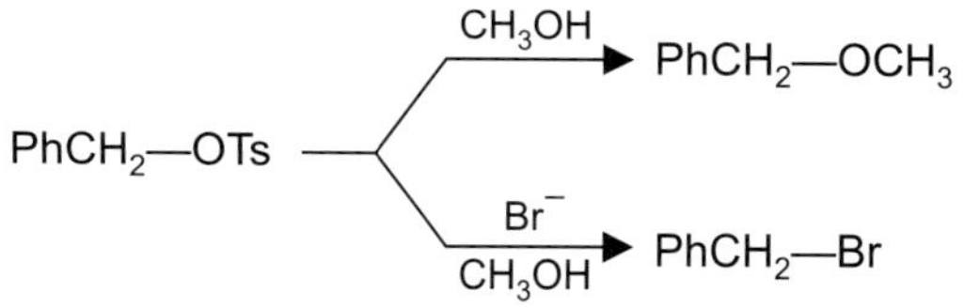

SCHEME 5.13a

than methanol and in the presence of this ion the product is benzyl bromide and not benzyl methyl ether.

Several aprotic solvents*e.g.*, dimethylsulphoxide (DMSO) and dimethylformamide (DMF) (Scheme 5.14) are used. An aprotic solvent is not a hydrogen bond donor since it does not have a hydrogen attached to an oxygen or to a nitrogen. Thus they do not solvate anions to any appreciable extent. Consequently with anionic nucleophiles, reaction rates are far greater in the polar aprotic solvents. There is lack of stabilization of these nucleophiles by hydrogen bonding, these are therefore, almost "naked" to be highly reactive as nucleophiles. Fluoride ion, therefore, is a better nucleophile in DMSO than in water during an S_N2 reaction.

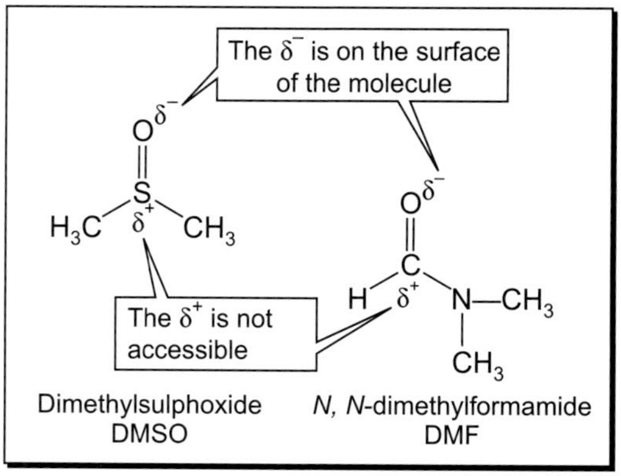

SCHEME 5.14

A polar aprotic solvent dissolves ionic compounds and it solvates cations, the way similar to protic solvents by orienting the negative end of its dipole around the cation (Scheme 5.14). It is however, unable to solvate the anion by H-bonding. Moreover, methyl groups in the case of DMSO shield the S which is the positive end of the dipole, this prevents the solvation of the anion, and same is the case with DMF.

Consider the S_N2 reaction on methyl iodide by a Br^- and an uncharged nucleophile Me_3N (Scheme 5.15) in an alcohol. In the reaction of Br^- the nucleophile is solvated by hydrogen

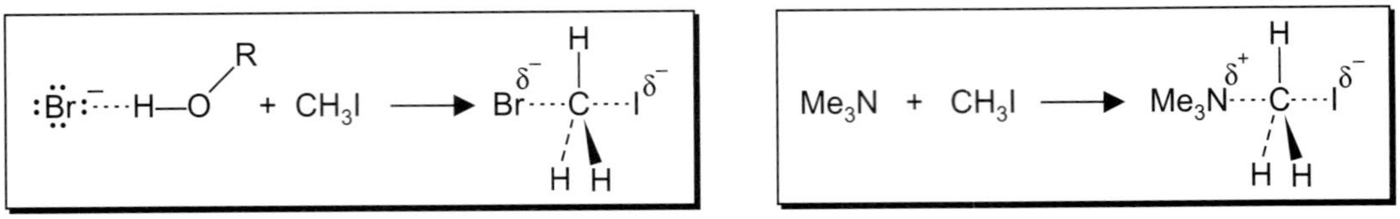

Slower reaction in a polar protic solvent, the nucleophile is solvated and not free.

Faster reaction in a polar protic solvent, both +ve and –ve charges are stabilized.

SCHEME 5.15

bonding. Thus the rate of an S_N2 reaction involving a negative nucleophile is slower in a more polar solvent. The transition state, however, behaves as a much larger ion and is thus much less stabilized by hydrogen bonding. In the case of neutral reactants a passage to the transition state involves the generation of opposite charges. Consequently the transition state is stabilized more than the reactants as the solvent is changed to more polar one. Thus the rate of an S_N2 reaction involving a neutral nucleophile is faster in a more polar solvent.

EXERCISE 5.4

(a) Explain whether the reaction of 1-chloropropane with NH_3 would be faster in $CH_3OH/80\%\ H_2O$ or in $CH_3OH\ 40\%/60\%\ H_2O$. (b) Whether the reaction of 1 chloropropane with CN^- will be faster in EtOH or DMSO.

ANSWER. *(a) This is an S_N2 reaction and the reactants are neutral, the transition state,* $CH_3—CH_2—CH_2Cl + NH_3 \longrightarrow CH_3CH_2\underset{\overset{\vdots}{NH_3^{\delta+}}}{\overset{\overset{Cl^{\delta-}}{\vdots}}{C}}H_2$ *is polar. The reaction will be faster in a more polar solvent. Since water is more polar than CH_3OH, so the reaction will be faster in 20% CH_3OH/80% H_2O.*

(b) Reaction is faster in DMSO.

(E) Leaving Groups

A good leaving group is the one which becomes a stable ion after its departure. As most leaving groups leave as a negative ion, the good leaving groups are those ions which stabilize this negative charge most effectively. The weak bases do this best, thus the best leaving groups are weak bases. In an S_N2 reaction the leaving group begins to gain negative charge as the transition state is reached. The more the negative charge is stabilized, the lower is the energy of the transition state; this lowers the energy of activation and thereby increases the rate of reaction.

The weak bases are the best leaving groups since weak bases are stable bases, since these can readily hold on to electrons which were earlier shared with a proton. Moreover, a weak base is not bonded as strongly to the carbon as would be the case with a strong base and consequently a weaker bond is more easily broken. When one considers S_N2 reactions on the similar reactants but with different leaving groups (Scheme 5.15*a*) one finds that I^- is the best leaving group while F^- is the worst. This is in keeping with the above discussion. One may also remember that larger atoms can better stabilize their negative charge.

Reactant	Leaving group	Relative rates of reaction
$HO^- + CH_3CH_2I \rightarrow CH_3CH_2OH$ +	I^-	30,000
$HO^- + CH_3CH_2Br \rightarrow CH_3CH_2OH$ +	Br^-	10,000
$HO^- + CH_3CH_2Cl \rightarrow CH_3CH_2OH$ +	Cl^-	200
$HO^- + CH_3CH_2F \rightarrow CH_3CH_2OH$ +	F^-	1

$I^- < Br^- < Cl^- < F^-$

Weakest base, most stable base, best leaving group

Strongest base, least stable base, worst leaving group

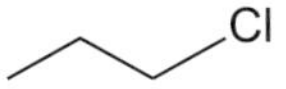

(I)

(II)

SCHEME 5.15a

An alkyl chloride or bromide reacts with NaI in acetone under S_N2 mechanism due to the presence of a strong nucleophile I^- and an aprotic solvent acetone. NaCl/NaBr formed is not soluble in acetone and therefore, the appearance of a precipitate in a test for alkyl chloride or bromide. As expected the compound (II, Scheme 5.15*a*) gives a precipitate immediately since bromide ion is a better leaving group than chloride ion.

Sulphonic acids, R SO_2OH are similar to sulphuric acid in acidity and the sulphonate ion RSO_3^- is a very good leaving group. Alkyl benzenesulphonates, alkyl *p*-toluenesulphonates are therefore, very good substrates in S_N2 reactions (see Scheme 5.4).

The triflate ion ($CF_3SO_3^-$) is one of the best leaving groups known, it is the anion of CF_3SO_3H which is a strong acid much stronger than sulphuric acid.

Recall that an alcohol can be converted into an alkyl halide on treatment with a hydrogen halide (see Schemes 5.3 and 5.5). A hydrogen halide protonates the alcohol as well as provides the nucleophile for the reaction. Other alternative could be the conversion of an alcohol into its sulphonate ester followed by an S_N2 substitution employing a halide nucleophile (Scheme 5.15*b*).

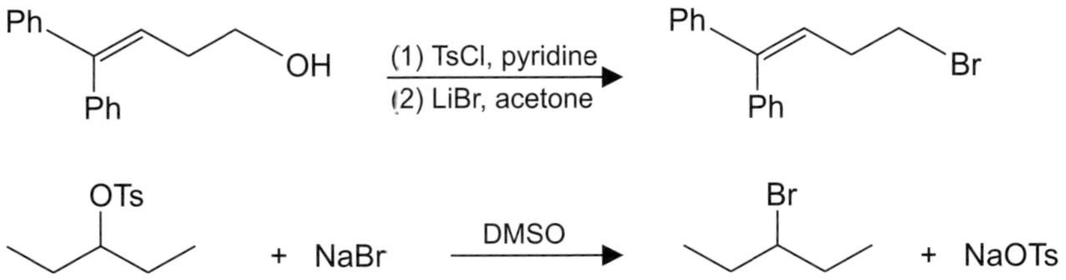

SCHEME 5.15b

The sulphonate ester method requires two steps for the conversion of an alcohol into an alkyl chloride. A one step method employs the use of a reagent thionyl chloride ($SOCl_2$) or a phosphorus trihalide *e.g.*, PBr_3. These reagents convert the OH group of an alcohol into good leaving groups chlorosulphite and bromophosphite groups respectively (Scheme 5.15*c*). These reagents replace the H of the alcohol group which makes the oxygen a weaker base and a better leaving group. Chloro-and bromoalkanes are generally used as reaction intermediates and since alcohols are widely available, the role of $SOCl_2$ and PBr_3 for their one step conversion into alkyl halides is synthetically very important. The mechanism may be S_N1 or S_N2 depending

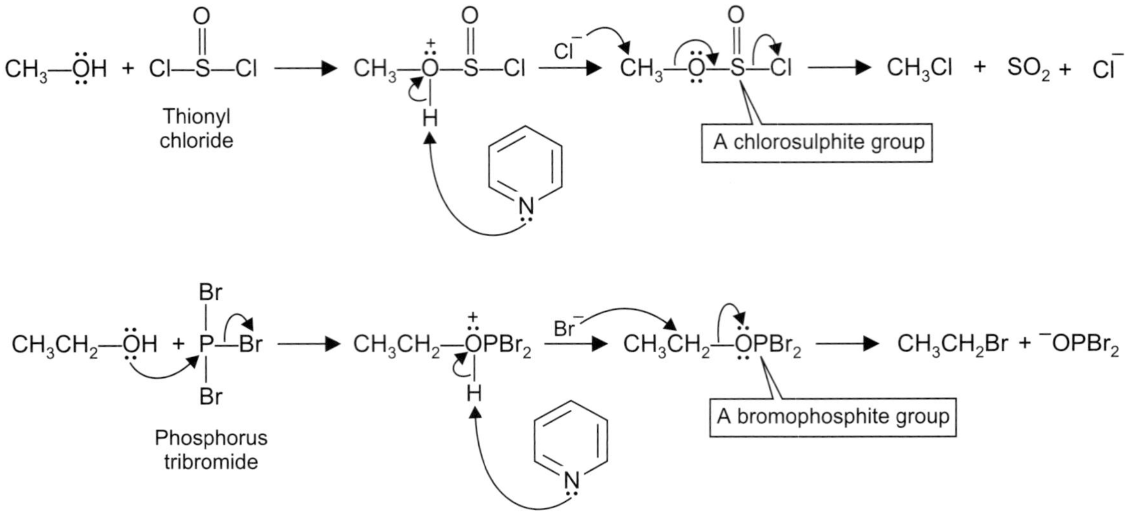

SCHEME 5.15c

on the structure of the compound. Pyridine is often used as a solvent to avoid the build up of HCl or HBr.

Role of triphenylphosphine and bromine in the conversion of alcohols into their bromides is more selective and occurs under milder conditions (see Scheme 7.18).

EXERCISE 5.5

Explain the stereochemical outcome of the reactions of the same chiral alcohol in reactions (I and II, Scheme 5.15d).

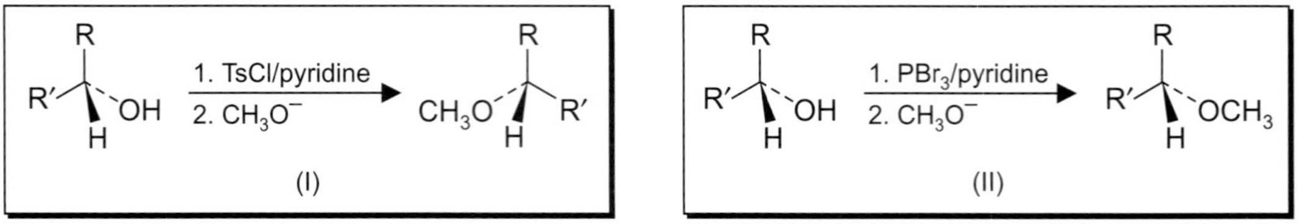

SCHEME 5.15d

ANSWER. *In reaction (I) only one S_N2 reaction is involved i.e., attack of CH_3O^- on the alkyl tosylate formed in the first step, thus the ether has the opposite configuration to that of alcohol. In (II) two S_N2 reactions are involved first is the attack of Br^- on bromophosphite followed by attack of CH_3O^- on the resulting alkyl halide, thus the parent configuration is retained.*

EXERCISE 5.6

What synthetic strategy one can use to obtain a high yield of 2-butanol from 2-bromobutane?

ANSWER. *One cannot use $\bar{O}H$ with a secondary reactant since it being a strong base, more of elimination will occur. One therefore, must modify the reagent suitably and replacing the H of $\bar{O}H$ by acetyl group (CH_3CO) will decrease the basicity in the acetate ion (resonance stabilization). One thus uses acetate anion in place of $\bar{O}H$ ion as the nucleophile in the S_N2 reaction using an aprotic solvent like DMSO. The acetate ester is formed in an S_N2 reaction in high yield which is then hydrolyzed $\bar{O}H/H_2O$ in a separate step (Scheme 5.15e).*

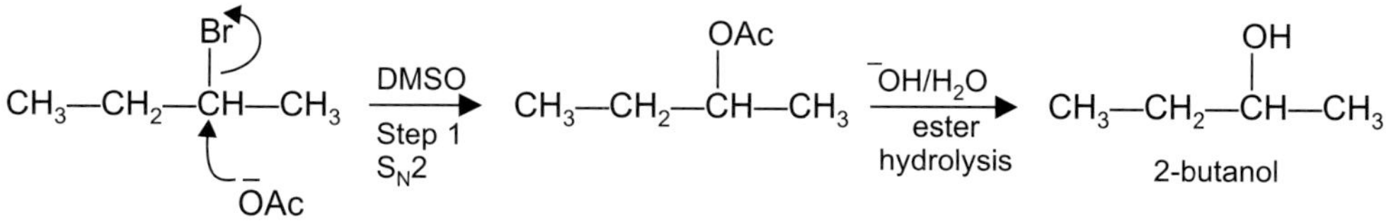

SCHEME 5.15e

(F) Electrophilic and Nucleophilic Catalysis

S_N2 reaction rates can be increased both by electrophilic and nucleophilic catalysis. A hydroxyl group can come off from an alcohol easily by adding a strong acid in a process termed electrophilic catalysis. In the case of alkyl halides, silver ion and mercury (II) ions are suitable, as these form strong bonds with halide ions. In case, when the nucleophile is weak and the leaving group is poor, reaction rates can be enhanced by the addition of iodide ion as a nucleophilic catalyst. Thus the S_N2 displacement (I, Scheme 5.16) is successfully catalysed since iodide ion is both a stronger nucleophile compared with pyridine and a better leaving group than acetate as shown (II, Scheme 5.16).

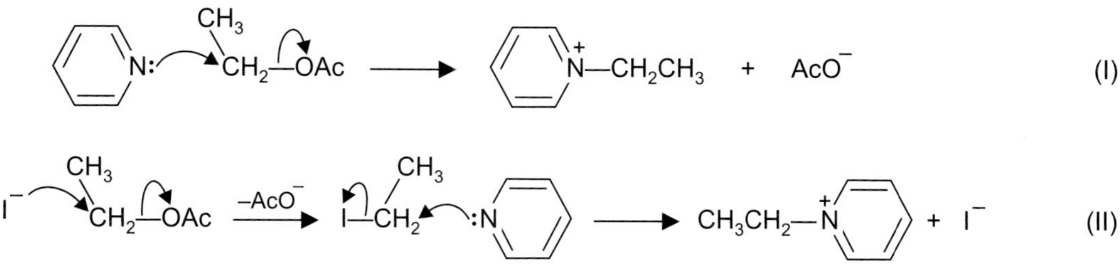

SCHEME 5.16

(G) The Role of Crown Ethers

Nucleophilic reactivity is increased in the presence of a suitable crown ether (see Scheme 1.20). The crown ether [18]-crown-6, *e.g.*, coordinates very effectively with potassium ions. Salts like KF, KCN and CH_3COOK which are otherwise insoluble in non-polar solvents like benzene, in the presence of [18]-crown-6 dissolve in it. Thus, the situation in the organic phase is the presence of relatively unsolvated anions to carry out nucleophilic substitution on an organic substrate. For example, KF which is both insoluble in benzene and unreactive to organic halides, brings about an efficient nucleophilic displacement on a halide in the presence of [18]-crown-6 (Scheme 5.17). The role of a crown ether here is of a catalyst which brings the anion into the organic phase, and the process is termed phase transfer catalysis.

$$C_6H_5CH_2Cl + K^+F^- \xrightarrow[\text{acetonitrile}]{\text{[18]-crown-6}} \underset{(100\%)}{C_6H_5CH_2F} + K^+Cl^-$$

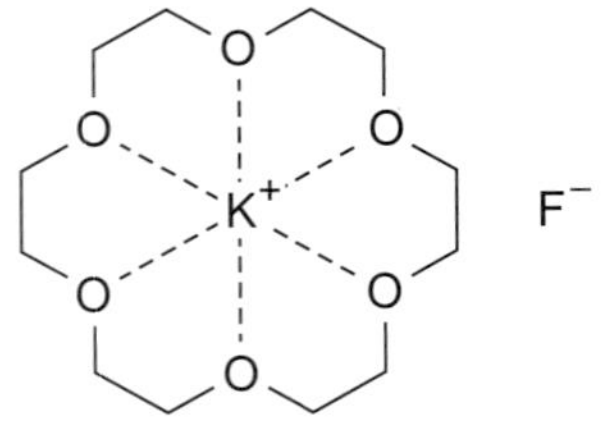

A crown ether complex of potassium ion (soluble in organic solvents)

SCHEME 5.17

(H) S_N2 Reactions with Allylic, Vinylic and Other Systems

The allylic systems are prone to react by S_N1 mechanism since these form delocalized allylic carbocations easily (see Schemes 2.11 and 2.18). However, the synthetic utility of these reactions is limited due to the allylic shift of the double bond. It is possible to have suitable reaction conditions under which allylic bromides react cleanly (without rearrangement) by way of the S_N2 mechanism. Thus allyl bromide undergoes bimolecular substitution about 40 times faster than *n*-propyl bromide. In the case of allylic system, the transition state receives resonance stabilization through conjugation with the *p* orbitals of the *pi* bond, (Scheme 5.18). The electronic structure of this transition state resembles the structure of the allyl anion. The stabilization of the transition state via the conjugation with the *p* orbital which is momentarily generated on the reacting carbon atom lowers the activation energy of the system, increasing the reaction rate.

Haloalkenes, in which the halogen is directly attached to the unsaturated carbon and phenyl halides display exceptionally low reactivity either by S_N1 or S_N2 mechanism. Thus while *n*-propyl chloride ($CH_3CH_2CH_2Cl$) undergoes rapid substitution with potassium iodide in acetone, 1-chloropropene ($CH_3CH{=}CHCl$) is inert. Simple alkenyl halides *e.g.*, vinyl chloride ($CH_2{=}CHCl$) also do not form carbocations readily. This is due to the increased strength of the vinyl halogen bond (see Scheme 2.15) in vinylic and phenyl halides. Moreover, the electrons of the double bond of benzene ring repel the approach of the nucleophile from the backside to be unreactive by S_N2 mechanism.

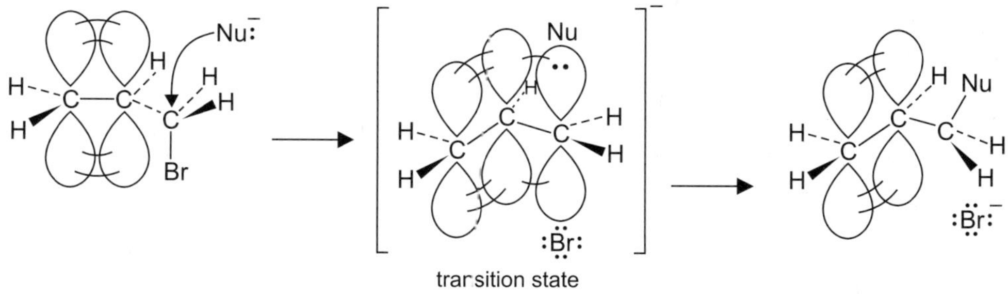

S_N2 reaction on allyl bromide

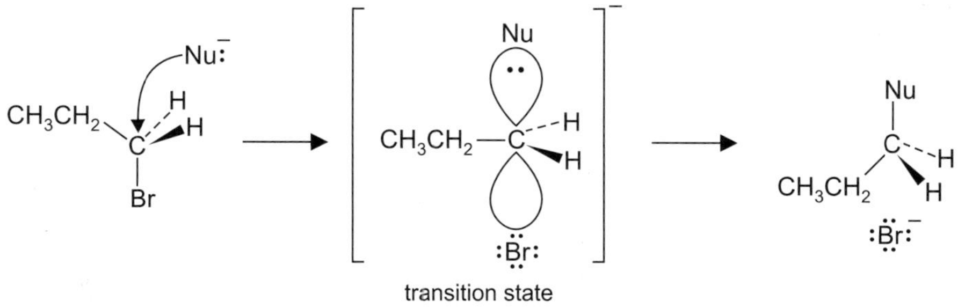

S_N2 reaction on *n*-propyl bromide

SCHEME 5.18

Thus S_N2 attack at sp^2 carbon does not occur and as said the failure is for the reasons that since the C—Br bond is in the plane of the ring the nucleophile shall have to be in the benzene ring to invert the carbon atom in an impossible way (Scheme 5.18*a*).

S_N1 type of solvolysis leading to vinylic cations can however, be carried out on suitable substrates provided one has an efficient leaving group like triflate anion OTf^- ($CF_3SO_2O^-$, which is a super leaving group) and the vinylic group contains electron releasing groups (Scheme 5.18*a*).

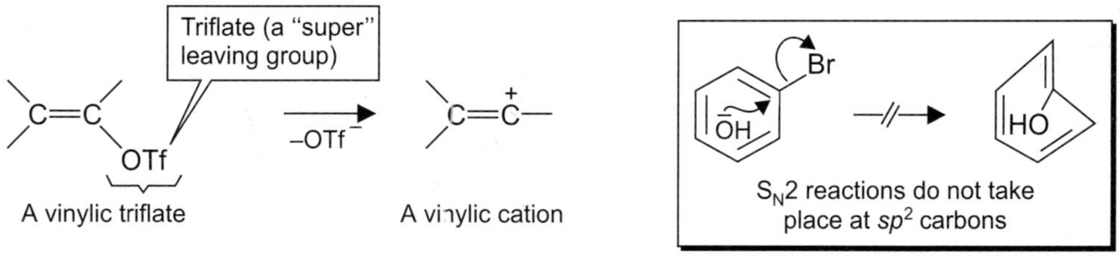

SCHEME 5.18a

Organocuparate Reagents and their Reactivity toward Alkenyl and Aryl Bromides

Grignards reagents and organolithium compounds (see Chapter 7) are reactive toward carbonyl compounds but do not react with organic halides to give new carbon-carbon bonds. The Gilman reagents-dialkylcuparates provide a useful *alternative. These reagents react with alkyl (primary or secondary) bromides and iodides by the displacement of the halide ion in a process known as coupling reaction. Significantly Gilman reagents also react with alkenyl and aryl bromides and iodides (Scheme 5.18b) to show that reactions are not S_N2 reactions. It seems that electron transfer process are involved.*

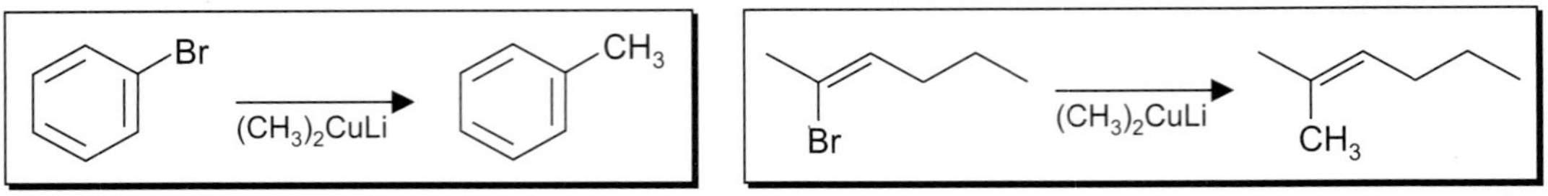

SCHEME 5.18b

5.3 SUBSTITUTION BY IONIZATION — S_N1 MECHANISM

(A) Mechanism — S_N1 Process

This reaction involves an ionization mechanism which proceeds by rate determining heterolytic dissociation of the substrate to a tricoordinate carbocation (Scheme 5.18*c*) and a leaving group. This dissociation is followed by a rapid combination of the highly electrophilic carbocation with a Lewis acid present in the medium. A free energy diagram for the S_N1 reaction of *tert*-butyl chloride and water is shown (Scheme 4.18).

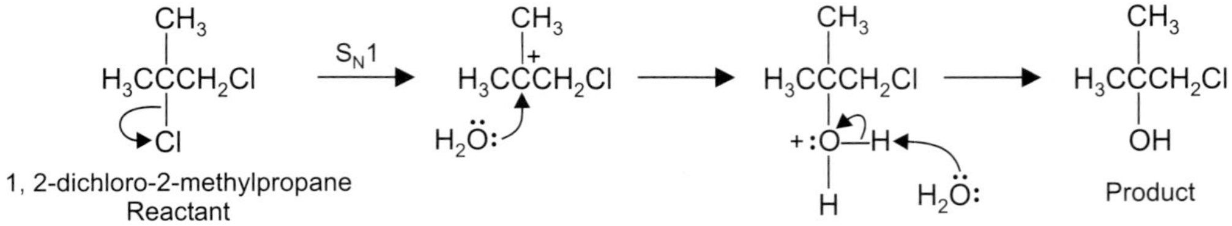

SCHEME 5.18c

Out of two chlorine atoms in the reactant (Scheme 5.18*c*), only one is replaced by the hydroxy group because it is an S_N1 reaction and only the chlorine atom bonded to a tertiary carbon is replaced. Water then acts as a nucleophile.

(B) The Effect of the Structure of the Substrate on S_N1 Reactions

The major structural features necessary for a substrate to undergo S_N1 substitution is the presence of substituents which stabilize the carbocation derived from it. These are the substituents which have *+I* and *+M* effects. Among alkyl halides, for all practical purposes only tertiary halides react by S_N1 mechanism. A tertiary carbocation being stabilized by three electron releasing group (see Scheme 4.28). Allylic and benzylic halides can also react by an S_N1 mechanism since these substrates can form relatively stable carbocations (Scheme 5.18*d*).

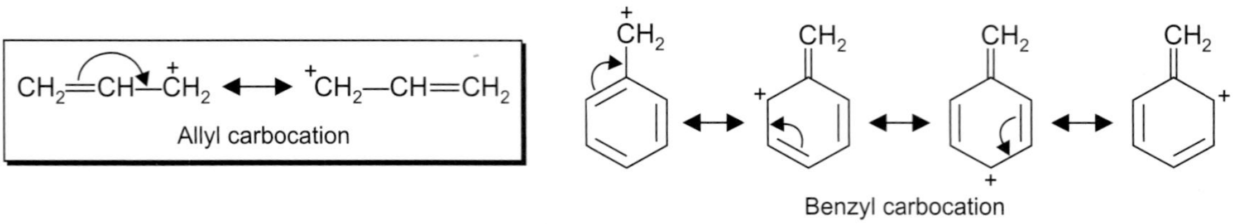

SCHEME 5.18d

The benzyl cation is almost as stable as an allyl cation and in benzyl cation as well, the positive charge is delocalized around the ring (Scheme 5.18*d*). Recall that an unsymmetrical allylic cation is attacked by the nucleophile at both ends and the regioselectivity is determined by steric hindrance (see Scheme 2.18). In the case of a benzyl cation the attack is almost always in the side chain.

Another point of interest is that due to high instability of a phenyl cation, phenyl halides show a retarded S_N1 activity, but for when the leaving group is very good as nitrogen in a diazonium salt (see Scheme 4.32*a*).

Methyl chloromethyl ether, with ether group of +*M* types is hydrolysed fast in water. The intermediate formed after heterolytic dissociation being the delocalized carbocation (oxonium ion, Scheme 5.19).

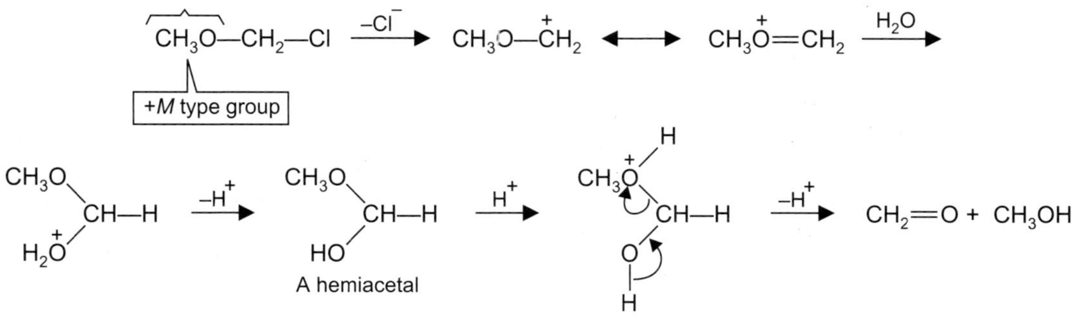

SCHEME 5.19

B strain and *I strain* effects are observed in many substrates and these effect the rate of S_N1 reactions. When *e.g.*, in a tertiary alkyl halide (R_3Cl), one or more *R* groups are highly branched like *e.g.*, *t*-butyl, the ionization is facilitated by relief of steric crowding in going from the tetrahedral ground state to the transition state for ionization and finally to the carbocation. This strain which may be present in a suitable substrate is called *B strain*[3] (for details see Scheme 1.36).

Similarly *I strain* effects S_N1 solvolysis rates in some cyclic compounds (see Scheme 1.36).

Nucleophilic S_N1 substitutions at bridgeheads is impossible or very slow, since a rigid bridged system prevents rehybridization to a planar sp^2 carbon. However, when such a structure is flexible the S_N1 reactions can take place, since now the bridgehead carbocation can be generated (see Scheme 4.32).

(C) Nucleophilicity

The rates of S_N1 reactions are independent of the nature or concentration of the nucleophile, since it does not participate in the rate determining step.

(D) Solvent Effects on S_N1 Reactions

The majority of the substrates in S_N1 reactions are neutral. In these cases, the more polar the solvent, the faster the reaction. There is a greater charge in the transition state than in the starting substrate (see Scheme 4.18). A polar protic solvent *e.g.*, will largely increase the rate of ionization of an alkyl halide, since it can solvate cations and anions. The solvation stabilizes the transition state (leading to the intermediate carbocation and halide ion) more than it does the reactants, consequently the free energy of activation is lower.

(E) The Nature of the Leaving Group

In either S_N1 or S_N2 reaction, the leaving group begins to acquire a negative charge as the transition state develops. Thus the effect of the leaving group is the same as studied earlier for S_N2 mechanisms.

(F) Stereochemical Outcome of S_N1 Reactions

The reactions at a stereocenter of an optically active alkyl halide lead to recemization due to the intermediate formation of a planar, achiral carbocation (see Scheme 4.22). However, the enantiomers are not normally formed in equal quantities, the major enantiomer has the opposite configuration to that of the substrate, *i.e.*, inversion predominates over retention. The carbocation R^+ and the leaving group X^- exist for a while as an ion pair $R^+ X^-$, consequently X^- part of the ion pair shields the side of the carbon atom to which it was attached. The incoming group therefore has a more chance to attack the other side.

5.4 S_N1 *VERSUS* S_N2 REACTIONS

S_N1 mechanism operates with those reactants which can form relatively stable carbocations. The use of weak nucleophiles and highly ionizing solvents favour S_N1 mechanisms. Thus, in the case of solvolysis of tertiary halides in the presence of highly polar solvents, S_N1 mechanisms are significant. During solvolysis, the nucleophile is weak, it being a neutral molecule (the solvent) rather than an anion.

S_N2 mechanisms operate with relatively unhindered alkyl halides, (tertiary halides do not react by S_N2 mechanism) by using strong nucleophiles, a polar aprotic solvent and a high concentration of the nucleophile.

Some Preparatively Useful S_N2 and S_N1 Reactions—Alkylations

- *Hydride nucleophiles from $LiAlH_4$ or $NaBH_4$ convert primary and secondary substrates into hydrocarbons via S_N2 mechanism (see Scheme 5.4).*
- *Carboxylate ion is the nucleophile during methyl ester formation from carboxylic acids from diazomethane (see Scheme 5.2a).*
- *O-Nucleophiles $Na^+ OR^-$ bring about Williamson ether synthesis –S_N2 reaction. Since the base is strong, a secondary substrate may lead mostly to elimination. Generally two routes may be available for the synthesis of the same product (see Scheme 5.2e). S_N1 conditions of solvolysis with substrates other than primary give good yield of ethers ($RL + HOR' \longrightarrow ROR'$).*
- *Carboxylate ion, a weak base gives good yield of esters when the substrate is secondary under S_N2 conditions (see Scheme 5.15e).*
- *Hal-nucleophiles convert alcohols in to alkyl halides $ROH + HX \longrightarrow RX$. The reaction takes an S_N1 path unless the substrate is primary. Reaction with HCl requires $ZnCl_2$ for primary and secondary alcohols (See Schemes 5.3 and 5.5). Conversion of the OH group of an alcohol into a good leaving group and subsequent reaction with a halide ion X^- gives an alkyl halide (S_N2 conditions, see Scheme 5.4)*
- *A one step procedure to convert an alcohol into an alkyl halide is reaction with thionyl chloride, phosphorus tribromide and phosphorus triiodide (see Scheme 5.15c).*

- *Amines can be prepared by using nucleophiles like NH_3, an amine, an azide ion or phthalimide anion (see Schemes 5.12a–5.12e).*
- *In subsequent chapters one will find the preparatively important S_N2 reactions e.g., ketones, esters and nitriles can be alkylated (alkylation at the α-carbon) to form a new C—C bond (Scheme 5.19a).*

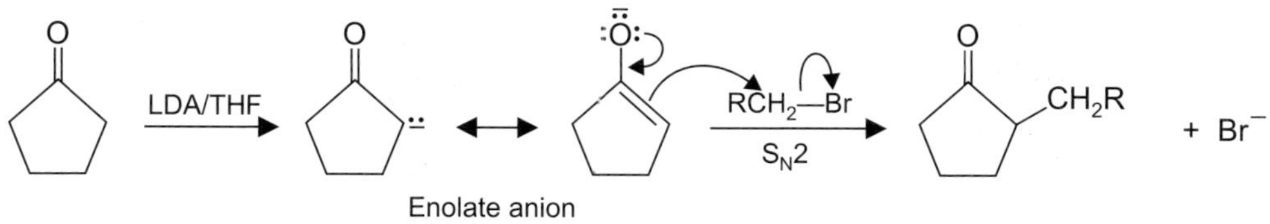

SCHEME 5.19a

The first step is the removal of a proton from the α-carbon of a carbonyl compound with a strong base LDA. Since the alkylation step is an S_N2 reaction it works best with primary alkyl halides.

Alkylation can also be carried out at the α-carbon via an enamine which is a very good carbon nucleophile like an enolate. Here again the alkylation step is an S_N2 reaction which therefore, works very well with methyl halides or primary alkyl halides (Scheme 5.19b).

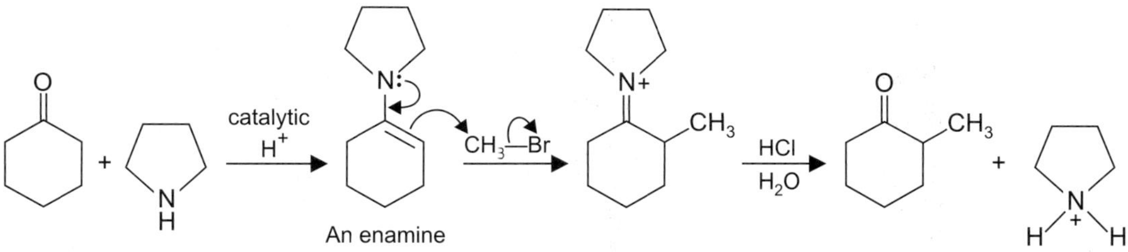

SCHEME 5.19b

The role of carbon nucleophiles CN^- and $RC{\equiv}C^-$ is explained (see Schemes 3.11b and 5.2) and these provide useful methods to make C—C bonds.

Phosphorus nucleophiles e.g., triphenylphosphine react via an S_N2 reaction with an appropriate alkyl halide to yield a phosphonium ylide needed during Wittig and related reactions. A proton on the carbon adjacent to the positively charged phosphorus atom is sufficiently acidic ($pK_a = 35$) for removal with a strong base like butyl lithium (Scheme 5.19c).

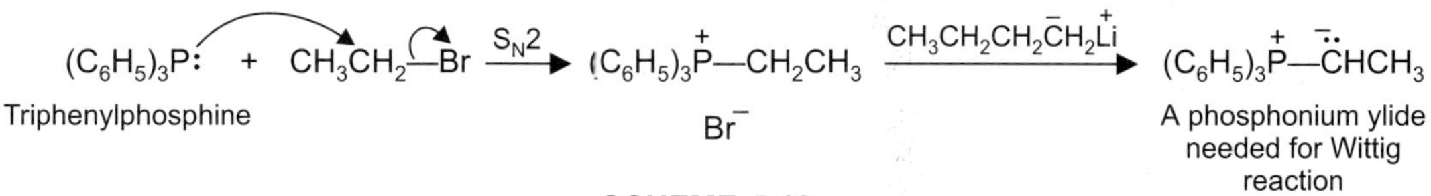

SCHEME 5.19c

Lastly mention may be made of sulphur nucleophiles HS^- or RS^- which are weak bases, but good nucleophiles and give good yields with primary and secondary substrates (Scheme 5.19 d)

$$(R—L + HS^- \longrightarrow R—S—H).$$

SCHEME 5.19d

5.5 OTHER ALIPHATIC SUBSTITUTION PATHWAYS

(A) The S_N2' and S_N1' Reactions

Allylic substrates undergo substitution with the migration of the double bond and is called S_N2' reaction. (see Schemes 2.20 and 2.21). These reactions normally take place when the S_N2 mechanism is sterically hindered. In the substrate (Scheme 5.20), the S_N2 process is suppressed by steric factors and S_N1 process is suppressed by using a reagent of high nucleophilic activity and a solvent of low ionizing power. The substrate, α, α-dimethylallyl chloride undergoes an S_N2' process with sodium thiophenoxide in ethanol to give the rearranged product in high yield. S_N1' reactions are discussed in Scheme 2.18.

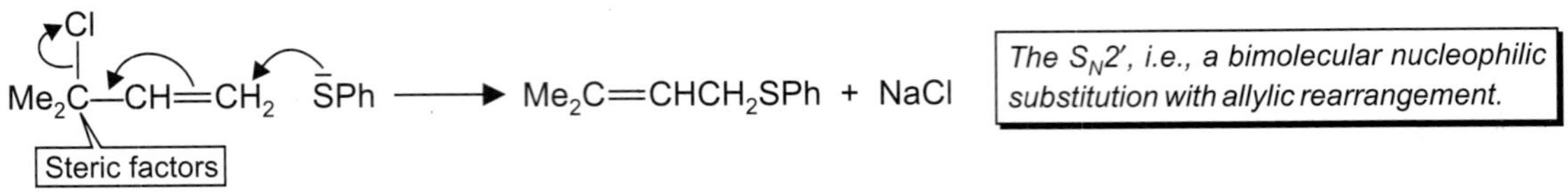

SCHEME 5.20

(B) The S_E2 Reaction

A carbon can undergo substitution electrophilic, bimolecular termed as S_E2 reaction when it is attached to strongly electropositive atoms *i.e.*, metals. Thus the bromination of an organo-mercurial occurs with retention of configuration at carbon (Scheme 5.21). This is in contrast to the inversion of configuration which is typical of the S_N2 reaction.

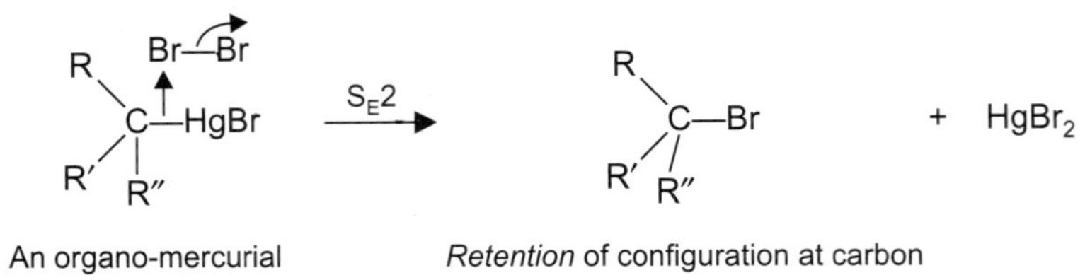

SCHEME 5.21

(C) The S_N1 Process

An alcohol may be converted into an alkyl chloride from the reaction of the alcohol with thionyl chloride either in ether solution or in the presence of pyridine. In the presence of ether, the chlorosulphite first formed decomposes *via* an intimate ion pair (Scheme 5.22). The ion pair collapses to give the alkyl chloride with the retention of configuration.

In the presence of pyridine, a chloride ion is liberated which then brings about an S_N2 displacement on the chlorosulphite and now the configuration at the sterocenter gets inverted (Scheme 5.23). In a solvent like diethyl ether most of HCl in lost as a gas and the chloride for substitution comes from chlorosulphite. In the presence of pyridine, pyridinium hydrochloride brings about an S_N2 reaction. This reaction shows as to how reaction conditions can change the stereochemical outcome of a reaction.

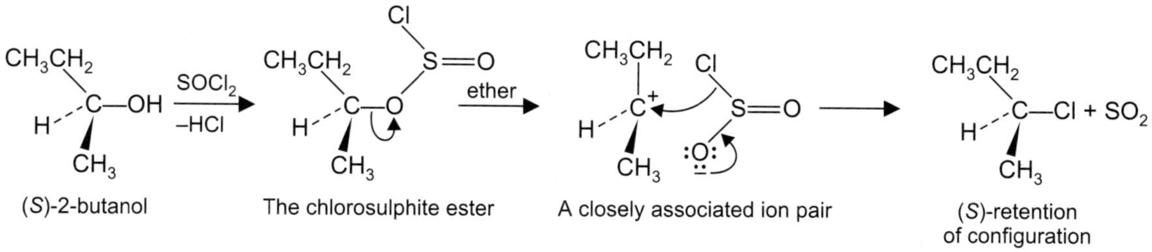

An S_N1 reaction (substitution, nucleophilic, internal) with retention of configuration

SCHEME 5.22

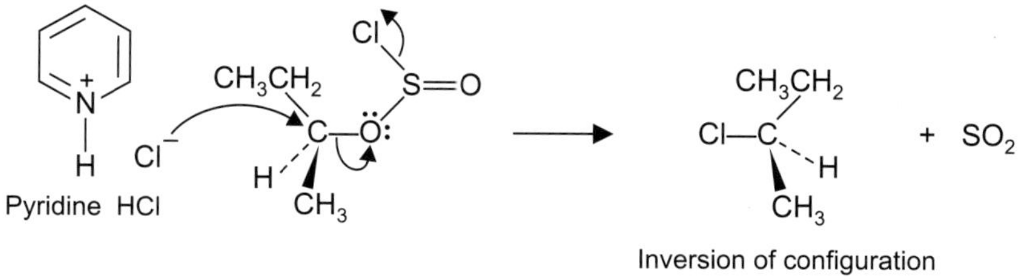

A normal S_N2 reaction

SCHEME 5.23

(D) Nucleophilic Substitution at a Bridgehead

When the leaving group *e.g.*, at [2.2.1] bridgehead is such that it cannot function as a nucleophile *i.e.*, to come back once it has gone, then a nucleophilic substitution can occur. Thus (I, Scheme 5.24) undergoes substitution with chlorobenzene as the nucleophile to give (II).

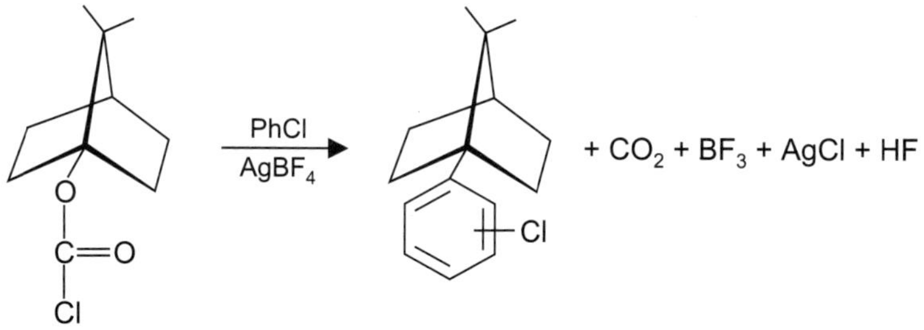

(I) (II)

SCHEME 5.24

(E) The SET Mechanism

In some nucleophilic reactions where S_N1 mechanism is highly probable, it has been proved by esr determination of intermediate that free radicals are infact involved. In such a mechanism a carbocation is a good electron acceptor and the nucleophile a good electron donor. The mechanisms are named SET (single electron transfer) mechanisms, *e.g.*, the reaction between trimethyl cation and *t*-butoxide ion (Scheme 5.25).

$$Ph_3\overset{+}{C} + t\text{-}BuO^- \longrightarrow Ph_3C\cdot + t\text{-}BuO\cdot \longrightarrow t\text{-}BuOCPh_3$$

SCHEME 5.25

5.6 THE ROLE OF ION PAIRS

The S_N1 mechanism involves the intermediate formation of a carbocation, it is planar and the nucleophile, therefore, should attack from either side of the plane with equal facility to give complete racemization. This result has been found in most first order substitutions (see Scheme 4.22), in many others, one finds that there is inversion to the extent of 2–20%. Thus, in some S_N1 reactions some of the product does not come from the carbocation but rather from ion pairs (Scheme 5.26).

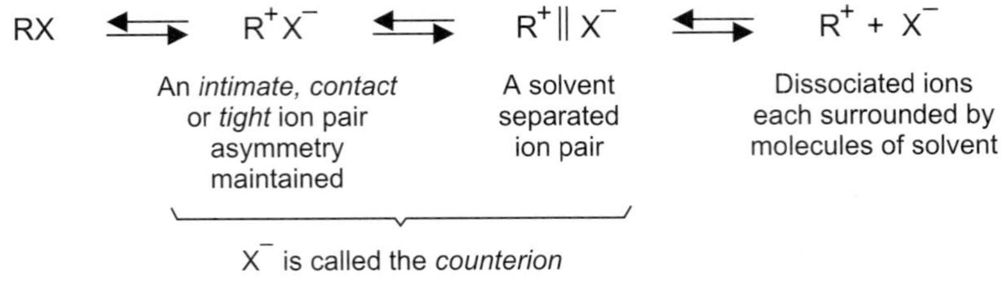

SCHEME 5.26

The reaction products can be formed by attack by the nucleophile at any stage. In case the products arise from tight ion pair one expects inversion of configuration. This is because in the case of tight ion pair R^+ is not completely free, there is still significant bonding between R^+ and X^- and asymmetry of the substrate is reasonably maintained to a considerable extent. As a consequence in tight ion pair, X^- solvates the cation on the side of its departure and therefore, it can only get solvation from the solvent molecules from the opposite side. This process will lead to inversion of configuration. In case the product arise from the solvent separated ion pair extensive racemization will result since here the stereochemistry is not maintained as tightly as in intimate ion pair. The product from the dissociated cation R^+ will give complete racemization. Thus in summary the following points may be noted:

Consider the reaction of optically pure 1-phenylethyl chloride of (*S*) configuration with water in aqueous acetone (Scheme 5.27) which as a whole shows first order kinetics (S_N1 reaction). The 2% net inversion of configuration of the substrate is due to the involvement of an ion pair mechanism (Scheme 5.28). In the initially formed intimate ion pair the carbocationic part is solvated on the side opposite to the leaving group and the product from these species will have inverted configuration. Thus many S_N1 reactions involve the formation of ion pairs.

- An ion pair is a closely associated cation and an anion and behaves as a single unit.
- In a tight ion pair, the individual ions retain their stereochemical configuration.
- A solvent separated ion pair has its ions separated by solvent molecules and the ions may and may not retain their stereochemical configuration.

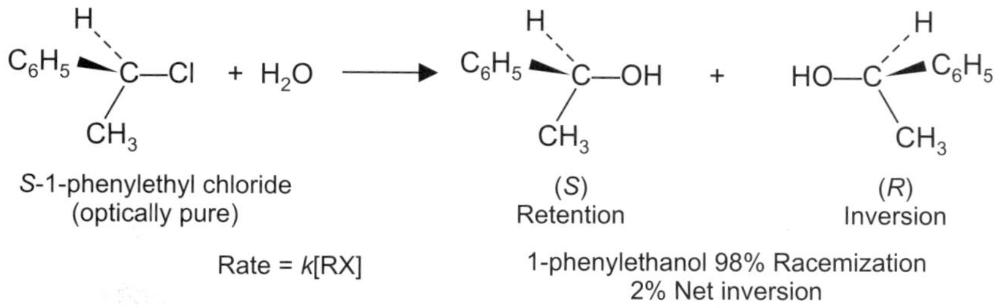

SCHEME 5.27

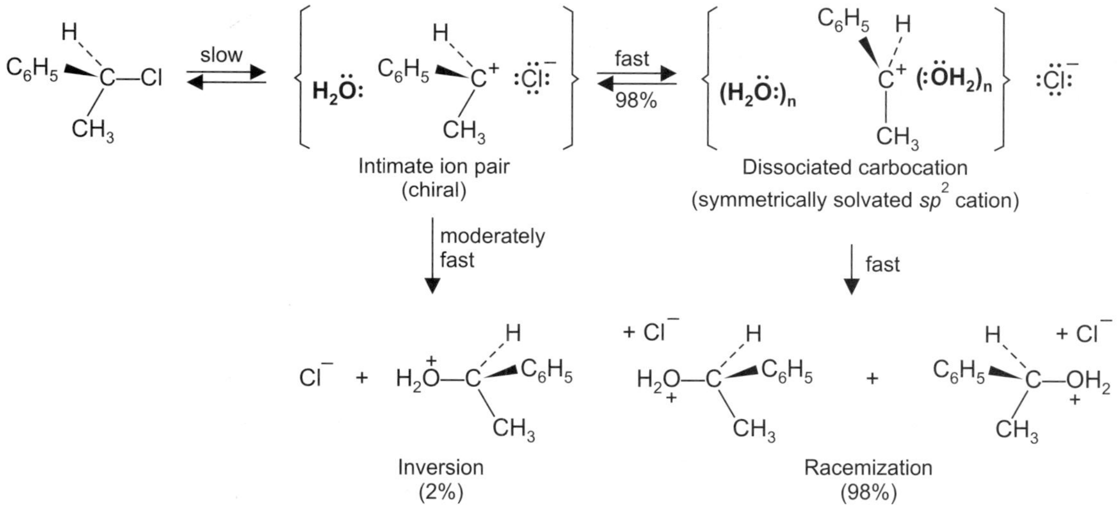

SCHEME 5.28

Other evidence also proves the formation of ion pairs. When 2-octyl brosylate (labelled at the sulphone oxygen with ^{18}O) was subject to solvolysis, the unreacted brosylate isolated from various stages of solvolysis had the ^{18}O considerably (but not completely) scrambled (Scheme 5.29).

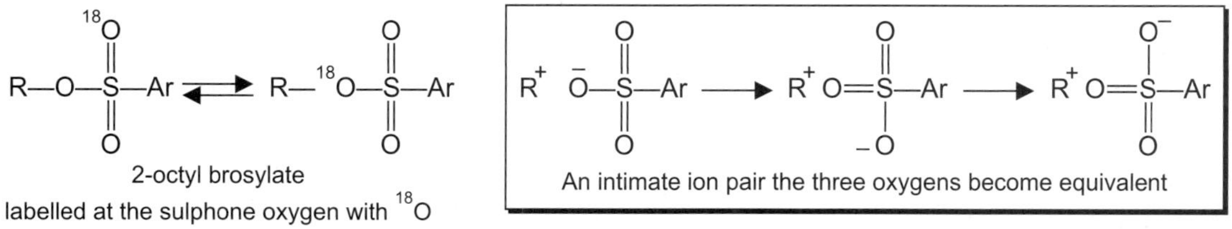

SCHEME 5.29

In this case the reaction involves the formation of an intimate ion pair, where the three oxygen atoms become equivalent. It may be remembered that an ion pair can recombine to afford the original substrate (an internal return, Scheme 5.29).

The addition of an inert salt like $NaClO_4$ or LiBr at a very low concentration to the solution of some substrate undergoing solvolysis leads to an initial large rate increase. This rate, subsequently falls off to become a normal ionic strength effect and the effect is called special salt effect. In the absence of the added salt, the solvolysis proceeds with intimate ion pair formation with considerable return to the starting substrate. The ClO_4^- or Br^- exchanges with the leaving group to prevent this return. Consequently the amount of the solvent separated ion pair that could have returned to the substrate is reduced. This leads to an overall increase in the reaction rate.

5.7 NEIGHBOURING GROUP PARTICIPATION AND NONCLASSICAL CARBOCATIONS

One has seen that nucleophilic substitution take place with racemization or with inversion of configuration. However, in several cases such reactions occur with overall retention of configuration. One factor which leads to retention of configuration during a nucleophilic substitution is neighbouring group participation. The neighbouring group is an electron rich substituent (Z:, Scheme 5.30) present in the proper position for backside attack *i.e.*, *anti* attack to the leaving group (X). The process infact is a two step process. In the first step (Scheme 5.30)

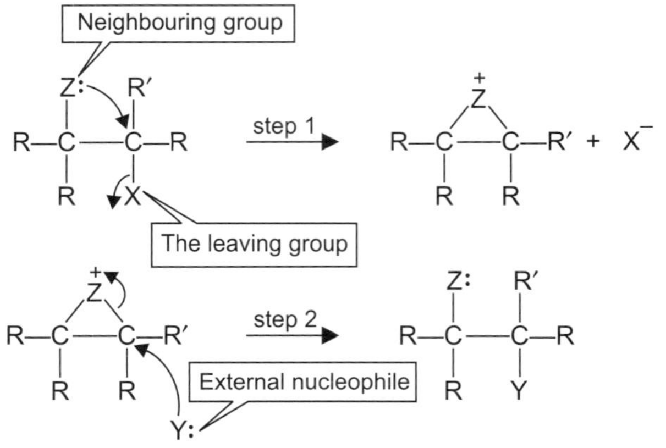

The neighbouring group mechanism two S_N2 substitutions, each brings about an inversion the configuration at a stereogenic carbon. Configuration is retained and not inverted or racemized.

SCHEME 5.30

the neighbouring group (acting as an internal nucleophile) attacks carbon at the reaction center (S_N2 attack) and the leaving group is lost to give a bridged intermediate. This is then attacked in the second step by an external nucleophile (Y:, another S_N2 attack) and the internal nucleophile goes back to where it came from, the net result is two consecutive S_N2 reactions leading to retention of configuration at the reacting carbon.

A graphic example of neighbouring group participation is found in the conversion of 2-bromopropanoic acid into lactic acid (Scheme 5.31). In the presence of concentrated sodium hydroxide, (*S*)-2-bromopropanoic acid (shown as its ion, Scheme 5.32) undergoes a bimolecular displacement with inversion of configuration as expected from the normal S_N2 reaction. The same reaction when carried out in the presence of Ag_2O and a low concentration of hydroxide ion, however, occurs with retention of configuration (Scheme 5.32). The reaction now involves two steps, in the first step the carboxylate group acts as a neighbouring group to displace bromide ion *via* backside attack on the stereocenter. The silver ion here acts as an electrophilic catalyst and aids the removal of bromine. In the second step, the α-lactone is attacked by a water molecule. Both the steps involve an inversion of configuration on the attacked carbon. Thus, the net result of two inversions in two steps is an overall retention of configuration.

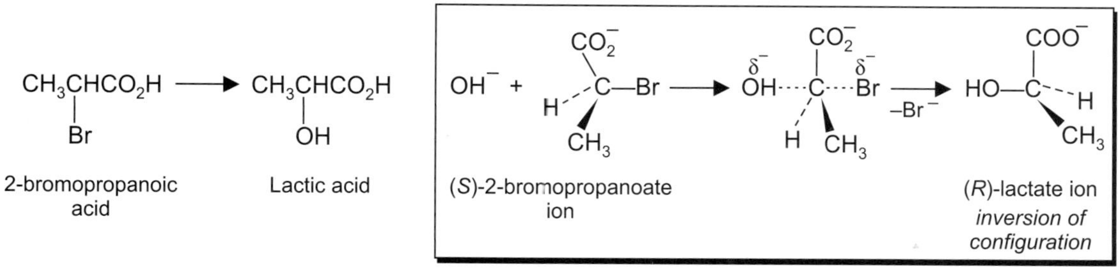

The normal stereochemical result for an S_N2 reaction

SCHEME 5.31

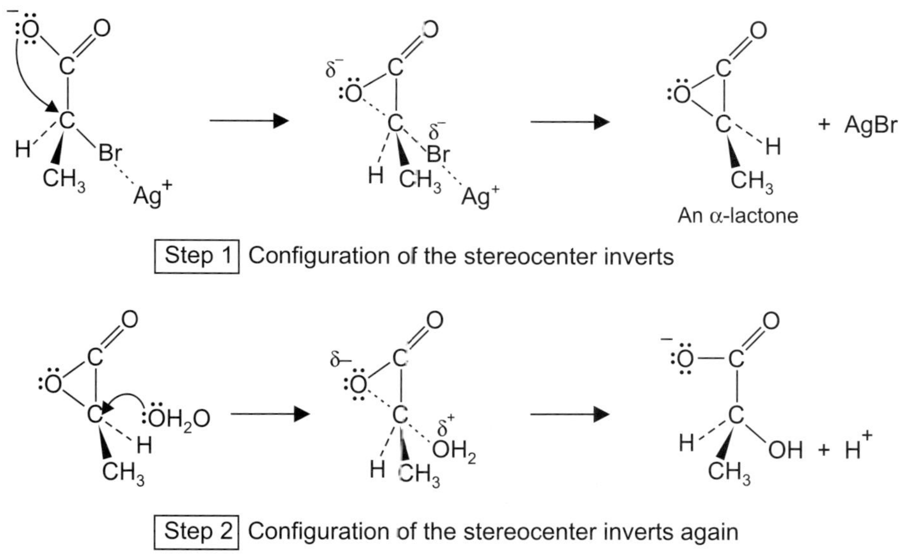

SCHEME 5.32

When the neighbouring group participation operates during the rate determining step of a reaction, the reaction rate is usually markedly increased. This effect is then termed anchimeric assistance. Sulphur atoms act as powerful nucleophiles and the participation of sulphur as a neighbouring group is common. On reaction with water both hexyl chloride (I, Scheme 5.33) and 2-chloroethyl-ethylsulphide (II) give their corresponding alcohols.

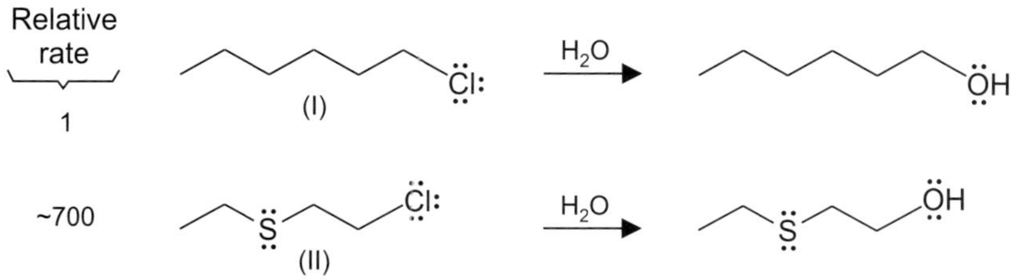

SCHEME 5.33

However, the rate of reaction of sulphur containing compound (II) is much greater than that of the alkyl chloride. The reaction in the case of (I) is a simple S_N2 displacement of chloride with water, while in the case of sulphide, it is the sulphur atom, which displaces the leaving group and acts as a neighbouring group. The intramolecular reaction (as expected) is much faster than the intermolecular reaction. The initial product from (II) is an episulphonium ion which is then opened by second S_N2 displacement (now intermolecular) to give the product (Scheme 5.34).

Thus a neighbouring group participation is the intramolecular involvement of one functional group in the reaction at other functional group. Anchimeric assistance is the increase in the reaction rate in the rate determining step of the reaction.

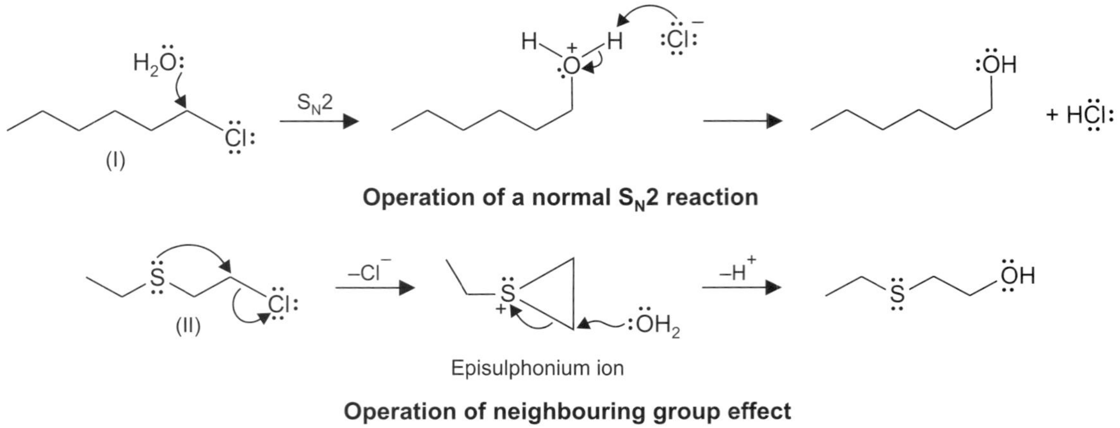

SCHEME 5.34

In a 1, 2-disubstituted cyclohexane derivative, for the neighbouring group participation to be operative the groups have to be *anti* to each other *i.e.*, diaxial as in (I, Scheme 5.35). A ring flip may be necessary to bring about such an arrangement of the groups. Consider the acetolysis of *cis* and *trans* isomers of 2-acetoxycyclohexyl tosylate (Scheme 5.36) which give the same product (I).

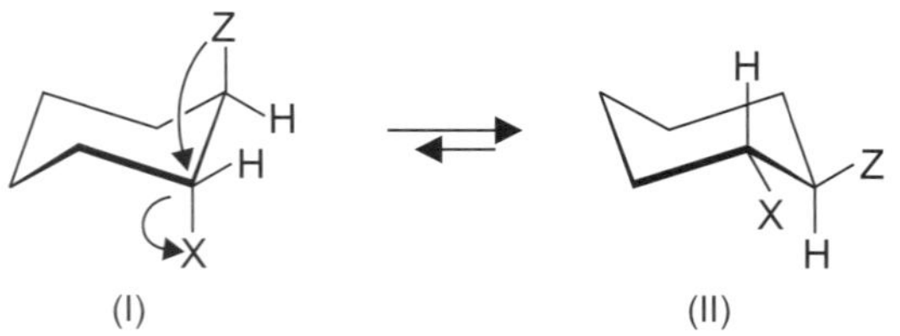

SCHEME 5.35

The *cis* isomer reacts *via* a direct S_N2 mechanism and the *trans* isomer reacts (about 700 times faster) *via* neighbouring group participation by involving an acetoxonium ion (A, Scheme 5.36). This acetoxonium ion (A, the resonance hybrid structure) from the *trans* isomer is, symmetrical achiral (Scheme 5.37) and can be attacked by the acetate ion at either of the two equivalent carbons shown by arrows. Thus, if one starts with an optically active *trans* isomer, the net result is the formation of a racemic mixture of diacetates.

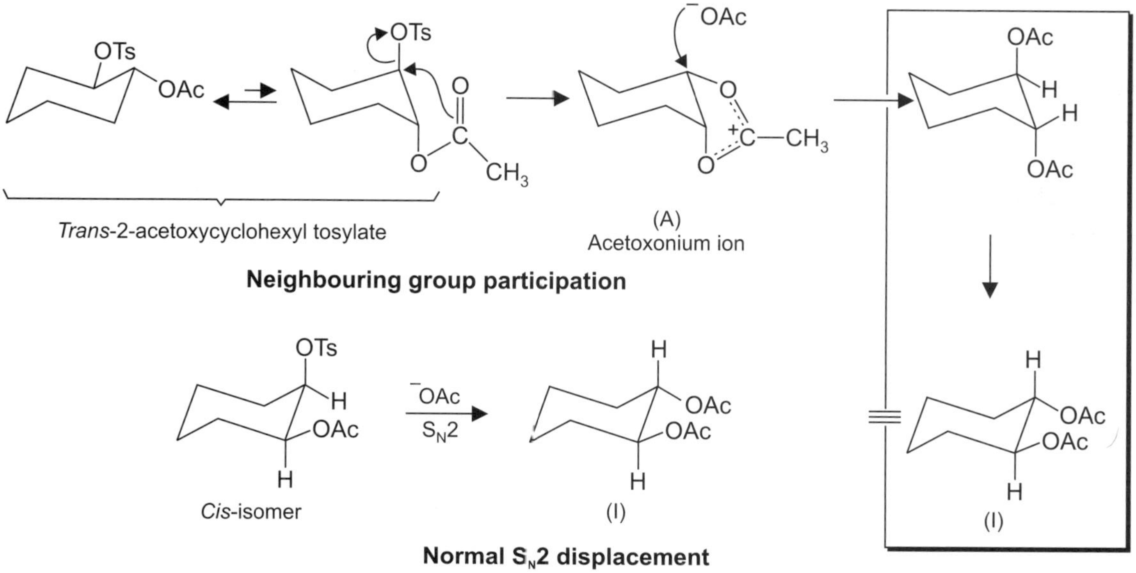

SCHEME 5.36

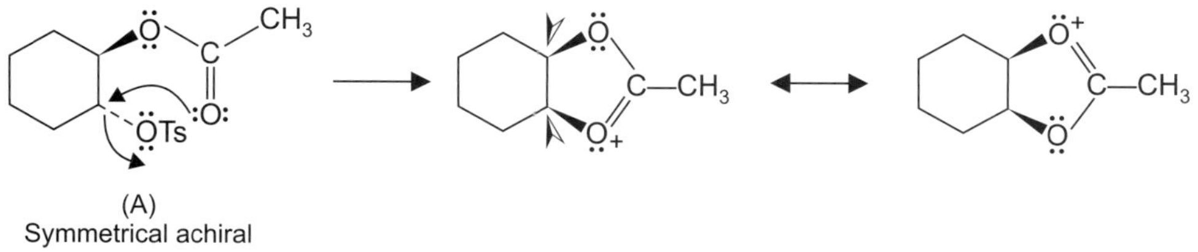

SCHEME 5.37

Among the norbornyl derivatives (on acetolysis) the *anti* tosylate (III, Scheme 5.38) reacts 10^{11} times faster than (I) while (II) has 10^4 times reactivity compared to (I).

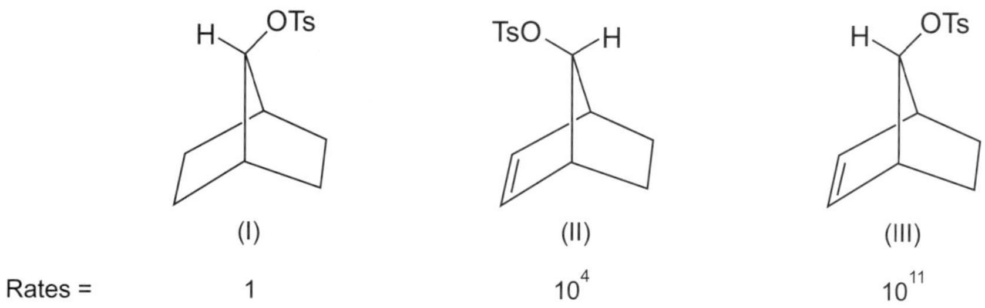

SCHEME 5.38

The fastest rate of acetolysis of *anti*-tosylate (III) compared to (I, Scheme 5.38) proves the removal of the tosyl group (the rate determining step) with strong anchimeric assistance by the double bond. The resulting non-classical carbocation *i.e.*, bridged ion can only react with acetate ion from the side opposite to the neighbouring group, with retention of configuration

(Scheme 5.39). In the *syn*-isomer (II, Scheme 5.38) the rate is slower because the double bond is not properly situated for participation. Thus this isomer dissociates without anchimeric assistance to give a homoallylic carbocation which rearranges to allylic carbocation (V, Scheme 5.40) and this reacts to give an acetate. The high reactivity (10^4 times) of (II) than (I) may be because of participation of σ electrons of two allylic 1, 6 and 4, 5 bonds.

Thus the bridged cation (Scheme 5.39) is an unusual situation which involves three-center two-electron bonding such species are called non-classical carbocations.

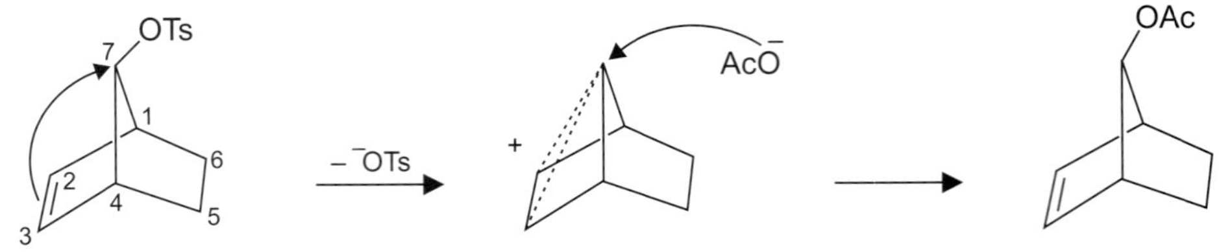

Anti-7-norbornenyl tosylate (III)

Bridged cation (Non-classical carbocation)

Anti-acetate

Involves three-centre two-electron bonding

SCHEME 5.39

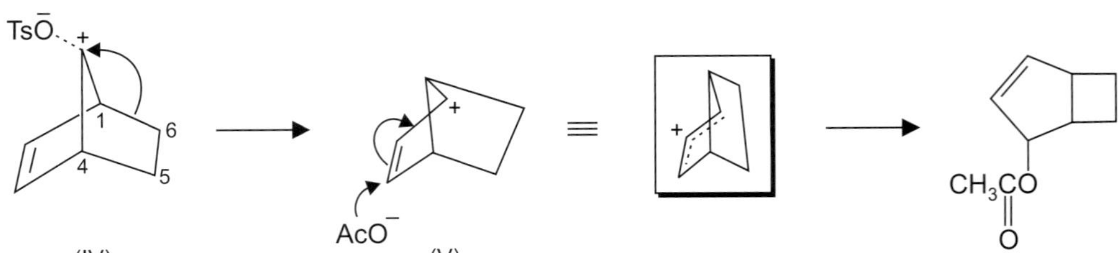

(IV) (V)

SCHEME 5.40

EXERCISE 5.7

A 2-thiosubstituted chlorocyclohexane reacts with aqueous solution of ethanol to give an alcohol and an ether due to presence of two nucleophiles water and alcohol. Explain why this rate of reaction is 70,000 times faster when the thio substituent is trans placed to the chloro substitutent?

ANSWER. *The thio substituent acts as an intramolecular nucleophilic catalyst and provides anchimeric assistance. It displaces the chloro substituent by the back side attack on the carbon with chloro substituent. Back-side attack requires both substituents to be diaxial. Subsequent attack by water or ethanol on the sulphonium ion is fast as the positively charged sulphur is a very good leaving group and cleavage of the three-membered ring releases strain (Scheme 5.41).*

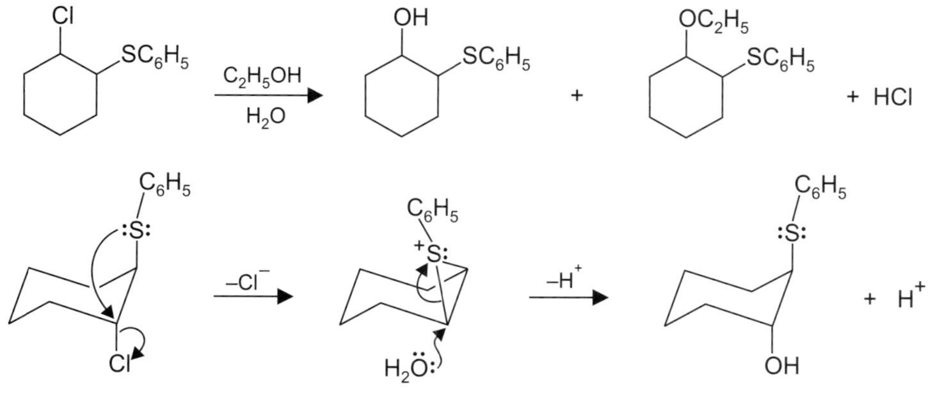

SCHEME 5.41

Evidence has been presented that C=C acts as a neighbouring group and that a non classical carbocation (a bridged cation) may be formed. Evidence is also available to show that a suitably located C—C in a substrate can also participate in the departure of the leaving group and non-classical carbocations may be involved.

During the acetolysis of *exo*- and *endo*- norbornyl tosylates (Scheme 5.42) it is found that (*a*), the solvolysis of *exo* isomer is 350 times faster than the *endo* isomer; (*b*), both the isomers give only the *exo* acetate; (*c*), and optically pure *exo*-tosylate gives 100% racemic product while an optically pure *endo* tosylate gives 93% racemic *exo*-acetate. These observations are explained:

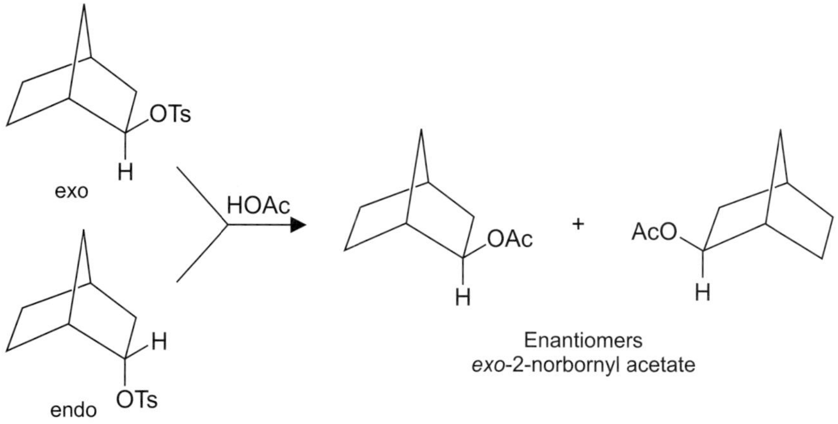

SCHEME 5.42

- In the *exo* isomer the 1, 6 σ bond is suitably located to act as a neighbouring group to lend anchimeric assistance *via* backside attack to give directly a non-classical carbocation (I, Scheme 5.43) which is more stable than the carbocation (II, Scheme 5.44) formed initially in the case of *endo*-isomer. The *endo*-isomer on the other hand first gives a carbocation (II, Scheme, 5.44) which subsequently forms the same non-classical carbocation.

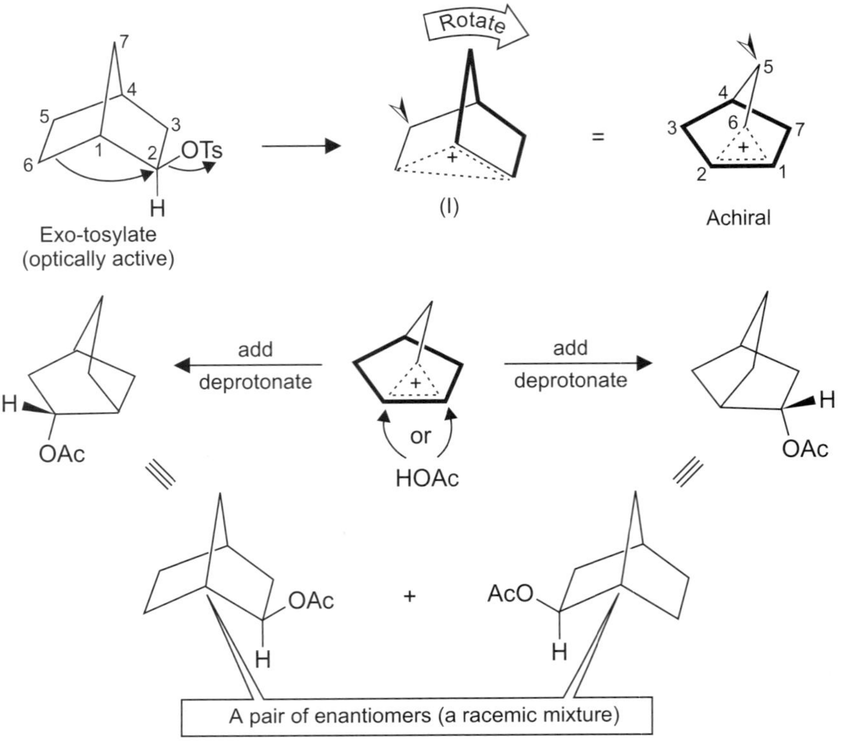

SCHEME 5.43

- The attack occurs from the *exo*-side due to the cage structure of the non-classical carbocation intermediate (I, Scheme 5.43). Moreover, the attack must occur from the direction opposite that of bridging interaction and this is *exo*-direction.
- Recall that a non-classical carbocation involves a three-center two-electron bonding. In a common case three carbon atoms are involved, two of which are bonded by a σ bond while the third is bonded to the other two by a two-electron three-center bond.
- The non-classical carbocation intermediate (I, Scheme 5.43) is achiral having a plane of symmetry passing through C-4, C-5, C-6 and midpoint of the C-1 and C-2 bond. The C-6 has two hydrogens and is pentacoordinate and is the bridging atom in the cation. Thus the attack at both C-1 and C-2 is equally likely which gives equal amounts of enantiomeric acetates—a racemic mixture.

In the case of *endo*-tosylate, till (II, Scheme 5.44) collapses to (I), it will give one enantiomer in excess to explain 93% formation of racemic acetate.

Bridging provides only stabilization. If other forms of stabilization are available, ions will then be open classical species. ^{13}C NMR spectroscopy is used to distinguish between equilibrating structures and bridged species. Thus 2-phenylnorbornyl cation (III, Scheme 5.44) has the classical structure. This benzylic cation is stabilized by π-electrons of the benzene ring and thus bridging is not involved. Firstly consider the resonance-stabilized carbocation (Scheme 5.45), the two carbons (shown by dots) which share the positive charge as expected are almost equivalent. In equilibrating ionic structures (two independent ions) such carbons differ by about 100 ppm.

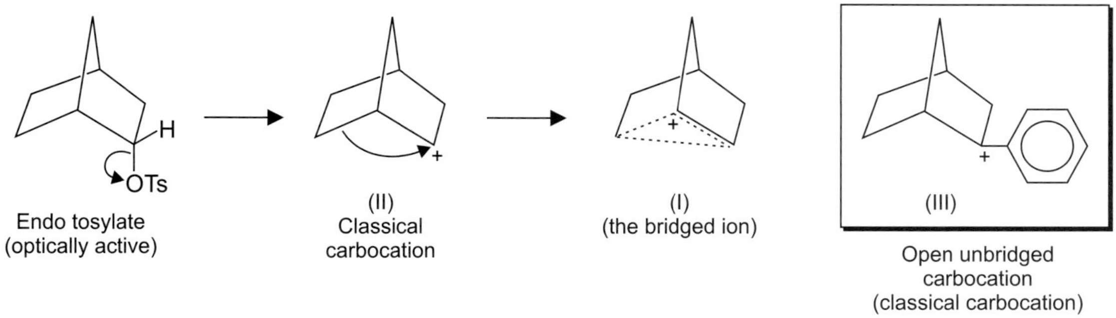

SCHEME 5.44

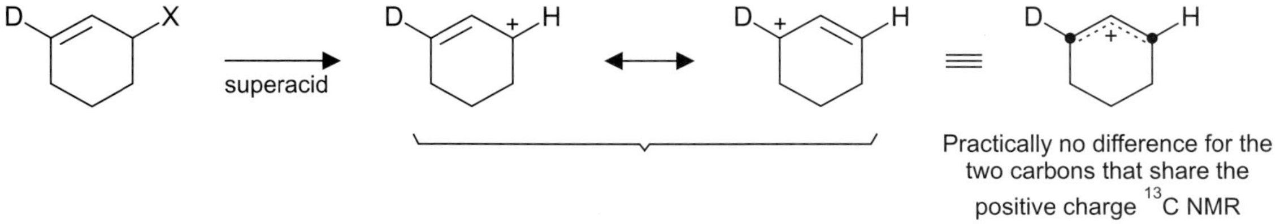

SCHEME 5.45

That 2-norbornyl cation is a bridged species has been shown by detecting it in a highly polar but non-nucleophilic solvent (super acid media, SbF_5-SO_2). The ^{13}C NMR showed very similar signals for both the deuterated as well as undeuterated positively charged carbons (Scheme 5.46). This evidence excludes the formation of an equilibrating pair of cations where the two boldly shown carbons should have displayed widely separated signals in ^{13}C NMR.

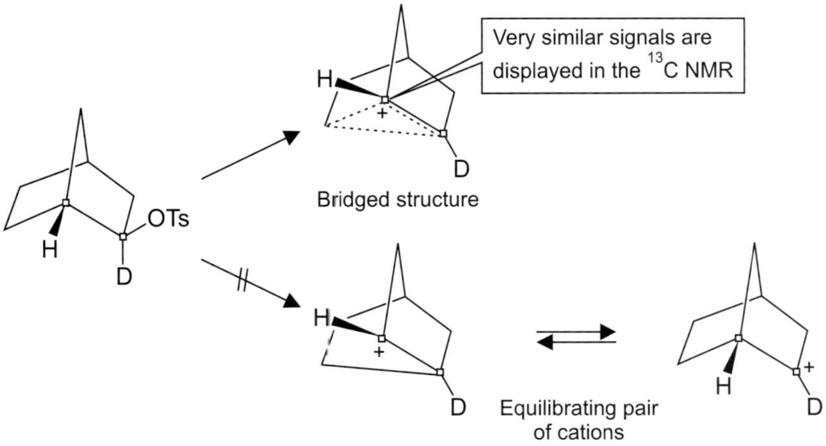

SCHEME 5.46

Certain properties of a cyclopropane ring are similar to that of an olefin. Cyclopropane rings in particular and cyclobutane ring generally display rate enhancements when these are suitably placed in a substrate. Thus cyclopropyl-methyl and cyclobutyl substrates undergo hydrolysis abnormally rapidly to yield the same products which include cyclopropyl methyl, cyclobutyl and homoallylic compounds (Scheme 5.47). Their formation is due to the intermediate formation of a non-classical cation (Scheme 5.47). In cyclic systems related to norbornyl system

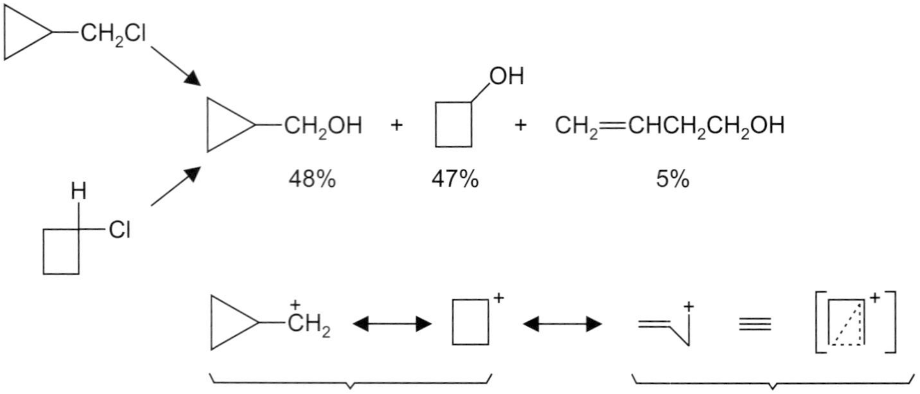

The formation of a common non-classical intermediate

SCHEME 5.47

the presence of a suitably placed cyclopropyl group acts as a neighbouring group. Thus (II, Scheme 5.48) reacts 10^{14} times faster than I. It has been suggested that in (II, Scheme 5.48) and other cyclopropyl derivatives which display rate enhancement, the developing *p* orbital of the carbocation is orthogonal to the participating bond of the cyclopropane ring.

(I) (II)

Ar = p-$NO_2C_6H_4$

SCHEME 5.48

5.8 NUCLEOPHILIC SUBSTITUTION AT SILICON

(For details see Scheme 7.36).

PROBLEMS

5.1. A factor which stabilizes an anion would be generally expected to reduce or enhance the rate of an nucleophilic substitution?

5.2. The free energies of activation for reaction of nucleophiles with CH_3I at 25°C in methanol and in DMF are given. How do you explain the relative nucleophilicities of the halide ions and thiocyanate ion?

Nu:$^-$	DMF	CH_3OH
Cl^-	16.9	25.0
Br^-	17.3	23.0
SCN^-	19.0	22.0
I^-	20.9	18.0

5.3. Why displacement of cyanide is never observed? Why azide and acetate ions are poor leaving groups?

5.4. Why an alcohol reacts with a halide ion only in the presence of a strong acid?

5.5. Why α-carbonyl substituted substrates like Br CH_2COCH_3 and $BrCH_2COO$ Et react more readily than the corresponding alkyl halides?

5.6. Why CH_3OCH_2Cl reacts with iodide ion in acetone several thousand times faster than CH_3Cl?

5.7. Why the inversion of configuration is much more during the solvolysis of $C_6H_{13}CH(CH_3)Cl$ than $C_6H_5CH\ (CH_3)Cl$?

5.8. Why the following allylic and benzylic halides react only by S_N1 process and not by S_N2 mechanism? Why both S_N1 and S_N2 mechanisms operate when R′=H or R=R′=H?

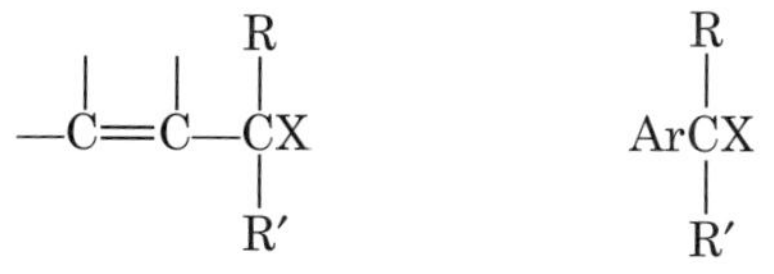

Allylic Benzylic

5.9. *Trans*-2-chlorocyclohexanol gives epoxycyclohexane in high yield on reaction with base however, the *cis*-isomer does not react this way. Explain.

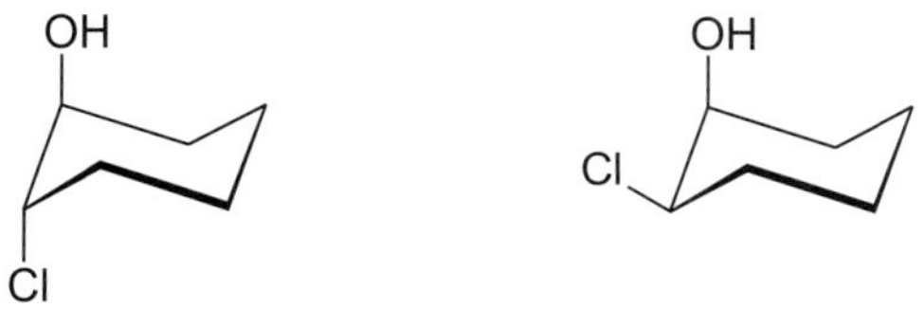

trans *cis*

5.10. Why the epoxide (I) reacts with acidic methanol to give a optically pure mixture of II and III (see under answer 5.10) and no racemization is observed?

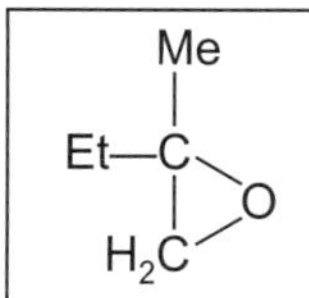

(I), (*R*)

5.11. Explain the outcome of the following reaction.

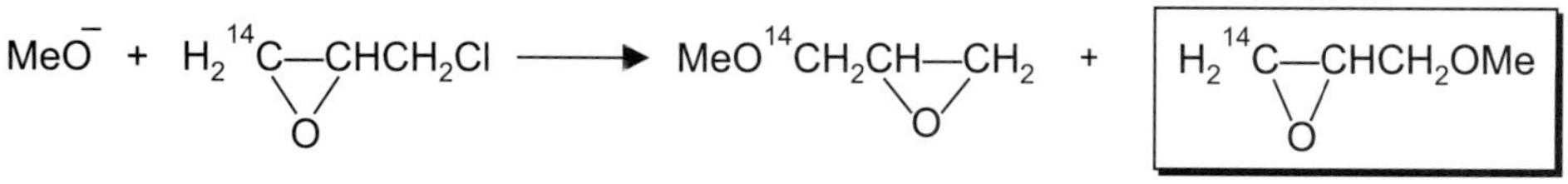

Not formed

5.12. Compound *I* reacts as shown. Explain.

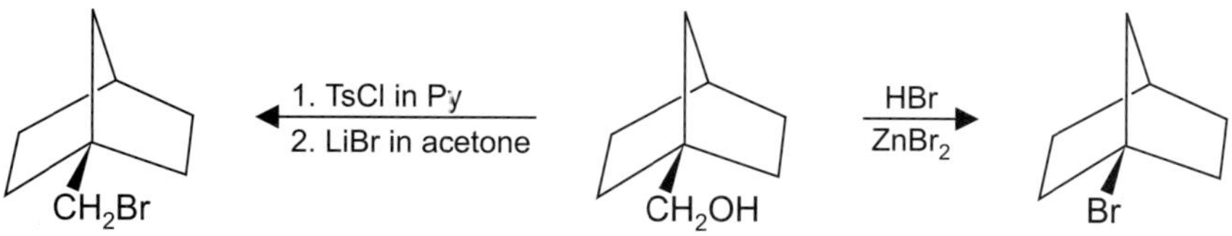

(II) (I) (III)

5.13. The following nucleophilic substitution reaction proceeds with a rearrangement. When the reactant is optically active the product is also optically active. Explain.

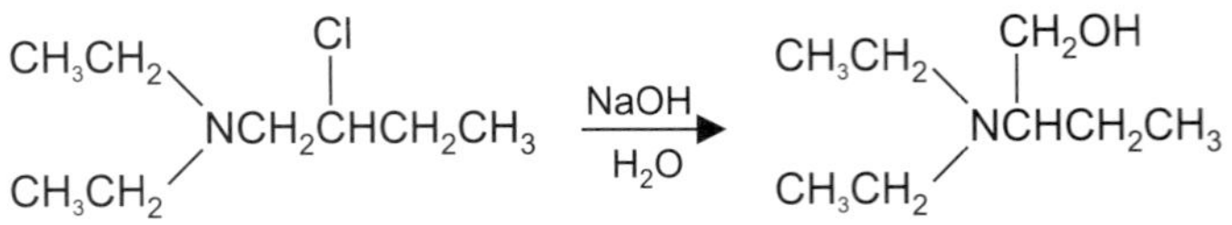

5.14. 2-Methyl but-3-en-2-ol reacts easily to yield almost exclusively one product through the unsymmetrical allylic carbocation intermediate. Explain.

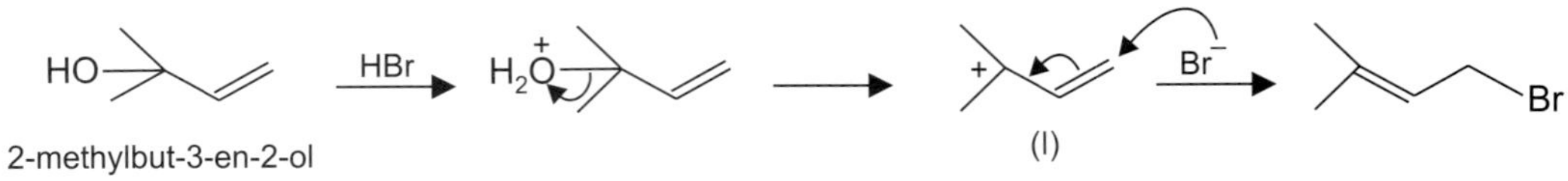

[**Hint:** Though the allylic carbocation (I) is unsymmetrical and resonance stabilized it is attacked by the nucleophile almost exclusively at the less substituted end predominantly. Compare with the explanation in Scheme 2.18.]

ANSWERS TO THE PROBLEMS

5.1. Reduce the rate.

5.2. Considering the halide ions, the smaller chloride is more hydrogen bonded in methanol compared to the larger halide ions. Generally nucleophiles (*e.g.*, SCN^-) with second row and larger atoms are less hydrogen bonded in hydroxylic solvents.

5.3. HCN is a weak acid ($pK_a = 10$). Similarly hydrazoic acid (HN_3) and acetic acid (CH_3COOH) are also weak acids (pK_a's of 5.8 and 4.8 respectively).

5.4. The hydroxide ion is a strong base and thus cannot be a leaving group. The acid protonates the hydroxyl group, and then the leaving group is water a much weaker base than a hydroxide ion.

5.5. In the case of carbonyl derivatives the transition states are stabilized by interaction of the *p* orbital on the carbon undergoing displacement with the adjacent *pi* system.

5.6. In case of the α-alkoxy compound the transition state (S_N2) is stabilized *via* the interaction of the *p* orbital on the carbon undergoing substitution with the (filled) oxygen 2*p* orbital.

5.7. It depends on the relative stability of the carbocation and subsequent ion-pair formation. In the case of phenyl containing substrate, the positive charge on the carbocation is delocalized on to the aromatic ring.

5.8. As with tertiary halides, the steric hindrance associated with the presence of three alkyl groups on the carbon having the halogen prevents these *tert* allylic and benzylic halides from undergoing S_N2 displacements. These, however, can react only by S_N1 process. However, with primary and secondary allylic and benzylic halides S_N2 displacements are possible. The primary and secondary allylic and benzylic halides react by S_N1 mechanism due to stability associated with the allylic carbocation.

5.9. In the conformation of a 1, 2-*trans* cyclohexane either the groups can be diaxial or diequatorial. In the diaxial conformation (*anti*-coplanar conformation) nucleophilic O^- displaces Cl *via* a backside attack (neighbouring group participation).

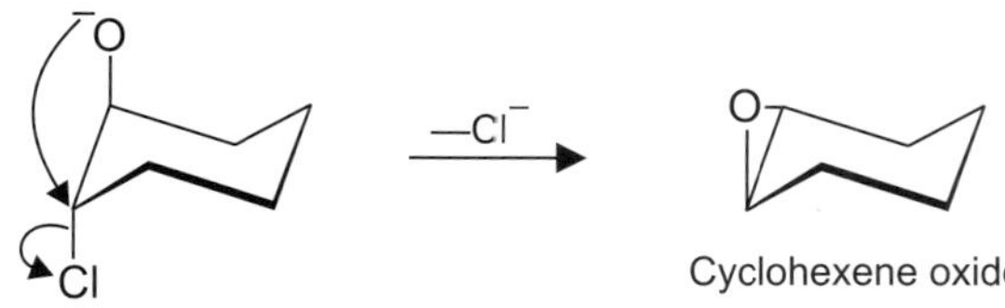

Cyclohexene oxide

In a 1, 2-*cis*-cyclohexane one group must be axial and the other equatorial. In one conformation Cl and H are diaxial and this situation can lead to elimination *i.e.*, dehydrochlorination to give a vinyl alcohol which finally gives a ketone (compare this situation with that in Scheme 12.15).

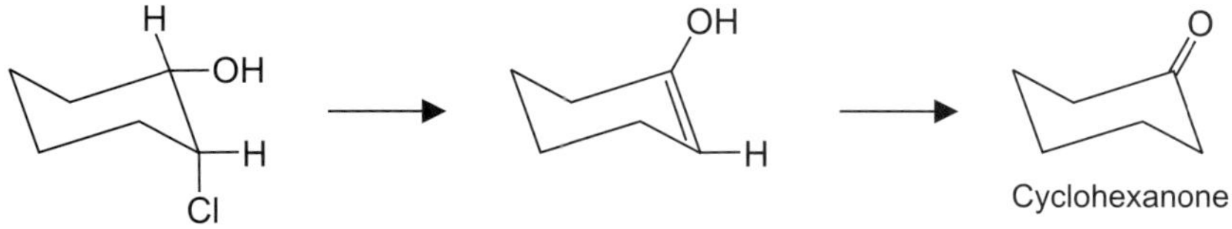

Cyclohexanone

5.10. The compound I arises from the S_N2 attack on the less substituted carbon of the protonated epoxide without disturbing the stereocenter.

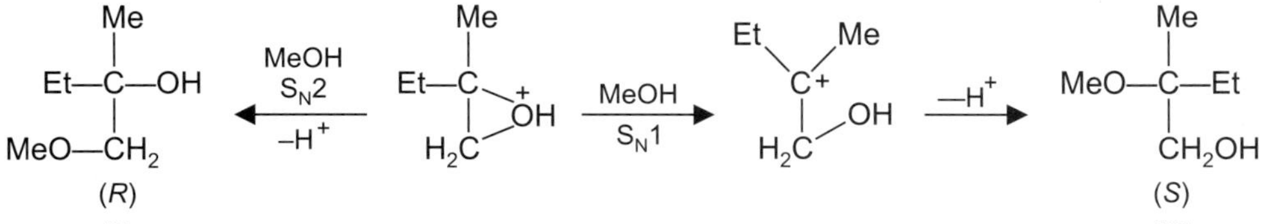

(*R*) (II) (*S*) (III)

The compound (III) is obtained *via* S_N1 attack on the intimate ion pair which occurs with inversion of configuration (compare with Scheme 5.28).

5.11. The reaction is initiated by an S_N2 displacement on less substituted carbon of the epoxide. The alkoxide ion then acts as a neighbouring group to displace Cl^- by another S_N2 reaction.

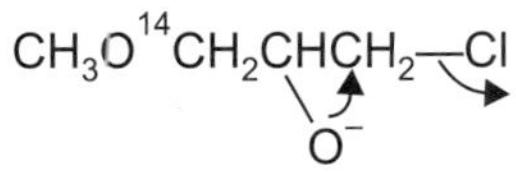

5.12. Neopentyl system is present in I, II is formed (S_N2) since (Br^-) does not face steric hindrance (ring residues on bridgehead carbon are tied back). The loss of OH^- as H_2O gives a 1° carbocation, the bridge methylene then migrates to give a 3° bridgehead carbocation (a flexible system) which picks up Br^- to give III.

5.13. An internal S_N2 reaction gives a three membered ring intermediate. This is then attacked by hydroxide at the less hindered site.

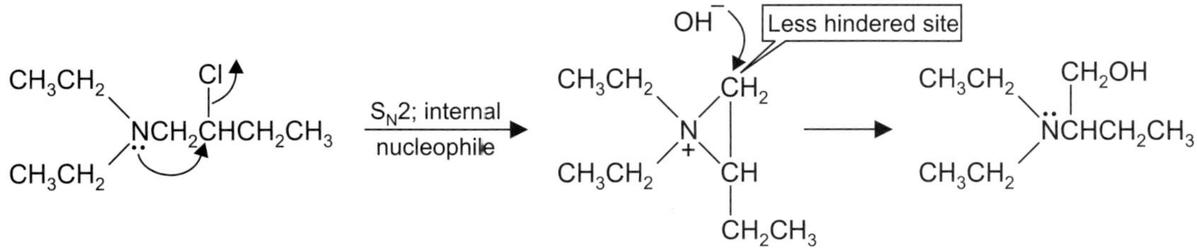

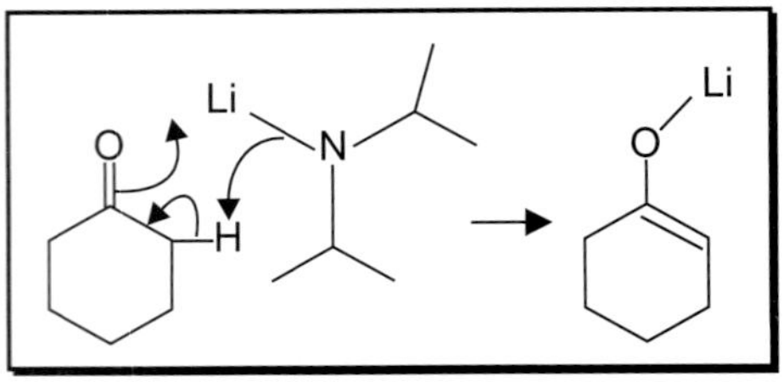

CHAPTER 6

Common Organic Reactions and their Mechanisms

6.1 BASE CATALYSED REACTIONS (FORMATION OF CARBON-CARBON BONDS)

(A) Enolates are Important Nucleophiles

In base catalysed reactions negatively polarized carbon reacts with the electrophilic carbon of carbonyl groups, alkyl halides and related compounds. The role of the base is to abstract a proton from a C—H bond which is adjacent to one or more groups of *–M* type to afford a carbanion. The group of *–M* type stabilizes the anion (see Schemes 3.25 and 3.26). Because of its relation to enol, the resonance stabilized anion is called an enolate ion which has two chemically different sites with a partial negative charge (an ambident nucleophile).

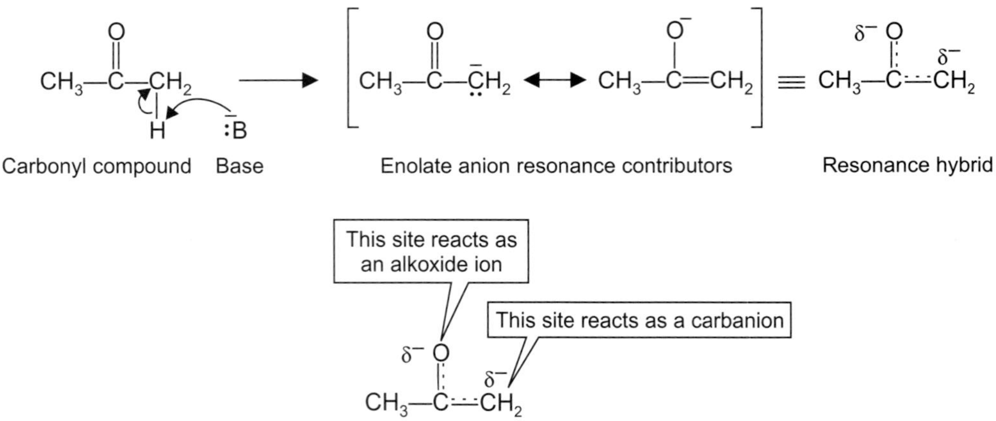

SCHEME 6.1

The enolate ions are capable of reacting at two sites (Scheme 6.1); as alkoxide ions and as carbanions. Thus when an enolate is treated with chlorotrimethylsilane, silylation occurs exclusively at the oxygen atom (Scheme 6.2). The silylation is a nucleophilic substitution at the silicon atom by the oxygen atom of the enolate. The formation of enol trimethylsilyl ether is a highly exothermic process. The oxygen-silicon bond thus formed is much stronger than a carbon-silicon bond. Consequently, the free energy of activation for reaction at the oxygen atom is lower than that for the reaction at the α-carbon. Silylation has its importance in several organic reactions. Thus electrophiles particulary silicon halides react at the oxygen atom to give silyl enol ethers (in a process called silylation).

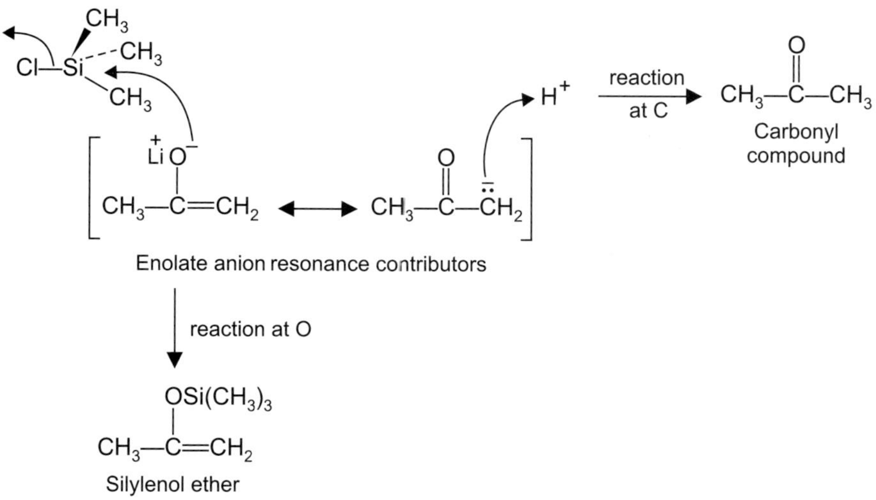

SCHEME 6.2

An enolate reacts as a carbanion with alkyl halides (Scheme 6.3) to give C-alkylation. One may appreciate that such S_N2 reactions are successful only with primary alkyl, primary

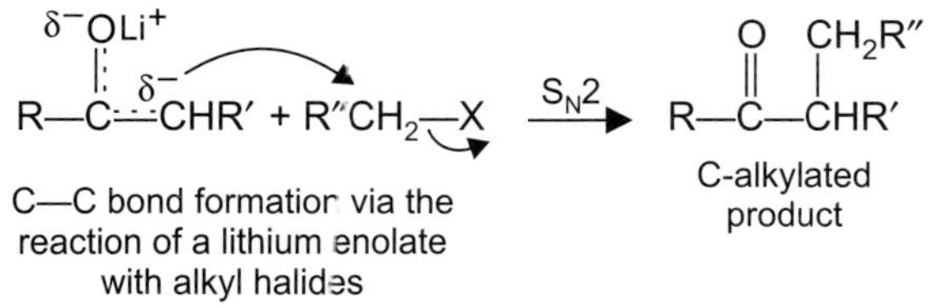

SCHEME 6.3

benzylic and primary allylic halides. As enolates are strong bases, with secondary and tertiary halides the major course taken by the reaction is elimination (Scheme 6.3*a*). Thus enolate ions are useful nucleophiles which are readily alkylated by primary alkyl halides (Scheme 6.3). The enolate ion being a relatively stronger base can involve itself into

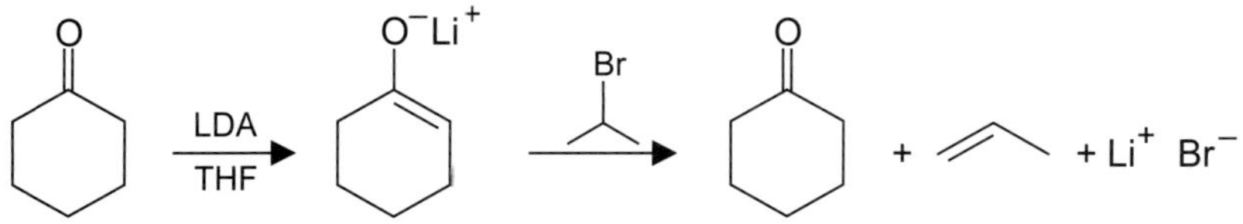

SCHEME 6.3a

E2 eliminations with secondary and tertiary alkyl halides (Scheme 6.3*a*). Another significant limitation is with aldehydes which cannot be alkylated as ketones (see Scheme 6.3). This aspect is discussed in detail in Schemes (6.46*a* and 6.46*b*).

The extent of enolate formation depends on the strength of the base used. When the base is weaker than the enolate itself, then the equilibrium lies to the left (eq. I, Scheme 6.4).

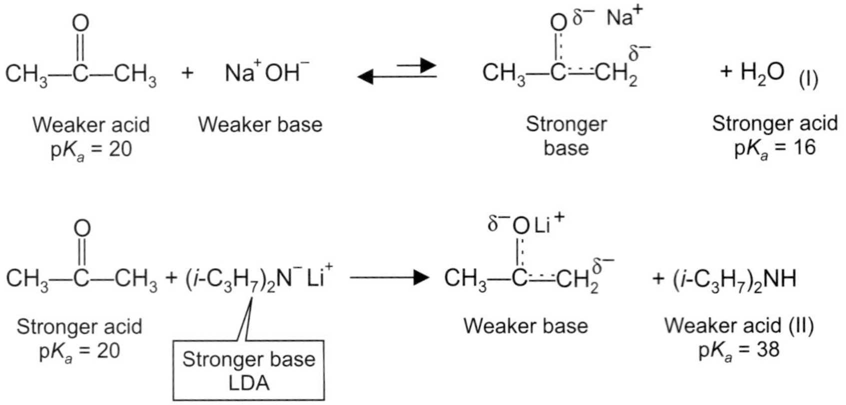

SCHEME 6.4

However, when a very strong base (LDA) is employed, the equilibrium lies far to the right (eq. II, Scheme 6.4, for this reasoning (see Scheme 3.42).

In an unsymmetrical ketone like 2-methylcyclohexanone there are two active sites and one can expect the formation of two possible enolates (I and II, Scheme 6.5). Of these, (I, thermodynamic enolate) with more substituted double bond is the thermodynamically more

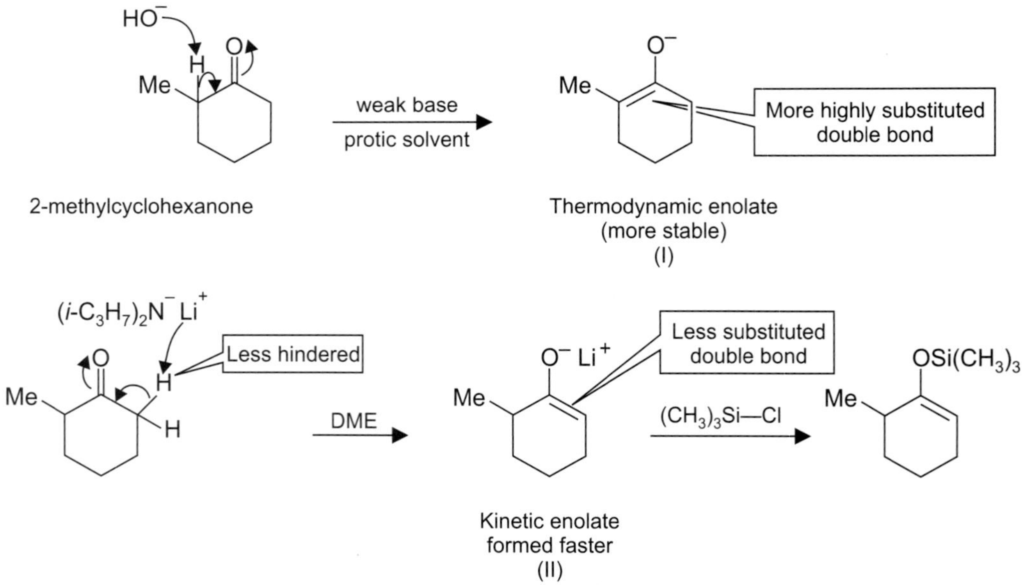

Kinetic versus thermodynamic deprotonation of unsymmetrical ketones

SCHEME 6.5

stable enolate. This enolate will be formed predominantly under conditions which allow the establishment of an equilibrium (use of a weak base in a protic solvent). The enolate (II, kinetic enolate, with less substituted double bond) is usually formed faster since the CH_2 group is sterically more accessible than the methyl substituted CH. Thus the kinetic enolate is formed predominantly when the reaction is kinetically controlled. This can be achieved by employing a sterically hindered base (LDA) which rapidly removes the proton from the less substituted α-carbon of the ketone. One can capture this enolate ion by reaction with chlorotrimethylsilane as the enol trimethylsilyl ether. This ether can be purified and converted back to the enolate (Scheme 6.6) by either reacting it with fluoride ions, making use of the very strong Si—F bond

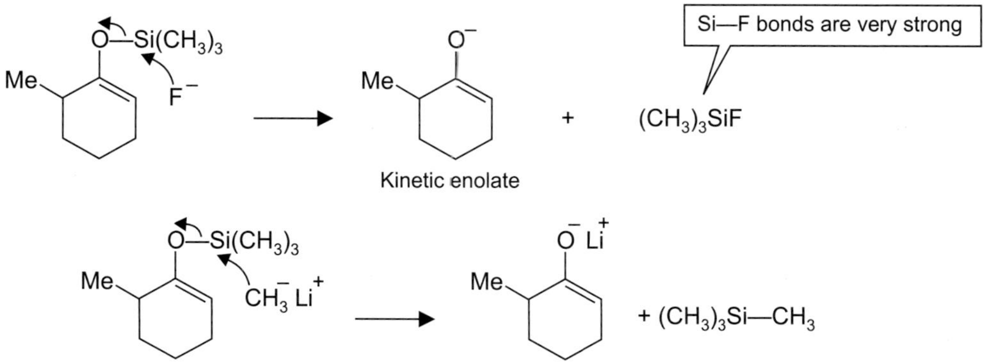

SCHEME 6.6

(594 kJ mol^{-1}), or by treatment with methyl lithium. Using this information one can, therefore, alkylate a ketone in a regioselective way (Scheme 6.7). The lithium enolate formed from 2-methylcyclohexanone *e.g.*, can be methylated with methyl iodide. Lithium enolates have their utility in directed aldol reactions (see Scheme 6.12).

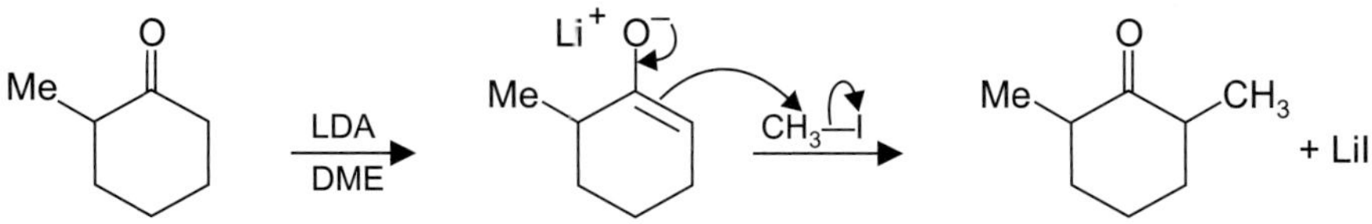

SCHEME 6.7

EXERCISE 6.1

How a β-dicarbonyl compound with pK_a ~ 10–14 or a nitro compound pK_a ~ 9–12 can be converted into its conjugate base ?

ANSWER. *With these substrates weaker bases e.g., alkoxides (pK_a ~ 17) can be employed to convert the material completely to its conjugate base. Aprotic conditions are thus no longer needed. However, a common practice to convert dicarbonyl compounds to their enolates in a clean, controllable manner is to use sodium hydride in dry THF (Scheme 6.7a).*

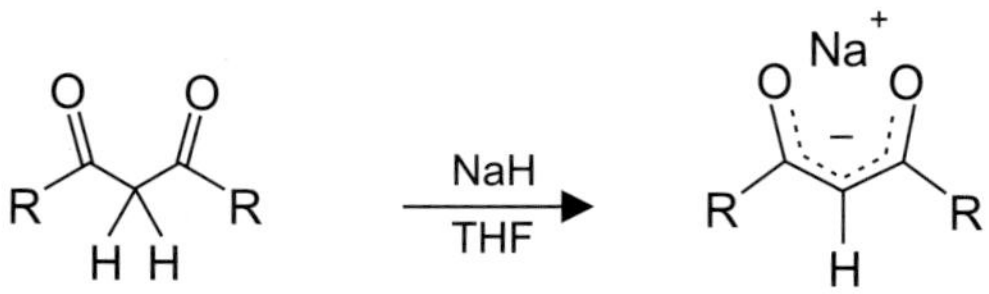

SCHEME 6.7a

(B) Regiospecific Generation of Enolates

Recall that one can either make a kinetic enolate or a thermodynamic enolate by choosing proper conditions. Advantage is taken of the fact that LDA removes preferentially a proton from the less substituted α-position to give a kinetic enolate. The most stable enolate is however, the thermodynamic enolate. The enolate formation may become reversible and the product formed then will be the one derived from thermodynamic enolate. Enolate formation may become reversible *e.g.,* by the use of a less hindered base like KH. The equilibrium between the enolates can also be established on using excess of ketone and if traces of protic inpurities are present. Thus carefully controlled conditions are required to prevent a possible equilibration (Scheme 6.7*b*).

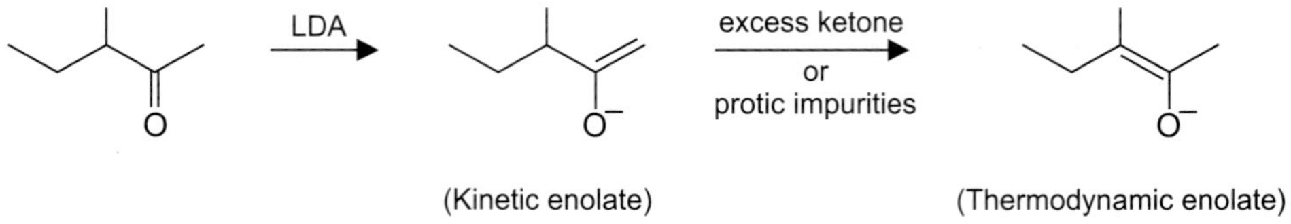

SCHEME 6.7b

Several alternatives are employed for the synthesis of compounds involving enolates which could avoid the reaction becoming reversible and one such alternative employs acetoacetic ester (see Scheme 6.18*a*). As a second approach, the unsymmetrical ketone is converted into its *N*, *N*-dimethylhydrazone and it is the geometry of this hydrazone which plays an important role in directing the base to least hindered α-position. The dimethylamino group points away from the more substituted carbon and coordinates with the lithium ion of butyl lithium ($Bu^- Li^+$) the base normally employed in this reaction. The hydrolysis of the hydrazone after the reaction regenerates the carbonyl product and thus the result is the same as working with the kinetic enolate (Scheme 6.7*c*).

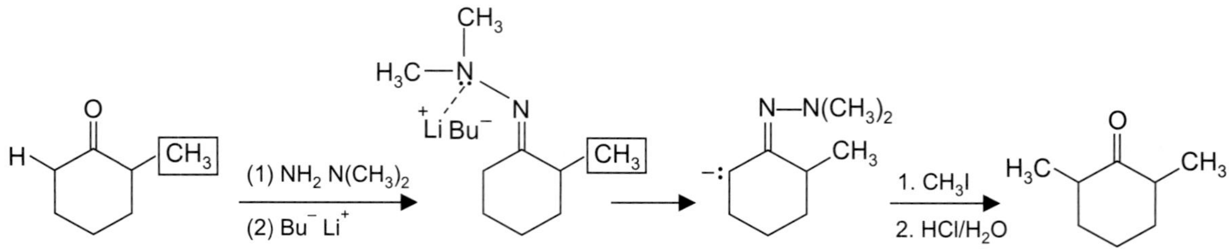

SCHEME 6.7c

(C) Aldol Condensation

When a dilute base is used, a condensation reaction involving two molecules of carbonyl compound occurs (Scheme 6.8). The process is the addition of the nucleophilic carbanion enolate (usually of an aldehyde) to the C=O of its parent compound and is termed an aldol condensation (Scheme 6.9). In a mixed aldol addition, the carbanion enolate adds to the C=O of the molecule

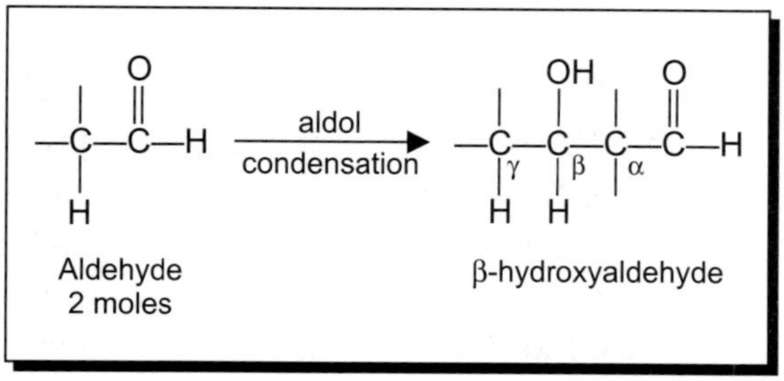

SCHEME 6.8

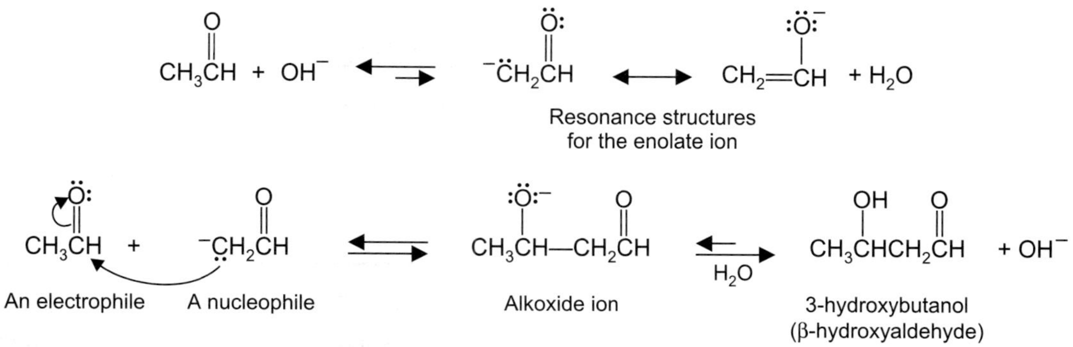

Mechanism of aldol condensation

SCHEME 6.9

other than its parent. Aldols as such are not always isolated from the condensation, and dehydration is brought about by base (Scheme 6.10). Intramolecular aldol reactions are used to make five and six membered cyclic enones (Scheme 6.10*a*). Three and four membered rings (strain) are not formed by this method due to strain and in these cases intermolecular reaction is favoured.

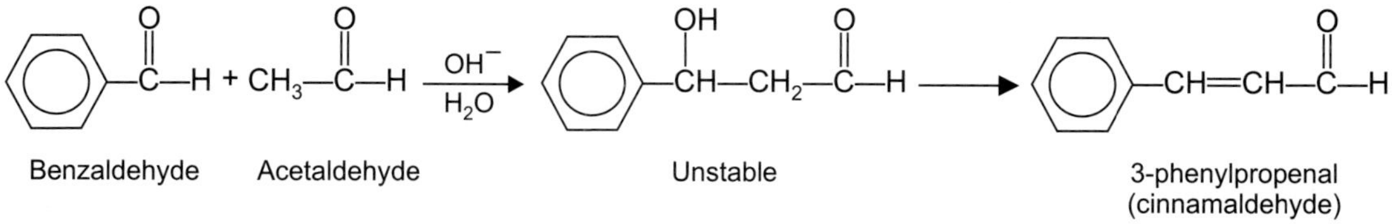

SCHEME 6.10

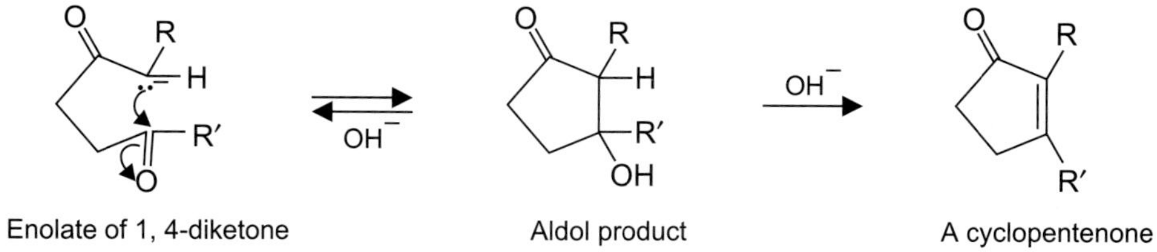

SCHEME 6.10a

(i) Aldol reaction under ordinary conditions

The aldol reaction, when conducted under ordinary conditions often gives a mixture of products, due to two reasons. Firstly *e.g.*, if each of the two aldehydes has an α-hydrogen atom, the condensation can give each of four products. This is due to the fact that each aldehyde can give an enolate and each can act as a carbonyl component. Secondly an aldol reaction can generate two stereocenters, thus it can lead to four possible stereoisomeric products (Scheme 6.11), two *syn* or *erythro* products and two *anti* or *threo* products. Thus it may be necessary to control the stereochemistry of the reaction. This requires both diastereoselection *i.e.,* if (±)-I (Scheme 6.11) or (±)-II (Scheme 6.11) will be the favoured product, as well as enantioselection *i.e.*, for a given diastereomer, whether the product will be (+)- or the (–)-epimer. The system of nomenclature is based on the main carbon chain which is drawn in the extended zig-zag form. The isomer, in which the two substituents at C-2 and C-3 are disposed in the same direction towards or away from the observer is called *syn* (or *erythro*) and the other *anti* (or *threo*).

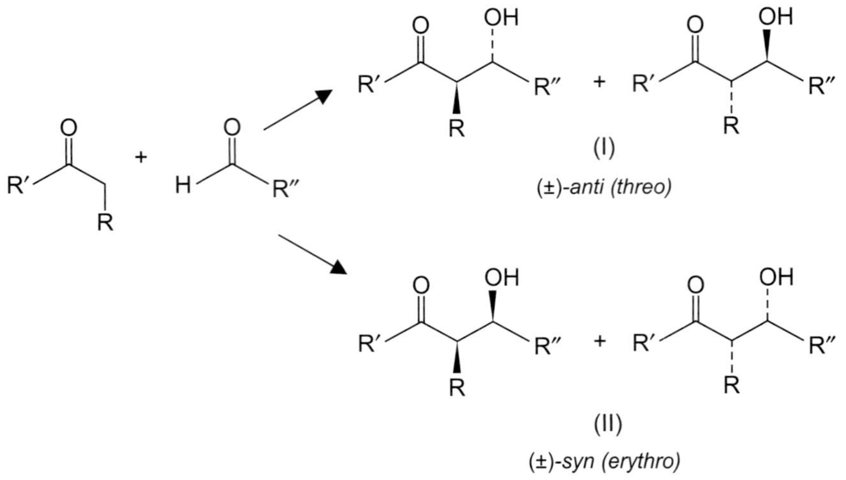

SCHEME 6.11

(ii) Use of preformed enolates in aldol reactions (directed aldol reactions)

A disadvantage while carrying out a crossed aldol reaction (*i.e.*, a mixed aldol reaction) is that more than one product is formed, when each of the carbonyl compound contains an active hydrogen. Under these situations one then directs an aldol reaction to follow a particular regioselectivity. One generates a kinetic enolate using LDA (see Scheme 6.5). This process ensures the complete formation of an enolate of one of the components (where the proton has been removed from the less substituted α-carbon, *i.e.*, from the position with larger number of α-hydrogens, Scheme 6.12). Thus in such a case, the mixed aldol reaction occurs specifically at one of the two α-positions and the ketone (I, Scheme 6.12) reacts as shown. Under ordinary conditions, this ketone would have reacted to afford a mixture of products (Scheme 6.13). When the addition of the preformed enolate to the second carbonyl component is rapid and the carbonyl component is added after the enolate formation the product is predictable and not a mixture (Scheme 6.12).

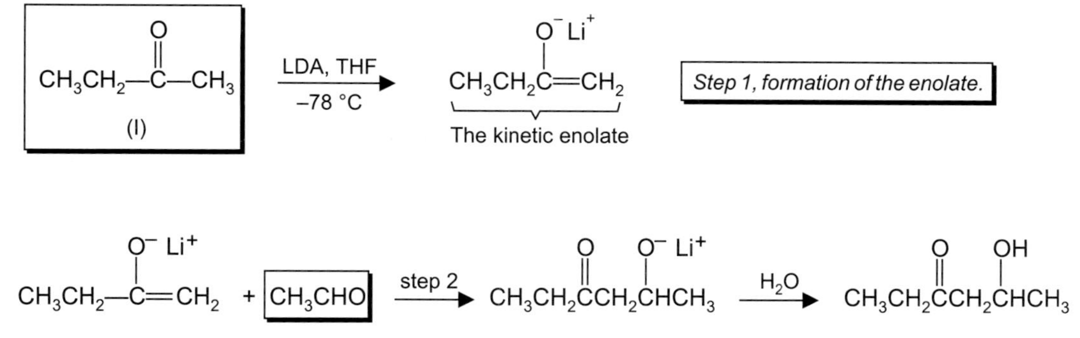

A directed aldol reaction

SCHEME 6.12

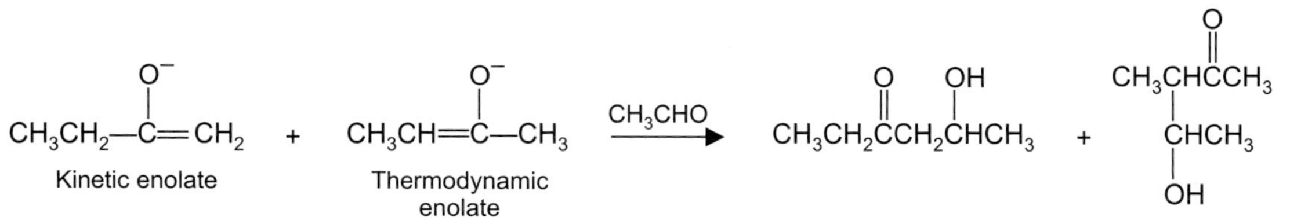

A crossed aldol reaction carried out in the classical way

SCHEME 6.13

(iii) The stereochemical course of the reaction—Diastereoselection

The following points may be considered:

- The aldol reaction creates two stereocenters from achiral starting materials and in a most general case, there are four stereoisomers of the aldol product (Scheme 6.11). Thus *syn* or *anti* diastereomers are produced, each as a pair of enantiomers.
- Regarding the stereochemistry of the reaction, therefore, one has to control diastereoselection *i.e.,* whether (racemic) *syn* or (racemic) *anti* product is formed as the major product. Secondly one has also to aim at enantioselection *i.e.*, formation of one of the four possible stereoisomers.
- Diastereoselectivity in the aldol reaction is achieved by employing the enolate of desired stereochemistry (*E* or *Z*).
- Enolates are generated *e.g.*, from a ketone and a base in the presence of chlorotrimethylsilane when the enolated are trapped as silyl enol ethers. These are separated and purified by chromatography and then converted into pure (*Z*)- or (*E*)-enolate with fluoride ion (see Schemes 6.5 and 6.6).
- Methods are available to produce either *E* or *Z* enolates in pure forms.
- *Z* enolates give mainly 2, 3-*syn* aldols while the *E*-enolates give the 2, 3-*anti* aldols (Scheme 6.14).

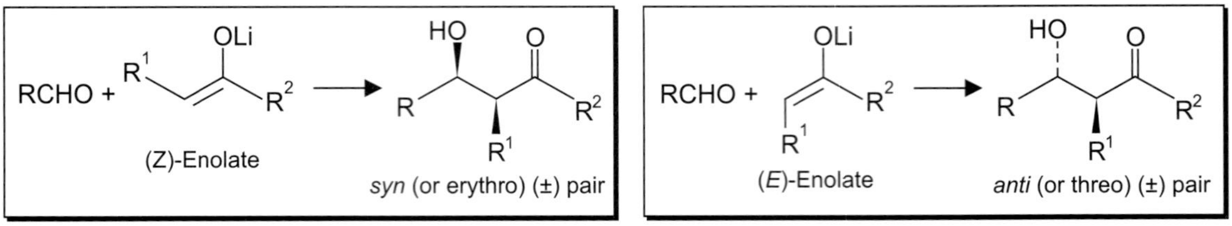

Specifically pure (E)-or (Z)-enolates can be made. Aldehydes react with (Z)-enolates to give 2, 3-syn aldols and (E)-enolates give the 2, 3-anti aldols.

SCHEME 6.14

- The diastereoselectivity is explained by involving a six-membered chair like transition state between the reactants *i.e.*, an enolate of defined geometry and the carbonyl compound *e.g.*, an aldehyde.
- Greater diastereoselectivity in aldol reactions is achieved by employing boron enolates as the carbon nucleophiles. The boron-oxygen bonds are shorter than lithium-oxygen bonds, and consequently the steric interactions in the chair like transition state are magnified to result in greater stereoselectivity. *Z*-vinyloxyboranes *e.g.,* are readily prepared by reacting ketones with a dialkylboron–trifluoromethanesulphonate (triflate) and a mild base diisopropylethylamine (in these enolates the boron atom is bonded to the oxygen atom of the ketone) and these react with aldehydes to give *syn* aldol in high yield (Scheme 6.14*a*). 3-Pentanone by this method gives *Z* and *E* enolates in a ratio of > 99 : 1 and subsequent reaction with benzaldehyde gives the *syn* and *anti* aldols in a ratio > 97 : 3 (The same condensation when carried out with lithium enolates gave a ratio of only 80 : 20).

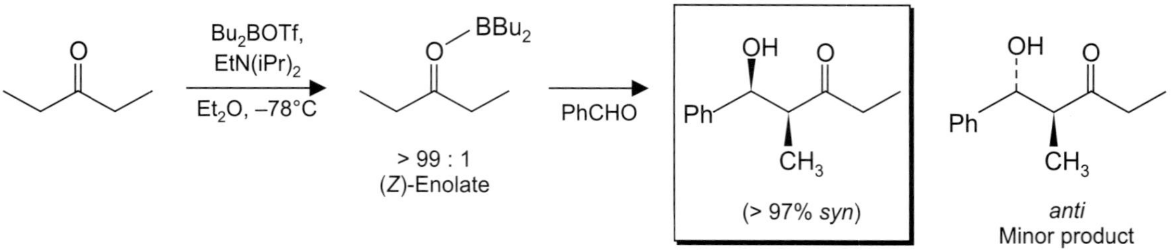

SCHEME 6.14a

- The diastereoselectivity is achieved by the reaction (*i.e.,* formation of the new C—C bond) proceeding *via* a chair like six membered transition state in which the ligated metal atom is bonded to the oxygen atom of the aldehyde as well to that of the enolate. If the geometry of the enolate is fixed, the only variable is the orientation of the aldehyde, and therefore, one deals with transition states of different stabilities. With *Z* enolate one of the transition states (III, Scheme 6.15) is disfavoured due to 1, 3-non-bonded interactions between the substituents and thus the reaction takes place largely *via* transition state (II, Scheme 6.15) to give *syn* aldol. Similar arguments show that the reaction with the *E*-enolate proceeds preferentially through the transition state (IV, Scheme 6.15).

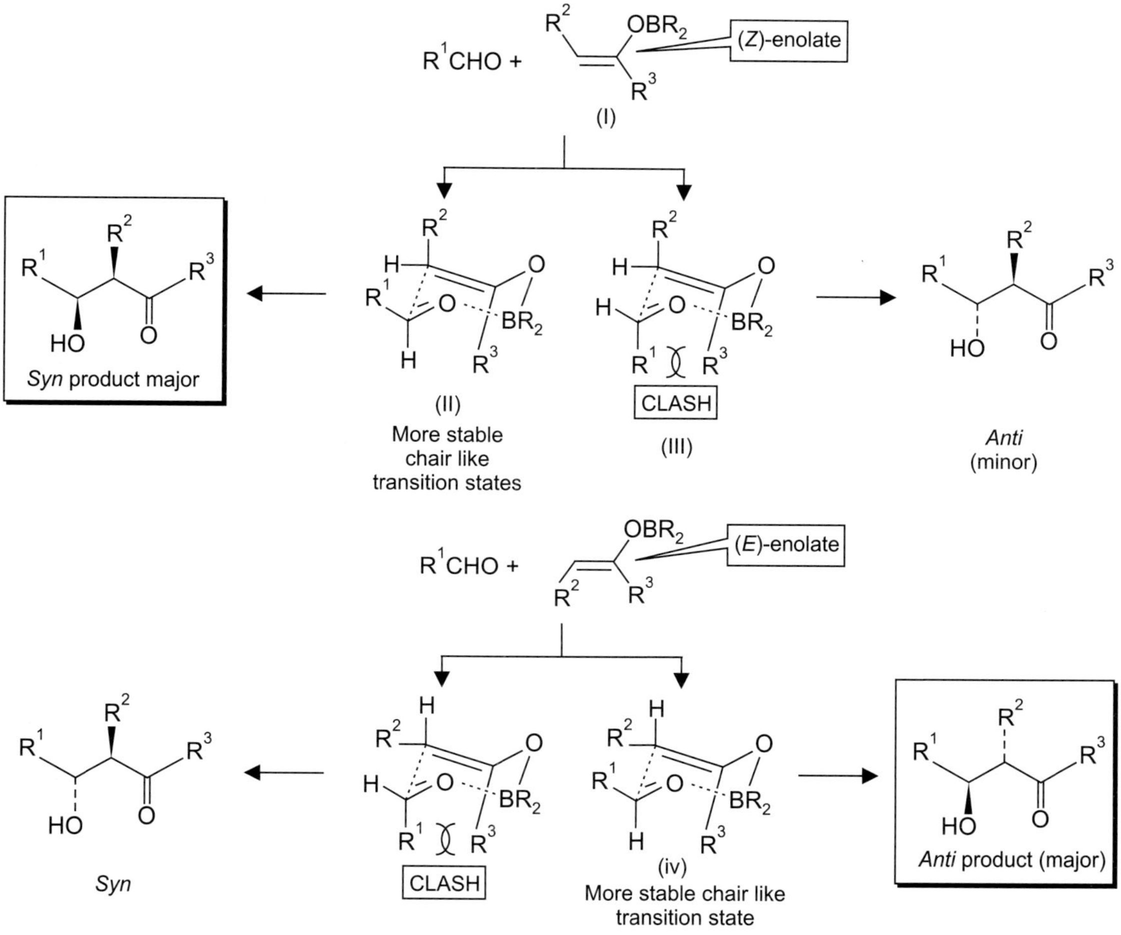

SCHEME 6.15

During reactions with lithium enolates the size of the substituents R^1 and R^2 explains selectivity, larger the size of R^2 (Scheme 6.14), the greater the selectivity. Thus in the case of a ketone CH_3CH_2COR, the *Z* enolate is generally formed the faster of the two, if the group R has reasonable size. The reason for this effect is that the methyl group has an eclipsing relationship with oxygen in the *Z* isomer, but with the bulkier R group in the *E*-isomer. This factor due to the steric strain is reflected in the transition states leading to their formation. Thus when R is very bulky *e.g.* *t*-butyl as in 2, 2-dimethyl-3-pentanone, the selectivity is very high (Scheme 6.16). The selectivity gets less pronounced when the size of group R is reduced, 3-pentanone under similar conditions gives a mixture containing 30% of the *anti* isomer.

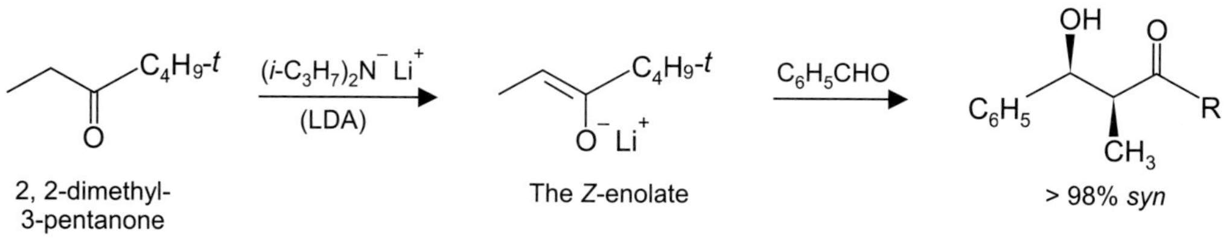

SCHEME 6.16

Strong Non-Nucleophilic Bases

Unlike in the past, now enolate nucleophiles are made by using very strong and non-nucleophilic bases (Scheme 6.16a). These bases have $pK_a > 35$ *and convert*

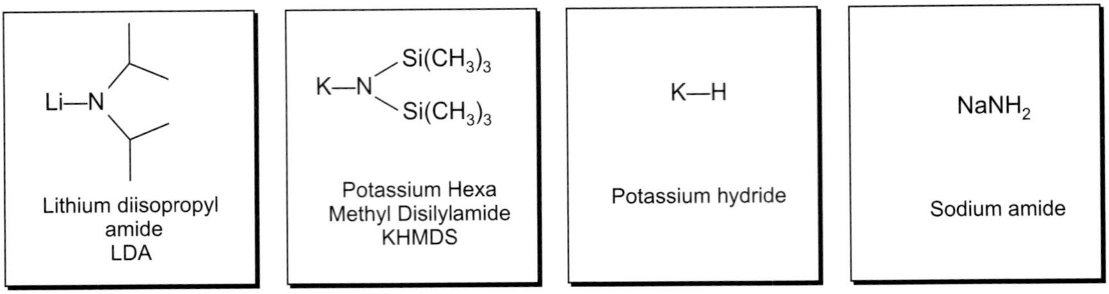

SCHEME 6.16a

carbonyl compounds with α protons that have pK_a *20–25 completely to their enolate anions. A carbonyl compound is completely converted into a stable nucleophile and thus cannot condense with itself. This is the key to success of a reaction e.g., aldol condensation (see Scheme 3.43d).*

(D) Other Reactions Resembling Aldol Condensations

(i) The Claisen-Schmidt reaction

Ketone enolates react with aldehydes effectively, however, aldehyde enolates donot undergo crossed aldol condensation with most ketones but on the other hand undergo self condensation.

This is a crossed aldol reaction when ketones are used as one component (Scheme 6.17) and bases like sodium hydroxide are used. Under these basic conditions ketones do not self condense appreciably since the equilibrium is unfavourable (see retroaldol reaction, Scheme 6.20).

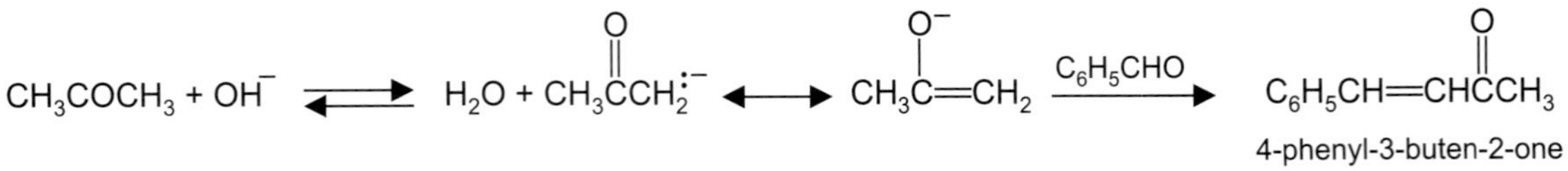

Claisen-Schmidt reaction

SCHEME 6.17

This difference can be understood by considering the following discussion. In the dimerization of acetaldehyde to give aldol (Scheme 6.9), the kinetics show that the first step *i.e.*, the generation of the enolate is rate determining. In the self condensation of acetone to yield diacetone alcohol (Scheme 6.20), the kinetics show, that in this closely related reaction, the rate determining step is instead the reaction of the enolate with a second molecule of acetone. The reason for this difference is that the carbonyl group in acetone is less rapidly attacked by nucleophiles (*i.e.*, the carbonyl carbon atom of a ketone is less positive) as compared to acetaldehyde. This is due to the fact that the methyl group due to its +*I* effect renders the adduct from acetone (and therefore, the preceding transition state) less stable. Moreover, the carbonyl group of acetone is more hindered. One may note that the C—C bond forming step in aldol condensations is facilitated by groups with –*I* effect on the carbonyl component and retarded by electron releasing groups.

An example of Claisen-Schmidt reaction is found in the synthesis of pseudoionone (Scheme 6.18) from the naturally occuring geranial and acetone. Pseudoionone is used in the commercial synthesis of vitamin A.

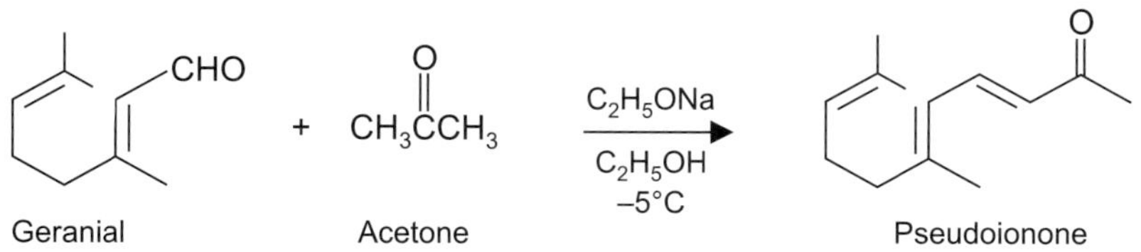

The Claisen-Schmidt reaction

SCHEME 6.18

Enolates and Other Nucleophiles—Alkylation of β-Dicarbonyl Compounds

- *Regiospecific formation of kinetic or thermodynamic enolates from ketones can be directly achieved by using a strong base which is a relatively poor nucleophile e.g., LDA, KHMDS or KH.*
- *For the removal of a proton attached to a carbon, a base with a higher pK_a than the pK_a of the proton to be removed is employed so that a complete conversion to carbanionic nucleophile is achieved (see Scheme 3.42).*
- *Thus one can either form a kinetic or a thermodynamic enolate from a ketone.*
- *The regiospecificity in the product can be forced by using an older but useful strategy. In order to make a methyl ketone, enolate anion could be generated from acetone itself (pK_a = 20) followed by the addition of an alkyl group from R—X via S_N2 reaction. Enolate anion of acetone requires the use of a very strong base to generate it and more importantly its high reactivity gives low yield of the desired product.*
- *One thus works with a different strategy, and uses a synthetic equivalent enolate anion of acetone (Scheme 6.18a) which is the enolate anion of ethylacetoacetate (ethyl acetoacetate pK_a 10, gives its enolate ion by using a moderate base e.g., NaOEt, however, the usual practice is to use NaH, pK_a > 35).*
- *The resulting nucleophile (as an enolate) is then used to form a new carbon-carbon bond.*

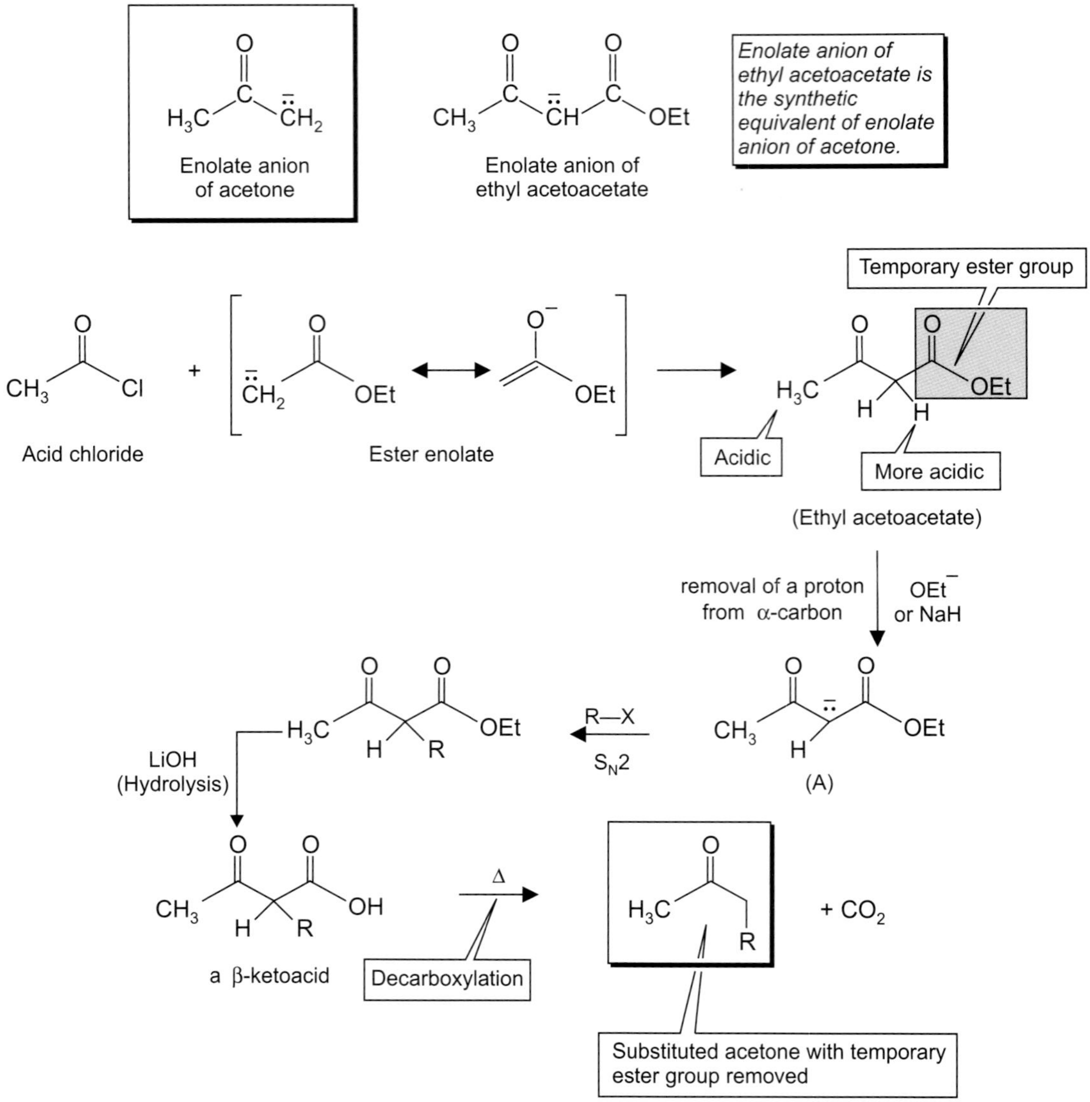

SCHEME 6.18a

- *Ethyl acetoacetate can be prepared by reacting an acid chloride with the ester enolate by Claisen type condensation. More acidic proton is removed and the enolate (A, Scheme 6.18a) thus formed can be used to create new carbon-carbon bond. After having guided the regioselectivity of the specific enolate anion formation, this ester grouping of the system is removed by hydrolysis followed by decarboxylation of the resulting β-keto acid (for mechanism of decarboxylation of a β-ketoacid (see) Scheme 6.52).*
- *One may recall the role of more stable acetate anion (OAc^-) as a synthetic equivalent of OH^- (see Scheme 5.15e) and the use of conjugate base of phthalimide as the synthetic equivalent of amide ion for the synthesis of primary amines (see Scheme 5.12e).*

(ii) Condensation with nitro alkanes and nitriles

The α-hydrogens of nitro alkanes are much more acidic (pK_a = 10) than those of aldehydes and ketones (see Scheme 3.30). Such nitro alkanes with α-hydrogens undergo base catalysed condensations with aldehydes and ketones in a way similar to aldol condensation (eq. I, Scheme 6.19). Similarly, nitriles (see Scheme 3.30) also display aldol type condensations (eq. II, Scheme 6.19).

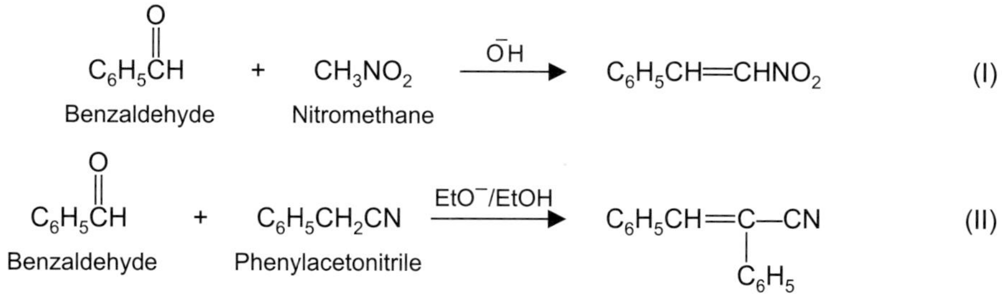

SCHEME 6.19

(E) The Retro-Aldol Reaction

The aldol reaction is reversible. Thus when diacetone alcohol (product of self condensation of acetone, Scheme 6.20) is heated with a strong base, the aldol addition is reversed. Infact one then gets an equilibrium mixture which consists largely of acetone. This type of reaction is called retro-aldol reaction.

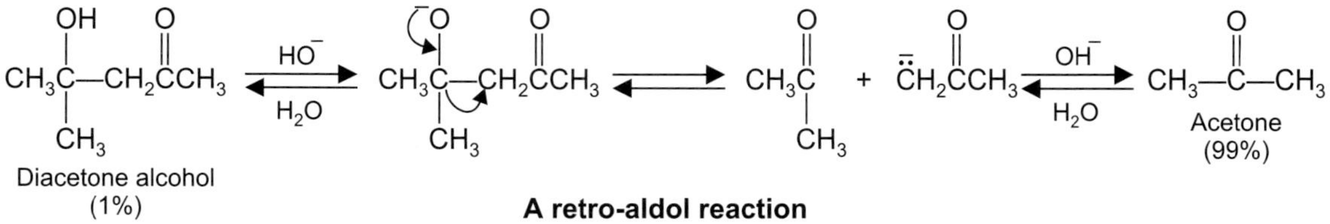

A retro-aldol reaction

SCHEME 6.20

(F) The Claisen Reaction

The Claisen reaction (different from Claisen ester condensation) is a base catalysed reaction between an aldehyde which does not have an active hydrogen and an ester which contains an active hydrogen. Thus *e.g.*, benzaldehyde condenses with ethyl acetate in the presence of sodium ethoxide to give ethyl cinnamate in high yield (Scheme 6.21).One may note that carbonyl in an aldehyde is more reactive towards nucleophiles when compared to the carbonyl group of an ester. The aldehydes with α-hydrogen atoms are not suitable partners for this reaction since these would prefer to undergo self condensation.

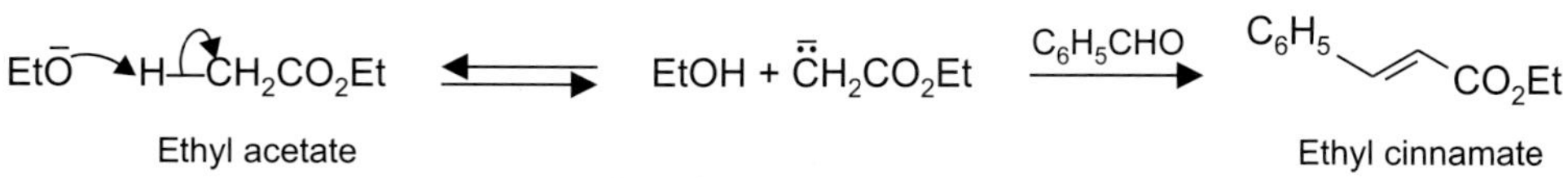

The Claisen reaction

SCHEME 6.21

(G) The Reformatsky Reaction

This is a crossed condensation reaction, and leads to aldol type products. The Reformatsky reaction involves the addition of an organozinc reagent to the carbonyl group of an aldehyde or ketone. This reaction extends the carbon skeleton of an aldehyde or a ketone and yields β-hydroxy esters. In this regard the mechanism of Reformatsky reaction (Scheme 6.22) resembles a Grignard reaction *e.g.,* with a ketone (Scheme 6.23). The organozinc reagent is less reactive than a Grignard's reagent, therefore a nucleophilic addition to the ester group does not occur.

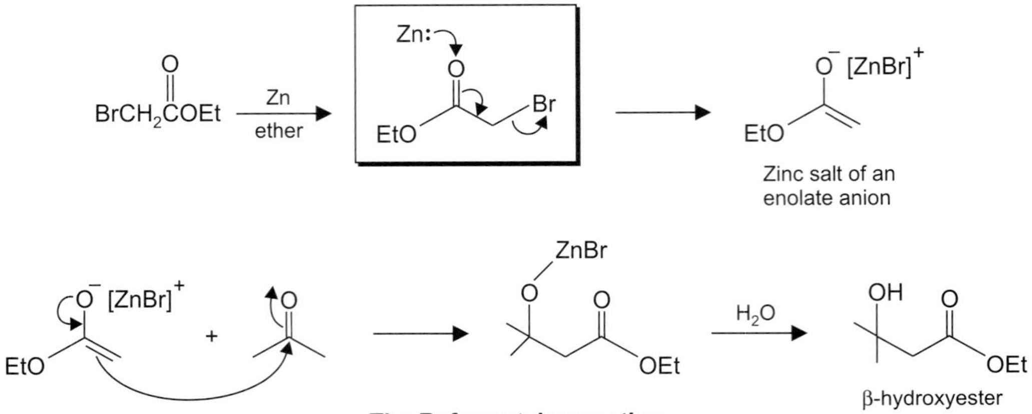

The Reformatsky reaction

SCHEME 6.22

On simply treating a mixture of a ketone and an ester with base does not lead to a synthetically useful product. A ketone is more acidic as well as more electrophilic than the ester. Thus, this combination leads only to the aldol condensation of the ketone. The problem is solved by having an ester enolate anion to act as a nucleophile to attack a ketone or an aldehyde. The ester enolate anion is formed first, in the absence of the ketone, by reduction of an α-bromoester with zinc (α-bromoacid is prepared by Hell Volhard Zelinski reaction is esterified to α-bromoester) Reformatsky reaction is not a base catalysed condensation. It resembles a Grignard reaction with a ketone (Scheme 6.23). The enolate attacks the ketone more rapidly compared to its ester precursor.

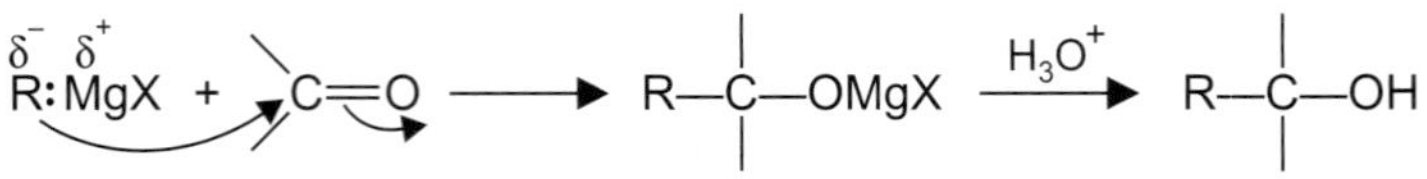

The addition of Grignard reagent to a carbonyl compound

SCHEME 6.23

(H) The Perkin Reaction

The reaction is used for the synthesis of α, β-unsaturated acids, and is an aldol type condensation between an aromatic aldehyde (ArCHO) and a carboxylic acid anhydride $(RCO)_2O$ catalysed by a carboxylate ion (the potassium salt of the carboxylic acid, RCOOK). The anhydride gives the enolate by reacting with basic carboxylate ion (Scheme 6.24).

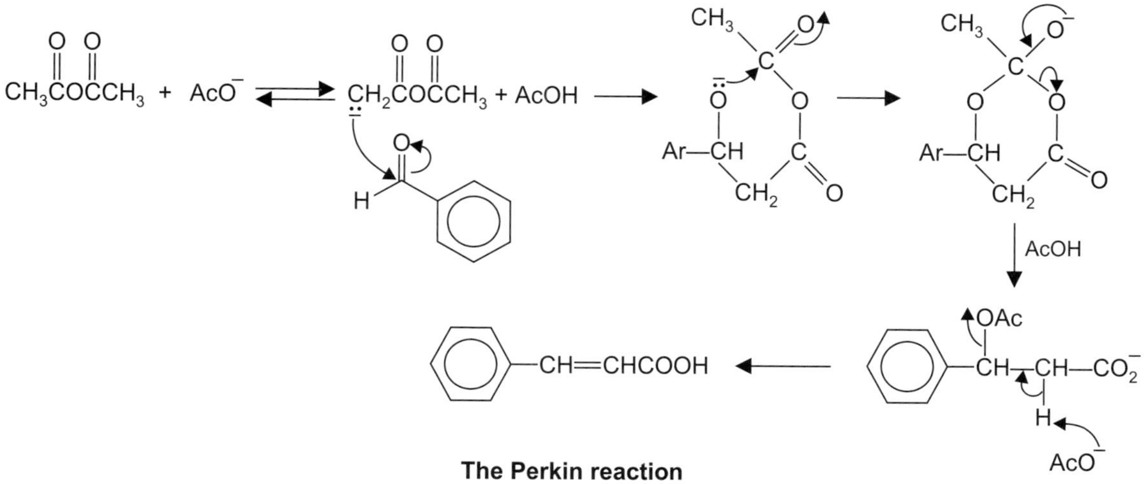

The Perkin reaction

SCHEME 6.24

(I) The Stobbe Condensation

This is a condensation between dialkyl succinates and ketones in the presence of bases like NaOEt. Mechanistically the enolate from the ester adds to the carbonyl group of the ketone (Scheme 6.25). One may compare this situation with that in Claisen ester condensation where in the presence of a base, the enolate from the ketone displaces alkoxide ion from the ester. In the Stobbe condensation, the condensation product (I, Scheme 6.25) undergoes cyclization to give a lactone intermediate, the oxygen anion of the adduct (I, Scheme 6.25) acting as an internal nucleophile facilitates the hydrolysis of one of the ester groups. The lactone then undergoes elimination (E1 or E2) to give a carboxylate salt. The net result is the attachment of a three carbon chain to the ketonic carbon atom.

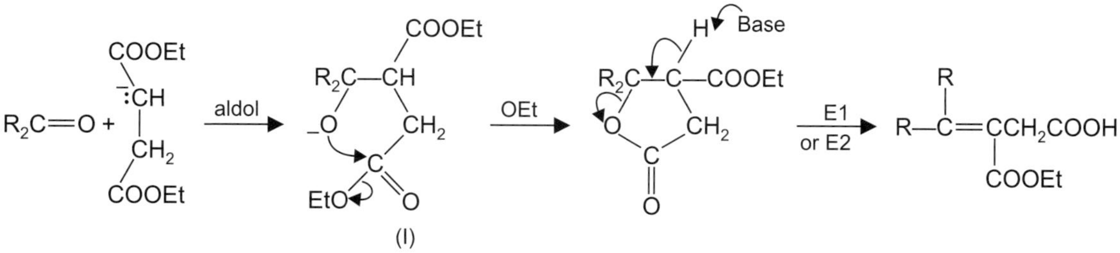

The Stobbe condensation

SCHEME 6.25

(J) Darzens Reaction

Aldehydes and ketones condense with α-haloesters in the presence of bases to give α, β-epoxy esters called (glycidic esters). The reaction called Darzens condensation involves the addition of the enolate to the carbonyl group to give an oxyanion (Scheme 6.26), this displaces the halide ion by an internal S_N2 reaction. On alkaline hydrolysis, these esters give glycidic acids which undergo a decarboxylative rearrangement when warmed in the presence of acids (Scheme 6.27) to give an aldehyde if R^3=H or a ketone if R^3=alkyl group. Thus the eventual outcome of such a reaction is to extend the chain by one carbon.

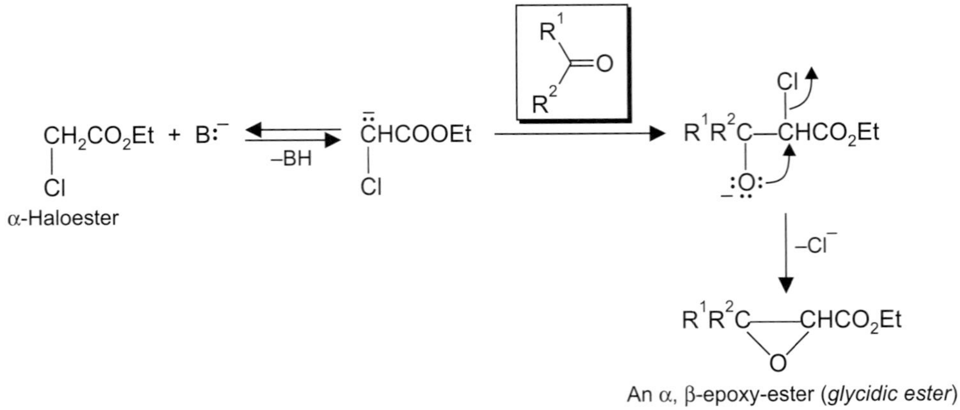

The Darzens reaction
Darzens reaction involves the reaction of an α-halo ester with an aldehyde or a ketone in an aldol type reaction to give a glycidic ester

SCHEME 6.26

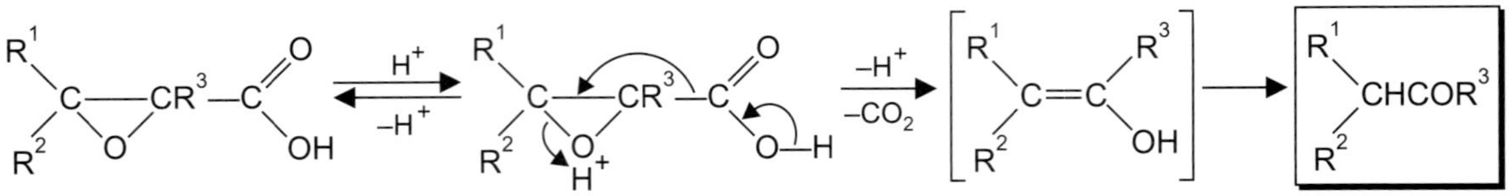

SCHEME 6.27

When t-butyl glycidates are used, the ester on pyrolysis eliminates isobutene to give an aldehyde or ketone (Scheme 6.28).

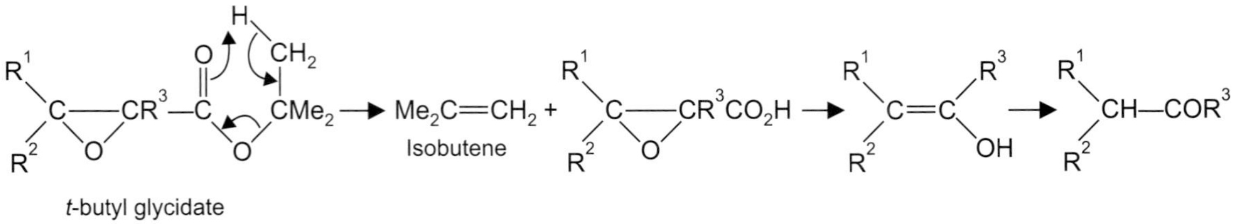

SCHEME 6.28

(K) The Knoevenagel Reaction

The condensation of aldehydes and ketones, usually not containing an α-hydrogen with compounds having an active methylene group (*i.e.*, a methylene bonded to two groups of $-M$ type) like malonic ester can take place even with a weaker base to give a sufficient concentration of the enolate ion. In reactions where amines like piperidine are used, are termed Knoevenagel condensation (Scheme 6.29). The initially formed condensation product (I), then undergoes a

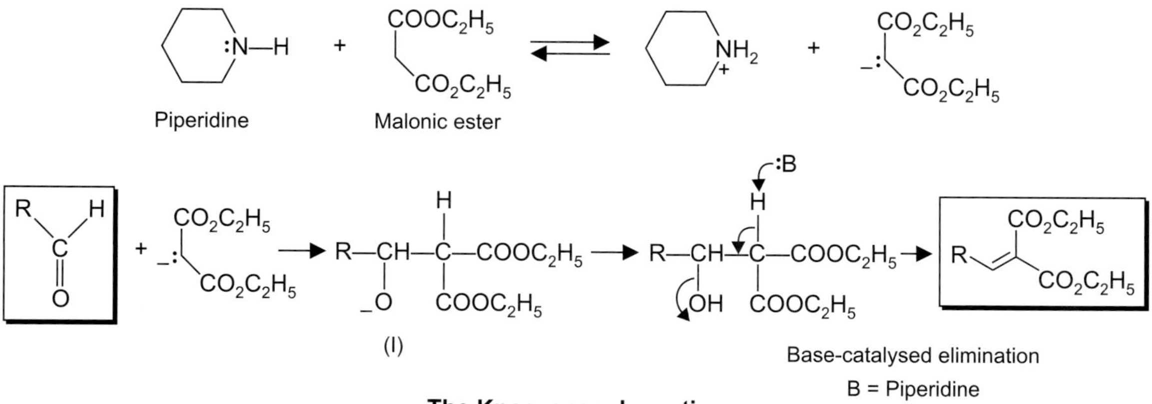

The Knoevenagel reaction

SCHEME 6.29

base catalysed elimination (Scheme 6.29). When malonic acid is employed, one of the carboxyl group gets eliminated. Thus using benzaldehyde and malonic acid, one ends up with cinnamic acid in high yield. (Scheme 6.30). In some cases it is possible for the second molecule of the

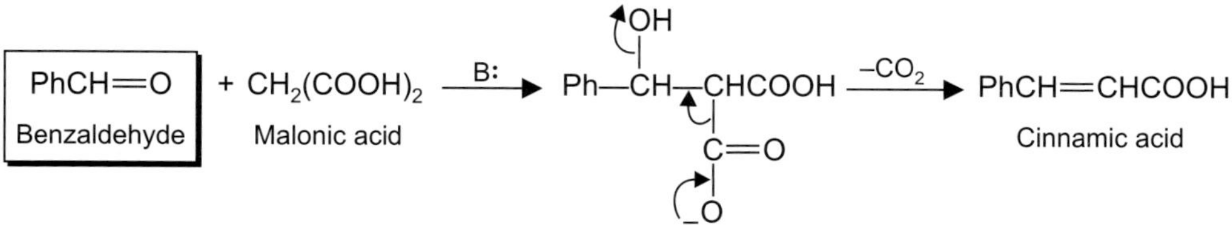

SCHEME 6.30

active methylene compound to add to the C=C bond of the product of Knoevenagel reaction (Scheme 6.31) via Michael reaction (for Michael reaction see Scheme 6.32). The Knoevenagel

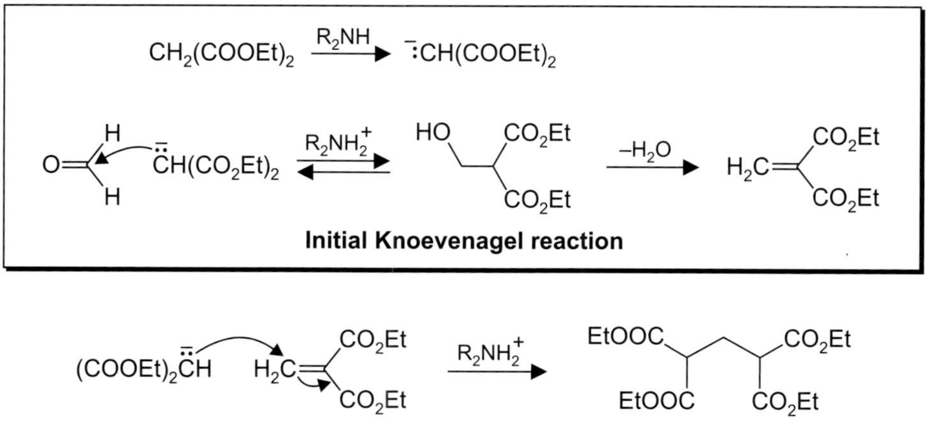

Initial Knoevenagel reaction

Michael addition process

SCHEME 6.31

reaction ends up with the elimination of water from the initially formed alcohol with a base. The mechanism of loss of water is of E1cb type. The Knoevenagel reaction has more synthetic value with aromatic aldehydes than aliphatic aldehydes. The addition to the C=C of the product from an aromatic aldehyde should be less likely due to loss of conjugation of the aromatic system.

(L) The Michael Reaction

This is a conjugate addition of enolate ions to α, β-unsaturated carbonyl compounds *i.e.*, to activated olefins (Scheme 6.32). Like other nucleophiles, the enolates do not react with simple olefins. The name Michael reaction is infact applied to a reaction between enolate forming component and an alkene which is not only activated by conjugation to a carbonyl group but to other groups of $-M$ type *e.g.*, ester, cyano, nitro and nitrile. With these structural features, the anion (I, Scheme 6.32) formed after addition (and thus the preceding transition state) is stabilized sufficiently by the delocalization of the charge on to an electronegative element and the addition therefore, occurs at a practicable rate.

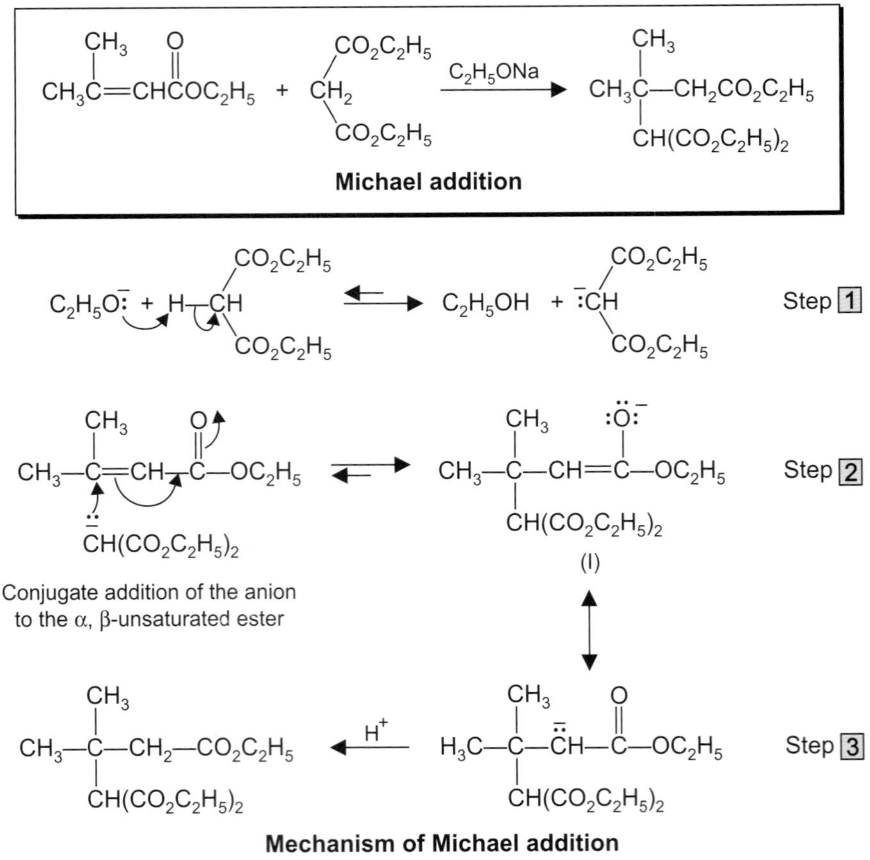

SCHEME 6.32

(M) Robinson Annulation

This ring forming reaction makes use of the Michael and aldol reactions and allows a six membered ring to be appended to a preexisting carbonyl group. The method is thus used to prepare cyclohexenone derivatives (Scheme 6.33).

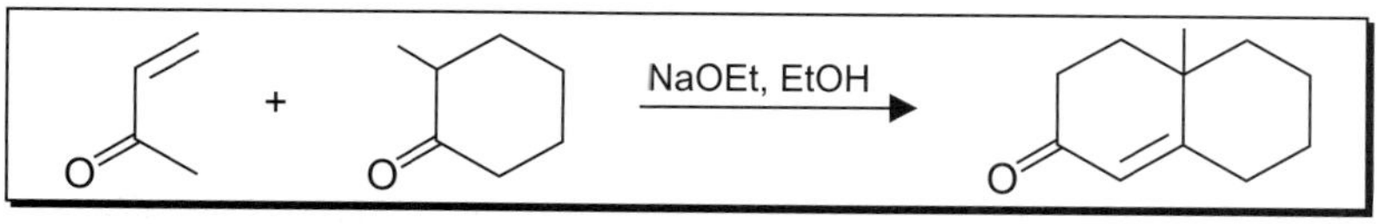

Robinson annulation

SCHEME 6.33

The first step is the formation of an enolate ion (II) from the ketone (I, Scheme 6.33*a*) which under the conditions of the reaction (use of OEt^-) is a thermodynamic enolate in this example. This enolate ion then undergoes a Michael reaction with an unsaturated ketone, *e.g.*, methyl vinyl ketone. The enolate ion (III) from the diketone intermediate thus formed undergoes an intramolecular aldol condensation to produce the six membered ring (Scheme 6.33*a*). Note that the enolate ion (III) is isomerized to its isomeric enolate. Recall that aldol reaction is reversible, the formation of a six membered ring exclusively is an example of kinetic as well as thermodynamic control since the more stable product is formed fastest.

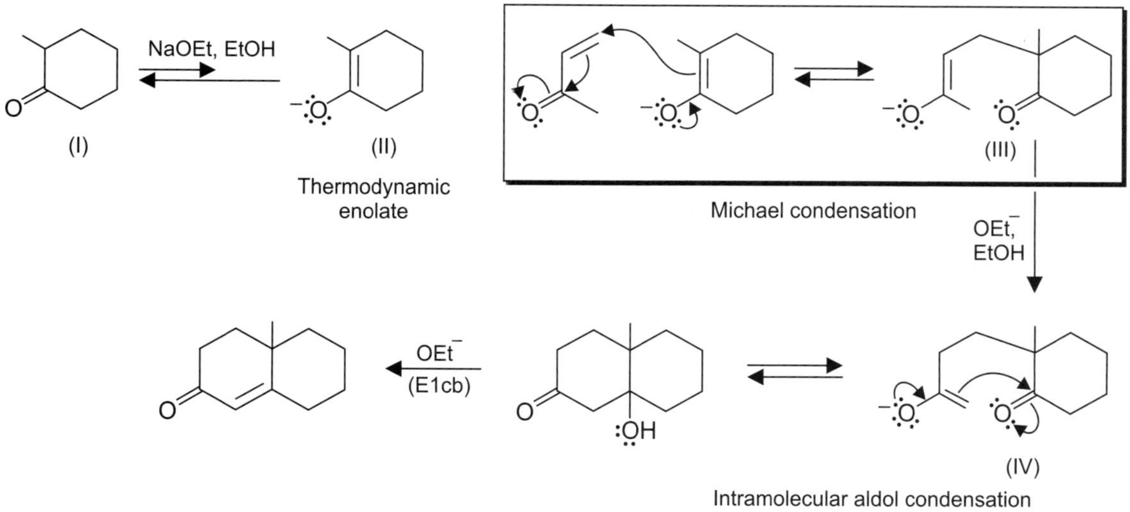

Mechanism of Robinson annulation

SCHEME 6.33a

EXERCISE 6.2

Why during the enolate formation from the diketone (I, Scheme 6.33b) out of several protons in the α-position, the removal of H_a is preferred ?

ANSWER. *It is only then that a stable six membered ring will be formed for other alternatives, the product would be either strained four membered ring or a strained bridged product.*

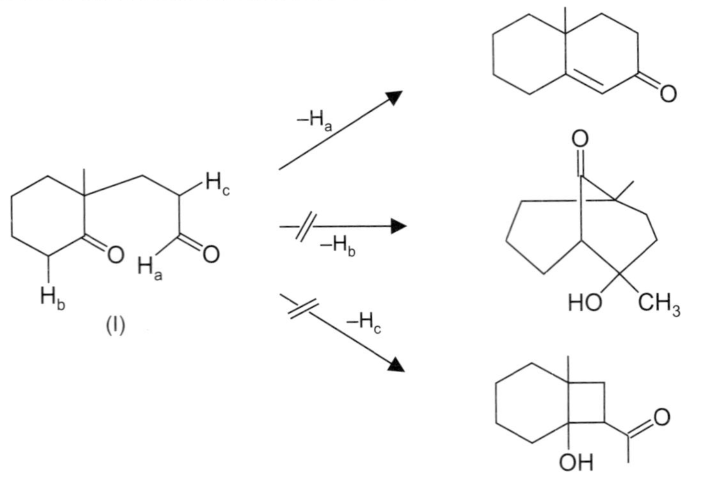

SCHEME 6.33b

(N) The Claisen (Ester) Condensation

On treatment with bases like sodium ethoxide, esters with α-hydrogen undergo self condensation; termed Claisen (ester) condensation (Scheme 6.34). The mechanism involves the conversion of one molecule of ester to a nucleophile by the base while the second molecule serves as a substrate (Scheme 6.34). The Claisen condensation reaction involves a series of equilibria *i.e.*, each step in the condensation is reversible. One may note that the formation of the new carbon-carbon bond is not thermodynamically favourable. The formation of the α-carbanion (ester enolate anion, Scheme 6.34) is not a favourable equilibrium reaction as alkoxide is a weaker base than the enolate and consequently only a low concentration of enolate forms. Attack of the α-carbanion on the second molecule of ester (acylation step), however, gives the product. This reaction is predicted to have an equilibrium constant around 1 and is reversible. In the final step 4, β-keto ester reacts with the alkoxide generated from substitution to give

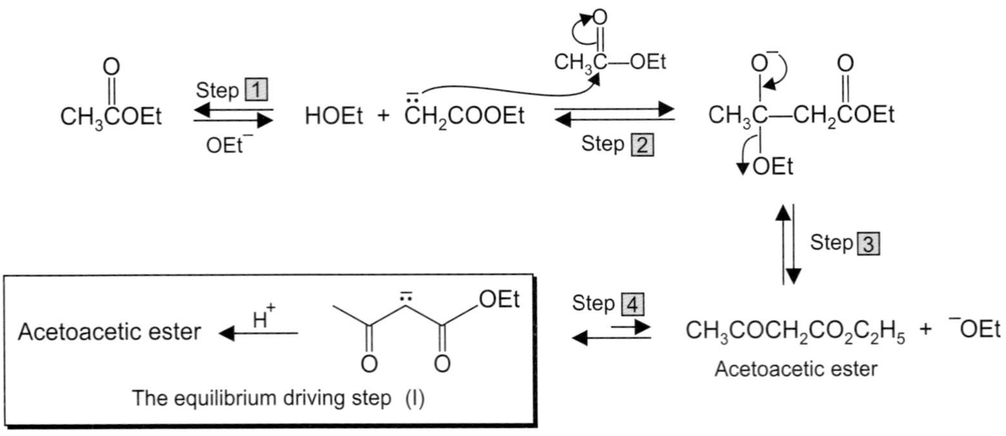

The Claisen condensation

SCHEME 6.34

the enolate anion of the product (I, Scheme 6.34). The equilibrium reactions are shifted toward this anion due to its stability since two carbonyl groups stabilize the common α-carbanion. The neutral β-keto ester product is isolated via acidification of the reaction mixture.

EXERCISE 6.3

Why the ester (A, Scheme 6.34a) does not give the expected Claisen ester condensation product ? How one can obtain this condensation product ?

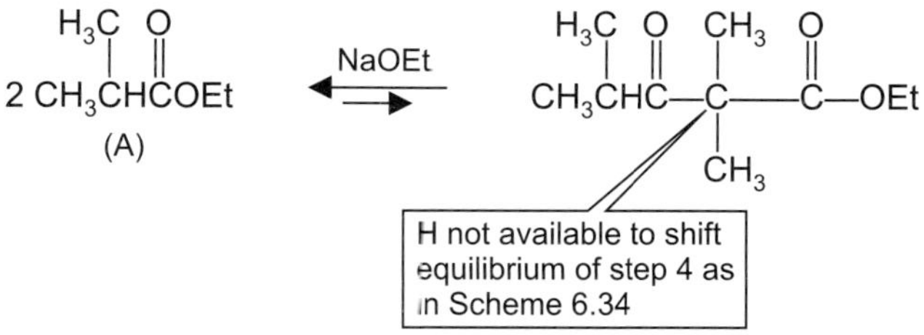

SCHEME 6.34a

ANSWER. *The alkoxide ion used as the base is much weaker than the enolate formed in step 1 of Claisen ester condensation. The equilibria for the first three steps (see Scheme 6.34) are not favourable. In fact step 4 is key to the success of the reaction. This condensation product (Scheme 6.34a) can be made if a powerful base KH is used, since now the enolate ion formed in the first step on the product side is weaker than the base on the reactant side of the equation.*

(O) The Dieckmann Reaction

It is an intramolecular Claisen condensation by way of which certain cyclic ketones can be obtained. It is most successful on diesters of C_6 and C_7 dibasic acids, where it gives the high yields of cyclic β-keto esters (Scheme 6.35). The primary driving force of the Dieckmann condensation (thermodynamically favourable) as with the Claisen condensation (see Scheme 6.34) is the

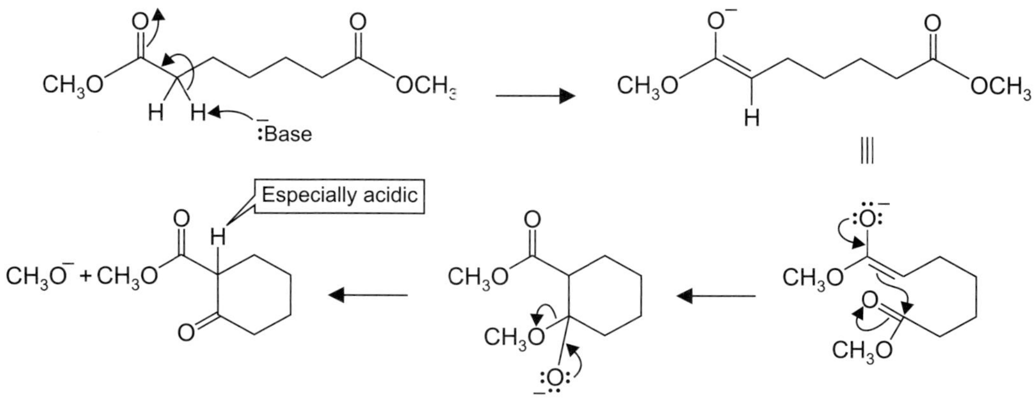

Dieckmann condensation

SCHEME 6.35

formation of the anion of the product β-ketoester (not shown) by deprotonation of the hydrogen bonded to the carbon between two carbonyl groups (shown by an arrow) of the product β-ketoester. The diesters of shorter chain dibasic acids due to the strain that would result in the formation of small rings react differently. Thus ethyl succinate undergoes an intermolecular condensation between two molecules to give a cyclohexanedione system (Scheme 6.36).

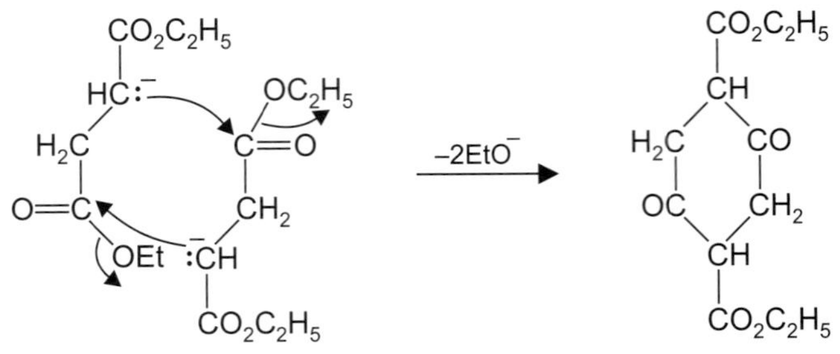

SCHEME 6.36

(P) The Thorpe Nitrile Condensation

In the Thorpe reaction, the α-carbon of one nitrile molecule (a carbanion source) adds to the CN carbon (acceptor site) of the another molecule. The C=NH bond that results can be hydrolyzed to get β-keto nitriles (Scheme 6.37).

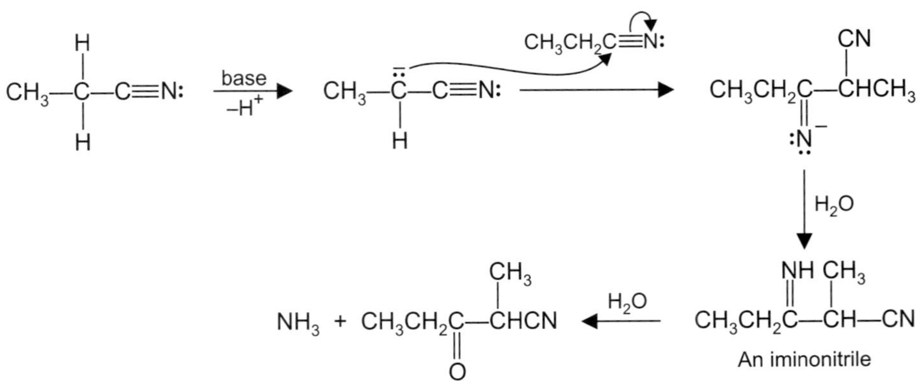

The Thorpe reaction

SCHEME 6.37

(Q) The Benzoin Condensation

This reaction is the cyanide ion catalysed intermolecular condensation of an aromatic aldehyde to give an acyloin. The condensation of benzaldehyde gives benzoin (Scheme 6.37*a*). The aromatic aldehydes do not form cyanohydrins, but the cyanide addition product, by base abstraction of a proton gives a carbanion. This reacts with a second molecule of the aldehyde. Thus benzoin condensation involves the reaction of two molecules of aldehyde (normally aromatic) in the presence of cyanide ion to give a benzoin (an α-hydroxyketone) which is also called an acyloin.

In benzoin condensation (recall that benzoin condensation resembles an aldol and Cannizzaro processes) cyanide ion is needed for the success of the reaction, however, it is regenerated at the end of the reaction (Scheme 6.37*a*).

$$\underset{\text{Benzaldehyde}}{2C_6H_5CHO} \xrightarrow{\text{KCN}} \underset{\text{Benzoin}}{C_6H_5CHOHCOC_6H_5}$$

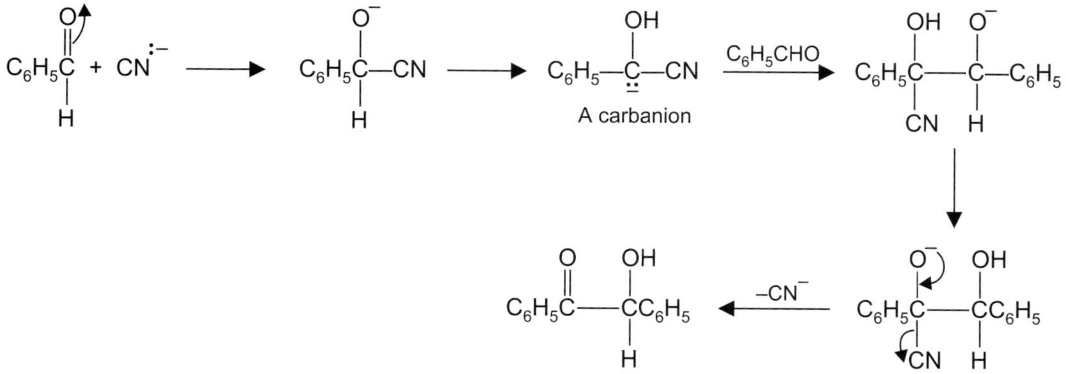

Mechanism of the benzoin condensation

SCHEME 6.37a

The success of the reaction is due to the cyanide ion. Firstly it is a reactive nucleophile and secondly it has the capacity to delocalize the negative charge on the carbanion. Thus the carbanion formation is assisted. Moreover, with aromatic aldehydes (unlike aliphatic aldehydes) the negative charge of the carbanion is further delocalized on the aromatic ring (Scheme 6.37*b*) and this factor provides the extra driving force for the reaction.

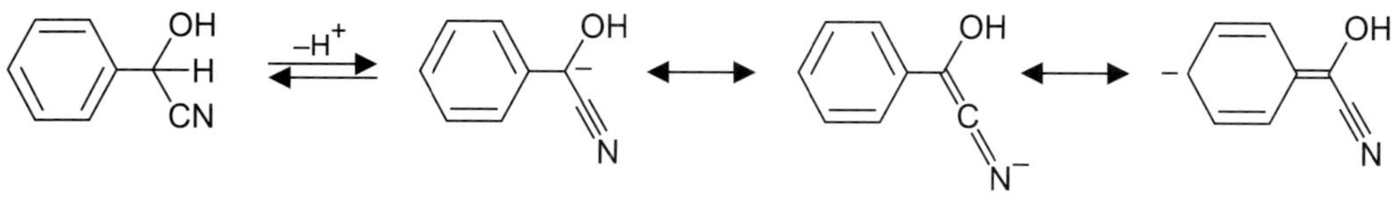

SCHEME 6.37b

6.2 STORK ENAMINE REACTIONS (FORMATION OF CARBON-CARBON BONDS)—REACTION OF AN ENAMINE WITH REACTIVE ELECTROPHILES

The name enamine usually refers to α, β-unsaturated amines. These can be prepared by reaction of an aldehyde or a ketone which contains at least one α-hydrogen atom and a secondary amine under acid catalysis (Scheme 6.38). More stable enamines are those derived from ketones rather than aldehydes. Moreover, the enamines derived from cyclic secondary amines (Scheme 6.39), are further more stable, than a acylic secondary amines.

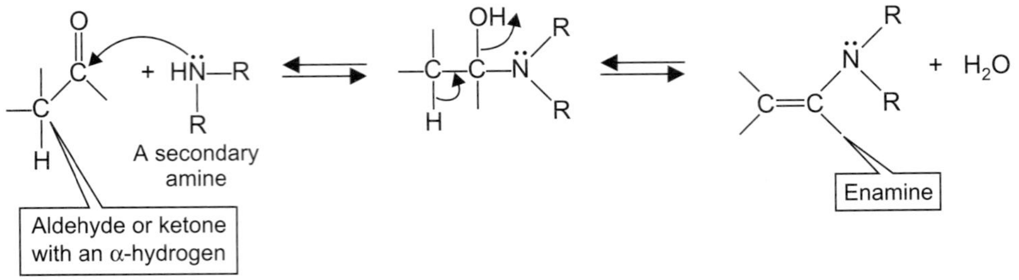

Enamine formation

SCHEME 6.38

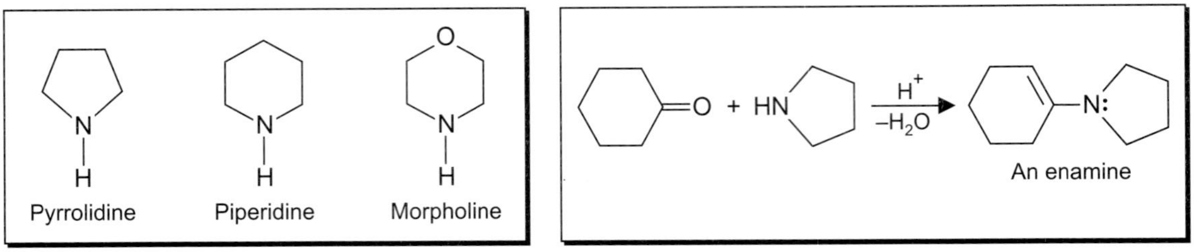

SCHEME 6.39

Primary amines also form enamines, however, in their case, enamine-imine tautomerism is possible. The equilibrium lies virtually completely on the imine side, it being more stable than enamine (Scheme 6.40). Thus an enamine from a primary amine cannot be isolated and used in synthesis.

The detailed mechanism of the formation of an imine and an enamine involving the nucleophilic attack of a primary and secondary amine respectively on the carbonyl carbon of an aldehyde or a ketone is detailed (see Scheme 11.15).

Enamines are represented as resonance hybrids of two canonical structures (Scheme 6.40) and are closely related to the enolates derived from ketones. In fact these are N analogs of an enol. An examination of the resonance structure (II, Scheme 6.40) shows that β-carbon is a nucleophilic center.

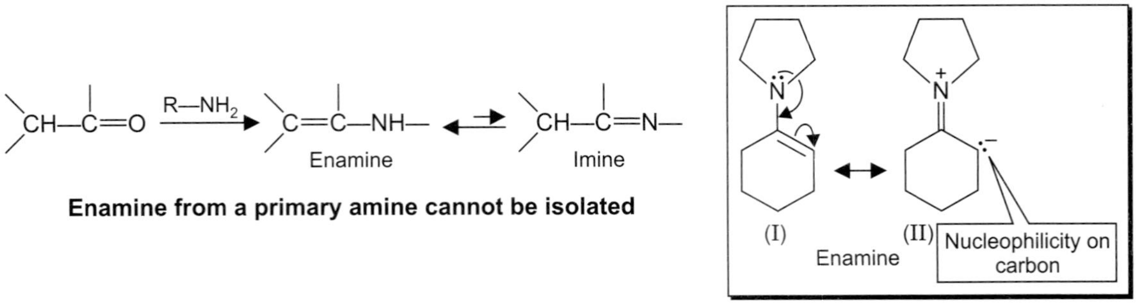

Enamine from a primary amine cannot be isolated

SCHEME 6.40

Thus enamines have a β-carbon with carbanionic character and can act as a nucleophile, and display alkylation and acylation reactions. The product iminium salt is hydrolyzed to the ketone. The net result of the entire sequence is the alkylation of the ketone in the α-position (Scheme 6.41). Enamines can thus be alkylated (Scheme 6.41) and these are S_N2 reactions, the alkylating agents have to be from, methyl, primary, allylic and benzylic halides.

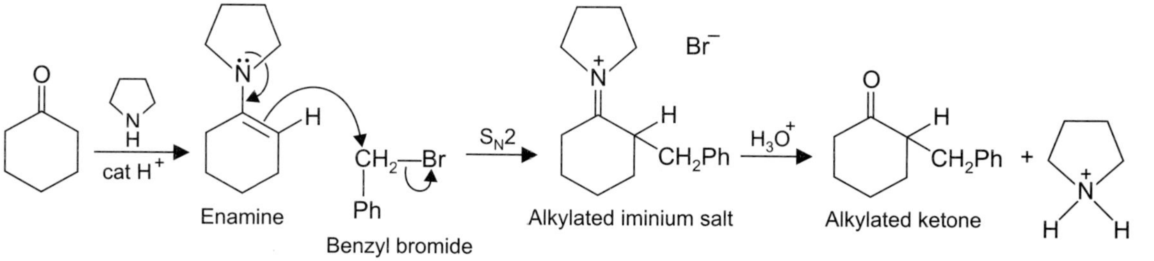

SCHEME 6.41

Acylation of an enamine is used for the synthesis of a β-diketone (Scheme 6.42) which can be further used as synthetic intermediates in more complicated compounds.

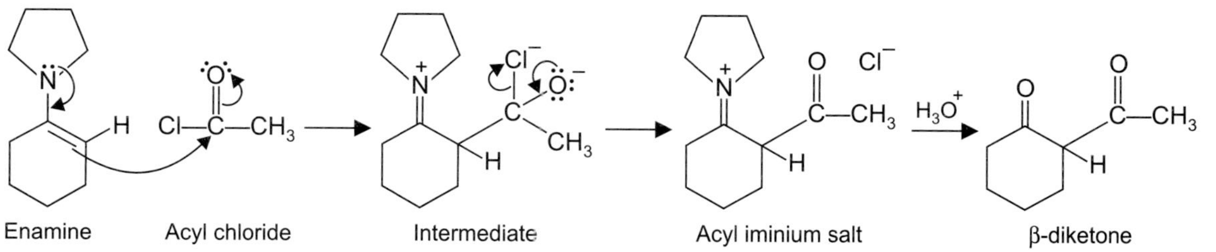

SCHEME 6.42

Enamines display Michael type condensation and behave as addenda to α, β-unsaturated ketones, esters and cyanides etc. (Scheme 6.43).

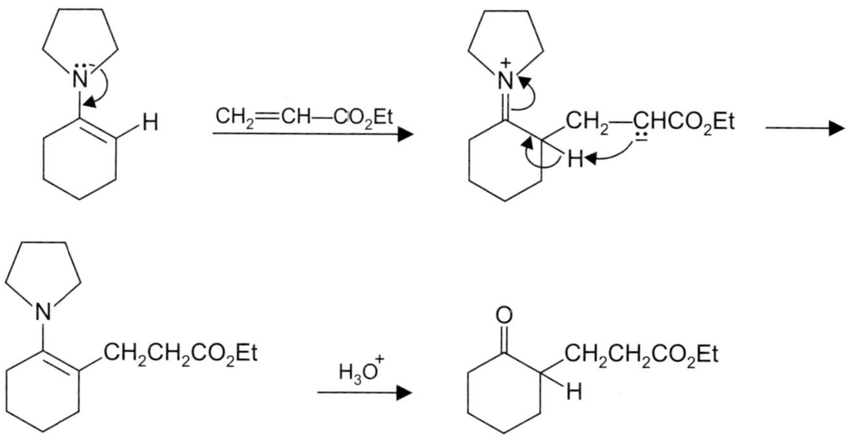

SCHEME 6.43

Some Problems of Working with Enamines—Enamines are Alkylated by Reactive Electrophiles

Among the three important enol equivalents mention may be made of (1) enamines, (2) silyl enol ethers and (3) azaenolates. The following points may be noted:

- *Alkylation of ketones is best carried out by employing lithium enolates (see Scheme 6.3).*

  Since enolates are very strong bases, secondary and tertiary halides undergo predominant E2 elimination (see Scheme 6.3a)
- *Lithium enolate formation is however, not satisfactory with aldehydes since aldehydes are highly electrophilic. Even the use of LDA (– 78°C) is not fast enough to avoid self condensation between the forming lithium enolate and unreacted aldehyde.*
- *Thus aldehydes (and ketones) are best alkylated via their enamines which are stable powerful nucleophiles and are specific enol equivalents (see Scheme 6.41).*
- *Specifically for alkylating aldehydes (and also for ketones) one uses aza-enolates (see Schemes 6.46a and 6.46b).*
- *Using a simple and less reactive alkylating reagent e.g., CH_3I, reaction with an enamine occurs at N instead of C to yield a quaternary ammonium salt which on hydrolysis gives back the starting reactant and therefore, the yields of the desired product are low (Scheme 6.43a). Thus more reactive alkylating agents e.g., benzyl halides, allylic halides and α-chlorocarbonyl compounds work best with enamines.*

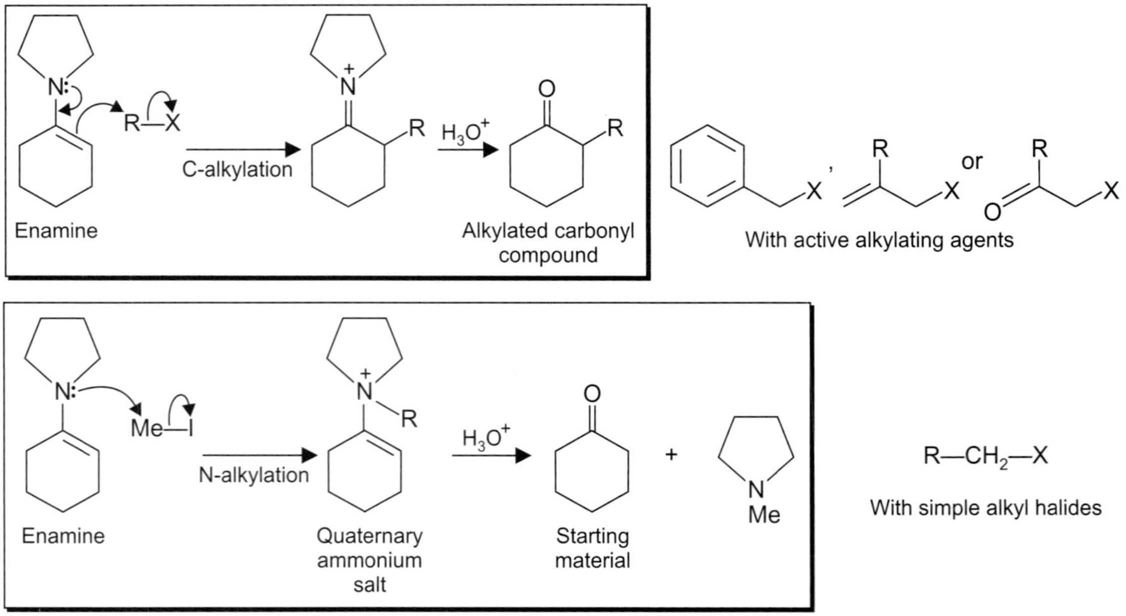

SCHEME 6.43a

Alkylation of Ketones via Silyl Enol Ethers—Silyl Enol Ethers are Alkylated with S_N1 Reactive Electrophiles

Compared to enamines which are potent neutral nucleophiles, silyl enol ethers are comparatively less reactive (less nucleophilic) and need a more powerful electrophile e.g., a tertiary carbocation. One has already seen (see Scheme 6.43a), that secondary and tertiary alkyl halides are not suitable alkylating agents for use with enamines or lithium enolates since elimination will be a predominant reaction. Thus tertiary alkyl halides work as very good alkylating agents for silyl enol ethers (Scheme 6.43b). The enolate is converted into its silyl enol ether which reacts with a carbocation generated from a tertiary alkyl halide in situ with a Lewis acid e.g., $SnCl_4$ or $TiCl_4$.

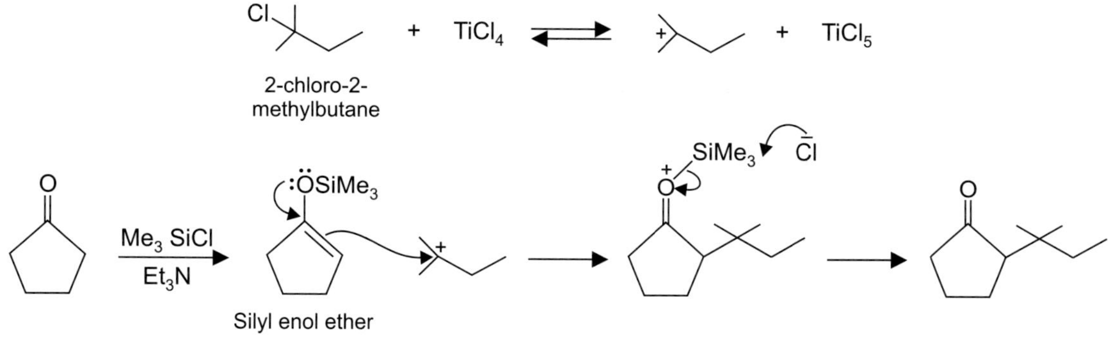

SCHEME 6.43b

An unsymmetrical ketone can react at two sites and the enamine formation occurs predominantly at the less substituted position (Scheme 6.44). The stabilization in an enamine is due to interaction of the alkene π-system with the unshared electron pair in a *p* orbital on nitrogen (compare with data in Scheme 3.35*b*). Consequently a coplanarity of the bonds on the unsaturated carbon atoms and those to nitrogen is favoured. This coplanarity shown by bold lines can be achieved in (I, Scheme 6.44) but not in II due to steric repulsions. In structure (I, Scheme 6.44), the methyl group adopts a quasi-axial conformation so that steric interaction with the amine moiety can be avoided. In structure (II), however, there is a serious steric clash ($A^{1,3}$ strain) which makes it unstable. Thus the enamine from 2-methyl cyclohexanone gives the C-alkylation predominantly on the less substituted carbon (Scheme 6.44). The formation of least substituted enamine minimizes steric interactions to maximize the planarity of N—C=C moiety and hence the overlap of the nitrogen lone pair with the alkene.

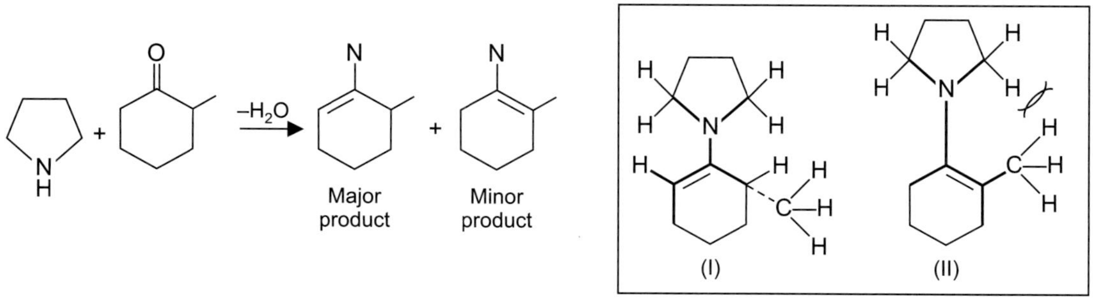

SCHEME 6.44

EXERCISE 6.4

Which of the two enamines will be formed preferentially from the tetralone (I, Scheme 6.45)?

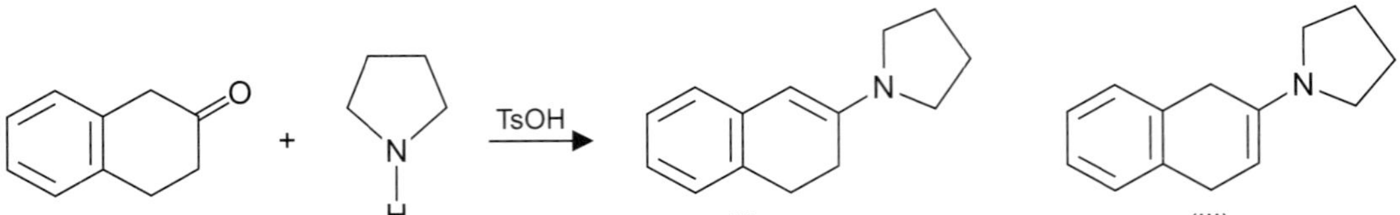

(I) (II) (III)

SCHEME 6.45

ANSWER. *Conjugated anamine (II) is formed exclusively at the expense of the non-conjugated isomer (III).*

EXERCISE 6.5

Show by drawing structures that enamines are closely related to the enolates derived from ketones.

ANSWER. *Enamines are the nitrogen analogues of enols (Scheme 6.46). Thus working with an alkylation of an enamine is the conversion a carbonyl compound*

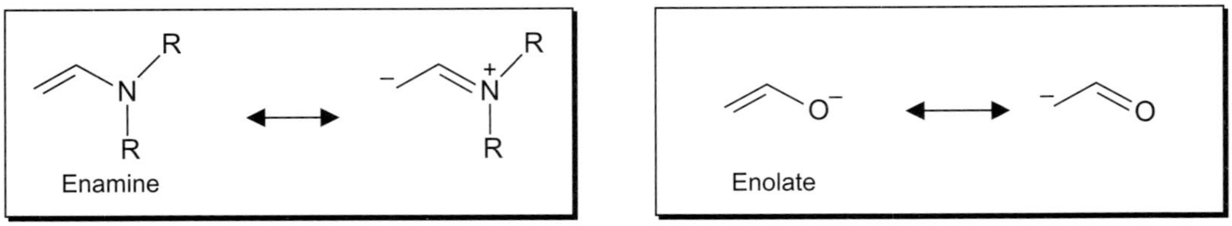

SCHEME 6.46

to another carbonyl compound and the overall process is as if an enolate is alkylated (see Scheme 6.41).The advantage of working with an enamine is that no strong base or an enolate is involved and the problem of self condensation is eliminated.

Aza-Enolates—A Summary of Enol Equivalents for Alkylation of Aldehydes and Ketone

Recall the following observations:

- *Ketones can be best alkylated by using their lithium enolates (see Scheme 6.7).*
- *Enamines are formed by reacting an aldehyde or a ketone with a secondary amine. Enamines can then react with S_N2 reactive haloalkanes (with reactive alkylating reagents i.e., allylic halides, benzylic halides and α-halocarbonyl compounds the yields are indeed high) to yield alkylated aldehydes and ketones (see Scheme 6.41).*

- *Silyl enol ethers derived from aldehydes and ketones can be alkylated with S_N1 reactive electrophiles e.g., tertiary, allylic or benzylic alkyl halides (see Scheme 6.43b).*
- *There is a difficulty with lithium enolate formation. Aldehydes are highly electrophilic and even a strong base like LDA (– 78°C) is unable to bring about fast deprotonation to rule out the aldol self-condensation between the lithium enolate and the unreacted aldehyde (Scheme 6.46a). Thus enamine of an aldehyde (where the aldehyde is present in the masked form during enolization) is instead employed for alkylating an aldehyde.*
- *As an other alternative one works with aza-enolates and aza-enolate alkylation of aldehydes is a successful procedure, which is also applicable to ketones. Recall that enamines can be made by reacting an aldehyde or a ketone with secondary amines. Reaction with a primary amine gives an imine which on reaction with a strong base LDA or Grignards reagent gets deprotonated to give an aza-enolate (Scheme 6.46a).*

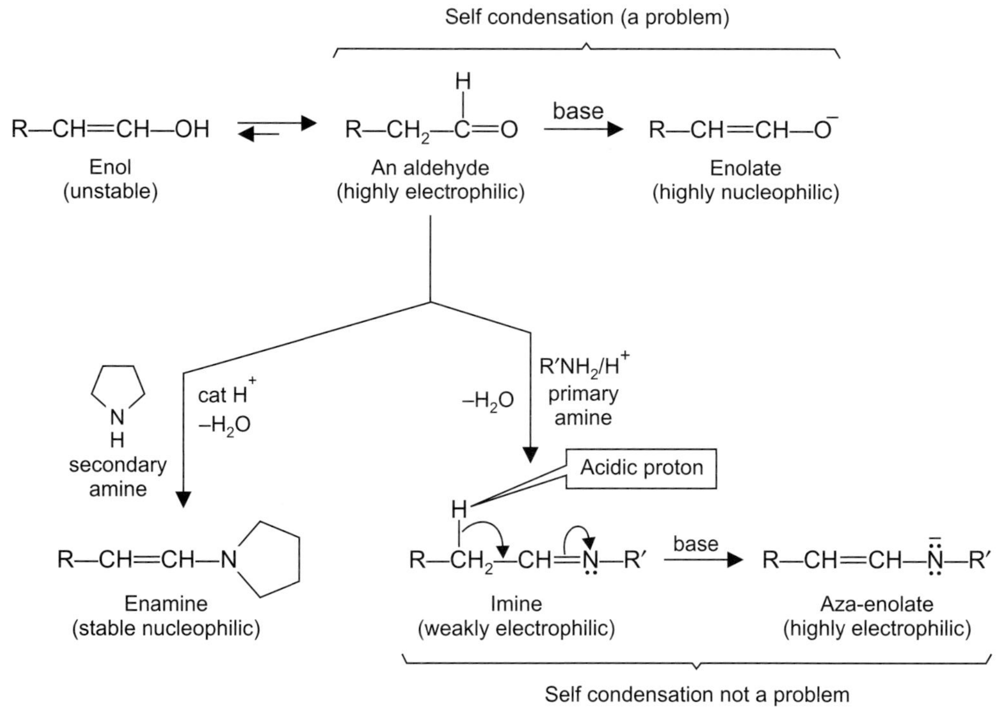

SCHEME 6.46a

- *Thus the alkylation of an aldehyde through an aza-enolate is presented (Scheme 6.46 b).*

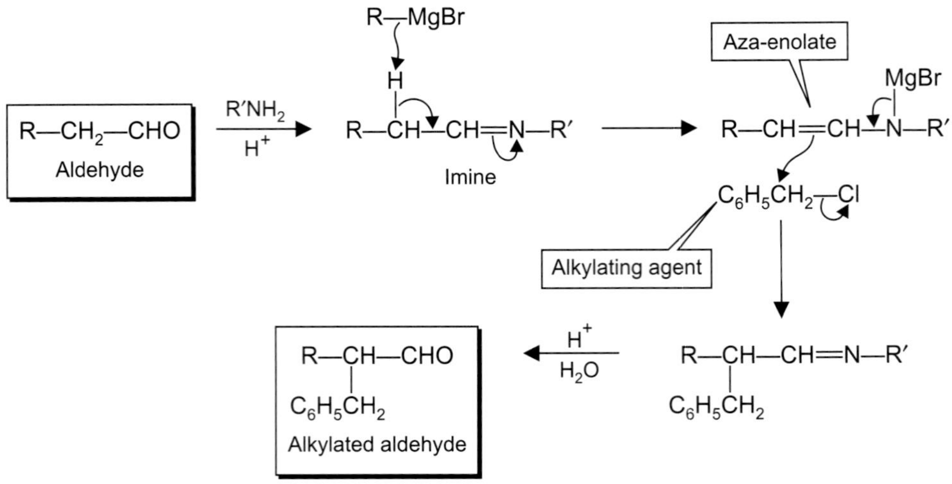

SCHEME 6.46b

6.3 ACID CATALYSED REACTIONS (FORMATION OF CARBON-CARBON BONDS)

(A) Acid Catalyzed Cyclization (Self Condensation of Alkenes)

The pseudoionone prepared by the Claisen-Schmidt reaction (see Scheme 6.18) can be elaborated to β-ionone during the synthesis of vitamin A. On treatment with BF_3 in acetic acid, the ring closure occurs in pseudoionone (Scheme 6.47).This is a typical acid catalysed cyclization of dienes, provided one can end up with stereochemically favoured five-or six-membered rings.

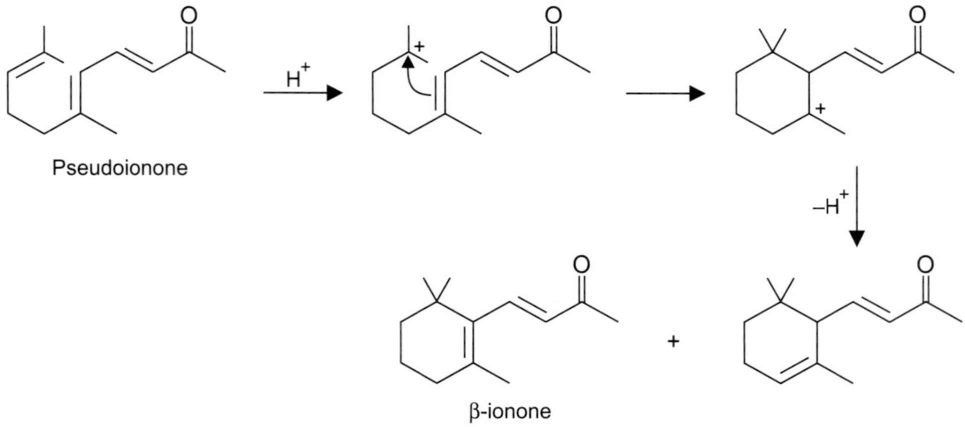

SCHEME 6.47

(B) Carbocations Derived from Aldehydes and Ketones

(i) Prins Reaction

The acid catalysed addition of an olefin to formaldehyde in the presence of an acid is called the Prins reaction. The net result being the addition of hydroxymethylene group to a double bond. The reaction gives a 1, 3-diol together with a cyclic acetal (1, 3-dioxan) which is formed from the diol by the addition of a second molecule of formaldehyde (Scheme 6.48). The mechanism involves electrophilic attack on both the double bonds. The acid protonates the C=O and the

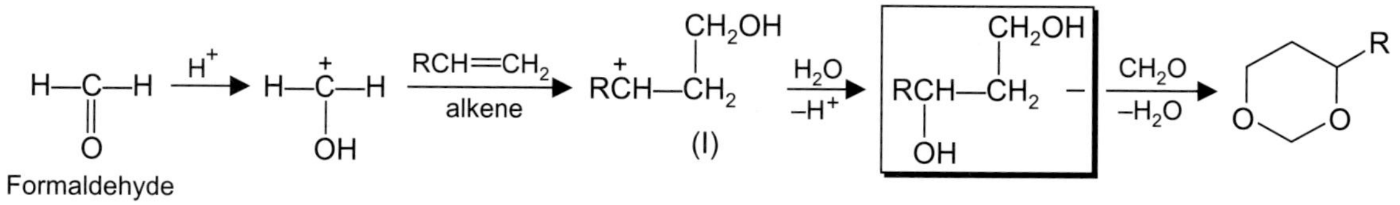

SCHEME 6.48

carbocation thus formed attacks the C=C. The resulting (I, Scheme 6.48) can add water to give a diol as shown or it can also undergo a loss of H^+ to give an olefin (Scheme 6.48*a*). Other examples of Prins reaction include the chloromethylation of phenol with formaldehyde and HCl (Scheme 6.48*a*).

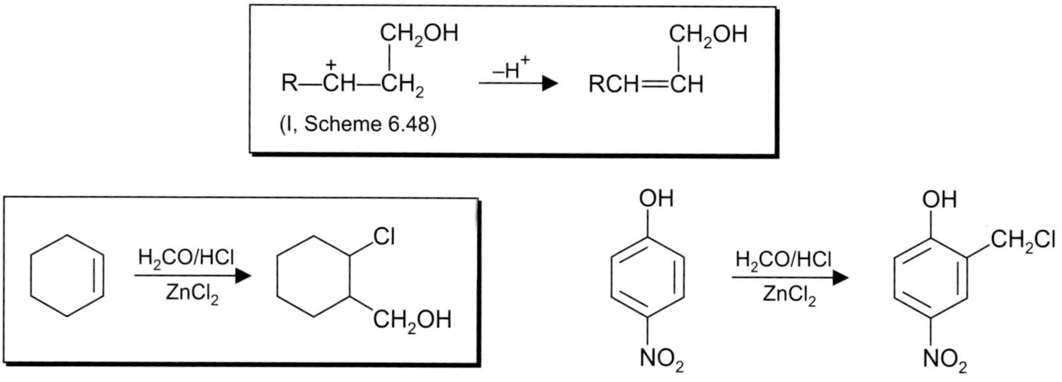

SCHEME 6.48a

(ii) Mannich Reaction

This reaction links together an amine an aldehyde (usually formaldehyde) and the enol of a ketone *i.e.*, a ketone with an α-hydrogen (Scheme 6.49). Thus the net result of Mannich reaction

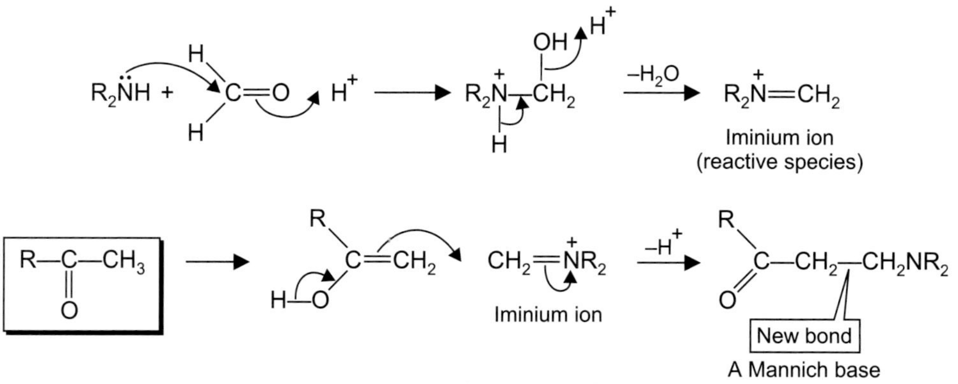

Mechanism of Mannich reaction

SCHEME 6.49

among CH_2O, RNH_2 and $R'COCH_3$ is the extension of the —CH_3 of the ketone by —CH_2NHR. The products of the Mannich reaction are known as Mannich bases and many are useful as intermediates in synthesis.

The reaction of acetophenone with diethylamine and formaldehyde gives the Mannich base which can be elaborated to a variety of compounds *e.g.*, on Hoffmann elimination (see Scheme 15.29) and so on (Scheme 6.50).

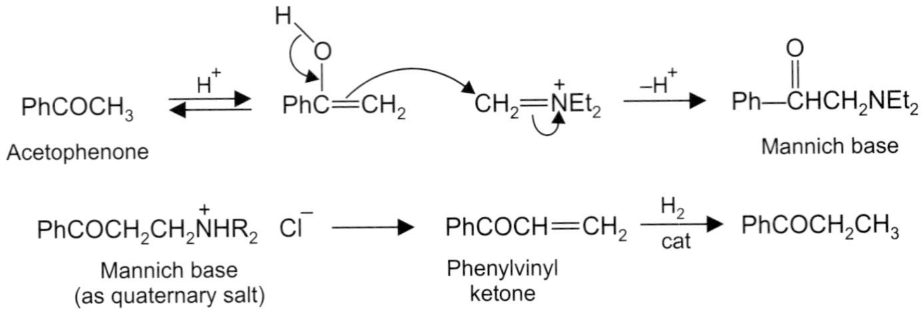

SCHEME 6.50

As seen above (Scheme 6.49) the electrophilic component in a Mannich reaction is the methyleneiminium ion generated from formaldehyde and a secondary amine. The nucleophile may be an enol, an activated arene or a π-excessive heteroarene (Scheme 6.50*a*).

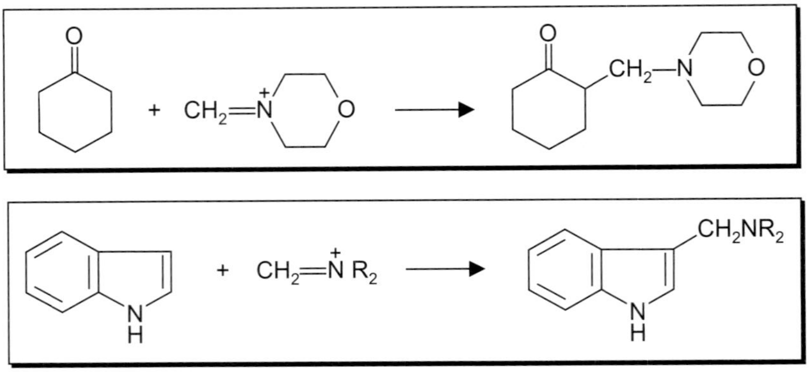

SCHEME 6.50a

EXERCISE 6.6

Why in indole (see Scheme 6.50a) the reactive species, the iminium ion reacts at the 3-position and not at position 2?

ANSWER. *The 2, 3-benzo derivatives of pyrrole, furan and thiophene react in the hetero ring. Substitution at the 3-position e.g., in indole generates a transition state (I, Scheme 6.50b) in which the stabilizing effect of N atom on the transition state is more and the appropriate resonance structure is benzenoid. The benzenoid system is disrupted on 2-substitution as shown (II, Scheme 6.50b).*

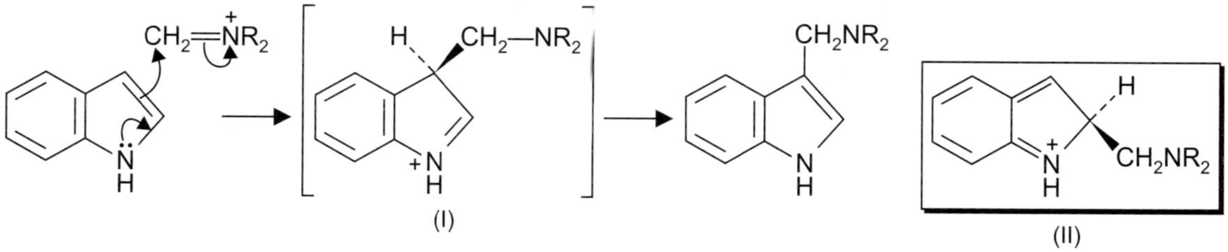

SCHEME 6.50b

The Mannich reaction is an important synthetic pathway in the biosynthesis of alkaloids, and many such reactions have been duplicated under the laboratory conditions. Thus tropinone can be synthesized from succindialdehyde, methylamine and acetone (Scheme 6.51).

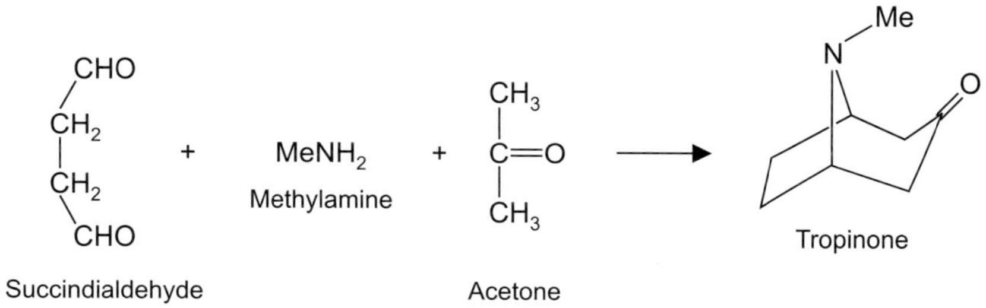

SCHEME 6.51

(iii) Building Heterocyclic Rings

When the ketonic carbonyl group of a β-keto aryl amide or an aryl ester is protonated, a ring closure leads to quinolines or coumarins (Scheme 6.51*a*). In this reaction the keto group is

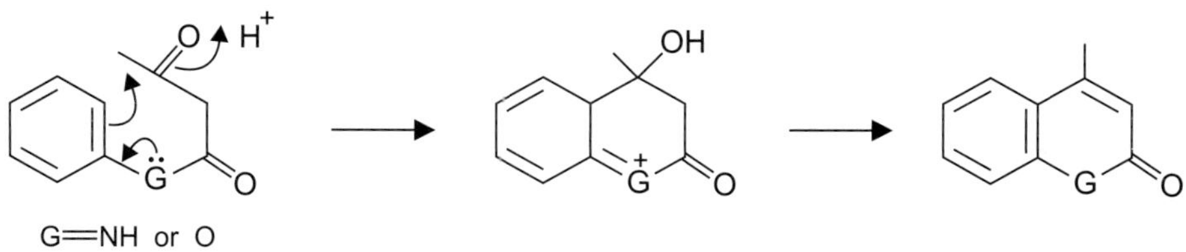

SCHEME 6.51a

acting as an alkylating agent (A Friedel-Crafts type alkylation) in the presence of a proton acid. The amidoketone (Scheme 6.51*a*) can be made by reacting aniline and acetoacetic ester (Scheme 6.51*b*).

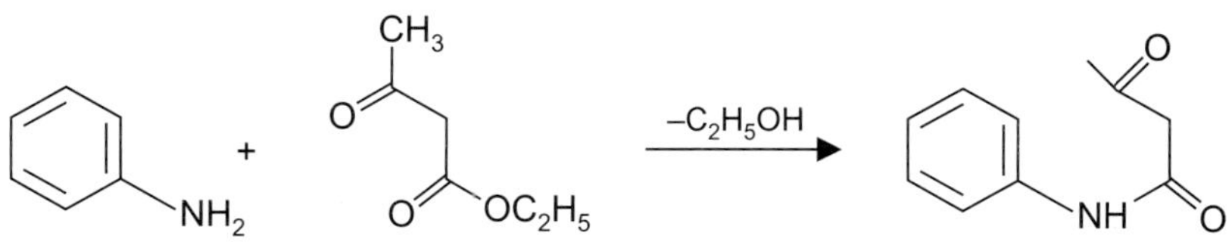

SCHEME 6.51b

6.4 REACTIONS OF CARBOXYLIC ACIDS AND THEIR DERIVATIVES

The carboxyl group (–COOH) containing compounds and their derivatives (*e.g.*, acyl chlorides, acid anhydrides, esters and nitriles) are one of the basic classes of compounds which are important both to organic and biochemistry. The derivatives of carboxylic acids are those functional groups that are converted to carboxylic acids by a simple acidic or basic hydrolysis. The mechanistic considerations of some of these reactions of these classes of compounds are discussed here.

(A) Decarboxylation of Carboxylic Acids

The carboxylate ions ($RCOO^-$) of simple aliphatic acids do not decarboxylate readily, however, the carboxyl radicals ($RCOO^\bullet$) undergo a ready decarboxylation. The one electron oxidation of a carboxylate ion ($RCOO^-$) produces a very unstable $RCOO^\bullet$ radical which decomposes with the loss of carbon dioxide and an alkyl radical. The unstable $RCOO^\bullet$ radical can be produced by electrochemical oxidation or by halogenation of a carboxylate ion. The alkyl radical formed on its decomposition can either be halogenated to the corresponding halo (normally bromo–) alkane in the Hunsdiecker reaction (see Scheme 16.22) or can couple to an alkane (Kolbe electrolysis, see Scheme 16.25).

When special groups are present in the molecule, the decarboxylation becomes rapid enough to be useful synthetically. One such structural feature is the presence of a keto group in the β-position to the carboxylic acid group. Thus, when such an acid decarboxylates it does so through a six-membered cyclic transition state (Scheme 6.52). This mechanism is consistent with the resistance of bridgehead bicyclic β-keto acids (Scheme 6.52) to decarboxylation as the product would be a highly strained bridgehead olefin.

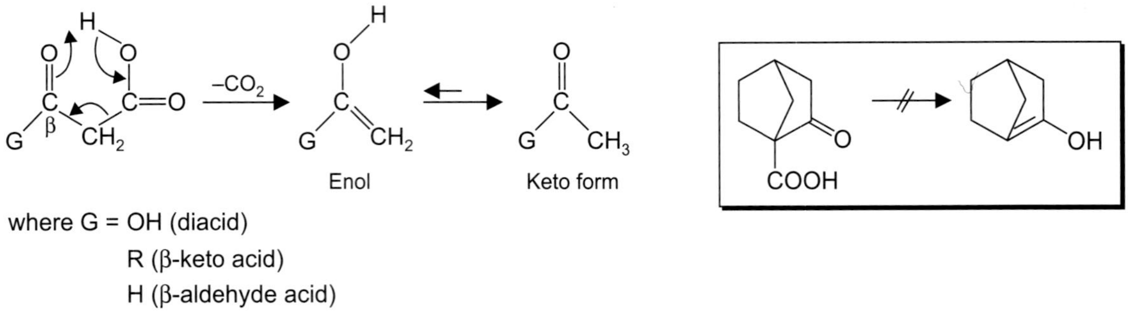

Decarboxylation of β-keto acids

SCHEME 6.52

Loss of a bromide ion from the β-position of a carboxylate group occurs readily under mild basic conditions. The process is decarboxylation accompanied by elimination (Scheme 6.53). Cinnamic acid dibromide is converted into β-bromostyrene on reaction with hot aqueous sodium carbonate solution. Dehydrobromination with potassium hydroxide gives the acetylenic compound phenylacetylene. Bromination of cinnamic acid (Ph CH=CH COOH) gives 2, 3-dibromo-3-phenylpropanoic acid in high yield as the starting compound.

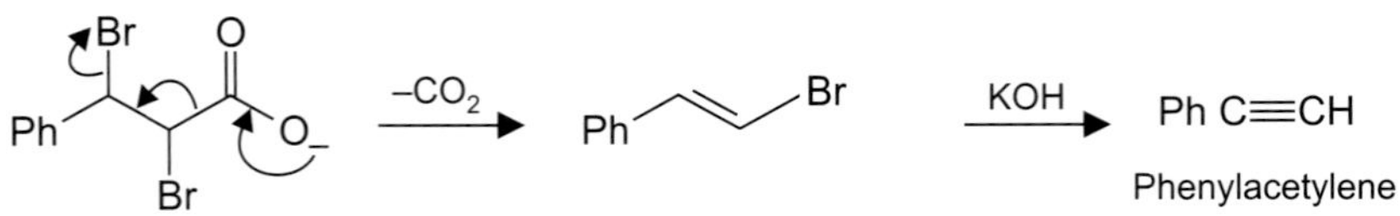

SCHEME 6.53

Decarboxylation of β-keto acids involves both the free acid as well as the carboxylate ion. When the carboxylate ion decarboxylates, it forms a resonance stabilized enolate anion (Scheme 6.54). This carbanion is much more stabler than the anion RCH_2^- which would be formed by the decarboxylation of an ordinary carboxylate ion.

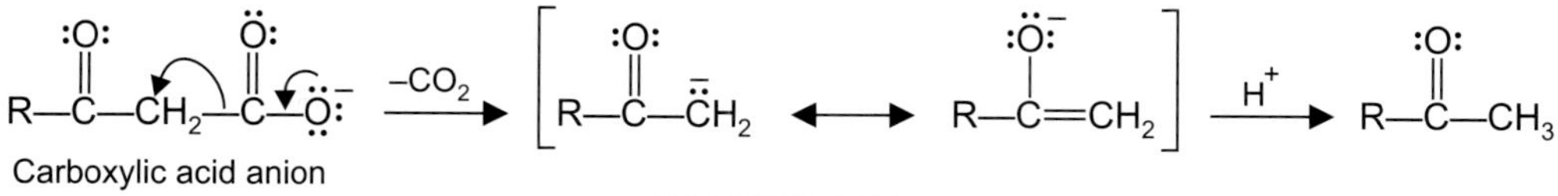

SCHEME 6.54

Pyrrole and furan carboxylic acids decarboxylate when heated (eq. I, Scheme 6.55), and this represents a reaction where during electrophilic aromatic substitution a group other than hydrogen is displaced. (The reaction starts *via ipso* attack). The decarboxylation which is operative when powerful activating substituents are present involves an internal electrophilic substitution by hydrogen, and a similar process is operative when phenolic acids are decarboxylated under acid catalysis (eq. II, Scheme 6.55).

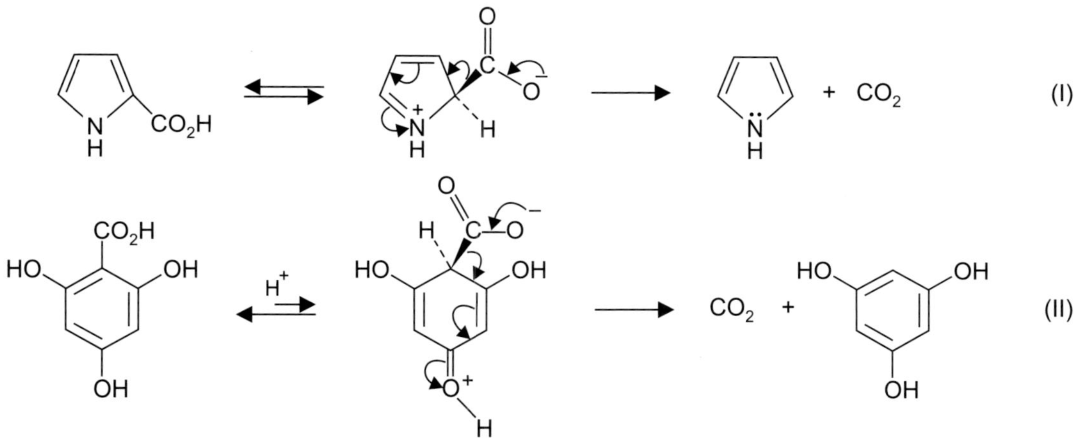

SCHEME 6.55

Aromatic carboxylic acids containing strongly electron-attracting groups, however, decarboxylate *via* the carbanions (Scheme 6.56).

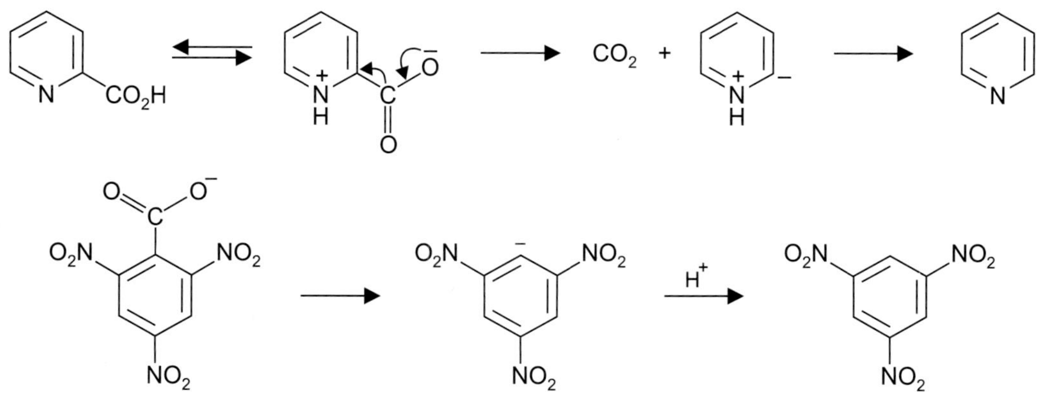

SCHEME 6.56

Oxidative decarboxylation of carboxylic acids has been discussed (see Scheme 13.58).

(B) Ester Hydrolysis and Esterification

The conversion of an ester into its acid and alcohol moieties (eq. I, Scheme 6.57) is termed ester hydrolysis. This hydrolysis can involve cleavage either at the acyl-oxygen or alkyl-oxygen bond (eq. II, Scheme 6.57). A variety of mechanisms may be involved both depending on the nature of R and R′ and the conditions of the reaction. The mechanisms of esterification are based on the principle of microscopic reversibility.

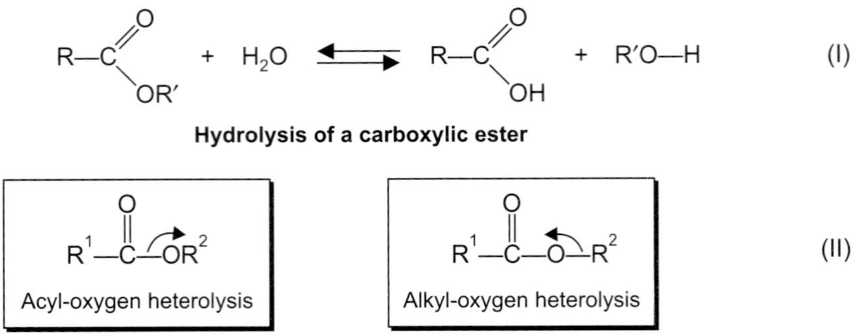

SCHEME 6.57

1. *$B_{AC}2$ mechanism (ester hydrolysis) [Base-catalysed (B), bimolecular (2) hydrolysis with acyl-oxygen (AC) cleavage]*

This is the most common mechanism for base-catalysed ester hydrolysis and involves nucleophilic attack by base on the ester to give an unstable tetrahedral intermediate which then decomposes to the products *i.e.*, an acid and alcohol. The following points proves this mechanism :

(*a*) The use of ^{18}O as an isotopic tracer has shown that the base promoted reaction generally proceeds by acyl oxygen cleavage and this is not dependent on the structure of the R and R′ groups (Scheme 6.58). Thus the alcohol produced (*e.g.*, ethanol if ethylpropionate labelled with ^{18}O, $C_2H_5—CO—^{18}O—C_2H_5$ is used) was found to be enriched in ^{18}O.

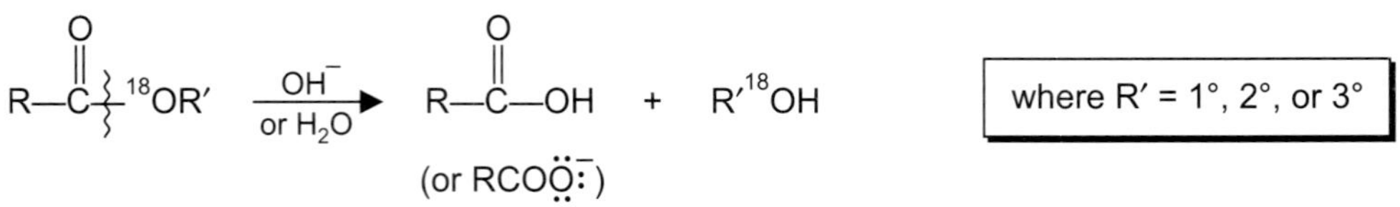

SCHEME 6.58

(*b*) Chemical kinetics of the alkaline hydrolysis of esters is second order, with the rate depending on the concentration of both ester and the base (Scheme 6.59). Thus, the reaction involves the attack on ester by the base.

Rate of reaction = k_2 [ester] [$:\ddot{O}H^-$]

SCHEME 6.59

(*c*) Further evidence for the acyl-oxygen fission during basic hydrolysis is stereochemical in nature. When an optically active ester is employed (Scheme 6.60) and if the $B_{AC}2$

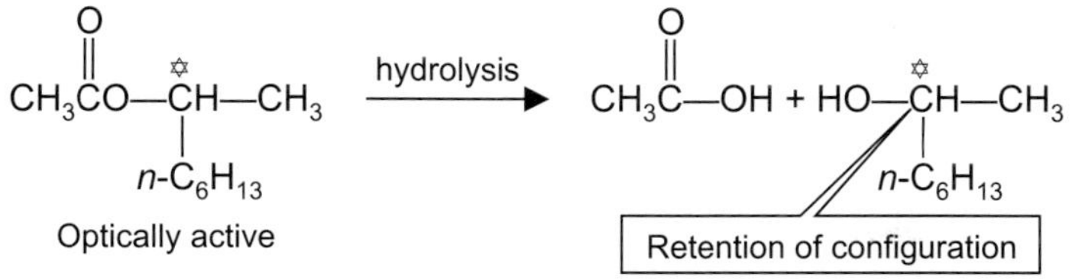

SCHEME 6.60

mechanism is operative then the complete retention of configuration of the alcohol formed is indicative of the formation of the ion RO^- (acyl-oxygen fission) *i.e.*, the bond R—O is never broken thus the mechanism (Scheme 6.61) involving an unstable tetrahedral intermediate is operative. This mechanism is consistent with the following

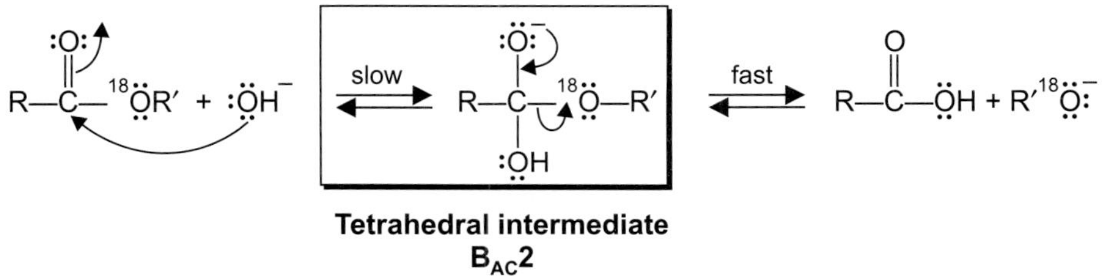

SCHEME 6.61

evidence. When the carboxyl labelled methyl benzoate (Scheme 6.62), undergoes alkaline hydrolysis in ordinary water and the reaction is interrupted after some time the unhydrolyzed ester is found to be a mixture (Scheme 6.62). The exchange of ^{18}O of the labelled ester for ordinary oxygen from the solvent occurs to give the unlabelled ester (*i.e.*, an exchange product is formed).

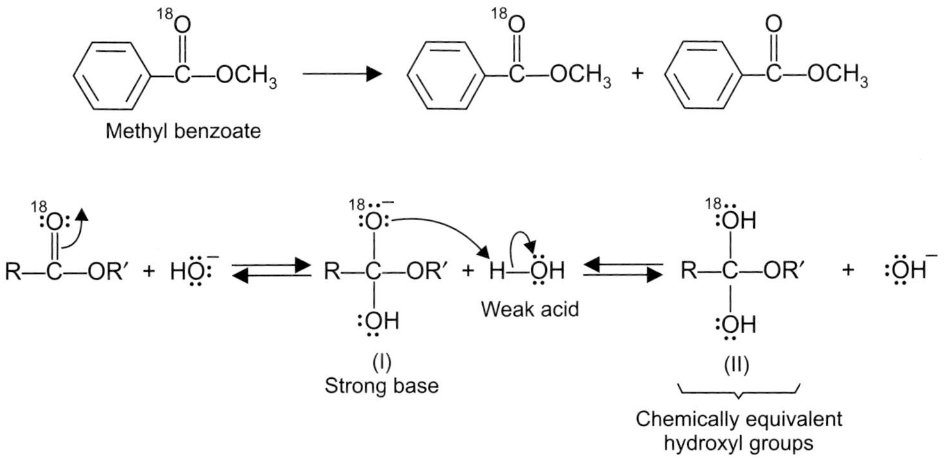

SCHEME 6.62

The initially formed intermediate (I, Scheme 6.62) *via* the addition of hydroxide ion to the starting ester is the alkoxide ion, a strong base. This intermediate reacts with water to give the intermediate (II, Scheme 6.62) having two chemically equivalent hydroxyl groups. Now the hydroxide ion can attack either of these two equivalent OH groups (Scheme 6.63). In case the base removes the proton from ^{18}OH group the original intermediate (I, Scheme 6.62) is reformed which then gives the starting ^{18}O labelled ester (or the hydrolysis products). In case the base removes the proton from the OH group of the intermediate (II, Scheme 6.63), then a new intermediate (III, Scheme 6.63) is formed which can either give the ester with no ^{18}O or could decompose to give hydrolysis products.

Thus in summary this mechanism of base catalysed ester hydrolysis involves nucleophilic attack by base on the ester to yield a tetrahedral ion which rapidly collapses to the products with consumption of one mole equivalent of base.

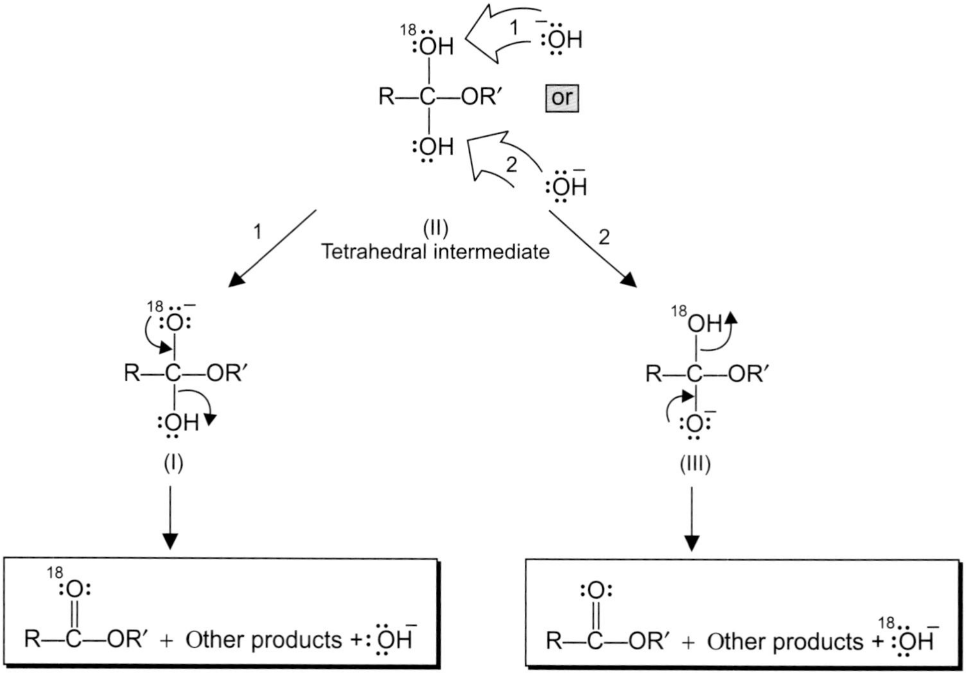

SCHEME 6.63

2. A_{AC} 2 [Acid-catalysed (A) bimolecular (2) hydrolysis with acyl-oxygen (AC) fission]

This is the most common acid catalysed hydrolysis mechanism and involves nucleophilic attack by water on the protonated ester to give a tetrahedral intermediate (infact, more like II, Scheme 6.64) which collapses to protonated acid (and then to acid) and alcohol. The evidence for this mechanism is on the same lines as in the alkaline hydrolysis. The oxygen –18 studies show that the acid catalysed hydrolysis is sensitive to the structure of alkyl group. For primary and most secondary alkyl groups, the acyl oxygen bond is broken (Scheme 6.64).

On the basis of principle of microscopic reversibility (Scheme 4.18*j*), the mechanism of reverse reaction *i.e.*, esterification is thus also contained in the same equilibria (Scheme 6.64).

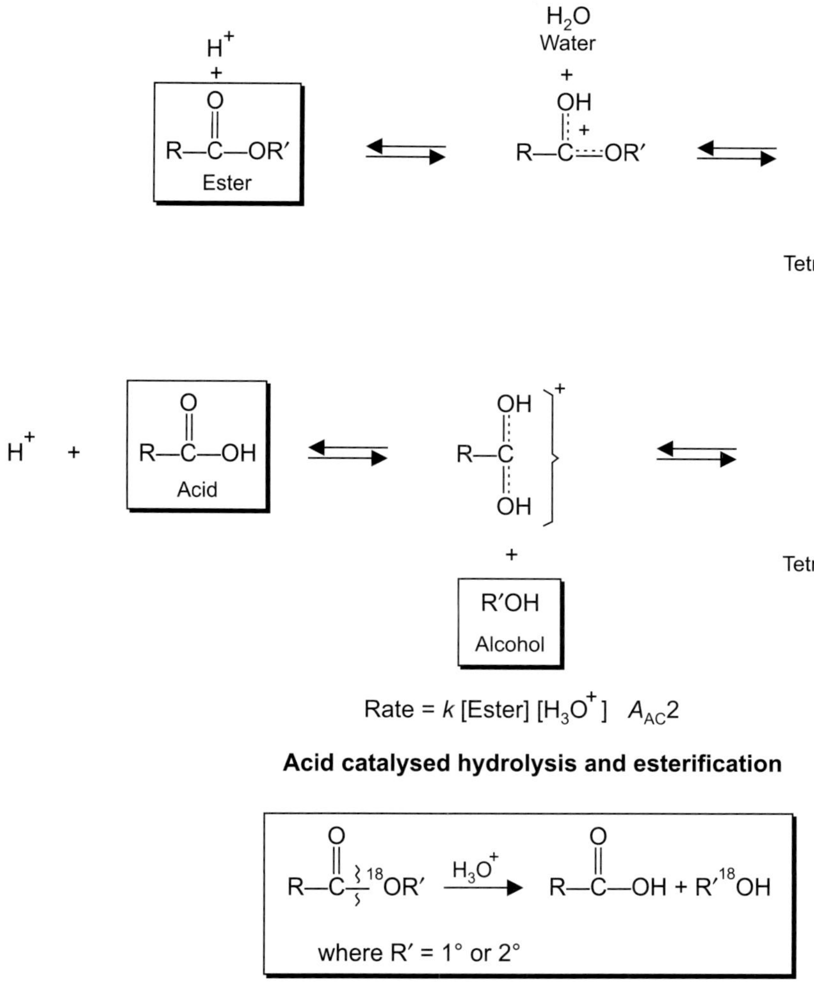

SCHEME 6.64

3. $A_{AL}1$ *[Acid-catalysed (A) unimolecular (1) hydrolysis with alkyl-oxygen (AL) cleavage]*

This hydrolysis involves the fragmentation of a protonated ester to give a carboxylic acid and a carbocation, which is rapidly trapped by water. This mechanism (S_N1 type) is operative only in those cases where R'^+ is a relatively stable carbocation. It is observed that for most tertiary alkyl groups, considerable ^{18}O ends up in the carboxylic acid, to indicate alkyl-oxygen fission (Scheme 6.65). Thus for esters derived from tertiary alcohols the reaction begins with the protonation of the ester (Scheme 6.66) and the leaving group is a stable tertiary carbocation. The carbocation subsequently reacts with water to give an alcohol. In keeping with this mechanism (Scheme 6.66), when the acid catalysed hydrolysis of *t*-butyl acetate is carried out in water enriched with ^{18}O, *t*-butanol containing ^{18}O was isolated.

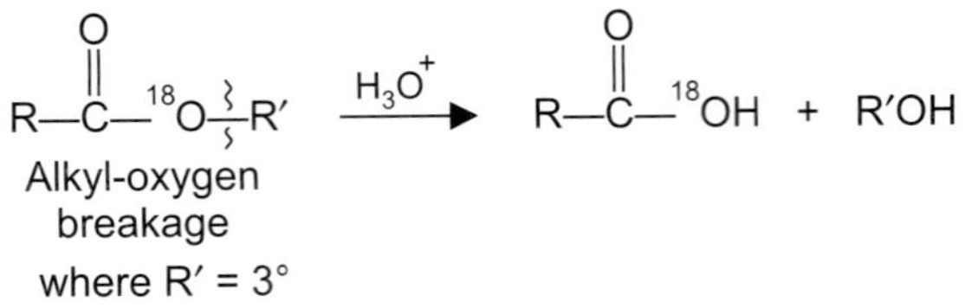

SCHEME 6.65

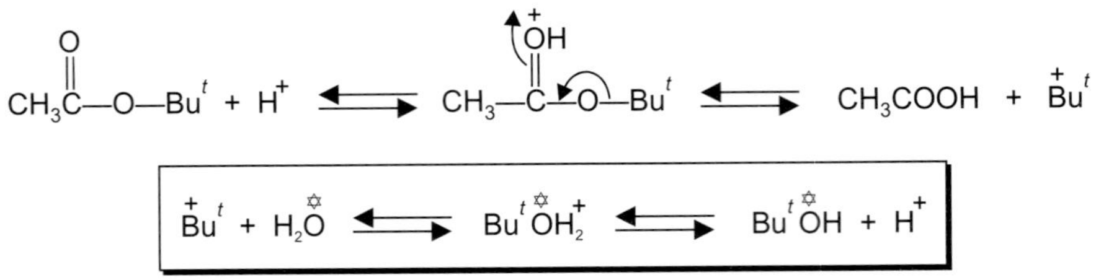

$A_{AL}1$
Common mechanism for *t*-alcohols

SCHEME 6.66

4. $B_{AL}2$[Base-promoted (B) bimolecular (2) ester hydrolysis with alkyl-oxygen(AL) cleavage]

This ester hydrolysis mechanism is operative in rare cases and involves nucleophilic displacement at the alcohol carbon (Scheme 6.67). The process infact is ester cleavage, thus methyl benzoate on reaction with sodium methoxide and methanol gives dimethyl ether and sodium benzoate (Scheme 6.67).

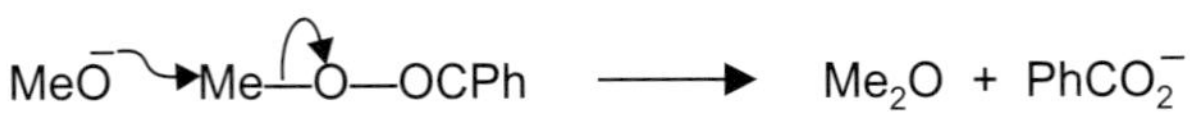

The $B_{AL}2$ mechanism

SCHEME 6.67

5. A_{AC} 1 Ester hydrolysis [Acid-promoted (A) unimolecular (1) ester hydrolysis with acyl-oxygen (AC) cleavage and esterification]

Esterification and hydrolysis of sterically hindered substrates (esters and acids *e.g.*, mesitoic acid or its ester) is carried out by dissolving them in concentrated sulphuric acid and subsequent reaction with water (for hydrolysis) or alcohol (for esterification). The mechanism operates *via* the initial addition of a proton to the substrate and subsequent heterolytic fission in the rate controlling step to give an acylium ion (Scheme 6.68) which then reacts further.

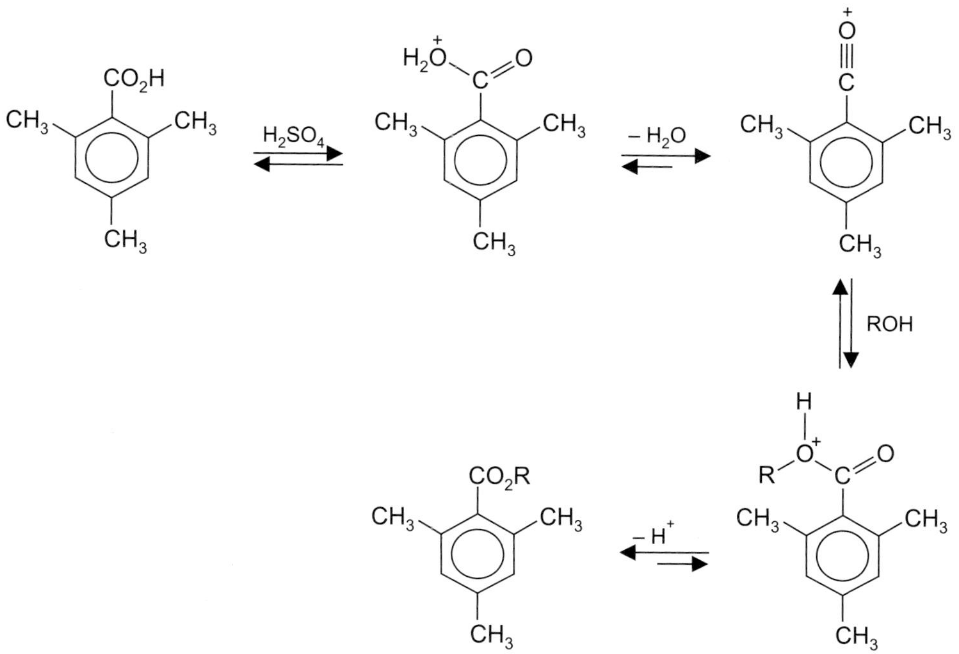

SCHEME 6.68

6.5 HYDROLYSIS OF AMIDES

1. Hydrolysis of an Amide in Aqueous Acid

The following points may be considered:

- The acid protonates the carbonyl oxygen increasing the susceptibility of the carbonyl carbon to nucleophilic attack by water (Scheme 6.69) to give a tetrahedral intermediate.

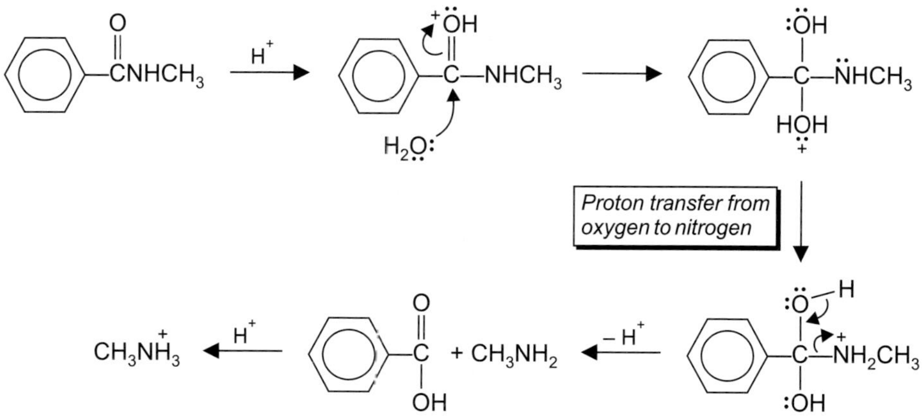

SCHEME 6.69

- Proton is transferred and an amine is expelled. This amine then reacts with H^+ to give an amine salt.

- The acid hydrolysis of amides follows the mechanism in analogy with ester hydrolysis. Under acidic conditions, the formation of the tetrahedral intermediate is promoted (attack by poor nucleophile water becomes feasible). The acid also changes the relative leaving abilities of the two groups on the tetrahedral intermediate. CH_3NH_2 is lost being a weaker base than OH^-.
- The formation of the amine salt at the end (NH_4^+) explains why H^+ is a reactant and not a catalyst and why the reverse reaction does not proceed—although R_2NH is a nucleophile, $R_2NH_2^+$ is not and this ion cannot attack the carbonyl group.

2. Hydrolysis of an Amide in Aqueous Base

The following points may be noted:

- In the hydroxide ion promoted hydrolysis of an amide; OH^- rather than water is the nucleophile. Since OH^- is a better nucleophile than water it is better to form the tetrahedral intermediate (Scheme 6.70).

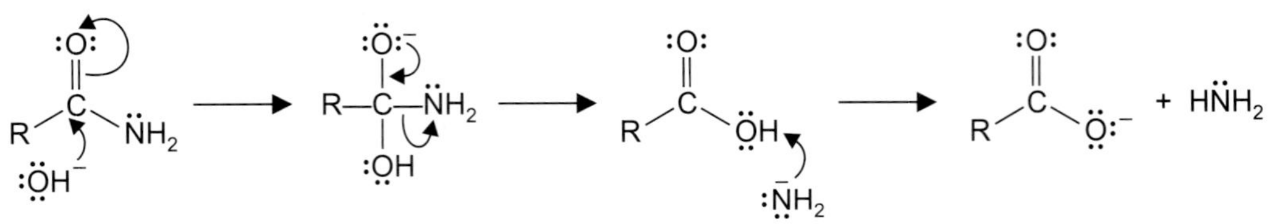

SCHEME 6.70

- The two leaving groups on the tetrahedral intermediate are OH^- and NH_2^- (since OH^- is the weaker base it is more easily eliminated, however, occasionally and more often NH_2^- is ejected).
- When, NH_2^- is ejected, the carboxylic acid thus formed immediately loses a proton. This step is irreversible (the carboxylate ion cannot be attacked by nucleophiles) and the reaction is driven towards products.
- Since hydroxide ion is consumed in the reaction, it is a reagent and not a catalyst.

PROBLEMS

6.1. On reacting acetaldehyde with a large excess of formaldehyde in the presence of a base, pentaerythrtcol, $C(CH_2OH)_4$ is formed. Suggest a mechanism.

6.2. (*a*) How one can extend the aromatic ring in (1) to obtain (A)? (*b*) How a combination of (II and III) can lead to (B).

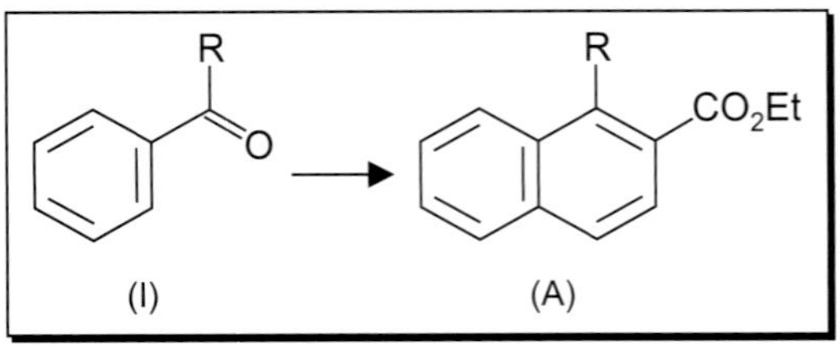

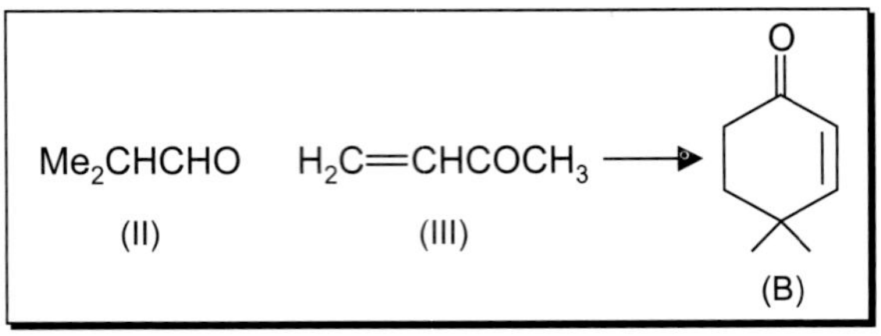

6.3. Ethyl 2-methylpropanoate (I) fails to undergo, the Claisen reaction with sodium ethoxide in ethanol and the possible product (II) is not isolated. Explain.

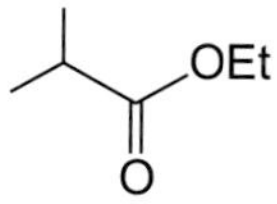

(I)

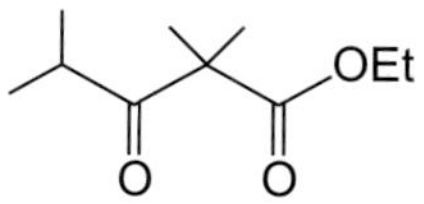

(II)

6.4. Claisen condensation of ethyl 2-methylpropanoate with ethanol and sodium ethoxide is unsuccessful. The expected Claisen product (see Problem 6.3) is obtained by using a much stronger base than ethoxide ion and the triphenylmethide ion (Ph_3C^-, added as sodium triphenylmethyl).

6.5. Intermolecular condensation between oxalic and glutaric esters under base catalysis gives five membered ring compounds. Give the mechanism.

6.6. How an α-halo ester reacts with an enamine? Give the mechanism.

6.7. Which major product would be formed by the reaction of enamine from 2-methylcyclohexanone and pyrrolidine and benzyl chloride?

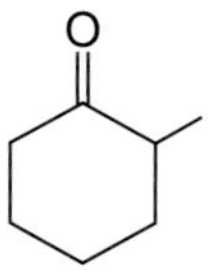

2-methylcyclohexanone

$PhCH_2Cl$

6.8. Explain the way, one may link together, phenol (C_6H_5OH), formaldehyde (HCHO), and a secondary amine (R_2NH)?

6.9. What evidence goes against the following one step mechanism for the basic hydrolysis of an ester ($B_{AC}2$ mechanism)?

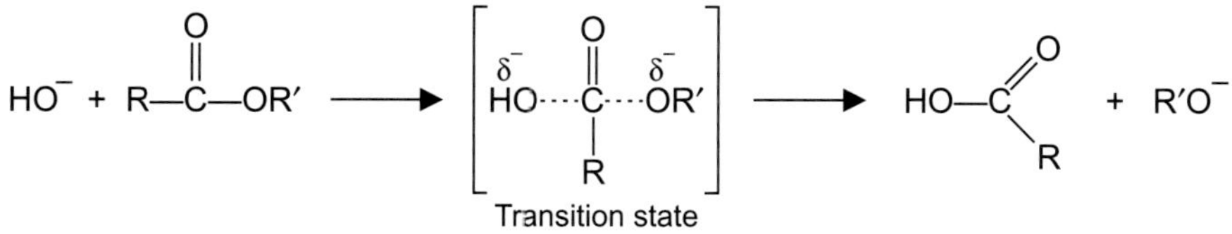

6.10. What is saponification?

6.11. The alkaline hydrolysis of an ester under $B_{AC}2$ mechanism has been presented (Scheme 6.61). The structure of the substrate particularly of the R group (Scheme 6.61) reflects on the different relative rates of hydrolysis. How one explains these differences?

	$\underbrace{Me}_{R}CO_2Me$	$\underbrace{ClCH_2}_{R}CO_2Me$	$\underbrace{CHCl_2}_{R}CO_2Me$	1
Relative rates	1	761	16000	
	$MeCO_2Et$	$EtCO_2Et$	$isoPrCO_2Et$	2
Relative rates	1.00	0.47	0.10	

Alkaline hydrolysis of esters, the $B_{AC}2$ mechanism

6.12. Write the reaction sequence for the conversion of diethylmalonate into mono and disubstituted acetic acids.

ANSWERS TO THE PROBLEMS

6.1. In the case only one of the two components *i.e.*, acetaldehyde has α-hydrogens, thus only two products could be formed. Moreover, it is only acetaldehyde that can form an enolate and the carbonyl group in formaldehyde is more reactive toward nucleophilic substitution; consequently β-hydroxypropionaldehyde predominates. However, this first condensation product still contains two acidic α-hydrogens and condensation continues until all the α-hydrogens in acetaldehyde have reacted. The final product (I) like formaldehyde has no active hydrogen and thus undergoes crossed Cannizzaro reaction (see Schemes 14.79 and 14.80) to give the final product.

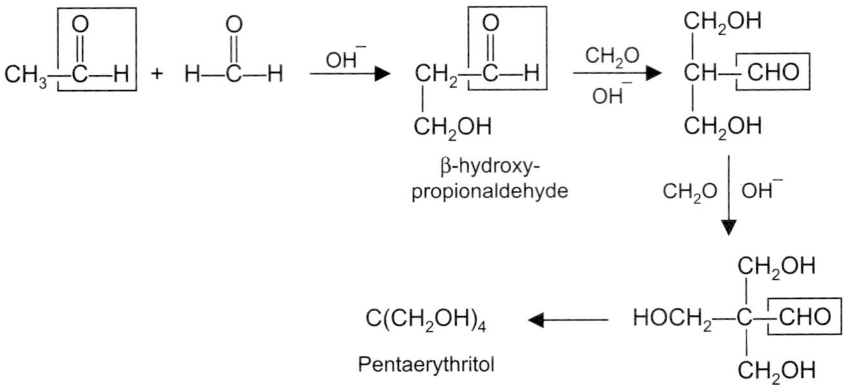

6.2. (*a*) The stobbe condensation on (1) leads to an attachment of a three carbon chain to the ketonic carbon atom. This undergoes ring closure *via* a Friedel-Crafts reaction. Its reduction and dehydration (gain in aromatic stabilization energy) gives (A).

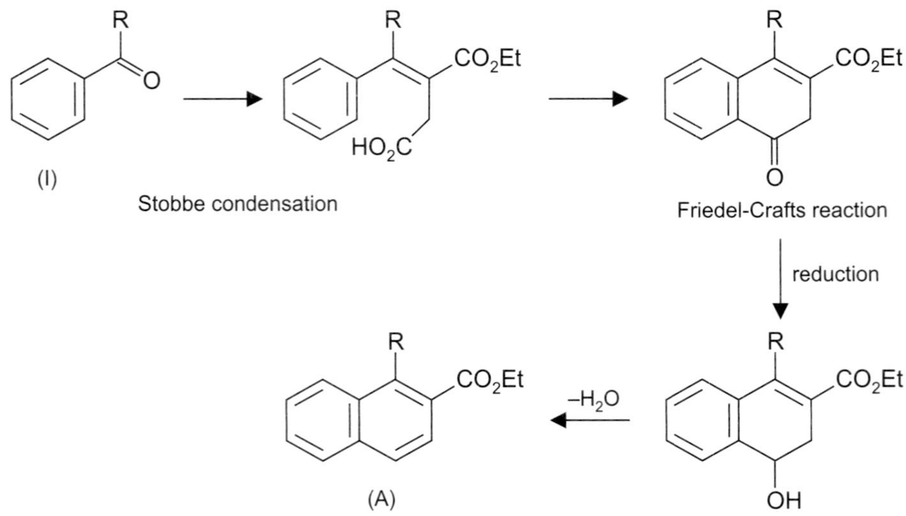

(*b*) The first reaction is the Michael addition of the enolate anion to methyl vinyl ketone followed by intramolecular aldol condensation.

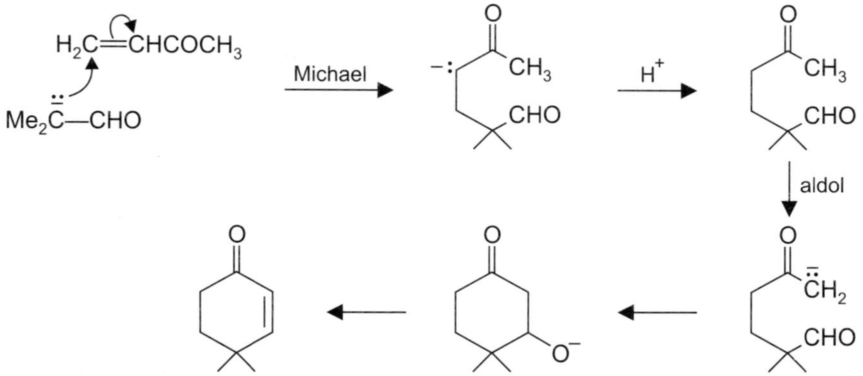

6.3. This failure reflects on the important of the final enolate anion which directs the equilibrium to the product. The ester (I) has only one α-hydrogen atom, and the expected β-keto ester product (II) which would result, is without any hydrogen atoms located between the carbonyl groups. Thus the final enolate cannot be formed under the reaction conditions.

6.4. The added Ph_3C^- ion leads to complete removal of ethanol formed in the reaction EtOH + $Ph_3C^- \rightarrow EtO^- + Ph_3CH$.

6.5.

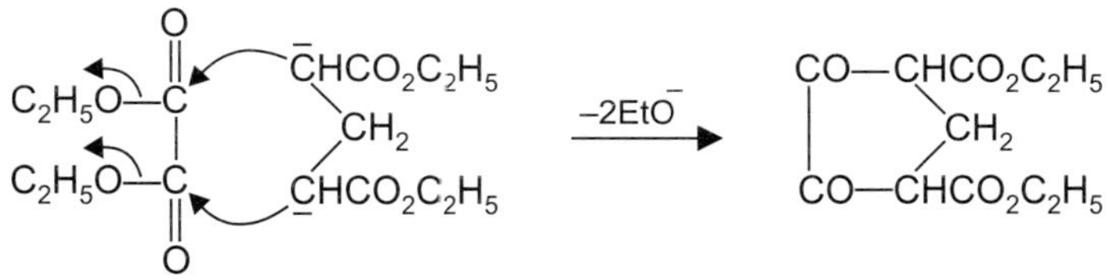

6.6. This reaction is similar to enamine alkylations and affords γ-keto esters.

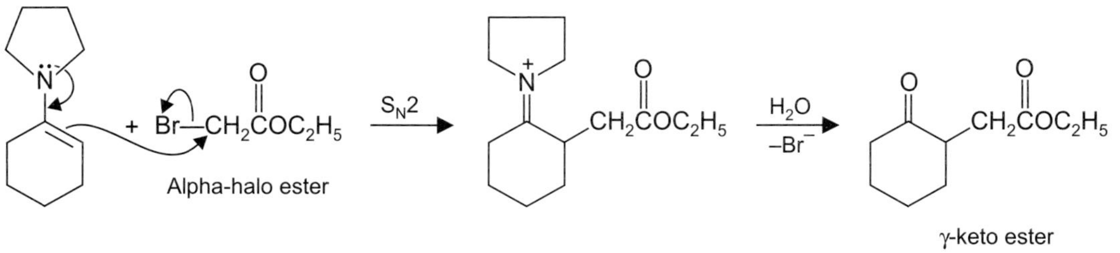

6.7. Of the two possible enamines (I and II), the enamine (I) will be formed as the major product (for reasoning see Scheme 6.44 which will lead to III as the final major product.

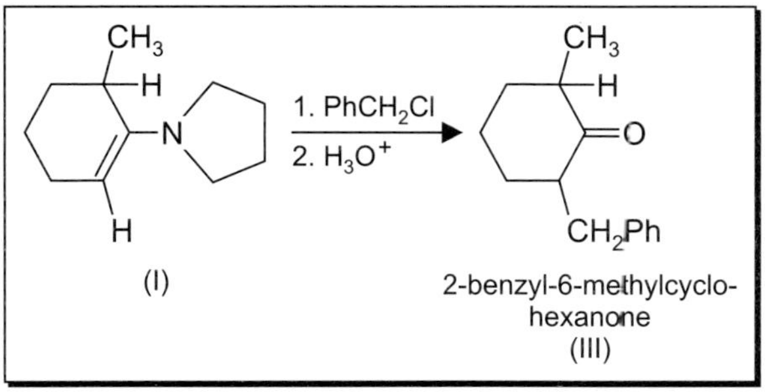

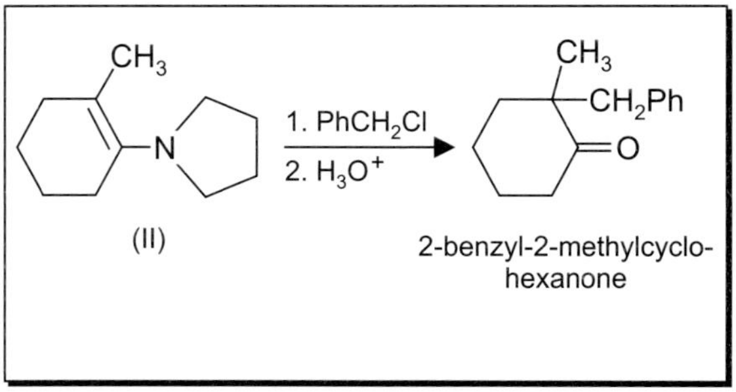

6.8. This is Mannich reaction, the active hydrogen compound being phenol (the hydrogens in the *o* and *p* positions are active) and the result is amino-alkylation in the *ortho*-position which is preferred.

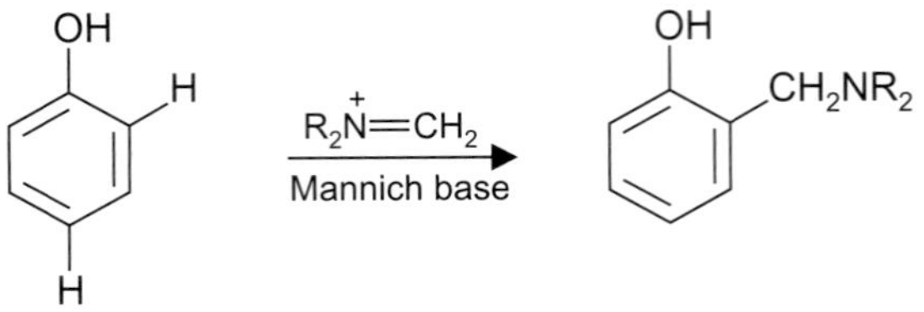

6.9. The observed oxygen exchange (Scheme 6.62) is not in keeping with this one step process (with a transition state as in S_N2 reaction).

6.10. This is an irreversible base (*e.g.*, OH^-) induced ester hydrolysis, the resonance-stabilized carboxylate anion shows little tendency to react with the alcohol.

6.11. In series 1, the increasing –*I* effect of group R accelerates the hydrolysis since a positive charge on the attached carbon facilitates the attack by hydroxide ion. Moreover, with the increase in the bulk of R one would expect (series 2), a steric retardation as well.

6.12. The nucleophilic ion after the removal of one of the α-methylene protons undergoes an S_N2 reaction with an alkyl halide to give a C–substituted malonic ester.

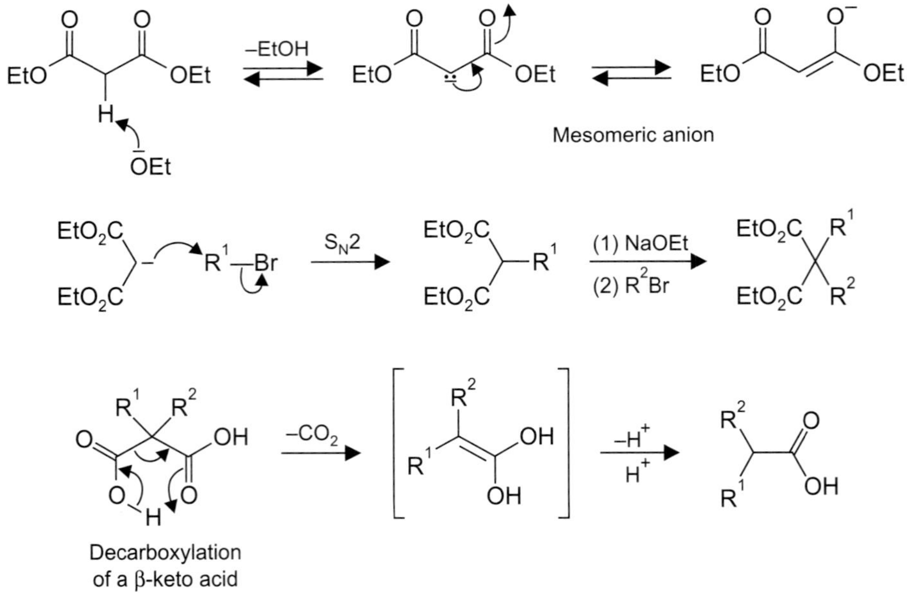

One may introduce similarly a second different alkyl group, or alternately two identical alkyl groups can be introduced in a one-step operation by employing appropriate amounts of reactants. On alkaline hydrolysis and careful acidification an alkyl malonic ester gives an alkyl malonic acid. The alkyl malonic acids (β-keto acids) undergo decarboxylation to give the mono- or disubstituted acetic acids.

CHAPTER 7

Reagents in Organic Synthesis and Relevant Name Reactions

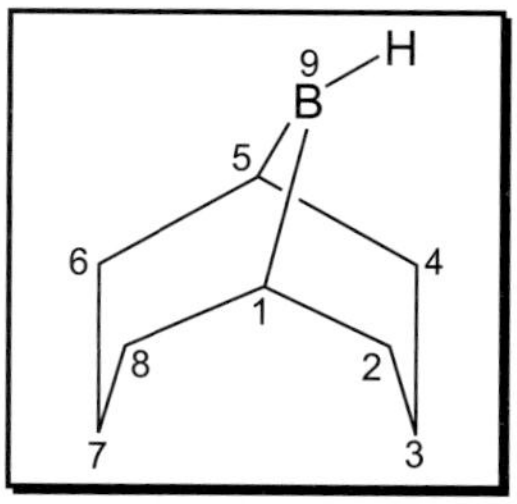

7.1 ORGANOTRANSITION METAL REAGENTS

(A) Introduction

The organoiron compound ferrocene (see Scheme 2.41) was prepared in 1951. Its unusual stability as well as bonding (holding iron to carbon) laid a strong foundation for the synthesis of organic complexes of the transition metals. Structure of ferrocene was elucidated by Wilkinson who received the Nobel Prize for this work (1973). Many of the transition metal complexes display extraordinary power as catalysts and selectivity.

The transition metals are defined as those elements which have partly filled *d* or *f*-shells, either in the elemental state or in their compounds. The transition metals react with a variety of groups or molecules called ligands (L) to yield transition metal complexes. Each ligand is bonded to the metal by overlap of an empty orbital on the metal with a filled orbital on the ligand. The bonding is thus covalent, but with considerable polar character which depends on the extent to which the positive and negative charges on the metal and ligand help to hold them together.

(B) The Catalytic Effect of Complexes

The ligands present on the metal are not directly involved in the reaction which is catalysed by a transition metal complex, however, their presence on the metal is indeed necessary. Their electronic or steric effects, lipophilicity or chirality largely determines the course of a reaction. Moreover, the ligands bring about the solubility of the complex in organic solvents. The catalytic effect is exerted by the metal by bringing the substrate and the reagent (*e.g.*, hydrogen and an alkene during hydrogenation with Wilkinson's catalyst, Scheme 14.16) close to each other via bond formation. Transition metal complexes can adopt a variety of geometries depending on the transition metal and the number of ligands present. Thus rhodium can form complexes with four ligands and is then square planar. When it forms a complex with five ligands, the geometry attained is trigonal bipyramidal while the complex holding six ligands is octahedral.

(C) Oxidation Number or Oxidation State (of Central Metal in Coordination Compounds)

This is the charge left on the central metal atom after all the ligands have been removed (theoretically) in their closed shell electronic configuration.

(D) The 18-Electron Rule

The transition metals have a partially filled *d*-orbital shell with a tendency to surround themselves with 18 electrons which is an electronic configuration corresponding to an "inert" rare gas (*e.g.*, $3d^{10}$, $4s^2$, $4p^6$ for Kr). When the metal of a transition metal complex has 18 valence electrons, it is called coordinatively saturated.

An exception is titanium which is low in available electrons and would be overcrowded if sufficient ligands were present to give a total of 18. The number of outer shell electrons in some of the metals are:

Outer shell electrons:	4	6	8	9	10
Metals:	Ti	Cr	Fe	Rh Co	Ni Pd

Thus chromium can bond to ligands which provide a total of 12 electrons.

(E) Total Valence Electron Count of the Metal in a Complex

The subtraction of the oxidation state of the metal in a complex from the total number of a valence electrons of the metal in the elemental state gives the *d* electron count (d^n). The total electron count of the metal in the complex is obtained by adding the number of electrons donated by all the ligands to d^n. The iron of ferrocene (see Scheme 2.41) has 18 valence electrons and is therefore, coordinately saturated. One calculates this number as follows:

- Iron has 8 valence electrons in elemental state
- Its oxidation state in ferrocene is +2
- Thus $d^n = 6$
- Each cyclopentadienyl ligand of ferrocene donates 6 electrons to the iron
- Thus for iron the valence electron count is $6 + 2 \times 6 = 18$

In the rhodium complex $Rh[(C_6H_5)_3P]_3\ H_2Cl$ (an intermediate in some alkene hydrogenations Scheme 14.16), the rhodium is coordinatively saturated since the total number of valence electrons is 18. The oxidation state of rhodium in the complex is +3. The two hydrogen atoms and the chlorine are each counted as – 1 and the charge on each of the triphenylphosphine ligands is zero. The removal of all the ligands would leave a Rh^{3+} ion. The d^n for the rhodium of the complex is $9 - 3 = 6$. Each of the six ligands of the complex donates two electrons to the rhodium in the complex to give the total 18.

(F) The Commonly Encountered Reaction Mechanisms

(i) Ligand Dissociation—Association (Ligand) Exchange

A complex which fulfils the 18 electron rule must necessarily dissociate (by the loss of a ligand with a pair of electrons to give a complex in which the metal has only 16 electrons and it becomes coordinatively unsaturated) before it can associated (coordinate) to become coordinatively saturated again.

In this process there is no change in the oxidation state of the metal, but only its electron count changes. A *ligand substitution* can take place by the dissociation of one ligand and the association of another (Scheme 7.1).

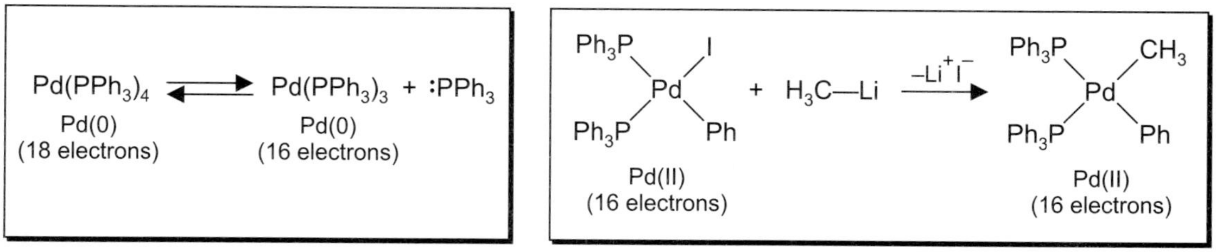

Ligand Dissociation-Association; Ligand Substitution

SCHEME 7.1

(ii) Insertion Reaction

It is the cleavage of a metal-ligand σ bond M—L, in complex A—M—L and the apparent insertion of a coordinated ligand A between M and L to give M—A—L. Thus the electron count of the metal decrease by two. These reactions are reversible, and the reverse reaction is called deinsertion (Scheme 7.1*a*).

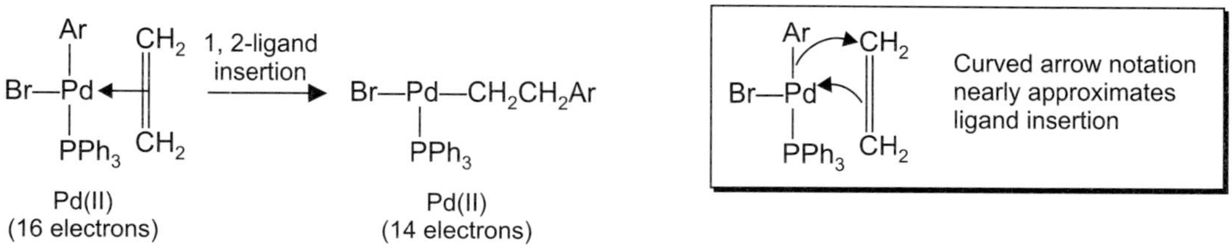

Ligand insertion

SCHEME 7.1a

(iii) Oxidative Addition—Reductive Elimination

This is the cleavage of a reactant molecule accompanied by addition of one or both its parts to the central metal atom of the complex. As a result of this, σ bonds are formed between the metal and the entering ligand(s). This requires formal electron donation from the metal, this process result in an increase in both the coordination number of the metal as well as in its oxidation number. Reductive elimination is the reverse of oxidative addition.

An example of oxidative addition is provided by the insertion of Pd into the carbon-halogen bond of iodobenzene (Scheme 7.1*b*). Electrons flow from the aryl halide to the metal

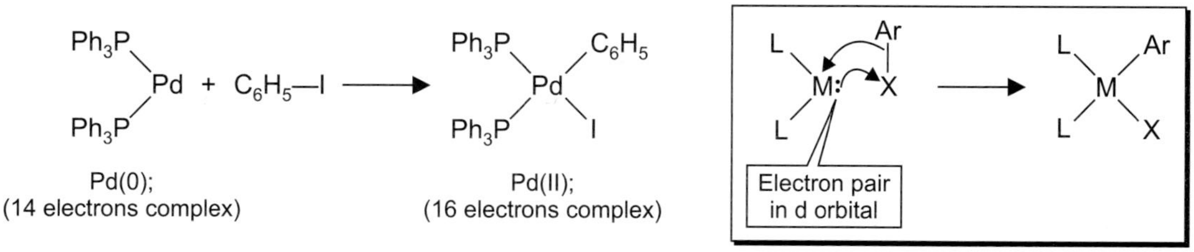

Oxidation addition

SCHEME 7.1b

and at the same time from the metal to the aryl halide. Reductive elimination can be exemplified by the formation of carbon-carbon double bond between two ligands within a nickel complex (Scheme 7.1*c*).

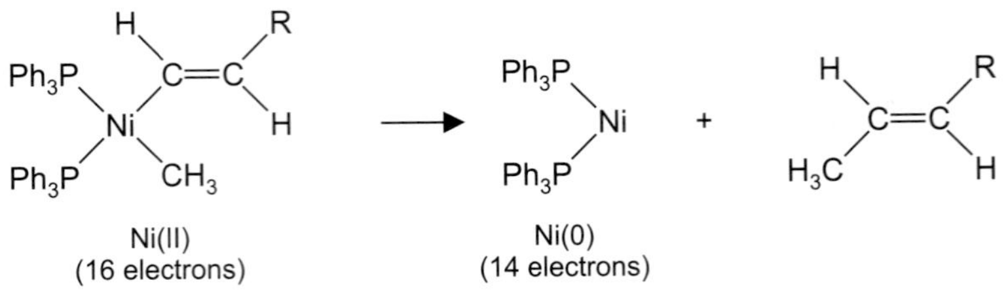

Reductive elimination

SCHEME 7.1c

7.2 SOME TRANSITION METAL ORGANOMETALLIC REACTIONS

(A) Rhodium Complexes—Wilkinson's Catalyst $(Ph_3P)_3$ RhCl

(i) Decarbonylation of Aldehydes

Wilkinson's catayst exists in a coordinatively unsaturated 16 electron state. The first step is therefore, oxidative addition which changes the electron count of the metal by +2. This reaction is of value when a carbonyl group is needed for activation in a step in a synthetic sequence and has to be removed subsequently. Decarbonylation with Wilkinson's catalyst goes through complexes of the type (I and II, Scheme 7.1*d*), is an intramolecular reaction and gives a retention of configuration at a chiral R.

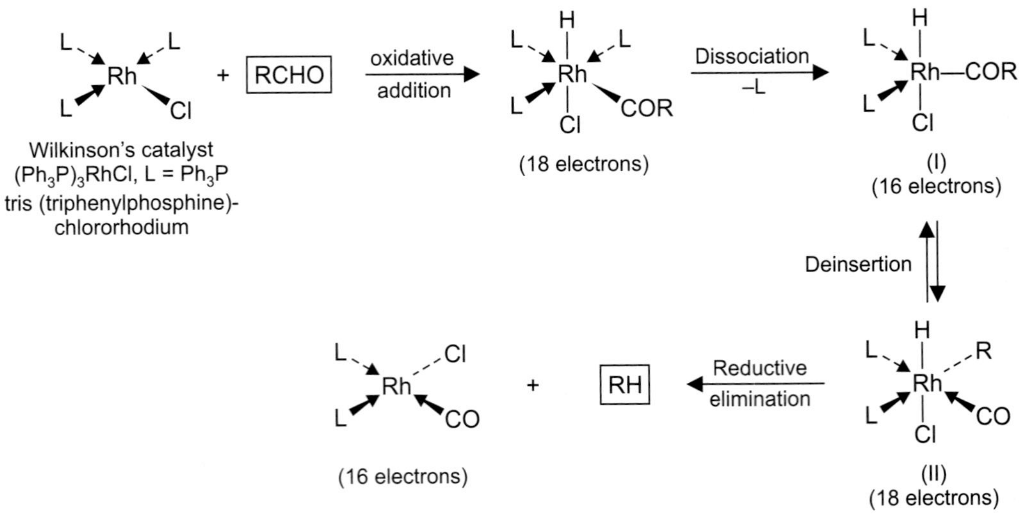

Decarbonylation of an aldehyde with Wilkinson's catalyst

SCHEME 7.1d

(ii) Formation of Carbon-Carbon Bonds

Rhodium complexes have also been used in the formation of carbon-carbon bonds (Scheme 7.1*e*).

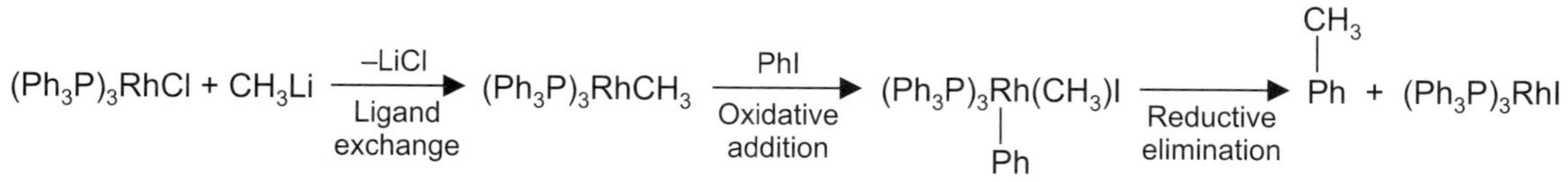

SCHEME 7.1e

In the example, the first step involves a ligand exchange by a combination of ligand association-dissociation step and a methyl group is inserted into the coordination sphere of rhodium. An oxidative addition then incorporates the phenyl group into the rhodium coordination sphere and subsequent reductive elimination then joins the methyl group and the benzene ring to yield toluene in the last step.

(iii) Hydrogenation of Alkenes using Wilkinson's Catalyst

For details see Scheme 14.16.

(B) Organo Palladium Compounds

(i) The Heck Reaction—Arylation and Alkylation of Olefins

It is a useful reaction for making carbon-carbon bonds. Aryl and alkenyl halides react with alkenes in the presence of catalytic amount of Pd when the halide is substituted by the alkenyl group. In this coupling reaction the R group in RPdX (X = halide or acetate) replaces hydrogen at the less hindered carbon atom of the alkene. R can be aryl or alkenyl and alkyl groups without a β-hydrogen (*e.g.*, CH_3, $PhCH_2$, $Me_3C—CH_2$, Scheme 7.1*f*). Ethylene is an effective olefin, with increased substitution the reactivity is lowered thus substitution takes place at the less substituted side of the double bond.

The species RPdX (*e.g.*, ArPdI) can be made *in situ* by the treatment of an aryl iodide with palladium acetate in the presence of a base such an tributylamine or potassium acetate (ArI $\rightarrow$ "ArPdI"). A reactant can be a simple olefin or it may carry a variety of functional

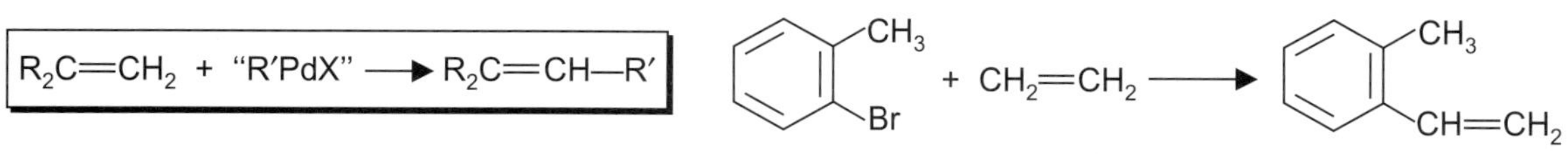

Heck reaction

SCHEME 7.1f

groups *e.g.,* ester, carboxyl, phenolic or cyano. Primary and secondary allylic alcohols afford aldehydes and ketones with double bond migration (Scheme 7.1*g*).

$$CH_2{=}CH{-}CH_2OH \xrightarrow{\text{"PhPdCl"}} PhCH_2CH_2CHO$$

SCHEME 7.1g

The Heck reaction is stereospecific and occurs by *syn*-addition of RPdX (Scheme 7.1*h*) to form a π-complex with the alkene and this rearranges to a σ complex by forming a carbon-carbon bond. The σ complex decomposes by *syn*-elimination (β-elimination). The sequence of reactions involves an inversion of configuration of the carbon where substitution takes place (Scheme 7.1*i*).

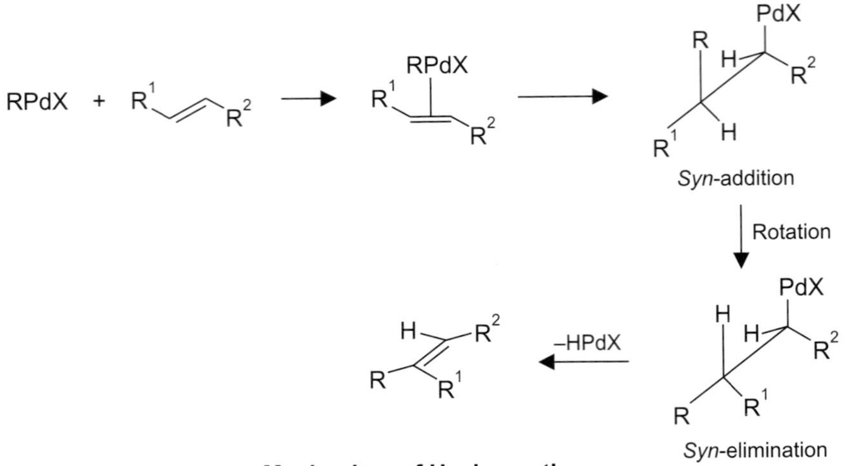

Mechanism of Heck reaction

SCHEME 7.1h

Thus when an acyclic alkene is employed in the Heck reaction, since internal rotation is possible, the hydrogen atom on the carbon at which insertion occurs gets eliminated.

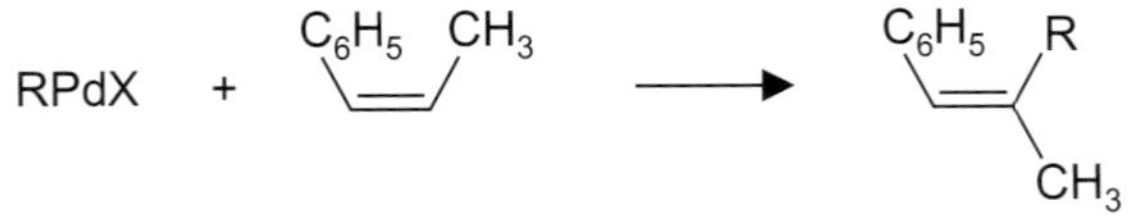

SCHEME 7.1i

EXERCISE 7.1

Explain the mechanism of the Heck reaction (Scheme 7.1j).

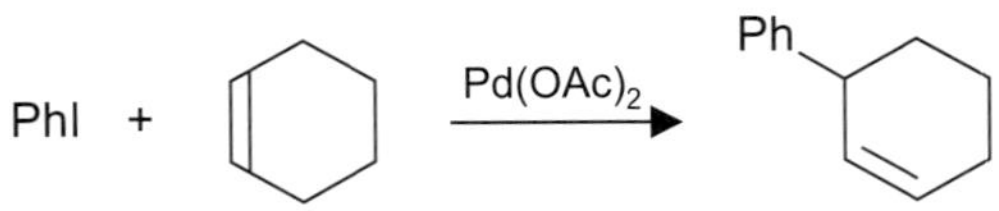

SCHEME 7.1j

ANSWER. *In this reaction the alkene double bond in the product is not at the site of coupling, but is one carbon removed i.e., an allylic shift has occured. The insertion reaction must be syn since Heck reaction is intramolecular. The insertion complex (Scheme 7.1k) has the Pd and the phenyl groups as cis-substituents (axial-equatorial* *relationship). As the subsequent β-elimination is a syn process, the hydrogen on the carbon at which the insertion had occured has trans relationship with Pd. Since internal rotation is prevented by the ring the cis relationship of Pd is available with a hydrogen on the other β-carbon.*

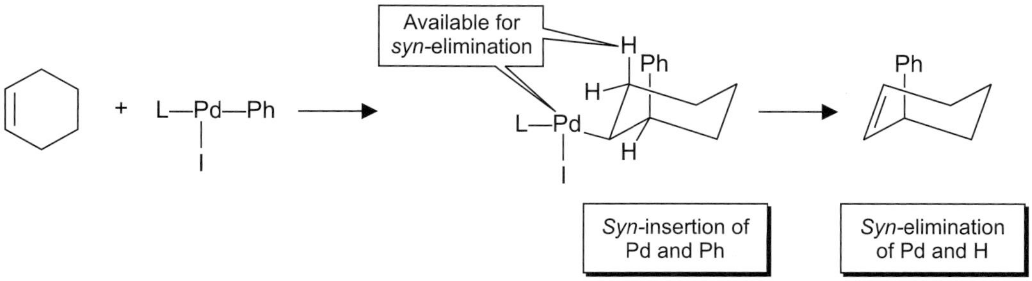

SCHEME 7.1k

(ii) Suzuki Reaction

Suzuki reaction is discussed in Scheme 8.26.

(iii) Oxidation of Alkenes to Aldehydes and Ketones—The Wacker Reaction

An alkene *e.g.*, ethylene forms a Pd-alkene complex and is followed by addition of water to this Pd(II) activated alkene. Subsequently elimination of Pd(0) and a proton affords the enol of acetaldehyde (Scheme 7.2). Further details are in (Scheme 13.45). The coreagents in this reaction are $CuCl_2$ and O_2 which reoxidize the Pd(0) to Pd(II).

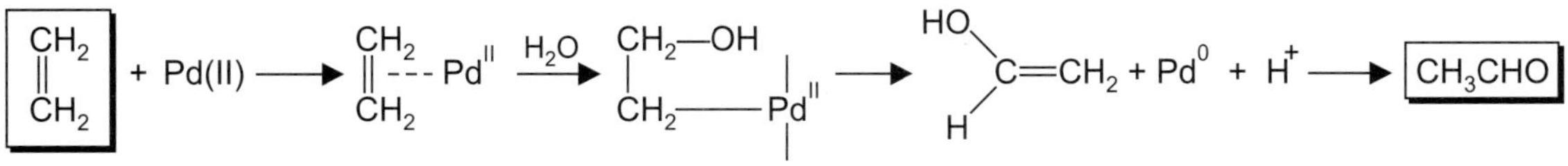

The Wacker reaction

SCHEME 7.2

(C) Octacarbonyldicobalt Complexes

(i) Hydroformylation (Oxo-reaction)

In this reaction an alkene reacts with carbon monoxide and hydrogen in the presence of a cobalt catalyst $HCo(CO)_4$ at high temperature and pressure to give an aldehyde (Scheme 7.3).

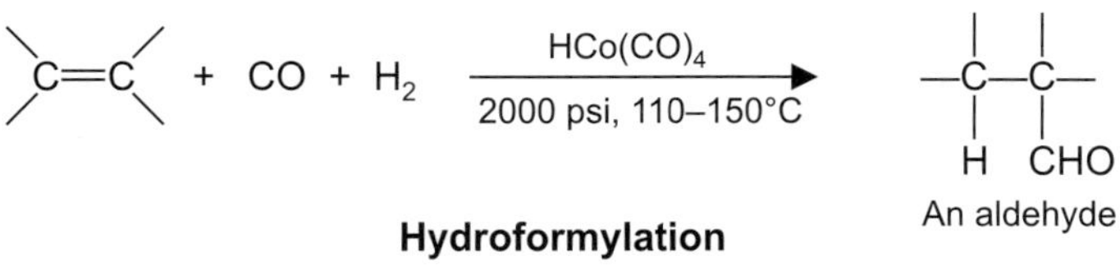

Hydroformylation

SCHEME 7.3

Metallic cobalt reacts with carbon monoxide to give octacarbonyldicobalt (Scheme 7.3*a*), which reacts with hydrogen to give hydridotetracarbonylcobalt which is the active catalyst. The following are some of the steps of oxo reaction.

1. The alkene replaces one molecule of carbon monoxide from $HCo(CO)_4$ to give the π complex (I, Scheme 7.3*a*).

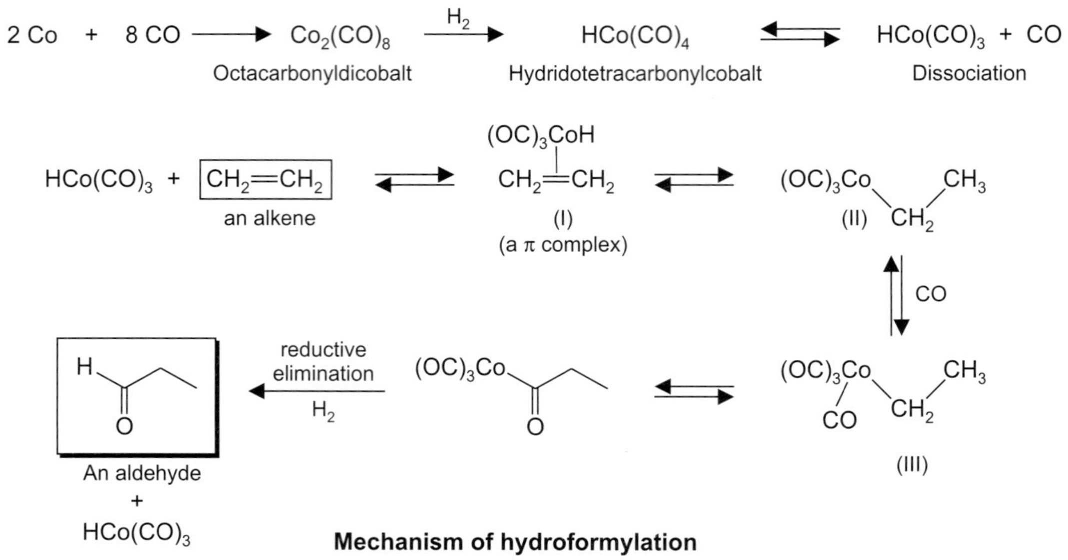

SCHEME 7.3a

2. In the insertion step, the hydrogen migrates from cobalt to one of the doubly bonded carbons, and the other carbon attaches to cobalt as in (II).
3. The metal alkyl formed above associates with an additional molecule of carbon monoxide to give (III). This is an important insertion step where a carbonyl group is inserted between the metal and the coordinated alkyl group. One can also consider this step as migration of the alkyl group from the metal to the carbon of a coordinated CO ligand and this is the key step where a carbon-carbon bond is formed.
4. Hydrogen is absorbed. One of the hydrogens migrates to the C=O group to give an aldehyde which leaves the coordinated sphere of the regenerated catalyst.

(*ii*) *Synthesis of Cyclopentenones—Pauson-Khand Reaction*

Alkynes react with octacarbonyldicobalt to give complexes where all four π-electrons are involved in bonding (I, Scheme 7.4). The complexes react with alkenes and carbon monoxide to give cyclopentenones. Reaction with the alkenes of the type $RCH{=}CH_2$ becomes highly regioselective (II).

In the Heck reaction, nucleophilic addition occurs to an unactivated olefin in the presence of catalytic Pd. In the Pauson-Khand reaction a five membered ring can be constructed from three components, an alkene, an alkyne and carbon monoxide in the presence of cobalt.

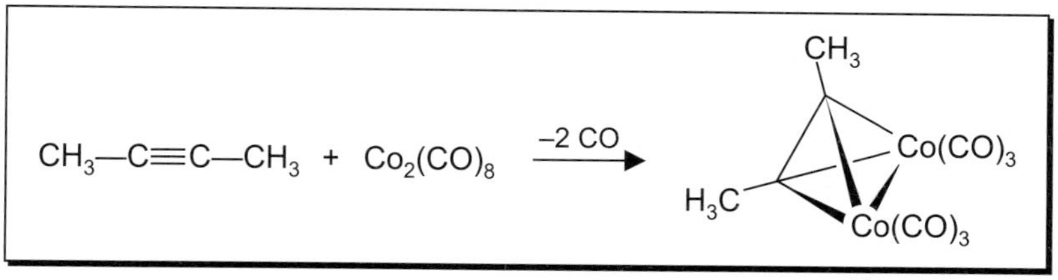

(I)

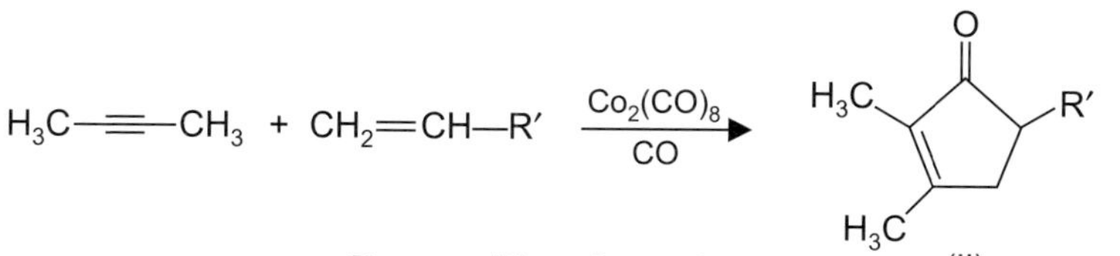

Pauson-Khand reaction (II)

SCHEME 7.4

(D) Compound Derived from Titanium

(i) Coordination Polymerization—Ziegler-Natta Catalysts

The catalytic system discovered by Ziegler and developed by Natta (both Noble Laureates, 1963) consists of a bimetallic coordination complex. These catalysts are made up of a transition metal salt particularly titanium tetrachloride and a metal alkyl *e.g.*, triethylaluminium ($Et_3Al : TiCl_4$).

The polymerization involves the insertion of the alkene *e.g.*, propylene into the C–Ti bond (I, Scheme 7.5). The process gives polypropylene where each carbon has the same configuration with all of the methyl groups on the same side of the polymer backbone. This

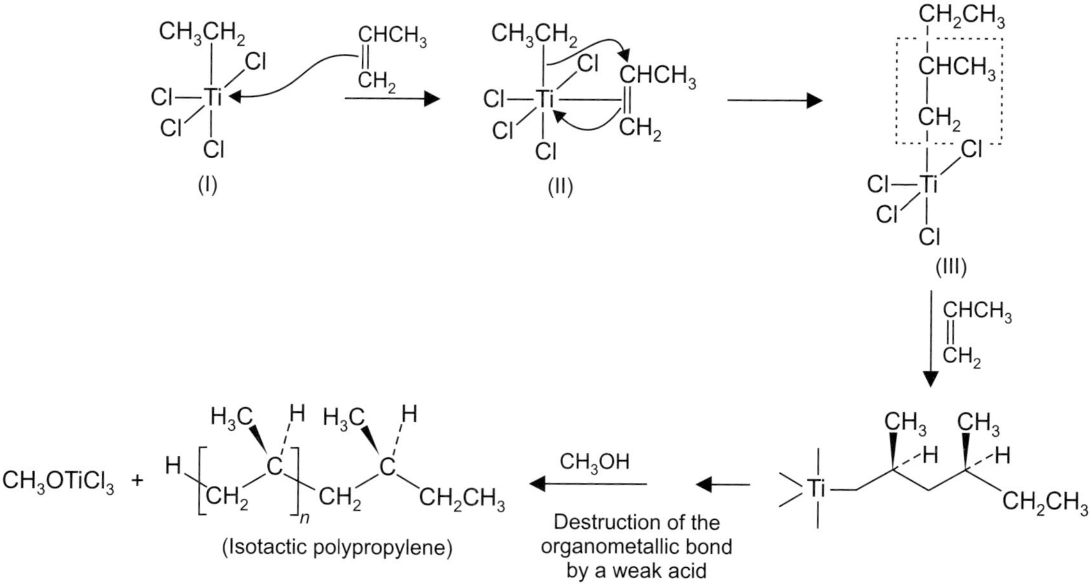

Ziegler-Natta polymerization-isotactatic polypropylene

SCHEME 7.5

regular and constant stereochemistry is called isotactic. This polymer has high density and greater strength compared to the one made from the radical polymerisation (atactic) which has a random orientation of substituents. One can modify the catalyst and make syndiotactic polypropylene with alternate carbons having opposite configuration (Scheme 7.6).

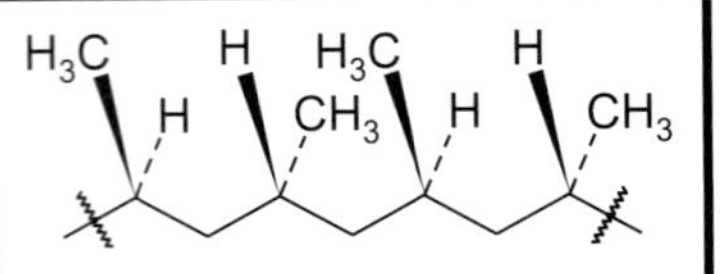

Syndiotactic stereochemistry of polypropylene

SCHEME 7.6

Mechanistically one views the transfer of an ethyl group from aluminium to form a sigma bond with titanium species (I, Scheme 7.5). Propylene forms a "pi-complex" (II) with the titanium by interaction of its pi MOs with a vacant coordination site on the metal. The ethyl group migrates to one carbon of the double bond of the coordinated propylene while the other carbon forms a sigma bond to the titanium (III). This creates a vacant coordination site on the metal for the process to occur again, which results in the formation of isotactic polypropylene—a stereoregular polymer which is linear without any short or long branches.

(ii) σ-Organotitanium compounds—Comparison with Grignards reagents

These compound of the type (I, Scheme 7.6*a*) are prepared from titanium tetraalkoxides. In these reagents the carbon atom bonded to titanium is strongly negatively polarized just like in a Grignard reagent and thus it acts as a nucleophile. The reagents (I, Scheme 7.6*a*) are far

$$3\ Ti(OR')_4 \xrightarrow{TiCl_4} 4\ ClTi(OR')_3 \xrightarrow[\text{or RLi}]{RMgX} \boxed{R\text{—}Ti(OR')_3}$$

(I)

SCHEME 7.6a

more selective when compared with Grignards reagents, and react with the carbonyl group of aldehydes and ketone and do not react with the ester group if present in the same molecule (Scheme 7.6*b*). Moreover, the reagent discriminates between an aldehyde and ketone and only the former is reduced when both are present in the same molecule. Moreover, these σ-organotitanium

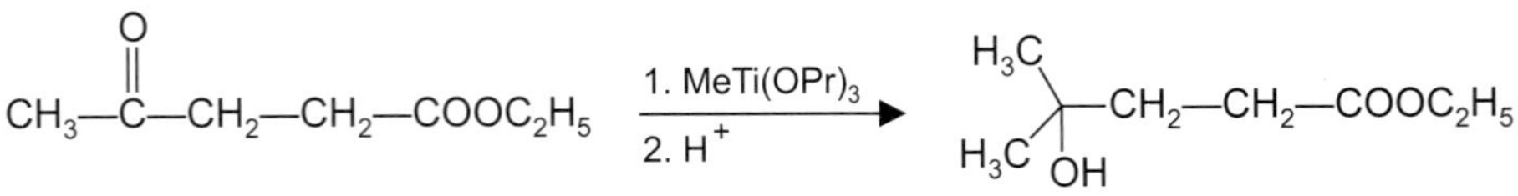

SCHEME 7.6b

compounds increase the diastereoselectivity observed with Cram's rule during nucleophilic addition with Grignards reagents to aldehydes with diastereotopic faces (Attack at the carbonyl group of an aldehyde with a neighbouring stereocenter, for details see Scheme 14.81).

(iii) Alkene Metathesis

On heating with different catalysts *e.g.*, tungsten, molybdenum, rhenium or titanium complexes, alkenes undergo disproportionation reactions. Thus under the influence of a suitable catalyst *e.g.,* Tebbe reagent, propylene give mainly *trans*-2-butene and ethylene (Scheme 7.6*c*).

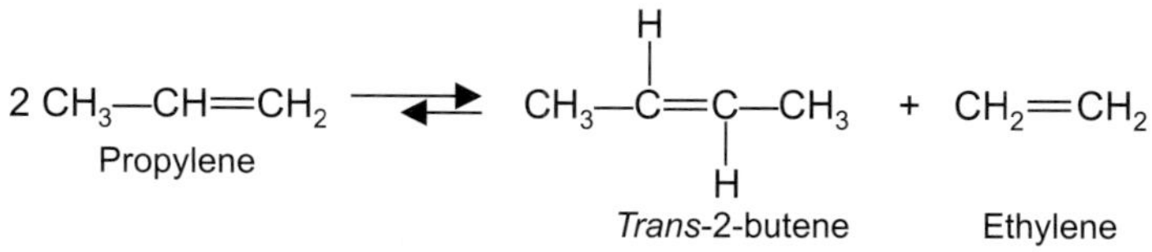

Alkene metathesis

SCHEME 7.6c

The Tebbe reagent contains titanium and is prepared from Cp_2TiCl_2 (Cp=Cyclopentadienyl) as shown (Scheme 7.6*d*). The reactive intermediate being a metallocarbene complex containing a carbon metal double bond ($Cp_2Ti{=}CH_2$).

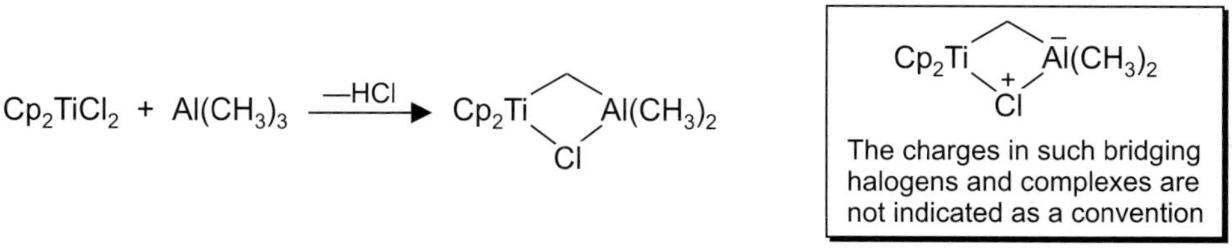

SCHEME 7.6d

Alkenes can form metallocycles as two metallocyclobutanes with Tebbe reagent in a reversible process (Scheme 7.6*e*). One may also use two different alkenes and *e.g.*, ethylene reacts with dimeric isobutylene to give neohexene which is a commercial process (Scheme 7.6*f*).

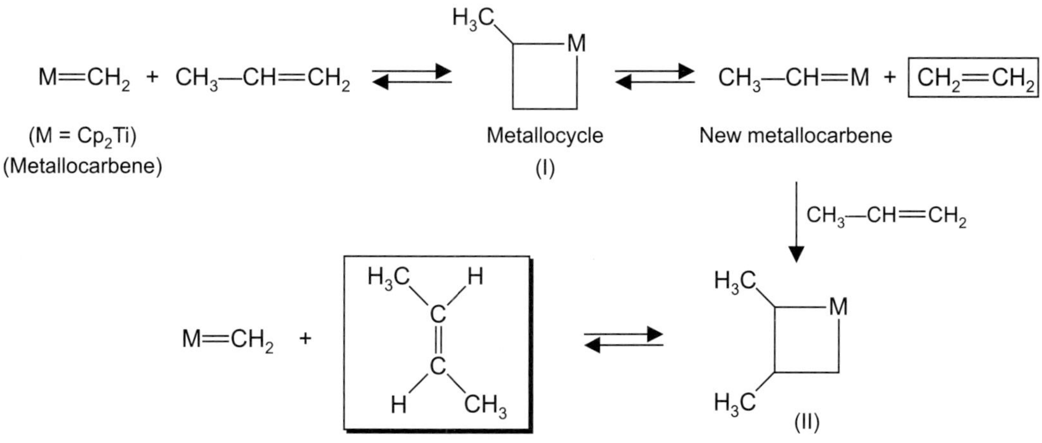

Alkene metathesis

SCHEME 7.6e

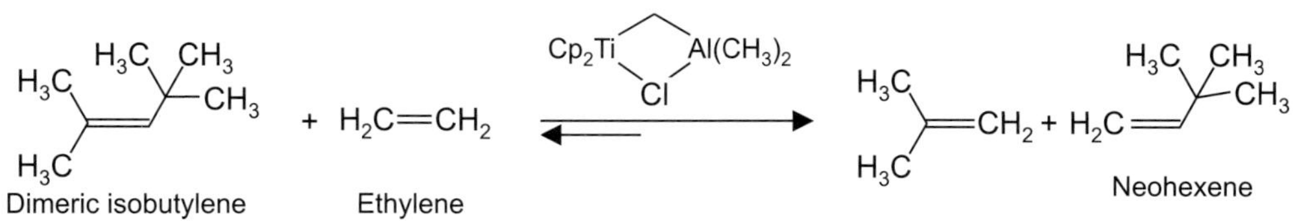

SCHEME 7.6f

The mechanism of the reaction involves the intervention of a metallocarbene and a four membered ring containing a metal (Scheme 7.6*e*). The metallocyclobutane is thought to undergo a cleavage across the ring by (A or B, Scheme 7.6*g*). Cleavage in (I, Scheme 7.6*g*) across the ring by (A) gives back the reactants and is thus, unproductive. The cleavage (B) in (I, Scheme 7.6*g*) gives the original alkene and a new metallocarbene. The new metallocarbene can add the starting alkene (propylene) to give a new metallocyclobutane(s) which can cleave to give a new alkene.

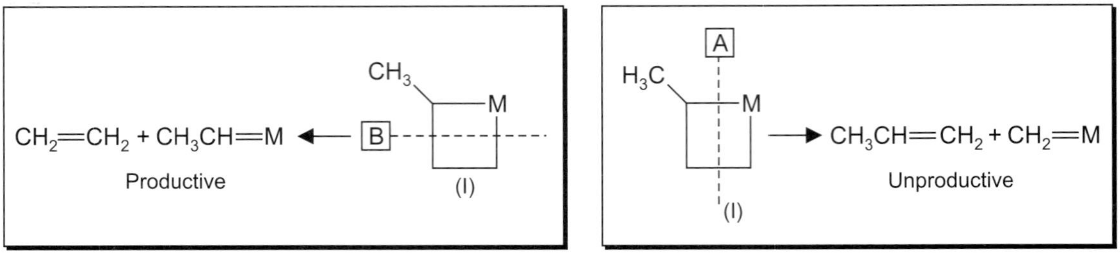

SCHEME 7.6g

Olefin metathesis represents a new exciting process which has been developed during the last 15 years for the creation of carbon-carbon double bonds. The process has a base in Ziegler-Natta catalysts for the polymerization of cyclic olefins. Olefin metathesis involves the breaking apart of alkenes at the double bond and the pieces get reassembled randomly. The name metathesis reflects the interchanging of the ends of carbon-carbon double bonds.

EXERCISE 7.2

A catalyst prepared from tungsten hexachloride and ethylaluminium dichloride is shown to be a metallocarbene complex, $(WCl_6 + C_2H_5AlCl_2 \longrightarrow Cl_n\,W = CHCH_3)$

On treatment with this catalyst, 2-pentene gives a mixture with 2-butene as one of the components. Propose a mechanism for this olefin metathesis.

ANSWER. *Olefin metathesis is a unique reaction which is successful only with catalysts prepared from transition metals. The metallocarbene can yield two metallocyclobutanes. Considering diffferent cleavages across the rings, one can see that one cleavage B′ in (II, Scheme 7.6h) is productive to yield the desired product.*

$$WCl_6 + EtAlCl_2 \longrightarrow Cl_nW{=}CHCH_3$$

Metallocarbene

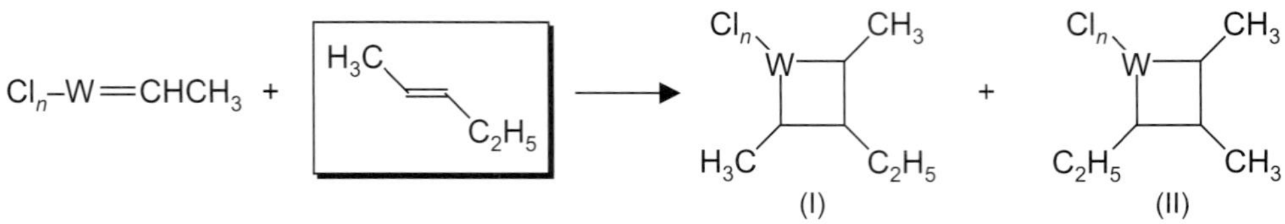

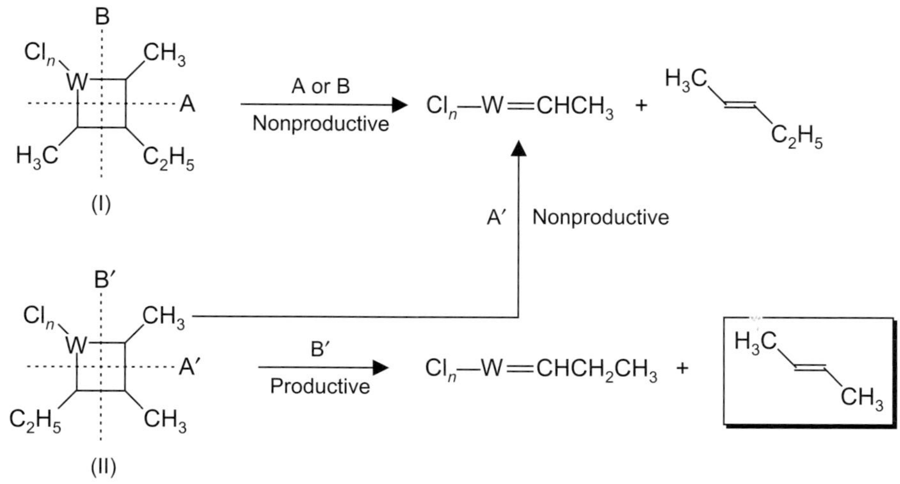

SCHEME 7.6h

(E) Role of Iron Compounds and Acyl-Iron Complexes

(i) Stabilization of Cyclobutadiene

Cyclobutadiene is an antiaromatic compound and thus highly reactive. It can, however be obtained as a complex with $Fe(CO)_3$ when the electron density is withdrawn from the ring by

> **Antiaromaticity and spectral behaviour of cyclobutadiene has been discussed in detail in sec. 2.3. Recall that cyclobutadiene is not a square arrangement, but it gets distorted to attain a rectangle form with short carbon-carbon double bonds and longer carbon-carbon single bonds. The distortion leads to minimize the connection *i.e.*, conjugation between the *p* orbitals (Scheme 7.6*i*).**

SCHEME 7.6i

the metal. This cyclobutadiene complex may be viewed as an aromatic duct and it thus undergoes electrophilic aromatic substitution (Scheme 7.7). This complex of cyclobutadiene from which cyclobutadiene is released via oxidation with cerium IV undergoes a Diels-Alder reaction (Scheme 7.7*a*) followed by photochemical [2 + 2] cycloaddition and base induced ring contraction (see under answer 10.8) to give cubane.

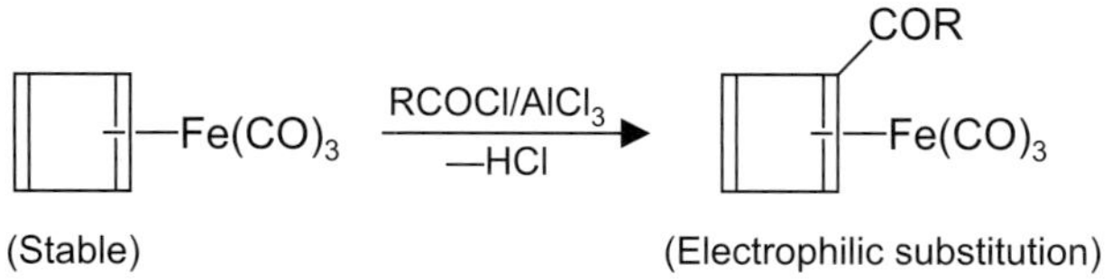

SCHEME 7.7

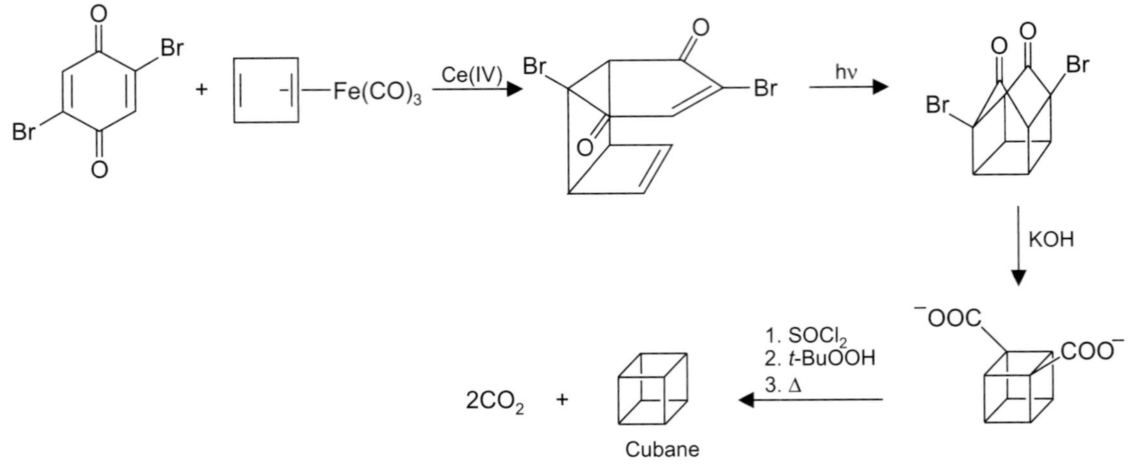

SCHEME 7.7a

(ii) Acyl-iron complexes—Direct Conversion of Alkyl Halides to Aldehydes and Ketones with Collman's Reagent

The Collman's regant $Na_2Fe(CO)_4$ is made from reacting iron pentacarbonyl $Fe(CO)_5$ with sodium amalgam in THF. The reagent reacts with an alkyl halide to give the ion $R\text{–}Fe(CO_4)^-$ (Scheme 7.7*b*). The reaction of Collman's reagent with CO in the presence of an alkyl halide yields an acylated iron complex which reacts with a second alkyl halide to give a ketone.

$$Na_2Fe(CO)_4 + RX \xrightarrow{-NaX} R\text{—}Fe(CO)_4^- \; Na^+$$

$$Na_2Fe(CO)_4 + RX + CO \longrightarrow R\text{—}\underset{\underset{O}{\|}}{C}\text{—}\overset{-}{Fe}(CO)_4 \xrightarrow{R'X} RCOR'$$

Acylated iron complex

SCHEME 7.7b

Interestingly acyl-iron anions can be converted into aldehydes by protonolysis and into ketones by alkylation. This technique, therefore, provides an alternative to the "reversed polarity" procedure (umpolung, see, Sec. 7.4*d*) for the conversion RX ⟶ RCHO and RCOR′ (Scheme 7.7*c*).

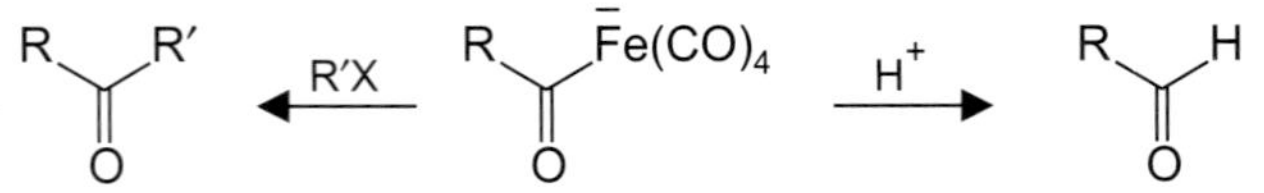

SCHEME 7.7c

The ion $RFe(CO)_4^-$ reacts with Ph_3P in the presence of CO (insertion and ligand exchange) to give a complex which reacts with alkenes and subsequent isomerization and hydrolysis yields ketones (Scheme 7.7*d*).

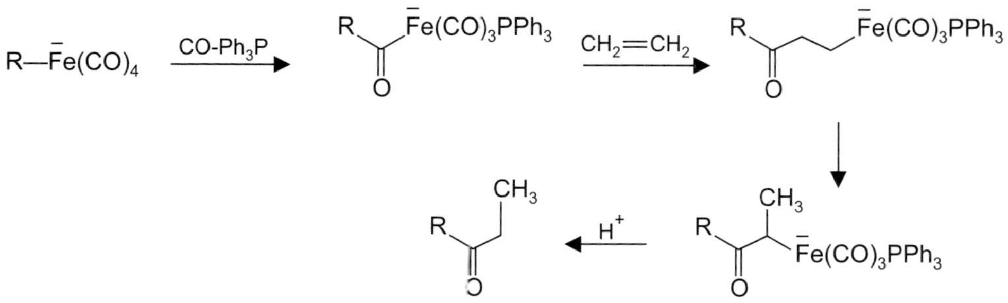

SCHEME 7.7d

(iii) C-Alkylation via Chiral Acyl-iron complexes (an Enantioselective Synthesis of an Acid)

The cyclopentadienyl triphenylphosphine iron complex with attached acyl units (I, Scheme 7.7*e*) is used. The following points may be noted.

- Butyl lithium transforms the acyl carbonyl group into the lithium enolate (II) with *E* geometry (methyl group on the same side as the enolate oxygen, atomic members of oxygen and iron are 8 and 28 respectively).
- One of the phenyl groups of PPh_3 unit effectively blocks the approach of the electrophile from CH_3I from the *Re* face at C2 of (II). The alkylation thus occurs from the *Si* face (closest to the observer.).
- On one electron oxidation with bromine in water gives an alkylated acid in an enantiomerically pure form (Scheme 7.7*e*).

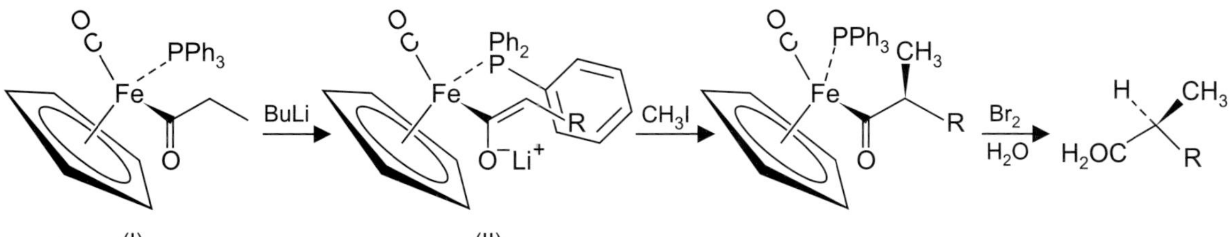

SCHEME 7.7e

(F) Chromium Arene Complexes

An arene on heating with chromium hexacarbonyl gives these complexes. The $Cr(CO)_3$ fragment is highly electron-withdrawing and activates an aromatic system towards nucleophiles. In these complexes the metal is bound to all six atoms of the aromatic ring and the complexation removes electron density from the π-cloud. The electron-withdrawing effect of chromium is comparable to that of a nitro group. The aromatic C—H bond of a Cr complex is acidic due to electron withdrawing effect of $Cr(CO)_3$ and thus treatment with MeLi or BuLi leads to deprotonation and the lithiated species then react with electrophiles (Scheme 7.7*f*). The product can be released from the complex by mild oxidation usually with cerium IV or also with I_2.

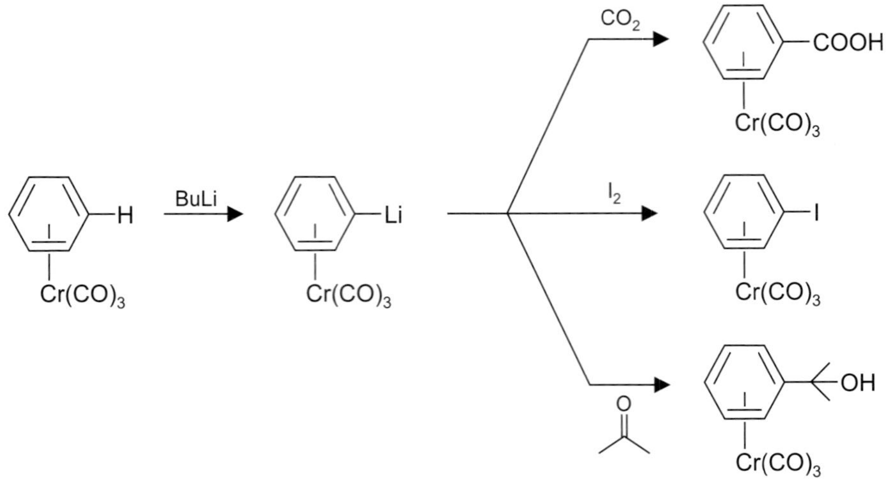

SCHEME 7.7f

Complexed aryl halides undergo displacement by nucleophiles *e.g.*, alkoxide, amines and stabilized anions (Scheme 7.7*g*) to give substituted benzenes.

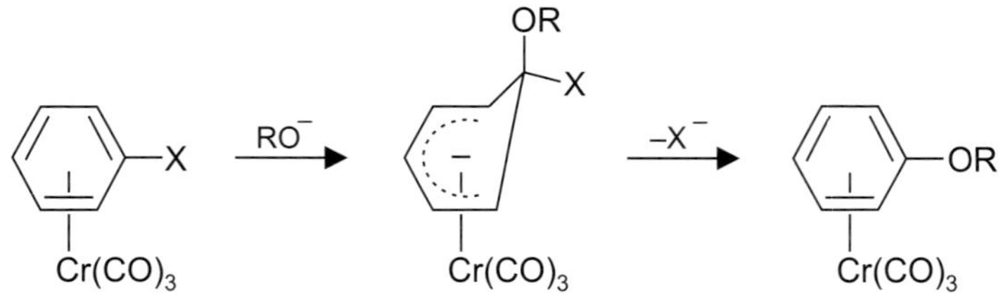

SCHEME 7.7g

> **Further details of nucleophilic substitution in arenechromium carbonyl complexes are in sec 9.6.**

(G) Use of Nickel and its Complexes

Reductive Coupling of Allyl Halides

Allylic halides can be symmetrically coupled on reaction with nickel carbonyl at room temperature in a solvent such as DMF, to give 1, 5-dienes (Scheme 7.8).

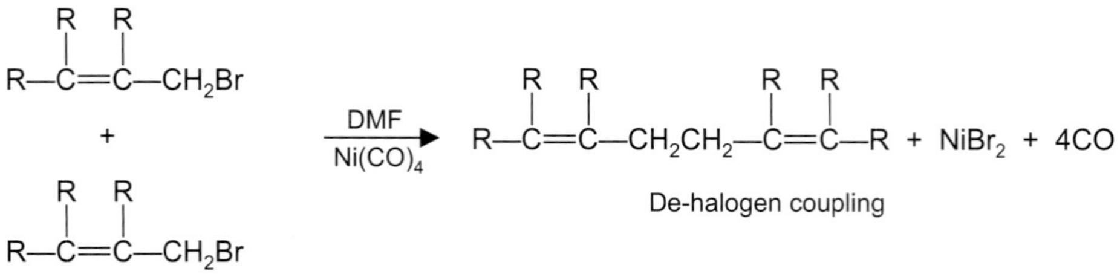

SCHEME 7.8

The order of halide reactivity is I > Br > Cl. With unsymmetrical allylic halides, coupling nearly always occurs at the less substituted carbon of the allylic group. A plausible mechanism may involve reversible dissociation of $Ni(CO)_4$, oxidative addition of the allyl compound to give a π-allyl complex of the type (I, Scheme 7.8*a*). The dimerization of this complex via loss of CO and halogen bridging gives a π-allylnickel bromide (II, Scheme 7.8*a*), which further reacts to give the coupled product. As the dimeric nickel complex (II, Scheme 7.8*a*) brings the coupling termini into proximity, the method is used intramolecularly for macrocylizations to build 11–20 membered rings in high yield.

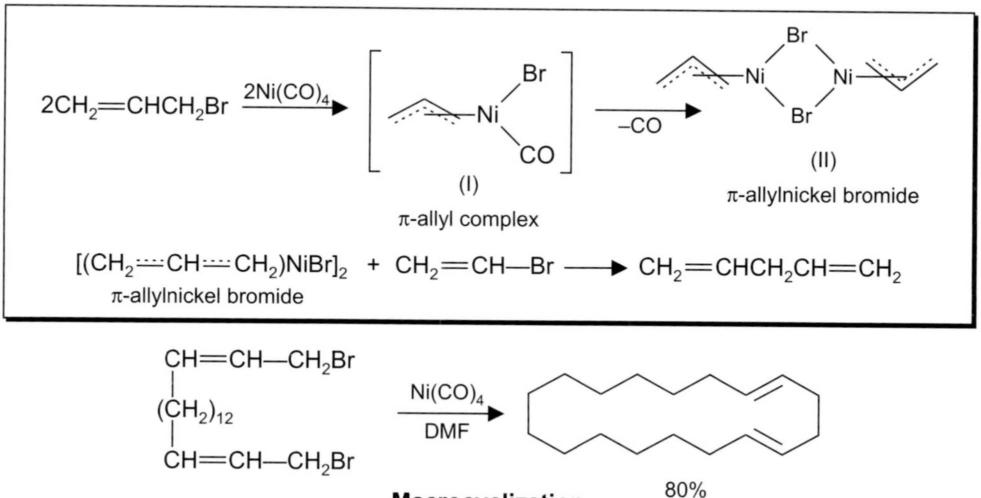

Macrocyclization

SCHEME 7.8a

When the reaction is carried out in a hydrocarbon solvent *e.g.*, benzene, the π-allylnickel bromide complex is formed which does not couple. Thus unsymmetrical (crossed) coupling can be achieved by treating an alkyl halide directly with (II, Scheme 7.8*a*) and even in this case unsymmetrical allylic groups couple at the less substituted end (Scheme 7.8*b*). The terpenoid

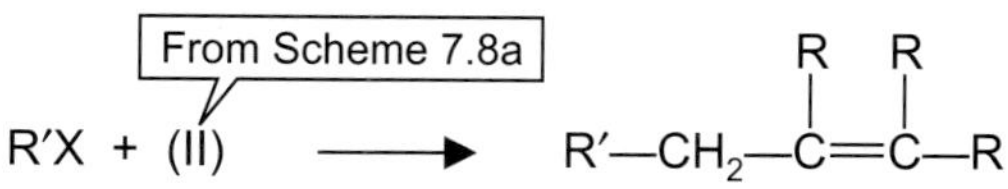

SCHEME 7.8b

geranyl acetate can be made this way (Scheme 7.8*c*). This reaction is of significance, since 1, 5-diene moiety is present in many natural products, consequently the method can be used in their synthesis.

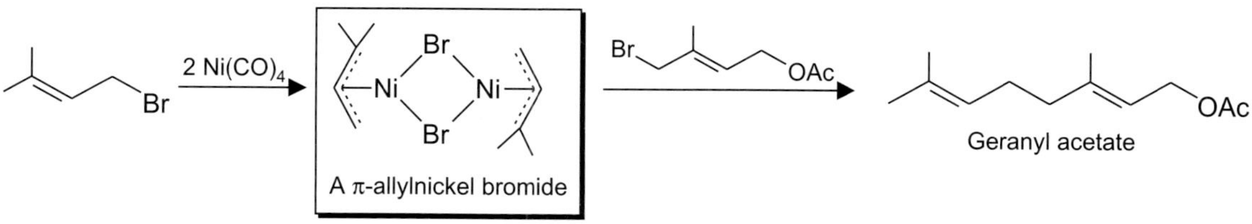

SCHEME 7.8c

(H) Organocuparates–Gilman Reagents

Transition metal organometallic compounds *e.g.*, organocuparates are derived from a Grignard or organolithium reagent in which the metal ion (Cu) is a transition metal. They are discussed in sec. 7.7D.

7.3 PHOSPHORUS CONTAINING REAGENTS

(A) Use of Phosphorus Ylides—Wittig Reaction and its Variants

The Wittig reaction (George Wittig, Nobel Prize 1979) is used primarily for the conversion of a carbonyl compound to an alkene (Scheme 7.9) using a special class of carbanion reagents

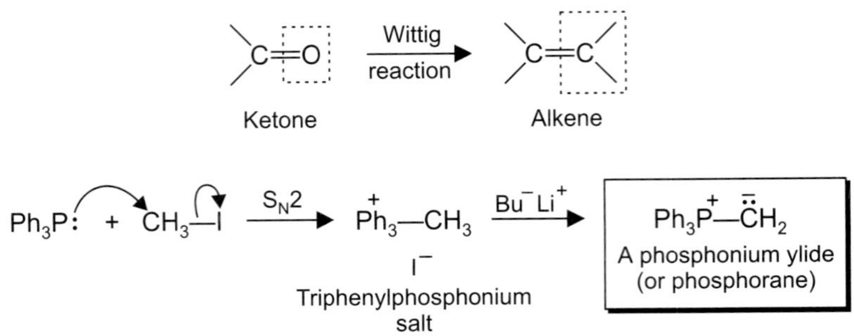

Wittig reaction-generating a phosphonium ylide

SCHEME 7.9

called *ylides*. A phosphorus ylide is prepared involving two steps, firstly an S_N2 reaction between triphenylphosphine and an appropriate alkyl halide gives a triphenylphosphonium salt. The proton on the carbon adjacent to the positively charged phosphorus which is sufficiently acidic (pK_a = 35) is then removed by a strong base like butyllithium (Scheme 7.9).

The ylide (nucleophilic carbanion) reacts with the electron-deficient centers like a carbonyl group to generate a cyclic 1, 2-oxaphosphetane. The cyclic intermediate is thought to get generated via a [2 + 2] cycloaddition reaction involving four electrons in the transition state. Decomposition of the cyclic 1, 2-oxaphosphetane eliminates phosphine oxide and brings about a regiospecific formation of an alkene (Scheme 7.10).

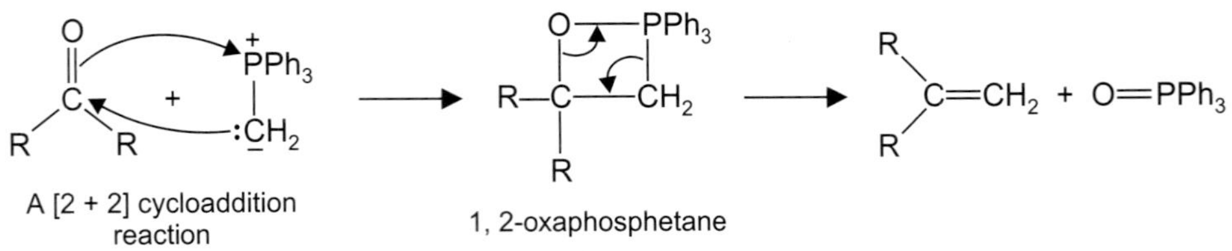

Mechanism of Wittig reaction involving a [2 + 2] cycloaddition

SCHEME 7.10

MECHANISM OF WITTIG REACTION

The negatively polarized carbon in the ylide is nucleophilic and can attack the carbonyl group (Scheme 7.10a). The result is a phosphorus betaine, a dipolar species. The betaine is short lived and may not be on the reaction pathway and rapidly forms a neutral oxaphosphacyclobutane (oxaphosphetane), characterized by a four-membered ring containing phosphorus and oxygen. This substance then decomposes to the product alkene. The consideration of a betaine structure is often invoked to explain the stereochemical outcome of Wittig reaction.

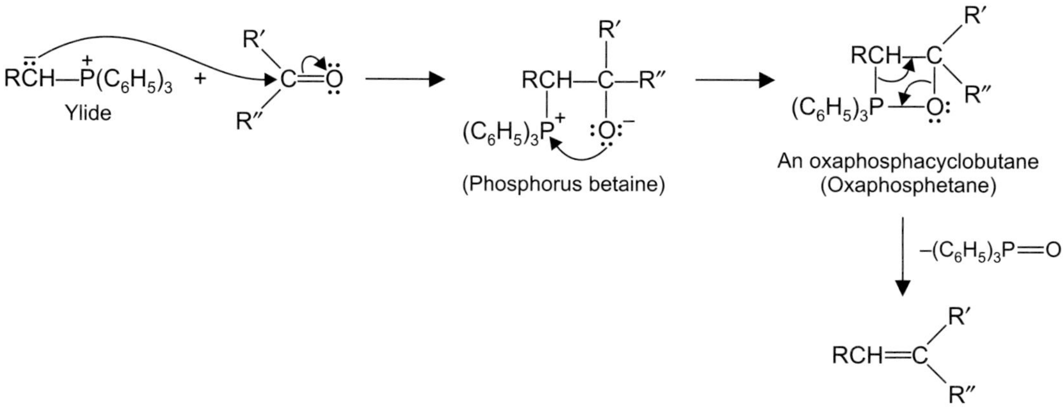

SCHEME 7.10a

(B) Stabilized Phosphoranes for use in Wittig Reactions

Simple phosphoranes (Scheme 7.9) are not only unstable but highly reactive. Thus reactions involving non-stabilized ylides have to be conducted under anhydrous conditions and in an inert atmosphere, since these ylides react both with water and oxygen.

The stabilized ylides have a group *e.g.*, a carbonyl group which can share the carbanions negative charge (II, Scheme 7.11). In such cases the phosphorane can be obtained by treatment of the phosphonium salt with less strong bases *e.g.*, sodium alkoxides, sodium hydroxide or even sodium carbonate may be used. Though, such phosphoranes are usually isolable crystalline compounds, but these have lower carbanionic activity. These react with aldehydes readily but ketones are frequently not attacked.

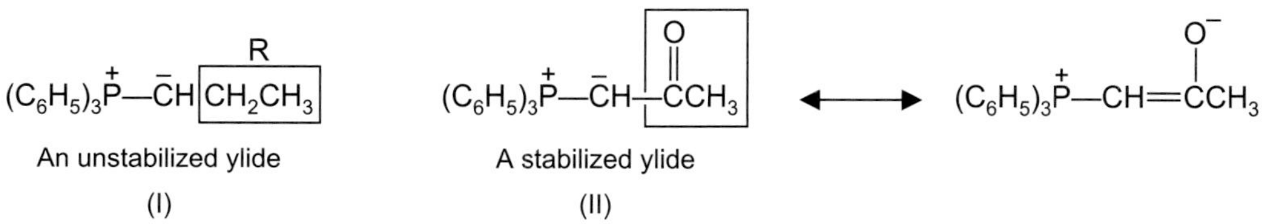

SCHEME 7.11

The Wadsworth-Emmons variant of the reaction uses carbonyl activated ylides in which triphenylphosphine is replaced by triethylphosphite $P(OEt)_3$. Reaction with a bromoester is a typical S_N2 reaction and the product further undergoes another S_N2 reaction to create a P=O bond and gives a phosphonate. This process called the Arbuzov reaction provides an important route to Wadsworth-Emmons reagents (Scheme 7.12).

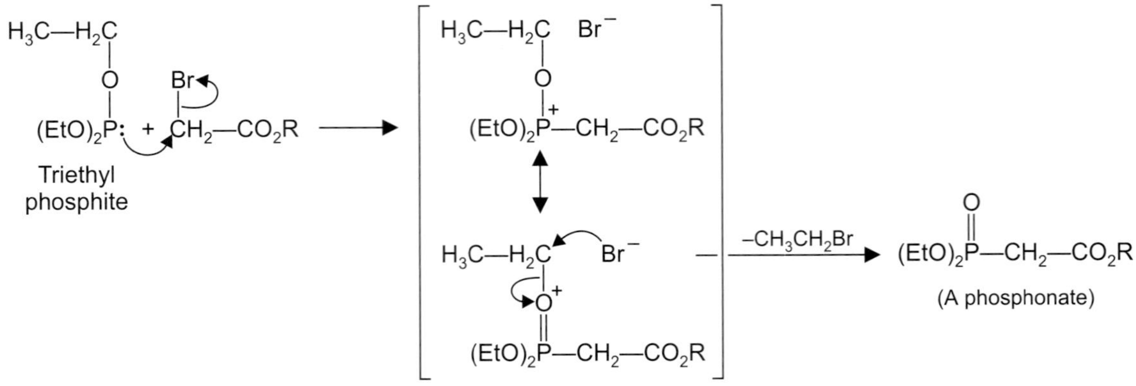

Formation of Wadsworth-Emmons reagents via Arbuzov reaction

SCHEME 7.12

The phosphonate obtained from the Arbuzov reaction is deprotonated in a separate step to give the corresponding carbanion. Much weaker bases compared to those employed in a Wittig reaction suffice since two electron withdrawing groups are attached to the methylene group. The resulting carbanion adds readily to the carbonyl groups of aldehydes and ketones and the steps of mechanism are similar to Wittig reaction (Scheme 7.13). The reactions where triphenylphosphine is replaced by triethylphosphite are termed Wadsworth–Emmons reactions.

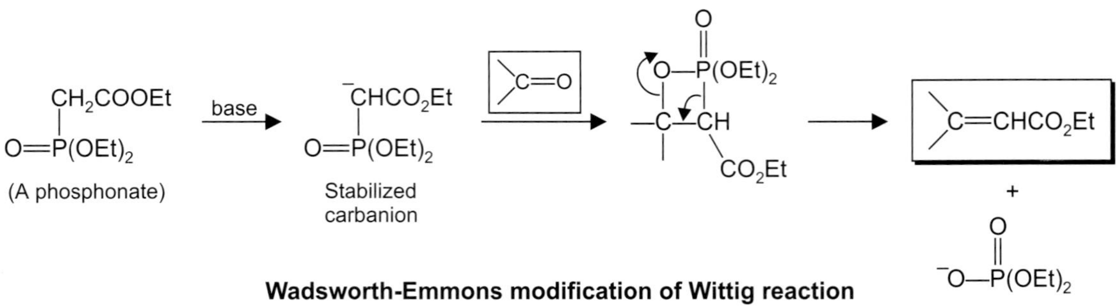

Wadsworth-Emmons modification of Wittig reaction

SCHEME 7.13

WITTIG REACTION AND ITS WADSWORTH-EMMONS VARIANT

Simple phosphoranes (phosphorus ylides) as used originally in a Wittig reaction are highly reactive and unstable (I, Scheme 7.14). More stable phosphoranes (II) are obtained when an electron withdrawing substituent is adjacent to the anionic carbon. However, these react with aldehydes but not with ketones. A futher modification is the Wadsworth-Emmons reagents in which triphenylphosphine is replaced by triethylphosphite to give reagents of the type (III) which easily add to ketones as well. These activated ylides are made via Arbuzov reaction, which is

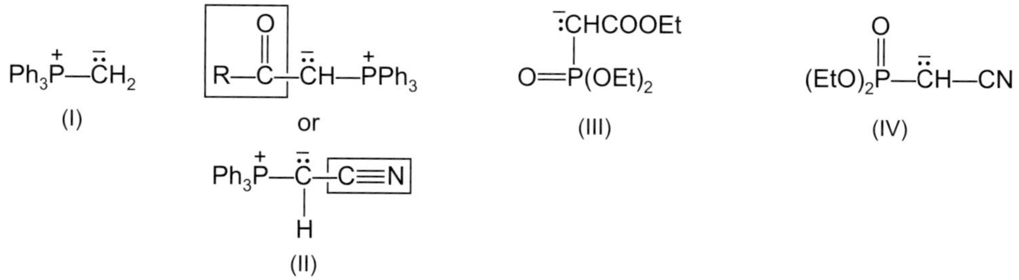

SCHEME 7.14

general method for making a variety of these activated ylides, other best substrates are benzylic and allylic halides and compounds with a halogen atom α- to the carbonyl, nitrile or sulphonyl groups (IV, Scheme 7.14).

EXERCISE 7.3

How can one prepare each of the compounds (Scheme 7.14a)?

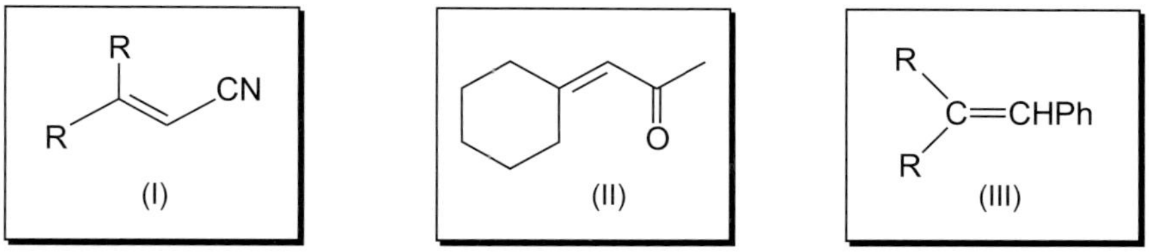

SCHEME 7.14a

ANSWER. *From retrosynthesis the disconnection for a Wittig reaction to make an alkene is the double bond between carbon atoms. Generally the ylide component is the end of the alkene double bond with the less highly substituted carbon atom. The phosphonate ylides for the synthesis of (I and II) are made by using Arbuzov reaction.*

(*a*) *Retrosynthesis:*

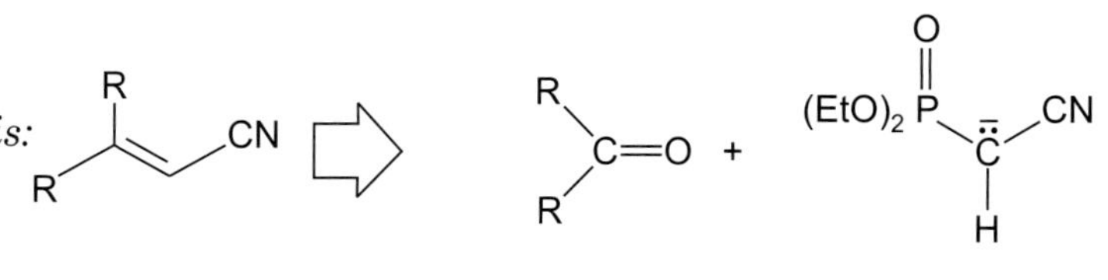

(I) *Synthesis*

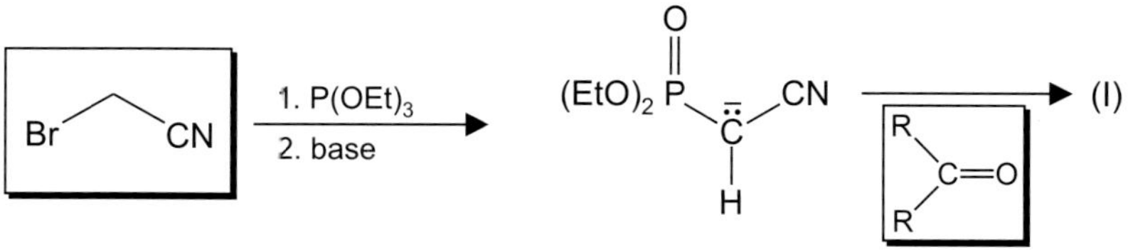

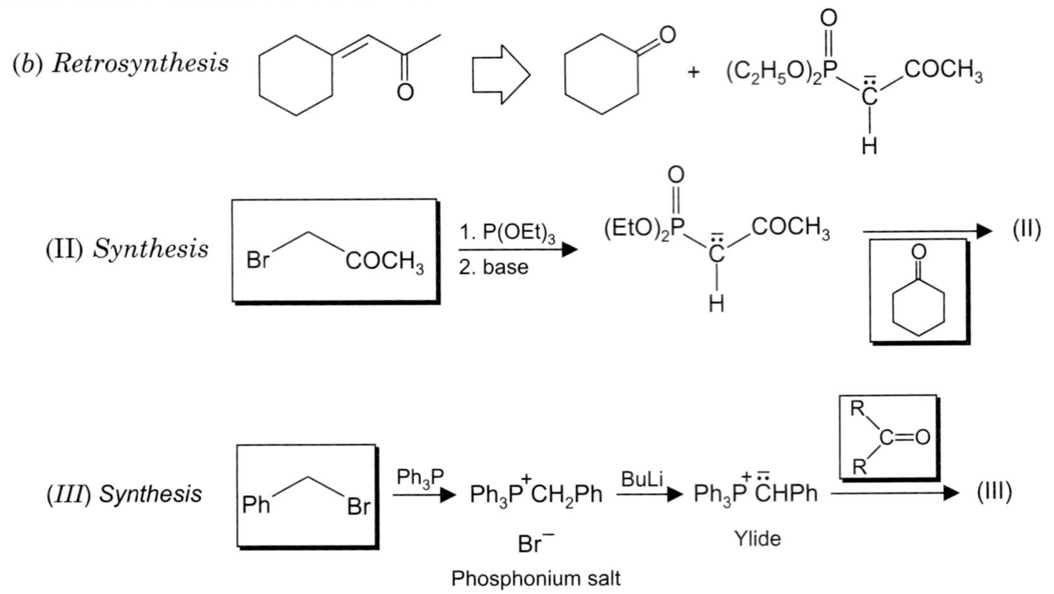

SCHEME 7.14b

MERITS OF WITTIG REACTION

The reaction is the best method to prepare less substituted double bond (the double bond is placed at the carbonyl group of an aldehyde or a ketone). Thus only one isomer is formed, the reaction is regiospecific and best method to make a terminal alkene (Scheme 7.14c), since other methods e.g., E2 reaction will give terminal alkene only as the minor product.

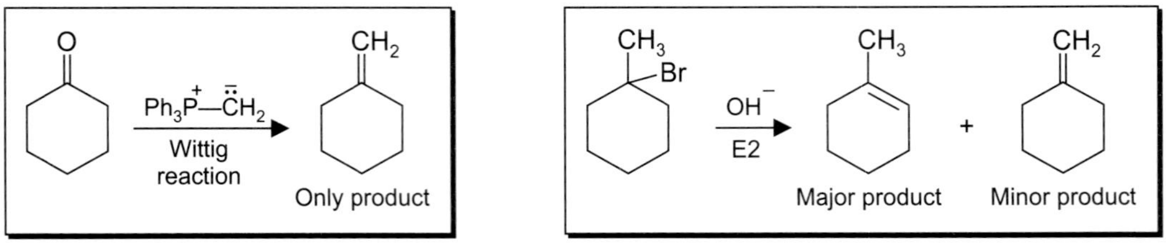

SCHEME 7.14c

When methoxymethylene is used as an ylide one can step up a ketone to an aldehyde with one more carbon atom. The reaction (Scheme 7.14d) involves an acid labile enol ether (see Scheme 14.37).

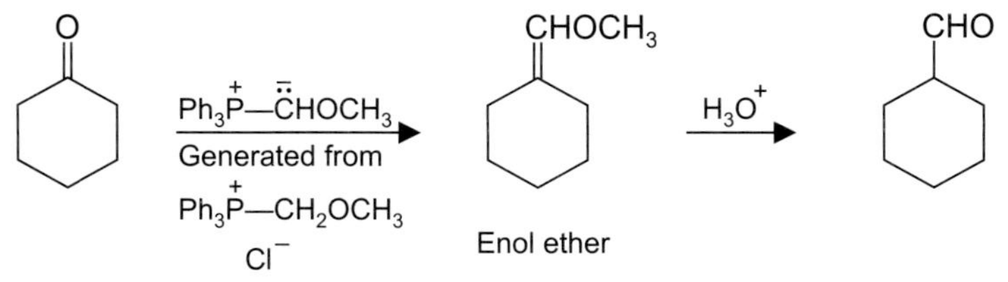

SCHEME 7.14d

(C) Stereoselectivity in Wittig and Wadsworth-Emmons Reactions

Sterechemistry of Wittig Reaction

The following are the general guidelines:

- *The Wittig reaction is E selective with stabilized ylides*
- *TheWittig reaction is Z selective with unstabilized ylides.*

These reactions are ideal for alkene synthesis, however, when the alkene product is capable of geometrical isomerism a mixture of *Z*- and *E*-isomers is often a product. Thus the reactions are not subject to steric control. It is found that the stereoselectivity depends strongly on both the structure of the ylide and reaction conditions. Generally, unstabilized phosphoranes give predominantly the *Z*-alkene particularly in the presence of polar aprotic solvents and in the presence of a salt. Stabilized phorphoranes (including Wadsworth-Emmons reagents) on the other hand give mainly the *E*-alkene.

The unstabilized phosphorane reacts with an aldehyde and the geometry of the oxaphosphetane is determined by the steric approach of the phosphorane (ylide). The first step in the addition is irreversible and the formation of the increased amount of *Z*-alkene is the result (Scheme 7.15).

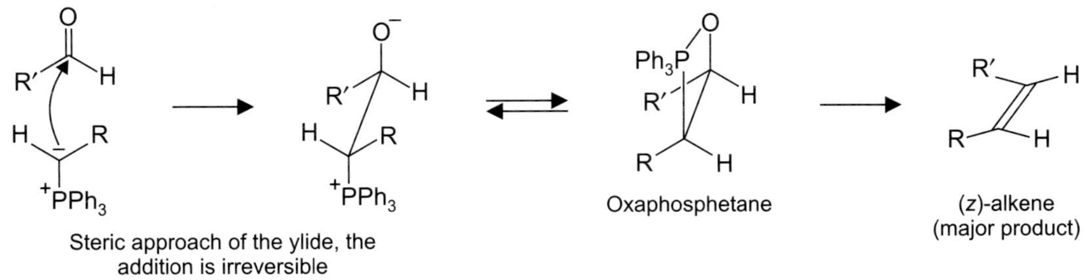

Stereoselectivity in Wittig reaction with unstabilized ylide

SCHEME 7.15

With stabilized phosphoranes, however, the first step in addition is reversible (Scheme 7.16) and this allows the formation of the more stable oxaphosphetane. Thus an interconversion to the more stable and thus more abundant isomeric form and a *syn*-elimination affords the *E*-alkene.

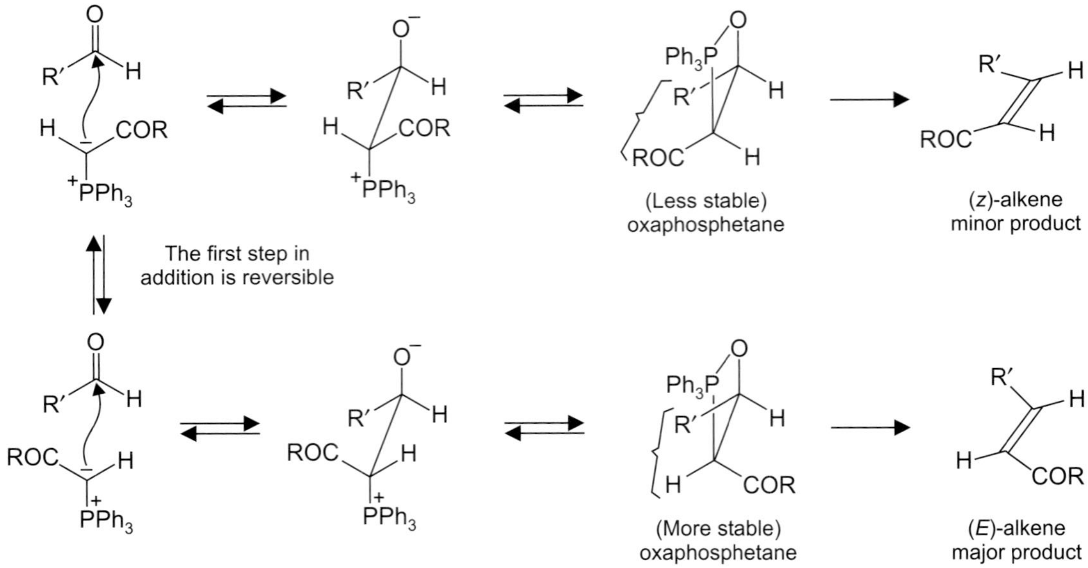

Stereoselectivity in Wittig reaction with stabilized ylides

SCHEME 7.16

Unlike the formation of Z-alkenes from unstabilized phosphoranes (see Scheme 7.15), by following the Schlosser modification of Wittig reaction one can end up instead with E-alkenes. Here the ylide is produced as a lithium halide complex (Scheme 7.17) and alllowed to react with an aldehyde at low temperature so that the betaine is stable. Treatment with a strong base such as phenyl lithium gives a β-oxido ylide (a new ylide). Addition of t-butanol protonates the β-oxido ylide stereoselectively to give the more stable betaine as a lithium halide complex. On warming, the complex collapses to the E-alkene.

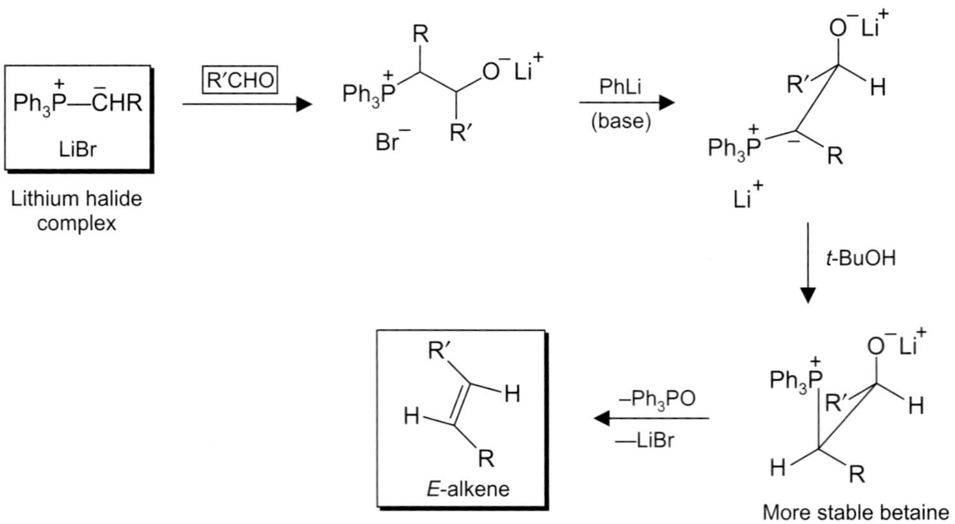

***trans*-Selective Wittig reaction-Schlosser variant**

SCHEME 7.17

(D) Conversion of Alcohols to Halides—Nucleophilic Substitution of Alcohols

> **Role of phosphorus tribromide (PBr_3) for the conversion of alcohols to bromides has been explained (see Scheme 5.15*c*).**

Use of thionyl chloride and phosphorus halides for the conversion of alcohols into halides is under vigorous and very acidic conditions. Milder conditions employ a 1 : 1 adduct from triphenylphosphine and bromine (Scheme 7.18) which converts alcohols to bromides. The alcohol displaces a bromide ion from the adduct which attacks the alkoxyphosphonium ion in an S_N2 reaction.

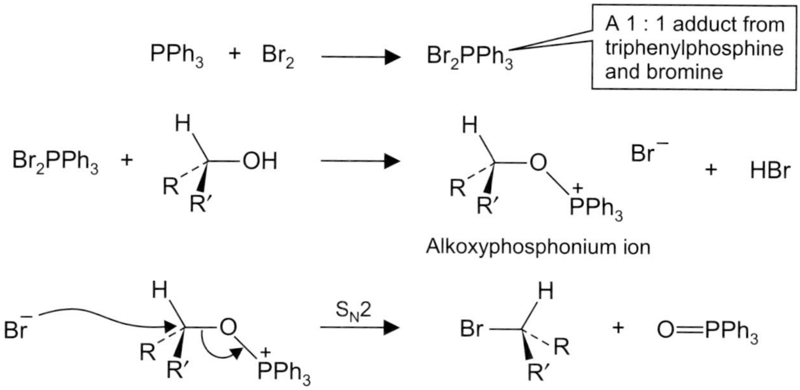

Conversion of an alcohol into bromide with Br_2PPh_3

SCHEME 7.18

Triphenyl dichloride displays the same reactivity to convert alcohols to chlorides. However, the most convenient method is to generate chlorophonium ions in *situ* by reacting trichlorophosphine with a suitable compound from which chlorine can be abstracted without its bonding pair so as to leave a stable carbanion *e.g.*, hexachloroacetone.

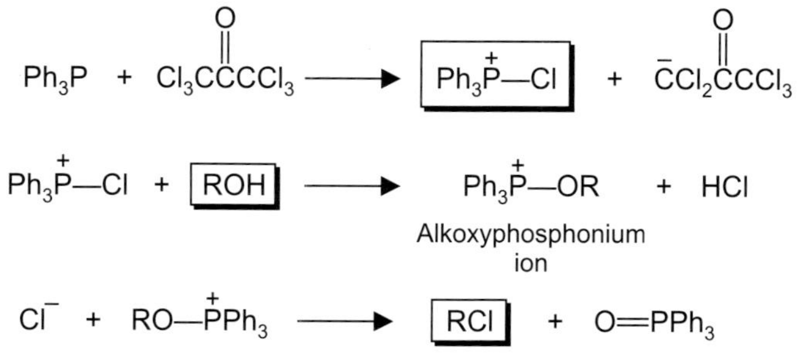

SCHEME 7.18a

A mild method to convert alcohols into iodides is via reaction with triphenylphosphine, diethyl azodicarboxylate and methyl iodide. The method generates the needed alkoxy phosphonium ion (which however, cannot be formed by methods discussed in (Schemes 7.18 and 7.18*a*). The ion then reacts with I^- to give alkyl iodide (Scheme 7.18*b*). Diethyl azodicarboxylate activates triphenylphosphine for a subsequent nucleophilic attack by the alcohol. After the initial reaction with Ph_3P, CH_3I captures the enolate ion (I, Scheme 7.18*b*) and generates the iodide ion for S_N2 reaction on the alkoxyphosphonium ion.

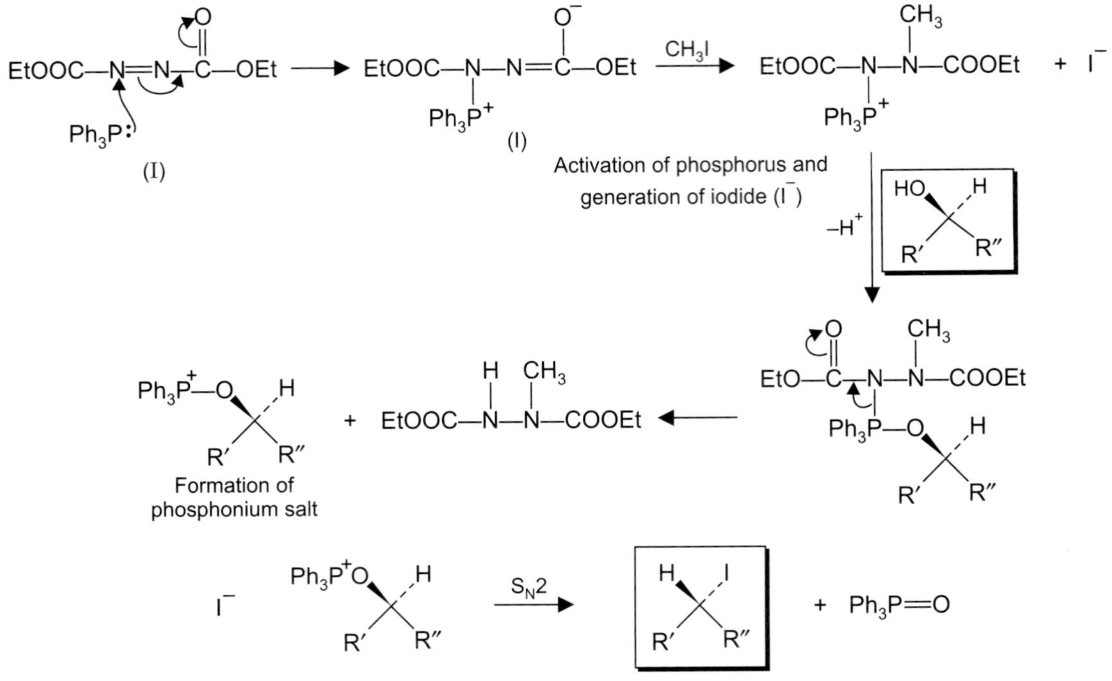

Nucleophilic substitution by I⁻, inversion of configuration

Mitsunobu reaction–conversion of an alcohol into an iodide

SCHEME 7.18b

EXERCISE 7.4

Write the products from the reactions (Scheme 7.18c) with stereochemistry.

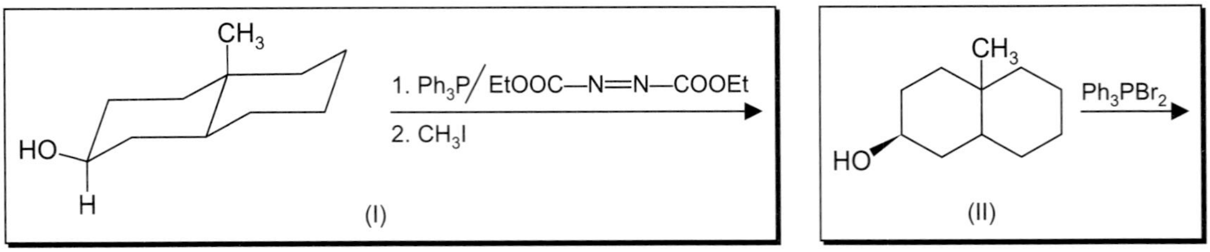

SCHEME 7.18c

ANSWER. *See Scheme 7.16d. Both reactions are S_N2 reactions leading to inversion of configuration.*

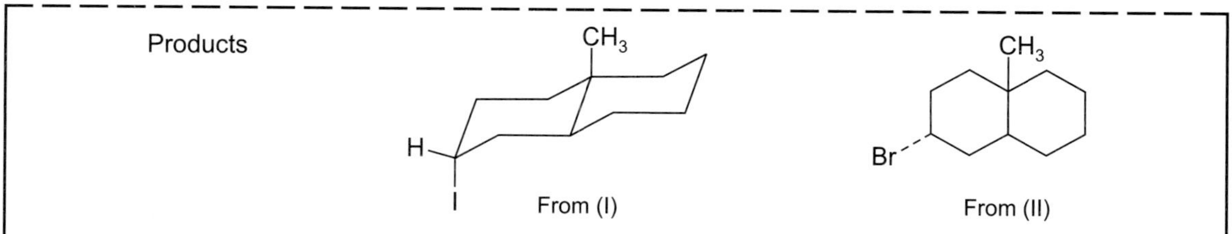

SCHEME 7.18d

(E) Conversion of 1, 2-Diols into Alkenes

A 1, 2-diol can be converted into its corresponding olefinic compound via a cyclic thiocarbonate. Its reaction with triethylphosphite gives the alkene (Scheme 7.18*e*). The success of the reaction lies in great affinity of phosphorus for sulphur.

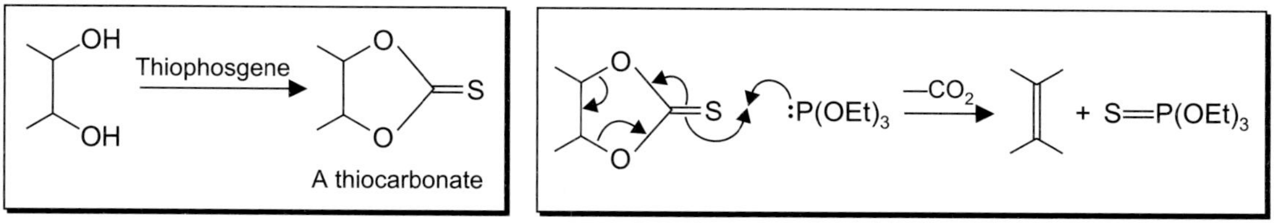

SCHEME 7.18e

(F) Cyclization of Aromatic Nitro Compounds

A great affinity of phosphorus for oxygen is the basis for the reductive cyclization of aromatic nitro compounds with triethylphosphite (Scheme 7.18*f*). In this cyclization reaction the oxygen

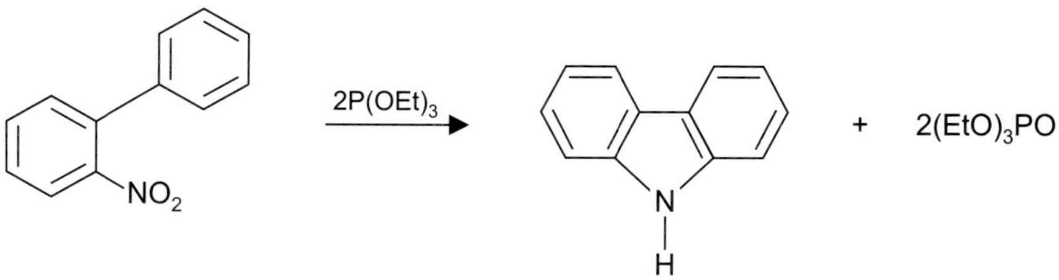

SCHEME 7.18f

atoms of the nitro group are transferred to phosphorus and the reaction is thought to proceed via the intermediate formation of a nitrene (ArṄ:).

7.4 ORGANOSULPHUR COMPOUNDS: SULPHUR YLIDES

(A) Introduction

The sulphur containing reagents are useful due to the ability of sulphur to utilize 3*d* orbitals for bonding and to occur in valence states higher than 2. The S—O bond is strong. Sulphur containing functional groups stabilize adjacent carbanions to provide carbanionic reagents for use in organic synthesis.

(B) Sulphur Ylides and their Formation

Just like phosphorus ylides sulphur ylides are also important synthetic reagents and are prepared from sulphonium salts which in turn are derived from sulphides by treatment with alkyl halides (Scheme 7.19).

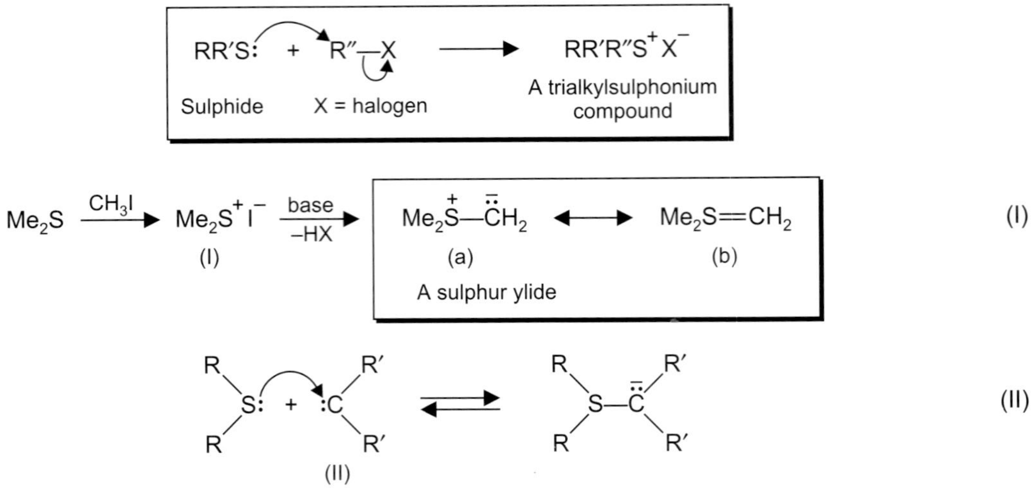

SCHEME 7.19

The abstraction of a proton from the corresponding sulphonium salt (I, Scheme 7.19) is the most common method, however, a direct method to make a sulphur ylide is via addition of a sulphide to a carbene (II, Scheme 7.19). The ylides are thought to be stabilized by resonance with the non-polar contributing forms (*i.e.,* bonding which involves a sulphur 3*d* orbital). In sulphur ylides the structure (*a*, Scheme 7.19) is the major resonance contributor *i.e.,* the sulphur ylides do not contain appreciable double bond character.

The stabilized ylides (like dialkylsulphoxonium ylides) are stabilized by the oxygen atom (Scheme 7.20). Nucleophilic dimethyl sulphoxide on reaction with methyl iodide gives the sulphonium salt which on reaction with a powerful base gives the ylide (sulphoxonium methylide).

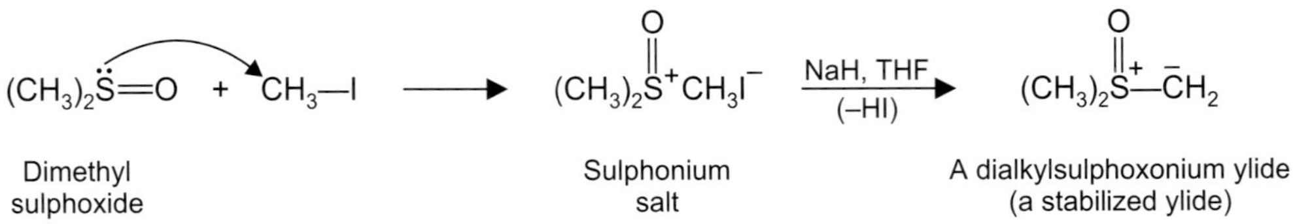

SCHEME 7.20

A highly basic sulphur ylide, dimsyl sodium is generated from dimethylsulphoxide (Scheme 7.21).

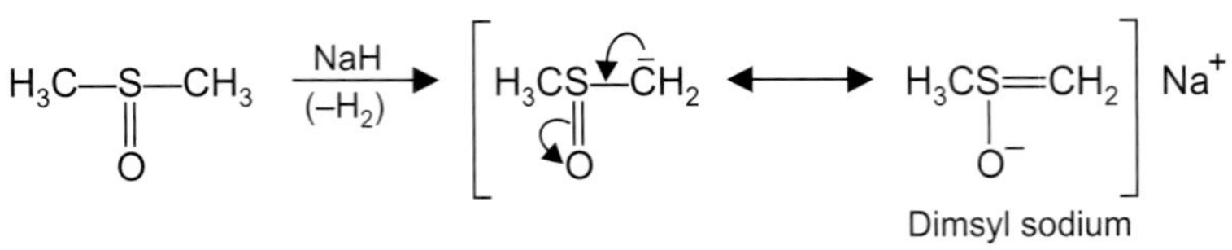

SCHEME 7.21

(C) Reactions of Sulphur Ylides

(i) Reaction with Carbonyl Compounds

Sulphur ylides react with aldehydes and ketones to give epoxides. In this respect they differ from phosphorus ylides which give alkenes. Sulphur ylides are also nucleophilic and attack the carbonyl carbon to give sulphur betaine. The first nucleophilic addition step is therefore, identical to the first step of the Wittig reaction. Sulfur has not that high affinity for oxygen as phosphorus thus the reaction after sulphur betaine formation follows a different path. The dimethyl sulphide functions as a very good leaving group, and is displaced by intramolecular nucleophilic attack of the alkoxide ion leading to the formation of an epoxide (Scheme 7.22). Thus, the sulphur ylide here (dimethylsulphonium methylide) acts as a methylene transfer reagent to adds a methylene group to the carbonyl double bond.

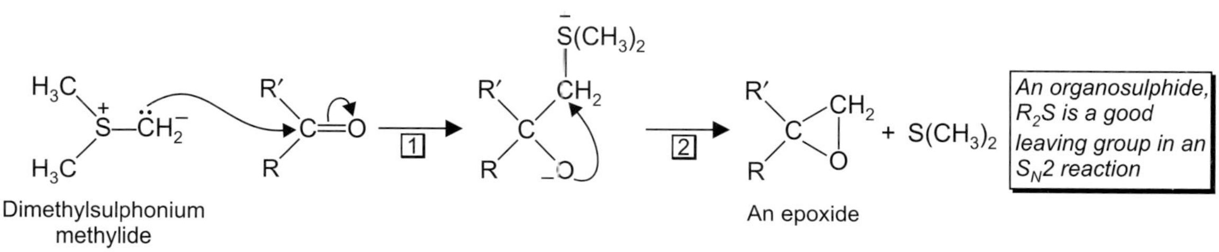

SCHEME 7.22

Epoxides provide an important group of synthetic intermediates and their synthesis (Scheme 7.23) is one of the most important applications of sulphur ylides. To give an example

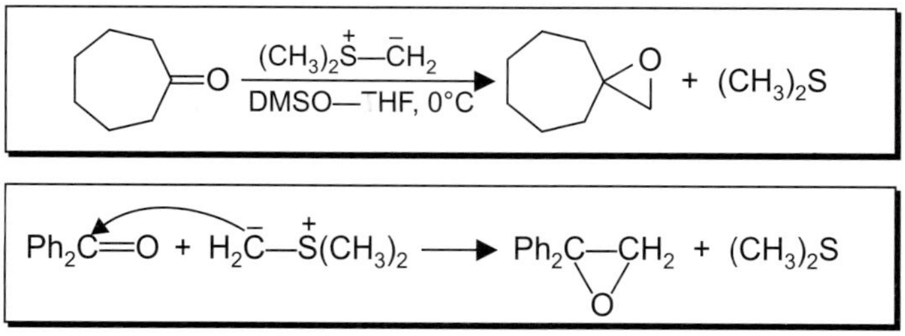

SCHEME 7.23

of this synthetic utility, the conversion ($>C{=}O \longrightarrow\ >CH{-}CHO$) can be brought about (Scheme 7.24) since epoxides rearrange to carbonyl compounds on reaction with BF_3 (see Scheme 13.29).

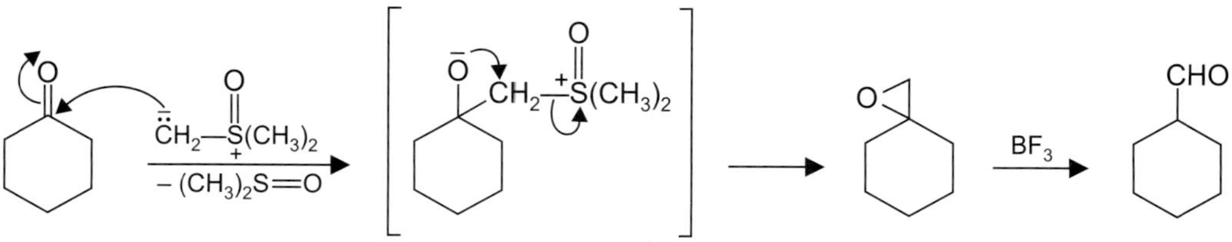

SCHEME 7.24

IMPORTANT SULPHUR YLIDES

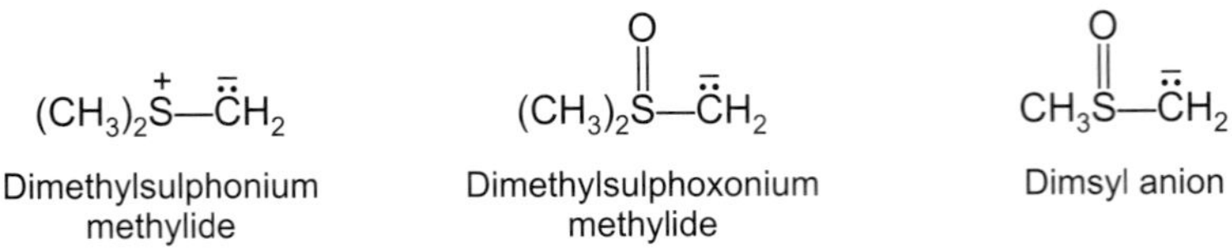

SCHEME 7.25

Unstabilized sulphonium ylides yield epoxides from α, β-unsaturated ketones, while stabilized ylides give cyclopropanes. In the absence of double bond both types of ylide give epoxides.

When a nucleophilic group is located close to the carbonyl group in a reactant then on reaction with the sulphonium methylide the intermediate epoxides are not isolated and instead heterocycles are generally obtained. Thus *o*-aminophenyl ketones (Scheme 7.26) react with dimethylsulphonium methylide to yield benzopyrroles via the epoxide formation and its subsequent opening.

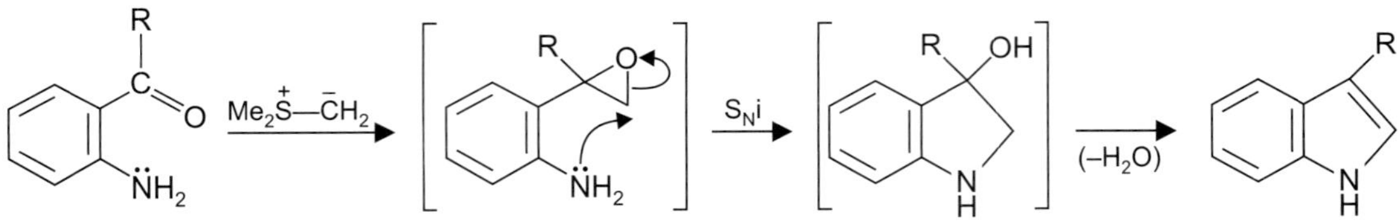

SCHEME 7.26

A synthetically useful sulphur ylide is cyclopropyl substituted (Scheme 7.26*a*). This reacts with carbonyl compounds to give oxaspiropentanes which are strained and undergo several ring opening reactions. Another example is in (Scheme 4.31*b*).

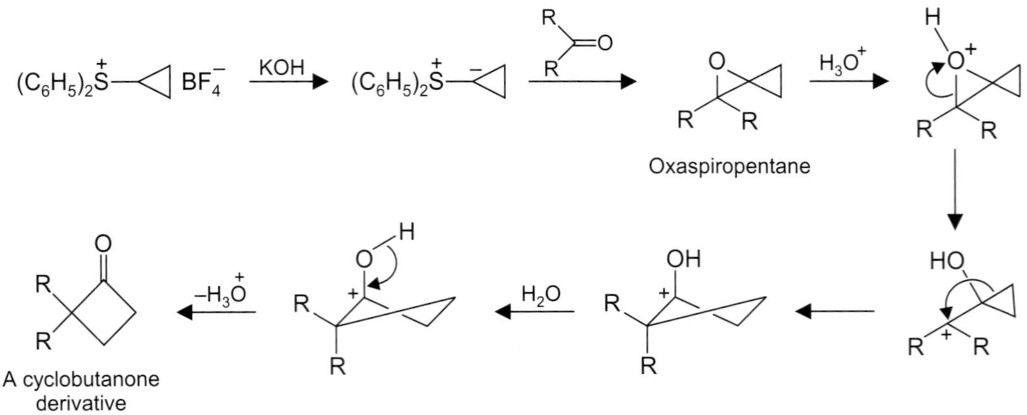

SCHEME 7.26a

EXERCISE 7.5

Write steps of the reaction of cycloheptanone with dimethylsulphonium methylide followed by $NaNH_2$ and HNO_2.

ANSWER. *Epoxide undergoes cleavage with amines to give aminoalcohols. The nitrous acid deamination will give a ring enlarged product (Tiffeneau-Demjanov reaction, Scheme 7.26b).*

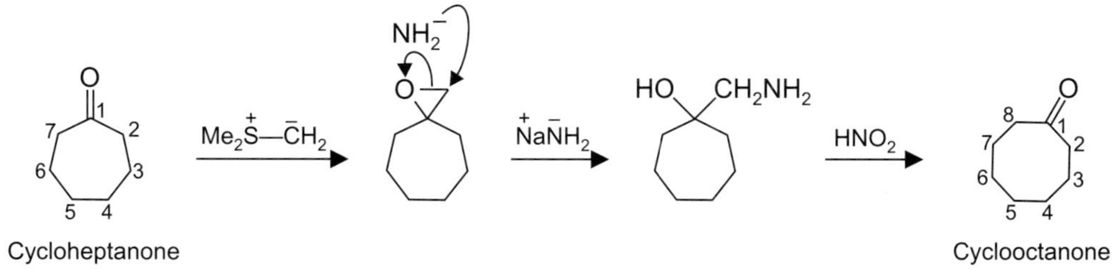

SCHEME 7.26b

(ii) Camparison in Reactivity Between Dimethylsulphonium Methylide and Dimethylsulphoxonium Methylide

These two types of sulphur ylides differ in their reactivity with α, β-unsaturated carbonyl compounds. The more reactive sulphonium ylides react rapidly by 1, 2-addition across the carbon-oxygen double bond and give epoxides. The less reactive sulphoxonium ylides react slowly by conjugate addition (1, 4-addition) to give cyclopropanes *via* Michael addition to the carbon carbon double bond (conjugate addition Scheme 7.27). The difference probably is due to two reactions the initially formed betaine can undergo: (*i*) reversal to starting materials or (*ii*) intramolecular nucleophilic displacement. Both reagent react most rapidly at the carbonyl

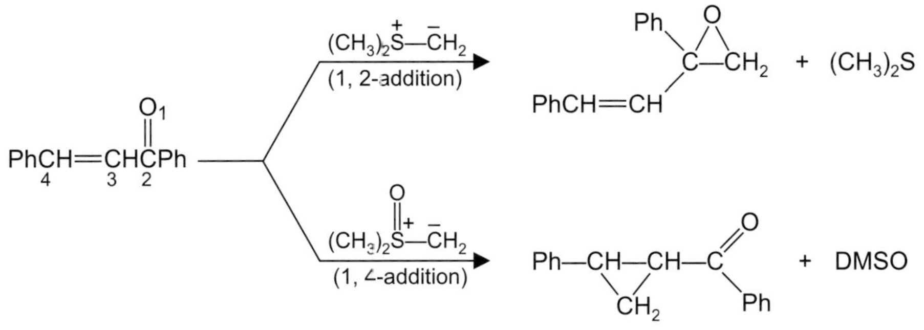

SCHEME 7.27

carbon. With dimethylsulphonium methylide, the intramolecular displacement step is faster than the reverse of addition leading to epoxide formation (I, Scheme 7.28). The reaction with more stable dimethylsulphoxonium methylide, the reversal is faster and the product formation occurs after conjugate addition (II, Scheme 7.28).

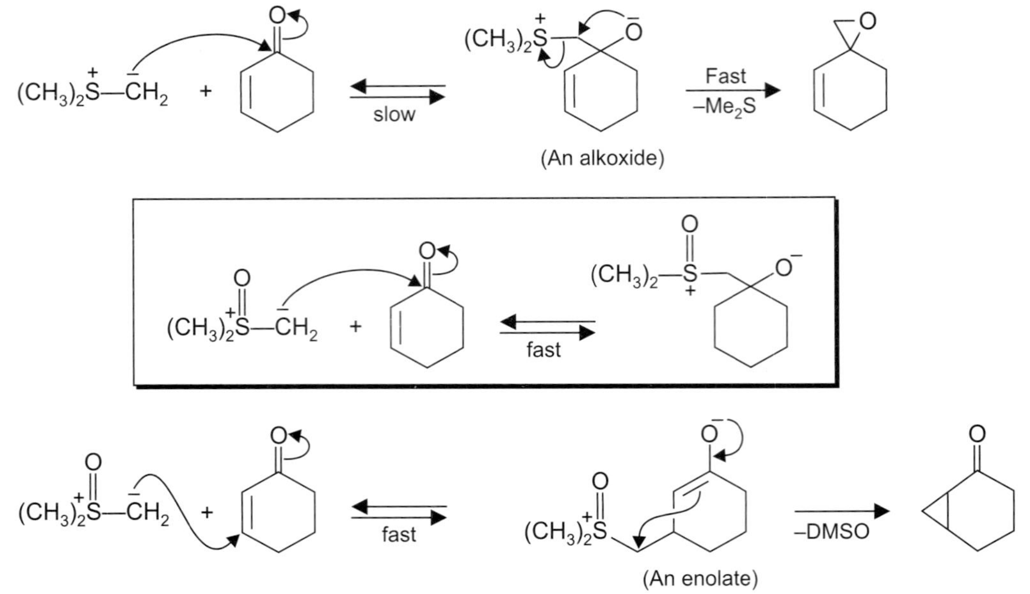

(I)

(II)

SCHEME 7.28

Another difference in these two types of sulphur ylides is the difference in the stereochemistry of epoxide formation with a cyclohexanone. Both dimethylsulphonium methylide as well as dimethyloxosulphonium methylide react the same way to give epoxides on reaction with non-conjugated aldehydes and ketones. These however, differ in their stereoselectivity with a cyclohexanone (Scheme 7.29). With the sulphonium ylide the epoxide has the new carbon-carbon bond axial, while with the oxosulphonium methylide this bond is equatorial. It is suggested that this difference may also be due to reversibility of addition in the case of the sulphoxonium methylide. The product from the sulphonium ylide is due to the kinetic preference for axial addition by small nucleophiles. In the case of reversible addition of the sulphoxonium ylide, product stereochemistry is determined by the rate of displacement which may be faster for the more stable epoxide.

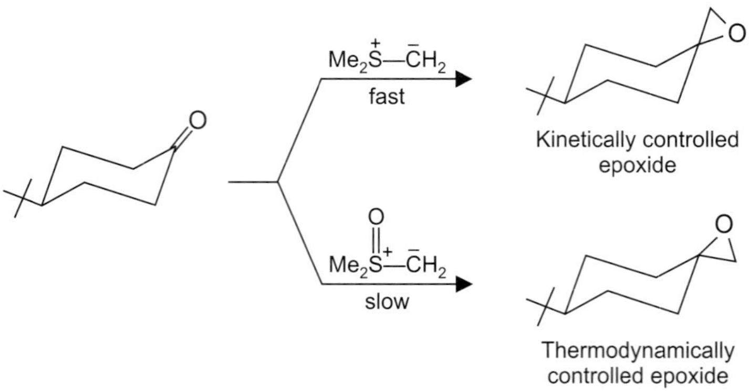

SCHEME 7.29

(iii) Reaction with Dimsylsodium—Synthesis of Complex Ketones

Dimsyl anion is very reactive as a nucleophile and reacts with aldehydes and ketones by nucleophilic addition to give epoxides, however reaction with esters is a nucleophilic substitution to give β-ketosulphoxides (I, Scheme 7.30).

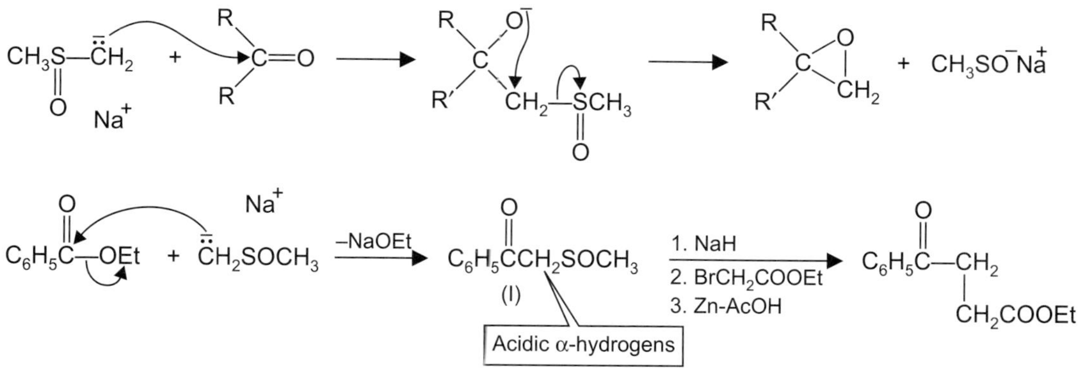

SCHEME 7.30

A β-ketosulphoxide contains a strongly activated methylene group and the acidic α-hydrogens can be removed to allow alkylations. Reductive desulphuration (Zn/AcOH) then gives ketones.

(D) 1, 3-Dithiane Anions—Umpolung (dipole inversion)

For introduction to 1, 3-Dithiane Anions see Scheme 4.36*a*.

The characteristic reactivity, nucleophilic or electrophilic of an atom or a group can be reversed temporarily in a process termed umpolung (German for polarity reversal, see Scheme 4.35). This *e.g.*, is seen in the case of aldehydes when these are converted into the anions of the corresponding 1, 3-dithiacyclohexanes (1, 3-dithianes). The otherwise electrophilic carbonyl carbon changes into nucleophilic center. The 1, 3-dithiane anions can be alkylated by a variety of reagents like, primary and secondary haloalkanes, aldehydes and ketone and epoxides.

A dithiane can be made by the reaction of an aldehyde with 1, 3-propane dithiol with a Lewis acid catalyst. The hydrogen on the carbon attached to two sulphur atoms can be removed by reaction with a strong base (Scheme 7.31). After the reaction, the 1, 3-dithiane system can be reconverted to the carbonyl compound by acid hydrolysis in the presence of mercury (II) ion.

Dithianes represent very important compounds in organic synthesis since as one goes from a ketone to a thioacetal the polarity at the functionalized carbon atom is inverted. Recall that aldehydes are electrophiles at the C=O carbon atom but a dithioacetal (a dithiane) after deprotonation to an anion become nucleophilic at the same carbon atom. This is an example of umpolung and dithianes represent the most important among the umpolung reagents.

(i) Synthesis of aldehydes (HCHO ⟶ RCHO)

The carbanion from 1, 3-dithiacyclohexane (I, Scheme 7.31) obtained by condensation of propane-1, 3-dithiol and formaldehyde can be alkylated by a suitable alkyl halide.

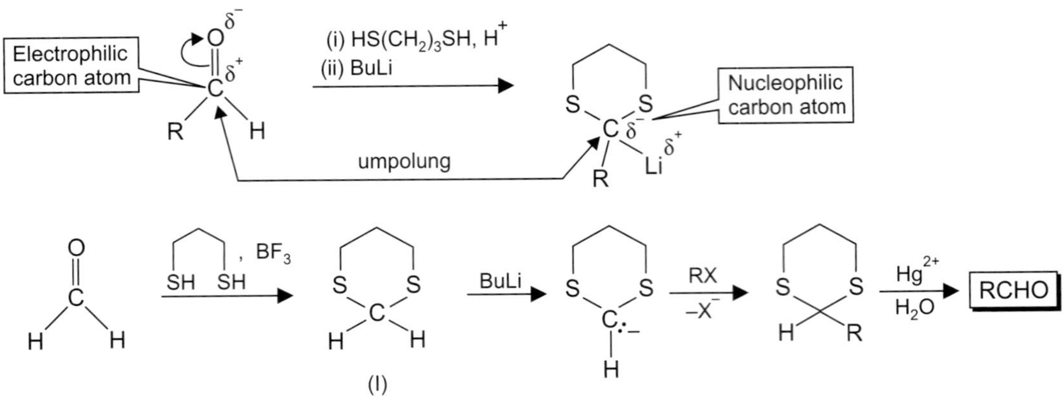

SCHEME 7.31

(ii) Synthesis of ketones (R—CHO ⟶ RR′CO)

This can be achieved by working with a dithiane obtained with an aldehyde other than formaldehyde and following the preocedure as above (see Scheme 4.36*a*).

(iii) Synthesis of Cyclic Ketones—Cyclobutanone

The carbanion derived from 1, 3-dithiacyclohexane (see I, Scheme 7.31) is alkylated with 3-chloroiodopropane (since iodine is the better leaving group it is selectively displaced) and the sequence of reactions (Scheme 7.32) gives cyclobutanone at the end.

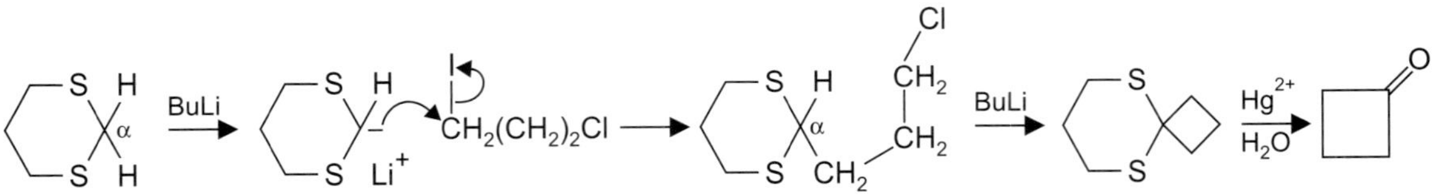

A dithiane
(cyclic thioacetal)

SCHEME 7.32

EXERCISE 7.6

A 1, 3-dithiane can be converted into a synthetically useful stabilized carbanion (Scheme 7.33). The corresponding 1, 3-dioxane which has even more electronegative oxygen atoms around the acetal carbon cannot be converted similarly to the anion by butyl lithium? Explain.

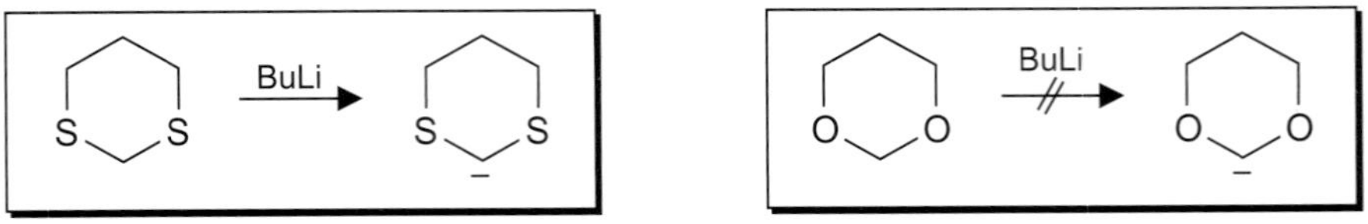

SCHEME 7.33

ANSWER. *The acidifying effect of sulphur is not based on electronegativity. The sulphur atoms have unfilled valence level d orbitals available which can accept electron density to stabilize an adjacent anion. One can consider the anion as being resonance stabilized with the negative charge being delocalized into the flanking sulphur atoms.*

(iv) Synthesis of a α- or β-hydroxy ketone

The nucleophilic 1, 3-dithianyl anions react with another source of electrophilic carbon in the form of an aldehyde, ketone or an epoxide. For example, reaction with an aldehyde gives an α-hydroxy ketone (Scheme 7.34) and the dithianyl anion is used as a nucleophile to open an epoxide ring, the hydrolysis of the resulting hydroxy dithiane gives a β-hydroxy ketone. In the reaction with an epoxide, it is the less hindered end of the epoxide that is preferentially attacked.

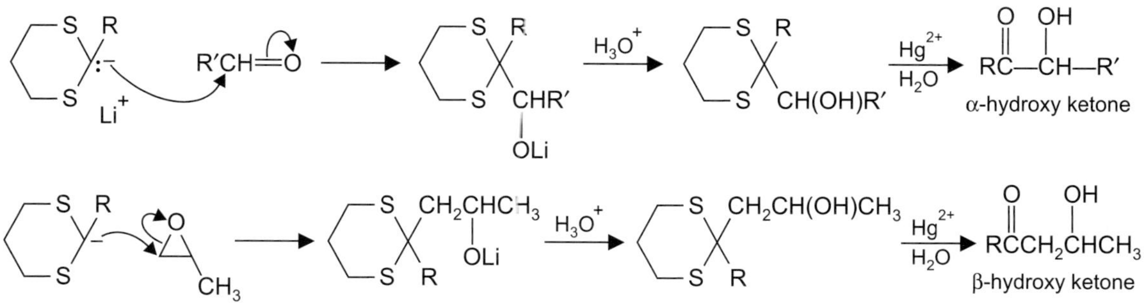

SCHEME 7.34

7.5 SILICON REAGENTS

(A) Introduction

Several organic reactions involving organosilicon compounds can take place since Si forms stronger bonds with O and F than does C but weaker bonds with C and H. Moreover the 3*p*-electrons of Si do not overlap effectively with the 2*p*-electrons of C or O. Multiple bonds C=Si and O=Si are not, therefore, commonly found in stable molecules.

Silicon is more electropositive than carbon so the carbon-silicon bond is strongly polarized as shown (I, Scheme 7.34*a*).This results in alkylsilanes being prone to attack by nucleophilic reagents. Silicon also has the ability to stabilize α-carbanions, (II, Scheme 7.34*a*) as well as β-carbocations (III).

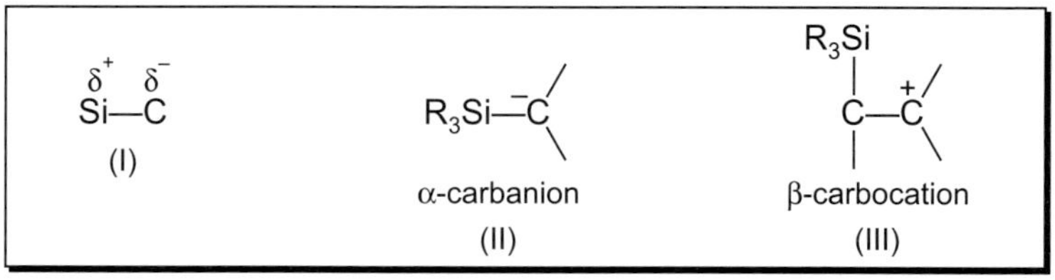

SCHEME 7.34a

Several chlorosilanes *e.g.*, $SiCl_4$ and $(Me)_3SiCl$ are readily available organosilicon reagents. These halides undergo a facile nucleophilic displacement to give a variety of useful synthetic intermediates (Scheme 7.35).

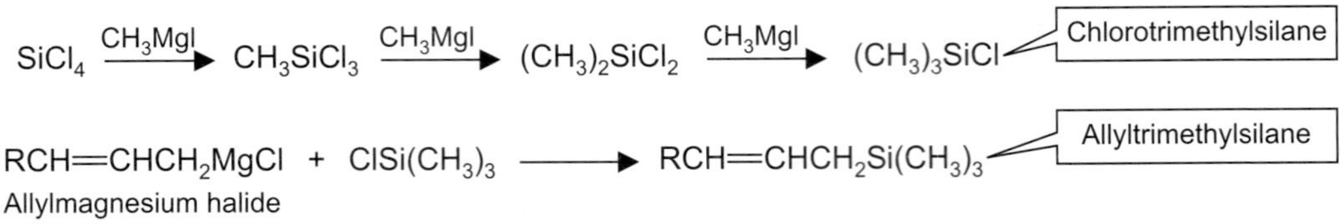

SCHEME 7.35

(B) Peterson Reaction—Synthesis of Alkenes

The Peterson silicon based alkene synthesis may be viewed as silicon version of Wittig reaction (Scheme 7.35*a*). Although the use of silicon stabilized carbanions is less common than those derived from phosphorus or sulphur, there are several significant advantages in steric terms.

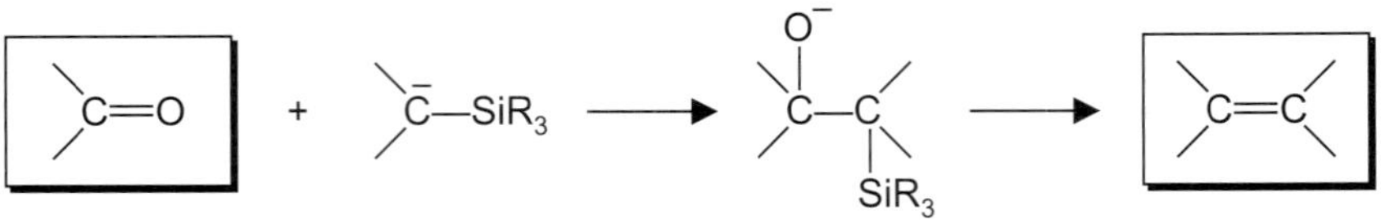

The Peterson synthesis—use of silicon stabilized carbanions

SCHEME 7.35a

The reaction can be carried out in two ways. In the first method the reactant *e.g.,* ethyl α-trimethylsilylacetate has a CH group which is adjacent to both a silicon containing moiety (which is normally $SiMe_3$) and a –*M* group displays a base induced reaction with an aldehyde or ketone to yield the alkene directly (Scheme 7.35*b*). The driving force for the reaction is the formation of strong silicon-oxygen bond, which converts the oxygen atom in to a much better silyloxy leaving group. In this case the more stable olefin isomer is formed because equilibration occurs in the enolate intermediate.

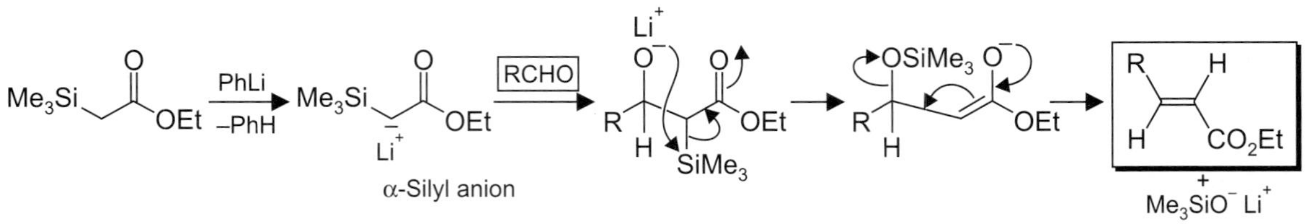

SCHEME 7.35b

In the second alternative when the group of –*M* type is absent in the reactant, one generates a C-metal bond adjacent to the $SiMe_3$ group. The reagent is therefore, trimethylsilylmethyl Grignard which adds to the carbonyl group of the reactant to form a β-hydroxysilane after hydrolysis. In this case the second step in needed to convert β-hydroxysilane into alkene (since the product alkene is not conjugated). These β-hydroxy silanes can be formed as diastereomers as *threo, erythro form*. These are separated and either of the diastereomers undergoes elimination of the trialkylsilyl group and the hydroxyl group to yield the alkene. The elimination is carried out under basic (KH) or acidic (H^+ or $BF_3.Et_2O$) conditions. Either diastereomer displays over 90% stereoselectivity and can give *E* or *Z* alkene, the elimination being *syn* under basic conditions and *anti* under acidic conditions (Scheme 7.35*c*).

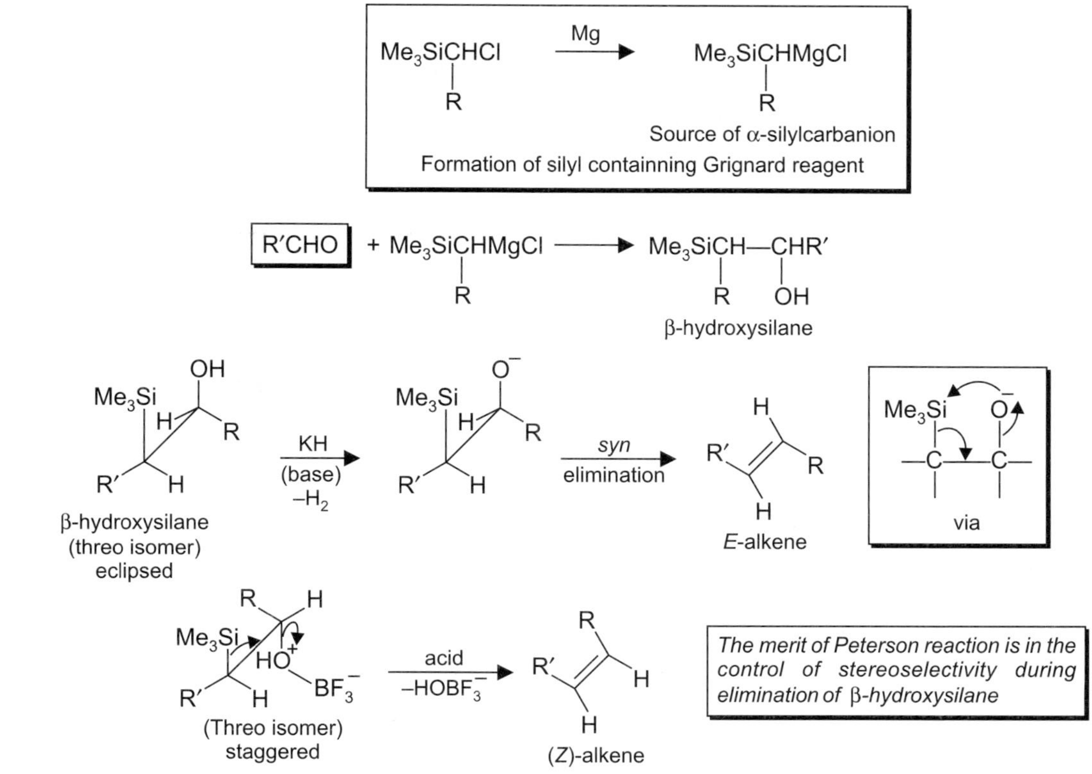

SCHEME 7.35c

(C) Silyl Epoxides and their Synthetic Applications to Make Methyl Ketones

A silyl epoxide may be made from ketones by the use of a useful reagent α-chlorotrimethylsilylmethyl-lithium (II, Scheme 7.35d) which is made from α-chloromethyltrimethylsilane (I) and *s*-butyllithium. This reagent (II,Scheme 7.35d) reacts with aldehydes and ketones to give α,β-epoxysilanes *via* a chlorohydrin (Scheme 7.35*d*). These epoxides undergo an acid catalyzed ring opening and the regioselectivity is dictated by the stabilizing effect of a β-Si-C bond on a carbocation (see III, Scheme 7.35*e*). An example is the conversion of an aldehyde RCHO to a methyl ketone RCH_2COCH_3 (Scheme 7.35*e*), which uses a related reagent α-methyl-α-chlorotrimethylsilylmethyl-lithium (A, Scheme 7.35*e*).

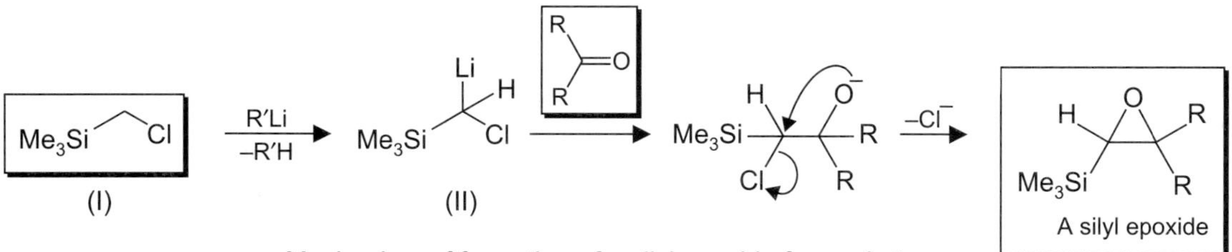

Mechanism of formation of a silyl epoxide from a ketone

SCHEME 7.35d

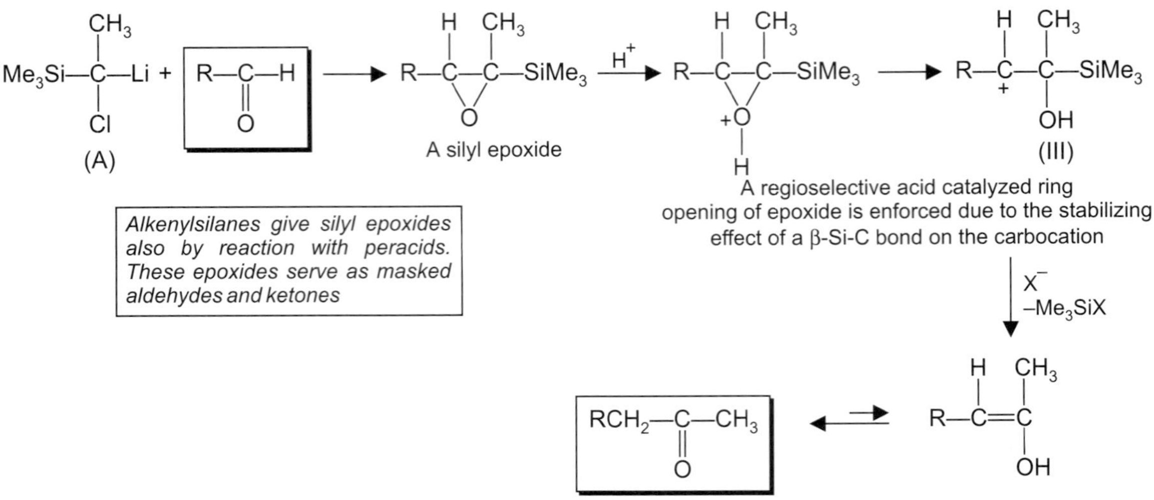

SCHEME 7.35e

(D) Synthetic Utility of Vinylsilanes (Alkenyl Silanes)

Vinylsilanes may be prepared by the nucleophilic displacement of halogen from a halosilane by an organometallic reagent (Scheme 7.36). An electrophile adds to the double

$$RCH{=}CHMgBr + (CH_3)_3SiCl \longrightarrow RCH{=}CHSi(CH_3)_3$$

A vinylsilane (an alkenylsilane)

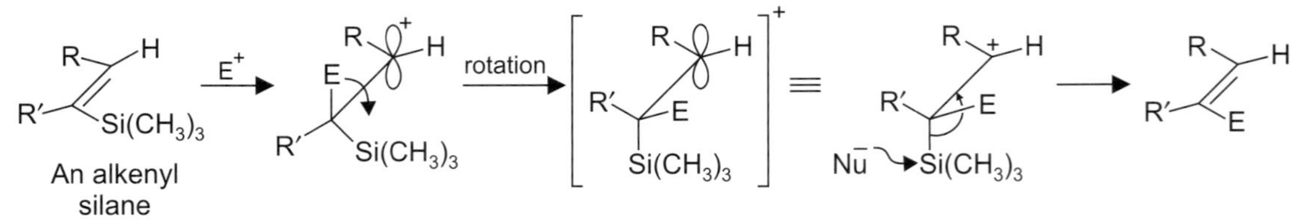

Mechanism of the reaction of a vinylsilane with an electrophile
SCHEME 7.36

bond of a vinylsilane regioselectively at the α-carbon due to stabilization provided by silicon to a carbocation β to it. This process results in a nucleophilic displacement at silicon to release an alkene in which the silyl group gets replaced stereospecifically by the electrophile. The reaction generally occurs with retention of configuration. It has been suggested that retention of configuration is due to the rotation around the C—C single bond in the carbocation to increase the stabilization interaction between the unoccupied 2*p* orbital and the C—Si bonding orbital (Scheme 7.36). Reaction of a vinylsilane with an electrophile normally requires a Lewis acid catalyst and some examples are in (Scheme 7.36*a*).

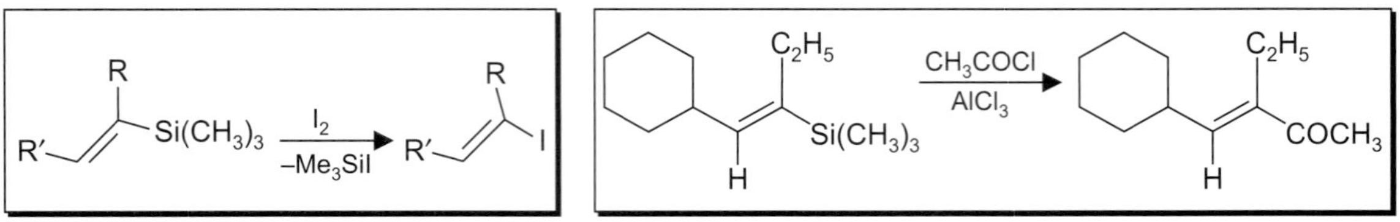

SCHEME 7.36a

(E) Reactions Involving Allylsilanes

Allylsilanes used for synthesis may be made from allyl halides via. Grignard reagents (Scheme 7.37). The dominant reaction of the addtion of electrophiles to these allylsilanes is again the attack of the electrophile at the double bond at the γ carbon due to the stability

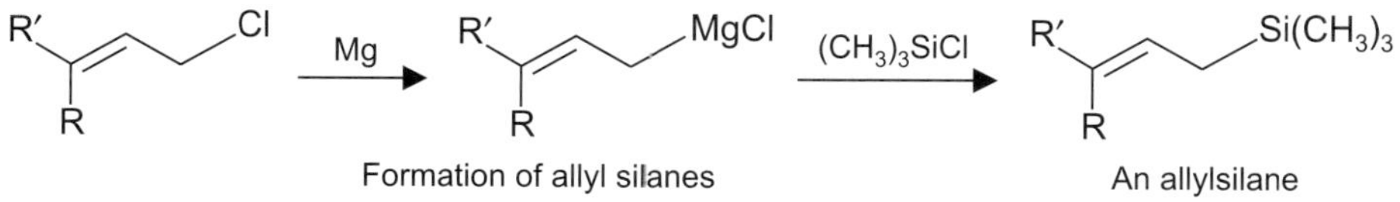

SCHEME 7.37

of the β-silyl carbocation. The overall process results in the replacement of the silicon substituent with an allylic shift of the double bond (Scheme 7.37*a*). The silyl group is removed via nucleophilic substitution at silicon. An example of the reaction involving electrophilic attack on allylic silanes is in (Scheme 7.37*b*), and probably involves acylium ion as the electrophiles.

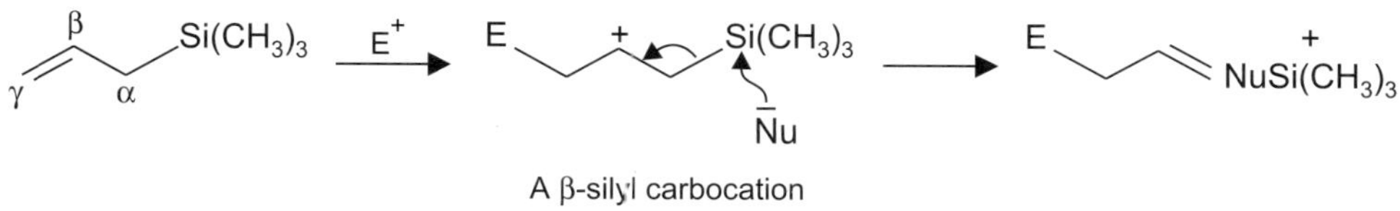

Addition of an electrophile to an allylsilane followed by double bond shift

SCHEME 7.37a

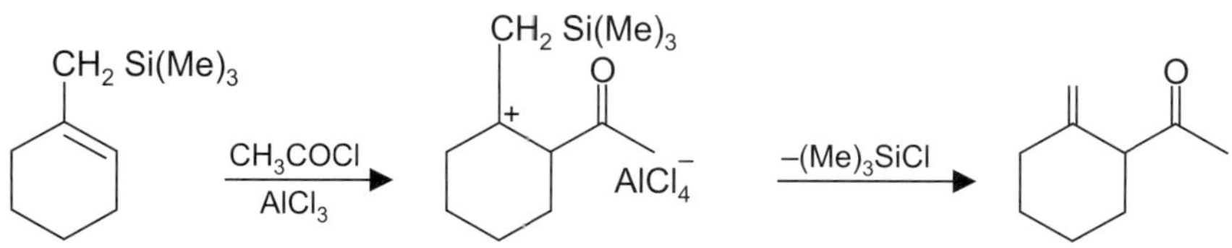

Acylation of an allylsilane to give a β, γ-unsaturated ketone

SCHEME 7.37b

(F) Trapping of Enolate Anions

When *e.g.*, chlorotrimethylsilane is included with the ketone and a base, the enolates (thermodynamic and kinetic) are trapped as silyl enol ethers, which can be separated by chromatography and converted to the parent enolates by treatment with fluoride ion (Scheme 7.38). Silylation, is a nucleophilic substitution at the silicon atom by the oxygen atom of the enolate, (oxygen-silicon bond that forms in the trimethylsilyl enol ether is very strong

For details of formation and purification of enolates see Schemes 6.5 and 6.6.

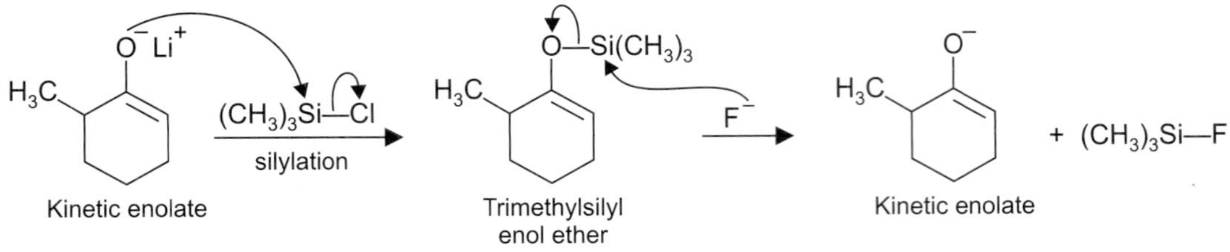

SCHEME 7.38

and is much stronger than a carbon-silicon bond). The trimethylsilyl enol ether is converted back to the enolate on treatment with fluoride ions (nucleophilic substitution) since Si—F bond is very strong.

A regiospecific α-alkylation of a ketone can be achieved by employing silyl enol ether intermediates (Scheme 7.38*a*). The source of fluoride ion is tetrabutylammonium fluoride $(C_4H_9)_4N^+F^-$ (TBAF).

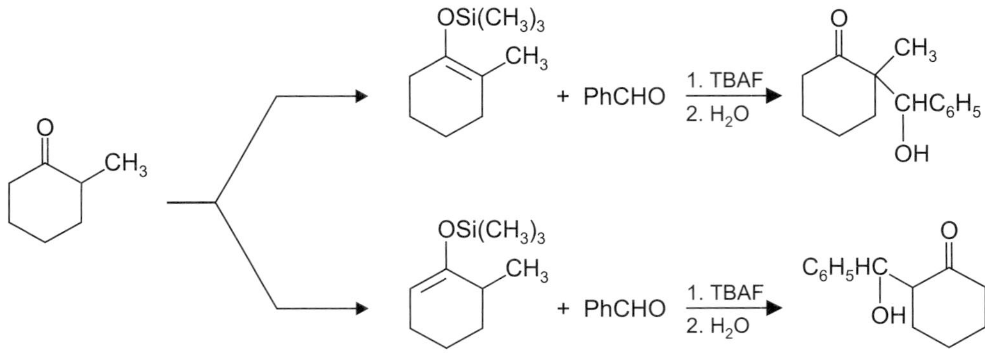

SCHEME 7.38a

Silyl enol ethers are alkylated by S_N1-reactive electrophiles with the use of Lewis acids (see Scheme 6.43 *b*).

(G) Silyl Ether Protecting Groups

A hydroxyl group may be protected by converting it to a silyl ether group. The most common is the *tert*-butyldimethylsilyl ether group [*tert*-butyl$(CH_3)_2$Si—O—R, or TBDMS—O—R], since it is stable over a pH range of about 4–12. A TBDMS group can be added by reacting the alcohol with *tert*-butyl-chlorodimethylsilane in the presence of an aromatic amine (a base such as imidazole or pyridine). This protecting group can be removed by treatment with fluoride ion (TBAF, Schem7.39). Conversion of an alcohol to its silyl ether makes it more volatile for application in analysis by gas chromatography and trimethylsilyl ethers are derivatives of choice. Trimethylsilyl ether group is however, too labile to solvolysis in protic media for use as a protecting group. In the case of R—O—TBDMS steric factors increase the stability and decrease the sensivity to hydrolysis. Reaction of an alcohol with the bulky chlorosilane is rather slow to be practicable, however imidazole is a far stronger nucleophile than alcohol and *N*-*t*-butyldimethylsilyl-imidazolyl ion (I, Scheme 7.39) thus formed provides an effective silylating agent and also provides a good leaving group (protonated form of imidazole).

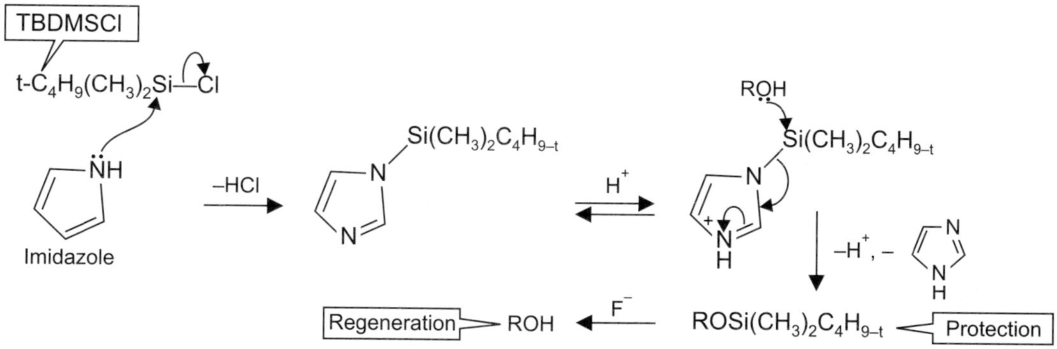

SCHEME 7.39

(H) Synthetic Uses of Iodotrimethylsilane

Trimethylsilyl iodide is an electrophilic reagent which forms strong Si—O bonds on reaction with oxygen nucleophiles and liberates strongly nucleophilic iodide ion. The reagent is often generated in *situ* by reacting trimethylsilyl chloride with sodium-iodide.

(i) Cleavage of Ethers

Benzyl and *t*-butyl systems get cleaved rapidly while secondary systems require longer times. The reaction involves the initial formation of a silyloxomium ion (I, Scheme 7.40). The direction

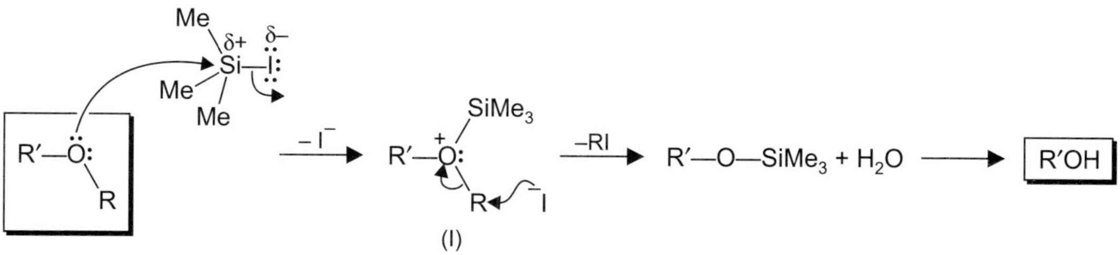

Cleavage of ethers with trimethylsilyl iodide

SCHEME 7.40

of cleavage in the case of unsymmetrical ethers is determined by the relative ease of O—R bond cleavage which may follow either S_N2 (methyl, benzyl) or S_N1 (*t*-butyl) processes.

(ii) Cleavage of Esters

Esters are also cleaved *via* the formation of trimethylsilyl esters which are easily hydrolyzed with water (Scheme 7.40*a*).

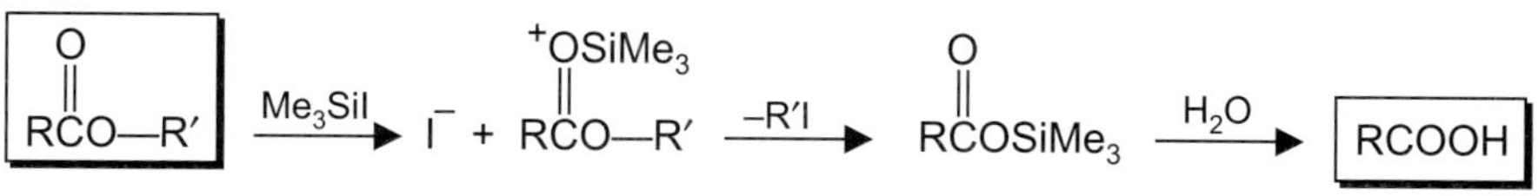

Cleavage of esters with trimethylsilyl iodide

SCHEME 7.40a

(iii) Reaction with Alcohols

Trimethylsilyl iodide converts alcohols into their iodides and even bridgehead alcohols gives good yields as seen in the case of adamantan-1-ol. Otherwise inversion of configuration (S_N2 mechanism) seems to be operative (Scheme 7.41).

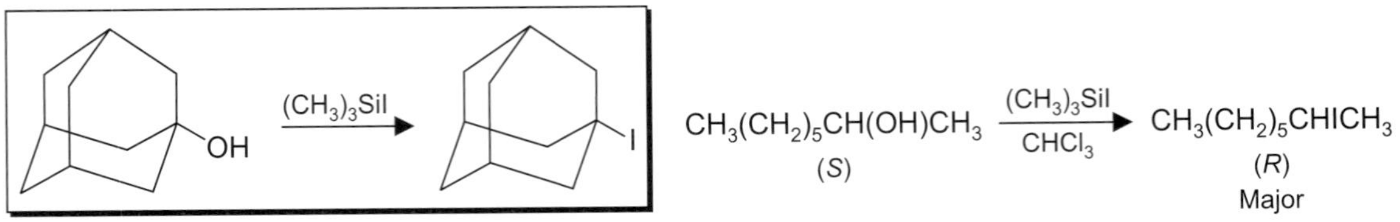

SCHEME 7.41

(iv) Reaction with Ketones

Enolizable ketones react with trimethylsilyl iodide to give silyl enol ethers. Iodine is introduced in the β or γ- position of the ketone when the reactant is α, β-unsaturated compound or a cyclopropyl compound (Scheme 7.42).

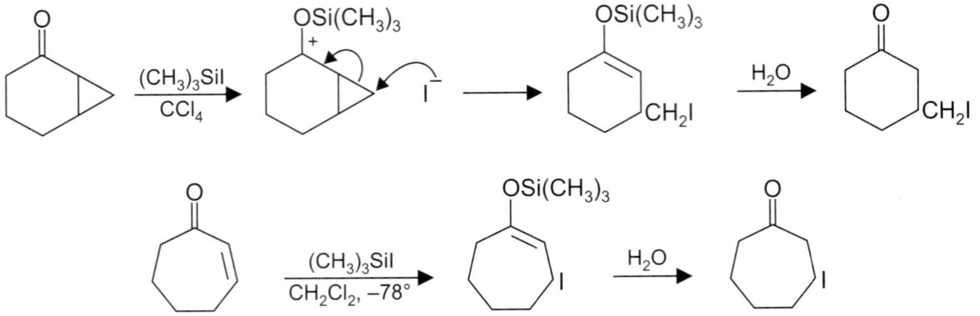

SCHEME 7.42

SELENIUM REAGENTS

SeO_2 in allylic oxidation of alkenes (Scheme 13.12), oxidation of C—H groups (Scheme 13.15), elimination from selenoxides (Schemes 12.32 and 12.33).

7.6 BORON CONTAINING REAGENTS

(A) Introduction

Borane (which exists as the gaseous dimer diborane B_2H_6) and organic boranes easily add to C=C bonds. The boron atom, subsequently can be removed from these adducts by using several reagents *via* reactions which use the ability of the boron atom (which is election poor) to accept an electron pair. These factors are responsible for synthetic applications of borane.

(B) Hydroboration

The boron-hydrogen bond *i.e.*, the B—H unit adds rapidly and quantitatively to many multiple bonds including carbon-carbon double bonds, a process known as hydroboration[10]. With simple alkenes, a trialkylborane is formed (Scheme 7.43). Diborane is commercially available in the form of complexes which it forms with ethers.

$$RCH{=}CH_2 \xrightarrow{BH_3} RCH_2CH_2BH_2 \xrightarrow{RCH=CH_2} (RCH_2CH_2{-})_2BH \xrightarrow{RCH=CH_2} (RCH_2CH_2{-})_3B$$

SCHEME 7.43

(C) Mechanism of Hydroboration

The π bond is electron rich and boron, electron poor. The reaction is initiated *via* the coordination of BH_3 with the π-electrons of the double bond (the participation of the empty *p* orbital on BH_3) followed by the formation of the carbon-hydrogen bond as in R—CH=CH_2 (Scheme 7.44), *via* a four center transition state. Both the new C—B and C—H are as a result formed from the same side of the double bond. The addition is dominated by steric considerations. Hydroboration is regioselective; unlike the electrophilic addition, steric and not electronic factors control the regioselectivity; the boron generally becomes attached to the less substituted and less sterically congested carbon. The alkylborane products are normally not isolated but are converted by subsequent reactions directly into desired products. The most important general reaction of alkyl-boranes is oxidation with alkaline hydrogen peroxide to give an alcohol (RCH_2CH_2OH).

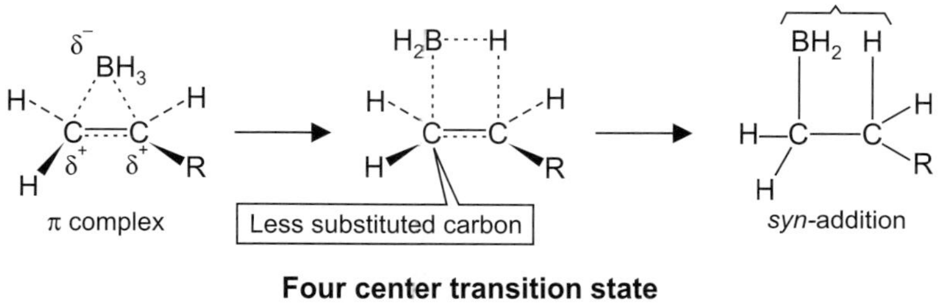

SCHEME 7.44

In case where stereochemistry may be defined, exclusive *syn* addition is observed. Thus addition occurs with *syn* stereospecificity. Consider 1-methyl-cyclopentene, since BH_3 has B—H bonds, it adds to the double bonds of three molecules of 1-methylpentene to afford a trialkyl borane (Scheme 7.45). The oxidation of trialkylborane with alkaline hydrogen peroxide replaces the boron atom with a hydroxyl group in the same stereochemical position. The net result of hydroboration and oxidation-hydrolysis is the addition of water (hydration) across a double bond with *anti*-Markovnikov orientation.

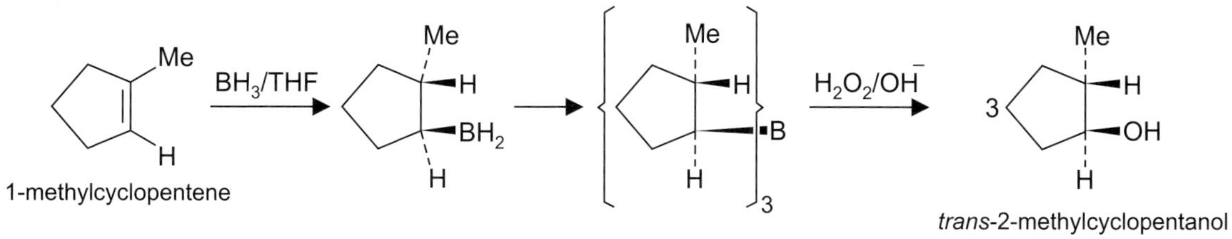

SCHEME 7.45

The alkene reacts at the less hindered side of the multiple bond (Scheme 7.46), there being a preference for approach of the borane from the less hindered side of the molecule.

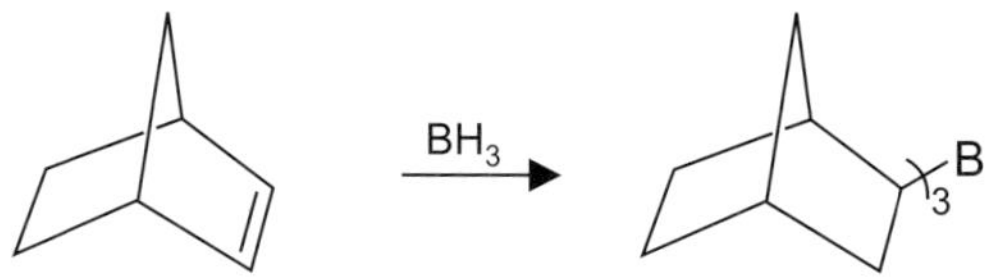

SCHEME 7.46

(D) Substituted (Sterically Congested) Boranes

Hydroboration of alkenes with borane have several limitations:

(*i*) Regioselectivity in the hydroboration of terminal alkenes is high, however, it is not complete (see Scheme 7.48).

(*ii*) In the case of 1, 2-disubstituted alkenes, there is little discrimination between the two termini of the double bond (see Scheme 7.48).

(*iii*) The rates of reaction of borane do not differ with differently substituted double bonds. Thus one may not achieve selective hydroboration of one double bond in the presence of other.

(*iv*) Hydroboration of terminal alkynes proceeds part the desired alkenylborane by the addition of second molecule of borane.

Most of these difficulties can be solved by using organoboranes formed from an appropriate alkene and diborane, using control of stoichiometry to terminate the hydroboration at the desired degree of alkylation. In practice, the number of hydroborations for a given alkene is strongly dependent on steric effects. The more hindered the alkene, the fewer the number of possible additions. For example, 2-methyl-2-butene hydroborates only twice, 2, 3-dimethyl-2-butene only once (Scheme 7.47). These reagents contain one or two B—H bonds; are sterically congested; less reactive and more selective than borane. Thus one can successfully achieve a desired regioselectivity between the two carbon atoms of the double bond in an alkene. Diborane *e.g.*, reacts with the internal alkene (Scheme 7.48) in a completely non-selective fashion, while with the same alkene disiamylborane (Sia_2BH) reacts almost exclusively at the methyl substituted unsaturated carbon. More example of regioselectivity of diborane and alkylboranes are given (Scheme 7.48).

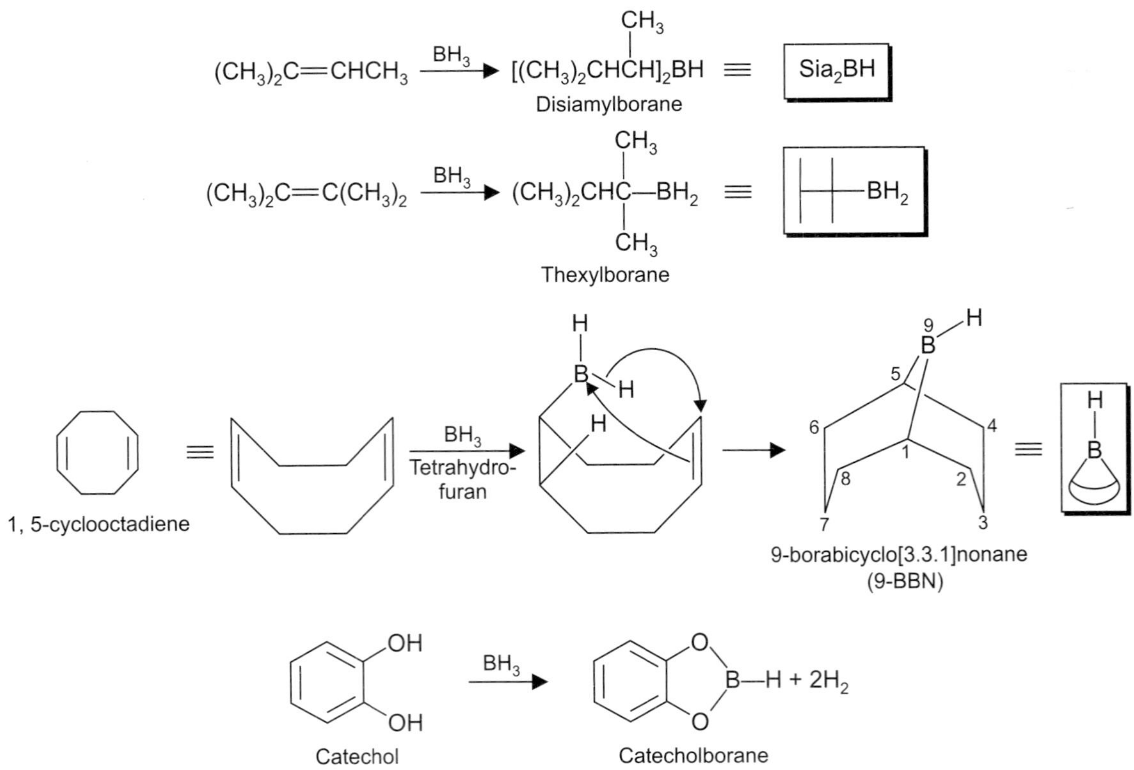

SCHEME 7.47

	1-hexene (a terminal alkene)	4-methyl-2-pentene (a 1, 2-disubstituted alkene)
Disiamylborane	99	97
9-BBN	99.9	99.8
Diborane	94	57 (43% adds at more substituted carbon)

% of Boron Added at Less Substituted Carbon

SCHEME 7.48

Often one succeeds in the selective hydroboration of one double bond in the presence of other. Vinyl cyclohexene can be monohydroborated with either disiamyl-borane or 9 BBN and oxidised to the corresponding alcohol (Scheme 7.49), also see Scheme 13.47. Catecholborane and 9-BBN are especially useful since these have only one B—H bond and thus one can control the reaction with an alkene or an alkyne.

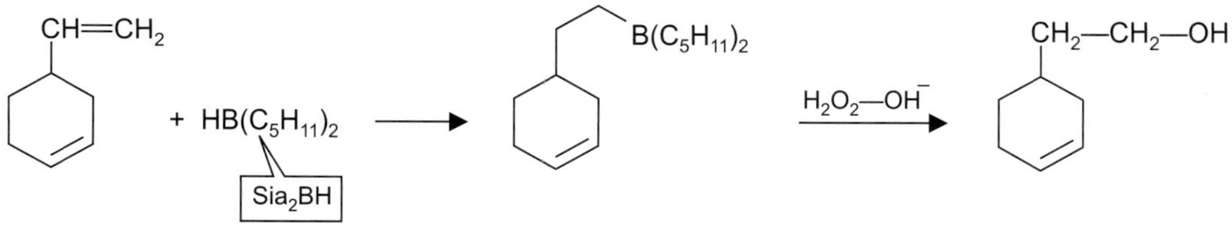

SCHEME 7.49

(E) Isomerization of Organoboranes

Hydroboration is thermally reversible. On heating around 150°C, organoboranes are isomerized to a mixture in which the major component has boron attached to the terminal carbon atom of the alkyl group (Scheme 7.50). The process involves a series of eliminations and additions and is catalysed by diborane. Using this techinque one can convert a readily available internal alkene to a primary alcohol.

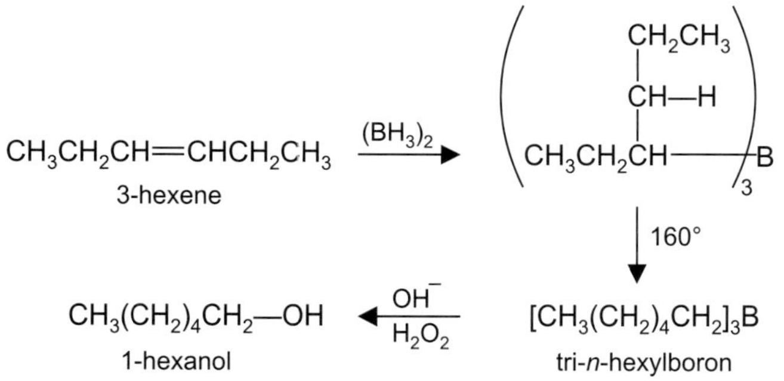

Isomerization of organoboranes

SCHEME 7.50

(F) Reactions of Organoboranes from Alkenes

(1) Oxidation

As seen above alkylborane oxidation with alkaline hydrogen peroxide solution gives alcohols. The reaction is initiated by addition of the HOO^- ion to boron (making use of boron's vacant $2p$ orbital) followed by boron to oxygen migration of alkyl groups (Scheme 7.51). When all the three alkyl groups have migrated, the resulting borate ester is hydrolyzed. The migrating group retains its configuration in the product.

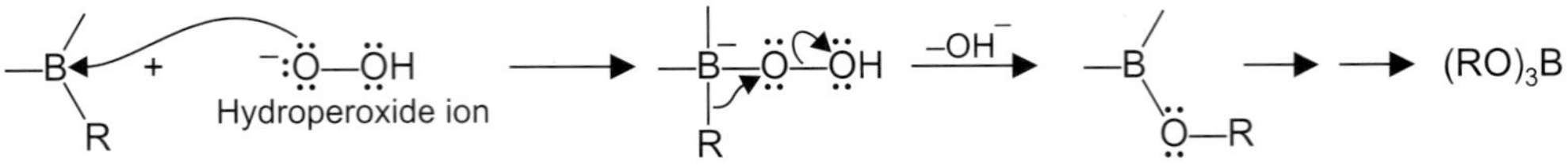

$$(RO)_3B + 3\ NaOH \xrightarrow{H_2O} Na_3BO_3 + 3\ ROH$$

Alkylborane oxidation with alkaline hydrogen peroxide

SCHEME 7.51

More vigorous oxidizing agents effect replacement of boron and oxidation to the carbonyl level (primary trialkylboranes to aldehydes and of secondary trialkyl boranes to ketones, see Scheme 13.47).

(2) Enantioselectivity

(i) Enantioselective Hydroboration of Alkenes

Optically active (asymmetric) boranes are prepared *e.g.*, by the reaction of borane with either (+) – or (–) – α-pinene (Scheme 7.52). These asymmetric boranes are of value in enantioselective synthesis. Optically active secondary alcohols of high optical purity are made from several disubstituted *Z*-alkenes by initial hydroboration with these reagents followed by oxidation (Scheme 7.53). Ipc_2 BH reacts very slowly with (*E*)-alkenes, but Ipc BH_2 can be used effectively.

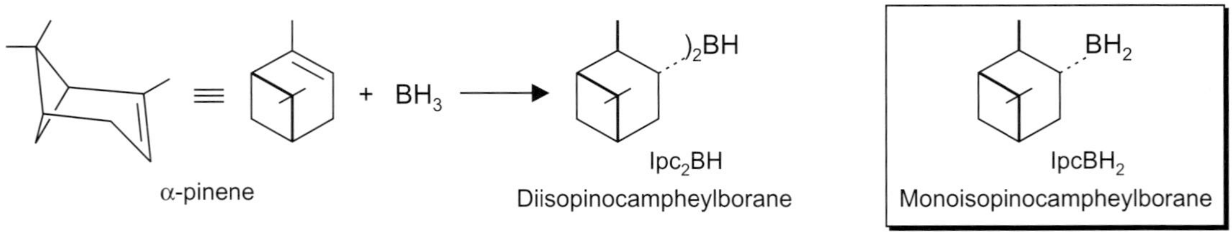

SCHEME 7.52

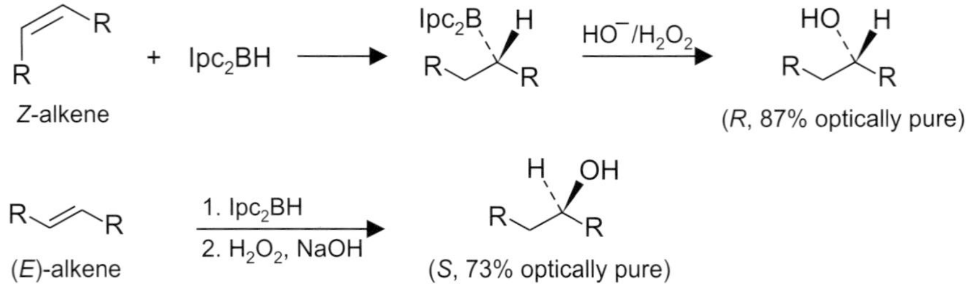

SCHEME 7.53

The chiral reagent $(Ipc)_2$ BH reacts with an alkene like any borane, but since it is chiral, the two faces of the alkene substrate react differently with the reagent producing tri (organo) boranes which are diastereomeric of which one predominates. Hydrolysis (oxidative) then gives a chiral alcohol is good yield and with high optical purity.

(ii) Enantioselective Reduction of Ketones

Boron derivative 9-BBN is prepared by reacting 1, 5-cycloctadiene with BH_3. (+) α-Pinene reacts with 9-BBN to give a chiral adduct (Scheme 7.54) which reduces ketones with a high degree of enantioselectivity, in some case 100%. It is thought that the adduct transfers a hydride to the

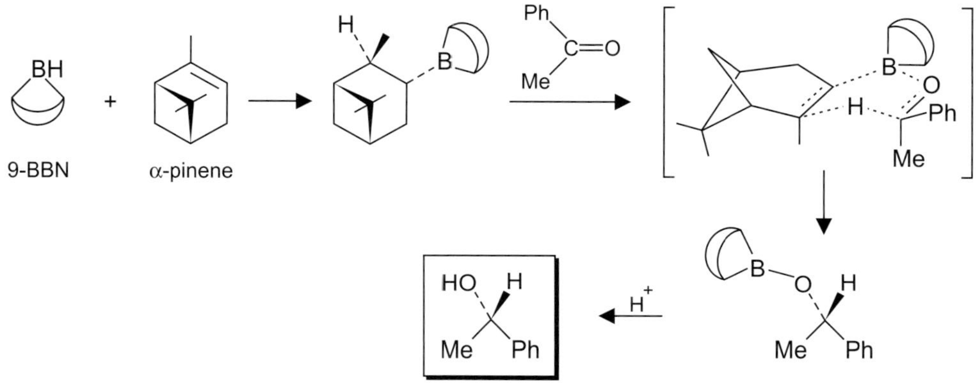

Enantioselective reduction of ketones

SCHEME 7.54

ketone via. a six-membered boat shaped cyclic transition state, where the larger group on the ketone (Ph in this example) preferentially lies-away from the α-pinene moiety to make the steric congestion minimum. The adduct of 9-BBN can be made with both (+)- and (–)- α-pinene to synthesize both the enantiomers individually.

(3) Protonolysis

The migration of R from boron to an oxygen atom during oxidation of an organoborane with alkaline hydrogen peroxide is a good way to make alcohols from alkenes via organoboranes. The R group can also be induced to migrate from boron to a proton. However, the proton cannot come from any source. Carboxylic acids react with organoboranes to cleave the C—B bond. When an organoborane is heated with a carboxylic acid, a Lewis acid–Lewis base reaction occurs between the boron atom (Lewis acid) and the carbonyl oxygen atom of the carboxylic acid (a Lewis base in this transformation Scheme 7.54*a*). The reaction proceeds through a six-membered transition state.

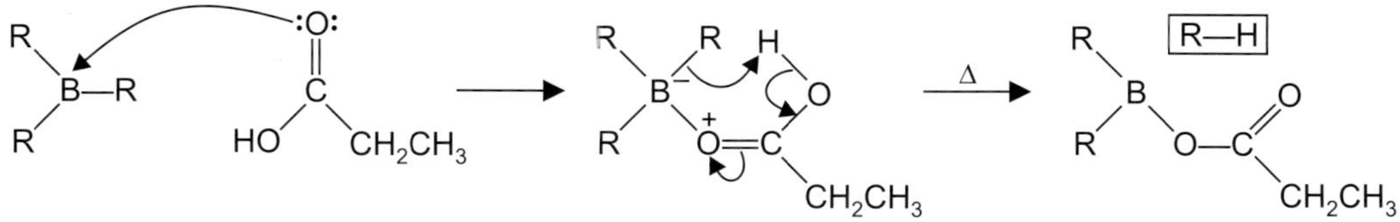

SCHEME 7.54a

The protonolysis of a carbon-boron bond is useful in two ways. First, it provides a good way to introduce deuterium (^{2}H, also denoted by D) into a molecule and an internal alkyne is converted into a *cis* alkene (Scheme 7.54*b*).

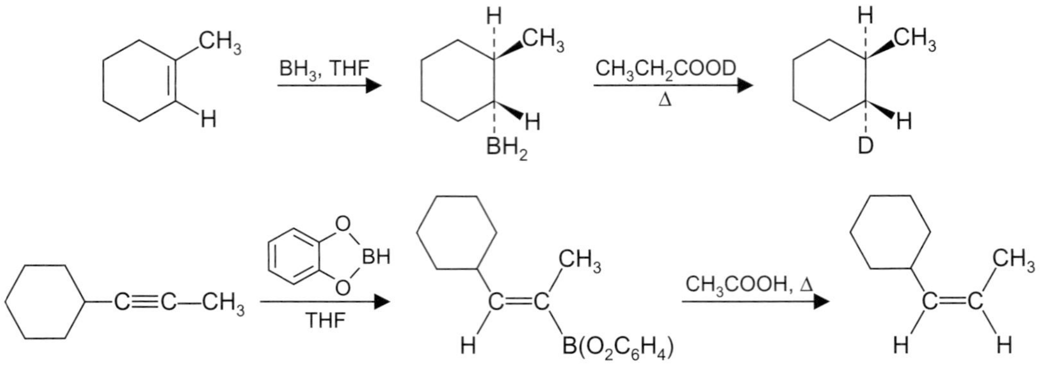

SCHEME 7.54b

(4) Carbonylation of Organoboranes

Carbon monoxide forms Lewis acid-base complexes with organoboranes, and these adducts undergo boron-to-carbon migration of the boron substituents.The reaction can be controlled and therefore, directed in the migration of one, two or all the three substituents on boron. Conditions can be defined so that carbonylation of organoboranes can lead to primary, secondary and tertiary alcohols, aldehydes and open chain, cyclic and polycyclic ketones.

(i) Conversion of Boranes to Tertiary Alcohols

Triallylboranes react with one molecule of carbon monoxide at 125°C in the presence of ethylene glycol to afford the 2-bora-1, 3-dioxolanes which on oxidation give tertiary alcohols in high yields. Ethylene glycol helps to intercept the boronic anhydride (Scheme 7.55), which otherwise forms polymers which are difficult to oxidize. The reaction pathway (Scheme 7.55) involves three successive intramolecular migrations.

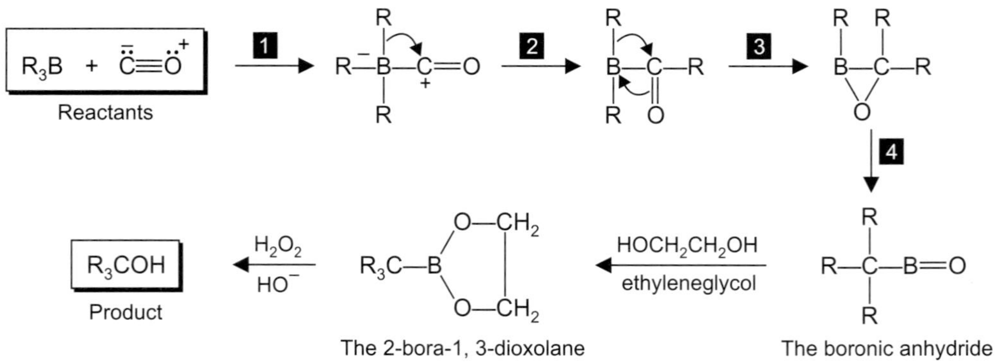

SCHEME 7.55

(ii) Conversion of Boranes to Ketones and Secondary Alcohols

When the carbonylation of a trialkylborane is conducted in the presence of water, the migration of the third alkyl group (step 4, Scheme 7.55) is intercepted. The hydrate is formed (Scheme 7.56), this can be oxidized ($^-$OOH) to a ketone or hydrolyzed to a secondary alcohol.

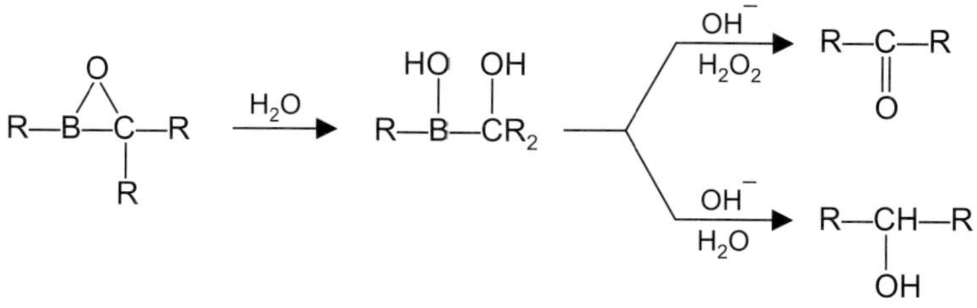

SCHEME 7.56

The reaction can be used to prepare unsymmetrical ketones by using 'mixed' organoboranes prepared from thexylborane (Scheme 7.57). The success of this method depends upon the thexyl group being noncompetitive with the other groups in migration steps. Thus the sequential introduction of two alkenes (Scheme 7.57) leads to the synthesis of unsymmetrical ketones.

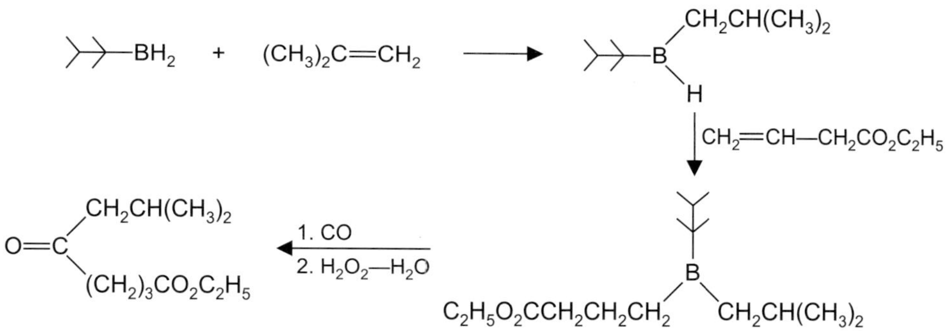

SCHEME 7.57

By working with appropriate dienes one can end up with cyclic or bicyclic ketones (Scheme 7.58).

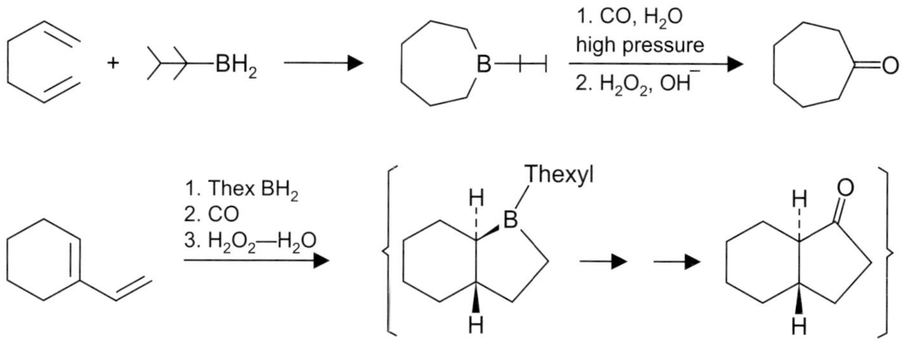

SCHEME 7.58

(iii) Conversion of Boranes to Aldehydes and Primary Alcohols

When the carbonylation of the trialkylborane is done in the presence of a reducing agent *e.g.*, lithium trimethoxyaluminium hydride, the reducing agent intercepts the intermediate after only one boron-to-carbon migration has taken place (see, Scheme 7.59). This on oxidation

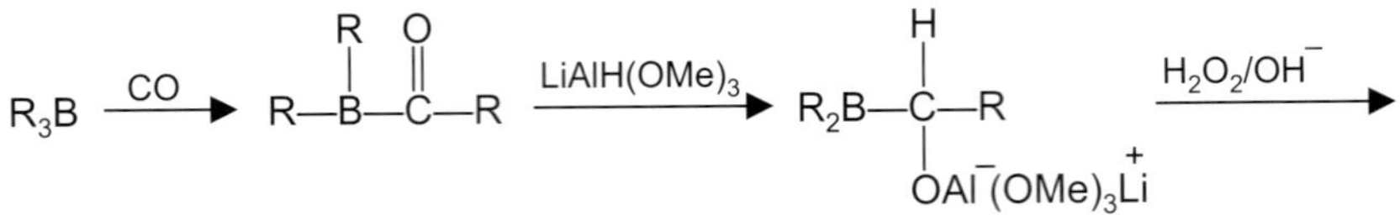

SCHEME 7.59

(route I, Scheme 7.59) gives aldehydes. An inspection of Scheme 7.59 shows that, the method as such has a disadvantage since only one of the three alkyl groups of the starting trialkylborane is converted into aldehyde, the others are wasted. This difficulty is solved by hydroboration of the alkene with 9-BBN. The B-alkyl derivative on reaction with CO in the presence of a reducing agent is attended with the preferential migration of the alkyl group (there is minimal tendency of the bicyclic ring to undergo migration) to give high yields of the aldehyde. While working with B-alkyl-9-BBN, since only the 9-alkyl group migrates, this method, thus converts (high yields) an alkene to an aldehyde containing one more carbon ($R—CH{=}CH_2 \longrightarrow RCH_2CH_2CHO$, Scheme 7.60).

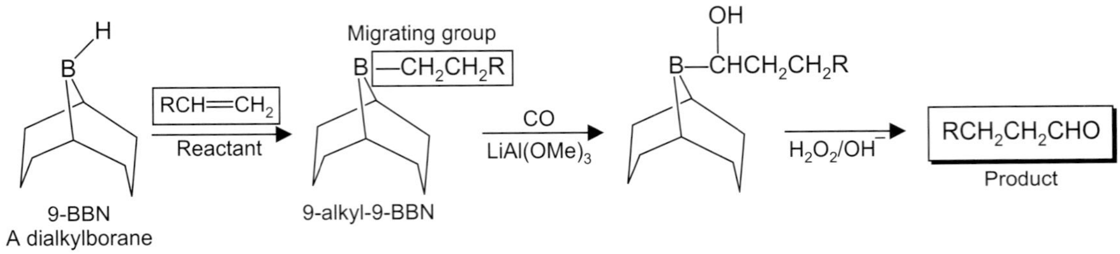

SCHEME 7.60

(5) Cyanidation of Organoboranes

Cyanide ion is isoelectronic with carbon monoxide and a borane reacts similarly to initially form an adduct with cyanide ion, on reaction with sodium cyanide (Scheme 7.61). Thus this is another reagent besides carbon monoxide which serves as the electrophilic migration terminus. The nitrogen atom in the initial adduct is not sufficiently electron attracting to induce migration. This ability is enhanced by acylation of the cyano group with trifluoroacetic anhydride. Two alkyl groups are transferred (Scheme 7.61) at low temperature to give ketones by oxidation. The method has a merit over carbonylation due to low temperature conditions of this reaction. Moreover, as in the carbonylation reaction, one can avoid the wastage of the alkyl groups by working with thexylborane. Also the thexyl group does not migrate and unsymmetrical ketones can be easily obtained.

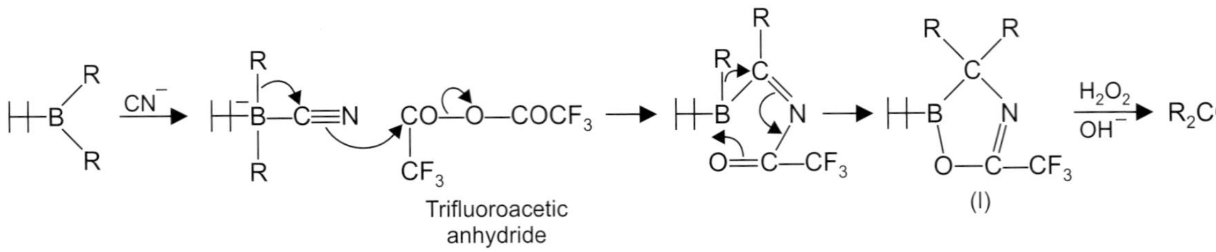

SCHEME 7.61

One can induce the migration of the third group also on (I, R_3B, Scheme 7.61) by using as excess of trifluoroacetic anhydride, to afford tertiary alcohols on oxidation (Scheme 7.62).

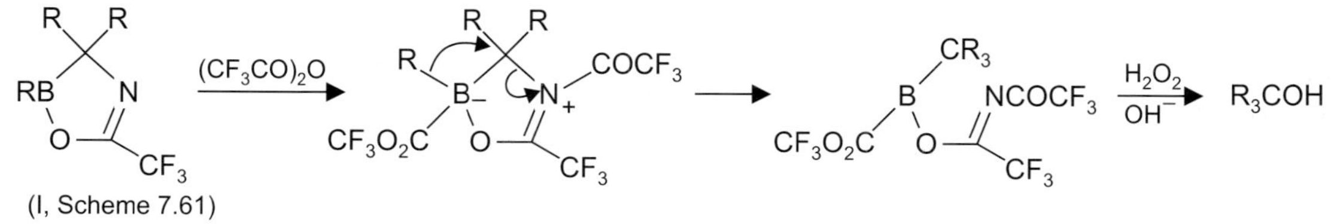

SCHEME 7.62

(6) Synthesis of Esters

Excellent yields of esters can be obtained by reacting trialkylboranes with *e.g.*, ethyl bromoacetate, (an α-haloester) in the presence of a base (eq I, Scheme 7.63). The enolate adds to the organoborane to give a tetracoordinate boron intermediate on which the migration of an alkyl group occurs with displacement of bromide ion. Hydrolysis of the rearranged product gives an ester. Similar reactions are displayed by α-haloketone and α-halonitriles. The key step in this reaction (Scheme 7.63) is again the migration of an alkyl group from boron to adjacent carbon atom. Only one of the three alkyl groups of the trialkylborane is used in this reaction. This problem is again solved by using an alkyl derivative of 9-BBN instead of the trialkylborane thus 9-BBN provides two of the alkyl groups in the borane R_3B. By following this method not only the alkyl group is fully utilized, but the 9-BBN takes no part in alkylation.

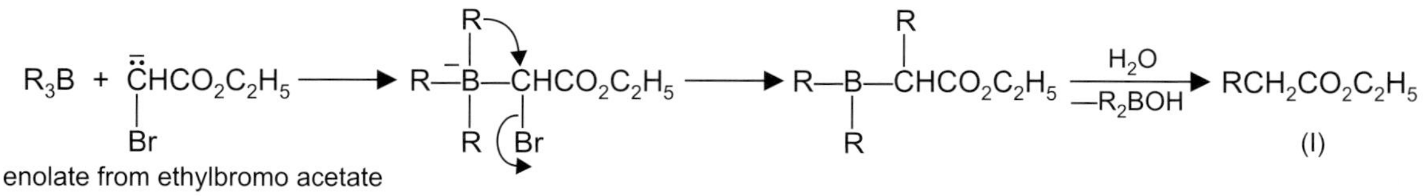

SCHEME 7.63

(G) Hydroboration of Alkynes and Reactions of Derived Organoboranes

(i) Synthesis of Z-alkenes

One can make both *E*-and *Z*-alkenes from a monosubstituted alkyne. A *Z*-alkene can be made by reacting a vinylborane (prepared from monohydroboration of an alkyne) with iodine.

Suzuki Reaction

Hydroboration of an alkyne with catecholborane followed by Pd(O) catalysed coupling with an aromatic bromide gives *cis* stereospecificity (also see, Scheme 8.26).

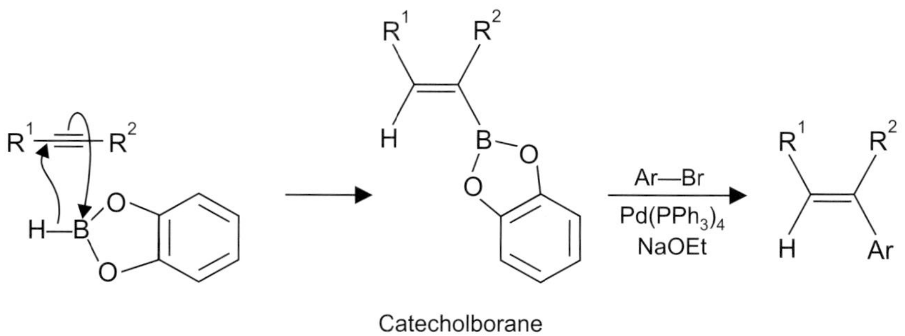

SCHEME 7.63a

On reaction with iodine an alkyl migration takes place from boron to carbon within an iodonium ion (Scheme 7.64). *Z*-alkene is formed *via anti* elimination after the alkyl group migration from an *anti*-periplanar transition state (I, Scheme 7.64). Thus this method, provides a pathway for the synthesis of a *Z*-alkene from a monosubstituted alkyne.

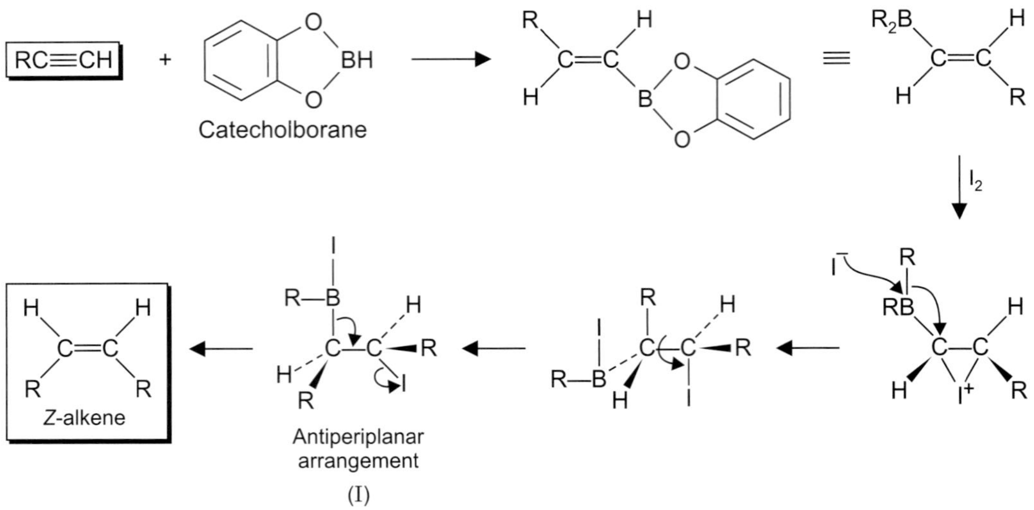

SCHEME 7.64

(ii) Synthesis of E-alkenes

For the synthesis of *E*-alkenes a I-haloalkyne is used. The addition of a dialkylborane gives an α-halovinylborane (*i.e.*, α-haloalkenylborane). On treatment with methoxide ion, this intermediate undergoes boron to carbon migration (Scheme 7.65) to afford an alkylated alkenylborane. Protonolysis gives an *E*-alkene.

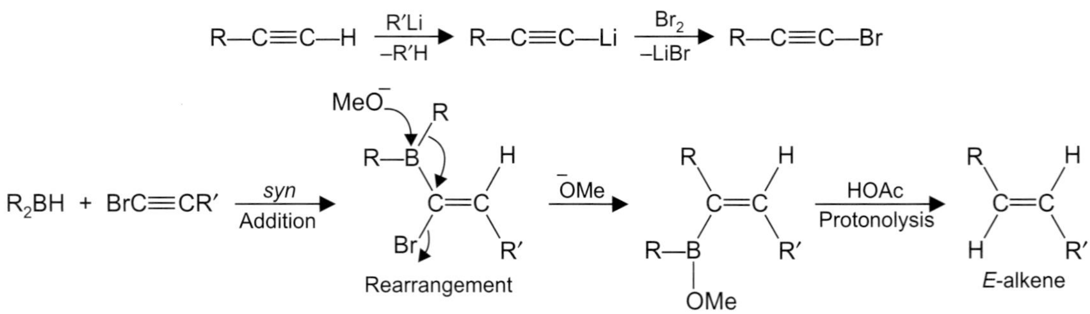

SCHEME 7.65

EXERCISE 7.7

Depict the outcome of reaction between both a terminal and an internal alkyne with 9-BBN followed by oxidation with alkaline hydrogen peroxide ?

ANSWER. *These are given (Scheme 7.65a).*

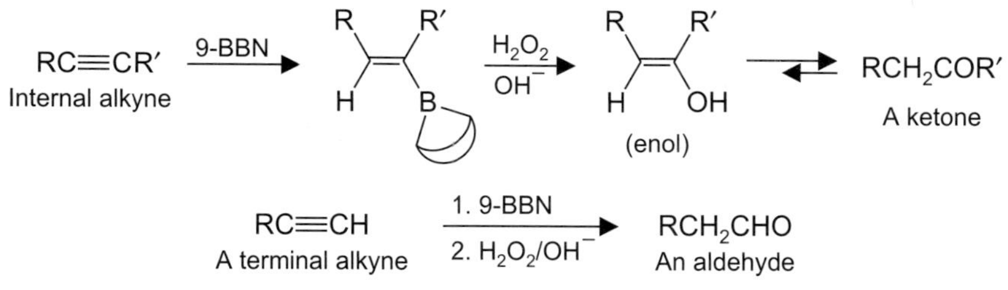

SCHEME 7.65a

(iii) Synthesis of Conjugated Dienes

The principles used in the synthesis of *E*- and *Z*-alkenes can be applied for an appropriate geometry of the double bond(s) in a conjugated diene. Thus *e.g.*, *E*, *E*-dienes can be made as shown (Scheme 7.66). Hydroboration of a 1-haloalkyne with thexylborane affords a thexyl-1-chloroalkenylborane. This reacts with another alkyne to give thexyldialkenylborane. Reaction with methoxide ion induces a rearrangement of alkenyl and protonolysis affords the diene.

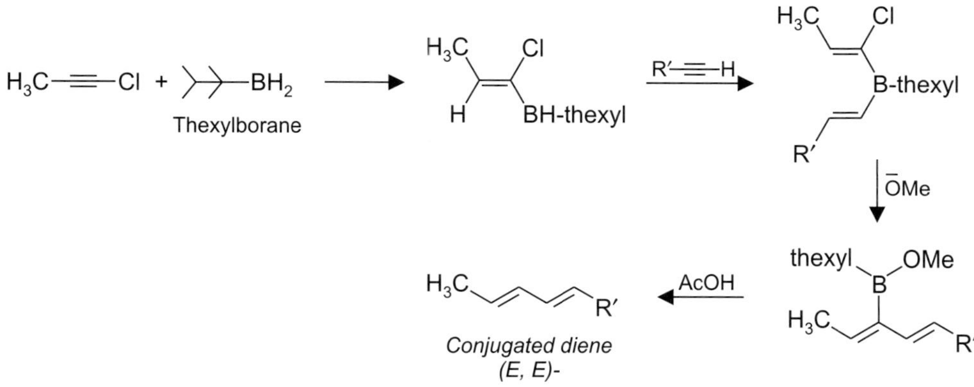

SCHEME 7.66

(H) Formation of Alkynes from Boranes and Acetylides

Organoboranes alkylate terminal acetylenes, adduct formation occurs between a lithium acetylide and a trialkyborane. Reaction with iodine involves an electrophilic attack of iodine on the triple bond thereby inducing a migration of an alkyl group from boron to carbon (Scheme 7.67). This is followed by elimination of dialkyliodoboron.

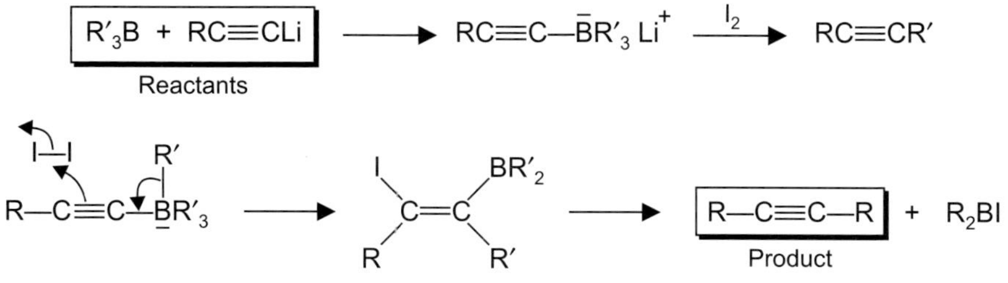

SCHEME 7.67

7.7 ORGANOMETALLIC REAGENTS

(A) Introduction

The compounds of lithium and magnesium are important organometallics, where the metals are highly electropositive. The polarity of the metal-carbon bond is such so as to place high electron density on carbon. Thus these reagents are strong sources of nucleophilic carbon.

(B) Grignard Reagents

Organometallic compounds RMgX are called Grignard reagents after the French chemist Grignard (Nobel Prize 1912). An alkyl halide (RX) with its electrophilic group is converted into its nucleophilic analog (I, Scheme 7.68) umpolung, reverse polarization as seen in 1, 3-dithianes (Sec 7.4D). Ether solvent is essential for its formation because the magnesium atom of a Grignard reagent is surrounded by only four electrons and it needs two more pairs of electrons to form an octet. The solvent molecules provide these electrons by coordinating (supplying electron pairs) to the metal. Coordination allows the Grignard reagent to dissolve in the solvent.

The organometallic compounds from when the metal (Li or Mg) donates its valence electrons to the partially positively charged carbon of the alkyl halide (Scheme 7.68).

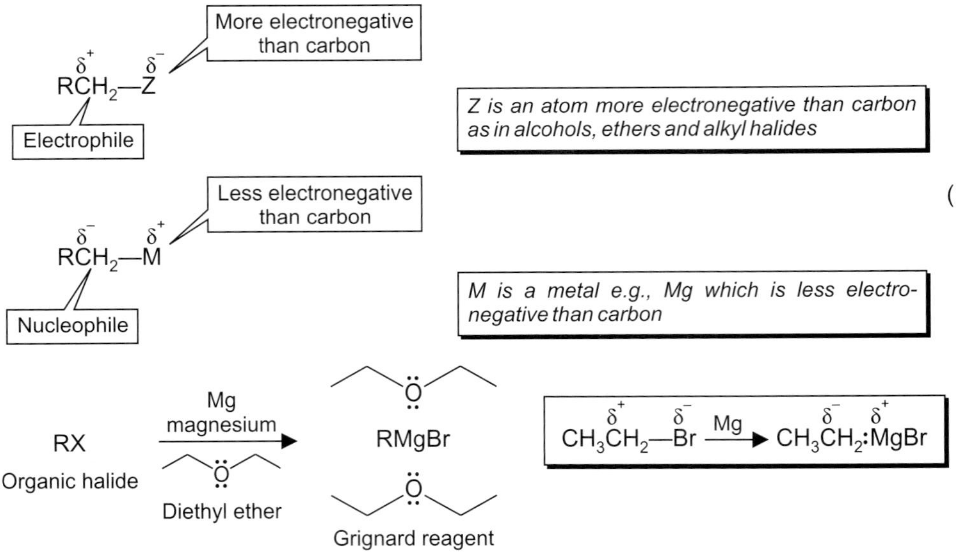

(I)

SCHEME 7.68

(1) Methods of Preparation

In addition to the general method described above, the Grignards reagents are prepared by metallation of hydrocarbons using a preformed Grignard reagent. When a C—H bond is significantly acidic *i.e.*, a stable carbanion can be formed, then such a C—H bond reacts with an organometallic reagent in which the carbon is less electronegative to yield the corresponding C—Metal derivative. Thus alkynyl Grignards and those from other acidic hydrocarbons *e.g.*, cyclopentadiene are made this way (see, Scheme 2.41).

$$R{-}C{\equiv}CH + C_2H_5MgBr \longrightarrow R{-}C{\equiv}C{-}MgBr + C_2H_6$$

SCHEME 7.69

(2) Reactivity

The Grignard reagent is a very powerful base, as a matter of fact this reagent contains a carbanion. One, therefore, fails to make a Grignard reagent from an organic compound that contains an acidic hydrogen (any hydrogen more acidic than the hydrogen atoms of an alkane or an alkene). Thus a Grignard reagent cannot be made from an organic compound containing an OH group, an NH group, an SH group, a COOH group, an SO_3H group —C≡CH group. The Grignard reagents are so sensitive to acidic compounds, that even during their preparation all moisture has to be excluded, otherwise, the Grignard reagent will react with the acidic group (Scheme 7.70). All OH and NH containing compounds react by replacement of hydrogen. In

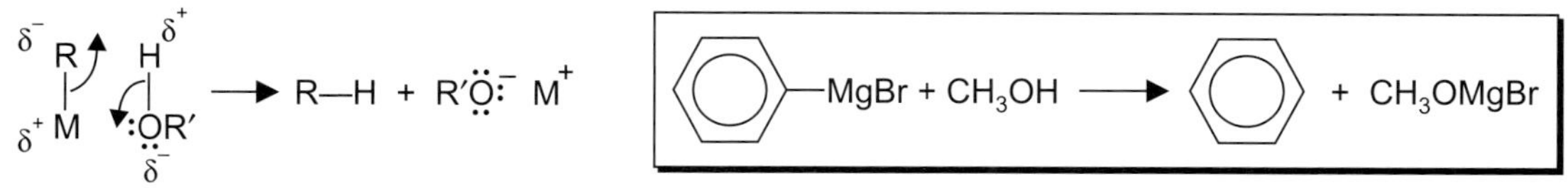

SCHEME 7.70

the reaction of Grignard reagents the direction of reaction is such that the magnesium atom is transferred to a more electronegative atom. This forms the mode of reaction of Grignard reagents, thus Grignards reagents react at the carbonyl group of aldehydes and ketones to afford the magnesium derivatives of alcohols, which on treatment with acids give alcohols (Scheme 7.71). Formaldehyde gives primary alcohols, other aldehydes give secondary alcohols, while ketones reacts to give tertiary alcohols.

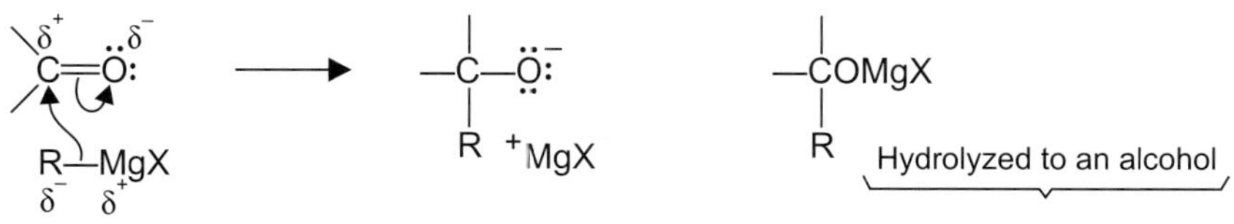

SCHEME 7.71

Alkenyl halides *e.g.*, vinyl bromide are unreactive in nucleophilic substitutions, but these can be metallated to organometallic reagents, and then these become functional (Scheme 7.71*a*).

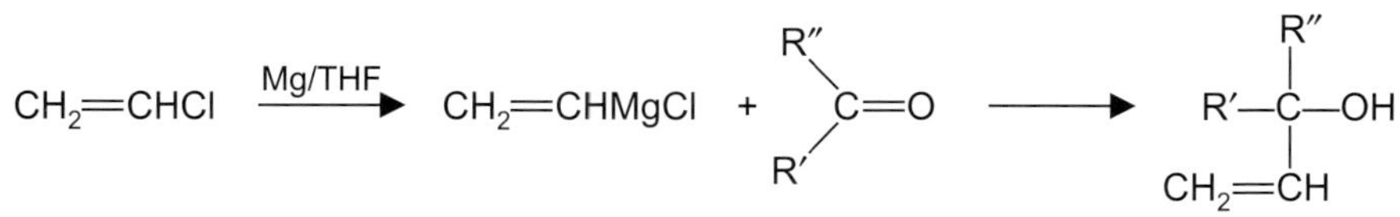

SCHEME 7.71a

(3) Reactions of Grignards Reagents

With carbonyl compounds, the addition of Grignard reagents as described above is the basis for the synthesis of a variety of alcohols. When the carbonyl group carries a substituent which can act as a leaving group (acyl halides, esters and anhydrides), the initially formed adduct can break down to regenerate a C=O bond and therefore, a second addition of the Grignard reagent can occur. Thus esters give tertiary alcohols (Scheme 7.72), the reaction probably begins with addition of the organometallic to the carbonyl function to give the magnesium salt of the hemiacetal. Rapid elimination regenerates a C=O bond for further addition.

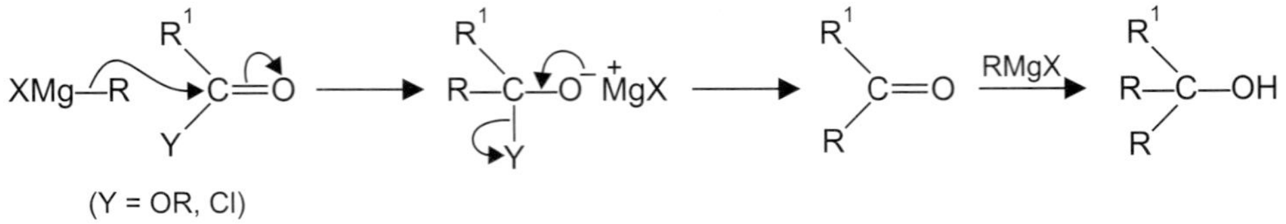

SCHEME 7.72

Acid chlorides are more reactive toward nucleophiles and their reaction with Grignard reagent can be effectively controlled to give ketones in high yield provided one equivalent of Grignard reagent is used at –78°C (Scheme 7.73). Grignard reagents also add to nitriles to give

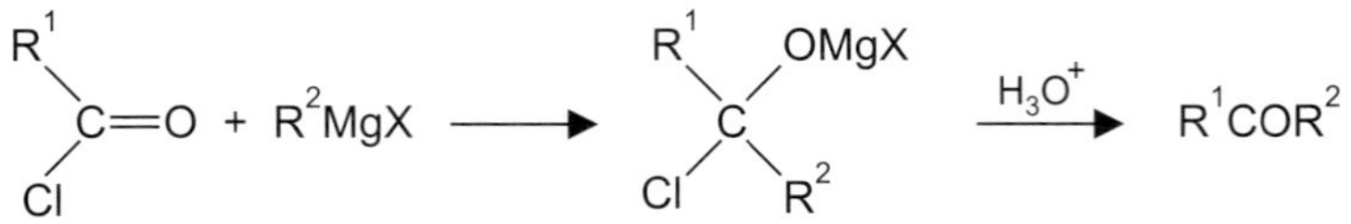

SCHEME 7.73

the magnesium derivatives which are unreactive to further addition and on hydrolysis give ketones *via* the unstable ketimines (Scheme 7.74). In these reactions the role of solvents to increase yield has been reported.

$$R^1CN + R^2MgX \longrightarrow R^1R^2C{=}NMgX \xrightarrow{H_3O^+} \underset{\text{Ketimine}}{[R^1R^2C{=}NH]} \xrightarrow{H_3O^+} R^1COR^2 + NH_3$$

SCHEME 7.74

EXERCISE 7.8

How primary, secondary and tertiary amides react with a Grignard reagent ?

ANSWER. *The principal raction with a primary and a secondary amide is the removal of acidic proton from nitrogen (Scheme 7.74a). The reaction with a tertiary amide, however, provides a useful synthesis of carbonyl compounds (ketones). Recall that –NR_2 is a poor leaving group and thus it does not depart from the initially formed adduct (Scheme 7.74a). The work up with acid provides a very good leaving group.*

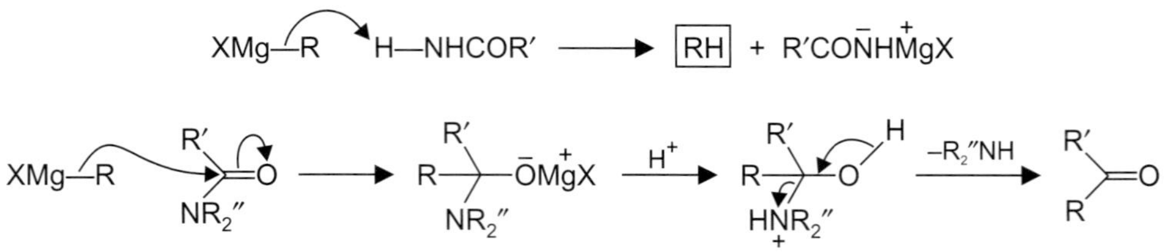

SCHEME 7.74a

The reaction of a Grignard reagent with ethylorthoformate leads to the formation of an acetal. The reaction begins with the elimination of one of the alkoxy groups (aided by magnesium ion acting as a Lewis acid) to generate an electrophilic carbon. The two remaining alkoxy groups tend to stabilize the resulting carbocationic species (Scheme 7.75). The acetal formed after the addition is hydrolyzed to an aldehyde.

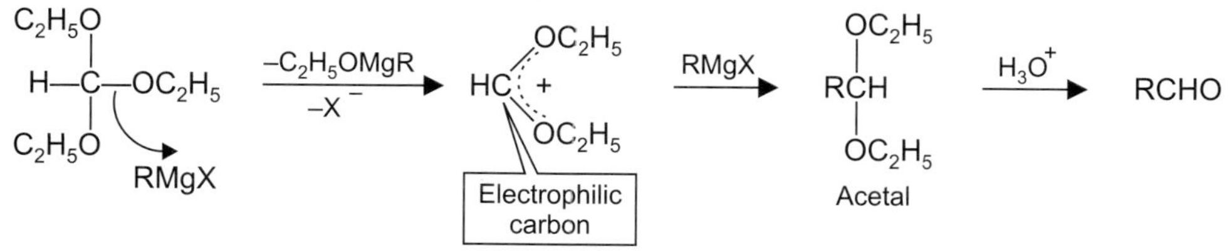

SCHEME 7.75

Carboxylic acids are formed by reacting a Grignard reagent with carbon dioxide (Scheme 7.76).

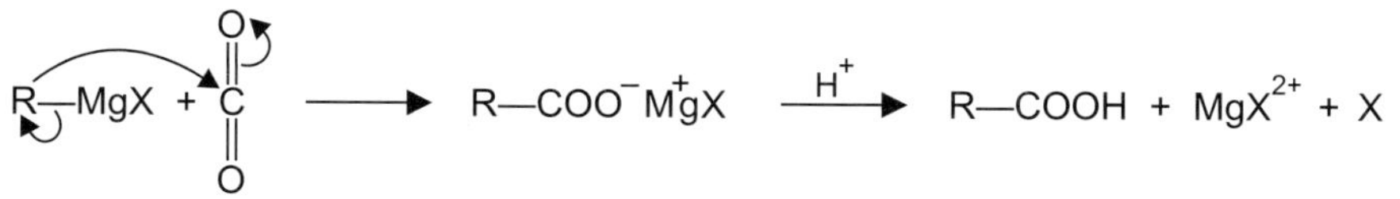

SCHEME 7.76

Like their alkyl analogs, allyl organometallics function as nucleophiles. Although structural rearrangements are not encountered with saturated Grignard reagents, allylic systems give products resulting from isomerization. Allyl Grignard reagents exist in solution as an isomeric mixture in rapid equilibrium (Scheme 7.77). Reaction of an allylic Grignard reagent with a carbonyl compound is attended with an allylic shift (shift of the double bond) and occurs through a six-membered transition stable (Scheme 7.77).

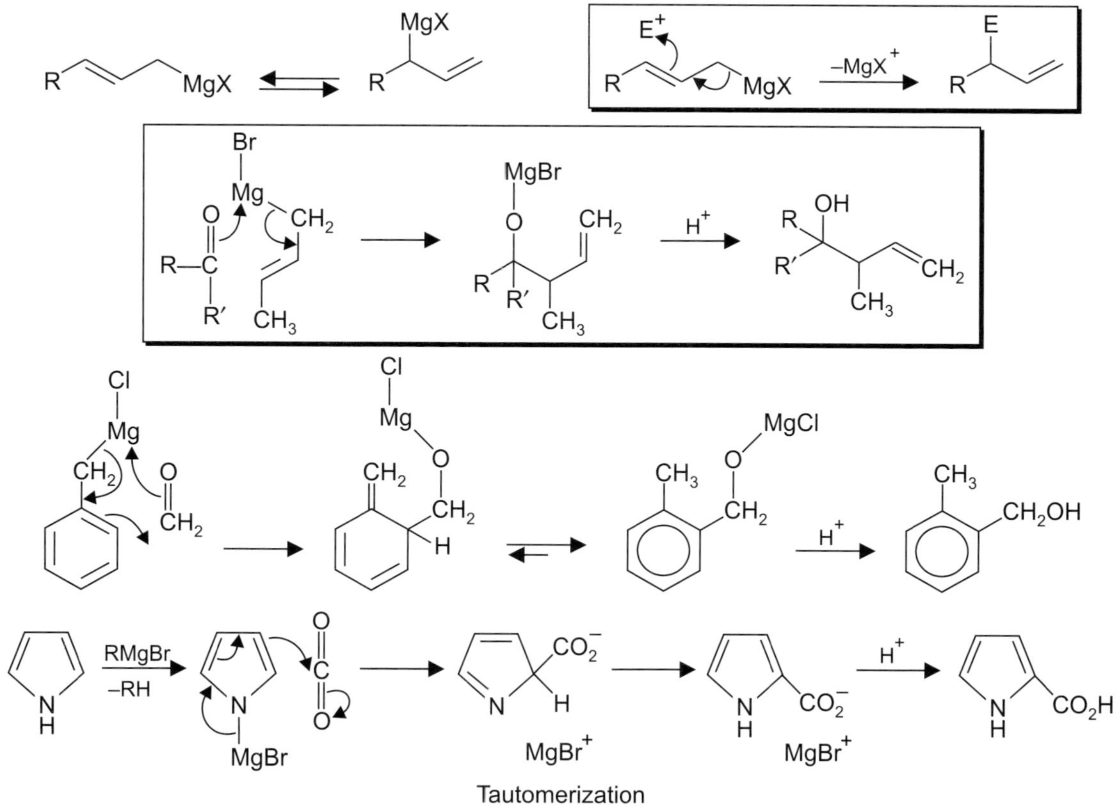

SCHEME 7.77

Same reactivity is observed with benzylic Grignards. Interestingly pyrrole reacts with a Grignard reagent at its NH group to yield an N-Mg derivative which then reacts with electrophiles at the 2-carbon atom (Scheme 7.77). One may note that in the case pyrrole (an allylic type Grignard) the reaction is however, not assisted by the formation of a six membered cyclic transition state.

Grignards reagents do not react with acyclic or strain free cyclic ethers. These however, react with epoxides and the less substituted ring carbon atom of the epoxide is attacked (Scheme 7.78), thus these act as an nucleophiles in S_N2 reactions with epoxides.

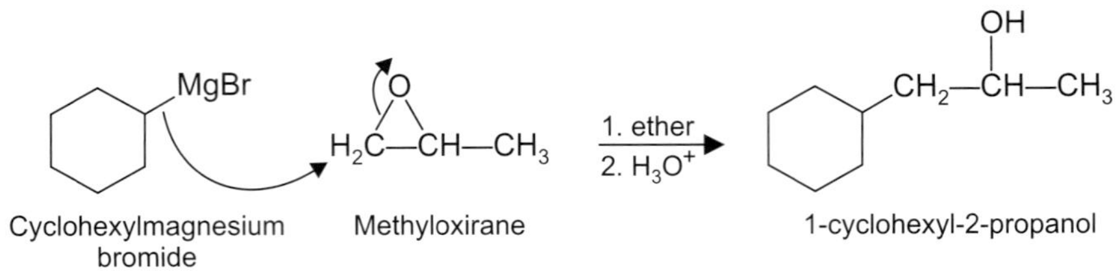

SCHEME 7.78

The reaction with epoxides is complimentary to the addition of organometallic reagents to carbonyl compounds. Thus reaction of formaldehyde with a Grignard reagent or an organometallic reagent extends the chain of the organometallic reagent by one carbon while the reaction with ethylene oxide, the chain is extended by two carbon atoms (Scheme 7.78*a*).

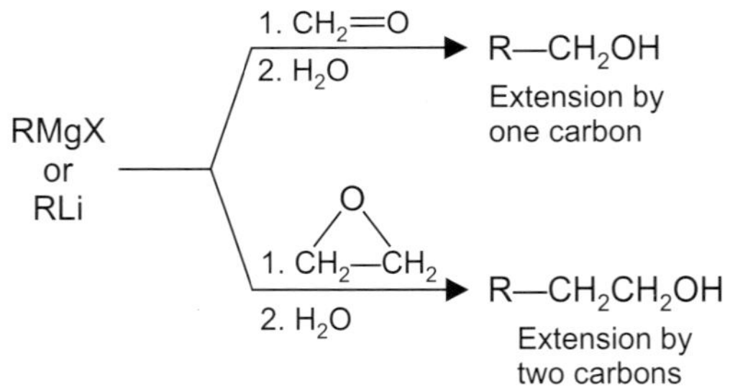

SCHEME 7.78a

The epoxide rings are also opened by lithium aluminium hydride, the hydride adds predominantly to the less hindered side of the epoxide. The initially formed alkoxide is protonated by water and this provides yet another method for the synthesis of alcohols (Scheme 7.78*b*).

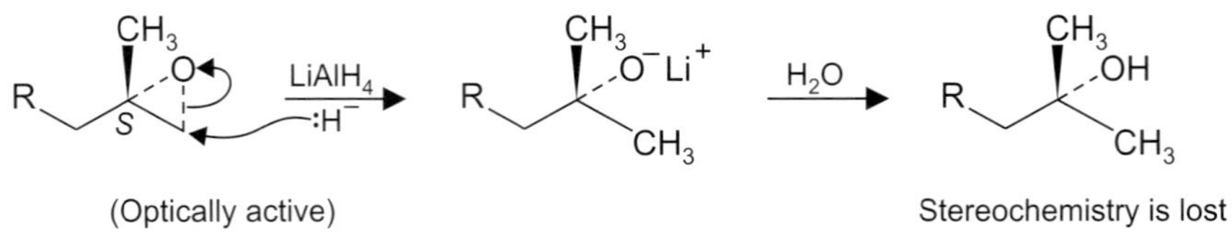

SCHEME 7.78b

(4) Stereoselectivity

The reaction of a Grignard reagent with a carbonyl group can create a stereocenter. When the carbonyl compound already has an adjacent stereocenter, there is predominance of one of the two possible diastereomers. The results are in keeping with the Cram's rule (see Scheme 14.81).

In the case of unhindered cyclohexanones, with Grignard reagents generally there is a preference for attack from the equatorial direction to give an axial alcohol as the major product (Scheme 7.79).

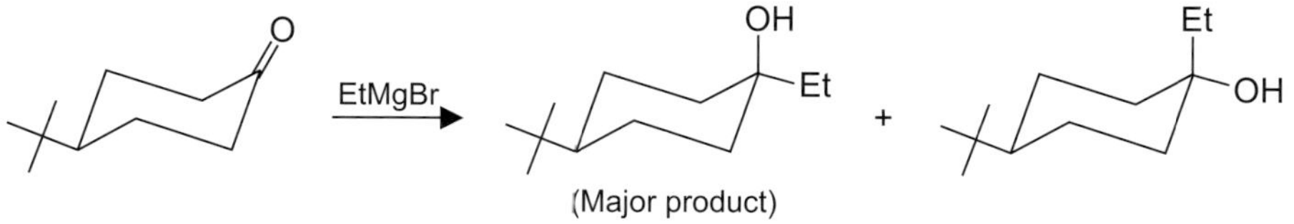

SCHEME 7.79

(5) Limitations—Steric effects

> **Grignard additions are sensitive to steric effects both in the reacting carbonyl compound and the Grignard reagent (see Scheme 1.33).**

(6) Reactions with elements other than carbon

(*i*) Thiols can be made by reaction with sulphur (Scheme 7.80), while sulphur dioxide reacts in a fashion similar to carbon dioxide (Scheme 7.80).

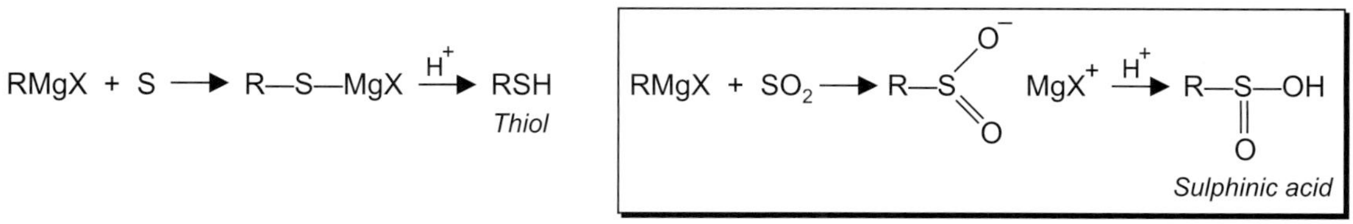

SCHEME 7.80

(*ii*) Hydroperoxides can be made by reaction with oxygen at low temperature and acidification of the magnesium derivative of the hydroperoxide (Scheme 7.81).

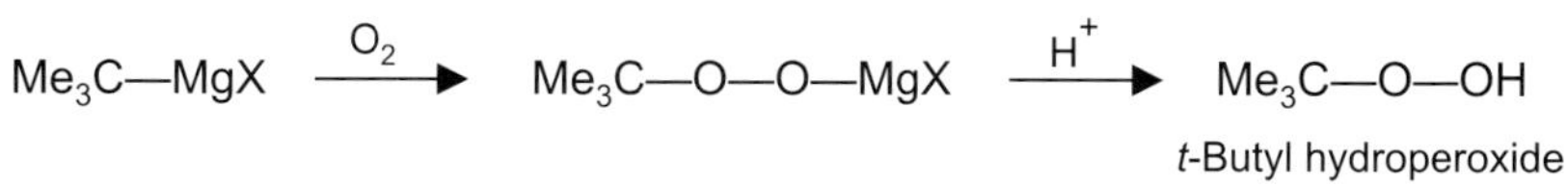

SCHEME 7.81

(*iii*) Iodides can be made by reacting with iodine and this provides an useful alternative to S_N2 reaction (see Scheme 5.8). Iodides are prepared from chlorides or bromides via S_N2 displacement (treatment with sodium iodide in acetone). This method, however, fails for the highly hindered neopentyl bromide (see Scheme 5.8). The Grignard reagent prepared *e.g.*, from neopentyl chloride on reaction with iodine gives neopentyl iodine in a good yield (Scheme 7.82).

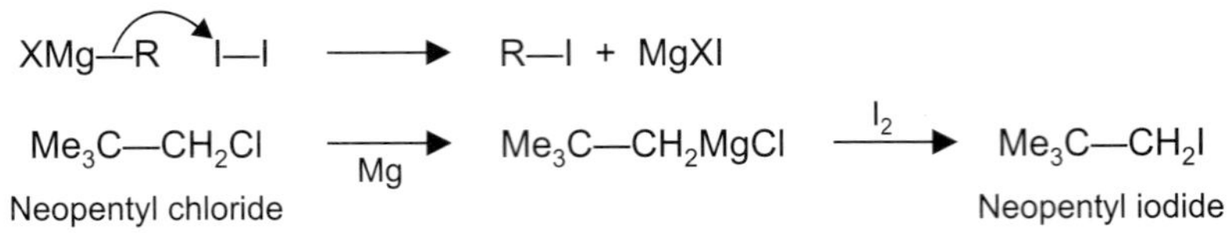

SCHEME 7.82

(*iv*) Derivatives of silicon can be made via Grignard reagents (see Schemes 7.35*c* and 7.37).

(C) Organolithium and Related Compounds

Organolithium compounds are more strongly nucleophilic than Grignard reagents. Although generally organolithium compounds show similar reactivity as Grignard reagents, however, in few instances these distinctly differ from their Grignard counterparts. Here a few such differences are outlined.

(1) Methods of Preparation

Organolithium compounds are generally prepared by reacting an alkyl halide with lithium metal. In several cases particularly with aryl and vinyl halides, the reagents are prepared by metal-halogen exchange (R—Br + BuLi ⟶ RLi + BuBr). The mechanism is similar to the formation of Grignards reagents.

(2) Reaction with α, β-unsaturated compounds—1, 2 or 1, 4-addition

Organometallic compounds may add either 1, 2 or 1, 4. In the case of Grignard reagents variable behaviour is observed depending on the structure of the conjugated system. The main reason seems to be steric hindrance. In several cases, where Grignard reagent gives predominant 1, 4-addition, the lithium compounds react predominantly by 1, 2-addition (Scheme 7.83). In fact if one wants to have maximum 1, 2-addition, an organolithium reagent is used. One may specifically achieve 1, 4-addition by using lithium dialkylcuparates.

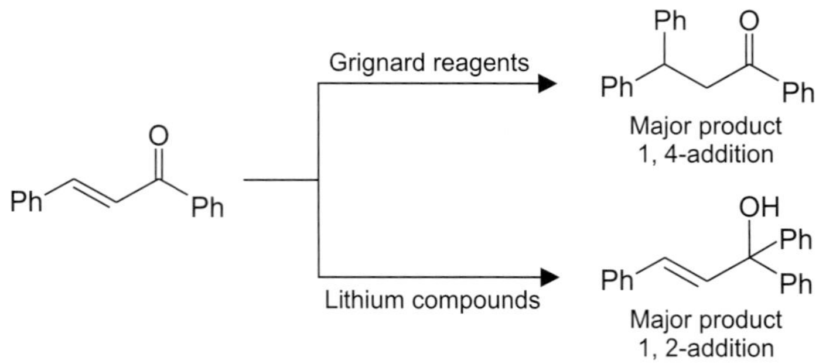

SCHEME 7.83

(3) Reaction with Carbon Dioxide

Unlike a Grignard reagent which gives carboxylic acid on reaction with carbon dioxide, the reaction with an organolithium compound gives ketones (Scheme 7.84). Organolithium compounds are much more nucleophilic than Grignard reagents, and therefore, can react with

SCHEME 7.84

the intermediate resonance stabilized carboxylate anion. Thus carboxylic acids themselves can be converted into ketone (Scheme 7.85).

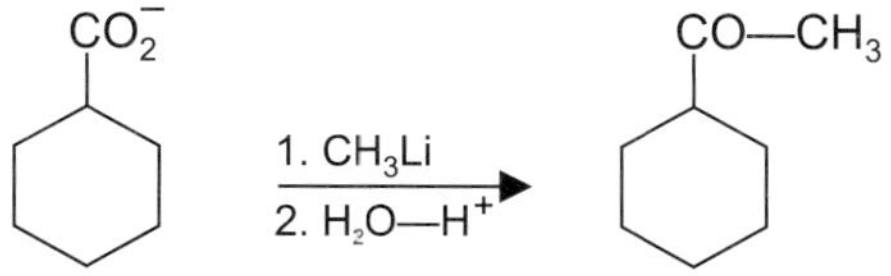

SCHEME 7.85

(D) Lithium Dialkylcuparates—Gilman Reagents

Lithium dialkylcuparates (referred to as Gilman reagents) are prepared by the reaction of two equivalents of the corresponding organolithium reagent with cuprous iodide (Scheme 7.86). These reagents are formally nucleophilic and have the same charge distribution as Grignard or organolithium reagents.

$$2CH_3I \xrightarrow{4Li} \underset{\text{Organolithium reagent}}{2CH_3Li} + 2LiI \qquad 2CH_3Li + CuI \xrightarrow[-LiI]{THF} \underset{\text{Gilman reagent}}{(CH_3)_2CuLi}$$

Preparation of an organocuparate (Gilman reagent)

SCHEME 7.86

The following reactions point out the synthetic importance of organocuparates

- Gilman reagents react with an alkyl halide (with the exception of alkyl fluorides) and one of the alkyl groups of the Gilman reagent replaces the halogen. Thus an alkane can be formed from two alkyl halides—one alkyl halide is used to form the Gilman reagent, which then reacts with the second alkyl halide (Scheme 7.87).

$$\underset{\text{Lithium dibutylcuparate}}{(CH_3CH_2CH_2CH_2)_2Cu^-\,Li^+} \xrightarrow[\text{1-iodopropane}]{CH_3CH_2CH_2—I} \underset{\text{Heptane}}{\boxed{CH_3CH_2CH_2CH_2CH_2CH_2CH_3}} + Cu(CH_2CH_2CH_2CH_3) + LiI$$

SCHEME 7.87

- Gilman reagents can be used to prepare compounds that cannot be made by using nucleophilic substitution reactions. For example, S_N2 reactions are not displayed by vinyl and aryl halides but these react with Gilman reagents to form carbon-carbon bonds (Scheme 7.88).

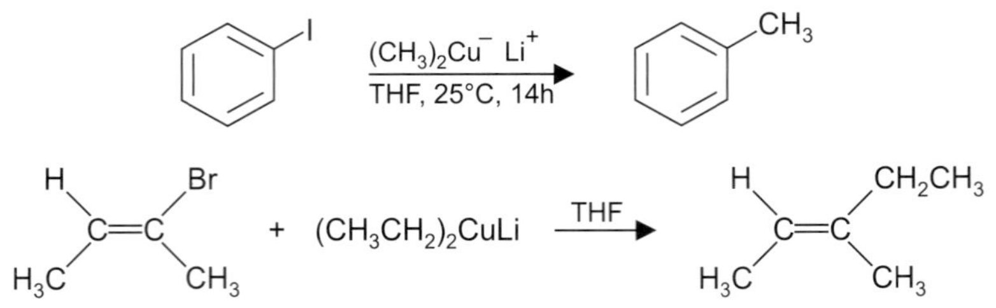

SCHEME 7.88

- Organocuparates generally do not react with carbon dioxide or other carbonyl compounds, the way Grignard or organolithium reagents (Scheme 7.89).

  Infact organocuparates react only with the most reactive carboxylic acid derivatives and are particularly useful for the conversion of acid chlorides to ketones (Scheme 7.89).

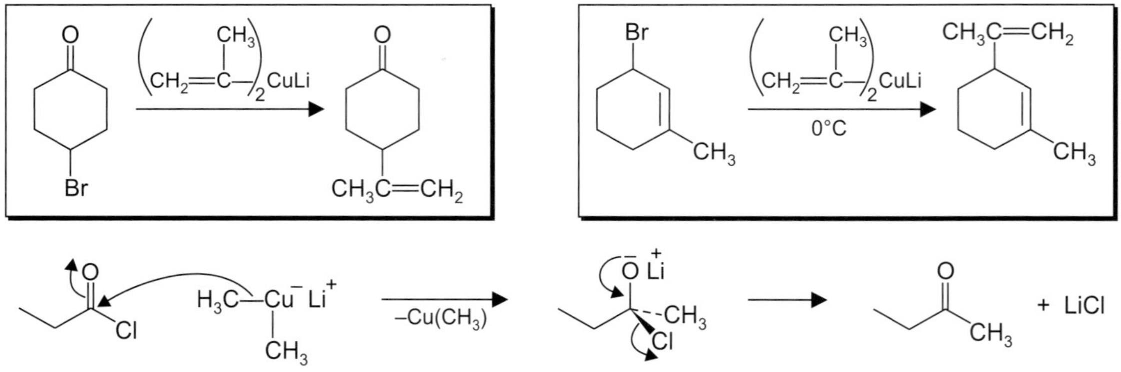

SCHEME 7.89

- Epoxides react with lithium organocuparates to give alcohols with inversion of configuration at the less substituted carbon atom of the epoxides as in an S_N2 process (Scheme 7.90).

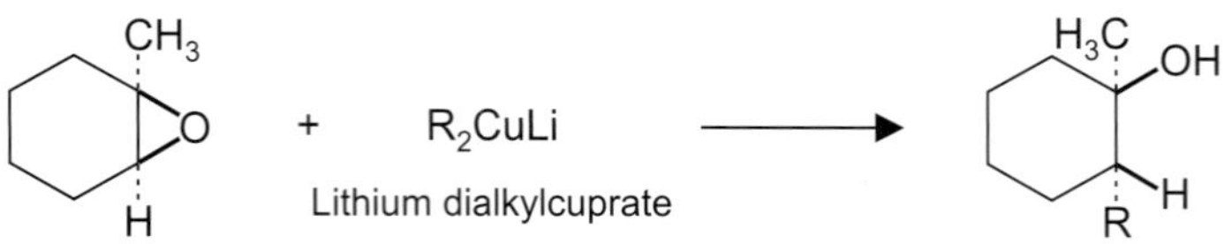

SCHEME 7.90

- Organocuparates are highly selective to bring about 1, 4-addition reactions (*i.e.*, conjugate addition) with α, β-unsaturated carbonyl compounds (Scheme 7.91). Thus methylvinyl ketone reacts cleanly to give 1, 4-addition reaction.

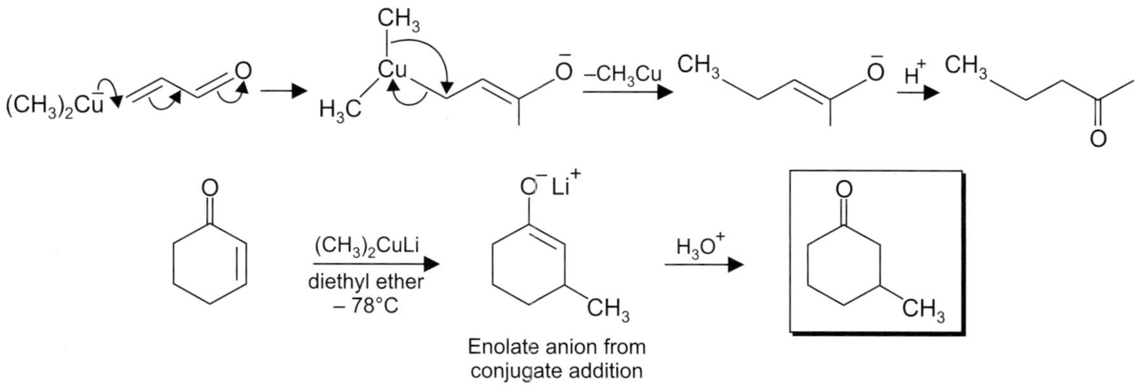

SCHEME 7.91

- Coupling reaction takes place in the presence of oxygen when organocopper reagents are heated or even at room temperature (Scheme 7.92).

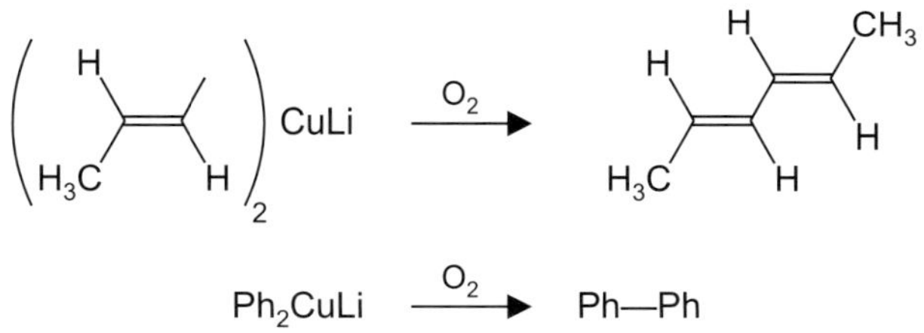

SCHEME 7.92

- Organocuparates involve oxidative addition and reductive elimination during carbon-carbon bond formation. The reaction between iodobenzene and an organocuparate *e.g.*, dimethylcuparate to give toluene occurs by the pathway that is not S_N2 in character. The overall mechanism probably consists of two steps (*i*) oxidative addition to the metal and the oxidation state of the copper ion increases from +1 to +3 (Scheme 7.93). Once the phenyl ring is bonded to the copper ion, the complex thus formed is unstable and undergoes reductive elimination forming toluene and a metal containing species.

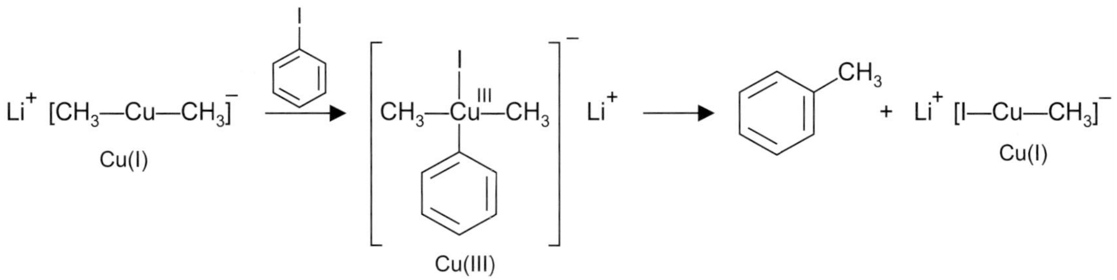

Mechanism with an organocuparate

SCHEME 7.93

PROBLEMS

7.1 How can one convert methylenephosphorane into keto-ylides?

7.2 Dichloromethylene (CCl_2) obtained by the reaction of base on chloroform $CHCl_3$ is trapped by triphenyl-phosphine to give an ylide. How this ylide can be converted into a vinylidene chloride ?

7.3 How can PhCHO be converted into PHCDO?

7.4 How can one transform (I) into (II) and (III) into (IV) by using Wittig reaction?

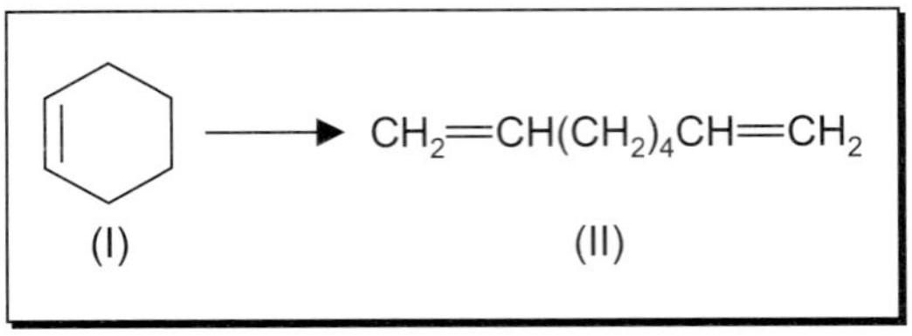

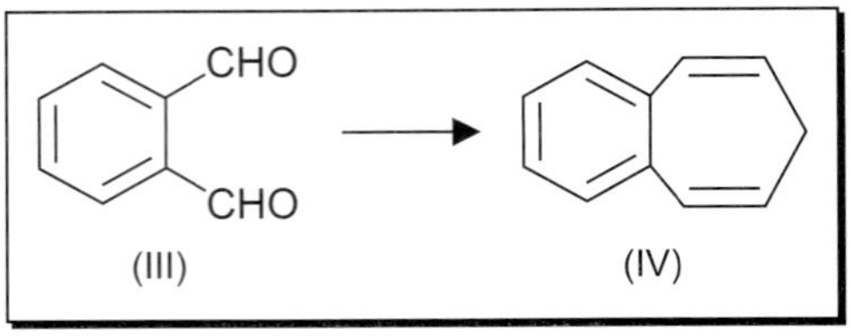

7.5 Write the structure of ylides which can bring about the following epoxidations:

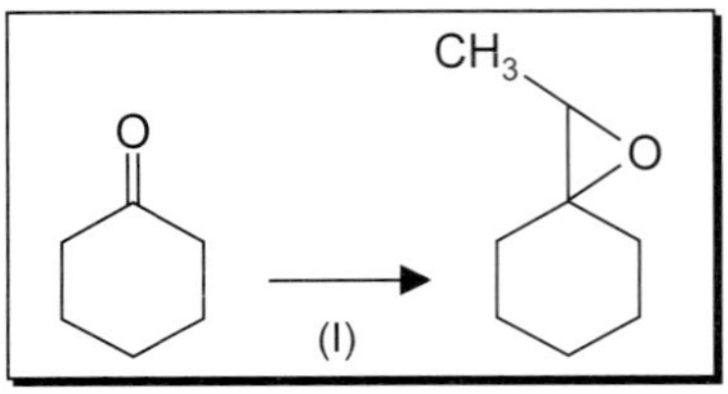

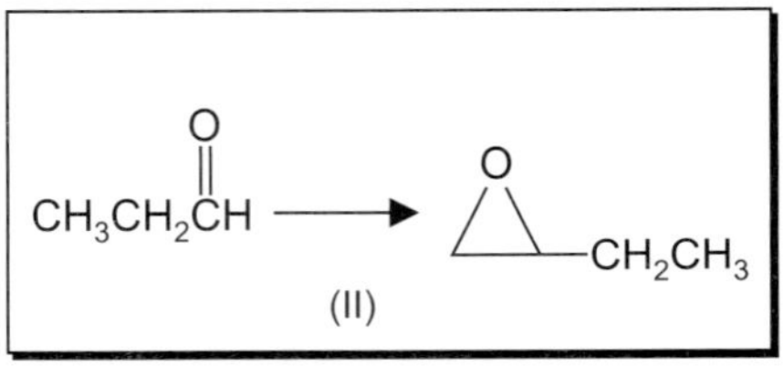

7.6 (*a*) Which reactants one can use to synthesize the following spirocyclic compound I?
(*b*) How can you convert II into III, and IV into V?

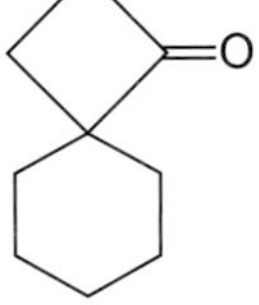

(I)

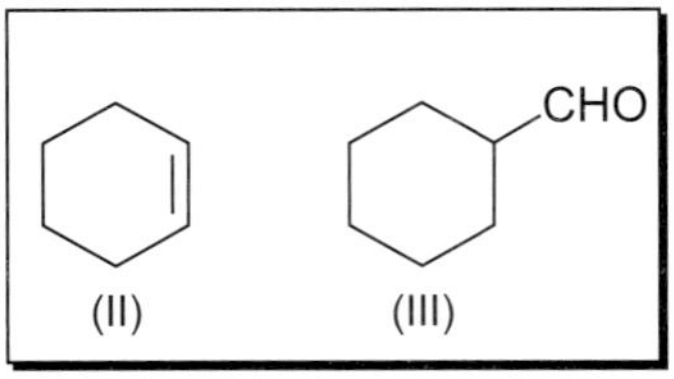

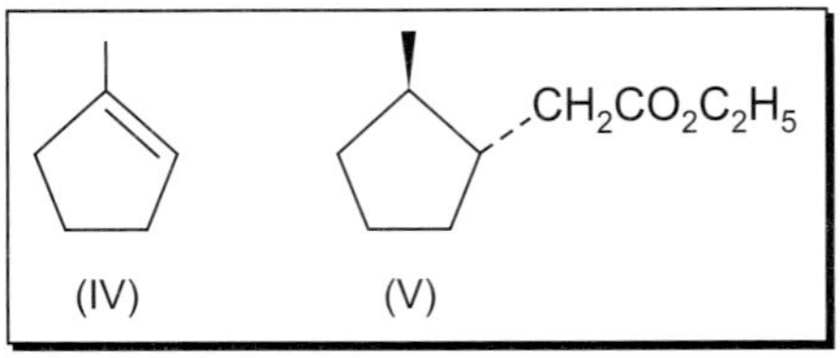

7.7 Using a Grignard reagent, describe a method to prepare 1-alkenes.

7.8 How one can prepare butane from chloroethane using the Corey-House Synthesis (see Scheme 7.86)?

7.9 How can one prepare 2-deuteropropane from isopropyl bromide?

7.10 How can one synthesize the following aldehyde $CH_3(CH_2)_4CH{=}O$

7.11 Write the product from the reaction. $2CH_2{=}CHCH_2Br + 2Ni(CO)_4$ 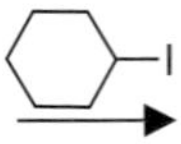

7.12 Write the products of reaction of cyclochexanone with (I) $Li{-}CH(Cl)\,Si(Me)_3/H^+$ and (II) with $(Me)_3\,SiC(Cl)(CH_3){-}Li/H^+$.

7.13 Write the product from reaction of each compound

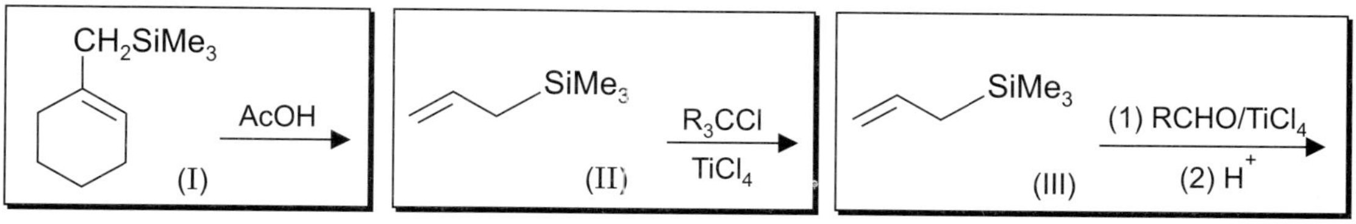

ANSWERS TO THE PROBLEMS

7.1 $Ph_3P{=}CH_2 \xrightarrow{RCOCl} [Ph_3\overset{+}{P}CH_2COR]Cl^- \xrightarrow{Ph_3P=CH_2} Ph_3P{=}CHCOR + [Ph_3\overset{+}{P}Me]\,Cl^-$

Keto-ylide

7.2 *Via* Wittig reaction with a ketone

$$CHCl_3 \longrightarrow [CCl_2] \xrightarrow{Ph_3P} Ph_3P{=}CCl_2 \xrightarrow{RR\,C=O} R_2C{=}CCl_2 + Ph_3P{=}O$$

Dichloromethylene; Dichloromethylenetriphenyl-phosphorane; A vinylidene chloride

7.3

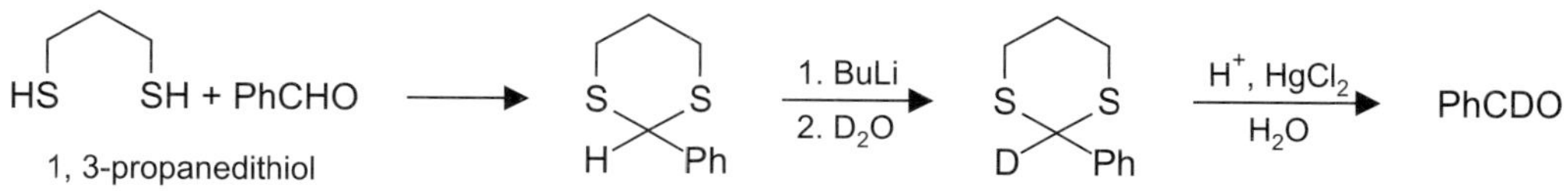

7.4 (I to II). The dialdehyde obtained after ozonolysis of (I) will undergo Wittig reaction to give yield (II), (III to IV). When (III) is reacted with a bifunctional Wittig reagent (V).

$$HC(=O)(CH_2)_4C(=O)H + (C_6H_5)_3P{=}CH_2 \longrightarrow (II)$$

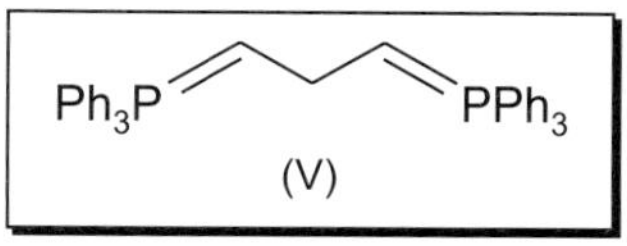

7.5 $CH_3CH{=}S(CH_3)_2$ (I) $\quad CH_2{=}S(CH_3)_2$ (II)

7.6 (*a*) Cyclohexanone and diphenylsulphonium cyclopropylide.

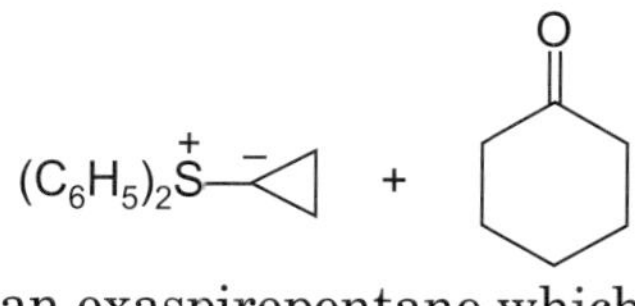

These will react to give an oxaspiropentane which will rearrange on treatment with acid (see Scheme 7.26*a*).

(*b*) (II to III). By the reaction of cyclohexene with 9-BBN to give 9-alkyl-9-BBN (where the alkyl group is cyclohexyl); followed by its reaction with CO and a reducing agent (see Scheme 7.60).

(IV to V). This is the synthesis of a substituted acetic acid and can be undertaken as shown, (see Scheme 7.63). Malonic ester route is another alternative for the synthesis of substituted acetic acids.

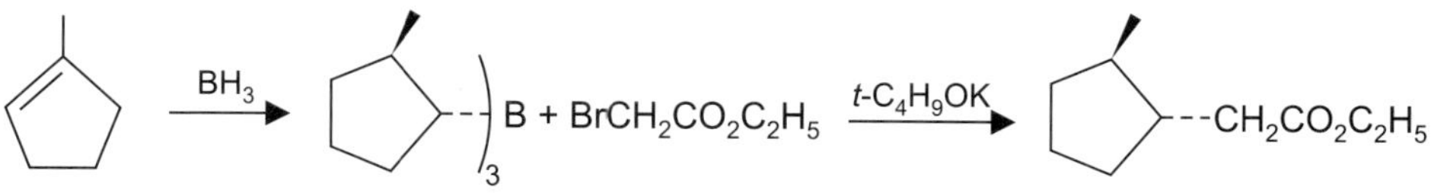

7.7 This can be done by using allyl halide with a suitable Grignard reagent or an organolithium compound.

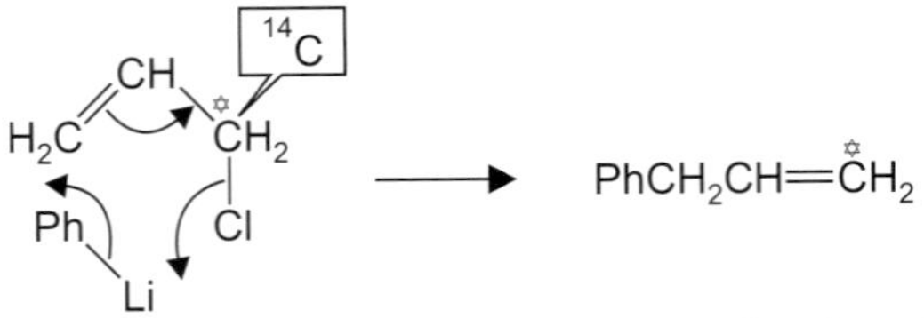

Alkylation by allylic halides often follows a cyclic mechanism. Thus allyl 1– ^{14}C chloride reacts with phenyl lithium to afford a major product with labelled carbon at the terminal methylene group.

7.8. $$\underset{\text{Chloroethane}}{CH_3CH_2Cl} \xrightarrow[\text{2. CuI}]{\text{1. Li}} \underset{\text{Lithium diethylcuprate}}{(CH_3CH_2)_2LiCu} \xrightarrow{CH_3CH_2Cl} CH_3CH_2CH_2CH_3$$

7.9. $$(CH_3)_2CHBr + \text{Mg/ether} \longrightarrow (CH_3)_2CHMgBr \xrightarrow{D_2O} \underset{\text{2-deuteropropane}}{(CH_3)_2CHD}$$

7.10. *Via* reaction of the appropriate Grignard reagent with ethyl orthoformate and by the hydrolysis of the resulting acetal $CH_3(CH_2)_4MgBr + HC(OC_2H_5)_3$

7.11. 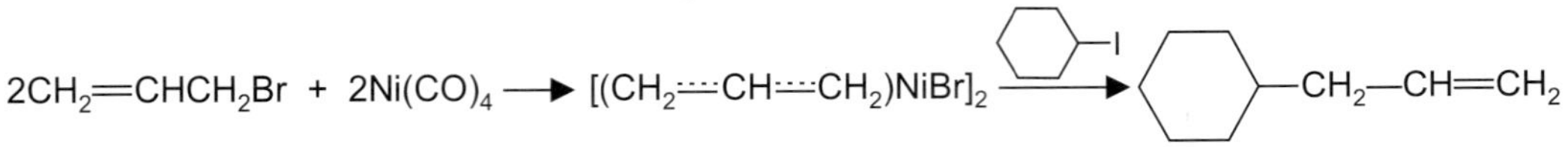

7.12 [Hint] For mechanism see schemes 7.35*d* and 7.35*e*

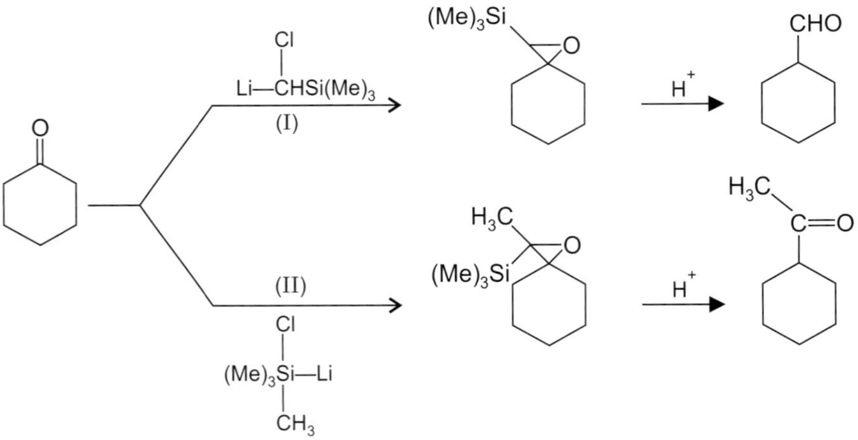

7.13. (I), 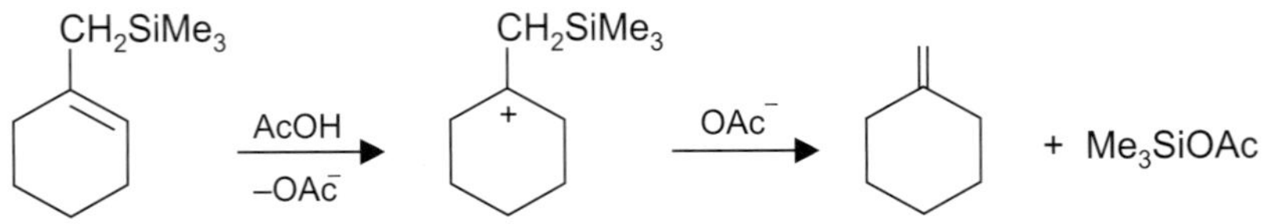

Since proton is the electrophile, only double bond shift occurs.

(II), 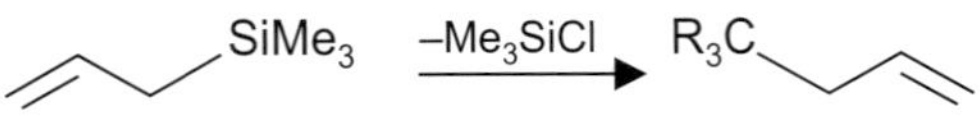

(III), 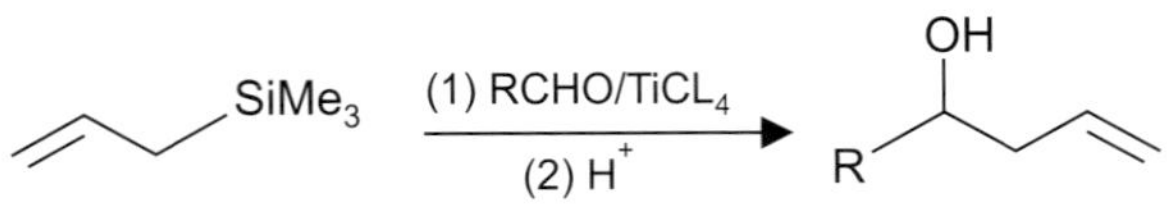

CHAPTER 8

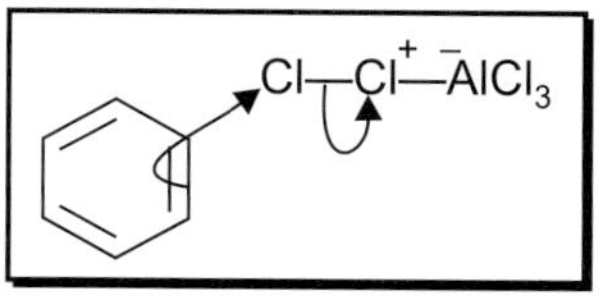

Electrophilic Aromatic Substitution

Both benzene and an alkene are susceptible to electrophilic attack primarily because of their exposed π-electrons (Scheme 8.1). However, benzene differs from an alkene in a very different way, since the closed shell of six π-electrons gives it a special stability. Thus unlike an alkene, where the carbocation formed after the initial attack of an electrophile undergoes an addition reaction (Scheme 8.1), the carbocation (arenium ion) from benzene undergoes substitution reactions by the loss of a proton. This loss of proton restores the aromatic sextet. Thus, *e.g.*, cyclohexene reacts to give *trans*-1, 2-dibromoc-yclohexane. This reaction is exothermic by about 29 kcal (121 kJ) per mole. A similar addition reaction on benzene is endothermic since it leads to loss of aromatic stability.

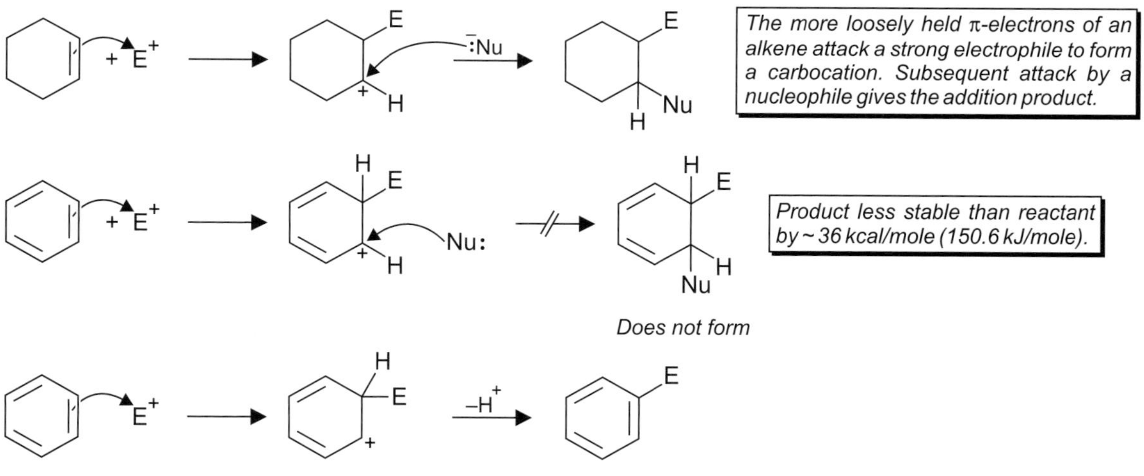

SCHEME 8.1

8.1 GENERAL VIEW—THE ARENIUM ION—THE ARENIUM ION MECHANISM—S_E2 REACTION

(A) Arenium Ion

In electrophilic substitution reactions, the benzene ring acts as an electron source. The reaction conditions employed are designed to form an electrophilic species (E^+, Scheme 8.2). Once formed, the electrophile (E^+) reacts with an arene *e.g.*, benzene itself to generate a carbocation. For benzene the structure of the ionic intermediate after attack by E^+ is a resonance hybrid of three resonance contributing structures which can be represented by a single structure showing the delocalization of charge (Scheme 8.2). This delocalized nonaromatic carbocation is termed an arenium ion, or often a sigma complex, since the electrophile is joined to the benzene ring *via* a new sigma (σ) bond. The sigma complex is not aromatic, however, because the sp^3 hybrid carbon atom interrupts the ring of *p*-orbitals (*i.e.*, the cyclic system of π-electrons is interrupted). The nature of electrophilic attack is thus highly endothermic (loss of aromaticity). The sigma complex, consequently regains aromaticity by the loss of the proton on the tetrahedral carbon atom with the help of a proton-accepting species, a base, *B* (*e.g.*, Nu^-:, to give a substitution product (Scheme 8.1).

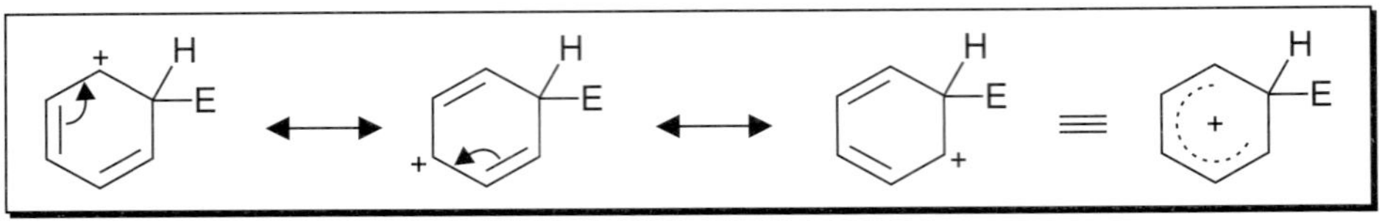

SCHEME 8.2

By using a Kèkule structure one can see that the arenium ion is a hybrid of three allylic type resonance structures (Scheme 8.2) and each has a positive charge on a carbon which is *ortho* or *para* to the site of electrophilic attack. The rate-determining step in this mechanism is the step in which the arenium ion is generated (the first step) and not the step in which a porton is lost from the arenium ion (second step). The breaking of C—D or a C—T bond is more difficult than the breaking of the C—H bond. If the second step in the reaction mechanism is the rate-controlling, electrophillic substitution reactions (*e.g.*, nitration) of aromatic compounds labelled with deuterium or tritium should be slower than those of the unlabelled compounds (Scheme 8.3). As a matter of fact, no significant change in reaction rate for the labelled compounds

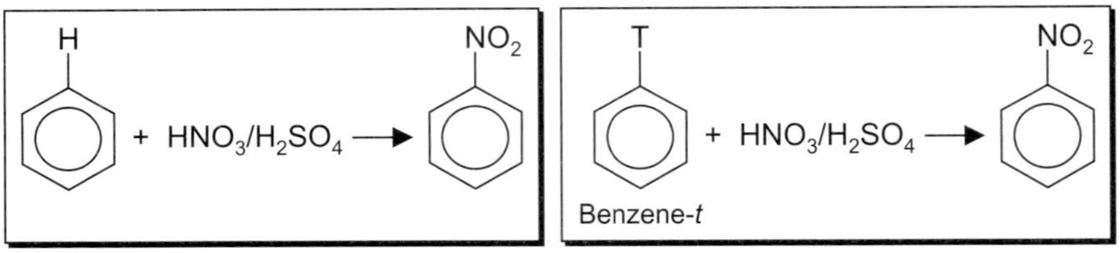

No difference in rate when nitro replaces hydrogen or tritium (no significant kinetic isotope effect).

SCHEME 8.3

(no significant isotope effect) is observed. Thus, the first step in the mechanism must be the slower, rate-determining step. Benzene is an extremely weak base and is only slightly protonated in concentrated H_2SO_4. Protonation of benzene (see Scheme 8.10) can be detected by carrying out hydrogen isotope exchange reactions in acid. Benzene in contact with 80% aqueous H_2SO_4 and tritium over longer periods gives benzene-*t* when the isotope distributes between the benzene and the aqueous acid.

The energy profile shown (Scheme 8.4) is in keeping with this type of mechanism. The arenium ion is a true intermediate lying between transition states 1 and 2. In transition state 1 the bond between the electrophile and one carbon of the benzene ring is only partially formed. In transition state 2 the bond between the same benzene carbon and its hydrogen is partially broken. In the slow, rate-determining (step), the aromatic nucleus (benzene) and the electrophile (E^+) come together to form a new bond between them. Because of this bimolecular attack, electrophilic aromatic substitution is often termed an S_E2 reaction, where S stands for substitution, E for electrophilic, and for the biomolecular nature of the reaction.

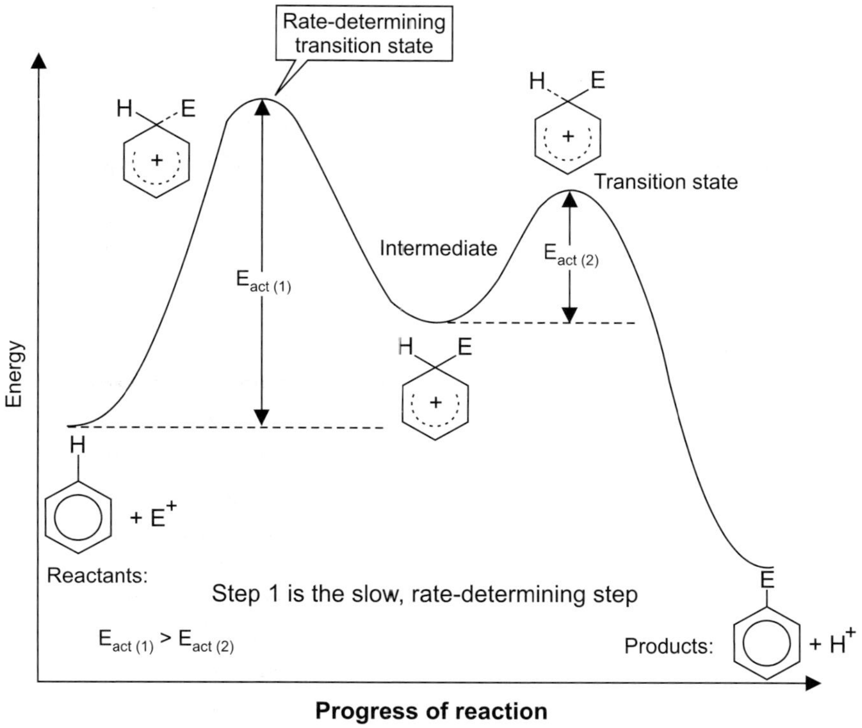

SCHEME 8.4

The arenium ion (I, Scheme 8.5) has been isolated as a crystalline compound (m.p. –15°C) by reacting mesitylene with ethyl fluoride with BF_3 as the catalyst at –80°C. On heating the arenium ion, the normal electrophilic substitution product was reached.

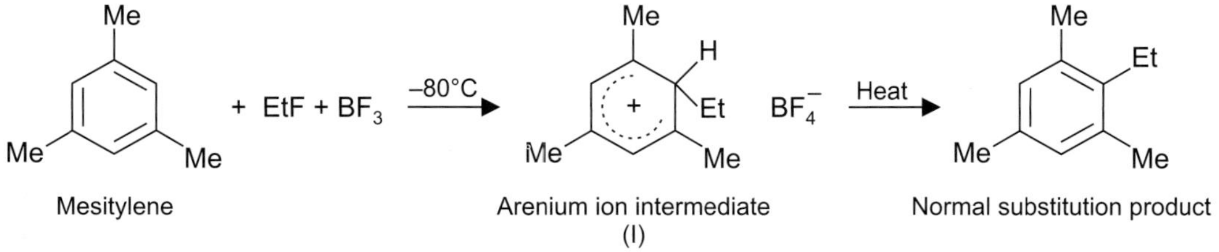

SCHEME 8.5

The simplest of the arenium ions *i.e.*, the benzenonium ion (Scheme 8.6) has been prepared by the protonation of benzene in HF—SbF_5—SO_2ClF—SO_2F_2 at –135°C and has been studied by ^{13}CNMR spectroscopy. The resonance stabilized carbocation, the benzenonium ion (σ complex) thus generated (Scheme 8.6) has about +1/3 charge at C-1, C-3 and C-5. In keeping with this fact these carbons have a greater chemical shift in ^{13}CNMR compared to C-2 and C-4 which remain uncharged.

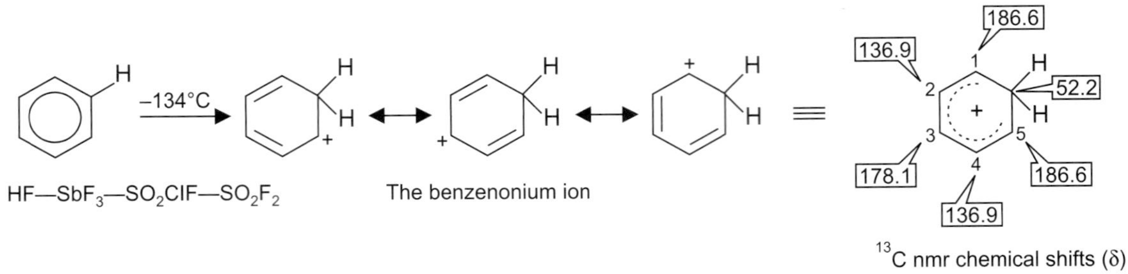

SCHEME 8.6

Another evidence for the formation of σ complexes as intermediates in electrophilic aromatic substitution is their trapping by nucleophiles. With suitable subtrates an addition is observed after the electrophile attacks a position which is already substituted (*ipso* attack), because now facile rearomatization by deprotonation is blocked. Thus nitration of the alkylated benzene (I, Scheme 8.7) at 0°C gives a diene (II), with acetate acting as the nucleophile. In several situations, however, atoms or groups other than hydrogen may be displaced from the aromatic ring (see, Scheme 8.48d).

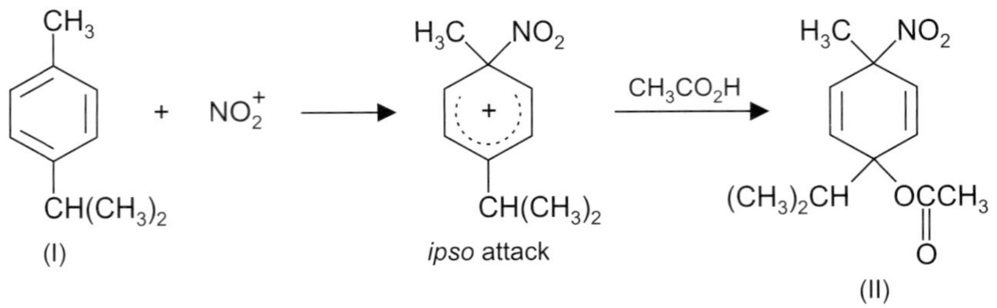

SCHEME 8.7

(B) Nitration of Benzene

Nitration of benzene with nitric acid requires sulphuric acid as the catalyst. The nitronium ion which is the electrophile is obtained when H_2SO_4 protonates HNO_3 (Scheme 8.8). H_2SO_4 is also the source of base HSO_4^- which removes the proton in the second step.

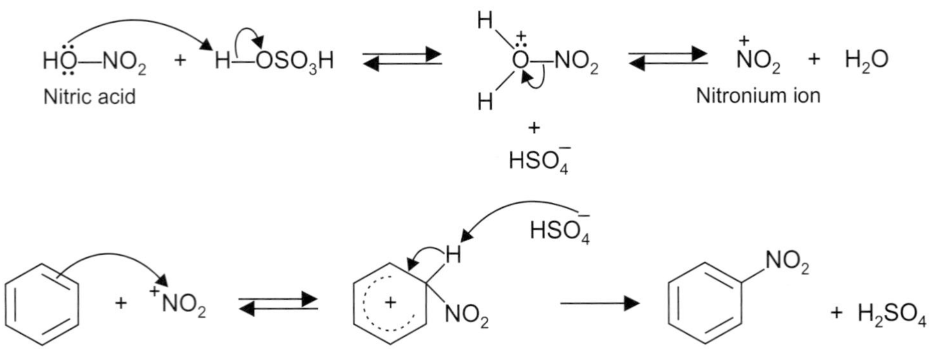

Nitration of benzene

SCHEME 8.8

As an evidence for the formation of nitronium ion as the electrophile, a salt *e.g.*, nitronium perchlorate, $NO_2^+ClO_4^-$ (which is a stable, isolable compound), reacts with benzene to give nitrobenzene in the absence of sulphuric or nitric acid. Nitric acid shows a peak in its Raman spectrum which disappears on the addition of sulphuric acid and two new peaks are displayed, one at 1400 cm^{-1} due to NO_2^+ and the other at 1050 cm^{-1} due to HSO_4^-.

(C) Sulphonation of Benzene

The sulphonation of benzene is normally carried out with fuming sulphuric acid (sulphuric acid containing sulphuric trioxide, SO_3), however concentrated sulphuric acid (95% H_2SO_4, 5% H_2O) works for "activated" rings. In either reaction the electrophile appears to be sulphur trioxide. In concentrated sulphuric acid, sulphur trioxide is generated in the equilibrium (Scheme 8.9) in which H_2SO_4 acts as both an acid and a base. Chemical kinetics show that the rate of sulphonation depends on the concentration of benzene and sulphur trioxide and not sulphuric acid.

$$2H_2SO_4 \rightleftharpoons H_3O^+ + HSO_4^- + SO_3$$

Generation of the electrophile, SO_3

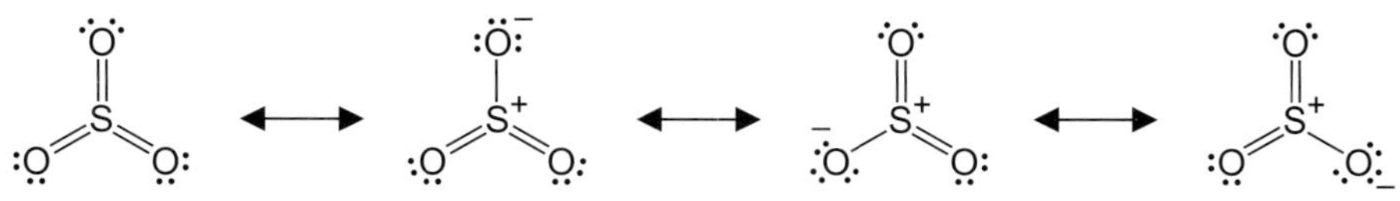

Sulphur trioxide powerful electrophile

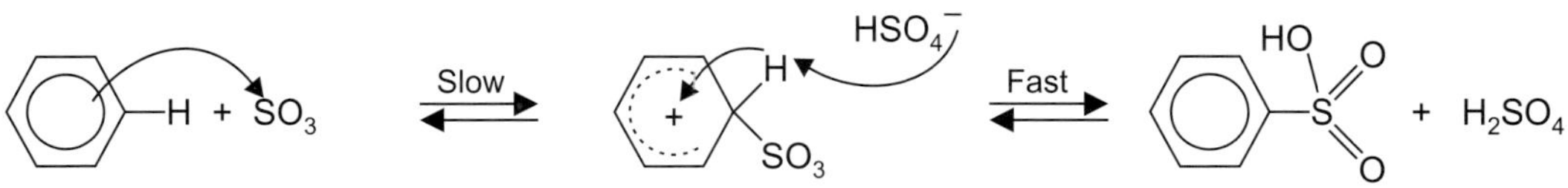

Attack of SO_3 on benzene ring   Sigma complex   Benzenesulphonic acid

SCHEME 8.9

(D) Protonation of Benzene

When a proton acting as an electrophile attacks benzene, the sigma complex can lose either of the two protons on the tetrahedral carbon. One can prove the formation of such a σ complex from benzene by using a deuterium ion (D^+) rather than a proton to show that the product contains deuterium atom in place of a hydrogen atom (Scheme 8.10). This is achieved by carrying out the reaction in the presence of deuteriosulphuric acid (addition of SO_3 to D_2O to give D_2SO_4).

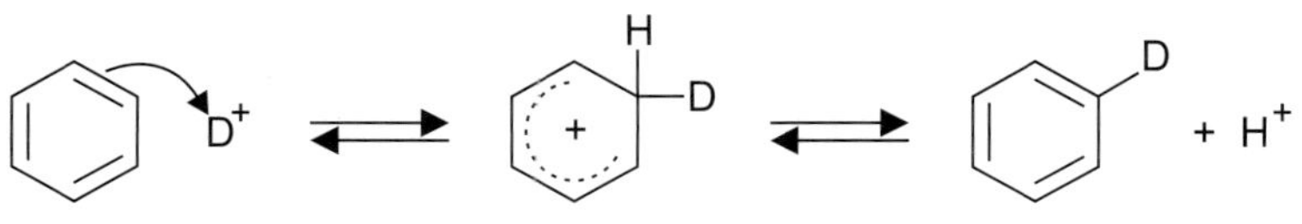

SCHEME 8.10

(E) Halogenation of Benzene

Bromine and chlorine molecules are not strong electrophiles and do not react with benzene. In the presence of a Lewis acid however, reaction occurs readily. The role of the catalyst is to accept a lone pair of electrons from the halogen molecule, which then weakens the Br—Br bond to provide Br^+ the electrophile for electrophilic aromatic substitution. The actual electrophile is probably the complex formed from the halogen and the catalyst, rather than a halonium ion *e.g.*, Br^+.

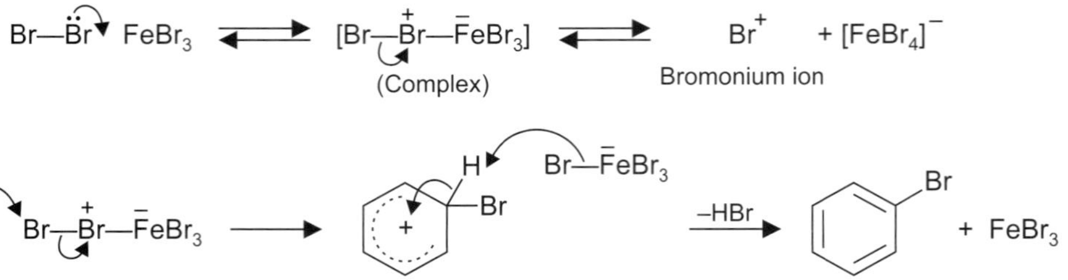

SCHEME 8.11

(F) Friedel-Crafts Alkylation—A Carbon-Carbon Bond Forming Reaction

In the Friedel-Crafts alkylation reaction a hydrogen is substituted for an alkyl group. A carbocation is generated during the first step of the reaction. There is not enough positive character on the carbon atom in alkyl halides for reaction with benzene; the catalyst increases the positive character. Aluminium chloride is the commonly employed Lewis acid.

The electrophile is a carbocation and alkyl fluorides, chlorides, bromides and iodides can all be used. Vinyl halides and aryl halides cannot be used since their carbocations would be too unstable if formed. As already said the electrophile may be a carbocation or perhaps more likely the complex itself (Scheme 8.12).

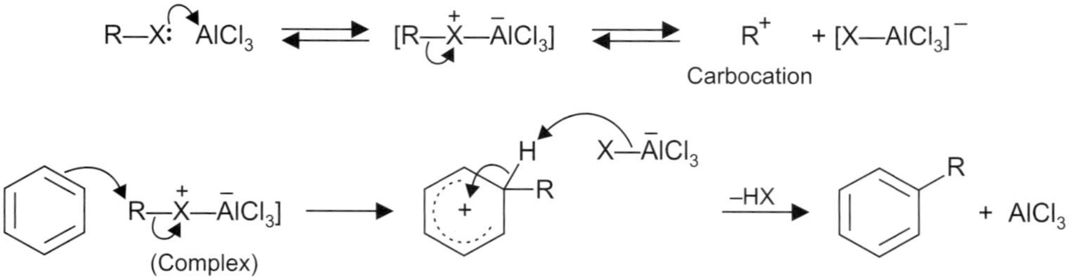

Friedel-Crafts alkylation of benzene

SCHEME 8.12

Alcohols and alkenes generate these carbocations on their reaction with an acid and thus these react analogously to alkyl halides. Alkenes can be protonated with HF, the fluoride ion being a weak nucleophile does not immediately attack the carbocation. Recall that carbocation is generated following the Markovnikov's rule (Scheme 8.13) in the protonation step.

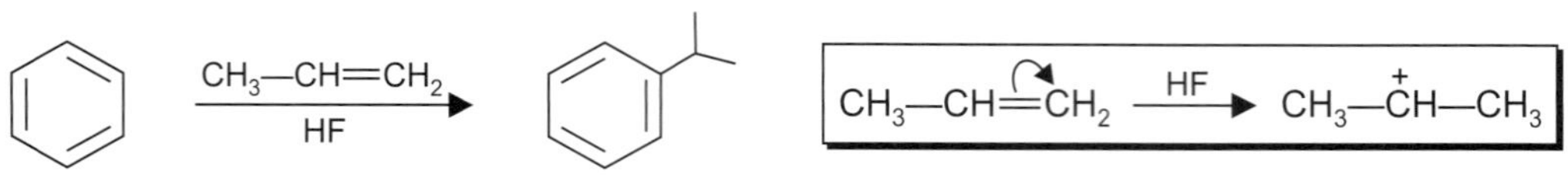

SCHEME 8.13

There are several drawbacks of the Friedel-Crafts alkylation reaction.

- The carbocations rearrange if such a rearrangement leads to a more stable carbocation. Thus it is not uncommon for mixtures to be produced. This is seen during alkylation of benzene with 1-chloropropane in the presence of $AlCl_3$ (Scheme 8.14).

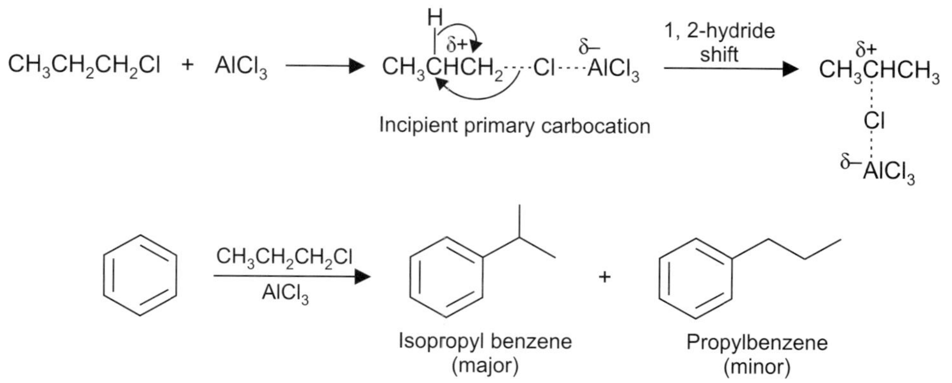

SCHEME 8.14

- Friedel-Crafts alkylations fail when the reactant contains more powerful electron withdrawing group than halogen. Nitrobenzene is an example which for this reason is used as a solvent for the reaction.
- Aromatic amines are reactive towards electrophilic attack, but do not undergo alkylation reactions. $AlCl_3$ forms a coordinate bond with the lone pair of electrons on the N atom of the amino group. This prevents the complexations of $AlCl_3$ with the alkyl halide and moreover the amino group is converted into a powerful electron-withdrawing group ($C_6H_5NH_2 + AlCl_3 \longrightarrow C_6H_5\overset{+}{N}H_2—\overset{-}{A}lCl_3$).
- The electron donating nature of an alkyl group, assists electrophilic attack on the benzene ring. After the initial alkylation, the product becomes more reactive than the starting reactant leading to mixed products of alkylation (Scheme 8.15).

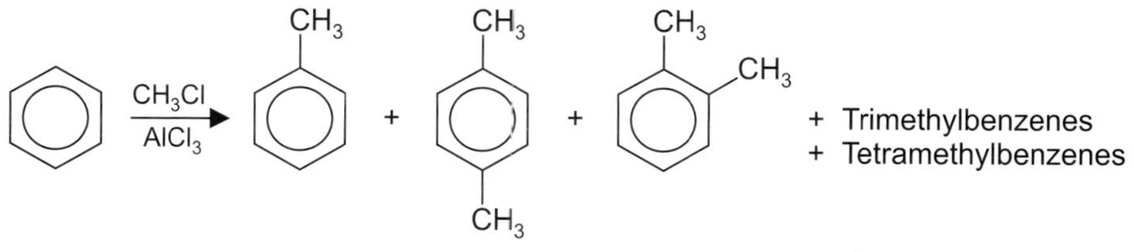

SCHEME 8.15

- Alkylation is reversible and an alkyl group can migrate from one molecule to other. This may lead to mixture of products. However, often advantage can be taken to transfer a *tert*-butyl group from one arene to another (Scheme 8.16).

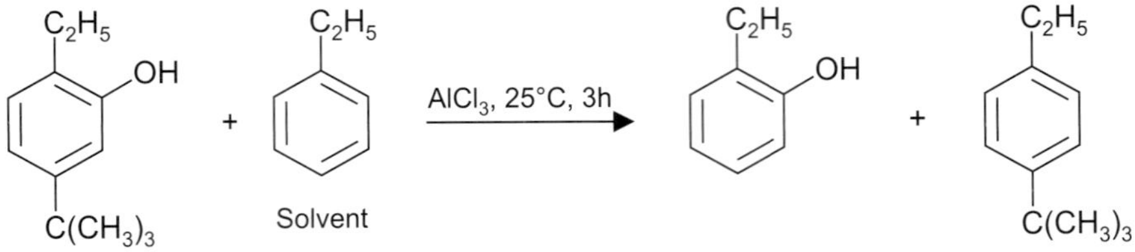

SCHEME 8.16

- Tertiary alkyl groups are most easily introduced during alkylation and depart as well most readily to give relatively stable carbocations. Thus, *t*-butyl group has been used in the protection of most reactive position in a reactant to instead initiate reaction elsewhere. After the protection, the *t*-butyl group is removed by adding excess of benzene to derive the equilibrium in the desired direction (Scheme 8.17).

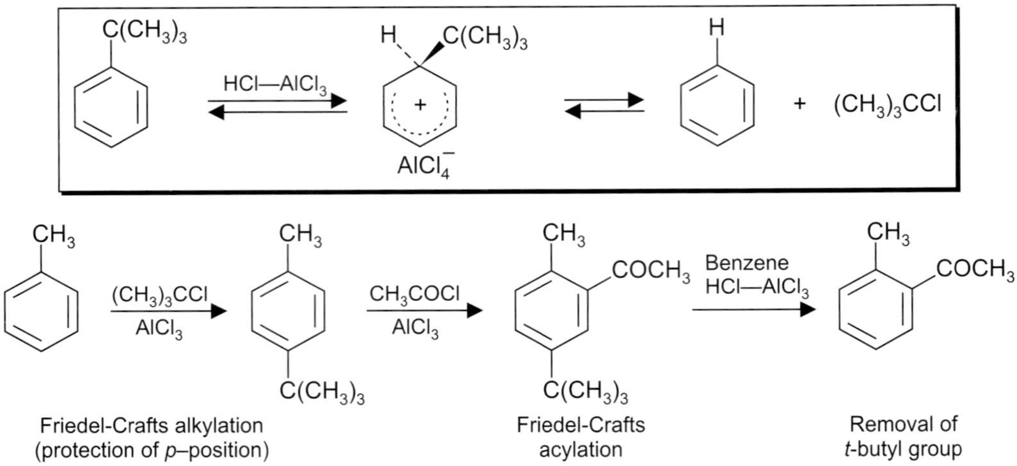

SCHEME 8.17

> *β-keto aryl amides or β-keto aryl esters act as alkylating agents in the presence of proton acids leading to the synthesis of quinolines or coumarins (see Scheme 6.51a).*

(G) Friedel-Craft Acylation Reaction

Acylium cations are generated by the reaction of acid halides with aluminium chloride. The Lewis acid initially coordinates to the carbonyl oxygen. This complex is in equilibrium with an isomer in which the aluminium chloride is bound to the halogen. Dissociation then gives the acylium ion, which is stabilized by resonance, most of the positive charge resides on the carbonyl carbon (Scheme 8.18). Reagents are not limited to acyl halides, but carboxylic acids, anhydrides and ketenes are also used.

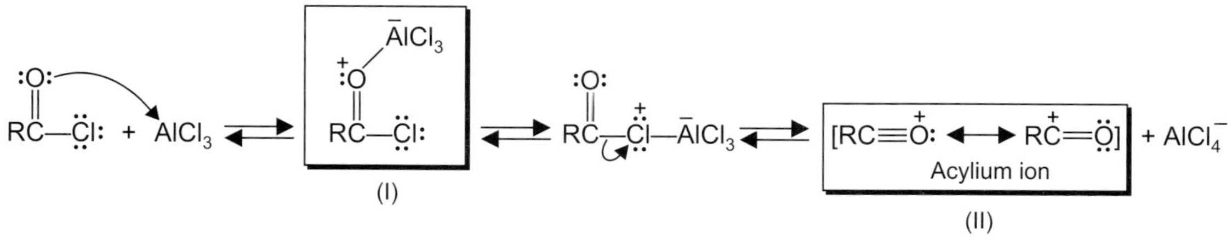

SCHEME 8.18

In the lewis acid catalysed acylation reaction two electrophiles are involved, one is the oxygen bound complex (I, Scheme 8.18) and the other is acylium ion which attack the benzene ring (Scheme 8.19). Which of the two is more effective in a particular case depends on the nature of R (for example, formation of the acylium ion is favoured when R is aromatic, since its positive charge can be delocalized on to the aromatic ring).

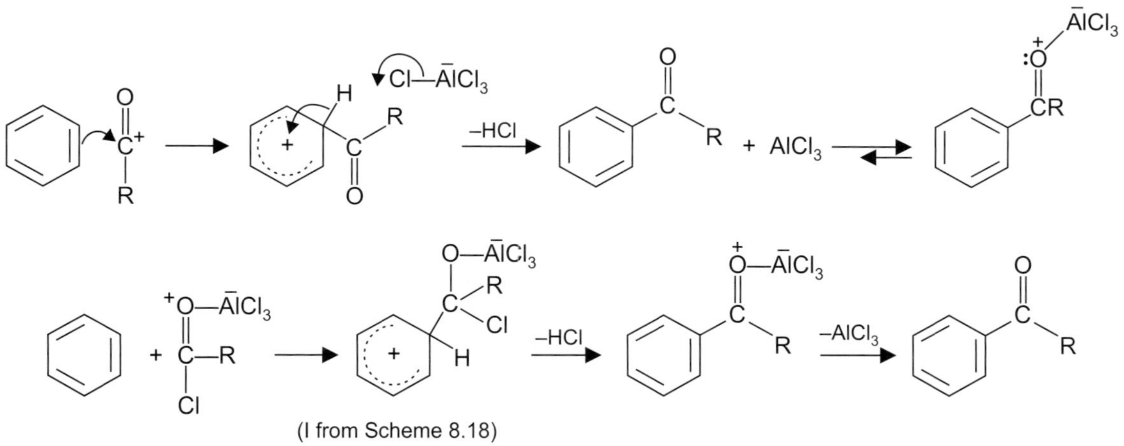

SCHEME 8.19

The major disadvantages seen in the case of Friedel-Crafts alkylation are not found here. Rearrangement in R is never observed and since RCO is a deactivating group, the reaction stops after one group is introduced. The effect is accentuated by the formation of a strong complex between the aluminum choride catalyst and the carbonyl function of the product ketone (Scheme 8.19). This complexation removes the $AlCl_3$ from the reaction mixture and requires the use of *at least one full equivalent* of the Lewis acid for the reaction to go to completion. Aqueous work-up is required to liberate the ketone from its aluminium chloride complex.

One has already seen that use of longer alkyl chains than ethyl is complicating due to carbocation rearrangements (see Scheme 8.14). Acylium ions, do not rearrange, however, thus straight chain alkyl groups can be placed on the benzene ring *via* Friedel-Crafts acylation and then reducing the carbonyl group to the methylene. Several methods can be used for reduction

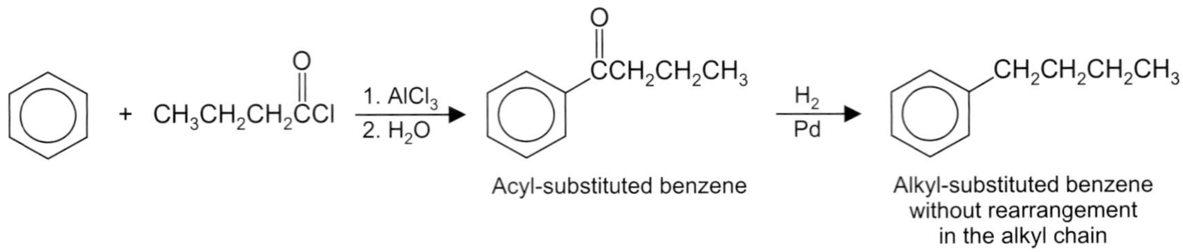

SCHEME 8.20

when a ketone carbonyl group is adjacent to a benzene ring it can be reduced to a methylene group by catalytic hydrogenation (H_2/Pd). Reduction of a carbonyl group to a methylene can be achieved by other methods namely by Clemmensen reduction (Zn/Hg/HCl) or via Wolff–Kishner reduction (heating with NH_2NH_2 and base).

Friedel-Crafts acylations have an important role to bring about ring closure provided the group introduced is in the proper position (Scheme 8.21). With cyclic anhydrides, the product contains a COOH group in the side chain which can be involved in cyclization by carrying out an intramolecular Friedel-Crafts acylation.

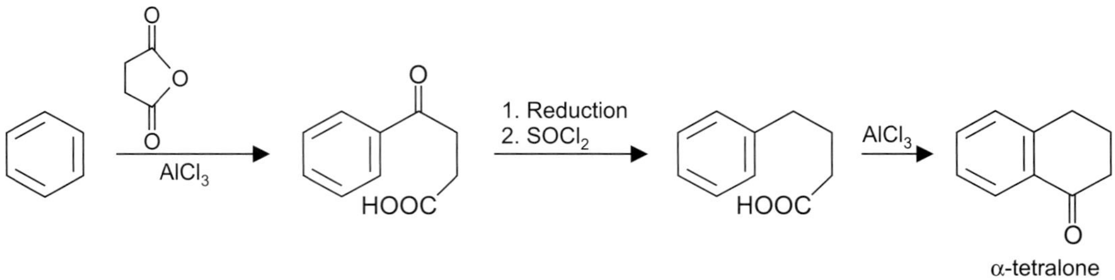

SCHEME 8.21

EXERCISE 8.1

Give the mechanism of formation of acylium ion from a carboxylic anhydride.

ANSWER. *It is in (Scheme 8.22).*

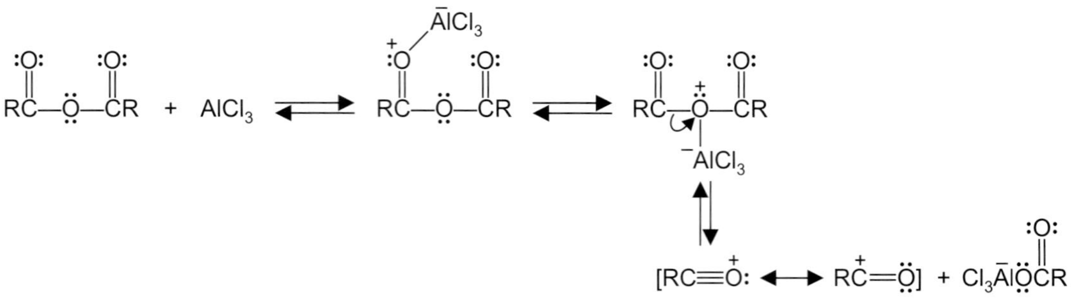

SCHEME 8.22

EXERCISE 8.2

Why benzene reacts with trimethylacetyl chloride in the presence of $AlCl_3$ to give t-butylbenzene, while anisole reacts to give the normal product $CH_3OC_6H_4CO(CH_3)_3$?

ANSWER. *Benzene is relatively less reactive and the acylium ion has the time to break down to CO and stable t-butyl carbocation (Scheme 8.23).*

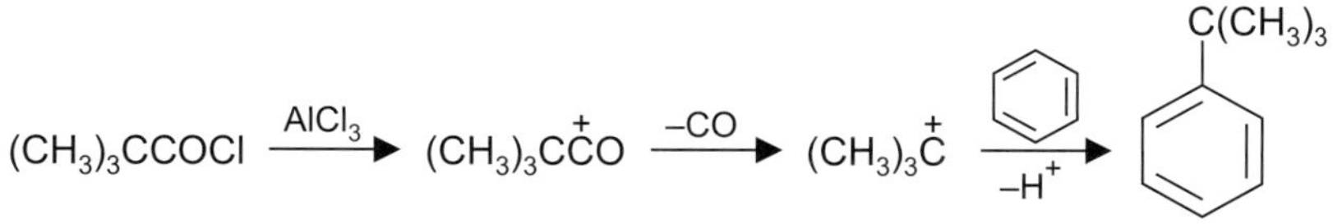

SCHEME 8.23

EXERCISE 8.3

Write the reagents for the conversion (Scheme 8.24).

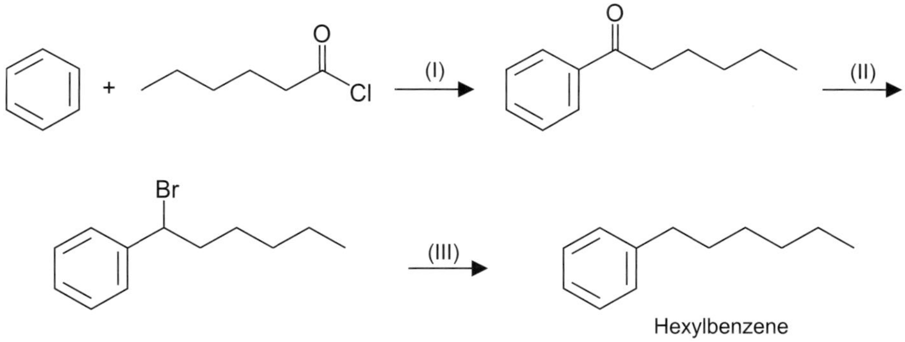

SCHEME 8.24

ANSWER. *I, $AlCl_3$ (–HCl); II, (1) $NaBH_4$; (2) HBr_2; III $LiAlH_4$.*

OTHER METHODS TO INTRODUCE STRAIGHT CHAIN ALKYL GROUPS ON A BENZENE RING

- *By using Gilman reagents (see Scheme 7.88) as shown (Scheme 8.25).*

SCHEME 8.25

- *By using Suzuki reaction. This is a versatile reaction for making carbon-carbon bonds and involves a palladium catalysed coupling of an organohalide and an organoborane. The desired organoborane can be made by reacting an alkene with catecholborane (Scheme 8.26).*

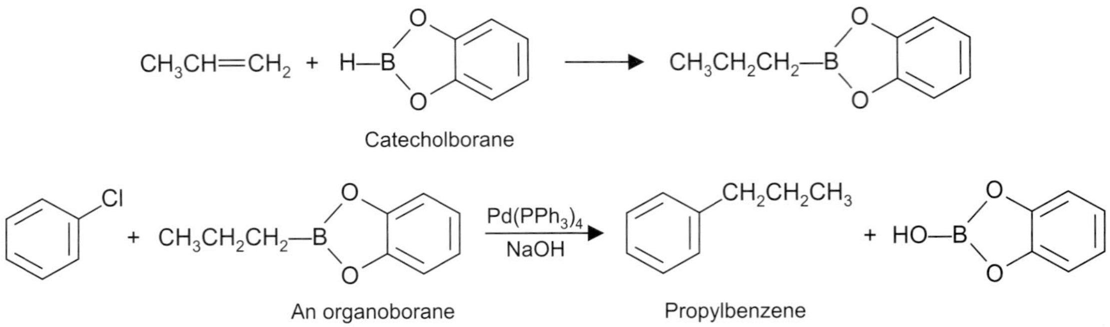

SCHEME 8.26

8.2 ELECTROPHILIC SUBSTITUTION ON MONOSUBSTITUTED BENZENES—ORIENTATION AND REACTIVITY

The reactivity of a monosubstituted benzene (C_6H_5—S) and the orientation of incoming substituent depends on the nature of the substituent (S) already present on the ring. A monosubstituted benzene on electrophilic substitution may give three possible disubstituted products (*ortho*, *meta*, and *para* isomers). From the yields of these isomers, it is possible to separate the substituents into two groups the *o*, *p*-directors which give predominantly *o*, *p* products and the *m*-directors which give mainly the *meta* products. The following points will help to understand both the reactivity *i.e.*, if the reaction will be slower or faster than with benzene and orientation of the incoming group in a monosubstituted benzene derivative, *i.e.*, the position (*o*, *m* or *p*) which the new group will take.

(A) Activation and Deactivation—Theory of Orientation

(i) Inductive Effect

The terms activation or deactivation are applied when the reactivity of a monosubstituted benzene (C_6H_5—S) in electrophilic substitution reactions is compared with benzene (C_6H_6) under the identical conditions. An electron-releasing group *i.e.*, *+I* effect of the substituent helps to stabilize the positive charge of the reaction intermediate (the σ complex) produced in the rate-determining step by furthering the delocalization of the charge (Scheme 8.27). This will serve to lower the activation energy in the rate-determining step relative to that of benzene. Thus, the compound (C_6H_5—S) where S has *+I* effect will undergo electrophilic substitution reactions more readily than benzene itself (*i.e.*, faster rate or milder conditions). Such a group is said to be an activating group. An example is of methyl group in toluene (Scheme 8.27). On the other hand, an electron-withdrawing group, (– *I* effect) will exert the opposite effect. It destabilizes the positive charge of the raction intermediate (the σ complex) in the rate-determining step by depleting electron density away from the ring, thus intensifying the positive character of the ring carbons compared with benzene to raise the activation energy of its rate-determining step, thus rendering the compound C_6H_5—S less reactive than benzene in electrophilic substitution reactions. Such a group is called a deactivating group and examples are of $-N^+Me_3$ group and trifluoromethyl (CF_3) group (Scheme 8.27).

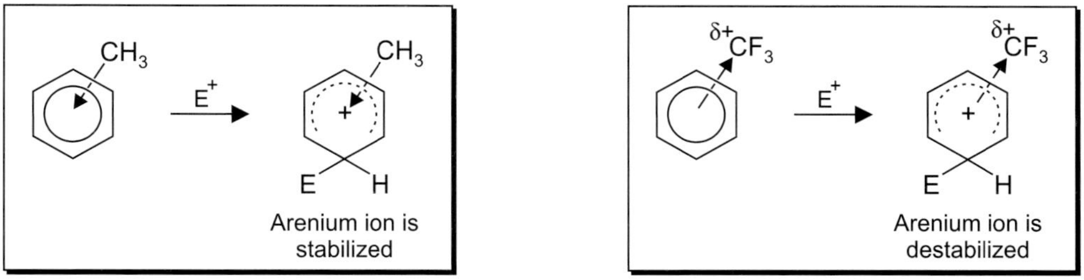

SCHEME 8.27

(ii) Mesomeric Effect

A substituent however, releases electron density to the aromatic ring or depletes electron density from it *via* both inductive (*I*) and mesomeric (*M*) effects. This is explained by taking following examples:

Toluene: The CH_3 group is activating and +*I* and +*M* effects (caused by hyperconjugation) work together to push electron density into the ring (Scheme 8.28), therefore, toluene is more susceptible to electrophilic attack than benzene.

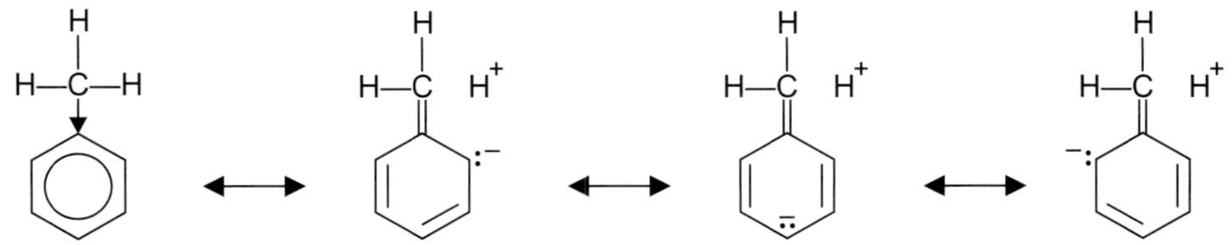

SCHEME 8.28

Anisole: When a substituent has a lone pair on the atom which is directly attached to the benzene ring, the lone pair can get delocalized into the ring; the substituent is said to donate electrons by resonance (mesomeric effect). Substituents *e.g.*, NH_2, OH, OR, and Cl donate electrons by resonance and these also withdraw electrons inductively since the atom attached to the benzene ring is more electronegative than a hydrogen (Scheme 8.29).

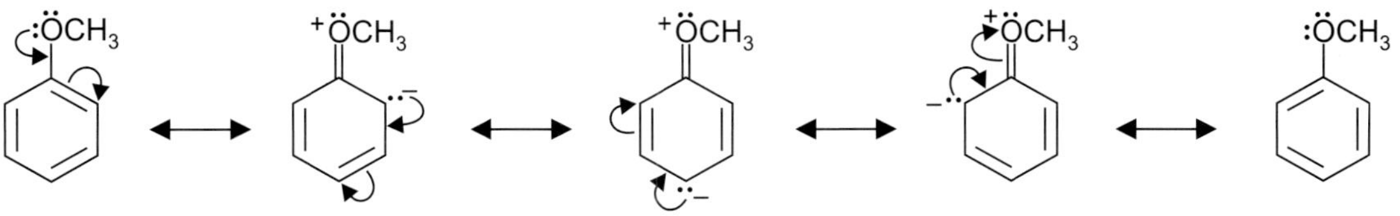

Electrons are donated to the benzene ring by mesomeric effect +*M* (resonance)

SCHEME 8.29

Nitrobenzene: When a substituent is attached to the benzene ring *via* an atom which is doubly or triply bonded to a more electronegative atom, the π-electrons of the ring can get delocalized onto the substituent. Thus C= O, C≡N, and NO_2 withdraw electrons by resonance. and these substituents also withdraw electrons inductively since the atom attached to the benzene ring has a full or partial positive charge and, is thus more electronegative than a hydrogen (Scheme 8.30).

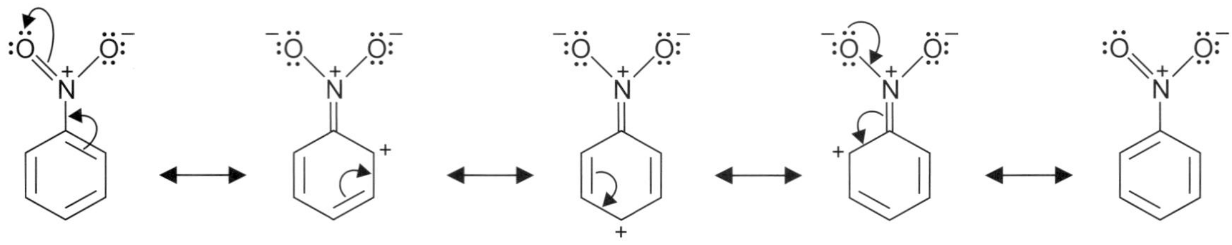

Electrons are withdrawing from the benzene ring by mesomeric effect (–*M*)

SCHEME 8.30

(B) An Introduction to Relative Reactivity of Substituted Benzenes and Orientation

The following points may be noted:

- Electron-donating substituents (activating substituents) activate the benzene ring to electrophilic attack to results in the formation of the *ortho*- and *para*- disubstituted benzene derivatives.
- Electron-withdrawing substituents deactivate the ring for attack by electrophiles, which occurs at the *meta* position (Scheme 8.31).

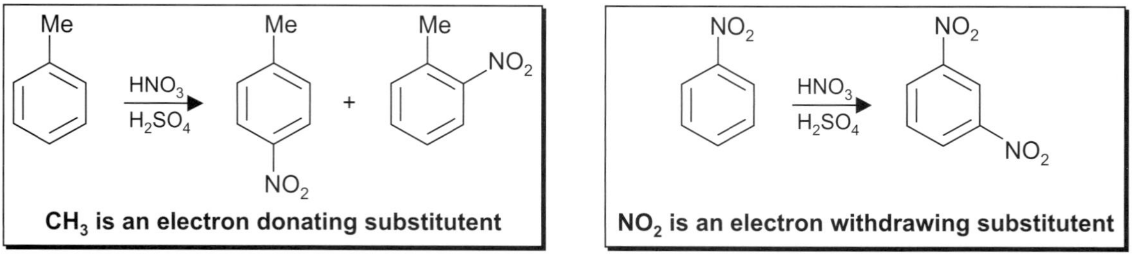

SCHEME 8.31

- All the *strongly activating substituents* (Scheme 8.32) donate electrons into the ring by resonance and withdraw from the ring inductively. These however, have been found experimentally to be strong activators to indicate that electron donation into the ring by resonance has more significance than inductive electron withdrawl from the ring (*i.e.*, +*M* effect dominates over the inductive effect –*I*).

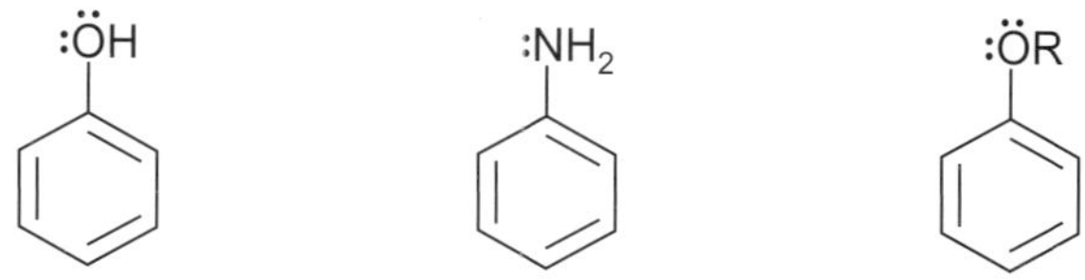

Strongly activating groups of –*I* and +*M* type—these are *o*, *p*-directing

SCHEME 8.32

- All strongly deactivating groups (Scheme 8.33) strongly withdraw electron density from the benzene ring both *via*; inductive (–*I*) and mesomeric (–*M*) effects (one may note that ammonium ions have no mesomeric effect, however, the positive charge on the nitrogen atom strongly withdraws electrons by inductive effect.

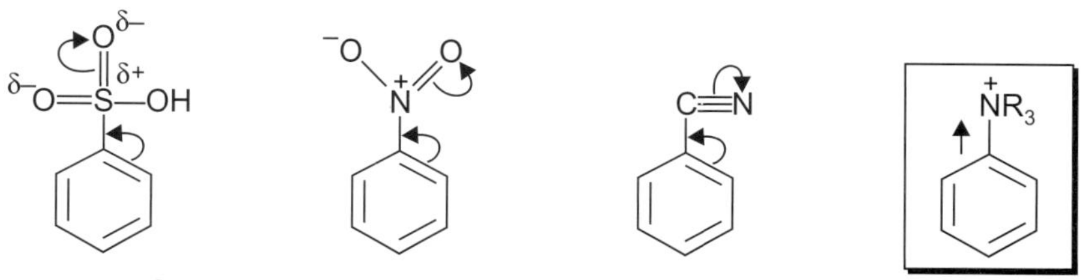

Strongly deactivating substituents–*meta* directing

SCHEME 8.33

- The relative stabilities of the three carbocations formed by the attack of the incoming electrophile (*ortho*-substituted carbocation, *meta*-substituted carbocation and *para*-substituted carbocation) help to predict the preferred pathway of a reaction. More stable the carbocation, the less energy will be needed to generate it and any feature which affects its stability will influence its ease of formation and thus the outcome of the reaction.

(C) Examples of *Ortho*-, *Para*-Directing Groups

Except for alkyl and phenyl substituents (which are *o*, *p*- directing) all of the *ortho*, *para* directing groups have atleast one non-bonding electron pair on the atom directly attached to the benzene ring (see Scheme 8.32). The resonance effect of the amino group is far more important than its inductive effect in electrophilic aromatic substitution, to make the amino group electron releasing. Consider the resonance structures of the arenium ions (carbocations) resulting from *ortho* and *para* attack (Scheme 8.34) on aniline. The structures (I and II, Scheme 8.34) apart from the same sets of three resonance structures are relatively stable.

In these structures the nonbonding pairs of electrons on nitrogen form an extra bond with the ring carbon. This extra bond and the fact that every atom in each of these structures has a complete outer octet of electrons renders these structures the most stable of all of the contributors. Because these structures are highly stable thus these make a large *and stabilizing* contribution of the hybrid. However, no such structure can be drawn following *meta* attack and thus the cation derived from this mode of attack is not additionally stabilized. This means that the *ortho*-and *para*-substituted arenium ions are for more stable than the arenium ion that is formed from *meta*-attack. The consequences of the involvement of the amino group are to stabilize the arenium ions formed after *ortho* and *para* attack which lowers the energy of

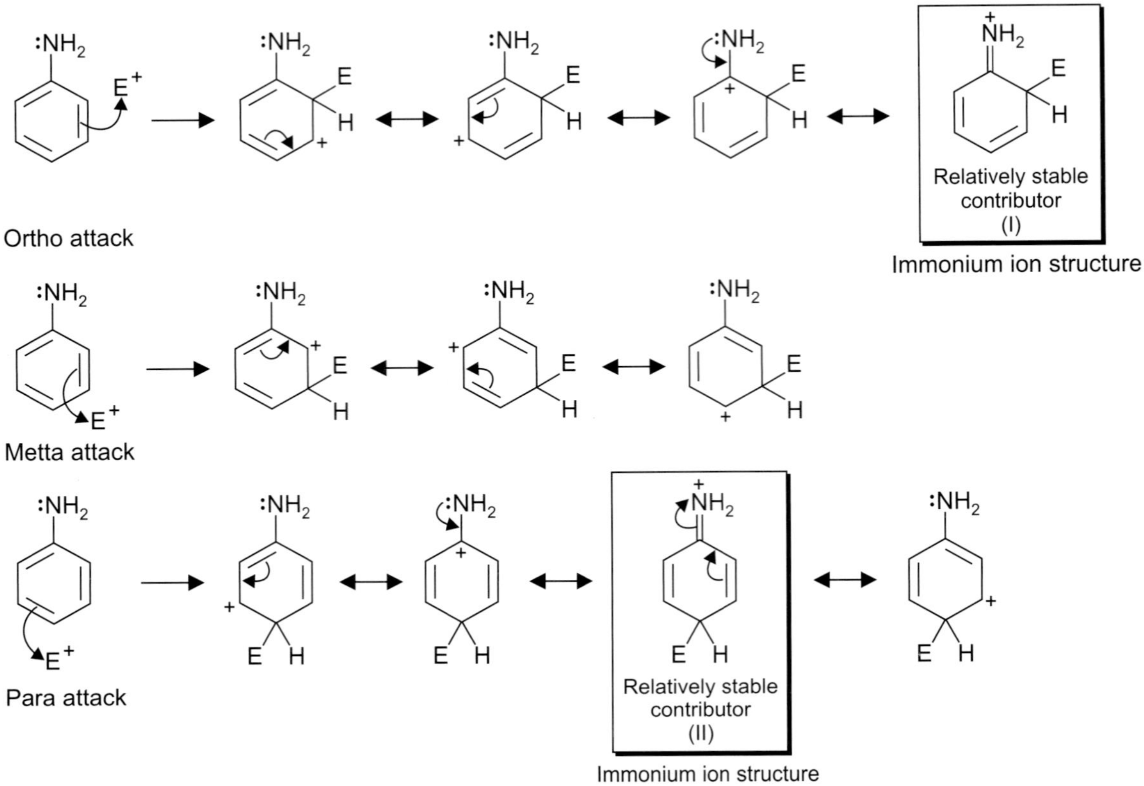

Attack of an electrophile $\overset{+}{E}$ on aniline at the three possible positions (*ortho, meta* and *para*)

SCHEME 8.34

activation for their formation (see Scheme 8.4). A point of further significance is that an additional (fourth) canonical form can be drawn for the arenium ions resulting from *ortho* and *para* attack. This fourth resonance structure confers extra stability on the intermediate and lowers the energy of the transition state leading to it. This however, is not so when the attack is *meta*. Thus just like immonium ion structures (Scheme 8.34), the oxonium ions from anisole and halonium ion structures from a halobenzene from attack on *o* and *p*-positions greatly stabilize the intermediate involved is each case (Scheme 8.34*a*).

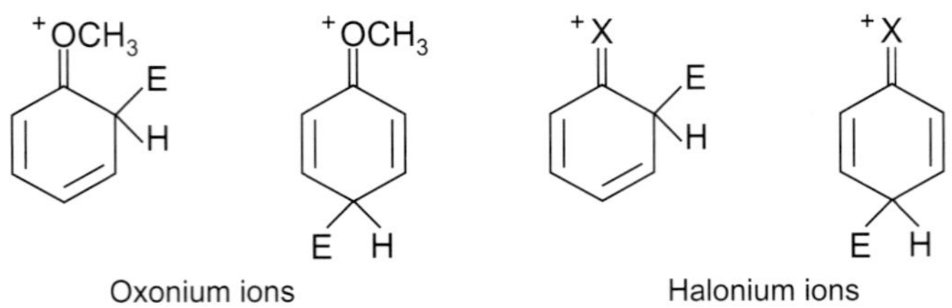

SCHEME 8.34a

Acyl derivatives of aniline and phenol are much less reactive with a smaller activating effect. The unshared pair of electrons on nitrogen or oxygen is readily delocalized within the substituent and is thus not readily available for π orbital overlap in the electron deficient transition state as shown for example in the case of an acyl derivative of aniline (Scheme 8.35).

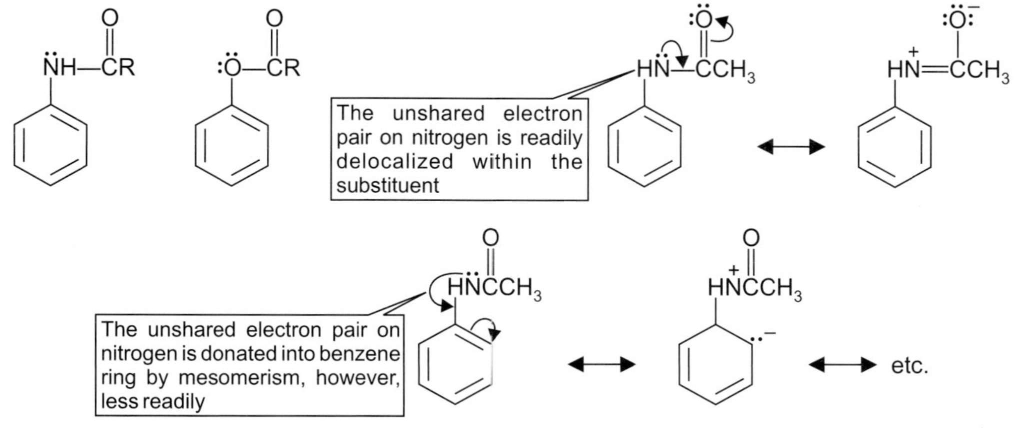

Acyl derivatives of aniline and phenol are moderately activating groups

SCHEME 8.35

EXERCISE 8.4

Why direct nitration of aniline is not a satisfactory reaction ? How it can be carried out ?

ANSWER. *Aromatic amines are highly susceptible to oxidation and nitric acid is a strong oxidizing agent. Nitration is therefore, carried out on its acyl derivative which reacts more slowly and then acyl group is removed by hydrolysis (Scheme 8.36). [Aniline is so reactive towards bromination that 2, 4, 6- tribromoderivative is formed instantaneously. Acetanilide, however reacts more slowly and the monobromoderivative (mainly para) is isolated. (Recall that both electrophiles and oxidizing agents seek electrons)].*

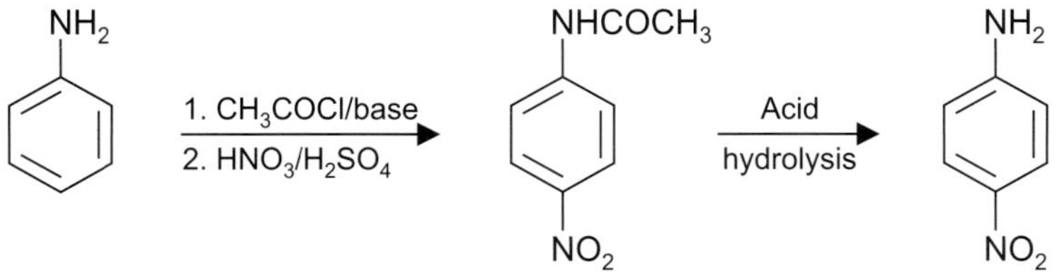

p-nitroacetanilide

SCHEME 8.36

MODERATING THE ACTIVATING POWER OF NH_2 AND OH GROUPS

Recall that NH_2 and OH are highly activating groups and consequentially it is difficult to stop electrophilic attack on aniline and benzene at the monosubstitution stage. Use of protecting groups, acetyl for aniline (as in acetanilide) and methyl for phenol as in anisole moderates their activating power. Deprotection can be brought about by hydrolysis.

In Alkylbenzenes and biphenyls the alkyl and aryl groups are somewhat weakly activating substituents (see Scheme 8.28), and are ortho-,para-directors. Consider the electrophilic substitution on toluene (Scheme 8.37). The arenium ions (I and II, Scheme 8.37) resulting from ortho and para attack are tertiary carbocations in which a methyl group is directly attached to a positively charged carbon of the ring making these still more stable. No such benefit results from attack at the meta position, which is therefore not a favoured position of attack.

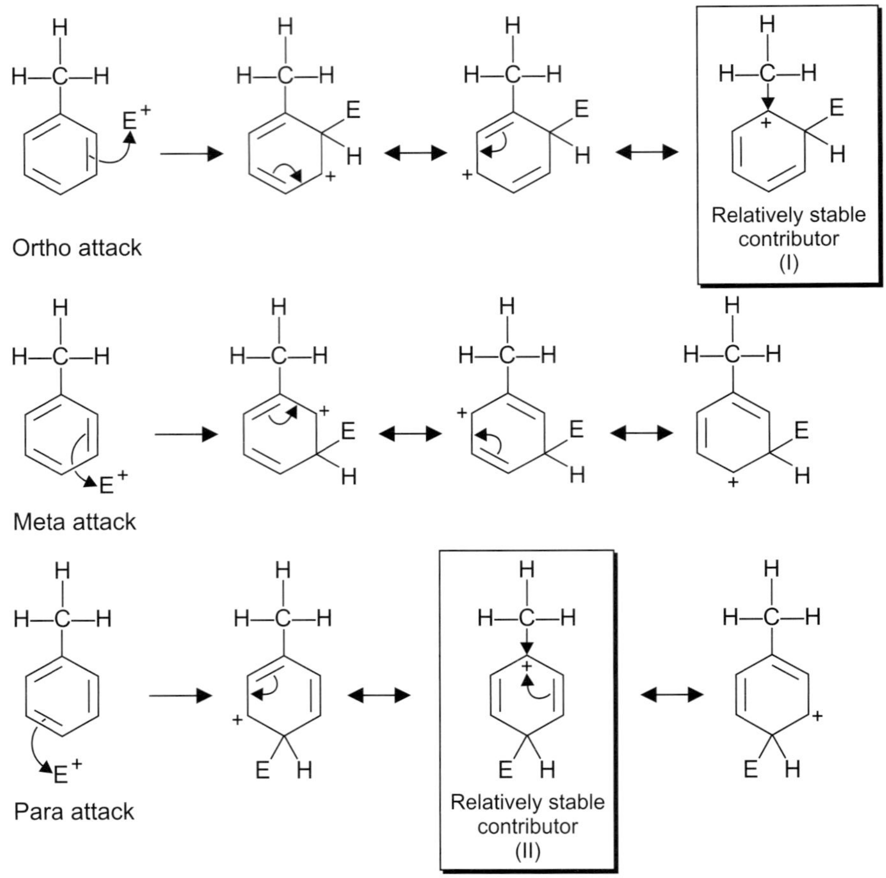

Electrophilic attack on toluene

SCHEME 8.37

EXERCISE 8.5

How biphenyl is attacked by an electrophile? How this reactivity is effected by the presence of electron—attracting or electron—releasing substituents on an aromatic ring.

ANSWER. *Biphenyl (a benzene with a phenyl substituent) is activated in the ortho and para positions and slightly deactivated in the meta-position. The deactivation in the meta-position is because of –I effect of the sp^2 hybridized carbon.*

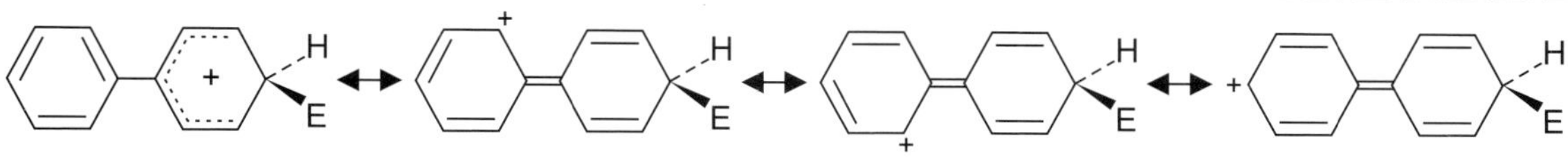

SCHEME 8.38

After the attack of E^+ (ortho or para-position, Scheme 8.38 only para attack is shown), the positive charge on the transition state is delocalized by the phenyl substitutent. The presence of an electron-attracting substituent slows the reaction and substitution occurs at the ortho and para positions of the unsubstituted ring. On the other hand, an electron-releasing substitutent increases the reactivity and causes reaction to occur in the substituted ring.

The halogens are regarded as weakly deactivating substituents and these direct an incoming electrophile to the ortho and para positions. If one considers chlorobenzene, the chlorine atom is highly electronegative which influences reactively. One finds that for electrophilic substitutions on halobenzenes, conditions comparable to those for benzene are required. Their electron donating mesomeric effect on the other hand governs orientation (Scheme 8.39). On electrophilic attack, chlorine atom stabilizes the arenium ion from ortho and para attack compared to the one from meta attack (this situation is similar to that seen in Scheme 8.34).

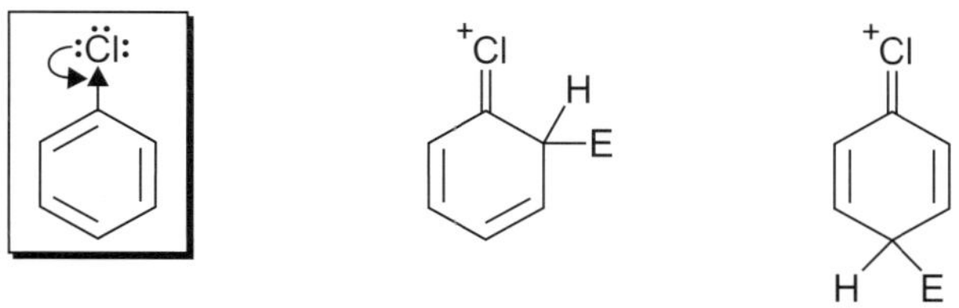

Chlorine atom stabilizes the arenium ion formed on *ortho* or *para* attack of the electrophile (E^+) relative to that from *meta* attack

SCHEME 8.39

(D) Examples of *Meta*-Directing Groups

One observes that in *meta*-directing groups, the atom directly attached to the ring either carries a partial positive charge or a full positive charge (see Scheme 8.33), these substituents are strongly deactivating and *meta*-directing. If one considers the electrophilic attack on nitrobenzene *e.g.*, nitration with HNO_3/H_2SO_4 it is more difficult than for benzene in keeping with the reduced electron density at the ring carbon atoms. Moreover, the reasonance contributors (I and II, Scheme 8.40) formed after attack of the electrophile (E^+) at the *ortho* and *para* positions are the least stable since these have a positive charge on each of the two adjacent atoms, thus the most stable carbocation is formed when the incoming electrophile is directed to the *meta*-position.

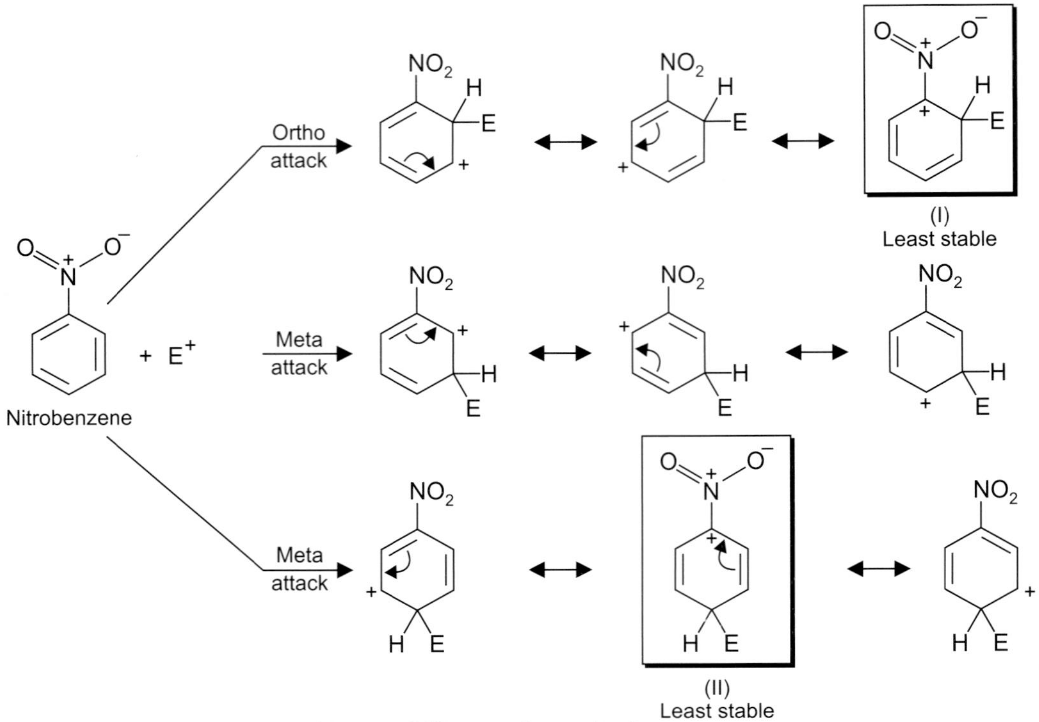

Electrophilic attack on nitrobenzene

SCHEME 8.40

META DIRECTING GROUPS

The arenium ion formed from ortho and para attack always leads to one contributing structure which is highly unstable relative to all others by having positive charges located on adjacent atom (the positive charge is located on the ring carbon which bears the electron withdrawing group). No such highly unstable resonance structure is involved from a meta attack, compare with Scheme 8.41.

The meta directing groups cannot undergo Friedel-Crafts reactions, since these reactions need a Lewis acid catalyst. In the presence of a meta director (deactivating group) the ring will become too unreactive to undergo Fridel-Craft reactions.

Recall that activating groups are o, p-directors these are:

$—\ddot{N}R_2$, $—\ddot{N}H_2$, $—\ddot{O}H$, $—\ddot{O}:^-$	$—\ddot{N}—C(=O)R$, $—\ddot{O}—C(=O)R$, $—\ddot{O}R$	R, Ar
Strongly activating	Activating	Weakly activating

Strongly deactivating groups are m-dir+ectors:

$—NO_2$, $—SO_3H$, $X{=}O$, $C{\equiv}N$, $—\overset{+}{N}R_3$, $—CF_3$, $—\overset{+}{S}R_2$	$—\ddot{X}:$, $—CH_2X$
Strongly deactivating	Weakly deactivating (*o*, *p*-directors)

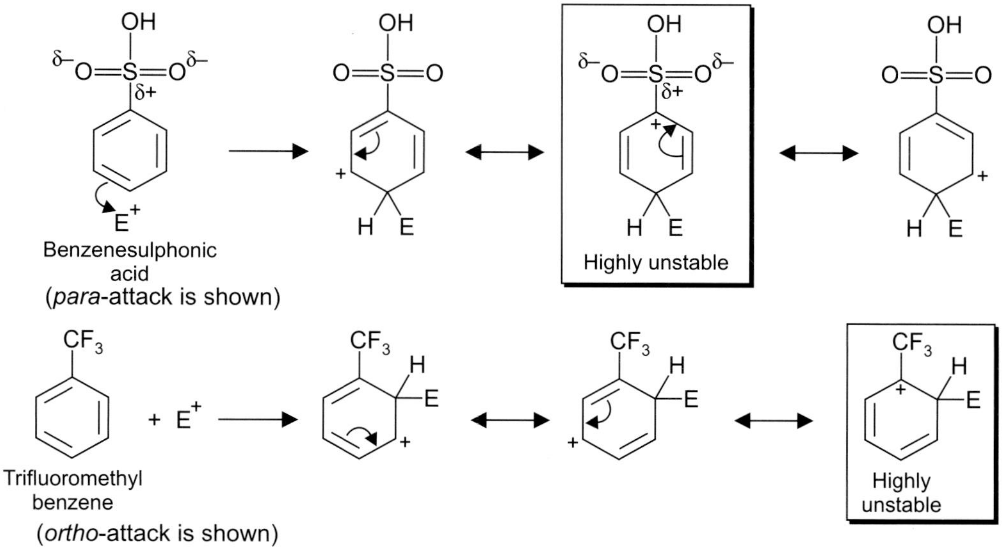

SCHEME 8.41

EXERCISE 8.6

Why aniline with its strongly activating substituent ($-\ddot{N}H_2$) does not undergo Friedel-Crafts reactions while phenol and anisole undergo these reactions?

ANSWER. *Due to the formation of a complex between the Lewis acid catalyst ($AlCl_3$) and the lone pair of the amino group (Scheme 8.42). The activating amino group is converted into a highly deactivating group.*

Phenol and anisole however, undergo Friedel-Crafts reactions because oxygen, being a weaker base than nitrogen, does not complex with the Lewis acid, and the expected o, p-orientation is observed.

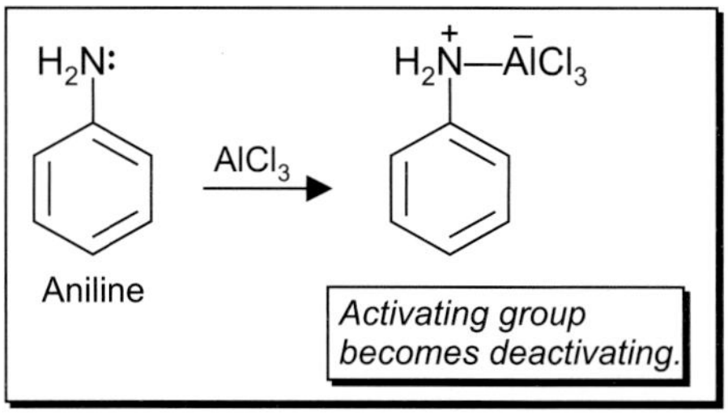

SCHEME 8.42

Other examples of moderately deactivating groups which are *meta* directing have a carbonyl group directly attached to the benzene ring (Scheme 8.43).

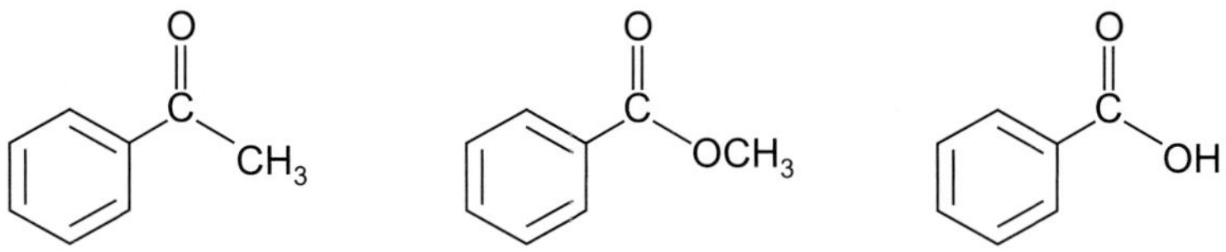

Moderately deactivating substituents

SCHEME 8.43

(E) Activating and Deactivating Substituents and their Synthetic Applications

(i) The Ortho, Para Ratio

One may expect more of the *ortho* product on electrophilic attack on a benzene ring with an *ortho, para*-directing substituent since two *ortho* positions are open for the incoming nucleophile. However *ortho* position is sterically hindered thus one would expect the preferential formation of para isomer if either the substituent on the ring or the incoming electrophile (E^+) is bulky.

(ii) Synthesis of Trisubstituted Benzenes

- When a disubstituted benzene has two different groups, the more powerful activating group generally determines the outcome of the reaction (Scheme 8.44). The acetamido group is far more stronger activating group compared to methyl, thus one observes predominant substitution at a position *ortho* to the acetamide group.

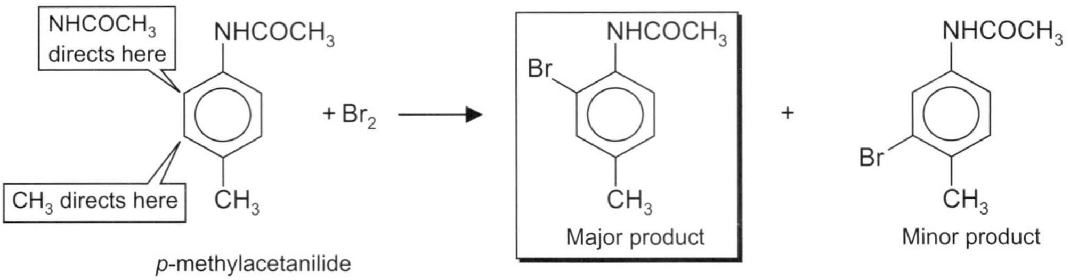

SCHEME 8.44

- Substitution does not occur between *meta* substituents since steric hindrance makes this position between the substituents less accessible (Scheme 8.45).

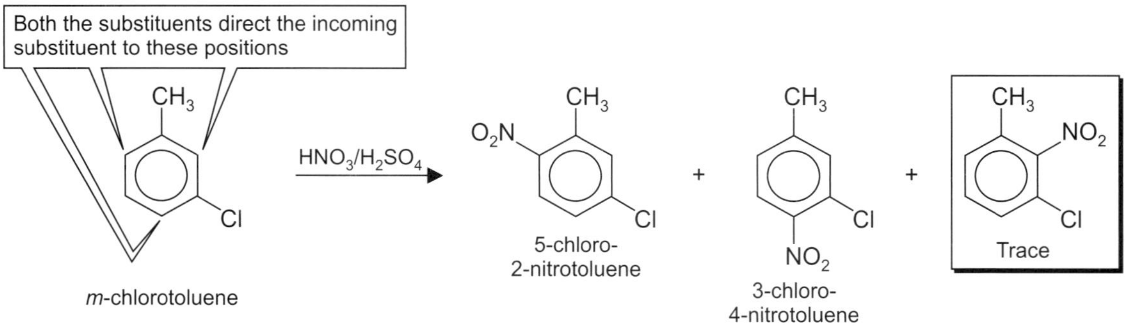

SCHEME 8.45

EXERCISE 8.7

Which product will dominate on electrophilic substitution of p-chlorotoluene?

ANSWER. *None, since both the substituents have almost similar activating effect (Scheme 8.46).*

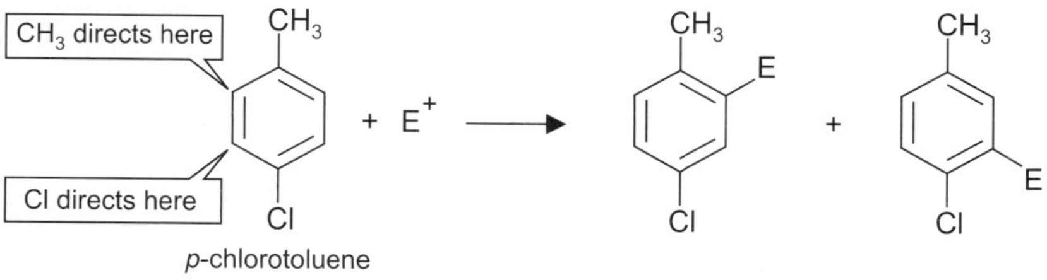

SCHEME 8.46

(iii) Synthesis of Meta-Substituted Toluenes e.g., Meta-Bromotoluene

One aims at the synthesis of *p*-acetamidotoluene. Since acetamido group is more strongly activating than the methyl group despite the steric hindrance to the entry of nucleophile at its *ortho* position bromination predominantly occurs *meta* to methyl. Removal of the acetamido group gives the desired product (Scheme 8.47).

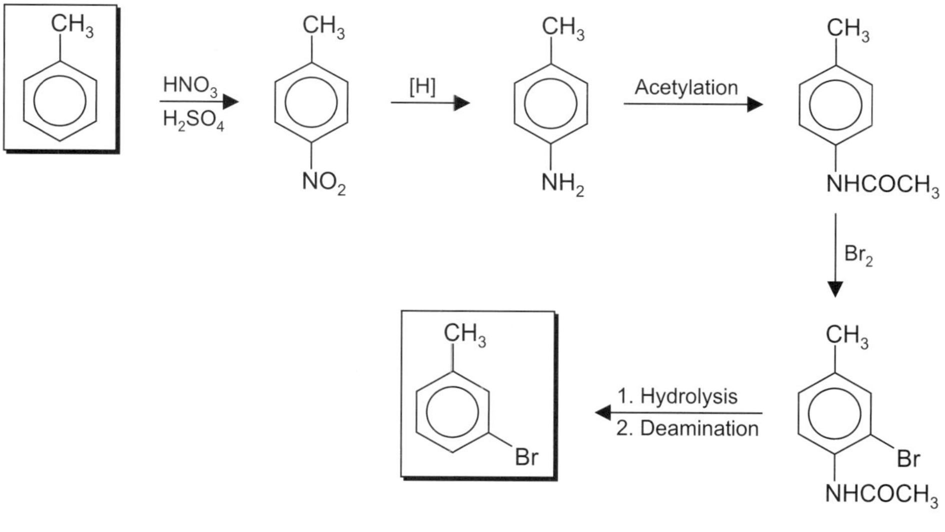

SCHEME 8.47

(iv) Synthesis of o-disubstituted Benzenes, the Role of Sulphonic Acid Group as a Blocking Group (Reversible Sulphonation as a Blocking Procedure)

Recall that nitration of aniline is carried out by first protecting (converting) the NH_2 group to $NHCOCH_3$ in order to moderate the high activating power of NH_2 group. One could thus prepare *p*-nitroaniline (see Scheme 8.36) in high yield, the acetamido group is purely a *para* director in many electrophilic substitution reactions). Sulphonic acid is used in many synthetic sequences as a blocking group which can be removed by heating in aqueous acid. Thus *o*-nitroaniline can be made (Scheme 8.48) in high yield.

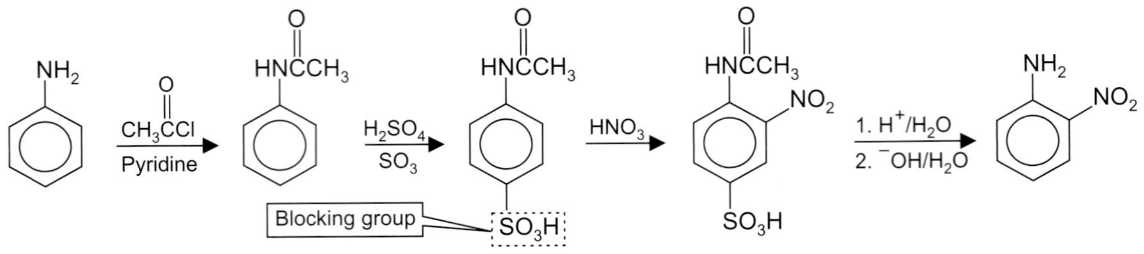

SCHEME 8.48

8.3 ELECTROPHILIC SUBSTITUTION IN NAPHTHALENE AND LARGER POLYCYCLIC AROMATIC HYDROCARBONS

Naphthalene is a fused bicyclic aromatic hydrocarbon (stabilization energy = 71 kcal/mole). Electrophilic substitution reactions on naphthalene and its derivatives occur at the α than β position due to the relative stability of the respective intermediates (Scheme 8.48*a*). The intermediate after the attack at α-position is stabilized by two resonance forms without disrupting the aromatic sextet in the other ring. Only one structure can be drawn for this intermediate after the attack at β-position without disrupting the aromatic sextet in the other ring.

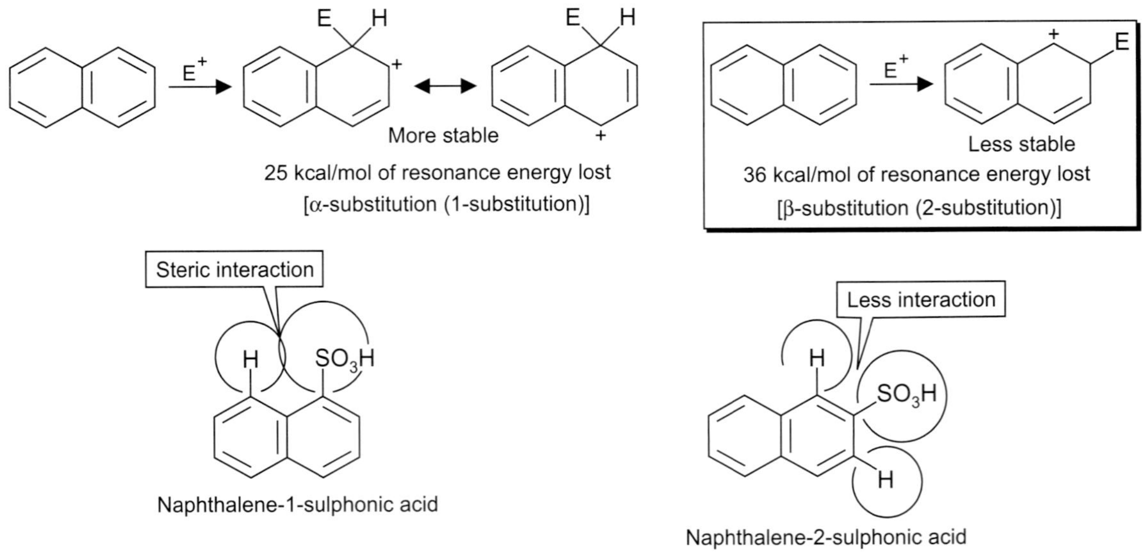

SCHEME 8.48a

Under certain conditions substitution at the β-position predominates, this is so when the reaction is thermodynamically controlled, as in sulphonation at high temperature. In the α-derivatives, there is steric repulsion between the substituent and the perihydrogen atom as shown (Scheme 8.48*a*) and it is thermodynamically the less stable. Another situation is when the reaction is kinetically controlled but the reagent is particularly bulky. The reaction then occurs at the α-position and is markedly hindered sterically by the peri-hydrogen. Naphthalene and many other polycyclic aromatic hydrocarbons are more reactive than benzene in electrophilic aromatic substitution. The activation energy for the formation of a complex in these cases is low than for benzene since more of the initial resonance stabilization is retained in the intermediated that have a fused benzene ring.

In naphthalene an activating group usually directs the incoming electrophile to the same ring and a deactivating group directs it away to the other ring preferentially at C-5 and C-8 (Scheme 8.48). Thus 1-naphthol (I, Scheme 8.48*b*) undergoes electrophilic substitution at C-2 and C-4 while 1-nitronaphthalene directs the incoming nucleophile *e.g.*, nitronium ion to C-5 and C-8 positions.

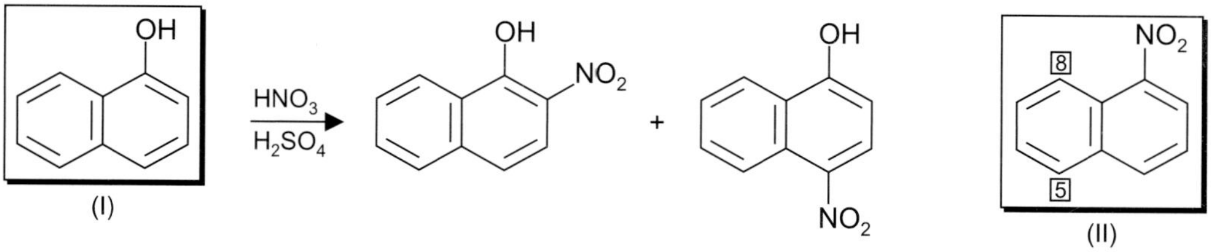

SCHEME 8.48b

The reactivies of larger polycyclic aromatic hydrocarbons are similar to those of naphthalene. Thus *e.g.,* the site of preferred electrophilic attack on phenanthrene is C-9 or C-10 since the major resonance contributor to the cation formed after electrophilic attack has two intact delocalized benzene rings (Scheme 8.48*c*).

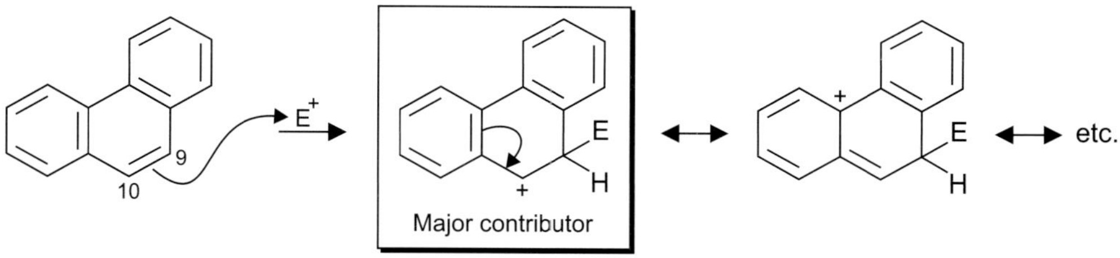

SCHEME 8.48c

8.4 ATTACK OF THE ELECTROPHILE AT A CARBON ALREADY BEARING A SUBSTITUENT (*Ipso* POSITION)—*Ipso* SUBSTITUTION

One has seen that during electrophilic aromatic substitution an incoming electrophile has only three choices in attacking a monosubstituted benzene: at the *o, m* or *p* positions. The attack of the electrophile at the carbon bearing the substituent is not competitve. Although this is generally true, however, in some cases, the electrophile may add to a position which is already substituted (*ipso* reaction). The carbocation formed after such *ipso* reaction can react in one of three ways:

(A) Formation of a Substitution Product by the Loss of a Substituent

Ipso substitution is observed in the protodealkylation of an alkylbenzene, a reaction that reverses the Friedel-Crafts alkylation. Tertiary alkyl groups are most easily removed (Scheme 8.48*d*).

The mechanism of this process probably begins with protonation by traces of HCl, followed by the loss of the 1, 1-dimethylethyl (*tert*-butyl) cation and its subsequent decomposition to regenerate the proton and 2-methylpropene (Scheme 8.48*d*). Thus *t*-butyl group is used to protect the most reactive position in a compound to effect reaction elsewhere (see, Scheme 8.17). For a related reaction (see Scheme 6.55).

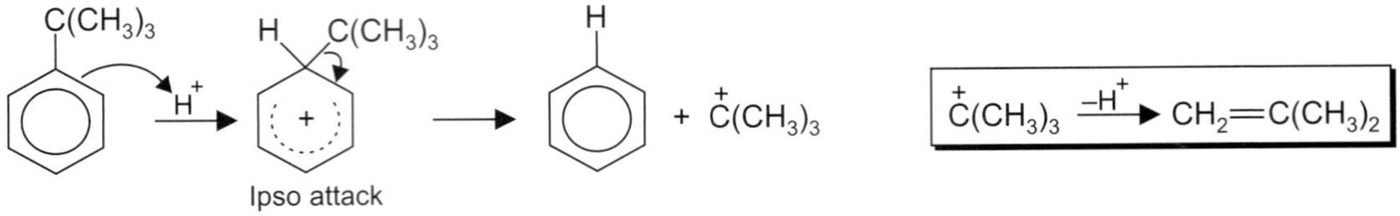

Electrophilic *Ipso* substitution

SCHEME 8.48d

(B) Formation of a Diene

When the substituent cannot depart this way (*e.g.*, a CH_3 group) a nucleophile in the system can attack the carbocation to give a diene (see Scheme 8.7). The carbocation formed after the *ipso* attack of the electrophile on a phenol can yield a dienone (Scheme 8.48*e*).

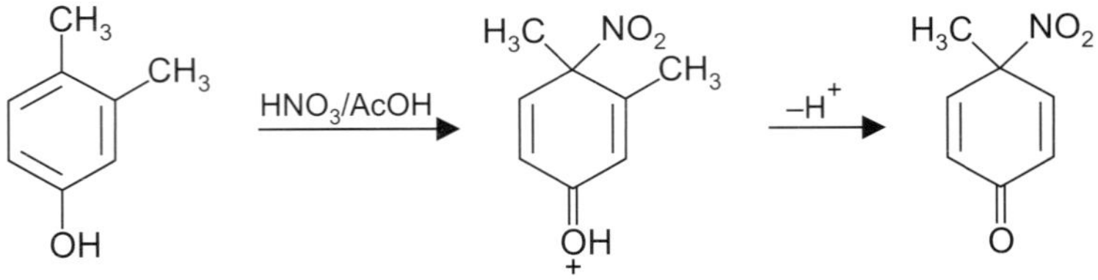

SCHEME 8.48e

(C) Rearrangement after *Ipso* Attack

The rearrangement of alkylbenzenes to their isomers has its base in *Ipso* attack. An alkyl group migrates from one carbon on the ring to another one, a situation so typical of an intermediate carbocations. Thus *o*-xylene rearranges to *m*-xylene (Scheme 8.48*f*). When the substituent already present on the ring is not a good cationic leaving group, the *ipso* addition product formed initially by the attack of an electrophile may undergo a rearrangement to provide a better leaving group. The end result of such a process may be the formation of the normal substitution product (Scheme 8.49), *e.g.*, during the nitration (with nitronium acetate) of *p*-xylene.

Silyl group has a strong tendency to direct the incoming electrophile to the position it occupies *i.e.*, "*ipso* attack".

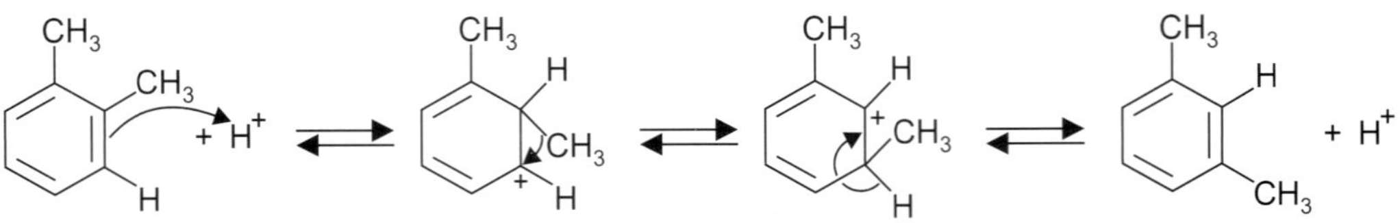

SCHEME 8.48f

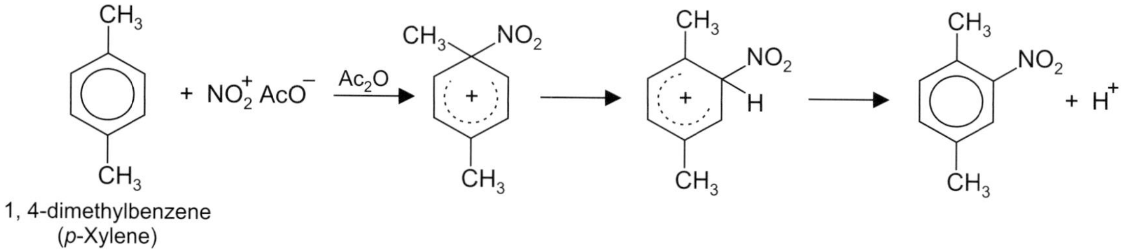

SCHEME 8.49

8.5 AROMATIC REARRANGEMENTS

(A) Fries Rearrangement

Phenolic esters can be rearranged by heating with Friedel-Crafts catalysts in a synthetically useful reaction known as the Fries rearrangement. The rearrangement amounts to intramolecular Friedel-Crafts acylation.The exact mechanism is still not known, but may follow the path shown (Scheme 8.50). The complex between the ester and the Lewis acid eliminates an acylium ion which gets substituted at the *ortho* and *para* positions, as in Friedel-Crafts acylation. As in Friedel-Crafts acylation in this case as well an initial complex is formed between the substrate and the catalyst, consequently one needs a substrate/catalyst molar ratio of atleast 1 : 1.

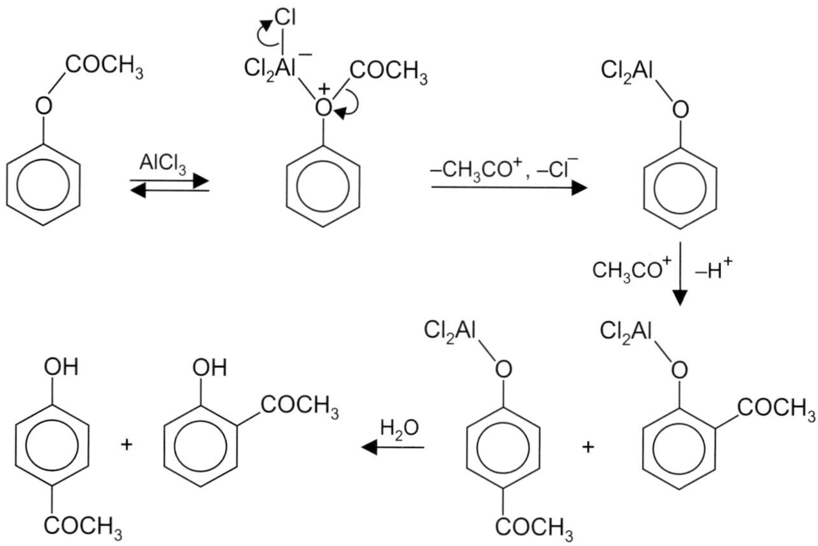

Fries rearrangement

SCHEME 8.50

(B) Photo-Fries Rearrangement (See Scheme 10.48)

(C) Reimer-Tiemann Reaction (Formylation)

This is a reaction of a phenol with chloroform in basic solution to give an aromatic aldehyde (Scheme 8.51). Although the yields are poor, the reaction is mechanistically interesting. The

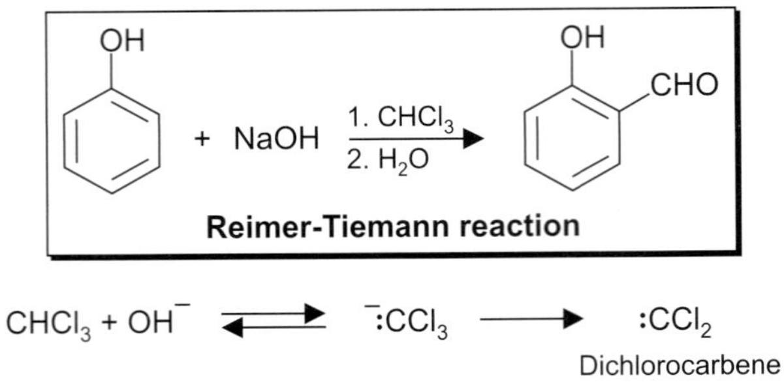

SCHEME 8.51

reaction occurs primarily in an *ortho-* position unless both are blocked. The electrophile in this reaction is dichlorocarbene produced by the reaction of chloroform with alkali. The reaction of phenolate ion with dichlorocarbene affords a dichloromethyl derivative which undergoes a rapid hydrolysis (Scheme 8.52). That the mechanism presented (Scheme 8.52) is essentially a

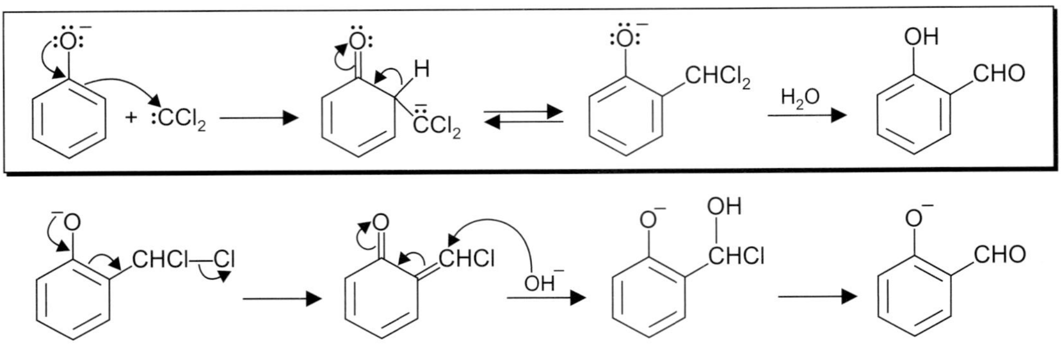

SCHEME 8.52

correct pathway is shown by the isolation of a byproduct from the Reimer-Tiemann reaction on *p*-cresol (Scheme 8.53). This can arise by attack of CCl_2 *para* to the OH group, as this position does not contain a hydrogen, a normal proton loss cannot occur and consequently the reaction ends when the CCl_2 moiety acquires a proton. *p*-Cresol represent a phenol with blocked *p*-position.

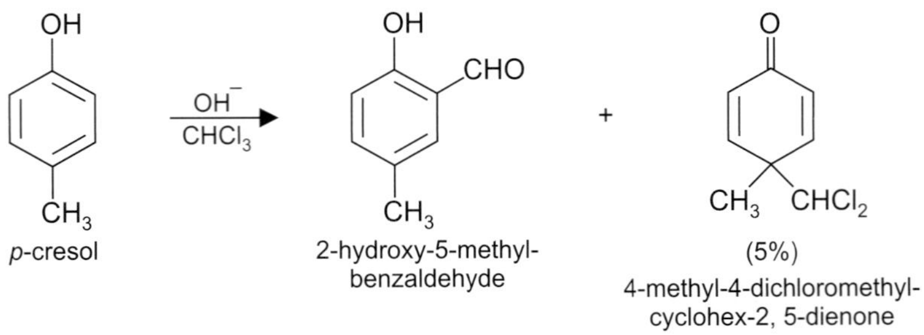

SCHEME 8.53

8.6 SOME NAME REACTIONS

(A) Gattermann-Koch Reaction (Formylation)

Formyl chloride HCOCl is unstable and decomposes to HCl and CO. Thus one cannot carry out Friedel-Crafts formylation of benzene. The formyl group (CHO) can be introduced into the benzene ring by treatment with CO under pressure in the presence of HCl and a lewis acid catalyst. It has been recently found that the electrophilic species is the formyl cation formed without the mediation of formyl chloride (Scheme 8.53*a*).

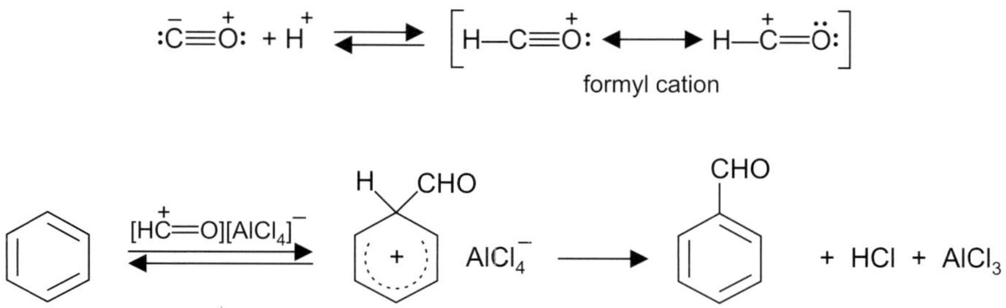

SCHEME 8.53a

Formylation fails with aromatic compounds of lower nuclear reactivity than the halobenzenes, so nitrobenzene is used as solvent. It is also unsuccessful with amines, phenols, and phenol ethers because of the formation of complexes with the Lewis acid.

(B) Gattermann Aldehyde Synthesis (Formylation)

This is an alternative to Gattermann-Koch reaction in which carbon monoxide is replaced by HCN. The ionic intermediate $\overset{+}{C}H{=}NH$, analogous to formyl cation is thought to be the electrophile (Scheme 8.53*b*). The reaction gives poor yields with benzene and halobenzenes and gives reasonable yields with aryl ethers and phenols.

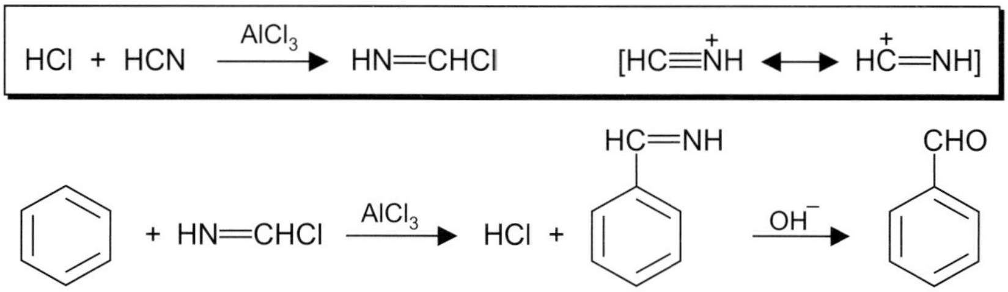

SCHEME 8.53b

(C) Vilsmeier Reaction (Formylation)

Activated aromatic compounds *e.g.*, *N*, *N*-dimethylaniline can be formylated by using a mixture of dimethyl formamide $HCON(CH_3)_2$ and phosphorus oxychloride $POCl_3$. The electrophilic species is chloroiminium ion $(CH_3)_2\overset{+}{N}{=}CHCl$ (Scheme 8.53*c*). The hydrolysis of imine formed as an intermediate completes the synthesis.

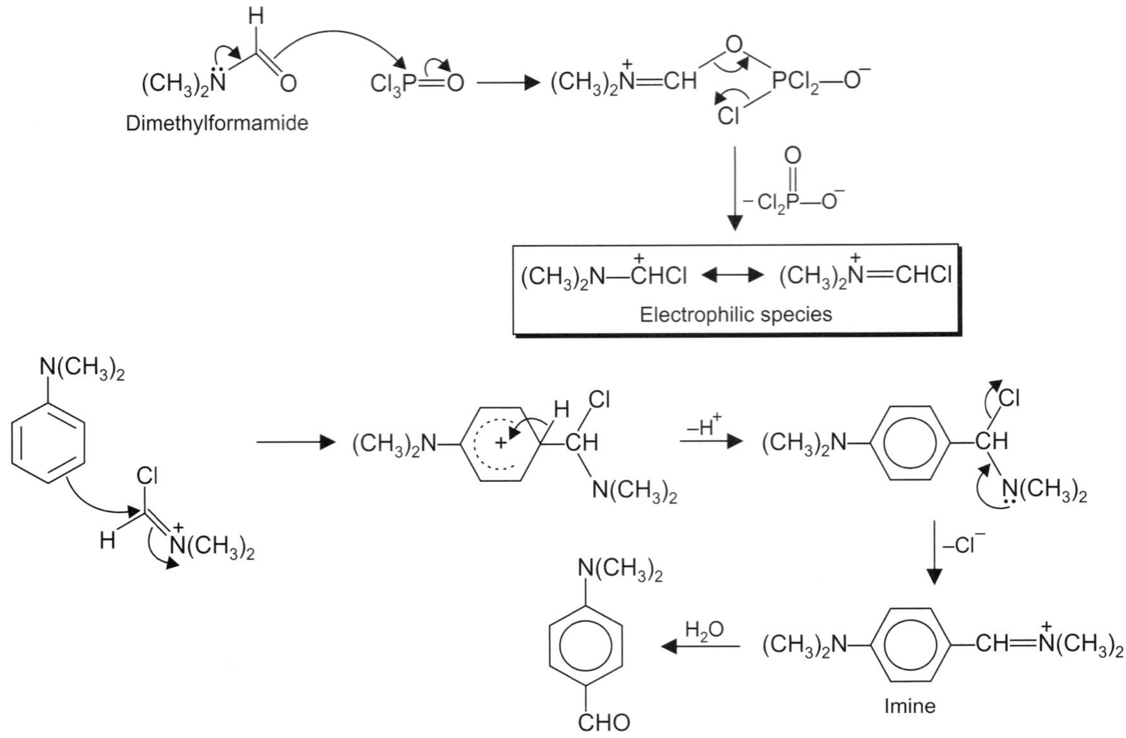

Mechanism of Vilsmeier reaction

SCHEME 8.53c

8.7 ELECTROPHILIC SUBSTITUTION ON HETEROAROMATIC COMPOUNDS

Several system like furan, pyrrole, thiophene and other heterocyclic compounds having the oxygen, nitrogen or sulphur atom in a structure in which it contributes two π-electrons belong to the π excessive group. This is shown by their resonance structure (see Scheme 2.40), and this donation of lone electron pair on heteroatom to the diene unit makes the carbon atoms in these systems electron rich and therefore, more susceptible to electrophilic aromatic substitution than those in benzene. The reactivity order is pyrrole > furan > thiophene (Scheme 8.54) and this reflects the order N > O > S for electron donating capacity.

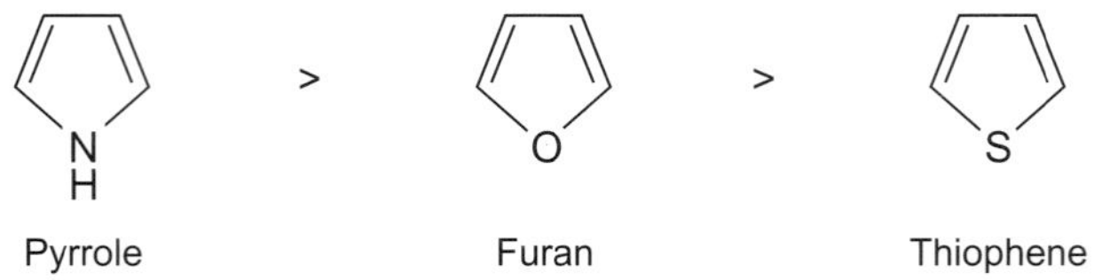

Electrophilic aromatic substitution

SCHEME 8.54

The order N > O is expected on the basis of electronegativity, while the order O > S points to the better overlap of the oxygen 2*p* orbital when compared to the sulphur 3*p* orbital, with the carbon 2*p* orbitals of the ring. During electrophilic substitution in these simple five membered heterocyclic rings (Scheme 8.54) position selectivity is usually 2 > 3. One accounts for this regioselectivity on the basis that the controlling step is the attachment of the electrophilic regent (E^+) to the aromatic ring in order to give the most stable intermediate carbocation. This approach explains the position selectivity (2 > 3) in these compounds and is explained by taking the example of pyrrole (Scheme 8.55).

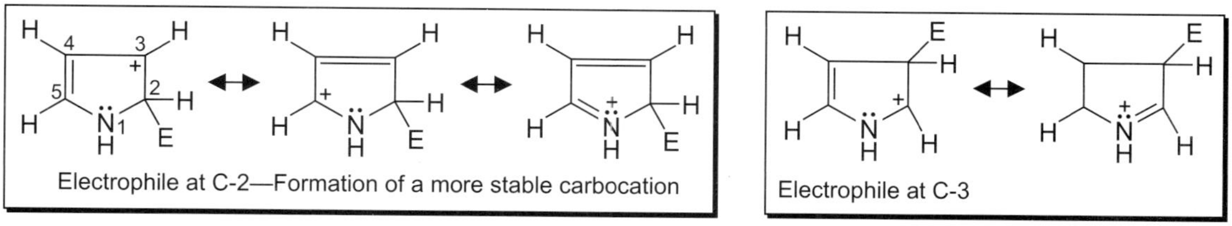

SCHEME 8.55

Both modes (attack at either C-2 or at C-3) benefit from the presence of the resonance contributing heteroatom, but attack at C-2 gives an intermediate with an additional resonance structure, *i.e.*, now the positive charge is delocalized over three atoms. Thus indicating this position to be the preferred center of substitution. As already explained, the carbon atoms in these five membered heterocycles are electron rich and therefore, undergo nitration, halogenation, sulphonation and Friedel-Craft acylation and are more reactive than benzene. These heterocycles *e.g.*, resemble the most reactive benzene derivatives like amines and phenols in undergoing reactions like nitrosation, Riemer-Tiemann reaction and coupling with diazonium salts. The reaction (I, Scheme 8.56) is an example of the Friedel-Crafts acylation. Furan and pyrrole are polymerized by acidic nitrating systems, and 2-nitro-derivatives are obtained by nitration by using acetyl nitrate CH_3COONO_2 formed by mixing acetic anhydride with nitric acid.

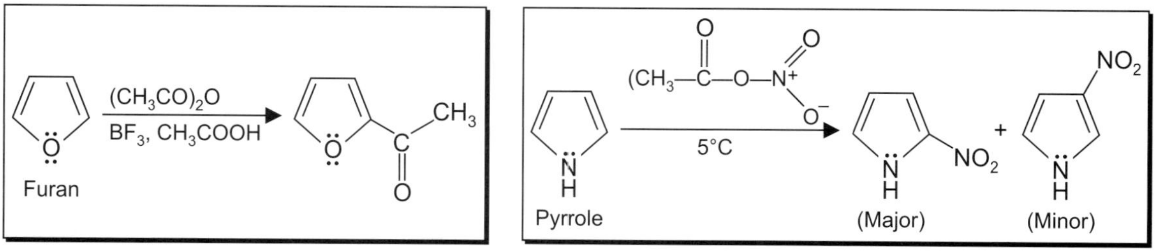

SCHEME 8.56

EXERCISE 8.8

Predict the poducts of Reimer-Tiemann reaction from pyrrole.

ANSWER. *Pyrrole is close to phenol in its reactivity with electrophiles. The reaction proceeds through pyrrolate anion and introduces CHO group at C-2 to yield pyrrole 2-aldehyde (Scheme 8.56a) 3-chloropyridine is also formed.*

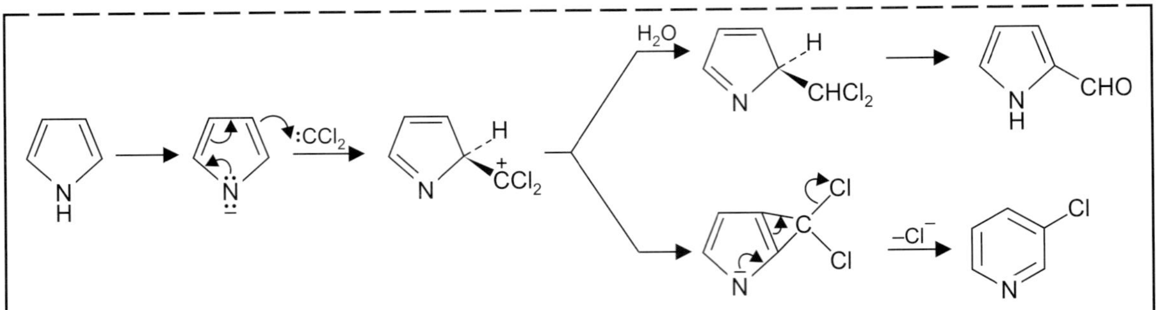

SCHEME 8.56a

A revealing trend in reactivities of the three heterocyclic compounds is found during their reaction with maleic anhydride. Pyrrole is significantly reactive toward electrophiles to take part as a nucleophile *e.g.*, in a Michael addition (see Scheme 6.32) (Scheme 8.57). Furan is less reactive toward electrophiles compared to pyrrole, which instead undergoes the Diels-Alder reaction (Scheme 8.57). Here it differs from benzene since less aromatic stabilization energy is lost on 1, 4-addition. Thiophene is not only less reactive to electrophiles but is less reactive as a conjugated diene as well, thus it does not react in a Diels-Alder reaction under normal conditions.

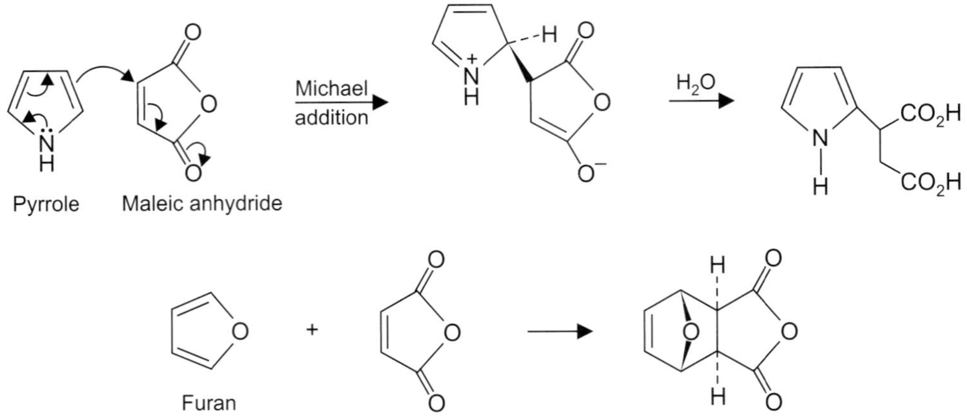

SCHEME 8.57

Compounds like pyridine incorporate the –N=CH– unit and are π-deficient and are deactivated to electrophilic attack. Pyridine has an sp^2 -hybridized nitrogen atom. In contrast to pyrrole, there is only one electron in the *p*-orbital that completes the aromatic π-electron arrangement of the aromatic ring, as is the case with phenyl anion, the lone electron pair is located in one of sp^2-hybrid atomic orbitals in the molecular plane (see Scheme 2.39*b*). Thus in pyridine the heteroatom does not donate excess electron density to the rest of the molecule. Moreover, in pyridine the –N=CH– unit is basic because the electron pair on the nitrogen does not from a part to the aromatic π system, consequently, nitrogen gets protonated or complexed with a Lewis acid under many of the conditions which are typical of electrophilic substitution reactions. This is another factor for the low reactivity of pyridine derivatives towards electrophilic substitution.

For pyridine, the reactivity towards electrophilic substitution is C-3 > C-4, C-2 and it is found that substitution (which needs vigorous conditions, like that in a highly deactivated benzene derivative) occurs mainly at the 3- or β- position. The ring nitrogen acts as a very strong destabilizing "internal" electron withdrawing substituent in the intermediates formed by attack at 2- and 4-positions (Scheme 8.58, attack only at C-2 is considered here, attack at C-4 position is similar to that at C-2 position just as *ortho* attack resembles *para* attack in the benzene series). The nitrogen also deactivates the C-3 position, but less than the 2- and 4-positions. When one considers the intermediates formed by electrophilic attack at C-2 or C-4 and compares these with formed by attack at C-3 (β-position) it is found that in the former case on structure (shown in box) is especially unstable because in it the electronegative nitrogen atom has only a sextet of electrons, thus substitution takes place predominantly at the C-3 position.

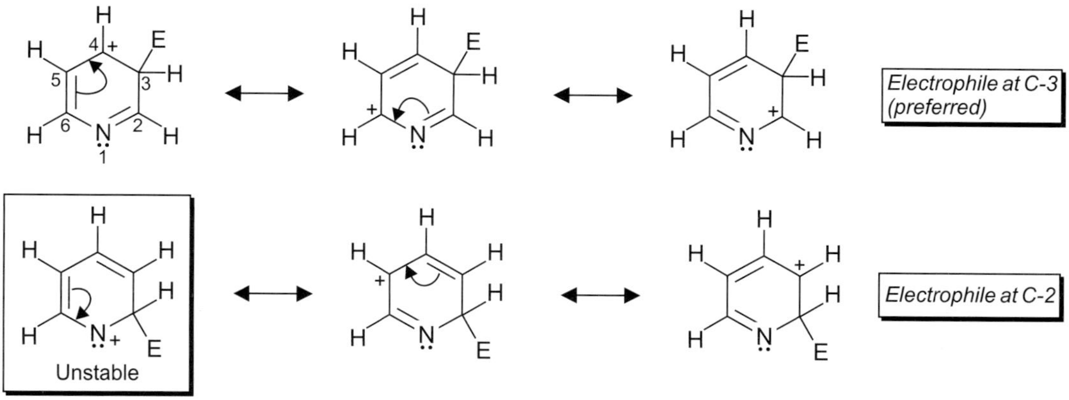

SCHEME 8.58

8.8 DIAZONIUM COUPLING

Diazonium ions are only weakly electrophilic and react particularly with only those aromatic reactants which are powerfully activated towards electrophiles *e.g.*, amines, phenols and heterocyclic systems such as pyrrole. The mechanism of electrophilic aromatic substitution using an arene diazonium ion is similar to electrophilic aromatic substitution with an electrophile (Scheme 8.59).

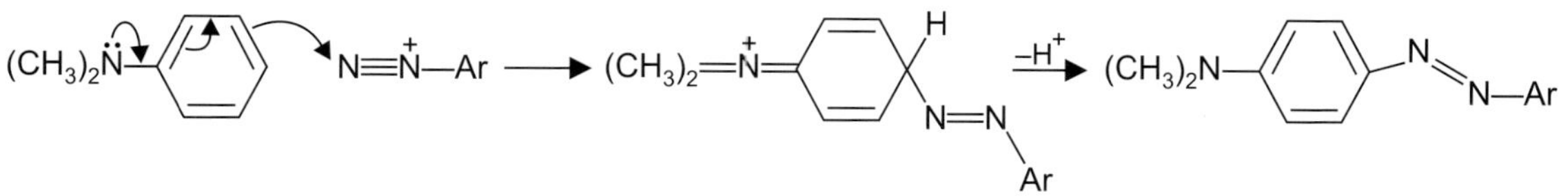

SCHEME 8.59

Azo benzenes are coloured compounds and are used commercially as dyes. Diazonium compounds are synthetically useful and may react to form an aryl cation by the loss of nitrogen in an S_N1 reaction, or the loss of nitrogen may generate a radical *via* one electron reduction (Scheme 8.60).

SCHEME 8.60

The utility of these is several synthetic reactions e.g., Gomberg reaction, Sandmeyer reaction etc. (see Chapter 16 on the free radicals)

PROBLEMS

8.1. Two procedures are available for the synthesis of 2-phenylethanol. Point out the route one would select and why?

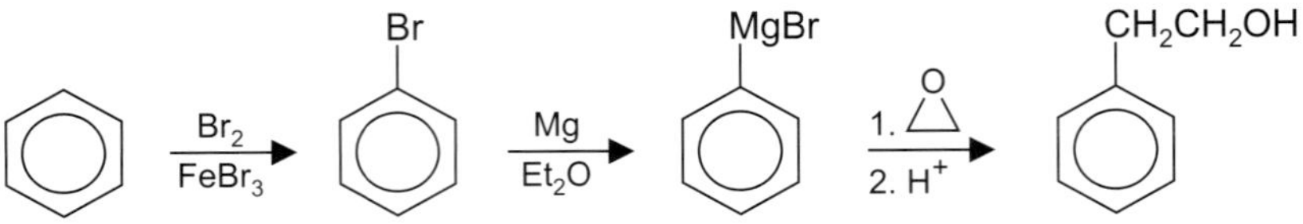

(I)

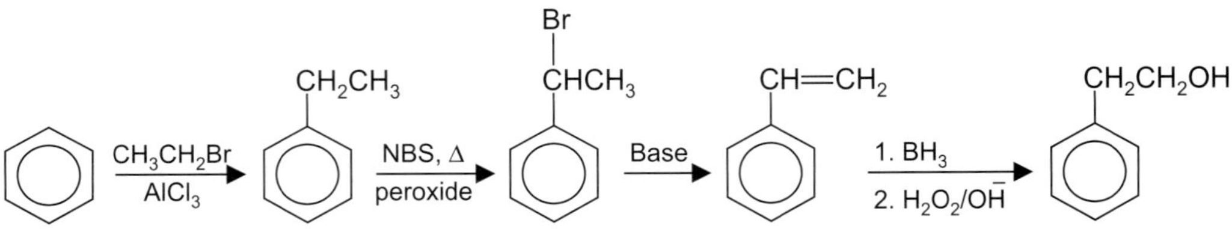

(II)

8.2. Direct nitration of *t*-butylbenzene to introduce a NO_2 group at the *ortho* position gives poor yields of the *o*-disubstituted product and the NO_2 group is introduced mainly at the *p*-position. How one can solve this problem?

8.3. Explain the result of nitration in the following alkylbenzenes:

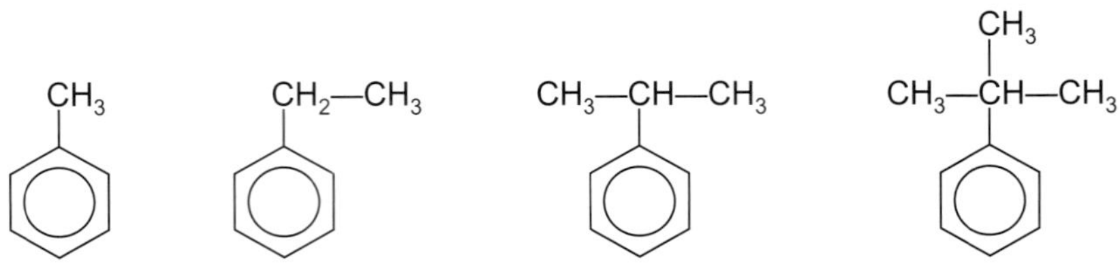

Toluene	Ethylbenzene	Isopropylbenzene	*tert*-Butylbenzene
1.57	0.93	0.48	0.22

The *ortho*/*para* ratios are for nitration

8.4. Which of the following products formed by the nitration of *p*-cymene with nitronium acetate can be assigned due to *Ipso* substitution?

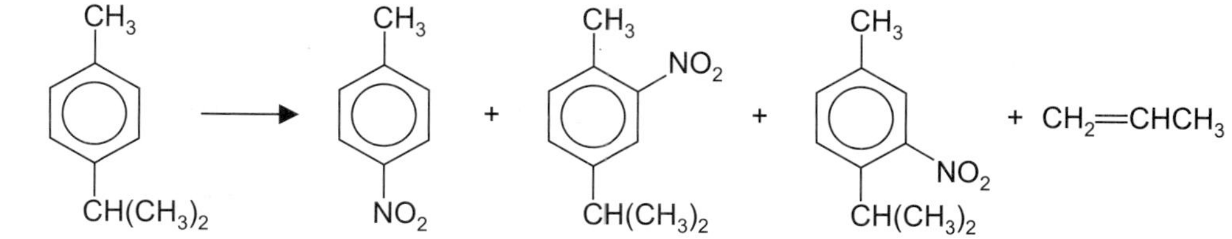

1-Isopropyl-4-methylbenzene (*p*-Cymene) (I) (II) (III) (IV)

8.5. The 2-3 benzo derivative of pyrrole *i.e.*, indole undergoes electrophilic substitution mainly at the 3-position. Explain.

8.6. Toluene on Friedel-Crafts acylation yields mainly the *para* acyl derivative. How can you obtain *ortho* derivative in high yield ?

8.7. The following phenol gives a dienone on nitration (HNO_3–AcOH). Explain.

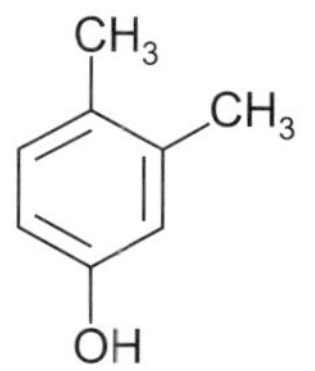

8.8. Nitration of pyridine mainly occurs at C_3 (see Scheme 8.58). How one can synthesize 4-nitroderivative?

8.9. Considering Scheme 8.56*a*, depict a synthesis of quinoline from indole.

8.10. *N*-alkyl-derivatives of aniline and *O*-alkyl derivatives of phenol have comparable reactivity with aniline and phenol, however, their acyl derivatives are much less reactive in electrophilic substitution. Explain.

8.11. For the synthesis of *m*-nitroacetophenone from benzene show the order in which the substituents are to be placed in the benzene ring.

ANSWERS TO THE PROBLEMS

8.1. One selects route (I) since the number of synthetic steps are less. Procedure (II) requires excess amount of the reactant (benzene) to prevent polyalkylation, free radical bromination (NBS, peroxide, Δ, see Scheme16.8) is likely to yield undesired side products.

8.2. By first blocking the *p*-position by carrying out sulphonation. Since both the electrophile (SO_3) and the substituent (*t*-butyl group) are sterically bulky, sulphonation occurs almost exclusively at *para* position. Now nitration can only occur in the *ortho* position to the *t*-butyl group and then the blocking group is removed.

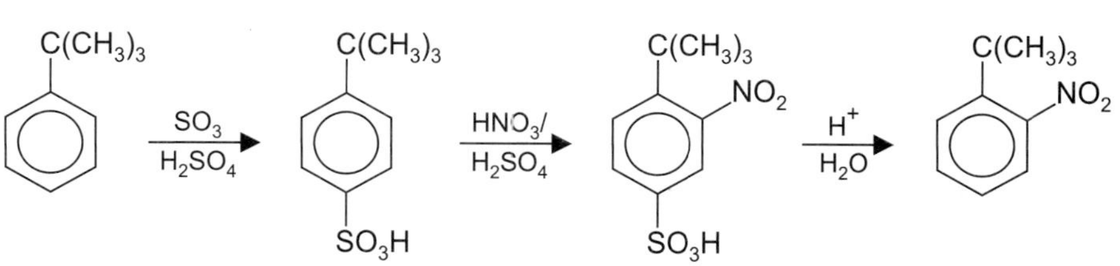

8.3. Among several factors, reaction temperature and steric hindrance, play a role in dictating the *ortho/para ratio*. The size of the substituent already present on the ring effects product distribution. Larger substituents "shields" to some extent the *ortho* position from attack by an electrophile, and when this occurs, more *para* substitution takes place.

8.4. The compounds (I and IV).

8.5. Indole is activated and reacts in the hetero ring. The stabilizing influence of the hetero atom on the transition state is more effective when the appropriate resonance structure is benzenoid compared to when the benzenoid system is disrupted.

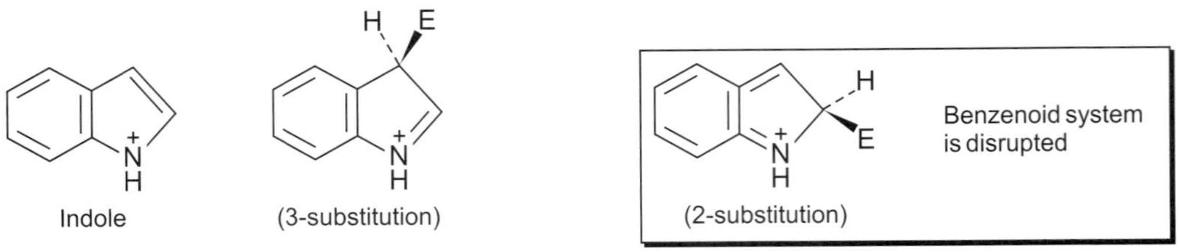

8.6. The *para* position in toluene is first protected by *t*-butylation.The acylation in then done on the *t*-butyl derivative and the *t*-butyl group is subsequently removed (see Schemes 8.17 and 8.47).

8.7. The electrophile adds to the *ipso* position and the dienone is formed by the loss of the hydoxylic proton.

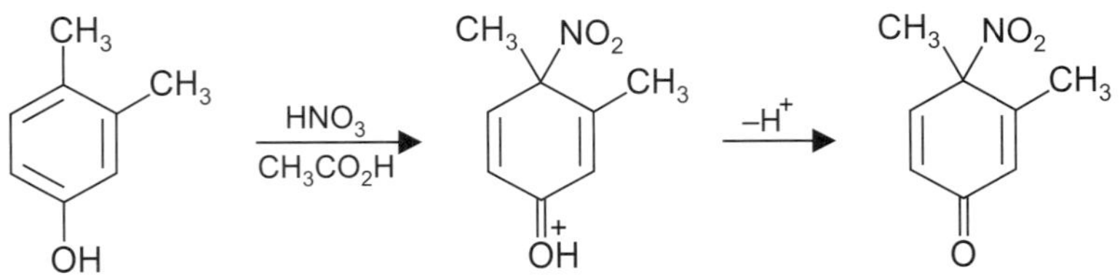

8.8. This can be achieved by nitrating pyridine *N*-oxide. The mesomeric electron-release from the oxide oxygen atom stabilizes the transition state for 4-substitution, as seen in the contribution of the resonance structure under box. (Here as expected 2-nitro-derivative is also formed but in small quantity.)

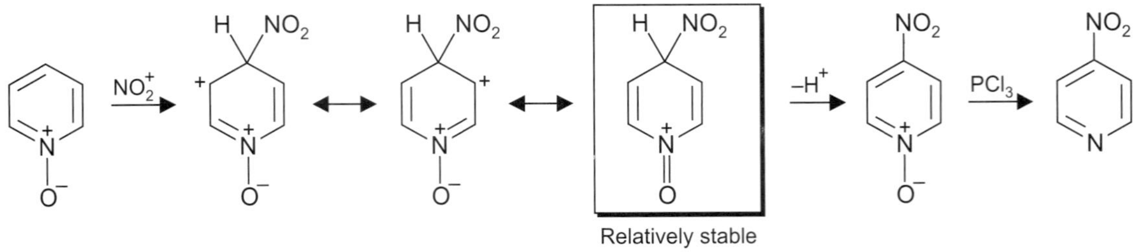

8.9. It can be achieved by adding methyl-lithium to indole in methylene dichloride solution, ring expansion occurs *via* the addition of chloromethylene.

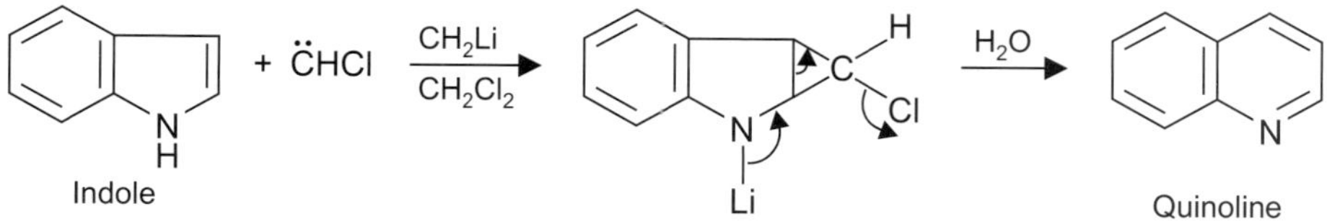

8.10. In the acyl derivative the unshared electron pair on nitrogen or oxygen is already delocalized within the substituent itself.

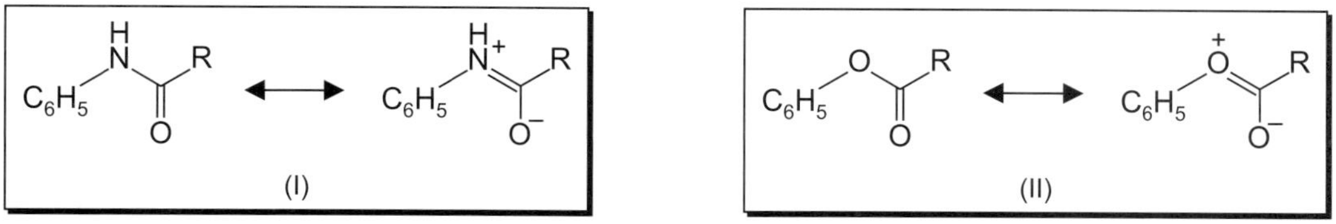

8.11. Both substituents of *meta*-nitroacetophenone are deactivating (meta directors). However, the Friedel-Crafts acylation reaction must be carried out first because the benzene ring of nitrobenzene would become too deactivated to undergo a Friedel-Crafts reaction.

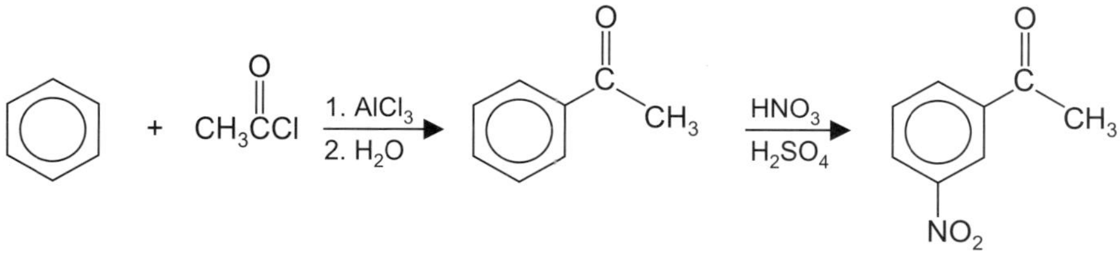

m-nitroacetophenone

CHAPTER 9

Aromatic Nucleophilic Substitution

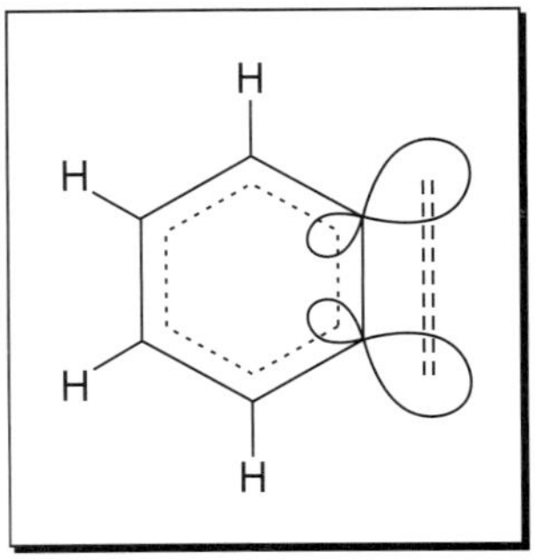

In electrophilic aromatic substitution, a strong electrophiles replaces a proton on the aromatic ring. In nucleophilic aromatic substitution, a strong nucleophile replaces a leaving group *e.g.*, a halide (Scheme 9.1). Several mechanisms for aromatic nucleophilic substitution are known and some of these are discussed. A nucleophile can be introduced into the ring provided it is sufficiently π-electron deficient due to the presence of an electron withdrawing group, *e.g.*, nitro (NO_2). The more effective leaving groups are the halogens. The nucleophile attacks the carbon atom to which the leaving group (a halogen atom) is attached (*ipso* attack). In the product the nucleophile is on the position of the original substituent. The aryl halides themselves undergo nucleophilic substitution with great difficulty unless strong electron-withdrawing groups are present in the *ortho* or *para* position to the halogen atom. It is only then that the mesomeric withdrawal of electrons is possible (Scheme 9.1) to make the cyclohexadienyl anion intermediate stable.

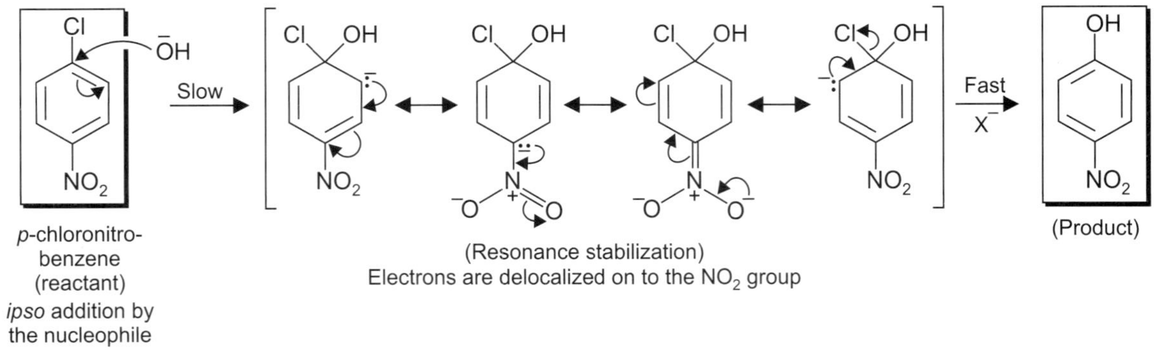

SCHEME 9.1

9.1 THE S_NAr MECHANISM—THE ADDITION—ELIMINATION MECHANISM—THE GENERAL NUCLEOPHILIC AROMATIC *Ipso* SUBSTITUTION

Reaction of *p*-chloronitrobenzene with nucleophiles *e.g.,* hydroxide ion, replaces the halogen with hydroxide ion (Scheme 9.1). This reaction is called nucleophilic aromatic substitution and the key to its success is the presence of at least one strongly electron withdrawing substituent, preferably more, on the benzene ring placed *ortho* or *para* to the leaving group. The presence of these substituents decreases the electron density in the benzene ring, making it more favourable for nucleophilic attack, and, these further stabilize the intermediate cyclohexadienyl anion by resonance. Unlike an S_N2 reaction of haloalkanes, substitution in these reactions takes place by a two-step mechanism, an addition-elimination sequence. This is the most important two step mechanism for nucleophilic aromatic substitution where the first step is usually (however, not always) the rate determining. A general representation of S_NAr mechanism is shown (Scheme 9.1).

In the first slow step, *ipso* addition by the nucleophile gives an anion with a highly delocalized charge. The important feature of this intermediate is the ability of the negative charge to be delocalized into the electron-withdrawing groups.

In the second step, the leaving group is eliminated to regenerate the aromatic ring.The reactivity of haloarenes in nucleophilic substitutions increases with the number of electron-withdrawing groups on the ring, particularly if they are in the *ortho* and *para* positions.

The following points may be noted:

- The reaction becomes facile when two or more nitro groups are present in the *ortho* and *para* positions as indicated by rate data (reaction with $\bar{O}CH_3/CH_3OH$, 50°C).

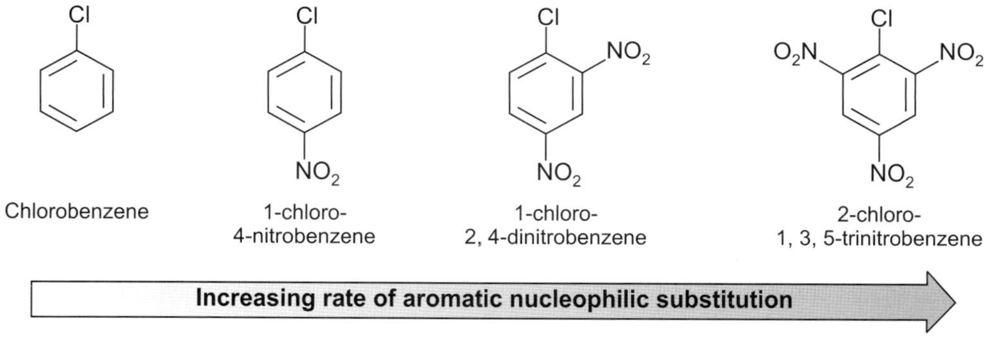

SCHEME 9.2

- Enough reactivity is generated in aryl halides when two nitro groups are present in favourable positions to delocalize the charge of the attacking nucleophile.

  Moreover, aryl halides which have *ortho-* or *para*-nitro groups become sufficiently reactive to display nucleophilic substitution even with neutral nucleophiles like NH_3. Thus 1-chloro-2, 4-dinitrobenzene (Scheme 9.2) on reaction with NH_3 gives 2, 4-dinitroaniline.

- The presence of electron-withdrawing groups is essential at the *ortho* and *para* positions to the site of nucleophilic attack, only then the charge of the nucleophile (electrons) can be delocalized (Scheme 9.3).

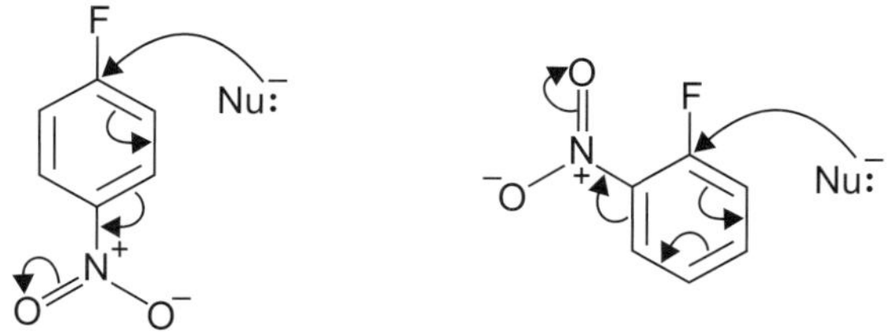

SCHEME 9.3

- 2, 4, 6-Trinitroanisole does not bear a halide leaving group thus its reaction (Scheme 9.4) with sodium ethoxide gives a Meisenheimer complex which corresponds to the product obtained by the nucleophilic addition stage of the S_NAr mechanism. This shows that nucleophilic attack on an aryl halide initially gives a resonance stabilized carbanion intermediate known as Meisenheimer complex by the initial *ipso* addition by the nucleophile'.

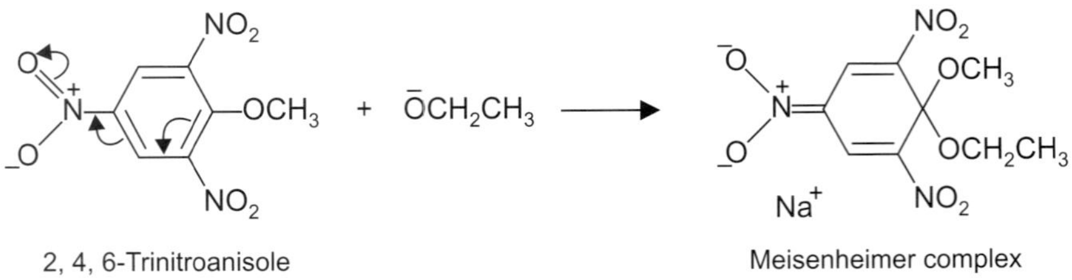

SCHEME 9.4

- In case this mechanism is similar to either the S_N1 or S_N2 mechanisms, the Ar—X bond should break in the rate-determining step. In the S_NAr mechanism however, the bond is not broken until after the rate-determining step. As expected a change in leaving group should not effect the rate of the reaction and this has been found to be the case. However, the rates cannot be expected to be identical, since the nature of leaving group X effects the rate at which nucleophile (Nu^-) attacks. When the electronegativity of the leaving group is more, there would be a decrease in the electron density at the site of attack leading to a faster attack by the nucleophile. The fact that from among halogens, fluoro is the best leaving group in several aromatic nucleophilic substitutions shows this mechanism to be different from the S_N1 and S_N2 reactions, where fluoro is the poorest of the leaving group from among the halogens.

The S_NAr mechanism differs from the S_N1 or S_N2 mechanisms in that fluoride ion is a good leaving group in the S_NAr substitution but is not a good leaving group in the S_N1 and S_N2 aliphatic substitution reactions. Aryl fluorides are actually better substrates than the corresponding aryl chlorides for the S_NAr reaction since the highly electronegative fluorine atom makes its attached carbon atom more reactive towards a nucleophile. The presence of a nitro group is still required and it should be present in the ortho or para position to the leaving group. Moreover, like vinyl halides, aryl halides cannot adopt the geometry necessary for a backside displacement since the ring shields the backside of the carbon-halogen bond (Scheme 9.4a).

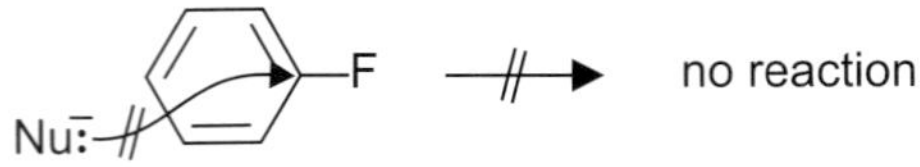

SCHEME 9.4a

EXERCISE 9.1

A fast reaction is observed with sodium methoxide with ortho and para isomer of fluoronitrobenzene as compared with m-fluoronitrobenzene. Explain.

ANSWER. *The mesomeric withdrawal of electrons is not possible from the anionic intermediate (Scheme 9.5). Direct conjugation of the negatively charged carbon with the nitro group is not possible in the cyclohexadienyl anion intermediate.*

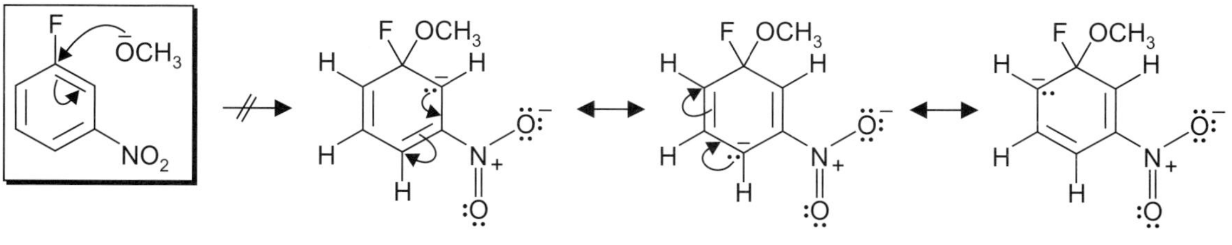

SCHEME 9.5

The most common type of reactants in nucleophilic aromatic substitutions are those which have *o*- or *p*- NO_2 substituents. However, highly fluorinated hydrocarbons fit the requirements for such a reaction and hexaflurobenzene undergoes substitution of one of its fluorines on reaction with a nucleophile (NaOR, Scheme 9.6).

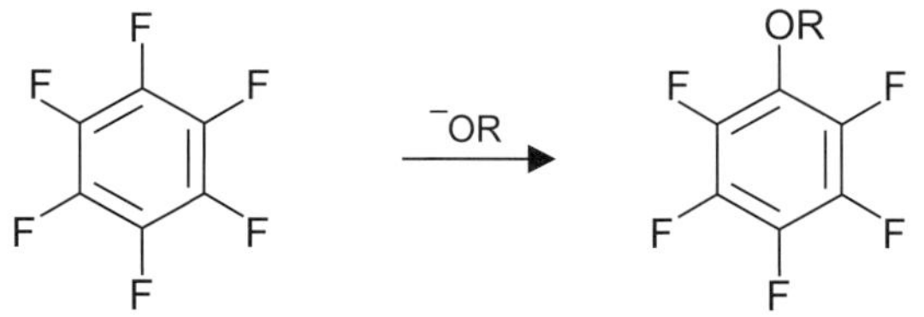

SCHEME 9.6

9.2 THE S_N1 MECHANISM IN NUCLEOPHILIC AROMATIC SUBSTITUTION—THE ARYL CATION MECHANISM—DIAZONIUM SALTS

When aryl diazonium salts are hydrolyzed in water these give the corresponding phenols. The net result of the reaction is substitution of the nucleophile (water) for the diazonium group (Scheme 9.7). There is sufficient evidence that this reaction proceeds through an aryl cation (I, Scheme 9.7). In the case of aryl halides and sulphonates a unimolecular mechanism has never been established with certainty. However, this mechanism is significant only in the case of diazonium salts, the driving force for formation of this reactive cation is departure of the good leaving group, nitrogen. The main evidences in support of aryl cation is the following:

- The rate of decomposition of a diazonium salt follows the first order kinetics and is independent of the concentration of nucleophile.
- On addition of halides salts in high concentration, the product formed is an aryl halide, however, the rate is independent of the concentration of the added salts.

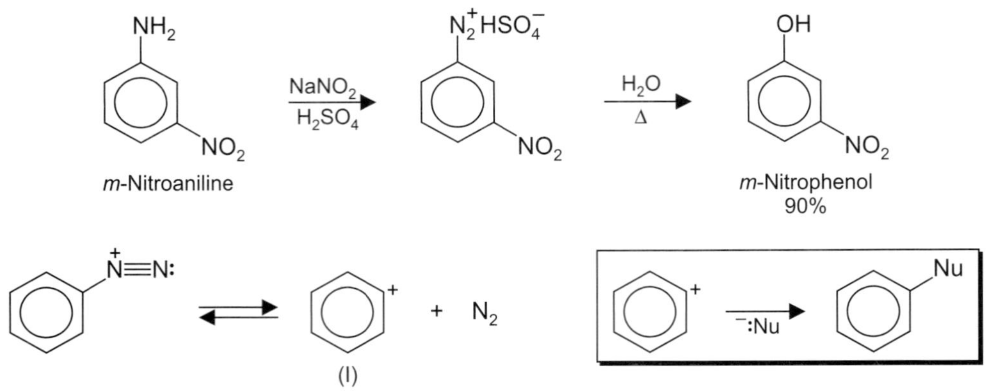

SCHEME 9.7

- The isomerization (Scheme 9.8) of isotopically labelled *p*-toluene-diazonium ^{15}N fluoroborate was observed during hydrolysis. This could arise if the nitrogen breaks aways from the ring and then returns.

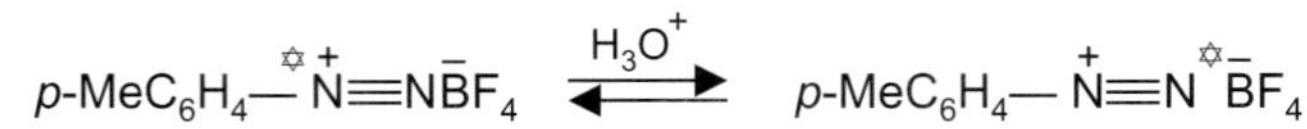

SCHEME 9.8

9.3 NUCLEOPHILIC AROMATIC SUBSTITUTION BY ELIMINATION–ADDITION–THE BENZYNE MECHANISM

Haloarenes do not undergo simple S_N2 or S_N1 reactions. However, at high temperature and pressure nucleophilic substitution can be achieved. Thus with hot sodium hydroxide followed by neutralizing work-up, chlorobenzene gives phenol. Similarly chlorobenzene reacts with sodium amide (a very strong base). One does not require high temperature in this case. It occurs in liquid ammonia at – 33°C (Scheme 9.9). Thus unlike the addition-elimination mechanism for nucleophilic aromatic substitution (Scheme. 9.1) which depends on strong electron-

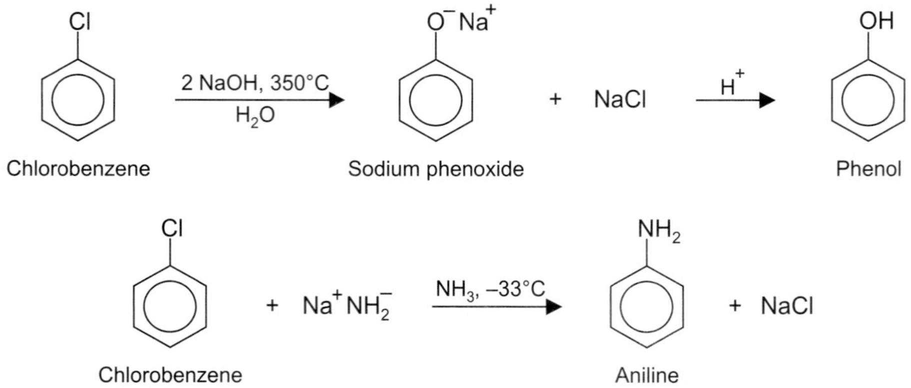

SCHEME 9.9

withdrawing substituents on the aromatic ring, under extreme conditions unactivated halobenzenes react with strong bases.

Direct substitution mechanisms are not operative on these compounds. Significantly the incoming nucleophile appears only at the *ipso* or at the *ortho* position relative to the leaving group. Thus, *p*-chlorotoluene reacts with sodium hydroxide to give a 50:50 mixture of *meta* and *para* products (Scheme 9.10). If at all the addition-elimination S_NAr mechanism was operative here, one would expect the formation of only the *para* product.

To show that the incoming nucleophile does not always end up at the position vacated by the leaving group (it appear either at *ipso* position or at the *ortho*-position relative to the

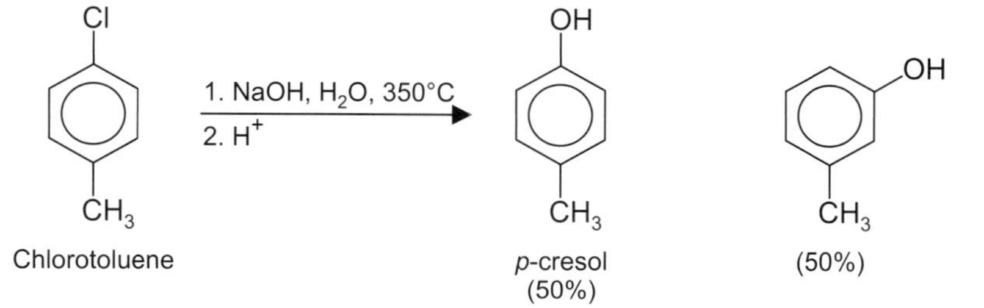

SCHEME 9.10

leaving group), the reaction of 1–^{14}C–chlorobenzene with sodium amide gave equal amounts of aniline labelled in the 1 position and in the 2-position (Scheme 9.11). This and other observations can be accounted for by an initial base induced elimination of HX from the benzene ring, a process reminiscent of the dehydrohalogenation of alkenyl halides to give alkynes. In the present case, step-by-step elimination through a phenyl anion intermediate gives a highly strained and reactive species called benzyne, or 1, 2-dehydrobenzene (Scheme 9.12).

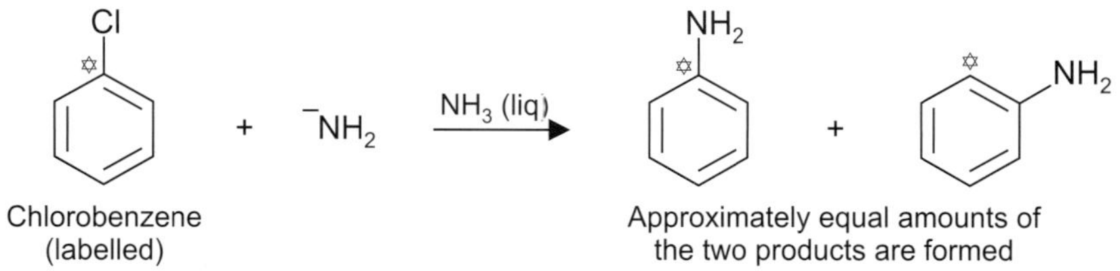

SCHEME 9.11

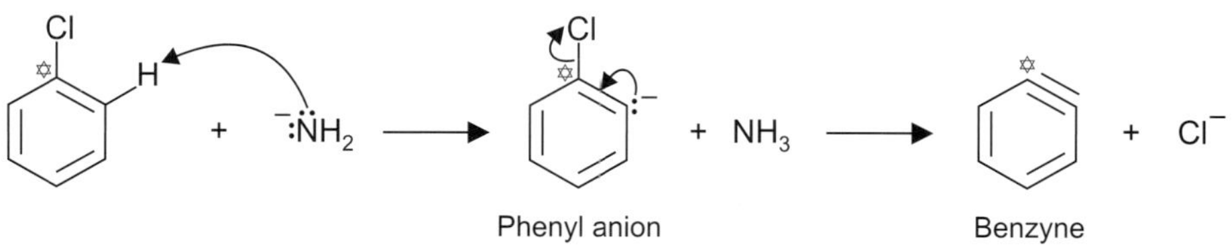

Formation of benzene intermediate

SCHEME 9.12

The benzyne intermediate is symmetrical and can be attacked by *e.g.*, NH_2^- at either of two positions. This formation of the intermediate benzyne, thus explains why half of the aniline produced from the radioactive chlorobenzene was labelled at the 2 position (Scheme 9.13).

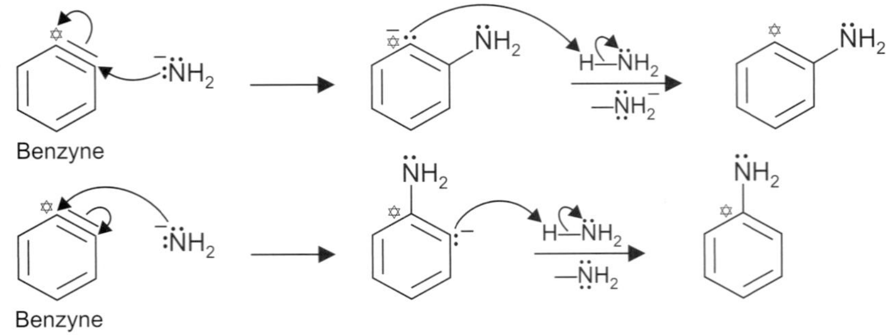

SCHEME 9.13

In some cases, the aromatic nucleophilic substitution occurs at an adjacent position and is called cine substitution. An important feature of benzyne mechanism is that, it can be attacked at two positions. The favoured position for nucleophilic attack is the one which leads to the more stable carbanion intermediate. Thus *e.g.*, in the case of–*I* groups, the more stable carbanion is the one in which the negative charge is closer to the electron-withdrawing substituent.The *ortho*-derivative (I, Scheme 9.14), on treatment with sodium amide gives *m*-(tri-fluoromethyl) aniline (III) as the exclusive product. This is explained on the basis of intermediate formation of the benzyne (II) which adds an amide ion in the way so as to form a more stable of the two carbanions (Scheme 9.14). Thus, both steric and electronic factors are operative. Based on electronic factors an anion (IV, Scheme 9.14) is more stable since CF_3 group is inductively electron-withdrawing. On steric grounds as well it is more favourable for the amide ion to attack away from CF_3 group.

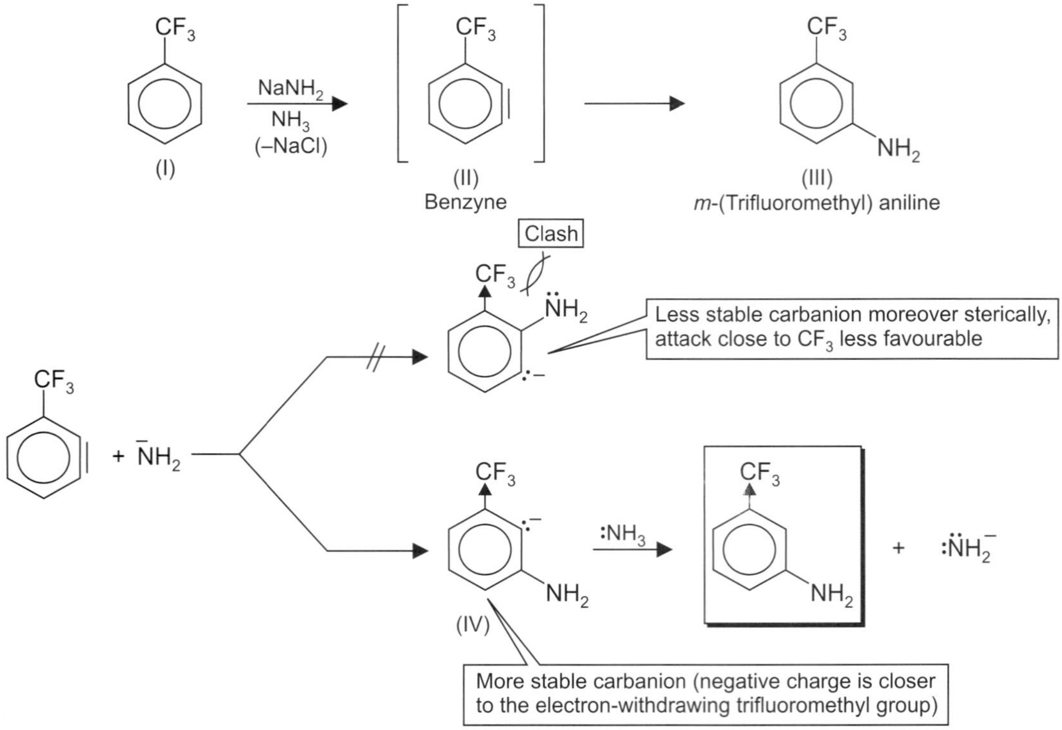

SCHEME 9.14

In some cases unactivated halides undergo nucleophilic substitution in a chain reaction involving anion radicals, where the initiation step is the electron transfer. This is $S_{RN}1$ process (see Scheme 16.30).

9.4 BENZYNE—A STRAINED CYCLOALKYNE

Benzyne is a highly reactive intermediate (an aryne intermediate). The formation of benzyne type of intermediate from a simple haloarene is favoured when the amide ion (NH_2^-) is used as a base. Due to its strong basic nature it successfully abstracts hydrogens from the aromatic ring as protons with the formation of ammonia. The halide ion X^- departs to give benzyne. One must have a hydrogen *ortho* to the halogen for benzyne formation and its formation is favoured with aryl halides which have electron-donating substituents. On the other hand, the presence of electron-withdrawing substituents on the ring causes the bimolecular nucleophilic aromatic substitution reaction to be favoured. The reason for the high reactivity of benzyne is the normal requirement for alkynes to adopt a linear rather than a bent structure. Due to the cyclic structure, benzyne cannot meet that requirement. The molecule has only a brief life and has never been isolated and can be easily trapped by any nucleophile present. Thus, *e.g.*, ammonia solvent adds to give the product benzenamine (aniline). As the two ends of the triple bond are equally reactive, nucleophilic addition may be at either carbon to explain the product mixtures observed from labelled chlorobenzene. The extra bond in benzene is a result of the overlap of sp^2 orbitals on adjacent carbons of the ring (Scheme 9.15). The aromatic framework of six π-electrons is arranged perpendicular to the two reactive and poorly overlapping additional hybrid orbitals which make up the distorted triple bond. There is a good deal of deformation in the molecular structure of benzyne from the regular hexagonal symmetry due to bond alteration imposed by the presence of triple bond (44 kcal/mole). Spectral data shows the contribution by a cumulated triene resonance structure (see Scheme 4.50).

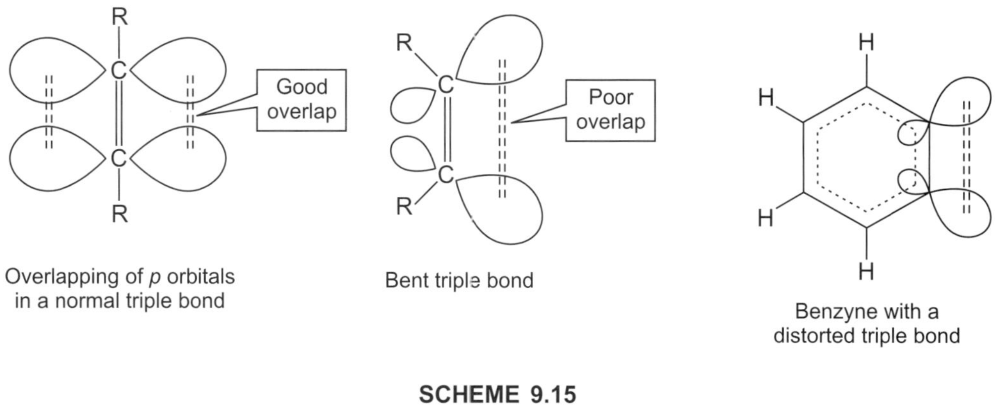

SCHEME 9.15

Benzyne has been observed spectroscopically under special conditions. One successful method is the irradiation of phthaloyl peroxide (see, Scheme 10.25) which gives species with IR and UV spectra typical of benzyne structure (formed by the elimination of two equivalents of CO_2). There are other methods to produce benzyne in addition to base catalysed elimination of hydrogen halide from a halobenzene. One of the convenient methods is the diazotization of *o*-aminobenzoic acids (Scheme 9.16). The benzyne formed in this method in the presence of other compounds *e.g.,* methanol reacts with these quickly.

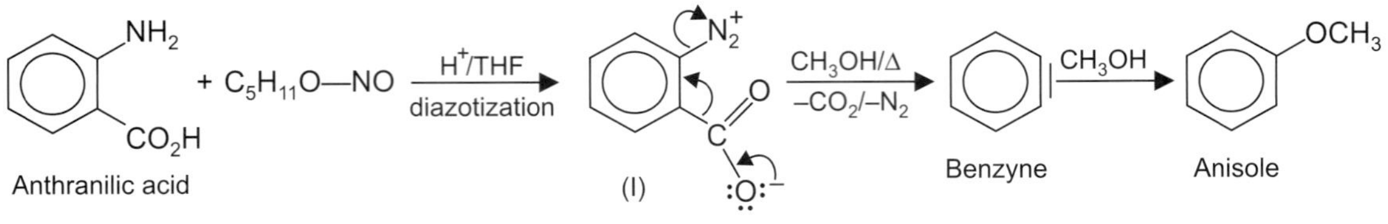

SCHEME 9.16

In the absence of any compound the zwitterion (I, Scheme 9.16) decomposes in an entropically favourable reaction to give CO_2, N_2 along with benzyne which has been detected mass spectrometrically. The peak at *m/z* 76 points to the presence of benzyne while the one at *m/z* 152 shows its dimerization to the dimer (Scheme 9.16*a*). Recall that the life time of a particle in a mass spectrometer is around 20 ns (nonasecond = 10^{-9} second) and during this period atleast benzyne can exist free in the gas phase.

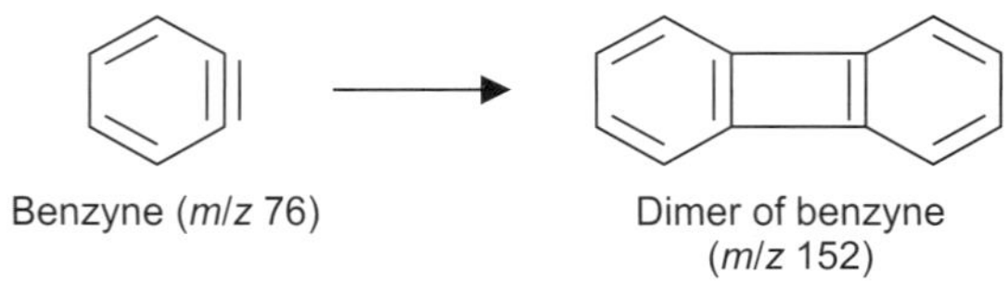

SCHEME 9.16a

One can prepare bis-Grignard reagents from aryl dihalides when the halogens are remote. However, when halogen atoms are present *ortho* to each other elimination occurs to give benzyne intermediates (Scheme 9.17). Benzyne is capable of dimerization to biphenylene when either a nucleophile or a reactive unsaturated compound is absent. One can trap benzyne by means of the Diels-Alder reaction, when benzyne is formed in the presence of a diene, the benzyne reacts as the dienophile. A diene often used for this purpose is anthracene, which provides the structurally interesting molecule triptycene (Scheme 9.18). Furans also react with benzyne to give Diels-Alder addition products (see Scheme 10.26).

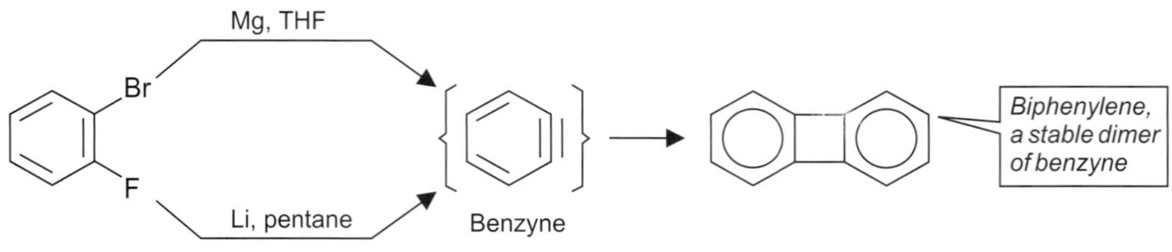

SCHEME 9.17

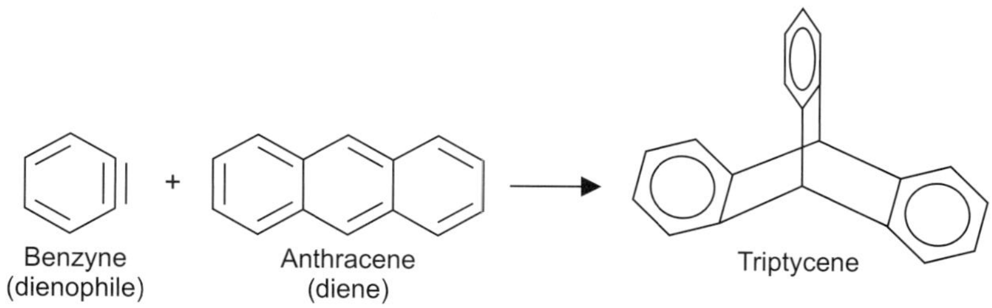

SCHEME 9.18

EXERCISE 9.2

Write the products from the reactions of (I and II, Scheme 9.18a)

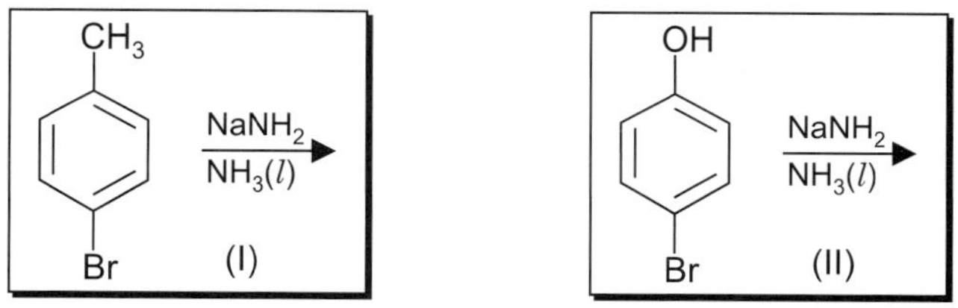

SCHEME 9.18a

ANSWER. *These are given (Scheme 9.18b).*

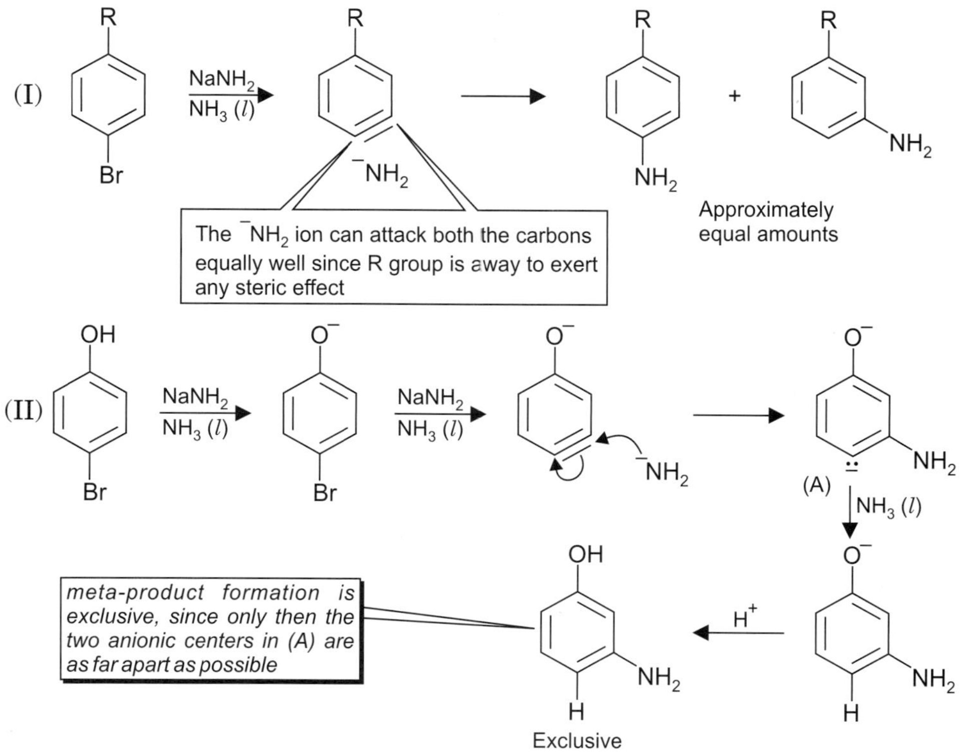

SCHEME 9.18b

9.5 NUCLEOPHILIC SUBSTITUTION OF PYRIDINE—THE CHICHIBABIN REACTION

Pyridine is less reactive than benzene towards electrophilic substitution (pyridine ring is electron poor) and this is due to the greater electronegativity of its ring nitrogen. Nitrogen in pyridine is less able to accommodate the electron deficiency in the transition state of the rate determining step in electrophilic aromatic substitution. Because the pyridine ring is relatively electron deficient, it undergoes nucleophilic substitution much more readily than benzene. A particularly useful and both unusual and remarkable example is the synthesis of aminopyridines by the reaction of a pyridine with an alkali metal amide (Chichibabin reaction). The reaction is initiated by attack by the nucleophile at C-2 or C-6. This is because the negative charge on the adduct formed by the addition of the nucleophile is stabilized by delocalization onto the electronegative nitrogen atom (Scheme 9.19). The second step is the loss of a hydride ion. Pyridine undergoes

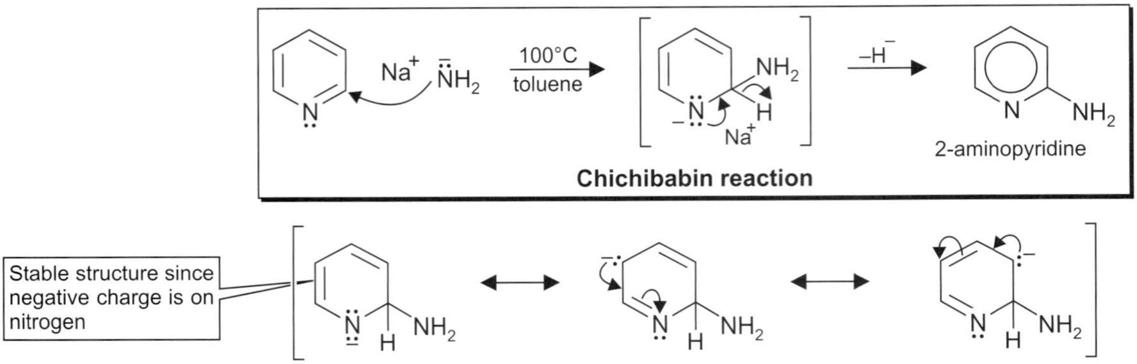

SCHEME 9.19

similar nucleophilic substitution reactions with *e.g.*, organolithium compounds and potassium hydroxide (Scheme 9.20).

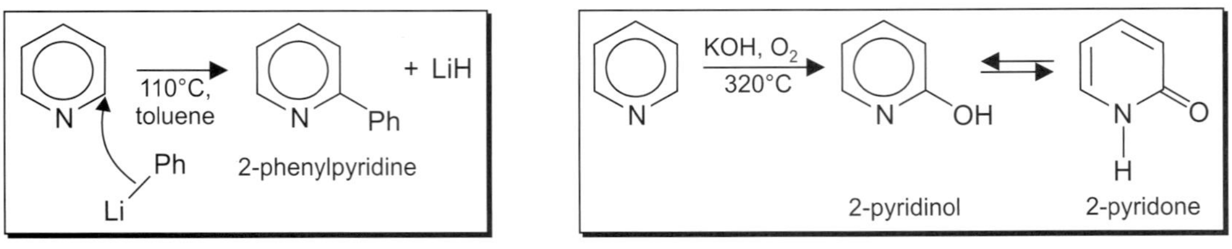

Chichibabin-like reactions

SCHEME 9.20

9.6 NUCLEOPHILIC SUBSTITUTION TO ARENECHROMIUM CARBONYL COMPLEXES

Ferrocene behaves as an electron-rich aromatic system and undergoes many electrophilic substitutions (Scheme 2.41). The most important π-complexes of aromatic compounds are the chromium complexes obtained by heating benzene or other aromatics with $Cr(CO)_6$

(Scheme 9.21). The $Cr(CO)_3$ unit in these compounds is strongly electron-withdrawing and activates the ring towards nucleophilic attack. Thus 1, 3-dithianyl anion reacts with the complex to form a carbon-carbon bond. The reaction of the nucleophile occurs on the aromatic ring away from the face with metal atom. The anion intermediate is stabilized by the transition metal atom which can accommodate extra electrons. Such an aryl-metal complex however,

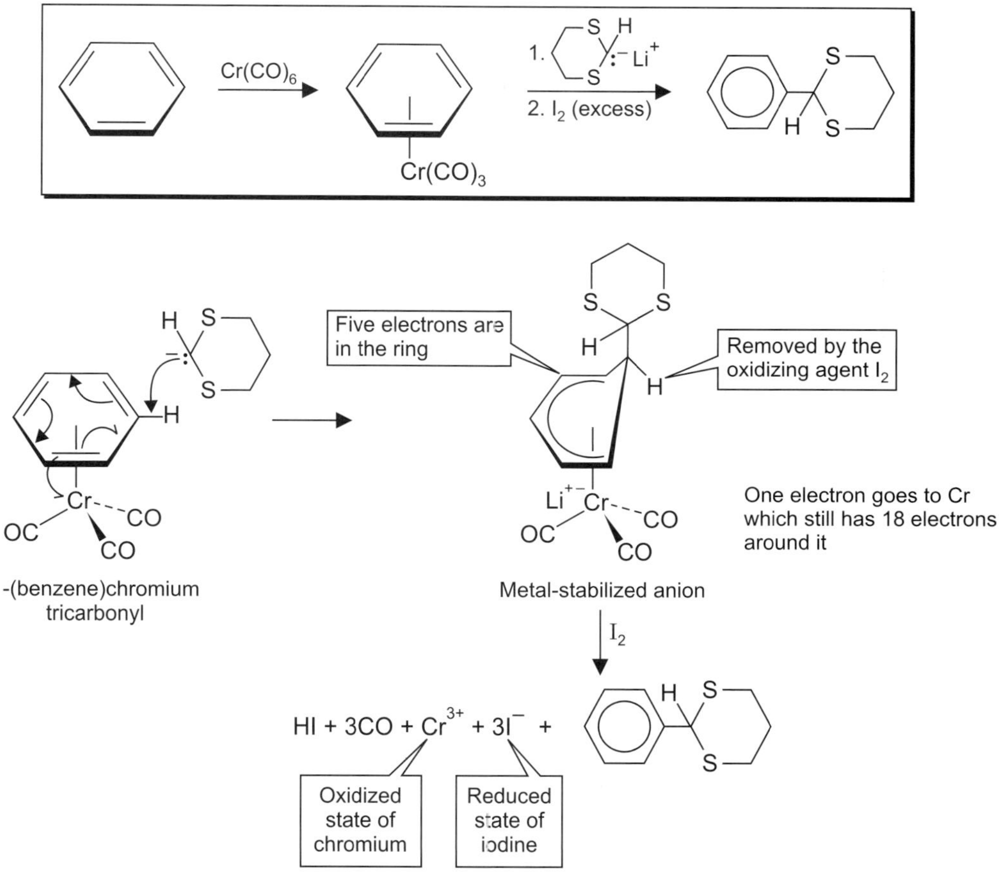

SCHEME 9.21

does not react with all carbon nucleophiles, it does not react with stabilized enolates *e.g.*, from diethyl malonate or with Grignard reagents or organocuparates. This aryl metal complex, however reacts well with ions stabilized by nitriles, and with anions of carbon acids with pK_a values greater than 20.

Complexes of aryl halides undergo nucleophilic substitution of the halogens just like halides with nitro groups (see Scheme 9.1). Thus chromium complex of fluorobenzene reacts via nucleophilic substitution by the enolate anion of diethyl malonate to yield diethyl phenylmalonate. This compound is an intermediate in the synthesis of barbiturates which cannot be prepared directly from diethyl malonate in any other way (Scheme 9.22).

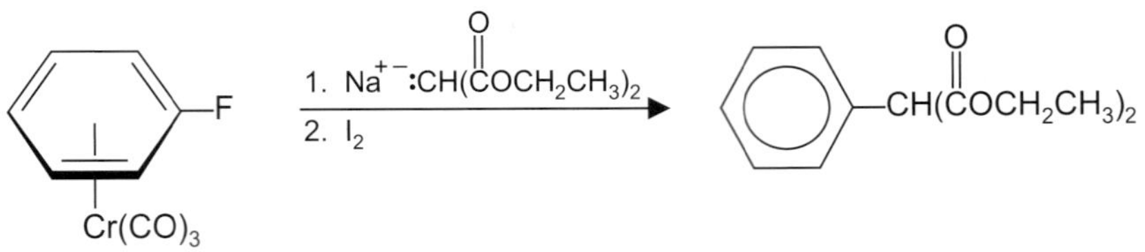

π-(fluorobenzene)-chromium tricarbonyl

Diethyl phenylmalonate 95%

SCHEME 9.22

Also see sec. 7.2

PROBLEMS

9.1. The reaction of *o*-bromoanisole gives only *meta*-aminoanisole on reaction with potassium amide in liquid ammonia. Explain this regioselectivity.

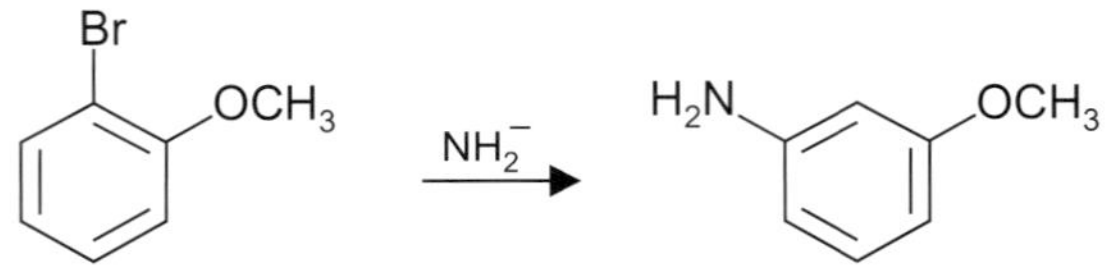

9.2. Nucleophilic aromatic substitution has been used for the identification of amino acids by Sanger (Nobel Prize 1958) using 2, 4-difluoronitrobenzene (2, 4-DNFB). Explain.

9.3. Show the nucleophilic attack at position 4 in pyridine?

9.4. A pyridinium ion *e.g.*, *N*-alkylpyridinium ion undergoes hydride addition with lithium aluminium hydride. Predict the nature of product. Explain if this type of reaction has biological significance.

ANSWERS TO THE PROBLEMS

9.1. The reaction proceeds through the intermediate formation of benzyne. The amide addition will occur only at the carbon that will lead to the more stable carbanion where the negative charge will be closer to the $-I$ effect of methoxy oxygen.

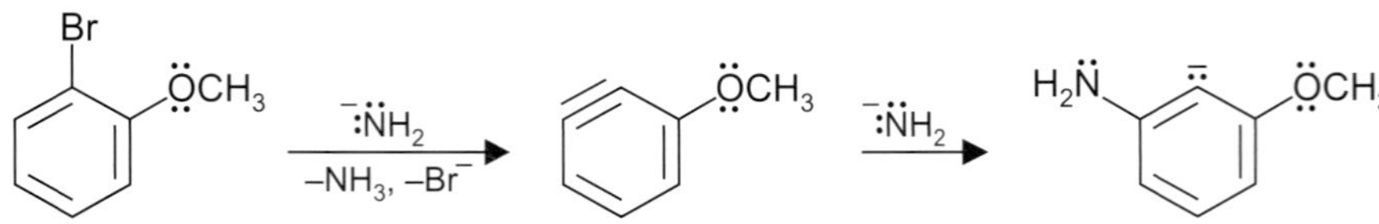

9.2. It is an addition elimination reaction. The amino group of the amino acid adds to the benzene ring to the position where fluorine is attached. Elimination of fluorine gives a derivatized amino acid which is highly coloured and identified by usual methods.

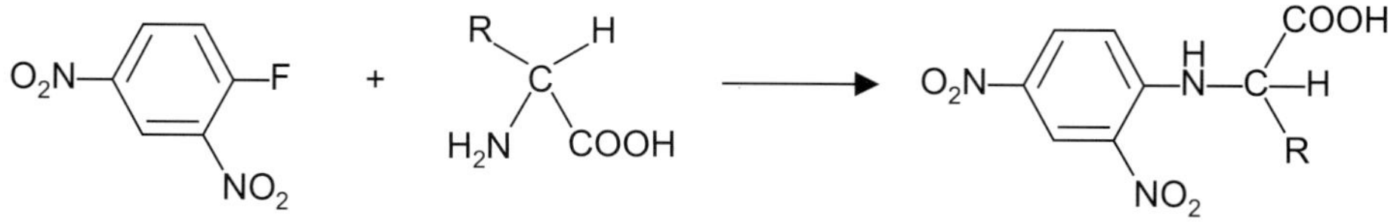

2, 4-DNFB An amino acid 2, 4-DNB derivative of amino acid

9.3.

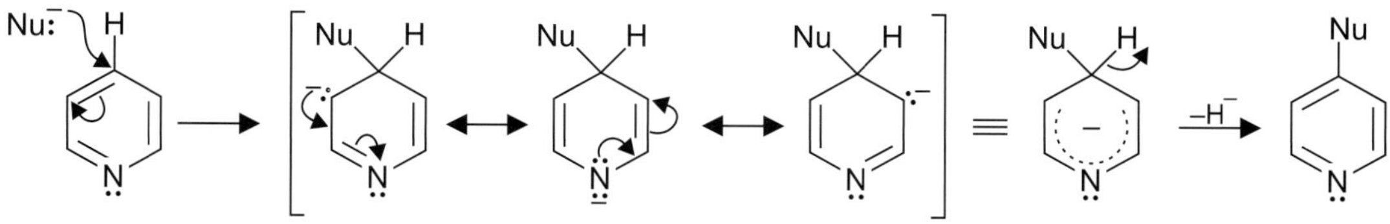

9.4. The nucleophilic attack by the hydride ion as expected occurs at position 2 and 4 and the product has a dihydropyridine skeleton. This reaction is the basis of the NADH—NAD+ oxidation-reduction process.

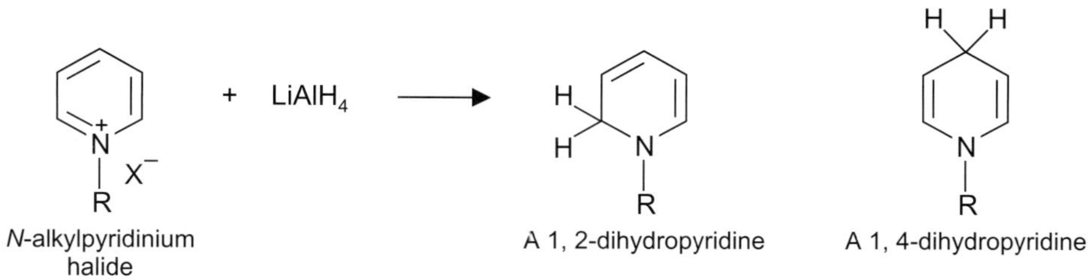

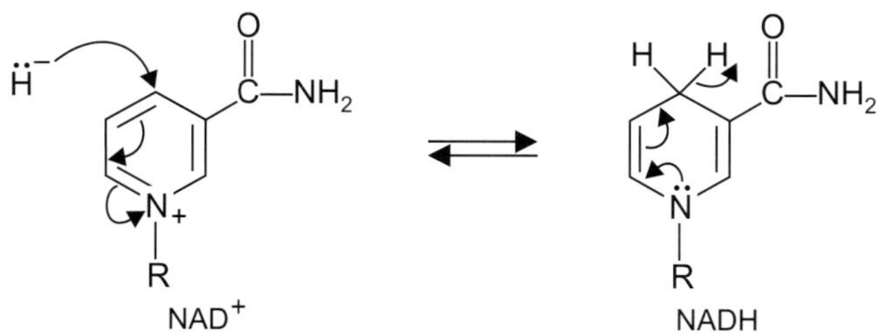

CHAPTER 10

Photochemistry

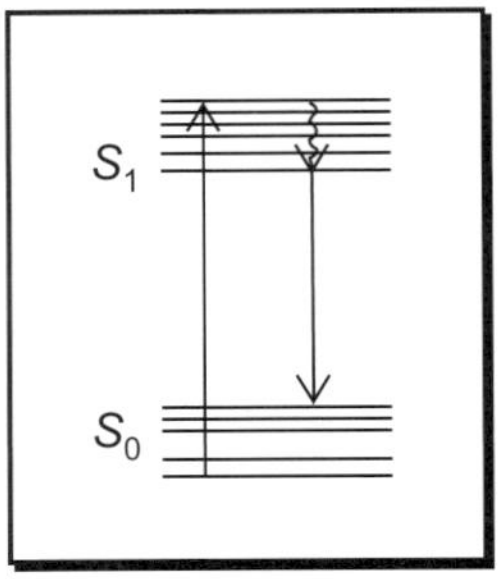

The reactions which take place when molecules absorb ultraviolet or visible radiation are called photochemical reactions.

10.1 ABSORPTION OF ELECTROMAGNETIC RADIATION—QUANTUM YIELD

In photochemistry light quanta are absorbed by individual molecules with the proper chromophore. The energy associated with ultraviolet and visible light is sufficient to excite electrons in molecules (photoexcitation). A molecule as a result is excited from its ground to electronically excited state. In several cases the energies imparted to molecules by photo-excitation are similar to covalent bond energies which can initiate chemical reactions. Covalent carbon-carbon bonds have energies around 100 kcal/mole. The absorption at 200 nm and 300 nm is equivalent to 143.0 and 95.0 kcal/mol respectively. Thus the typical upper limit of energy available for photochemical processes is near 143 kcal/mol (598 kJ/mol). This corresponds to a lower wavelength limit of about 200 nm for effective transmission of light through air. Below 200 nm strong absorption by oxygen in the air needs the use of vacuum ultraviolet apparatus if higher energies (shorter wavelengths) are to be employed. With soft glass as the reaction vessel, most of the UV-radiation below 360 nm is absorbed by the glass, consequently the practical energy maximum is near 80 kcal/mol (335 kJ/mol). Pyrex glass is a better light transmitter, while quartz is the most transparent of the common materials employed in the study of photochemical reactions.

The efficiency of a photochemical reaction is usually expressed in terms of quantum yield Φ which is the relation between the number of the molecules that undergo a particular photochemical reaction and the number of photons absorbed.

$$\Phi = \frac{\text{Number of molecules undergoing a particular process}}{\text{Number of photons absorbed}}$$

In summary, the radiations in the infrared region of the spectrum correspond to 1–10 kcal/mole of energy which produces only vibrationally or rotationally excited molecules. The light in the visible and ultraviolet region however, has sufficient energy to cover the range of chemical bond energies and can induce chemical changes by exciting molecules to higher electronic states. The energy of visible light varies from 38 kcal/mole (750 nm) to 71 kcal/mole (400 nm), whereas ultraviolet light is still more effective as it provides energy up to 143 kcal/mole (200 nm).

10.2 EXCITED STATES

The following points may be noted:

- In most stable compounds the electrons are paired and their spins (represented as + 1/2 or – 1/2) cancel each other.
- When the electron spins in a molecule cancel, regardless of whether the electrons are all paired in orbitals, the molecule is then said to be in the singlet state (S). Thus a molecule with all electrons paired is said to be in a singlet state (S).
- When the electrons are paired in their lowest energy orbitals, the molecule is in the ground state (S_0), and the molecule is said to be in the lowest energy singlet state (S_0).
- The energy associated with ultraviolet and visible light is sufficient to excite electrons in molecules. An electron is promoted from the ground state orbital to an excited state orbital, a process termed photoexcitation.
- Initially, almost all molecules are in the lowest vibrational level of the ground state, S_0. When a ground state singlet absorbs a photon of sufficient energy, it is converted to an excited singlet state. The photoexcitation process is very rapid (10^{-15} second), it is even faster than a molecular vibration. This excited state in which there is no spin inversion in called excited singlet state S_1, S_2 or higher which depends on the energy of the excited state.
- When two electrons occupy different orbitals, according to Hunds rule, the lowest energy state will be that where these two electrons have parallel spins and are unpaired. A triplet state (T_1, T_2 etc.) of a molecule is that in which two electrons are unpaired. This may be represented by considering a diatomic molecule in which the configurations in its ground and lowest excited states are presented (Scheme 10.1). Thus a triplet (T) represents an excited state where the spin states of electrons in a molecule do not cancel as the spin state of one electron in the molecule is changed.

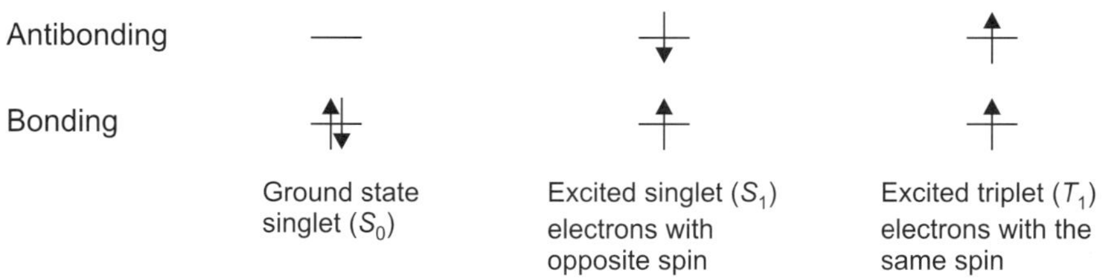

SCHEME 10.1

- In organic photochemistry the transitions $n \longrightarrow \pi^*$ and $\pi \longrightarrow \pi^*$ are of significance. The $n \longrightarrow \pi^*$ transition is the lowest energy transition *e.g.*, in the case of most ketones and consequently the $S_0 \longrightarrow S_1$ transition. The lower energy $n \longrightarrow \pi^*$ transition occurs at longer wavelengths compared to the $\pi \longrightarrow \pi^*$ transition. The $\pi \longrightarrow \pi^*$ transition is the $S_0 \longrightarrow S_2$ transition.

10.3 THE FATE OF THE MOLECULE IN S_1 AND T_1 STATES (JABLONSKI DIAGRAM)

The S_1 state of a molecule (lowest vibrational level) has a longer lifetime, about 10^{-8} to 10^{-7} sec. and its energy is rapidly dissipated through molecular collisions and one of the following processes takes place (explained by Jablonski diagram).

- An excited molecule with electrons in S_1 state can return to ground state S_0 by emitting radiation ($h\nu$), a process called fluorescence (Scheme 10.2).

  The fluorescent light has longer wavelength than the light needed for the original excitation.
- Immediately after promotion (~ 10^{-11} sec.), the molecule returns to the lowest excited singlet state S_1 by giving up it energy as heat. This process is called internal conversion (Scheme 10.2).

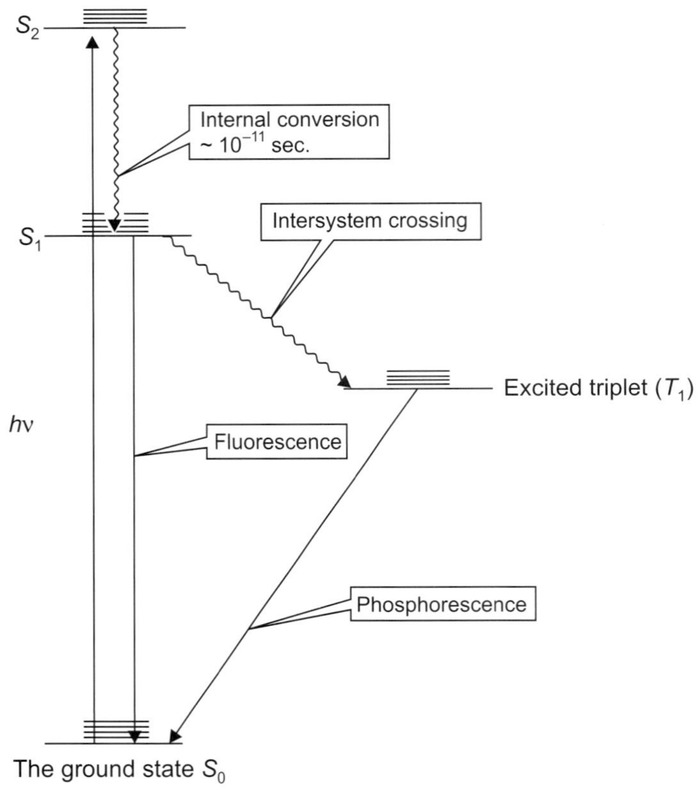

Jablonski diagram—photochemical reactions

SCHEME 10.2

- S_1 can undergo intersystem crossing involving a spin inversion to the triplet state T_1. A triplet state is reached when one of the unpaired electrons in the excited molecule undergoes an inversion of its spin and gives rise to a lower-energy state described as a triplet, T_1.
- The S_1 and T_1 are the major reactive states in photochemical reactions since nondissociative chemical reactions are more probable in long lived excited states. The triplet state T_1 has even a longer lifetime compared to S_1.
- The triplet state may return to the ground state with a further spin inversion, either by emitting radiation (phosphorescence) or by giving up energy as heat. It may also

take part in a chemical reaction. It may transfer its energy to another molecule which is thereby raised to the triplet state. The triplet state is paramagnetic since it has two parallel unpaired electrons, while the single state is diamagnetic. Thus the photochemical reactions involving triplet states can be quenched by paramagnetic salts and by free radical scavengers like oxygen.

- Phosphorescence is the light emission from the triplet state as it returns to the ground state. Phosphorescence and internal conversion of T_1 to S_0 are spin forbidden processes. An intermolecular reaction is particularly favoured due to the longer life of T_1 state compared to S_1 state.

10.4 ENERGY TRANSFER

Excitation energy of an electronically excited molecule can be transferred to the ground state of another molecule. The method provides alternative means to generate electronically excited molecules. The method is normally employed to produce triplet excited states.

10.5 ENERGY TRANSFER AND PHOTOSENSITIZATION

A molecule in an excited state (S_1 or T_1) may transfer its excess energy all at a time to a different compound in the environment. This process is termed photosensitization. Thus, the excited molecule is converted to S_0 while the acceptor molecule becomes excited. A triplet excited state generates another triplet and a singlet generates a singlet.

Photosensitization is very important aspect of photochemistry. It is an alternative process for creating excited states which are otherwise difficult to attain by direct irradiation. Thus a photochemical reaction can carried out on a molecule which cannot be brought into the desired excited state by direct absorption of light by the use of a photosensitizer. An example is of simple (unconjugated) alkenes which absorb only in the far UV region which is difficult to reach experimentally. The photochemistry of such compounds, is thus studied with the help of photosensitization.

The role of a sensitizer is to transfer the excitation energy from an electronically excited molecule (the sensitizer itself) to the ground state of another molecule to give a triplet excited state. Singlet excitation energy can be transferred, however, the lifetimes of singlet excited states are very short (10^{-8} sec.) compare to triplet excited states ($> 10^{-6}$ sec.). The triplet energy transfer requires that the triplet energy of the donor be higher compared to the acceptor molecule by 3 kcal/mol.

A common triplet sensitizer is benzophenone (Scheme 10.3), with a triplet energy level of 69 kcal/mol (289 kJ/mol). The triplet energy of naphthalene (61 kcal/mol *i.e.,* 255 kJ/mol) is lower than that of this sensitizer. Naphthalene does not absorb appreciably at 345 nm. When however, a mixture of benzophenone and naphthalene in ethanol-ether is irradiated at 345 nm (at low temperature) one observes the phosphorescence of naphthalene. Thus the requisite triplet excitation must come from excited triplet of the sensitizer benzophenone. In conclusion, since naphthalene is photosensitized by triplet benzophenone. The triplet energy of naphthalene must be lower than that of benzophenone which indeed is so.

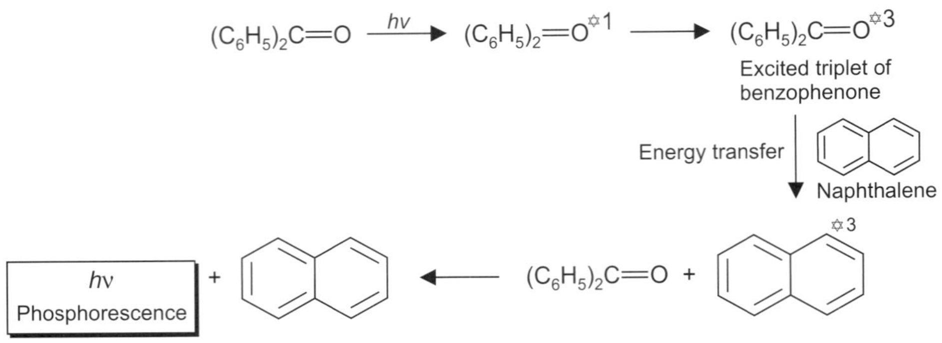

SCHEME 10.3

10.6 FORBIDDEN TRANSITIONS—INTERSYSTEM CROSSING

The following points may be noted:

- Intersystem crossing $S_1 \longrightarrow T_1$ is although energetically gainful but it is formally a spin forbidden process. The average energy of the lowest-energy triplet excited state T_1 is normally greater than S_0, however less than that of S_1 (Scheme 10.3*a*).
- If the singlet state is sufficiently long lived as is so in aromatic and carbonyl systems, the $S_1 \longrightarrow T_1$ change called intersystem crossing occurs with almost 100% efficiency.
- The efficiency in intersystem crossing depends, among other factors on the energy gap between S_1 and T_1 (the $S_1 - T_1$ energy gap). When this energy difference is small, the intersystem crossing is efficient. Generally intersystem crossing efficiencies for ketones is maximum. A typical example is of benzophenone for which intersystem crossing has 100% efficiency since the energy gap is small (5 kcal/mol, Scheme 10.3*a*). Aromatic compounds have intermediate to high, while olefins have low intersystem crossing efficiencies. Thus when the energy gap is large, intersystem crossing efficiency may be low or zero and spin forbiddenness becomes important.

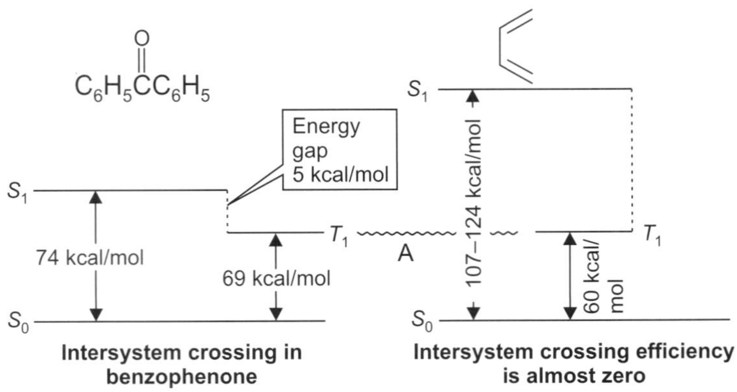

A represents 100% efficiency $S_1 \rightarrow T_1$ intersystem crossing for a sensitized reaction

SCHEME 10.3a

For other examples of sensitization in photochemistry (see Schemes 10.39*a* and 10.39*b*).

A Summary of Photochemical Processes

- $S_0 \xrightarrow[\text{Excitation}]{h\nu} S_1 \xrightarrow[\text{or internal conversion}]{\text{Fluorescence}} S_0$

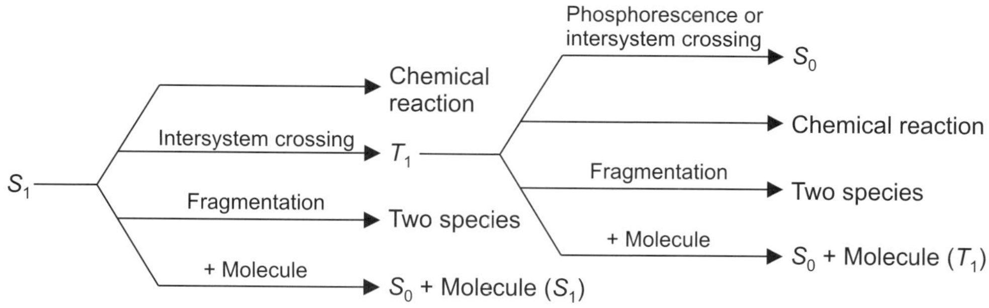

SCHEME 10.3b

- *S_1 and T_1 are main reactive states in photochemical reactions.*
- *Chemical reactions are favoured by the longer lifetime of T_1 state relative to S_1.*
- *S_1 on intersystem crossing goes to the triplet state T_1. In the triplet state the spin of one electron has been changed so that now the molecule has two electrons which cannot pair.*
- *Intersystem crossing ($S_1 \longrightarrow T_1$) in many compounds is an improbable process which involves switching of an electronic spin, thus triplet states assume no importance in the photochemistry of such compounds. In most other compounds e.g., $\pi \longrightarrow \pi^*$ states of polycyclic aromatic hydrocarbons and $n \longrightarrow \pi^*$ of several ketones, the process of intersystem crossing occurs with high efficiency.*
- *The efficiency of intersystem crossing depends largely on the difference in energy between singlet and triplet excited states (i.e., $S_1 - T_1$ energy gap). Excited benzophenone provides an example which converts completely from S_1 to T_1.*
- *Triplet state is long-lived with lifetime greater than 10^{-5} sec., since the conversion to S_0 would require a switching an electronic spin. The energy difference between S_1 and T_1 is far less than between T_1 and S_0 so the latter intersystem crossing is far less probable.*

10.7 PHOTOCHEMICAL REACTIONS

(A) Photoreduction (Hydrogen Atom Abstraction)

Many aromatic ketones react by hydrogen atom abstraction from solvent or some other hydrogen donor to give diols as the stable products formed by the coupling of the resulting α-hydroxybenzyl radicals (Scheme 10.3*c*). This is the common reaction of a photoexcited carbonyl group and the details are in (Scheme 10.4).

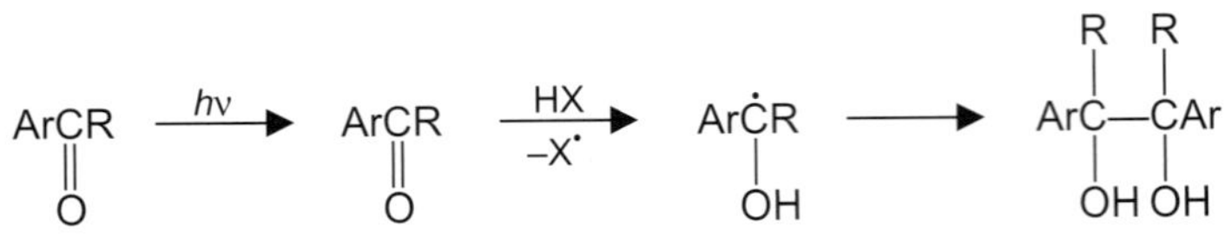

SCHEME 10.3c

An aryl ketone *e.g.*, benzophenone on irradiation in the presence of a hydrogen donor like a secondary alcohol, brings about a reductive coupling reaction. The initial excited singlet state (S_1) goes to a triplet state (T_1). The triplet contains two unpaired electrons and thus is a diradical. This radical can abstract a hydrogen atom from some other molecule in solution. In the photoreduction, the diradical abstracts a hydrogen atom from the secondary alcohol and forms two benzhydryl radicals that can dimerize to benzopinacol.

In this reaction the quantam yield for disappearance of benzophenone is 2.0. This is so since, the radical after hydrogen atom abstraction from 2-propanol transfers a hydrogen atom to benzophenone. Thus two molecules of benzophenone are reduced by absorption of only one photon of light.

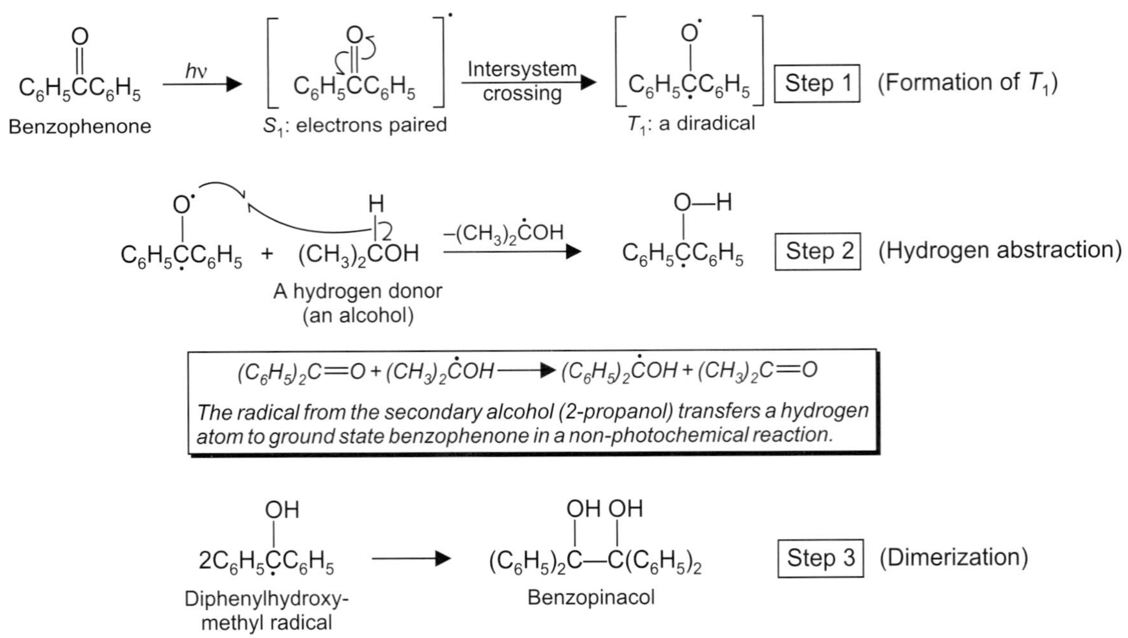

SCHEME 10.4

PROBLEM 10.1

The photoreduction of benzophenone is carried out in the presence of toluene. The benzyl radical formed after the triplet excited state of benzophenone abstracts a hydrogen atom from toluene dimerizes to bibenzyl. In the same reaction in the presence of 2-propanol the intermediate free radical 2-hydroxy-2-propyl radical instead gives up a hydrogen atom to benzophenone. Explain and give the mechanism and other products formed if any.

ANSWER. *The benzyl radical is resonance stabilized (Scheme 10.5).*

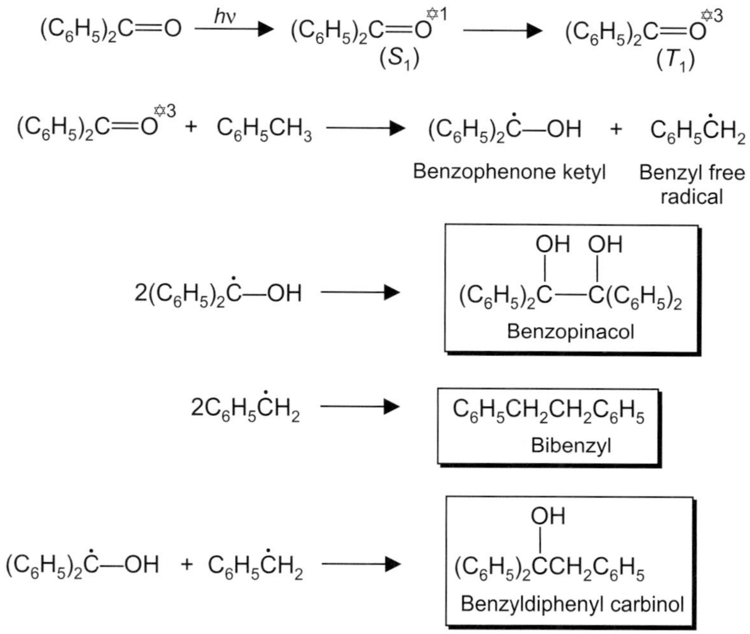

SCHEME 10.5

In the case of reduction of benzophenone, pinacol could also be formed *via* the dimerization of the hydroxyisopropyl radicals, however, pinacol is not formed. This shows that the lifetime of the hydroxyisopropyl radical is too short for it to combine with another radical to form the dimer, it however, transfers its hydroxylic proton too rapidly (see box, Scheme 10.4) and is converted to acetone. The diphenylhydroxy methyl radical has a relative long lifetime due to resonance stabilization, and is sufficiently stable not to attack Me_2CHOH.

Photoreduction may be described as light initiated reduction of different type of functional groups in the presence of an electron donor.

(B) Photoenolization

Benzophenone derivatives which have an *ortho* alkyl substituent do not display photoreduction. These derivatives however, display a different photoreaction in which the intramolecular hydrogen abstraction occurs from the benzylic position. This yields an enol which gives the more stable starting benzophenone without any photoreduction. This photoenolization can be detected when photolysis is carried out in deuterated solvents (Scheme 10.6). Photoenolization

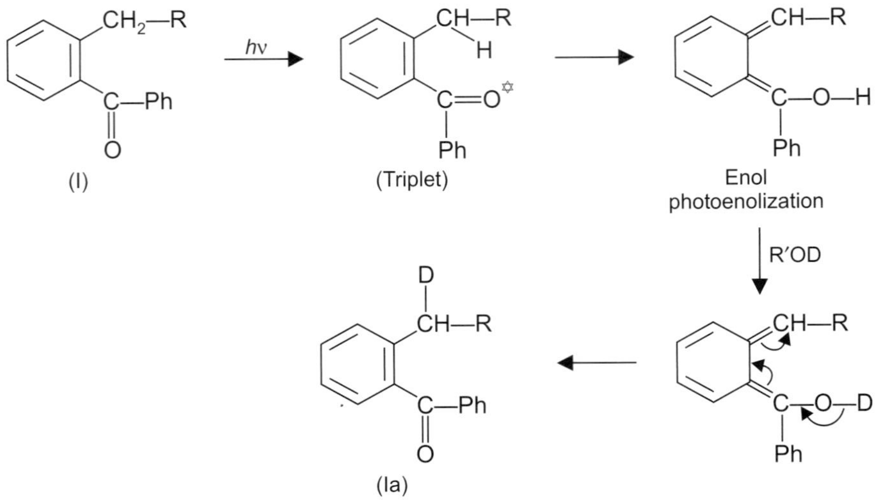

SCHEME 10.6

can as well be detected when an enol is trapped as its Diels-Alder adduct (Scheme 10.7).

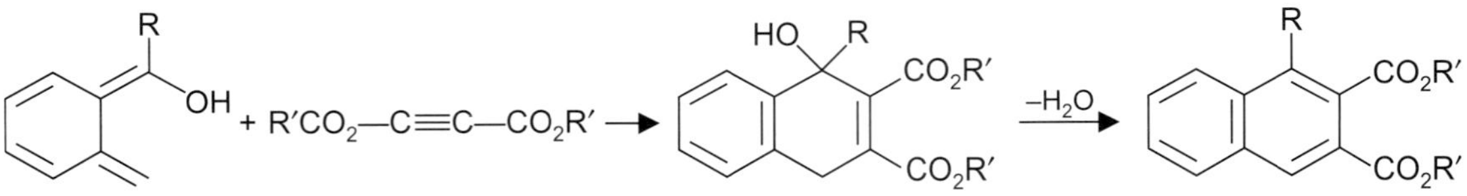

SCHEME 10.7

(C) Photooxidation (Formation of Peroxy Compounds)

Molecular oxygen is a triplet in its ground state (T_0) and it does not add to cyclic dienes. It is promoted to an excited state, known as singlet oxygen with all electrons paired. An interesting cycloaddition reaction takes place on irradiation of some dienes and polyenes with oxygen in the presence of a triplet sensitizer such as the dye methylene blue (Scheme 10.8). The sensitizer in its triplet state transfers its energy to the triplet oxygen molecule *i.e.*, the ground state of oxygen to give singlet oxygen (*i.e.*, an excited oxygen molecule). The sensitizer subsequently returns to its ground state (S_0). Singlet oxygen behaves as a dienophile and adds to the diene to give an endoperoxide by a typical Diels-Alder reaction (Scheme 10.8).

The synthetic utility of this reaction is based in the reduction of peroxides to diols (Scheme 10.8). Olefins with an allylic hydrogen atom give hydroperoxides (Scheme 10.9) which can be subsequently reduced to allylic alcohols. When an olefin does not have an allylic hydrogen, it then gives dioxetans (Scheme 10.9) which on heating afford carbonyl compounds.

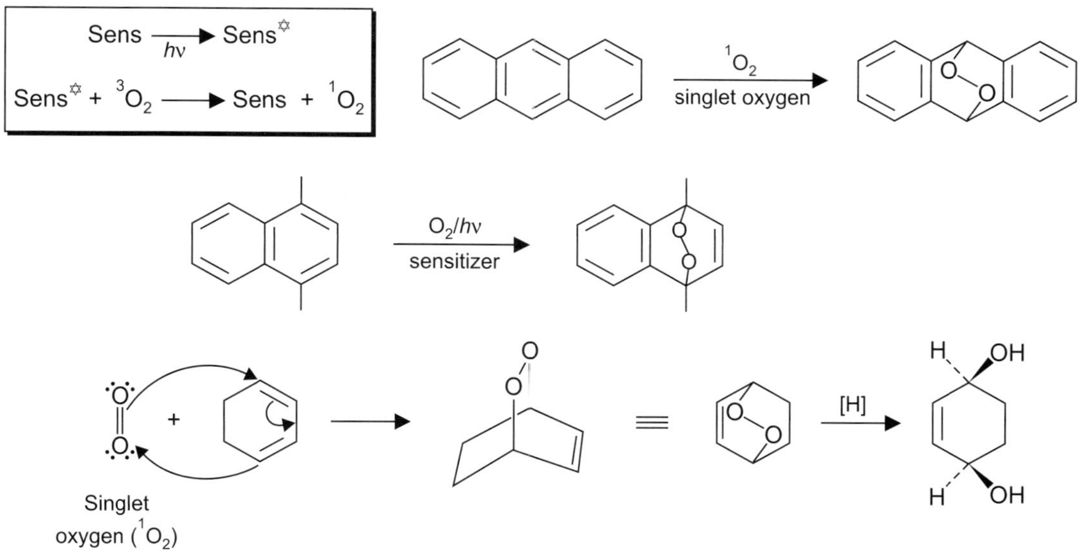

SCHEME 10.8

In summary the peroxy compounds are formed when one irradiates the substrate in the presence of oxygen and a sensitizer. The sensitizer is excited to the triplet state which then

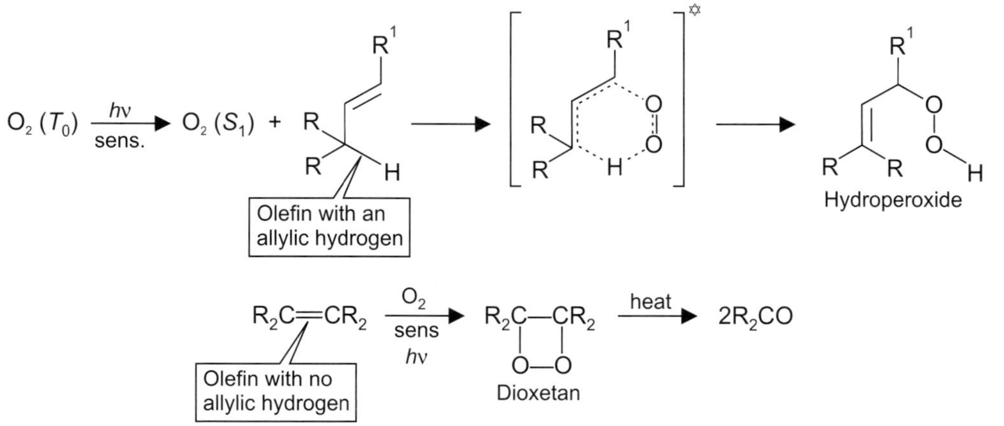

SCHEME 10.9

activates the oxygen molecule which then reacts with a variety of unsaturated hydrocarbons (Schemes 10.8 and 10.9). The triplet sensitizer can as well abstract a hydrogen atom from the substrate to generate a radical (Scheme 10.10) which then reacts with oxygen. Thus a secondary alcohol can be oxidized to a hydroxy-hydroperoxide by using benzophenone as the sensitizer. The triplet benzophenone generates a carbon radical by reacting with alcohol (Scheme 10.10). This adds oxygen and a chain reaction is propagated. The hydroxy hydroperoxides tend to eliminate hydrogen peroxide to give carbonyl compounds.

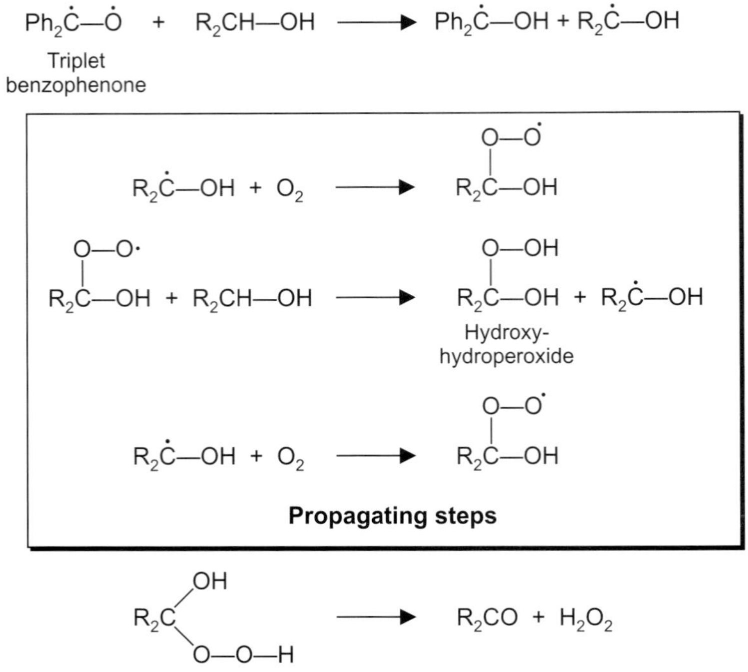

SCHEME 10.10

(D) *Cis-trans* Isomerization (Photoisomerization)

Alkenes display a characteristic photochemical interconversion of *cis* and *trans* isomers (Scheme 10.11). Generally, the *trans* isomer is thermodynamically more stable and photolysis gives a mixture which however is richer in the *cis* isomer. Irradiation, is therefore, a useful method of converting a *trans* alkene to its *cis* isomer. The reaction is promoted by direct irradiation of the substrate and also by photosensitized energy transfer. In several alkenes the *E* isomer absorbs energy more effectively (it has a larger molar absorptivity ε) and at a slightly different wavelength compared to the *Z* isomer. It is sometimes possible to convert an *E* isomer to its thermodynamically less stable *Z* from—a technique called optical pumping (Scheme 10.11). The isomerization takes place, since the excited states (both S_1 and T_1) of many olefins have a perpendicular rather than a planar geometry (Scheme 10.12).

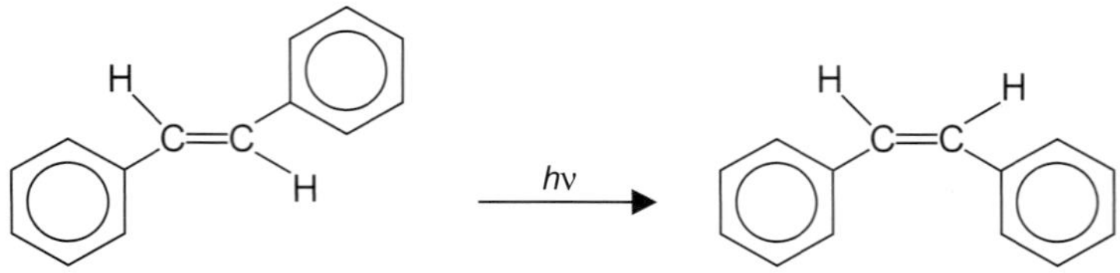

trans-stilbene
λ_{max} = 295 nm
ε = 27,000

cis-stilbene
λ_{max} = 280 nm
ε = 13,500

Photosensitized *cis-trans* isomerization

SCHEME 10.11

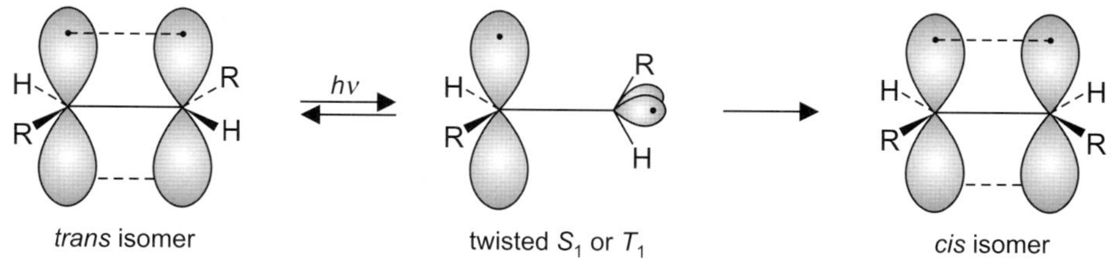

SCHEME 10.12

The isomerization occurs because the π bond which normally prevents it, is lost in passage to the excited state, in which the two sets of substituents now occupy mutually perpendicular planes (Scheme 10.12). Thus, *cis-trans* isomerism disappears on excitation. When the excited molecule drops back to the S_0 state either isomer can be formed. In case a wavelength of light is selected at which the *trans* isomer absorbs while the *cis*-isomer doesnot than the *trans* isomer can be converted completely to the *cis*.

The photochemical isomerization of vitamin A aldehyde (11 *Z*-retinal to 11 *E*-retinal) on absorption of light is the primary event in vision processes (Scheme 10.12*a*). The natural form

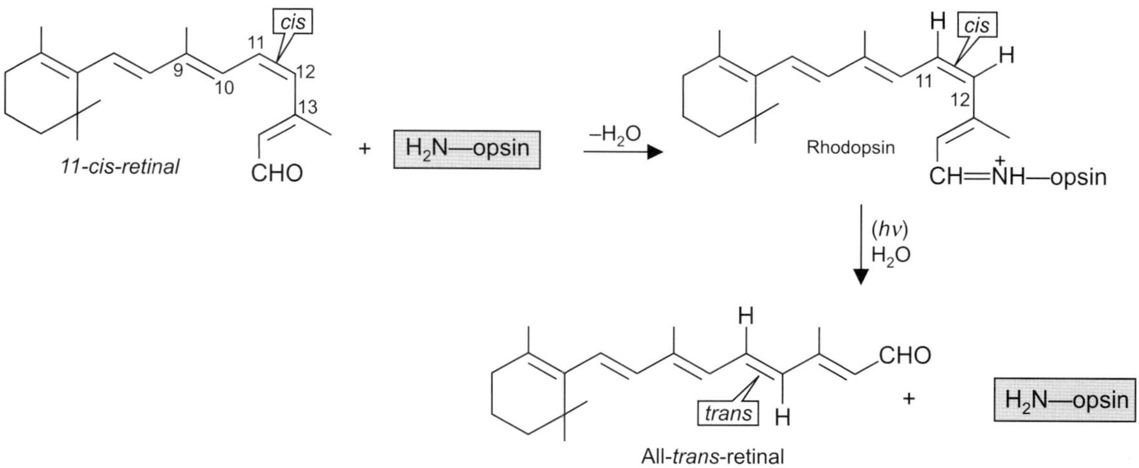

SCHEME 10.12a

of vitamin A aldehyde is all -*trans*. A protein opsin in the retina of human eye has a pocket where only 11 *Z*-retinal fits and is bonded with it via an imine linkage. The complex molecule which is formed is called rhodopsin which absorbs light energy in the visible region of the spectrum. On absorption of light 11 *Z*-retinal is isomerized to the all *trans*-from 11 *E* retinal. In this form the fit into the opsins pocket is disturbed and dissociation to 11 *E*-retinal and opsin takes place, neither of which absorbs light in the visible region of the spectrum 11 *E*-retinal is reconverted into the *cis*- form now by an enzyme and the visual cycle begins again.

Photoexcited cyclic olefins add hydroxylic solvents provided the size is appropriate to accommodate a *trans* double bond. These additions involve the formation of *E* isomer of the alkene as the key intermediate which may be strained as *e.g.*, in the case of cyclohexene.

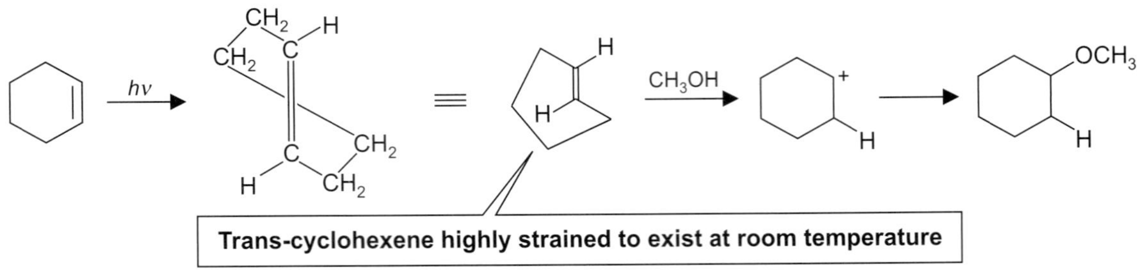

SCHEME 10.12b

The *trans*-cycloalkenes can be protonated very easily since strain is relieved on protonation. Norbornene does not display this addition of methanol, since the *trans* isomer of norbornene would be highly strained if formed. This is again the case with cyclopentene. It may be mentioned that rings of cycloalkenes containing five carbon atoms or fewer exist only in the *cis* form. The introduction of a *trans* double bond into these rings would introduce large strain. *Trans*-cyclohexene can be formed as a very reactive short lived intermediate in some reactions. *Trans*-cycloheptene has a very short lifetime and has not been isolated, *trans*-cyclooctene has however, been isolated and is stable at room temperature. In this case the ring is large enough to accommodate the geometry required by a *trans* double bond, and thus these add methanol.

Cycloheptenone and larger rings also undergo a similar photoisomerization (see Scheme 10.36a).

Cyclohexene, cycloheptene and cyclooctene on photoexcitation all give products derived from ring contraction and carbene insertion products (Scheme 10.12c).

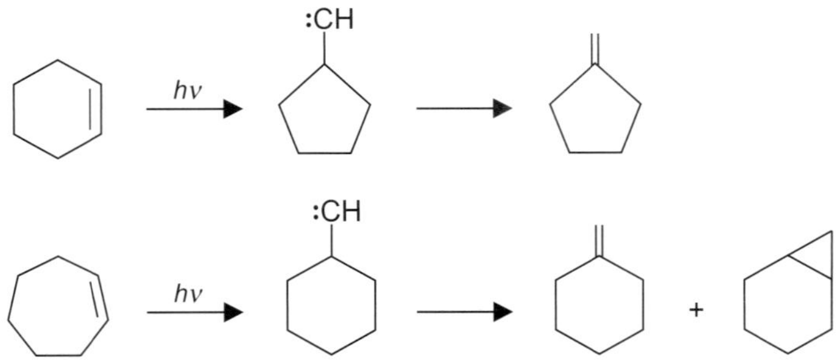

SCHEME 10.12c

(E) Photoisomerization Followed by Oxidative Coupling

Many olefinic compounds undergo geometrical isomerization which is their typical photoreduction. The reaction occurs under direct irradiation of the reactant as well as by photosensitized energy transfer.

On irradiation *cis*- or *trans*-stilbene gives only an equilibrium mixture of the two (Scheme 10.11). In the presence of oxygen, however, *cis*-stilbene photocyclizes reversibly to give a small proportion of dihydrophenanthrene which is oxidized to phenanthrene irreversibly with oxygen (Scheme 10.13).

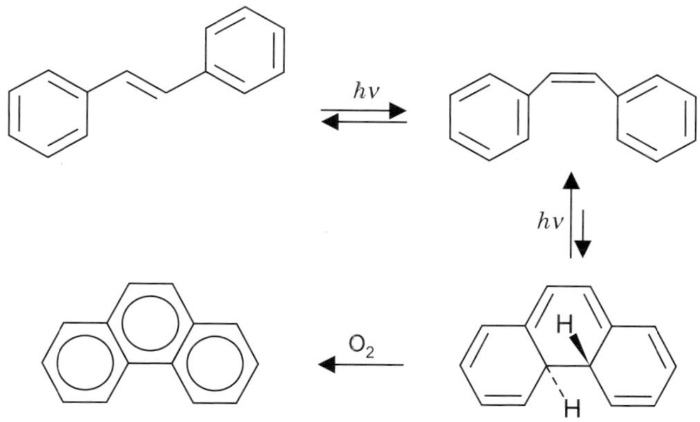

SCHEME 10.13

(F) Photolysis (Photochemical Fragmentation)

(i) Photolysis of carbonyl compounds

It was seen during photochemical reductions (Sec. 10.7, A) that the excited states of the carbonyl groups of many aldehydes and ketones are very efficient hydrogen abstractors from solvent or some other hydrogen donor (see Scheme 10.3*c*).

The prominant photochemical reaction displayed by ketones in the gas phase is the fission of the carbon-carbon bond adjacent to the carbonyl group and is called *α-cleavage* (Scheme 10.13*a*).

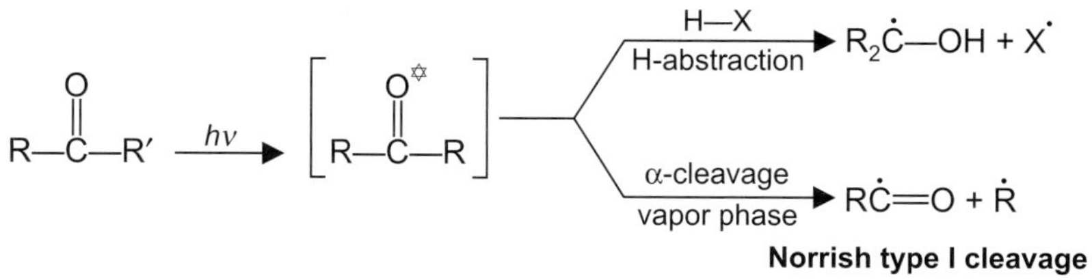

SCHEME 10.13a

Norrish Type I Process

A molecule on irradiation usually leads to homolytic bond cleavage to generate free radical intermediates. The cleavage at the carbon-carbon bond α- to the carbonyl group is often termed Norrish Type I cleavage.

The Norrish type I process dominates in the vapor phase than in solution, however, under solution conditions the two radicals generally recombine within the solvent cage where these are generated. This process becomes significant when the relatively stable carbocation (tertiary or benzylic type) is formed.

(a) Norrish type I reaction of carbonyl compounds

The process 'Norrish type I' cleavage (α-cleavage) originates from the carbonyl *n*, π* state and involves the homolytic cleavage of one of the carbonyl alkyl groups (*i.e.*, photochemical cleavage at the carbon-carbon bond α to the carbonyl group) as the primary photochemical reaction

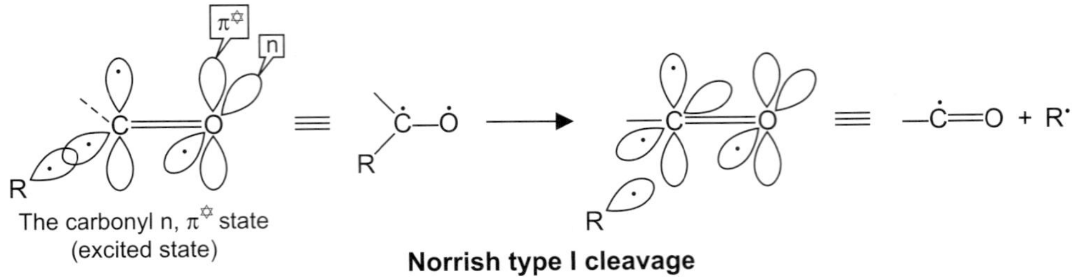

SCHEME 10.14

(Scheme 10.14). This is followed by decarbonylation and subsequent reactions of alkyl free radicals (Scheme 10.15).

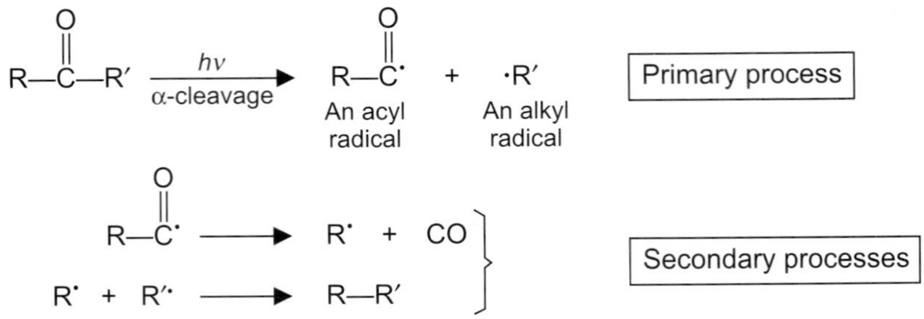

SCHEME 10.15

The nature of the products depends on the structure of the radicals. This is shown by Norrish type I reaction on irradiation of gaseous acetone at 313 nm (Scheme 10.15*a*).

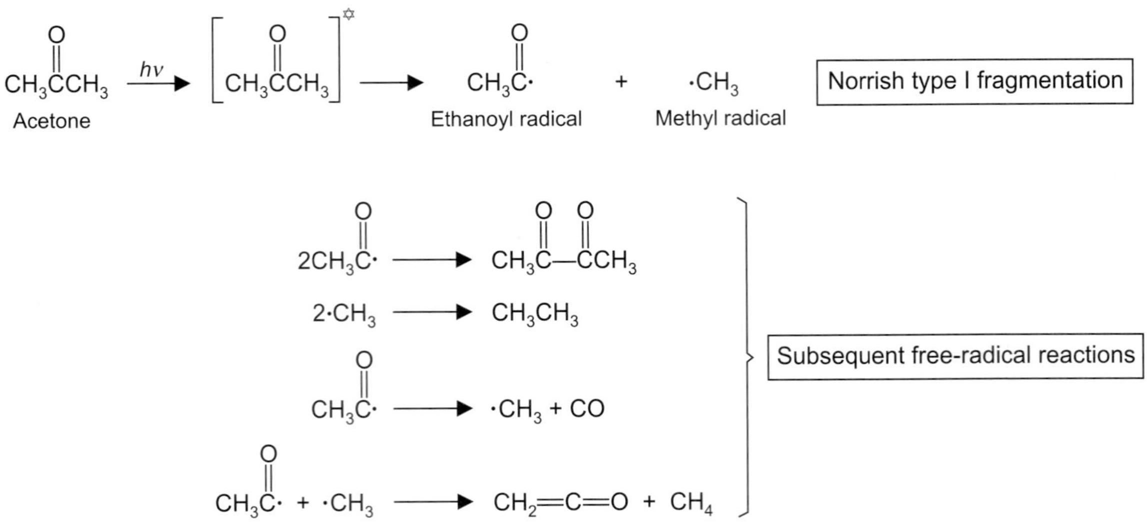

SCHEME 10.15a

With unsymmetrical ketones, the α-cleavage occurs so as to give the more stable of the two possible free radicals. This is followed by reactions like disproportionation or intermolecular hydrogen atom abstraction by the acyl radical (Scheme 10.15*b*).

SCHEME 10.15b

Smaller ring ketones *e.g.*, cyclohexanone on α-cleavage (Norrish type I process) yield a diradical and subsequent intramolecular hydrogen abstraction gives an aldehyde (Scheme 10.16), while loss of CO gives cyclopentane.

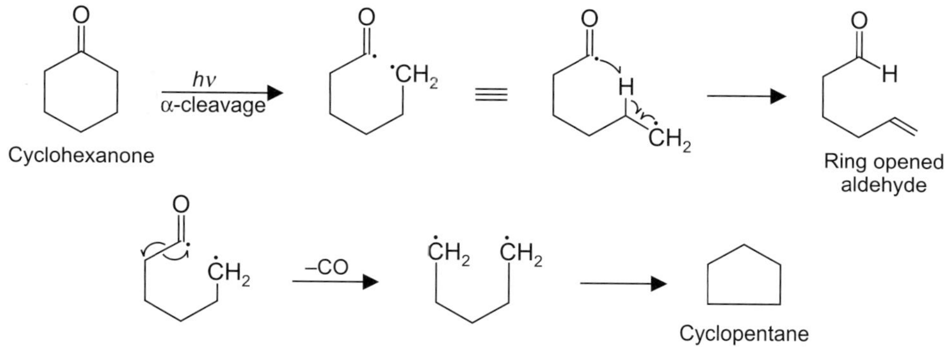

SCHEME 10.16

The Norrish type I cleavage (Ronald Norrish shared the Nobel Prize in Chemistry in 1967) is useful for ring cleavage of cyclic ketones.

EXERCISE 10.2

Predict the products from the compounds (Scheme 10.17) on photochemical reactions via α-cleavage.

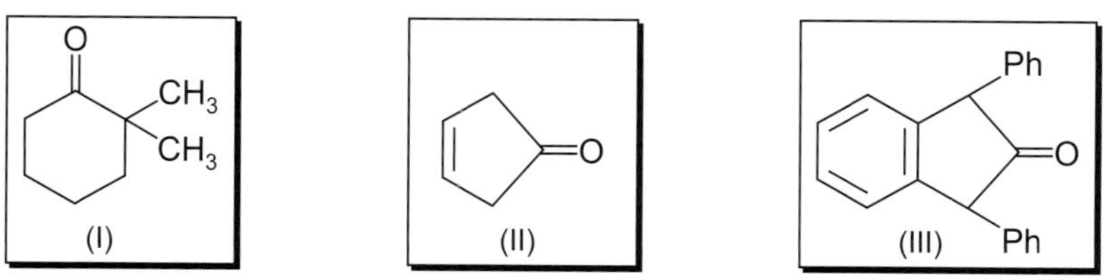

SCHEME 10.17

ANSWER. (*I*) *It cleaves to give the most stable tertiary radical rather than primary radical (Scheme 10.18), intramolecular hydrogen transfer reactions then gives different products.*

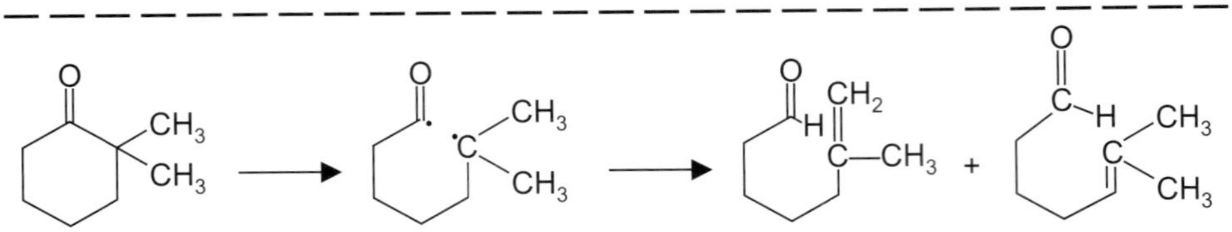

SCHEME 10.18

In (II) and (III), decarbonylation is the major reaction pathway (Scheme 10.19).

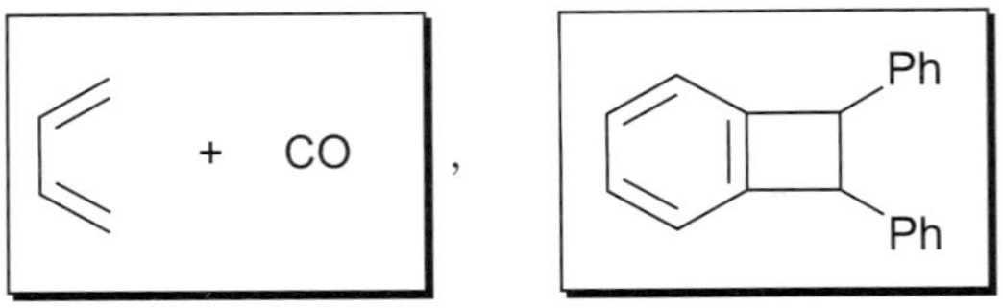

SCHEME 10.19

EXERCISE 10.3

Often a ketene is observed as one of the products during α-cleavage (Norrish Type 1 process) of small ring compounds. How this is formed ?

ANSWER. *Via disproportionation (Scheme 10.19a).*

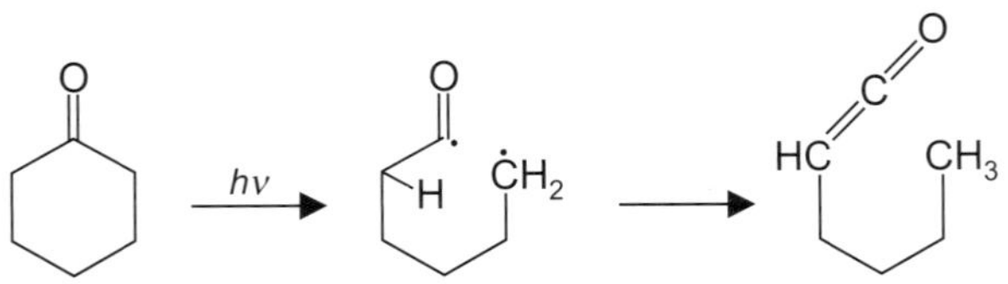

SCHEME 10.19a

(b) Norrish type II, cleavage

Norrish Type II Process

In this photolysis reaction, the carbonyl n, π^* state is involved leading to photoelimination which proceeds through a six centered fragmentation initiated by γ-hydrogen abstraction in the primary photochemical step. Secondary reaction of the diradical leads to cleavage to yield an alkene and a new ketone.

This reaction (a photoelimination) is observed in ketones in which a hydrogen atom attached to the γ-carbon atom is available. The photoexcited carbonyl group (n, π^* state) abstracts the γ-hydrogen *via* a six-membered cyclic pathway (Scheme 10.19*b*) to give a diradical which leads to fission between C_α and C_β (referred to as Norrish type II, photoelimination) to an alkene and an enol which tautomerizes to the carbonyl compound (a new ketone). A similar reaction occurs in the mass spectral fragmentation of carbonyl compounds where it is identified as the Mc Lafferty cleavage. In addition to the fragmentation process between α- and β-carbon atoms (Scheme 10.19*b*), the Norrish type II process is also followed by ring closure (Scheme 10.20).

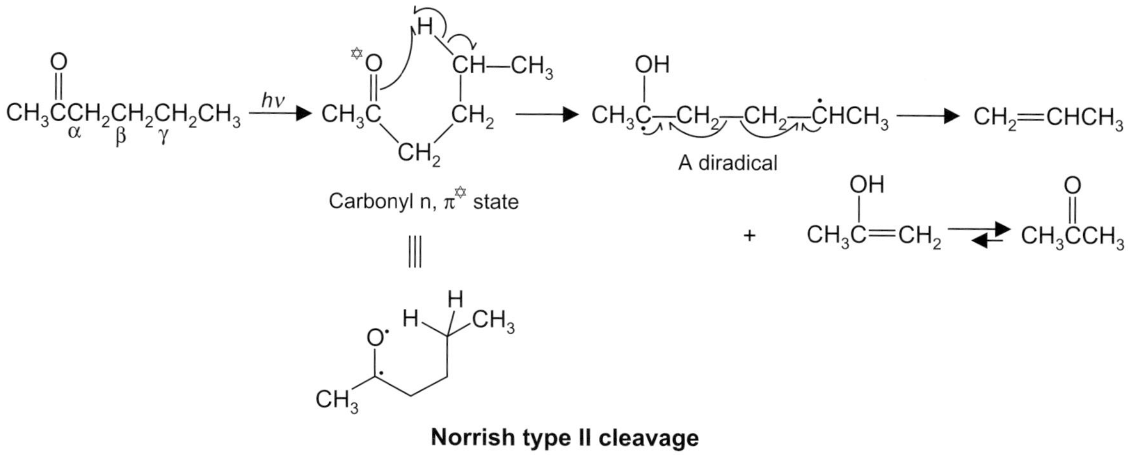

SCHEME 10.19b

which involves the formation of a cyclobutanol by ring closure of the diradical.

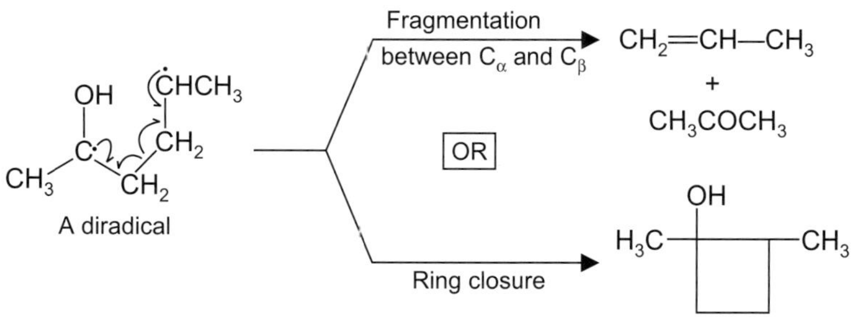

SCHEME 10.20

In larger ring compounds a Norrish type process is observed to give ring closure as shown in the case of cyclodecanone. This gives a bicyclic compound in which both rings are six-membered (Scheme 10.20*a*).

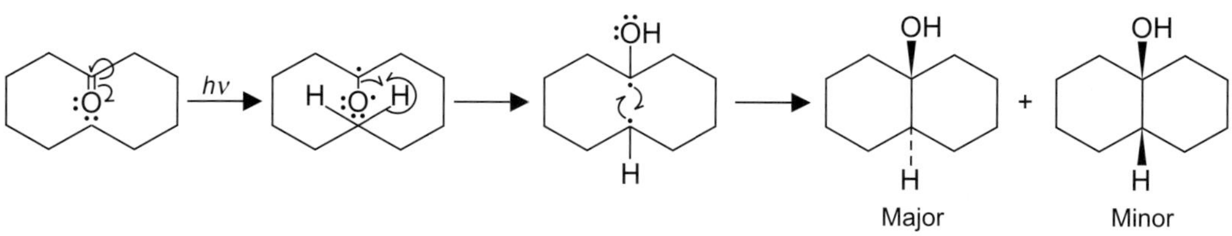

SCHEME 10.20a

EXERCISE 10.4

Predict the product from the compound (Scheme 10.20b) involving intramolecular hydrogen abstraction by the excited carbonyl group followed by ring formation.

ANSWER. *The reaction can be considered as Norrish type II process (Scheme 10.20c).*

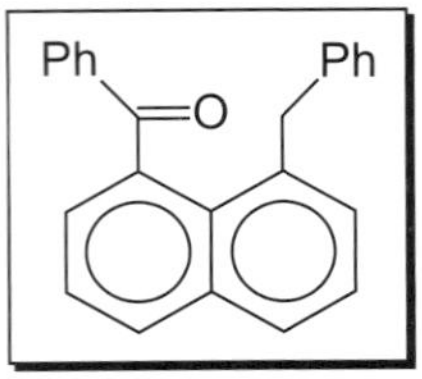

SCHEME 10.20b

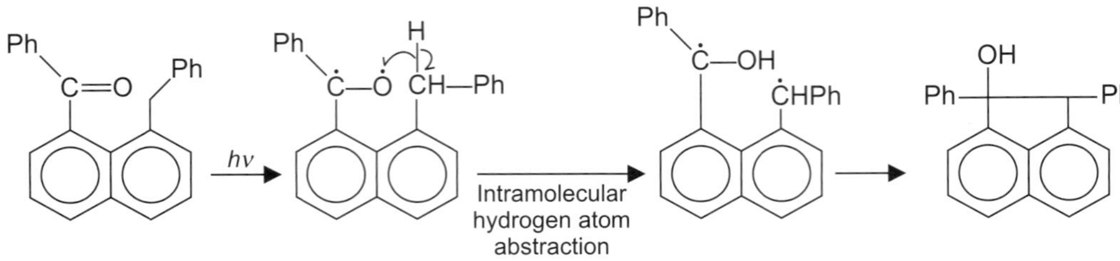

SCHEME 10.20c

(ii) Photolysis of compounds containing the $=\overset{+}{N}=\overset{-}{N}$ *group*

On irradiation, diazomethane gives the simplest carbene *i.e.*, methylene (Scheme 10.21). This reactive intermediate can exist in either a singlet or triplet state. In singlet methylene

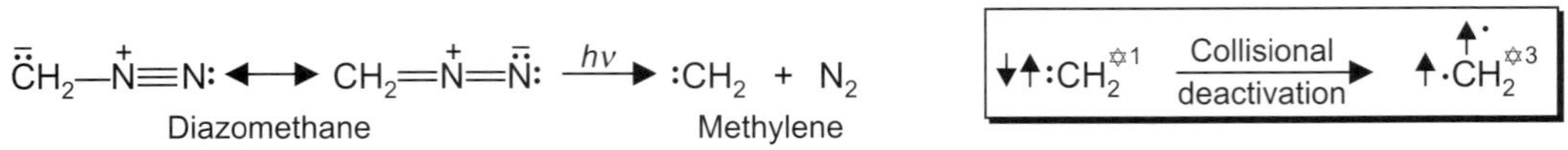

SCHEME 10.21

the nonbonding electron spins are paired and this is usually formed first in the photolysis reaction. Its triplet state has however, lower energy of the two electron configurations and the electrons are not spin paired.

When diazomethane is irradiated it generates carbene in the singlet state. The deactivation of singlet to triplet state (the triplet state of methylene has the lower energy of the two electron configurations) occurs on collision with other molecules if these are present in the reaction medium. Carbene in triplet state can be generated *via* photolysis in the presence of a triplet sensitizer.

Diazomethane on photolysis in the presence *cis*-2-butene is largely stereospecific concerted *syn* addition to the double bond to give *cis*-1, 2-dimethylcyclopropane. While this stereospecificity decreases when an inert gas or a liquid is introduced in the medium (Scheme 10.21*a*) and both *cis*- and *trans*-cyclopropanes are formed. In the latter case the insertion of carbene is not concerted but is a stepwise addition. Where the initial adduct, a triplet has sufficient time to allow rotation about the central C—C bond (more details are in Schemes 11.13*a* and 11.13*b*).

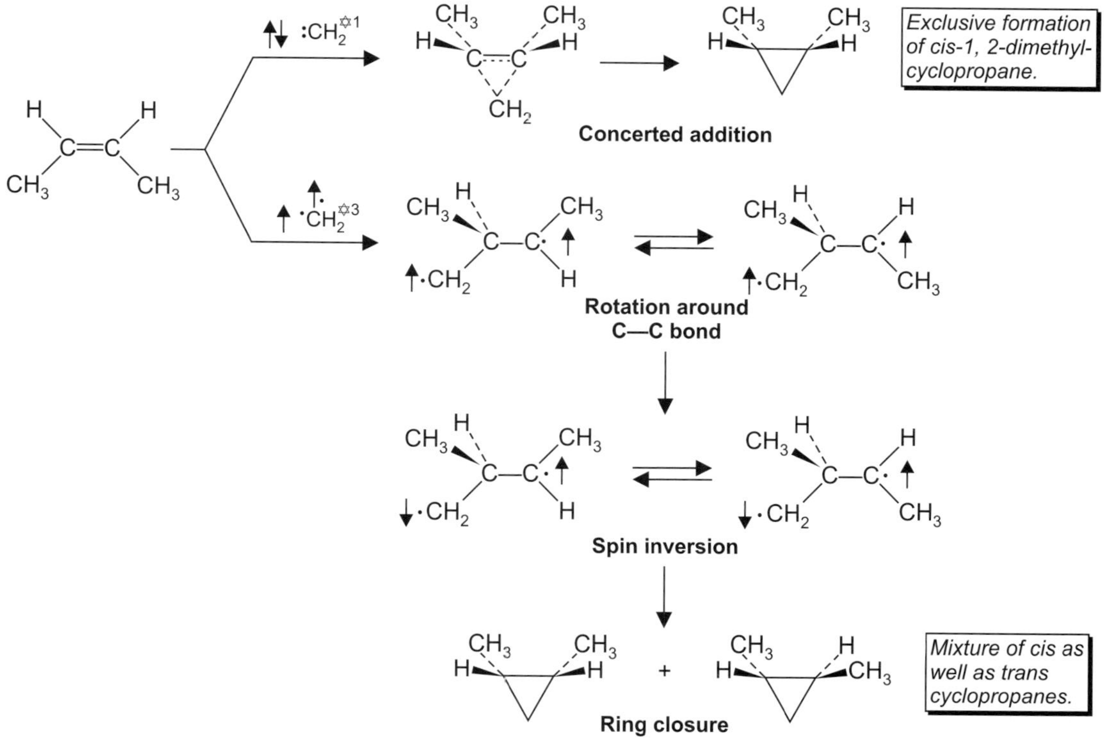

SCHEME 10.21a

Carbenes are so reactive that they add to the "double bonds" of aromatic rings. The products of initial addition are generally not stable and rearrange to give ring expansion. Thus irradiation of diazomethane in benzene leads to addition of methylene across C=C, followed by a spontaneous Cope rearrangement (Scheme 10.22).

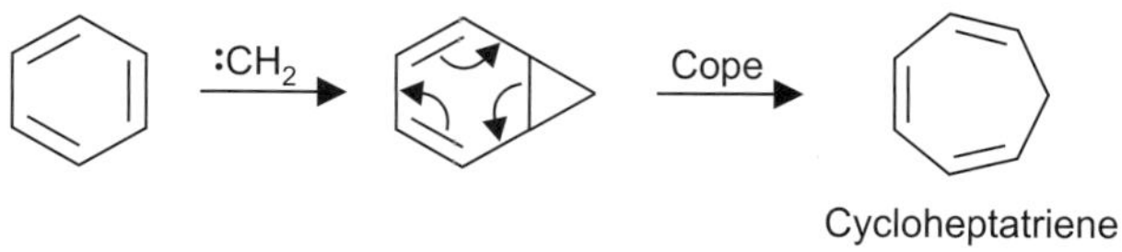

SCHEME 10.22

The α-diazoketones on photolysis give carbenes which undergo a fast rearrangement to a ketene (Scheme 10.23) which can be subsequently trapped by an olefin and the techniques form a useful method for the synthesis of four membered rings (Scheme 10.23). The method forms a good technique for ring contraction starting from cyclic. α-diazoketones (Scheme 10.24).

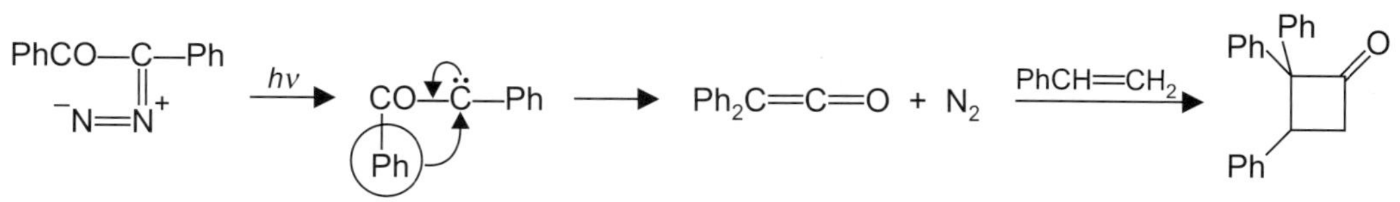

SCHEME 10.23

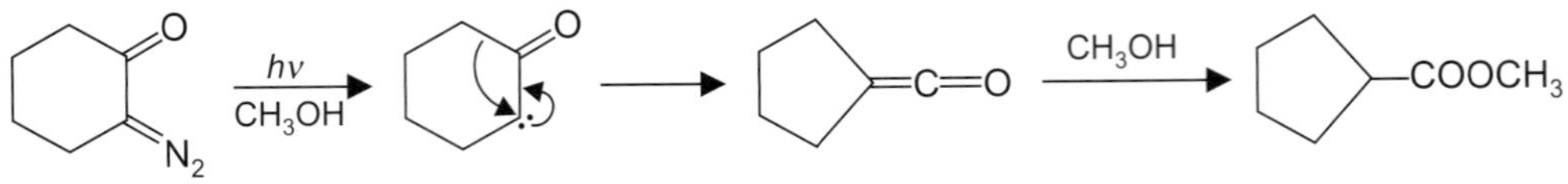

SCHEME 10.24

(iii) Photolysis to produce reactive intermediates

Photolysis to generate some highly reactive intermediates under the condition where these are stable has been successfully carried out on suitable substrates. The process for benzyne formation involves a photochemical elimination of carbon dioxide (Scheme 10.25) from 1, 2-benzenedioyl (phthaloyl) peroxide. It may be mentioned that benzyne intermediate can be

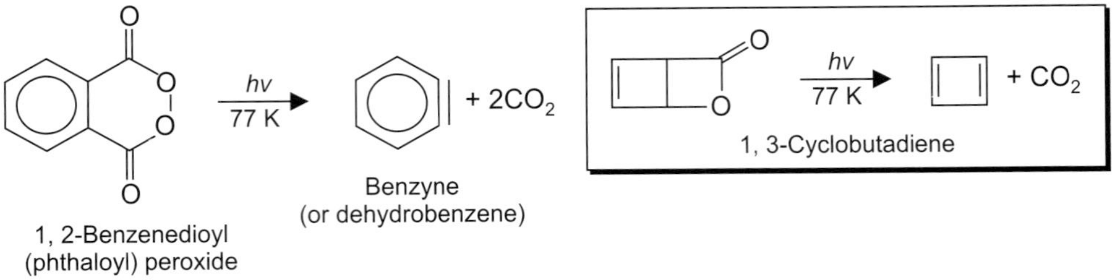

SCHEME 10.25

trapped as its Diels-Alder adduct with furan (Scheme 10.26) and as other compounds, (see Schemes, 4.50, 9.17 and 9.18).

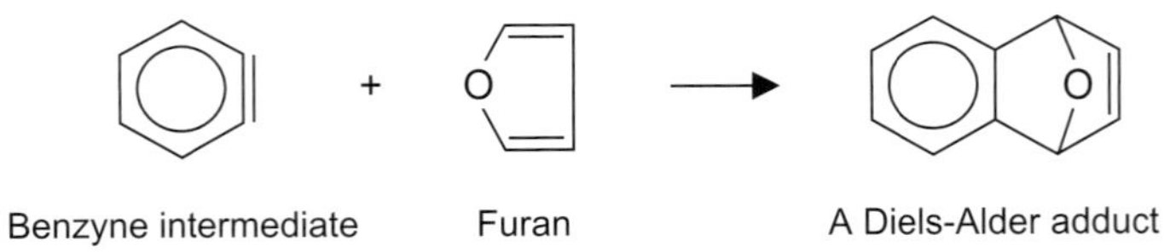

SCHEME 10.26

(G) Photoaddition

Most of the typical photoadditions involve the formation of a 1 : 1 adduct *via* reaction of an excited state of one molecule (generally *e.g.*, a carbonyl compound, aromatic compound or another molecule of the same olefin) with the ground state of another (generally an olefin) with the formation of a ring compound.

(a) Photoaddition of olefins to carbonyl compounds—Paterno-Büchi reaction

Paterno-Büchi reaction normally involves the reaction (photoaddition) of the triplet state of the carbonyl compound with the ground state of an alkene (Scheme 10.27) to give an oxetane. The regiochemistry of the reaction is based on the preferential formation of the most stable

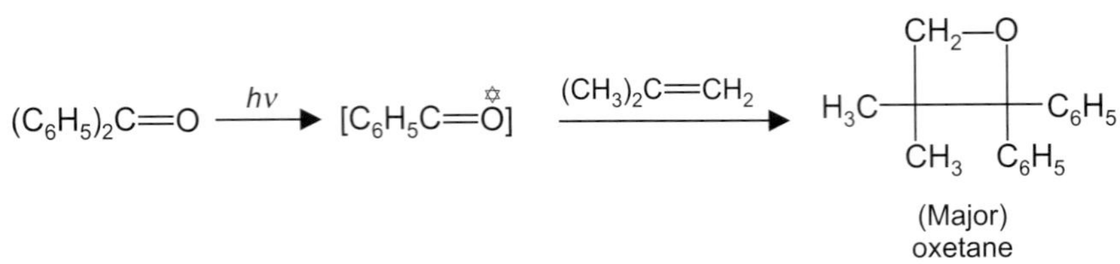

SCHEME 10.27

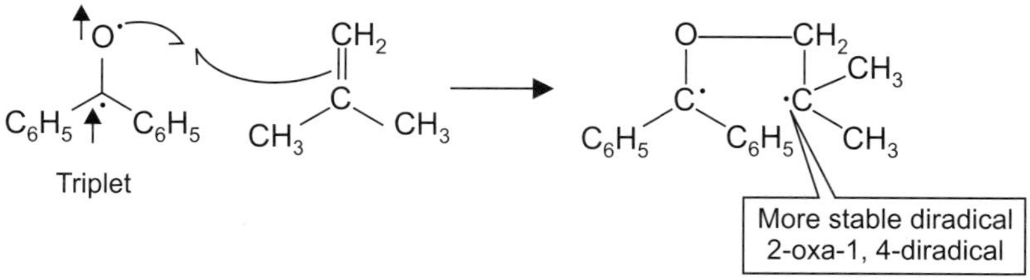

SCHEME 10.28

2-oxa-1, 4-diradical intermediate as explained from the reaction between benzophenone and trimethyl ethylene as another example (Scheme 10.28). The ring formation to generate an oxetane occurs in two stages. The excited carbonyl compound as its triplet adds *via* its oxygen atom to the alkene, as the tertiary radical is more stable. Thus the mode of addition from (I, Scheme 10.29) is preferred.

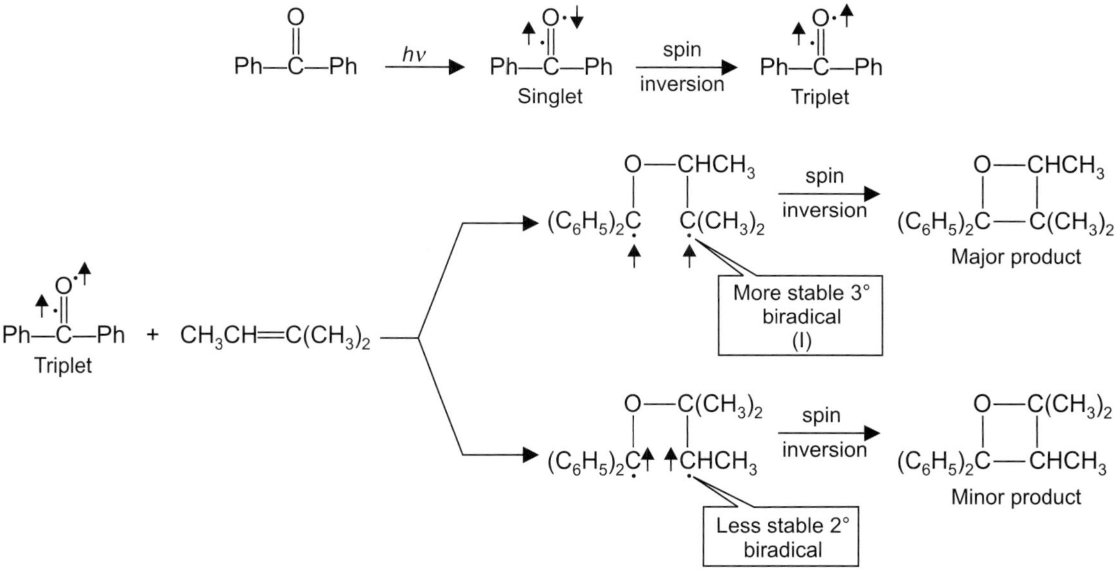

SCHEME 10.29

As discussed above the reaction is regioselective, however, it is not necessarily stereospecific since *cis*- as well as *trans*-2-butene give the same mixture of products of *cis*- and *trans*-oxetanes when reacted with benzophenone (Scheme 10.29*a*). This shows that the intermediate diradical intermediate is long lived to make the rotation around single bonds possible before the final spin-inversion and subsequent cyclization.

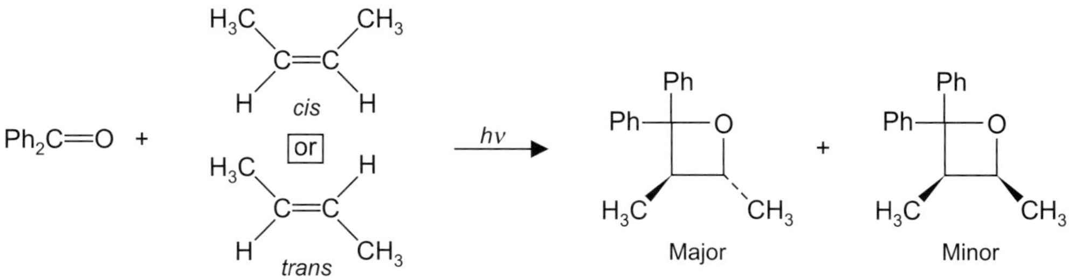

SCHEME 10.29a

Stereospecificity of Paterno-Büchi Reaction

Evidence has been provided to show that Paterno reaction though regioselective but is not stereospecific. Recent work with cis- and trans-isomers of cyclooctene has shown that at low temperature this photochemical [2 + 2] cycloaddition becomes almost stereospecific (Scheme 10.29b).

W. Adam, V.R. Stegmann and S. Weinkotz, J. Am. Chem. Soc., 123, 2452 2001; W. Adam and V.R. Stegmann, J. Am. Chem. Soc., 124, 3600 (2002).

In this case it is suggested that the intermediate 1, 4-diradical undergoes final spin inversion faster and there is not enough time for rotation to occur about single bonds.

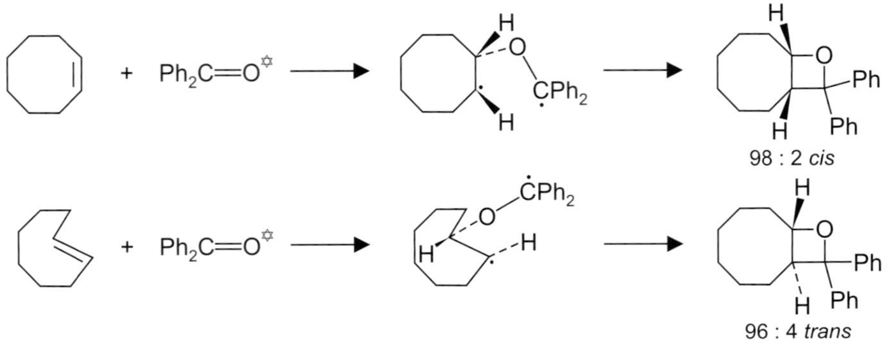

SCHEME 10.29b

EXERCISE 10.5

Write the structure of the product from each of the photochemical cycloaddition reactions (Scheme 10.29c).

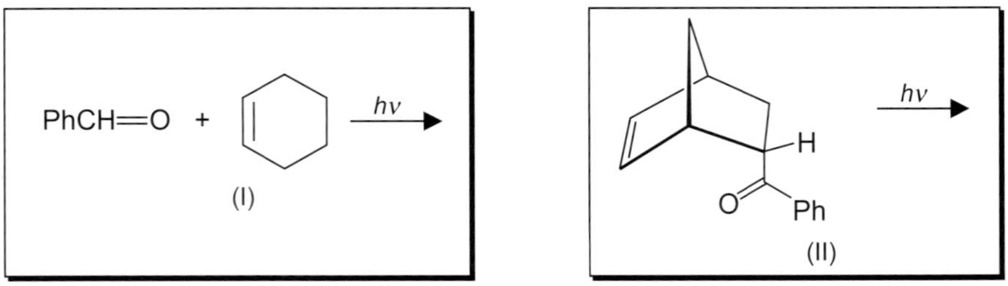

SCHEME 10.29c

ANSWER. *These are given (Scheme 10.29d).*

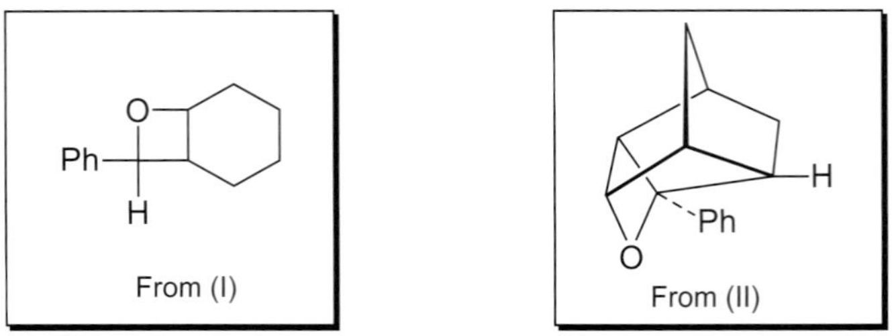

SCHEME 10.29d

An essential condition for the success of Paterno-Büchi reaction is that the triplet energy of the alkene is comparable or higher than the carbonyl compound. If this is not so then the energy transfer from the excited carbonyl group to the ground state alkene will take place and the carbonyl compound will return to its ground state leading to triplet state olefin. In case this happens, one will not get an oxetane, but dimerization of the olefin. Since the energy of the triplet from benzophenone is less than that from norbornene, an oxetane is obtained by the expected Paterno-Büchi reaction (Scheme 10.30) on irradiation of benzophenone in the presence of norbornene. Since the energy of the triplet from acetophenone is more as compared with norbornene, the irradiation of acetophenone in the presence of norbornene gives norbornene dimerization.

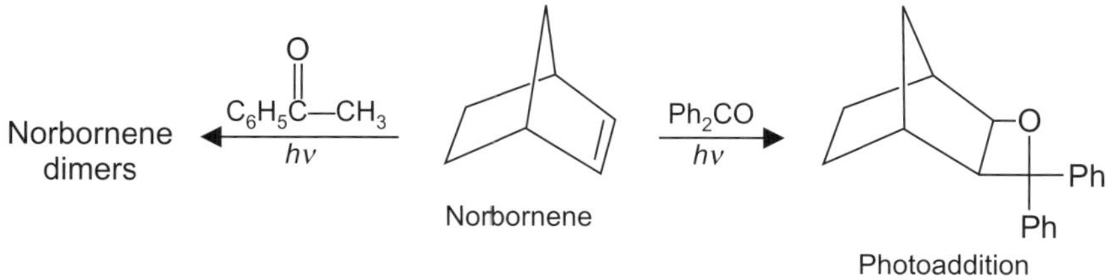

SCHEME 10.30

(b) Photoaddition of alkenes, alkynes and amines to aromatic compounds

Interesting products of cycloaddition are formed when both olefins and acetylenes undergo photochemical addition to benzene. These [2 + 2] cycloadditions are common photochemical reactions, and do not normally take place thermally. Similarly, the important [4 + 2] thermal Diels-Alder cycloadditions usually do not occur on irradiation. These types of contrasting reaction pathways contributed to the development of the theories of pericyclic reactions. The photochemical 1, 2-addition to the benzene ring (initiated by the $n \rightarrow \pi^*$ excitation of the maleic anhydride, Scheme 10.31) immediately undergoes a 1, 4-addition of Diels-Alder type to the 1, 3-diene. An extremely strained cyclobutene ring formed (Scheme 10.32) opens up by a spontaneous electrocyclic reaction.

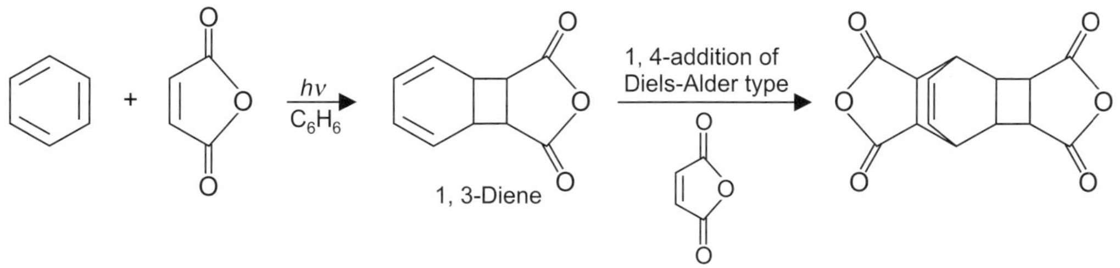

SCHEME 10.31

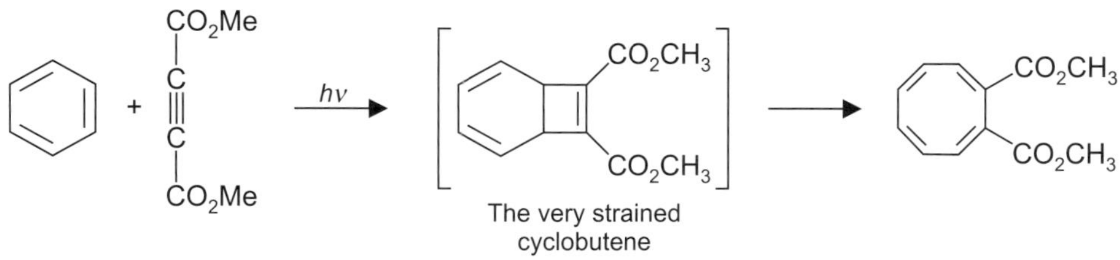

SCHEME 10.32

1, 3- and 1, 4-Addition to benzene ring are known reactions. Irradiation of solutions of alkenes in benzene (or substituted) benzenes gives primarily 1 : 1 adducts where the alkene bridges *meta* positions of the aromatic ring. Thus, the key step in the 1, 3-addition of olefins with benzene is the formation of the intermediate prefulvene (Scheme 10.33) which is formed from a high vibrational level of the first excited singlet of benzene. When butadiene adds to benzene, the part of the product structure derived from butadiene has *trans* geometry (Scheme 10.33). During the 1, 4-addition of a primary or a secondary amine to benzene (via irradiation), the T_1 state of benzene is involved (Scheme 10.34).

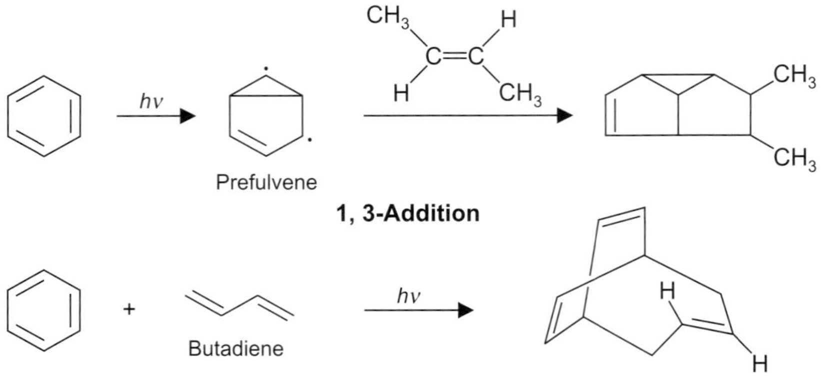

SCHEME 10.33

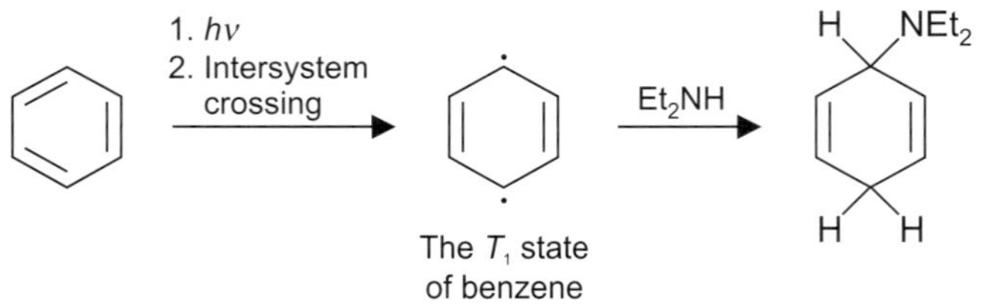

SCHEME 10.34

(H) Photochemistry of α, β-Unsaturated Ketones

When the carbonyl group is conjugated with a carbon-carbon double bond to give an α, β-unsaturated ketone, the energy of the highest π orbital is raised and the energy of the π* orbital is lowered. The energy gap between π and π* states becomes lesser, and less energy (light of longer wavelength) is needed for the π → π* transition in which the entire C=C—C=O group is involved. There is no change in the energy level of the *n* orbital, however the lowering of the level of π* orbital shifts the *n* → π* transition to a longer wavelength. Generally an α, β-unsaturated ketone shows two absorption maxima at 220 nm (π → π*) and at 310 nm (*n* → π*). The irradiation of an α, β-unsaturated ketone can induce either of these transitions, and the excited state, normally has more π, π* than *n*, π* triplet character but an *n*, π* excitation, may be involved initially.

Among the typical reactions initiated by the π, π* states of α, β-unsaturated ketones are (*i*) photocyclodimerization and (*ii*) olefin addition across the double bond to yield cyclobutane derivatives.

(a) Photocyclodimerization

Both cyclopentenone and cyclohexenone dimerize to give four membered ring compounds (Scheme 10.35).

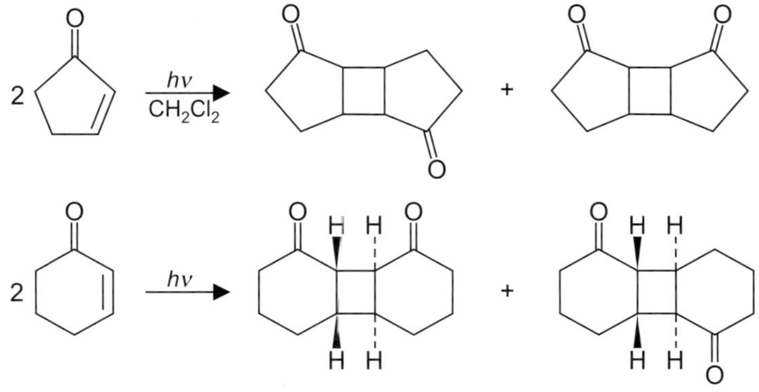

SCHEME 10.35

(b) Olefin addition

The triplet state of an α, β-unsaturated carbonyl compound is a delocalized system and can be represented as a hybrid (Scheme 10.36). Consequently, addition to an olefin can occur both at C=C as well as at C=O. However, the former path is normally favoured (Scheme 10.36).

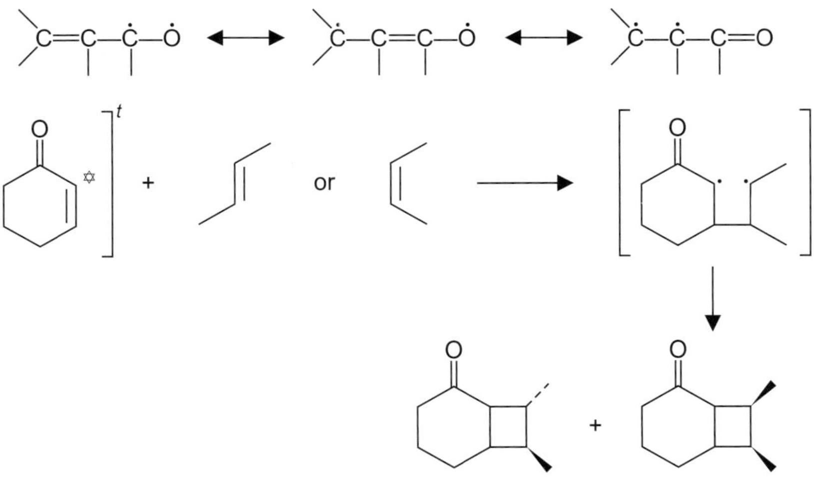

SCHEME 10.36

(c) Cycloheptenone and larger rings

When the enone system is part of the seven or eight membered ring then photoisomerization takes place rather than dimer formation. Cycloheptenone and larger rings, undergo photoisomerization to the *trans*-cycloalkenones. In seven- and eight-membered rings, the *trans* double bonds are highly strained and in nucleophilic solvents addition takes place (Scheme 10.36*a*).

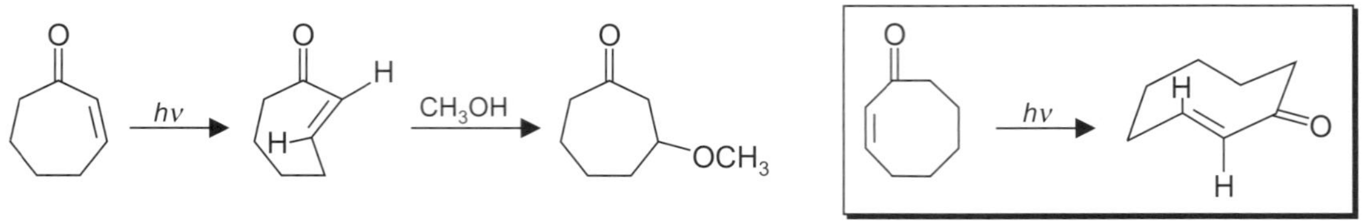

SCHEME 10.36a

(d) 4, 4-Dialkylcyclohexenones

4, 4-Dialkylcyclohexenones display a photochemical rearrangement which involves the formal shift of the C-4—C-5 bond to C-3 with the formation of a new C-2, C-4 bond (Scheme 10.36*b*).

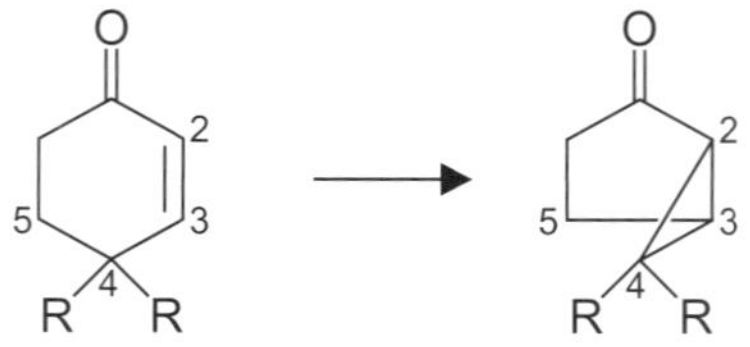

SCHEME 10.36b

(e) Cyclohexadienones

These reactants undergo deep seated rearrangements through a triplet excited state (Scheme 10.36 c).

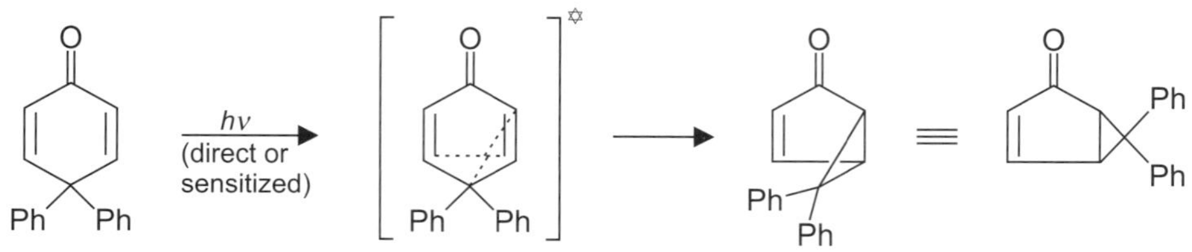

SCHEME 10.36c

(I) Photochemistry of Alkenes, Dienes and Aromatic Compounds

(a) Cis-trans isomerization (see Schemes 10.11 and 10.12)

(b) Photodimerization

This process is common with olefins and aromatic compounds, whereby the photoaddition, the formation of a 1 : 1 adduct (photodimerization) by the reaction of an excited state (singlet or triplet) of one molecule of a reactant with another molecule of the same reactant in its ground states gives dimeric products. Non-conjugated olefins absorb only in the difficulty accessible region around 200 nm and below. Thus their unsensitized photodimerizations are not common, however these are concerted reactions and occur through the excited singlet, S_1.

The photochemical [2 + 2] cycloaddition is symmetry allowed process (Scheme 10.37). The Diels-Alder reaction is however, an example of a Thermal cycloaddition reaction. 2-Butene adds to itself to give a four membered ring when it is irradiated with UV radiation at a wavelength of 214 nm. The product retains the stereochemistry of the starting alkene. The major photochemical reaction (Scheme 10.37) is the stereochemical isomerization of alkenes; the 1-butene isomer is formed by the isomerization of the position of double bond and these reactions are not concerted reactions.

Olefin photodimerization is easy, through the triplet state, which is generated with a photosensitizer. Acetophenone with λ_{max} 270 nm undergoes intersystem crossing from its S_1 state to T_1 efficiently. In the presence of an olefin a singlet sensitizer and a triplet olefin are formed. The olefin in its triplet state then adds through one carbon atom to a carbon atom of a second molecule of olefin and subsequently after a spin-inversion ring formation is completed.

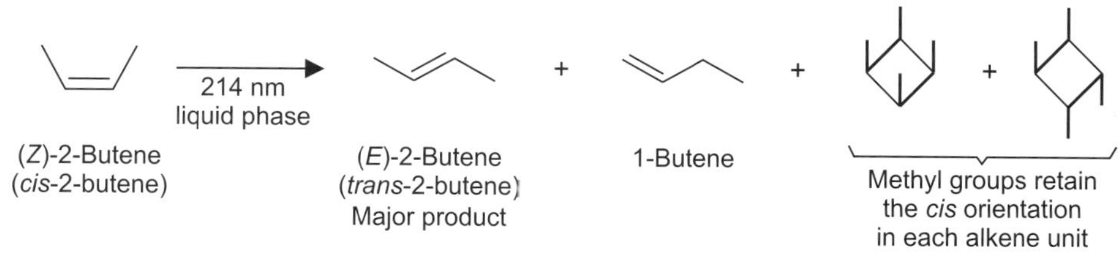

SCHEME 10.37

Conjugated dienes do not display photodimerization through the excited singlet but instead undergo photocyclization. Photodimerization of conjugated dienes can be brought about *via* triplet sensitization as in the case of cyclopentadiene with benzophenone (Scheme 10.38). It may be mentioned that thermal dimerization of cyclopentadiene, gives only *endo*-dicyclopentadiene (Scheme 17.45), unlike in the present case where both *endo*- as well as *exo*-dicyclopentadiene is formed.

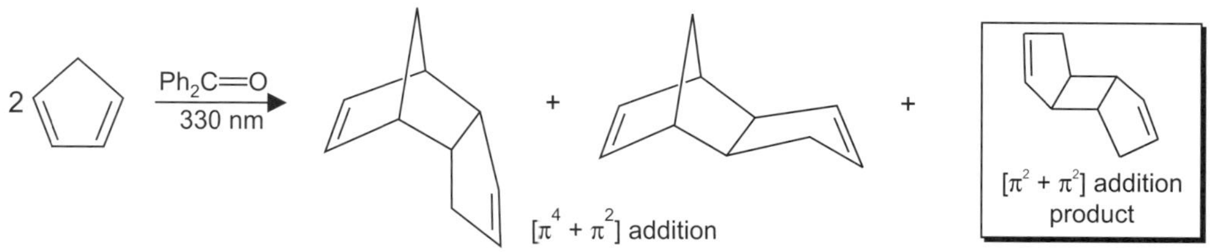

The dimerization of cyclopentadiene

SCHEME 10.38

Anthracene dimerizes (Scheme 10.39) *via* its excited singlet (π, π*) state. Conjugated dienes display a variety of photoreactions depending on if the excitation is direct or photosensitized. When the photochemical dimerization of butadiene is carried out in ether solution in the absence of a photosensitizer two products are formed by intramolecular cyclization (Scheme 10.39*a*). These arise from the excited singlet state, intersystem crossing efficiency in 1, 3-butadiene is almost zero and as a consequence the triplet derived products are not observed.

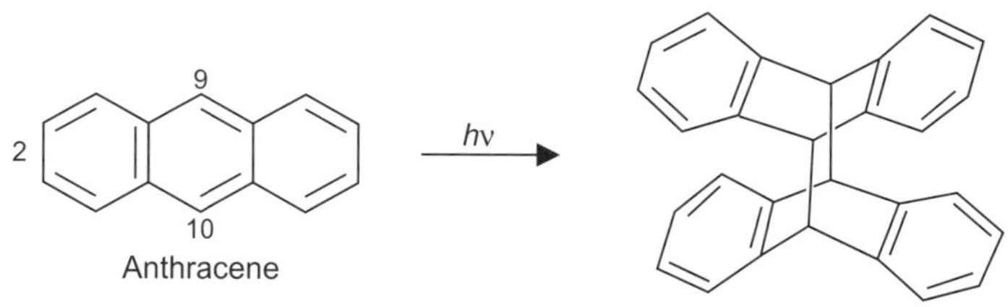

SCHEME 10.39

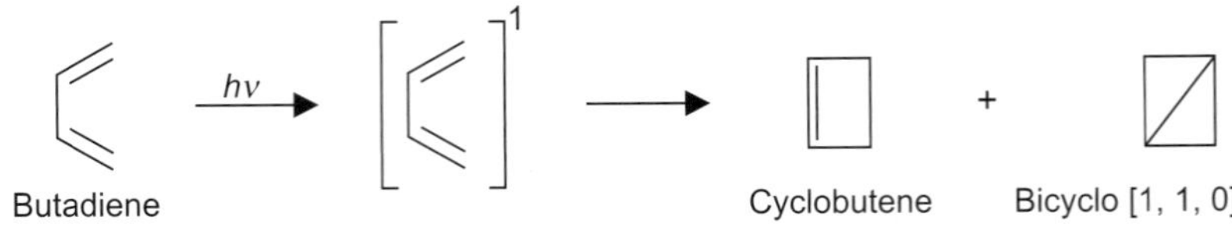

Butadiene Cyclobutene Bicyclo [1, 1, 0] butane

SCHEME 10.39a

However, when the photolysis is carried out in the presence of a sensitizer *e.g.*, triplet excited benzophenone, only dimers are formed involving triplet-excited 1, 3-butadiene formed by energy transfer from triplet excited benzophenone (Scheme 10.39*b*). The following points may be noted:

- On irradiation a mixture of butadiene and benzophenone at 366 nm (78.1 kcal/mol) light is only absorbed by benzophenone.
- Benzophenone has a small energy gap ($S_1 \rightarrow T_1$), therefore, intersystem crossing ($S_1 \rightarrow T_1$) has 100% efficiency.
- The triplet energy of benzophenone (69 kcal/mol) is more than adequate for diffusion controlled energy transfer to butadiene since triplet energy of butadiene is about 60 kcal/mol.
- $S_0 \rightarrow S_1$ energy gap for benzophenone is lower than that of butadiene, therefore transfer of singlet energy from excited singlet benzophenone to butadiene does not occur (see Scheme 10.3*a*).

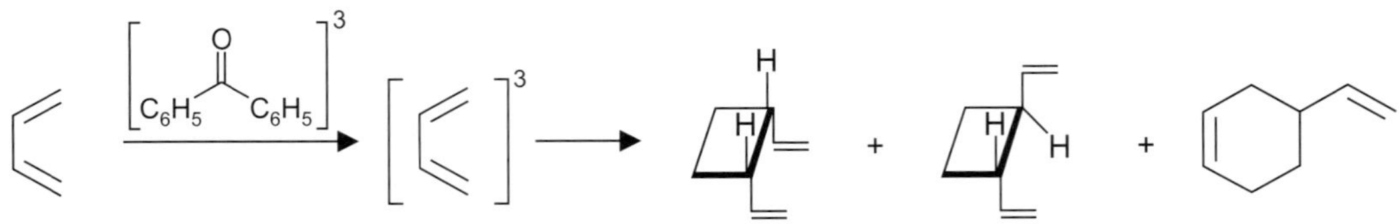

cis-1, 2-divinyl-cyclobutane *trans*-1, 2-divinyl-cyclobutane 3-vinylcyclohexene

SCHEME 10.39b

The geometry of the reacting species *i.e.*, orientations reflect in determining the course of the reaction in terms of the formation of cyclobutane or cyclohexene derivatives (Scheme 10.39*c*).

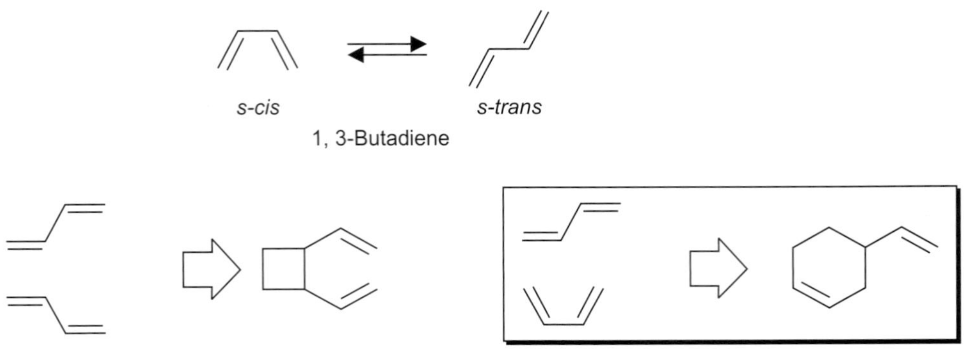

SCHEME 10.39c

(c) Intramolecular photocyclization of 1, 4-dienes

A 1, 4-diene system in which the two π components are separated by an sp^3 carbon displays a photochemical rearrangement known as di-π-Methane Rearrangement (Scheme 10.39*d*) to give a vinylcyclopropane derivative. In this rearrangement either a singlet or a triplet excited state is involved and may proceed *via* a diradical formed *via* bonding between C-2 and C-4. When in a substrate the central sp^3 carbon is unsubstituted, the rearrangement becomes unfavourable as the second step, would lead to the formation of a unfavourable primary radical.

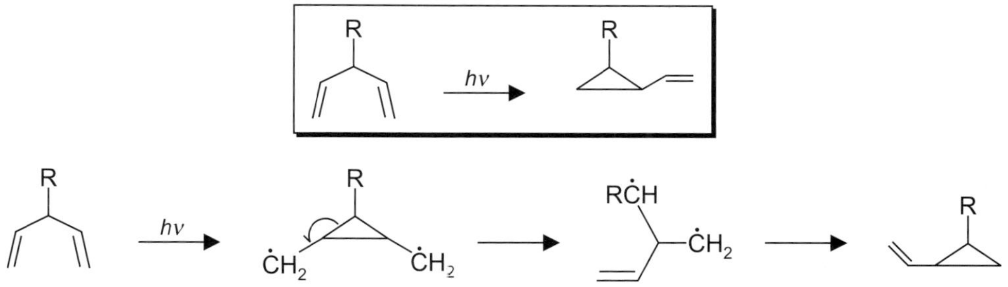

Di-π-methane rearrangement

SCHEME 10.39d

An analogous rearrangement occurs with β, γ-unsaturated aldehydes and ketones (Scheme 10.39*e*) and the rearrangement is then called oxa-di-π-methane rearrangement and gives cyclopropyl ketones.

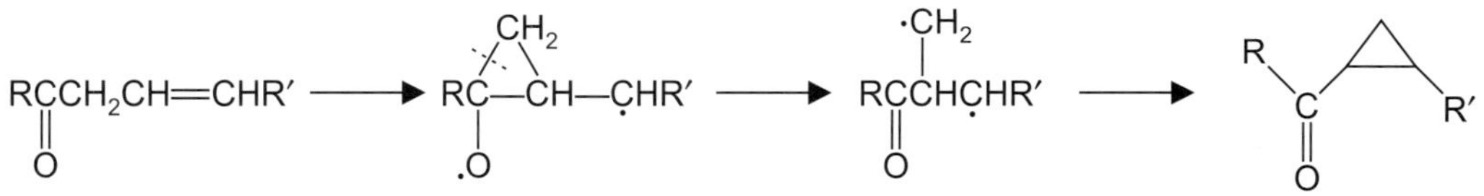

SCHEME 10.39e

EXERCISE 10.6

Write the structure of the product from a photochemical intramolecular cyclization (Scheme 10.39 f).

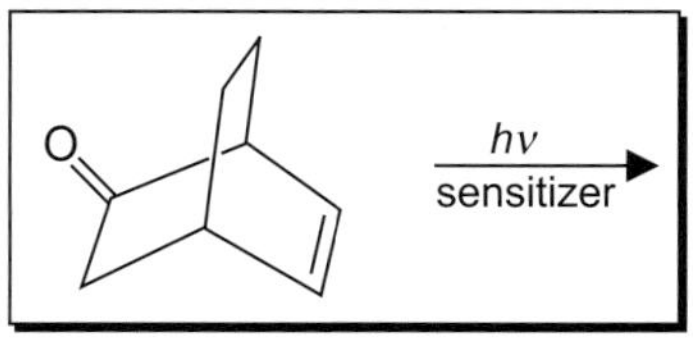

SCHEME 10.39f

ANSWER. *It is a system containing a β, γ-unsaturated moiety, and will undergo oxa-di-π-methane rearrangement (Scheme 10.39g)*

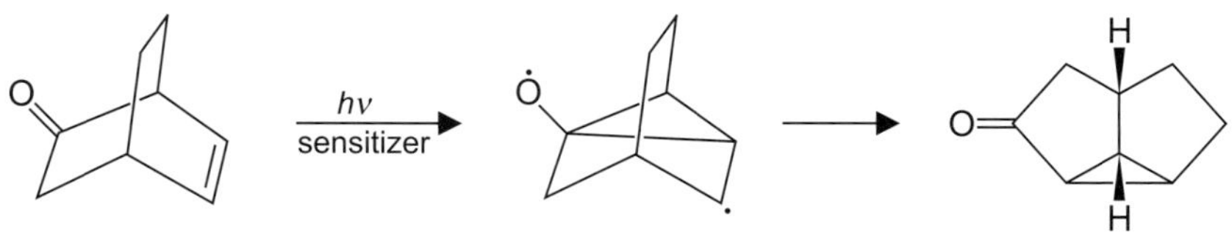

SCHEME 10.39g

Several dienes and polyenes can be photochemically converted into cyclic isomers. The

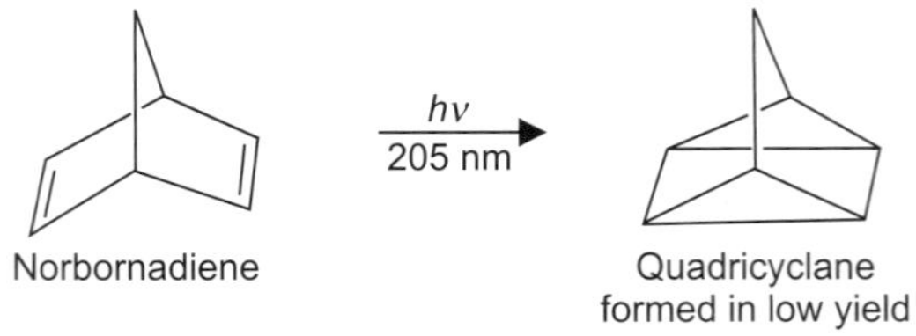

SCHEME 10.40

direct irradiation of norbornadiene (205 nm) gives quadricyclane slowly in low yield (Scheme 10.40). However, the reaction becomes highly efficient by triplet energy transfer by using a triplet sensitizer. Norbornadiene is transparent to light of 313 nm wavelength.

The triplet excited norbornadine thus produced (by energy transfer from triplet excited acetophenone) gives quadricyclane (Scheme 10.41).

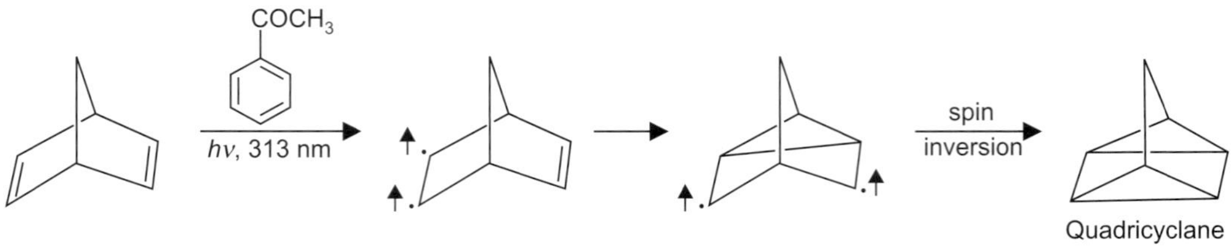

SCHEME 10.41

(d) Conjugated dienes

One has seen the formation of four membered rings during photocyclodimerization (see Scheme 10.35). Other examples include the formation of four membered rings from dienes (Scheme 10.42).

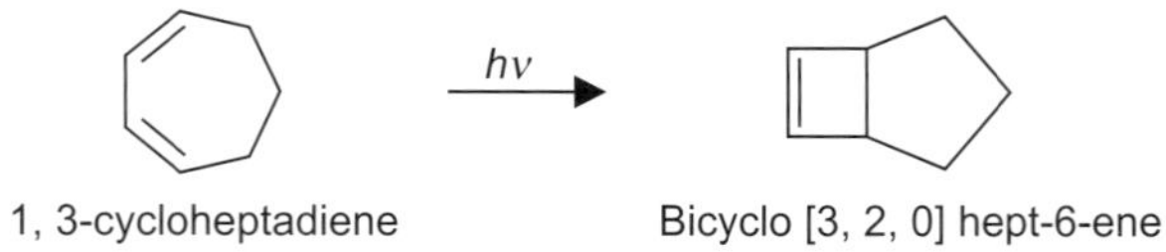

SCHEME 10.42

The different stereochemical outcome of thermal and photochemical reactions is often exploited when reactions are reversible, to bring about isomerization between stereoisomers. The thermal conrotatory ring opening of trans-3, 4-dimethylcyclobutene (I, Scheme 10.42a) to trans, trans-hexa-2, 4-diene (II) is an efficient reaction (the relief of the ring strain). The photochemical ring closure is also very effective: the conjugated diene absorbs light at longer wavelengths than the mono-ene product cis-3, 4- dimethylcyclobutene (III) which is photochemically inert to normal UV radiation. The photon of light provides lot of energy to make the endothermic reaction proceed. By this sequence of reactions, geometric isomers can be interconverted.

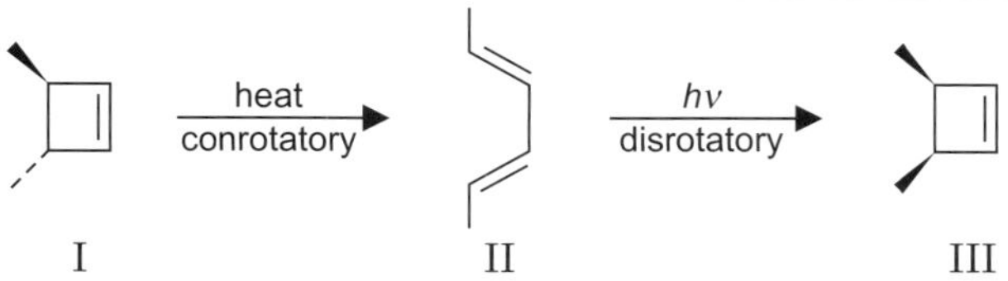

SCHEME 10.42a

Thus in the case of conjugated dienes when the excited singlet state is involved, the reaction is concerted, electrocyclic type and stereospecific. Orbital symmetry principles (see Chapter 17) are employed to rationalize their outcome.

(e) Cage compounds

Several interesting cage compounds can be made from appropriate reactants on irradiation (Scheme 10.42*b*).

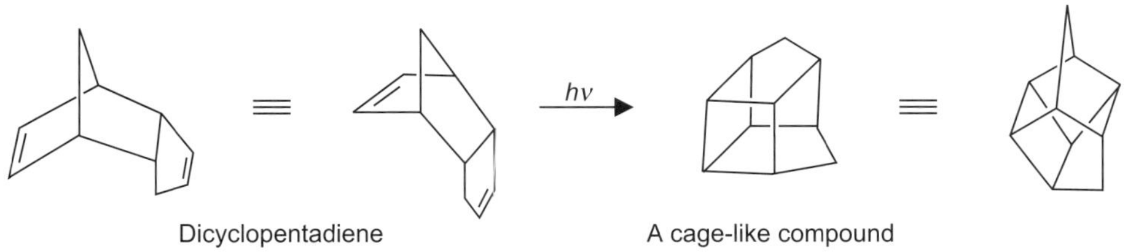

SCHEME 10.42b

(f) Photoisomerization of benzenoid compounds

Benzene on irradiation with light of 254 nm gives both fulvene and benzvalene (Scheme 10.43), *via* prefulvene which are all non-aromatic. Several strained systems like that in Dewar benzene and a prismane have been reached by the irradiation of substituted benzenes (Scheme 10.44).

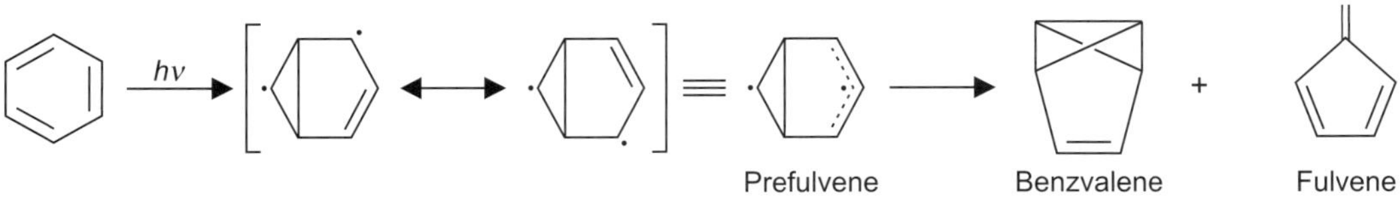

SCHEME 10.43

All these photoproducts are valence isomers of the normal benzenoid structure. These products (Scheme 10.44) are reached from the excited state, however a precise mechanism is not known. Probably the skeletons of Dewar benzene and a prismane arise from the bent triplet state (Scheme 10.45) drawn by ignoring the substituents, for simplicity.

The photochemistry of 1, 3, 5-tri-*t*-butylbenzene and 1, 2, 4-tri-*t*-butylbenzene is rather complex and the photostationary state is established which involves carbon frameworks of a benzvalene, a prismane and a Dewar benzene (Scheme 10.44).

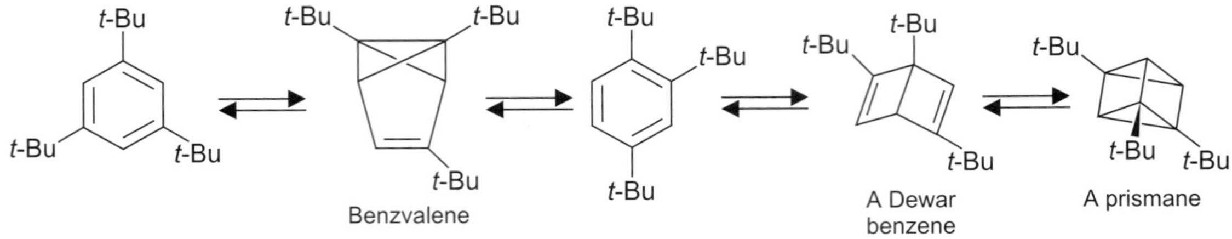

SCHEME 10.44

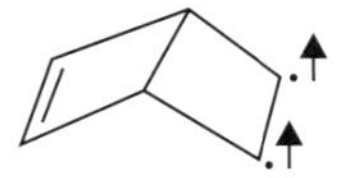

SCHEME 10.45

All these strained systems *i.e.*, Dewar benzene, prismane, benzvalene (as well as their derivatives) are thermally labile and ultimately end up in benzenoid compounds. These strained species are thought to be intermediates in photochemical isomerization reactions of alkyl benzenes to other alkyl benzenes. Thus *o*-xylene on irradiation ends up in mixtures containing *m*- and *p*- isomers (Scheme 10.46).

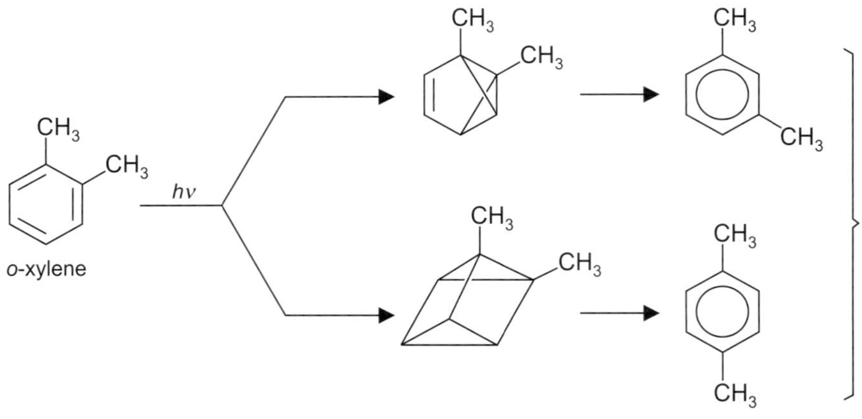

SCHEME 10.46

(J) Photorearrangements

Several photorearrangements (photocyclizations) leading to intramolecular photocyclized products, structural isomers or valence bond isomers have already been discussed, leading to the synthesis of highly stained compounds. Under this section some typical photorearrangements are discussed.

(a) Rearrangement of acyclic α, β-unsaturated ketones—conversion of conjugated substrate to unconjugated product

Acyclic α, β-unsaturated ketones with a γ-hydrogen undergo a double bond migration on irradiation (Scheme 10.47). This process is observed if energy of light is not that high as to bring about dissociation. The γ-hydrogen abstraction and its transfer to the oxygen atom occurs *via* a six-membered cyclic transition state (Scheme 10.47). The substrate changes from conjugated to non-conjugated system without any equilibrium (as is the case with *cis-trans* isomerisation). The reason for this is that irradiation is carried out at a wavelength absorbed by the conjugated substrate and since this is longer than that absorbed by the non-conjugated product, the latter remains unaffected after it is formed. Isomerization therefore, yields a less stable isomer and this isomerization is another example of optical pumping.

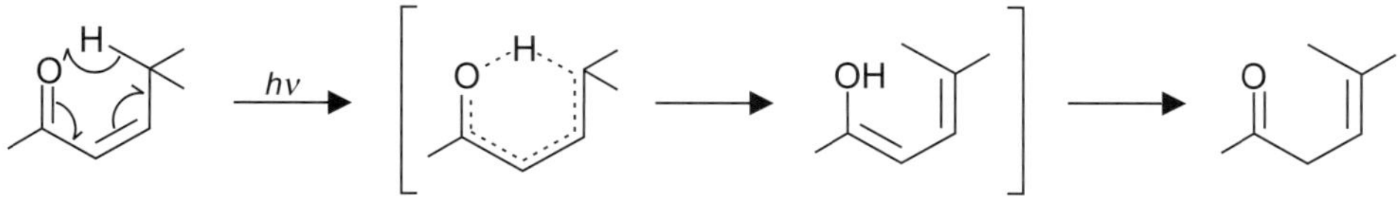

SCHEME 10.47

(b) Photo-Fries reaction

Phenol esters in solution on photolysis give a mixture of *o*- and *p*- acylphenols (Scheme 10.48). In the gas phase, where solvent cage effects are not operative many other products are formed.

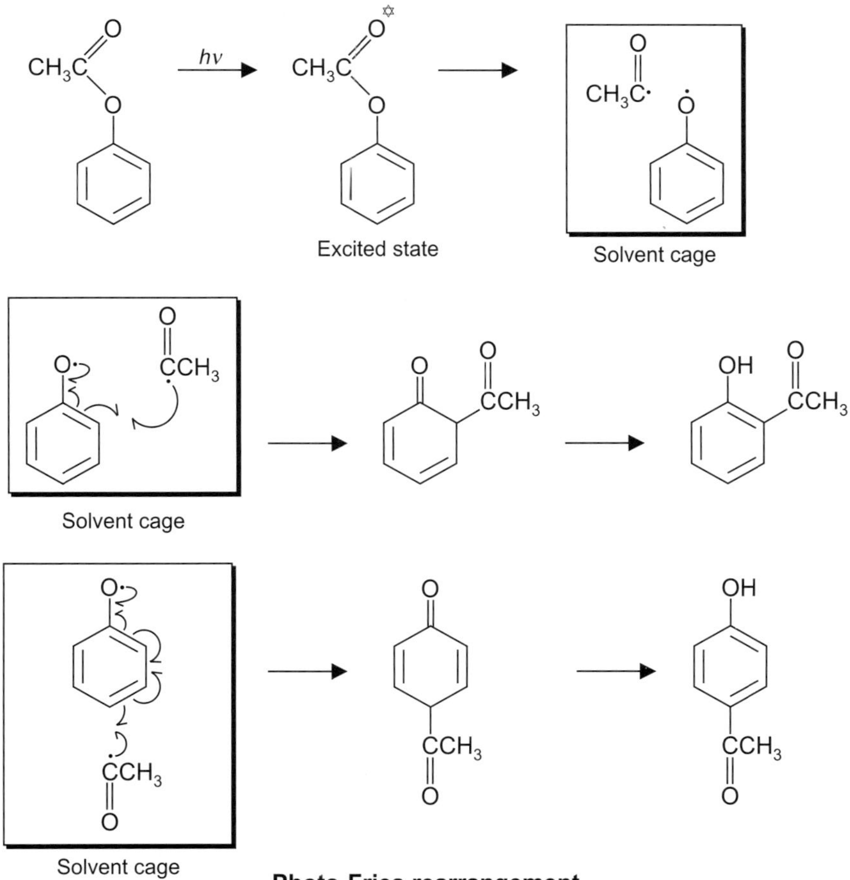

Photo-Fries rearrangement

SCHEME 10.48

In Fries rearrangement (Scheme 8.50), however, such products are formed in the presence of a catalyst. Photo-Fries reaction does not need a catalyst and is normally an intramolecular free radical process. Significantly, the phenol (ArOH) is always a side product which arises from some ArO• which leaks from the solvent cage and can abstract a hydrogen atom from a neighbouring molecule.

(c) Photorearrangement of 2, 5-cyclo hexadienones

The excited singlet produced from (I, Scheme 10.49) by an $n \rightarrow \pi^*$ transition in the excited singlet on intersystem crossing gives triplet state (III, Scheme 10.50). Reorganization of bond, as shown in (IV, *via* free radicals) gives another triplet state molecule (V, Scheme 10.49) which contains a three membered ring. The relaxation yields a singlet state and a ground state zwitterion (VI). This carbocation then undergoes rearrangements which are typical of such an ion to give the observed product (VII).

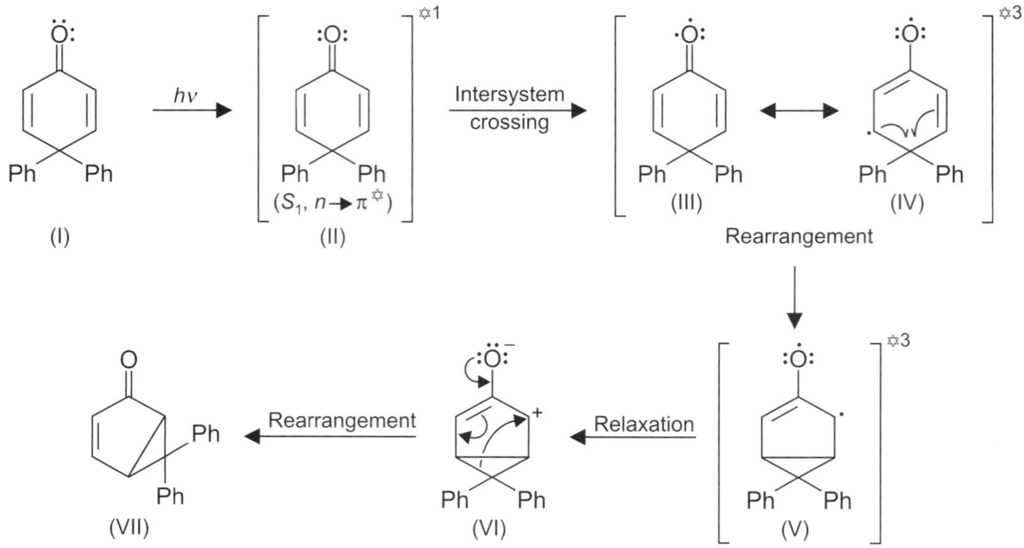

SCHEME 10.49

The cyclohexadienone system in (I, Scheme 10.49) absorbs around 300 nm ($n \rightarrow \pi^*$) and 240 nm ($\pi \rightarrow \pi^*$) and the rearrangements are observed by irradiation at either wavelength. Thus, one may assume that a lower energy transition ($n \rightarrow \pi^*$) is involved initially. Moreover, energy transfer is found to occur when (I, Scheme 10.49) is irradiated with triplet energy sensitizers *e.g.*, acetophenone. These data therefore, reasonably suggest that the photorearrangement of 2, 5-cyclohexadienones occurs *via* an $n \rightarrow \pi^*$ triplet state.

(K) Photochemical Aromatic Substitution

Compared to the ground state, the distribution of charge in an excited species can be entirely, different. Several examples are available where one can control the positional selectivity in nucleophilic aromatic substitutions by working under photochemical conditions. Thus, the reaction of 3, 4-dimethoxynitrobenzene on heating with OH^- ion leads to the replacement of 4-methoxy group, while at room temperature under UV irradiation, instead the 3-methoxy substituent is substituted (Scheme 10.50). On heating, the nitro group through its –*M* effect makes the *ortho* and *para* positions positive compared with the *meta* position. Under irradiation, however, the *ortho* and *meta* positions are rendered positive compared to the *para* position.

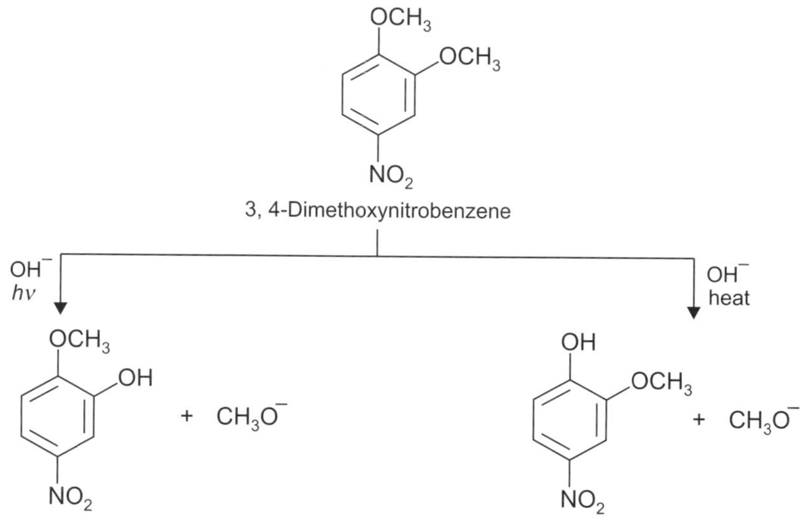

SCHEME 10.50

PROBLEMS

10.1 Plan a synthesis of the following compound:

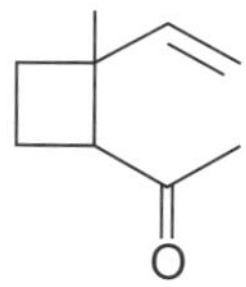

10.2 Write the structures and mechanism of formation of products from the irradiation of cyclopentanone.

10.3 Write the structure of the products from the photochemical reactions of the following compounds:

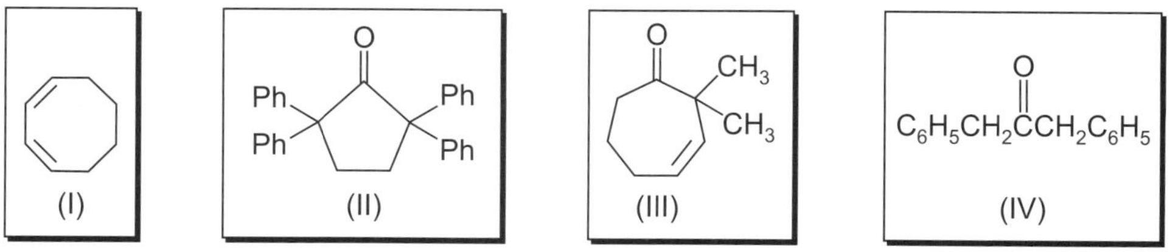

10.4 Considering the photocyclization of norbornadiene to quadricyclene (Scheme 10.40), write the structure of the product of irradiation of cholesta-3, 5-diene.

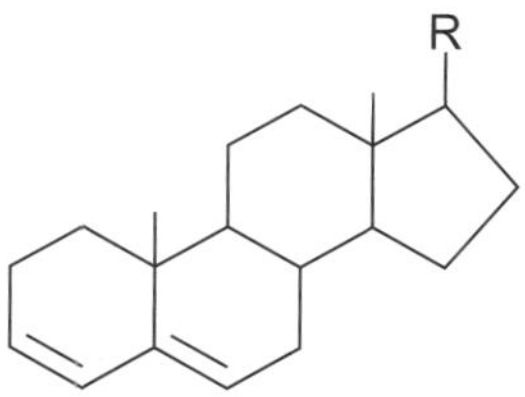

10.5 Write the structure of the product (with the name of reaction) formed from the irradiation of the following compound:

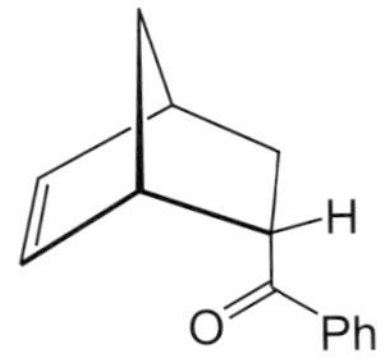

10.6 Rationalize the following reaction:

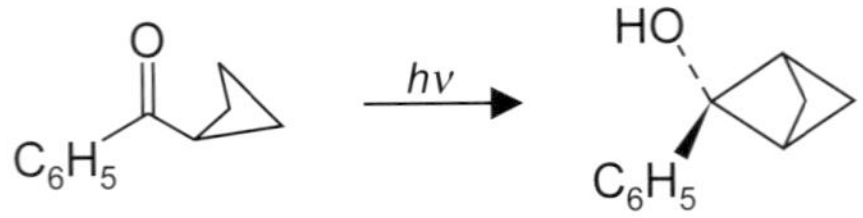

(I) (II)

10.7 Cyclopentadiene adds readily to *p*-benzoquinone in a [4 + 2] manner on heating. What product will be formed on the irradiation of product (I) of this reaction?

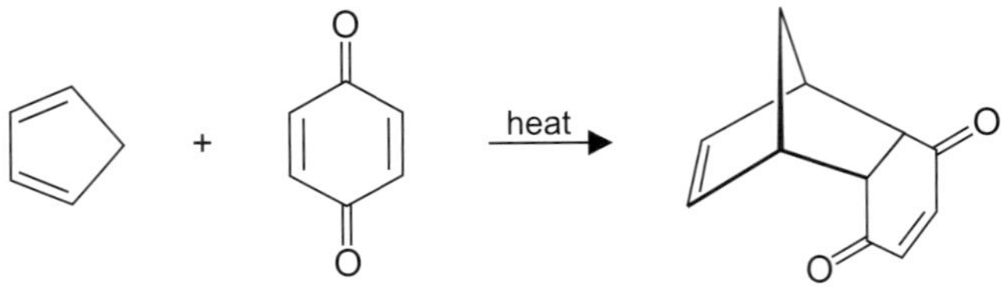

(I)

10.8 Considering the reaction sequence (Scheme 10.42), plan a synthesis of cubane.

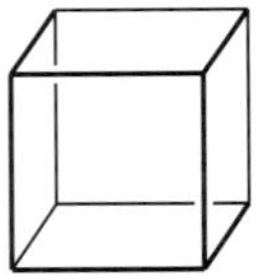

Cubane

10.9 A diazo compound on irradiation in heptane at –78°C gives rise to a saturated hydrocarbon C_4H_6. This displays three signals in ^{1}HNMR and two signals in ^{13}C NMR spectra in the aliphatic region. Suggest a structure to this compound.

$$CH_2{=}CHCH_2CH{=}\overset{+}{N}{=}\ddot{N}:^-$$

10.10 1, 4-Cineole and ascaridole are the naturally occurring oxide and peroxide respectively. There are three isomeric terpinenes which out of these is a suitable isomer for photochemical conversion into either 1, 4-cineole or ascaridole?

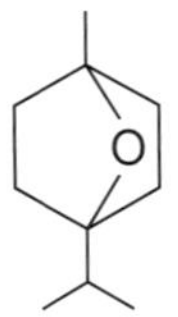

1, 4-cineole

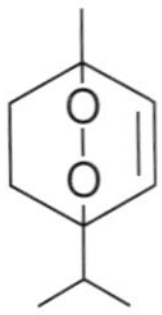

Ascaridole

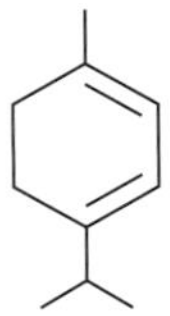

α-terpinene

β-terpinene

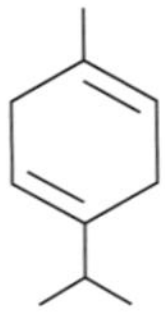

γ-terpinene

10.11 The sesquiterpenoids caryophyllene and its geometrical isomer occur naturally in the oil of cloves. Plan a synthesis of a suitable starting material containing a four-membered ring.

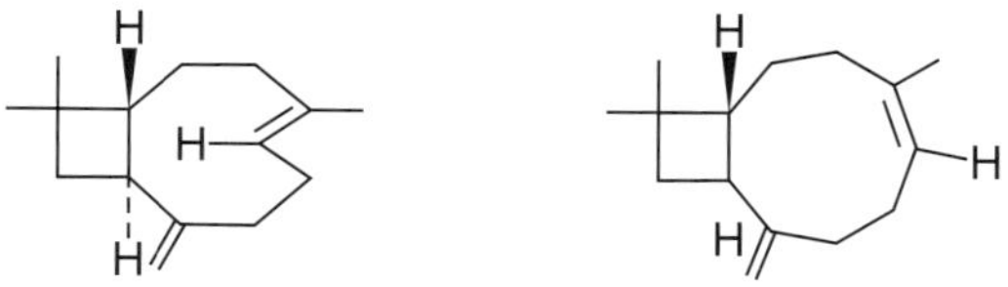

Caryophyllene Isocaryophyllene

10.12 Write equations for the outcome of Paterno-Büchi reaction of ArCHO with furan and dihydrofuran.

ANSWERS TO THE PROBLEMS

10.1 The cyclobutane ring can be constructed *via* photochemical cycloaddition of the alkene using acetophenone sensitizer.

10.2 The path followed is the n, π^* excitation of the carbonyl group, α-fission, a hydrogen atom transfer from the γ-position to the carbonyl group (formation of aldehyde) or the formation of cyclobutane *via* the loss CO and cyclization of the diradical.

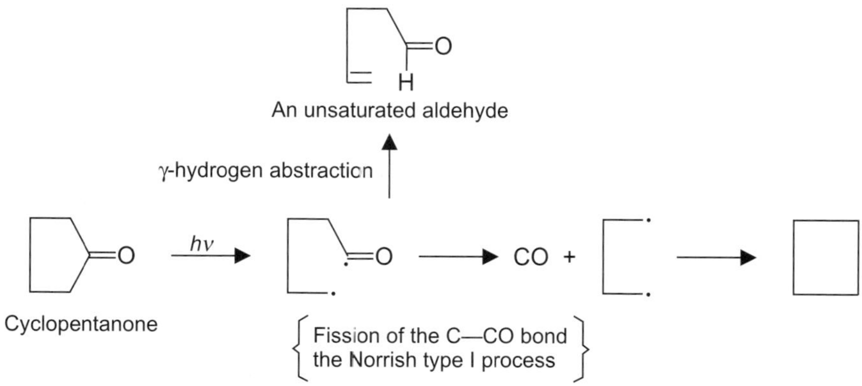

10.3 The initial α-cleavage product gets decarbonylated fast if the radical is stabilized (as in the case of II). The compound (II) thus reacts *via* the diradical to give either the cyclized product or the fragmentation product. In the case of (III) which is a β, γ-unsaturated cyclic ketone, the α-cleavage gives a resonance stabilized allyl radical on one side. The diradical recombines to give another isomerized ketone.

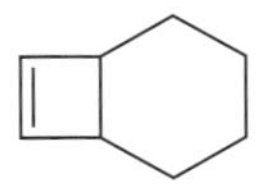

From I

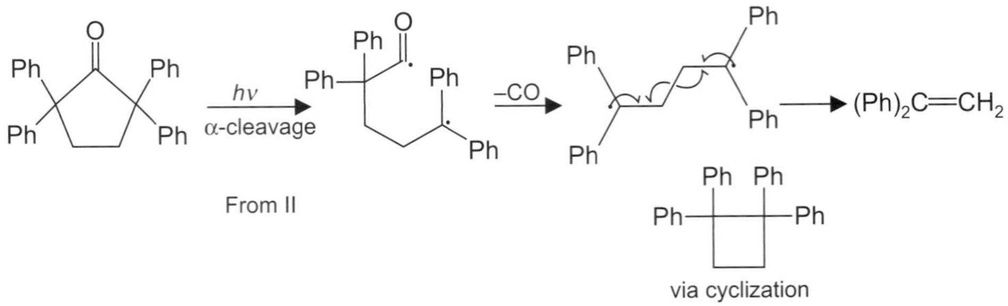

From II

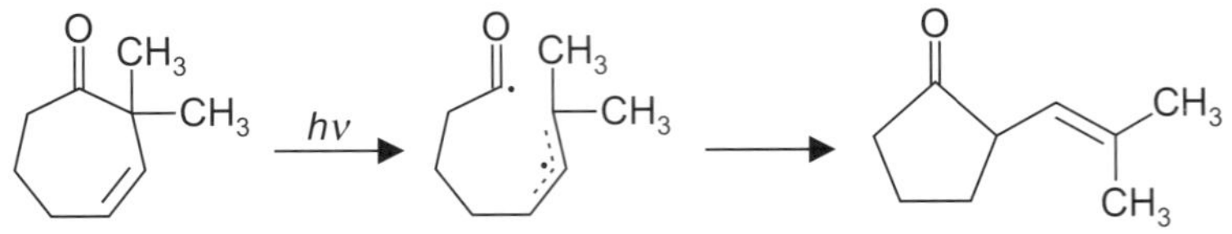

From (III)

$C_6H_5CH_2—CH_2C_6H_5$

From (IV)

10.4

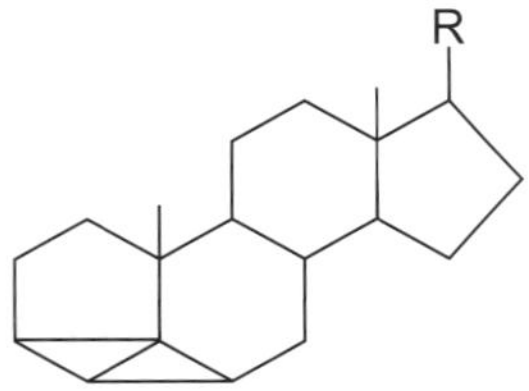

10.5 It is Paterno-Büchi reaction which leads to the formation of a tetracycloxetane.

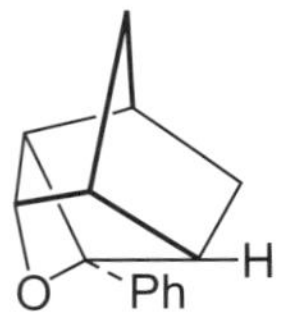

10.6 The photochemical conversion of (I to II) involves the coupling rather than the fragmentation of the Norrish Type II biradical to the cyclobutanol.

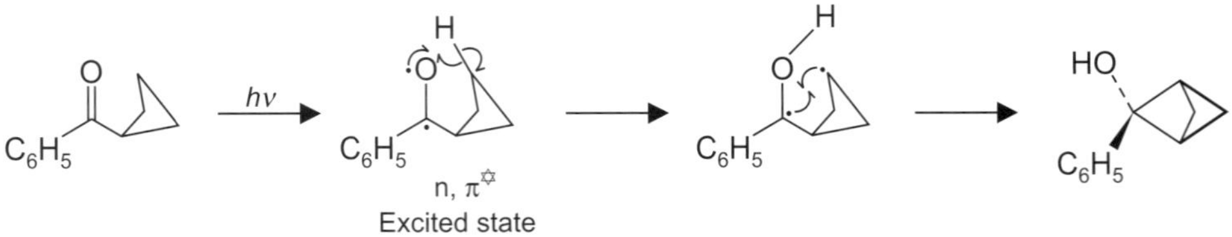

10.7 The compound will undergo ring closure to give a cage compound.

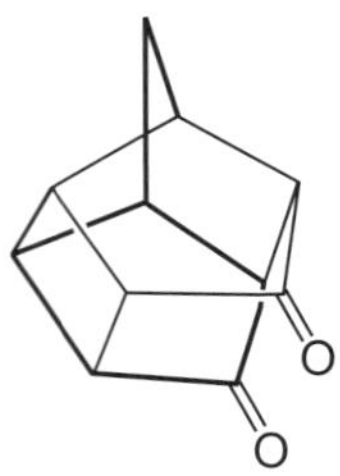

10.8 The potential cage structure of cubane is reached by the dimerization of 2-bromocyclopentadienone which occurs spontaneously. The irradiation step was best performed on a ketal derivative of the dimer from which the keto group was subsequently regenerated by hydrolysis. The successive base induced ring contractions are reactions of Favorskii type. The cubanedicarboxylic acid has been converted into cubane itself by heating the *t*-butyl perester (from the acid chloride with *t*-butyl-hydroperoxide) in cumene; homolysis of the peroxide bond is followed by decarboxylation and uptake of a hydrogen atom from the solvent.

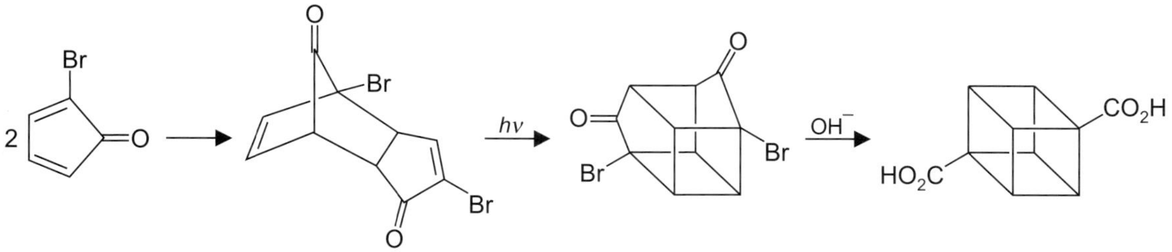

$$RCO_2H \longrightarrow RCO{-}O{-}O{-}C(CH_3)_3 \xrightarrow{heat} RCO{-}\dot{O} + \dot{O}C(CH_3)_3$$

$$RCO_2^{\bullet} \xrightarrow{-CO_2} R\cdot \xrightarrow{PhCHMe_2} RH + Ph\dot{C}Me_2$$

10.9

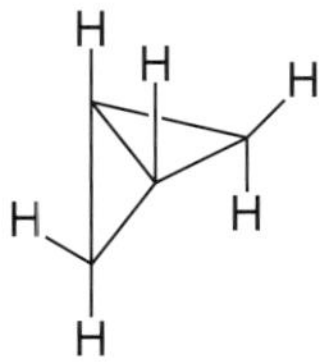

10.10 Formation of cyclic peroxides by conjugated dienes is a general reaction, and although ultraviolet light often initiates the reaction, better results are achieved by carrying out the irradiation of α-terpinene in the presence of sensitizers, *e.g.*, chlorophyll, dyes, etc.

10.11

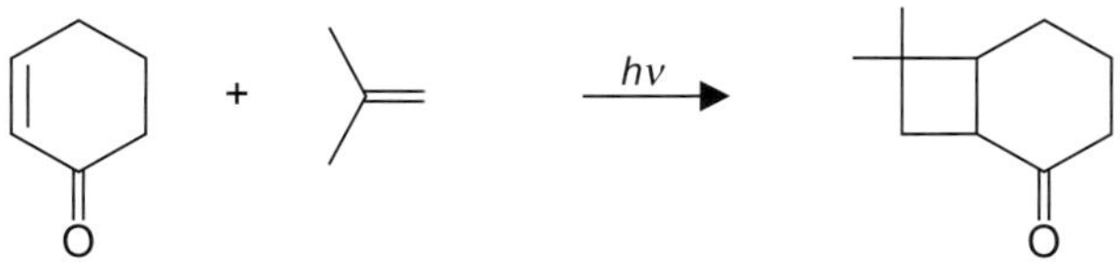

cis + trans

10.12 The regioselectivity of the reaction can be explained by invoking the formation of the most stable intermediate-2-oxa-1, 4-diradical. For this reason one finds the reversal of orientation during the photochemical addition of ArCHO to furan and dihydrofuran.

A.G. Criesbeck and S. Stadtmuller, *Chem. Ber.* 123, 357 (1990).

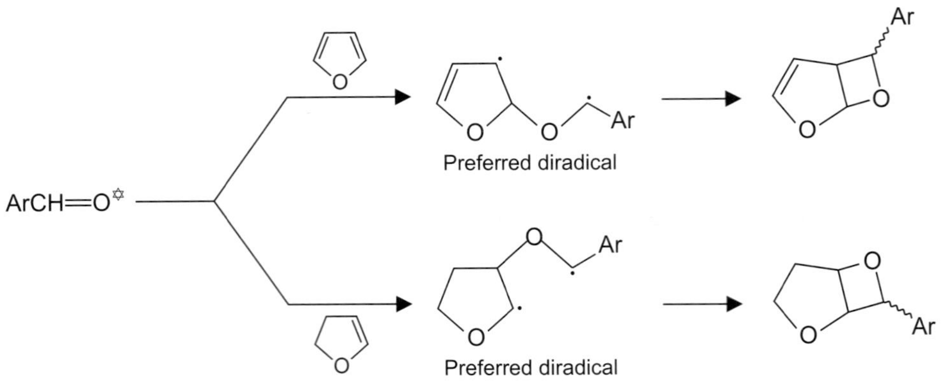
ArCH=O*
Ar
Preferred diradical
Ar
Ar
Preferred diradical
Ar

CHAPTER 11

Addition to Carbon-Carbon and Carbon-Hetero Multiple Bonds

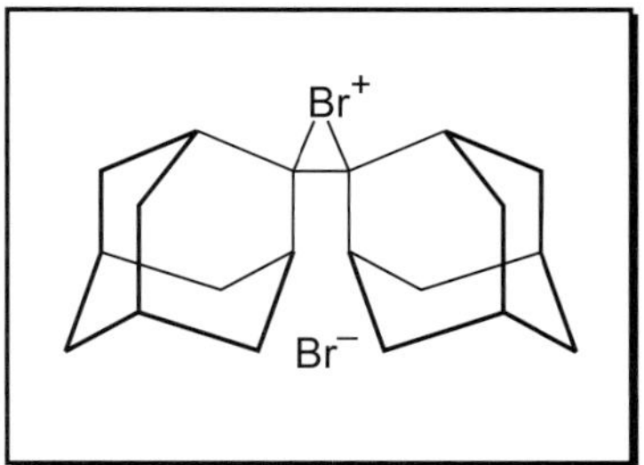

In this chapter the broad mechanistic and stereochemical aspects of addition reactions involving electrophiles, nucleophiles and free radicals to a multiple bond are described. The other specific topics and reactions are described elsewhere as indicated:

- *HYDROGENATION*
- *HYDROBORATION*
- *MICHAEL REACTION*
- *SHARPLESS ASYMMETRIC EPOXIDATION*
- *METAL HYDRIDE REDUCTIONS*
- *ADDITION OF ORGANOMETALLIC REAGENTS TO CARBONYL COMPOUNDS*
- *WITTIG REACTION*
- *CONDENSATION REACTIONS INVOLVING ENOLATES*
- *HYDROLYSIS OF ESTERS, AMIDES AND AMMONOLYSIS OF ESTERS*

11.1 ADDITION OF ELECTROPHILES TO A MULTIPLE BOND

(A) Introduction

A proton and other electrophiles *e.g.*, Br^+ readily add to one end of a C═C or C≡C bond, and to the more electronegative atom in C═N, C≡N or C═O bonds. During addition to a C═C or C≡C bond, the addition involves the formation of the most stabilized cation. Alkyl groups stabilize a carbocation by inductive and hyperconjugative effects; unsaturated and aromatic groups stabilize a cation by delocalization of the positive charge onto remote atoms. Propene reacts with HBr in the first step to give the secondary prop-2-yl cation, rather than the less stable primary prop-l-yl cation (Scheme 11.1).

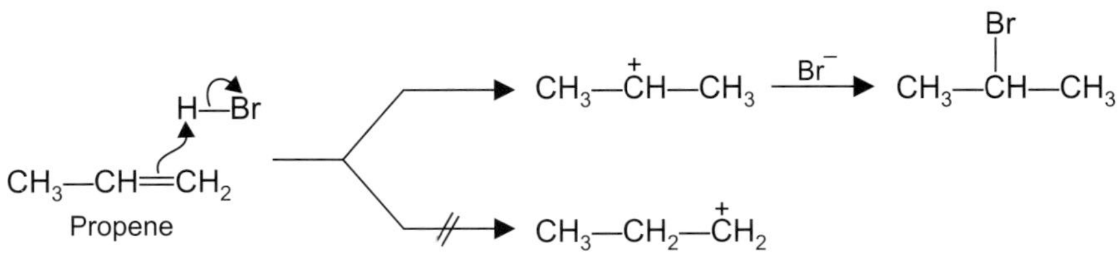

SCHEME 11.1

The addition of HCl to 3, 3, 3-trifluoropropene occurs in the opposite direction, the proton adds to the central carbon instead of the terminal carbon as seen in the case of propene. In this case, the secondary carbocation would be destabilized by the electron withdrawing CF_3 group (Scheme 11.1*a*).

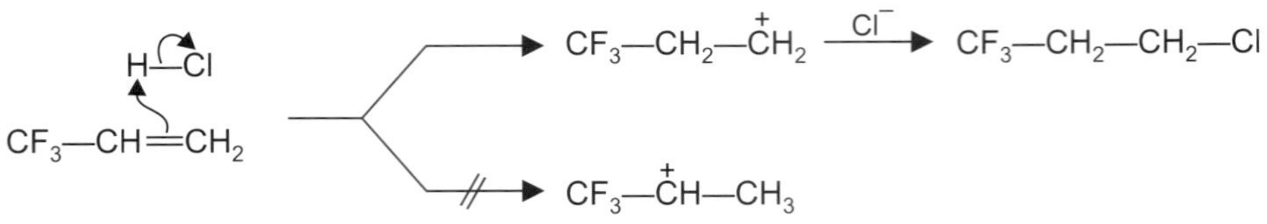

SCHEME 11.1a

EXERCISE 11.1

Why unlike the hydration of a mono-alkyl-substituted ethene, the hydration of alkoxysubstituted alkene (a vinyl ether) with aqueous acid occurs under much milder conditions and with complete regioselectivity ?

ANSWER. *The substituents which stabilize the intermediate carbocation by an electron donating resonance interaction will not only largely accelerate the reaction but also control the regioselectivity (Scheme 11.1b).*

$$RO—CH{=}CH_2 \xrightarrow{H_3O} RO—\overset{+}{C}H—CH_3 \longleftrightarrow R\overset{+}{O}{=}CH—CH_3$$

$$RO—CH{=}CH_2 + HCl \longrightarrow RO—CHCl—CH_3$$

SCHEME 11.1b

A vinyl ether reacts rapidly with hydrogen chloride, however in the presence of water i.e., reaction with hydrochloric acid, the intermediate carbocation reacts with water to give a hemiacetal (see, Scheme 14.37).

EXERCISE 11.2

Why ethylene reacts with HCl faster than vinyl chloride ?

ANSWER. *Consider the carbocations involved during the addition of HCl (Scheme 11.1c). The cation (III) is less stable than (I, –I effect of chlorine). The cation (II) has the halogen nearer the positive charge so considering –I effect of chlorine, this cation is more strongly destabilized and this factor overweighs the +M effect of chlorine* ($CH_3—\overset{+}{C}H—Cl \longleftrightarrow CH_3—CH{=}Cl^+$)*. Thus the ease of formation of the ions (I-III) decreases in the order* $CH_3—\overset{+}{C}H_2 > CH_3—\overset{+}{C}HCl >$

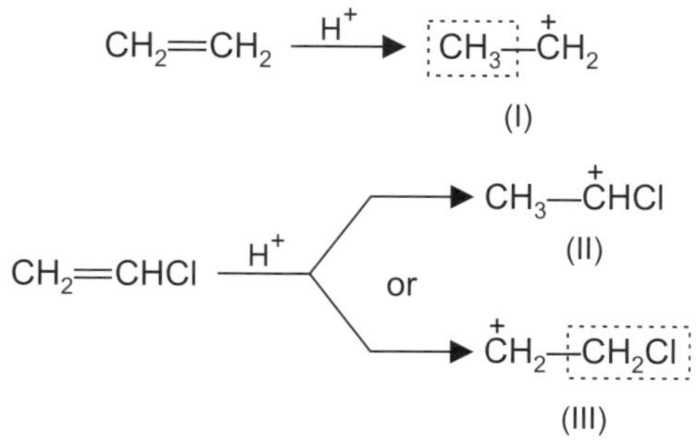

SCHEME 11.1c

$\overset{+}{C}H_2$—CH_2Cl. *Thus vinyl chloride reacts at a slower rate than ethylene but orients in the same way as propylene.*

The carbonyl group gets protonated to give a delocalized cation. This reversible reaction is the first step of several acid catalysed reactions on carbonyl compounds (Scheme 11.1*d*).

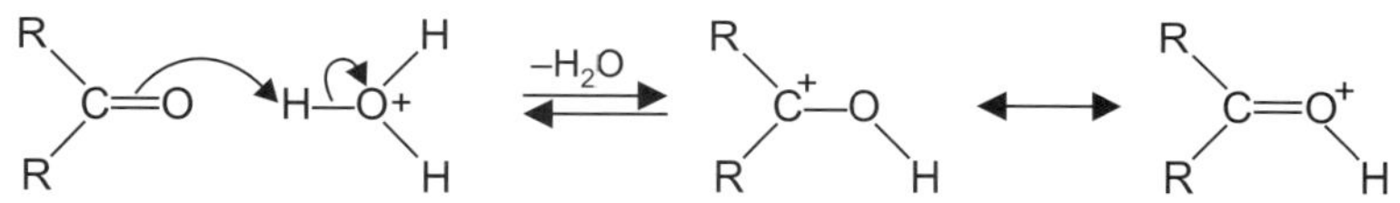

SCHEME 11.1d

(B) Addition of Halogens

Chlorinations and brominations of alkenes are electrophilic additions and involve a discrete positively charged intermediate which may be a bridged cyclic halonium ion (*e.g.*, bromonium ion I, Scheme 11.2) or a carbocation. This positively charged intermediate is formed after the addition of Br^+ to the alkene (see, also Scheme 4.23).

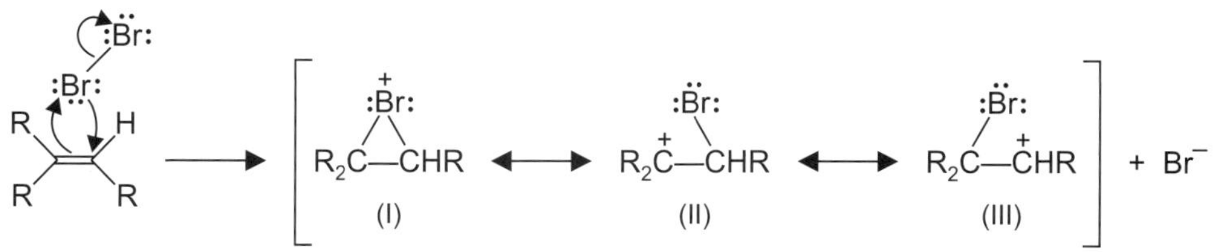

SCHEME 11.2

The bridged bromonium ion is best represented by the hybrid (Scheme 11.2) where a tertiary carbocation (II) is a far more important contributor to the hybrid than the secondary carbocation (III, Scheme 11.2). Consequently the carbon with an alkyl substituent(s) will be more electrophilic. The regioselectivity is explained on this stability feature and the addition of bromine to propylene in water (water competes with bromide ion as the nucleophile) gives in addition to dibromo compound, 1-bromo-2-hydroxypropane as the major product than 2-bromo-1-hydroxypropane (Scheme 11.2*a*). Thus regioselectivity of addition of BrOH to an unsymmetrical olefin is explained on the charge distribution within the delocalized reactive intermediate (bromonium ion) (See, Scheme 11.2). In fact the three membered ring of the halonium ion is symmetrical provided the original double bond is symmetrical. Greater stabilization of the 2° carbocation makes the bridged bromonium ion infact an unsymmetrical species (I, Scheme 11.2*a*).

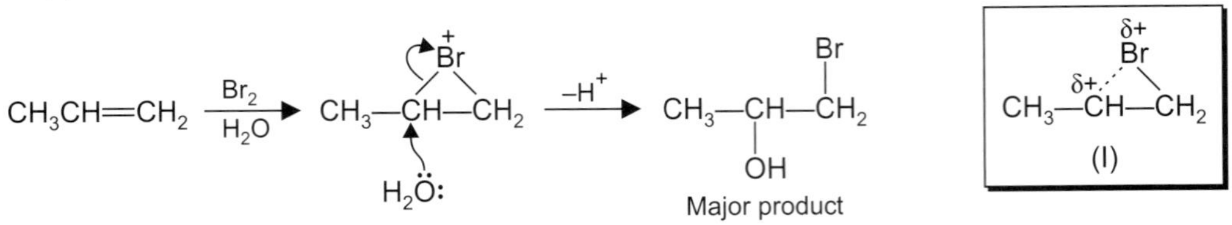

SCHEME 11.2a

The formation of a bromonium ion intermediate ensures the formation of *anti* product. The bridging will prevent free rotation around the C—C bond which will ensure the formation of a mixture of diastereomers. Note that the opening of the cyclic bromonium ion is an S_N2 reaction in which the incoming nucleophile attacks on the opposite side of the leaving group. Moreover, when bridging is strong, the *anti* product is formed as a major or exclusive product.

The *anti*-addition of bromine is observed for alkenes which do not have substituent groups that would stabilize a carbocation intermediate (Scheme 11.2). Thus, addition of bromine to *cis*- and *trans*-2-butene is stereospecific (Scheme 11.3). The first step is the formation of a

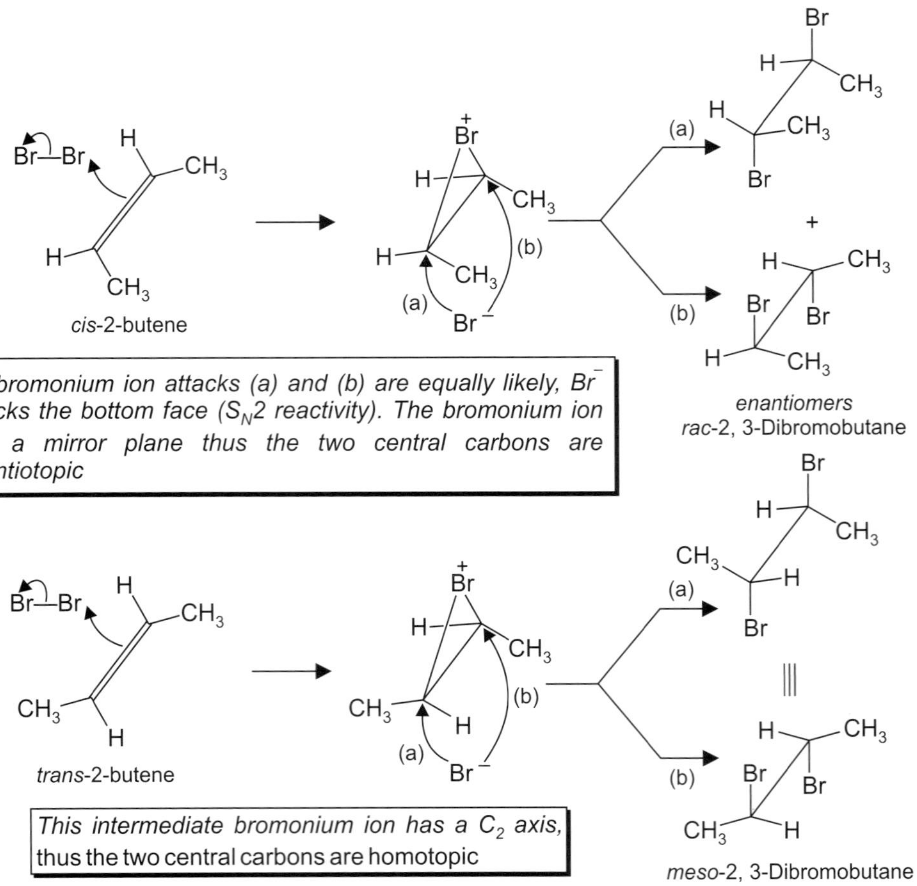

SCHEME 11.3

positively charged bridged bromonium ion and in the second step the nucleophile, Br^- adds to the face away from the bridging group to give the overall *anti* addition.

When the alkene has phenyl group on the double bond, the selectivity becomes less and both *anti* and *syn* adducts are formed. This is so, because now the positive charge in the intermediate is delocalized on the aromatic ring (Scheme 11.3*a*). The presence of a phenyl group, therefore, provides sufficient stabilization to allow carbocation formation (Scheme 11.2). This situation reduces the strength of bromine bridging and allows rotation to occur as shown (Scheme 11.3*a*). The freely rotating open carbocation would give both *syn* as well as *anti* addition

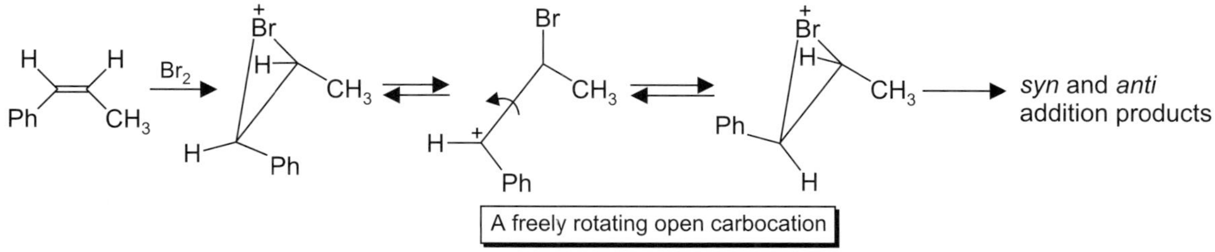

SCHEME 11.3a

products. Thus *syn* and *anti* addition is observed with both *Z*-1-phenylpropene (Scheme 11.3*a*) as well as with *E*-1-phenylpropene

> **ADDITION OF HALOGENS TO ALKENES**
>
> *A weakly bridged bromonium ion can lead to the loss of stereospecificity during halogen addtion. The unconjugated alkenes involved in strong bridging leading to predominant anti-stereospecificity. The presence of a phenyl group on the double bond generates a cationic character at the benzylic site leading to the formation of more syn addition. Chlorine has a smaller size and lesser polarizability than bromine, consequently chloronium ion is far less bridged. Thus overall bromination gives more pronounced anti addition than chlorination.*

Cyclohexene and its derivatives add chlorine and bromine to yield *trans* diaxial product since only axial positions on adjacent carbons in a cyclohexane are *anti* and coplanar. The initial *trans* diaxial product undergoes a conformational change and is in equilibrium with the more stable *trans* diequatorial conformation (Scheme 11.3*b*). When a *t*-butyl group is introduced on the cyclohexene ring, then the molecule exists almost exclusively in a conformation in which the *t*-butyl group is equatorial. Thus when 4*t*-butylcyclohexene is brominated, the product has *t*-butyl group equatorial with bromine atoms in the axial positions in the favoured product.

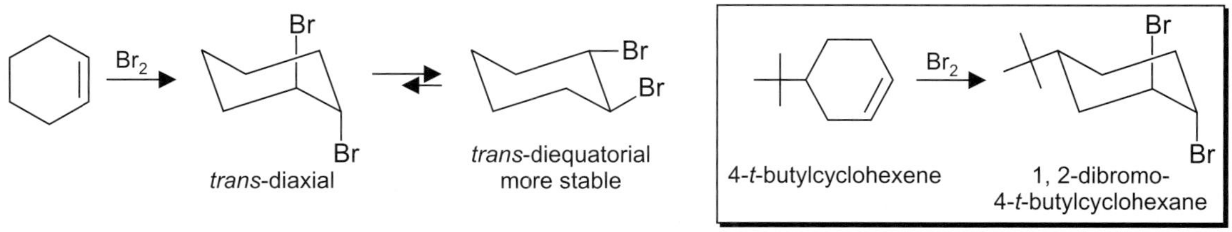

SCHEME 11.3b

In several situations bromonium ions have been observed. Under superacid conditions 1-bromo-2-fluoropropane (Scheme 11.4) gives a cation, which infact is a bromonium ion related to propene as shown by NMR spectroscopy. The highly hindered alkene adamantylideneadamantane gives a bromonium ion (Scheme 11.5). An X-ray crystal structure determination on its derivative has shown the cyclic nature of the bromonium ion. In this case the bromonium ion is not attacked by Br^-, the attack is completely prevented by the steric hindrance offered to the backside approach of the bromide ion by the extremely bulky cage like structure.

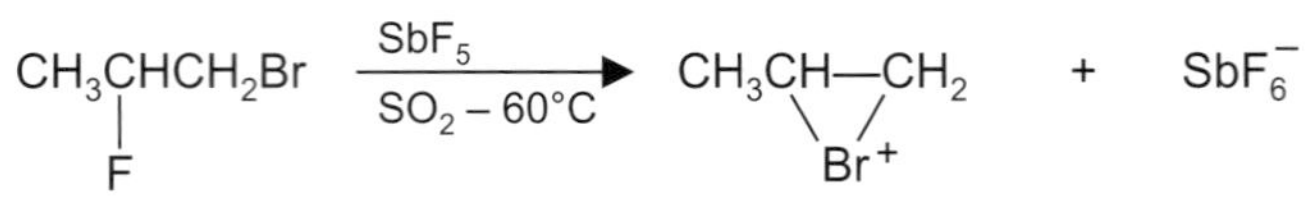

SCHEME 11.4

SCHEME 11.5

Whether the intermediate is a halonium ion or an open carbocation the mechanism is termed Ad_E2 *i.e.*, electrophilic addition, bimolecular.

One may recall that allylic bromination (instead of addition) of an alkene is the reaction on treatment with NBS in CCl_4 in the presence of peroxides or light (see, Scheme 2.19, Problem 2.3). In summary, in a non-polar solvent and a very low concentration of bromine, the reaction is not addition of bromine but it reacts to substitute an allylic hydrogen atom. NBS provides a very low concentration of bromine by reacting with HBr. To understand the reason for allylic substitution at high temperature over addition a consideration of entropy changes is important. Addition of bromine combines two molecules into one, thus the reaction has a substantial negative entropy change (Sec. 4.4). However, at low temperature the $T\Delta S°$ term in $\Delta G° = \Delta H° - T\Delta S°$ is now not large to offset the favourable $\Delta H°$ term, at higher temperatures, the $T\Delta S°$ term gains more significance, consequently $\Delta G°$ becomes more positive and the equilibrium becomes more unfavourable. (In the addition of bromine, the bromonium ion formation is a reversible step).

Bridged ions

The term bridged ions is infact synonymous with nonclassical cations. A nonclassical carbocation involves three-center two electron bonding, a bromonium however, does not involve this. The term bridged ion as used here for a bromonium ion is thus ambiguous.

EXERCISE 11.3

Why acenaphthylene (Scheme 11.5a) on reaction with bromine gives the expected trans dibromide (anti addition) alongwith a large amount of cis dibromide (syn addition).

ANSWER. *Due to the formation of a resonance stabilized open carbocation. S_N2 reaction on cyclic bromonium ion will give only the trans product, while the open cation will give both cis and trans products (Scheme 11.5b)*

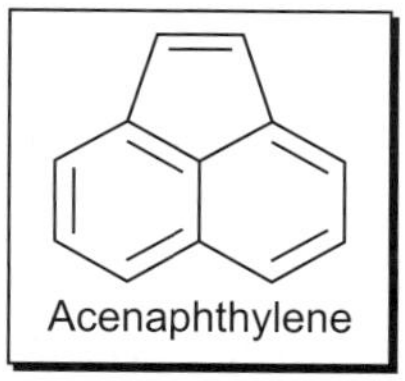

SCHEME 11.5a

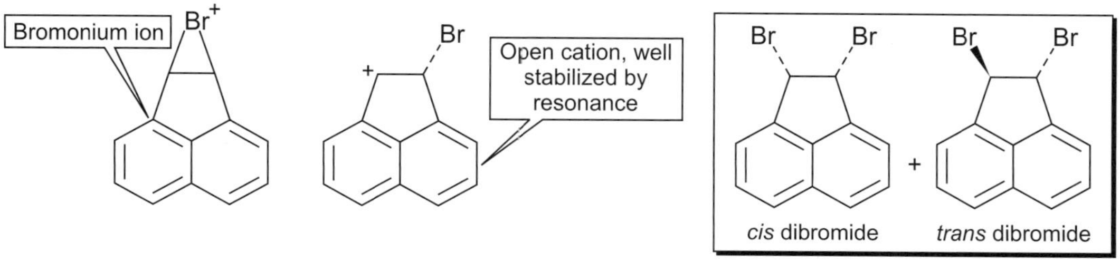

SCHEME 11.5b

(C) Addition of Hydrogen Halides to Alkenes

The first aspect of mechanism of addition of hydrogen halides to alkenes is the observed regioselectivity controlled by Markovnikov rule *i.e.*, the relative ability of the carbon atoms to accept positive charge. The regioselectivity of addition of HBr can be complicated when a free radical chain addition occurs (formation of *anti*-Markovnikov addition product) in competition with the ionic addition (see, Scheme 16.13).

The addition of hydrogen halides to alkenes yields mostly *anti* product. An unsymmetrical alkene will follow the Markovnikov rule during addition of hydrogen halides, since the partial positive charge, which develops will reside largely at the carbon which is most able to

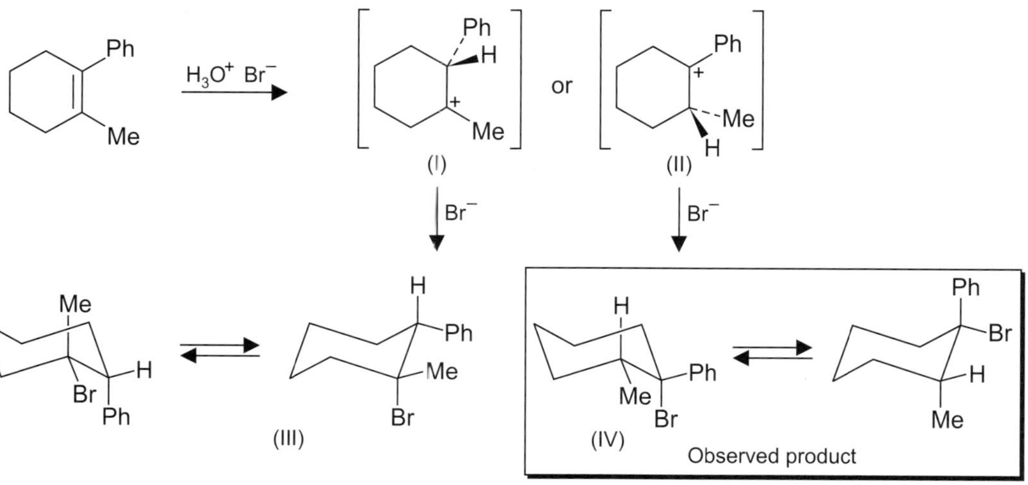

Regioselectivity in the addition of HBr to 2-methyl-1-phenylcyclohexene using

SCHEME 11.6

accommodate it, *i.e.*, the more substituted of the unsaturated carbons. Consider the addition of HX (*e.g.* HBr) to an unsymmetrical alkene 2-methyl-1-phenyl-cyclohexene (Scheme 11.6). Considering the normal *anti* addition, the regioselectivity can be determined by the following argument. If the protonation occurs from above the plane of cyclohexene derivative, attack at C1 gives (I), while at C2 give II. The carbocation (II, Scheme 11.6) is far more stable, since in its case extensive π-delocalization of the positive charge is available. The carbocation II is captured by bromide from the face opposite to the one holding the proton to give (IV) the product of normal *anti* addition. While the other route to give III is not followed.

Hydrogen halides and other acids do not form bridged ions with alkenes. The contribution of a structure of the type (I, Scheme 11.7) *i.e.*, a complex may be there on the mechanistic pathway, which may be attacked from backside to give an *anti* product (II, Scheme 11.7).

Many additions of HBr and HCl to alkenes follow a third order rate expression and the stereochemistry of addition to unconjugated alkenes is largely *anti*. These observations of rate expression and stereochemistry point to the formation of a complex (I, Scheme 11.7) the *anti* product being formed by the backside attack on the complex, *i.e.*, the transition state involves proton transfer to the alkene from one hydrogen halide molecule and capture of the halide ion from the second (Scheme 11.7).

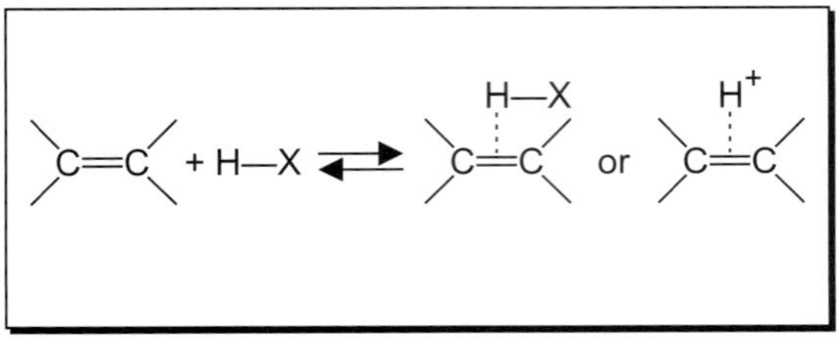

(I)

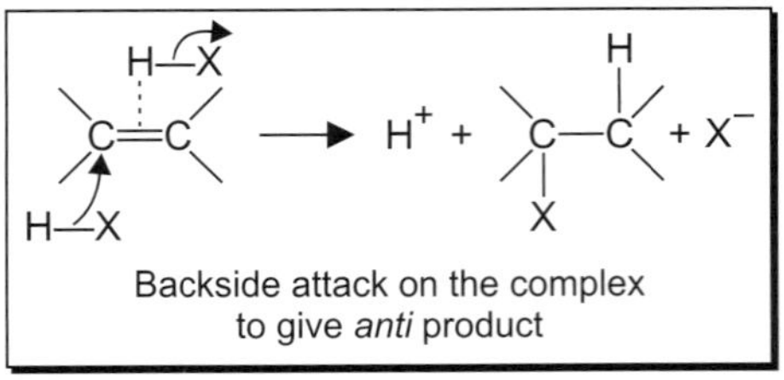

(II)

SCHEME 11.7

When the double bond is conjugated with an aryl group the *syn* adduct predominates. As in the case of bridging, the stabilization of the carbocation intermediate by the aryl group reduces the effectiveness of the complex formation and an ion pair may be the key intermediate (Scheme 11.8). When the ion-pair (formed by the initial alkene protonation), collapses to the product faster than the rotation, the result would be *syn* addition because the proton and the halide ion initially remain on the same side of the molecule.

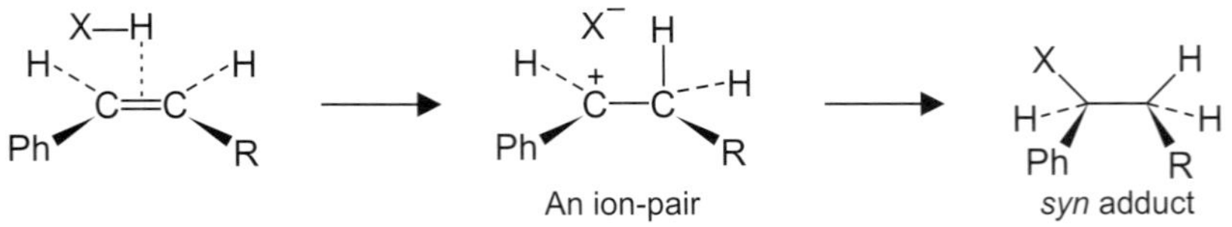

SCHEME 11.8

An example where both *syn* and *anti* addition of DBr is observed is of acenaphthylene (Scheme 11.8*a*).

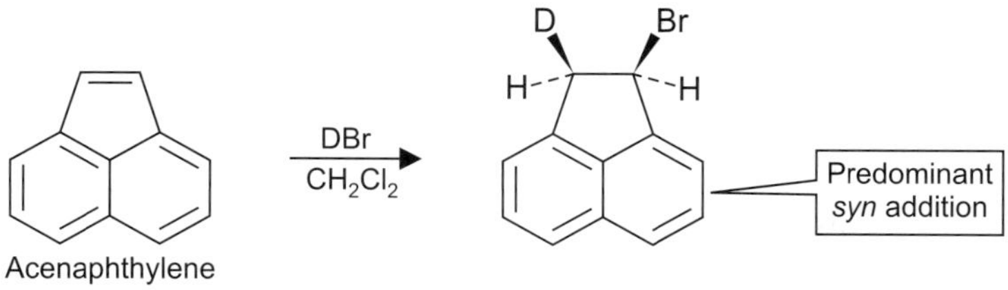

SCHEME 11.8a

> **Other examples of electrophilic addition reactions to alkenes include, epoxidation (A, Sec. 13.4), bis-hydroxylation (D, Sec. 13.4) and hydroboration-oxidation (Sec. 7.6)**

(C) Electrophilic Addition to Alkynes

The mechanism for electrophilic addition to alkynes appears similar to those for alkenes. The first step of the reaction of HBr to terminal alkyne generates formally a vinyl carbocation. Markovnikov regiochemistry is observed for this reaction. The more stable 2° vinylic carbocation is formed in preference to a less stable 1° vinylic carbocation (Scheme 11.9). The alkenyl bromide then reacts with the second equivalent of HBr, and the product again with Markovnikov regiochemistry is the geminal dibromide, which has both bromine atoms attached to the same carbon atom.

The vinyl carbocation formed as the first intermediate is not equally stable as a secondary or tertiary carbocation. Thus, the free energy of activation is higher for addition to a triple bond than that for a double bond. Consequently the reaction of a second equivalent of the reagent with an alkyne is usually more facile than reaction of the first equivalent. One may add only one equiv. of a reagent to a triple bond with controlled experimental conditions.

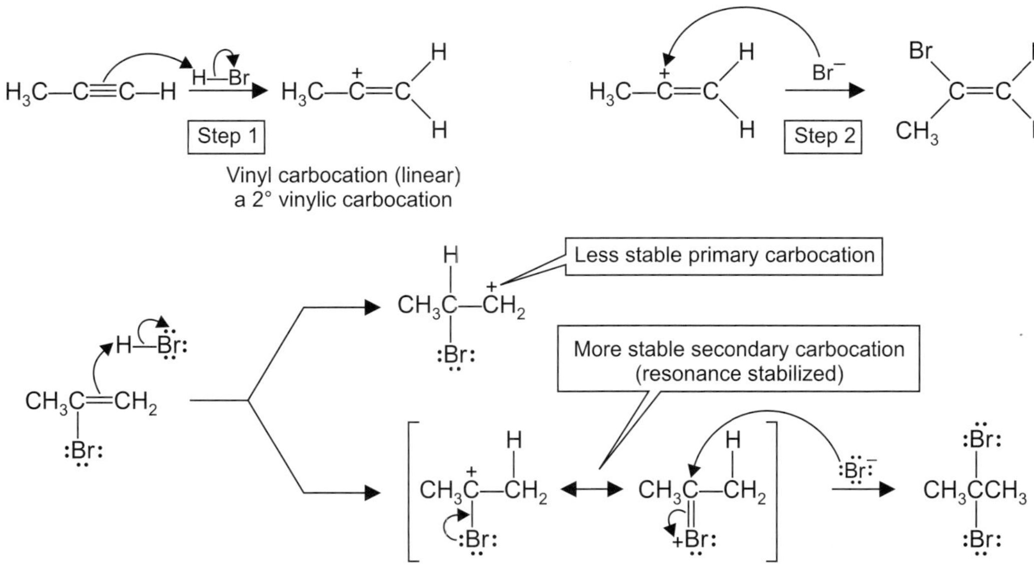

SCHEME 11.9

Compared to regioselectivity, stereoselectivity is much lower than seen for alkenes and in less polar solvents there is some preference for *syn* addition (Scheme 11.9*a*).

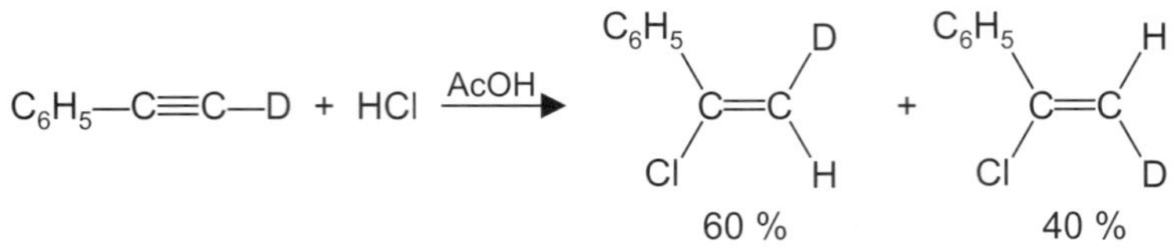

SCHEME 11.9a

EXERCISE 11.4

During addition of bromine to acetylenedicarboxylic acid, the reaction is easily stopped after the addition of the first mole of Br_2. Explain.

ANSWER. *Addition of one mole of Br_2 to an alkyne gives a dibromoalkene via the bridged bromonium ion. The reaction is stereoselective leading to anti addition. After the addition of first mole of bromine to acetylene-dicarboxylic acid, the initial product (Scheme 11.9b) has four deactivating groups*

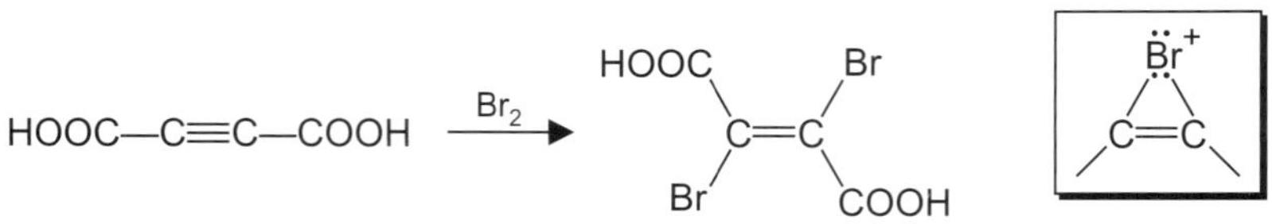

SCHEME 11.9b

(D) Some Other Aspects of Electrophilic Addition Reactions to Alkenes and Alkynes

- The intermediate carbocation may not react with the nucleophile to complete the addition, but may lose a proton. This situation arises when there is steric hindrance (see, Scheme 1.35).
- The enols also lose a proton (see, problem 11.2).
- Though aromatic carbon-carbon double bonds react with electrophiles by substitution, however, addition may compete with substitution as in *e.g.*, anthracene (see, Scheme 2.41*c*).
- Like alkenes, alkynes react with carbene to form cyclopropenes or bicyclobutanes depending on the amount of carbene present (Scheme 11.9*c*).

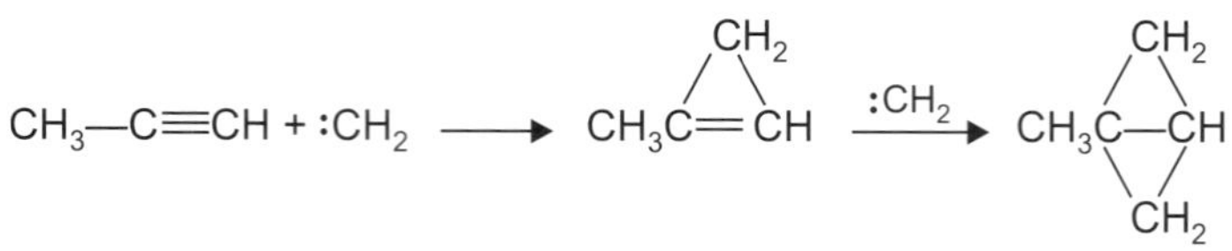

SCHEME 11.9c

- The initially formed carbocation can undergo molecular rearrangement, particularly when after rearrangement a more stable carbocation is formed. This is seen *e.g.*, during the addition of HBr to 3, 3-dimethyl-1-butene (Scheme 11.9*d*).

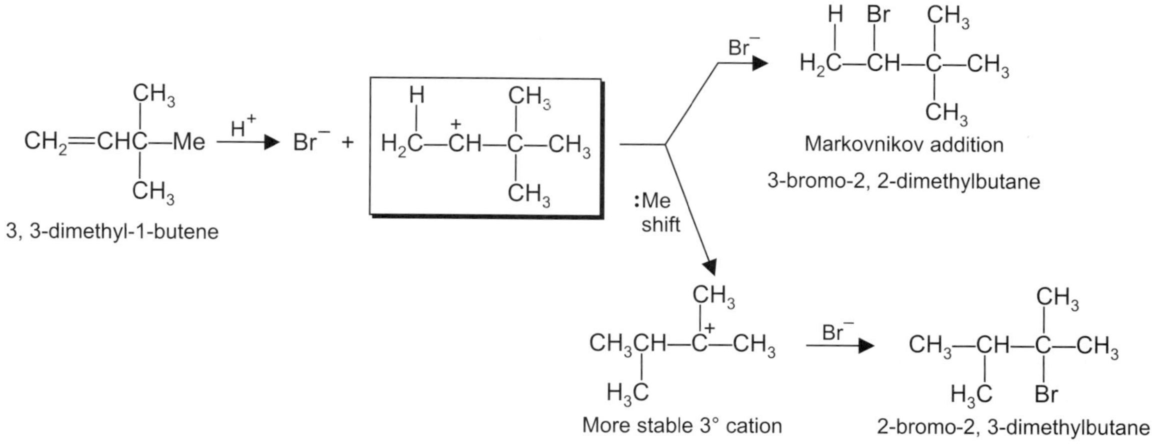

SCHEME 11.9d

(E) Electrophilic Additions Involving Metal Ions (Mercuration—Reduction)

Metal cations promote addition by electrophilic attack on alkenes in a reaction *e.g.*, oxymercuration which involves the addition of OH from water and HgOAc group from a solution of mercuric acetate $Hg(OAc)_2$ in water/THF. The reaction gains its importance due to the ease with which the mercury substituent is removed with sodium borohydride. Oxymercuration—demercuration yields the same product as obtained by hydration. The reaction is generally stereospecific (*anti*) and regioselective. Thus in a conformationally biased cyclic alkenes *e.g.*, 4-*t*-butylcyclohexene, the reaction yield exclusively the product of *anti*-addition and reflects on the intermediate formation of a cyclic mercurinium ion intermediate (Scheme 11.10) to give *anti*-diaxial addition.

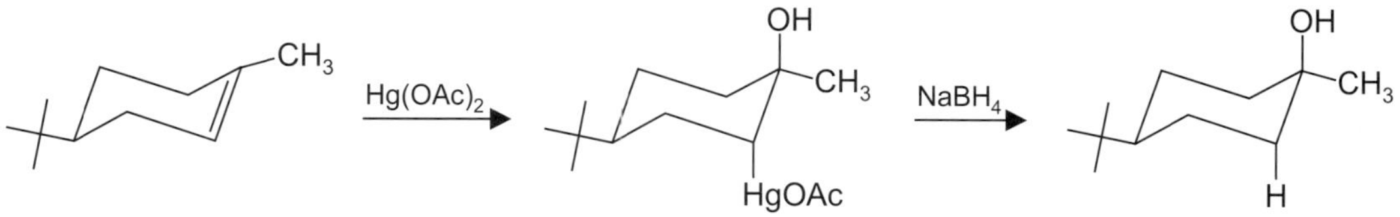

SCHEME 11.10

The mechanism of the reaction may be studied by considering the following points:

- This is a two step process and in the first step, the electrophilic attack by $Hg(OAc)^+$ at the less substituted carbon of the double bond gives a cyclic (bridged) mercurinium ion intermediate. As in the case of bromonium ion the cationic intermediate may be bridged (mercurinium ion) or open which is dictated by the structure of the olefin (Scheme 11.10*a*).
- The addition is completed by the attack of a nucleophile with the highly substituted carbon atom which is more positive.
- After the first step of addition the mercury is replaced by hydrogen by the use of a reducing agent.

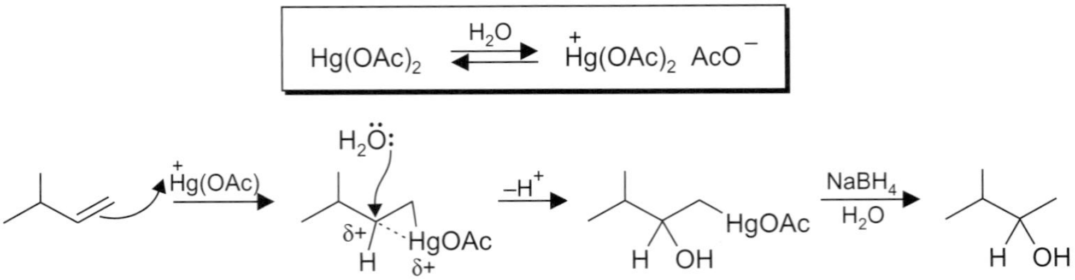

Oxymercuration reduction

SCHEME 11.10a

- The two step procedure thus generally involves the formation of a mercurinium ion intermediate and avoids the formation of an open carbocation. Therefore, rearrangements are not observed as is the case during acid catalysed hydration (Scheme 11.10*b*).

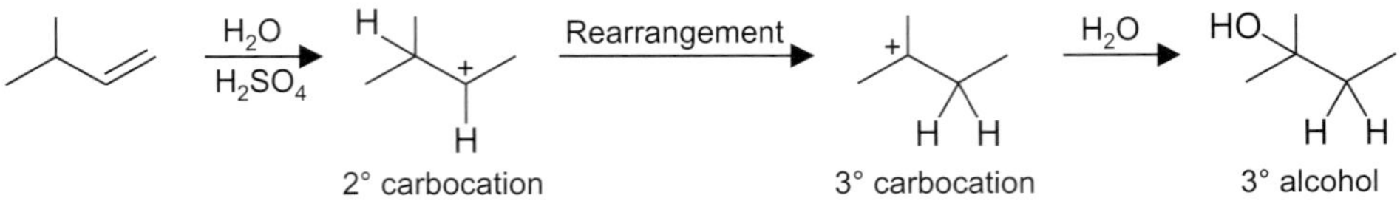

Acid-catalysed hydration

SCHEME 11.10b

11.2 NUCLEOPHILIC ADDITIONS TO ALKENES AND ALKYNES

Nucleophiles do not attack alkenes, however, if the carbon-carbon double bond is conjugated to a group of $-M$ type, they readily attack the alkene. Examples of this are seen during Michael additions and epoxidation of α, β-unsaturated aldehydes and ketones with alkaline hydrogen peroxide.

Powerful nucleophiles *e.g.*, an alkoxide ion (in an alcoholic solution) react with alkynes (Scheme 11.11). A terminal alkyne on base catalysed nucleophilic addition of ethanol gives a substituted vinyl ethyl ether (Scheme 11.11). A simple alkene however, does not undergo nucleophilic addition of an alkoxide. In this case as well the regioselectivity as observed in alkenes is followed *i.e.*, the nucleophile adds to that carbon which gives more stable of the possible carbanion intermediates.

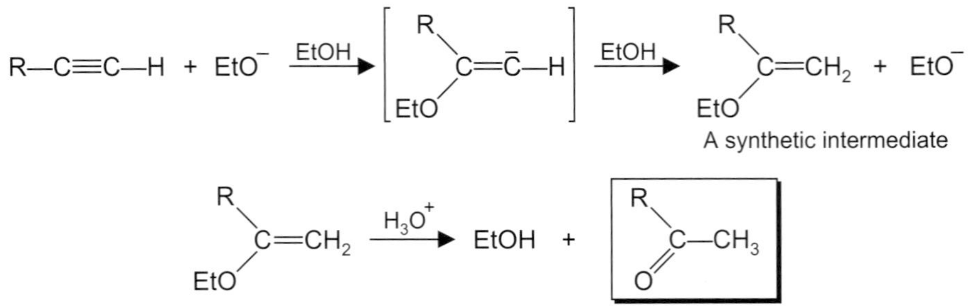

SCHEME 11.11

However, with less powerful nucleophiles like water a catalyst like mercury (II) ion is needed. This ion complexes and consequently draws electrons from the triple bond. Thus water adds to an alkyne in the presence of mercury (II) sulphate and dilute sulphuric acid to initially give an enol which tautomerizes to a ketone (Scheme 11.12).

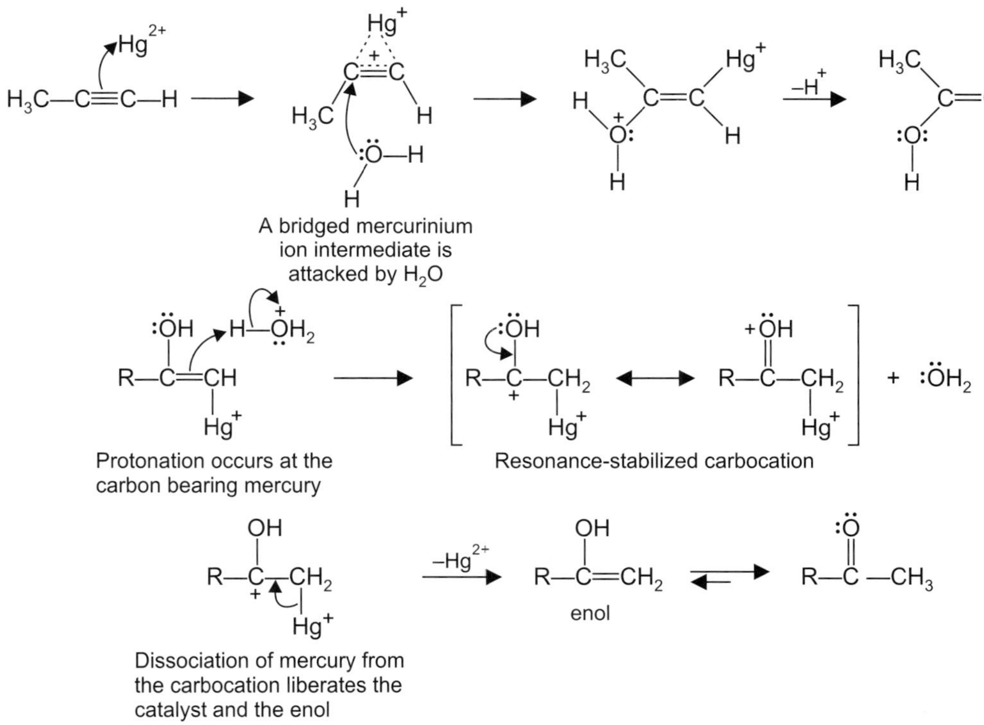

SCHEME 11.12

11.3 ADDITION OF CARBENES

Carbenes add to alkenes to give cyclopropane derivatives. Singlet methylene *e.g.*, reacts with an alkene is stereospecifically, the addition occurs in one step (concerted addition) and the stereochemistry of the alkene is preserved in the product (Scheme 11.13). The electrons in the

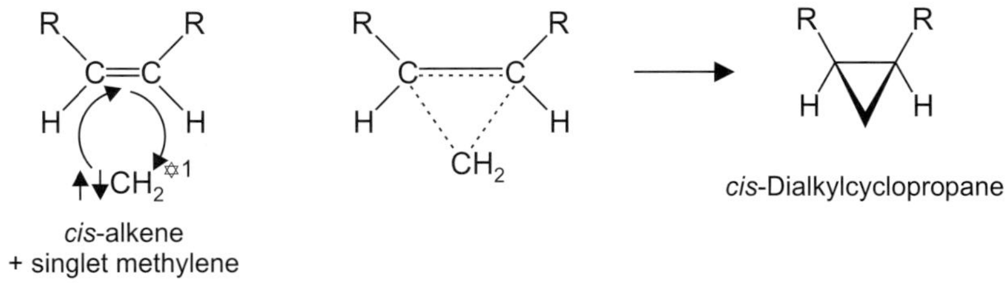

SCHEME 11.13

triplet methylene are not paired and consequently it reacts with an alkene in a stepwise process (Scheme 11.13*a*). The initial reaction is the formation of a biradical and this has sufficient lifetime to allow rotation. The addition is therefore, non stereospecific.

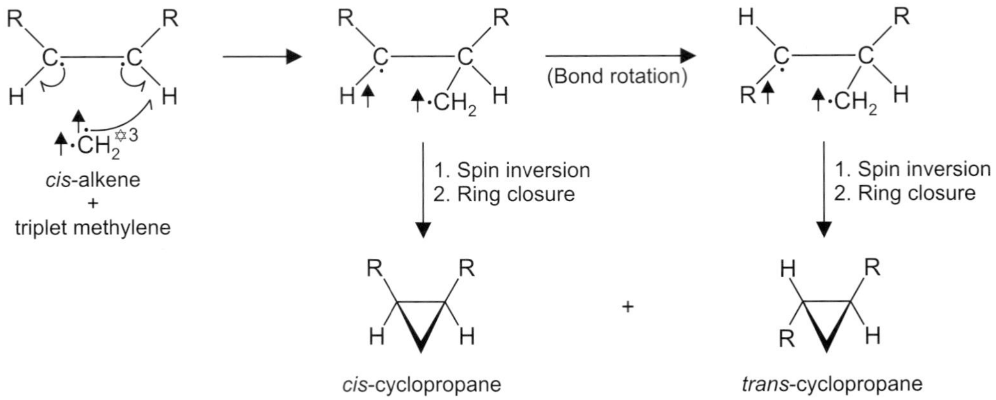

SCHEME 11.13a

Carbene addition to a double bond is complicated due to too many side products. The Simmons-Smith procedure is superior and leads to same results. The reaction involves reaction of a double bond compound with diiodomethane (CH_2I_2) and a Zn-Cu couple, the attacking species is an organozinc intermediate (ICH_2ZnI), a carbene like species called a carbenoid (Scheme 11.13*b*).

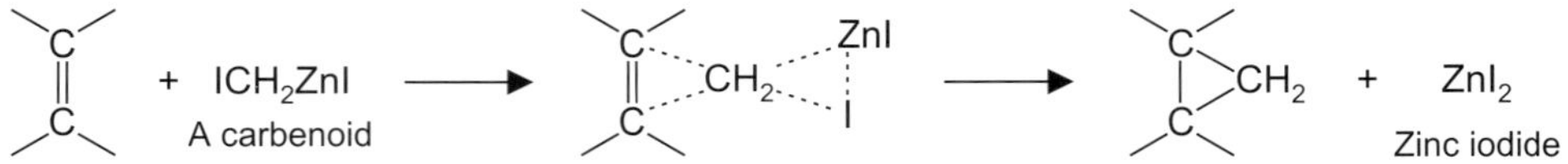

The Simmons–Smith synthesis

SCHEME 11.13b

11.4 NUCLEOPHILIC ADDITIONS TO CARBONYL COMPOUNDS (ALDEHYDES AND KETONES)

Aldehydes and ketones display nucleophilic addition reactions when the nucleophile is either a hydride ion or a carbon nucleophile *e.g.*, CN^-. In their case since the carbonyl group is attached to a group which is too strong a base to be eliminated (H^- or R^-), thus substitution does not occur (Scheme 11.14). When the nucleophile that adds to the aldehyde or ketone is one in which Nu is *not* electronegative (Nu is an H or a C nucleophile), the tetrahedral addition product will be stable. The alkoxide ion will be protonated to yield the final product.

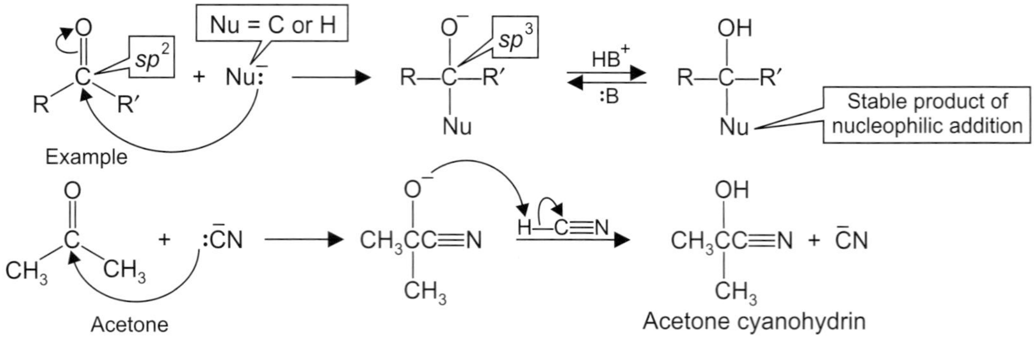

SCHEME 11.14

In case however, the nucleophile Nu that adds to the aldehyde as ketone is a nitrogen or an oxygen nucleophile, than the tetrahedral intermediate is not stable and from it water will be eliminated and then the process is termed nucleophilic addition-elimination reaction. The example is of the formation of an imine and an enamine when a primary or a secondary amine adds respectively as a nucleophile to the carbonyl group of an aldehyde or a ketone (Scheme 11.15). Note that when the nucleophile is a secondary amine, the proton loss is from α-carbon rather than nitrogen in the last step.

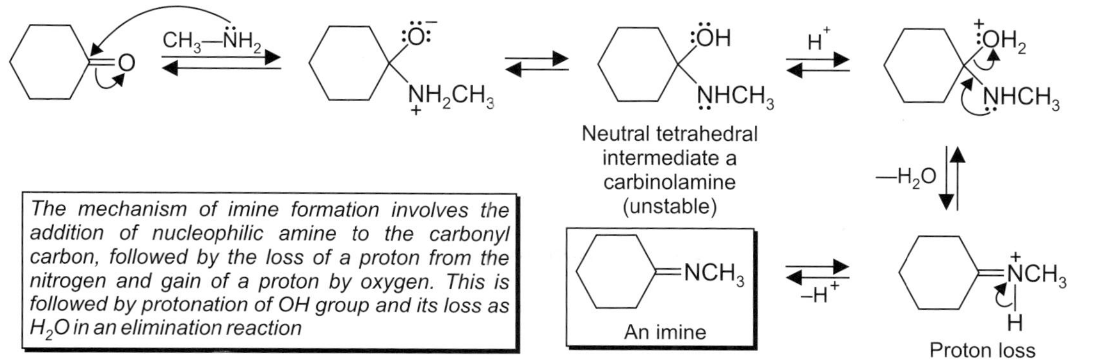

Mechanism of imine formation

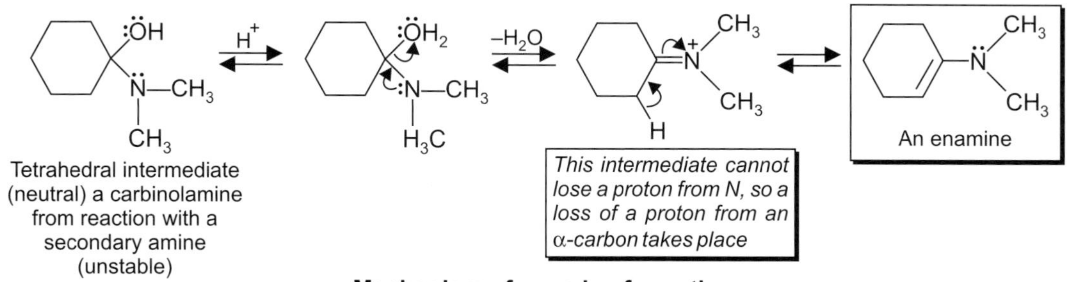

Mechanism of enamine formation

SCHEME 11.15

11.5 NUCLEOPHILIC ADDITIONS TO THE CARBONYL CARBON OF A CARBOXYLIC ACID DERIVATIVE

These derivatives form a tetrahedral intermediate, since both Y and Nu are electronegative atoms. This intermediate is unstable (Scheme 11.16) the relative basicities of Nu^- and Y^- determine as to which will be lost from the tetrahedral intermediate.The weaker base is lost easily (*weaker the base, the better it is as a leaving group*). Because a weak base does not share its electrons that well as a strong base. Thus, a weaker base forms a weaker bond which is to break. If Nu^- is a much weaker base than Y^- then Nu^- will be expelled and thus no new product will be obtained. When on the other hand Y^- is a much weaker base compared to Nu^- then Y^- will be expelled leading to the formation of a new product, this mechanism is called nucleophilic acyl substitution reaction. The mechanism of hydroxide ion promoted hydrolysis of an ester (see Scheme 6.61) is an example.

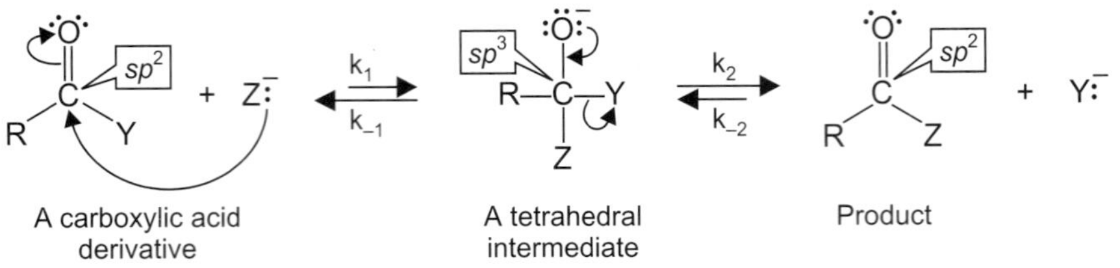

SCHEME 11.16

11.6 RADICAL ADDITIONS TO ALKENES

For a detailed discussion see Chapter 16.

11.7 NUCLEOPHILIC ATTACK ON CARBON-NITROGEN TRIPLE BOND

The nitriles are attacked by powerful nucleophiles *e.g.*, $^{-}$OH ion in water (Scheme 11.17) while acid is needed for attack by weaker nucleophiles like water and alcohols. Nitriles are treated as carboxylic acid derivatives since their hyrdrolysis yields carboxylic acids.

Alkaline hydrolysis occurs by nucleophilic attack on the partially positive carbon of the nitrile group.

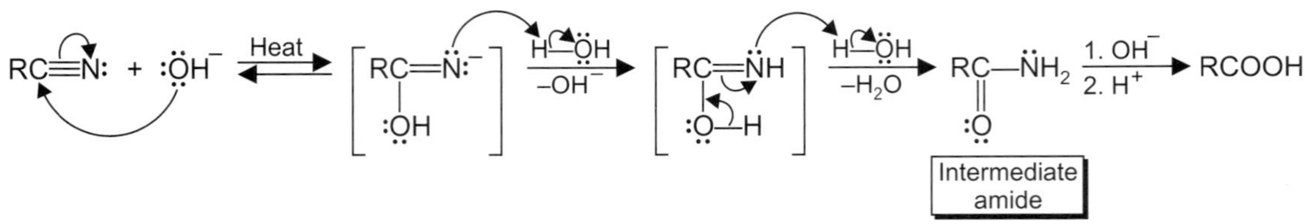

SCHEME 11.17

During acid hydrolysis the weakly basic nitrogen is protonated and then a weaker nucleophile *e.g.*, water attacks the electropositive carbon atom (Scheme 11.18).

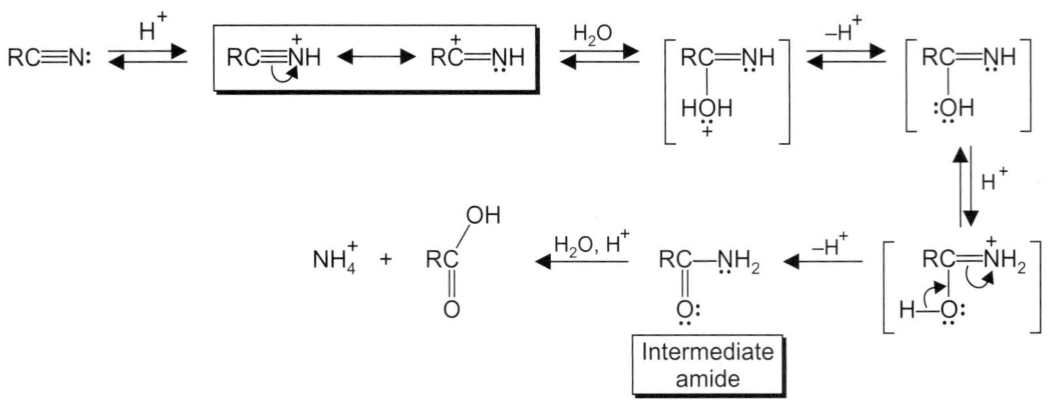

SCHEME 11.18

PROBLEMS

11.1 How ICl will add to $Me_2C{=}CH_2$?

11.2 Write the mechanism for the formation of a ketone on addition of bromine to the enol (I), and the addition of HCl to (II).

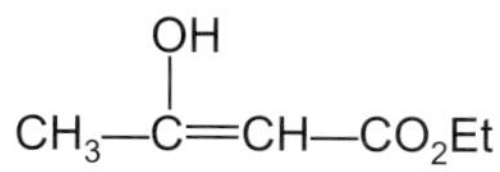

(I)

$O_2N—CH{=}CH_2$

(II)

11.3 Cyclohexene adds bromine to yield only a *trans*-1, 2-dibromo product (see, Scheme 4.23). Dihydropyran on a similar addition however, gives both *cis* and *trans* isomers. Explain.

11.4 Why alkynes are less reactive compared to alkenes toward addition of bromine?

ANSWER TO THE PROBLEMS

11.1 As chlorine is more electronegative than iodine, ICl contains the electrophilic iodine. Thus according to Markovnikov rule the product is 2-chloro-1-iodo-2 methylpropane ($Me_2CCl\ CH_2I$).

11.2 Electrophilic addition to enols takes this course *i.e.*, the loss of a proton after the addition of an electrophilic reagent.

(*i*)

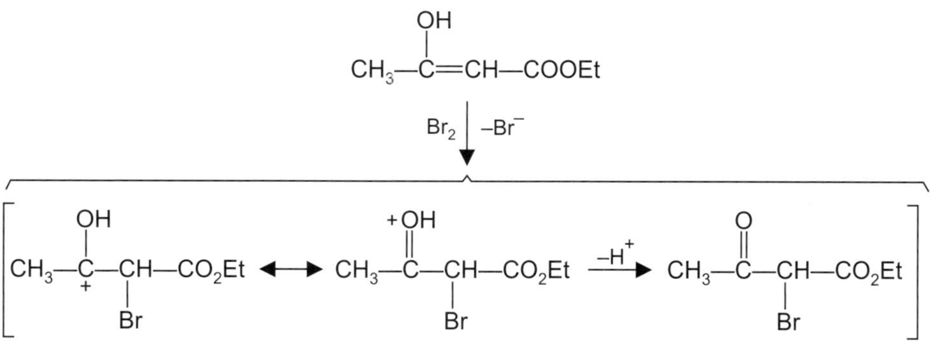

(*ii*) $O_2N—CH{=}CH_2 + HCl \longrightarrow O_2N—CH_2—\overset{+}{C}H_2 \xrightarrow{Cl^-} O_2N—CH_2—CH_2Cl$

11.3 The initially formed product of electrophilic addition is a resonance stabilized carbocation and not a bromonium ion.

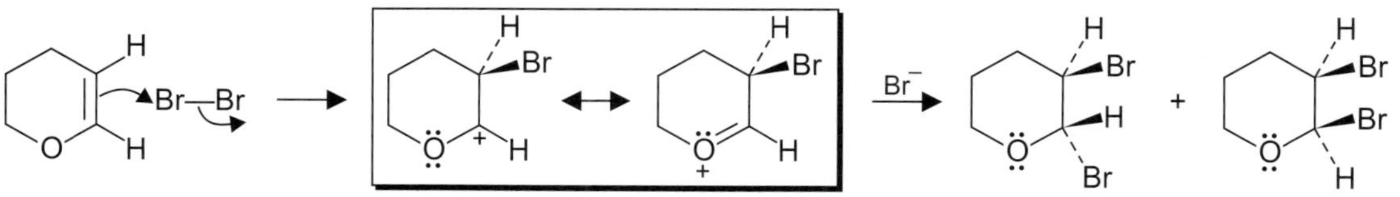

11.4 Alkynes undergo the same type of reactions as alkenes, however, since alkynes contain two π bonds, they undergo addtion with two moles of the reagent. The bromonium ion from an alkyne has a double bond, thus it is more strained and less stable than that from an alkene. Moreover, the carbon atoms in I have more *s* character than II, making I less stable than II.

CHAPTER 12

Elimination Reactions

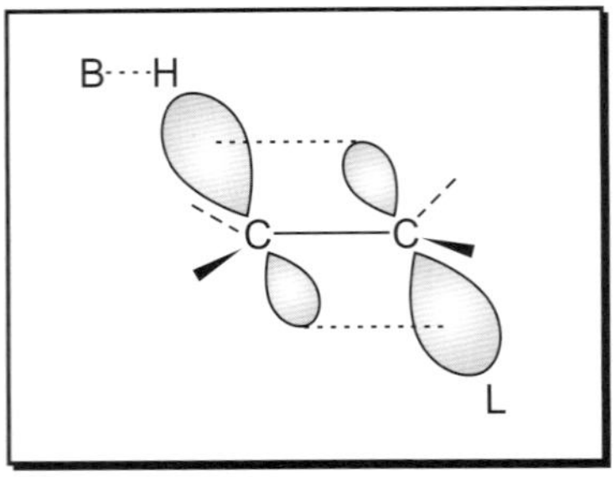

Elimination is one of the major classes of reactions displayed by organic compounds. Elimination reactions involve the loss of two groups or atoms from a molecule. Elimination reactions are classified under two general headings, β-eliminations [(1, 2 eliminations) are the most common elimination reactions] in which groups on adjacent atoms are eliminated with the formation of an unsaturated bond (Scheme 12.1). The β-eliminations include acid catalysed dehydration of alcohols, solvolytic and base induced elimination reactions from sulphonates, alkyl halides, and the Hofmann eliminations from quaternary ammonium salts (see, Scheme 15.29).

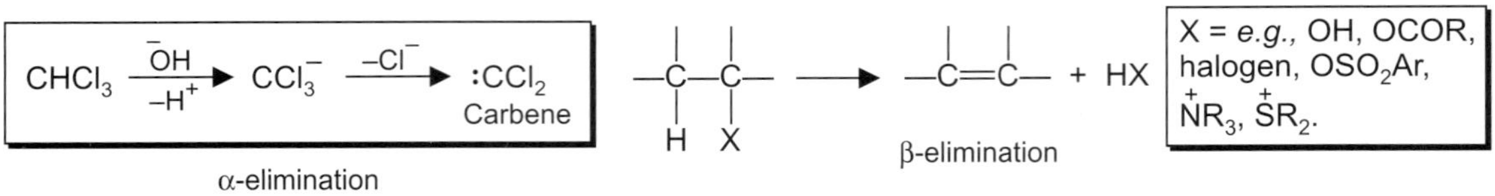

SCHEME 12.1

The second mode of elimination involves two groups departing from the same atom. These 1,1- or α-eliminations are used for generating the reactive intermediates called carbenes (for further details see, Scheme, 4.44).

β-elimination reactions can occur by a variety of mechanisms and three mechanistic pathways are normally distinct routes (Scheme 12.2). Of these *E*2 and *E*1 (see, Schemes 4.3 and 4.6) are the most common. These two processes are closely related to the S_N2 and S_N1 mechanisms of substitution. A third mechanism is designated as E1cB (elimination, unimolecular of the conjugate base) which is less common. The E1cB mechanism involves a carbanion intermediate and the substrate must contain substituents which stabilize it. The substrate undergoing E1cB elimination has a leaving group which is β placed to a carbanion stabilizing group *e.g.*, C=O, NO_2, cyano etc. A good example is of Knoevenagel reaction (see, Scheme 6.29). Another example is in (Scheme 12.2*a*). One may note that if a substrate undergoing elimination has a poor leaving group (*e.g.*, OH^-), the transition state of otherwise E2 elimination gains E1cB character.

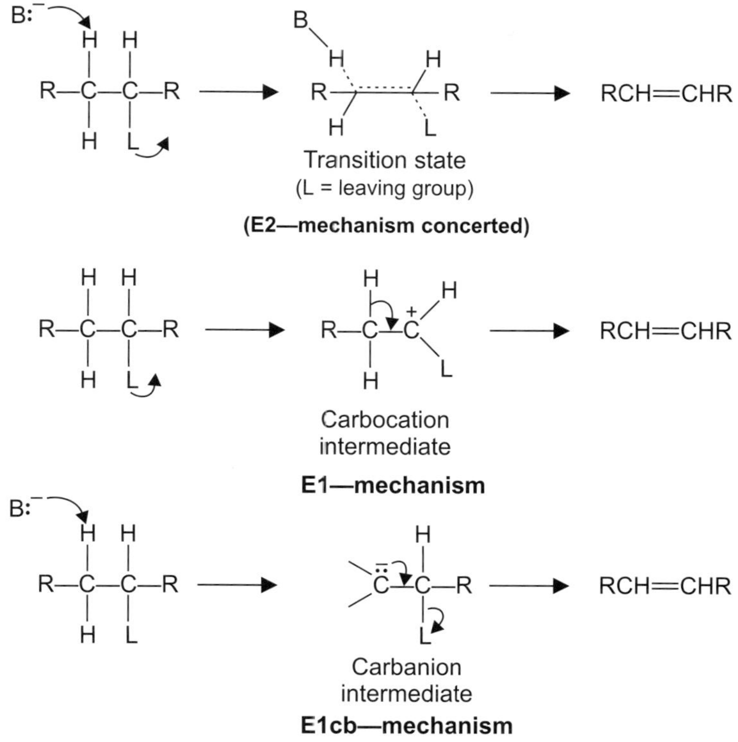

SCHEME 12.2

Consider the example of elimination of (OH^-) as water from a β-nitroalcohol after it is deprotonated with base. The resulting carbanion is stabilized by resonance from which OH^- group eliminates in the second slower step (Scheme 12.2*a*).

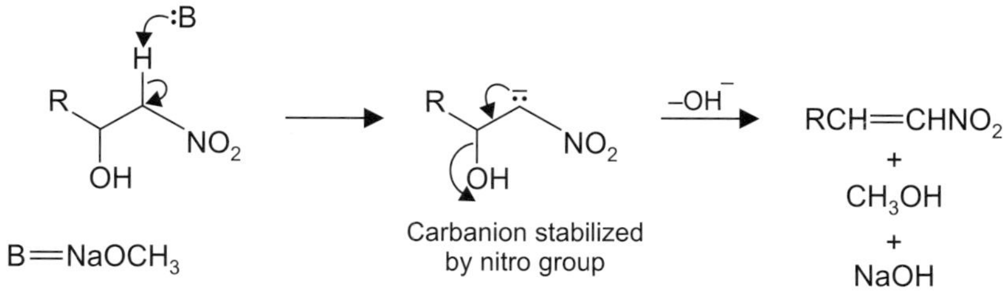

SCHEME 12.2a

12.1 THE BIMOLECULAR MECHANISM FOR ELIMINATION—E2 PROCESS

(A) E2 Mechanism

In this process both substrate and the base participate in the single step in the bimolecular transition state from which the removal of a proton β to the leaving group is concerted with the leaving group (Scheme 12.2).

The structure of the organic compound undergoing elimination and the strength of the base used for the $E2$ elimination reflects on the extent of these three bond changes at the transition state. One may thus study the transition state for $E2$ reaction as a hybrid of structures (I-IV, Scheme 12.3). These structures gain individual importance which are based on several factors:

- Structure (III, Scheme 12.3) would gain importance when L^- is the poor leaving group, or if the negatively charged carbon is adjacent to group with $-I$ effect.

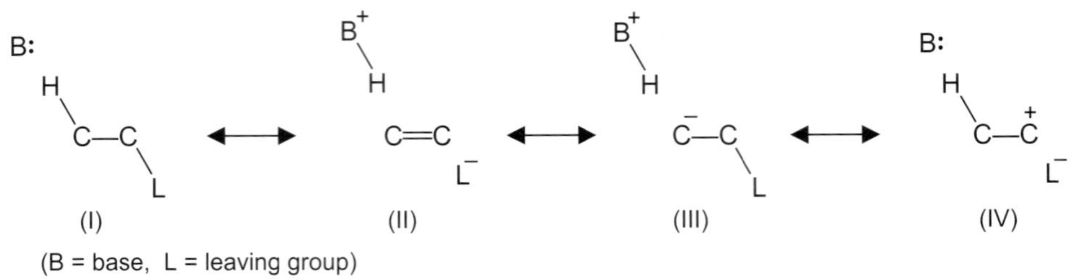

The spectrum of E2 transition states (variable transition state theory of elimination reactions)

SCHEME 12.3

- Structure (IV) gains importance, when on the other hand, L^- is a good leaving group, while B: is a weak base.
- The structure (II, Scheme 12.3) *i.e.*, the alkene like character of the transition state becomes significant when L^- is a good leaving group and B: is a strong base.
- The transition state (III) will have considerable carbanion character when the leaving group is poor *e.g.*, NMe_3 in Hofmann elimination.

(B) Direction of Elimination in E2 Reactions

A study of the spectrum of E2-transition states (Scheme 12.3) helps to explain the orientation of the double bond. With several substrates the elimination can take place in more than one way. Generally the more substituted alkene is formed as the major product (Scheme 12.4).

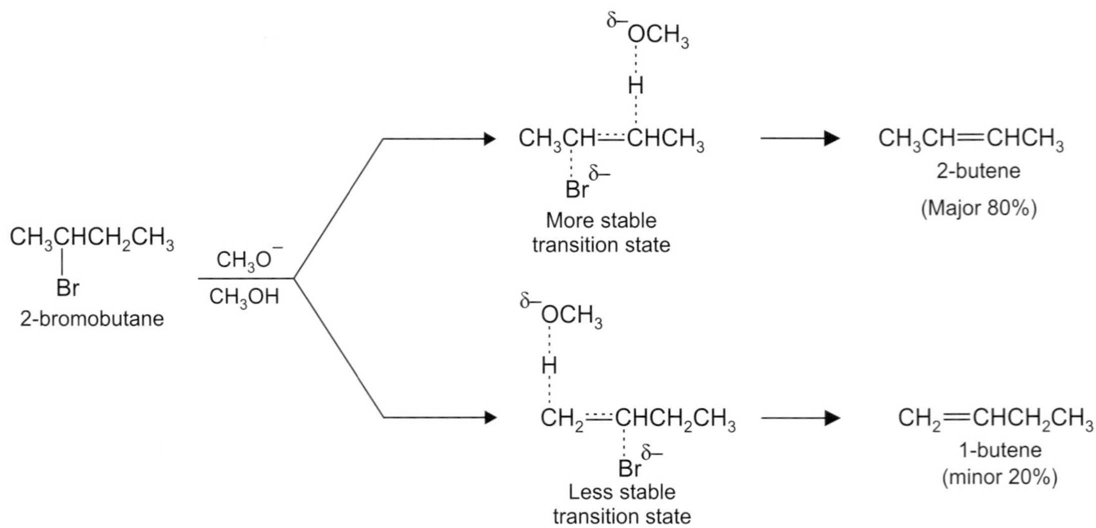

SCHEME 12.4

This generalization is known as the Saytzeff rule. When 2-bromobutane reacts with a base, one expects two elimination products (Scheme 12.4), since in the transition state both the C—H as well as C—Br bonds are breaking. The transition state has alkene like structure and the factors which stabilize an alkene (the larger number of alkyl substituents bonded to the sp^2 carbons) also stabilize the transition state. Thus 2-butene is formed as the major product.

The relative reactivities of alkyl halides in an E2 elimination follows the order:
tertiary alkyl halide > secondary alkyl halides > primary alkyl halides
This is due to the predominant formation of a more substituted alkene (Scheme 12.4*a*).

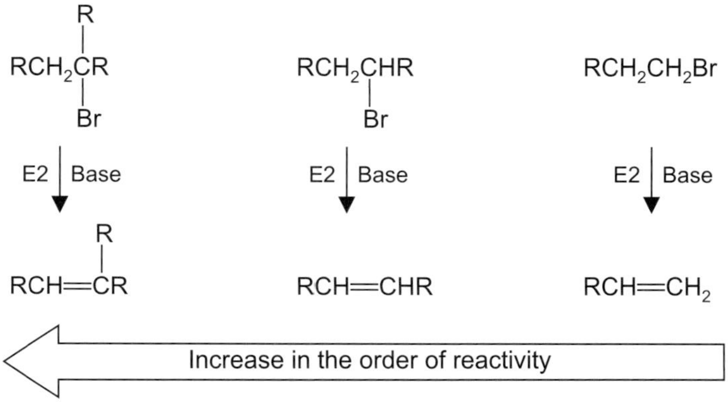

SCHEME 12.4a

Exceptions to Saytzeff rule are observed from base induced eliminations from quaternary ammonium salts and from sulphonium salts which give predominantly the less substituted alkene (Hofmann rule, Scheme 12.5).

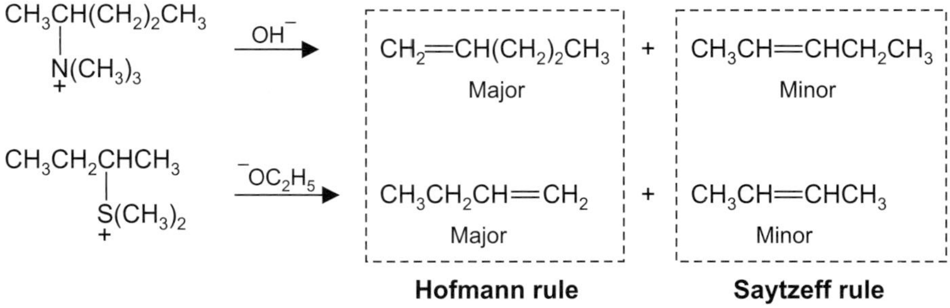

SCHEME 12.5

One can understand this difference between the eliminations from an alkyl bromide (see Scheme 12.4, Saytzeff rule) and from a quaternary ammonium ion (Scheme 12.5, Hofmann rule) on the basis of the poor leaving group tendency of amine compared with Br^- (Scheme 12.5*a*).

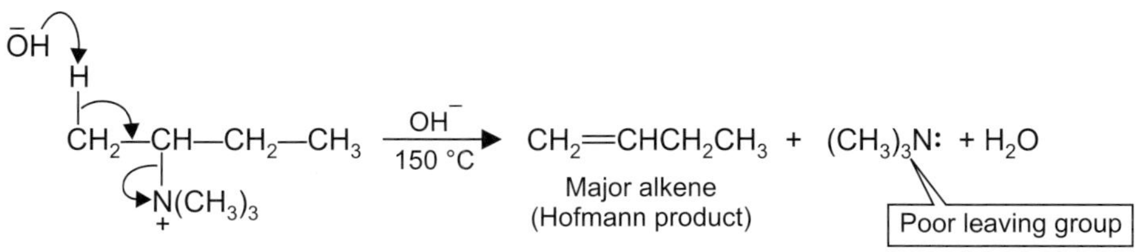

SCHEME 12.5a

Thus in the E2 elimination of a quaternary ammonium ion, the C—H bond is almost fully broken in the transition state and consequently the structures (I or II, Scheme 12.6) may be considered as important contributors to the transition state *i.e.,* the transition state is "carbanion-like", but between (I and II), structure (I, Scheme 12.6) is much more significant

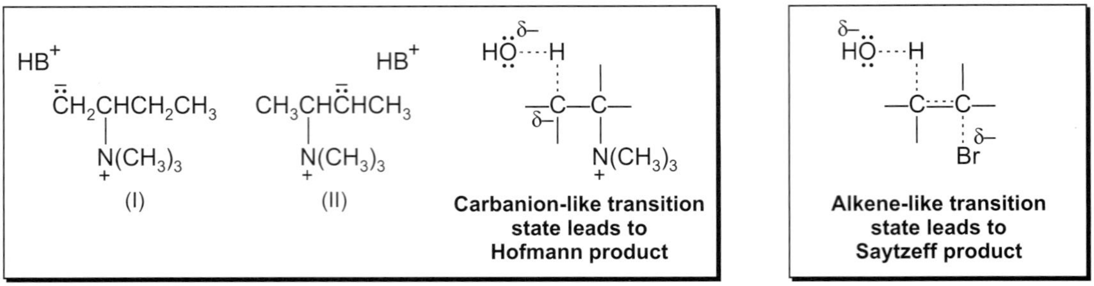

SCHEME 12.6

since an alkyl group destabilizes an adjacent negative charge, therefore, the less substituted alkene predominates. In the case of an alkyl bromide the transition state, will be more alkene like. Further, it is seen that with an alkyl bromide itself, the Hofmann orientation predominates in case the proton removed the Saytzeff orientation is in an sterically hindered environment. In such a case the use of sterically hindered base may lead to Hofmann orientation (Scheme 12.7).

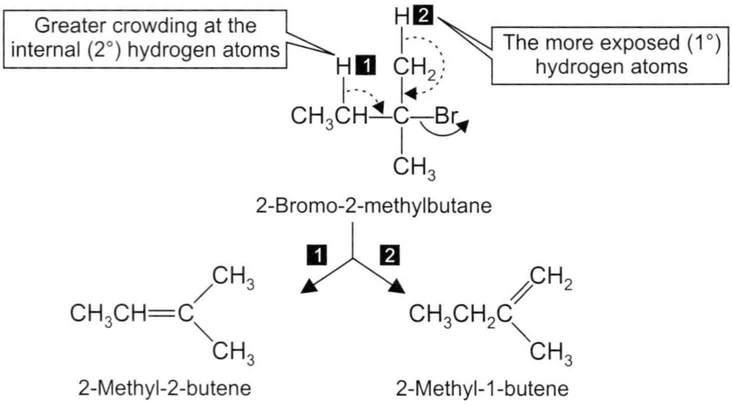

Base	2-Methyl-2-butene %	2-Methyl-1-butene %
EtO^-	70	30
Me_3CO^-	28	72
Et_3CO^-	12	88
	Saytzeff	Hofmann

SCHEME 12.7

A consideration of the transition state as the hybrid of structures (Scheme 12.3) helps in explaining the orientation observed in the four 2-halohexanes (Scheme 12.8). With X=F; the orientation is largely Hofmann while with X=I, the orientation is predominantly Saytzeff. Two factors may be considered, firstly the bond strengths lie in the order C—I < C—Br < C—Cl < C—F, and secondly the electron withdrawing effect of X follows the order F > Cl > Br > I.

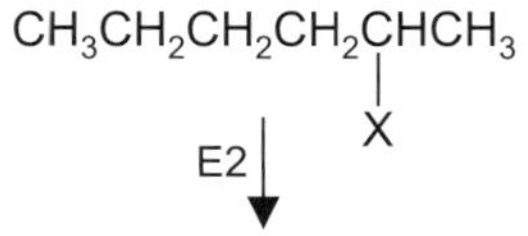

	2-hexene +	1-hexene
X = I	81%	19%
Br	72	28
Cl	67	33
F	30	70
	Saytzeff	Hofmann

SCHEME 12.8

Thus iodide is the best leaving group of the series and fluoride is the worst. With fluorine therefore, one has predominant C—H bond breaking with little alkene character but considerable carbanion character in the trasition state (*i.e.*, it resembles III, Scheme 12.3). In the fluorine case, the primary hydrogen is preferentially abstracted by base, since it allows the negative charge to develop on a primary carbon which can best accomodate it to give Hofmann orientation.

(C) The Rate of E2 Reactions and other Aspects

The reaction rate increases with the increasing strength of the base (I, Scheme 12.9). The rate also increases with a good leaving group and the leaving group ability parallels the stability of the anion. Ethers and alcohols do not undergo *E*2 elimination reaction since alkoxide and hydroxide ions are relatively high energy species, while a sulphonate (II, Scheme 12.9) displays this reaction readily, since sulphonate anion is very stable (the conjugate base of a strong acid.)

$$CH_3CO_2^- < HO^- < EtO^- < Me_3CO^- < NH_2^- \qquad \text{(I)}$$

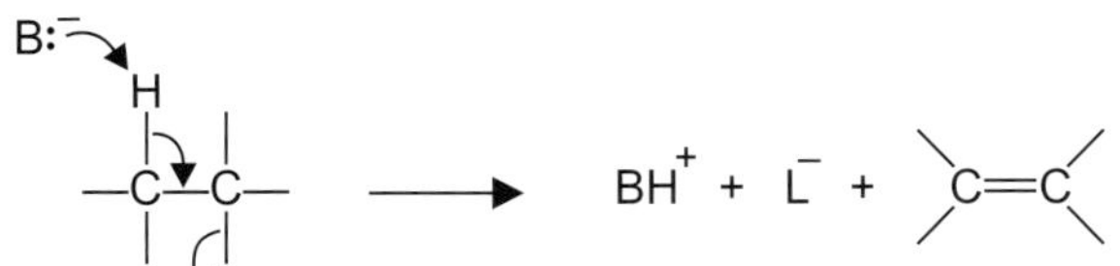

(II)

(Sulphonate L = $-OSO_2R$

SCHEME 12.9

Elimination is facile when the new double bond can come into conjugation with an existing unsaturated bond. The stabilization energy associated with conjugation in the product formed is partly developed at the transition state. Thus $CH_2{=}CH{-}CH_2{-}CH_2Br$ eliminates HBr to give butadiene more readily when compared with 2-bromobutane.

One has seen that normally during E2 elimination, Saytzeff rule predicts the formation of a more substituted alkene and the exceptions are when the leaving group is poor (see Scheme 12.5). In this case negative charge will build up on the carbon from which the proton is lost and therefore, the carbanion stability determines the major alkene product. Moreover, in several eliminations, the less stable alkene predominates *e.g.*, when the base is bulky and sterically hindered (see Scheme 12.7). As the last example, it is conjugation which determines the regiochemistry of $E2$ reactions. A conjugated alkene is more stable even though it may not be the most substituted alkene. A conjugated product is more stable than a nonconjugate product due to resonance. The transition state in such cases has a partial development of conjugation which provides it enough stabilization. Thus the elimination (Scheme 12.10) gives the conjugated product as the major alkene though it is not highly substituted.

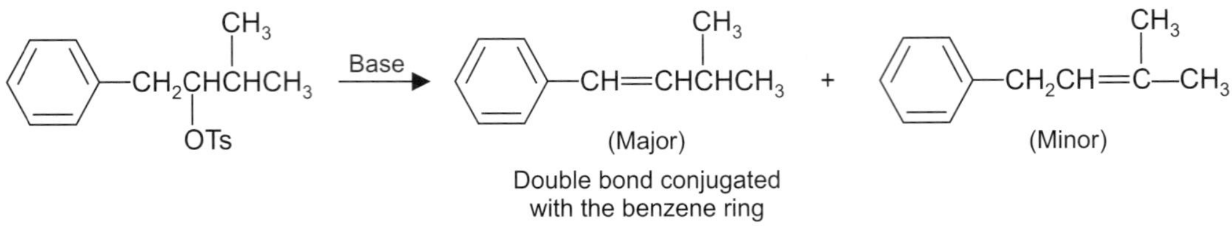

SCHEME 12.10

EXERCISE 12.1

Write the products of elimination (Scheme 12.11)

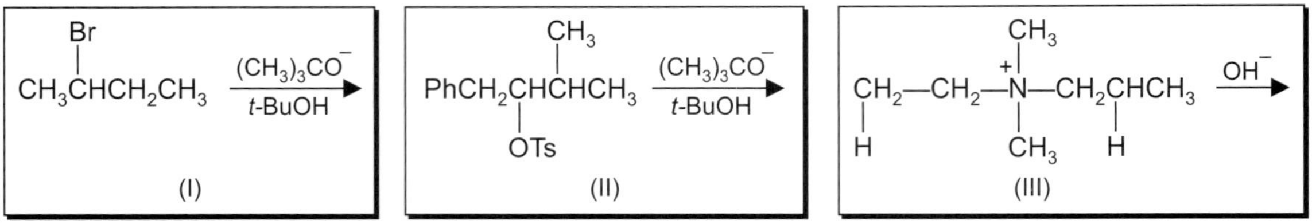

SCHEME 12.11

ANSWER. (I) *More of 1-butene ($CH_2{=}CHCH_2CH_3$) would be formed than 2-butene ($CH_3\,CH{=}CHCH_3$) since the base is strong and sterically hindered.*

(II) *The major product is conjugated less highly substituted ($PhCH{=}CH\underset{}{C}H(CH_3){-}CH_3$).*

(III) *In this case the alkyl groups bonded to nitrogen are larger than methyl thus elimination from these groups is also to be considered. In previous example (see Scheme 12.5a) the elimination could occur only in the butyl group. In the present case, β hydrogen could be lost either from ethyl or propyl group, since Hofmann's rule predicts the formation of less highly substituted product, ethene would be formed as the major product (Scheme 12.11a) while propene ($CH_3{-}CH{=}CH_2$) would be a minor product.*

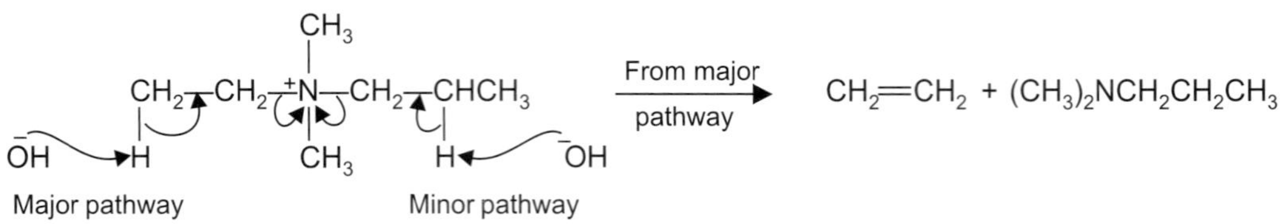

SCHEME 12.11a

(D) Stereochemistry of E2-Elimination

In an $E2$ elimination the carbon sp^3 orbital of the C—H σ bond and the carbon sp^3 orbital of the *C—L* σ bond (L is the leaving group) must begin to overlap to form a π bond. Consequently these two sigma bonds have to be coplanar and two arrangements are *syn*-periplanar and *anti*-periplanar. In *syn*- periplanar conformation these two bonds are on the same side of C—C bond while in *anti*- periplanar conformation these are on opposite sides of the C—C bond (Scheme 12.12).

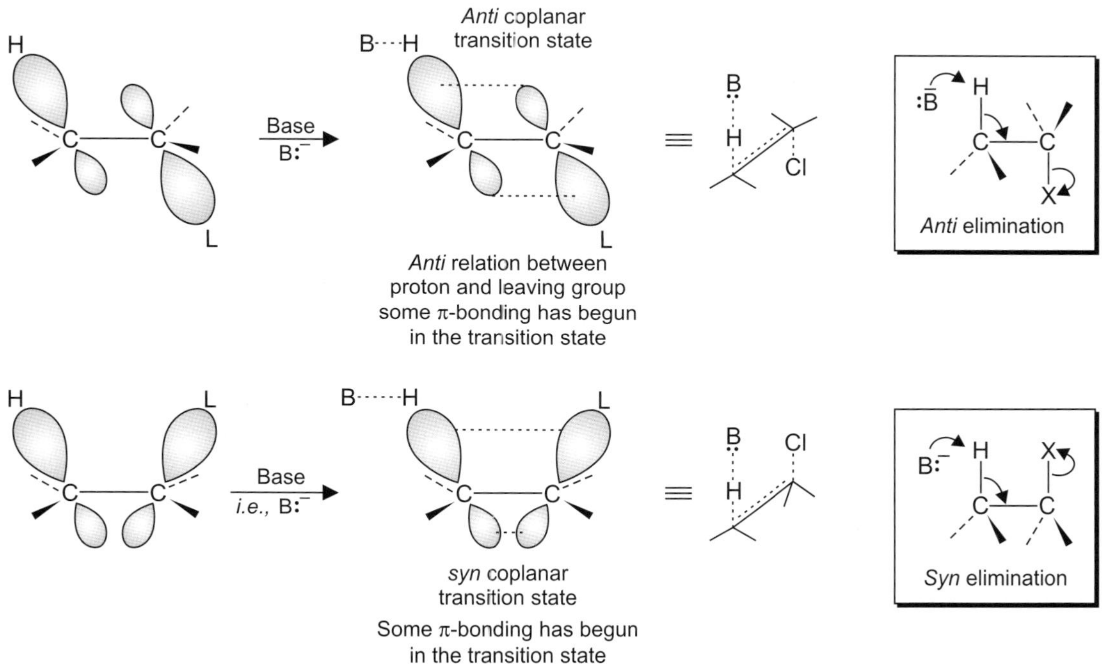

SCHEME 12.12

The evidence that a partial alkene structure exists at the transition state is shown by the predominant formation of more of the *trans* alkene (sterically less congested) when geometrical isomers are possible in the $E2$ elimination. Consider $E2$ elimination from 2-bromopentane which is regioselective and as expected the Saytzeff product 2-pentene is formed as the major product (Scheme 12.12*a*). $E2$ reaction is also stereoselective since 2-pentene which can exist as a mixture of stereoisomers, contains more of E stereoisomer than Z isomer. In the two conformations (I and II, Scheme 12.12*a*) of 2-bromopentane, conformation (II) is more stable than (I), since in the former $-CH_3$ is in gauche relationship with H, in (I), however, $-CH_3$ is gauche placed with respect to the bulkier $-CH_2CH_3$,

The $E2$ eliminations occur very fast if the two eliminated groups are *anti* to each other and they as well as the connecting carbon atoms are coplanar. This geometry is termed *anti*-periplanar stereochemical arrangement for concerted elimination reactions. Thus the dihedral angle between the proton and the leaving group in the *anti*-periplanar conformation is 180°. In this arrangement *i.e.*, when the hydrogen and the leaving group are at a dihedral angle of 180°, their orbitals are aligned.

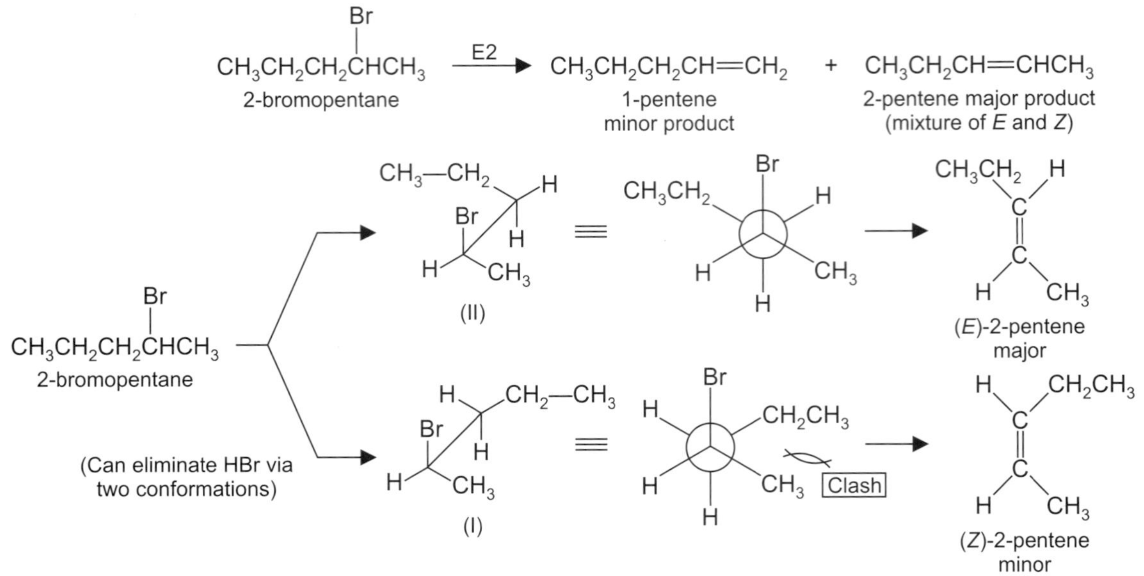

SCHEME 12.12a

The other arrangement for the concerted transition state of the *E*2 raction where the orbitals of the hydrogen atom and the leaving group are aligned so that they can begin to form a *pi* bond in the transition side is *syn* coplanar (*i.e.*, the departing groups are coplanar on the same side of the molecule). In the *syn*-coplanar arrangement the hydrogen and the leaving group are eclipsed (dihedral angle 0°).

The transition state for the *anti*-coplanar arrangement (Scheme 12.12) represents a staggered conformation and the base (nucleophile) is far removed from the leaving group and generally this transition state is of lower energy. The transition state for the *syn*-coplanar elimination is an eclipsed conformation. This is higher in energy due to eclipsing interactions and due to an interference between the attacking base and the leaving group.

A strong preference is observed for *anti* elimination in cyclohexyl systems where the departing groups on the adjacent carbons occupy axial positions *(anti-* coplanar arrangement).

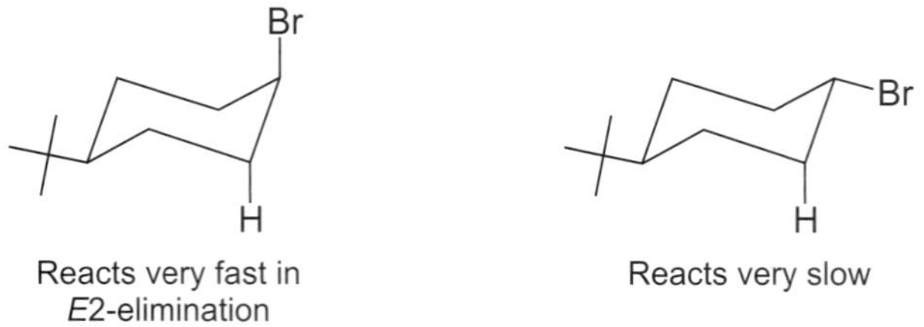

SCHEME 12.12b

(E) Examples of *Anti* and *Syn* E2 Eliminations—Conformation and Chemical Reactivity

1. Reactions in acyclic systems

Consider an alkyl bromide, 2-bromobutane (Scheme 12.13). It has one stereocenter and therefore, only two stereoisomers exist *i.e.*, the compound has an enantiomeric pair. Elimination of HBr from either enantiomer gives the same result: the mixture of *cis-* and *trans-* isomers of butene (Scheme 12.13). This is an example of a compound which reacts stereoselectively. The *E*2 elimination reaction must involve only those conformations of bromobutane molecule in which the ligands to be eliminated attain an *anti-* periplanar arrangement. One can write three possible staggered conformations of 2-bromobutane (I—III, Scheme 12.13). The conformations I and II satisfy the coplanarity condition. Thus I gives rise to the *cis* olefin while II to its *trans* isomer.

In conformation (III, Scheme 12.13) *anti* relationship between the departing ligands H and Br is not available and therefore, it cannot undergo *E*2 elimination of HBr. Moreover, it is found that mainly *trans* alkene is obtained to prove that elimination from rotamer (II) is more facile than (I). The reaction from (II) is favoured due to larger stability of transition state from (II, Scheme 12.13) than a trasition state from (I).

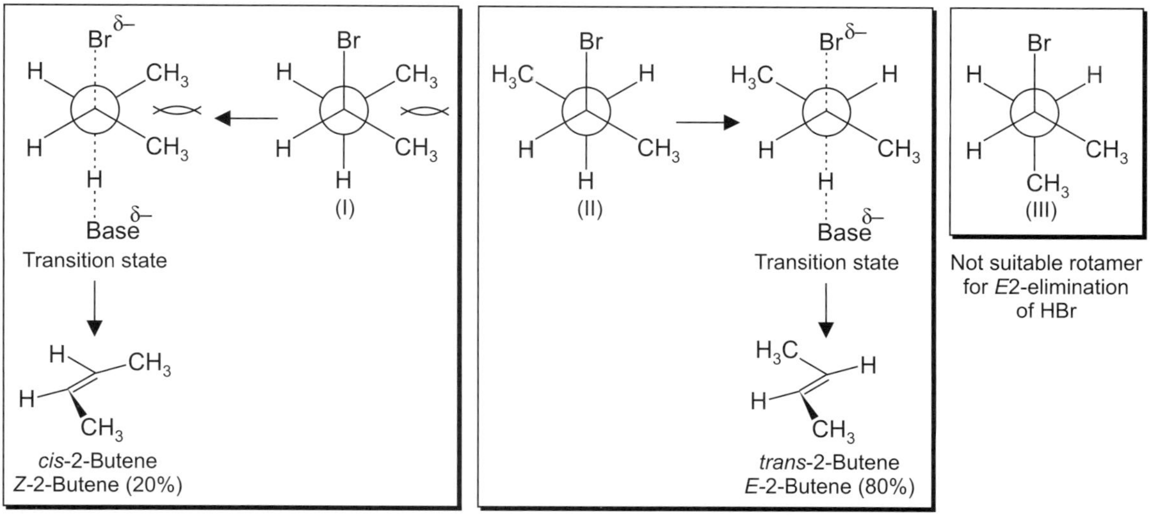

anti-Periplanar arrangement is available between H and Br in both I and II, the staggered conformations of 2-bromo-butane. Steric interactions between the methyl groups in (I), the resulting transition state and the product cis-2-butene destabilize each of these compared to (II). Thus the formation of trans-2-butene (major product) is more favourable.

SCHEME 12.13

Stereospecificity and Stereoselectivity

A stereoselective reaction is one in which a single starting material can give two or more stereoisomeric products but one of these in greater amounts (or even to the exclusion of the other). Thus 2-bromobutane undergoes a stereoselective base induced elimination of hydrogen bromide (Scheme 12.13).

In a stereospecific reaction stereoisomeric starting materials yield products which are stereoisomers of each other. The dehalogenation of meso and (±)-2, 3-dibromobutane is thus a stereospecific reaction.

Recall that Curtin-Hammett principle relates the product distribution to relative energies of the transition states (see, Scheme 4.18*i*).

Dehalogenation of *vic*-dihalides with I^- (iodide ion) has an *E*2-like transition state and has similar mechanistic requirements *e.g.*, in 2,3-dibromobutane (Scheme 12.14). The molecule has two stereocenters and therefore, four stereoisomers can exist. The compound has, however, only three stereoisomers, a (±) pair and a *meso* form. Elimination of two bromines with iodide ion from the *meso*-isomer (I, Scheme 12.14) gives *trans*-2-butene while the (+) or (–) or the (±)–pair gave *cis*-2 butene. This is so, because elimination can occur only *via a* comformation of the starting compound in which the two bromine atoms are in an *anti*-periplanar arrangement, regardless of the fact if or not this is the most stable conformation. The reaction on (±) isomer (II, Scheme 12.14) involves a less stable transition state compared to the *meso* isomer (I). In B two methyl groups are gauche placed, as a result, the elimination is slower by a factor of about two for the (±)- than for the *meso*-isomer.

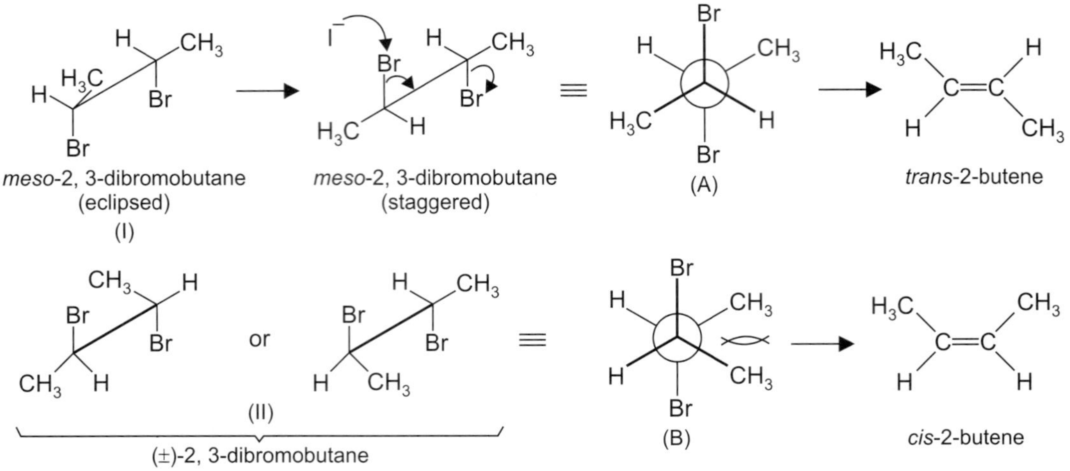

SCHEME 12.14

Preparing to understand the outcome of E2 Eliminations—Conformation and Reactivity

- *Understand the stereochemistry of the reactant*
- *Start by drawing the Fischer projection*
- *Convert the Fischer projection to a staggered conformer*
- *Consider only the conformers in which the groups to be eliminated are anti coplanar.*

Consider the reactant 1, 2-dibromo-1,2-diphenylethane with two stereocenters. As the groups on each stereocenter are the same one has a meso stereoisomer. Consider E2 elimination with base on this stereoisomer (Scheme 12.14a) which gives only the cis- isomer of the product alkene. One may alternately draw the meso form of the compound on a wedge and dash structure (Scheme 12.14b) which is easy since the projection has a plane of symmetry (a consideration of models will help). Rotation (by only 60°) to bring H and Br anti periplanar to each other followed by E2 reaction leads to cis- alkene (see, Scheme 12.14b).

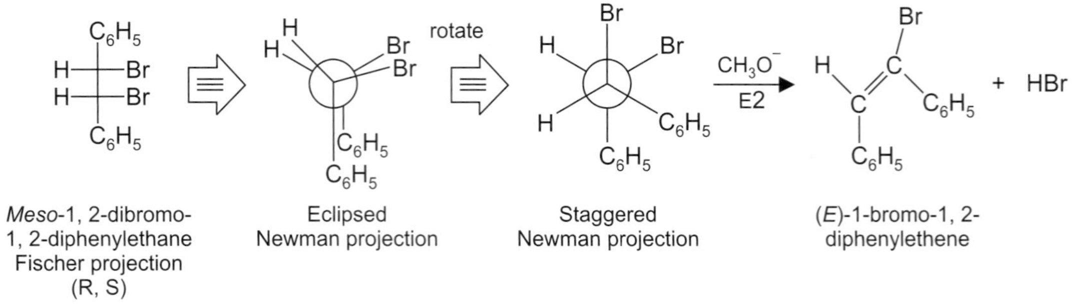

SCHEME 12.14a

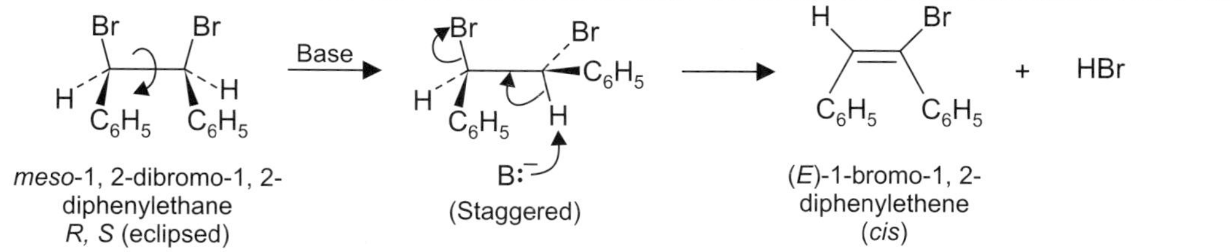

SCHEME 12.14b

Apart from the meso- stereoisomer other two stereoisomers represent a pair of enantiomers. Either of these enantiomers or the racemic mixture give the same product trans alkene (Scheme 12.14c). By switching (interchanging) two groups around one stereocenter on the meso- stereoisomer gives one of the enantiomers (Scheme 12.14c) while the third stereoisomer would be its mirror image.

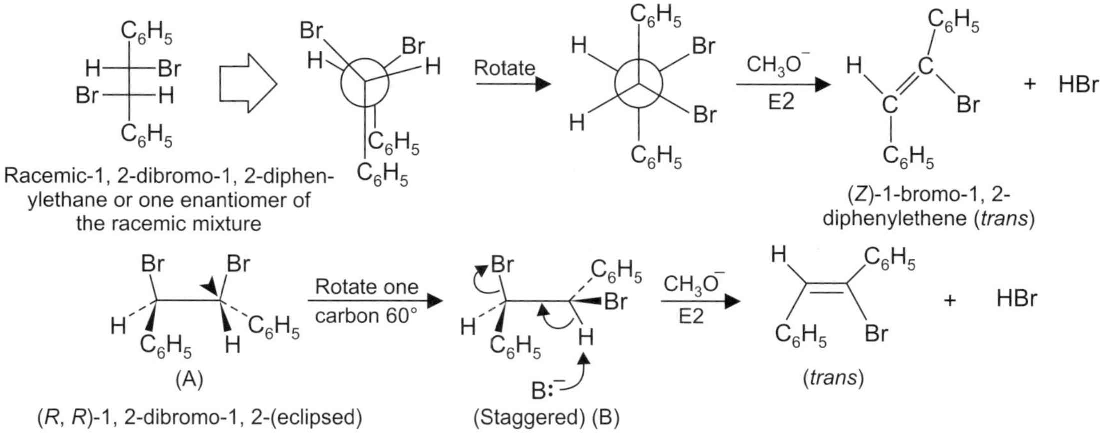

Rotation of the carbon shown by arrow in (A) by 60° brings H and Br anti coplanar. After rotation H comes in the plane of paper (shown on a continuous line) similarly Br rotates from the plane of paper towards ones eyes to be put on a thick wedge. The orientation of C_6H_5 remains same i.e., away from ones eyes as seen in (B).

SCHEME 12.14c

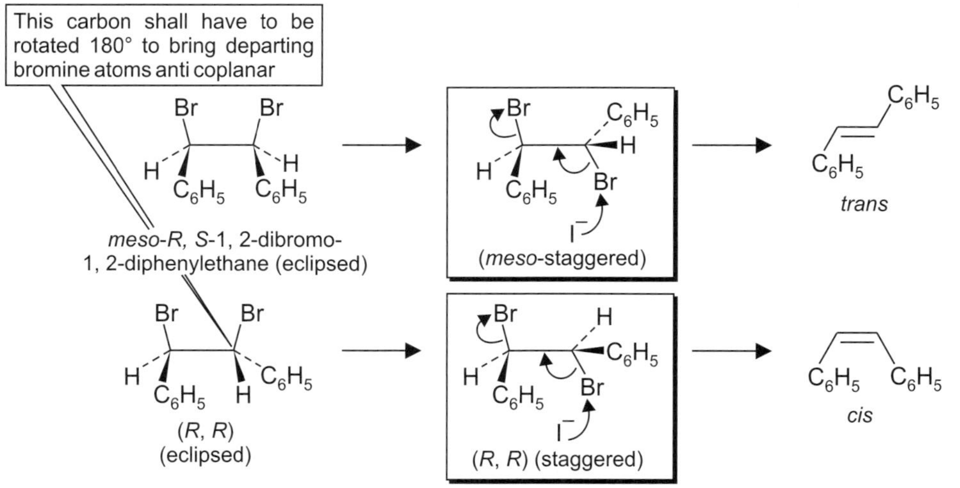

SCHEME 12.14d

Now consider, the elimination of Br_2 with an iodide ion instead of HBr from the molecule of 1, 2-dibromo-1, 2-diphenylethane (Scheme 12.14d). Considering the meso stereoisomer written as eclipsed conformation has to be properly staggered (180° rotation of one carbon with respect to other) to bring two bromine atoms anti coplanar. Unlike the previous elimination of HBr (see, Scheme 12.14b) elimination of bromine now gives trans alkene. Similarly either of the pure enantiomers alone or the racemic mixture gives the cis alkene.

2. Reactions in cyclohexane systems

Generally all cyclohexanes are most stable in their chair conformations with all the carbon-carbon bonds staggered. In a chair cyclohexane any two adjacent axial bonds (*trans-* diaxial) are in an *anti-* periplanar conformation. A chair cyclohexane molecule can flip to bring about an *anti*-coplanar relationship between the departing groups for the success of an *E*2 reaction. When HBr eliminates from bromocyclohexane to give cyclohexene, the initial more stable conformation, in which the bromine atom is equatorial undergoes a chair-chair interconversion (Scheme 12.15). This process inverts the axial and equatorial relationships. This flip can now bring the leaving group in the desired axial position to have an *anti*-coplanar relationship with an axial hydrogen on the adjacent carbon.

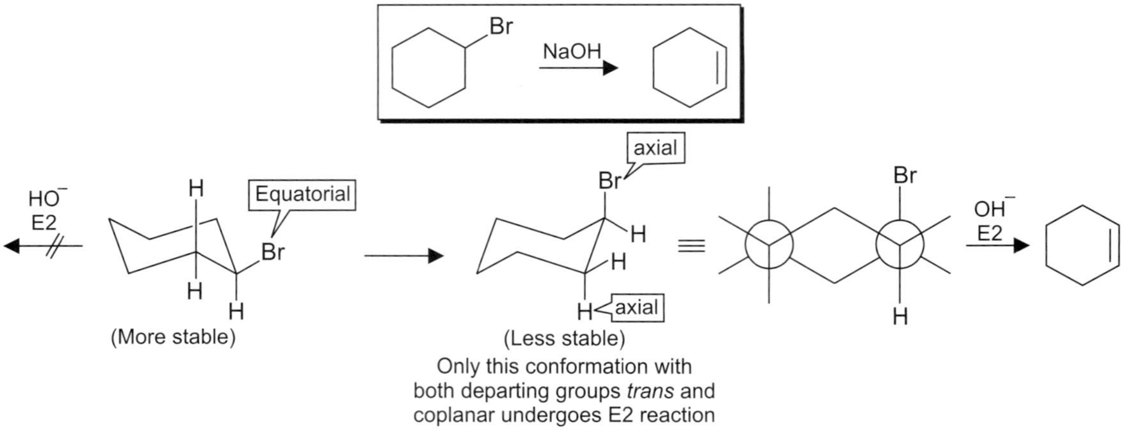

SCHEME 12.15

Another example to show that the *E*2 elimination in a six membered ring proceeds best when the adjacent *trans* groups can adopt an *anti*-periplanar conformation (1,2-diaxial) even if this is a higher energy conformation is of menthyl chloride. Menthyl chloride can have two conformations (I and II, Scheme 12.16). In (I, all the three substituents equatorial) the chlorine is equatorial and as such *anti-* periplanarity with an adjacent hydrogen cannot be achieved. For the E2 elimination to occur menthyl chloride undergoes a ring flip to give a high energy and therefore, unfavourable conformation (II, Scheme 12.16) which has an axial hydrogen on one side. The reaction occurs through an unfavourable conformation. Consequently the elimination of HCl is very slow. In neomenthyl chloride (Scheme 12.17) the chlorine is axial (methyl and isopropyl being equatorial) and axial hydrogens for E2 elimination are available on neighbouring carbons on both sides. Thus a facile elimination occurs to give two olefins. The olefin (I, Scheme 12.17) being the Saytzeff product predominates.

3. Eliminations in bridged compounds—Syn Elimination

In most of these systems *syn*-eliminations are more common. In these compounds, *anti*-elimination is disfavoured both by conformational and steric reasons. Deuterated norbornyl

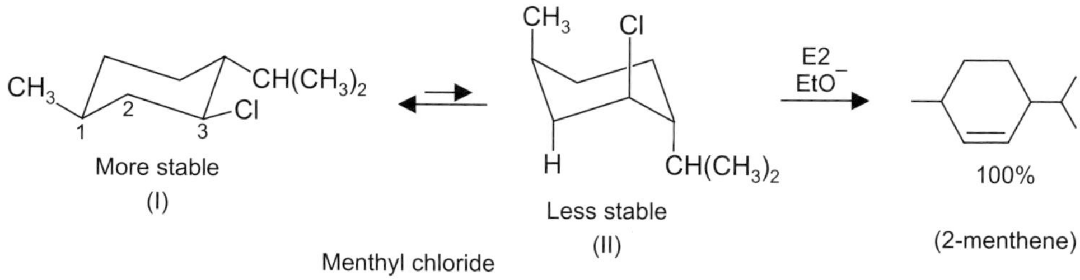

SCHEME 12.16

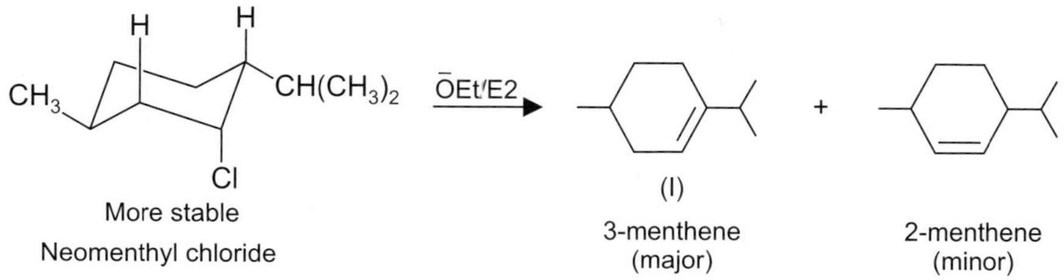

SCHEME 12.17

bromide (A, Scheme 12.18) on *E*2 elimination gave almost exclusive product containing no deuterium. Firstly the rigid ring system prohibits attainment of an *anti-* elimination process *i.e.*, the leaving *exo* Br group cannot achieve a dihedral angle of 180° with *endo* hydrogen (the angle being only 120°). Thus the leaving groups (D and Br) prefer *syn* elimination (*via* a planar transition state) with a dihedral angle of about 0° to *anti*-elimination.

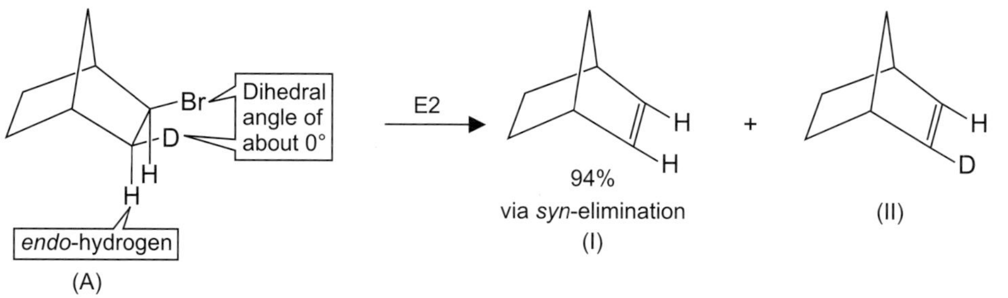

SCHEME 12.18

12.2 THE UNIMOLECULAR MECHANISM FOR ELIMINATION—E1 PROCESS

Some Guidelines for E1 and E2 Reactions

- *E1 reaction is favoured by a protic polar solvent and when no good base or nucleophile is present.*
- *During E1 reactions since a carbocation is formed, the carbon skeleton can rearrange before the loss of a proton.*

- *High concentration of a strong base in the presence of an aprotic polar solvent will favour an E2 reaction*
- *Primary substrates (R–CH_2–L) with low steric hindrance generally undergo S_N2 reaction with any nucleophile. These can be forced to display E2 elimination by using a hindered strong base like potassium t-butoxide. Primary allylic or benzylic substrates may display E1 reactivity (or S_N1 substitution).*
- *When the substrate is secondary, a weak basic nucleophile e.g., CH_3COO^-, CN^- will bring about S_N2 substitution, however, E2 elimination will be predominant with a strong base e.g., OH^-, OR^- etc. E1 elimination (along with S_N1 substitution) will be observed with a polar solvent in the absence of a good nucleophile or base. An example is the use of an alcohol both as a solvent and a nucleophile.*
- *Tertiary substrates R_3CL give high yields of elimination product by E2 mechanism using a strong base.*
- *An E2 reaction is regioselective, the more stable of the alkenes being formed as the major product. When the leaving group is poor or the base is sterically hindered then the less stable alkene may be the major product. An E2 reaction also displays stereoselectivity both E and Z products are formed from suitable reactants. The product alkene with bulky groups on the opposite side of the double bond being more stable will be formed in greater proportion.*
- *An E1 reaction is also stereoselective. The major alkene is the one with bulkiest groups on opposite side of the double bond.*

(A) E1 Mechanism

This elimination takes place (without the participation of a base) in two steps, unimolecular ionization, being rate determining (see Scheme 4.6 and 12.2)

(B) Direction of Elimination

The *E*1 eliminations from the intermediate carbocation (see Scheme 12.2) give rise to the more substituted alkene as the principal product (Saytzeff rule, Scheme 12.19). 2-Bromo-2-methylbutane on reaction with water in ethanol gives substitution as the major product when both water and ethanol can act as nucleophiles to give an alcohol or an ether (Scheme 12.19). The major alkene formed is more highly substituted.

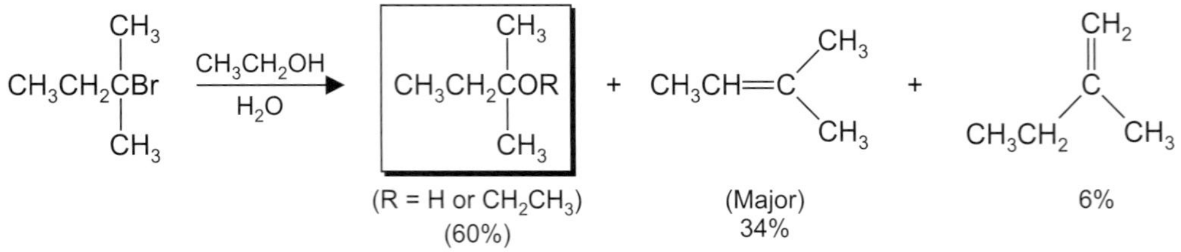

SCHEME 12.19

As a last point recall that weak bases are good leaving groups, thus from among alkyl halides with the same alkyl groups the alkyl fluorides display least reactivity in *E*1 elimination reactions (Scheme 12.20).

RF < RCl < RBr < RI

Increasing reactivity in an E1 reaction →

SCHEME 12.20

(C) E1 Elimination from Cyclic Compounds

Since a carbocation is formed in the first step of the $E1$ reaction, (unlike a $E2$ reaction) the relative stereochemistry of the leaving groups (*anti* coplanarity) is not important. When menthyl chloride (see Scheme 12.16) undergoes $E2$ reaction only one alkene is formed in 100% yield due to the need for the departed groups to attain diaxial positions. When menthyl chloride is subjected to $E1$ reaction conditions two alkenes are formed, the major product is in accord with Saytzeff rule (Scheme 12.21).

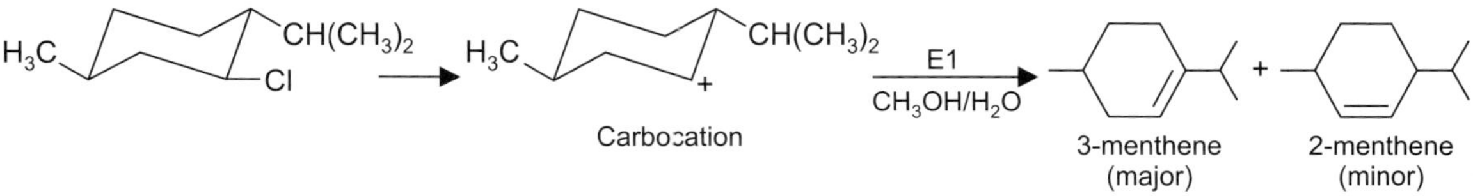

SCHEME 12.21

EXERCISE 12.2

Predict the product from the E1 elimination of compound (Scheme 12.22).

ANSWER. *The reaction conditions of E1 elimination (CH_3CH_2OH/H_2O) generate a carbocation. A carbocation rearrangement may involve the formation of a more stable carbocation and the major/exclusive product may be derived from this (Scheme 12.23).*

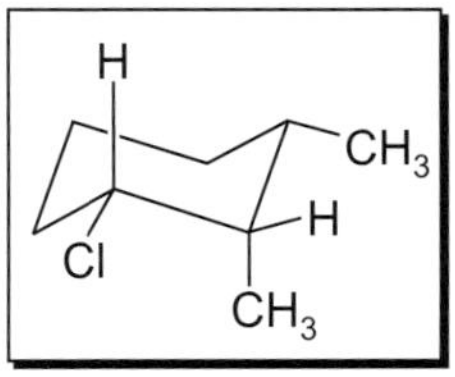

SCHEME 12.22

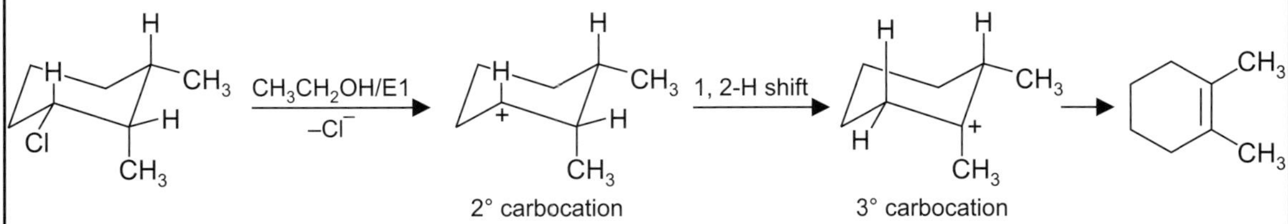

SCHEME 12.23

EXERCISE 12.3

Recall the normal E2 anti pathway followed during the elimination of a bromine molecule from a 1, 2-bromide with iodide ion. The meso-1,2-dibromo-1,2-diphenylethane gives the trans alkene (see, Scheme 12.14d).

Why compound (Scheme 12.24) instead eliminates bromine to give a cis alkene.

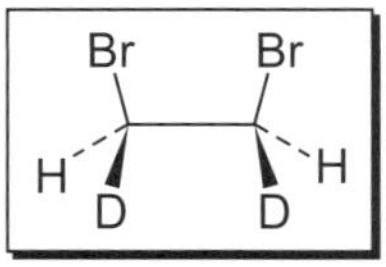

SCHEME 12.24

ANSWER. *In this case now bromine is attached to a primary carbon. One may explain the outcome by first invoking an S_N2 reaction followed by E2 elimination (Scheme 12.25).*

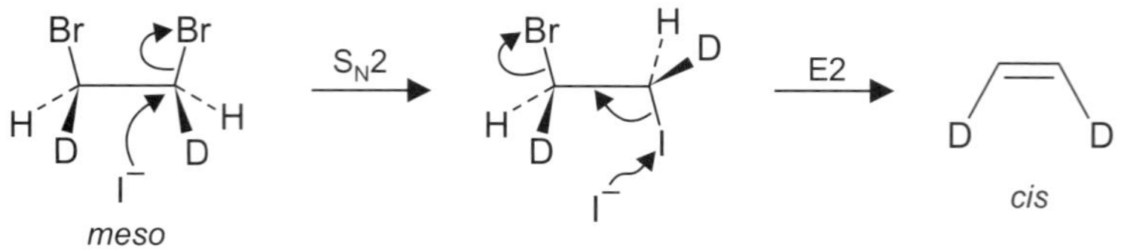

SCHEME 12.25

(D) Curtin-Hammett Principle

The Curtin-Hammett principle applies to a conformationally heterogenous reactant where the products must be non-equilibrating. (IUPAC Commission, Gold 1983). The Curtin-Hammett principle implies that in a chemical reaction which give one product from one conformer and a different product from another conformer (provided the products do not interconvert while the two conformers are rapidly interconverting relative to the rate of product formation). The product composition is not determined by the relative populations of the ground state conformers but largely depends on the relative energies of the corresponding transition states involved (Scheme 12.26).

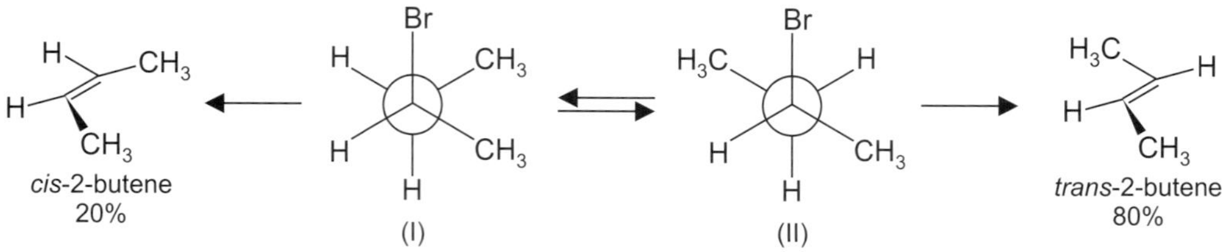

Curtin-Hammett principle relates the product distribution to relative energies of the transition states from e.g., the conformers of 1-bromobutane (conformationally heterogeneous starting material) rather than the relative populations of the conformers [the products must be non-equilibrating and the conformers are rapidly interconvertable relative to the rate of product formation].

SCHEME 12.26

12.3 PYROLYTIC *SYN* ELIMINATION—*Ei*—ELIMINATION INTERNAL

These thermal eliminations occur within a small family of compounds, like acetate esters, methyl xanthate esters and tertiary amine oxides with the formation of olefins and eliminating acetic acid, COS and CH_3SH and dimethylhydroxyl amine respectively (and are the reverse of ene-reaction, see, Scheme 17.97). These eliminations have a common mechanistic feature: a concerted reaction *via* a cyclic transition state within which an intramolecular proton transfer is accompanied by elimination to form a new carbon carbon double bond. The cyclic transition states dictate the *syn* eliminatation *i.e.*, the hydrogen atom and the leaving group depart from the same side of the incipient double bond (Scheme 12.27). These eliminations do not involve acidic or basic catalysts. There is a wide variation in temperatures at which these eliminations proceed. The pyrolysis of carboxylic esters and xanthates provide a useful alternative for dehydration of alcohols without rearrangement.

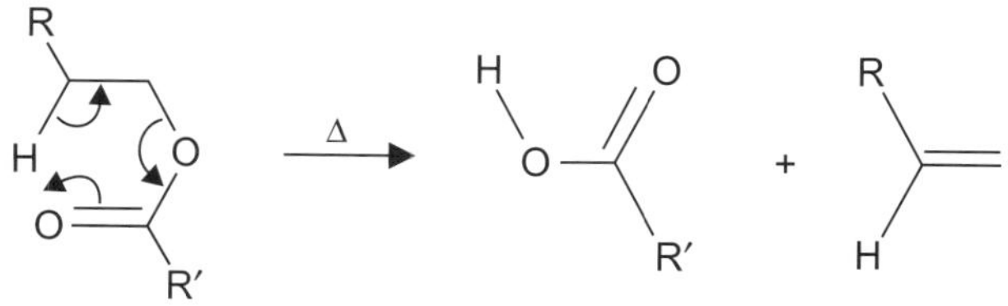

Pyrolytic *syn* elimination

SCHEME 12.27

(A) Stereochemistry of Thermal Eliminations

As already said these eliminations occur with *syn* stereochemistry and the reactions therefore, on the whole are often referred to as thermal *syn* eliminations. On pyrolysis, *erythro* and *threo* isomers of 1-acetoxy-2-deutero-1,2-diphenylethane gave in each case *trans-* stilbene (Scheme 12.28). The *syn* elimination is proved by retention of deuterium in the product from *erythro* isomer (A, Scheme 12.25) and its loss from the *threo* isomer B. For this *syn* process either the hydrogen or the deuterium could be *syn* placed to the acetoxy group. However, in the preferred conformations have the phenyl groups as far apart as possible. The molecule has two stereocenters and therefore, four stereoisomers are possible *i.e.*, a pair of enantiomers in each *erythro* and *threo* isomer. In eclipsed conformations (I and II, Scheme 12.28) Ph/Ph θ = 0°. The elimination, thus takes place from conformations with Ph/Ph far removed. In the transition state from A, a H is *syn* placed with the acetoxy group. In B now it is D which is *syn* placed with it.

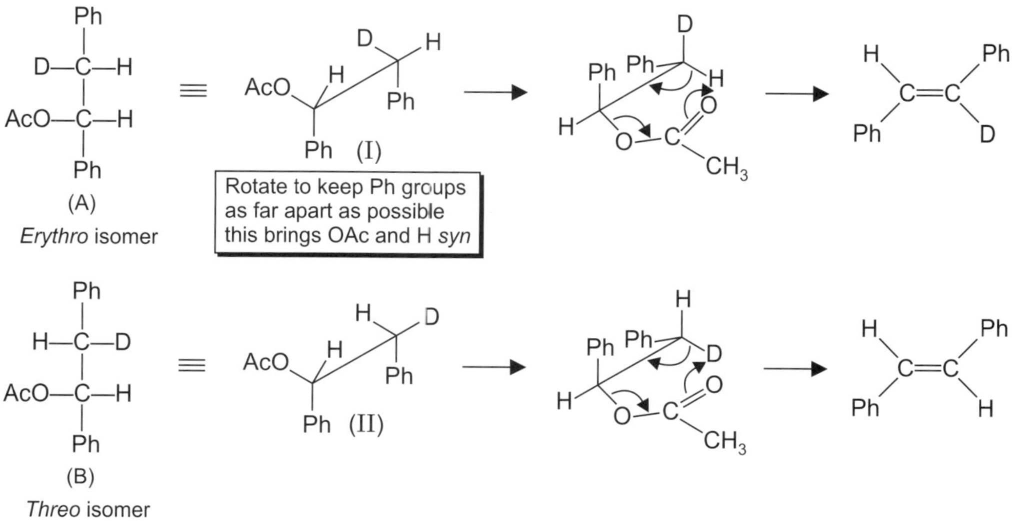

Pyrolysis of 1-acetoxy-2-deutero-1, 2-diphenylethane

SCHEME 12.28

(B) Product Composition

When a conjugating group is present in the β position, the elimination takes place in that direction to yield the conjugated olefin. In the absence of such a situation, the composition of the product is mainly determined by the number of hydrogen atoms available on each of the β-carbons. Thus 2-butyl acetate on pyrolysis affords a mixture containing 57% 1-butene and 43% 2-butene (Scheme 12.29).

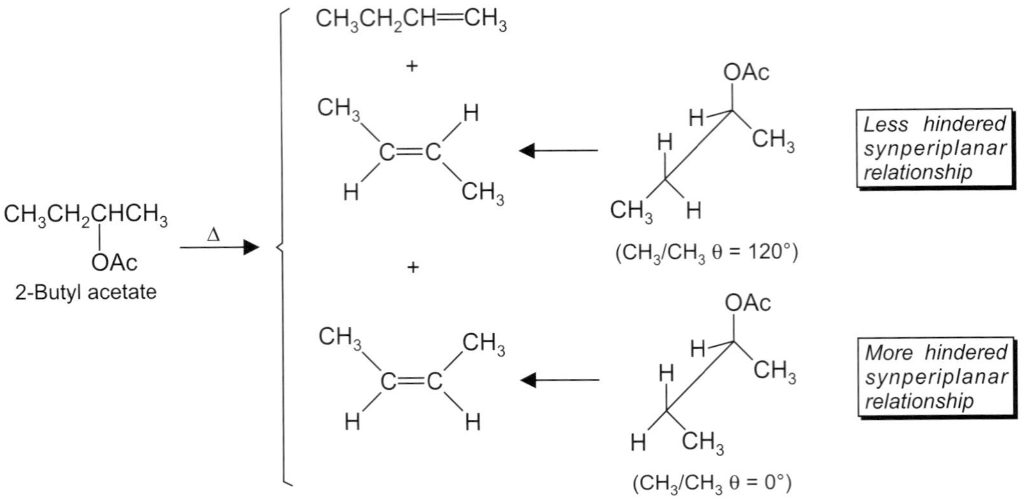

SCHEME 12.29

This product formation closely agrees with the 3:2 ratio of β-hydrogens. Moreover, of the 2-butenes, the *E*-isomer is formed as the major component. This may be due to less steric crowding between the two bulky methyl substituents in the transition state which leads to the *E*-alkene (Scheme 12.29).

(C) Thermal Elimination Reactions in Cyclohexanes

In cyclohexane systems if a leaving group is axial then for the *cis* relationship a β-placed equatorial hydrogen on the ring should be available for *syn* elimination. This is explained by

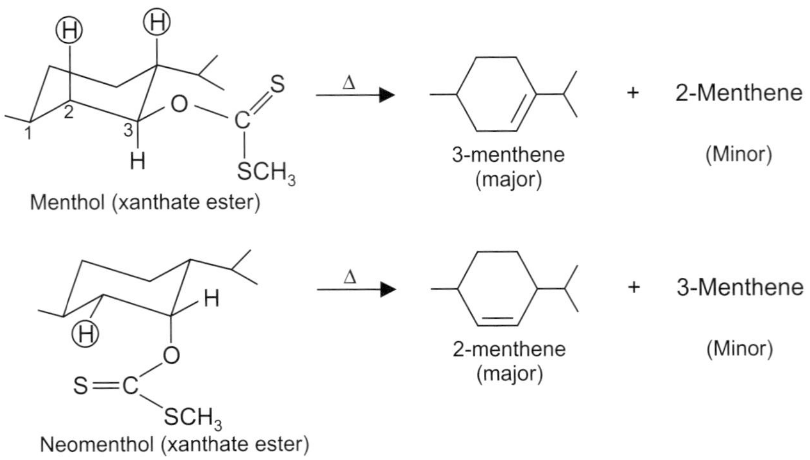

SCHEME 12.30

considering the methyl xanthate esters from menthol and neomenthol (Scheme 12.30). Recall that during $E2$ reaction the system of menthol and neomenthol (*trans* coplanarity condition) give opposite results.

The Cope reaction involves the pyrolysis of amine oxides (Scheme 12.31) having a hydrogen atom β to the amine group. The *syn* elimination affords an alkene and dialkylhydroxylamine.

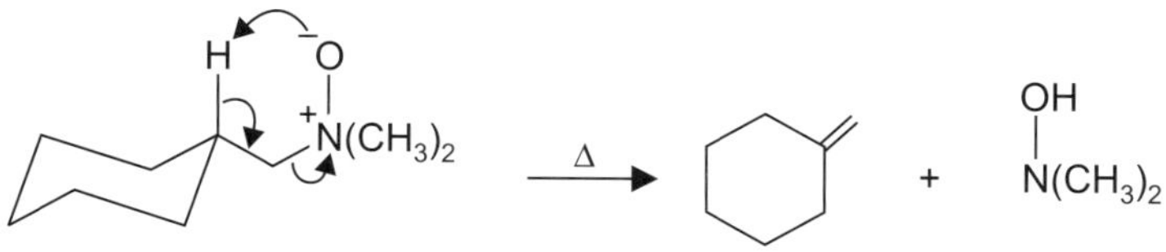

SCHEME 12.31

(D) Elimination from Selenoxides

Selenoxides are more reactive than amine oxides toward β elimination and many undergo thermal *syn* elimination to form alkenes under milder conditions (room temperature). The selenoxides can be easily made from the corresponding selenides by oxidation with hydrogen peroxide. A selenides is formed *e.g.*, by the nucleophilic substitution on a tosylate by selenol (Scheme 12.32). The orientation is satistical *i.e.,* depending on the number of β-hydrogens available on each side and Hofmann product predominates.

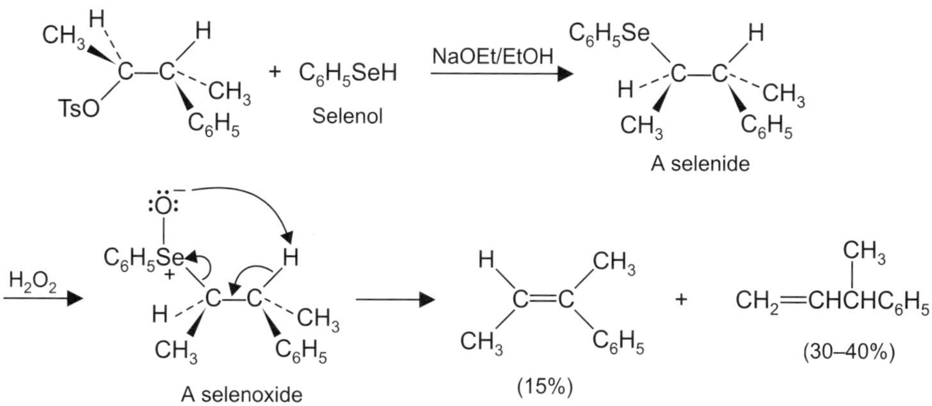

SCHEME 12.32

Selenoxide eliminations find many synthetic uses and an example is the formation of α, β-unsaturated ketones. Lithium enolates from ketones (made by employing LDA) react with benzene selenide bromide (C_6H_5SeBr) to give selenides in which C_6H_5Se-group is attached in the α-position. Treatment of the selenide with H_2O_2 (room temperature) gives α, β-unsaturated ketone (Scheme 12.33).

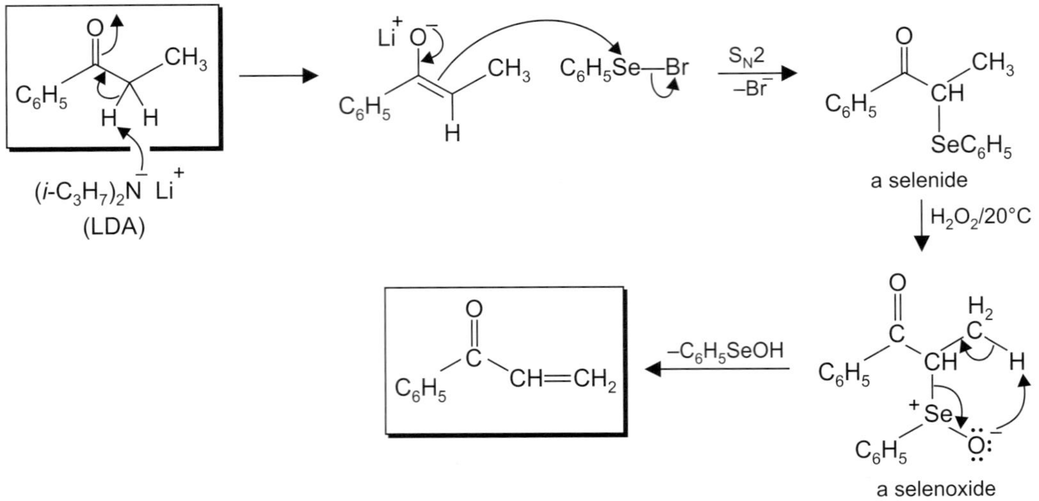

SCHEME 12.33

PROBLEMS

12.1 Why in *syn* thermal eliminations no skeletal rearrangements are observed?

12.2 What products are expected on E_2 elimination and in what yield from the following ammonium and sulphonium substrates?

(I) (II)

12.3 Which of the *threo-* or *erythro-* 1,2-diphenyl-propylamine will react faster under *E*2-elimination reaction conditions (sodium ethoxide and ethanol)?

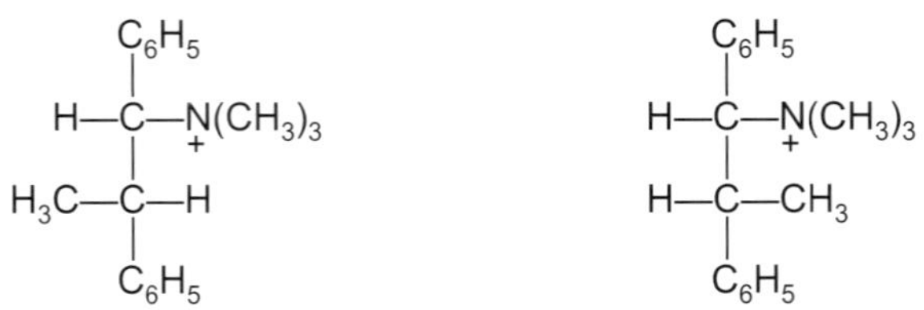

(*threo*) (*erythro*)

12.4 On pyrolysis of the following ester gave in addition to acetic acid, the mixture of olefins I and II in a nearly statistical ratio of 3:2. Explain.

$$H_3COOCH(CH_3)CH_2CH_3 \longrightarrow \underset{(I)}{H_2C{=}CHCH_2CH_3} \quad \underset{(II)}{CH_3CH{=}CHCH_3}$$

12.5 In several situations the *E*2 elimination gives a complex mixture of regio- and stereoisomers of alkenes. Explain giving a suitable example.

12.6 Sodium alkynides acting as nucleophiles, displace a halide ion from primary alkyl halides (alkylation of alkynide ion). With secondary or tertiary halides, the alkyne from which sodium alkynide was orginally derived is isolated. Explain.

12.7 Predict the outcome of the following reactions:

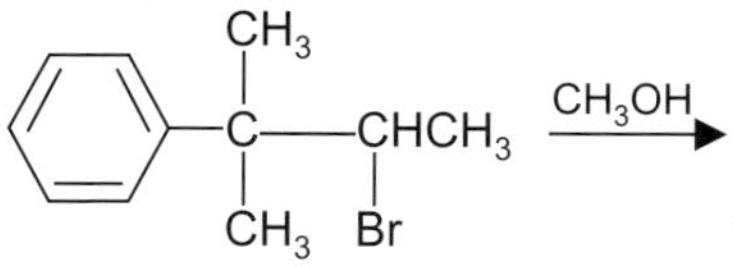

(I)

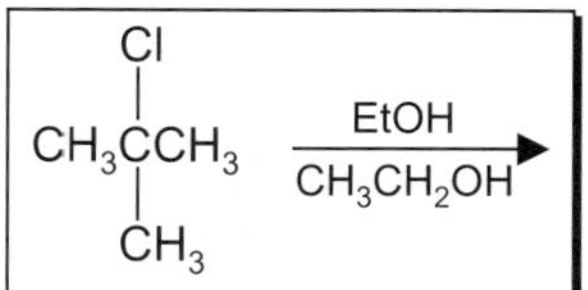

(II)

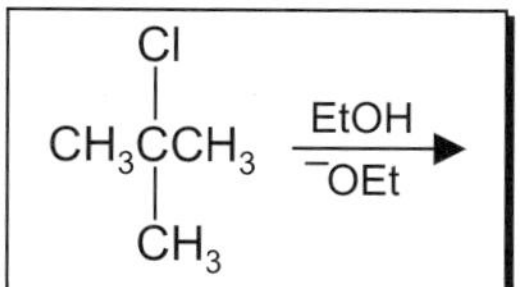

(III)

12.8 Select one product from each of the following reactions giving arguments.

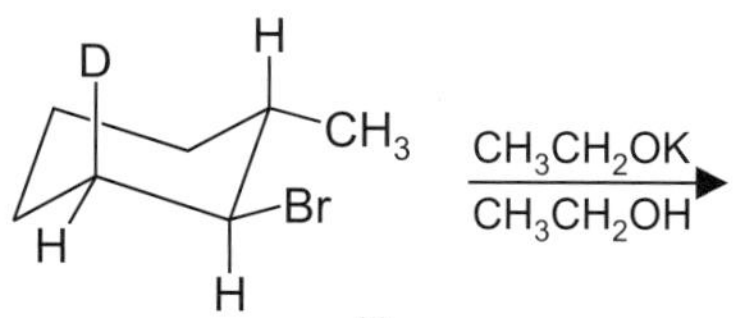

(I)

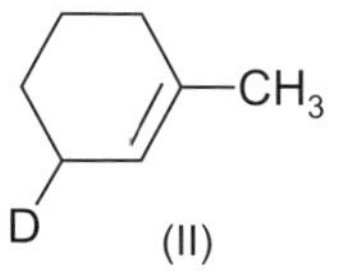

(II)

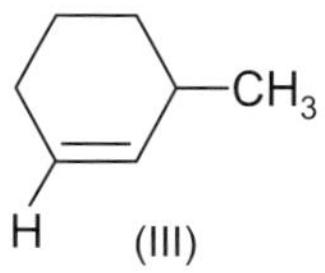

(III)

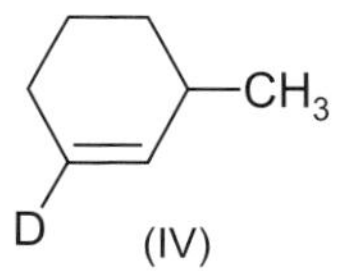

(IV)

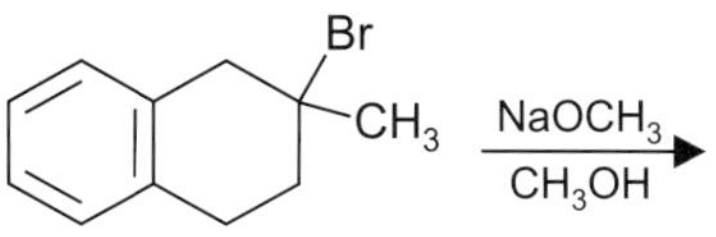

(V)

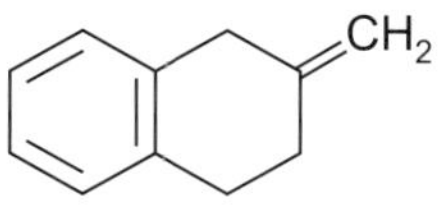

(VI)

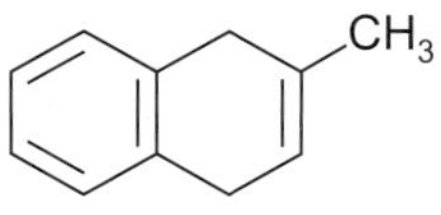

(VII)

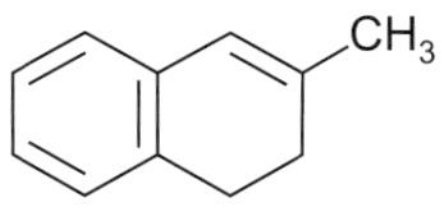

(VIII)

ANSWERS TO THE PROBLEMS

12.1 Since no intermediates that are prone to such rearrangements are involved.

12.2 The elimination will largely involve that β position, the loss of a proton from which will end up with a stabler carbanion (involvement) of a carbanion like transition state).

(I) ⟶ $CH_2{=}CH_2$ (Major product) + $MeCH{=}CH_2$ (Minor product)

(II) ⟶ $MeCH_2CH{=}CH_2$ (Major product) + $MeCH{=}CHMe$ (Minor product)

12.3 These isomers will undergo stereospecific elimination. Due to the rigid requirements of the transition state (*trans*-coplanarity) the *erythro*-isomer will react slowly since the desired conformation has crowding of the two phenyl groups.

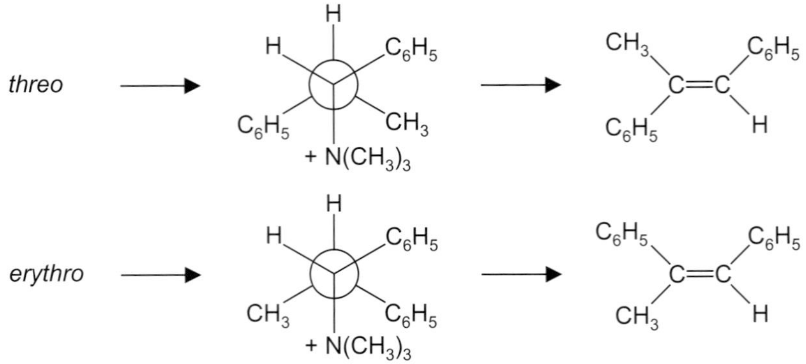

12.4 There are two β-positions from which H can be lost. Product (I) is obtainable by the loss of 3 β-hydrogens while II by the loss of only two β-hydrogens.

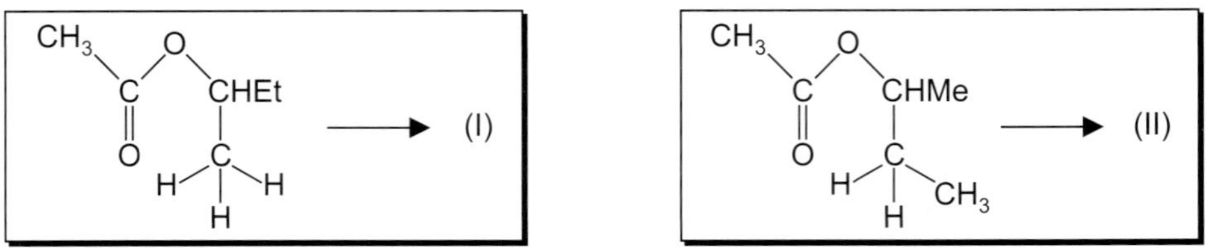

12.5 It is due to several possible *E*2 elimination routes *e.g.*, in 3 chlorohexane.

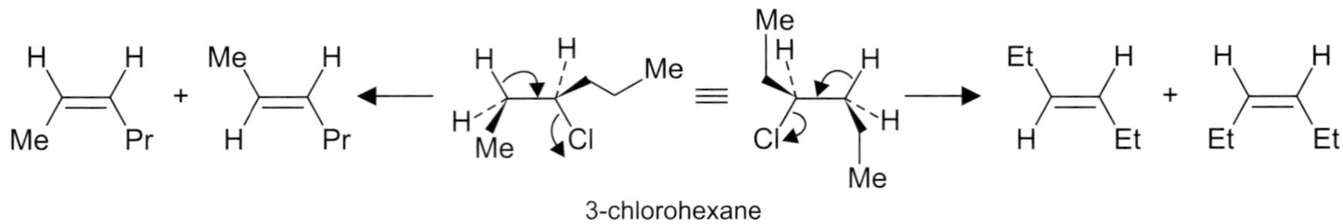

12.6 The alkynide ion, acting as a nucleophile displaces a halide from the primary alkyl halide in an S_N2 reaction to give a substituted alkyne. With secondary and tertiary halides, the alkynide ion acting as a base (rather than a nucleophile) brings about an *E*2 elimination.

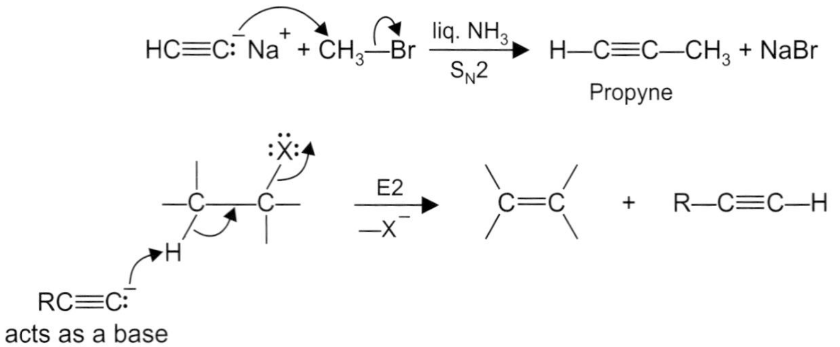

12.7 It is an *E*1 reaction, the initially formed secondary carbocation from (I) undergoes a 1, 2-methyl shift to give a more stable tertiary benzylic cation and the final product after the loss of a proton. The substrate II is tertiary and strong base is absent. The reaction will give mainly substitution, with some elimination. The reactant III is a tertiary halide, strong base gives elimination by *E*2 pathway in a high yield.

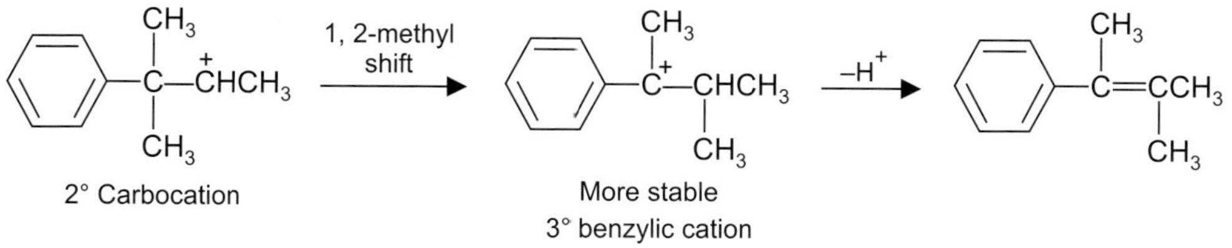

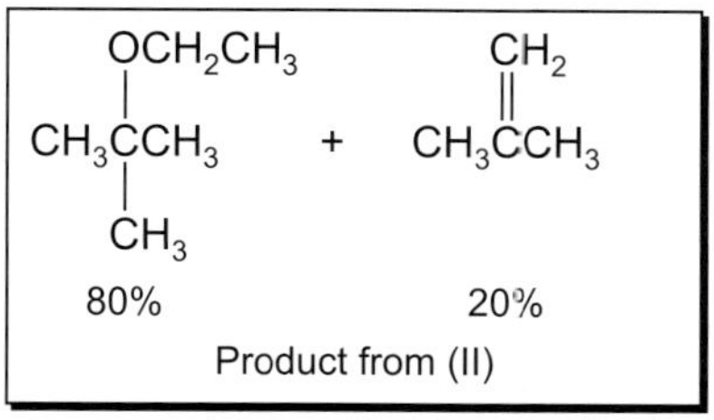

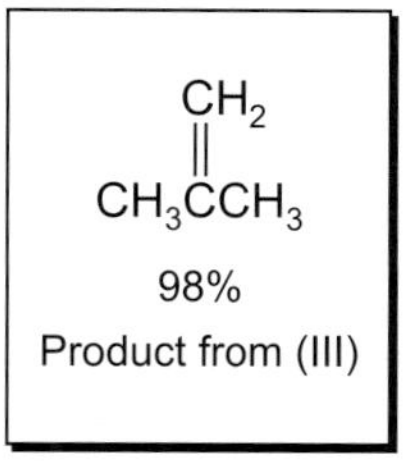

12.8 Generally $E2$ eliminations require a conformation in which the β-hydrogen atom and the leaving group are *anti*. In a cyclohexane ring of (I) the hydrogen and the leaving group have to be diaxial. To place the bromine in axial orientation ring flips and *anti* elimination from it will give (IV) as the only alkene that can be formed. The products (II and III) could arise via a *syn* elimination which is not the case.

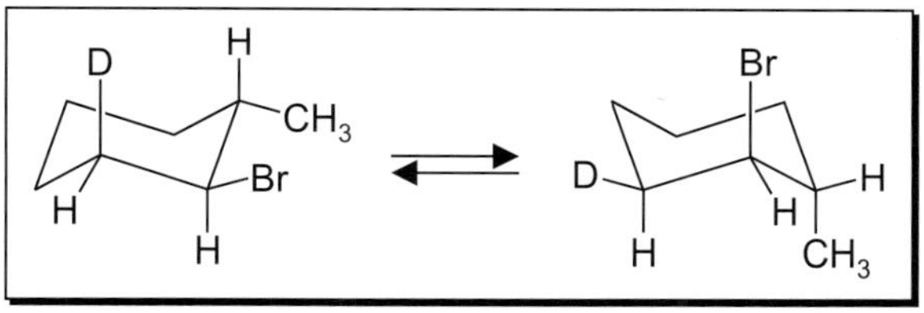

The substrate (V) has got three types of β-hydrogens. The compounds (VII and VIII) are trisubstituted, while (VI) is the least stable. The compound (VIII) is the alkene formed since the new double bond is in conjugation with the aromatic ring.

CHAPTER 13

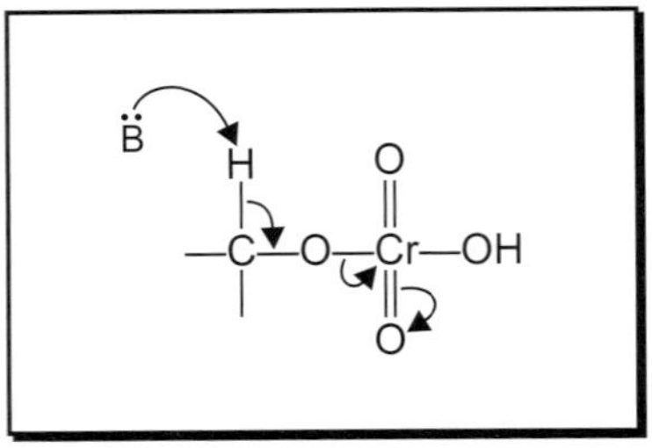

Oxidation Methods

An oxidation reaction is defined as addition of oxygen to an organic compound, removal of hydrogen or removal of one electron as *e.g.*, in the conversion of phenoxide anion to phenoxy redical (see Schemes 14.2 and 14.3).

13.1 OXIDATION OF ALCOHOLS TO ALDEHYDES, KETONES OR CARBOXYLIC ACIDS

(A) Chromium (VI) Oxide/Chromium Trioxide (CrO_3)

From among a variety of transition metal oxidants, the Cr (VI)— derived reagents are highly useful. The oxidation state of Cr in these reagents is (VI), and these are powerful oxidizing species. The CrO_3 based oxidants convert an alcohol to the corresponding ketone or aldehyde (oxidation of a primary alcohol proceeds through the aldehyde to carboxylic acid) and the general mechanism is given (Scheme 13.1). The most commonly used reagent is chromic acid H_2CrO_4, which is usually prepared by adding chromium VI oxide (CrO_3) or sodium dichromate ($Na_2Cr_2O_7$) to sulphuric acid. When the oxidations are carried out in aqueous acetone the process is termed Jones oxidation (or oxidation by the Jones reagent, CrO_3—H_2SO_4—H_2O.

The mechanism involves the formation of a chromate (VI) ester and a subsequent fragmentation (an *E*2 elimination reaction) to form a ketone (by the loss of H^+ and $HCrO_3^-$).

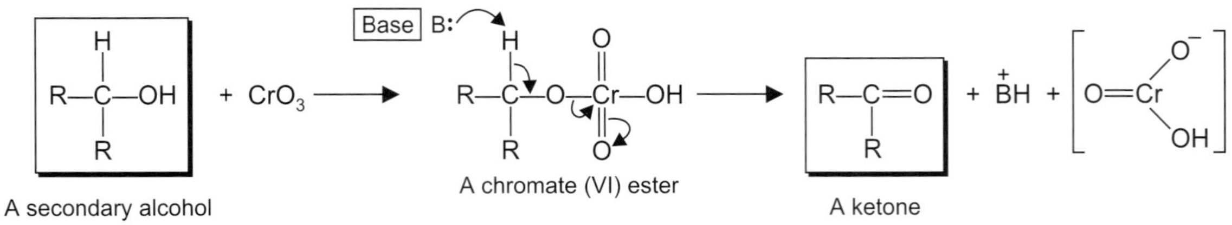

SCHEME 13.1

A marked isotope effect (~ 6) was observed with propan-2-ol. Thus when chromic acid is used in the oxidation of $CH_3CDOHCH_3$ and isopropanol, $CH_3CHOHCH_3$, the deuterated compound reacted about six times slower, so that $^{kH}/_{kD}$ is ~ 6 to prove that the rate determining step involves the cleavage of a C—H bond.

EXERCISE 13.1

A useful laboratory test for aldehydes is the silver mirror test when silver (I) oxide (Ag_2O) brings about the oxidation of an aldehyde to a carboxylic acid. Give a mechanism of the reaction.

ANSWER. *This is in Scheme 13.1a and the mechanism involves the nucleophilic addition of silver oxide to the carbonyl group of the aldehyde.*

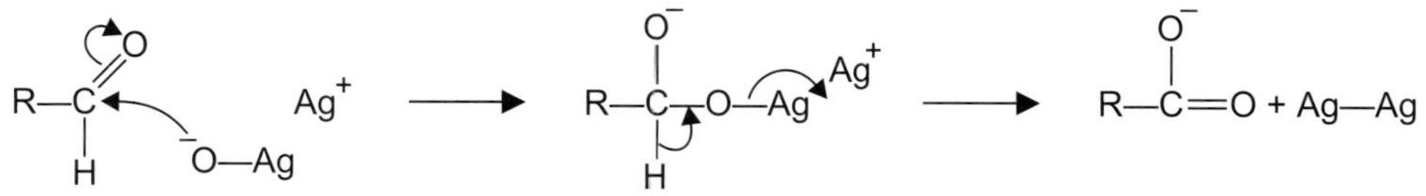

SCHEME 13.1a

A very useful Cr (VI) reagent is pyridinium chlorochromate (abbreviated PCC), which is prepared by dissolving CrO_3 in hydrochloric acid and by subsequent treatment of the solution with pyridine (Scheme 13.2). When PCC is dissolved in DMF or is used as a suspension in CH_2Cl_2, it oxidizes the secondary alcohols to ketones and allylic primary alcohols to the corresponding aldehydes. Saturated primary alcohols are oxidized to an aldehyde or the carboxylic acid depending on the conditions.

Collins reagent is obtained by adding chromium (VI) oxide to pyridine and is chromium (VI) oxide-pyridine complex, (CrO_3–2 pyridine) which is used for the oxidation of alcohols containing acid-sensitive functional groups. Pyridinium dichromate ($C_5H_5\overset{+}{N}H_2$) Cr_2O_7 abbreviated (PDC) is obatined by the reaction of chromium trioxide with pyridine in water. All these reagents are readily soluble in organic solvents and PCC and PDC are conveniently available as stable solids. Irrespective of which one is used these perform the same function. Other complexes *e.g.*, chromium (VI) oxide dimethylpyrazole complex provide an added advantage as the internal base is available for the decomposition of the intermediate chromate (VI) ester (see, Scheme 13.1). Examples of oxidation with these reagents are in (Scheme 13.2*a*).

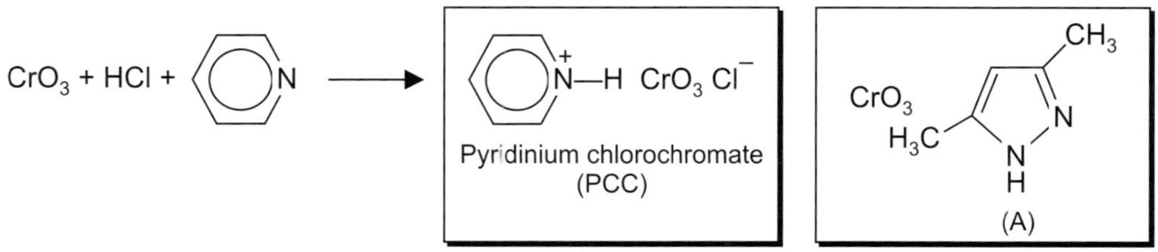

SCHEME 13.2

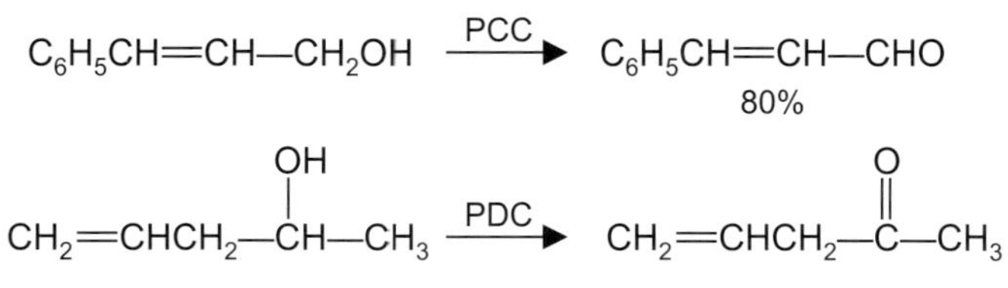

SCHEME 13.2a

(B) Use of Dimethyl Sulphoxide (DMSO)

Dimethylsulphoxide in combination with an electrophilic molecules like dicyclohexylcarbodiimide, acetic anhydride, oxalyl chloride and sulphur trioxide brings about the oxidation of alcohols.

The following points may be noted:

- DMSO on reaction with an electrophile gives a species which is activated and thus an alcohol adds to the sulphur atom and the adduct thus formed has a good leaving group as well.
- The alcohol is converted into its halide or tosylate and then reacted with DMSO. In either case there is an intermediate formation of an alkoxysulphonium ion from which carbonyl compound is formed by base catalysed elimination.

(i) By the use of DCC

In this method (Scheme 13.3) DMSO is activated by using DCC and an acid is used to catalyse the first step to give an intermediate (A, Scheme 13.3) which is attacked by alcohol. This oxidation can be specifically used for the oxidation of primary alcohols to aldehydes.

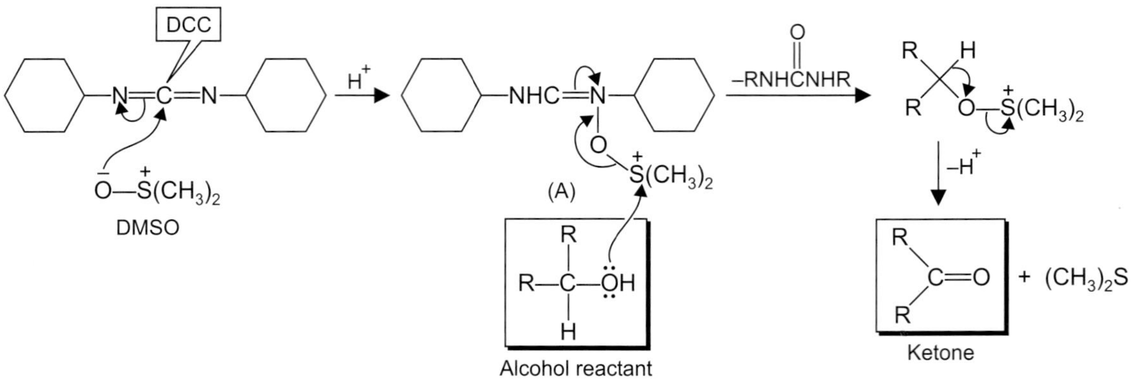

SCHEME 13.3

Further uses of DCC-Formation of Amide Bond in Peptide Synthesis (formation of aliphatic carbon-nitrogen bonds)

Carboxylic acids react with amines only under very vigorous condition. It is thus necessary to convert the acid into a derivative which is more reactive towards the nucleophiles. The requirement is that the group X in the derivative RCOX should be a good leaving group. Conversion to an acyl chloride is a common way to accomplish this for normal organic reactions. However, acyl chlorides are quite reactive and do not give high enough yields in peptide synthesis due to side reactions, thus milder methods for the amide bond formation are to be employed. The most common method to form an amide bond in peptide synthesis makes use of DCC as the coupling reagent.

The two C=N bonds in DCC (Scheme 13.3a) make the central carbon electrophilic and highly reactive towards nucleophiles. The carboxylic acid (nucleophile) first reacts with DCC to form an intermediate that resembles an anhydride but with a carbon-nitrogen double bond in place of one of the carbonyl groups (—CO—O—C=N—R).

The amino group then reacts with the intermediate (I, Scheme 13.3a) to give a tetrahedral intermediate (II). The C—O bond of the tetrahedral intermediate breaks easily as these bonding electrons are delocalized to form dicyclohexyl urea—a stable diamide. One may recall that weaker (more stable) the base, the better is the tendency for it to act as a leaving group.

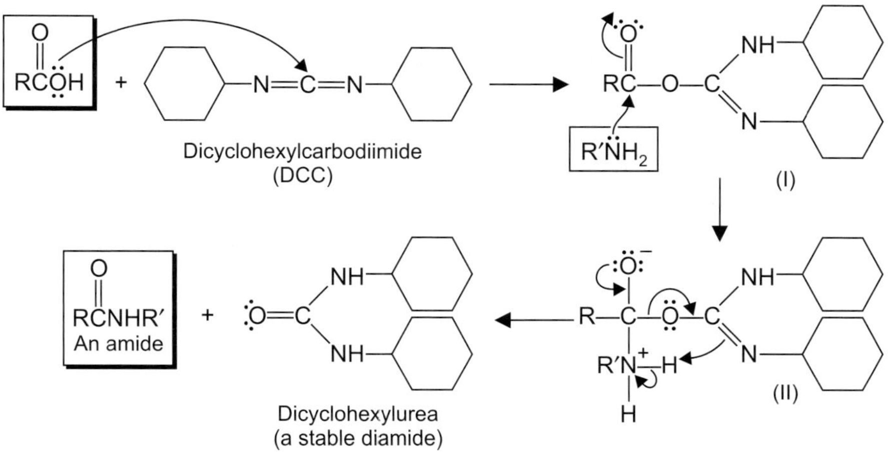

SCHEME 13.3a

(ii) Swern oxidation—use of DMSO and oxalyl chloride

Swern oxidation (Scheme 13.4) DMSO is activated towards the addition step by oxalyl chloride. The resulting alkyloxy-sulphonium salt (A, Scheme 13.4) on treatment with a base (usually triethylamine) gives the corresponding carbonyl compound in high yields under mild conditions. Thus a primary alcohol gives an aldehyde and a secondary alcohol gives a ketone.

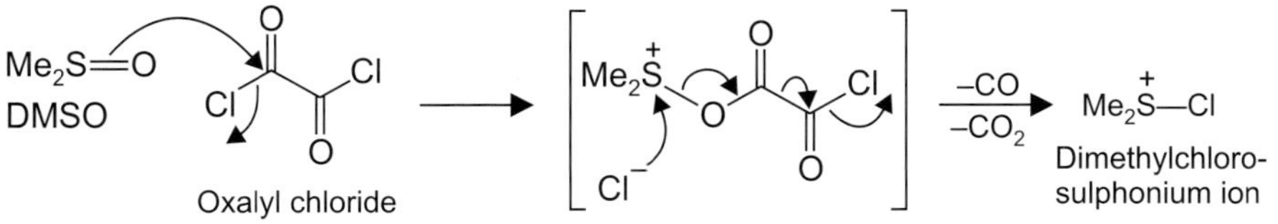

Activation of DMSO—formation of oxidizing agent

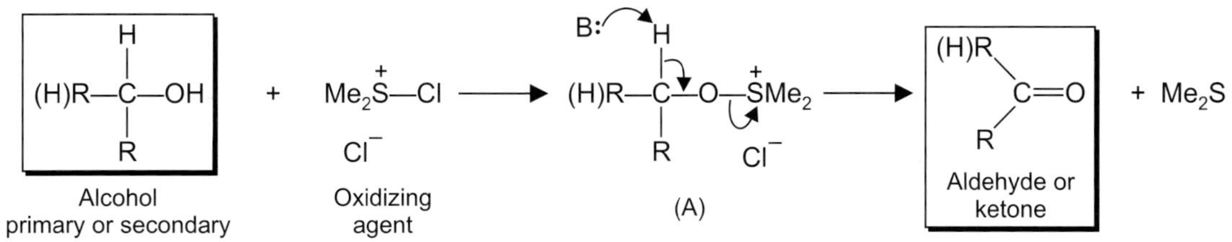

Oxidation of an alcohol with dimethylchlorosulphonium ion–Swern oxidation

SCHEME 13.4

(iii) Use of substrates with good leaving groups

Substrates with good leaving groups *e.g.*, the alcohols toluene–*p*–sulphonates usually derived from an alcohol and the sulphonyl chloride give the corresponding aldehydes or ketones on treatment with DMSO (Scheme 13.5). The presence of a base facilitates the reaction by removing the proton. Reaction occurs *via* an initial S_N2-displacement followed by base catalysed elimination on the resulting sulphonium salt (Scheme 13.5).

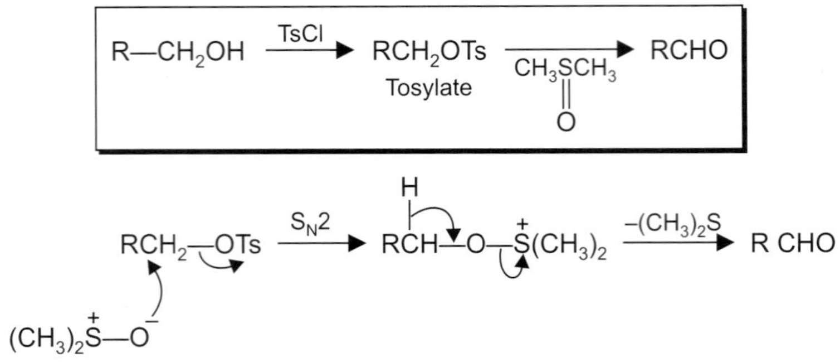

SCHEME 13.5

Specific Oxidation of Primary Alcohols to Aldehydes

Methods are available e.g., use of DMSO to specifically oxidize primary alcohols to aldehydes by arresting over-oxidation to carboxylic acids. A variety of amine oxides e.g., pyridine N-oxide react in a way similar to DMSO and are useful reagents to oxidize primary alcohols to aldehydes (Scheme 13.6). In this reaction pathway, nothing is in the reaction sequence which can oxidize the initially formed aldehyde to an acid.

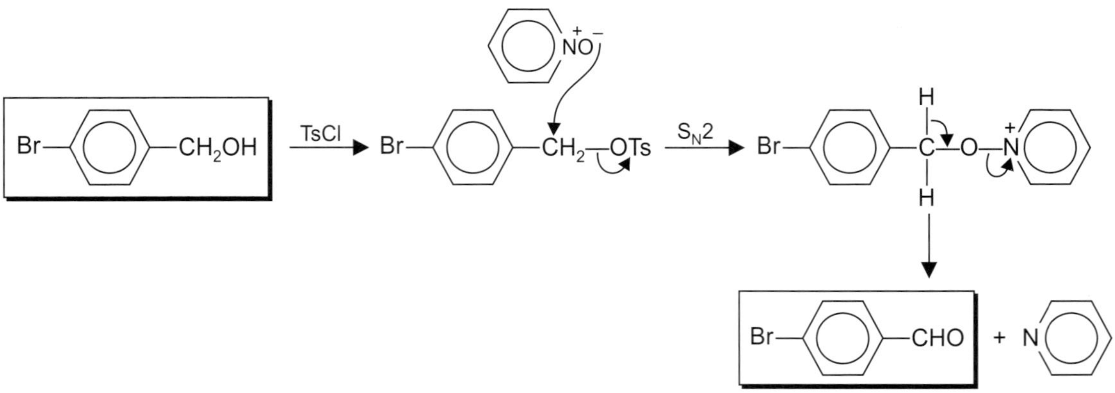

SCHEME 13.6

Periodinane Oxidation (Use of Dess-Martin Reagent) DMP

This is a very mild oxidation via which a primary alcohol can be converted into an aldehyde. The reagent periodinane is made from 2-iodoxybenzoic acid and acetic anhydride (Scheme 13.6a). An alcohol reacts with it to displace an acetate moiety and the intermediate formed undergoes a fragmentation as in Swern oxidation.

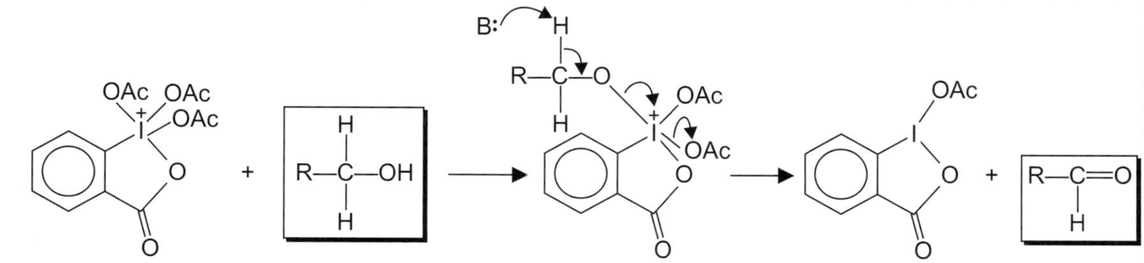

Dess-Martin Periodinane (DMP)

SCHEME 13.6a

One may recall that DMP has a high valent iodine atom bonded to several oxygen atoms.

(C) Oppenauer Method

The developments in CrO_3-pyridine and DMSO based methods for oxidation have pushed several classical methods into background. One such method which is interesting mechanistically, is Oppenauer oxidation, this is the reverse of the Meerwein– Pondorff – Verley reduction. The technique involves heating the alcohol to be oxidized with an aluminium alkoxide in the presence of a carbonyl compound usually acetone in large excess. The carbonyl compound acts as the hydrogen acceptor within a cyclic complex (Scheme 13.7).

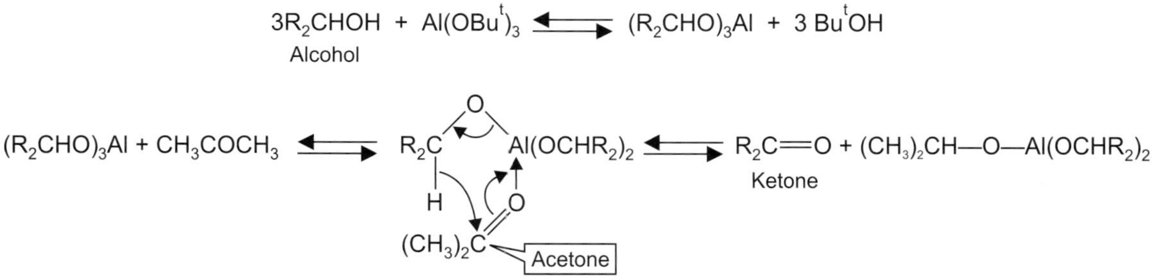

SCHEME 13.7

(D) Oxidation with Oxoammonium Ions (*A recent method N. Merbouh, J.M. Bobbitt and C. Brueckner, Org. Prep. Proced. Int. 36, 3 (2004)*).

This oxidation method uses an oxoammonium ion, generally derived from the stable nitroxide—tetramethylpiperidine nitroxide (A, Scheme 13.8) abbreviated TEMPO. It is regenerated in a catalytic cycle by the use of hypochlorite ion as the stoichiometric oxidant. During oxidation an alcohol forms an intermediate adduct with the oxoammonium ion.

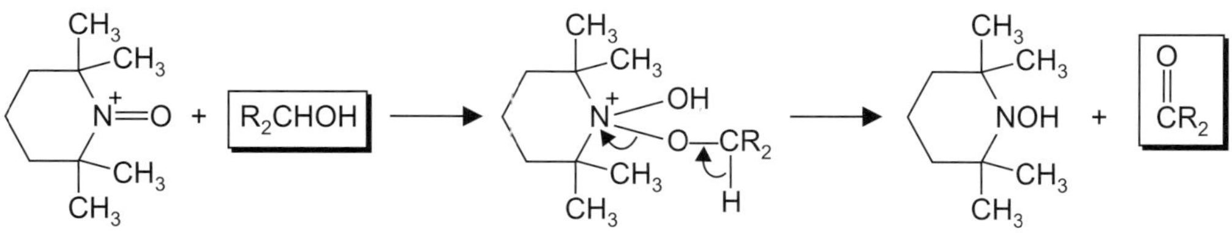

(A)

SCHEME 13.8

This oxidation system has been selectively used to oxidize primary alcohols in the presence of secondary hydroxyl groups (Scheme 13.9). The reagent has been further reported to form carboxylic acids from primary alcohols by using sodium chlorite as a cooxidant.

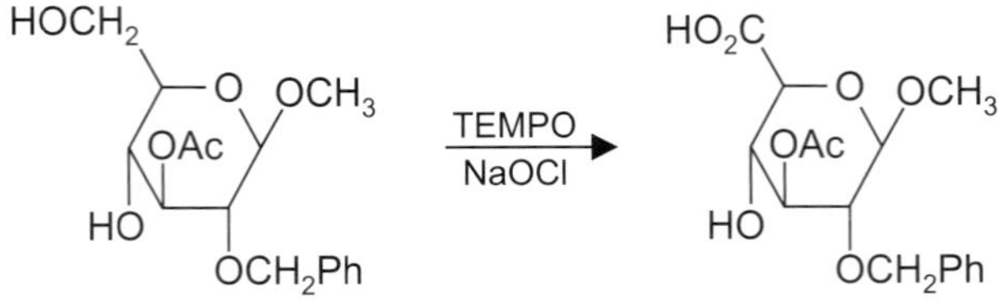

SCHEME 13.9

(E) Oxidation of Allylic Alcohols—Active Manganese Dioxide and Quinones

Manganese dioxide specifically oxidizes allylic and benzylic hydroxyl groups to give α, β-unsaturated carbonyl compounds. The reagent does not attack carbon-carbon double and triple bonds and saturated hydroxyl groups (eq. I, Scheme 13.10).

Several high potential quinones for example, chloranil are capable of oxidizing allylic, benzylic and propargylic alcohols. Mechanistically, the reaction proceeds *via* the formation of resonance stabilized carbocations which are generated by hydride loss from the reactant (eq. II, Scheme 13.10).

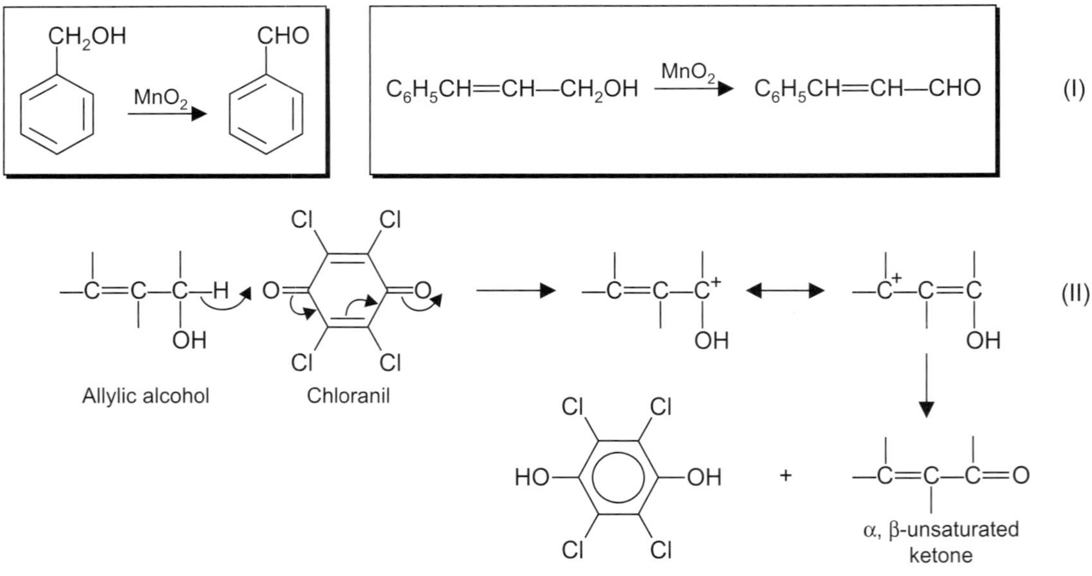

SCHEME 13.10

(F) Benzylic Alcohols—Sommelet Reaction

In this method a halide from benzyl alcohol is reacted with hexamethylenetetramine and the salt thus obtained (Scheme 13.11) is subjected to hydrolysis in the presence of more hexamethylenetetramine. (Recall Oppenauer oxidation, in this case as well, the equilibrium is diplaced with the excess of amine.) The mechanism involves the tranfer of a hybride–ion from benzylamine to methyleneimine.

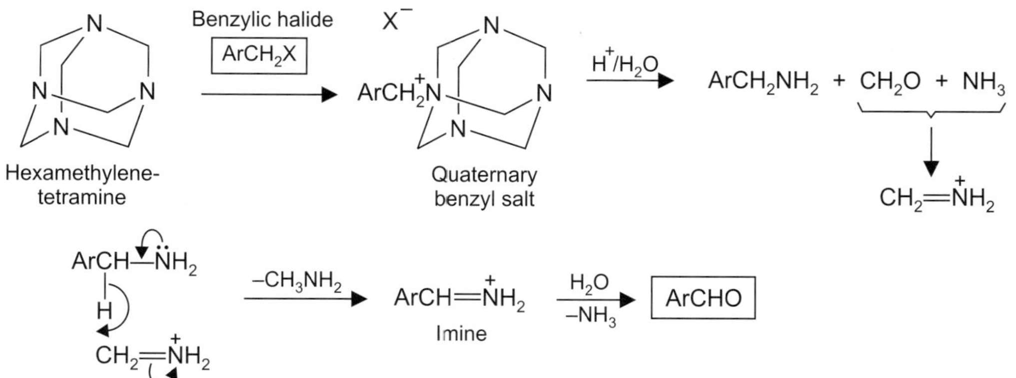

Sommelet reaction

SCHEME 13.11

13.2 ALLYLIC OXIDATION OF ALKENES

Allylic oxidation of alkenes can be brought about with selenium dioxide to give carbonyl compounds, allylic alcohols, or esters depending on conditions. This oxidation is thought to be an ene reaction and subsequent [2, 3] sigmatropic rearrangement. The hydrolysis of the Se(II) ester finally leads to an allylic alcohol (Scheme 13.12). The sigmatropic rearrangemant brings the double bond in the original position (A, Scheme 13.12). The alcohols formed originally are further oxidized to the carbonyl compound.

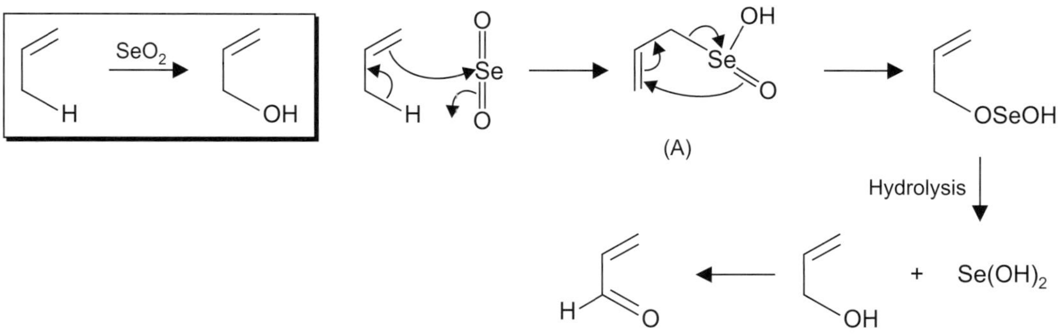

SCHEME 13.12

Allylic alcohols are the initial products of oxidation and these are further oxidized to carbonyl groups with selenium dioxide. Essentially it is the carbonyl compound that is isolated. If one wants to end up with an alcohol, the oxidation is carried out in acetic acid as the solvent and then acetate esters and formed. The oxidation is carried out with catalytic amount of selenium dioxide and *t*-butyl hydroperoxide (TBHP) which reoxidizes the used catalyst.

In summary, the allylic oxidation of alkenes follows the following steps :

- The initial reaction is the formation of allylic selenic acid (A, Scheme 13.12)
- Allylic rearrangement on (A) gives an unstable compound which rapidly gives an allylic alcohol
- The oxidation continues to yield an aldehyde or ketone.

A useful alternative technique which is equivalent to allylic oxidation of alkenes, but now with a shift in the position of the double bond involves the intermediates formation of a β-hydroxy-selenide. The hydroxyselenide is obtained by the addition of phenyl selenenic acid to an alkene (Scheme 13.13). The hydroxyselenide is oxidised with *t*-butylhydroperoxide (TBHP) to unstable selenoxide which immediately eliminates phenylselenenic acid (PhSeOH) to give the *E*-allylic alcohol. With trisubstituted alkenes (I, Scheme 13.13) the addition of phenyl selenenic acid is highly regioselective, the hydroxyl group gets attached at the more substituted end of the carbon-carbon double bond. Thus this addition follows Markovnikoff's rule where "Ph Se^{+}", behaves as the electrophile. The elimination always specifically proceeds away from the oxygen functionality to give the allylic alcohol. Eliminations from selenoxides are discussed (see, Schemes 12.32 and 12.33).

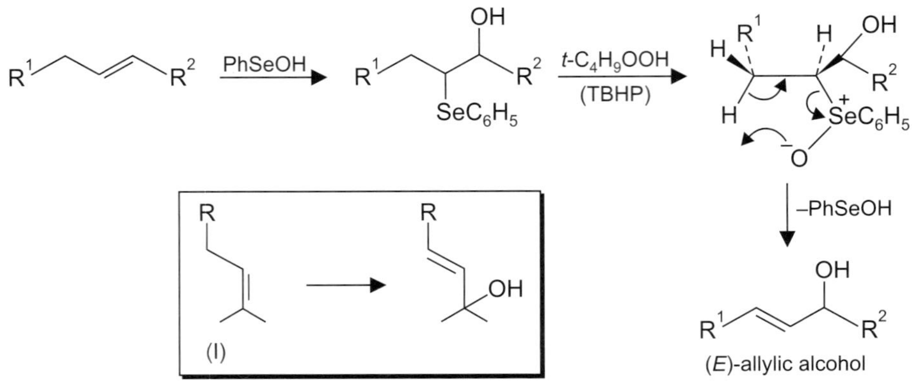

SCHEME 13.13

13.3 OXIDATION OF SATURATED C—H GROUPS

(A) Selenium Dioxide Oxidation

Methylene groups adjacent to cabonyl (–CH_2–CO–) can be oxidized with selenium dioxide to give α-dicarbonyl compounds (Scheme13.14). In the case of unsymmetrical ketones, oxidation generally occurs at that CH_2 group which is easily enolized.

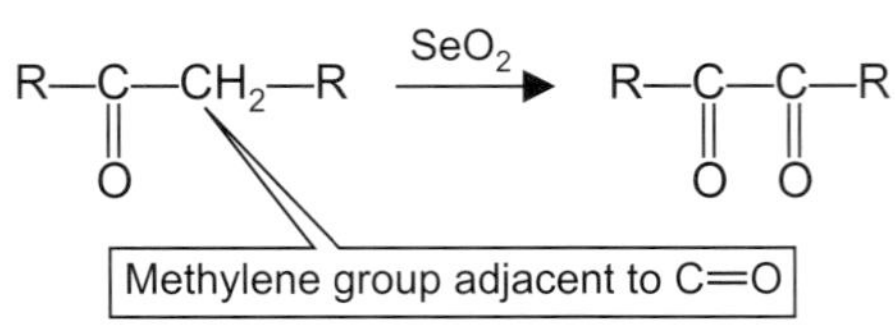

SCHEME 13.14

This conversion uses selenium dioxide in the presence of a base. The mechanism involves the attack by the enolate on the selenium atom to yield the selenate ester of the enol (I, Scheme 13.15). The selenate ester rearranges to regenerate the carbonyl group with the transfer of the oxygen atom to the α-carbon. The removal of the remaining α-hydrogen atom with base initiates the fragmentation shown (II, Scheme 13.15) to yield the α-diketone.

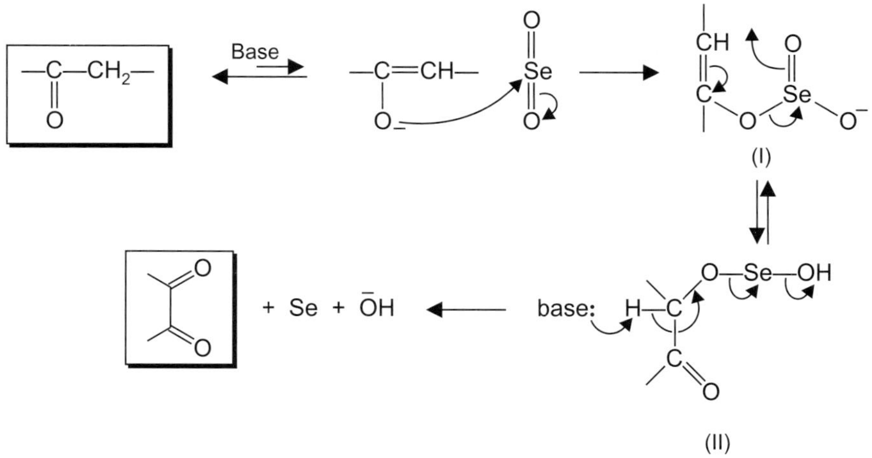

SCHEME 13.15

Another proposal suggests the intermediate formation of a β-ketoselenic acid (Scheme 13.16). An example of its application is the conversion of acetophenone to phenylglyoxal.

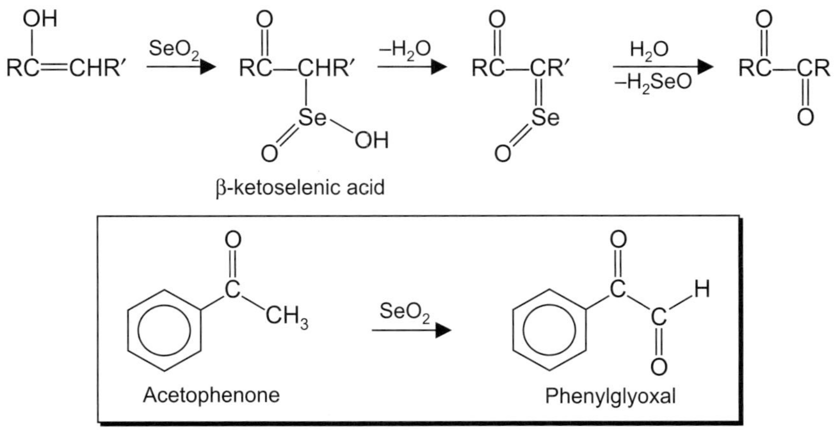

SCHEME 13.16

(B) Monohalogenation and Subsequent Reaction with DMSO

It is substitution-elimination reaction (Scheme 13.17) similar to the one observed during oxidation of *e.g.*, secondary alcohols to ketones with chromium (VI) oxidants.

SCHEME 13.17

It is also possible to selectively oxidize an unactivated C—H group in the molecule which has alternative centers. An example is the Barton reaction (see, Scheme 16.31) where an intramolecular free radical abstraction reactions occur specifically through six membered cyclic transition states and it is possible to oxidize selectively δ C—H bonds.

One can also introduce a C=C bond in a substrate, *e.g.*, a steroid with high regioselectivity *via* specific dehydrogenation. Thus in 3 α-cholestanol (Scheme 13.18) a double bond at a specific position is introduced remote from any functional group. One starts with irradiation of an ester of 3 α-cholestanol of suitable geometry so that the ketones triplet (in the benzophenone portion) is now within bonding distance only with the hydrogen atom at C-14. The abstraction of this hydrogen followed by another internal abstraction affords the unsaturated ester from which the alcohol is generated by hydrolysis. In case the nature (size) of the esterifying acid is changed, other regioselective dehydrogenations can be carried out.

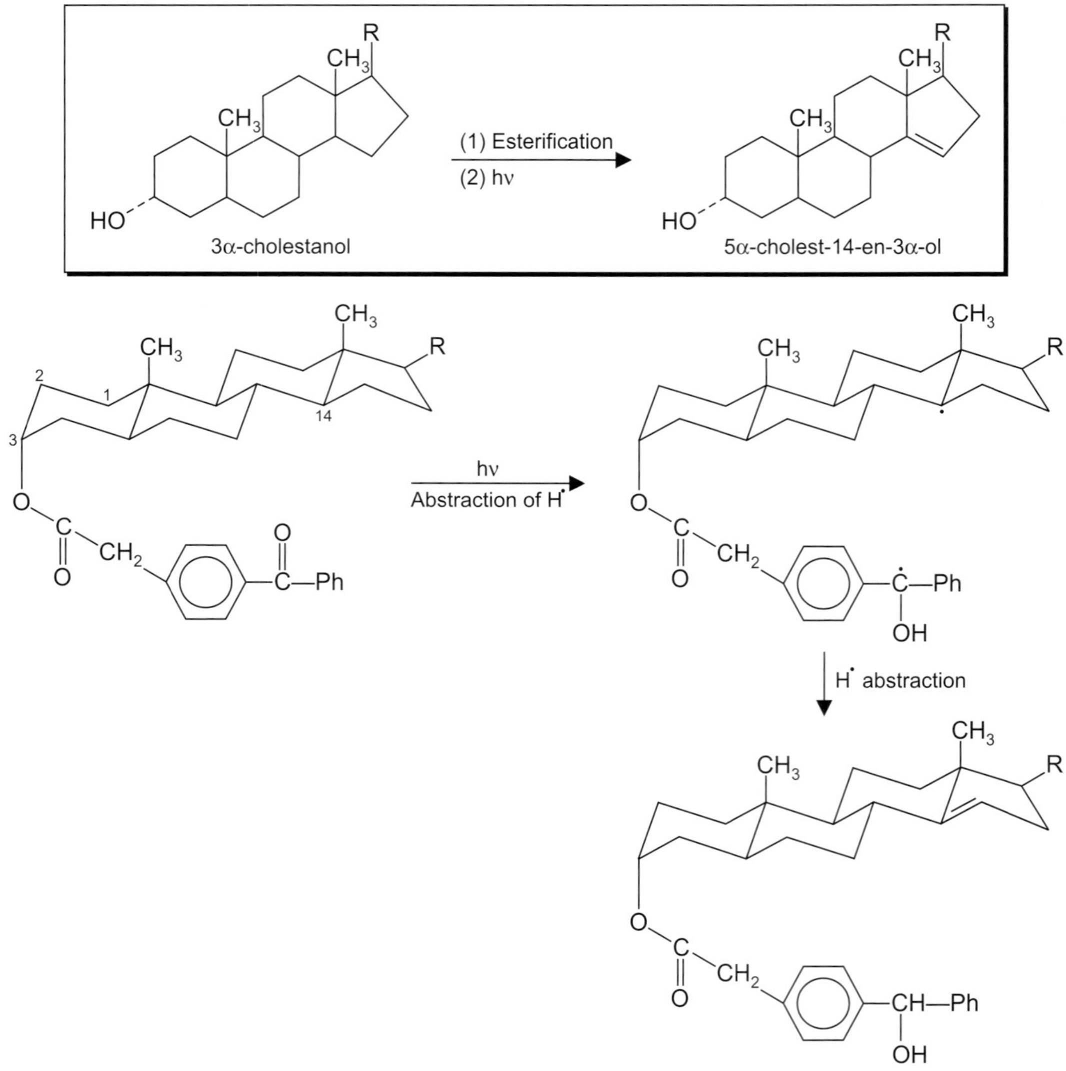

SCHEME 13.18

Alicyclic compounds (the reduction products of aromatic systems *i.e.*, hydroaromatic compounds) can be dehydrogenated to their corresponding aromatic systems in several ways.

This can be done by heating the compounds with those catalysts which are used for their hydrogenation or by using selenium (Scheme 13.19). Quinones which are easily reduced to the corresponding hydroquinones can also be used for aromatizations. Chloranil (2, 3, 5, 6-tetrachloro-1, 4-benzoquinone) and DDQ (2, 3-dichloro-5, 6-dicyano-1, 4-benzoquinone) are often used. The mechanism with quinones involves a transfer of hydride from the substrate to

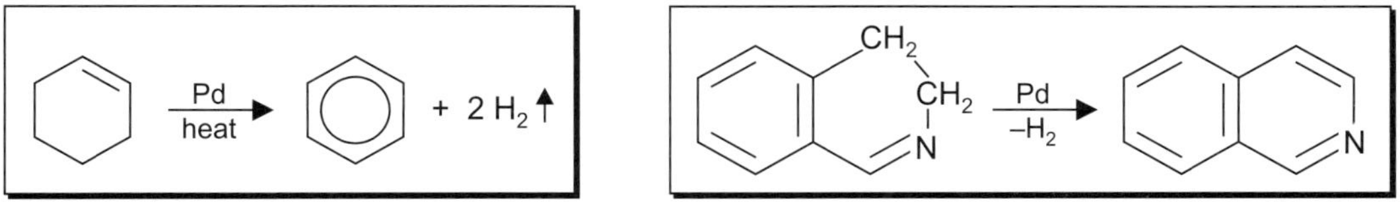

SCHEME 13.19

the quinone oxygen and subsequent loss of a proton (Scheme 13.20). Loss of a hydride ion from the reactant gives a carbocation which can only be formed readily provided it is stabilized *e.g.*, in case it is allylic or benzylic.

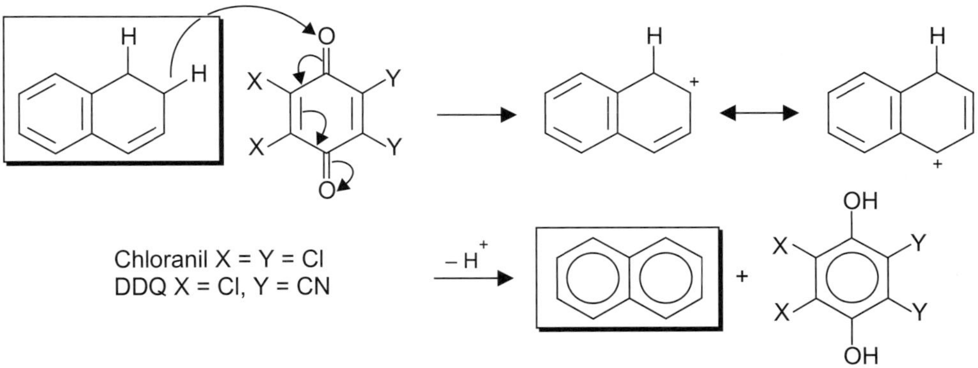

SCHEME 13.20

EXERCISE 13.2

Predict the reagents to bring about each of the oxidations (Scheme 13.20a).

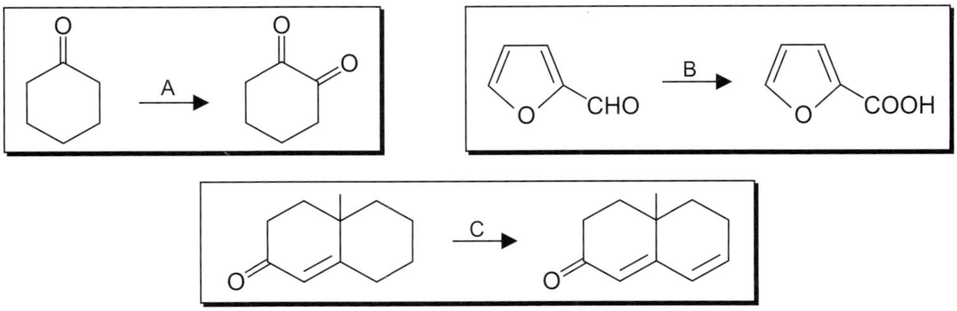

SCHEME 13.20a

ANSWER. *A, SeO_2 ; B, Ag_2O ; C, DDQ.*

Oxidation of side chains on aromatic rings is a useful reaction. Benzylic site is activated to oxidation since the intermediate radical or carbocation intermediates are stabilized by resonance. Moreover, since the aromatic ring is resistant to attack by Cr (VI) and Mn (VII) reagents, these reagents are used widely for these oxidations (Scheme 13.21).

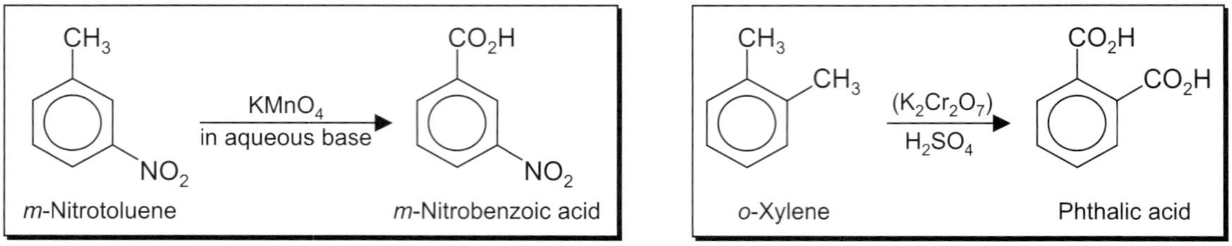

SCHEME 13.21

13.4 ADDITION OF OXYGEN AT CARBON-CARBON DOUBLE BONDS

(A) Epoxidation

Peroxycarboxylic acids *e.g.*, perbenzoic acid (PBA) or *m*-chloroperoxybenzoic acid (MCPBA) are used to convert alkenes to epoxides. The ionic intermediates (*e.g.*, carbocations) are not on the reaction pathway which excludes the formation of one bond at a time. In that case, however, a rotation about the C-C single bond will convert *e.g.*, a *cis*-alkene to a mixture of both *cis*-and *trans-* epoxides. Epoxidation is thus a stereospecific *syn*-addition *e.g.*, *cis*-2-butene gives only the *cis*-product (Scheme 13.22). It is a concerted process (no intermediates) where the two bonds are formed at the same time and thus, the stereochemical relationships of the groups in the starting alkene (see, Scheme 13.22) *e.g.*, *cis*-2-butene does not change and it gives only the *cis* product (Scheme 13.22). The oxidation thus follows a concerted pathway (Scheme 13.22). The peroxy acid acts as an electrophile since the rate of epoxidation is increased by alkyl groups and other electron donating substituents on the alkene (electron rich alkenes). Electron poor alkenes *e.g.*, α, β-unsaturated aldehydes and ketones can be epoxidized only by alkaline solutions of hydrogen peroxide involving nucleophilic addition of HO_2^- which is facilitated by the C=O group (see, Michael addition).

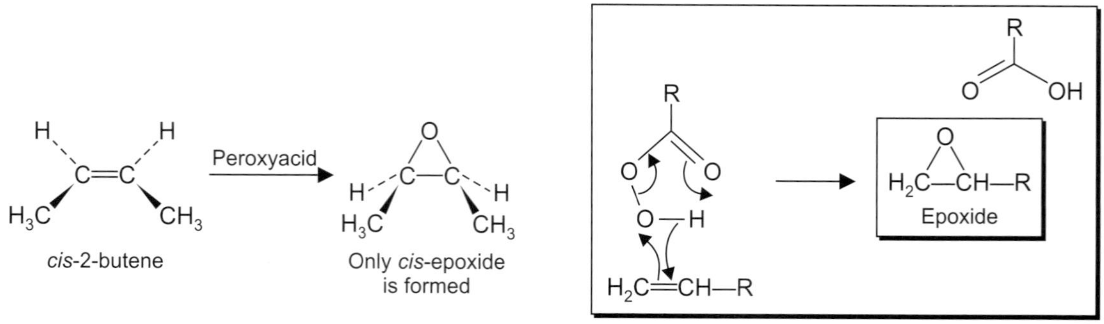

SCHEME 13.22

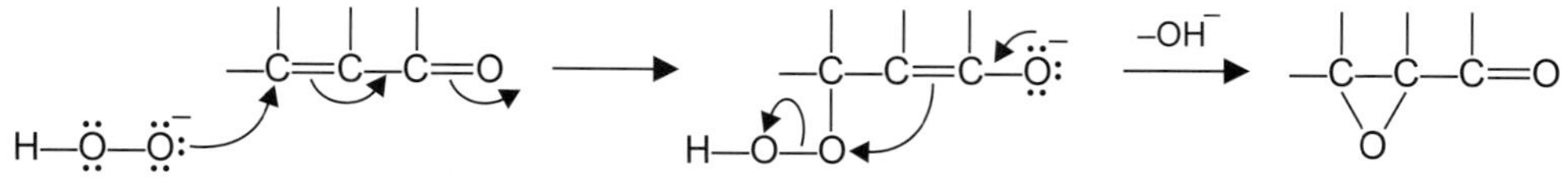

SCHEME 13.23

Stereochemically the addition of oxygen occurs preferentially from the less hindered side of the molecule, norbornene gives almost exclusively the *exo*-product (Scheme 13.24). In such molecules where the alternative modes of approach are not largely different, mixture of products are formed; the unhindered exocyclic double bond in 4-*t*-butyl-methylene-cyclohexane yields both stereoisomeric epoxides (Scheme 13.24).

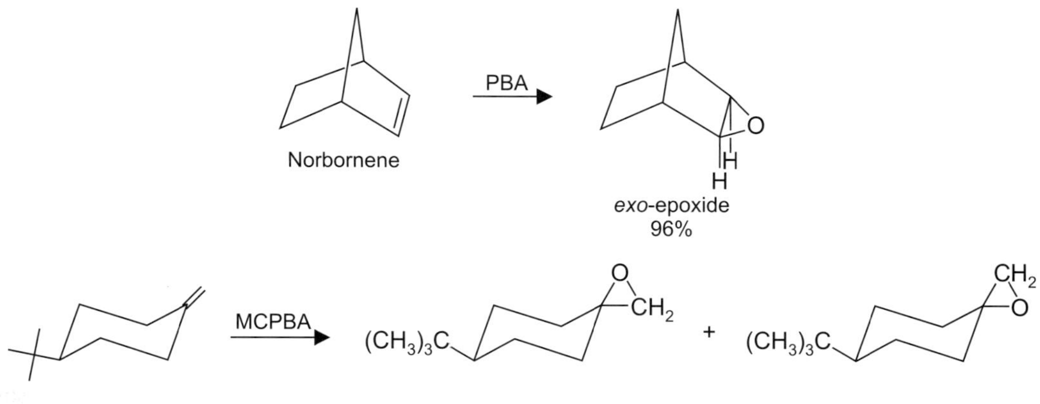

SCHEME 13.24

As expected, 2-cyclohexenyl acetate yields the *trans*-epoxide from an attack by the peracid on the less hindered side of the double bond (Scheme 13.25). However, with the free alcohol (I, Scheme 13.25) *i.e.*, with the OH group in the allylic position, the epoxidation is on the more hindered face. It is suggested that hydrogen bonding between the hydroxyl group and the peracid stabilizes the transition state for *cis*-epoxidation.

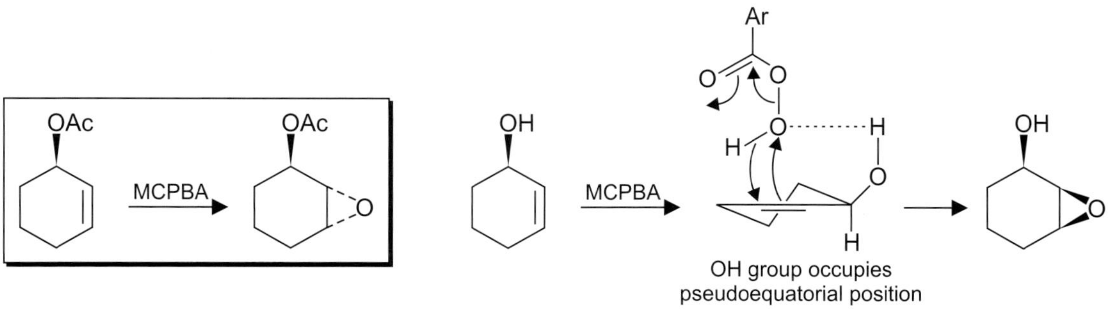

SCHEME 13.25

Sharpless epoxidation reaction is a highly useful stereoselective reaction for carrying out asymmetric epoxidation of allylic alcohols. Oxidation of allylic alcohols with *t*-butyl hydroperoxide in the presence of either (+) – or (–) – diethyl tartrate (DET) and titanium tetraisopropoxide affords the corresponding enantiomer in high optical yield. Thus oxygen is delivered from the bottom face in the presence of (+)–tartrate or from the top face in the presence of (–)–tartrate (Scheme 13.25*a*).

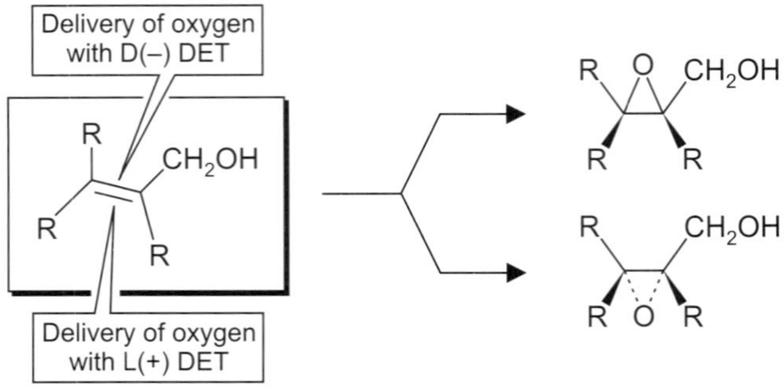

SCHEME 13.25a

(B) Epoxidation with Dioxirane and its Derivatives

[D.Yang, *Acc. Chem. Res.*, 37, 497 (2004)]

Dimethyldioxirane, DMDO can be obtained from acetone and hydrogen persulphate ion HSO_5^- (Scheme 13.25*b*). DMDO and other dioxiranes derived from other ketones can bring about the epoxidation of double bonds and often react with less reactive compounds (A, Scheme 13.25*b*).

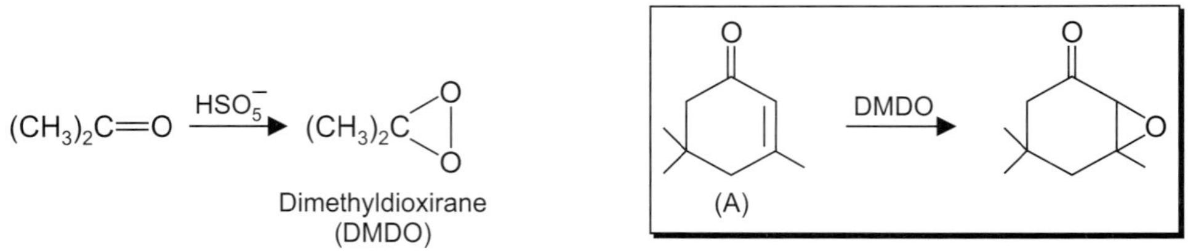

SCHEME 13.25b

Chiral dioxiranes derived from fructose derivatives (Scheme 13.25*c*) display good enantioselectivity with several alkenes (Scheme 13.25*d*).

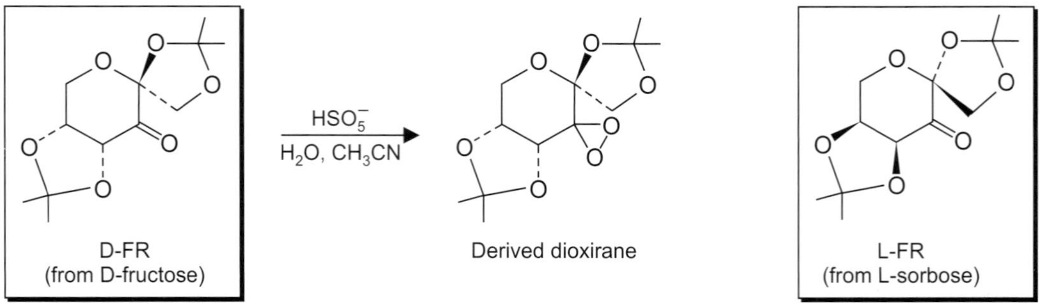

SCHEME 13.25c

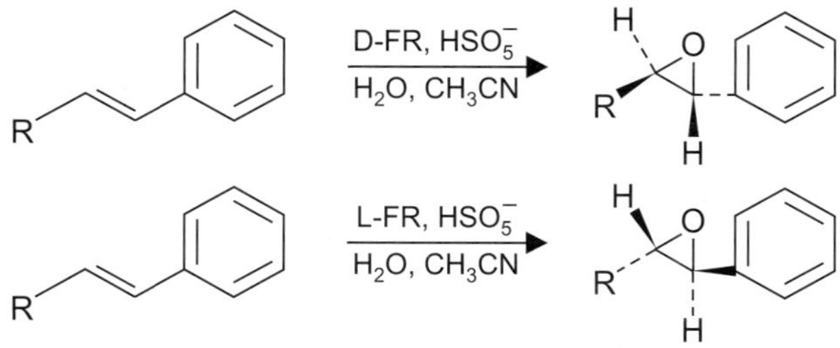

SCHEME 13.25d

(C) Transformations of Epoxides—Use in Synthesis

Epoxides are highly reactive and useful synthetic intermediates, since these react with a variety of nucleophiles with opening of the epoxide ring, which relieves the strain of the three membered ring.

Epoxidation of an alkene followed by the acid-catalysed hydrolysis gives a method for *anti-hydroxylation* with the general mechanism (*eq.* I, Scheme 13.26). This is well compared with *syn-hydroxylation* carried out with an alkene with osmium tetroxide.

Acid catalysed hydrolysis of 1, 2-epoxy-cyclopentane proceeds by water attacking as the nucleophile on the protonated epoxide from the side opposite the epoxide group. The carbon atom being attacked (shown by an arrow) undergoes an inversion of configuration (the attack similarly could occur at the other carbon as well on the symmetrical system and will give the enantiomeric form of *trans*- 1, 2-cyclopentanediol (eq. II, Scheme 13.26).

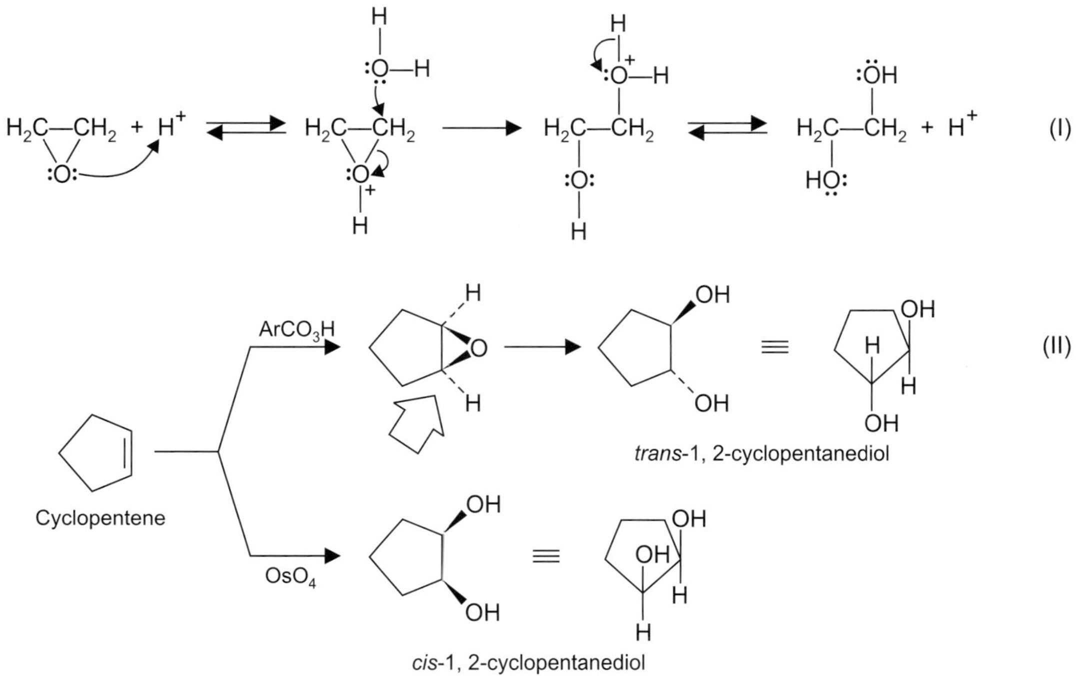

SCHEME 13.26

The reactions of epoxides leading to cleavage of one carbon-oxygen bond in a nucleophilic substitution reaction that follows generally the S_N2 pathway. Under neutral or basic conditions, as expected of S_N2 mechanism the nucleophile attacks the less sterically congested carbon (Scheme 13.27).

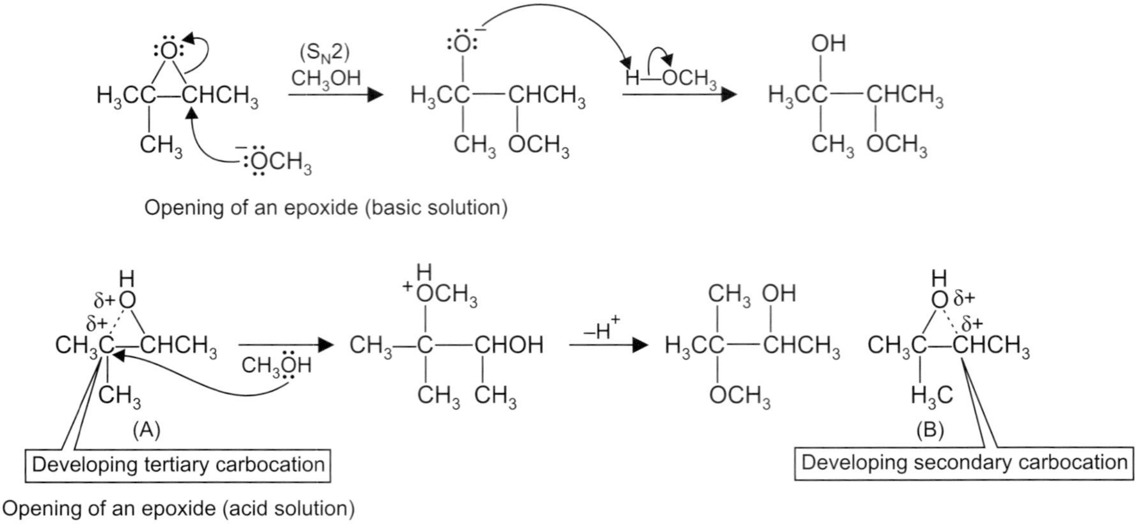

SCHEME 13.27

In the acid solution, the protonation of the epoxide weakens the C—O bonds. If the C—O bonds are largely intact at the transition state the nucleophile will attack the less substituted carbon (steric requirements of S_N2 reaction). However, if the C—O rupture is significant when the transition state is reached, the attack by the nucleophile is on the more substituted carbon (Scheme 13.27). This change, in regioselectivity is due to the ability of the more substituted carbon to stabilize the developing positive charge so that nucleophile attacks this carbon. One may note that the protonated epoxide opens preferentially so as to generate the partial positive charge on the more substituted carbon. Thus the other opening pathway (B, Scheme 13.27) is far less favourable since a tertiary carbocation (A) is more stable than a secondary carbocation. Although this is typically true of S_N1 reactions, however, the nucleophile approaches the protonated epoxide from the side opposite the leaving oxygen (S_N2 characters). One would call such a reaction which is partially S_N1 and partially S_N2. Thus working with a chiral epoxide (Scheme 13.27) the ring opening under two conditions will give optically active products which will be constitutional isomers.

In cyclohexanes the ring opening gives the axial alcohol (Scheme 13.27*a*).

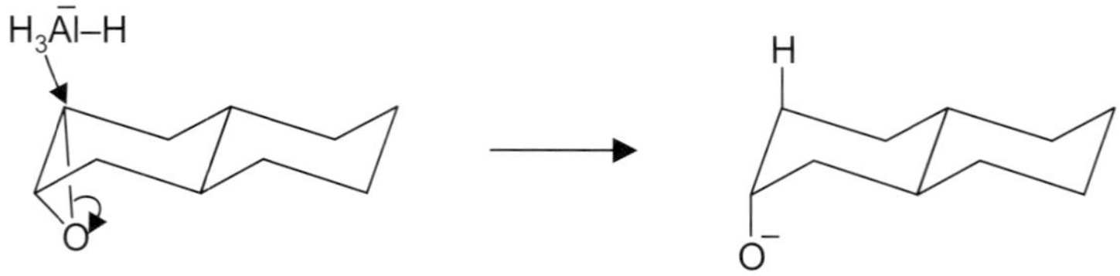

SCHEME 13.27a

Epoxides undergo a rearrangement on treatment with strongly basic reagents to give allylic alcohols. A proton abstraction takes place from a carbon adjacent to the epoxide ring and the proton which is *cis* to the epoxide is selectively removed *i.e.*, the reaction involves a concerted *syn* elimination (Scheme 13.27*b*).

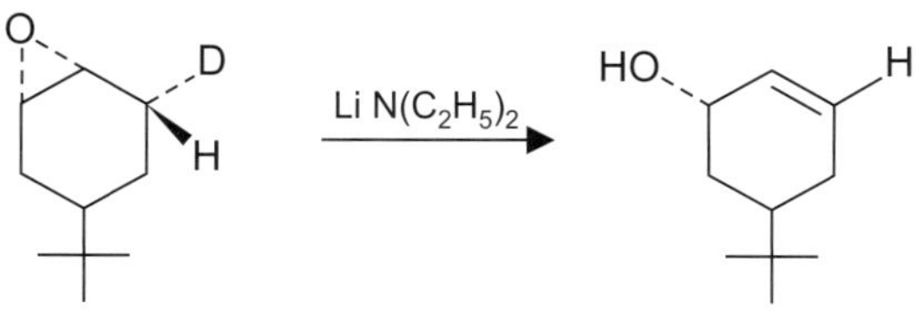

SCHEME 13.27b

Lithium aluminium hydride acts as a nucleophilic reagent in the reduction of epoxides to give saturated alcohols. The hydride is added to the less substituted carbon atom of the epoxide ring (Scheme 13.28).

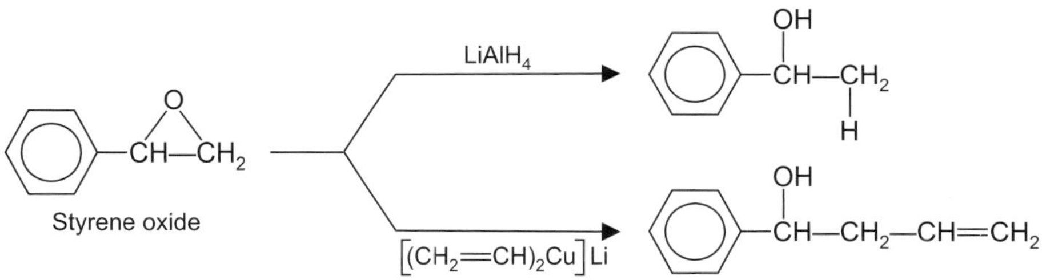

SCHEME 13.28

The reaction of epoxides with lithium diorganocopper (Gilman) reagents also brings about regioselective ring opening at the less substituted carbon which is also in agreement with S_N2 reactivity.

EXERCISE 13.3

Depict the reaction of trans-2-butene with meta-chloroperbenzoic acid followed by the reaction of the product with CH_3OH/H_2SO_4.

ANSWER. *It is a stereospecific epoxidation leading to only the trans epoxide. The ring opening in acid solution is an S_N2 reaction occurring with inversion of configuration (Scheme 13.28a).*

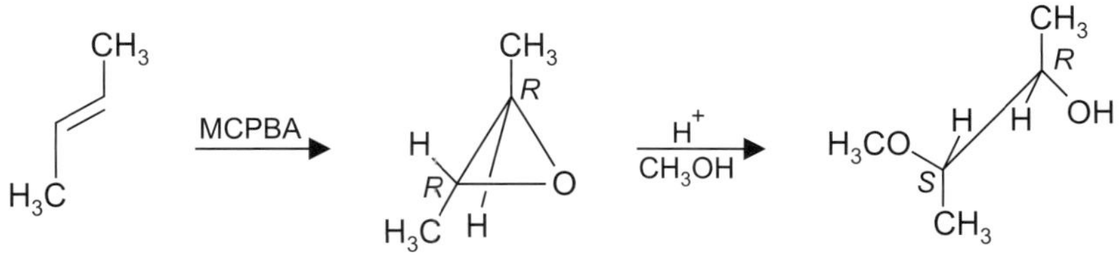

SCHEME 13.28a

EXERCISE 13.4

Predict the reaction of optically active (S)-1, 2-epoxybutane with HCl which gives two products. Explain the stereoisomerism / constitutional isomerism of the reaction.

ANSWER. *It is an "S_N2 like S_N1" reaction (Scheme 13.28b). The protonated epoxide exists in equilibrium with a structure (III, Scheme 13.28b) which places the positive charge on the carbon that can best stabilize it (position b). The structure (III) is infact a bridged carbocation which is blocked on one face by the oxygen, therefore, the nucleophile attacks it from the opposite side. The nucleophile Cl^- can attack either at the less substituted carbon atom (less hindered) or the carbon atom with a positive charge, but in both cases from the face opposite to oxygen. The products are optically active constitutional isomers.*

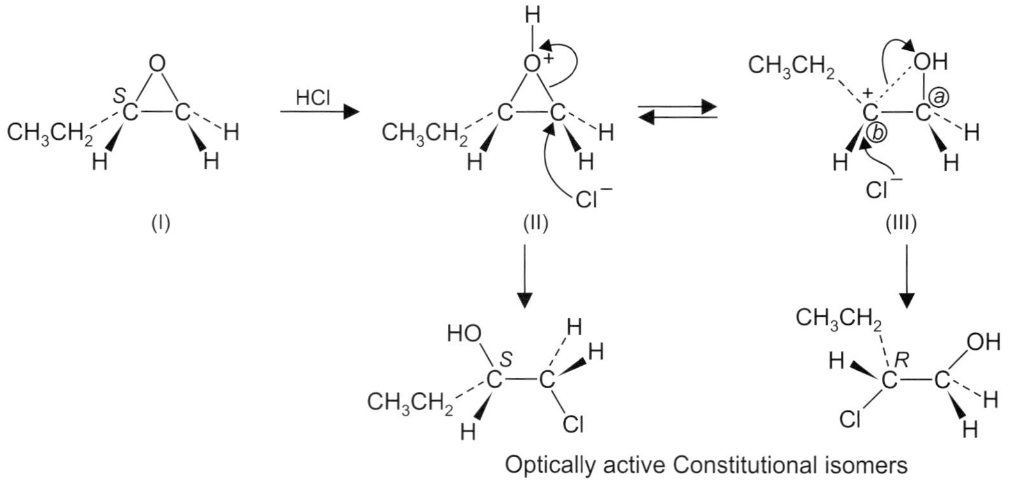

SCHEME 13.28b

Epoxides undergo acid catalysed rearrangement to carbonyl compounds. Lewis acids *e.g.*, boron trifluoride etherate, mineral acids acids or magnesium bromide are often used as catalysts (Scheme 13.29).

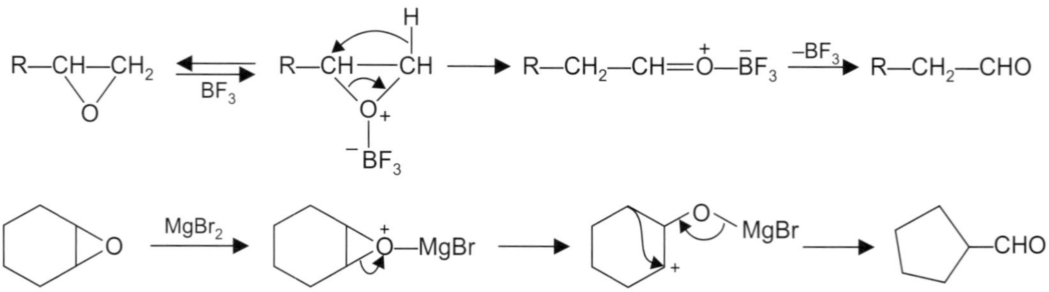

SCHEME 13.29

Reaction with dimethyl sulphoxide yields an α-ketol (Scheme 13.30).

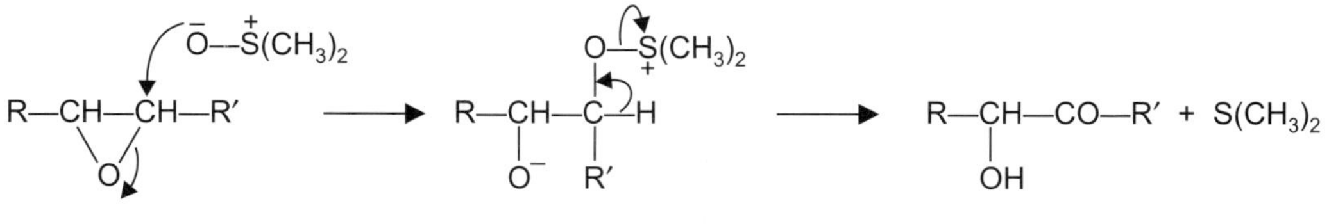

SCHEME 13.30

Alkenes are epoxidized with retention of configuration (see Scheme 13.22). Methods are available for the inversion of configuration of alkenes and in one process involves epoxidation deoxygenation. The nucleophilic attack by a phosphorus reagents *e.g.*, triphenylphosphine at the oxirane carbon leads to inversion of configuration and gives a charge-separated intermediate (a betaine). This undergoes *syn* elimination *via* a four center cyclic transition state which requires a 180° rotation around the C—C bond to establish the appropriate geometry. The stereochemical outcome of such a reaction is to put the R groups attached to the oxirane carbons in a different stereochemical relationship in the oxirane and alkene (Scheme 13.31).

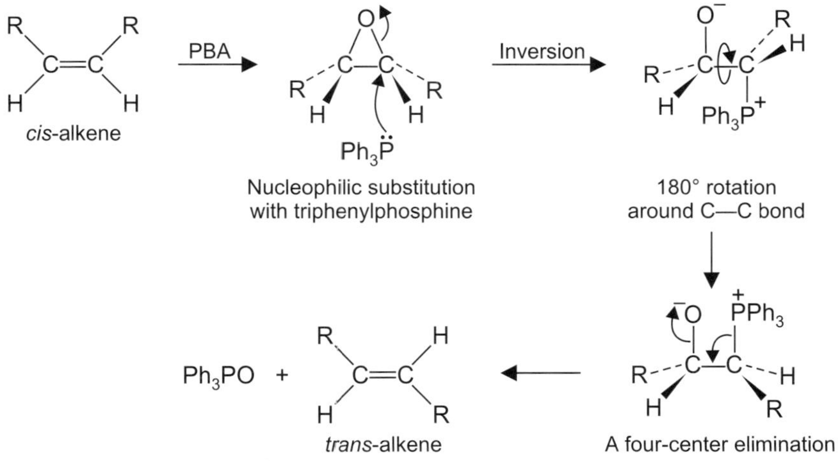

SCHEME 13.31

(D) Diol Formation

Two reagents oxidize an alkene to a 1, 2-diol: osmium tetroxide OsO_4 and potassium permanganate $KMnO_4$. Both osmium and manganese are in a highly positive oxidation state and therefore, both attract electrons. Both MnO_4^- and OsO_4 form cyclic intermediates on reaction with an alkene. Both add *syn* to the π bond of an alkene and as a result both the oxygens are added to the same side of the double bond to give only a *cis*-diol after the hydrolysis of the intermediate cyclic ester and thus, the reaction is stereospecific (*syn*-addition, Scheme 13.32).

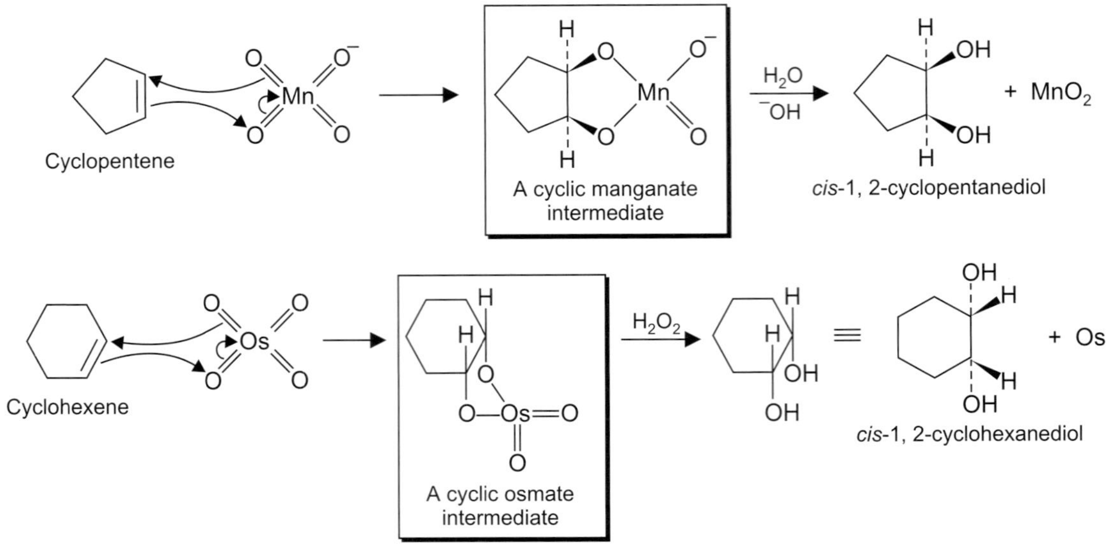

SCHEME 13.32

The attack of the reagent can either be from top or bottom face which gives both enantiomers (as a racemic mixture). When there is a choice the reagents predominantly attacks from the less hindered side of the double bond (Scheme 13.32*a*).

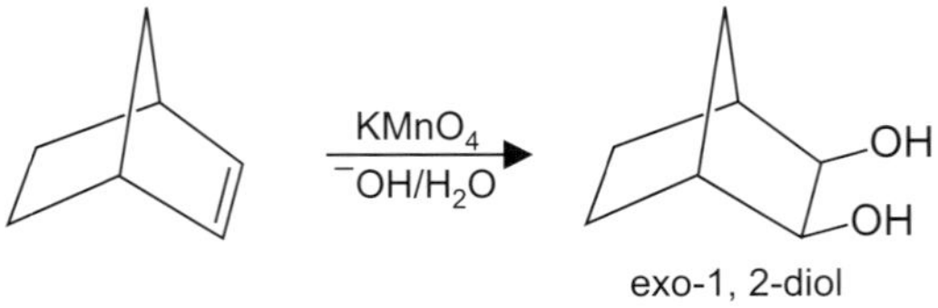

SCHEME 13.32a

The permanganate ion is a more powerful oxidizing agent than osmium tetraoxide and yields of glycols are generally low due to overoxidation. $KMnO_4$ is used in acid solution to cleave a double bond of an alkene (Scheme 13.32*b*).

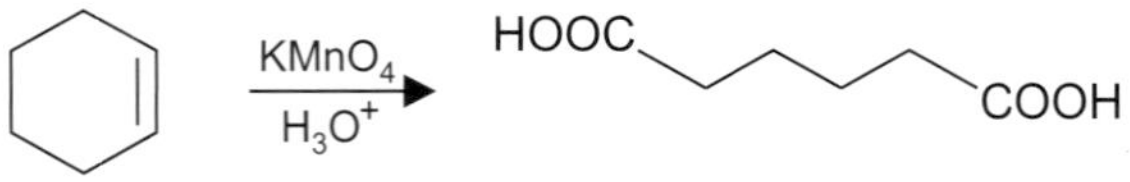

SCHEME 13.32b

Alkenes are oxidized to 1, 2-diols with a basic solution of $KMnO_4$ which are then cleaved with HIO_4 to form carbonyl compounds. Under acidic conditions or if the basic solution of $KMnO_4$ is heated, the reaction does not stop at the diol and the alkene is cleaved. A solution of $KMnO_4$ in benzene containing [18]-crown 6 (pink benzene, see Scheme 1.20) cleaves the alkenes to gives the products in high yield (Scheme 13.32*c*).

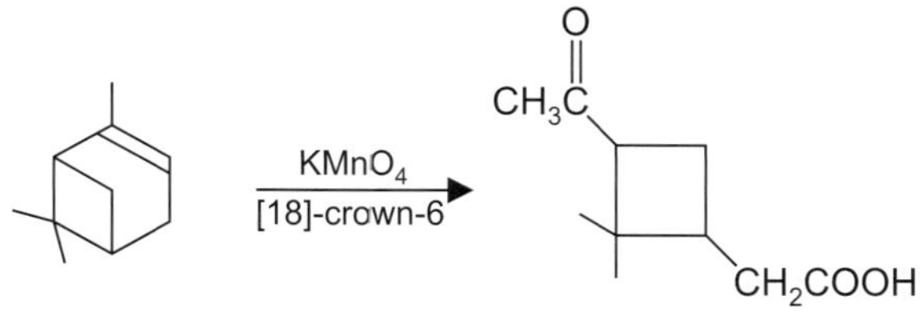

α-pinene Pinonic acid 90%

SCHEME 13.32c

EXERCISE 13.5

How maleic and fumaric acid can be converted into meso-tartaric acid and fumaric acid into its enantiomeric pair by using two different methods.

ANSWER. *Maleic acid (cis-isomer) adds $KMnO_4$ by syn addition and the original configuration of the reactant is retained in the product to give meso-tartaric acid. Meso-tartaric acid can also be obtained from fumaric acid (the trans-isomer) by epoxidation and then opening the epoxide with aqueous solution of hydroxide. Since this reaction involves one inversion of configuration fumaric acid also gives meso-tartaric acid (Scheme 13.32d).*

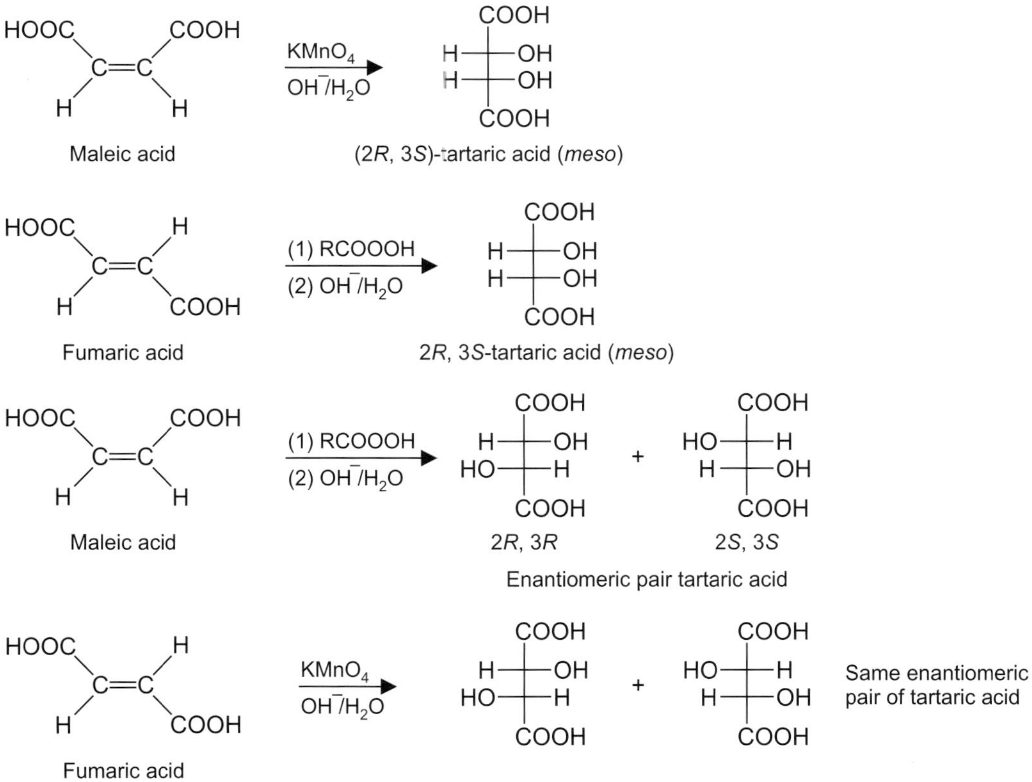

SCHEME 13.32d

Woodward and Prevost method involves the reaction of an olefin with iodine and silver acetate. Under dry condition (absence of water) this method leads to a *trans*-1, 2-diacetate (Scheme 13.33) from which the *trans*- diol is obtained by hydrolysis. Woodward reaction is carried out in the presence of water to yield the monoester of the *cis*- diol, and the final hydrolysis gives the *cis*- diol (Scheme 13.33). The olefin is reacted with iodine in the presence of silver acetate. Iodine reacts with the double bond to give an iodonium ion which undergoes displacement by acetate in the S_N2 type reaction, giving a *trans*-iodo-acetate. Anchimeric assistance by the acetate group, together with the powerful bonding capacity of silver ion for iodide, gives a cyclic acetoxonium ion (I, Scheme 13.38) which gives a *trans*-1, 2-diacetate. The acetoxonium ion, under wet conditions traps water and reacts to yield a *cis*-hydroxyacetate.

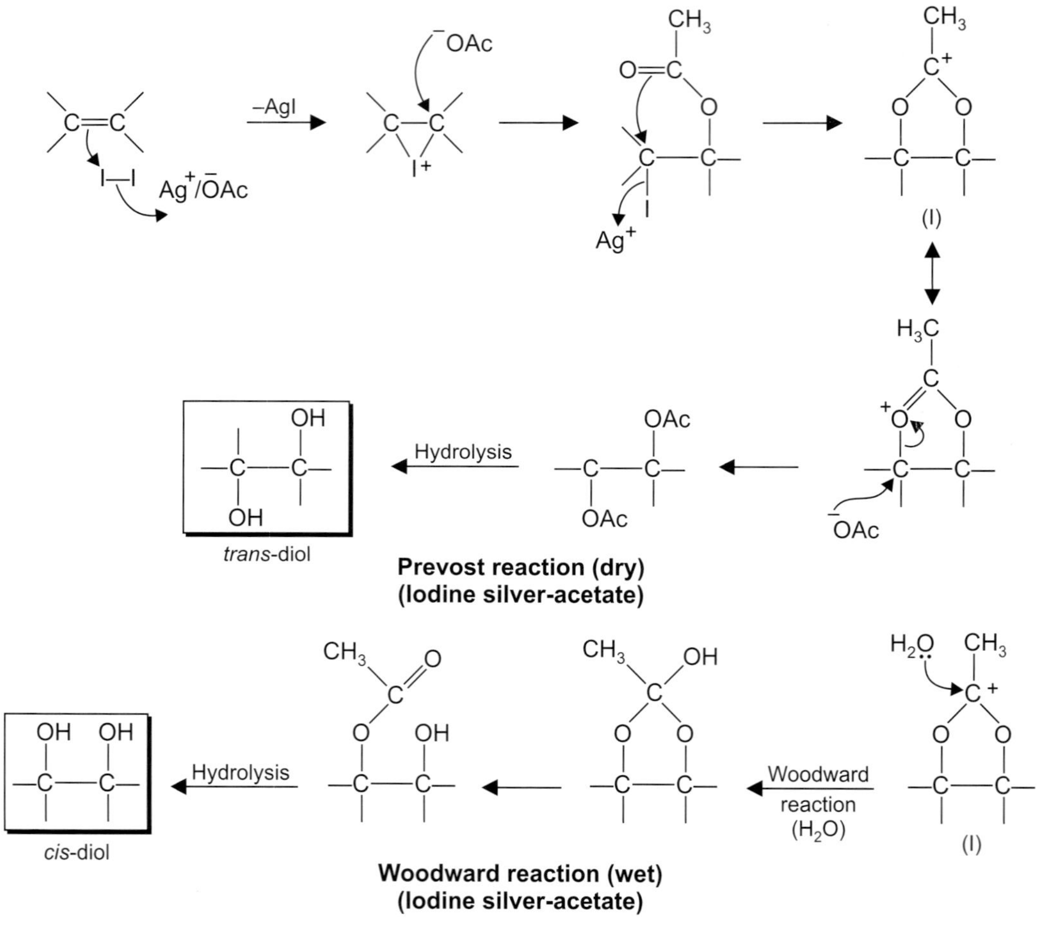

SCHEME 13.33

The reaction (Prevost) is depicted (Scheme 13.34) to show the stereochemical consequences of the reaction more clearly with the chair conformation of cyclohexane system.

Thus in summary, both Woodward and Prevost reactions a 1, 2-glycol is formed from an olefin by reacting it with iodine and silver acetate. In the presence of an aprotic solvent (Prevost-method) *trans*-1,2-glycols are formed. In the presence of water (Woodward-method) the product is a *cis*-diol.

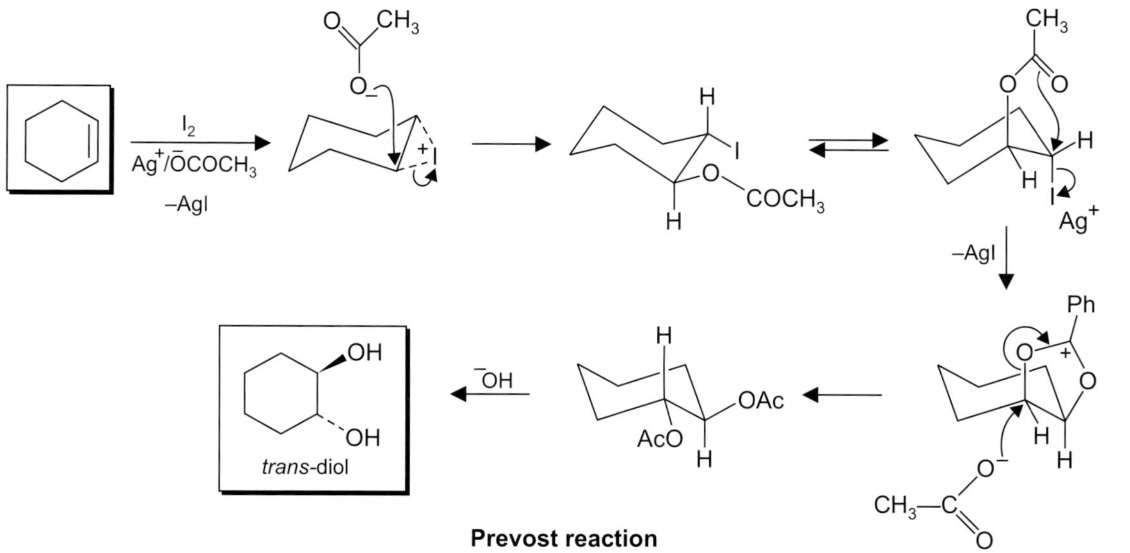

Prevost reaction

SCHEME 13.34

13.5 OZONOLYSIS

Ozonolysis provides a method for the fission of carbon-carbon double bonds. The process involves two key intermediates, an initial ozonide (1, 2, 3-trioxolane) and the ozonide (1, 2, 4-trioxolane). The first step of the mechanism involves a cycloaddition to give the initial ozonide, subsequent fragmentation and recombination yields the isomeric product, ozonide. The first step is a 1, 3-dipolar cycloaddition reaction (Scheme 13.35), ozone being highly electrophilic 1, 3-dipole (resonance structures, Scheme 13.35).

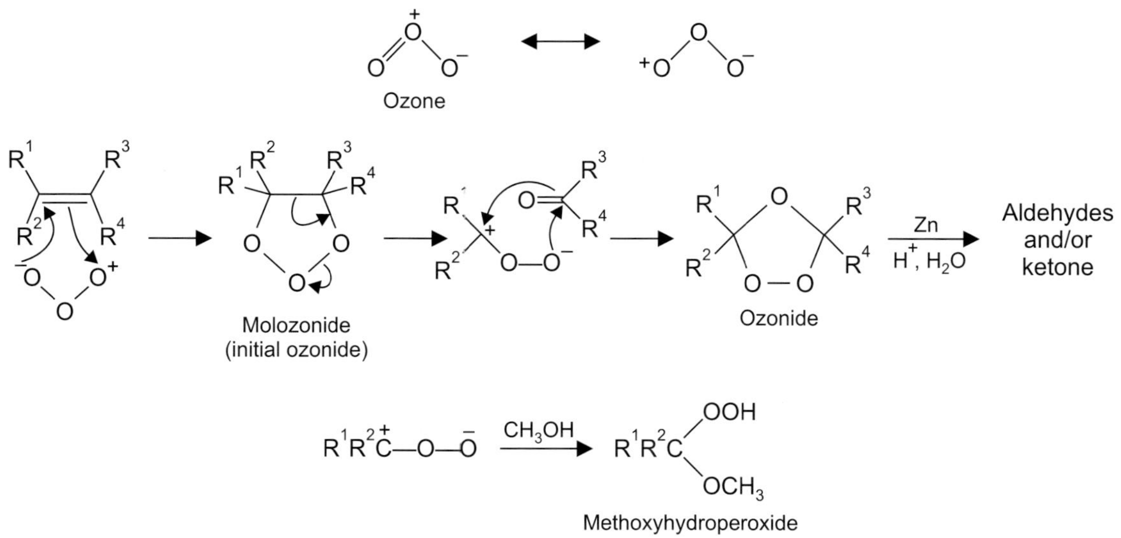

SCHEME 13.35

Methanol (the nucleophilic solvent) when cooled to – 20°C is not attacked by ozone and is used in many cases as the ozonisation solvent. In these cases the main product of addition of ozone to the alkene is not the usual ozonide but, is the methoxyhydroperoxide. Direct solvolysis of ozonides gives ketones and/or acids, depending on the structure of the alkene (Scheme 13.36).

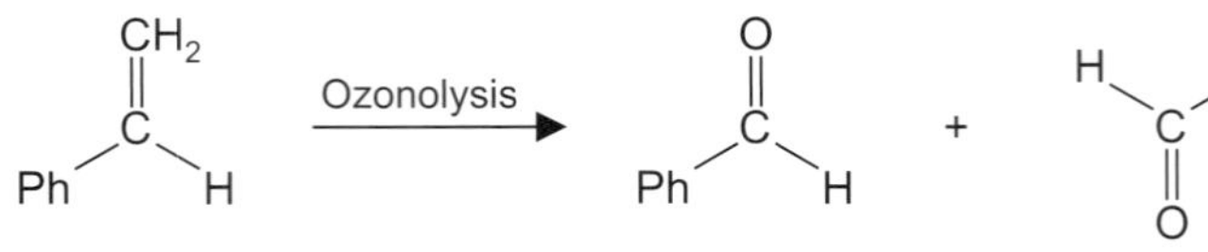

SCHEME 13.36

Ozonolysis of α, β-unsaturated carbonyl compounds *e.g.*, ketones and acids give abnormal products of ozonolysis with fewer than the expected number of carbon atoms. With α, β-unsaturated compounds the loss of a carbon involves the formation of a hydrated form of the ozonide (Scheme 13.37).

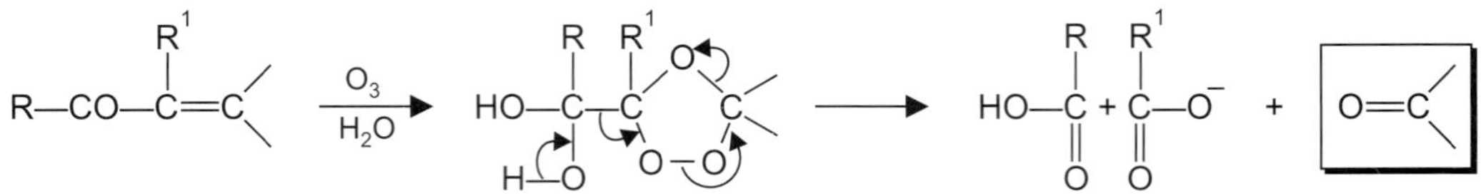

SCHEME 13.37

13.6 CLEAVAGE OF GLYCOLS AND RELATED COMPOUNDS

Carbon-carbon double bonds are cleaved *via* the glycols (1, 2-diols) by reaction with lead tetra-acetate. The fragmentation is believed to occur within a cyclic adduct of the glycol and the oxidant (Scheme 13.38).

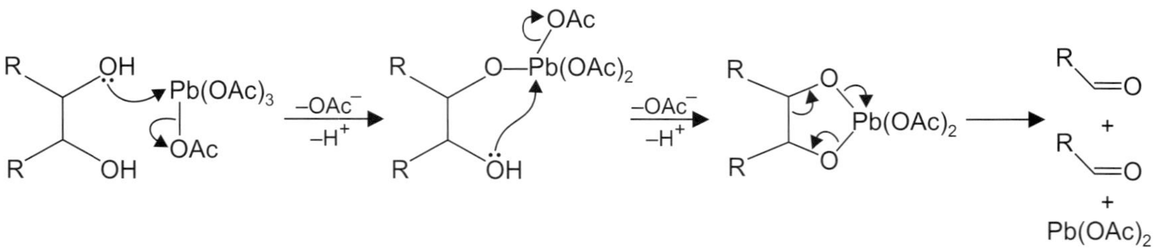

SCHEME 13.38

Glycol cleavage with periodic acid (HIO_4) involves a similar cyclic intermediate (Scheme 13.38*a*). The reaction, therefore, seems to have stereochemical requirements and to form a cyclic intermediate, the two hydroxyl groups of a cyclic diol should be positioned appropriately to form this intermediate. In the case of a 1, 2-cyclohexane diol the two OH groups can be both

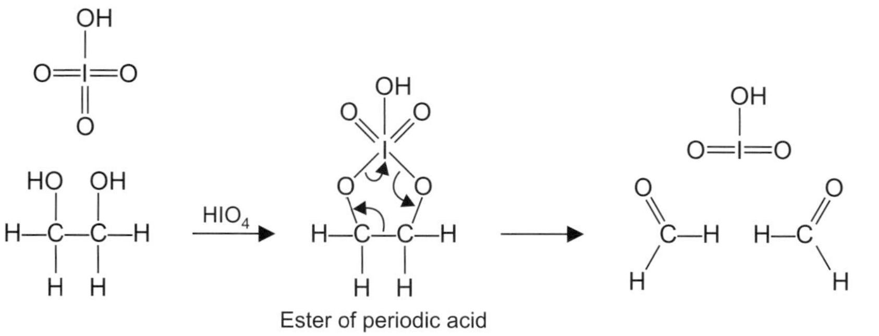

SCHEME 13.38a

equatorial, both axial or one axial and one equatorial. In one of these situations when the two hydroxyl groups occupy axial positions, these are too far away from each other to form the cyclic intermediate and such a conformer will not react with HIO_4. Thus between (I and II, Scheme 13.39) with *t*-butyl group in equatorial position, (II) does not react with HIO_4.

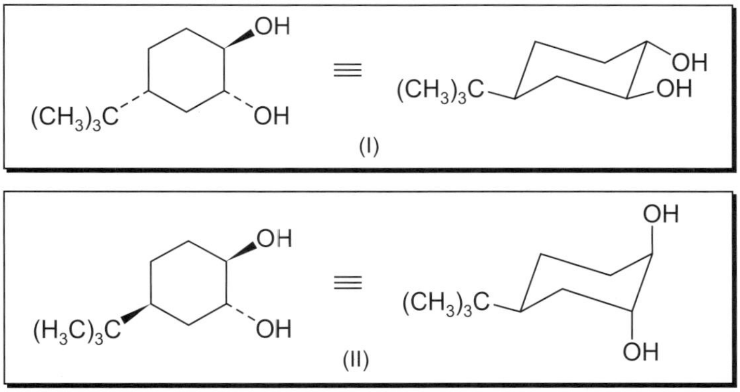

SCHEME 13.39

Oxidative Cleavage of 1, 2-glycols

Some glycols which however, cannot form a cyclic intermediate have been found to undergo cleavage. To explain this an acyclic pathway may be followed (Scheme 13.40).

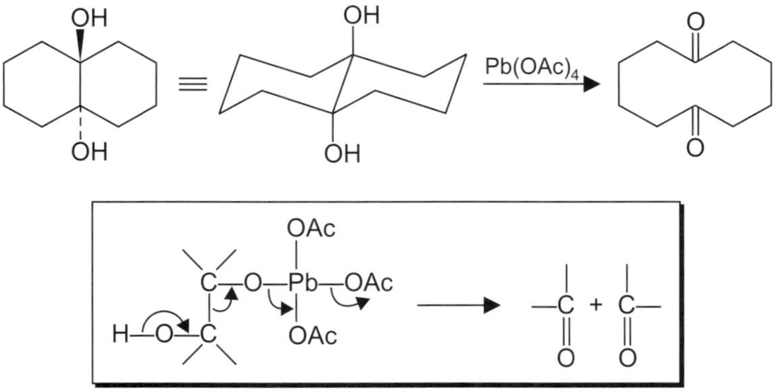

SCHEME 13.40

Several other combinations of adjacent functional groups are also cleaved with these reagents (Scheme 13.41). The cleavage of diketones with IO_4^- involves a reactive cyclic intermediate formed *via* the nucleophilic attack on the diketone (Scheme 13.42). Compounds with carboxyl groups on adjacent carbons (succinic acid derivatives) can be bisdecarboxylated with lead tetra-acetate (Scheme 13.43) to give alkenes.

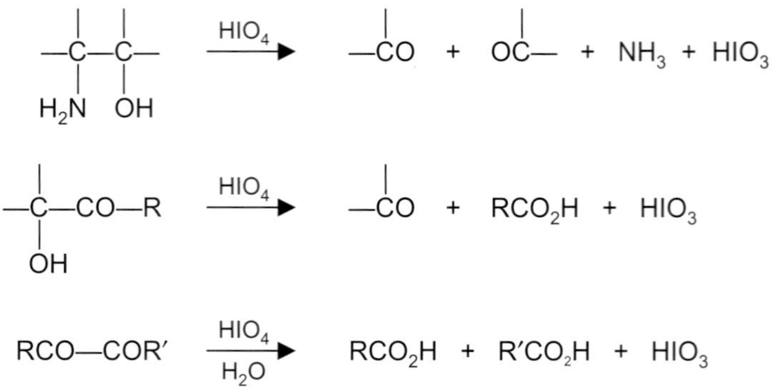

SCHEME 13.41

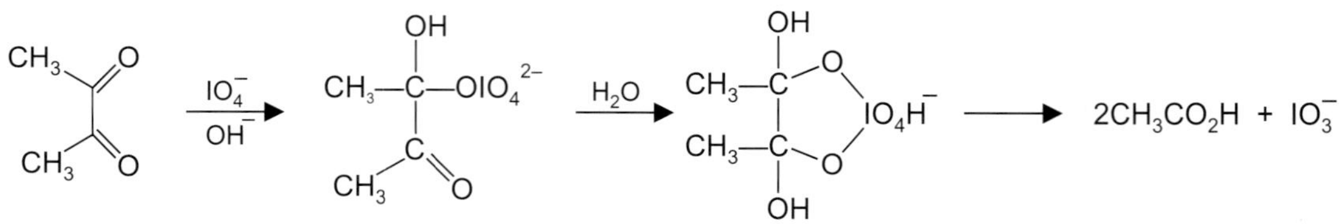

SCHEME 13.42

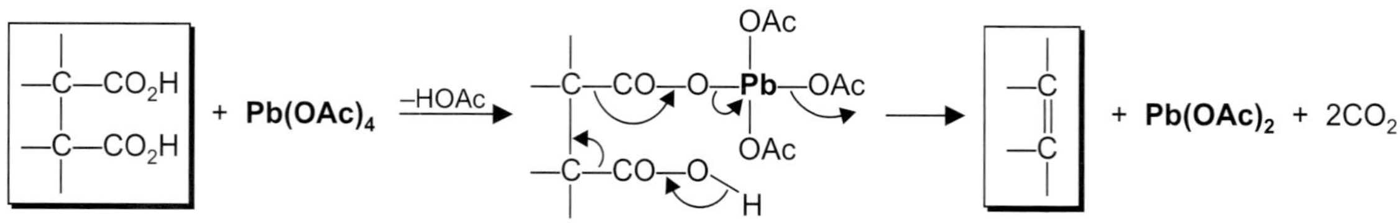

SCHEME 13.43

13.7 OXIDATION OF ALKENES TO ALDEHYDES AND KETONES CATALYSED WITH PALLADIUM AND OXIDATION OF ALKYLBORANES

Oxidation of mono-substituted and 1, 2-disubstituted olefins can be carried out to give aldehydes and ketones with palladium chloride and similar salts of noble metals (Scheme 13.44). The oxidation of ethylene to accetaldehyde with this reaction is an industrial process (the Wacker process). The palladium chloride is reduced to the metal, however the reaction is made catalytic with a co-oxidant ($CuCl_2$), whereby the palladium Pd(0) is re-oxidized to Pd(II). The mechanism of the reaction (Scheme 13.45) involves π-comlexes from Pd(II) and an alkene and these complexes are activated to nucleophiles. The mechanism of reaction for ethylene and water as the nucleophile is presented (Scheme 13.45).

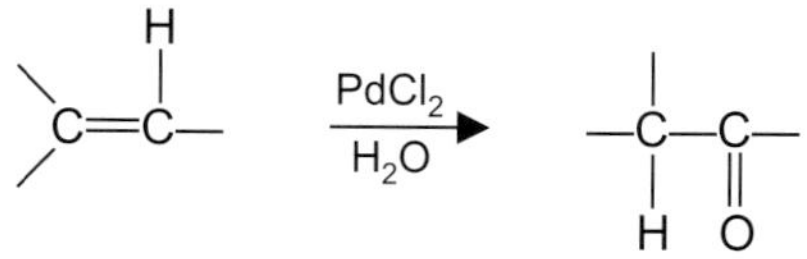

SCHEME 13.44

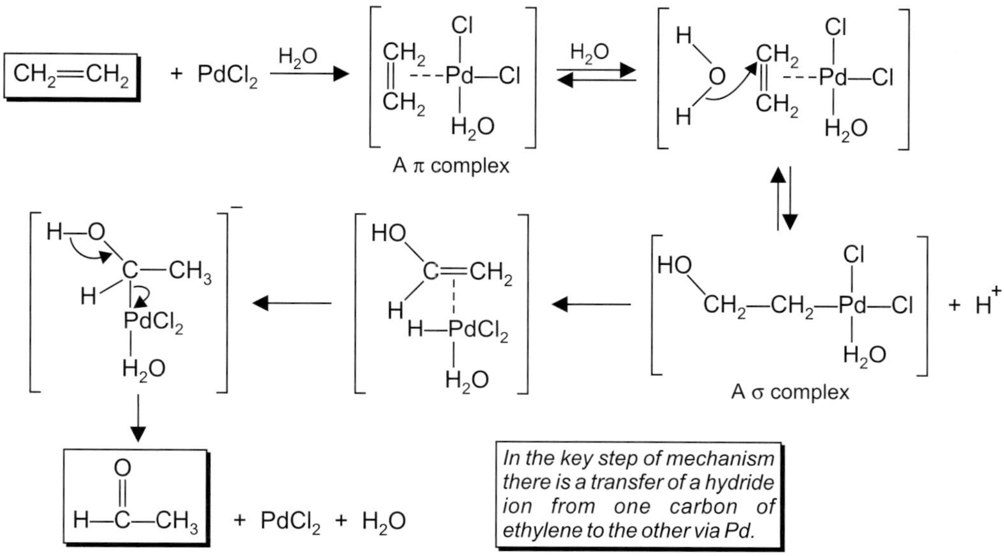

SCHEME 13.45

Ethylene reacts with Pd(II) to afford a π complex (Scheme 13.45) which is attacked by nucleophile (water) to give a σ complex.

The following steps may be involved (also see Scheme 7.2)

1. *Trans* hydroxypalladation of ethylene to an unstable complex.
2. β-Elimination within the complex, with transfer of hydride ion from one carbon of ethylene to the other *via* palladium (see, Scheme 7.1).
3. When the oxidation is conducted in deuterium oxide, acetaldehyde formed does not contain deuterium, to show that all the four hydrogens of the acetaldehyde come from the original ethylene and none from the solvent. This explains the hydride migration step of the mechanism (Scheme 13.45).

With monosubstituted alkenes bonding to palladium is through the unsubstituted carbon atom therefore, the reaction of a terminal alkene with water gives a methyl ketone (Scheme 13.46). 1, 2-Disubstituted alkenes (internal alkenes) are oxidized only slowly compared to terminal bonds and this enables the reaction to become selective.

Oxidation with PCC of an alkylborane formed by the hydroboration of a terminal alkene with Sia_2BH provides a useful alternative route to an aldehyde (Scheme 13.47). Disiamyl borane is highly regioselective, a very small amount of the alkyl methyl ketone (< 1%) is formed. Moreover, with a non-conjugated diene, with a terminal and non-terminal carbon-carbon double bonds, the terminal bonding system reacts preferentially to give at the end an unsaturated aldehyde as in the case of limonene.

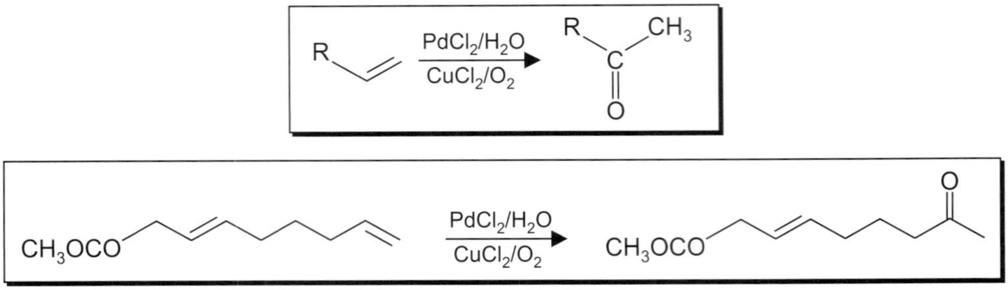

SCHEME 13.46

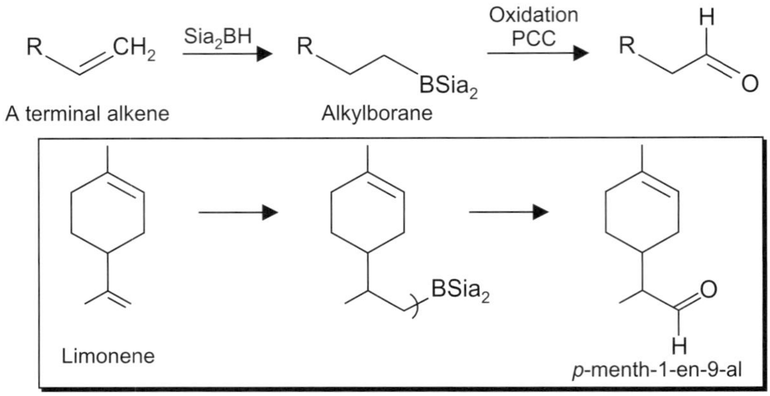

SCHEME 13.47

13.8 OXIDATION OF KETONES

1. Cyclohexanone gives adipic acid (Scheme 13.48) and is an important industrial process. Several reagents can be used for the purpose, chromic acid or alkaline potassium

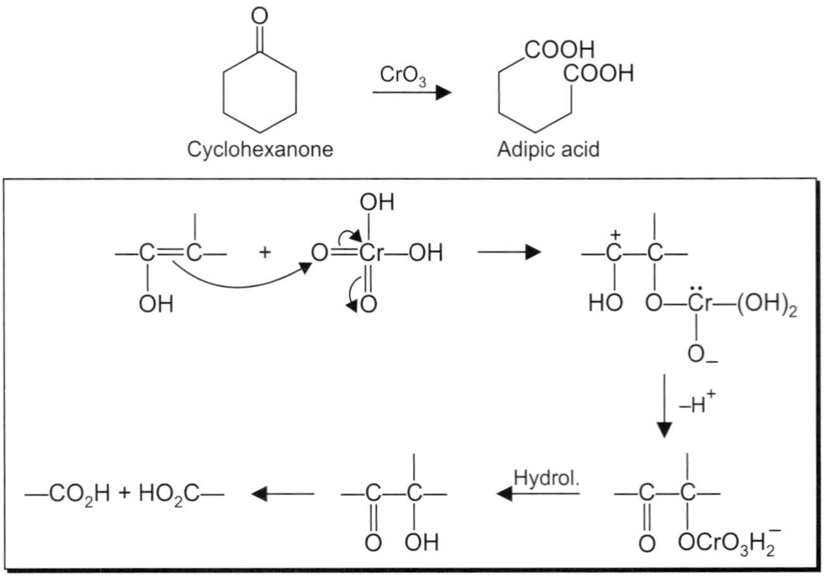

SCHEME 13.48

permanganate. These strong oxidizing agents give carboxylic acids, reaction occurring through the enol (acid solution) or the enolate anion (basic solution). The process for alkaline potassium permanganate is also given (Scheme 13.49).

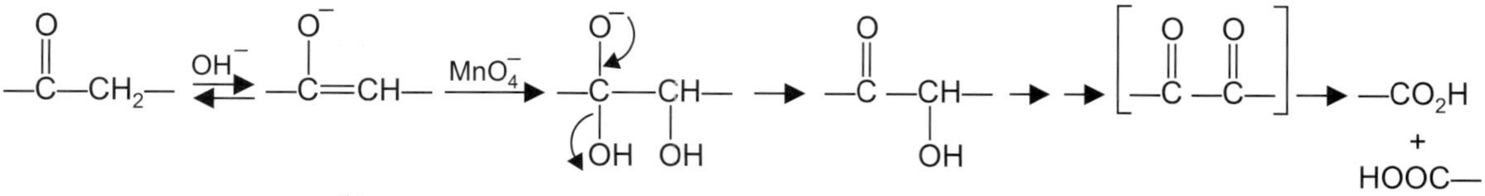

SCHEME 13.49

2. Methyl ketone are oxidized by chlorine, bromine or iodine in alkaline solution to yield acids and the corresponding haloform. The process is a base catalysed halogenation followed by elimination of the conjugate base of the haloform (Scheme 13.50). The process provides a useful route to aromatic acids, a methyl ketone which may be prepared by Friedel Crafts reaction, is oxidized with iodine in sodium hydroxide solution (Scheme 13.50).

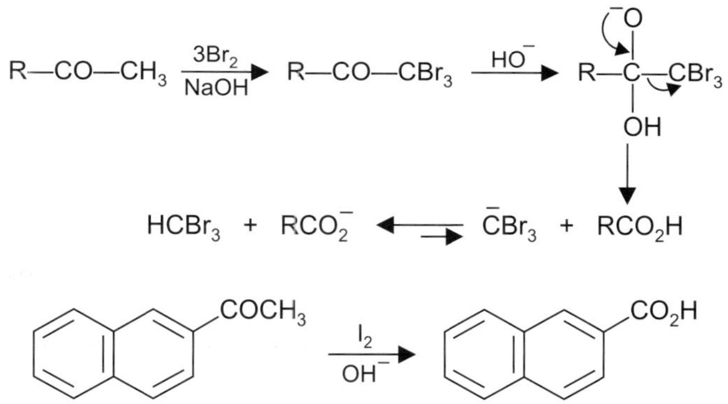

SCHEME 13.50

3. Ketones on treatment with peracids like perbenzoic acid or peracetic acid in the presence of acid catalyst give related esters by insertion of oxygen. The mechanism of this reaction (Baeyer-Villiger reaction) involves a rearrangement to electron deficient oxygen and is related to the pinacol rearrangement. Nucleophilic attack of the peracid on the carbonyl group gives an intermediate which rearranges with the loss of the anion of an acid (Scheme 13.51).

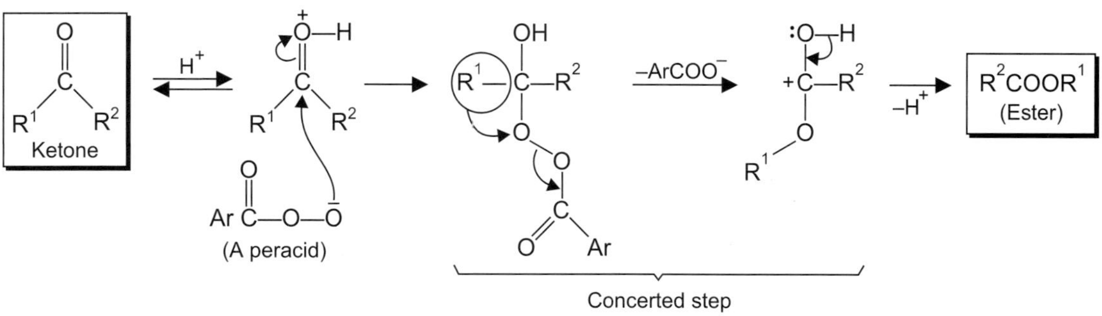

The Baeyer-Villiger reaction

SCHEME 13.51

In an unsymmetrical ketone, that group migrates which is more nucleophilic of the two better able to supply electrons. Thus among alkyl groups the ease of migration is tertiary > secondary > primary > methyl (Scheme 13.52). Aryl groups migrate in preference to primary alkyl groups. The intramolecular concerted nature of the reaction has been proved, since the chiral migrating groups retain their configuration in the rearranged product. Tertiary hydrocarbons form hydroperoxides readily by autoxidation. Cumene can be prepared by Friedel Crafts alkylation of benzene with propylene and it gives cumene hydroperoxide readily. Cumene hydroperoxide undergoes an acid catalysed rearrangement (Scheme 13.53), like Baeyer-Villiger reaction to give a hemiacetal which is hydrolyzed under acid conditions (see Scheme 14.37).

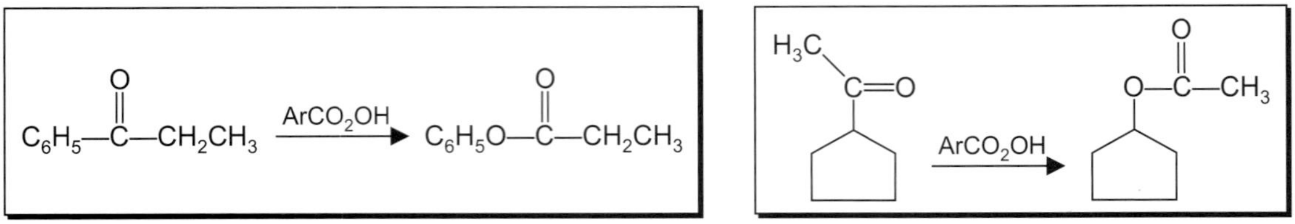

SCHEME 13.52

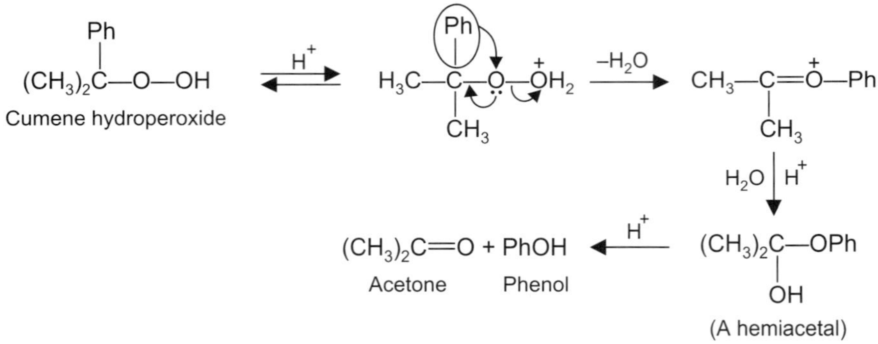

SCHEME 13.53

Aromatic aldehydes and ketones are converted to phenols and quinones on treatment with alkaline H_2O_2, provided there is an OH or NH_2 group in the *o*– or *p*–position (Scheme 13.54). This is Dakin reaction which proceeds by a mechanism similar to that of Baeyer-Villiger reaction.

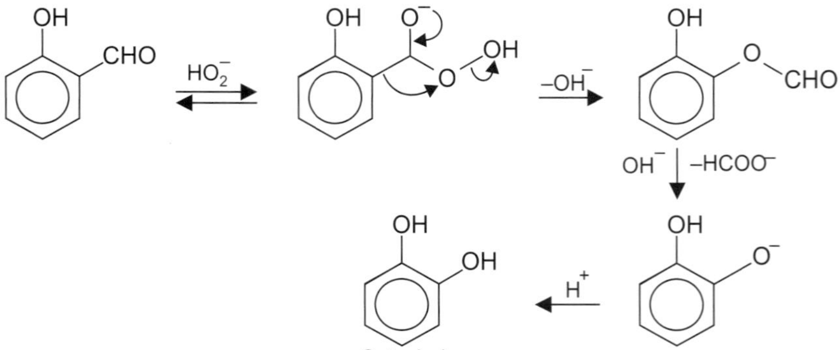

Dakin reaction

SCHEME 13.54

EXERCISE 13.6

What product one expects from the Baeyer-Villiger oxidation of the ketone (Scheme 13.55). Give a mechanism.

ANSWER. *The Baeyer-Villiger oxidation of ketones gives esters and therefore, cyclic ketones give lactones (cyclic esters). Considering the migratory aptitudes t-alkyl > s-alkyl > aryl > n-alkyl > methyl, the more highly substituted group (as expected) migrates i.e., oxygen is inserted toward the more highly substituted carbon (Scheme 13.56).*

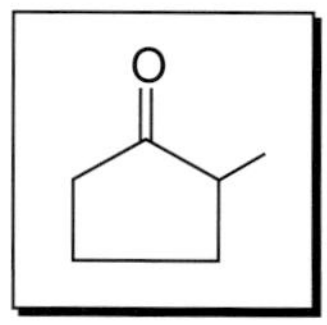

SCHEME 13.55

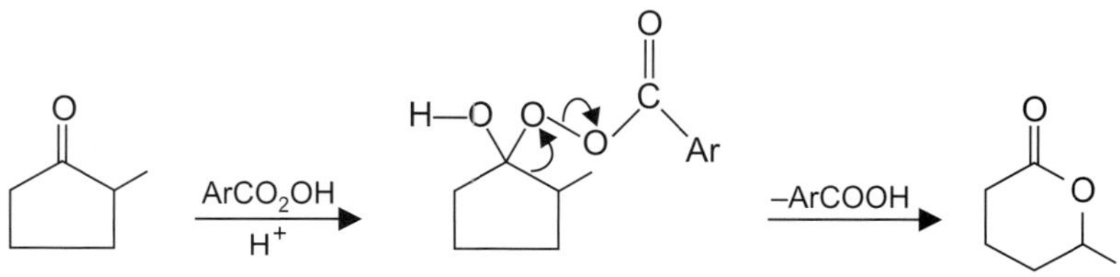

Lactone

SCHEME 13.56

13.9 OXIDATION OF α-KETOLS

These systems are oxidized easily by using one electron oxidants in basic solution to give α-dicarbonyl compounds. On mechanistic grounds, the carbanion formed on reaction with base can donate one electron to the oxidant to yield a delocalized radical (Scheme 13.57) subsequent loss of a second electron completes the oxidation. Thus benzoin can be oxidized to give benzil in 90% yield.

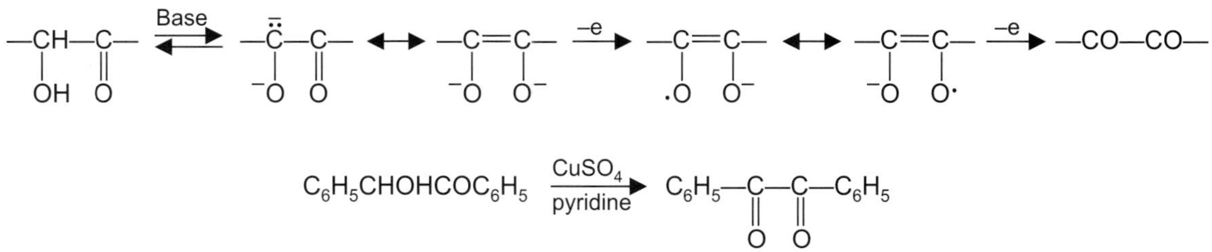

Benzoin Benzil

SCHEME 13.57

13.10 OXIDATIVE DECARBOXYLATION OF ACIDS

Carboxylic acids can be decarboxylated with lead tetraacetate and if a β-hydrogen is present, the alkene is formed by the elimination of H and COOH. High yields of alkenes are formed on heating carboxylic acids with lead tetraacetate in the presence of a catalytic amount of copper (II) salt (Scheme 13.58).

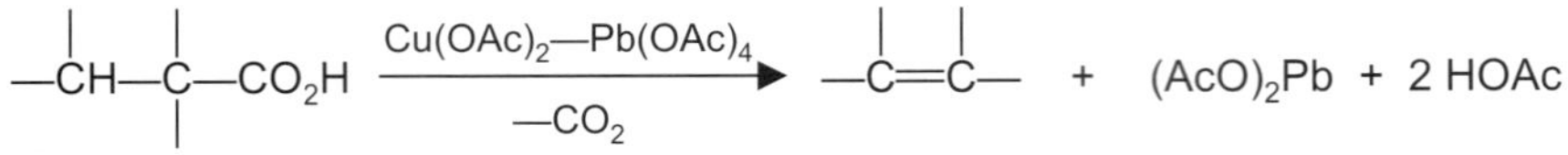

SCHEME 13.58

The mechanism is of free radical type and occurs *via* the homolysis of the lead carboxylate followed by a free radical chain mechanism (Scheme 13.59). The effect of Cu^{2+} ions is to oxidize the radicals to alkenes.

$$RCH_2CH_2CO{-}O{-}Pb(OAc)_3 \longrightarrow RCH_2CH_2CO{-}O^{\bullet} + {}^{\bullet}Pb(OAc)_3$$

$$RCH_2CH_2CO{-}O^{\bullet} \xrightarrow{-CO_2} RCH_2\dot{C}H_2$$

$$RCH_2\dot{C}H_2 + Cu^{2+} \longrightarrow RCH{=}CH_2 + H^{+} + Cu^{+}$$

SCHEME 13.59

13.11 AROMATIC RINGS OF PHENOLS—COUPLING

The aromatic rings of phenols are highly prone to oxidation by one electron oxidants, since the removal of a hydrogen atom gives a delocalized aryloxy radical (Scheme 13.60). The phenoxy radical displays several reactions depending on its structure. One important mode is by coupling *i.e.*, dimerization (Scheme 13.60) which is predominantly of the C—C type (*o–o, p–p,* or *o–p*).

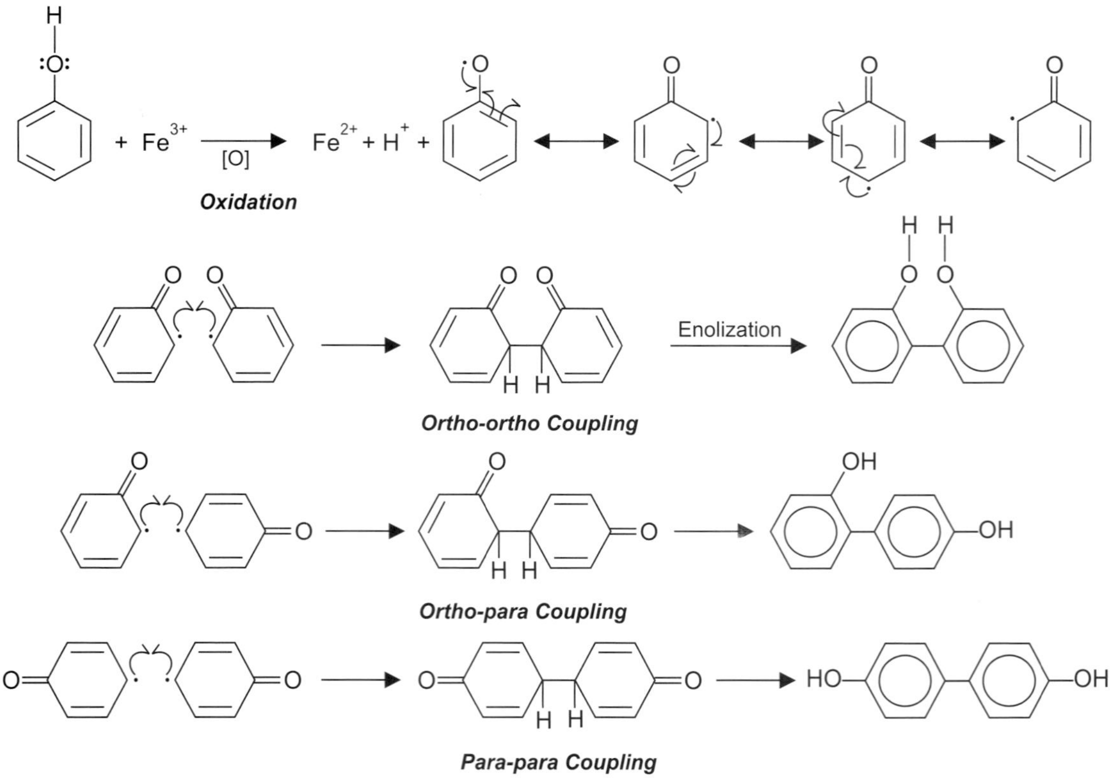

SCHEME 13.60

The *ortho* and *para* dihydric alcohols can be oxidized to yield the corresponding quinones by the use of one electron oxidants like Fe^{3+} (Scheme 13.61).

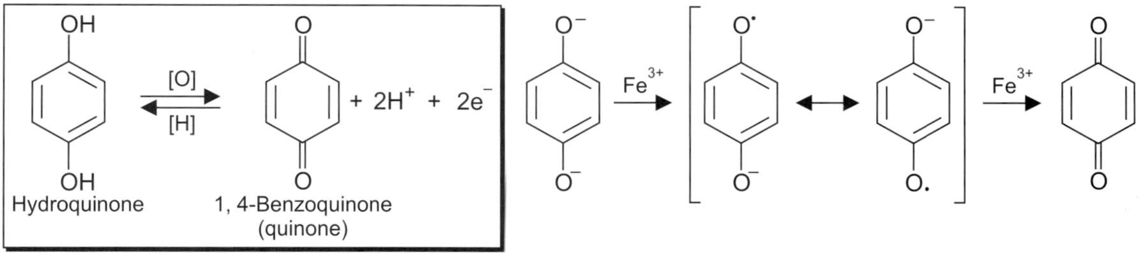

SCHEME 13.61

Substituted quinones and hydroquinones form an important part of the electron transport system in biological organisms. These compounds are involved in the cellular interconversions of Fe^{3+} to Fe^{2+} reactions which are essential for the utilization of oxygen gas.

13.12 OXIDATION OF AMINES

Dioxirane prepared by the reaction of acetone with persulphate ion (see Scheme 13.25*b*) is a highly useful reagent for the oxidation of amines. Primary amines give nitro compounds while secondary amines give the corresponding hydroxylamine (Scheme 13.62).Tertiary amines on reaction with a peracid or H_2O_2 give an *N*-oxide.

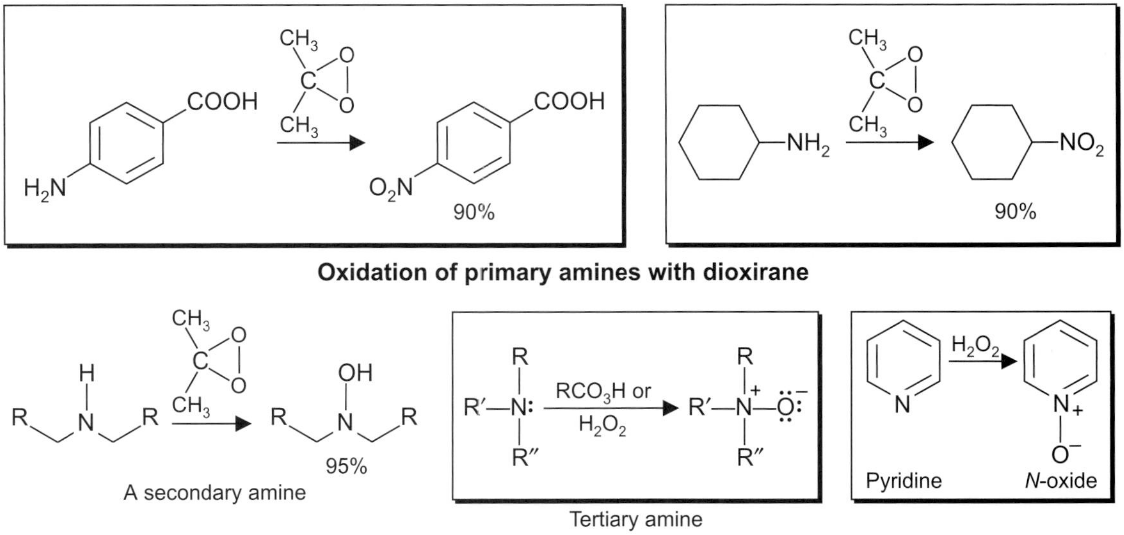

SCHEME 13.62

13.13 PHOTOOXIDATION OF ALKENES

This is discussed (chapter 10) and offers another method to give allylic hydroperoxides which can be reduced to allylic alcohols. Singlet oxygen is the excited state of molecular oxygen, the ground state of molecular oxygen is a triplet with unpaired electrons (Scheme 13.63). Singlet oxygen (a dienophile) reacts *e.g.*, with a 1, 3-diene to form a peroxide *via* cycloaddition–Diels-Alder reaction. These peroxides can be reduced to diols (see Scheme 10.8).

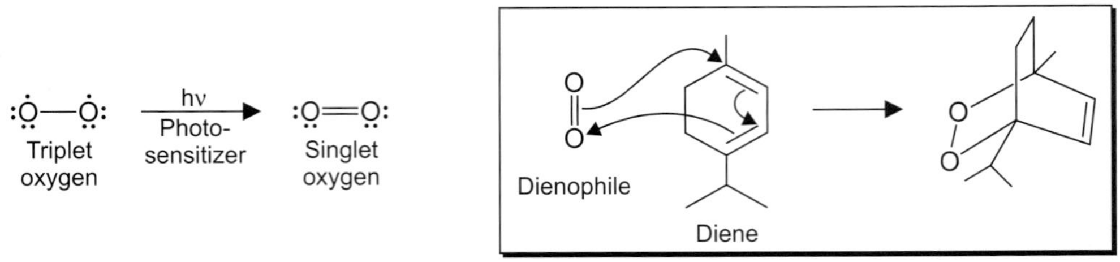

SCHEME 13.63

PROBLEMS

13.1 Write the structure of the product of monoepoxidation with a peroxy acid from the following compounds:

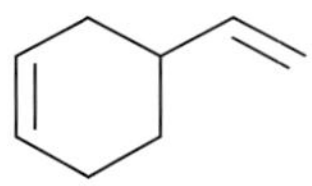

(I)

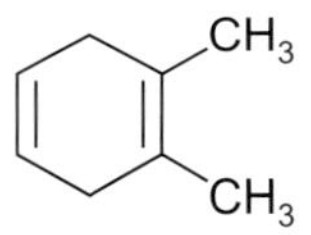

(II)

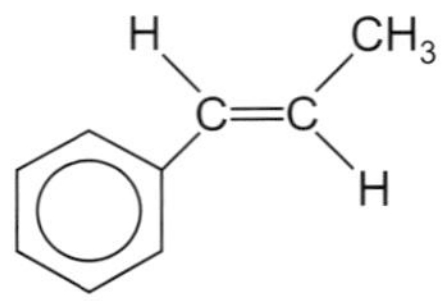

(III)

13.2 Why trifluoroperacetic acid is an effective reagent compared to peracetic acid for epoxidation of alkenes?

13.3 Write the stereostructure of the product of epoxidation from the following olefin.

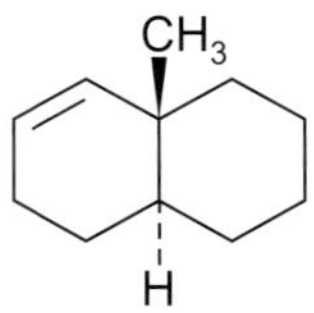

13.4 Which of the 1, 2-diols is expected to react faster with lead tetraacetate?

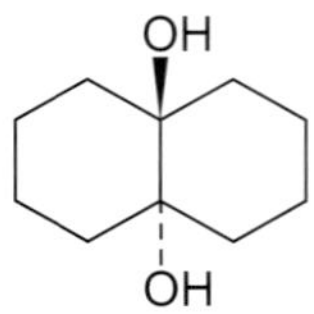

(I)

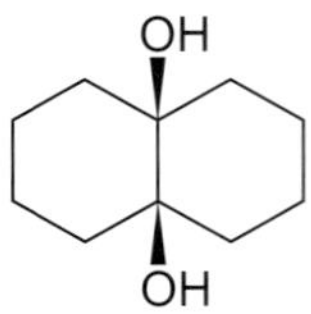

(II)

13.5 How peracids react with carbon–carbon double bond and carbon–oxygen double bond? Give the mechanism of each pathway.

13.6 Write the stereostructures of the epoxide ring opening.

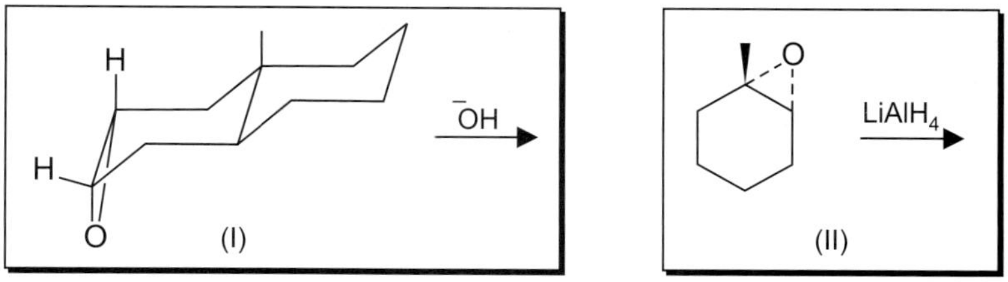

13.7 How can hexanedial (adipaldehyde) be prepared starting from cyclohexene?

13.8 Write the mechanism of ozonolysis of an α, β-unsaturated carboxylic acid which proceeds by the loss of CO_2.

13.9 For the success of Dakin reaction (see Scheme 13.54) there must be an OH or NH_2 in the *ortho* or *para*-position of the aromatic aldehyde or ketone. Explain.

13.10 Why in rigid ring systems, axial alcohols are oxidized (CrO_3) faster than equatorial ones?

Relative rates: 3.2 1

13.11 How can (II) be obtained from (I)? What product is expected from the ozonolysis of (II), using dimethyl sulphide for reductive cleavage?

(I) (II)

13.12 Predict the products from each of the unsymmetrical ketones on Baeyer-Villiger reaction.

$Me_3C—CO—CH_3$ (I) $Ph—CO—CH_3$ (II)

13.13 α-Hydroxycarboxylic acids undergo an easy oxidative decarboxylation with lead tetraacetate to give ketone. Suggest a mechanism.

13.14 Which of the following 1, 2-glycols is expected to react faster with HIO_4 to bring out the oxidative cleavage.

(I) (II)

ANSWERS TO THE PROBLEMS

13.1 Epoxidation of an alkene with a peroxy acid takes place by electrophilic attack of the peroxy acid on the double bond. In this case the reaction will, therefore, preferentially occur on the electron rich double bond.

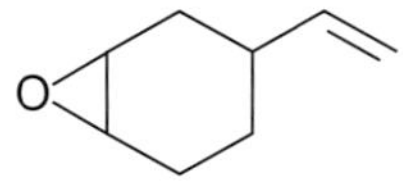

(I)

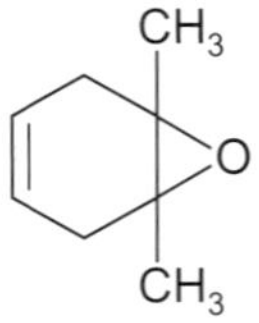

(II)

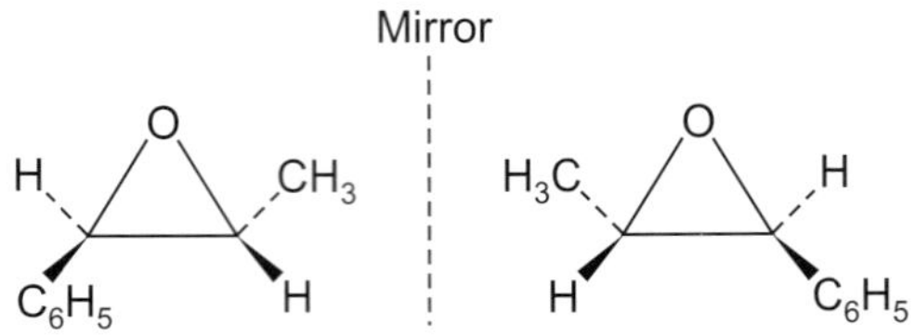

(III)

In the case of (III) two stereogenic carbons are created and since epoxidation is a *syn* addition, only two of the four possible stereoisomers are formed.

13.2 The rate of epoxidation is expected to increase with electron withdrawing groups in the peroxy acid.

13.3 The reagent will approach predominantly from the less hindered side of the substrate *i.e.*, α-side in this case.

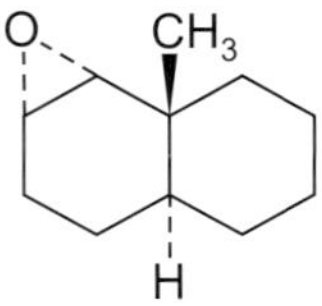

13.4 Compound (II) will cleave faster since the preferential mechanism involving a cyclic adduct can be easily reached.

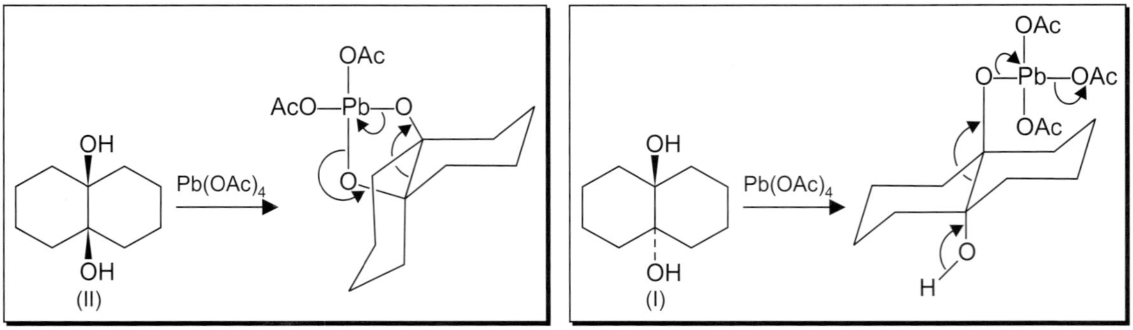

13.6 Like the action of aqueous acid on epoxides, base also gives a 1, 2-diol. The product again has the *anti*-stereochemistry as a result of the stereo-specificity of S_N2-displacements. The ring-opening of epoxides from rigid cyclohexenes as in the present case yields *trans*-diaxial products.

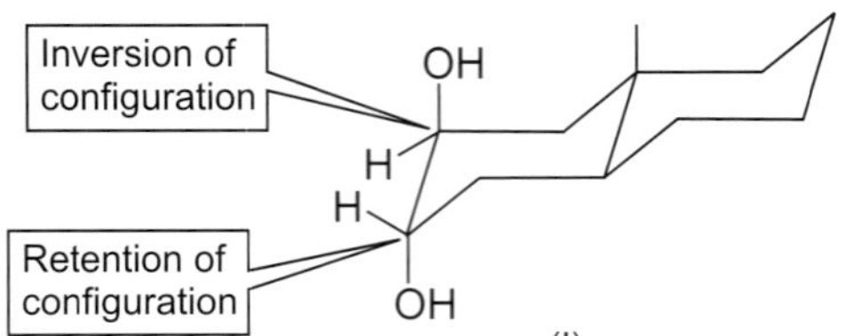

(I)

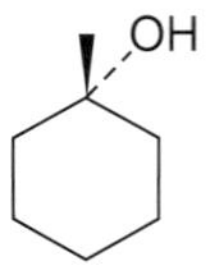

(II)

13.7 By its ozonolysis and a careful reductive cleavage of the ozonide.

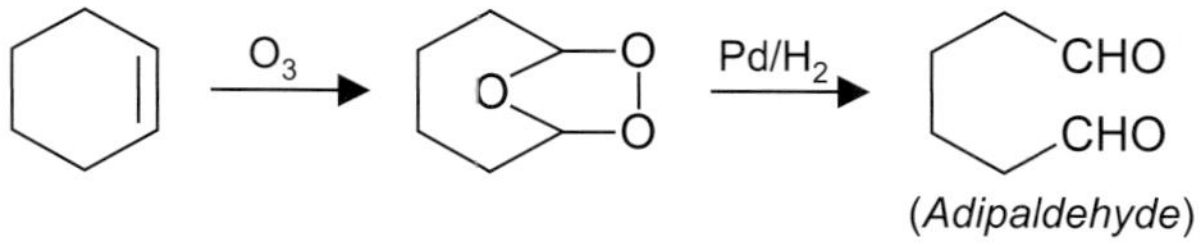

13.8

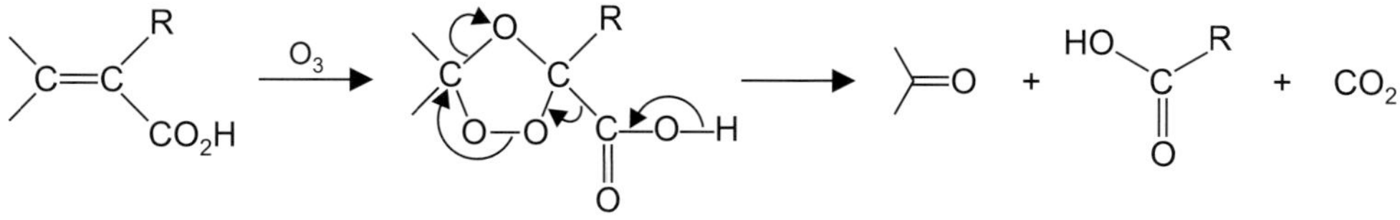

13.9 The hydroxide ion is not a good leaving group. The function of the aromatic substituent may be to provide powerful anchimeric assistance to bring about heterolysis of the *o–o* bond.

13.10 Since the rate determining step in this reaction involves the breaking of the C–H bond the more exposed equatorial C–H in reacts faster than axial C–H in rigid ring systems so that axial alcohols react faster than equatorial ones.

13.11 *p*-Methoxytolune (I) can be converted into diene (II) by Birch reduction. The presence of methoxyl group activates the double bond to which it is attached for the attack by electrophilic reagents.

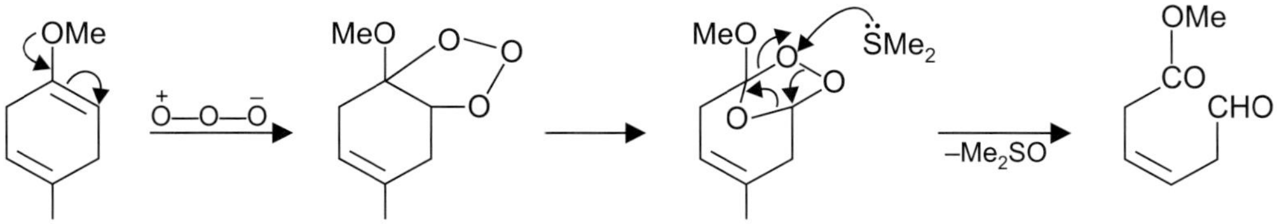

13.12 That group migrates which is better in supplying electrons.

$Me_3C—O—CO—CH_3$	$Ph—O—CO—CH_3$
t-butyl acetate	Phenyl acetate
(I)	(II)

13.13

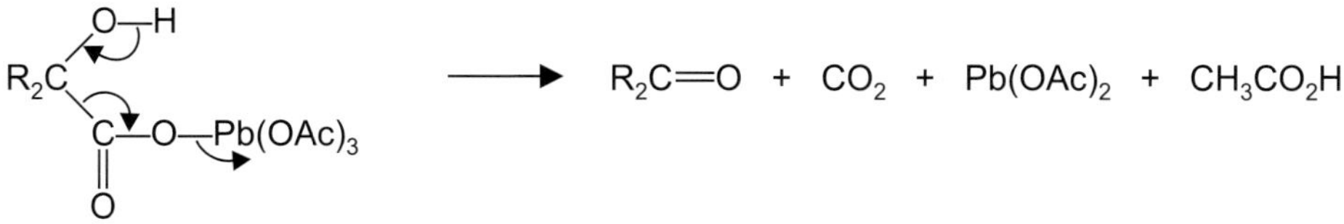

α-hydroxycarboxylic acid
(undergoes oxidative decarboxylation)

13.14 The cyclic intermediate formation during oxidative cleavage with HIO_4 is easy when the two OH groups are on the same side of the molecule. Thus (II) will be cleaved faster than (I).

CHAPTER 14

Reduction Methods

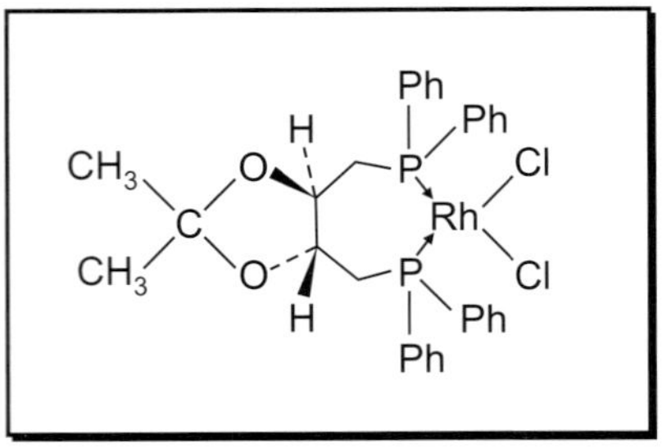

An increase in hydrogen content or a decrease in oxygen content of an organic compound is usually described as its reduction (Scheme 14.1). The opposite of reduction is oxidation *i.e.*, it is either increasing the oxygen content or decreasing the hydrogen content of an organic molecule (Scheme 14.2). Generally, the oxidation of an organic molecule is defined as a reaction which leads to an increase in it of any element which is more electronegative than carbon (Scheme 14.3). The replacement of hydrogen atoms in a compound by chlorine atoms is an oxidation reaction.

When an organic compound is reduced the "reducing agent" is oxidized, whereas when an organic compound is oxidized the "oxidizing agent" is reduced.

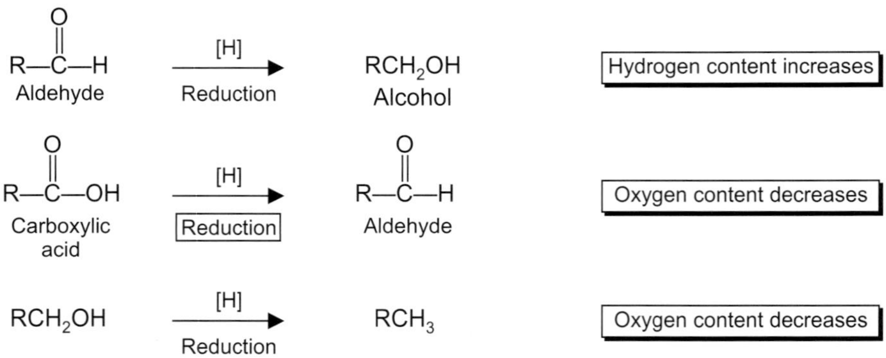

SCHEME 14.1

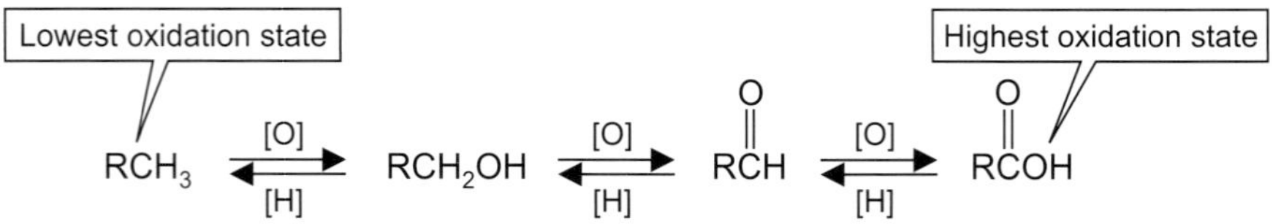

SCHEME 14.2

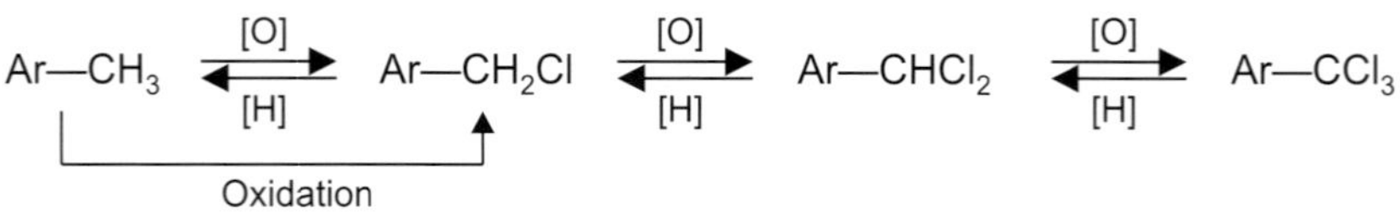

SCHEME 14.3

Most of the oxidizing and reducing agents are generally inorganic compounds. The reductions are generally carried out either by catalytic hydrogenation or chemically. Three main pathways for reduction may be considered:

1. The catalysed addition of molecular hydrogen (Scheme 14.4, eq. I). Hydrogenolysis process involves, the addition of hydrogen followed by bond rupture (Scheme 14.4, eq. II).

$$\underset{\text{Propene (Propylene)}}{CH_3CH{=}CH_2} + H_2 \xrightarrow{Ni} \underset{\text{Propane}}{CH_3CH_2CH_3} \quad (I)$$

Hydrogenation

$$ArCH_2{-}NMe_2 + H_2 \xrightarrow{\text{Catalyst}} ArCH_3 + HNMe_2 \quad (II)$$

Hydrogenolysis

SCHEME 14.4

2. By the transference of hydride ion *e.g.*, in the reduction of a carbonyl group by lithium aluminium hydride (Scheme 14.5).

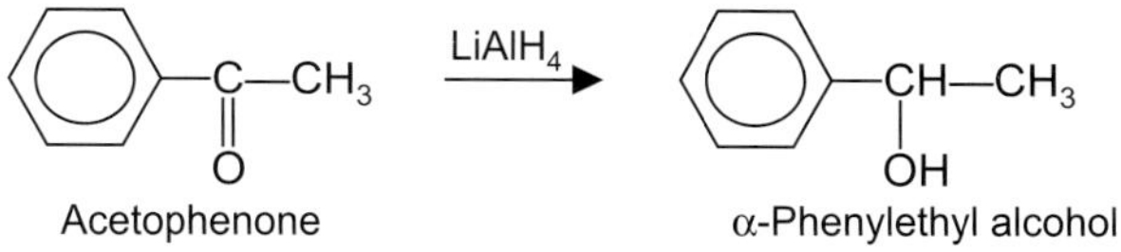

The transference of hydride ion

SCHEME 14.5

3. By the addition of electrons, followed by coupling, as in the reduction of ketones to pinacols (Scheme 14.6).

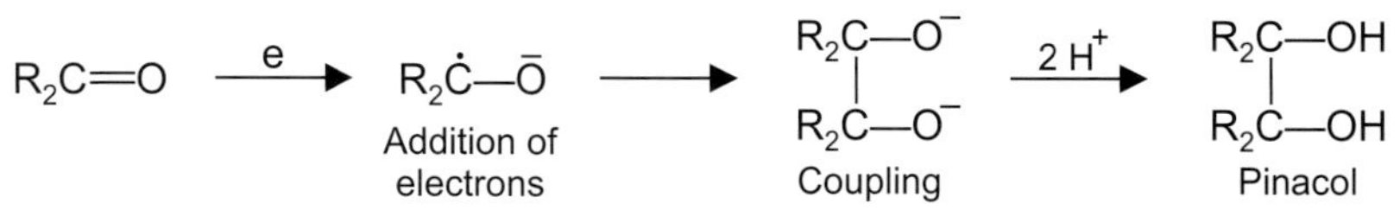

SCHEME 14.6

14.1 CATALYTIC REDUCTION-REDUCTION WITH DIIMIDE AND HYDROBORATION

(A) Heterogeneous Catalytic Hydrogenation

(i) The role of catalyst

Alkanes, which are strained can be reduced catalytically by rupturing C—C bonds. The C—C cleavage relieves the strain (Scheme 14.7). Normally almost all olefins can be reduced by treatment with hydrogen in the presence of a catalyst. The process of hydrogenation is thought to proceed via a mechanism involving the adsorption of hydrogen molecules on the surface of

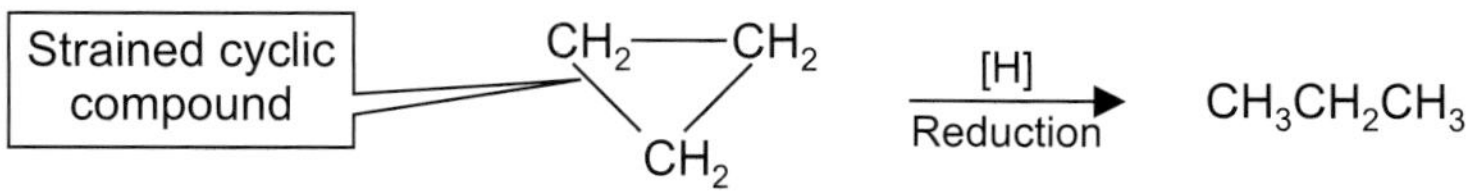

SCHEME 14.7

the catalyst to form metal hydrogen bonds. The surface bearing adsorbed hydrogen causes adsorption of the alkene as well and a subsequent stepwise transfer of hydrogen atom occurs. This process yields an alkane before the organic molecule leaves the catalyst surface. As a result, both hydrogen atoms usually add from the same side of the molecule, a process termed *syn* addition (Scheme 14.8).

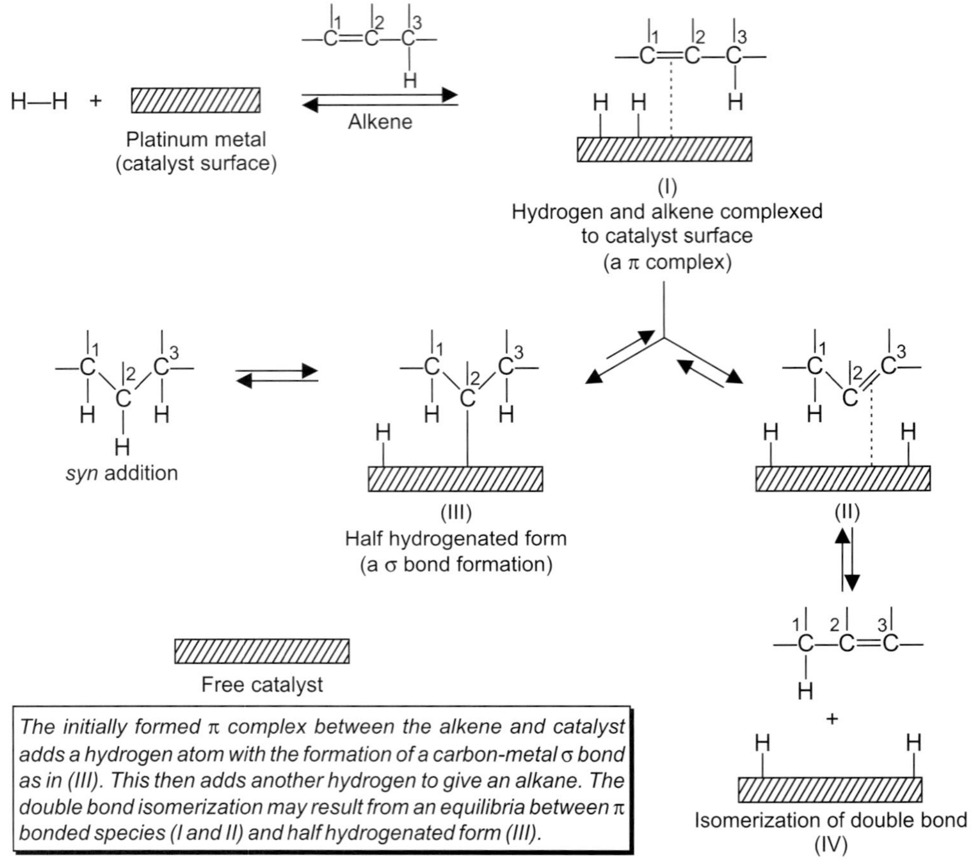

The initially formed π complex between the alkene and catalyst adds a hydrogen atom with the formation of a carbon-metal σ bond as in (III). This then adds another hydrogen to give an alkane. The double bond isomerization may result from an equilibria between π bonded species (I and II) and half hydrogenated form (III).

SCHEME 14.8

Thus as expected *Z*-1, 2-dimethylcyclohexene gives a *meso* product on catalytic hydrogenation and a racemic mixture results from (I, Scheme 14.8*a*).

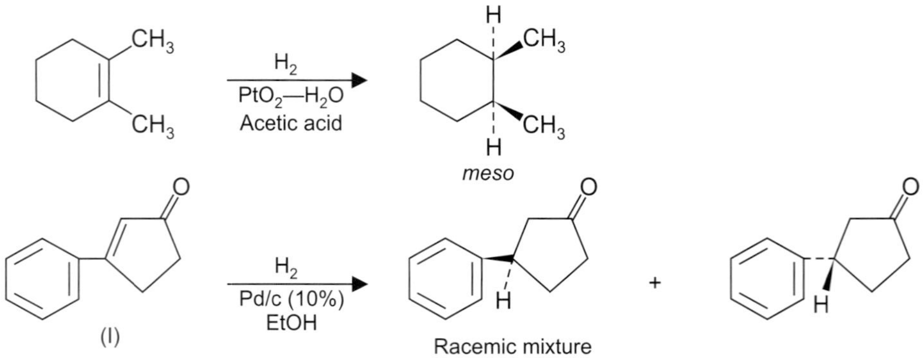

SCHEME 14.8a

The hydrogenation is thought to involve an equilibria between (II and IV, Scheme 14.8), and (III) which is the half hydrogenated form. The half hydrogenated form then either picks up the second hydrogen atom to afford the hydrogenated product or it can give back the starting alkene, or it can afford an isomeric alkene (IV). The formation of isomeric alkene can explain the *anti* addition of hydrogen which is often the mode of addition particularly with Pd as the catalyst (see Scheme 14.12).

(ii) Evidence for syn addition

Stereochemically, the hydrogenation of an olefin generally occurs in a *syn-* fashion (Scheme 14.8). Thus, when one considers the hydrogenation of the *E*-stilbene derivative (I, Scheme 14.9), the *syn* addition gives the (±)-dihydro-derivative while the *syn*-addition to the Z isomer (II) gives the *meso* isomer.

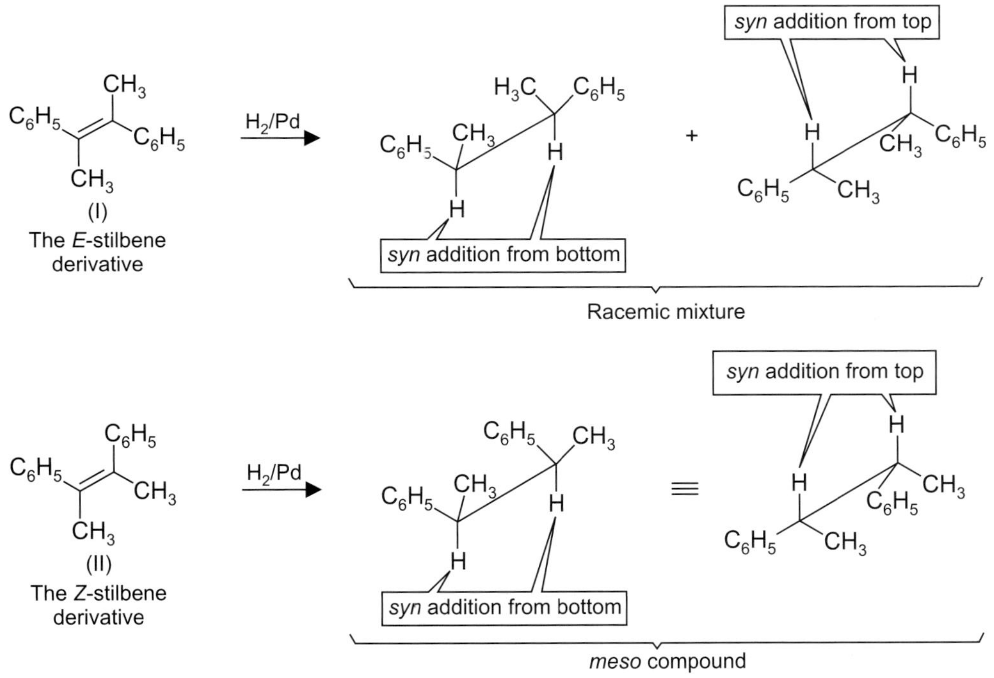

A *syn* stereospecificity:
Both hydrogen atoms add from the same side of the molecule

SCHEME 14.9

(iii) Syn addition from less hindered side

Generally the hydrogenation occurs by *syn* addition of hydrogen and from the less hindered side of the unsaturated center. This is seen in the case of cholesterol (Scheme 14.10), which on hydrogenation gives mainly, *trans-* ring-fusion (5α), since the angular methyl group hinders the fit of the catalyst on the β-face. In cholesterol the 3-OH group is equatorial, however if in such a system a 3-substituent is axial (A, Scheme 14.10) the fit to the catalyst is hindered on both sides and hydrogenation then gives a mixture of *cis* and *trans* decalin systems in almost equal amounts.

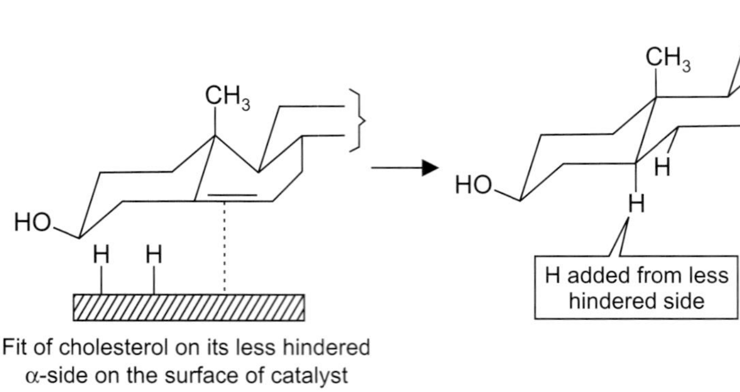

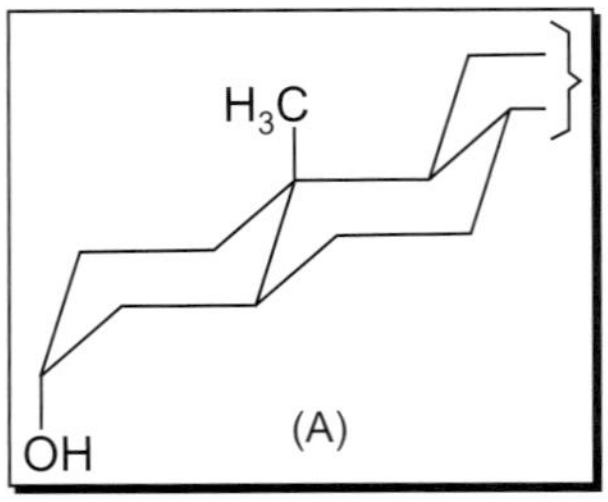

SCHEME 14.10

(iv) The facial stereoselectivity of hydrogenation

In some instances, the affinity of a particular substituent group for the catalyst surface may force addition of hydrogen from the same side of the molecule occupied by the substituent irrespective of steric effects. The hydroxy methylene group (– CH_2OH) is in particular effective in this regard. Thus, in the hydrogenation of the tetrahydrofluorene derivative (Scheme 14.11) there is excess of *syn* addition of hydrogen. However, the *syn*-addition is reduced to only 10% when this group is replaced by – $CONH_2$ in the same system (Scheme 14.11). The *syn* directive effect indicates that the hydroxyl group interacts with the catalyst surface. Thus the alcohol yields a predominant product with *cis* ring juncture, the same system with –$CONH_2$ group gives as expected a major product with *trans*-stereochemistry. (Scheme 14.11).

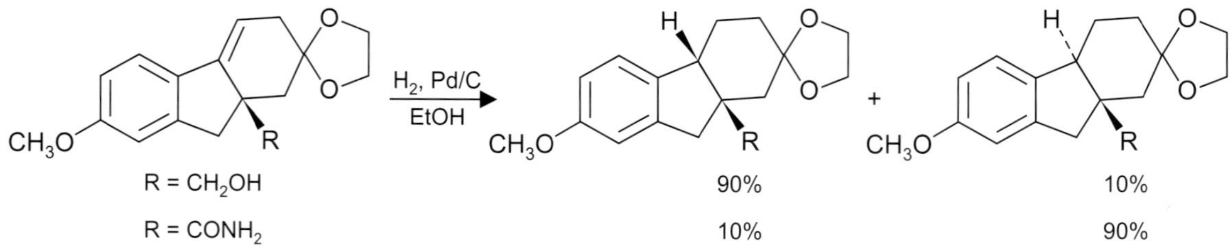

SCHEME 14.11

(v) Stereoselectivity for unhindered alkenes—the nature of a catalyst

Generally hydrogen adds to an alkene in a *syn* fashion, *anti*-addition is often observed particularly with palladium catalyst. Thus 1, 2-dimethylcyclohexene (Scheme 14.12) gives different mixtures of *cis*- and *trans*- 1, 2-dimethylcyclohexene depending on catalyst. This *anti*- addition has been explained as a result of the migration of the methyl substituted double bond (Scheme 4.13).

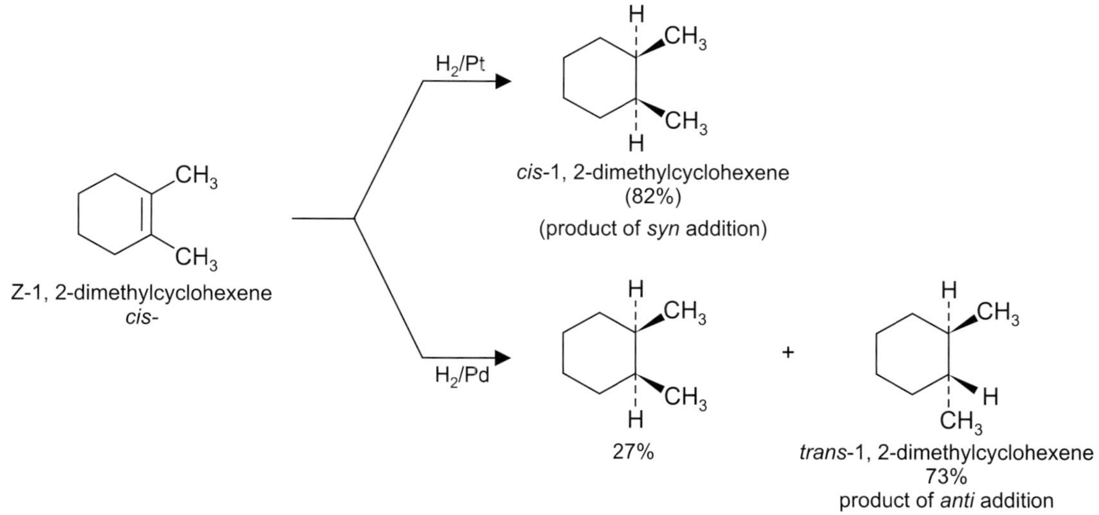

SCHEME 14.12

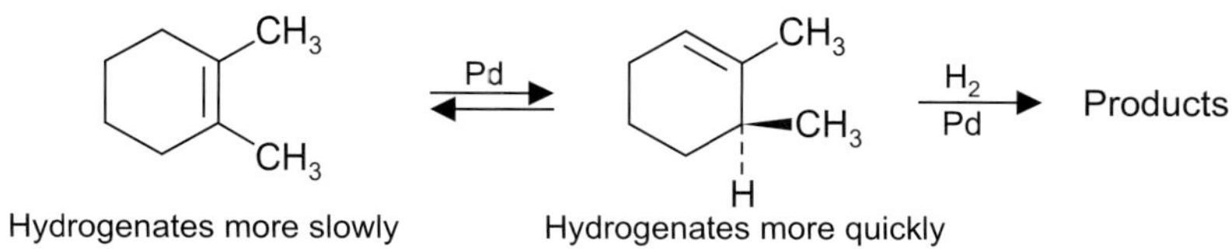

SCHEME 14.13

The hydrogenation of a carbon-carbon double bond can also be carried out by the supply of hydrogen by a donor *e.g.*, cyclohexene or hydrazine (Scheme 14.14). The driving force in the case of cyclohexene is due to the gain in aromatic stabilization energy on the formation of benzene while with hydrazine, the strongly bonded N_2 molecule is formed. The advantage of this method is that no special apparatus is needed.

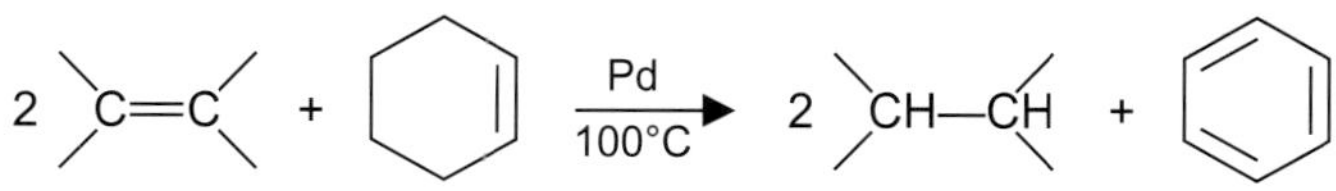

SCHEME 14.14

(B) Homogeneous Catalytic Hydrogenation Using Wilkinson's Catalyst

Heterogeneous catalytic hydrogenation suffers from several disadvantages and these are:

1. No selectivity is observed when more than one unsaturated center is present.
2. Double bond migration occurs.
3. Several groups suffer an easy hydrogenolysis (see Scheme 14.18).
4. One may not be able to predict, the stereochemical outcome, despite a number of rules, since heterogeneous catalytic hydrogenation depends on chemisorption and not on reaction between molecules.

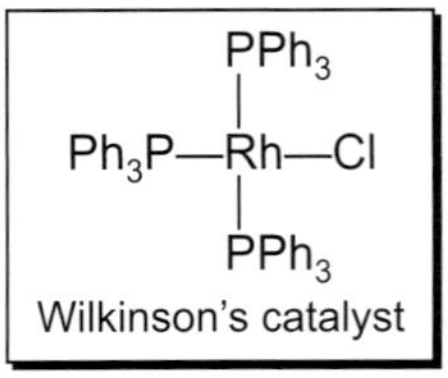

SCHEME 14.15

Some of these difficulties have been solved by homogeneous catalytic hydrogenation in which the metal is replaced by a soluble complex of rhodium or ruthenium and by carrying out the reduction in homogeneous solution. Several soluble catalysts have been used and the more effective are the ones derived from rhodium and ruthenium (Scheme 14.15). The rhodium complex used is $(Ph_3P)_3RhCl$ (called Wilkinson's catalyst, see, Sec. 7.1E), is prepared by heating rhodium chloride $RhCl_3$, $3H_2O$ with excess of triphenylphosphine in boiling ethanol and with this catalyst, hydrogenation occurs in a single phase — in solution.

The Wilkinson's catalyst is thought to act by exchanging one phosphine ligand for a solvent molecule (*S*) to afford a complex which can then bind two hydrogen atoms on the metal (Scheme 14.16). Displacement of the solvent molecule by the alkene is followed by step-wise stereospecific *syn*-transfer of the two hydrogen atoms, and the saturated molecule then leaves the metal center.

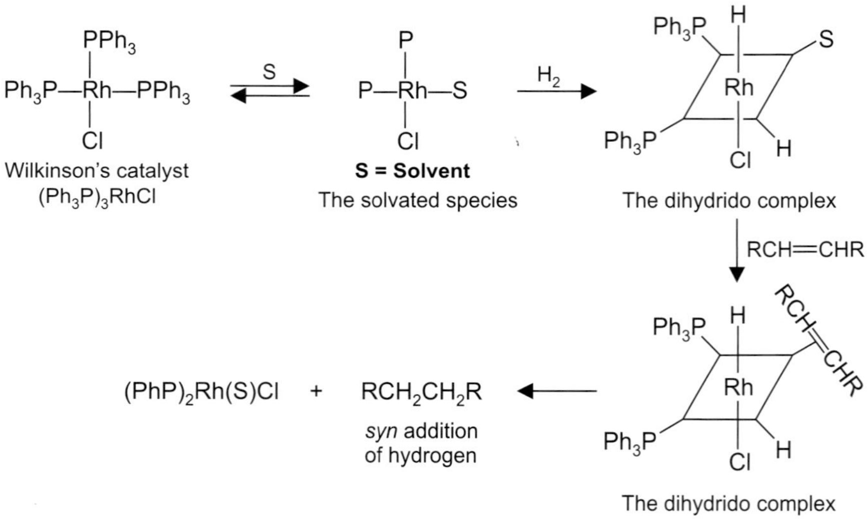

Mechanism of hydrogenation with Wilkinson's catalyst

SCHEME 14.16

The Wilkinson's catalyst has its merit that whereas olefins and acetylenes are reduced, other common groups like C=O, C≡N and NO_2 are not attacked and selective reductions can be carried out (Scheme 14.17). Moreover, *mono-* and disubstituted double bonds are reduced more rapidly compared to more substituted types *i.e.*, *tri-* and *tetra-* substituted double bonds.

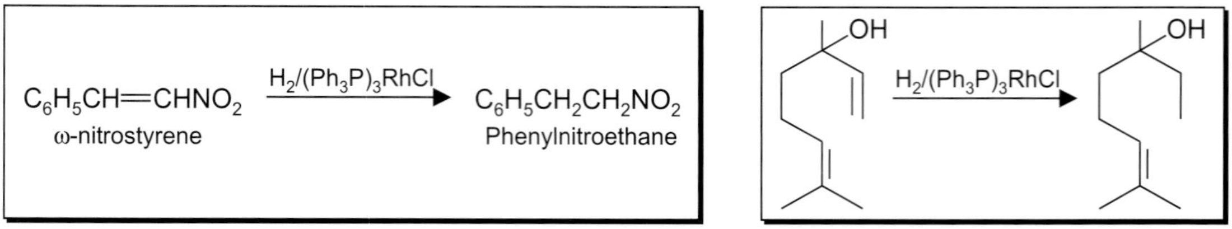

SCHEME 14.17

Hydrogenolysis implies a process of cleavage of a carbon-heteroatom bond by catalytic hydrogenation. Aliphatic alcohols *e.g.*, are inert toward catalytic hydrogenation, but benzylic alcohols are reduced to the corresponding hydrocarbons (Scheme 14.18). This type of a reaction where reduction is used to cleave a single bond *e.g.*, C—OH in the case of a benzylic alcohol is called hydrogenolysis.

Thus hydrogenolysis is the cleavage of a C—X single bond during addition of hydrogen and represents one of several reactions hydrogen brings about over metal catalysts.

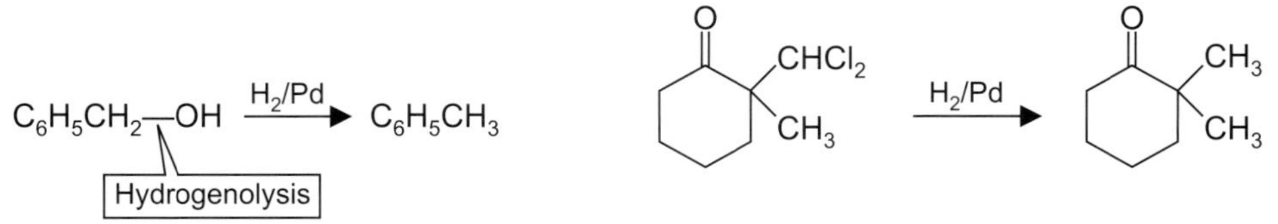

SCHEME 14.18

An advantage of working with homogeneous catalytic hydrogenation is that hydrogenolysis does not take place. Thus with Wilkinson's catalyst, *e.g.*, benzyl cinnamate affords cleanly the dihydro–derivative while, hydrogenation using a metal catalyst results (in addition to hydrogenation) in cleavage of the O-benzyl bond (Scheme 14.19).

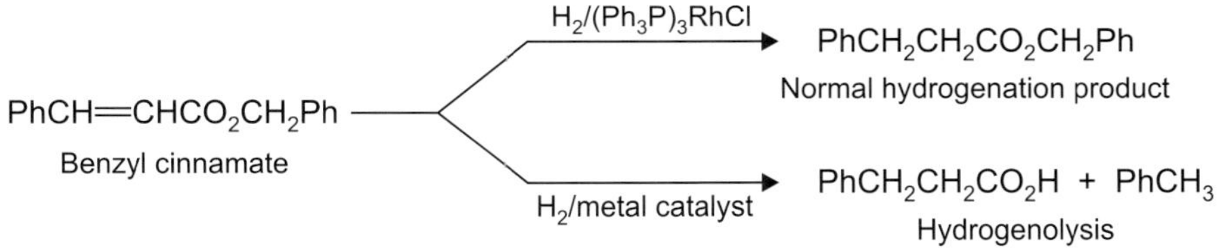

Hydrogenolysis is the cleavage of C—X single bond with hydrogen over metal catalyst

SCHEME 14.19

Rhodium has a strong affinity for carbon monoxide, the Wilkinson's catalyst thus brings about decarbonylation of an aldehyde group. Olefins therefore, with aldehyde groups are degraded (*e.g.*, cinnamaldehyde gives styrene, Scheme 7.1*d*).

The hydrogenation of an achiral substrate with optically inactive Wilkinson catalyst will always produce a racemic modification. Thus an alkene with enantiotopic faces on hydrogenation with Wilkinson's catalyst gives racemic product since there is equal probability for hydrogen to add to the either face. However, when the Wilkinson's catalyst is made chiral (optically active) by attaching optically active ligands, then the hydrogenation occurs in a chiral medium and can give optically active products. A large number optically active ligands have been developed where the chirality is due to either stereogenic carbon or stereogenic phosphorus.

It may be mentioned that catalytic hydrogenation is the second important class of reduction methods after hydride reductions.

DIOP is a complex of rhodium with a chiral ligand which has been used to synthesize biologically active (–) form of dopa. As the catalyst is chiral, the transition states which would lead to two enantiomers are diastereotopic, and have different energies, the transition state which leads to the one of the enantiomers is of lower energy and is favoured (Scheme 14.19*a*).

Rhodium $(DIOP)Cl_2$—a chiral catalyst for enantioselective reduction =

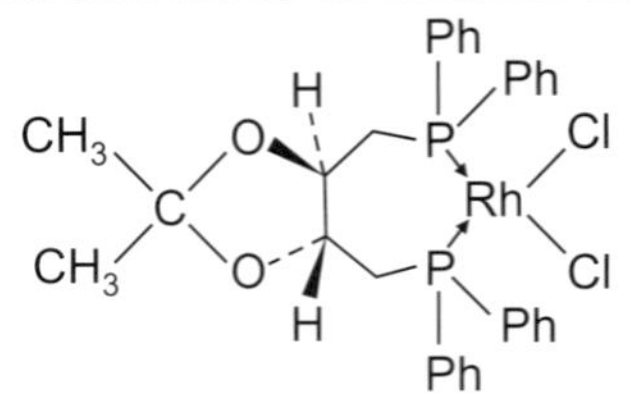

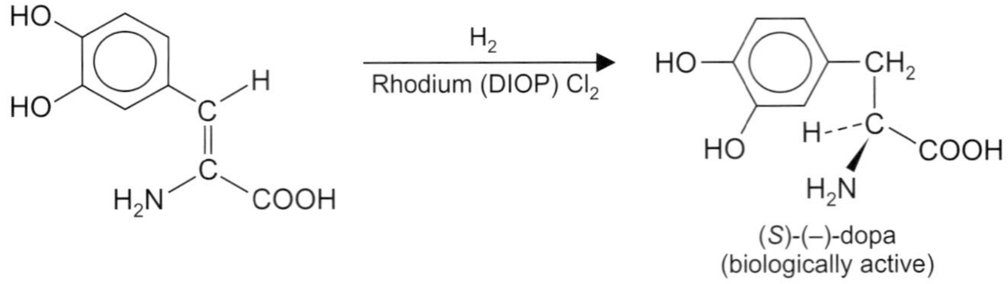

SCHEME 14.19a

(C) Other Catalysts and Specially Conditioned Catalysts

A catalyst which brings about hydrogenation of an alkyne to an alkene is the nickel boride catalyst (P-2 catalyst). This catalyst can be prepared by the reduction of nickel acetate with sodium borohydride (Scheme 14.20). Hydrogenation of alkynes with P-2 catalyst leads to *syn*-addition of hydrogen and gives an alkene with (*Z*) or *cis* configuration (Scheme 14.20).

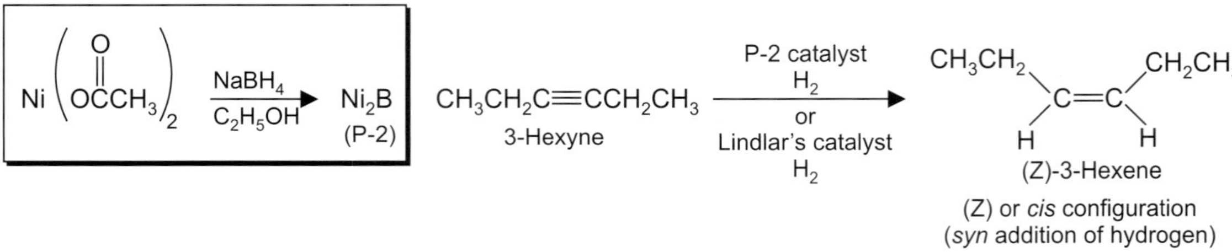

SCHEME 14.20

Specially modified catalysts lead to partial hydrogenation of alkynes to (*Z*) alkenes. Metallic palladium on calcium carbonate after it has been partially deactivated with lead acetate and quinoline is called Lindlar's catalyst. Reductions of alkynes with these type of catalysts proceed with high stereoselectivity. Normally the reduced product is composed largely of the thermodynamically less stable (*Z*)-alkene (Scheme 14.20).

In Rosenmunds reduction of acid chlorides to aldehydes (Scheme 14.21) with hydrogen on a palladium catalyst, barium sulphate it is often helpful to poison the catalyst.

$$RCOCl + H_2 \longrightarrow RCHO + HCl$$

Rosenmunds reduction

SCHEME 14.21

A variety of functional groups are prone to catalytic hydrogenation and acid chlorides are the most reactive and the arenes are the least reactive. Since C=C is weaker than C=O, catalytic hydrogenation can be used to carry out selective reduction of C=C in the presence of a carbonyl group. Other groups which are reduced during catalytic hydrogenation are in (Scheme 14.21*a*).

SCHEME 14.21a

(D) Reduction of Double Bonds via Hydroboration of Alkenes Followed by Reaction with Organic Acids

Alkanes from Alkenes Via Hydroboration

The organoboranes made from alkenes on reaction with organic acids give alkanes. Details are in Schemes 7.54a and 7.54b.

(E) Hydrogen Transfer from Diimide

Diimide (HN=NH, obtained by the copper (*II*)–catalysed oxidation of hydrazine) is unstable and decomposes to nitrogen and hydrogen (in the absence of additive). When diimide is liberated in the presence of an alkene a rapid *syn*- stereospecific reduction occurs (Scheme 14.22).

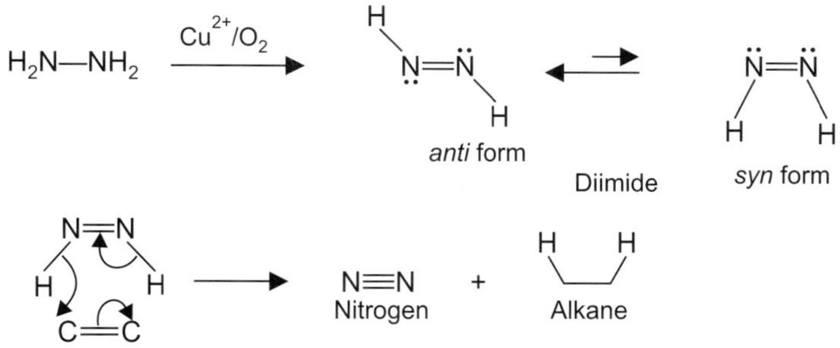

SCHEME 14.22

Although both the *syn* and *anti* forms of diimide are produced by the oxidation of hydrazine, only the *syn* form reduces the double bonds by a cyclic mechanism. The addition is therefore, stereospecifically *syn* (Scheme 14.22). More strained double bond reacts selectively in the presence of a *cis* double bond (Scheme 14.23).

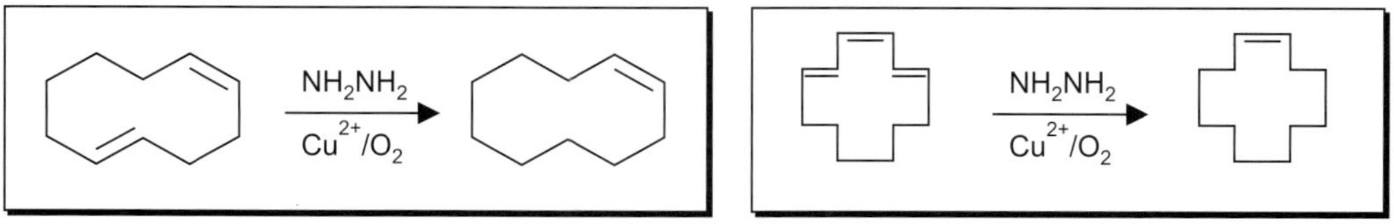

SCHEME 14.23

Diimide is a highly selective reagent which generally reduces readily the symmetrical bonds like C≡C, C=C, N=N, and carbonyl containing groups, nitro groups and sulphoxides are not reduced (Scheme 14.24). Significantly the otherwise highly reducible disulphide bond remains unaffected with diimide.

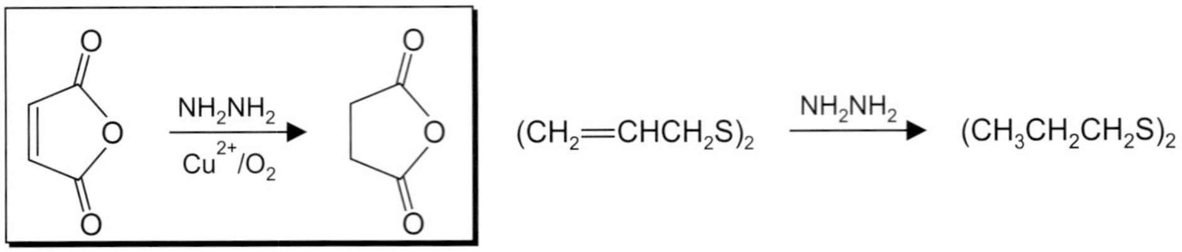

SCHEME 14.24

14.2 REDUCTION BY DISSOLVING METALS–METAL AND AMMONIA

(i) Reduction of a double bond conjugated with a carbonyl group

Conjugate reduction of α, β-unsaturated ketones is carried out by adding electrons from a metal *e.g.*, lithium dissolved in liquid ammonia (a dissolving metal reduction). An alcohol is often added to the reduction medium to provide protons (Scheme 14.25). The stereochemical outcome of the reduction is established by the proton transfer to the β-carbon. When one stereocenter is created a racemic mixture is formed (I, Scheme 14.26). With a chiral enone system of $\Delta^{1,9}$-2-octalone, the product has a *trans-* ring junction.

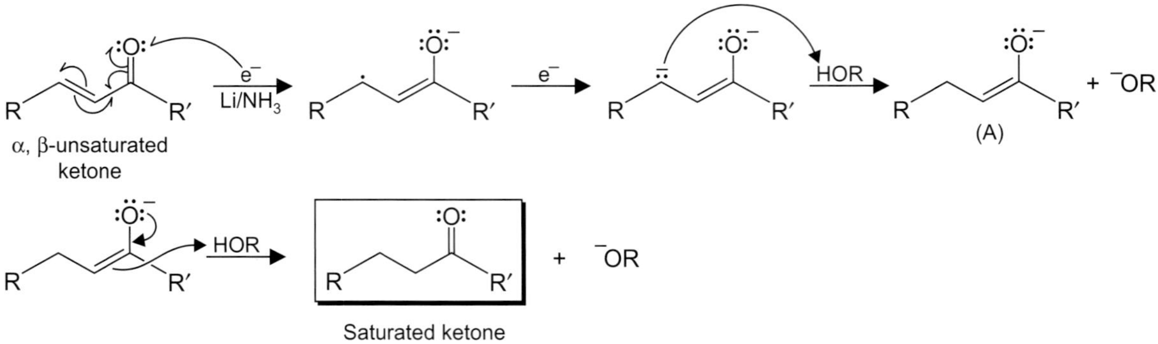

SCHEME 14.25

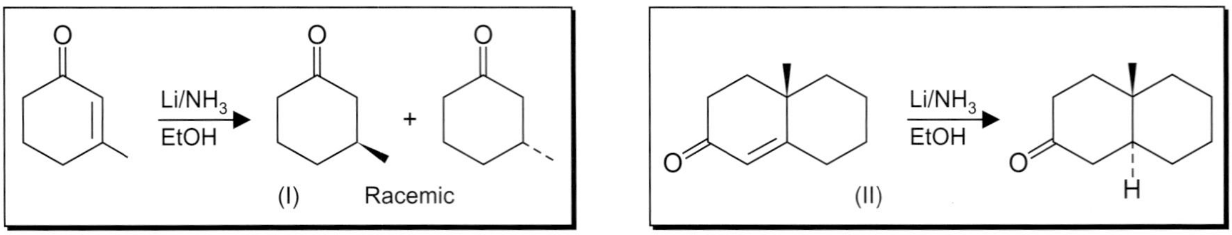

SCHEME 14.26

(ii) Utility in carbon-carbon bond formation

If during reduction of an α, β- unsaturated ketone, only one equivalent of the proton donor (ROH) is used the enolate (A, Scheme 14.25) remains available for a suitable subsequent reaction. Thus if the enolate (I, Scheme 14.27) is not protonated it can be employed in alkylation reactions.

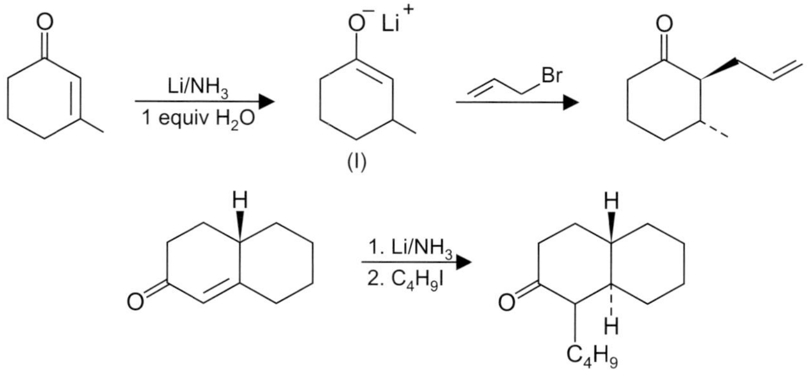

SCHEME 14.27

Generating a Specific Enolate of a Ketone

The dissolving metal reduction of an enone is method of high merit to generate a specific enolate of a ketone regioselectively. Use of LDA with a ketone (Scheme 14.28) will however, give a mixture of enolate ions.

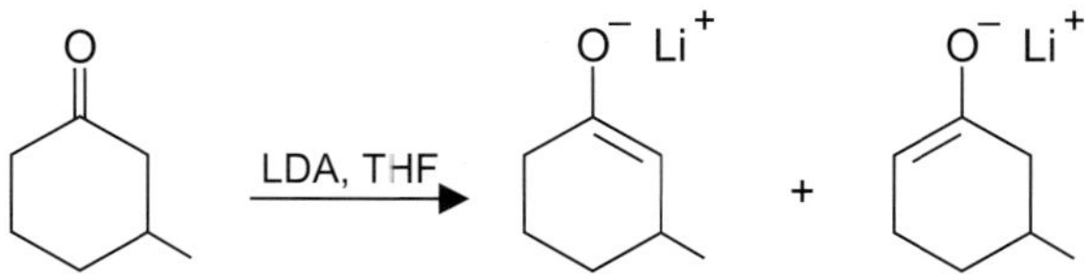

SCHEME 14.28

Thus the following points regarding reduction of α,β-unsaturated ketones in liquid ammonia may be noted:

- Li/NH_3 in the presence of excess protons from a source like *t*-butyl alcohol will reduce the C=C bond of enones to give saturated ketones.
- Li/NH_3 without an added proton source/or with 1 equivalent of proton source will give an enolate ion which can undergo alkylation in the α-position of the ketone.

(iii) Reduction of alkynes with metal and liquid ammonia

Electrons are transferred to acetylenes more readily than to alkenes because of greater reactivity of acetylenes towards nucleophiles. The reagents of choice are sodium or lithium in liquid ammonia. The reaction stops at the alkene stage and internal alkynes are converted into *trans* alkenes using lithium in liquid ammonia (Scheme 14.29). A *cis* alkene can however, be made by the catalytic hydrogenation on a deactivated Lindlar catalyst. With a metal catalyst only *syn* addition occurs (Scheme 14.29).

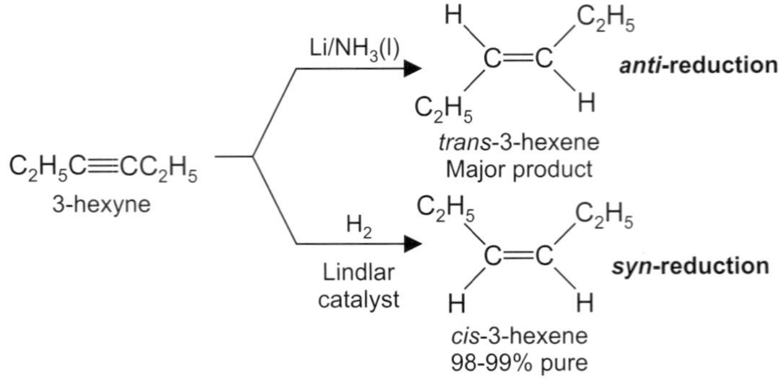

SCHEME 14.29

The mechanism of the reaction with Li/NH_3 is thought to involve successive electron transfer and protonation steps (Scheme 14.30).

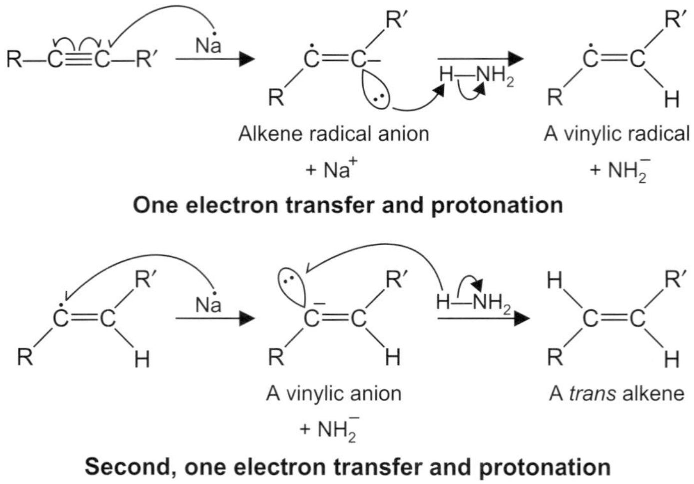

SCHEME 14.30

The following points may be noted:

- The transfer of the *s* orbital electron from sodium or lithium is an easy process. Sodium and lithium have a strong tendency for losing the single electron of their outer-shell *s* orbital.
- The radical anion formed after the first electron transfer (Scheme 14.30) is a very strong base and can therefore, remove a proton from NH_3.
- The vinylic anion can adopt either the *cis* or *trans* configuration, however, the equilibrium favours the more stable *trans* configuration with bulky groups as far apart as possible (Scheme 14.31).

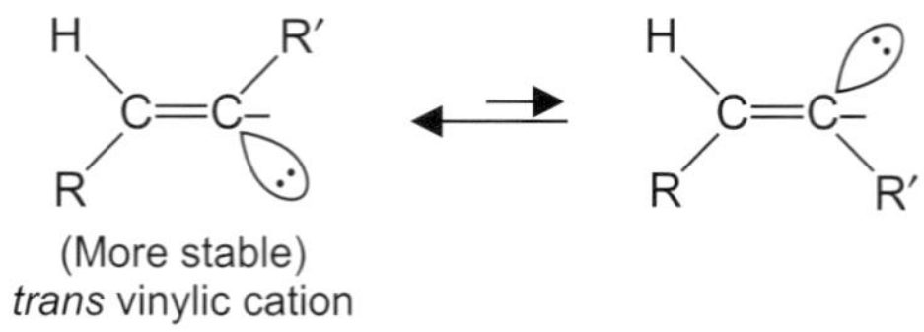

SCHEME 14.31

Thus in summary an alkyne can be reduced to a *cis*-alkene by employing a poisoned catalyst. Treatment of an alkyne with Na/NH_3 brings about reduction to give a *trans*-alkene. Na/NH_3 reduction proceeds involving a series of electron and proton-transfer steps.

EXERCISE 14.1

Write the outcome from the reactions (Scheme 14.32).

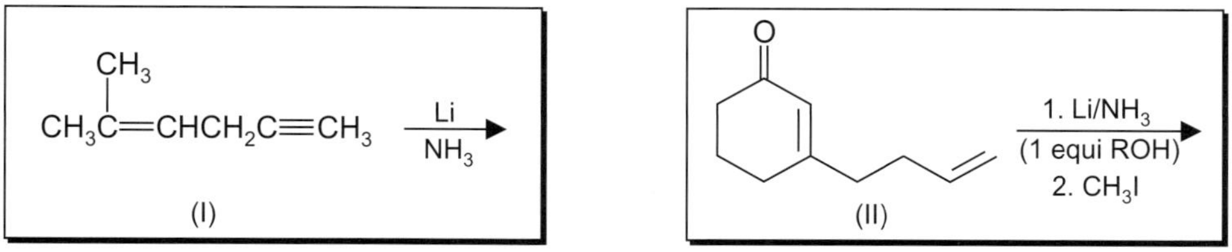

(I) (II)

SCHEME 14.32

ANSWERS. *(I) A carbon-carbon double bond is inert towards lithium in liquid ammonia and is not reduced. Alkyne is reduced to a trans alkene the product formed is (III, Scheme 14.33).*

(II) The conditions of the reaction will give a specific enolate ion which subsequently undergoes alkylation with CH_3I. These reactions normally form the trans isomer as the predominant product provided the alkyl substituents are on adjacent carbons, thus the product formed is (IV).

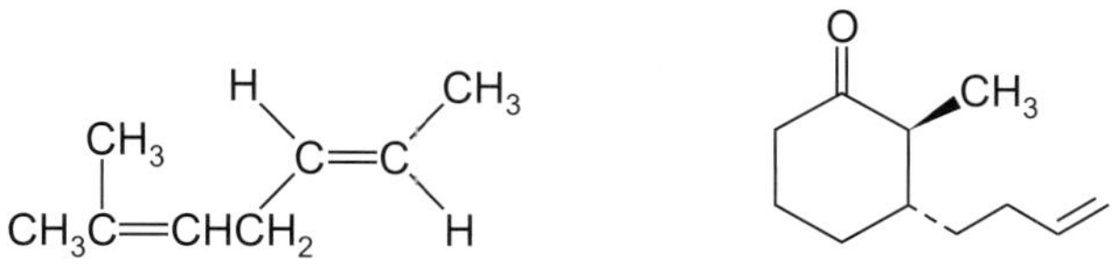

(III) (IV)

SCHEME 14.33

EXERCISE 14.2

Convert 1-butyne (I, Scheme 14.33a) in to the cis-epoxide (II). How the same epoxide with trans- geometry (III) could be prepared?

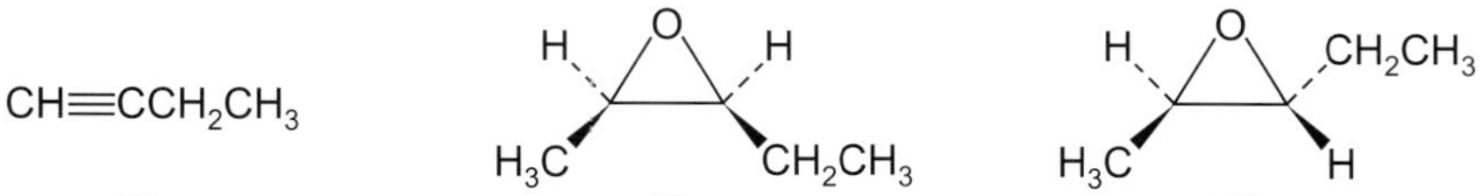

(I) (II) (III)

SCHEME 14.33a

ANSWER. *This is given in Scheme 14.33b.*

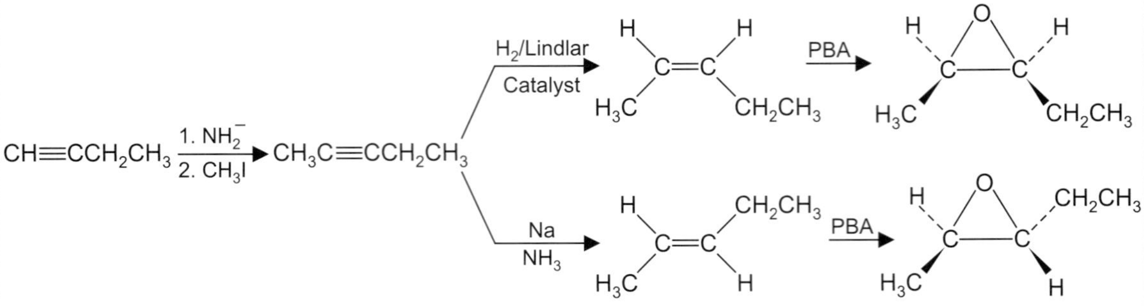

SCHEME 14.33b

(iv) Reduction of aromatic rings—two successive electron transfer/protonation steps
Partial reduction of aromatic rings is carried out by dissolving metal systems and the reaction is called Birch reduction. The reduction medium usually consists of lithium or sodium in liquid ammonia and an alcohol is usually added to serve as a proton source. The mechanism of Birch reduction begins with the addition of the electron to form a radical ion (Scheme 14.34). Protonation of the radical ion gives a neutral radical. Addition of another electron gives a carbanion, which is protonated to yield 1, 4-cyclohexadiene. The isolated double bonds in the dihydro product are much less easily reduced than the aromatic ring and thus reduction stops at the dihydro stage.

$$NH_3 + Na \longrightarrow (NH_3)\ ...e'(NH_3) \text{ (deep blue solution)} + Na^+$$

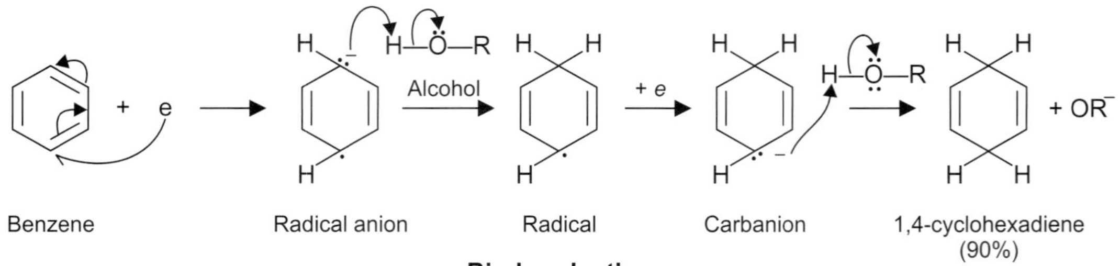

Birch reduction
A two-electron reduction of aromatic rings using a solution of alkali metal in liquid ammonia

SCHEME 14.34

The two carbon atoms which are reduced to give the dihydro product pass through the carbanionic intermediates (see Scheme 14.34). The electron withdrawing substituents (COOH) stabilize them while electron donating substituents (OCH_3) destabilize them. Reduction thus takes place on carbon atoms attached with electron withdrawing substituents and not on carbon atoms on which electron releasing substituents are present (Scheme 14.35).

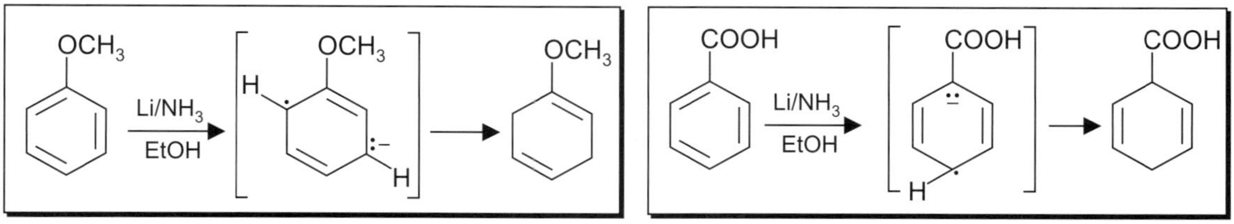

SCHEME 14.35

The reduction of methoxybenzenes is used in the synthesis of cyclohexenones via hydrolysis of the intermediate enol ethers (Scheme 14.36) vinyl ethers react with aqueous acids *e.g.*, with aqueous hydrochloric acid to give an intermediate carbocation, which reacts with water to give a hemiacetal which is hydrolyzed (Scheme 14.37).

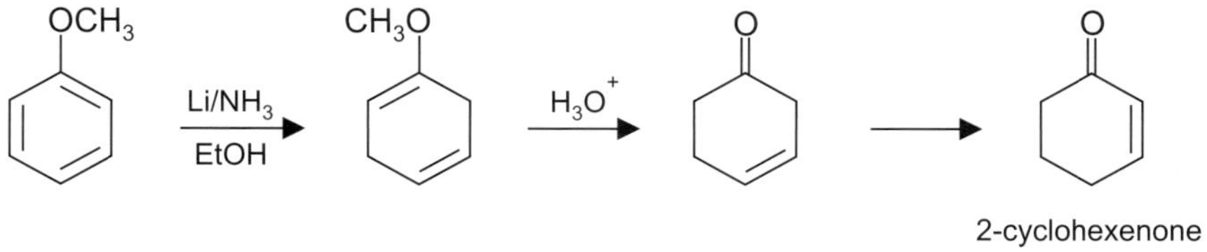

SCHEME 14.36

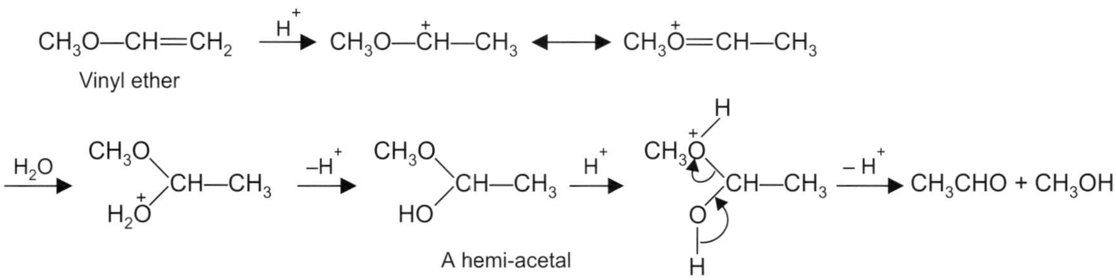

SCHEME 14.37

Recall that unlike catalytic hydrogenation of benzene which is a difficult process, sodium in liquid ammonia and a little alcohol brings about its reduction. However, there is an important difference, during Birch reduction the ring is not totally reduced and the reaction steps at the diene stage.

14.3 ADDITION OF HYDROGEN AND REDUCTIVE COUPLING OF CARBONYL COMPOUNDS—DISSOLVING METAL REDUCTIONS

One has already studied the role of a metal as the reducing agent where the metal (Na or Li) provides one or more electrons to the organic reactant. A ketone forms a ketyl radical after the transfer of a single-electron by a metal to it. The radical may be protonated to give alcohols or may get dimerized to a pinacol by the dimerization of anion radicals in the absence of a proton-donor (Scheme 14.38).

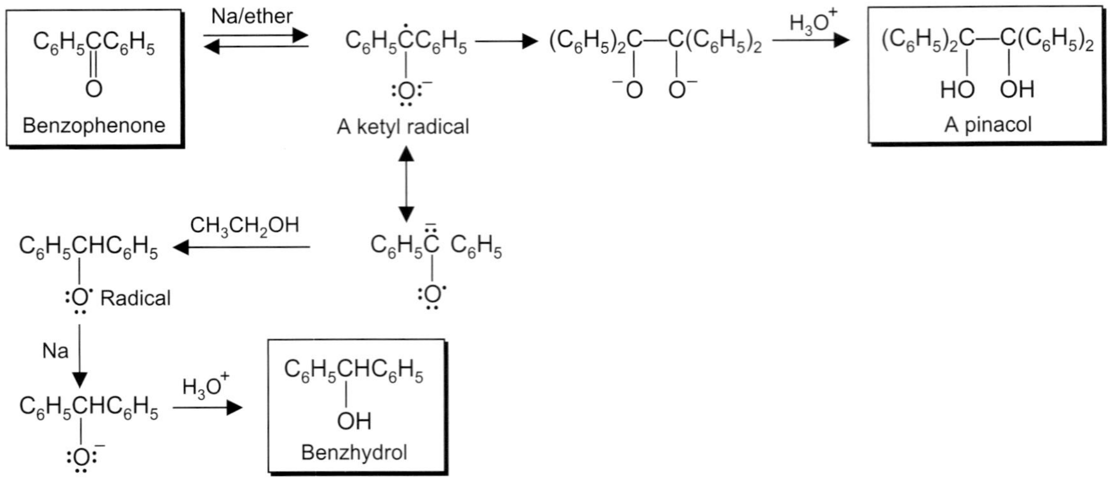

SCHEME 14.38

Acetone on reduction with dissolving metals in the absence of proton donors gives pinacol (Scheme 14.39). The common reagents are magnesium, and magnesium amalgams. Currently use of titanium tetrachloride along with magnesium amalgam gives improved yields.

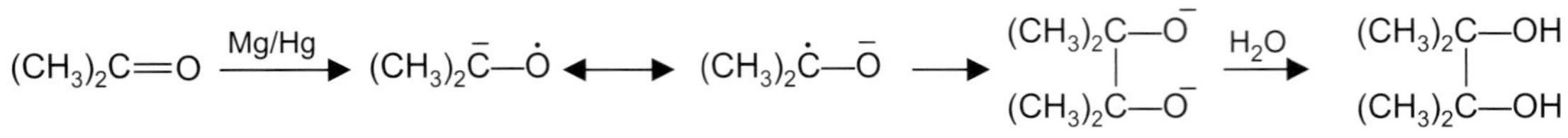

SCHEME 14.39

14.4 REDUCTIVE REMOVAL OF FUNCTIONAL GROUPS AND REDUCTIVE FISSION—HYDROGENOLYSIS

Role of a variety of reagents for this purpose under different experimental conditions are known. Here a few typical examples are discussed.

(A) Zinc as the Metal

Compared to alkali metals, metallic zinc is a milder reducing agent which can selectively remove many oxygen functional groups from α-substituted ketones. The reaction is thought to be a two electron reduction followed by the loss of the substituent as an anion. This reaction is thought to be a concerted process, since the isolated functional groups do not react. The reaction is shown with a cyclohexanone derivative (Scheme 14.40) which proceeds with zinc and acetic acid.

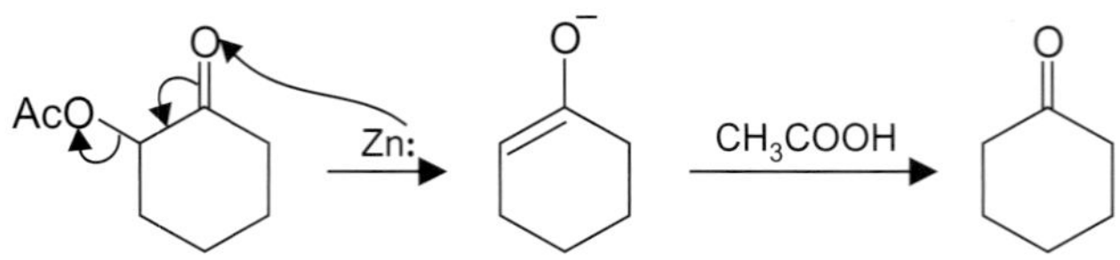

SCHEME 14.40

Zinc as electron-transfer agent reacts with 1, 2-dihalides in a different manner; both halogen atoms are eliminated and an olefinic bond is introduced (Scheme 14.41). Allene may be obtained in 80% yield by the treatment of 2, 3-dichloropropylene with zinc in aqueous ethanol.

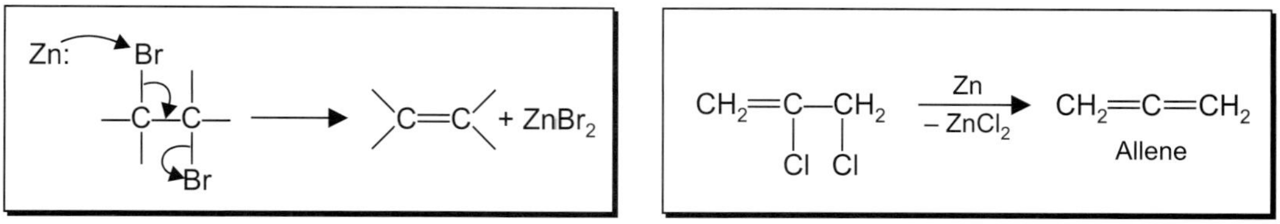

SCHEME 14.41

(B) Conversion of Alcohols into Their Parent Hydrocarbons

This can be achieved by the reductive elimination of the tosylate group from an alcohol tosylate with lithium aluminium hydride (Scheme 14.42) and is an S_N2 type of reaction.

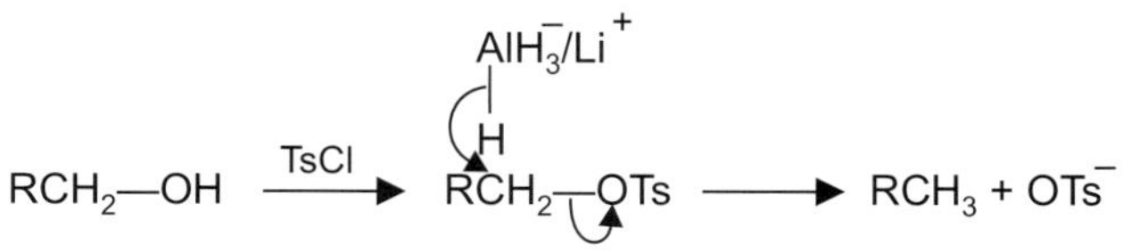

Alcohol Tosylate derivative

SCHEME 14.42

Lithium aluminium hydride and sodium borohydride also reduce primary and secondary alkyl halides to hydrocarbons (Scheme 14.42*a*) via a similar S_N2 process, however other functional groups may, also be reduced. Sodium cyanoborohydride is a weaker and more selective reagent (see, Scheme 14.85).

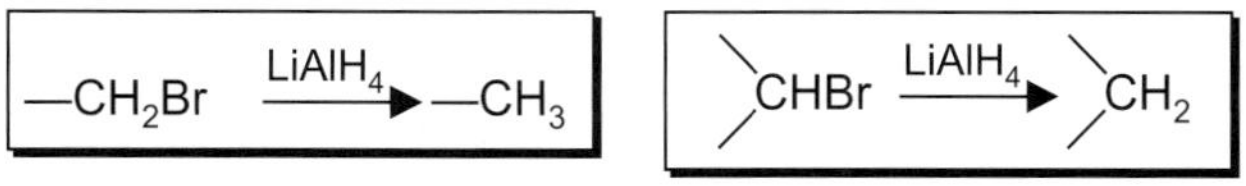

SCHEME 14.42a

(C) Conversion of Primary Amines to Their Parent Hydrocarbons

Primary amines can be reduced to their corresponding hydrocarbons on reacting their derived toluene-*p*-sulphonamide with a large excess of hydroxylamine-*O*-sulphonic acid in alkaline ethanol. Use is made of the strong leaving group potential of sulphates and sulphinates and the instability of the –N=NH grouping (Scheme 14.43).

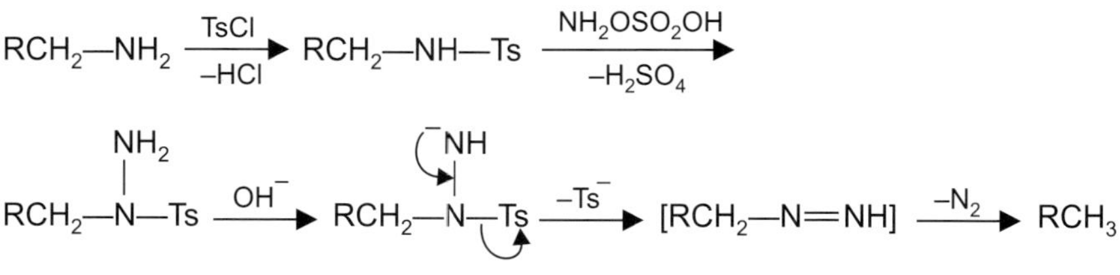

SCHEME 14.43

(D) Synthesis of Peptides—Amino and Carboxyl Group Protection and Removal of Protecting Groups via Hydrogenolysis

The following points may be considered:

- In order to obtain a correct sequence in peptide synthesis the *N*–terminal end of one of the amino acids (or peptides) is protected, the *C*–terminal end of the second amino acid (or peptide) is also protected.
- In case this is not done a random peptide bond formation takes place *e.g.*, a reaction of phenylalanine with glycine would result in the formation of four dipeptides as products (Scheme 14.43*a*).

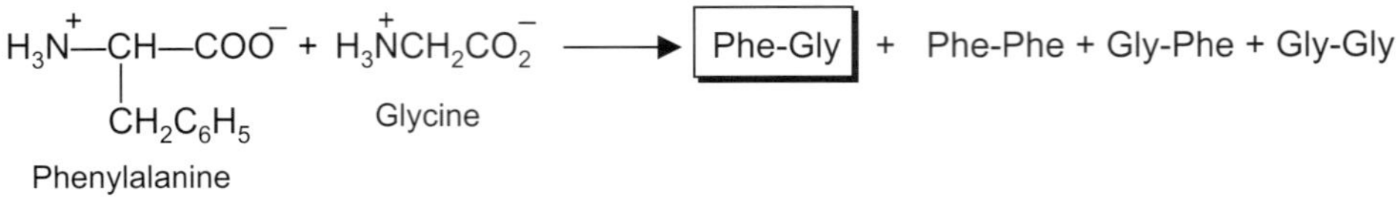

SCHEME 14.43a

- In order to form Phe-Gly, the *N*-protected phenylalanine has to be reacted with *C*–protected glycine followed by the removal of protecting groups (Scheme 14.43*b*). The

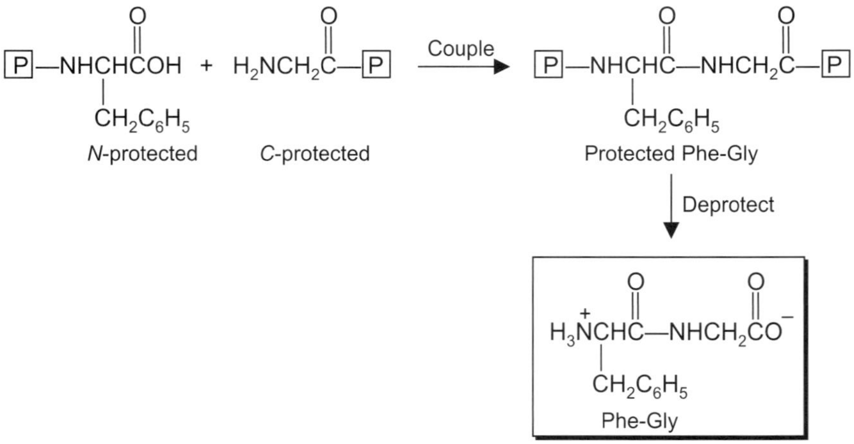

SCHEME 14.43b

amino group is protected by acylation with benzyloxycarbonyl group. This group is introduced on acylation of an amino acid with benzyloxycarbonyl chloride (Scheme 14.43*c*). The benzyloxycarbonyl protecting group is removed easily by hydrogenolysis in the presence of Pd or by reduction with Na/NH_3.

Recall that hydrogenolysis is the cleavage of a C—X bond with hydrogen over metal catalyst (See, Scheme 14.19). Hydrogenolysis represents one of the several reactions which hydrogen brings about over metal catalysts.

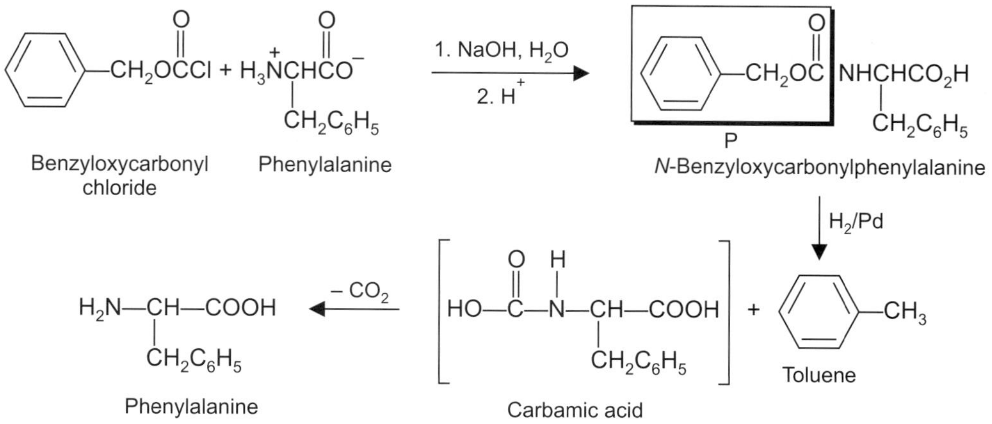

SCHEME 14.43c

- Carboxyl groups of amino acids are generally protected as esters and benzyl esters have a choice since these can also be removed by hydrogenolysis (Scheme 14.43*d*)

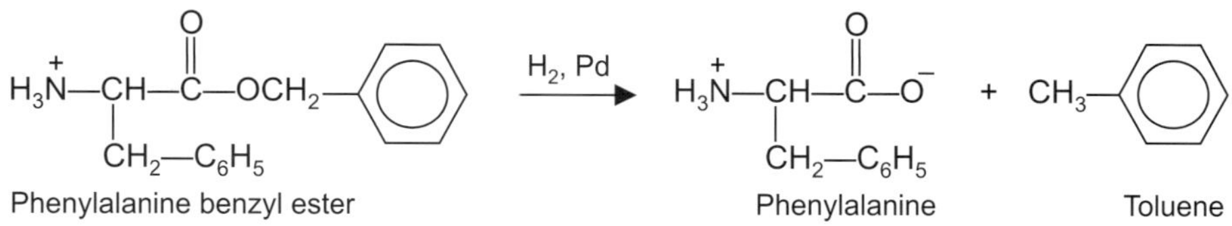

SCHEME 14.43d

Hydrogenolysis

When the benzylic group is attached to OH, OR, OCOR, NR_2, SR or a halogen, then it is sensitive to catalytic reduction, electron transfer agents and nucleophilic reducing agents. From the protected thiol groups the benzyl residue is removed by electron transfer reduction (Scheme 14.43e). Here hydrogenolysis with Pd is not carried out since sulphur will poison the catalyst.

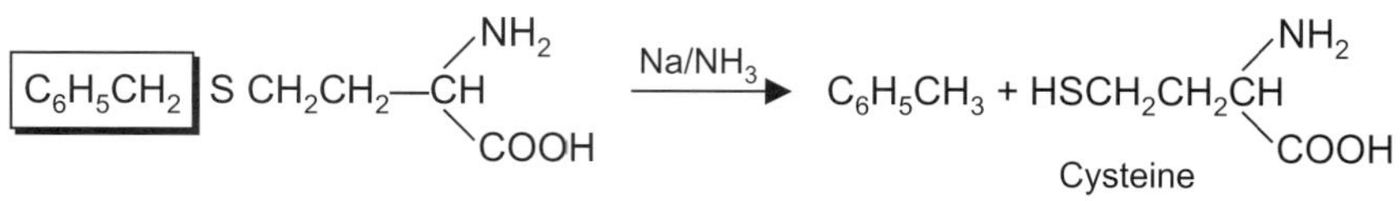

SCHEME 14.43e

14.5 REDUCTIVE DEOXYGENATION OF CARBONYL GROUPS

The reductive removal of carbonyl groups from organic molecules can be carried out by several methods and these lead to complete reduction to methylene groups or conversion to alkenes. The choice of a method depends on the sensitivity of the substrate under reducing conditions.

(A) Clemmensen Method

Amalgamated zinc and hydrochloric acid is a classical reagent combination for conversion of carbonyl groups to methylene groups. The mechanism is not known in detail and probably takes the course shown (Scheme 14.44). The mechanism may involve carbon-zinc bonds at the metal surface. Transfer of electrons from the metal surface to the carbonyl carbon atom has been suggested. The concentrated acid is probably needed to bring about the initial protonation; amalgamation of the zinc raises its hydrogen-overvoltage so that molecular hydrogen is not generated. Thus it survives as a reducing agent in the acid solution and is not used up in reaction with the acid to give molecular hydrogen. Only halogen acids are effective, probably because by complexing the initial $-Zn^+$ species, they provide a medium for the reduction of this species by a second atom of zinc.

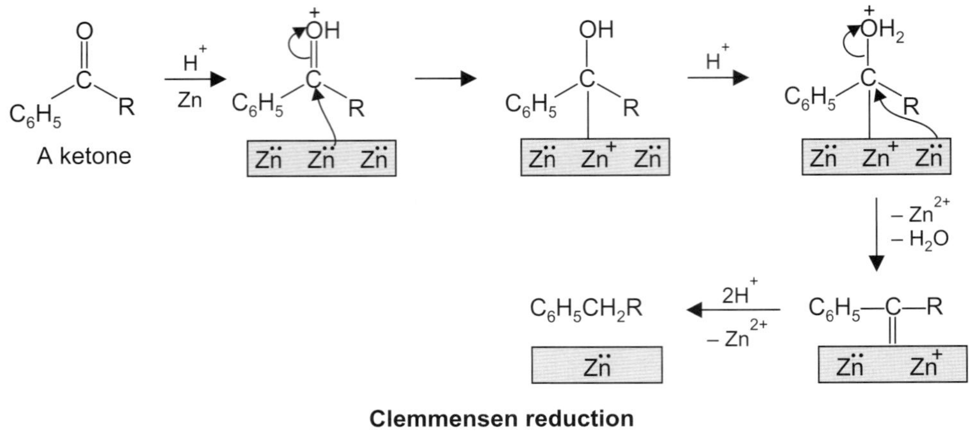

SCHEME 14.44

(B) The Wolff-Kishner Reduction

The reaction involves the reduction of carbonyl group to a methylene by base catalysed decomposition of the hydrazone of the carbonyl compound. Alkyldiimides are believed to be formed which collapse with loss of nitrogen (Scheme 14.45). The loss of the especially stable molecule of nitrogen provides the driving force for the reaction.

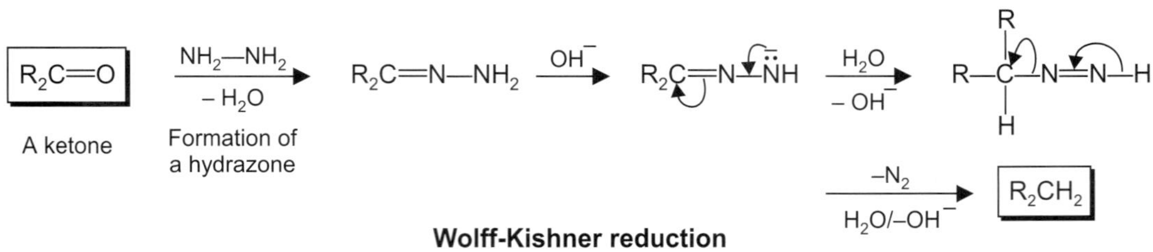

SCHEME 14.45

A modification employs potassium *t*-butoxide as the base and dimethyl sulphoxide as solvent. Alkoxide bases are very much more powerful in this solvent than in water or hydroxylic solvents and reduction occurs at room temperature.

(C) Tosylhydrazone Method

A carbonyl compound reacts with toluene-*p*-sulphonylhydrazine to give its corresponding tosylhydrazone (Scheme 14.46). The reduction of tosylhydrazones with sodium borohydride converts the carbonyl group to a methylene group. The mechanism (Scheme 14.47) probably involves the formation of a diimide as in the case of Wolff-Kishner reduction.

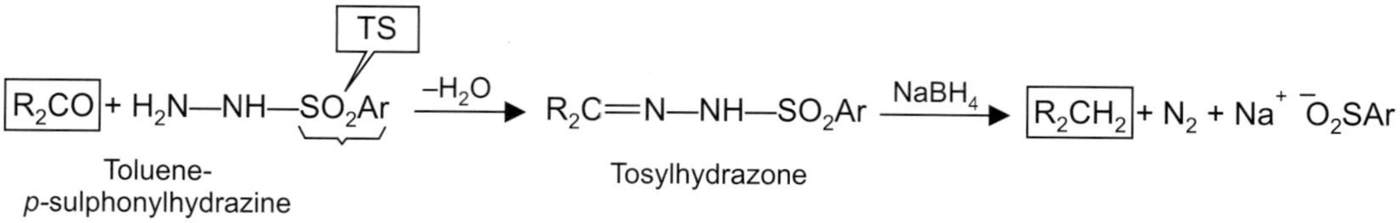

SCHEME 14.46

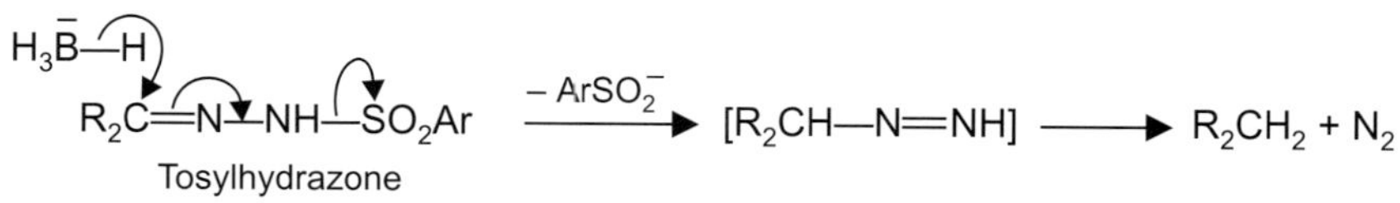

SCHEME 14.47

The conversion of ketone *p*-toluenesulphonyl hydrazones to alkenes occurs on treatment with a strong base *e.g.*, an alkyl lithium called Shapiro reaction (Scheme 14.48). This reaction proceeds via the anion of a vinyldiimide which subsequently decomposes to a vinyl lithium reagent. This intermediate on contact with a proton source gives the alkene (Scheme 14.48).While dealing with unsymmetrical acyclic ketones, one has to consider both regiochemistry and stereochemical aspects. Thus, 2-octanone gives exclusively 1-octene (Scheme 14.49) and the observed regiospecificity is dictated by the stereochemistry of the C=N bond of the starting hydrazone. There is preference for the removal of a proton which is *syn* to the arenesulphonyl group since the arrangement allows chelation with the lithium ion (Scheme 14.49).

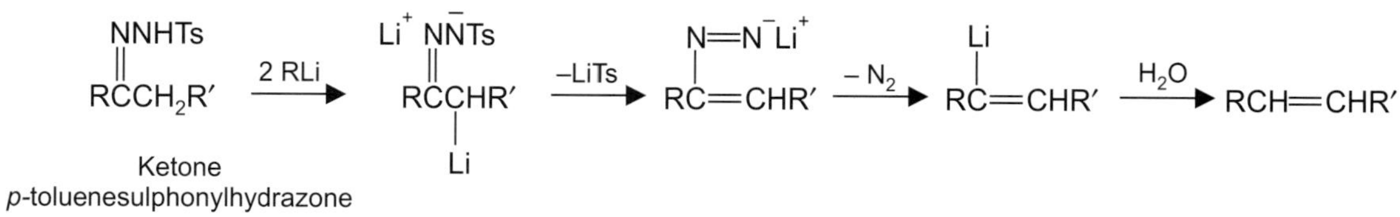

The Shapiro reaction

SCHEME 14.48

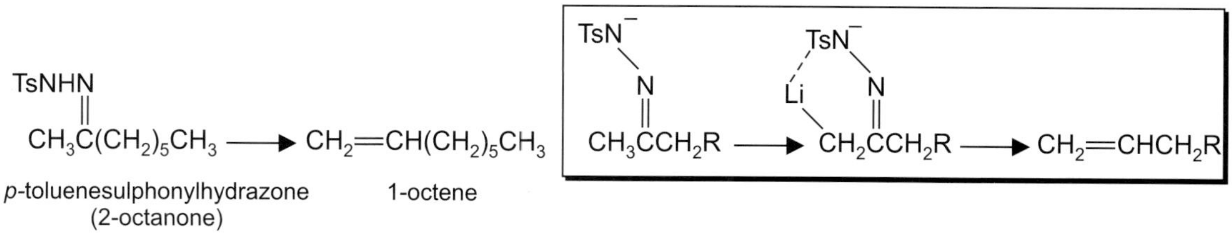

SCHEME 14.49

(D) Desulphurization of Thioketals (Mozingo Reaction)

A carbonyl group of a ketone can be converted into a methylene group by desulphurization of its thioketal. The carbonyl compound an aldehyde or ketone is reacted with ethylene dithiol in the presence of a Lewis acid to its thioacetal or ketal. Reaction with excess Raney nickel causes hydrogenolysis of both C—S bonds (Scheme 14.50). Freshly prepared Raney nickel has enough hydrogen to reduce the thioacetal or thioketal without added hydrogen.

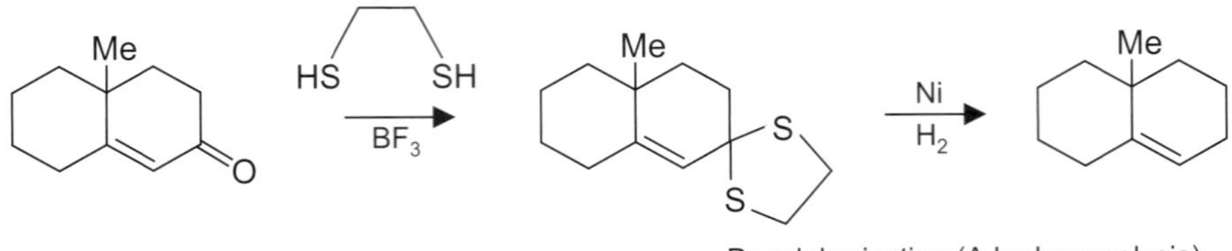

Desulphurization (A hydrogenolysis)

SCHEME 14.50

Recall that C—O bonds are strong while C—S bonds are very weak and one can remove a keto group of an aldehyde or ketone *via* its conversion into a thioacetal or thioketal which is hydrogenolized over Raney nickel. This reaction is sometimes called Mozingo reaction.

Reduction of aldehydes and ketones to hydrocarbons

The Mozingo reaction offers a useful alternative for the reduction of carbonyl compounds which are sensitive to acids and bases under Clemmensen and Wolff-Kishner reduction conditions respectively.

Acetals and ketals are used as protecting groups

One can reduce an ester in the presence of a ketone when both are present in the same compound by first protecting the ketone group to a function which would be stable to the reducing agent (Scheme 14.50a). The ketone group is protected as a ketal and after the reduction of the ester the protecting group is removed to regenerate the ketone.

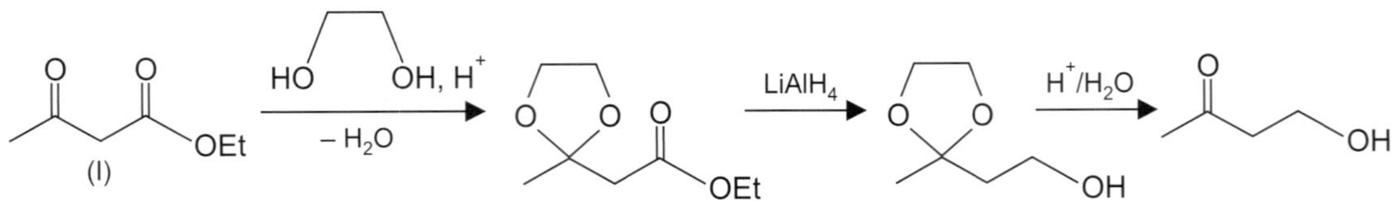

SCHEME 14.50a

14.6 REDUCTION BY HYDRIDE TRANSFER REAGENTS

(A) Lithium Aluminium Hydride and Sodium Borohydride—Reduction of Aldehydes and Ketones to Alcohols

Carbonyl componds are reduced with reagents which transfer a hydride from boron or aluminium. A variety of reagents of this type are available to achieve a considerable degree of chemoselectivity and stereochemical control. Sodium borohydride and lithium aluminium hydride are the most widely used of these reagents. Sodium borohydride is a mild reducing agent which reacts rapidly with aldehydes and ketones but only slowly with esters. Lithium aluminium hydride is a more powerful hydride-donor reagent (Scheme 14.51). It reduces esters, acids, nitriles, and amides, as well as aldehydes and ketones. Neither sodium borohydride nor lithium aluminium hydride react with isolated carbon-carbon double bonds.

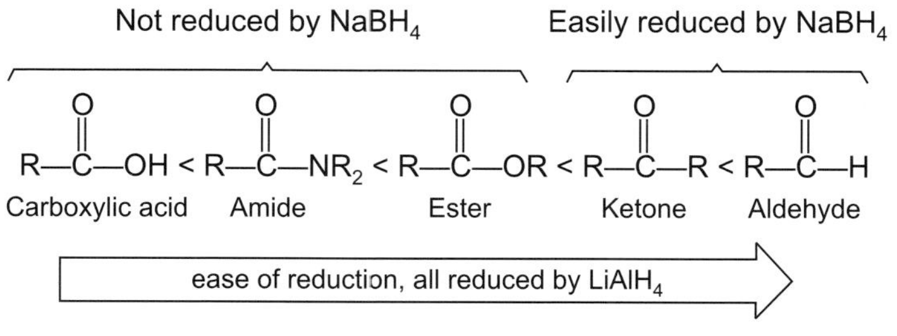

SCHEME 14.51

Sodium borohydride is easy to handle than lithium aluminium hydride and is also far more selective. Thus $NaBH_4$ will reduce the nitroketone only at the carbonyl group while $LiAlH_4$ will reduce at the nitro group as well (Scheme 14.52). $NaBH_4$ dissolves in water while $LiAlH_4$ reacts with water violently and catches fire. Reductions with $LiAlH_4$ are thus carried out in anhydrous solvents (ether or THF). This is due to the fact that lithium is a stronger Lewis acid compared with sodium and also AlH_4^- is far more reactive hydride donor than BH_4^-.

SCHEME 14.52

The most commonly used reagent for the reduction of aldehydes and ketones is sodium borohydride ($Na^+ \ ^-BH_4$). Since a BH_4^- ion contains four hydrides, it is capable of reducing four molecules of aldehyde or ketone (Scheme 14.53). The mechanism with both $NaBH_4$ and $LiAlH_4$ involves the activation of the carbonyl group with the metal cation and a subsequent hydride transfer (Scheme 14.53). As all the four hydrides are transferred, several distinct reducing agents may be involved during reduction with these reagents. For example during reduction with $NaBH_4$ in ethanol an alkoxyborohydride anion $EtOBH_3^-$ formed after the first hydride transfer, may reduce three more molecules of carbonyl compound and transfer all of its three

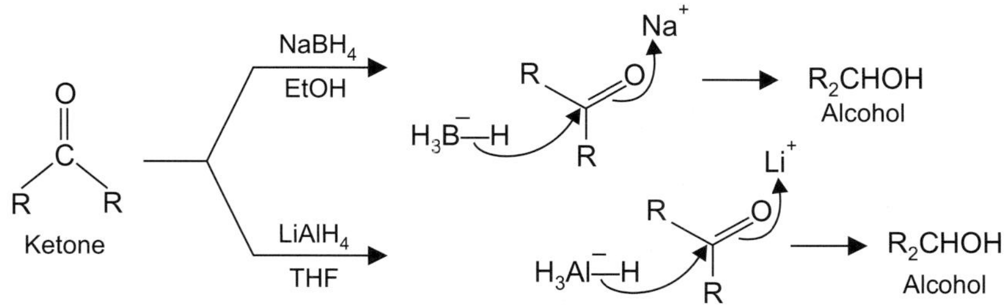

SCHEME 14.53

hydrogen atoms stepwise. The mechanism with $LiAlH_4$ is same (see, Scheme 14.53). As $LiAlH_4$ catches fire when in contact with water, the reduction with it are carried out in aprotic solvents *e.g.*, THF. At the end of reaction the product is isolated by the hydrolysis of the aluminium alkoxide. The reduction of the four molecules of a carbonyl compound with $LiAlH_4$ is depicted (Scheme 14.54).

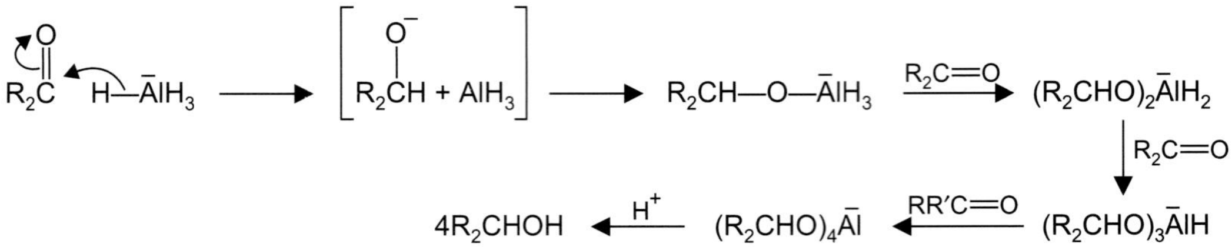

SCHEME 14.54

(B) Role of Lithium Aluminium Hydride Reduction of Esters to Alcohols

Often $LiAlH_4$ is the reagent of choice for the reduction of esters to alcohols (Scheme 14.55). This reduction involves the collapse of tetrahedral intermediate (A, Scheme 14.55) after the first hydride transfer which involves an elimination step. The aldehyde gets reduced to an alcohol by a second reduction.

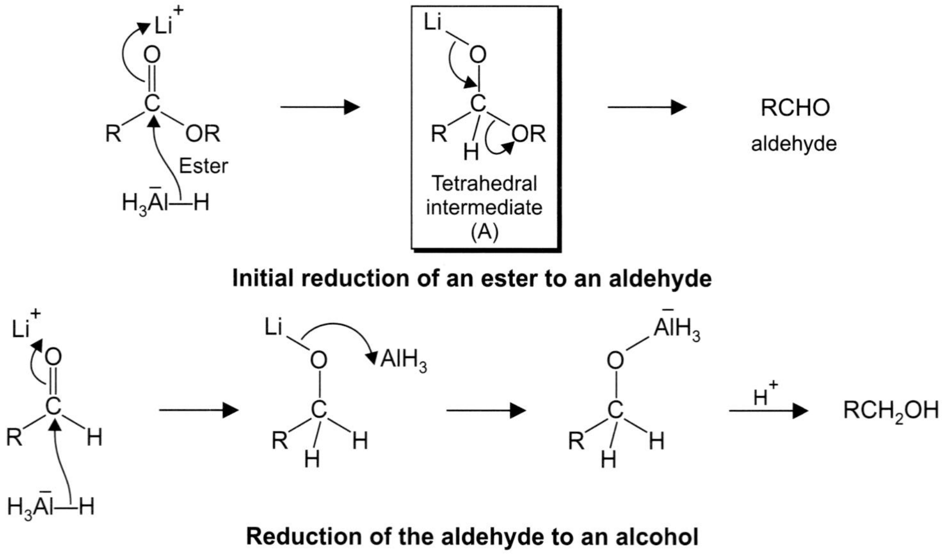

SCHEME 14.55

(C) Role of Lithium Borohydride—Reduction of Esters to Alcohols

A milder alternative to lithium aluminium hydride is lithium borohydride which reduces esters and not acids or amides (Scheme 14.56). Sodium borohydride reduces esters but very slowly if at all. Thus the chemoselectivity of hydride donors is reflected in the nature of metal cation and the ligands present. $LiBH_4$ has enhanced activity over sodium borohydride which is due to greater Lewis acid strength of Li^+ over Na^+. $LiBH_4$ also reduces lactones in a facile manner.

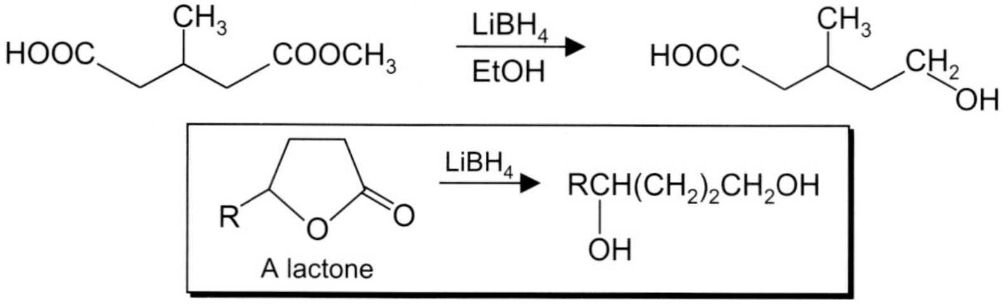

SCHEME 14.56

(D) Lithium Aluminium Hydride Reduces Amides to Amines

The mechanism of reduction of amides to amines with $LiAlH_4$ is similar to that of reduction of esters to alcohols. However, the initially formed tetrahedral complex collapses to give an iminium ion and does not eliminate NR_2^- since nitrogen is a poorer leaving group than oxygen (Scheme 14.57).

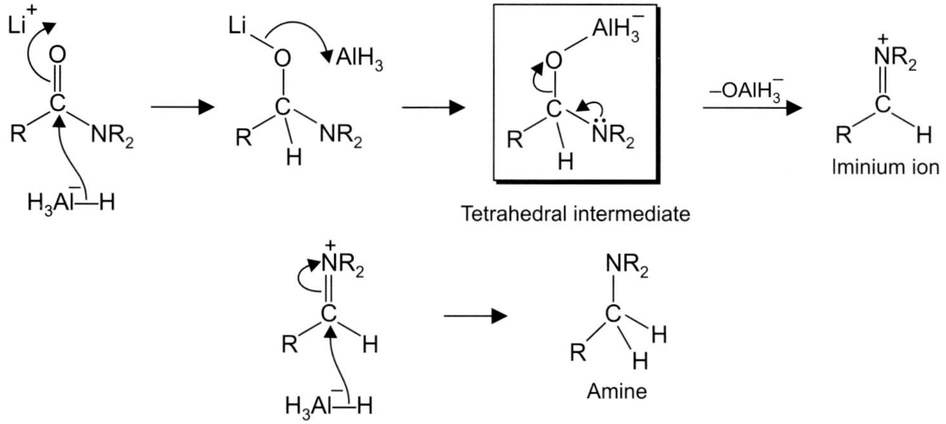

Reduction of amides to amines with $LiAlH_4$

SCHEME 14.57

Diborane Reduces Amides to Amines

Neutral borane BH_3 can also transfer its hydrogen as a hydride. Borane, however, is an electrophilic species which has a strong tendency to accept an electron pair in its vacant p orbital, BH_3 is therefore, expected to reduce electron rich carbonyl groups. Sodium borohydride ($NaBH_4$) is, however, nucleophilic and reacts by addition of a hydride ion to the more positive end of the polarized bond. Thus although $NaBH_4$ quickly reduces an acyl halide to a primary alcohol, borane does not react with an acyl halides, the carbonyl group of an acyl chloride is electron poor due to the electron—withdrawing effect of halogen (Scheme 14.58). Similarly, the reaction of borane with esters is also sluggish.

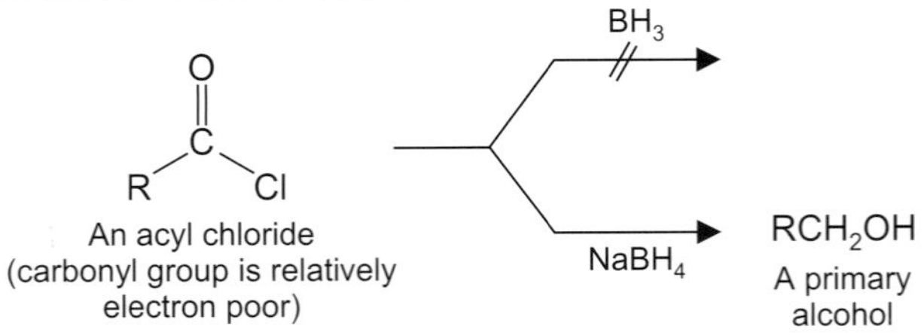

SCHEME 14.58

Reduction of carbonyl groups with borane occurs via addition of electron deficient borane to the oxygen atom followed by irreversible transfer of a hydride ion from boron to carbon (Scheme 14.59).

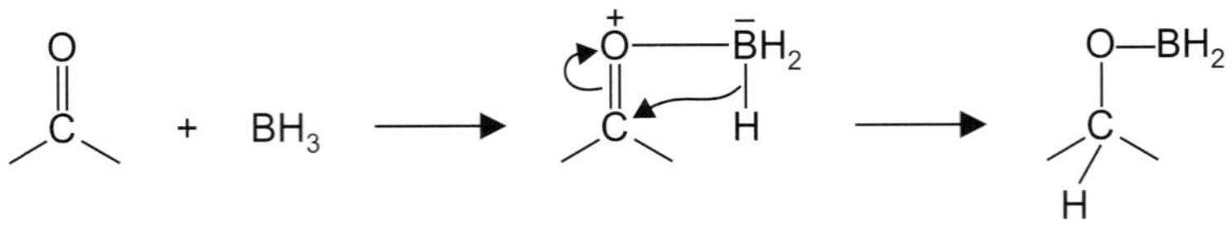

SCHEME 14.59

Amides require vigorous conditions for reduction with $LiAlH_4$ to amines so that chemoselectivity may be lost. Diborane offers a useful alternative for this reduction and involves the electrophilicity of borane (Scheme 14.60). The process involves a complex formation between carbonyl group (Lewis base) and the empty p orbital of borane (Lewis acid). A transfer of hydride from anionic boron to electrophilic carbon gives a tetrahedral intermediate which collapses to an iminium ion. The iminium ion is again reduced by BH_3.

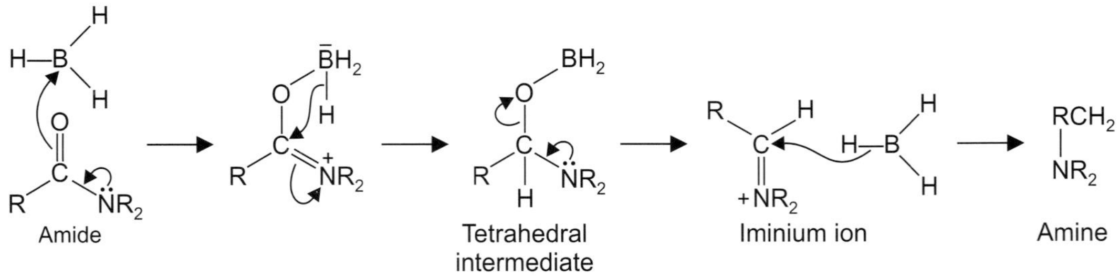

SCHEME 14.60

Borane—a Reagent of Choice for the Reduction of Acids to Alcohols

Borane is a choice reagent for the reduction of carboxylic acids to primary alcohols. Moreover, it is highly stereoselective and reduces carboxylic acids to primary alcohols without affecting other unsaturated groups e.g., nitro and cyano groups and even other reducible groups like esters are left unchanged. The reduction of a carboxylic acid proceeds through the formation of the triacyloxyborane intermediate by the protonolysis of the B—H bonds (Scheme 14.61) and the carbonyl groups are subsequently reduced by further reaction with borane.

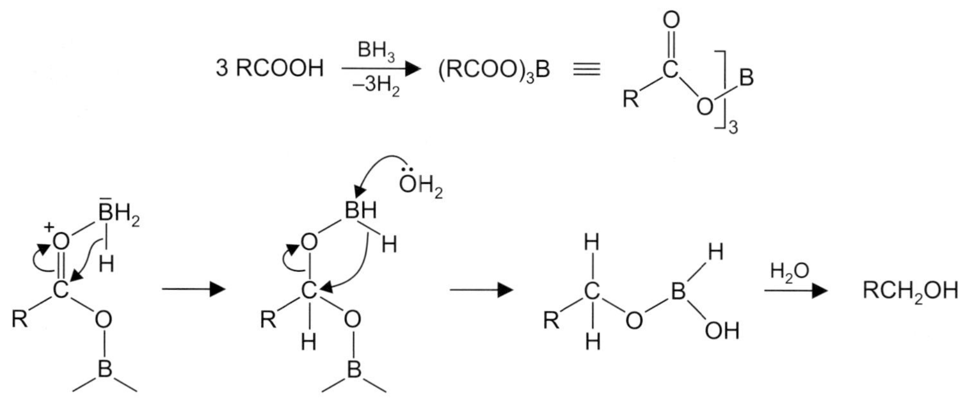

SCHEME 14.61

EXERCISE 14.3

Comment on the reaction outcome (Scheme 14.62)

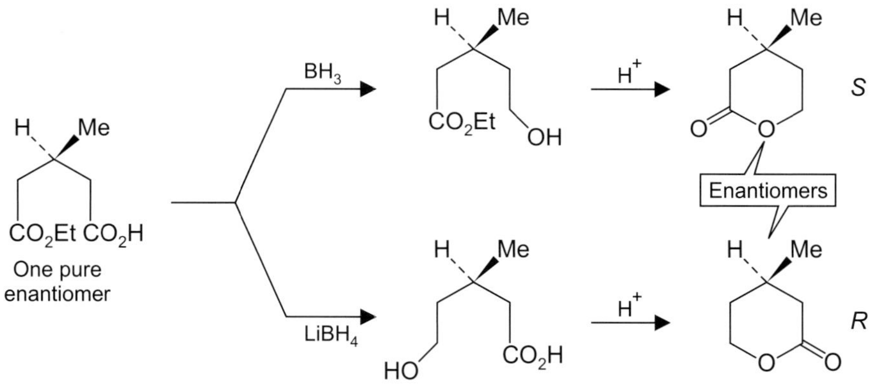

SCHEME 14.62

ANSWER. *Borane and lithium borohydride have opposite relative reactivities. One can either reduce the ester group or carboxylic acid by reaction with either of the reagents. The same chiral substrate can therefore, be converted into either R or S lactone.*

(E) Further Uses of Lithium Aluminium Hydride and Sodium Borohydride

The following points may be noted:

- **Reduction of carboxylic acid with $LiAlH_4$**

  Recall that carboxylic acid derivatives are converted into alcohols or amines by the addition of hydride ion to the carbonyl group during reduction with $LiAlH_4$. The tetrahedral intermediate thus formed then collapses to give the product after the elimination of the better leaving group (see, Schemes 14.55 and 14.57).

  Carboxylic acids are reduced with $LiAlH_4$ to the corresponding alcohols with a different mechanism than involved in the reduction of carboxylic acid derivatives since a leaving group is not present (Scheme 14.63). The aluminium atom rather than the lithium ion bonds with the second oxygen atom. This gives a good leaving group $H_2Al—O^-$ which is the conjugate base of $H_2Al—OH$ which is a strong acid. (Recall that alone AlH_3 like BH_3 can transfer a hydride ion and is also an electrophilic species due to a vacant *p* orbital and is a Lewis acid).

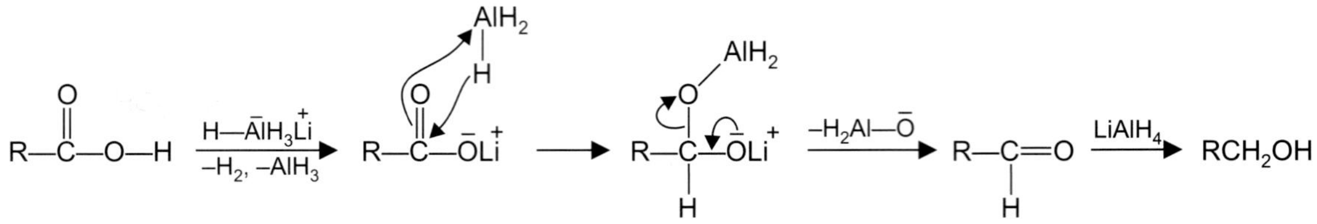

SCHEME 14.63

- **Reduction of nitriles with $LiAlH_4$**

Nitriles on reduction with $LiAlH_4$ give a primary amine *via* the formation of the imine salt (Scheme 14.64). When the reduction is carried out with an hindered hydride the reaction stops at the imine stage and hydrolysis gives an aldehyde.

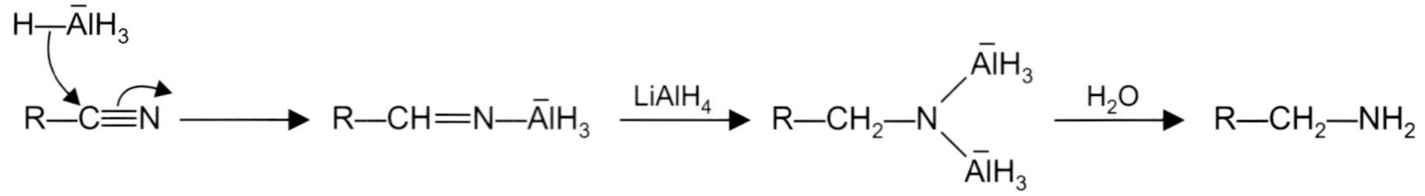

SCHEME 14.64

- **Reduction of α, β-unsaturated aldehydes and ketones**

Both $NaBH_4$ and $LiAlH_4$ react preferentially with the carbonyl group of an α, β-unsaturated aldehyde or ketone, however the reduction of both π bonds is common (Scheme 14.65). When 1, 2-addition is desired, the reduction is carried out in the presence of a lewis acid cerium chloride. The cerium ion seems to activate the carbonyl group (I, Scheme 14.65) toward nucleophilic attack and directs hydride reduction by complexing with the carbonyl group.

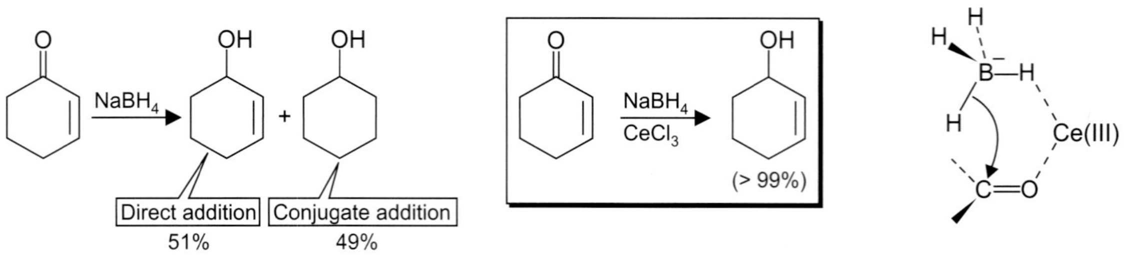

SCHEME 14.65

Reduction of Alkene Portion of an α, β-Unsaturated Compound

Recall that C=C is weaker than C=O. Catalytic hydrogenation reduces the C=C double bonds and leaves the carbonyl group of an α,β-unsaturated ketone unaffected (Scheme 14.66).

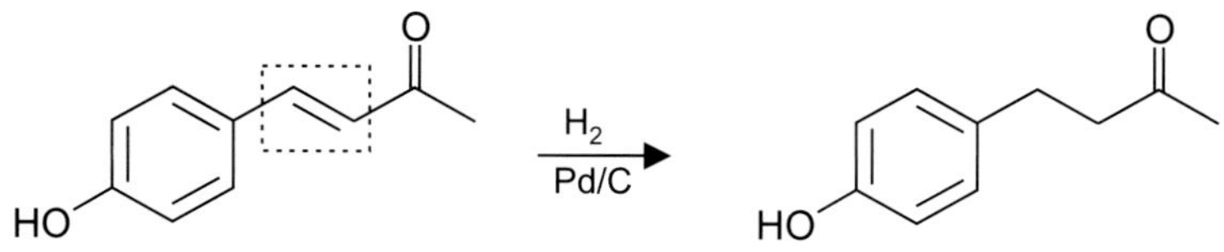

SCHEME 14.66

14.7 STEREOSELECTIVITY OF REDUCTION WITH SMALL HYDRIDE DONORS

In rigid cyclic ring ketones, the stereochemistry of hydride reductions appears to be determined normally by the relative importance of competing influences:

1. Stability of the final product.
2. Steric hindrance to the approach of the reagent (steric approach control).

The two faces of the carbonyl group of camphor (*i.e.*, *exo* and *endo*) have different accessibility to nucleophilic reagents. The approach of the reagent to the bottom (*endo*) face is hindered due to the U-shaped cavity of the molecule, but the top (*exo*) face is strongly hindered by the overhanging C-7 methyl group. Thus reduction with $LiAlH_4$ (or with bulky hydrides and Grignards reagents) occurs exclusively by approach of the AlH_4^- from the *endo* side (Scheme 14.67). In the case of norcamphor, however, the *exo* side now becomes free of hindrance but the hindrance to approach from the *endo* side still remains. Thus, hydride reduction occurs by attack of the AlH_4^- now from the *exo* side (Scheme 14.68).

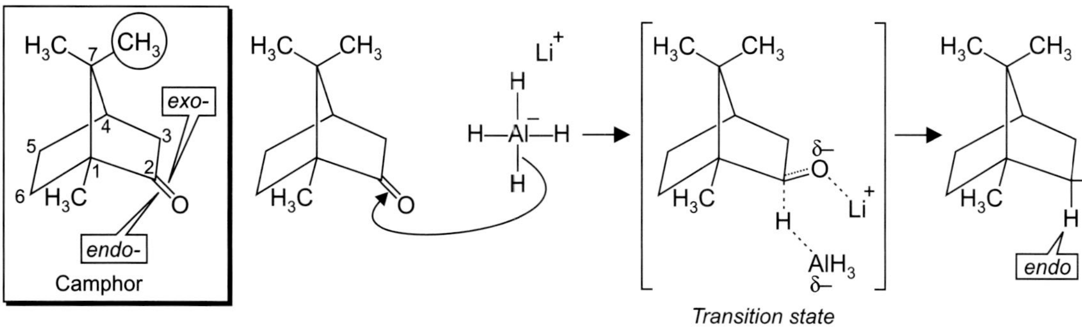

SCHEME 14.67

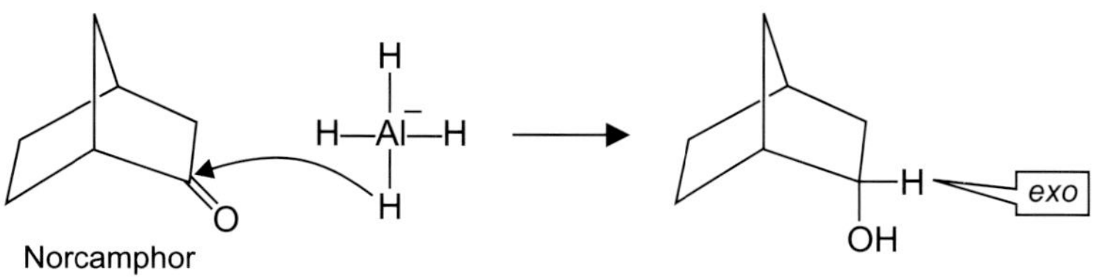

SCHEME 14.68

The stereochemical outcome of the reduction of cyclohexanes *e.g.*, with lithium aluminium hydride is difficult to predict. With comparatively unhindered ketones, the more stable equatorial alcohol generally predominates. According to Felkin's model, in cyclohexanones the outcome of reduction (and addition of other nucleophiles) is a balance of both steric and torsional strain. Formation of an equatorial alcohol (*e.g.*, by an axial attack of hydride) requires a staggered transition state (I, Scheme 14.69) which could suffer from steric strain between the nucleophile and the β-axial substituents (substituents at C-3). Thus with less bulky nucleophile *e.g.*, AlH_4^- and with no axial substituent at C-3 *i.e.*, with only a tiny hydrogen at C-3, the energy of I is lower and an equatorial alcohol predominates. In case the nucleophile is sterically demanding (see Scheme 14.72) and/or a bulky axial substituent is present at C-3 (Scheme 14.70) the energy of transition state (I, Scheme 14.69) increases and then an axial alcohol formation predominates (Scheme 14.70). Formation of an axial alcohol *i.e.*, attack by nucleophile *e.g.*, a hydride at equatorial position requires a partially eclipsed transition state (II, Scheme 14.69) with torsional strain between the nucleophile and the axial α-hydrogen substituents.

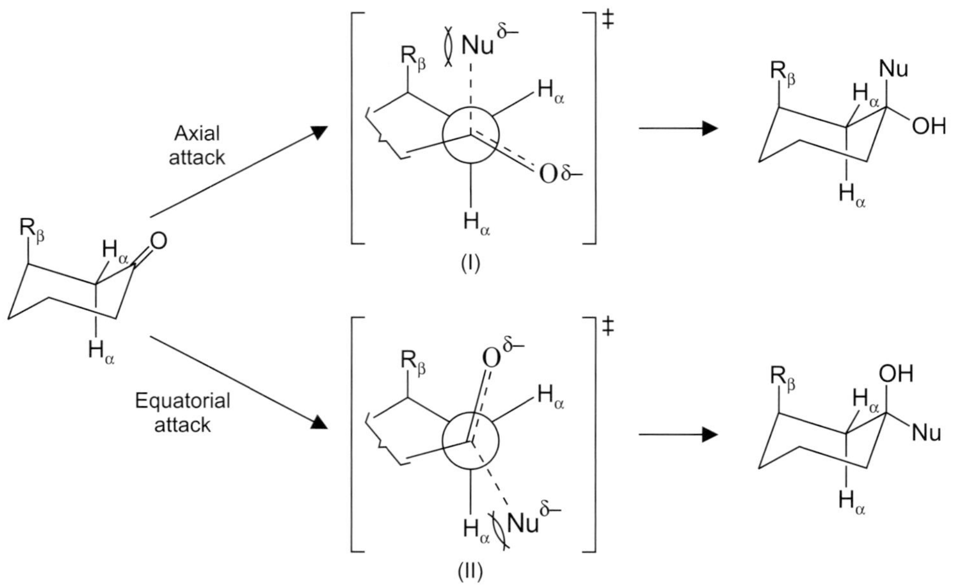

SCHEME 14.69

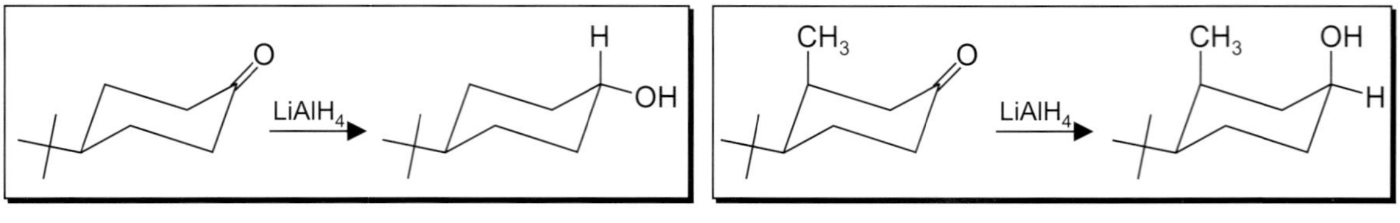

SCHEME 14.70

14.8 STEREOSELECTIVITY OF REDUCTION WITH HINDERED HYDRIDE DONORS-SELECTRIDE

The stereochemistry of reduction of ketones by bulky hydride donors have been studied. The steric approach control becomes significant when the hydride reagent becomes more highly substituted and thus hindered. The alkyl substituted borohydrides have particularly high selectivity for the least hindered direction of approach.

The reagent lithium tri-*sec*-butylborohydride trade name as 'Selectride (Scheme 14.71) is the reagent of choice for the reduction of cyclic and bicyclic ketones in a highly stereoselective manner, giving the less stable isomer. This is in contrast to the usual reagents *e.g.*, lithium aluminium hydride and sodium borohydride. Thus 4-*t*-butylcyclohexanone (Scheme 14.61) which gives equatorial alcohol on reduction with $NaBH_4$ (80%) or $LiAlH_4$ (92%) as the major product, with lithium tri-*sec*-butyl-borohydride selectride give the axial alcohol as the major product. In summary the introduction of bulky alkyl groups into the borohydride anion dramatically alters the direction of attack on a cyclic ketone, and cyclohexanones are attacked predominantly from the equatorial side to afford the axial alcohol.

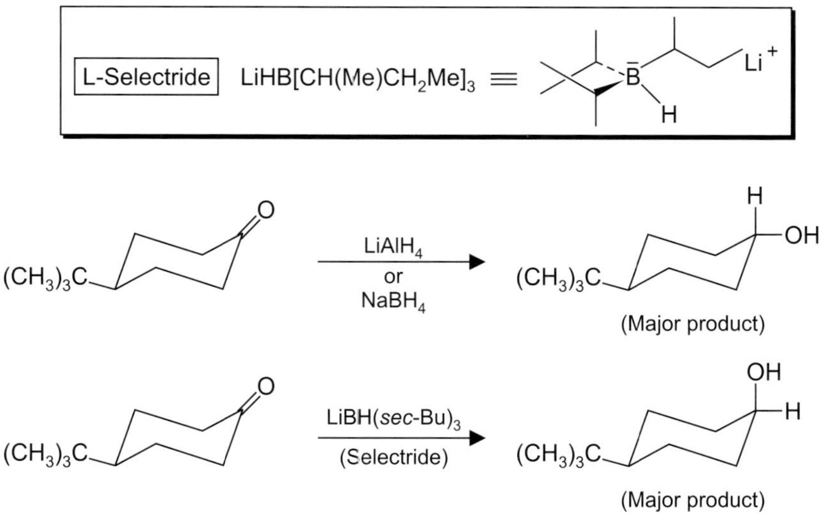

SCHEME 14.71

Consider the delivery of the hydride ion *via* the axial approach *i.e.*, the top face of the cyclohexanone derivative. This axial approach of the bulky reagent is severely restricted by interaction between the large groups —$CH(CH_3)CH_2CH_3$ on the boron and the vertical axial H atoms on the chair conformation (Scheme 14.72).

Thus the ketone (Scheme 14.72) during reduction with a bulky reducing agent is under kinetic control and the selectride is facially selective. The major product being formed by the delivery of the hydride *via* equatorial face which involves a stable and less crowded transition state as compared to the one involved *via* axial approach by the bulky reducing agent.

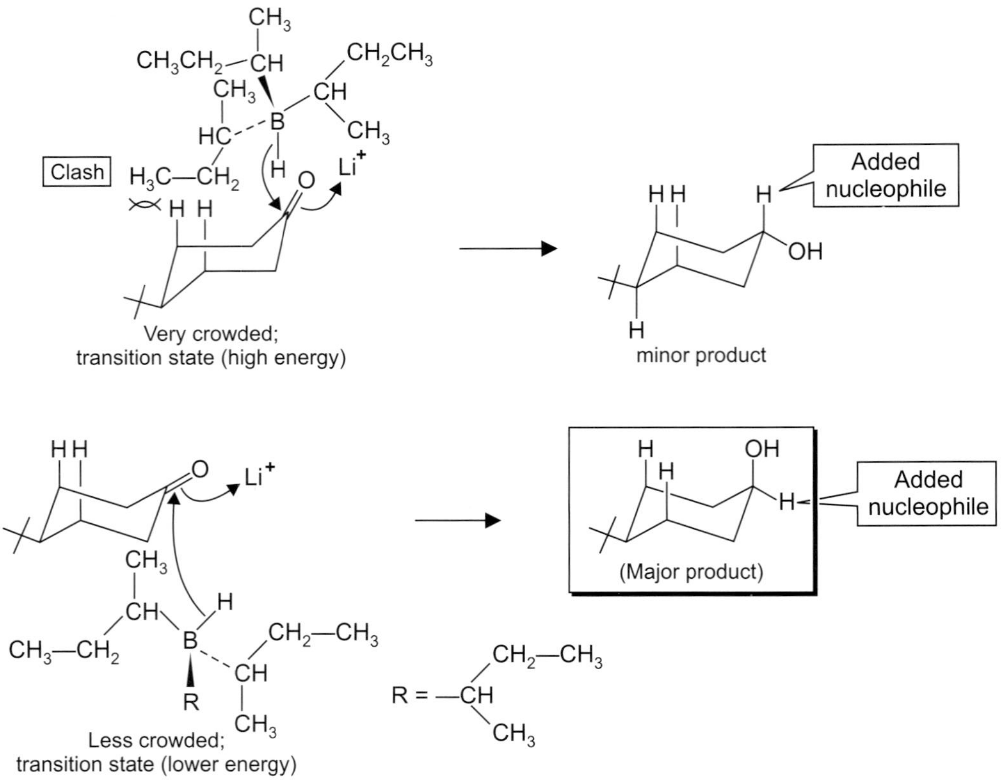

SCHEME 14.72

Another hindered reagent is diisobutylaluminium hydride –DIBAL (or DIBAL—H) which is used as the hydride donor at low temperature. Using DIBAL, the reduction of an ester can be stopped after one equivalent of hydride ion has been added (Scheme 14.73). This however, is not the case with $LiAlH_4$ and the aldehyde formed in the first reduction is reduced further to an alcohol (see, Scheme 14.55) DIBAL is also a good reagent for reducing nitriles to aldehydes (Scheme 14.74).

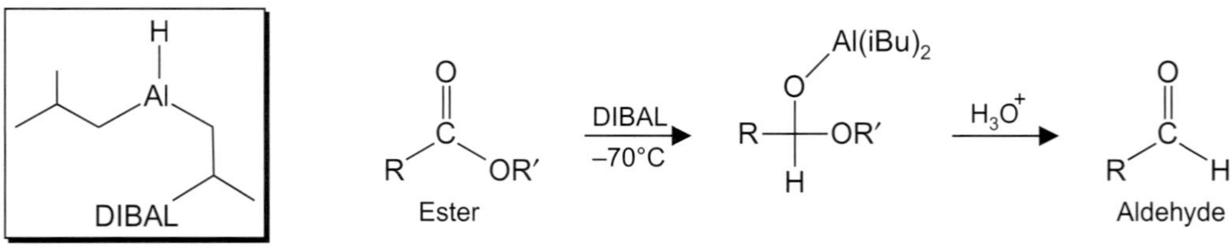

SCHEME 14.73

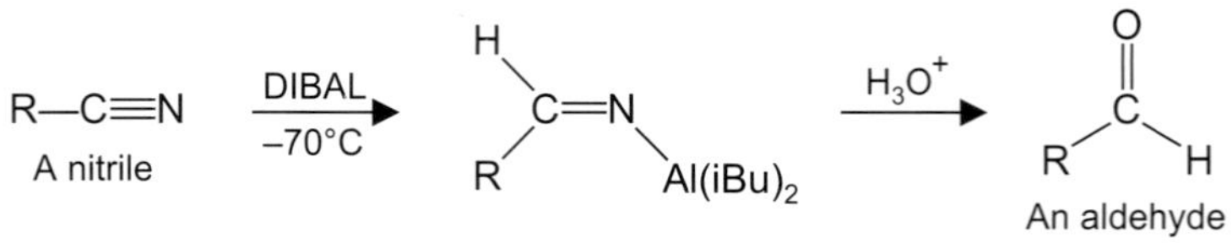

SCHEME 14.74

Use of DIBAL

Diisobutylaluminium hydride DIBAL is a reagent of choice for the preparation of aldehydes (see, Schemes 14.73 and 14.74) by working at low temperatures. At room/ordinary temperatures ketones, esters (Scheme 14.74a) and epoxides are reduced to alchols while nitriles give amines. At ordinary temperatures this reagent brings about the conversion of α, β-unsaturated compounds to allylic alcohols (Scheme 14.74b).

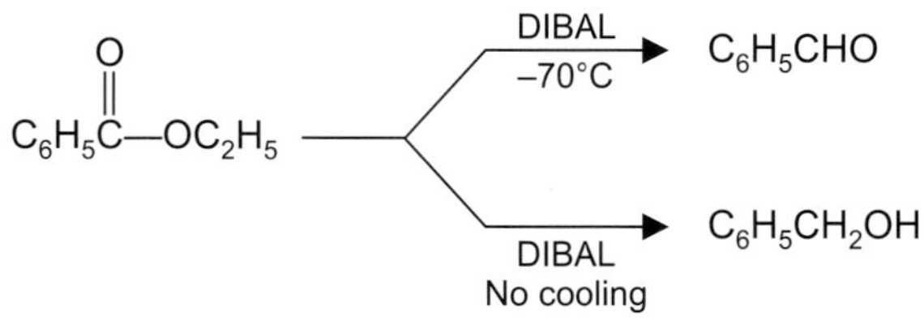

SCHEME 14.74a

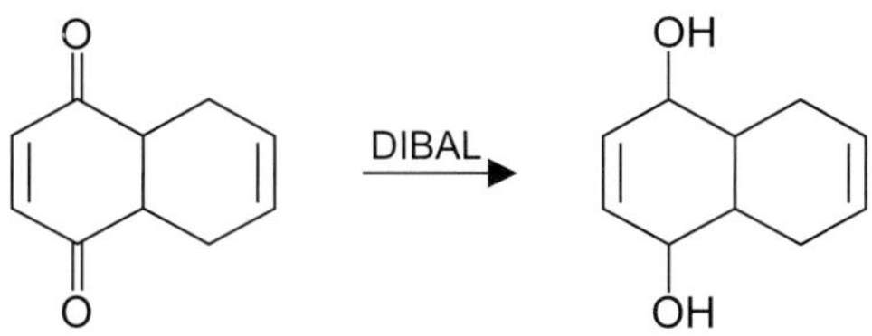

SCHEME 14.74b

When some of the hydrogens of $LiAlH_4$ are replaced by OR groups, the original reactivity of the metal hydride is significantly reduced. One such reagent is lithium tri-*tert*-butoxyaluminium hydride which reduces an acyl halide only up to the aldehyde stage whereas with $LiAlH_4$ the product is an alcohol (Scheme 14.75).

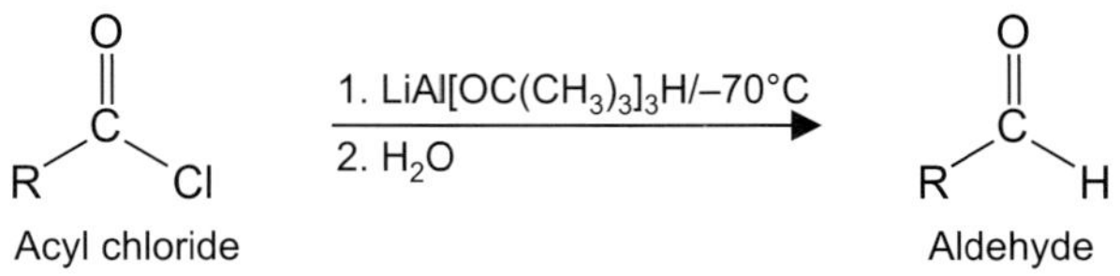

SCHEME 14.75

14.9 CHIRAL BORANES—ENANTIOSELECTIVE REDUCTION OF CARBONYL COMPOUNDS

A chiral, optically active reducing agent, *e.g.*, Ipc. BBN (alpine-borane, Scheme 14.76 brings about asymmetric reduction of unsymmetrical (prochiral ketones R—CO—R′ R ≠ R′) to optically active secondary alcohols with a very high degree of enantioselectivity even when the ketone itself is achiral. The chiral reagent is prepared from 9-BBN and α-pinene. It is suggested that hydride is transferred by the adduct (Ipc. BBN) by adopting a boat shaped six-membered cyclic transition state (Scheme 14.76).

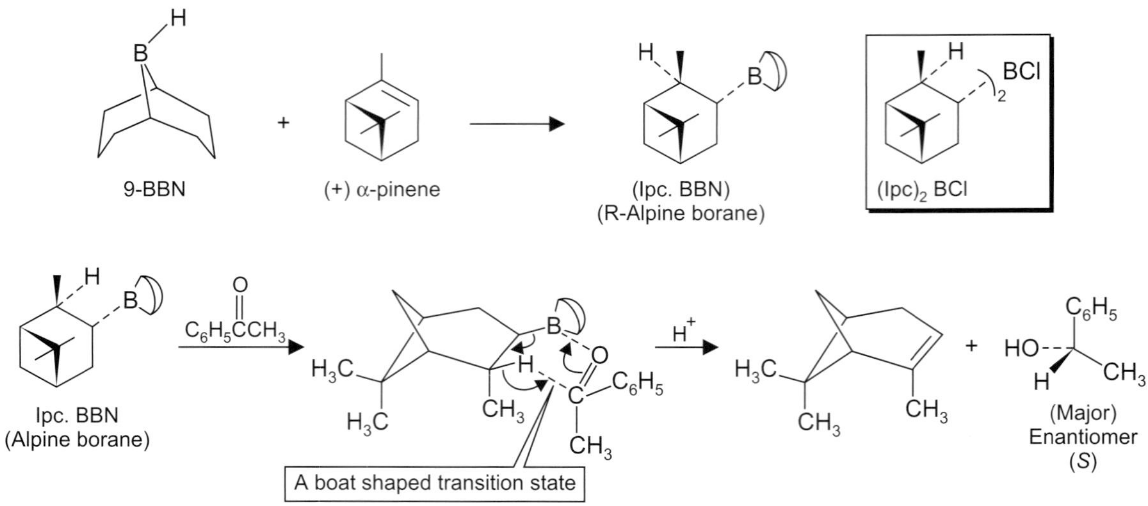

SCHEME 14.76

The larger group on the ketone (phenyl group in this case) directs away from the terpenoid moiety so that steric hindrance is kept at minimum thus this arrangement leads to the major enantiomer. The other alternative arrangement (Scheme 14.77) would give the minor enantiomer of the product.

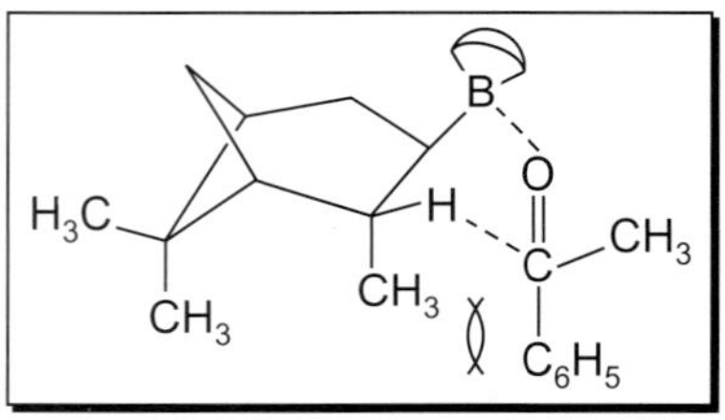

SCHEME 14.77

Both the enantiomers of α-pinene can be made to react with 9-BBN and thus the *S* enantiomer of the secondary alcohol (Scheme 14.76) can also be synthesized.

Diisopinocampheylchloroborane $(Ipc)_2$ BCl (Scheme 14.76) is another reagent of choice which brings about an enantioselective reduction of a wide variety of ketones.

14.10 MEERWEIN-PONNDORF REDUCTION—THE HYDRIDE TRANSFER REACTION

The reduction of carbonyl compounds to alcohols with aluminium isopropoxide has long been known and is called Meerwein-Ponndorf-Verley reduction. The reduction is done by heating the components together in solution in isopropanol. An equilibrium is set up and the product is obtained by using an excess of the reagent or by distilling off the acetone as it is formed. The reaction involves a transfer of hydride ion from the isopropoxide to the carbonyl compound through a six-membered cyclic transition state (Scheme 14.78).

Thus in summary Meerwein-Ponndorf-Verley reduction is carried out by aluminium isopropoxide when the carbonyl group of an aldehyde or a ketone gives a primary or a secondary alcohol respectively. In this reduction, the hydride ion is donated by carbon unlike a metal hydride. The reverse reaction, oxidation of a primary or a secondary alcohol to the carbonyl function in the presence of an aluminium alkoxide is the Oppenauer oxidation.

> *Instead of using a complex metal hydride as a source of hydride ion, the hydride ion may be donated by a carbon atom in Meerwein—Ponndorf Verley reduction.*

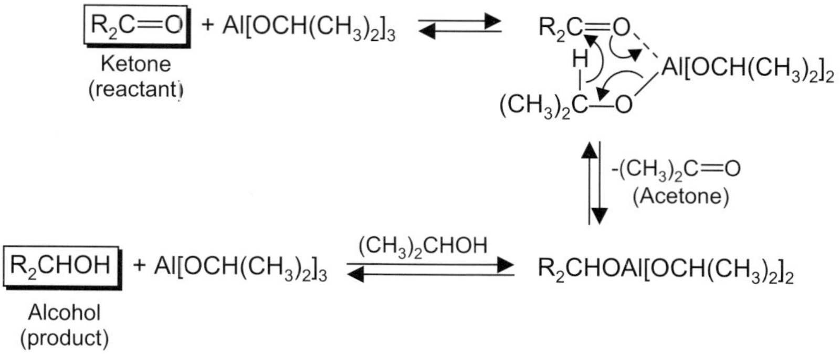

SCHEME 14.78

14.11 CANNIZZARO REACTION

Aldehydes without α-CH groups cannot undergo base-catalysed condensation, however, they react with bases by disproportionation involving the transfer of hydride ion (Scheme 14.79).

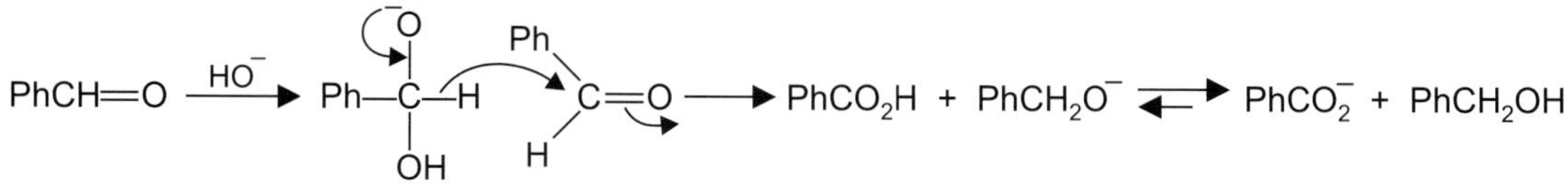

SCHEME 14.79

Crossed Cannizzaro reaction between one molecule of such aldehyde and formaldehyde leads to reduction of the former and the oxidation of the latter, since formaldehyde is more reactive than other aldehydes towards nucleophiles and rapidly gives a high concentration of the donor anion (Scheme 14.80). This fact has been used for reductions, thus benzaldehyde is reduced by formaldehyde in the presence of potash in refluxing methanol to give benzyl alcohol in high yield.

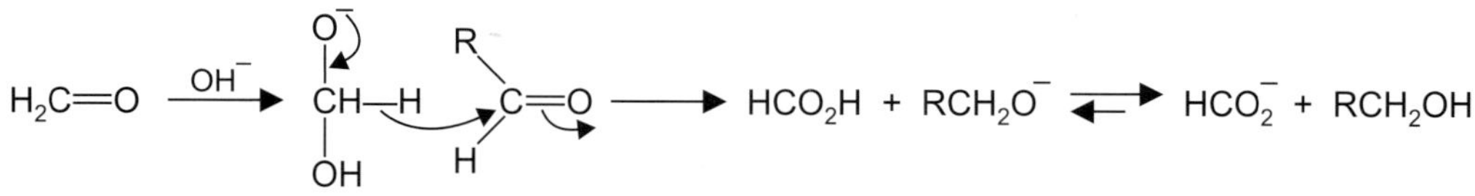

SCHEME 14.80

14.12 REDUCTION OF ALDEHYDES AND KETONES WITH AN ADJACENT STEREOCENTER (ASYMMETRIC INDUCTION)

The stereochemistry of reduction of acyclic aldehydes and ketones is a function of the substitution on the adjacent stereogenic carbon. The formation of the major product can be predicted on the basis of a conformational model of the transition state (Scheme 14.81, Crams rule). According to this model, that diastereomer will be formed as the major product which involves minimal steric interaction of the nucleophile with the groups L and M (S, M and L are the relative sizes of the substituents). Thus the reagent (*e.g.*, hydride ion $H:^-$ during reduction with lithium aluminium hydride I, Scheme 14.70) approaches the less hindered side of the carbonyl group when the rotational conformation of the molecule is the one in which the carbonyl group is flanked by the two least bulky groups on the adjacent stereocenter (Scheme 14.81).

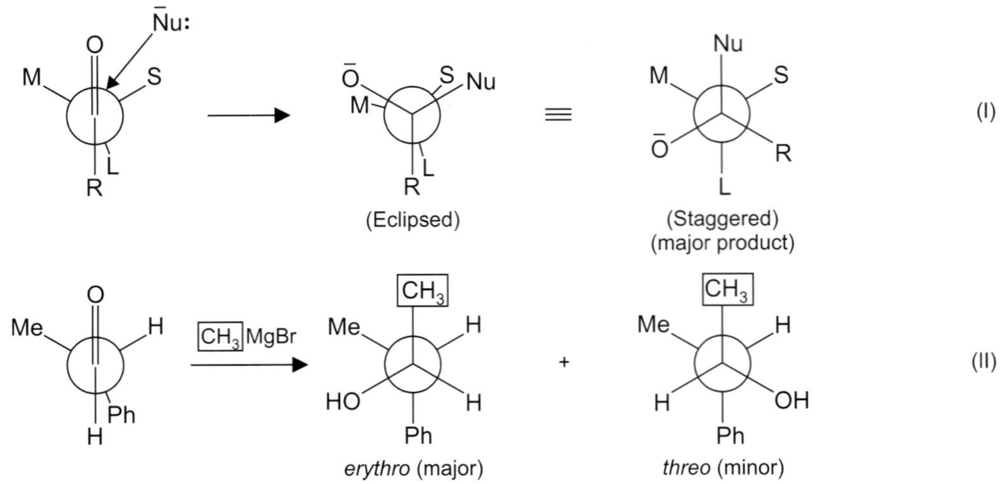

Application of Crams rule

SCHEME 14.81

Erythro and *Threo* Nomenclature

A compound with two stereocentres and with two substituents in common C_{abc} — C_{abd} are given shorthand nomenclature erythro and threo. When written in Sawhorse or Newman projection one can recognize these. In erythreo configuration it is possible to eclipse each of the like substituents and also the unlike substituents as in (Scheme 14.81). In a threo configuration one can only eclipse at a time either any two of the like substituents or the unlike substituents.

Crams rule can thus be used to determine the stereochemistry of the major product. The rule has been now rationalized in the form of Felkin–Anh model. The Newman projection (in the lowest energy arrangement) is the one with the C—L bond perpendicular to the carbonyl group (L, M and S represent the largest, medium-sized and smallest substituents on the stereogenic carbon). The nucleophile is delivered to the carbonyl carbon in a plane perpendicular to the C=O fragment from the side opposite the C—L bond so as to make an obtuse angle with C=O corresponding nearly to the tetrahedral angle of Nu—C—O in the product. Of the two possible arrangements (I, Scheme 14.81*a*) is of lower energy due to less steric interference (between the Nu and the smallest group). The major product is thus (II).

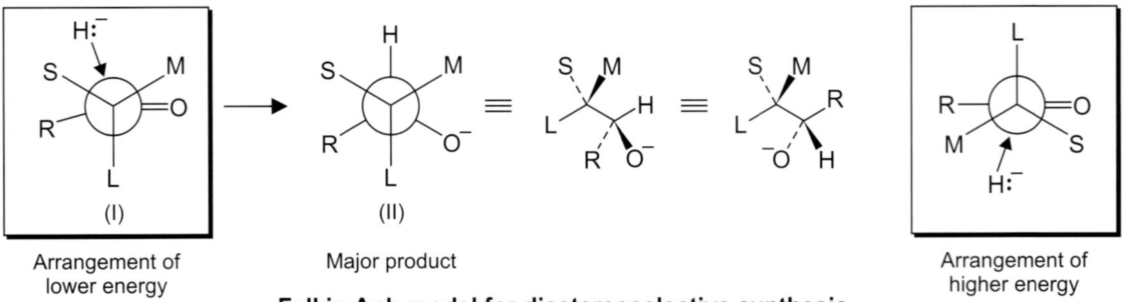

Felkin Anh model for diastereoselective synthesis

SCHEME 14.81a

The diastereoselectivity as predicted by Cram's rule (Felkin Anh modification) during the nucleophilic addition with Grignards reagents is further increased by the use of σ organotitanium compounds (see Scheme 7.6*a* the yield of predicted diastereomer is for greater) as shown (Scheme 14.81*a*).

Cram's rule may not be followed when the conformation of the carbonyl compound in the transition state does not depend on steric factors. In such examples one of the substituents on the stereocenter is an alkoxy, hydroxyl or other complexing group. Lithium cations *e.g.*, coordinate effectively with oxygen atoms and thus reduction of ketones of the type (I, Scheme 14.81*b*) with lithium aluminium hydride involves a rigid chelate system of type (II). Thus the reagent brings about the reduction by first locking the conformation by coordination with both the methoxy oxygen and the ketone oxygen. A high degree of stereoselectivity is due to the attack of the chelate from the less hindered side (*i.e.*, away from the phenyl group, in this case it is the *Re* face which offers the least resistance to the approach of the reagent) and this may not be in accordance with Cram's rule.

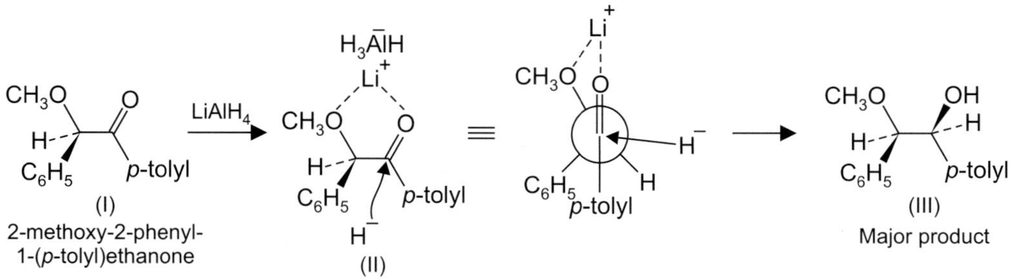

SCHEME 14.81b

The formation of one product stereoisomer over the other depends on the difference in size between S and M. In case S and M are very similar, their interactions with the incoming nucleophile (as well as R) would be small and the diastereoselectivity of such molecules will be poor. In cases when one of the substituents is strongly electronegative (*e.g.*, chlorine) the preferred transition state corresponds to conformation (Scheme 14.81*c*) due to the tendency of the negatively polarized oxygen and chlorine atoms to be as far apart as possible and in these cases Cram's rule may not be followed.

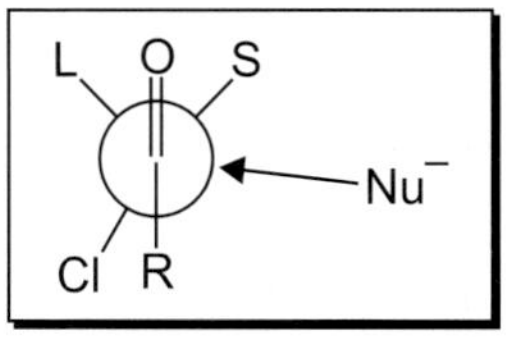

SCHEME 14.81c

14.13 REDUCTION OF EPOXIDES

A detailed study of reaction of epoxides is presented in chapter 13 (Schemes 13.27-13.28). Here some examples of reduction of epoxides with $LiAlH_4$ and borane are given.

Epoxides are converted to alcohols by $LiAlH_4$. Since epoxides can be easily prepared from olefins, the overall reaction provides a method to hydrate an olefin. The reaction proceeds by S_N2 substitution by hydride ion and a new C—H bond is formed. With unsymmetrical epoxides therefore, the hydride addition occurs at the less hindered carbon of the epoxide (eq. I, Scheme 14.82). Moreover, as expected of nucleophilic S_N2 attack of an epoxide by the hydride reagent, inversion of configuration occurs at the carbon atom which is attacked (eq. II, Scheme 14.82).

Moreover, substituted cyclohexene epoxides are preferentially reduced in such a direction as to give an axial alcohol (eq. III, Scheme 14.82), *i.e.*, the reaction has the *trans* stereochemistry characteristic of S_N2 reactions.

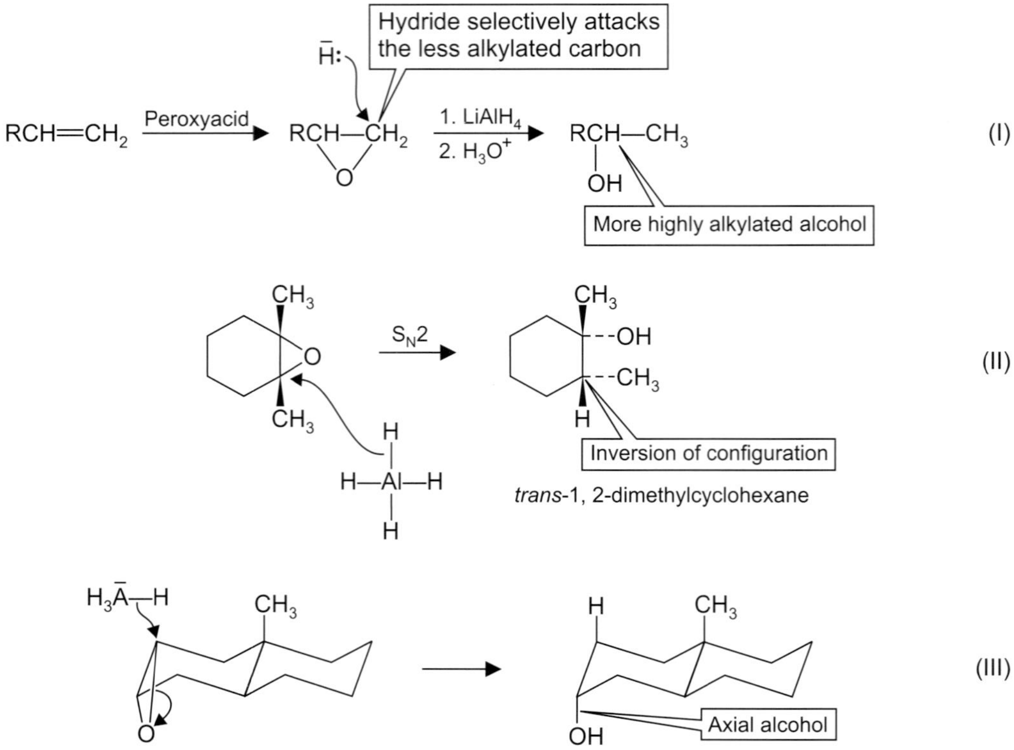

SCHEME 14.82

Epoxides on reduction with borane (in contrast with lithium aluminium hydride) give rise to less substituted alcohol and the reaction is thus complementary to the reduction with $LiAlH_4$ (Scheme 14.83). This reduction is catalysed by lithium borohydride.

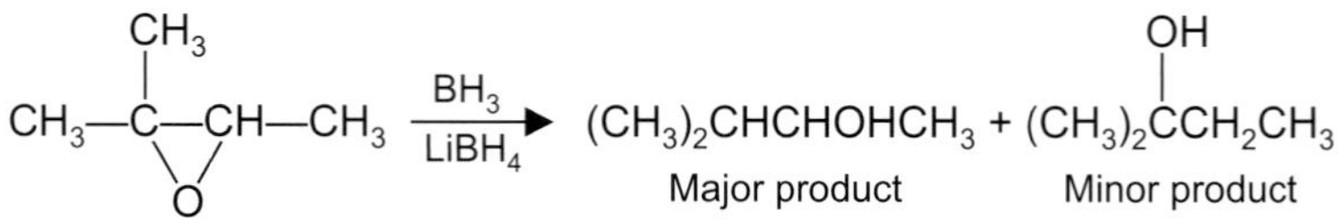

SCHEME 14.83

14.14 REDUCTIONS WITH ENZYMES—BAKERS YEAST

Reductions can be catalysed by enzymes *e.g.*, a prochiral ketone is reduced to optically active secondary alcohol. The enzyme is only the chiral catalyst which, however, does not provide the hydrogen atom for reduction. The hydrogen atoms, instead are provided by the *relevant coenzyme e.g.*, NADH. One may carry out the reduction using whole cells, like bakers yeast where both the enzyme and the coenzyme are provided by the organism. Thus ethyl acetoacetate (Scheme 14.84) is reduced selectively to ethyl (*S*)-(+)-3-hydroxybutanoate using one of the reducing enzymes found in Baker's yeast alcohol dehydrogenase. It is one of the enzymes which are involved in the metabolism of D-glucose to ethanol. In this reaction enantiomeric excesses ranging from 70–97% have been achieved particularly when oxygen is excluded during the fermentation (anaerobic). The best enantioselectivities are obtained provided the ketone has a β-ester group. However, a dependence on the size of the two groups has been discovered. Reduction of ethyl acetoacetate (Scheme 14.84) with Baker's yeast yielded the (*S*) alcohol but reduction of ethyl β-ketovalerate instead gave the (*R*) alcohol.

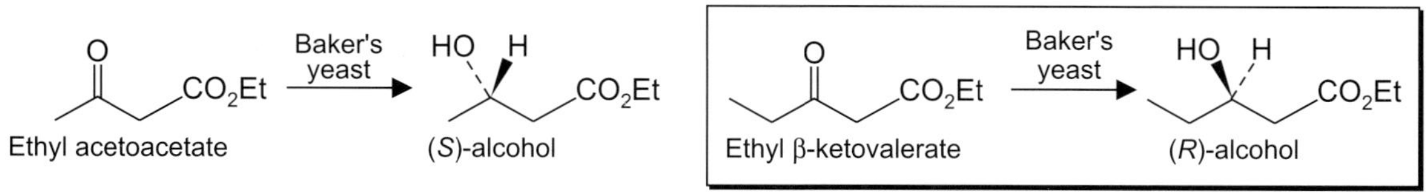

SCHEME 14.84

14.15 LESS REACTIVE MODIFIED BOROHYDRIDES—SODIUM CYANOBOROHYDRIDE AND SODIUM TRIACETOXYBOROHYDRIDE

Role of several modified and hindered hydride donors has been discussed. These are:

- Lithium *tri-sec*-butylborohydride (selectride)

$$\left(\begin{array}{c}CH_3CH_2CH—\\ |\\ CH_3\end{array}\right)_3 \bar{B}H\ Li^+ \text{ (see, Schemes 14.71, 14.72)}$$

- Diisobutylaluminium hydride (DIBAL) AlH $[CH_2CH(CH_3)_2]_2$ a derivative of alane ($AlCl_3$), (see Schemes 14.73, 14.74 and 14.74*b*)
- Lithium *tri-tert*-butoxyaluminium hydride $LiAlH[OC(CH_3)_3]_3$ (see Scheme 14.75). Sodium cyanoborohydride ($NaBH_3CN$) is yet another useful reagent which is weaker than sodium borohydride, but is more selective. It readily reduces iodides, bromides and tosylates to the hydrocarbons (just like $LiAlH_4$ see Scheme 14.42*a*) in neutral solution in hexamethylphosphoramide (HMPA) as the solvent even in the presence of carbonyl and other reducible groups like epoxides (Scheme 14.85).

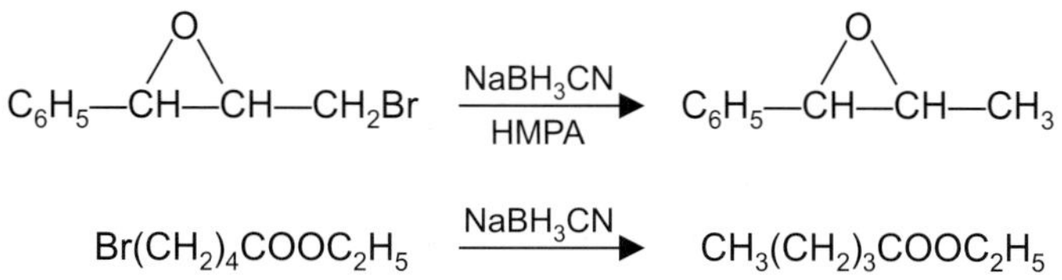

SCHEME 14.85

Due to the electron withdrawing effect of the cyano group, the reagent becomes weaker (less nucleophilic and thus more selective than $NaBH_4$. This reagent is stable at pH 3–4 and in this acid medium a carbonyl group is converted to its protonated form $C{=}\overset{+}{O}H$ which is readily reduced to the corresponding alcohol. Thus aldehydes and ketones are reduced in acidic media and remain unaffected around a neutral pH.

Aldehydes and ketones at pH = 6 are not disturbed by sodium cyanoborohydride, therefore, amines are made through reduction of an imine derivative (reductive amination Scheme 14.86).

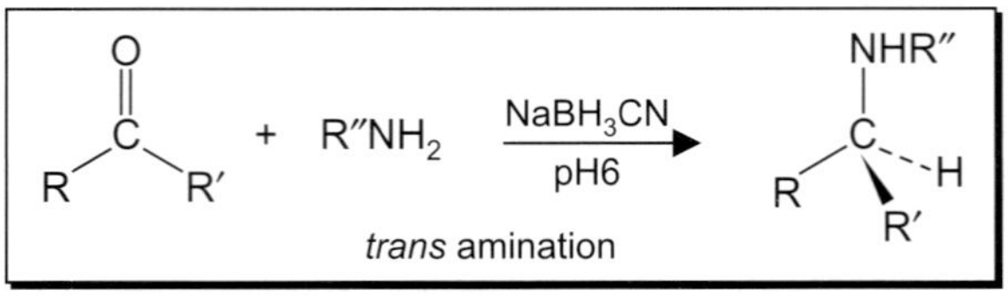

SCHEME 14.86

Recall that C=N double bond reacts with nucleophiles in a way similar to a C=O double bond. Moreover, an imine is a better base than its carbonyl precursor so it is readily protonated near pH=6 and the mechanism of transamination is presented (Scheme 14.87).

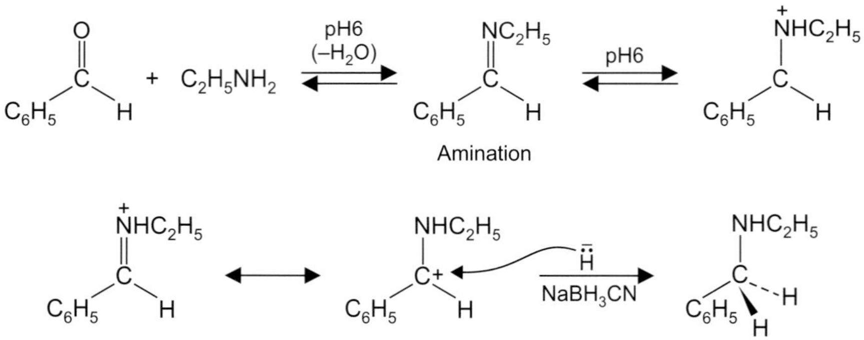

SCHEME 14.87

Sodium acetoxyborohydride $NaB(OAc)_3$ H is less reactive than sodium borohydride and can reduce selectively an aldehyde in the presence of a ketone (Scheme 14.88).

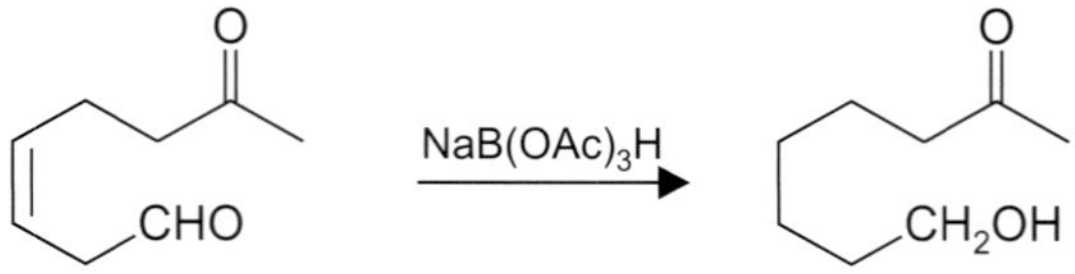

SCHEME 14.88

The use of $NaBH_4$ may reduce both the groups. $NaB(OAc)_3H$ is generated by dissolving $NaBH_4$ in acetic acid.

PROBLEMS

14.1 Predict the stereochemistry of the product which would be obtained on hydrogenation of α-pinene over platinum catalyst.

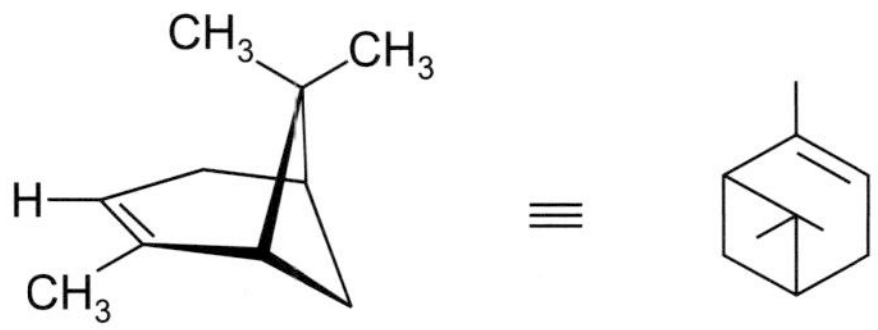

α-Pinene

14.2 When acetone is treated with magnesium in benzene, a strongly exothermic reaction occurs. This on aqueous work up gives pinacol. What is the mechanism of this reaction?

14.3 Write the structure of the product when the following compound is treated with (Na/Hg).

$$PhCH{=}CHCH_2\overset{+}{N}(CH_3)_3\overset{-}{I}$$

14.4 Sodium in liquid ammonia, containing ammonium acetate reduces amides to aldehydes. Write the mechanism of the reaction. What end product is expected in the presence of ethanol?

14.5 Write the outcome of the following reactions:

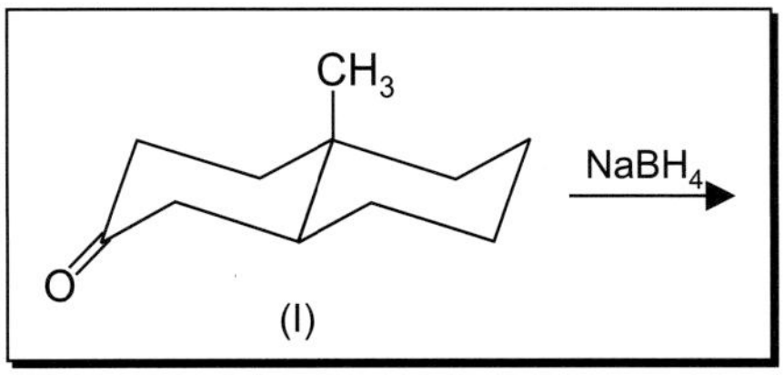

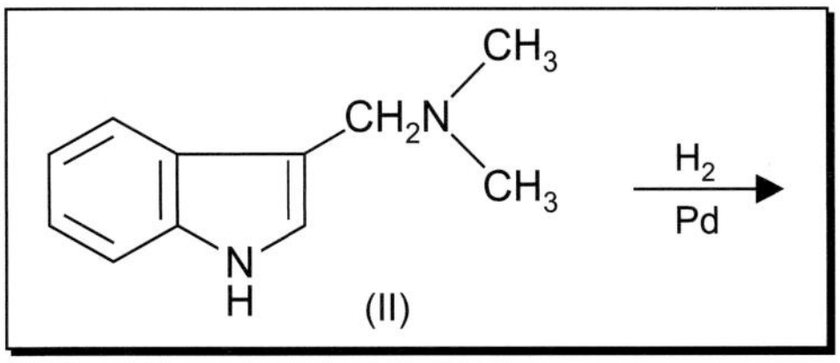

14.6 Write the structure of the major products from following reactions:

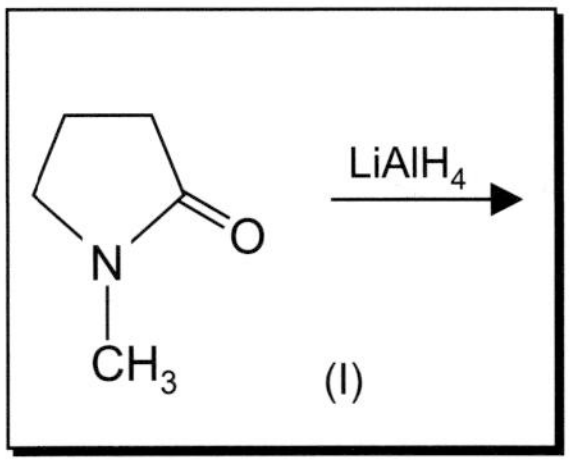

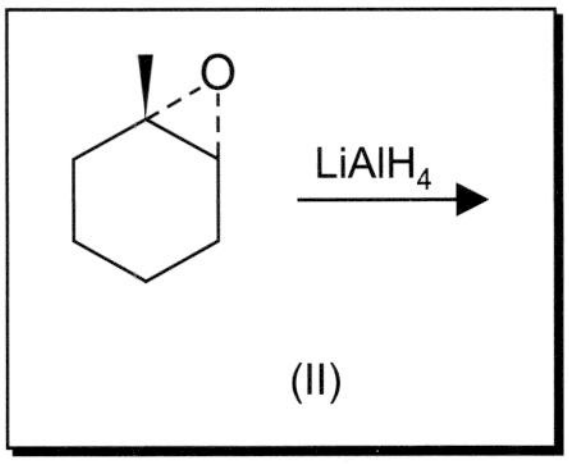

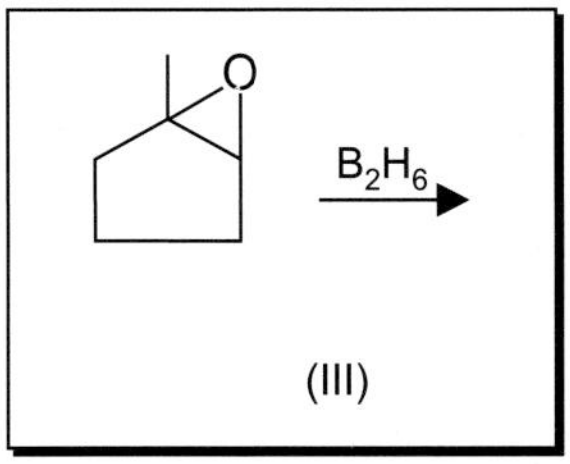

ANSWERS TO THE PROBLEMS

14.1 The major product will be the one formed by addition of hydrogen from the more accessible side of the double bond. Thus in the product the added hydrogen occupies the less hindered face *i.e.*, *trans* to the bridge with gem dimethyl group.

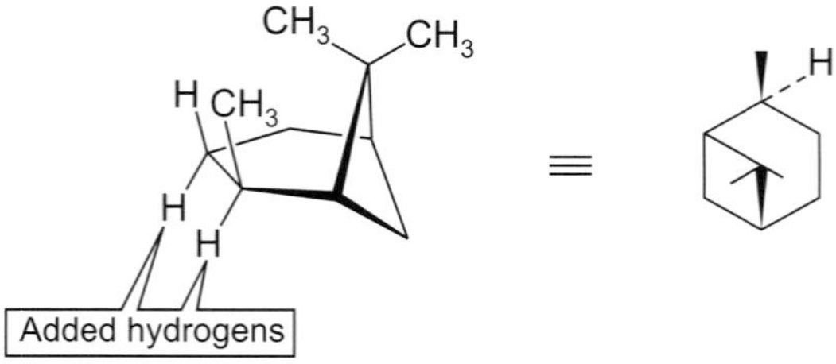

14.2 It is the one electron reduction of acetone with magnesium metal. The carbonyl group accepts an electron to give a radical anion. In the subsequent reaction, two of the radicals couple to give a dianion which is protonated on aqueous work up.

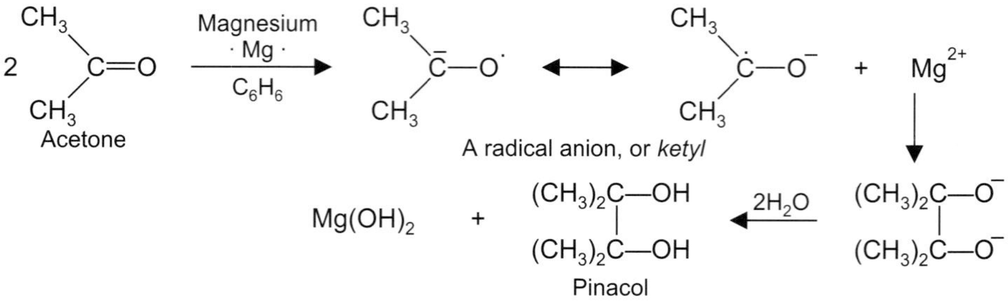

14.3 The compound will undergo a hydrogenolysis reaction *i.e.*, a reductive fission of a benzylic type system. The activating effect of the benzene ring is transmitted through conjugated double bonds.

$$PhCH{=}CHCH_3 + N(CH_3)_3$$

1-Phenylpropene

14.4 Ammonium acetate is acting as a weak proton donor. In the presence of a stronger proton-donor *e.g.*, ethanol further reduction will occur to give an alcohol.

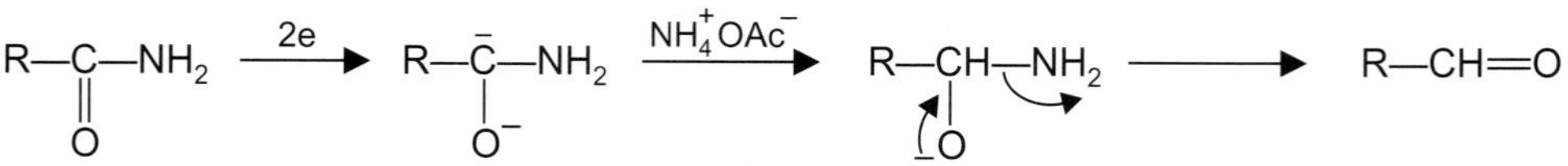

14.5 The steric effects around the carbonyl group in (I) are not severe, thus the product will be the more stable equatorial alcohol. In the case of (II) hydrogenolysis occurs.

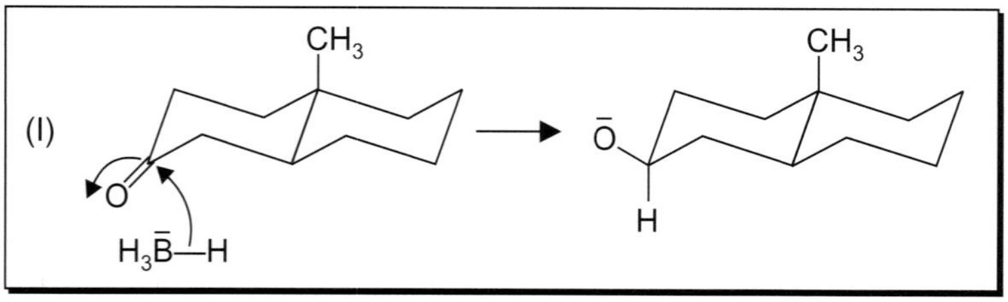

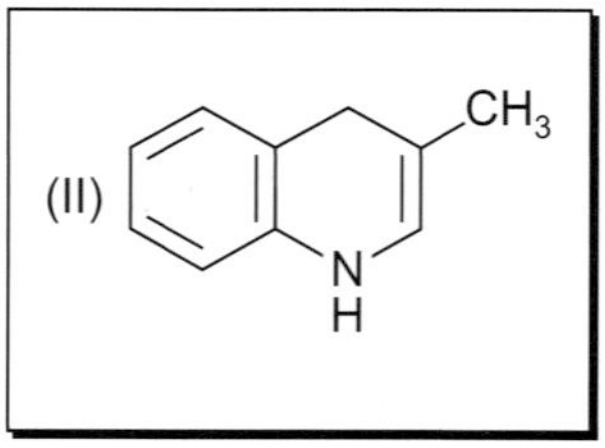

14.6

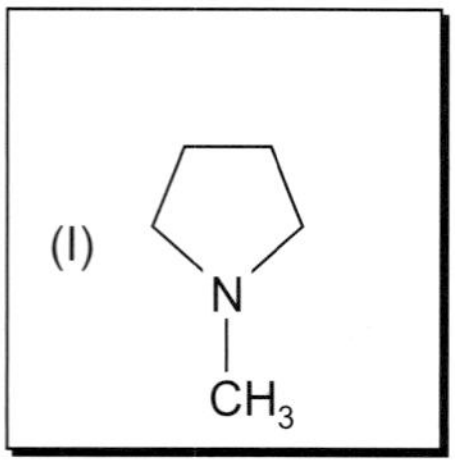

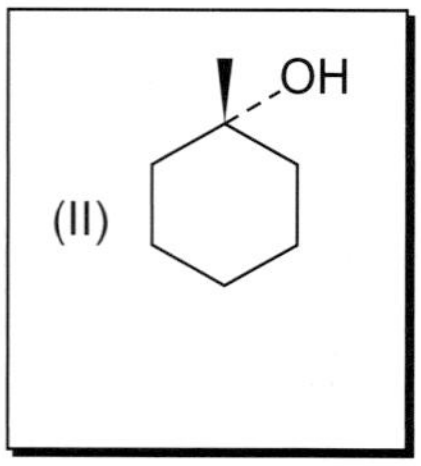

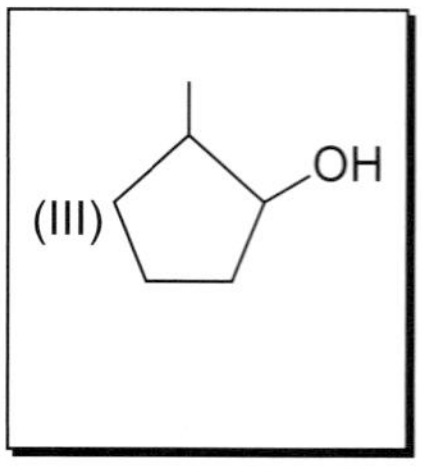

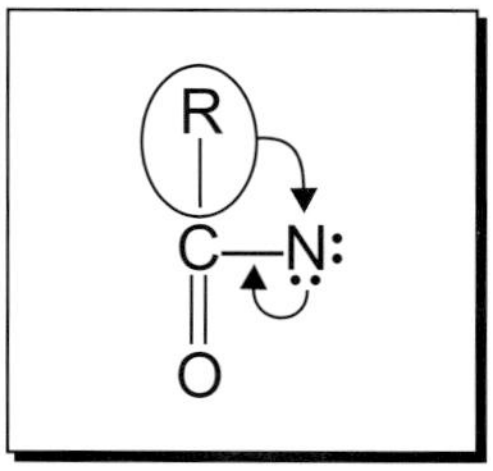

CHAPTER 15

Molecular Rearrangements

When in a reaction on an organic compound, the basic skeleton of the molecule undergoes a change, one then speaks of the occurrence of a molecular rearrangement. The majority of the rearrangements involve a migration from an atom (migration origin) to an adjacent one (migration terminus) and are termed 1, 2 shifts (Scheme 15.1). Most of these rearrangements are intramolecular processes and may occur in:

1. Electron deficient systems

Migrating group moves with its bonding pair of electrons to the migration terminus which is electron deficient with an open sextet. This is seen in the formation of a nitrene (electron deficient nitrogen) during the decomposition of an acyl azide. Once formed the nitrene undergoes a molecular rearrangement (Scheme 15.1). These rearrangements are also called nucleophilic or anionotropic rearrangements.

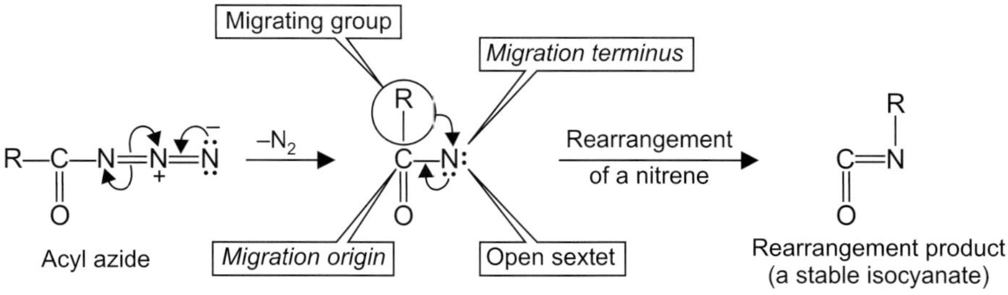

SCHEME 15.1

The migrating group or atom (M) *e.g.*, M = halogen, RO^-, RS^-, R_2N^-, (—C— or H) migrates from one atom A to an adjacent atom B along with a pair of bonding electrons (Scheme 15.1*a*). The essential requirement for the 1, 2-shift involved in this rearrangement is

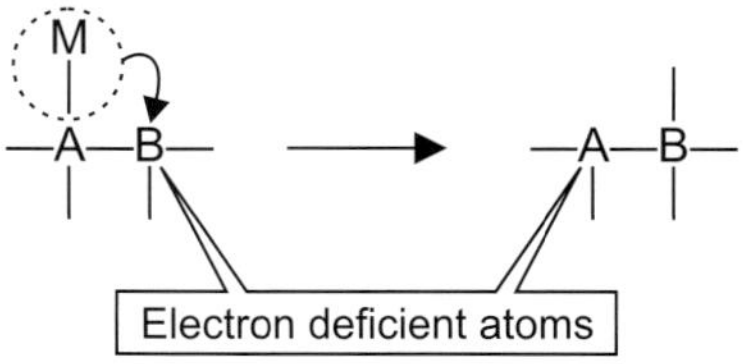

SCHEME 15.1a

the formation of the atom B which has only a sextet of electrons. The electron sextet may be generated in neutral species for examples nitrenes (see, Scheme 15.1) or carbenes and also in cationic species like carbocations, and in electron deficient oxygen and nitrogen.

2. Electron rich systems

In these rearrangements (which are less common), the migration group migrates without its electron pair. An example is prototropic rearrangement where the migrating group is hydrogen. These rearrangements are also termed electrophilic or cationotropic rearrangements. These rearrangements include reactions which are initiated by the formation of an anion (Scheme 15.1*b*) and are usually called anionic rearrangements. Most anionic reactions begin with the removal of a proton by a strong base and these rearrangements may proceed by ionic or free radical pathways.

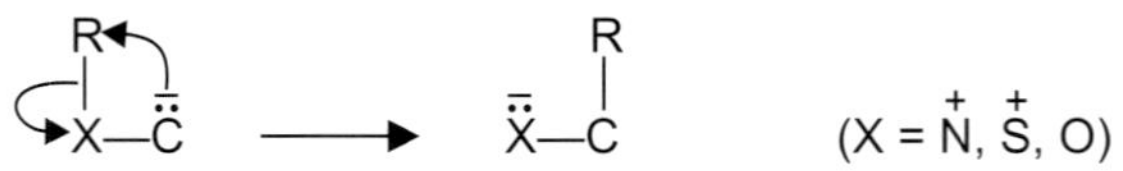

R migrates without its bonding pair

SCHEME 15.1b

3. Free radical rearrangements

The least common of the rearrangements are free radical rearrangements where the migrating group moves with just one electron.

The 1, 2-shifts (rearrangements) via carbocations are much more common than the similar rearrangements that involve carbanions or free radicals. Only in the transition state for the carbocation case (I, Scheme 15.2) one has two electrons which both can go into the bonding orbital, consequently one has a low energy transition state. However, in the free radical or carbanionic migrations, one has, electrons in excess of two which have to be accommodated in antibonding orbital of much higher energy. Thus as one would expect, during 1, 2 shifts of free radicals or 1, 2 shifts involving carbanions the migrating group is generally an aryl or a similar group which can accommodate the extra electrons and therefore, effectively removes these from the three membered transition state.

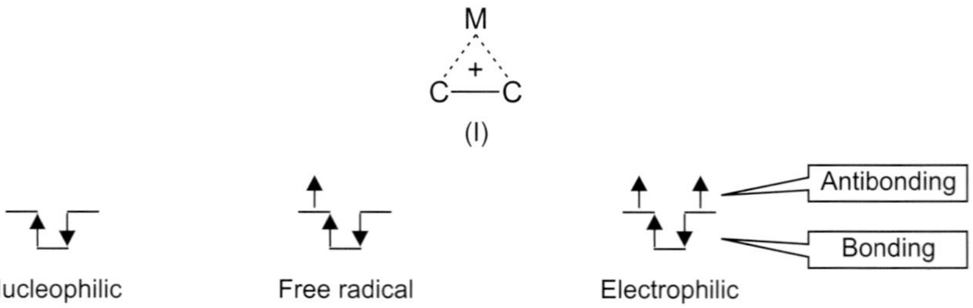

SCHEME 15.2

15.1 REARRANGEMENTS TO ELECTRON DEFICIENT CARBON

(A) Wagner-Meerwein Rearrangement (Carbon to carbon migration of R, H and Ar)

1. A migration of an alkyl group in neopentyl system

These rearrangements involve the intermediate formation of a carbocation which is an electron deficient open sextet in which a 1, 2 shift (rearrangement) occurs when such a shift leads to the formation of a carbocation of greater stability and this is the driving force of the rearrangement. A further requirement which is often necessary for this rearrangement to occur is the presence of two or more alkyl groups (high degree of substitution) on the migration origin. One of the simplest systems within which carbon (as a methyl group with its bonding pair of electrons) migrates to an electrons deficient carbon atom is the neopentyl system.

Migrations to an electron deficient carbon atom are broadly known as Wagner-Meerwein rearrangements. As an example consider the solvolysis of neopentyl iodide (H_2O, Ag^+). In the absence of Ag^+ ion no reaction is observed at all. Neopentyl iodide cannot display S_N1 reaction since the primary carbocation if formed would be highly unstable and S_N2 reaction is also not observed since the substrate is too hindered. In fact on solvolysis a rearranged alcohol 2-methyl-2-butanol is formed (Scheme 15.3 and 15.3*a*).

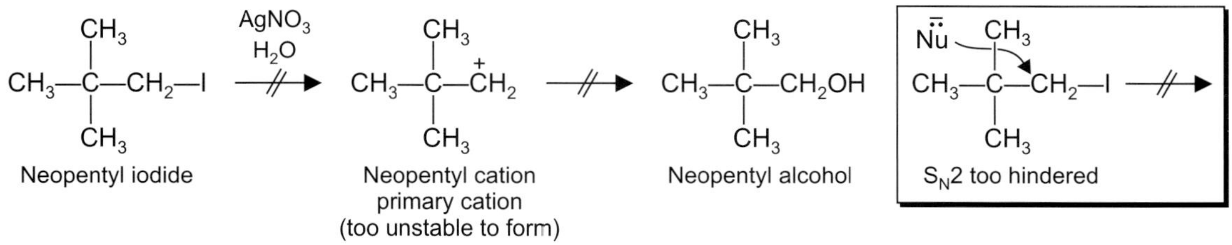

SCHEME 15.3

A shorthand method to depict this Wagner-Meerwein rearrangement is in (Scheme 15.3*a*). However, recall from (Scheme 15.3) that the presence of a primary carbocation is highly unsettling and this being highly unstable may not be at all on the reaction pathway.

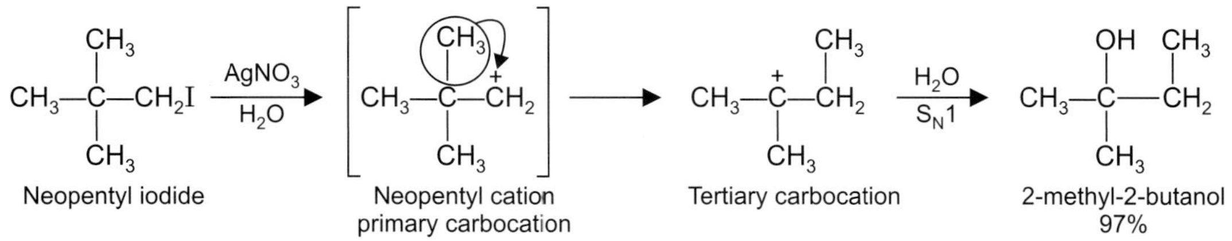

Wagner-Meerwein rearrangement on a neopentyl system

SCHEME 15.3a

Keeping this in mind a variant of mechanism (Scheme 15.3*b*) is presented to display as if methyl group migrates as iodide departs and by doing so the migrating alkyl group allows the formation of a stable tertiary carbocation. Infact it is suggested that the migration is concerted with the departure of the leaving group. The migrating group infact is never free but keeps on attached with the substrate in some way. It has been demonstrated that if the migrating group is chiral, its configuration is retained in the rearranged product. In summary it is suggested that the migrating group must bridge the migration origin and terminus. Much effort has been spent in determining if such bridged ions are transition states or intermediates and recall this information from chapter 5 on nonclassical carbocations.

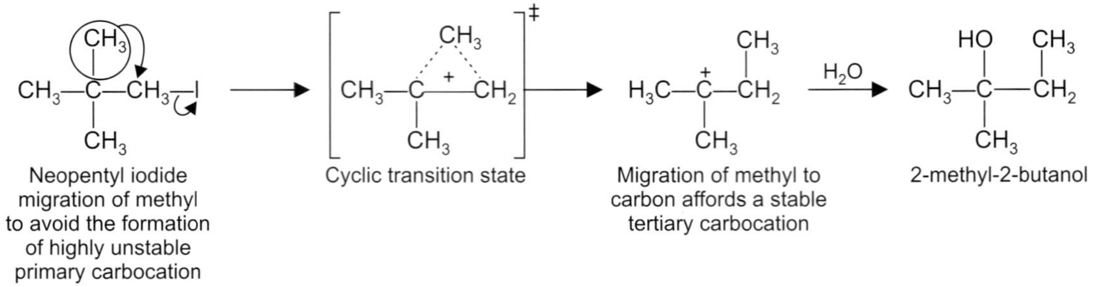

More realistic representation of a 1, 2-shift during Wagner-Meerwein rearrangement

SCHEME 15.3b

2. *Relief of strain via ring expansion provides a driving force for Wagner-Meerwein shift-shift of a carbon-carbon bond*

α-Pinene with a strained four membered ring on reaction with hydrogen chloride gives a strained carbocation (I, Scheme 15.3*c*) in which the ring expands to give a less strained five-membered

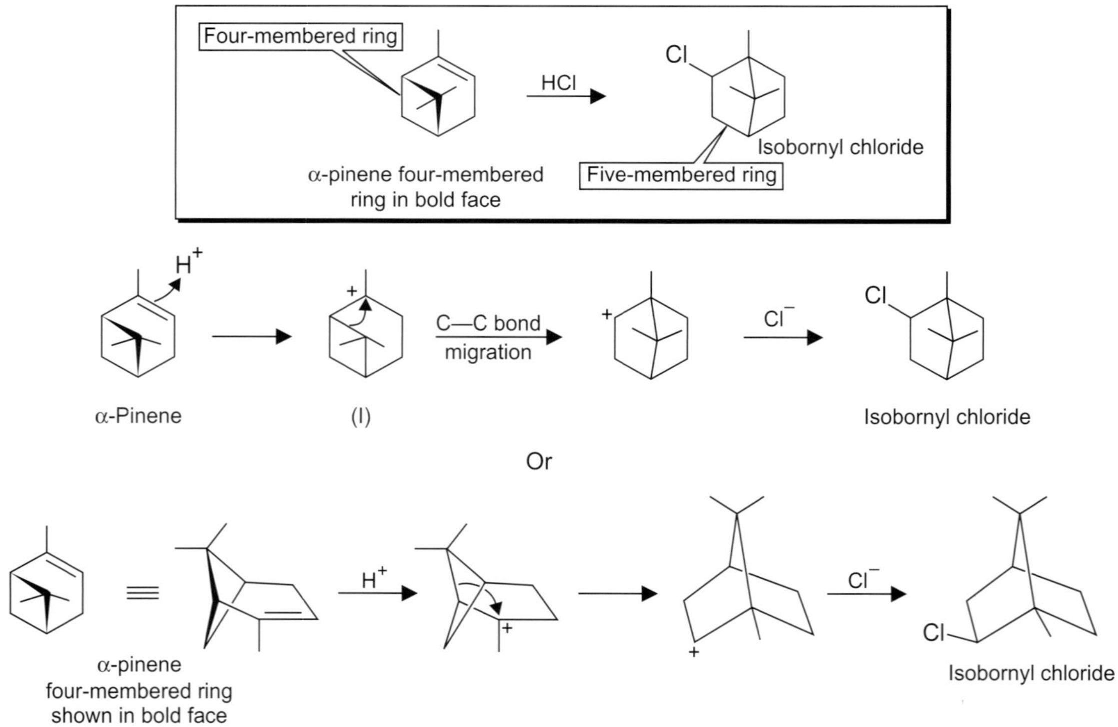

SCHEME 15.3c

analogue. In this case despite the fact that the rearrangement transforms the initially formed stable tertiary carbocation to a less stable secondary carbocation, relief of ring strain makes the rearrangement favourable.

3. *The migrating group to be anti-periplanar to the leaving group—a general stereochemical requirement in rearrangements*

Consider the conversion of camphene into isobornyl chloride (Scheme 15.3*d*). Camphene on protonation gives a tertiary carbocation followed by its capture by the chloride ion. The initial product (I, Scheme 15.3*e*) loses chloride assisted by nicely poised C—C bond which acts as a neighbouring group to lend anchimeric assistance via backside attack to give (II) which is then captured by chloride to give isobornyl chloride.

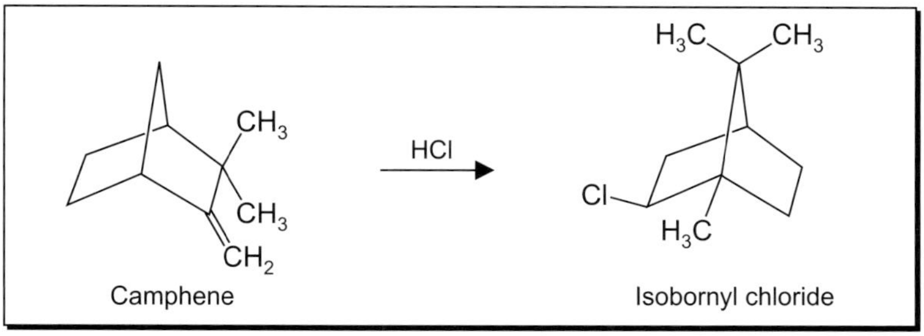

SCHEME 15.3d

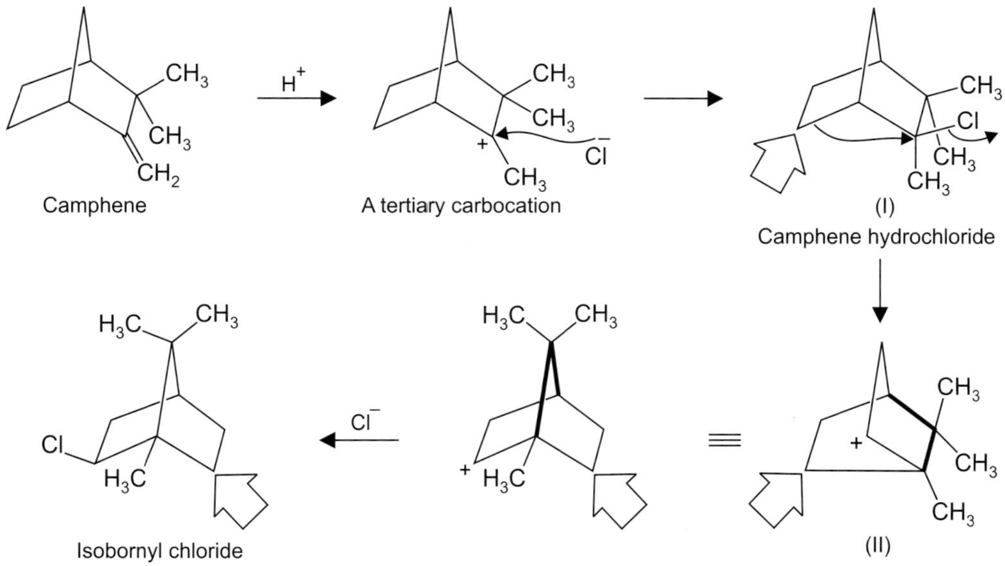

SCHEME 15.3e

This is the case of rearrangement since the participating C—C bond ends up bonded to a different atom.

PROBLEM 15.1

Predict the product of the reaction (Scheme 15.3f) and give the mechanism.

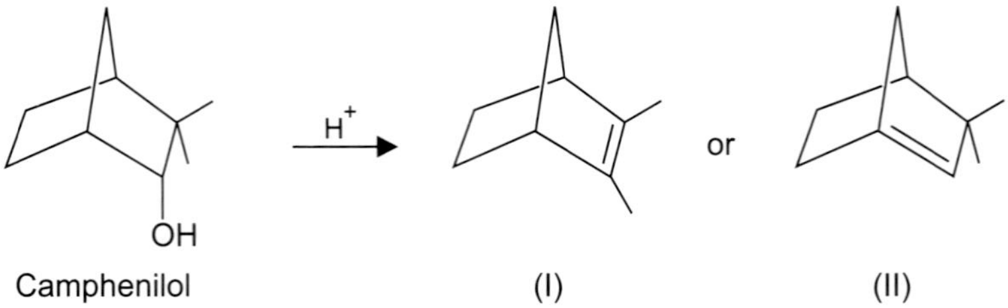

SCHEME 15.3f

ANSWER. Santene will be formed. Recall that in (III, Scheme 15.3*d*) the loss

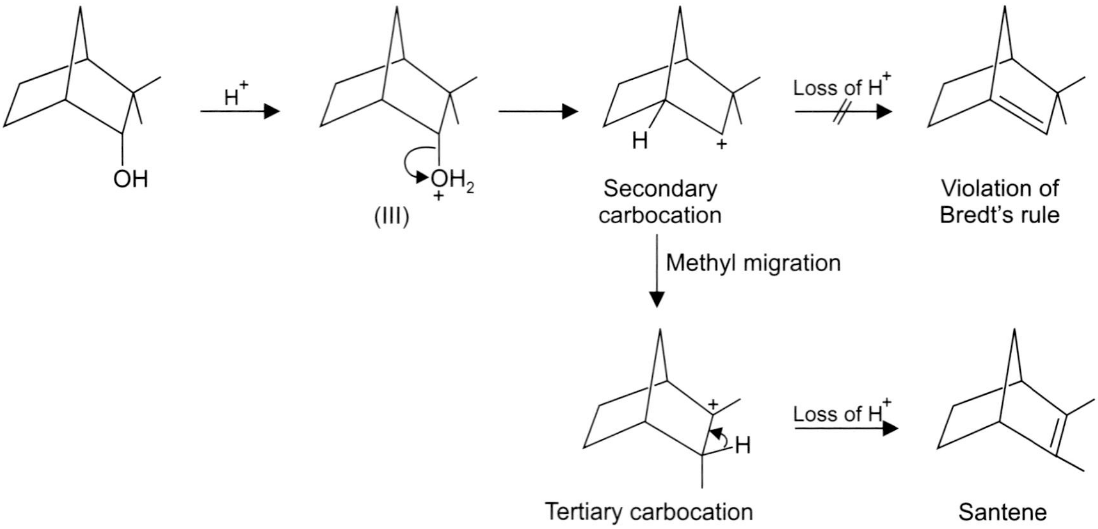

SCHEME 15.3g

of water cannot be assisted by the C—C bond as in (Scheme 15.3*e*) since now it is not in a *anti* position. Thus unassisted loss of water gives a carbocation on which a methyl migration gives the product after proton loss.

PROBLEM 15.2

Why the trans decalin system in (I, Scheme 15.3h) undergoes ring contraction on treatment with acid?

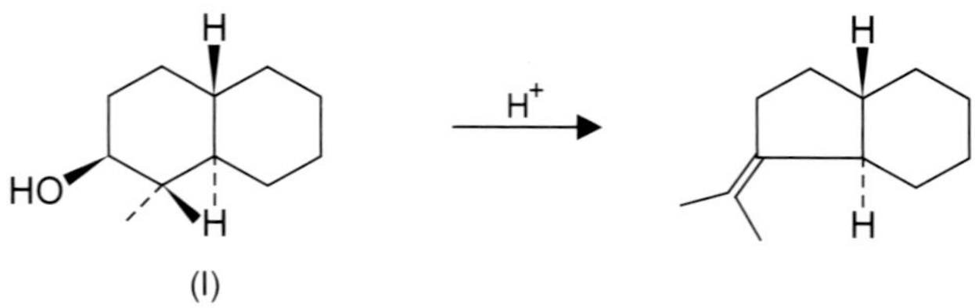

SCHEME 15.3h

ANSWER. *This is a result of Wagner-Meerwein rearrangement. The equatorial OH group is antiperiplanar to a C—C ring residue and the loss of H_2O from (II) is assisted by this C—C bond to give a ring contracted carbocation from which loss of proton gives the product. Recall that this is a molecular rearrangement since the participating C—C bond has ended up bonded to a different carbon atom. In (I, Scheme 15.3i) there is no anti-hydrogen available to the hydroxyl group so as to exclude the possibility of simple dehydration.*

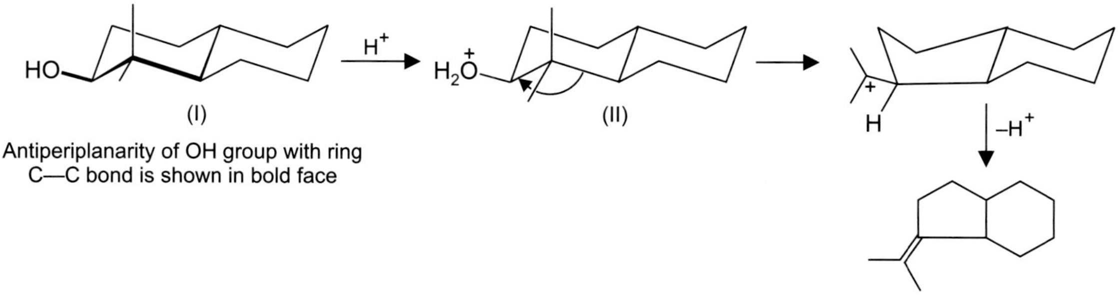

SCHEME 15.3i

4. *Aryl groups are prone to migrate and have a far greater migratory aptitude than alkyl groups or hydrogen*

Compared to neopentyl system, in neopentyl iodide (see, Scheme 15.3), the halide (I, Scheme 15.4) undergoes solvolysis with rearrangement several thousand times faster. The aryl group provides anchimeric assistance to the loss of halogen and involves the formation of a phenonium ion.

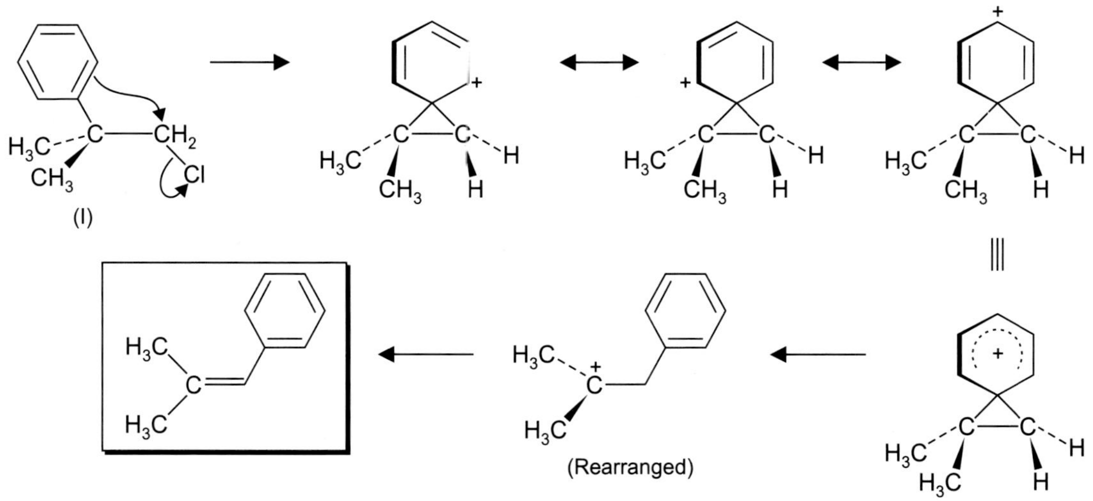

SCHEME 15.4

(B) Pinacol and Semipinacol Rearrangement

Pinacols (1, 2-diols) on treatment with acids display a rearrangement to ketones (pinacolones). The rearrangement is similar to Wagner-Meerwein shift, but for the fact that here the rearranged intermediate carbocation, the conjugate acid of the ketone is more stabilized than the rearranged carbocation formed in the Wagner-Meerwein shift (Scheme 15.5). Thus in pinacol rearrangement a shift still takes place eventhough the migration terminus may be a tertiary carbocation.

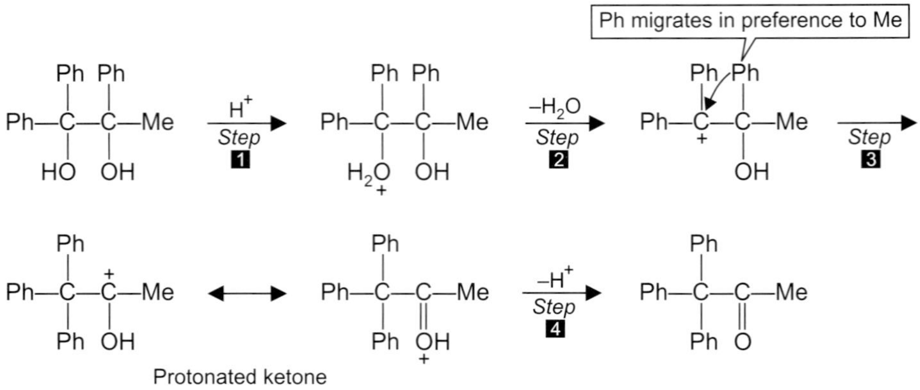

Pinacol rearrangement

SCHEME 15.5

A further observation is that with unsymmetrical glycols the product formation depends mainly by which OH is lost to leave behind the more stable of the two carbocations, and thereafter by which is the better migrating groups (order of migratory aptitude, is Ar > H > R). Thus pinacol (Scheme 15.6) reacts as shown *i.e.*, initial formation of a more stable carbocation (benzylic) and then migration of hydrogen rather than methyl. Another example is the rearrangement (Scheme 15.7).

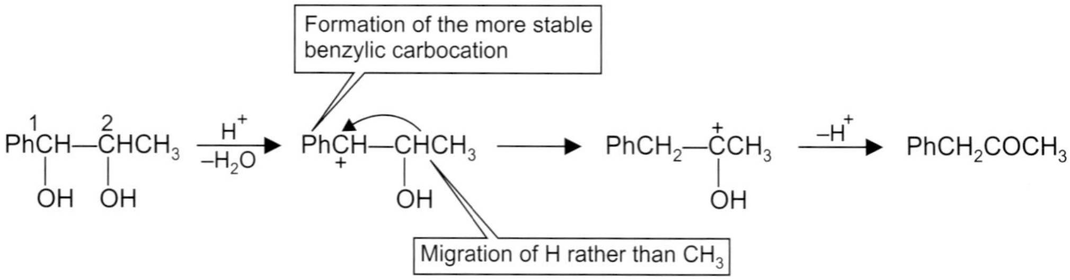

Order of *migratory aptitudes* is Ar >> H > R

SCHEME 15.6

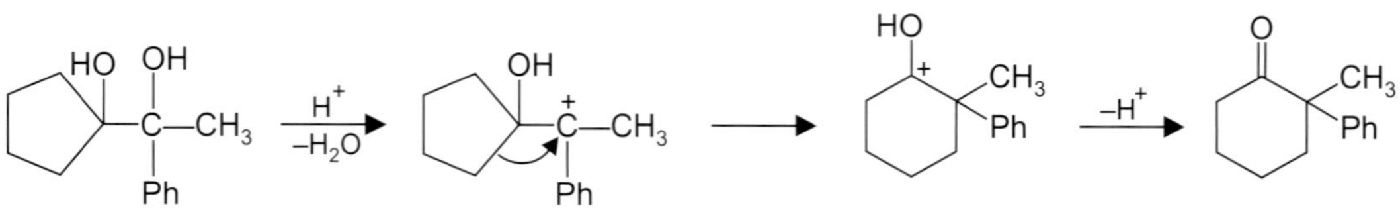

SCHEME 15.7

Deamination of α-amino alcohols is closely related with pinacol rearrangement and is called semipinacol rearrangement. These rearrangements are typical where a hydroxyl group provides the electrons to migrate a group and the driving force is provided by the loss of the leaving group other than water (Scheme 15.8).

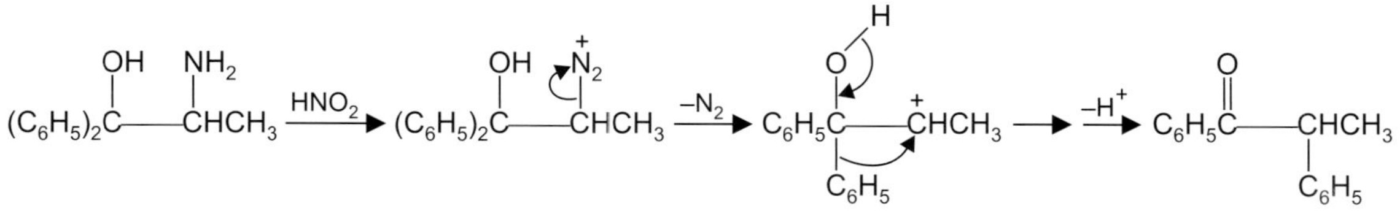

Semipinacol rearrangement

SCHEME 15.8

The rearrangement has been used to bring about both ring contraction as well as ring expansion (Scheme 15.9).

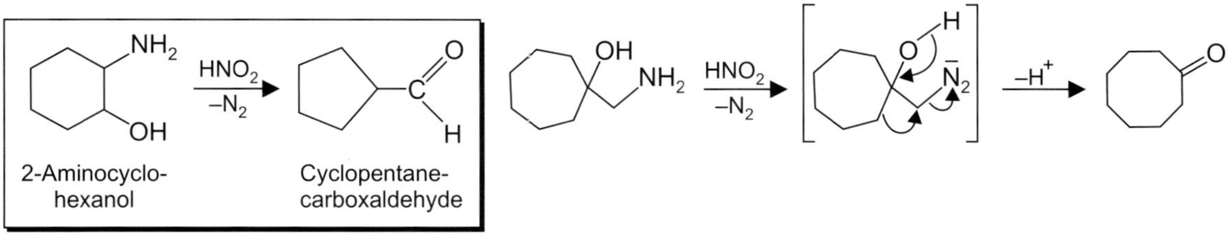

SCHEME 15.9

PROBLEM 15.3

Write the product with mechanism on reaction with HNO_2 from each of the four diastereomers with mechanism (Scheme 15.10).

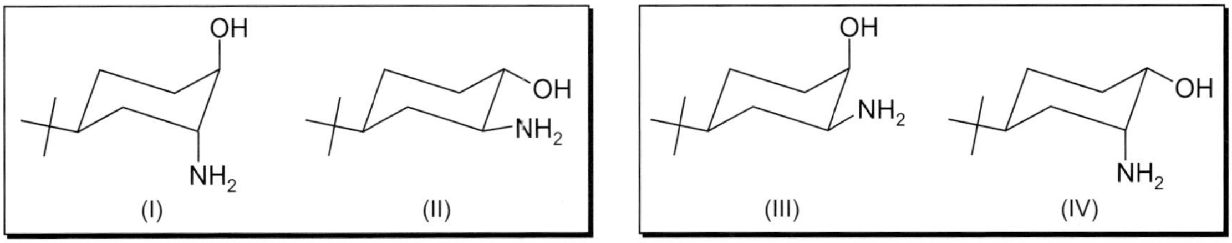

SCHEME 15.10

ANSWER. *In each case t-butyl group locks the ring in the conformation shown. The stereochemical requirement requires an antiperiplanar arrangement between the migrating and leaving group ($-N_2^+$). In each of the reactions the OH group provides the electronic push (Scheme 15.11).*

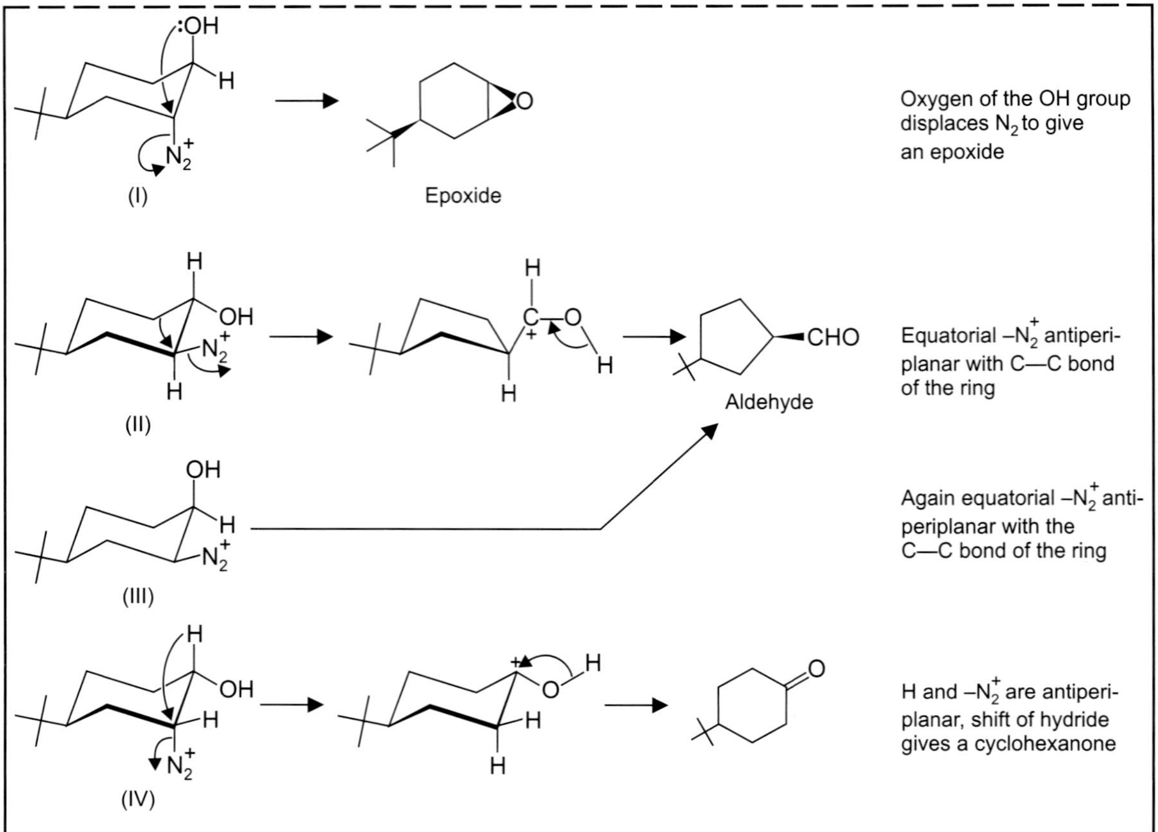

SCHEME 15.11

(C) Benzil-Benzilic Acid Rearrangement

1, 2-Diketones (α-diketones) on treatment with hydroxide ion undergo a rearrangement to give α-hydroxy acids *e.g.*, benzil gives benzilic acid (Scheme 15.11*a*). The arrangement has its driving force in the removal of the product by ionization of the carboxyl group.

One may note here, that the migrating phenyl (with its bonding pair of electrons) is not migrating to a carbon with open sextet, but to a carbon from which the π electrons shift from the carbonyl bond to the oxygen (I, Scheme 15.11*a*).

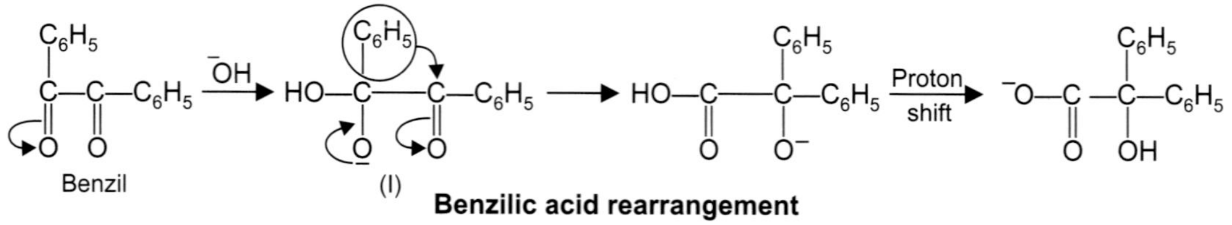

SCHEME 15.11a

(D) Rearrangements Involving Diazomethane

Wolff rearrangement is the loss of nitrogen from a α-diazoketone by the action of silver oxide or irradiation with light to yield an intermediate carbene. The Arndt-Eistert synthesis is a method to convert a carboxylic acid to its next higher homolog by the operation of Wolff rearrangement. The carboxylic acid is converted into its acyl halide which on reaction with diazomethane gives a diazoketone (I, Scheme 15.12).

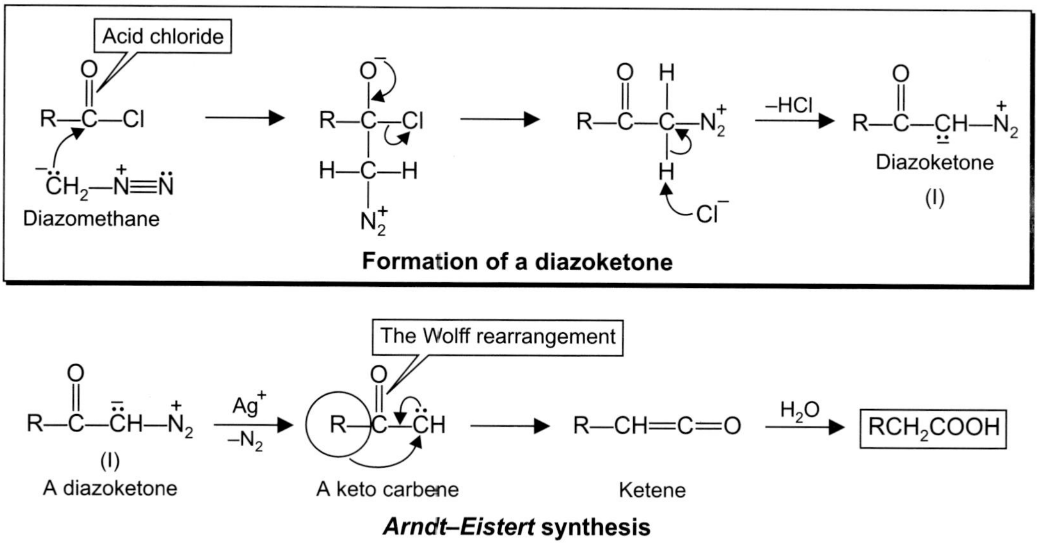

SCHEME 15.12

When diazomethane adds to an aldehyde or a ketone a product of CH_2 insertion is obtained. Thus cyclohexanone is converted into cycloheptanone (Scheme 15.12*a*). The intermediate is similar to the one involved in semipinacol rearrangement (See, Scheme 15.8).

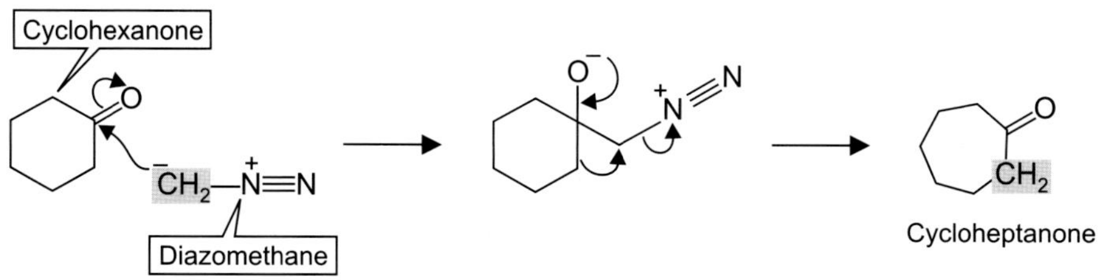

SCHEME 15.12a

A disadvantage of this reaction is that an epoxide is formed as a by-product and in some cases it may be the only product (Scheme 15.12*b*).

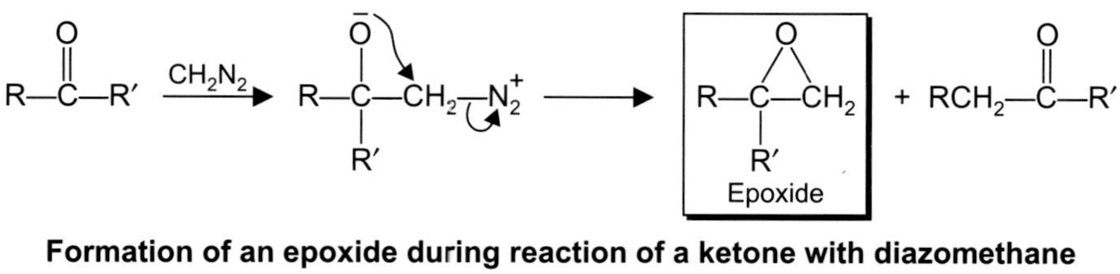

SCHEME 15.12b

The mechanism (Scheme 15.12) involves the formation of a carbene (electron deficient carbon) to which the migrating group brings its electron pair to afford a ketene and finally a carboxylic acid.

(E) Rearrangement of Alkanes

Saturated hydrocarbons display skeletal rearrangements when treated with a Lewis acid in the presence of a catalytic quantity of an organic halide (Scheme 15.13). The rearrangement is of Wagner-Meerwein type and involves the intermediate formation of a carbocation whose formation is initiated by the organic halide. The carbocation then initiates the rearrangements by abstracting hydride ion from the alkane (Scheme 15.13).

Significantly in the rearrangement of alkanes the product is derived from the less stable carbocation. Unlike Wagner-Meerwein rearrangement which is kinetically controlled (immediate reaction of the rearranged carbocation with nucleophile), the alkane rearrangements are under thermodynamic control. Practically no nucleophile is present (the concentration of $AlCl_4^-$ is negligible), the reactions are thus reversible. As a consequence the relative proportions of the alkanes formed are under thermodynamic control.

An interesting applications of this method is the conversion of all tricyclic alkanes to adamantanes. The success of this reaction (Scheme 15.13) is due to high thermodynamic stability of adamantane. Dimeric cyclopentadiene (which is readily available *e.g.*, *via* Diels-Alder reaction on cyclopentadiene) gives adamantane on reaction with $AlCl_3$ around 180°C.

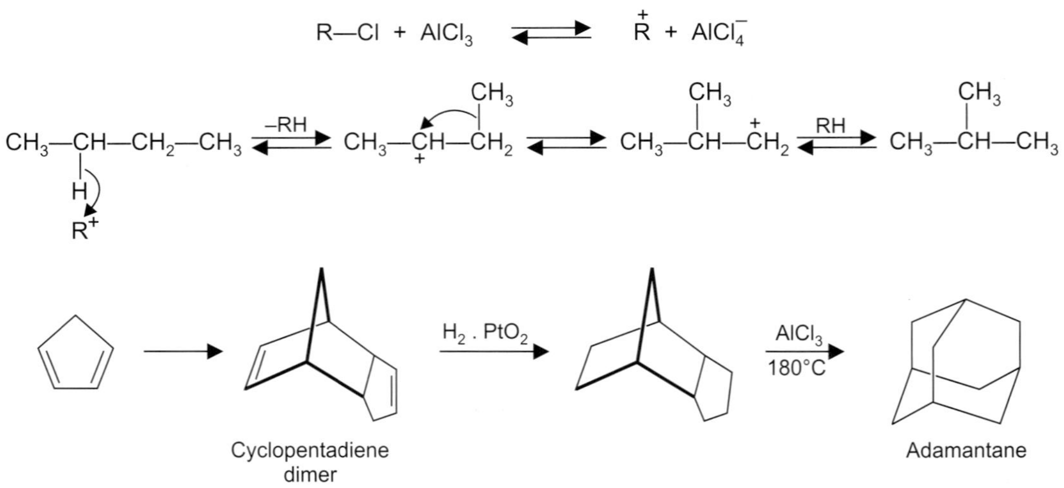

Rearrangement of alkanes are thermodynamically controlled

SCHEME 15.13

15.2 REARRANGEMENTS TO ELECTRON DEFICIENT NITROGEN

(A) Hofmann Rearrangement

Amides which do not have a substituent on the nitrogen display a molecular rearrangement (Hofmann rearrangement) on treatment with alkaline aqueous solutions of bromine or chlorine to give primary amines (Scheme 15.14). In this rearrangement, the carbonyl carbon atom of the amide is lost and the *R* group of amide gets attached to the nitrogen of the amine (Scheme 15.15). The mechanism involves the following steps:

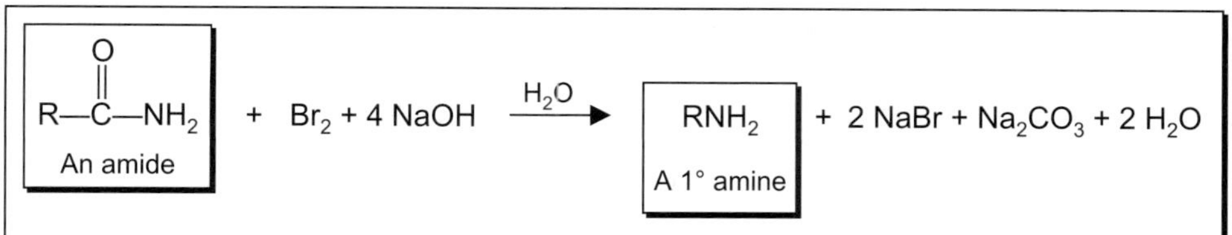

Hofmann rearrangement

SCHEME 15.14

1. It is a base promoted bromination of an amide, just like the base promoted halogenation of a ketone.
2. The base abstracts a proton from the nitrogen to give an unstable bromamide anion. The anion spontaneously rearranges by the migration of *R* to the electron deficient nitrogen to give an isocyanate.
3. Lastly the isocyanate undergoes hydrolysis to give an amine.

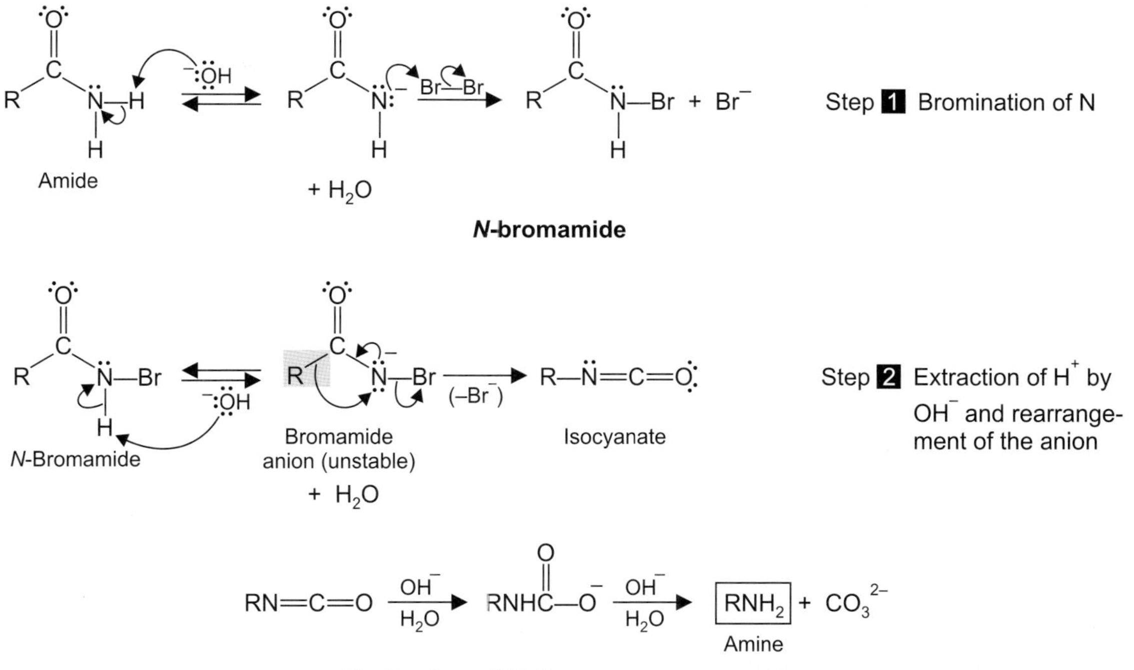

Mechanism of Hofmann rearrangement

SCHEME 15.15

An interesting stereochemical observation is, that if the migrating group (*R*) is chiral its configuration is retained in the product amine (I, Scheme 15.16). Thus this arrangement is intramolecular, the migrating group does not become free, but remains attached with the substrate in some way *e.g.*, via a bridged transition state. Thus if one starts from (*R*)-2-methylbutanamide, the end product is (*R*) *sec*-butyl amine (Scheme 15.16). The intramolecular nature of the rearrangement was shown by carrying out the reaction with a mixture of 3-deuteriobenzamide and ^{15}N-benzamide. Mixed anilines were not formed to indicate that the migrating group does not separate during the rearrangement (Scheme 15.16*a*).

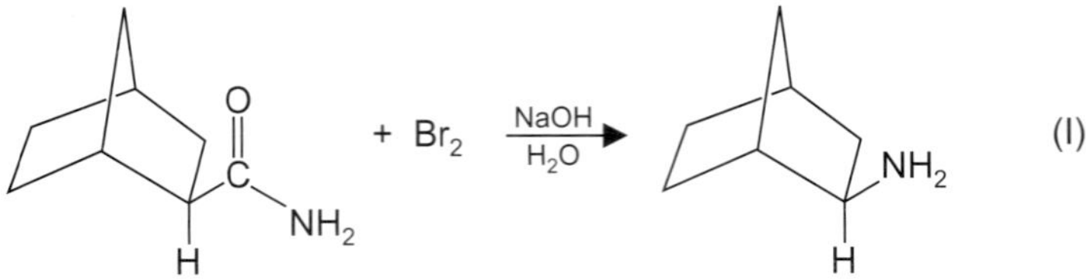

Hofmann degradation proceeds with retention of configuration of the migrating group.

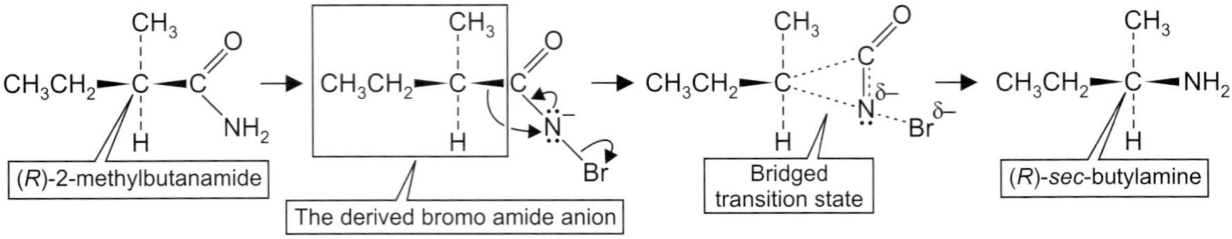

SCHEME 15.16

$$C_6H_4DCONH_2 + C_6H_5CO^{*}NH_2 \longrightarrow C_6H_4DNH_2 + C_6H_5^{*}NH_2$$

$C_6H_4D^{*}NH_2$ or $C_6H_5NH_2$
Not formed

SCHEME 15.16a

(B) Curtius, Schmidt and Lossen Rearrangements

These rearrangements and also Hofmann rearrangement are closely similar in that a carbon migrates from carbon to nitrogen to give an isocyanate. In these 1, 2-shifts the migrating group (carbon) is an alkyl or aryl group and the leaving group may be Br (Hofmann

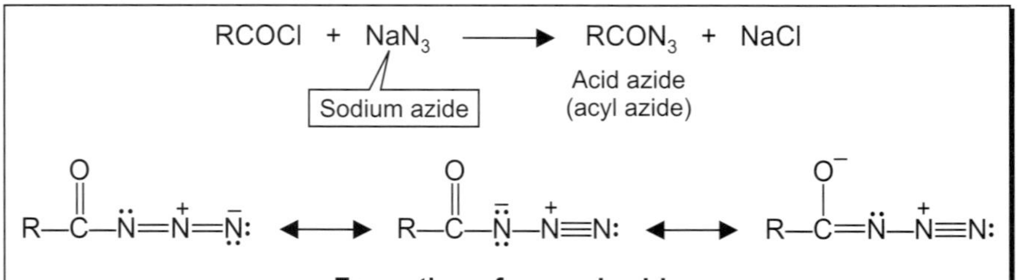

Formation of an acyl azide

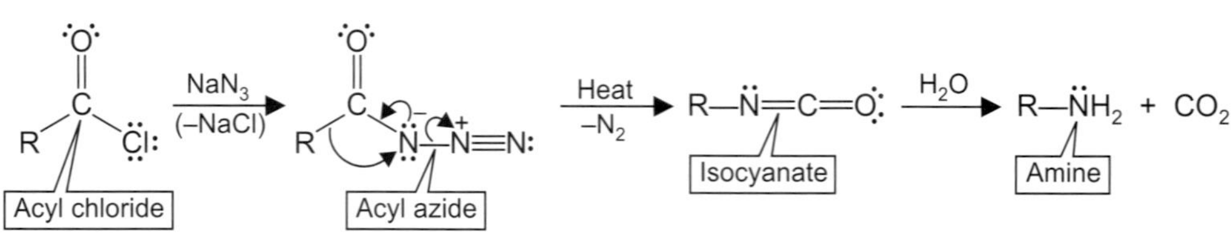

The mechanism of Curtius rearrangement

SCHEME 15.17

rearrangement); N_2 (Curtius and Schmidt rearrangements) or $RCOO^-$ (Lossen rearrangement). The Curtius rearrangement involves the pyrolysis of acyl azides (acid azides) to give isocyanates. Acid azides are prepared from acid chlorides on treatment with sodium azide, and best represented as a resonance hybrid (Scheme 15.17).

Schmidt reaction involves the treatment of a carboxylic acid with hydrazoic acid (HN_3, hydrogen azide) which yields amines via the isocyanate when catalysed by an acid *e.g.*, sulphuric acid (Scheme 15.18). The protonated azide undergoes the rearrangement.

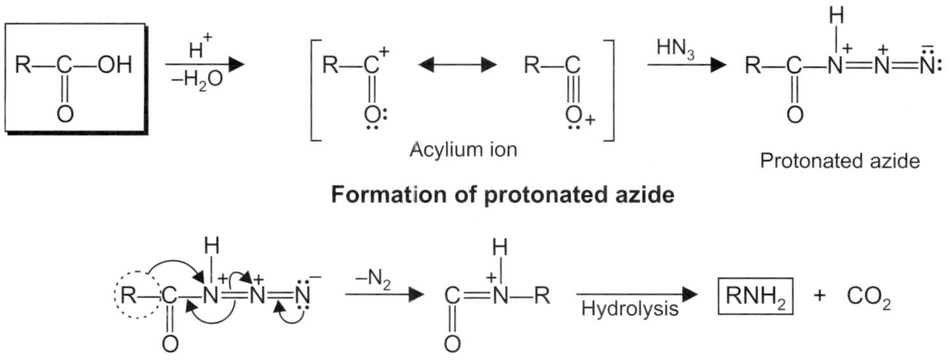

Formation of protonated azide

Schmidt reaction

SCHEME 15.18

There are many variations of Schmidt reaction and can be applied to ketones to give amides which is a general procedure (Scheme 15.18*a*).

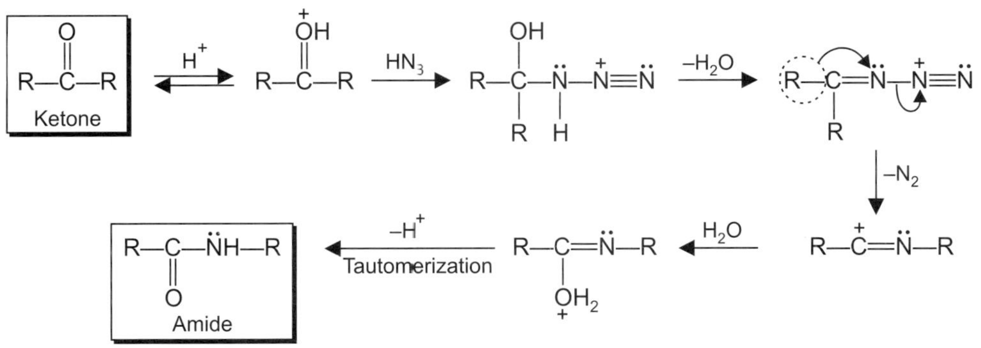

Conversion of ketones to amides via Schmidt reaction

SCHEME 15.18a

The Lossen rearrangement occurs when *O*-acyl derivatives of hydroxamic acid are heated with bases (Scheme 15.19). The hydroxamic acids display tautomerism, the keto form is termed hydroxamic form while the enol form is called hydroximic acid. Hydroxamic acid is prepared by the action of hydroxylamine on acid chloride (Scheme 15.19).

Thus in summary all the three reactions give an amine as the end product with one less carbon. Curtius reaction starts with a carboxylic acid through an acyl azide, Schmidt reaction starts with a carboxylic acid through an azide and Lossen reaction is the dehydration of a hydroxamic acid.

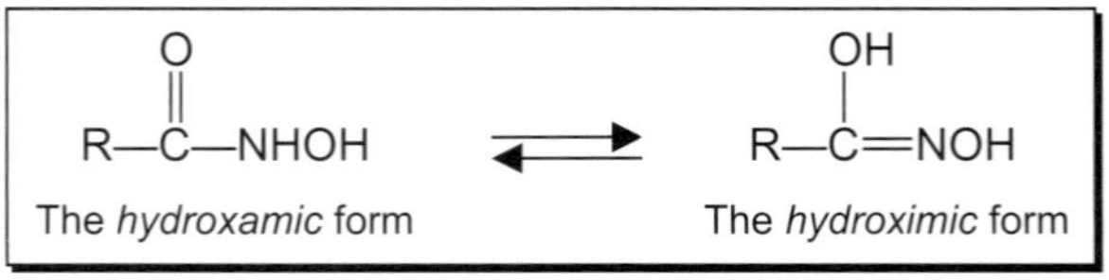

Hydroxamic acids exhibit tautomerism

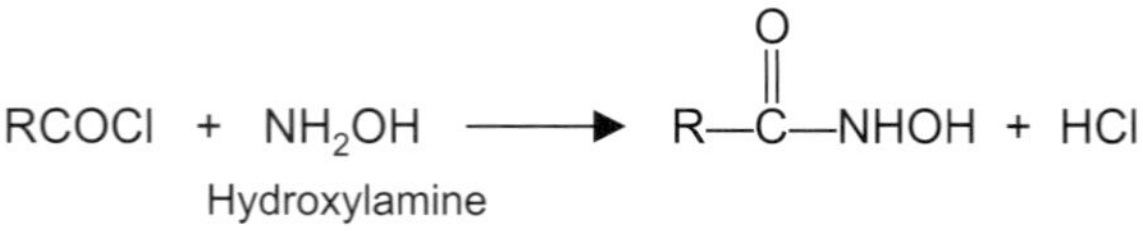

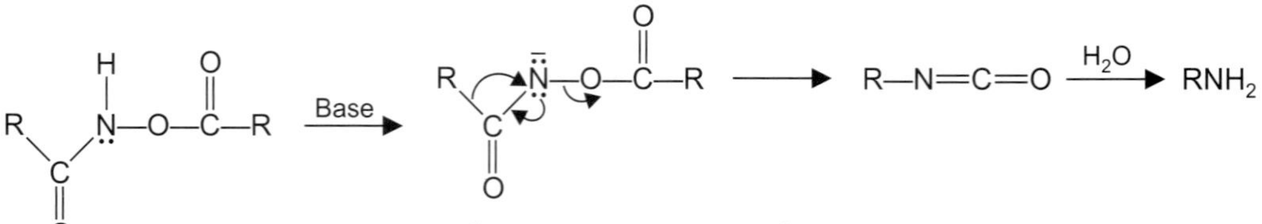

Lossen rearrangement

SCHEME 15.19

(C) The Beckmann Rearrangement

On treatment with acids, oximes rearrange to substituted amides and this reaction is termed as Beckmann rearrangement (Scheme 15.20). The rearrangement is stereospecific and the group that normally migrates is the one that is *anti* placed with respect to the hydroxyl. The method is often used to determine the configuration of the oxime. Acetophenone oxime gives only acetanilide (Scheme 15.21). The intermediate formation of nitrilium iron (Scheme 15.20) has been confirmed by spectral methods. The rearrangement is intramolecular, since if the migrating group is chiral it retains its configuration in the product.

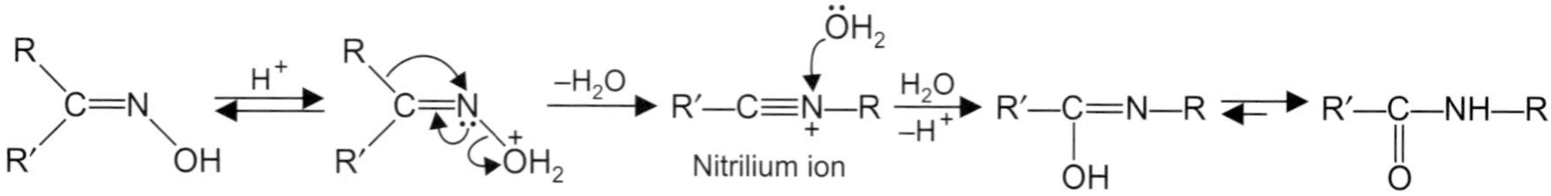

The Beckmann rearrangement

SCHEME 15.20

The function of the acidic reagents is to convert the hydroxyl group to a better leaving group. Thus *e.g.*, with H_2SO_4, this group is converted into OH_2^+ and with PCl_5, it is the phosphate which is the leaving group (Scheme 15.21).

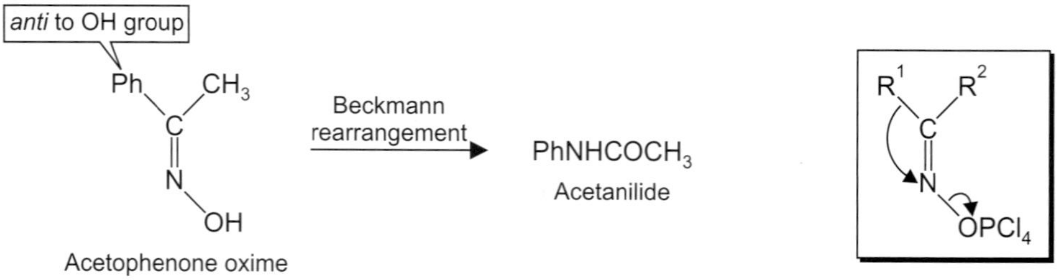

SCHEME 15.21

The oximes of the cyclic ketones give ring enlargement (Scheme 15.22). Caprolactam gives a polymer of the nylon group when heated.

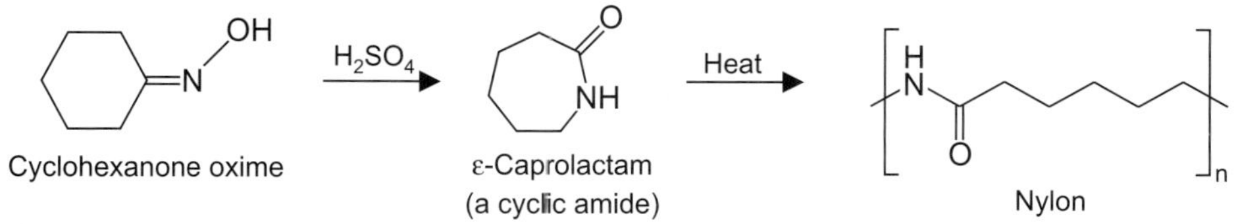

SCHEME 15.22

15.3 REARRANGEMENTS TO ELECTRON DEFICIENT OXYGEN

(A) Baeyer-Villiger Rearrangement

Baeyer-Villiger oxidation has been discussed (see, Schemes 13.51–13.53) and involves the insertion of oxygen atom next to the carbonyl group. The migratory aptitudes have been discussed and it is seen that methyl group has the least tendency to undergo migration and consequently methyl ketones end up with acetates and retention of stereochemistry if the migrating group is chiral is the rule (Scheme 15.22*a*).

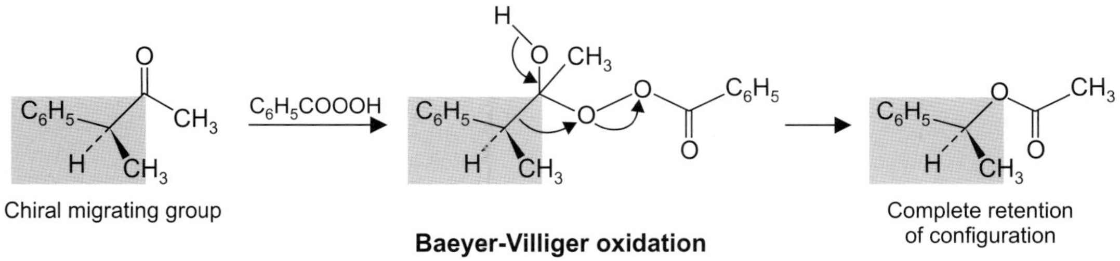

SCHEME 15.22a

Often unsaturated ketones are not good candidates for Baeyer-Villiger oxidation since an alkene will get epoxidized. However, if a double bond is not so electron rich as a disubstituted double bond in (Scheme 15.22*b*) it will remain unattacked.

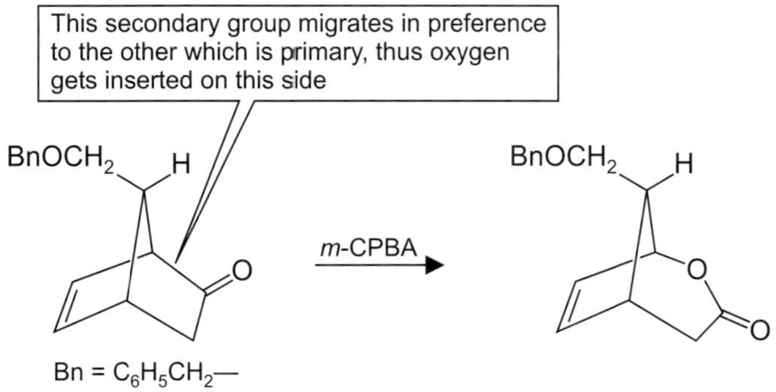

SCHEME 15.22b

Recall that during hydroboration as well during the oxidation of an organoborane with alkaline H_2O_2 gives alcohols by the migration of carbon from boron to oxygen. The migrating group retains its configuration (Scheme 15.22*c*). The stereochemistry of the product is dictated during the hydroboration step itself and is retained in the product.

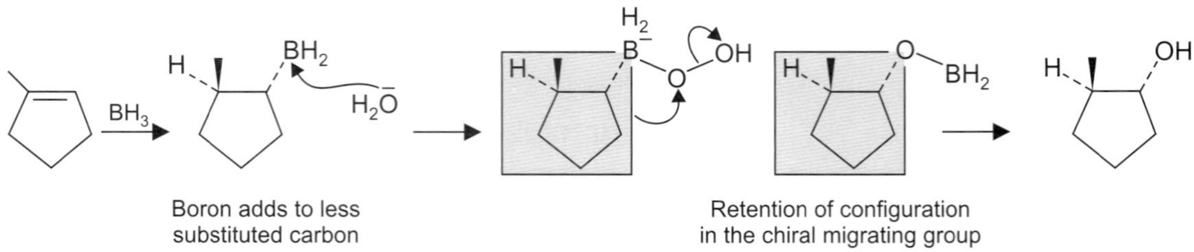

SCHEME 15.22c

Identical product can be made using a Baeyer-Villiger oxidation by starting with a substrate with appropriate stereochemistry and keeping in mind that this configuration will be retained during migration step (Scheme 15.22*d*).

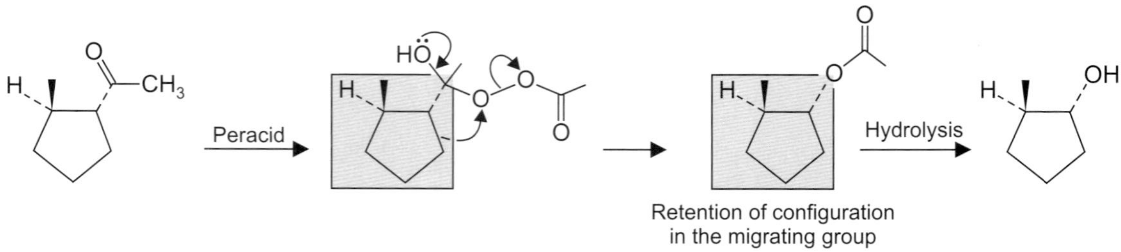

SCHEME 15.22d

(B) Dakin Reaction

(See Scheme 13.54)

15.4 REARRANGEMENT TO ELECTRON RICH CARBON

(A) Favorskii Rearrangement

The reaction of α-halo ketones (containing an α-hydrogen on the non halogenated side of the carbonyl) with alkoxide ions give rearranged esters, a process termed Favorskii rearrangement (use of hydroxide ions gives the free acid).

The mechanism of Favorskii rearrangement involves a cyclopropanone derivative (II,Scheme 15.23) which is formed by the "internal S_N2" displacement of the halide ion by an initially formed α-carbanion (I). Base then adds to the cyclopropanone carbonyl and the ring opening takes place in a way (IV, Scheme 15.23) so as to gives more stable of the carbanions. Thus in the absence of resonance stabilization, the less substituted alkyl anion is formed (Scheme 15.24).

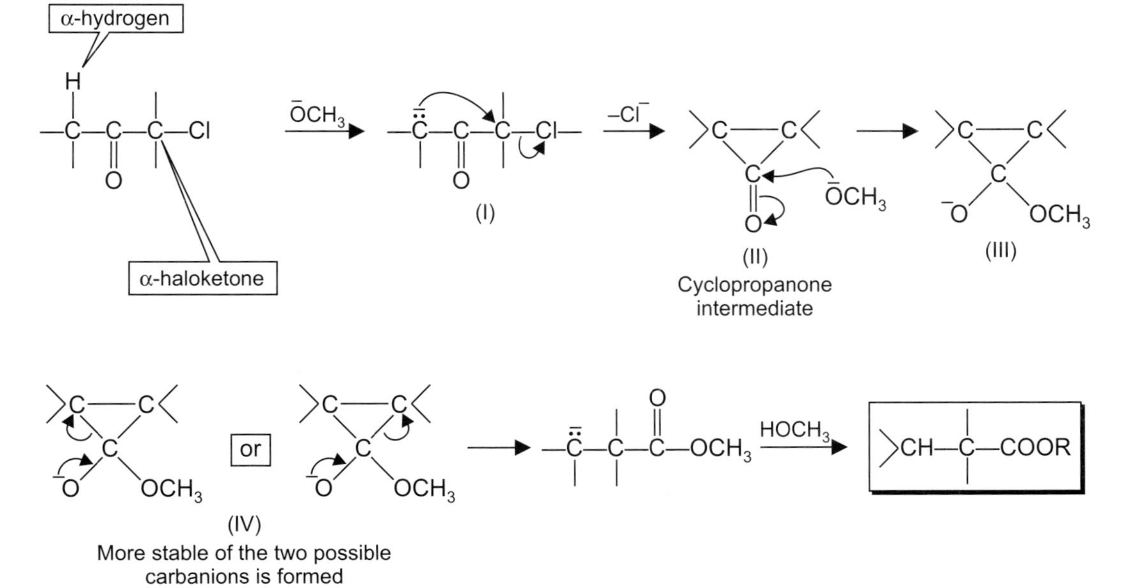

Mechanism of Favorskii rearrangement

SCHEME 15.23

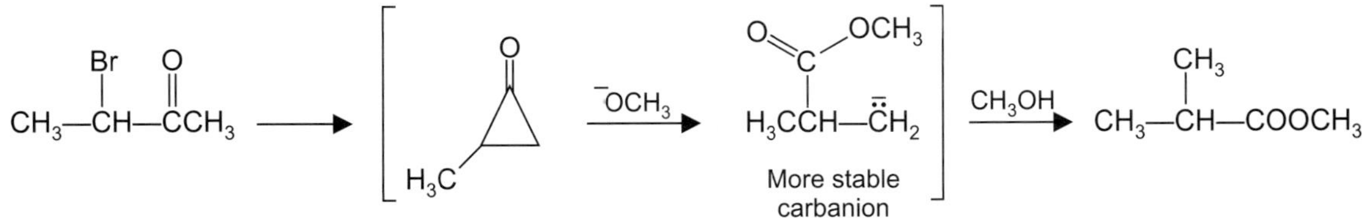

SCHEME 15.24

Favorskii reaction involves the formation of cyclopropanones as the intermediates. The formation of such intermediates has been clearly established *e.g.*, 2, 3-di-*tert*-butylcyclopropanone (Scheme 15.24*a*) which is somewhat unreactive towards base has been isolated from the reaction of the corresponding α-chloroketone with potassium hydroxide (Scheme 15.24*a*).

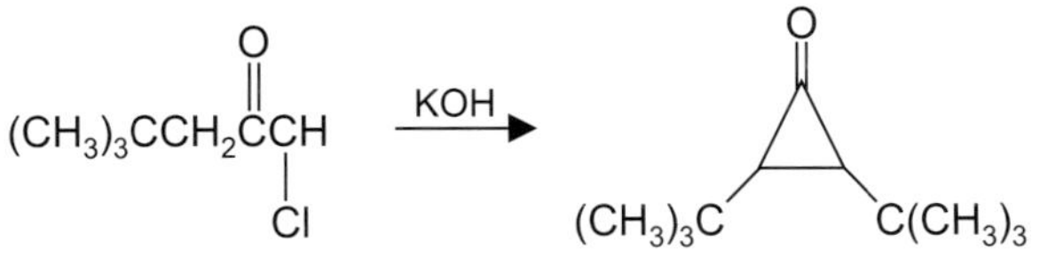

SCHEME 15.24a

Alkyl groups destabilize carbanions while aryl groups stabilize them (delocalization of the negative charge). Both the isomers (I and II, Scheme 15.25) give the same product and this requires the formation of the same common cyclopropanone intermediate from both the reactants which opens only in one direction to give the more stable resonance stabilized benzylic carbanion (Scheme 15.25).

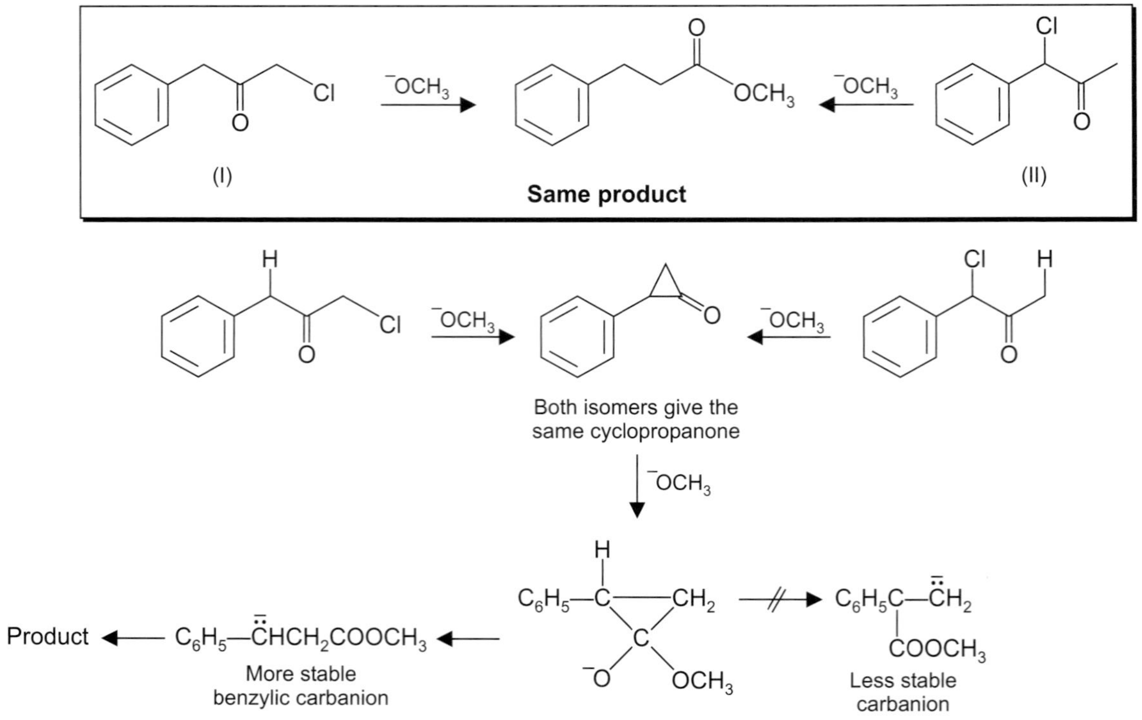

SCHEME 15.25

One can thus bring about a ring contraction by working with cyclic systems (I, Scheme 15.26). Moreover, by working with 2-chlorocyclohexanone in which C-1 and C-2 were equally labelled with ^{14}C the product contained 50% of the label on the carbonyl carbon and 25% each on C-1 and C-2 (II, Scheme 15.26). These results are further in keeping with the formation of symmetrical cyclopropanone derivative.

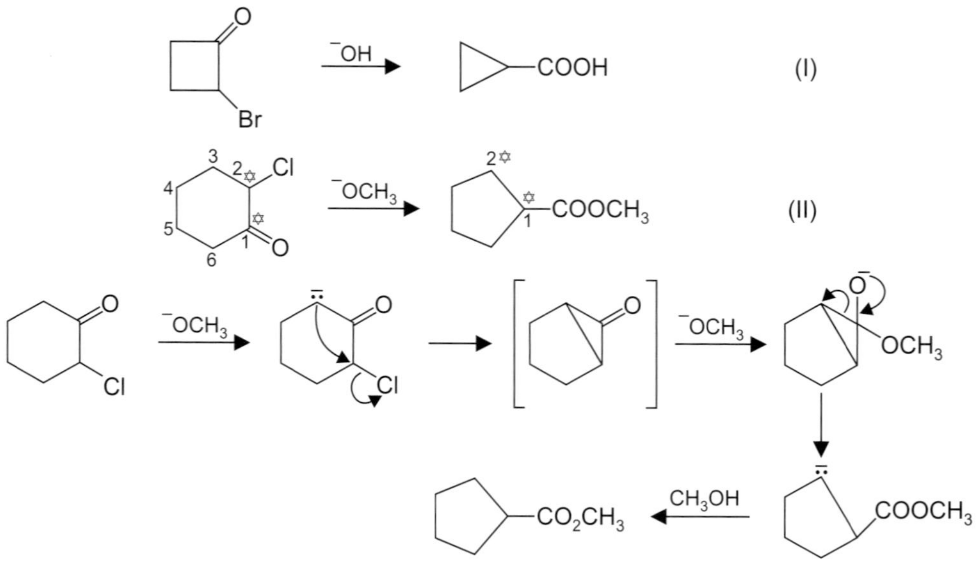

SCHEME 15.26

(B) Wittig Rearrangement

Alkyl benzyl and related ethers rearrange when reacted with a strong base to give alcohols *via* a rearrangement called Wittig rearrangement (Scheme 15.27). A radical pair mechanism is suggested after the base removes the proton from the ether. As part of the evidence for this mechanism (Scheme 15.27) it is seen that migratory aptitudes in this 1, 2-shift follow the order of free radical stabilities and not the carbanion stabilities.

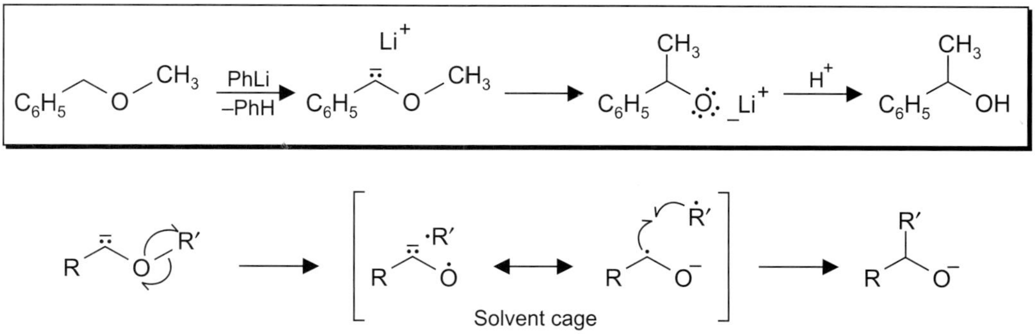

Wittig rearrangement

SCHEME 15.27

(C) The Neber Rearrangement

Ketoxime tosylates on reaction with base rearrange to give α-amino ketones in a reaction called Neber rearrangement. The mechanism of the rearrangement (Scheme 15.28) involves the intermediate formation of an azirine.

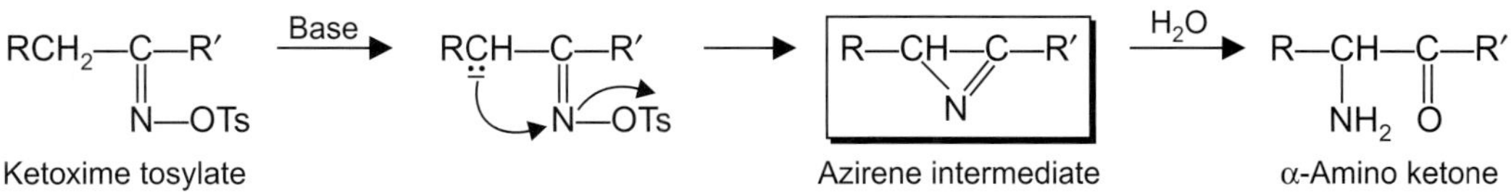

The Neber rearrangement

SCHEME 15.28

(D) Stevens Rearrangement

Quaternary ammonium salts which have β-hydrogen atoms undergo E2 (Hoffmann) elimination on reaction with base (Scheme 15.29). However, if in a quaternary ammonium salt none of the alkyl groups have a β-hydrogen, but one has an electron withdrawing group in this position *e.g.*, a β-carbonyl group then an ylide is formed by the loss of an α-hydrogen with a base. The ylide rearranges to a tertiary amine and is termed Stevens rearrangement (Scheme 15.30). The role of the electron withdrawing carbonyl group in this example, is to assist the formation of the ylide by stabilizing the charge. This charge is neutralized by a 1, 2-shift.

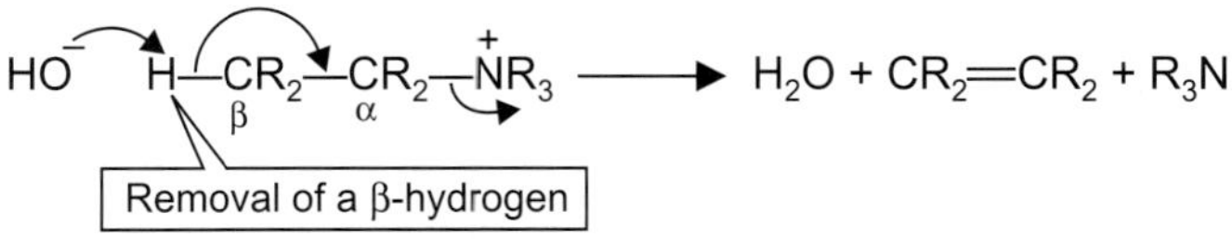

E2 (Hoffmann) elimination with base

SCHEME 15.29

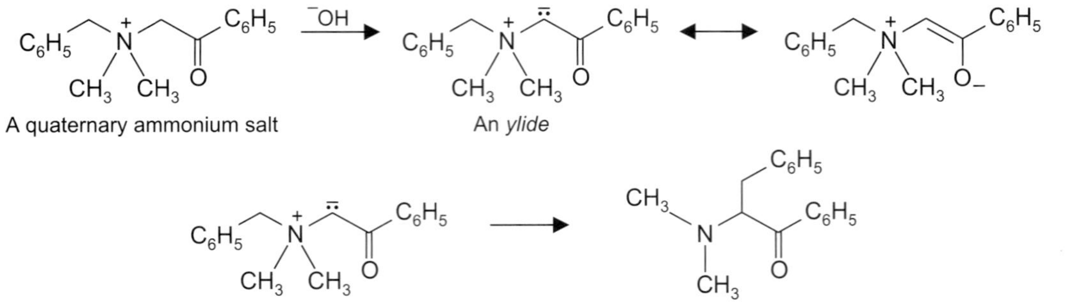

The Stevens rearrangement

SCHEME 15.30

(E) The Sommelet-Hauser Rearrangement

This rearrangement is typical of benzyl quaternary ammonium salts which on reaction with alkali metal amides give benzyl tertiary amines (*o*-methylbenzylamines, Scheme 15.31). Thus compared to the substrates in the Stevens rearrangement, in Sommelet Hauser rearrangement, a strongly electron withdrawing group (*e.g.,* a β-carbonyl) is absent. The α-hydrogen is too weakly acidic for the rearrangement to be induced by hydroxide ion. Stronger base, is thus required and the rearrangement then follows a different path instead of a 1, 2-shift. The Sommelet rearrangement is in fact a [2, 3]-sigmatropic rearrangement.

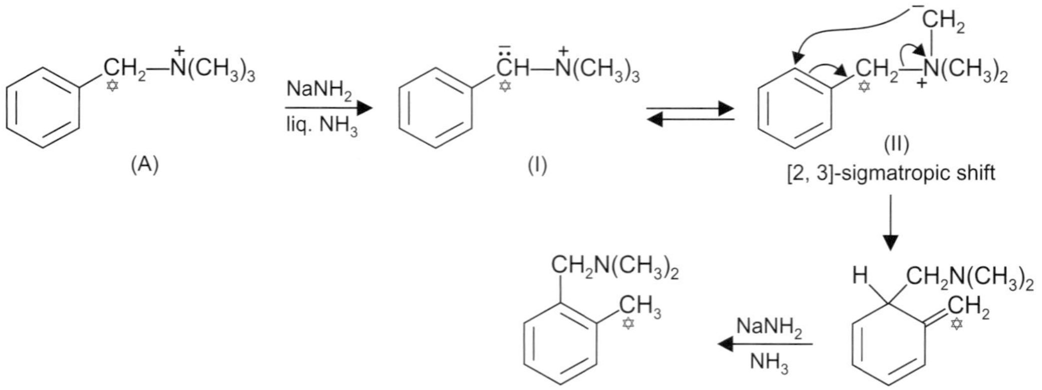

Sommelet-Hauser rearrangement

SCHEME 15.31

The mechanism involves the loss of a proton from the benzylic position to give the ylide (I, Scheme 15.31) which is in equilibrium with the second ylide (II). The ylide (II, Scheme 15.31) then undergoes a [2, 3]-sigmatropic rearrangement followed by aromatization to give the product.

The evidence that indeed it is a [2, 3] migration on the ylide (II, Scheme 15.31) rather than a [1, 4] shift on the benzylic ylide (I) was provided by working with a substrate (A, Scheme 15.31) in which benzylic carbon was labelled (^{14}C). As expected of [2, 3] sigmatropic shift on (II, Scheme 15.31), the label was found on the methyl group of the aromatic ring. In case this rearrangement instead involved a [1, 4]-shift on the benzylic ylide (I, Scheme 15.31) the label will not be on the methyl group of the aromatic ring (Scheme 15.32).

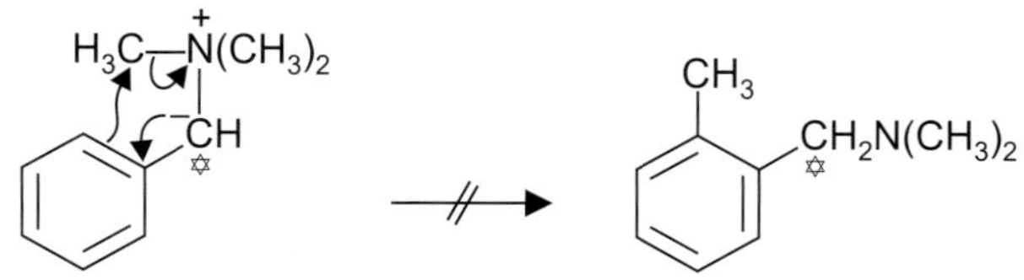

SCHEME 15.32

PROBLEM 15.4

When dibenzyldimethylammonium salt (Scheme 15.32a) is reacted with a strong base PhLi, both Steven's and Sommelet-Hauser rearrangements were observed. Write the products from each rearrangement.

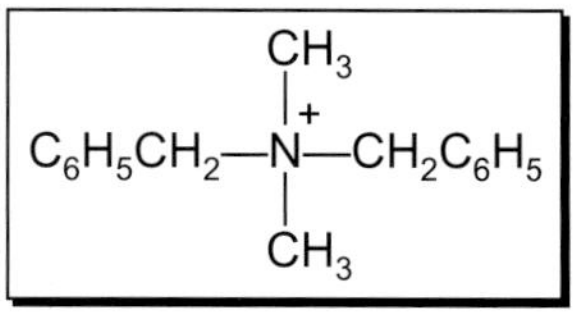

SCHEME 15.32a

ANSWER. *These are in Scheme 15.33.*

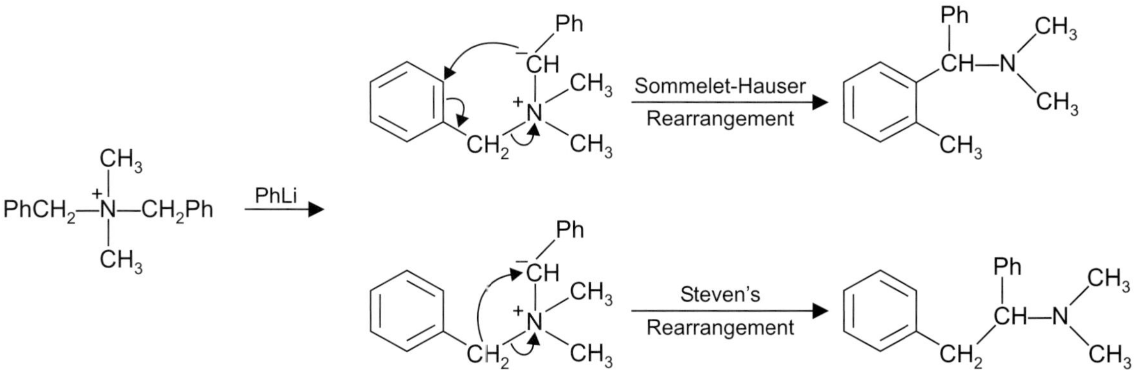

SCHEME 15.33

15.5 AROMATIC REARRANGEMENTS

The rearrangement of alkyl benzenes to their isomers is observed after *Ipso* attack. The alkyl migration in these cases are typical of carbocation intermediates. The following aromatic rearrangements have already been discussed.

A. Fries Rearrangement (see Chapter 8).

B. Photo-Fries Rearrangement (see Chapter 10).

C. Reimer-Tiemann Reaction (see Chapter 8).

D. Benzidine Rearrangement.

In this remarkable rearrangement (Scheme 15.34) an intramolecular migration from nitrogen to carbon takes place, and hydrazobenzenes give benzidines on treatment with acid. The 4, 4′-compound is formed by a [5, 5]-sigmatropic rearrangement.

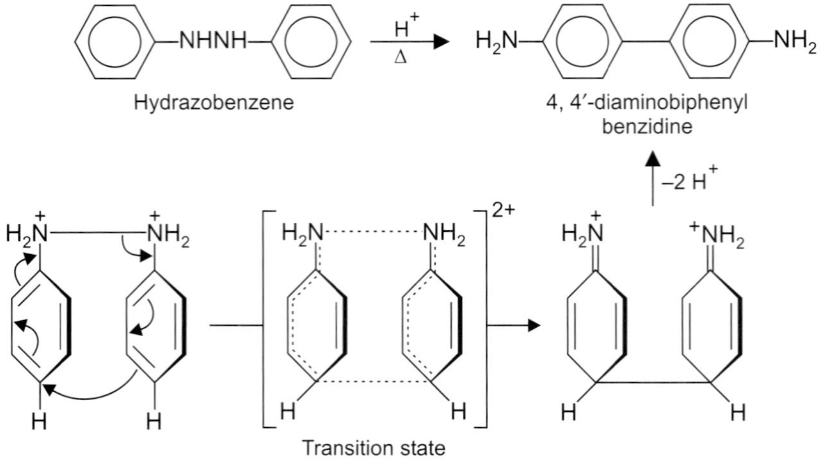

Benzidine rearrangement

SCHEME 15.34

15.6 FREE RADICAL REARRANGEMENTS (See Sec. 16.9)

PROBLEMS

15.1 How the following compound will undergo a pinacol-pinacolone rearrangement?

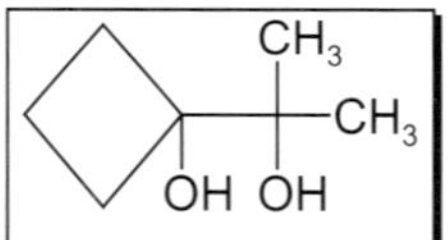

(I)

15.2 A ketone reacts with diazomethane to give either the next higher homologue or an oxirane. Give mechanisms.

15.3 Give the mechanism of the reaction of a ketene with diazomethane to give cyclopropanone.

15.4 Write the structure of the reaction product from succinimide with bromine and aqueous potassium hydroxide.

15.5 Preparation of β-amino pyridine *via* electrophilic nitration of pyridine which occurs chiefly at the β-(or 3-position) gives poor yields. Pyridine resembles a highly deactivated benzene derivative. How one can plan the synthesis of β-aminopyridine from naturally occuring nicotinamide.

15.6 Write the products from the reaction of a peracid with cyclohexanone, reaction of diazomethane with cyclohexanone and reaction of cyclohexanone with HN_3/H_2SO_4.

ANSWERS TO THE PROBLEMS

15.1 The OH group attached to the cyclobutyl group cannot be lost to give a cyclobutyl carbocation since this would involve an increase in ring strain in going from 90° to 120°. The formation of the carbocation at the other carbon is followed by the migration of the ring residue (equivalent to alkyl group).

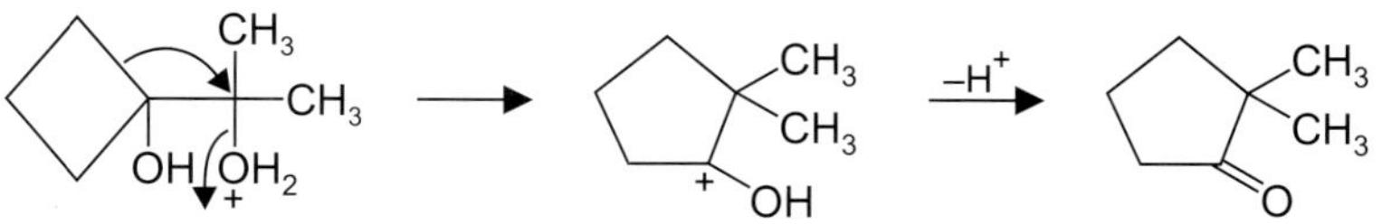

15.2 Diazomethane is a resonance hybrid and acts first as a carbon nucleophile.

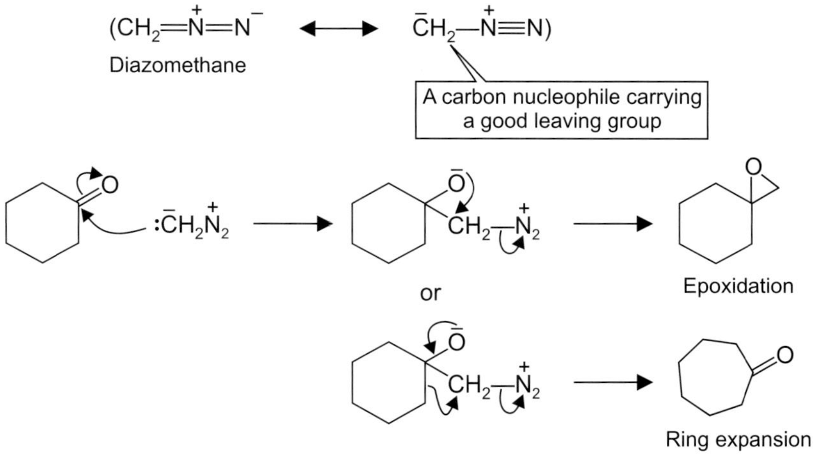

15.3 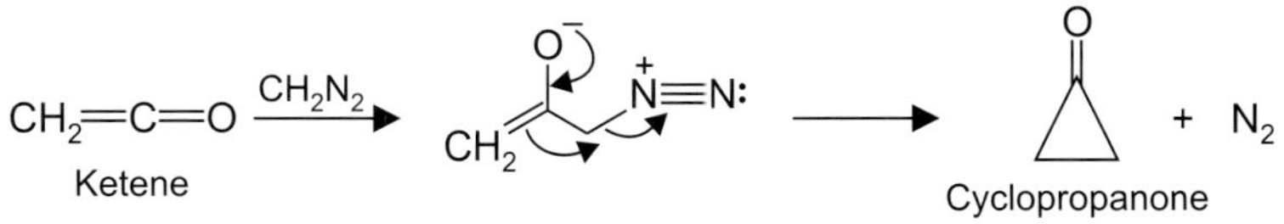

15.4 The product formed will be β-alanine *via* the half-amide of succinic acid involving Hoffmann rearrangement.

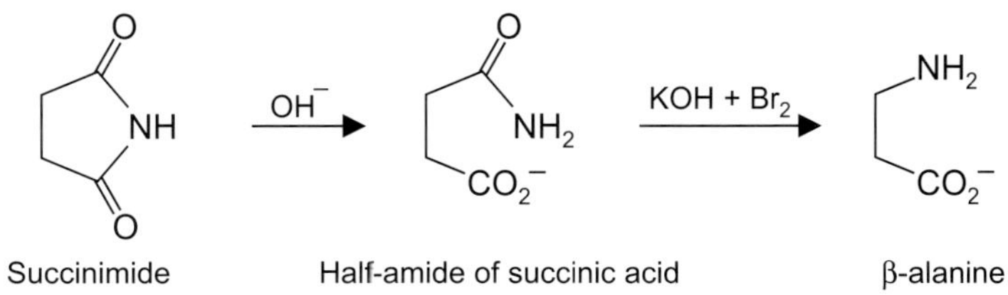

15.5 This can be done via Hoffmann rearrangement.

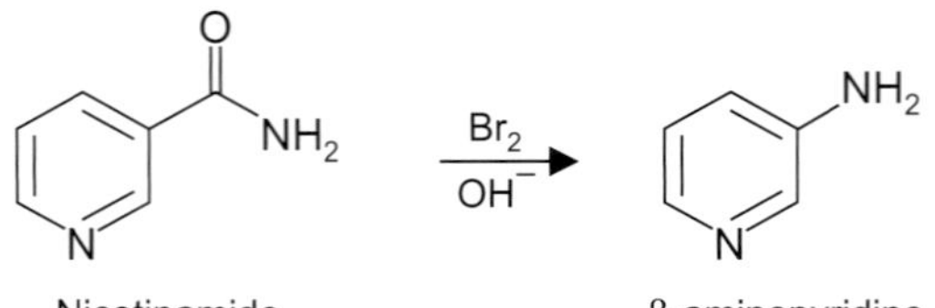

Nicotinamide β-aminopyridine

15.6 The first two reactions are insertion reactions of oxygen and CH_2 group, the nucleophilic group in a peracid and diazomethane carry a good leaving group. The last reaction is Schmidt reaction, which with ketones gives amides.

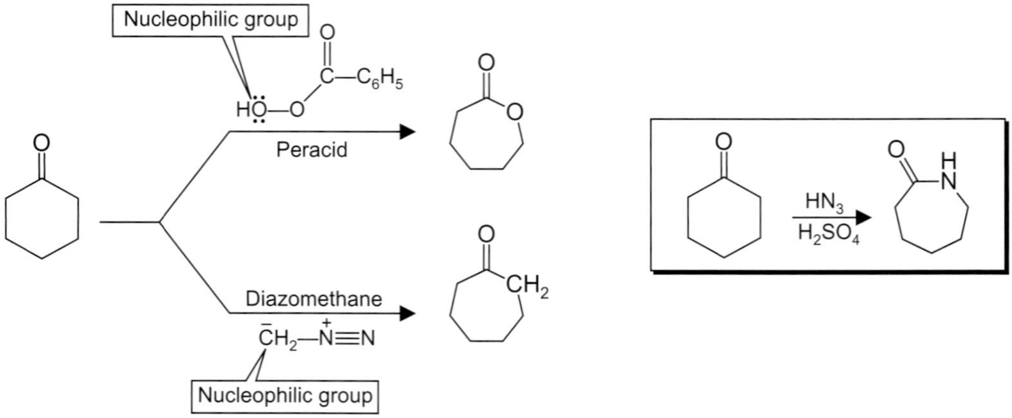

CHAPTER 16

Free Radical Reactions

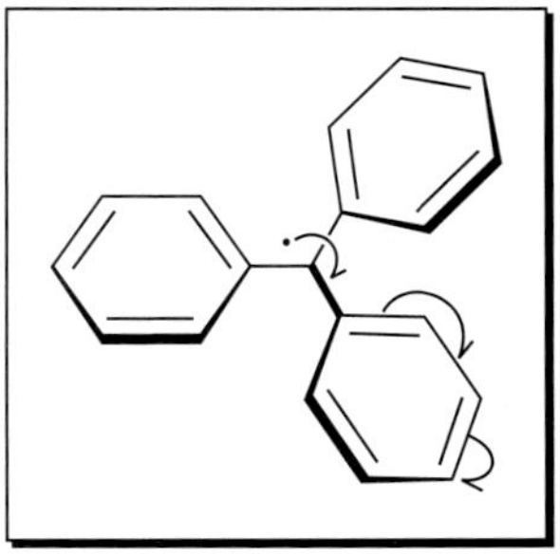

Free radical is any atom or group that possesses one or more unpaired electrons. Elements such as halogen atoms ($Cl^{\bullet}$) and alkali metals ($Na^{\bullet}$) are free radicals, since they have odd numbers of electrons. Molecular oxygen (O_2) exists as a diradical molecule which is its triplet state (two unpaired electrons, Scheme 16.1) instead of having all of its nonbonded electrons paired. A species with paired electrons is termed singlet oxygen. Singlet oxygen can be obtained by the photochemical excitation of O_2, generally in the presence of a photosensitizer. It is singlet oxygen which reacts with unsaturated species by pericyclic mechanisms. One or two electron reduction of molecular oxygen gives superoxide and peroxide ions which are harmful to several living organisms and several enzymes eliminate these from a living system.

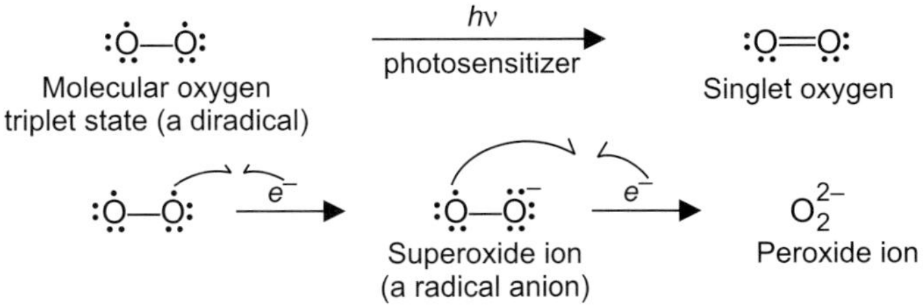

SCHEME 16.1

16.1 STRUCTURE, STABILITY AND GEOMETRY

(i) Structure (see Schemes 4.38 and 4.39)

(ii) Stability and Geometry (see Schemes 4.40–4.42)

Some obvious trends have already been discussed to show that weakest bonds have most stable radicals. A C—H bond decreases in strength in R—H when R goes from primary to secondary and then to tertiary. Thus tertiary alkyl radicals are the most stable while methyl radicals are the least stable (see Table 1.2). Similarly since the benzylic C—H bond and allylic C—H bond are particularly weak (DH° = 87 kcal/mol), therefore allyl and benzyl radicals are more stable. These are discussed under section 4.16*c*. Bond dissociation energies reflect on the stability of free radicals (see Schemes 1.25, 2.19 and 4.40).

16.2 PREPARATION

Initiation of Radicals

Weak bonds in molecules undergo homolytic cleavage to form free radicals. The bond homolysis of even weak bonds is endothermic and therefore, energy in the form of heat (Δ) or light (hν) is normally needed to generate free radicals for initiation step of a free radical chain reaction.

(*i*) Triphenylmethyl Radical—A Stable Radical in Solution

Gomberg (1900) prepared triphenylmethyl radical (Scheme 16.2) by treating triphenylmethyl chloride with silver metal when a yellow solution developed. The solution was rapidly decolourized by oxygen and iodine, reagents known to react with carbon free radicals. The dimeric structure displayed UV (λ_{max} = 313 nm) and ^{1}HNMR spectra (determined in 1970) are characteristic of a cyclohexadiene structure (Scheme 16.2). The ^{1}HNMR spectrum showed a typical 1H, singlet at δ 5.0 ppm due to the proton on the cyclohexadiene ring. The other expected signals are at δ 7.4 ppm (m, 25 protons on the five Ph groups), and a double doublet 5.8 – 6.4 for the four olefinic protons of the cyclohexadiene moiety.

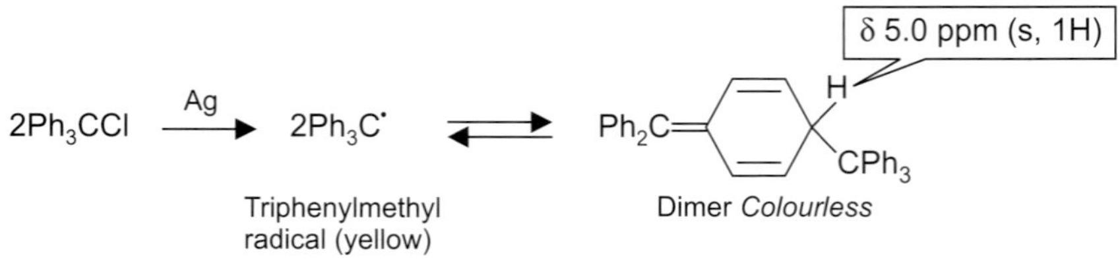

SCHEME 16.2

The triphenylmethyl radical exists in solution and is stable enough to account for about 2% of the equilibrium mixture. Its stability is due to steric reasons. In fact the phenyl groups in triphenylmethyl radical are not coplanar but have a propeller shape with phenyl groups twisted out of the plane by about 30° (Scheme 16.3). The delocalization of the unpaired electron is thus far less than ideal. The central carbon accommodates most of the radical character, the steric shielding by the twisted phenyl groups inhibits its reactions. The dimerization is obviously via its least hindered carbon atoms (Scheme 16.3). Radicals like triphenylmethyl which can maintain their concentration almost indefinitely (or atleast for a few hours) are called persistent radicals.

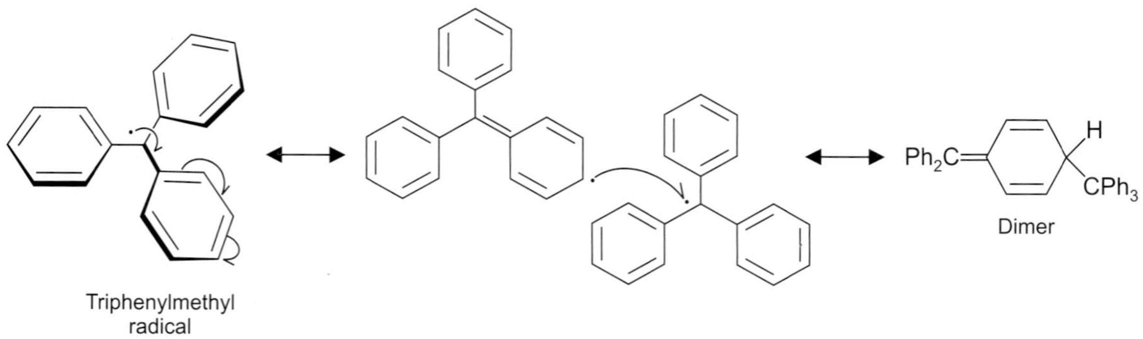

SCHEME 16.3

(*ii*) Other Stable and Persistent Radicals

A good example of another persistent radical is of a nitroxide TEMPO (2, 2, 6, 6-tetramethyl piperidine *N*-oxide) a commercial product which can even be sublimed (Scheme 16.3*a*). Steric hindrance is one factor which leads to its exceptional stability. Moreover, there is an interaction between the lone pair orbital on nitrogen and the singly occupied orbital on oxygen which results in the formation of π bond (Scheme 16.3*a*). With three electrons available, one has to occupy the π^* orbital. As a consequence the N—O bond order is $1\frac{1}{2}$ and in case of abstraction or addition reaction of the oxygen atom this strength shall have to be reduced to 1 which is an unfavourable situation.

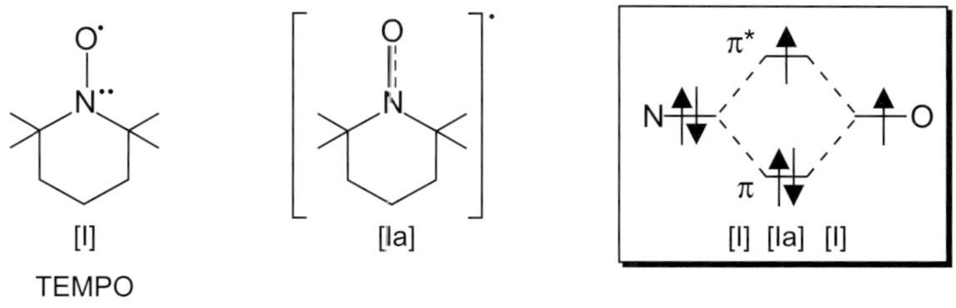

SCHEME 16.3a

(*iii*) By Thermal Homolysis of Weak Bonds

Homolytic cleavage of sigma bonds can be successful with any compound provided the temperature is sufficiently high. However, this thermal method is useful with selective weak σ bonds within a molecule which dissociate at temperatures below about 200°C. Bonds of peroxy and azo compounds have bond dissociation energies below 40 kcal/mol (170 kJ/mol) and are therefore, sufficiently weak to be good sources of radicals under typical reaction conditions (Scheme 16.4). AIBN is another good radical initiator. In this case the carbon radical which is formed is stabilized by cyano group and due to the strong nitrogen-nitrogen triple bond of the N_2 product. Here unlike the peroxy compound, the C—N bond is not so weak (see Scheme 4.42*a*). In the thermal decomposition of a peroxide (Scheme 16.4) the reaction involves the fission of the oxygen-oxygen bond and the initially formed free radical then decarboxylates to gave the fragmentation products. A fragmentation is thus a process where some initially formed radical loses a small stable molecule.

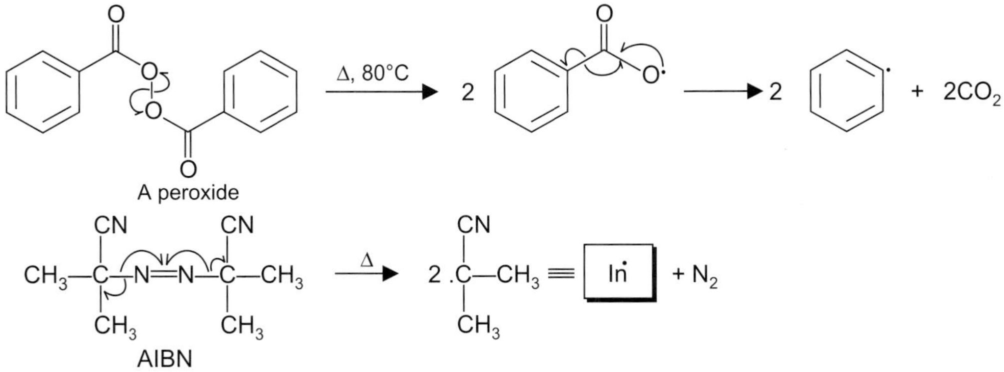

SCHEME 16.4

(*iv*) By Photochemically Induced Homolytic Bond Cleavage

When a compound is excited by the absorption of a photon of light an electron gets promoted to an unoccupied orbital. Since this orbital is normally antibonding in character, a bond in the excited molecule gets weak and subsequently cleaves in a homolytic fashion. The halogens are readily homolyzed by light to generate radicals which can be useful in chemical reactions (Scheme 16.5).

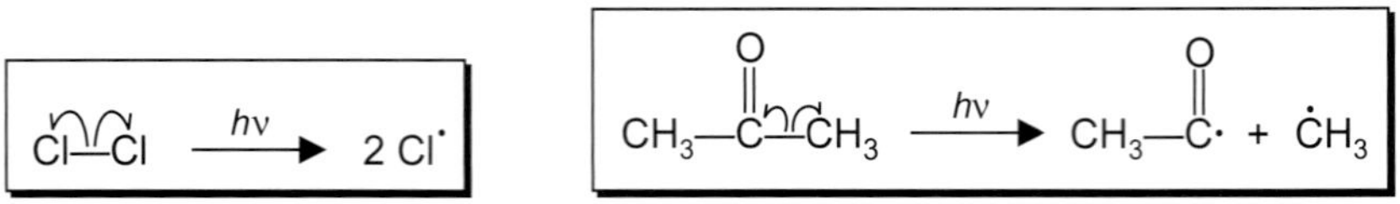

SCHEME 16.5

(*v*) By Abstraction

Recall that the H—Br bond is almost as strong as a C—C bond yet Br• radicals can be obtained from H—Br in the presence of alkoxy radicals generated by the homolysis of the weak O—O bond of a peroxide *e.g.*, dimethylperoxide ($CH_3O—OCH_3$, Scheme 16.6). In this generation of Br• from HBr the peroxy radical RO• abstracts a H• from HBr to yield a new radical Br• and gives ROH.

SCHEME 16.6

(*vi*) By Addition Reactions

A Br• radical for example adds to a double bond to generate a new carbon-centered radical (Scheme 16.6*a*).

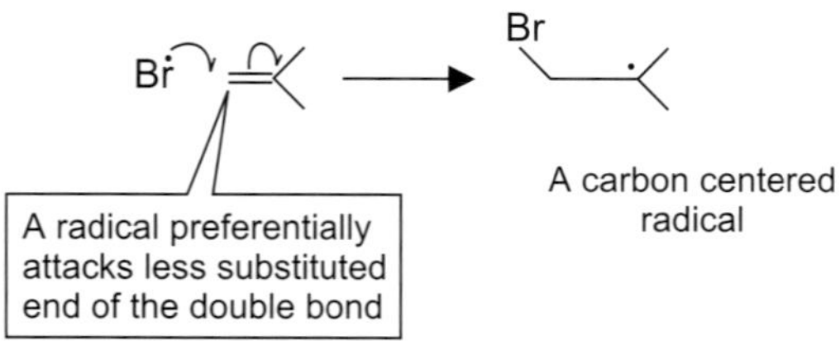

SCHEME 16.6a

(*vii*) Redox Generation

Free radicals can also be formed from oxidation—reduction (redox) processes. Transfer of electrons to or from metal atoms and ions is yet another common method to initiate radical reactions (Scheme 16.6*b*).

$$Fe^{2+} + H_2O_2 \longrightarrow Fe^{3+} + \bar{O}H + H\dot{O}$$

SCHEME 16.6b

(*viii*) Generation of Specific Carbon Centered Radicals within a Molecule

One of the most significant development in organic chemistry of radicals during the past 20 years is the use of tributyltin hydride as a reagent to generate specific carbon centered radicals in a molecule. Such specific free radical species would undergo a single reaction and will then undergo termination. The termination step leads to a single product and regeneration of that radical to continue the chain. Tributyltin radical is one such species obtained by initiation with AIBN (Scheme 16.6C). Tin forms strong covalent bonds to halogens but very weak bonds to hydrogen. This feature has been exploited to initiate a radical chain reaction. Thus if the substrate has a halogen, bromide or iodide (but not fluorine) one can generate a free radical specifically on that position (Scheme 16.6*d*).

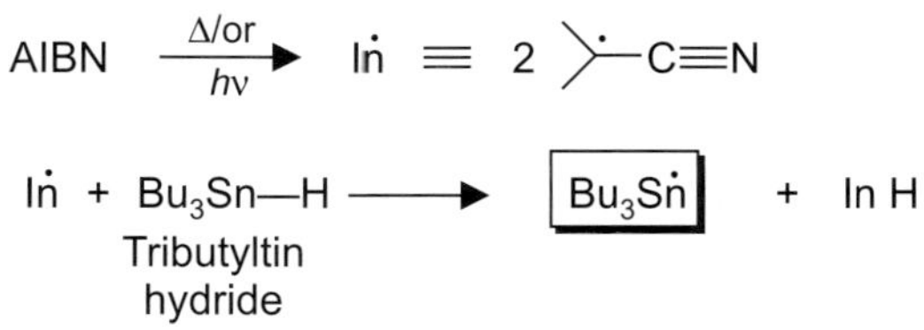

SCHEME 16.6c

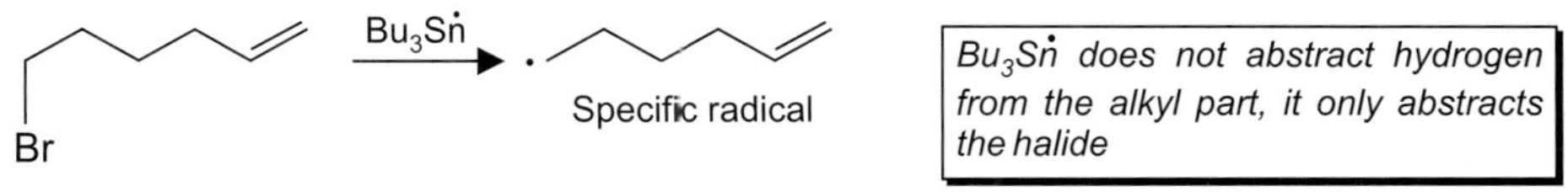

SCHEME 16.6d

In another alternative method, esters of *N*—hydroxypyridine-2-thione are used to generate radicals. The key to radical formation is via decarboxylation of an adduct formed with tributyltin radical and *N*—hydroxypyridine-2-thione due to the tendency of tin radicals to add to carbon-sulphur bonds (Scheme 16.6*e*). The net result of this sequence is the decarboxylation and reduction of the original acyloxy group.

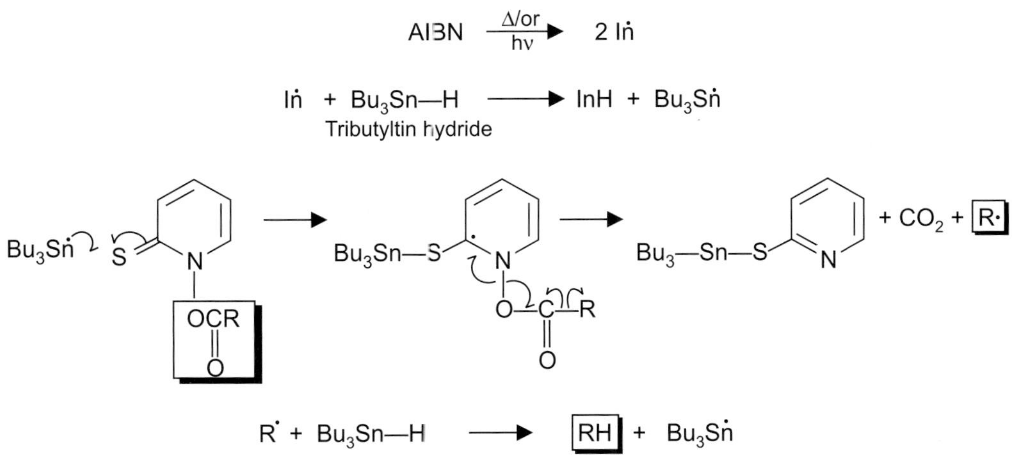

SCHEME 16.6e

Lastly mention may be made of alkyl radicals which can be derived from alkyl mercury halides (Scheme 16.6*f*). Alkyl mercury halides are stable compounds, but these on reduction with $NaBH_4$ give highly unstable alkyl mercury hydrides, which give alkyl radicals.

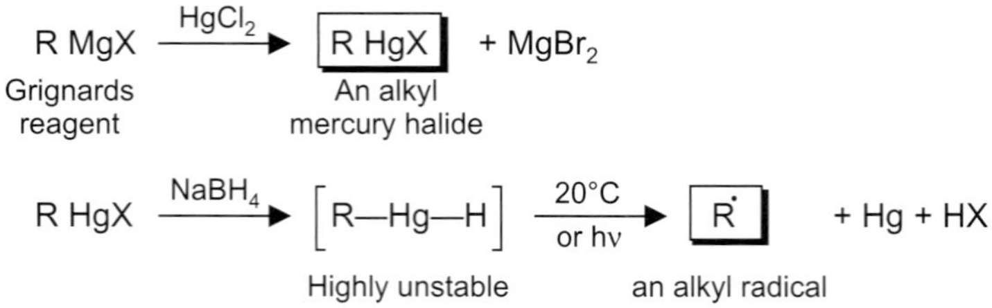

SCHEME 16.6f

16.3 PROPERTIES OF FREE RADICALS

A general review of properties of free radicals is given (see, Scheme 4.42*b*).

(A) Abstraction

1. Introduction

Free radicals react with saturated organic compounds by abstracting an atom, normally hydrogen, from carbon. A major factor that determines the selectivity of a free radical towards C—H bonds of different types is bond dissociation energy. Firstly, the rate of abstraction increases as the bond dissociation energy decreases (Scheme 16.7), and in general the order is tertiary C—H > secondary C—H > primary C—H. Recall that weakest bonds have most stable radicals. In cyclohexene (Scheme 16.7), the bromine atom could abstract a vinylic, an allylic or the methylene hydrogen, however, since it is only the allylic radical (II, Scheme 16.7) which is resonance stabilized, it will be formed far more easily than (I or III). Similarly due to low bond

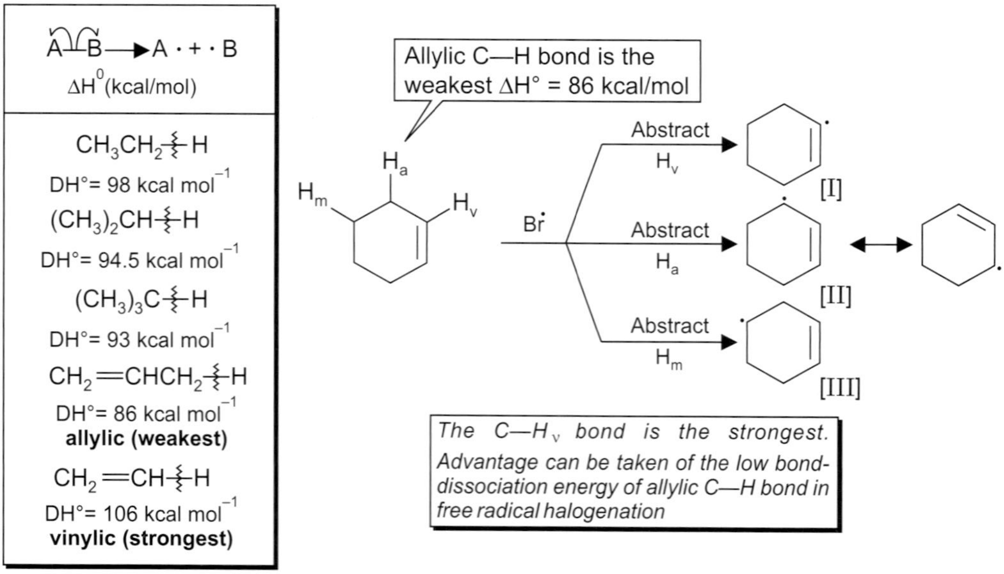

SCHEME 16.7

dissociation energy of the benzylic C—H bonds, the benzyl radical is especially stable since the delocalization spreads out the odd electron into the ring (for details see Scheme 4.40).

Advantage is taken of low bond dissociation energy of allylic (as well as benzylic) carbon-hydrogen bonds in free radical halogenation, but only under special experimental conditions to avoid addition to the double bond. Allylic bromination is carried out with *N*-bromosuccinimide (Scheme 16.8) in CCl_4 in the presence of light.

The reaction is initiated by the formation of small amount of $Br^{\bullet}$ (possibly formed by the homolytic cleavage of N—Br bond of NBS). NBS provides a constant but very low concentration of bromine. It does this by rapid reaction with HBr formed in the substitution reaction. Each molecule of HBr is replaced by one molecule of Br_2. With these conditions, in a non polar solvent and a very low concentration of bromine, it is then involved in the radical chain bromination of the allylic hydrogen and no significant addition of bromine to double bond occurs (also see, Scheme 2.19).

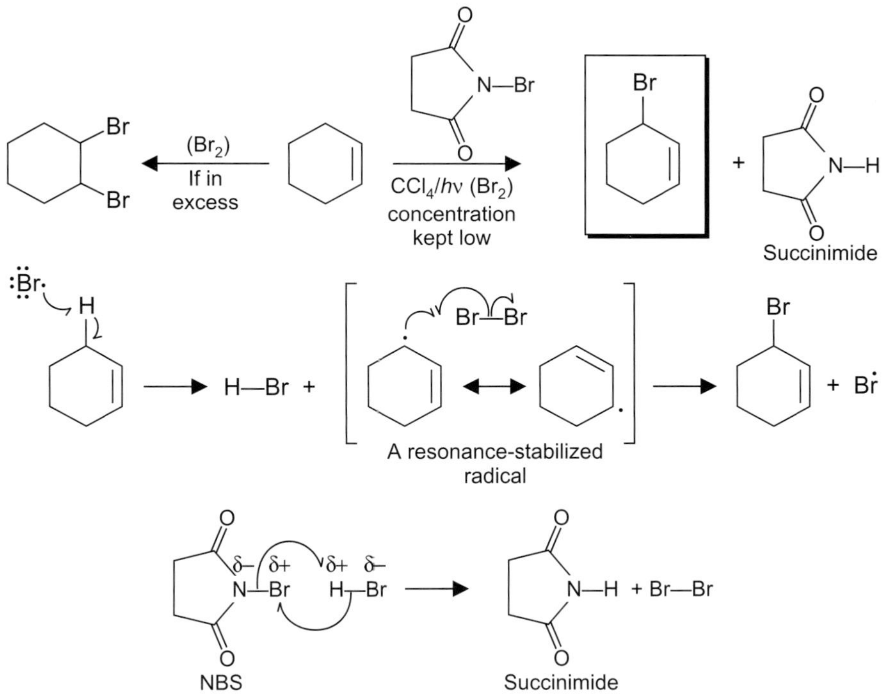

SCHEME 16.8

Allylic substitution of H by Br

Radical allylic bromination is an important method to functionalize an unfunctionalized center. There are few methods available to an organic chemist for this purpose and a well known method is Barton reaction (Scheme 16.32). Allylic bromides can be made to undergo nucleophilic substitutions to generate other functional groups and in (I, Scheme 16.8a) the regioselectivity is to remove the less sterically hindered H atom.

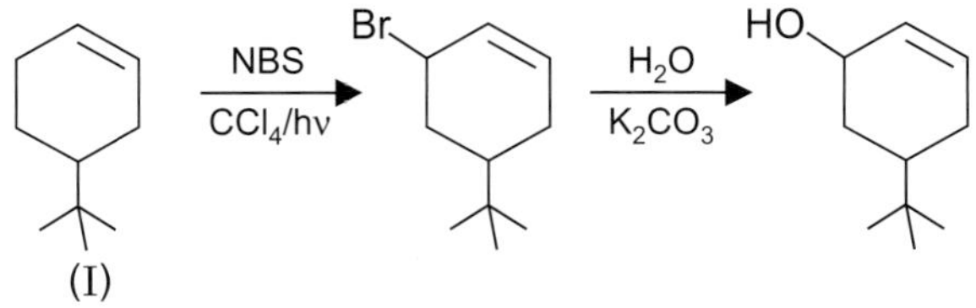

SCHEME 16.8a

PROBLEM 16.1

What product will be formed from monobromination reactions (Scheme 16.9)

(I) (II) (III)

SCHEME 16.9

ANSWER. (I) $C_6H_5CHBrCH_2CH_3$ via $C_6H_5\dot{C}HCH_2CH_3$

(II) Of the several allylic hydrogens the tertiary allylic hydrogen would be abstracted easily (Scheme 16.9*a*)

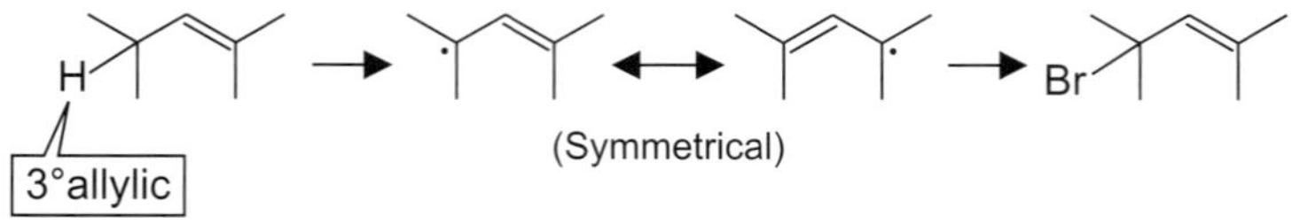

SCHEME 16.9a

(III) A hydrogen which is both benzylic as well as allylic is the most reactive (Scheme 16.9*b*)

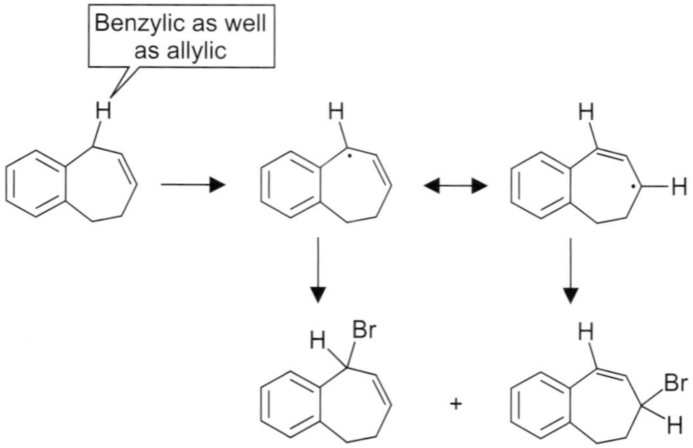

SCHEME 16.9b

The allylic and benzylic C—H bonds are significantly weaker than those in saturated systems, since the unpaired electron in the resulting radicals is delocalized. Thus compounds containing these systems not only react readily with free radicals, but also react selectively (Scheme 16.7) at the benzylic position (or allylic position).

2. Bromine atoms are significantly more selective than chlorine atoms

When chlorine atom abstracts a hydrogen atom *e.g.*, from cyclohexane then only one product is formed as all the 12 hydrogen are equivalent. A chlorine atom is far less selective than a bromine atom. In the case of cyclohexene, radical bromination is highly selective and gives the product of allylic bromination almost exclusively. However, the allylic position in cyclohexene is only slightly more reactive toward a chlorine atom as compared to other methylene positions

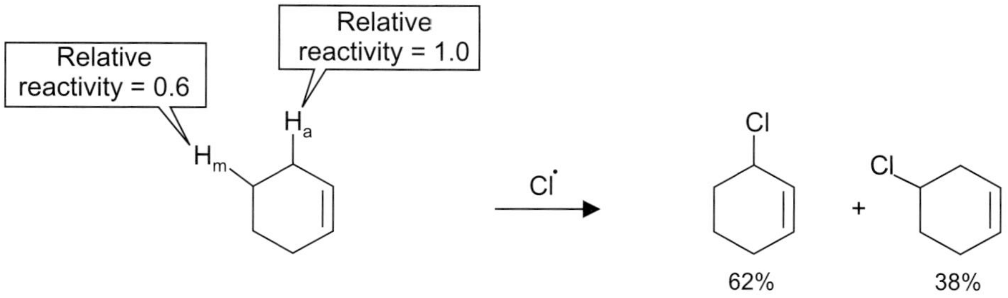

SCHEME 16.10

(Scheme 16.10). Similar is the situation with substitution on saturated compounds. Isobutane is chlorinated to give comparable quantities of isobutyl chloride and *t*-butyl chloride while bromination gives *t*-butyl bromide almost exclusively (Scheme 16.10*a*). This selectivity is explained on the basis of Hammonds postulate (see Scheme. 4.18*b*).

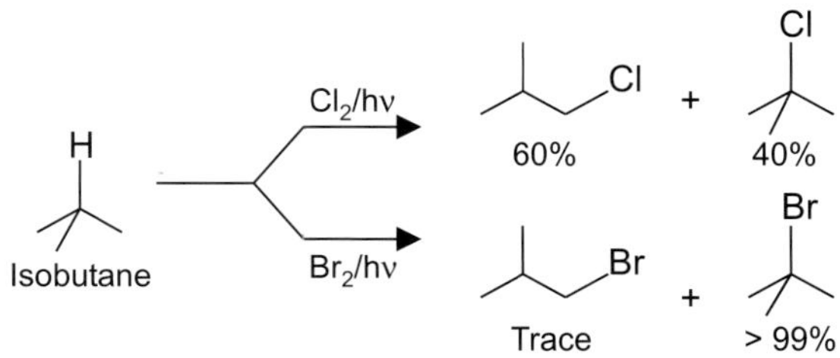

SCHEME 16.10a

The following points may be noted:

- Hammonds postulate can be applied well to a multistep reaction.
- An exothermic step has a transition state which looks more like the reactants of that step.
- The transition state of an endothermic step of a reaction looks more like the products of that step.
- Consider the rate-limiting step of the radical halogenation of alkanes—the abstraction of a hydrogen atom by a halogen radical, *e.g.*, during chlorination and bromination of ethane (Scheme 16.10*b*).

			ΔH^0 [kcal/mol (kJ/mol)]
$CH_3CH_3 + \cdot Cl \longrightarrow CH_3CH_2\cdot$	+	HCl	– 5.0 (21 kJ/mol)
+ 98		– 103	(exothermic)
$CH_3CH_3 + \cdot Br \longrightarrow CH_3CH_2\cdot$	+	HBr	+ 10.0 (41.8 kJ/mol)
+ 98		– 88	(endothermic)

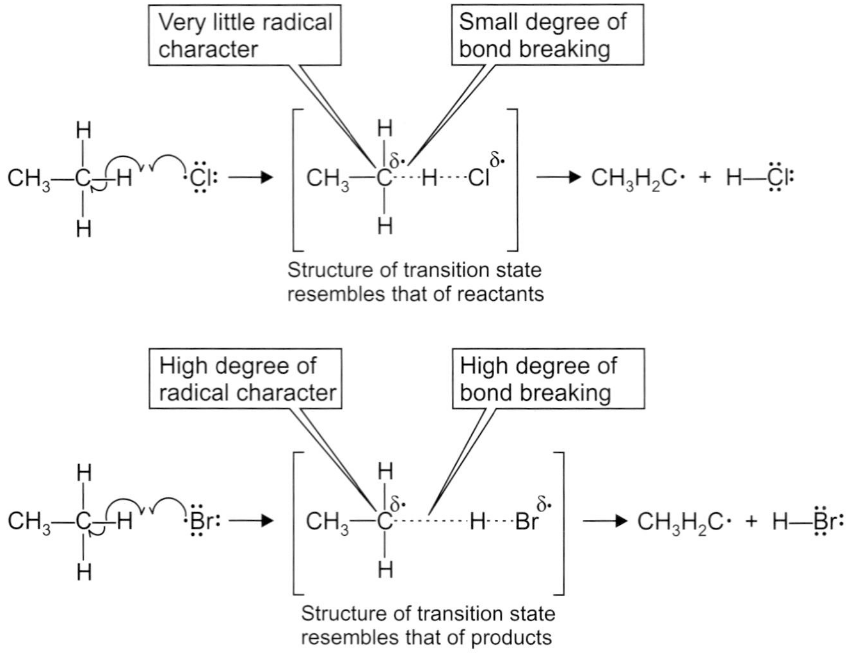

SCHEME 16.10b

- The abstraction of hydrogen by chlorine is exothermic, the transition state will thus resemble a typical exothermic reaction (Scheme 16.10*c*) and resembles starting materials more than the products (the transition state is reached early). Thus with less radical character on the carbon in the transition state, the stability of the radical intermediate does not reflect much in the product.
- However, when bromine abstracts hydrogen, it is an endothermic reaction and the transition state is reached later during the reaction and has a structure which must be similar to product radical (Scheme 16.10*c*).

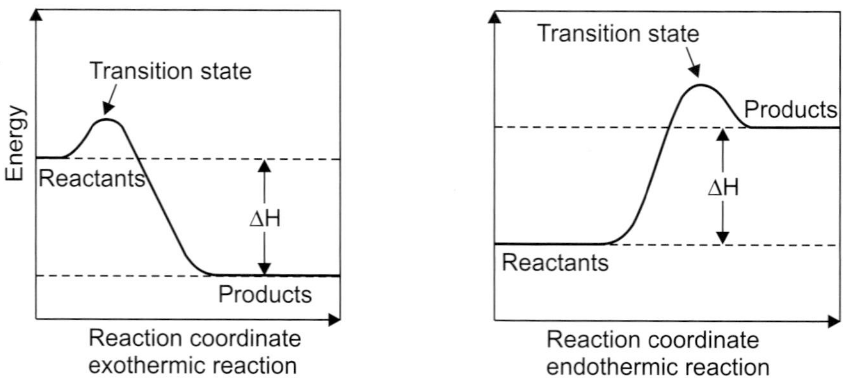

SCHEME 16.10c

Therefore, the radical stabilities 3° > 2° > 1° are markedly reflected in the transition state stabilities and regioselectivities in the same order.

3. Neighbouring group assistance in free radical substitution reactions

Photolytic halogenation generally gives a mixture of products although bromine often shows far better selectivity (*see, Scheme 6.10a*). However, the bromination of alkyl bromides *i.e.*, bromination of carbon chains with a bromine atom displays high regioselectivity. This is explained by invoking a neighbouring group effect by bromine atom during abstraction (Scheme 16.10*d*). Normally during abstraction, $\dot{Br}$ abstracts a hydrogen from R—H to give $R^{\bullet}$,

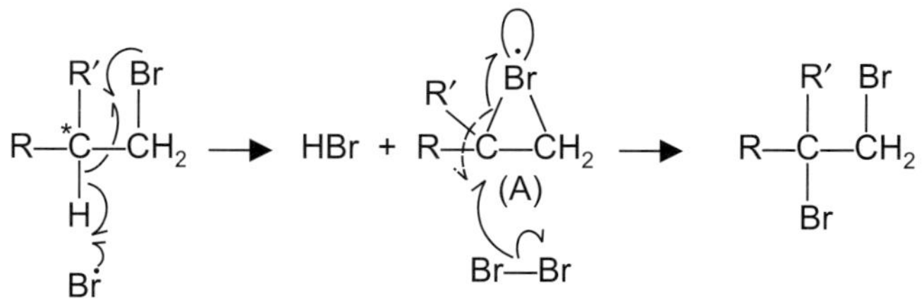

Neighbouring-group assistance in free radical reactions

SCHEME 16.10d

in this case suitably located bromine assists the abstraction process to afford a cyclic intermediate *i.e.*, a bridged free radical (A, Scheme 16.10*d*). This cyclic free radical intermediate is similar to the one invoked in S_N2 neighbouring group participation. The involvement of bridged free radical intermediate is shown by the retention of configuration of the stereogenic carbon in the substrate.

(B) Addition Reactions of Hydrogen Halides

1. Orientation of addition—A summary

Free radicals add to the common unsaturated groupings to give new radicals. The most important of the unsaturated groups in free radical synthesis is the C=C bond, addition to which is markedly selective. In particular, addition to the olefins of the type CH_2=CHX occurs almost exclusively at the methylene group, whatever the nature of X (Scheme 16.11). Three factors seem to control this selectivity. Firstly the steric hindrance between the radical and X avoids reaction at the substituted carbon atom. Secondly the grouping X stabilizes the radical (tertiary relative to primary or secondary Scheme 16.11). Moreover, if X is an alkyl group, the stabilization will result due to hyperconjugation.

2. Chain reactions are involved during addition to multiple bonds

Recall that most useful radical reactions are chain reactions and a radical chain reaction of the addition of HBr to isobutene is presented (Scheme 16.11). In each step of such a cycle a radical is consumed while a new radical is generated.

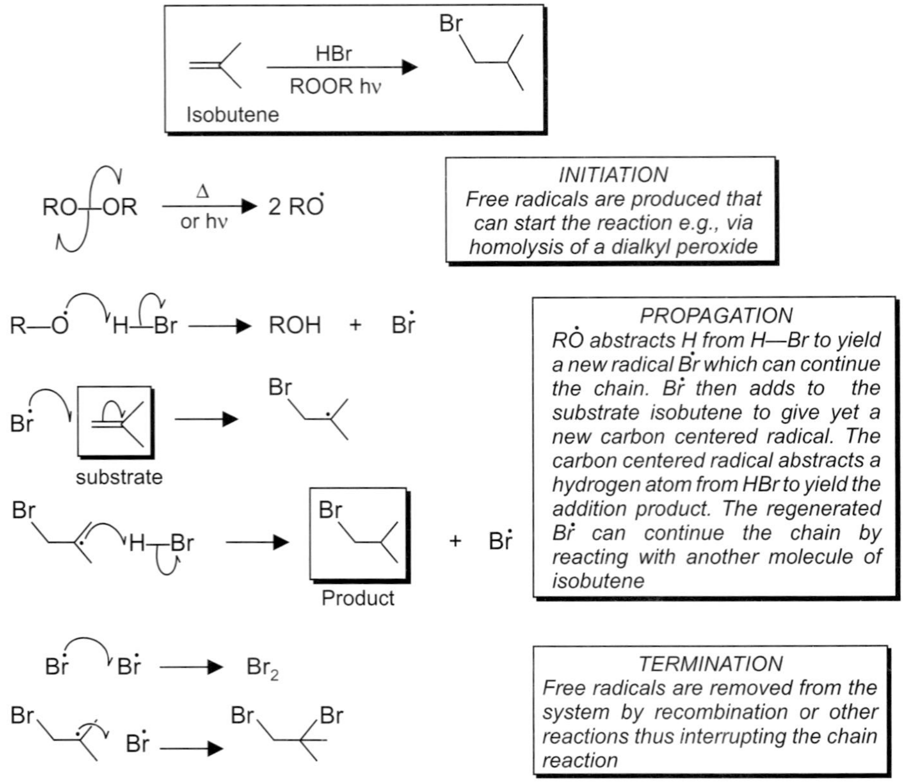

SCHEME 16.11

3. Formation of carbon-carbon bonds via addition of some carbon radicals to alkenes
Here (Scheme 16.12), a classic example of addition of a halomethane (chloroform $CHCl_3$ or bromoform $CHBr_3$) to olefins initiated by benzoylperoxide is presented (the reaction can also be initiated by photolysis).

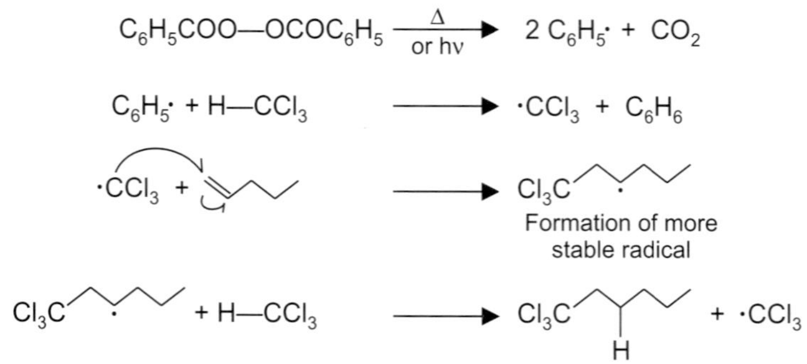

Addition of a haloform to an alkene

SCHEME 16.12

PROBLEM 16.2

Write the product of the reaction (Scheme 16.12a)

$$C_6H_5CH{=}CH_2 \xrightarrow[h\nu]{Br{-}CCl_3}$$

SCHEME 16.12a

ANSWER. *Under the influence of light the weakest C—Br bond (rather than C—Cl bond) undergoes homolysis and* $\dot{C}Cl_3$ *radical then adds to the less hindered (unsubstituted end of the alkene to yield the more stable secondary benzylic radical. The secondary benzylic radical (I, Scheme 16.12b) then abstracts a* $B\dot{r}$ *from* $BrCCl_3$ *to give the observed product.*

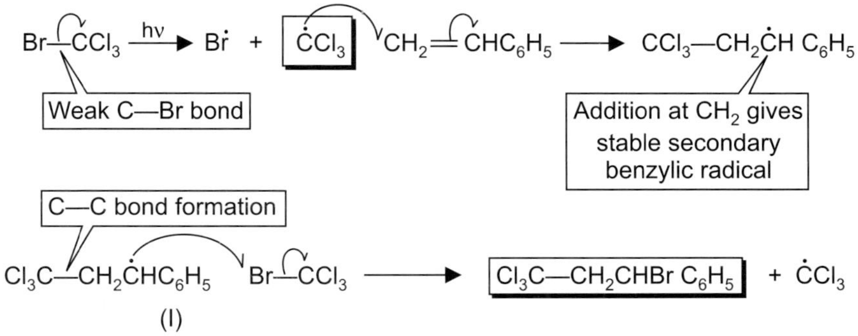

SCHEME 16.12b

Aldehydes can also add to alkenes to give ketones in a chain reaction (Scheme 16.12*c*). Recall that organic free radicals are often formed when other radicals or atoms abstract hydrogen from C—H bonds of a substrate. However, often this procedure is of less value to form radicals with specific structures, since in a substrate a hydrogen can be abstracted from several positions. Free radicals react with aldehydes by a selective attack at a hydrogen bonded to the carbonyl group to generate acyl radicals. The aldehydic C—H bond is far weaker than other bonds of CH_3 group.

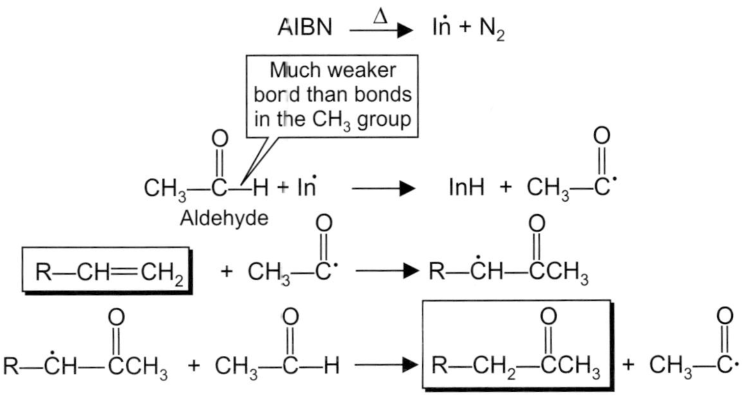

Addition of an aldehyde to an alkene

SCHEME 16.12c

4. *Role of tributyltin hydride $Bu_3Sn—H$ to reduce carbon-halogen bonds to carbon-hydrogen bonds and in carbon-carbon bond formation*

Carbon-centered free radicals role of organotin hydride

Recall the carbon-carbon bond formation reaction e.g., during the addition of actaldehyde to an alkene (see, Scheme 16.12c). In acetaldehyde the aldehydic C—H bond is far weaker than C—H bonds of CH_3 group. Thus the abstracting radical selectively attacks at the hydrogen bonded to the carbonyl group. In other cases there may be however, little selectivity between different C—H bonds leading to mixture of products to be of real synthetic value. Thus, one may look for an abstracting radical which abstracts an atom other than hydrogen and organotin radical [trialkyltin radical $(Bu)_3$ $Sn^\bullet$] is one such radical. A trialkyltin radical readily abstracts bromine or iodine (but not fluorine) from carbon to yield a carbon center radical (Scheme 16.12d). This radical however, does not abstract a hydrogen from C—H bond since Sn—H bond is weak.

$$\boxed{Bu_3Sn^\bullet} + Br—R \longrightarrow \boxed{R^\bullet} + Bu_3Sn—Br$$

SCHEME 16.12d

Tin hydrides have played a highly significant role in organic synthesis involving radical reactions. Tin forms strong covalent bonds to halogens, but the covalent bond with hydrogen is very weak. This aspect has been exploited in a radical chain reaction particularly in following reactions:

- Reduction of haloalkanes to alkanes. During the reduction of an organohalide molecule with tributyltin hydride an iodine atom is more easily transferred than a bromine atom or a chlorine atom. When one has a choice, an organoiodide or a bromide can be selected as the substrate. The substrate reactivity also follows the expected order, the more stable radical is generated faster in the first propagation step (allyl, benzyl > 3° > 2° > 1° > vinyl, phenyl).
- Carbon-carbon bond formation
- Intramolecular addition—Formation of five membered rings.

Reduction of haloalkanes to alkanes (Scheme 16.12*e*) is an excellent synthetic method for this purpose. The mechanism starts with the homolysis of Bu_3SnH with the initiator AIBN. Use of AIBN is sufficient and a stronger peroxide initiator is not needed since the Sn—H bond of $Bu_3Sn—H$ to be cleaved is very weak and a comparatively less reactive nitrile stabilized radical generated from AIBN would be sufficient. The $RO^\bullet$ radicals from peroxides are highly reactive and will thus lead to problems by abstracting hydrogens from other positions as well. The organotin radical preferentially abstracts halogen atoms from the haloalkane (except fluorine). To generate an alkyl radical, the organotin radical, however, does not abstract a hydrogen from the C—H bonds of the alkyl halide since the Sn—H bond thus formed would be very weak. In the next step the alkyl radical preferentially abstracts a hydrogen atom from the Sn—H bond which is the weakest. Thus in the reaction (Scheme 16.12*e*) one has generated a specific radical which abstracts a halogen efficiently and undergoes a single reaction process and is then terminated. The termination step gives only a single product and the reactive radical to continue the chain. This reaction can be applied to substrates which have other

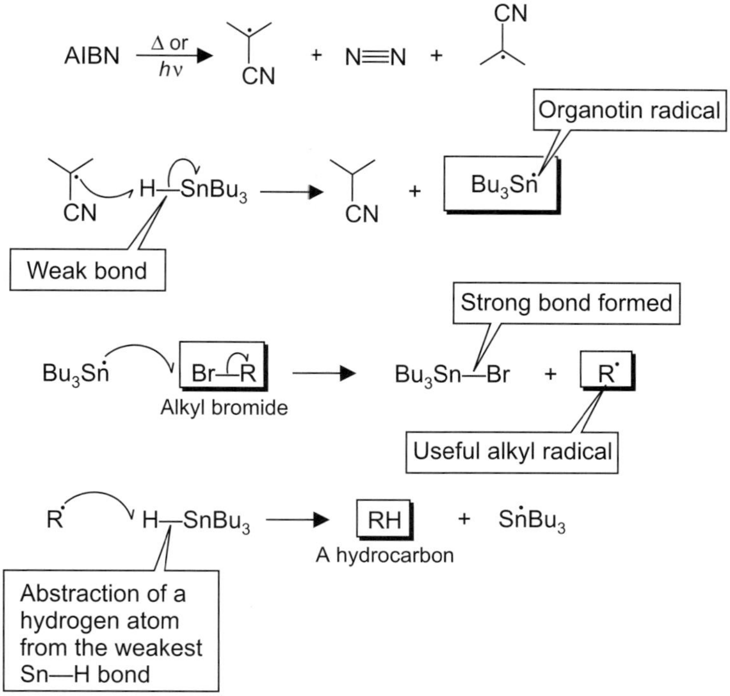

Trialkyl tin hydride reagents convert organohalides to the corresponding hydrocarbons

SCHEME 16.12e

groups *e.g.*, carbonyl which would be readily reduced by reagents like lithium aluminium hydride.

The method is used for C—C bond formation and the net addition of an alkyl group to a reactive double bond follows the halogen abstraction by an organotin radical. In this reaction an organic radical and a hydrogen atom add to the C—C double bond. This is a successful reaction since the new C—H and Sn—I bonds are far stronger than the previous bonds Sn—H and C—I which are cleaved.

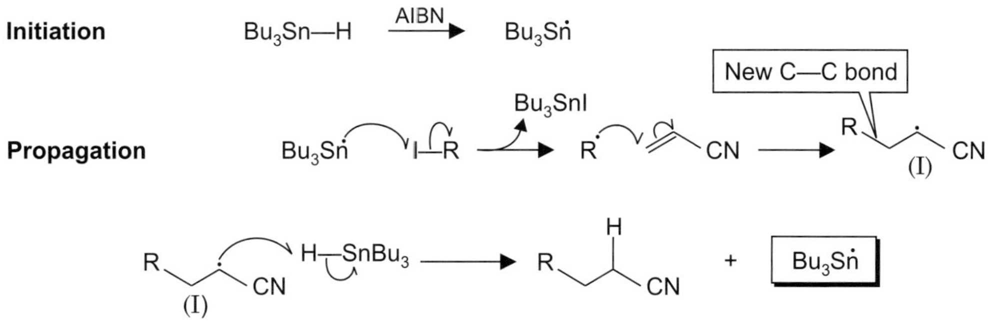

Role of tributyltin hydride in carbon-carbon bond formation

SCHEME 16.12f

Carbon-carbon bond formation by using radical reactions

The carbon-carbon bond forming reactions (Schemes 16.12b and 16.12f) using radicals have preparative value since the product radicals (I, Schemes 16.12b and 16.12f) can abstract an atom (Br• in the first case from Br—CCl_3 and Ḣ from Bu_3Sn—H in the second case) by the cleavage of unusually weaker bonds. The efficiency of this reaction indeed depends on this step since polymerization can compete with the atom abstraction step in the chain mechanism (Scheme 16.12g).

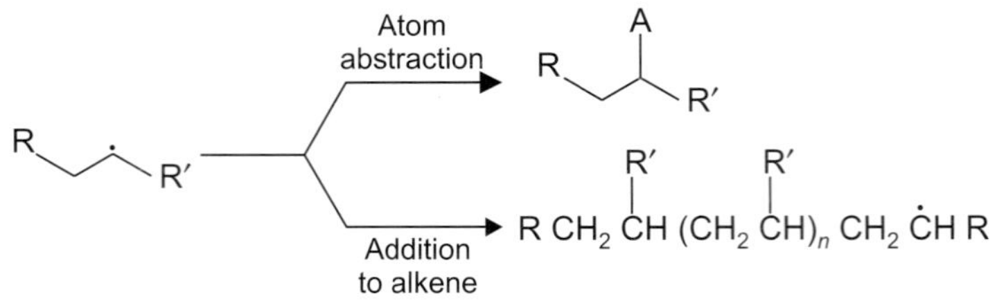

SCHEME 16.12g

When this method of free radical generation is applied to an unsaturated alkyl halide *e.g.*, 6-bromo-1-hexene (Scheme 16.12*h*), the reaction can lead to two pathways. Intramolecular addition of carbon radical to the C=C bond produces a ring or the carbon radical abstracts a hydrogen atom from tributyltin hydride to give a reduction product (Scheme 16.12*h*).

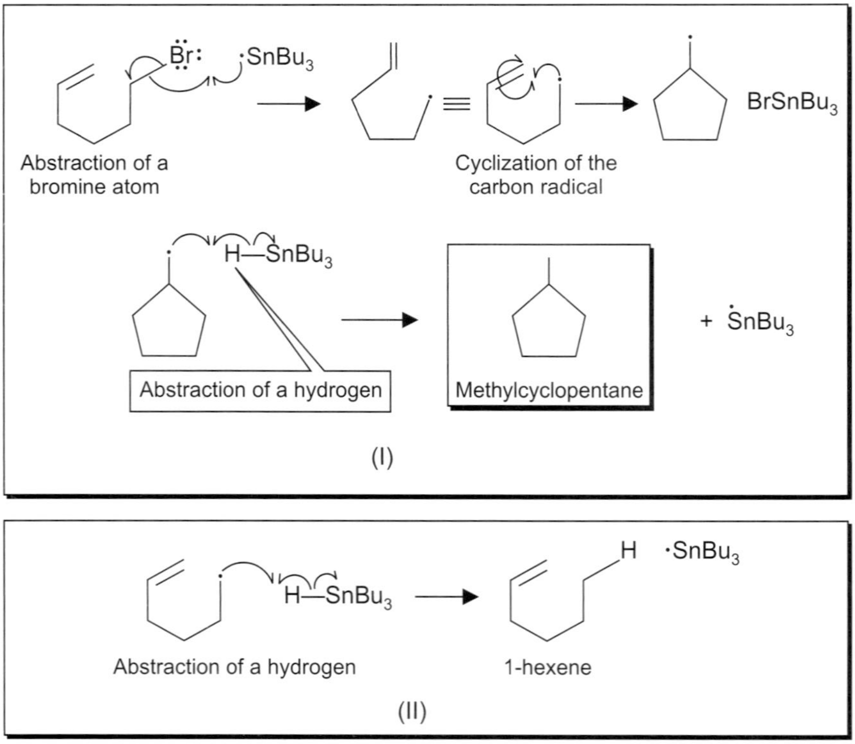

SCHEME 16.12h

These two pathways *i.e.,* substitution (reduction) and addition compete with each other. The course of reaction can be changed by changing the concentration of tributyltin hydride. The substitution reaction is a bimolecular process, thus high concentration of tributyltin hydride will favour it, while a lower concentration of tributyltin hydride will favour the ring formation.

It is known that the rates of ring-forming free radical cyclizations are 5 > 6 > 7. It was found that reaction (I, Scheme 16.12*h*) gives the five-membered ring product exclusively. Thus the regioselectivity of ring formation is not controlled by thermodynamic considerations, but rather by kinetic control of the cyclization. It is found that bond formation between the radical and the alkene stereoelectronically requires an approach angle of about 110° between the free radical center and the olefinic plane. This is because the free radical addition results via donation of the unpaired electron on the radical into the π antibonding orbital of the olefin, which coincidentally makes an angle of about 110° with the olefinic plane (Scheme 6.12*i*).

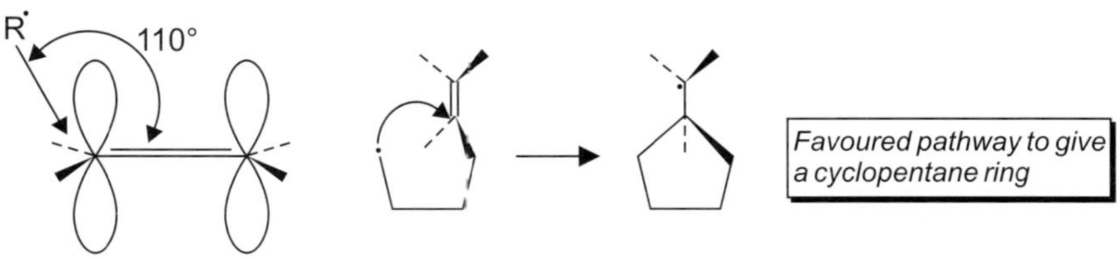

SCHEME 16.12i

It is easy to achieve this approach angle during cyclization via attack on that end of the double bond which is closest to the radical centre (favourable entropy factors) leading to a five membered ring. For attack on the other olefinic carbon the radical shall have to reach across the double bond to achieve the proper approach angle. This indeed would be a higher energy path and is kinetically not favoured, thus six-membered ring formation is not favoured.

One may appreciate that while Diels Alder reaction is one of the methods to form fused six-membered rings, radical cyclization is superior to synthesize a fused ring system containing a five membered carbocycle. The power of the method for the synthesis of fused ring system with a five membered carbocycle is presented (Scheme 16.12j). In this case the reaction is induced at a bridgehead position where radical reactions are normally difficult since the radical cannot achieve planar geometry. Thus when the halogen is separated by four carbons from the double bond, cyclization to give a five membered ring can occur.

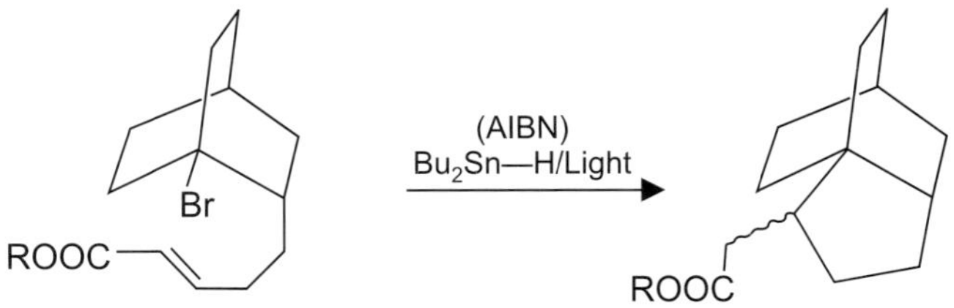

SCHEME 16.12j

PROBLEM 16.3

Write the product of the reactions (Scheme 16.12k).

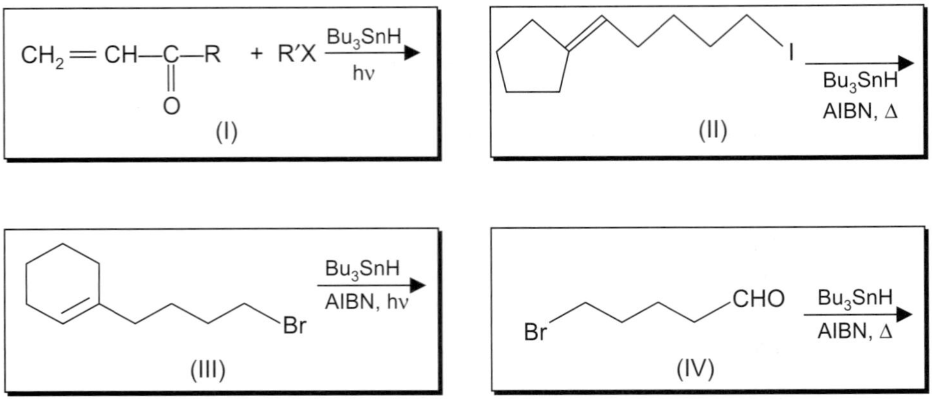

SCHEME 16.12k

ANSWER. *This is in (Scheme 16.12l)*

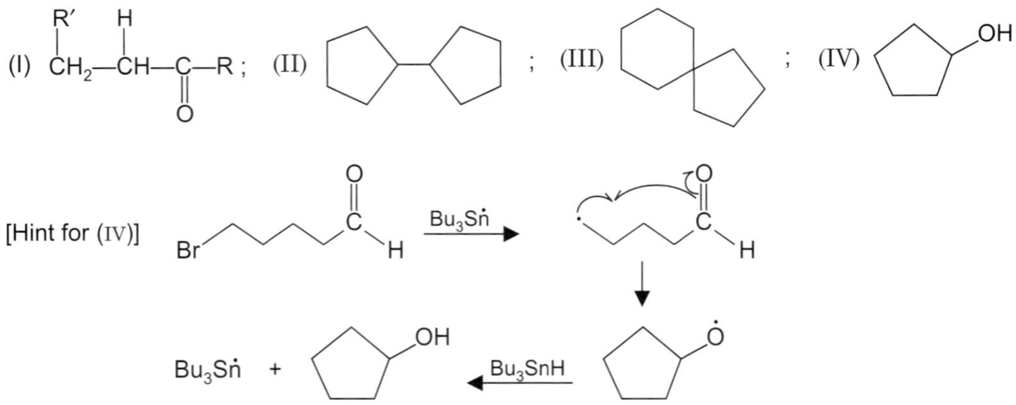

SCHEME 16.12l

5. *Peroxide effect (Hydrogen bromide adds to alkenes via a radical pathway and with apparent anti-Markovnikov regiochemistry*

Peroxide effect (Kharasch 1933). The presence of oxygen or peroxides which are formed when an alkene is exposed to the air, or added peroxides causes the addition of hydrogen bromide to take place in the direction opposite to that predicted by Markovnikov's rule (Scheme 16.13). This departure from the rule is termed as the 'abnormal' reaction, and was shown to be due to the 'peroxide effect' (Kharasch *et al.,* 1933). Hydrogen chloride, hydrogen iodide and hydrogen fluoride do not exhibit the abnormal reaction. It has been found that the addition of hydrogen bromide is 'abnormally' effected photochemically as well as by peroxide catalysts.

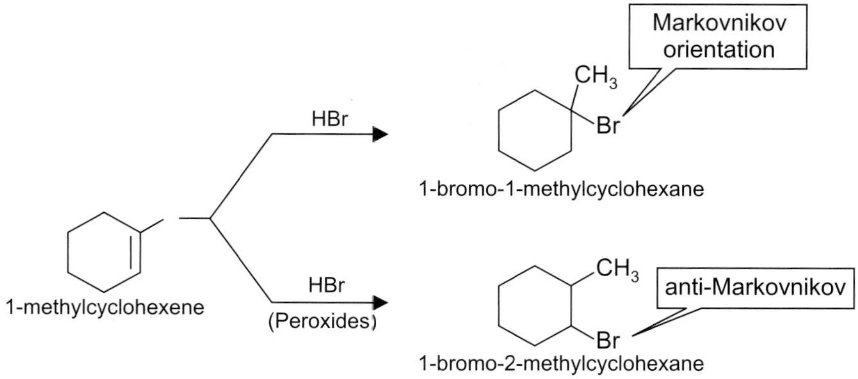

SCHEME 16.13

Peroxide is a radical initiator, thus addition of HBr to an alkene involves radicals as intermediates, the more stable the radical the easier it is to form—the energy barrier is lower for its formation. Thus the bromine radical adds to that sp^2 carbon in the alkene (I, Scheme 16.14) that is bonded to the most hydrogens to from a stabler (tertiary) of the two possible free radicals

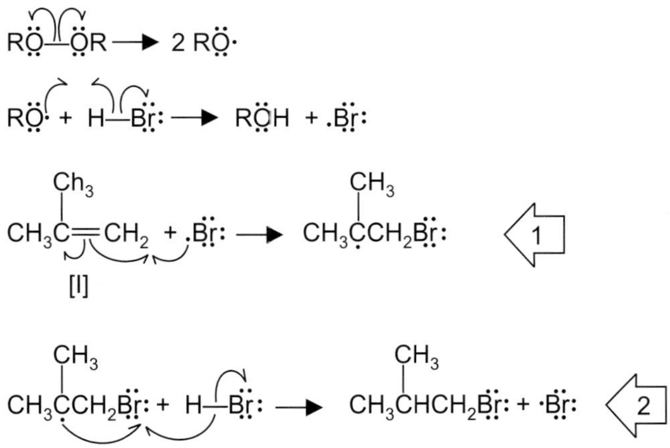

SCHEME 16.14

(tertiary and primary). In the reaction of HBr (absence of peroxides) to an alkene the addition is ionic, initiated by the addition of H^+ to give the most stable carbocation (Markovnikov addition

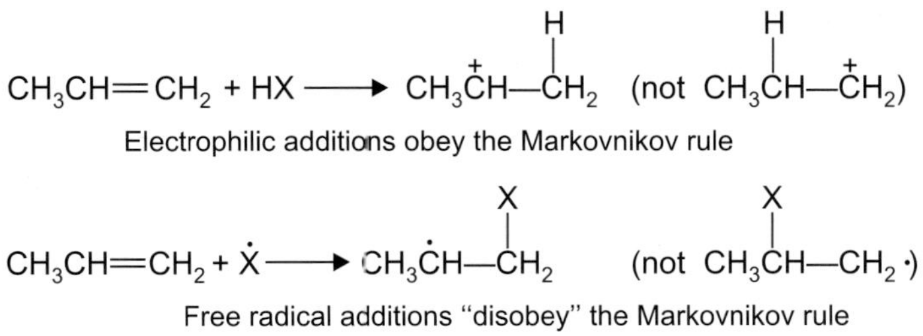

SCHEME 16.15

Scheme 16.15). In the presence of peroxides the addition of HBr to an alkene is initiated by $\dot{Br}$ to produce the most stable free radical (anti Markovnikov rule, Scheme 16.13). Peroxide has no effect on the addition of HCl or HI to an alkene, which, however, occurs according to Markovnikov rule only (heterolytic addition). In a radical reaction, the steps which propagate the chain reaction compete with the steps which terminate it. Termination steps are always exothermic since only bond making occurs. Thus, when both propagation steps are exothermic only then these compete with termination. When one considers the energetics of the two propagating steps (1 and 2, Scheme 16.16) only for hydrogen bromide these steps are exothermic. For HCl and HI one of these two steps is on the other hand endothermic.

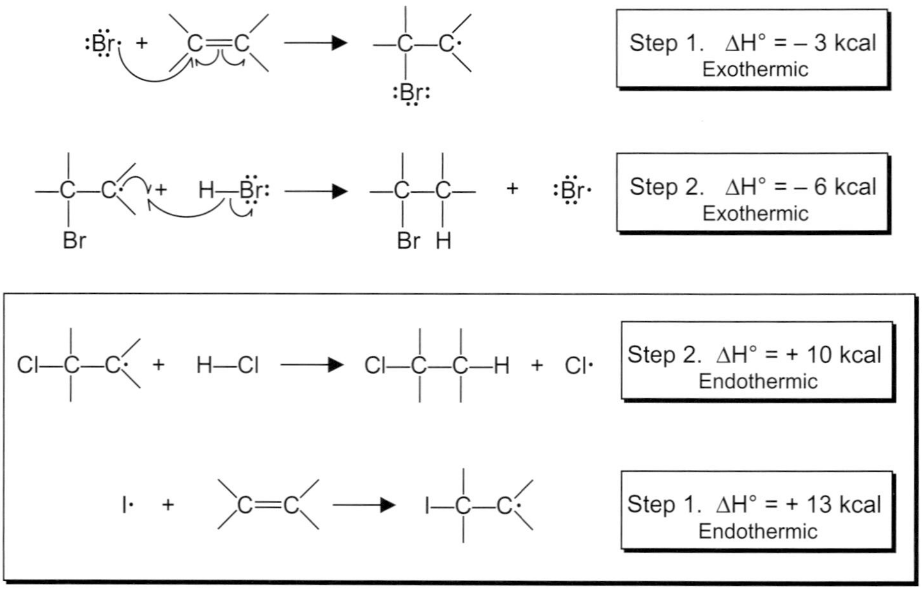

SCHEME 16.16

16.4 AROMATIC NUCLEOPHILIC SUBSTITUTION—$S_{RN}1$ SUBSTITUTION

Now several substitution reactions are known which occur by electron transfer processes and are designated $S_{RN}1$ reactions (R for radicals). These refer to a nucleophilic substitution *via* a radical intermediate which proceeds *via* a unimolecular decomposition of a radical anion derived from a substrate. The general mechanistic details may be explained by with an aryl halide as the substrate.

1. Electron transfer to the substrate gives a radical anion, which then expels the leaving group.
2. The radical thus obtained reacts with the nucleophile to afford a radical anion which should be capable of continuing the chain reaction (Scheme 16.17). The similarity of the first two steps to those in an S_N1 reaction, has lead to the designation $S_{RN}1$ (R for radicals).

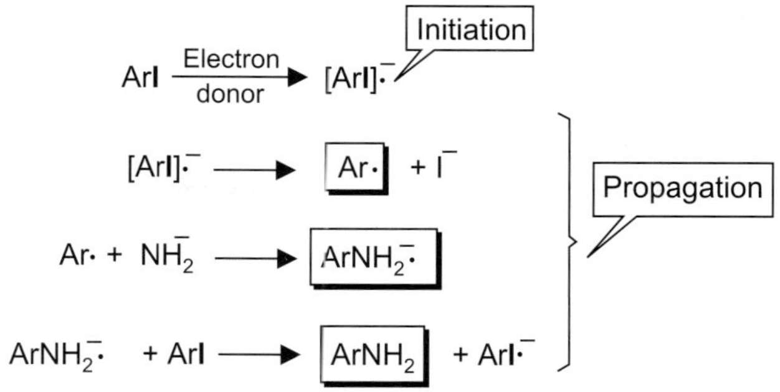

The $S_{RN}1$ Mechanism

SCHEME 16.17

16.5 HOMOLYTIC AROMATIC SUBSTITUTION

Both alkyl as well aryl radicals substitute in aromatic nuclei by an addition-elimination sequence. The free radical adds to the aromatic ring to give a reactive cyclohexadienyl intermediate (I, Scheme 16.18). Alkylation can be carried out on heating the compound with *e.g.*, diacylperoxide. The yields of the product (II, Scheme 16.18) tend to be low since alkylbenzenes display enhanced reactivity toward free radicals at the benzylic carbon.

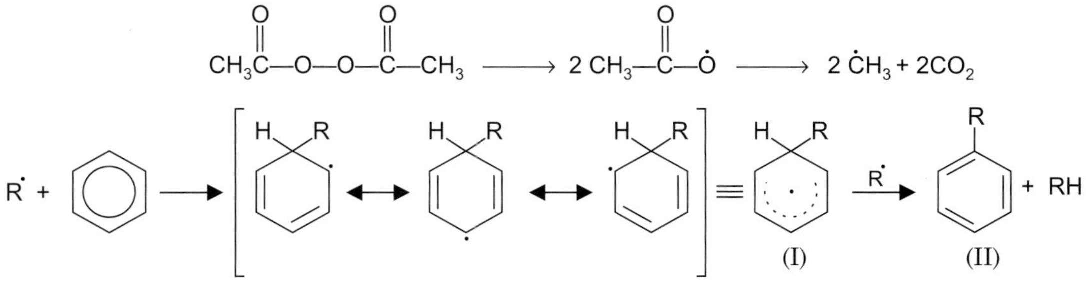

SCHEME 16.18

For arylation, diacylperoxides like dibenzoylperoxide are employed. Thus biphenyl can be made in a high yield from benzene and dibenzoyl peroxide and dimeric products are also formed (Scheme 16.19).

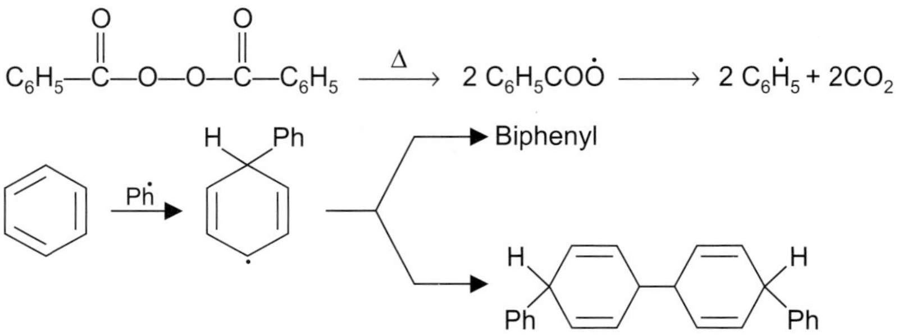

SCHEME 16.19

16.6 SOME NAME REACTIONS

(A) Gomberg Reaction

This reaction involves the synthesis of biaryls by radical reaction. When the acidic solution of a diazonium salt is made alkaline, the aryl portion of the diazonium salt couples with another aromatic ring (Scheme 16.20).

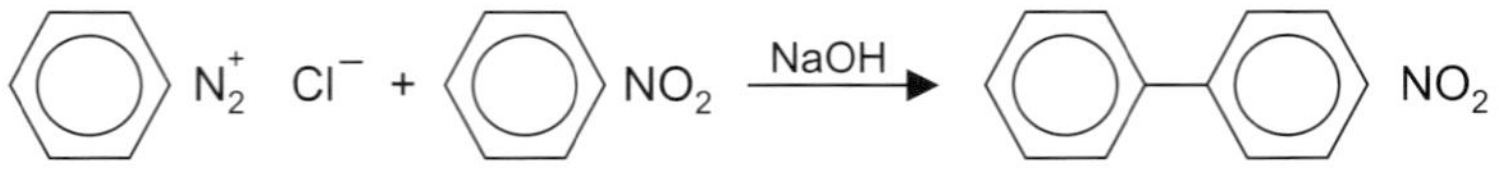

Gomberg-Bachmann reaction

$$C_6H_5\overset{+}{N_2}\,Cl^- \xrightarrow{OH^-} C_6H_5{-}N{=}N{-}OH \longrightarrow C_6\dot{H}_5 + N_2 + \dot{O}H$$

$$C_6\dot{H}_5 + C_6H_5NO_2 + \dot{O}H \longrightarrow p\text{-}C_6H_5{-}C_6H_4NO_2 + H_2O$$

a free radical mechanism

SCHEME 16.20

(B) Pschorr Ring Closure

When the Gomberg-Bachmann reaction is carried intramolecularly either by alkaline solution or with copper powder the procedure is called Pschorr ring closure (Scheme 16.21). Fluorene is prepared starting with *o*-aminodiphenylmethane.

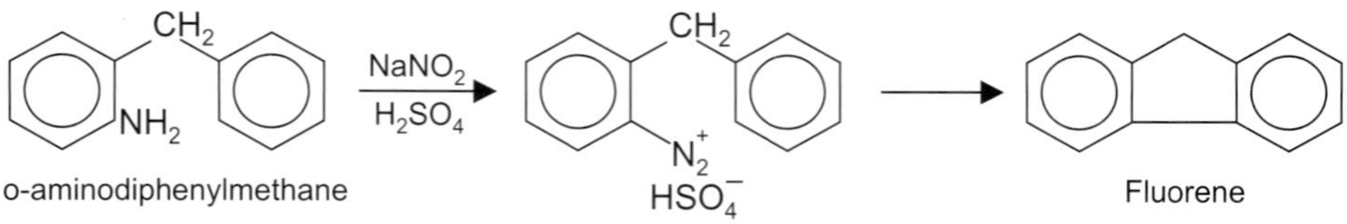

The Pschorr synthesis

SCHEME 16.21

(C) Hunsdiecker Reaction (Dearboxylative bromination)

This is a free radical substitution reaction, where the reaction of a silver (or mercury) salt of carboxylic acid with bromine gives a bromo compound and is a method of decreasing the length of the carbon chain by one unit (loss of carbon dioxide, Scheme 16.22). The reaction involves the formation of an acyl hypobromite (step 1, Scheme 16.22 which is not a free radical reaction). The acyl hypobromite then undergoes homolysis (step 2). The acyloxy radical loses carbon dioxide (step 3) and the resulting radical abstracts bromine from a second molecule of the hypobromite.

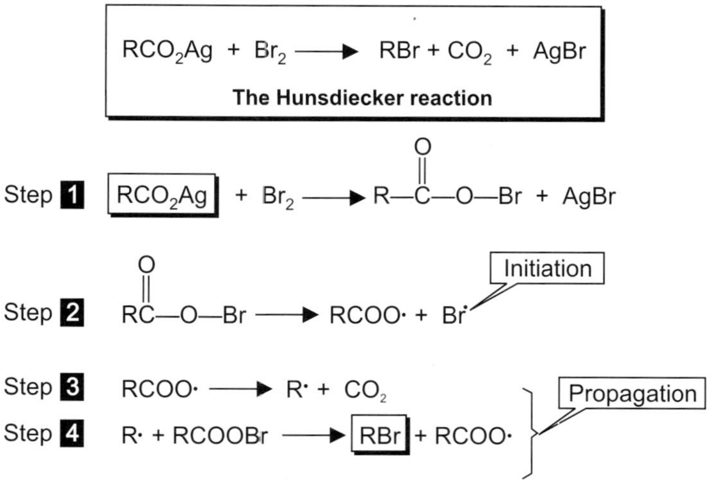

The mechanism of Hunsdiecker reaction

SCHEME 16.22

A free radical has been generated at bridgehead position via the Hunsdiecker reaction (Scheme 16.23) to show that a free radical need not be planar.

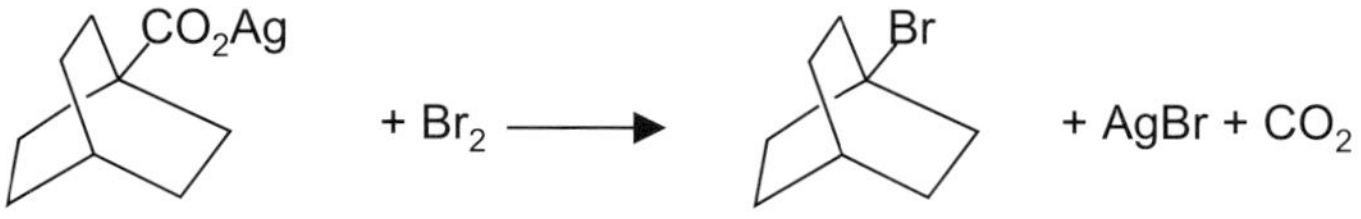

SCHEME 16.23

(D) The Ullmann Reaction (Dimerization of aryl radicals)

This reaction is a useful technique for the synthesis of biaryls through the reaction of two moles of aryl halide in the presence of copper (Scheme 16.24). Thus, *o*-nitrochlorobenzene on heating to around 220°C in the presence of copper gives 2-2′-dinitrobiphenyl. The reaction may involve the formation of free radicals.

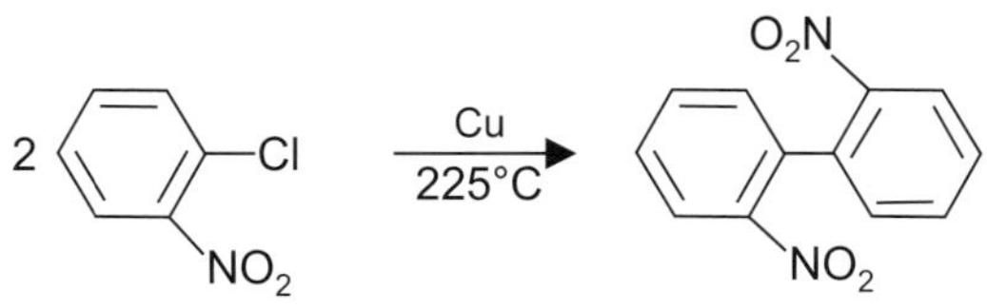

o-nitrochlorobenzene 2, 2'-dinitrobiphenyl

The Ullmann reaction

SCHEME 16.24

(E) The Kolbe Electrolytic Reaction (Dimerization of alkyl radicals)

This reaction is a free radical substitution. On electrolysis the alkali metal salt of a carboxylic acid, RCOOH gives the coupled product R—R, *via* decarboxylation and combination of the resulting radicals (Scheme 16.25). The alkyl radicals are liberated at the anode. The sodium liberated at the cathode reacts with the solvent (the electrolysis is done between platinum foil electrodes, of a solution of the acid in methanol containing enough sodium methoxide to neutralize about 2% of the acid) to generate more carboxylate anion.

$$C_2H_5CO_2^{\cdot -} \longrightarrow C_2H_5CO_2^{\cdot} + e \qquad C_2H_5CO_2^{\cdot} \longrightarrow C_2H_5^{\cdot} + CO_2$$

Propionate ion discharges at the anode; Propionate free radical

$$2C_2H_5^{\cdot} \longrightarrow C_4H_{10}$$

n-butane

Electrolysis of sodium propionate

SCHEME 16.25

(F) The Acyloin Condensation

It is the intermolecular, sodium promoted condensation of two moles of ester or the intramolecular condensation of a diester to yield an α-hydroxyketone (acyloin, Scheme 16.26).

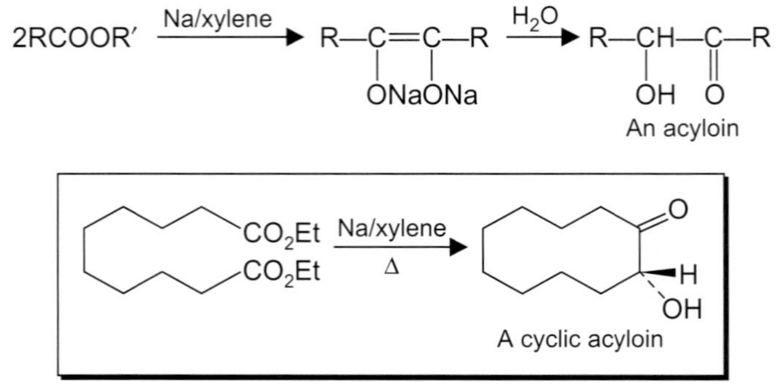

SCHEME 16.26

The formation of cyclic acyloins is a useful method to synthesize medium size ring compounds. The mechanism of the reaction (Scheme 16.27) involves the electron transfer to the carbonyl group of the ester. The radicals thus formed undergo dimerization and the loss of alkoxide

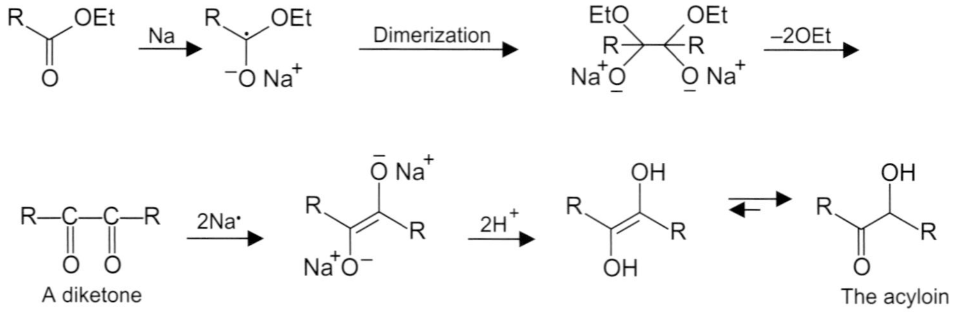

SCHEME 16.27

groups gives a diketone (RCOCOR) intermediate (small amounts of a diketone have been isolated as side products). Further electron transfer affords the disodium derivative of the acyloin. In the acyloin reaction ethoxide ion is generated, and therefore, base catalysed ester condensations can compete (*e.g.*, Dieckmann reaction). Thus the ethoxide ions are trapped by including chlorotrimethylsilane (eq. I, Scheme 16.28) while carrying out this condensation. This reagent also reacts with the acyloin dianion (eq. II, Scheme 16.28) to give the bis-silyl enol ether (A, Scheme 16.28) from which the acyloin is liberated with acid.

$$(CH_3)_3SiCl + EtO^- \longrightarrow (CH_3)_3Si{-}OEt + Cl^-$$

(I)

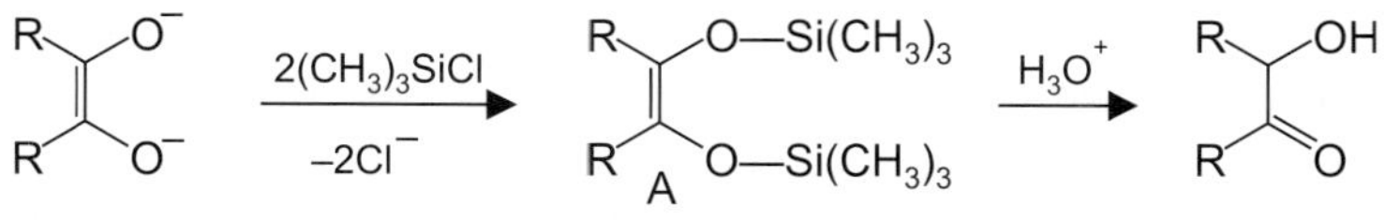

(II)

SCHEME 16.28

The first reported catenane (see, Scheme 1.23) was made using acyloin condensation (Scheme 16.29). This involved the following steps:

1. An acyloin condensation on the diethyl ester of the C_{34} dicarboxylic acid gave the cyclic acyloin (I, Scheme 16.29).
2. This was reduced under Clemmensen reduction conditions using DCl instead of HCl to incorporate deuterium in the reduced product (for detection via spectroscopy) and the C_{34} cycloalkane (II) was formed.
3. A repetition of the reaction in the presence of (II, Scheme 16.29) was anticipated to give (IV), since there was a chance that some molecules of ester will get threaded through (II, Scheme 16.29) before the cyclization of III, and indeed the catenane IV was isolated.

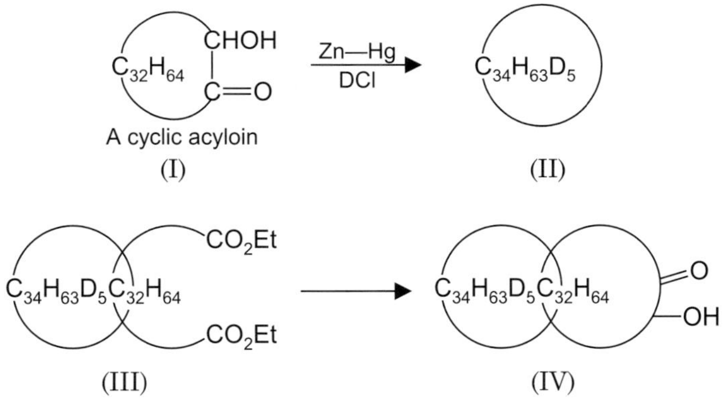

SCHEME 16.29

(G) The Hofmann-Loffler-Freytag Reaction (Intramolecular free radical reaction)

N-haloamines undergo a photochemical decomposition in acid solution and lead to an introduction of functionality at a carbon atom remote from the functional group already present. The following points may be considered.

- The initial product of the reaction is a δ-haloamine.
- The key step during its formation involves the migration of the halogen in a *N*-chloroamine (I, scheme 16.30) to the δ-position to give (IV). In this process the nitrogen cation radical (II, scheme 16.30) abstracts an internal hydrogen atom from the δ-carbon via a six membered cyclic transition state.
- The abstraction of a chlorine atom from the chloroamine by the alkyl radical (III) gives (IV, Scheme 16.30). Treatment with a base generates the amino group which displaces chlorine intramolecularly to a pyrrolidine (V).

(H) The Barton Reaction (Photolysis of nitrites)

This is another non-chain radical reaction which oxidizes a methyl group in the δ-position to an OH group to a CHO group. The alcohol is first converted to its nitrite ester (I, Scheme 16.31) with nitrosyl chloride, on irradiation the oxyradical (II) abstracts a hydrogen from δ—CH bond. The alkyl radical (III) thus formed combines with nitric oxide liberated in the photolysis to give a nitroso compound which tautomerises to the corresponding oxime. This technique has been used for the synthesis of aldosterone (as its 21-acetate) from corticosterone acetate. The nitrite ester of the 11 β-hydroxyl group was made by reaction with nitrosyl chloride. After photolysis the resulting oxime was hydrolysed. Aldosterone exists in part as its hemi-acetal (Scheme 16.32).

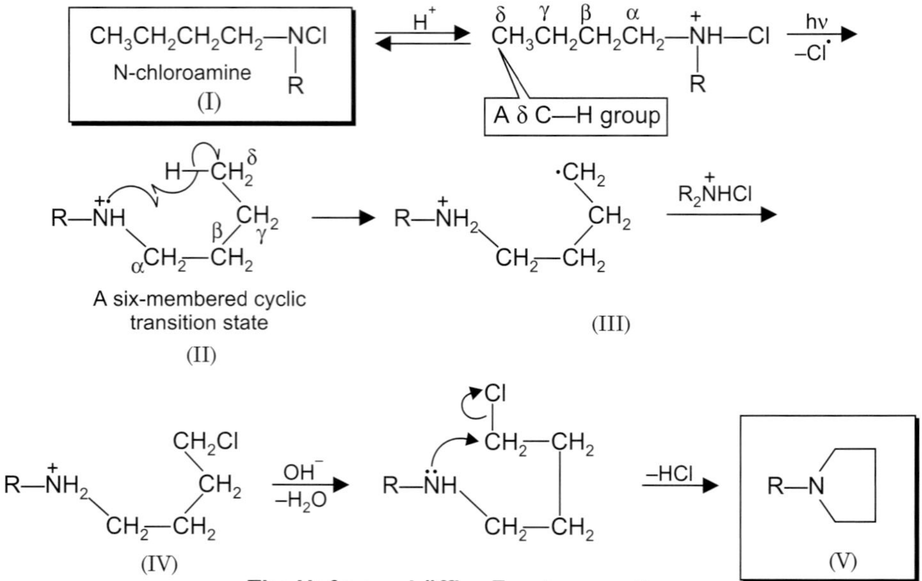

The Hofmann-Löffler-Freytag reaction

SCHEME 16.30

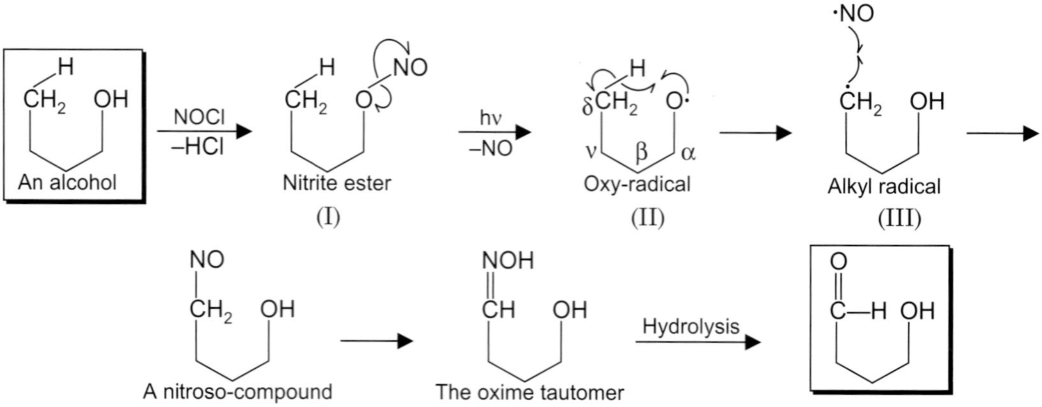

Photolysis of nitrites: the Barton reaction

SCHEME 16.31

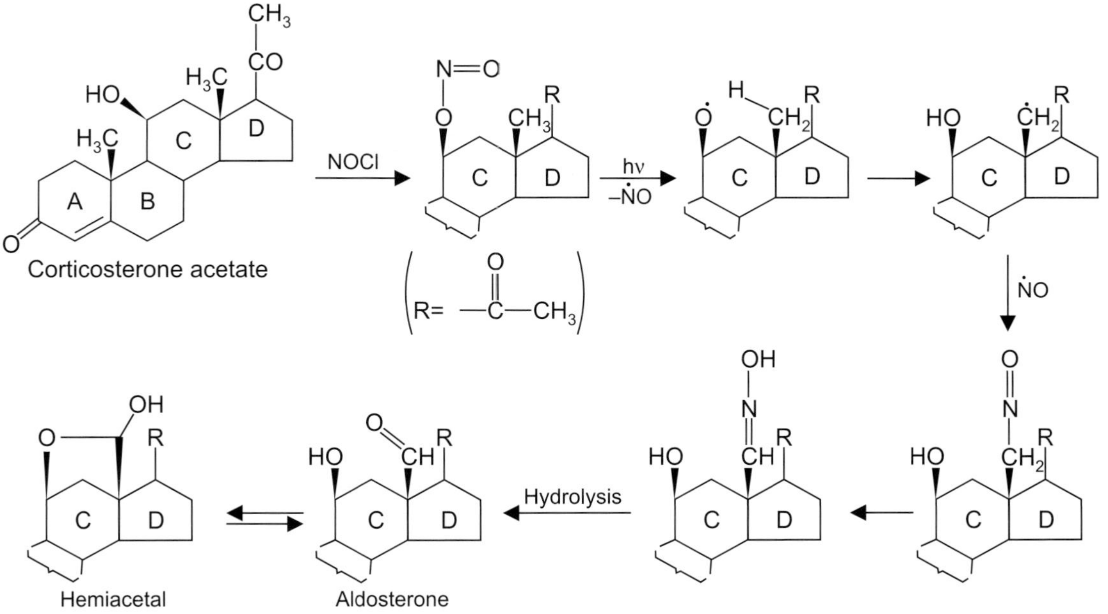

Application of Barton reaction

SCHEME 16.32

16.7 THE COUPLING OF ALKYNES

An oxidative coupling of terminal alkynes gives diynes and the reaction is catalysed by Copper (II) acetate in pyridine or similar bases.

The loss of an acidic proton of acetylene by the base, and one electron oxidation of the acetylide anion by copper (II) ion is followed by the dimerization of the resulting acetylide radicals (Scheme 16.33).

$$R{-}C{\equiv}C{-}H \rightleftharpoons R{-}C{\equiv}\ddot{C}^{-} \xrightarrow{Cu^{2+}} R{-}C{\equiv}C\cdot + Cu^{+}$$

$$2R{-}C{\equiv}C\cdot \longrightarrow R{-}C{\equiv}C{-}C{\equiv}C{-}R$$

SCHEME 16.33

In a different procedure an alkyne is reacted with an aqueous mixture of copper (I) chloride and ammonium chloride in air. Under these acidic conditions the proton is removed by complexing between C≡C bond and (Cu^+, copper (I) ion). Some of the copper (I) ions get oxidized with air to copper (II) ions which in turn oxidizes the acetylide ion to the radical and then get back to the copper (I) state. A related coupling method consists of treating a monosubstituted acetylene with a 1-bromoacetylene in the presence of copper (I) ion (Scheme 16.34). The method is used to synthesize [18] annulene from 1, 5-hexadiyne. The cyclic trimer (Scheme 16.35) is treated with base to make the system fully conjugated involving prototopic shifts. A partial hydrogenation of the remaining triple bonds gives [18] annulene.

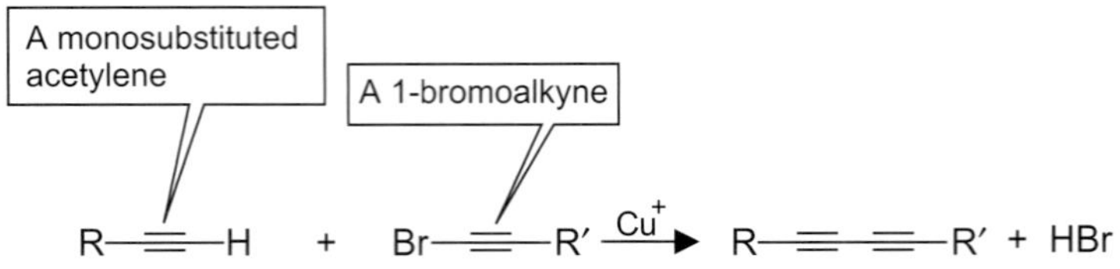

SCHEME 16.34

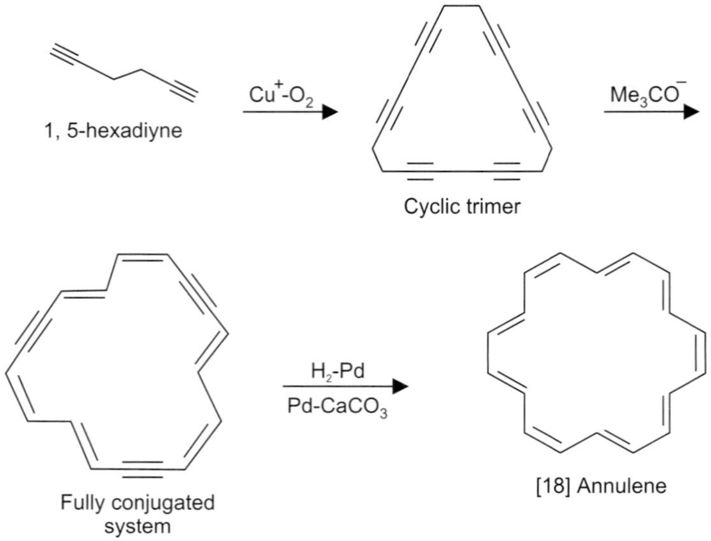

SCHEME 16.35

16.8 REACTIONS INVOLVING ELECTRON TRANSFER STEPS

One has already come across a reaction which involves, electron transfer and decarboxylation of acyloxy radicals in the Kolbe electrolysis (see Scheme 16.25). Many transition metal ions have two or more relatively stable oxidation states which differ by one electron. Consequently transition metal ions frequently take part in electron transfer reactions. The strongly catalysed decomposition of peroxy esters by Cu(I) is due to the oxidation of copper to Cu(II) (Scheme 16.36). Consider the reaction of cyclohexene with *t*-butyl perbenzoate

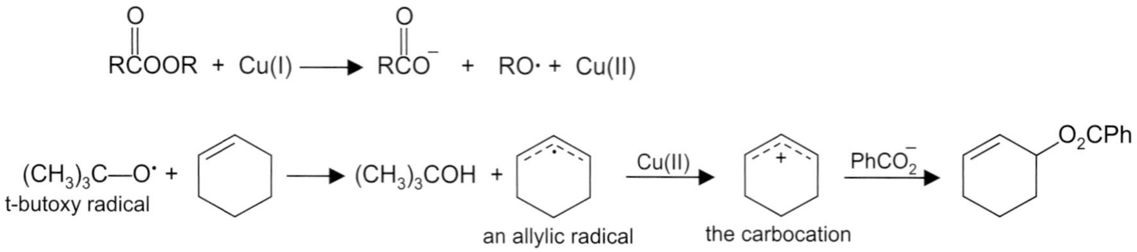

SCHEME 16.36

mediated by Cu(I). The *t*-butoxy radical thus formed (*via* the reductive cleavage of the peroxy ester) abstracts a hydrogen from cyclohexene to afford an allylic radical. The oxidation of the radical is brought about by Cu (II) and the intermediate carbocation formed then reacts with the benzoate ion. This is therefore, a reaction where both the intermediates a radical and a carbocation are involved.

16.9 MOLECULAR REARRANGEMENTS

Rearrangements which are very common with carbocation intermediates are rare in the case of free radicals (Scheme 16.37). Infact the migration of a saturated group is highly unlikely.

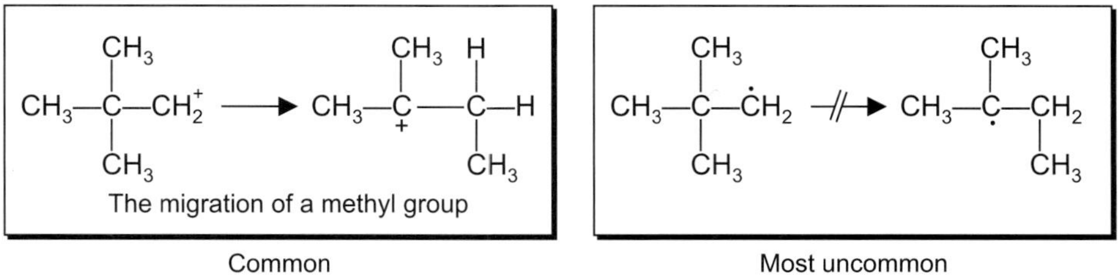

SCHEME 16.37

In the case of cationic intermediates, migration occurs through a bridged transition state (or intermediate) which involves a three center two electron bond (I, Scheme 16.38). In the case of a free radical there is a third electron in the system. It however, cannot occupy the same orbital as the two other electrons. It shall then have to be in an antibonding level. Thus the transition state for migration is less favourable compared to that in a carbocation.

In the case of free radicals, migrations however, can occur with aryl, vinyl, acyl and other unsaturated groups. The migration of an aryl group (Scheme 16.38) *e.g.*, involves the formation of bridged intermediate by an addition process and the intermediate is a cyclohexadienyl radical. The substrate (I, Scheme 16.39) adds an acyl radical (acyl radicals are formed by the abstraction of the formyl hydrogen from an aldehyde). The phenyl group migration then gives another free radical (II, Scheme 16.39) which abstracts a hydrogen atom from the

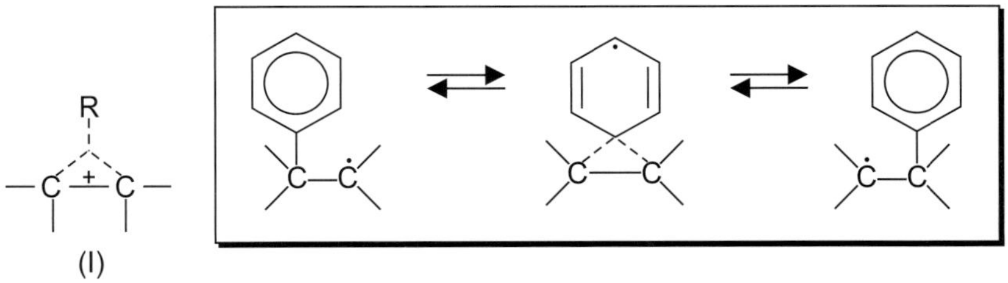

SCHEME 16.38

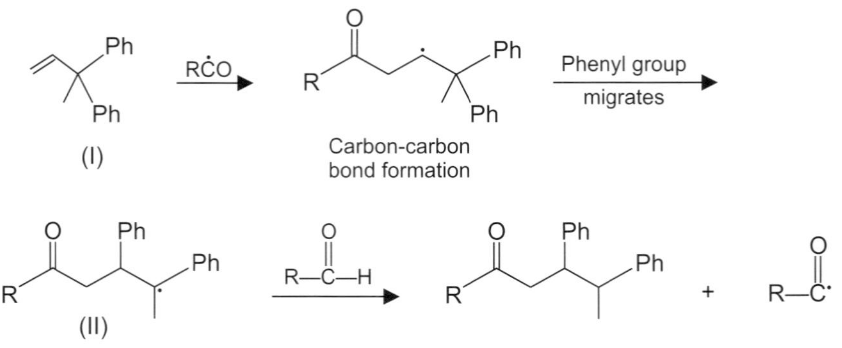

SCHEME 16.39

aldehyde. Migration of phenyl groups have been observed during decarbonylation of suitable aldehydes under free radical conditions. Consider the example of 3-methyl-3-phenylbutanal (Scheme 16.40) and consider the following points:

- Free radicals react with aldehydes by selective attack at the hydrogen bonded to the carbonyl group to form acyl radicals (I, Scheme 16.40).

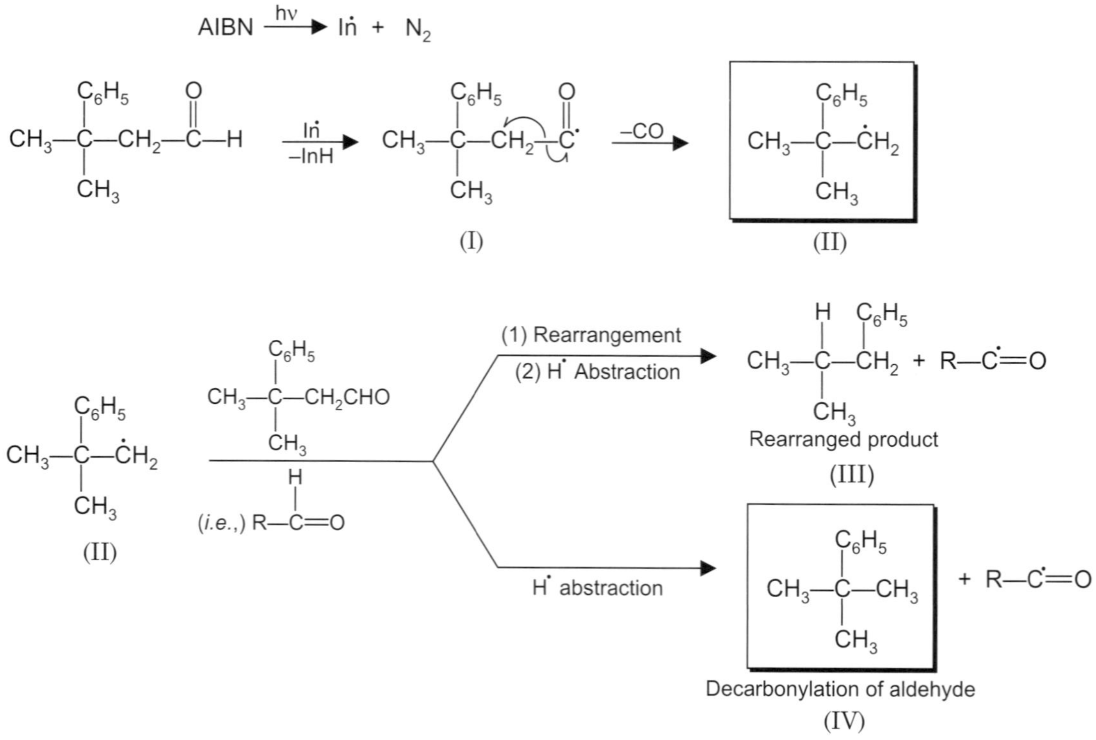

SCHEME 16.40

- AIBN or organic peroxides can be used as radical initiators.
- The acyl radicals usually lose CO to form alkyl or aryl radicals (II, Scheme 16.40).
- Migration of a phenyl group followed by abstraction of hydrogen atom from the aldehyde gives the rearranged product (III). One may note that only phenyl but not methyl migrates.
- When the radical (II, Scheme 16.40) abstract only a hydrogen atom one get the product of decarbonylation of the reactant aldehyde.

PROBLEM 16.4

Write the product formed from the reaction (Scheme 16.41) with mechanism

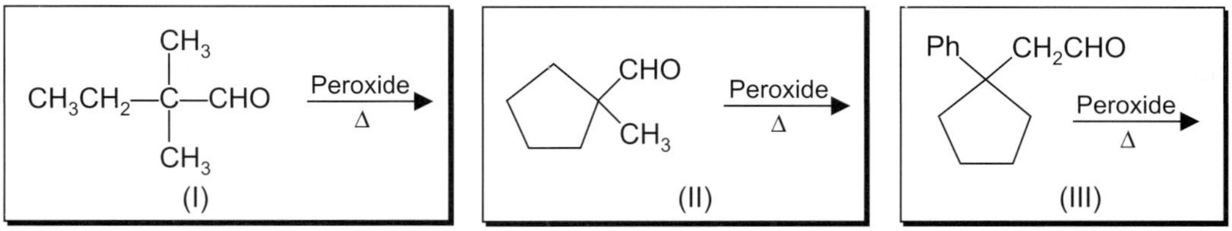

SCHEME 16.41

ANSWER. *From (I), $CH_3CH_2CH(CH_3)_2$ due to decarbonylation, no rearrangement possible; (II),* 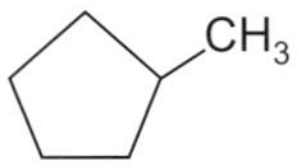 *due to decarbonylation and no rearrangement; (III),* 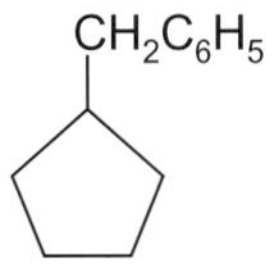 *due to free radical formation on methylene group in the chain, Ph migration and hydrogen atom abstraction.*

Migrations have also been observed for chloro groups. Thus in the reaction of (I, Scheme 16.42) with bromine in the presence of peroxides a rearrangement to (II) was observed, which was formed along with the normal addition product $Cl_3CCHBrCH_2Br$. Migration of a halogen could occur via a transition state in which the odd electron is accommodated in a vacant *d* orbital of the halogen.

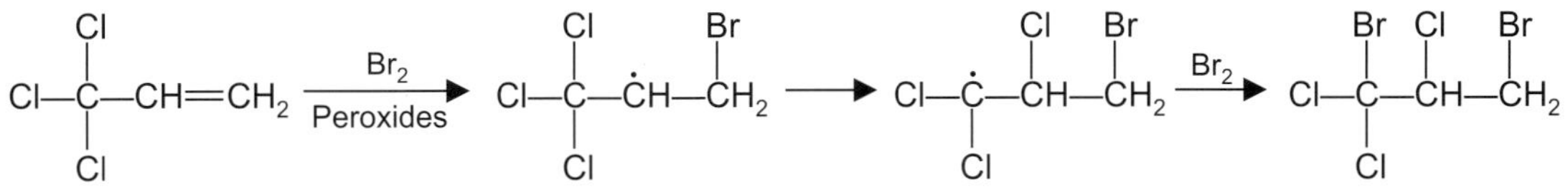

(I) (II)

SCHEME 16.42

16.10 SOME FURTHER SUBSTITUTION AND OTHER REACTIONS

(i) Nitration of Alkanes (Formation of carbon-nitrogen bonds)

This is carried out in the gas phase around 400°C. It has been shown that nitric acid itself does not react at a saturated carbon unless nitrogen dioxide (this is a stable free radical) is present. Thus the purpose of nitric acid is the supply of nitrogen dioxide and the key steps of nitration are (Scheme 16.43).

$$\boxed{RH} + \cdot NO_2 \longrightarrow R\cdot + HNO_2$$

$$R\cdot + \cdot NO_2 \longrightarrow \boxed{RNO_2}$$

Nitration of alkanes

SCHEME 16.43

(ii) Cyclic Ether Formation

Alcohols which have δ-hydrogens get cyclized with lead tetracetate to give tetrahydrofurans in high yields (Scheme 16.44). The reaction is carried out by heating (~ 80°C) or at room temperature on irradiation with UV light. Evidence has been given to show that the last step in cyclic ether formation is through a carbocation intermediate. The four and six membered cyclic ethers (oxetanes and tetrahydropyrans) are not formed by this method.

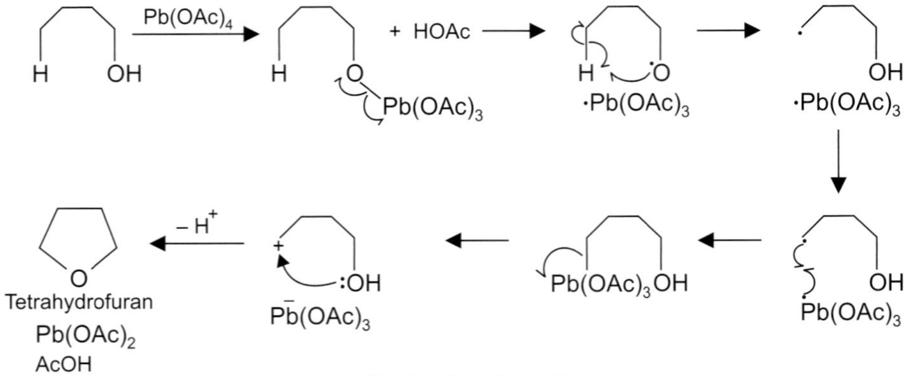

Cyclic ether formation

SCHEME 16.44

(iii) $S_{RN}1$ Mechanism For Substitution on Hindered Alkyl Halides

The $S_{RN}1$ mechanisms are now known to occur with a variety of substrates (for example on aryl halides, see Scheme 16.17). The operation of this mechanism with hindered alkyl halides reflect both on the synthetic utility of the reaction and geometry of free radical intermediates (Scheme 16.45).

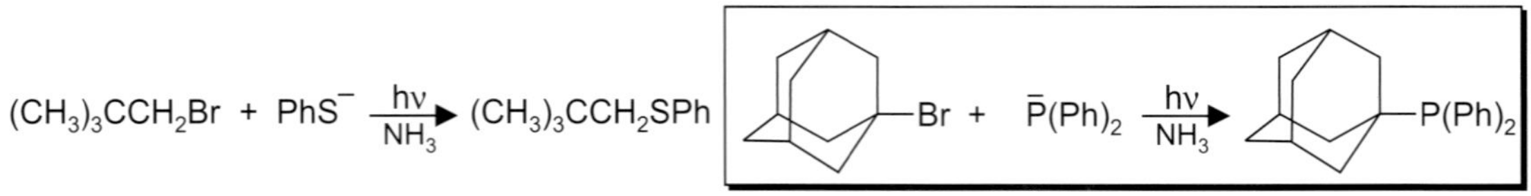

SCHEME 16.45

The mechanism involves similar steps as in Scheme 16.17, *i.e.*, initiation by the supply of electrons to the alkyl halide; formation of an alkyl free radical and its reaction with nucleophile etc.

(iv) Polymerization of Olefins

In the presence of free radical initiators (a peroxide or an azo compound) several olefins are polymerized (Scheme 16.46). The chains continuously grow and termination occurs by dimerization or disproportionation. Coordination polymerization of olefins is gainfully brought about using Ziegler-Natta catalysts which gives isotactic (stereoregular) polymers (Chapter 7).

$$Y\cdot + CH_2{=}CH_2 \longrightarrow Y{-}CH_2{-}\dot{C}H_2$$

$$YCH_2\dot{C}H_2 + CH_2{=}CH_2 \longrightarrow YCH_2CH_2CH_2\dot{C}H_2 \xrightarrow{\text{etc.}} Y(CH_2CH_2)_nCH_2\dot{C}H_2$$

SCHEME 16.46

In free radical polymerization chain branching occurs when the growing end of the polymer chain abstracts a hydrogen atom from its own chain. A new chain brand then starts growing at this point (see arrow Scheme 16.47).

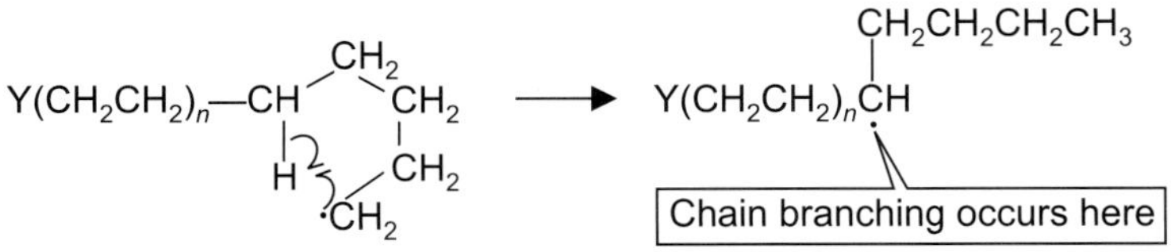

SCHEME 16.47

Nitrosation of cyclohexane with nitrosyl chloride (NOCl) is a photochemical reaction of great commercial significance for the manufacture of nylon. The NO radical is not reactive, however chlorine readily abstracts a hydrogen atom from cyclohexane to yield cyclohexyl radical. This radical combines with NO radical to give nitrosocyclohexane (Scheme 16.48). Note that it is not a chain reaction, and nitrosocyclohexane in other steps gives oxime. The cyclohexanone oxime undergoes a Beckman rearrangement to the ε-lactam which is then converted into nylon (see, Scheme 15.22).

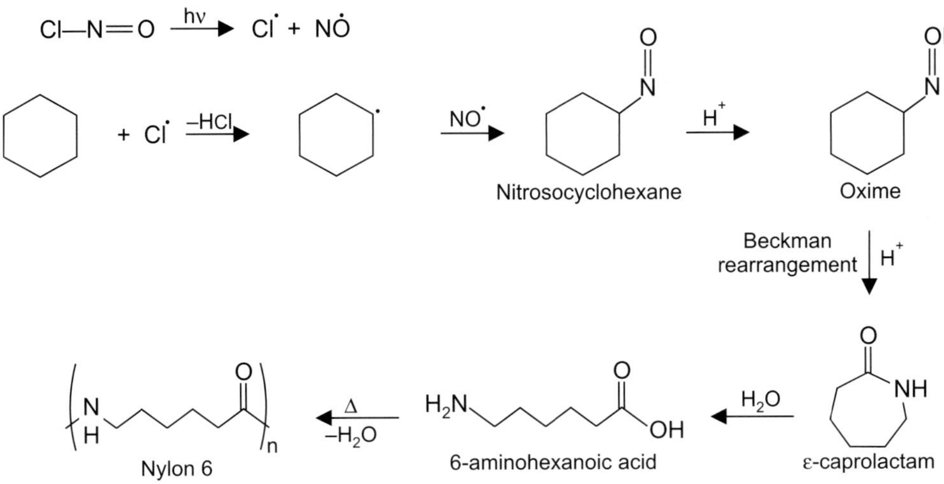

SCHEME 16.48

(v) *Radical Inhibitors*

Some compounds are highly reactive towards radicals and react with these immediately on their generation to give inactive products. A low concentration of these inhibitors is sufficient, to arrest a free radical reaction. These inhibitors may be stable free radicals like nitric oxide (NO reacts with organic radicals *e.g.*, $R^{\bullet}$ to give nitroso compounds R—NO) or these may be compounds which react with radicals to generate radicals of a stability so that these do not perpetuate the chain. An example of this is quinol (Scheme 16.49). Thus to achieve a good efficiency from a radical catalysed reaction it is necessary to carry out radical reactions with reactants of high purity. One must appreciate that a chain propagating step cannot be achieved till all the inhibitor (if present) is consumed. In case an inhibitor is present, the reaction mechanism then instead may take up the ionic pathway which is not effected by an inhibitor.

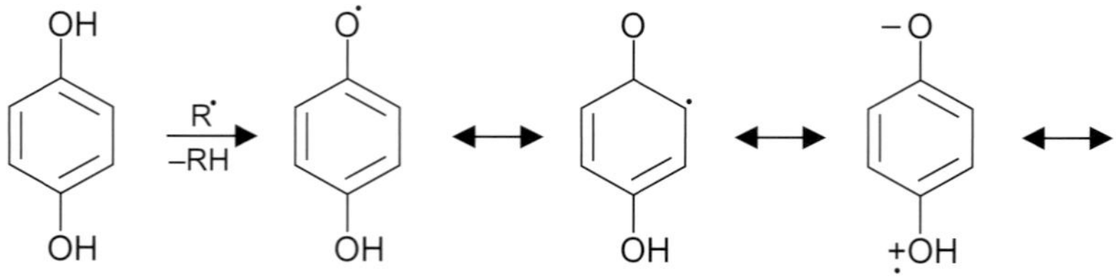

SCHEME 16.49

Inhibitors are required for storage of some organic compounds which are highly susceptible to radical catalysed polymerisation. The inhibitor will then act to eliminate efficiently any stray radicals which may originate through the action of light or oxygen.

PROBLEMS

16.1 Explain the following reactions in terms of yield of the products.

$$(CH_3)_3CH \xrightarrow[-HCl]{Cl_2} (CH_3)_2CHCH_2Cl + (CH_3)_3CCl$$

1-chloro-2-methylpropane (Isobutyl chloride) 64% ; 2-chloro-2-methylpropane (*tert*-Butyl chloride) 36%

$$(CH_3)_3CH + Br_2 \longrightarrow (CH_3)_3C\text{—}Br + HBr$$

t-butyl bromide over 99%

16.2 Devise a method for the synthesis of α-tetralone from tetralin. How can one synthesis the compounds (I and II)?

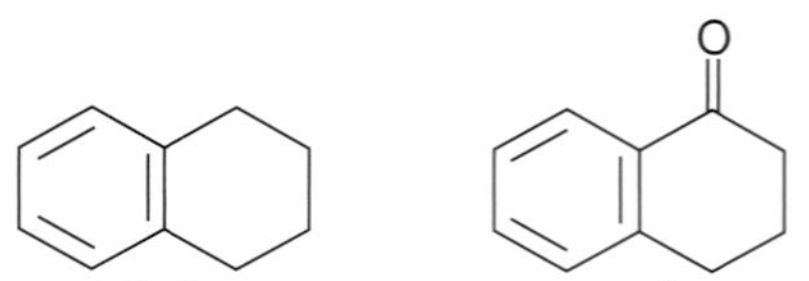

tetralin α-tetralone

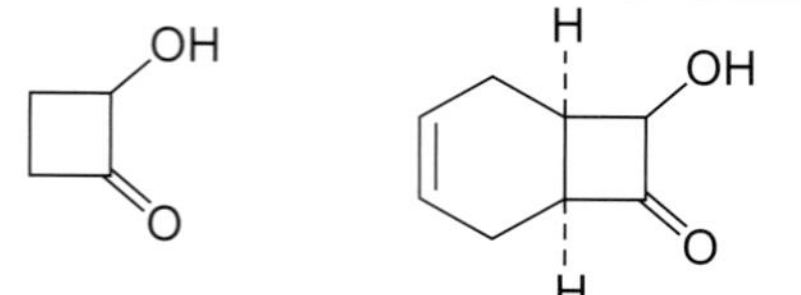

(I) (II)

16.3 Bromotrichloromethane adds to cyclooctene according to expectations. However, the addition of carbon tetrachloride gives the unusual product (II) as well as the expected product (I). Explain.

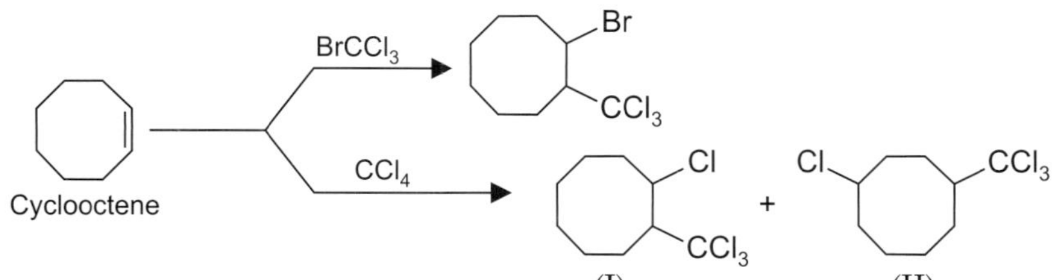

(I) (II)

16.4 Why in the free radical rearrangement of the vinyl group in (I) to give (III) one invokes the formation of the intermediate cyclopropyl species?

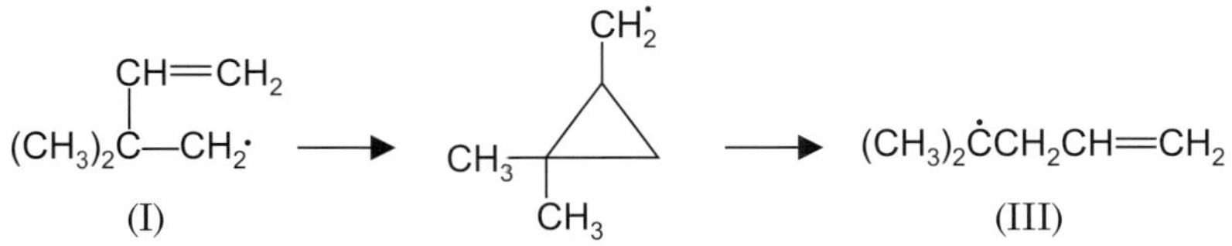

(I) (III)

16.5 Why the radical initiated decarbonylation of optically active (I) leads to a racemic product? Give mechanism.

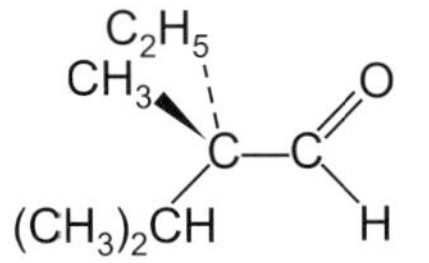

Optically active
(I)

16.6 Predict the radical catalysed addition of carbon tetrachloride to β-pinene.

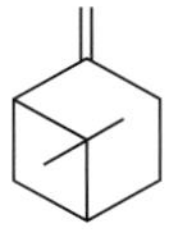

β-pinene

ANSWERS TO THE PROBLEMS

16.1 Bromine atoms are significantly more selective than chlorine atoms in their reactions at primary, secondary and tertiary C—H groups.

16.2 It can be achieved via hydroperoxide and its subsequent E2 elimination with base. The cyclic acyloins (I and II) can be made from the corresponding esters (III and IV) by treatment with sodium, chlorotrimethylsilane followed by acid hydrolysis.

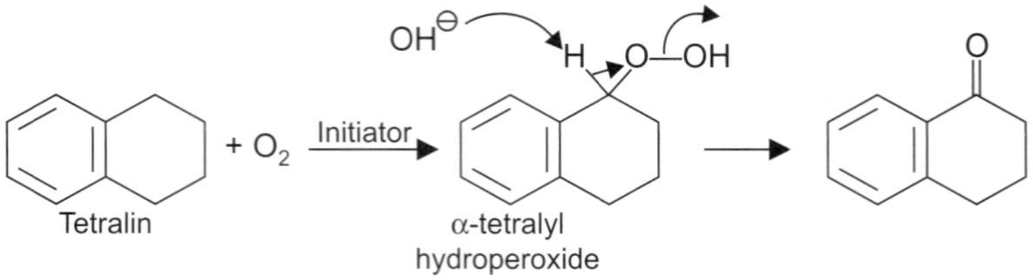

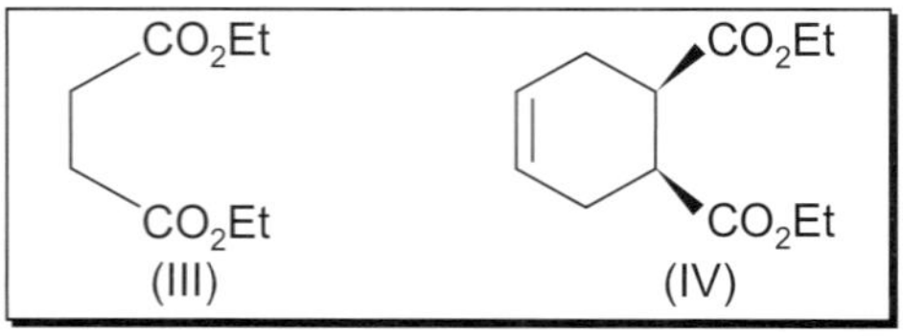

16.3 The radical intermediate from addition of (CCl_4) faces two competing reactions *i.e.*, intramolecular hydrogen abstraction (compare with intramolecular free radical reactions *e.g.*, Barton reaction) as well as abstraction of a chlorine atom from carbon tetrachloride. In the medium sized ring under study, transannular hydrogen abstraction is not observed during the addition of $BrCCl_3$, as the bromine atom abstraction is rapid and this prevents a competition by the intramolecular hydrogen abstraction.

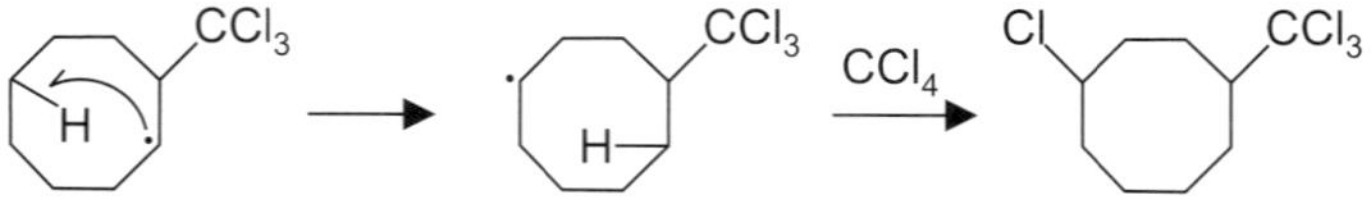

16.4 The migration occurs through addition to give a cyclopropyl species. The involvement of the usual transition state for a radical 1, 2 shift will have three electrons and electrons in excess of two can be accommodated only in an *anti* bonding molecular orbital of much higher energy. This in fact is a cyclization fragmentation reaction and the overall driving force is the conversion of a primary to a tertiary radical. D.A. Lindsay, J. Lusztyk, and K.U. Ingold, *J. Am. Chem. Soc.* **106,** 7087 (1984).

16.5 The process involves the formation of a free radical at the stereogenic carbon.

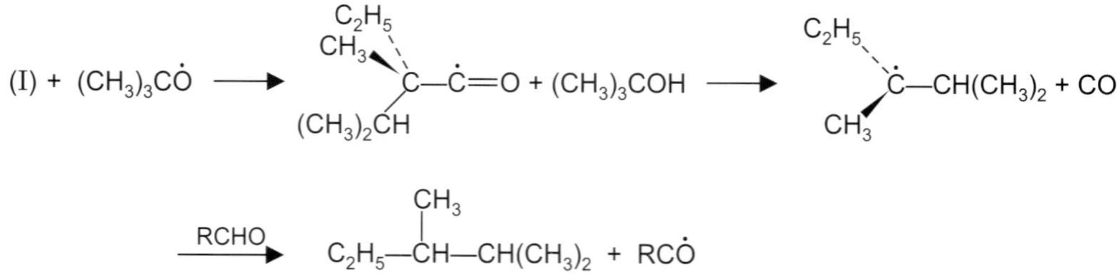

16.6 The initial addition affords a radical which undergoes molecular rearrangement whereby the ring strain associated with the cyclobutane ring is relieved.

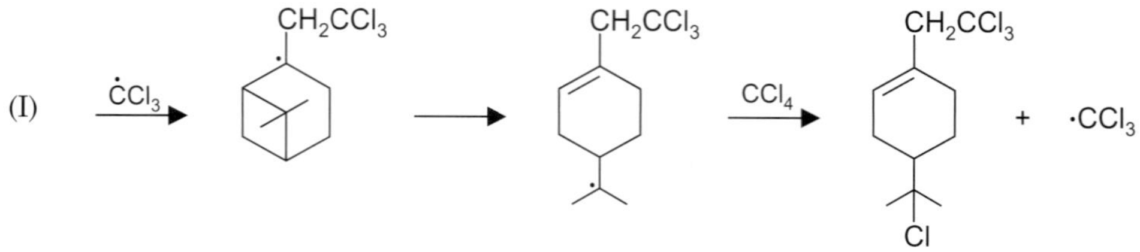

CHAPTER 17

Pericyclic Reactions

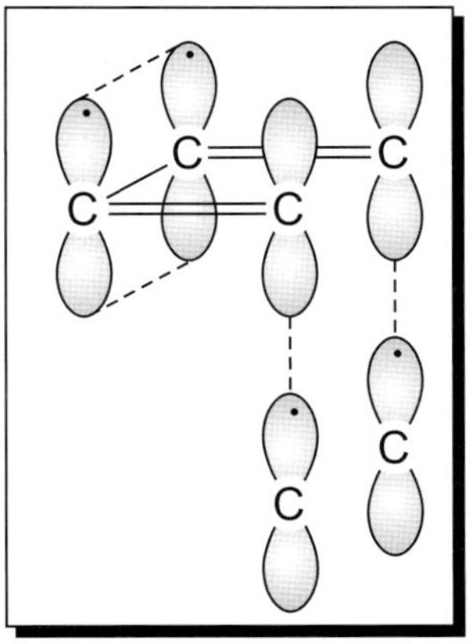

In a pericyclic reaction, there is a concerted bond reorganization and the essential bonding changes occur within a cyclic array of the participating atomic centers. These reactions do not involve the intermediate formation of either ions or radicals. Pericyclic reactions are also largely unaffected by polar reagents, solvent changes, radical initiators etc. These can however, be influenced only thermally *i.e.,* reactants are in their ground state or photochemically, *i.e.*, the excited state of a reactant is involved in the reaction.

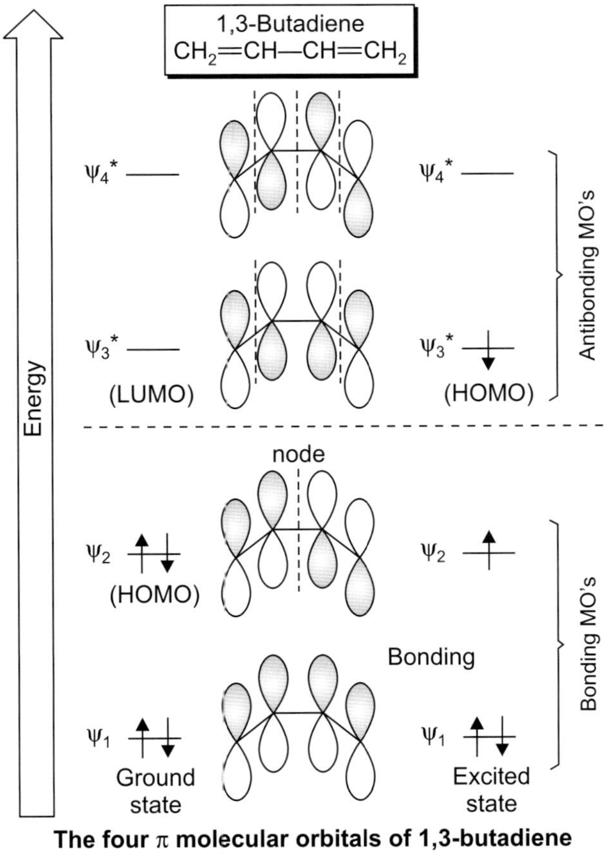

The four π molecular orbitals of 1,3-butadiene (asterisk denotes an antibonding orbital)

SCHEME 17.1

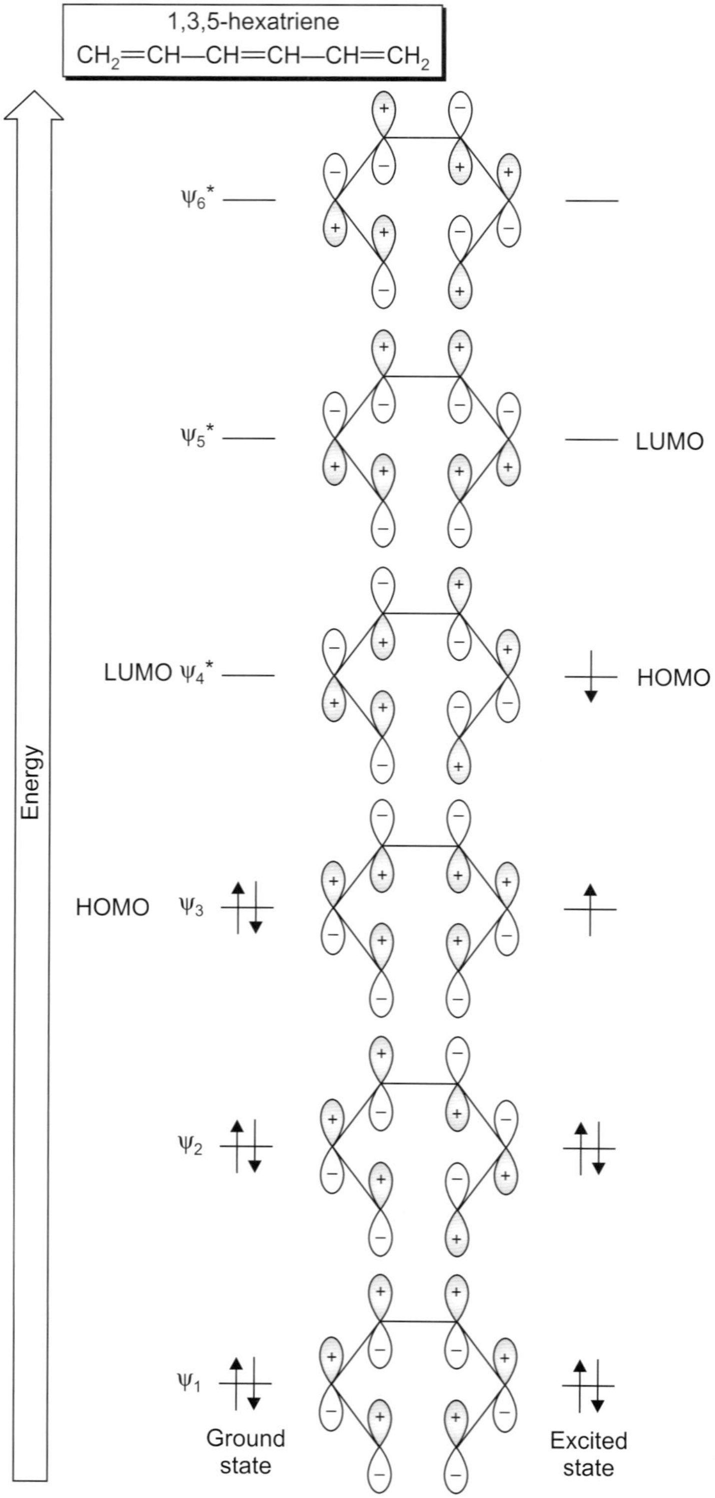

The six π molecular orbitals of 1, 3, 5-hexatriene

SCHEME 17.2

A consideration of the phase of orbitals of 1, 3-butadiene and 1, 3, 5-hexatriene are presented (Schemes 17.1 and 17.2) has significance, since only orbitals of the same phase will overlap to result in bonding. The orbitals of the different phase lead to a repulsive anti-bonding situation.

The following points may be noted:

- The normal electronic configuration of a molecule is called its ground state. When one combines *four* adjacent *p* atomic orbitals *e.g.,* in 1, 3-butadiene a set of four π molecular orbitals, two of which are bonding and two of which are antibonding are obtained. The four π electrons occupy the two bonding orbitals to leave the antibonding orbitals vacant.
- The lowest-energy π molecular orbital (ψ_1, Greek psi) is a fully additive combination with no nodes between the nuclei and is thus bonding. The π MO of the next lowest energy ψ_2, with one node between nuclei is also bonding. Above ψ_1 and ψ_2 in energy there are two antibonding π MO's, ψ_3* and ψ_4*. (The asterisks indicate antibonding orbitals.) One may note that the number of nodes between nuclei increases as the energy level of the orbital increases. The ψ_3* orbital has two nodes between nuclei while ψ_4*, the highest-energy MO, has three nodes between nuclei.
- When one considers the ground state of 1, 3-butadiene (Scheme 17.1) the highest occupied molecular orbital (HOMO) is ψ_2 and the lowest unoccupied molecular orbital (LUMO) is ψ_3*.
- Ultraviolet irradiation of a polyene excites an electron which is promoted from its ground state HOMO to its LUMO *i.e.*, from ψ_2 to ψ_3*. The molecule is then in an excited state. In the excited state the HOMO of 1, 3-butadiene is ψ_3*.
- The electronic excitation changes the symmetries of HOMO and LUMO and it also changes the reaction stereochemistry.
- In a thermal reaction the reactant is in its ground state while in a photochemical reaction the reactant is in its excited state.
- To show the different phases of the two lobes of a *p* orbital one phase may be shaded and other left as such (Scheme 17.1) or different phases may be represented by mathematical signs (+) or (–). Thus the ground state HOMO and excited state HOMO of a conjugated diene are depicted (Scheme 17.3).

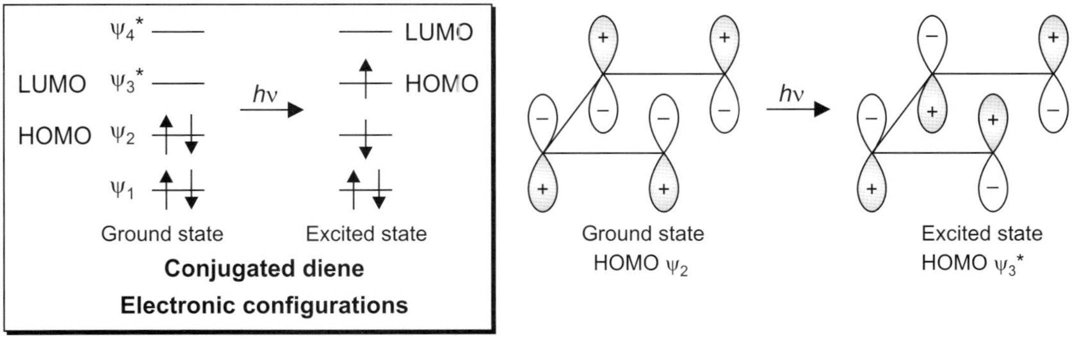

Ground state and excited state electronic configurations of a conjugated diene

SCHEME 17.3

A consideration of the six π molecular orbitals of 1, 3, 5-hexatriene (Scheme 17.2) shows that ψ_3 is the HOMO in the ground state and $\psi_4{*}$ is the LUMO. In the excited state of 1, 3, 5-hexatriene $\psi_4{*}$ is the HOMO and $\psi_5{*}$ is the LUMO.

17.1 CONSERVATION OF MOLECULAR ORBITAL SYMMETRY

The resulting stereochemistry of the concerted pericyclic reactions depends on whether the reaction is thermal or photochemical. Woodward and Hofmann in 1965 pointed out that symmetry of the molecular orbitals which participate in the chemical reaction determines the course of the reaction and they proposed the principle of the conservation of orbital symmetry in concerted reactions. A pericyclic reaction can take place only provided the symmetry of all reactant molecular orbitals is the same as the symmetry of the product molecular orbitals (symmetry allowed reaction). In other words the lobes of the reactant molecular orbitals must be of correct algebraic sign for bonding overlap to take place in the transition state to give the product.

In a concerted reaction, the symmetry present in the reactants is maintained during the course of the reaction and is present in the product as well (principle of conservation of orbital symmetry). For example, in the Diels-Alder reaction, the reactants, the diene and the dienophile each has a plane of symmetry which is maintained in the *transition* state as well as in the product cyclohexene (Scheme 17.4).

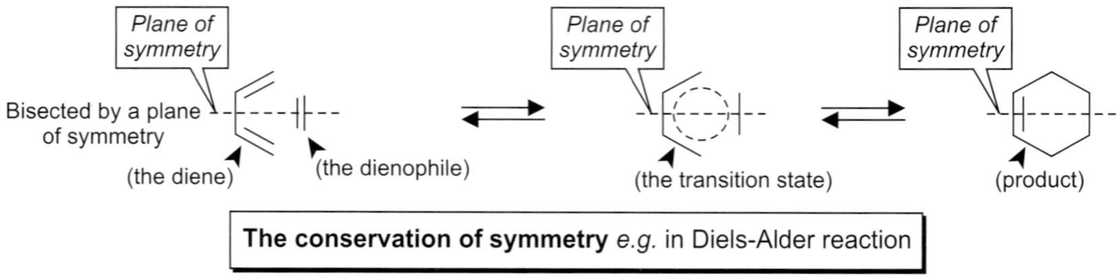

SCHEME 17.4

17.2 METHODS TO EXPLAIN PERICYCLIC REACTIONS

Three approaches have been employed to explain the pericyclic reactions and these are:

1. Woodward-Hofmann Rules—Correlation Diagrams

These rules require the smooth passage of the participating molecular orbitals of the reactants into the molecular orbitals of the product. The process is described by a correlation diagram. In case the conversion of the reactant orbitals into the product orbitals is favoured in terms of energy and if orbital symmetry is conserved in the process, the reaction is called symmetry allowed (The symmetries of the orbitals to be maintained are around the mirror plane and the two fold axis of rotation as a symmetry element). In case either of these conditions (energy and orbital symmetry) is not met the reaction is called symmetry-forbidden.

2. HOMO-LUMO Method—FMO Approach

The Fuki's method concentrates on the so called frontier molecular orbitals, the HOMO (highest occupied molecular orbital) and the LUMO (lowest unoccupied molecular orbital). In the ground state of 1,3-butadiene ψ_2 is HOMO and ψ_3^* is the LUMO (Scheme 17.1). The Fuki's FMO approach examines as to how the orbitals of HOMO or in some cases, the orbitals of HOMO of one component and the LUMO of other overlap to form new bonds. If the overlaps are favourable (bonding overlaps) then the reaction is allowed and if not favourable (antibonding overlaps) then the reaction is forbidden.

3. Möbius-Hückel Analysis (PMO Method)

The idea behind this method is that a pericyclic reaction which proceeds through a *transition* state which has aromatic characteristics (electron interactions is energetically favourable) is allowed process. For system of $4n + 2$ electrons Hückel transition states are aromatic; for systems of $4n$ electrons Möbius transition states are aromatic.

17.3 ELECTROCYCLIC REACTIONS (FMO-APPROACH)

These are pericyclic reactions (intramolecular) which under the influence of heat or light involve either the formation of a ring, with the generation of one new sigma-bond and the consumption of one *pi*-bond or the reverse (Scheme 17.5). The reverse reaction, ring opening proceeds by the same mechanism, but in reverse.

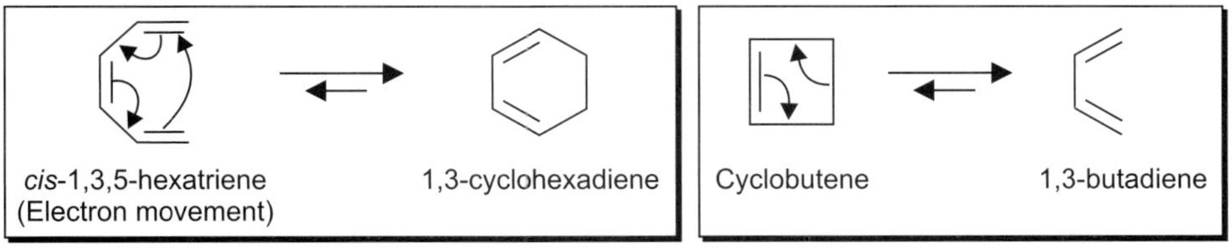

Electrocyclic reactions

SCHEME 17.5

The stereochemistry of electrocyclic reactions can be studied by using suitably substituted molecules. An intriguing feature of electrocyclic reaction is the stereochemistry of the product which depends on whether the reaction is thermally induced or photo-induced.

> *Electrocyclic reactions are brought about by heat or light and are concerted and stereospecific. A symmetry-allowed pathway requires in-phase orbital overlap. When the HOMO is symmetric which has the end orbitals identical e.g., (II, Scheme 17.6) the rotation will have to be disrotatory to achieve the in-phase overalap (a symmetry allowed process). When however, the HOMO is asymmetric e.g., (I, Scheme 17.6) the rotation must be conrotatory to achieve an in-phase overlap. A symmetry-allowed pathway requires an in-phase orbital overlap.*

Regarding Electron Movement

In electrocyclic reactions conjugated polyenes close to give rings or rings open to polyenes shown by having electrons "chase each other's tails" around in a circle. These are among the conceptually simplest organic reactions. (It doesn't matter in which direction the curved arrows depicting electron motions are drawn in electrocyclic reactions, as long as they all move in the same direction (Scheme 17.5).

The following points may be noted:

- All the electrocyclic reactions are accounted for by orbital-symmetry arguments (FMO approach) by looking only at the symmetries of the two outermost lobes of the polyene. Thus the inner lobes of a reactant may not be shown (Scheme 17.6) and if shown, these may not be labelled +ve or –ve. The lobes of like sign can be either on the opposite side or on same side of the molecule (I and II respectively; Scheme 17.6). For bond formation the outermost lobes must rotate—a positive lobe overlapping a positive lobe or a negative lobe overlapping a negative lobe. When the two lobes of like sign are on the same side of the molecule (symmetric arrangement) the two orbitals (on the ends of the *pi* system) must rotate in different direction (clockwise and counterclockwise) and this motion is termed Disrotatory (Scheme 17.7). When, however, the lobes of like sign are on opposite side of the molecule (asymmetric arrangement) both orbitals must rotate in the same direction (both clockwise or both counter-clockwise) and this motion is termed Conrotatory.

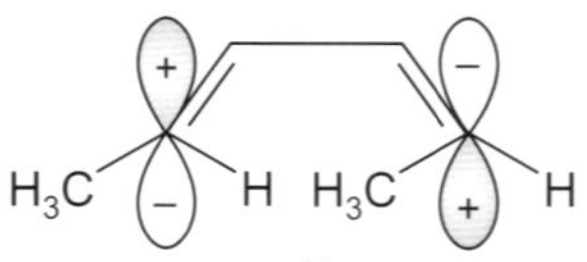

(I)
Ground state HOMO ψ_2 of a substituted butadiene (2*E*, 4*Z*-hexadiene)

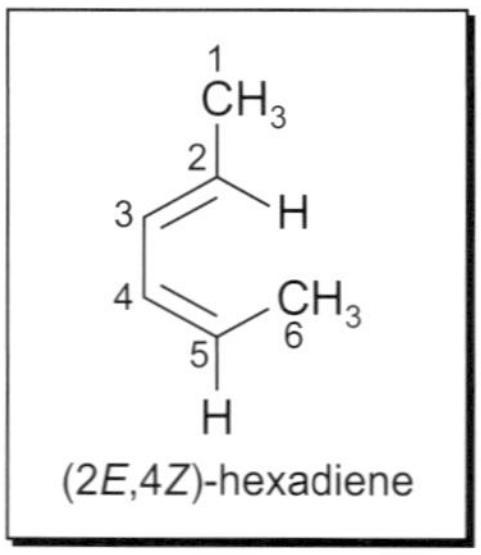

(2*E*,4*Z*)-hexadiene

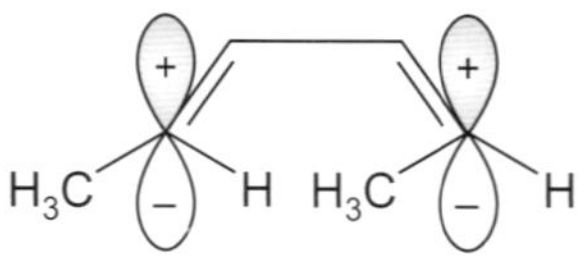

(II)
Excited state HOMO ψ_3^* of a substituted butadiene (2*E*, 4*Z*-hexadiene)

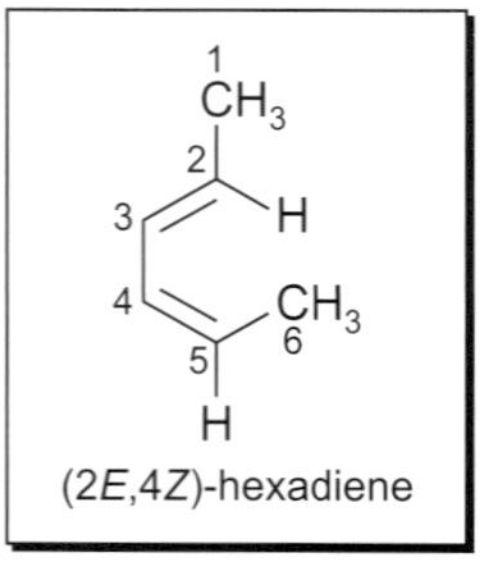

(2*E*,4*Z*)-hexadiene

SCHEME 17.6

- The stereochemical outcome of an electrocyclic reaction depends on the number of double bonds in a polyene and on whether the reaction is thermal or photochemical. A thermal electrocyclic reaction involving 4*n* *pi* electrons (n = 1, 2, 3,...) proceeds with conrotatory motion (*i.e.*, a motion in which the bonds rotate in the same direction) while the photochemical reaction involves disrotatory motion (a motion, in which the bonds rotate in opposite directions).

A thermal reaction involving $(4n + 2)$ *pi* electrons (where $n = 0, 1, 2, ...$) proceeds with disrotatory motion while the photochemical reaction proceeds with conrotatory motion.

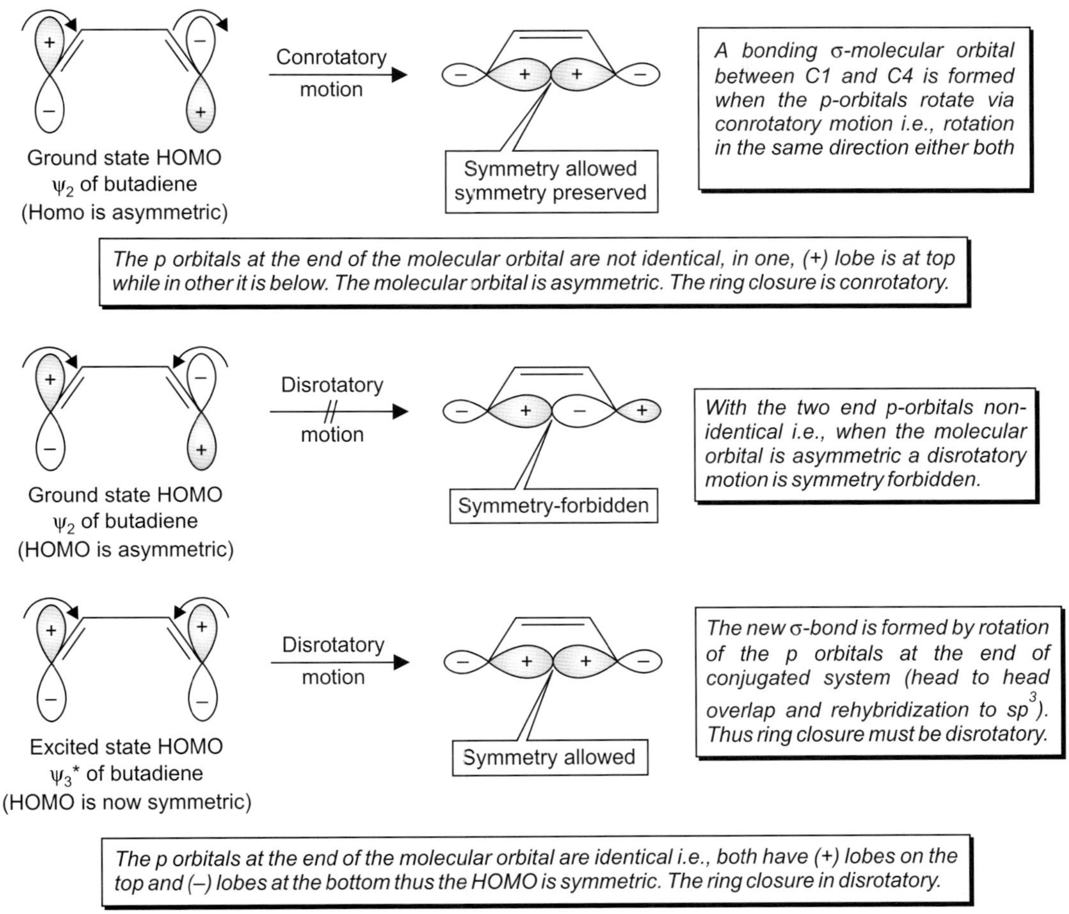

SCHEME 17.7

- The direction taken by an electrocyclic reaction is dependent on the relative stabilities of the ring and open-chain reactants. In the case of cyclobutanes the open chain structure is favoured because of the strain in the ring, during the thermal reaction.

A. Thermally Induced Interconversion of a Conjugated Diene and a Cyclobutene

The following points may be noted:

- The symmetry of the HOMO of a conjugated polyene undergoing ring closure determines the outcome of the electrocyclic reaction. All conjugated polyenes with asymmetric HOMO's undergo conrotatory ring closure. The ground state HOMO of a conjugated compound with an even number of double bonds *e.g.*, in a conjugated diene is asymmetric and should undergo conrotatory ring closure (Scheme 17.8). Moreover the reaction is remarkably stereospecific (Scheme 17.8). One may recall that there are always two conrotatory modes clockwise and anticlockwise and both are equally probable. It is obvious that neither of the two possible disrotatory modes can be a favourable process each being an antibonding situation.

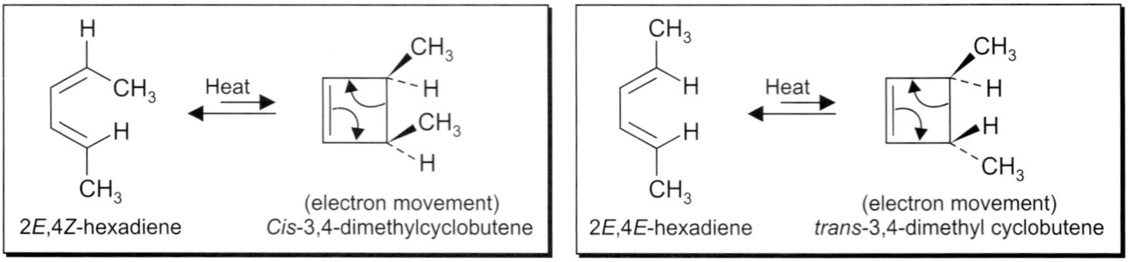

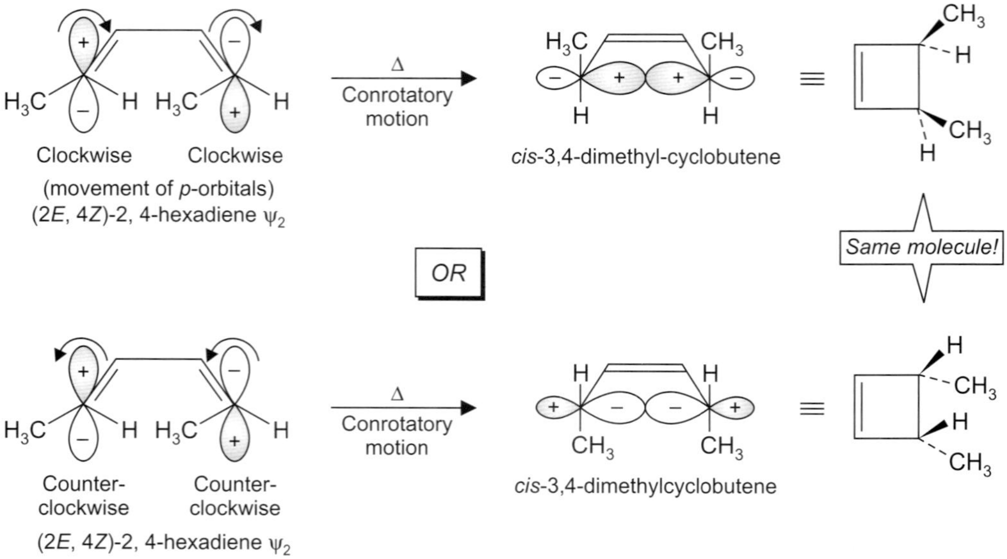

SCHEME 17.8

EXERCISE 17.1

Depict the conrotatory ring closure of 2E, 4E-hexadiene and predict the stereochemistry of the product.

ANSWER. *Conrotatory ring closure is depicted in the counterclockwise fashion, the product is trans-3, 4-dimethylcyclobutene (Scheme 17.9). Same product would result by conrotation via clockwise motion.*

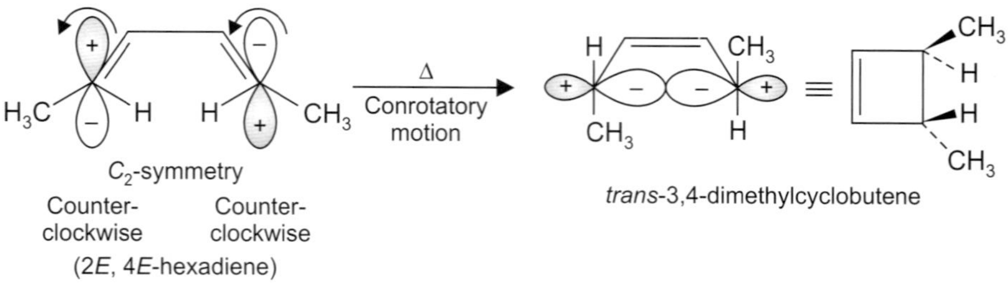

SCHEME 17.9

More on Electrocyclic Reactions

- *The diene must assume a s-cis-conformation for the terminal carbons p orbitals overlap.*
- *The alkyl substituents (as an approximation) do not effect the π molecular-orbital structure of a conjugated alkene. Thus the π molecular structure of 2, 4-hexadiene is very close to that of parent 1,3-butadiene.*
- *The relative orbital phase at the terminal carbon atoms of the HOMO (the orbital symmetry) is what that determines if the reaction would occur by conrotation or disrotation.*

B. Thermal Ring Opening of Cyclobutenes

In keeping with the principle of microscopic reversibility the reverse process of thermal ring opening takes exactly the same path.

Due to conrotatory motion (Scheme 17.10). A σ bond will open so as to give the resulting *p* orbitals which will have the symmetry of the highest occupied π orbital of the product. Since in the case of cyclobutenes the HOMO of the product (*i.e.*, a butadiene) in the thermal reaction is ψ_2 therefore, the cyclobutene must open so that on one side the positive lobe lies above the plane, which on the other side it is below it. This process also forces the stereochemistry in the product from a substituted cyclobutene (Scheme 17.10).

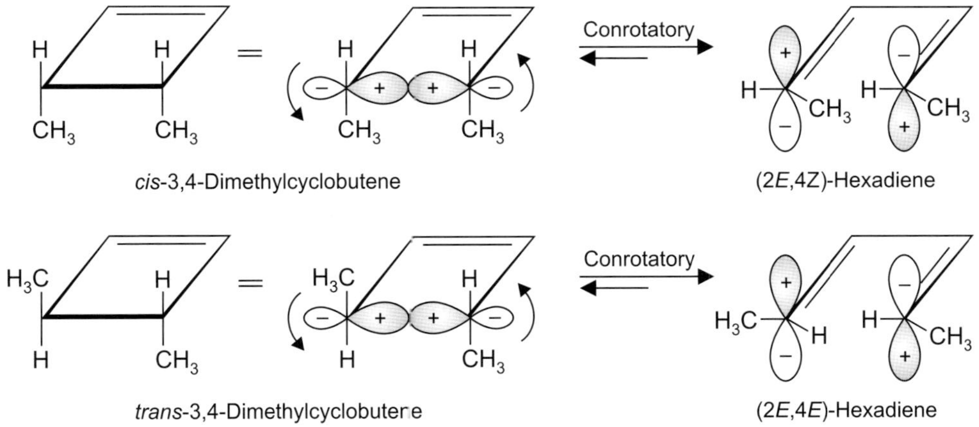

Thermal ring opening of *cis*- and *trans*-dimethylcyclobutene involves conrotatory motions

SCHEME 17.10

One may depict the opening of cyclobutane rings directly along with the stereochemistry. Consider the thermal electrocyclic ring opening in *cis*-3, 4-dimethylcyclobutene to 2*E*, 4*Z*-hexadiene. This is a concerted 4*n*-electron reaction in which two of electron pairs are involved ; the sigma bond and the *pi* bond of the reactant are converted to the two *pi* bonds of the product. (Scheme 17.11). Recall that a conrotation involving 4 electrons is thermally allowed. The conrotation in anticlockwise fashion will yield the same compound as already seen (see, Scheme 17.8).

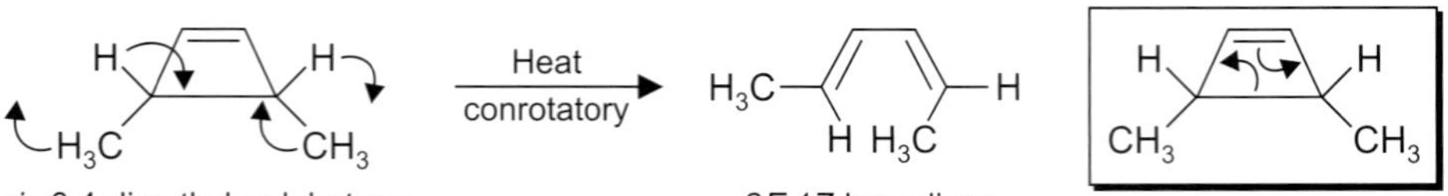

cis-3,4-dimethylcyclobutene — 2*E*,4*Z*-hexadiene — Movement of electrons

Movement (rotation) of end carbon atoms (conrotation) clockwise

SCHEME 17.11

EXERCISE 17.2

Depict two possible conrotatory (thermal) modes of ring opening in trans-3, 4-dimethylcyclobutene.

ANSWER. *The two conrotatory modes lead to different products. (The two disrotatory modes on cis- and trans-disubstituted cyclobutenes are in Scheme 17.20a).*

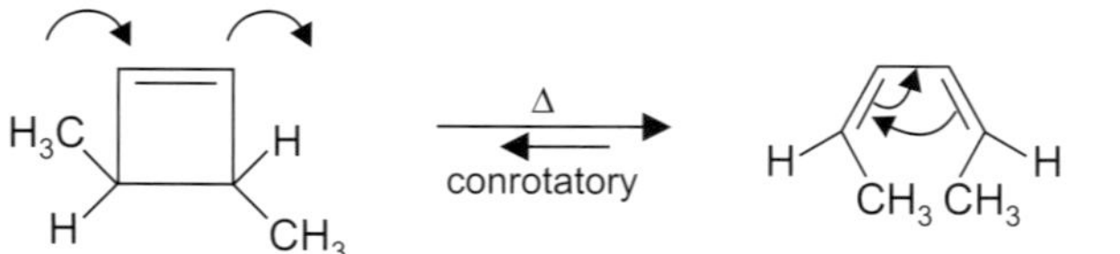

trans-3,4-dimethylcyclobutene (rotation of *p*-orbitals clockwise) — *cis, cis* (*Z, Z*) (movement of electrons)

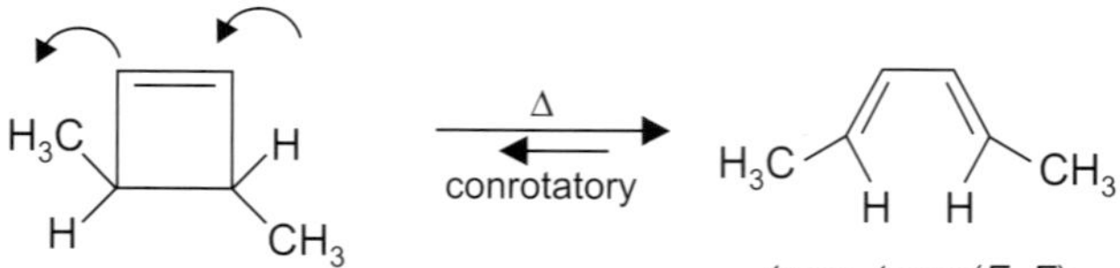

trans-3,4-dimethylcyclobutene (rotation of *p*-orbitals) Anticlockwise — *trans, trans* (*E, E*)

SCHEME 17.12

Thus one can also easily depict these conrotations (and also disrotations) directly without drawing molecular orbitals (Scheme 17.13). When a polyene undergoes an electrocyclic ring closure to give a cycloalkene, the terminal carbons of the polyene chain must rotate about 90° to convert the p orbitals on these carbons into the sp^3 orbitals forming the new σ bond. The substituents on these carbons are to be rotated into a plane which is approximately at right angles to the newly formed ring. Conversely, during the ring opening of a cycloalkene, the substituents on the atoms forming the bond undergoing cleavage will rotate into the plane of the new double bonds.

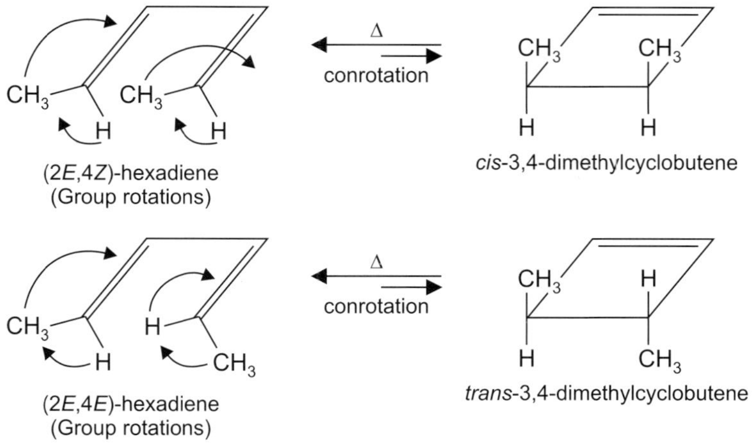

SCHEME 17.13

C. Thermal Electrocyclic Interconversion of 1, 3, 5-Hexatriene and 1,3-Cyclohexadiene

> ***Thermal Electrocyclic Ring Closure or Ring Opening—Conrotatory or Disrotatory Motions***
>
> *One has already seen that the ground state HOMO of a compound with an even number of conjugated double bonds is asymmetric and involves conrotatory motion e.g., in interconversion of a conjugated diene and cyclobutene (involvement of 4n pi electrons).*
>
> *On the other hand the ground state HOMO of a compound with an odd number of conjugated double bonds is symmetric and involves disrotatory motion e.g., in interconversion of 1, 3, 5-hexatriene and 1, 3-cyclohexadiene (involvement of 4n + 2 electrons).*

A thermal electrocyclic reaction involving $(4n + 2)$ π electrons where $(n = 0, 1, 2,...)$ proceeds with disrotatory motion.

The HOMO for the ground state of a hexatriene is ψ_3 and when compared with the HOMO of the ground state of butadiene *i.e.*, ψ_2 one finds that the relative symmetry about the terminal carbons is opposite (Scheme 17.14). Thus unlike the thermal opening of a cyclobutene (or the reverse reaction—the ring closure) which requires conrotatory motion, in the thermal opening of a 1,3-cyclohexadiene and likewise the ring closure requires a disrotatory motion. Based on these arguments 2*E*, 4*Z*, 6*E*-octatriene gives specifically *cis*-5,6-dimethylcyclohexadiene (Scheme 17.15).

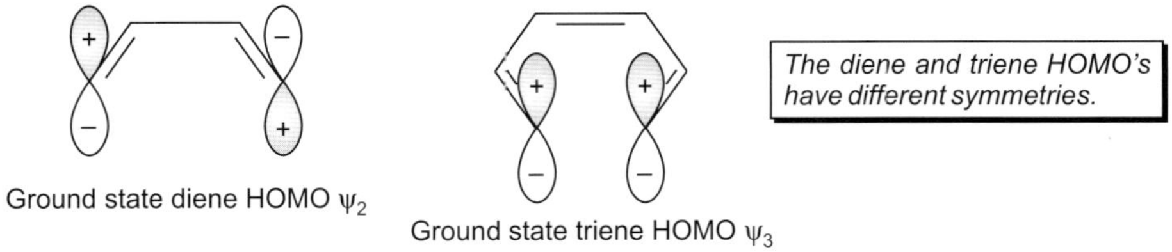

A conjugated diene and a conjugated triene react in opposite stereochemical senses. The diene opens and closes by conrotatory motion while a triene opens and closes by disrotatory motion.

SCHEME 17.14

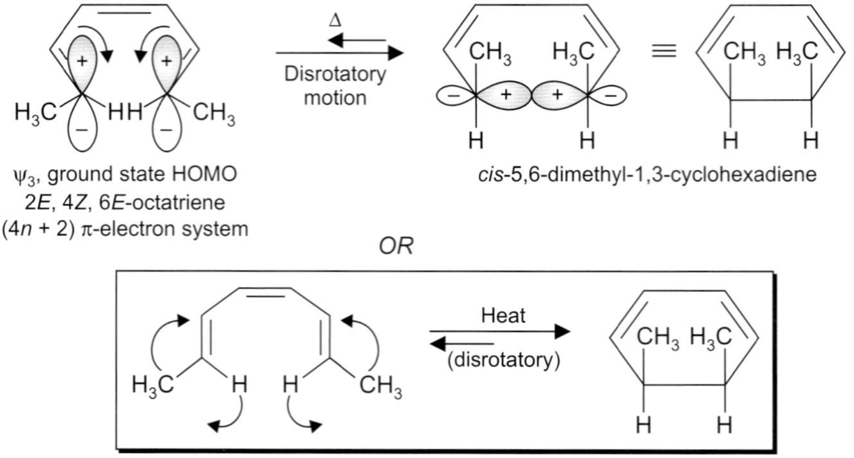

SCHEME 17.15

Similarly one finds that in the thermal cyclization of (2*E*, 4*Z*, 6*Z*)-octatriene as well the methyl groups rotate in the disrotatory fashion (Scheme 17.16).

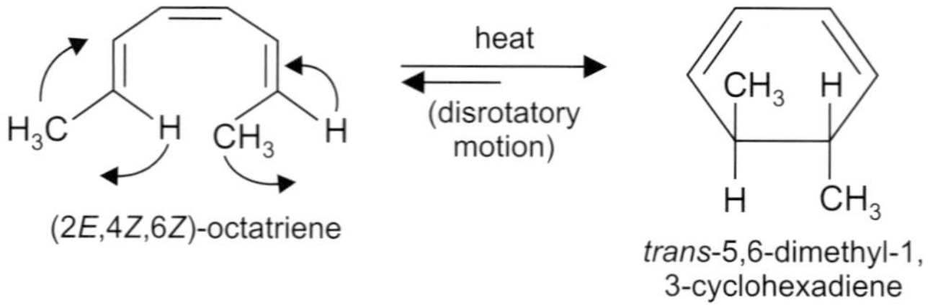

SCHEME 17.16

A thermal ring opening *e.g.*, in the case of *cis*-5, 6-dimethyl-1, 3-cyclohexadiene (see Scheme 17.15) must also be disrotatory. The ground state HOMO of the derived triene is to be ψ_3 (Scheme 17.17).

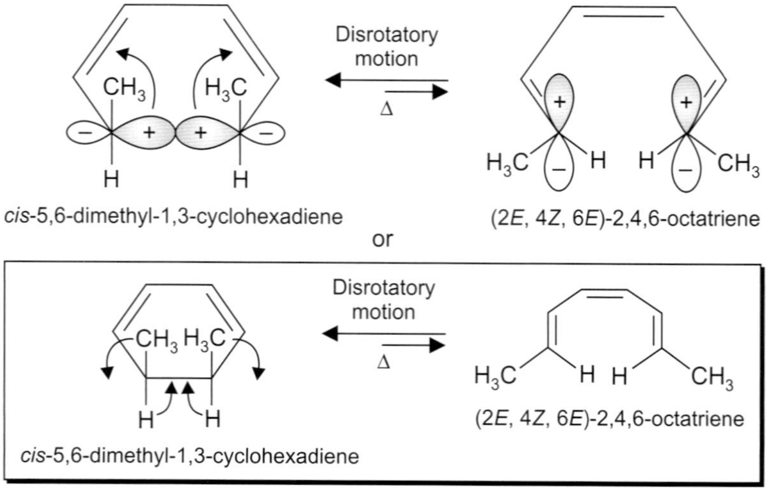

SCHEME 17.17

EXERCISE 17.3

What stereochemistry of the allyl carbocation is expected under thermal conditions from cis-dimethylcyclopropyl carbocation (Scheme 17.18)?

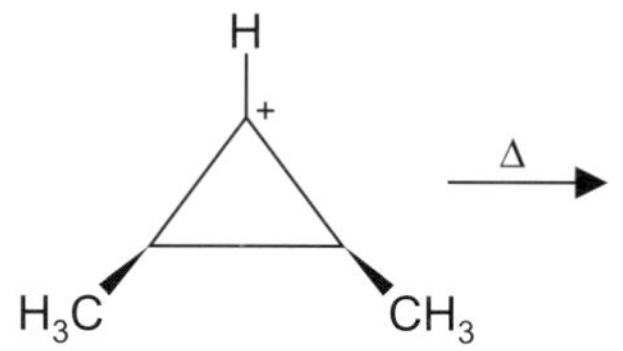

SCHEME 17.18

ANSWER. *The allyl carbocation will be formed by electrocyclic ring opening. Since the reactant involves two electrons it is a (4n + 2, n = 0, 1, ...) system and thus disrotatory opening is thermally allowed (Scheme 17.19).*

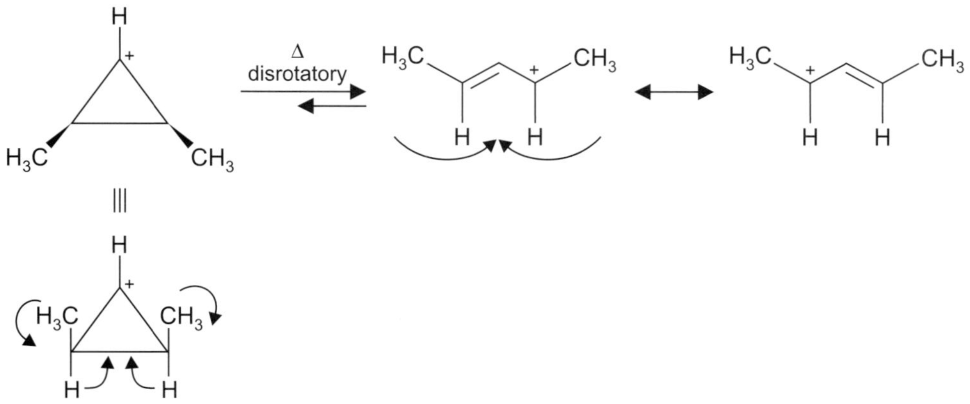

SCHEME 17.19

D. Summary—Electrocyclic Thermal Reactions

Electrocyclic closure of a conjugated diene is conrotatory, while that of a conjugated triene is disrotatory. This is due to the difference is the phase relationships within the HOMO at the terminal carbons of these π systems. In the diene the HOMO has opposite phase at these two carbons while in the triene the HOMO has the same phase.

Conjugated alkenes with 4n π electrons (n = any integer) have antisymmetric HOMOs and undergo conrotatory ring closure while those with (4n + 2) π electrons have symmetric HOMOs and undergo disrotatory ring closure. Thus a conrotatory ring closure is allowed for systems with 4n π electrons and it is forbidden for systems with (4n + 2) π electrons. Conversely, disrotatory ring closure is allowed for systems with (4n + 2) π electrons and is forbidden for systems with 4n π electrons.

E. Excited State (Photochemical) Electrocyclic Reactions

On absorption of light a molecule reacts through its excited state. Recall that the HOMO of the excited state is different from the HOMO of the ground state and therefore, has different symmetry (thermal electrocyclic reactions occur through electronic ground states).

Consider for comparison the thermal or photochemical cyclization of (2*E*, 4*Z*) hexadiene which gives *cis* or *trans* 3,4-dimethylcyclobutene respectively. Under thermal cyclization the ground state HOMO is ψ_2 which is asymmetric, the reaction has to be conrotatory in order to achieve in-phase overlap (Scheme 17.20).

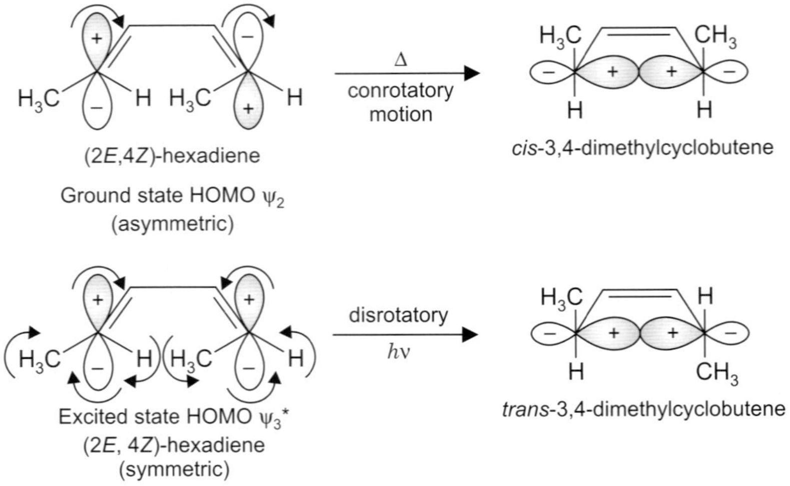

SCHEME 17.20

Similarly when (2*E*, 4*E*)-hexadiene on heating cyclizes to form *trans*-3, 4-dimethylcyclobutene, none of the *cis*-isomer is formed. In the reverse reaction the cyclobutene opens to produce only the (*E*, *E*)-isomer of the hexadiene. When the same isomer is photolyzed rather than heated, only *cis*-3, 4-dimethylcyclobutene is produced as the only product (Scheme 17.20*a*). Thus the reaction is completely stereospecific and involves conrotation under thermal conditions and disrotation on photochemical cyclization.

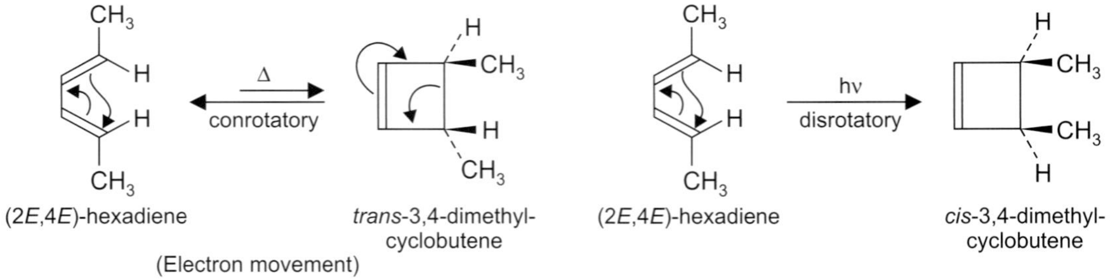

SCHEME 17.20a

EXERCISE 17.4

Depict the stereochemistry of the product from two disrotatory modes involved in the photochemical reaction of cis and trans-disubstituted cyclobutene.

ANSWER. *The products will be formed by disrotation, a cis-disubstituted cyclobutene opens to give cis, cis and trans, trans isomers of butadiene (Scheme 17.20b), while a trans-disubstituted cyclobutene gives the same product (in which there is one cis and one trans double bond).*

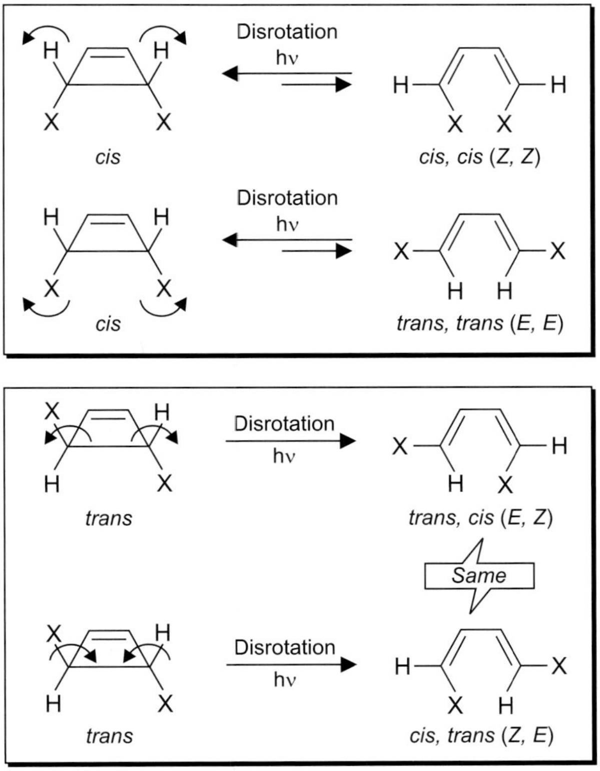

SCHEME 17.20b

When photocyclization is carried out the excited state HOMO of (2*E*, 4*Z*)-hexadiene is now ψ_3^*, because this HOMO is symmetric (this HOMO has the same phase at each end of the π system) thus the bonding overlap can occur only provided the ring closure is disrotatory (Scheme 17.20).

One quickly gets to the same stereochemical out come by avoiding writing of the *p* orbitals at the ends of the conjugated system (Scheme 17.21).

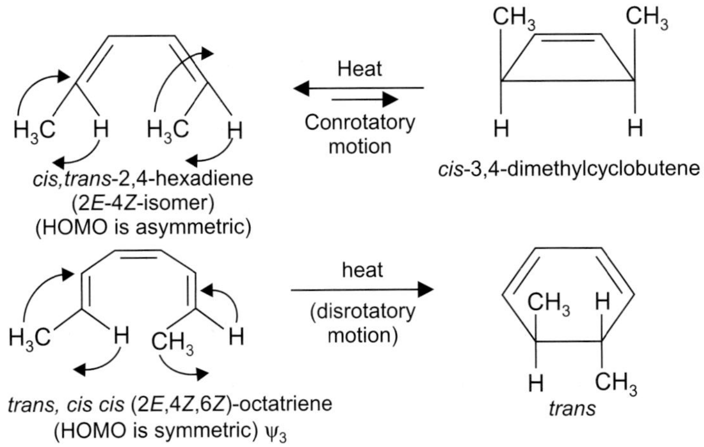

The ground state HOMO of a compound with an even number of conjugated double bonds is asymmetric (4n) π-electron system (conrotatory ring closure). The ground state HOMO of a compound with odd number of conjugated double bonds (4n + 2) π-electron system is symmetric (disrotatory ring closure).

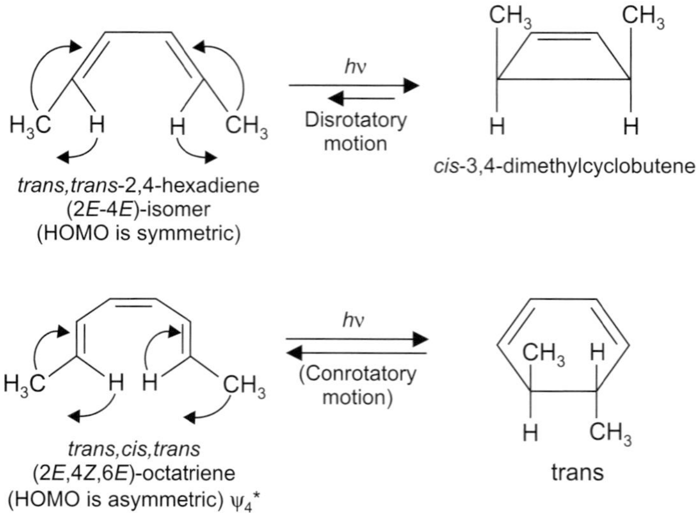

In reactions under photochemical conditions every thing is reversed. The ground state and excited state HOMO's have opposite symmetries. If the ground state HOMO is symmetric the excited state HOMO is asymmetric.

SCHEME 17.21

EXERCISE 17.5

Predict the stereochemical outcome from the electrocyclic reactions (Scheme 17.22)?

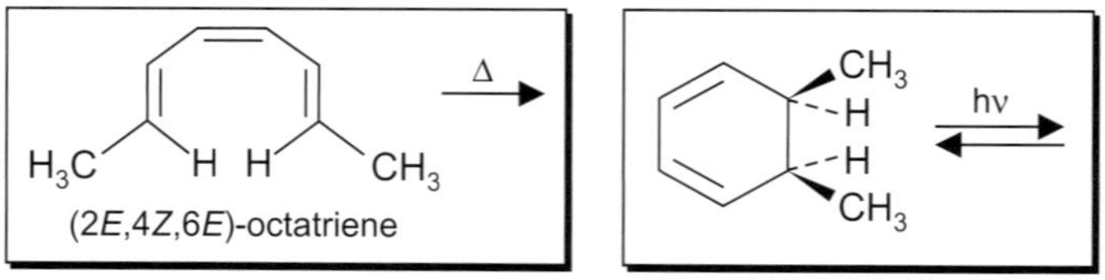

(I) (II)

SCHEME 17.22

ANSWER. (I) *Reactant has odd number of double bonds (4n + 2) π electron system, ground state HOMO of the triene (ψ_3) is symmetric. Ring closure is disrotatory.*

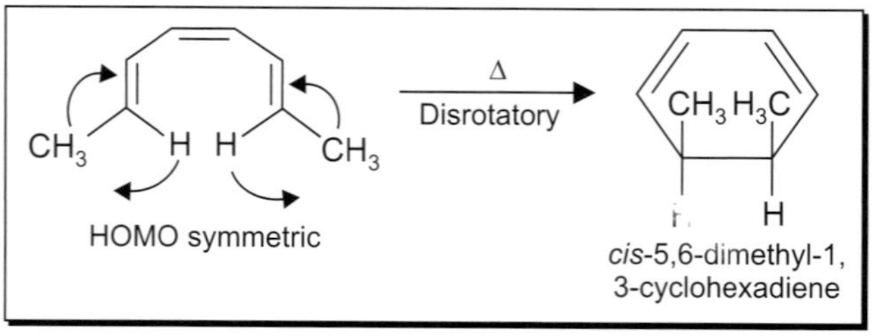

SCHEME 17.23

(II) *Reactant is a (4n + 2) π electron system, photochemically it involves a conrotatory motion.*

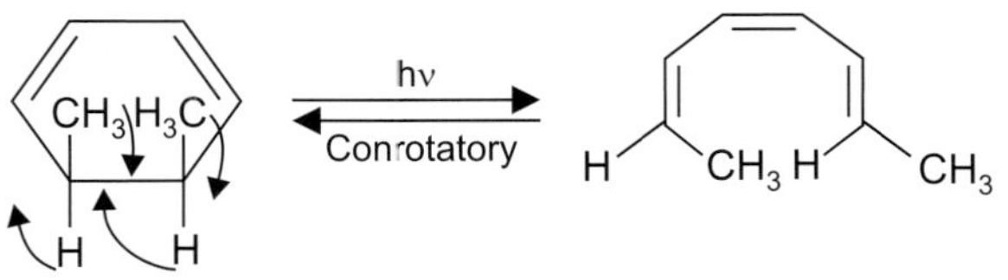

SCHEME 17.24

The selection rules for electrocyclic reactions which are given again (Table 17.1) will thus help to know the outcome of an electrocyclic reaction. Compound (I, scheme 17.25) is a $4n\pi$ system, therefore, a thermal cyclization *via* a conrotatory motion is an allowed process.

Table 17.1. Selection Rules for Electrocyclic Reactions

Number of electrons*	Mode of activation	Allowed stereochemistry
$4n$	thermal photochemical	conrotatory disrotatory
$4n + 2$	thermal photochemical	disrotatory conrotatory

*n = an integer. These selection rules are based on the orbital symmetry of the open-chain (conjugated alkene) reactant.

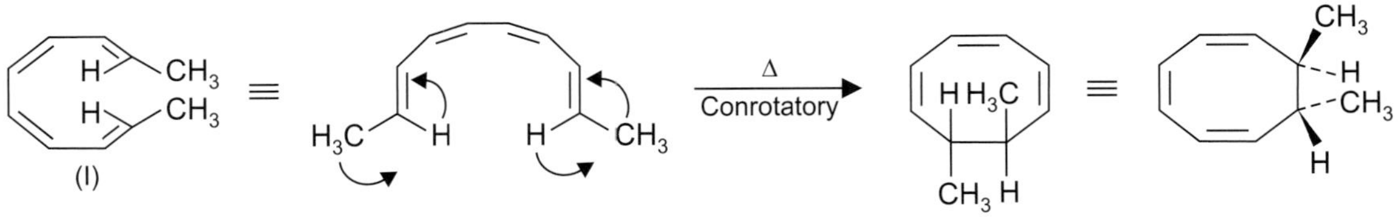

(I)

SCHEME 17.25

F. Stereochemical Rules—Electrocyclic Reactions—Problem Solving Hint

A conjugated diene and a conjugated triene react in opposite (alternating) stereochemical senses during thermal reaction. The diene opens and closes by a conrotatory path while the triene opens and closes by a disrotatory path. These results are due to different symmetries of the HOMO of a diene and a triene.

On electronic excitation (photochemical reactions) the symmetries of HOMO and LUMO change and with it changes the reaction stereochemistry (which is reversed) and stereochemical rules for electrocyclic reactions are given (Table 17.1).

Problem Solving Hint 1

If in a reactant the bonds to the substituents point in opposite directions then these substituents will have cis stereochemistry in the product provided the motion is disrotatory (Scheme 17.25a)

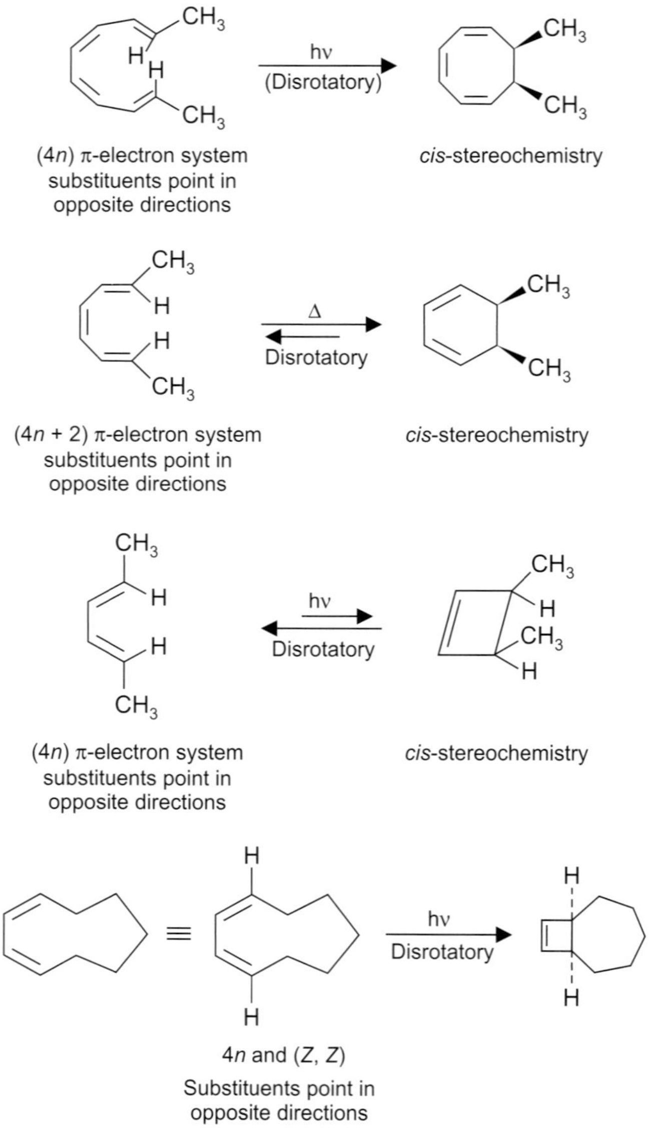

SCHEME 17.25a

Problem Solving Hint 2

If in a reactant the bonds to the substitutents point in opposite directions then these substituents will have trans stereochemistry in the product provided the motion is conrotatory (Scheme 17.26).

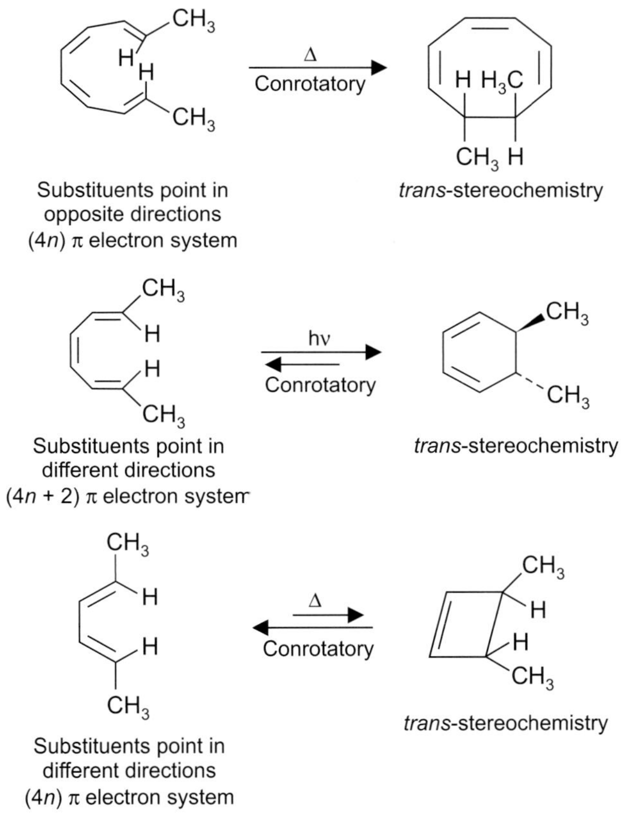

SCHEME 17.26

Problem Solving Hint 3

If in a reactant the bonds to the substituents point in the same direction then these substituents will have cis-stereochemistry in the product provided the motion is conrotatory (Scheme 17.27).

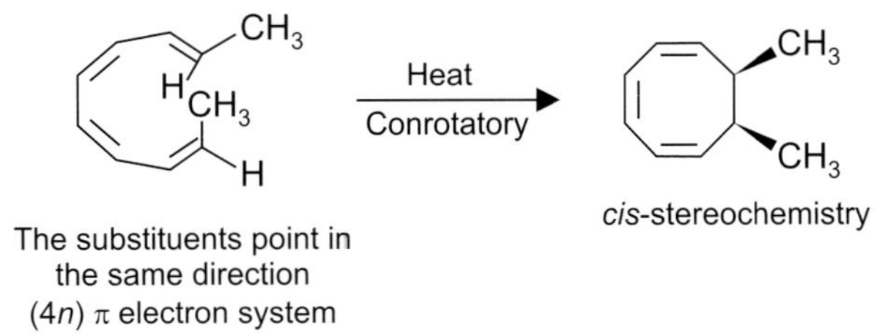

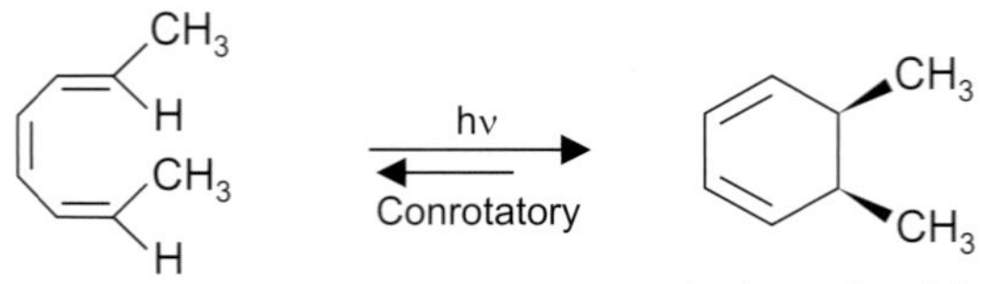

The substituents point in the same direction
$(4n + 2)$ π-electron system

cis-stereochemistry

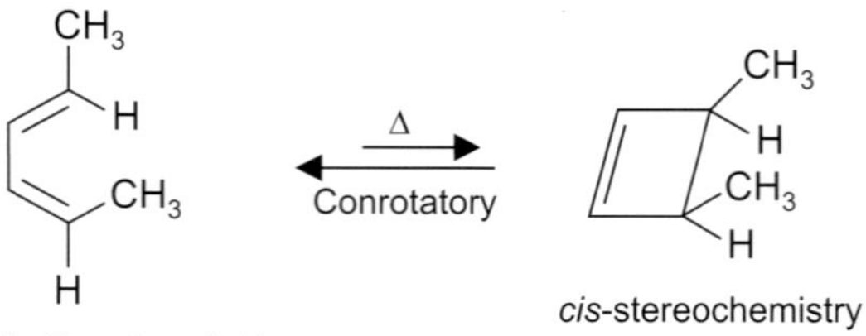

The substituents point in the same direction
$(4n)$ π-electron system

cis-stereochemistry

SCHEME 17.27

Problem Solving Hint 4

If in a reactant the bonds to substituents point in the same direction then the substituents will have trans-stereochemistry in the product provided the motion is disrotatory (Scheme 17.28).

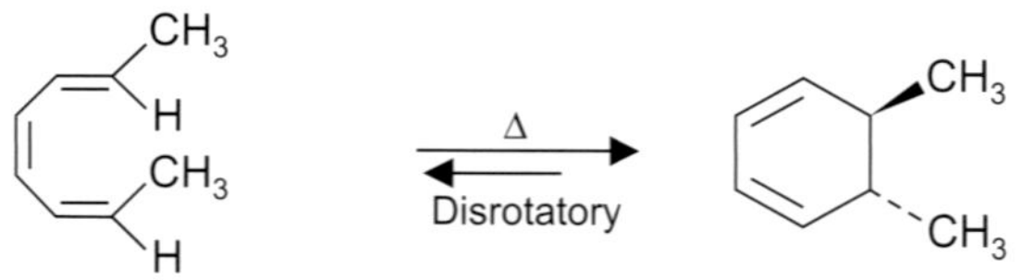

The substituents point in the same direction $(4n + 2)$ π electron system

trans-stereochemistry

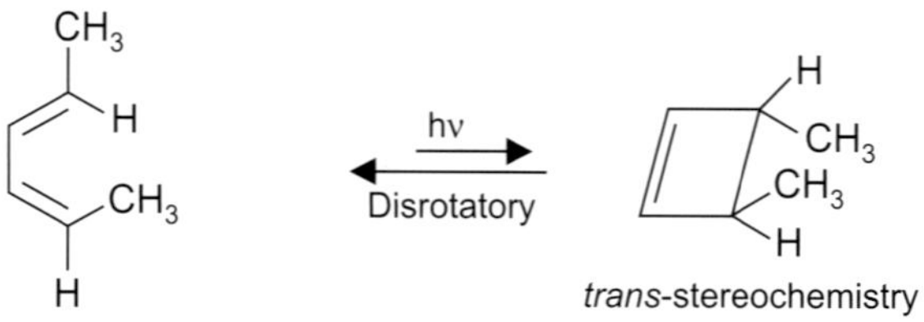

The substituents point in the same direction $(4n)$ π electron system

trans-stereochemistry

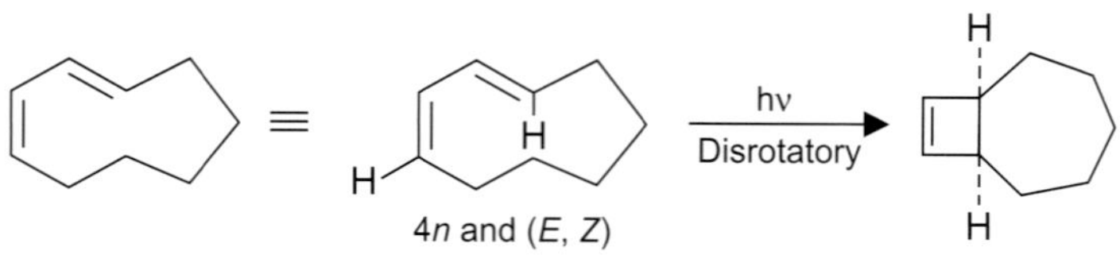

$4n$ and (E, Z)

The substituents point in the same direction

SCHEME 17.28

Example 1: Consider, firstly the photochemical ring closure of precalciferol which gives ergosterol and lumisterol (I and II respectively; Scheme 17.29) in which hydrogen and methyl substituents are *trans* to one another.

The reactant precalciferol has three conjugated π bonds $(4n + 2)$ π electron system thus ring closure under photochemical conditions is conrotatory (see Table 17.1). The methyl and hydrogen at C-10 and C-9 respectively point in opposite directions in precalciferol (the H atom at C-9 is not shown which if drawn is in opposite direction to CH_3 at C-10, note that ring residue at C-9 and CH_3 substituent at C-10 are in the same direction).

Thus in precalciferol a conrotatory ring closure will cause the substituents which point in opposite directions in the reactant to be *trans* in the product. On disrotatory ring closure of precalciferol, however, these substituents will assume a *cis* relationship. The reason for the formation of two *cis* and two *trans* products is that *e.g.,* in the case of disrotatory ring closure (Scheme 17.29) two *cis* products arise due to the outward disrotatory or inward disrotatory motion.

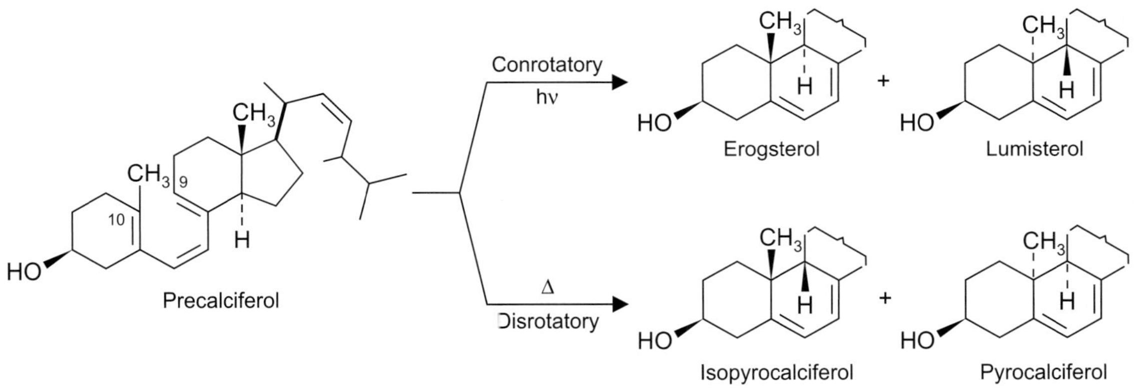

SCHEME 17.29

Example 2: On heating 1,3,5-cyclononatriene (I, Scheme 17.30) gives a bicyclic product with *cis*-ring fusion. In (I) when the hydrogens are drawn, these point in opposite directions (II). The reactant has three π bonds $(4n + 2)$ π electron system thus under thermal conditions the ring closure is disrotatory (table 17.1) and therefore, the hydrogen atoms in the product will have a *cis* relationship.

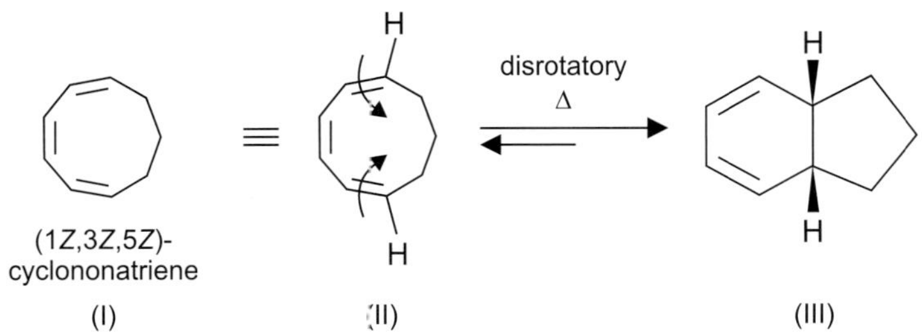

SCHEME 17.30

Example 3: Consider the photochemical ring opening electrocyclic reaction of (I, scheme 17.31). The product (II) undergoing ring closure has three conjugated double bonds and thus under photochemical conditions (see Table 17.1) ring closure or ring opening is conrotatory. To get a product with *cis* hydrogens in (I, Scheme 17.31) the substituent hydrogens have to point in the same direction (II). Thus compared to 1,3,5-cyclononatriene (I, Scheme 17.30) in which all the three double bonds have *Z* geometry in (II, Scheme 17.31) one of the double bonds has *E* configuration.

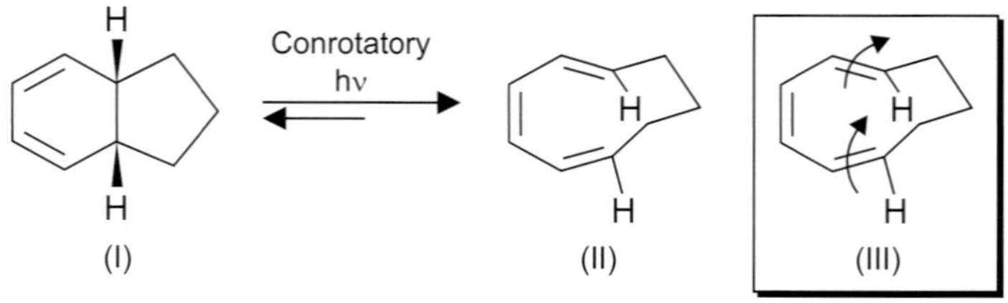

SCHEME 17.31

Example 4: 1,3-Cycloheptadiene (I, Scheme 17.32) closes to the cyclobutene by a disrotatory motion under photochemical conditions with expected *cis* ring fusion. A conrotatory opening of cyclobutene (II) is thermally allowed, since the hydrogens in (II, Scheme 17.32) are *cis*, these must point in the same direction in the ring opened product (III) *i.e.,* one of the double bonds in III must be *cis* and the other must be *trans*. However, a *trans*-double bond cannot be accommodated in a seven membered ring and thus (II, Scheme 17.32) is stable under thermal conditions. Similar arguments prove that compound (IV, Scheme 17.32) does not undergo ring opening under thermal conditions while (V) does.

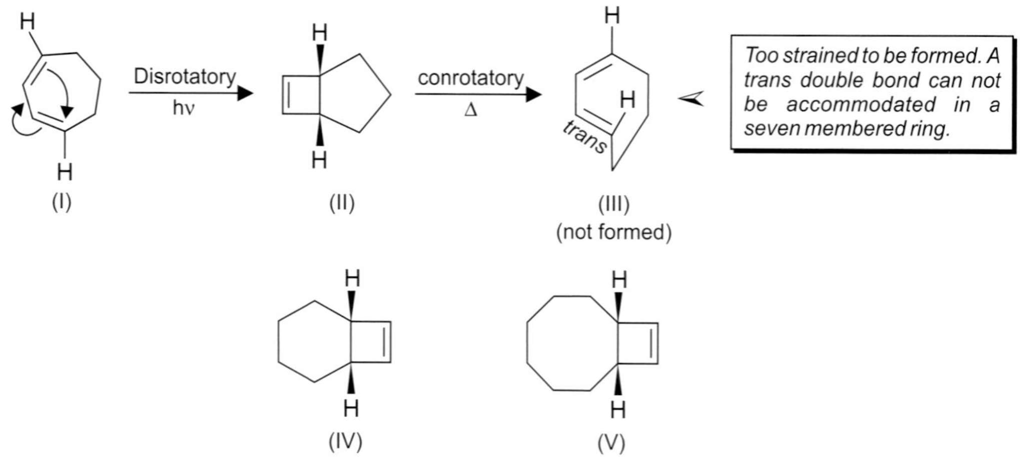

SCHEME 17.32

EXERCISE 17.6

Predict if the conversion shown (Scheme 17.33) is allowed or forbidden?

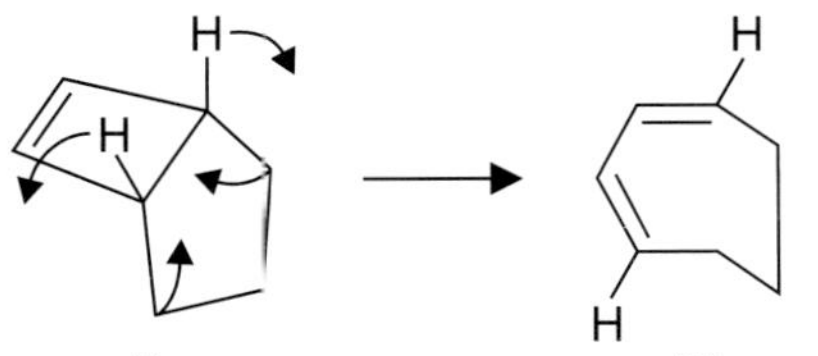

(I) (II)

SCHEME 17.33

ANSWER. *I t is a concerted 4n electron reaction. The ring opening shall have to be conrotatory (Table 17.1). Thermal process is allowed but the product is strained and is therefore, not formed (see Scheme 17.32). However, the conversion (Scheme 17.33) to unstrained all cis-diene is a disrotatory process as shown and is forbidden by the selection rules (Table 17.1).*

EXERCISE 17.7

One of the cyclobutenes (Scheme 17.34) on heating reacts very fast while the other reacts at an extremely slow rate and at much higher temperature. Explain.

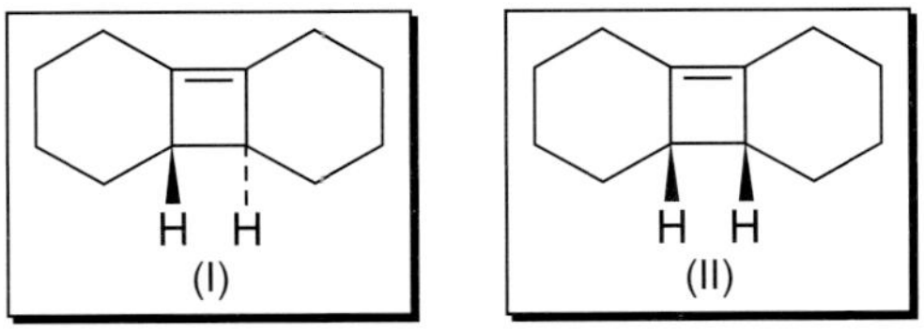

SCHEME 17.34

ANSWER. *Recall problem solving hints 1–4. Compound (I) reacts faster.*

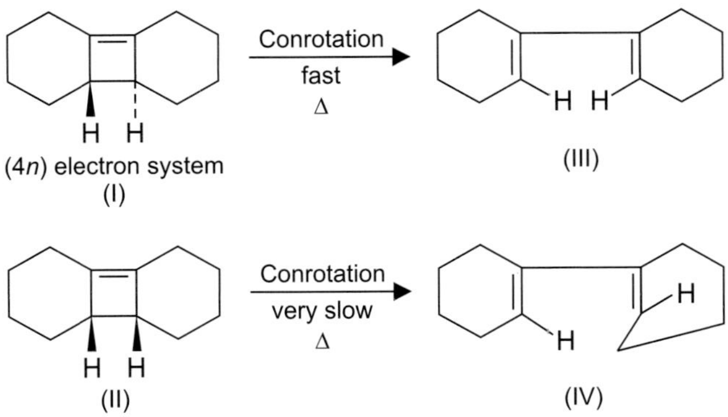

Both the H atoms point in opposite directions, these are to be trans in the reactant.

Both the H atoms point in the same direction, these are to be cis in the reactant, however (IV) has a trans double bond in a six membered ring. Its formation is very slow if at all.

SCHEME 17.35

The compound (II) can however, undergo an easy photochemical opening to the butadiene (III see Problem 17.9), involving disrotation. One may recall that e.g., on ring closure a butadiene under thermal or photochemical conditions gives cyclobutenes with opposite configurations. (see, Scheme 17.20a)

EXERCISE 17.8

How one can convert trans-5, 6-dimethyl-1, 3-cyclohexadiene into its cis isomer?

ANSWER. *This may be achieved by electrocyclic ring opening followed by electrocyclic ring closure under appropriate conditions (Scheme 17.36).*

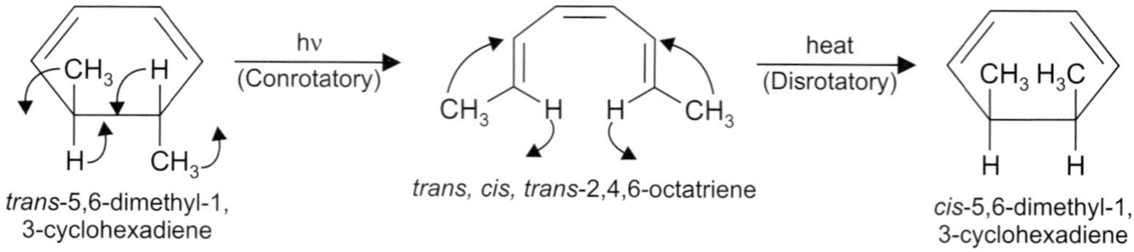

SCHEME 17.36

Example 5: Dewar benzene has been synthesized and is stable at room temperature (at 25°C, the half life for its conversion to benzene is two days). It is much less stable than benzene due to angle strain and no stabilization due to aromaticity. Dewar benzene could therefore, easily isomerize to benzene. However, this conversion is an electrocyclic reaction (Scheme 17.37), it involves two pair of electrons one pair of π electrons and one pair of sigma electrons of the Dewar benzene, the third pair of electrons is located in exactly the same place both in the reactant and the product and is thus not involved in the reaction).

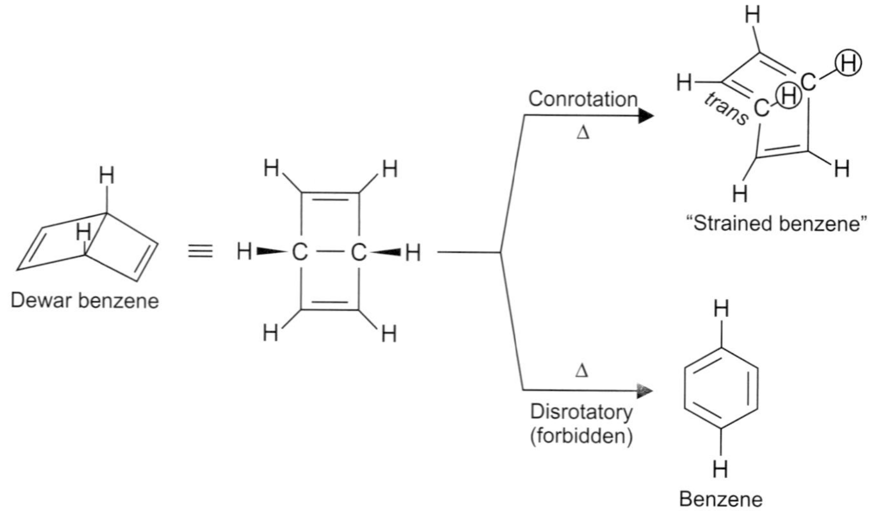

SCHEME 17.37

A thermally allowed electrocyclic reaction with two pairs of electrons ($4n$) π electron system must be conrotatory and the opening of the cyclobutene ring in Dewar-benzene (Scheme 17.37) would result in strained isomer of benzene ("strained benzene") with a *trans* double bond in six membered ring. (In strained benzene two hydrogens on the newly created diene moiety which point to the same side are circled. Thus the otherwise thermally allowed conversion of Dewar benzene into benzene is geometrically impossible since low energy pathway for this conversion is not available.

Example 6: The tetraene (Scheme 17.38) has an even number of π bonds ($4n$) π electron system and therefore, under thermal conditions (– 10°C) it will undergo conrotatory ring closure. Since the two methyl substituents on the tetraene point to opposite directions these will be *trans* in the ring closed cyclooctatriene. In cyclooctatriene one now has three double bonds in conjugation (an odd number, $4n + 2$ system) and therefore, the second thermal ring closure will now be disrotatory. Since the hydrogen substituents at the ends of the triene system (which are not drawn) are in opposite directions (Compare with Scheme 17.30), these must be *cis* in the final bicyclo product.

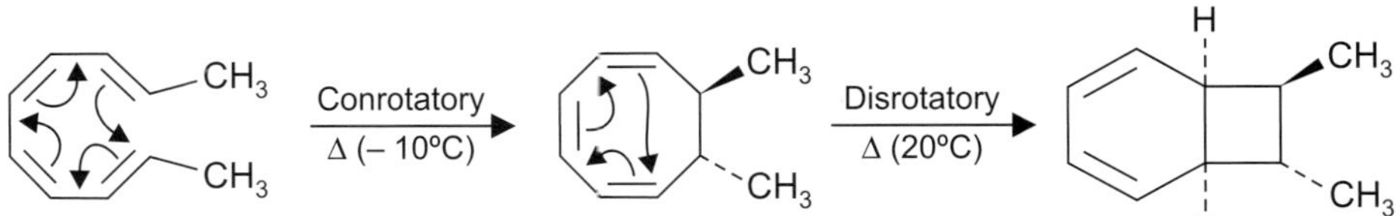

(2*E*, 4*Z*, 6*Z*, 8*E*)-decatetraene (electron movement) — *trans*-dimethyl-cyclooctatriene — *cis* ring junction

SCHEME 17.38

Example 7: A remarkable distinction between photochemical and thermal reactions is displayed by all *cis*-cyclodecapentaene (Scheme 17.39). One may consider its reacting system with three conjugated π bonds. A conrotatory ring closure under photochemical conditions with hydrogen substituents at the end of the considered triene system in opposite directions will end up in a *trans* ring junction. Thermal ring closure of a three π bond system is disrotatory, and with hydrogen atoms in opposite directions will end up in a *cis* ring junction.

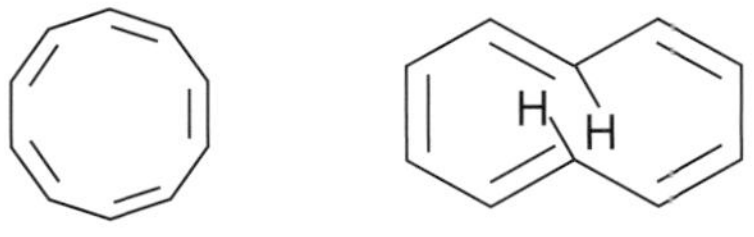

cis,cis,cis,cis,cis-Cyclodecapentaene (angle strain)

trans,cis,trans,cis,cis-Cyclodecapentaene (strain caused by the inside hydrogens)

Two cyclodecapentaenes are shown. All cis isomer (known) is not aromatic. Infact it is a tub shaped molecule. All double bonds cannot be considered as conjugated.

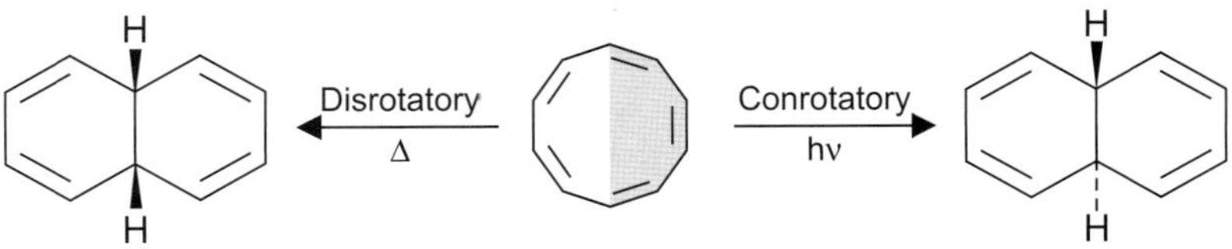

SCHEME 17.39

17.4 CYCLOADDITIONS (FMO-APPROACH)

The reactions of alkenes (the dienophiles) and polyenes (conjugated dienes) in which two molecules react to form a cyclic product, with π electrons being used to form two new σ bonds are called cycloaddition reactions. These reactions are classified on the basis of π electrons involved, in each component, the [4 + 2] cycloaddition being the well known Diels-Alder reaction (Scheme 17.40). The reaction of two alkenes to form a cyclobutane derivative is termed a [2 + 2] cycloaddition reaction (Scheme 17.40). A cycloaddition reaction requires only heat or light for initiation, radical and ionic intermediates are not involved.

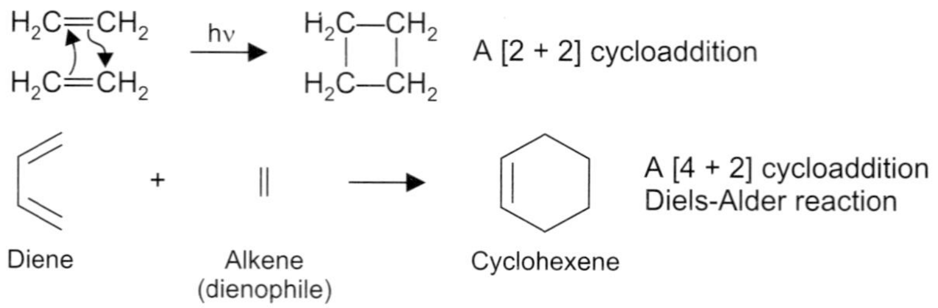

SCHEME 17.40

Molecular Orbitals of Ethylene

On heating ethylene its π electrons are not promoted, but remain in the ground state ψ_1. Irradiation with UV light excites an electron from ψ_1, the ground-state HOMO, to ψ_2^*, which becomes the excited-state HOMO. Interaction between the excited-state HOMO of one alkene and the LUMO of the second alkene indicates that a photochemical [2 + 2] cycloaddition reaction can occur by a suprafacial pathway (Scheme 17.41).

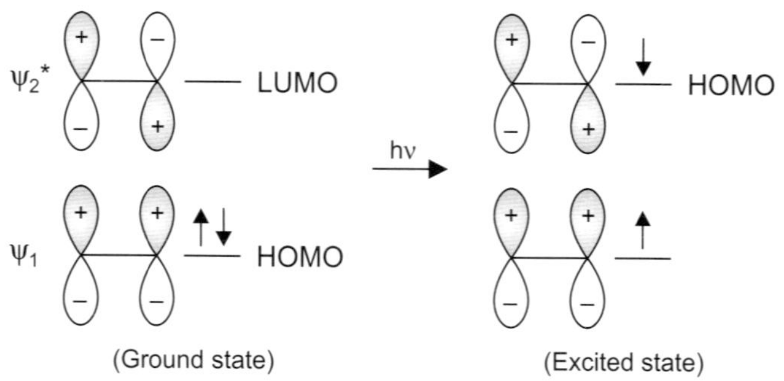

Molecular orbitals of ethylene

SCHEME 17.41

A. Diels-Alder Reaction—[4 + 2] Cycloadditions

These are concerted, thermal [4 + 2] cycloadditions. A consideration of orbital interactions (two combinations) accounts for this (Scheme 17.42), *i.e.*, the overlap can take place between the HOMO of one component and the LUMO of the other and *vice versa*.

As with electrocyclic reactions in cycloadditions as well, one is only concerned wih the terminal lobes. The simplest [4 + 2] system involves the cycloaddition of 1, 3-butadiene (the diene) and ethylene (the dienophile) which is a thermally induced reaction. This thermally allowed reaction involves *e.g.*, the HOMO of 1, 3-butadiene (ψ_2) with the LUMO of ethylene ψ_2* (one could equally well use the diene LUMO and the alkene HOMO).

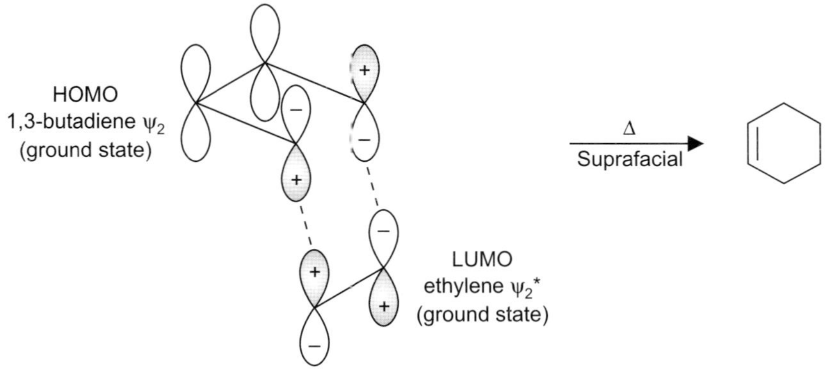

Symmetry-allowed thermal [4 + 2] cycloaddition : 1,3-butadiene and ethylene

SCHEME 17.42

In either case, the overlap brings together lobes of the same phase. Addition to the lobes on the same side of a π system is called suprafacial addition, while addition to lobes on opposite sides of a π system is termed antarafacial addition (for an example of antarafacial addition see Scheme 17.57). These modes of addition are identified by the symbols *s* and *a* respectively. Thus cycloaddition of two π bonds each reacting suprafacially would be called [$\pi^2 s + \pi^2 s$] reaction.

EXERCISE 17.9

Explain by orbital drawings that [4 + 2] cycloaddition is photochemically forbidden.

ANSWER. *Usually the absorption of a photon will promote an electron from HOMO to LUMO. In the case of a photochemical Diels-Alder reaction (which is the most uncommon) the lower energy HOMO–LUMO gap is in the diene partner. Thus on absorption of light a new photochemical HOMO for the diene (ψ_3*) is generated and now the HOMO–LUMO interaction with the dienophile partner involves one antibonding overlap. Thus the new bonds cannot be formed at the same time and the photochemical Diels-Alder reaction is forbidden by orbital symmetry (I, Scheme 17.43). However, one may note that photoinduced [4 + 2] cycloaddition cannot occur if either the diene or the dienophile is excited (II, Scheme 17.43).*

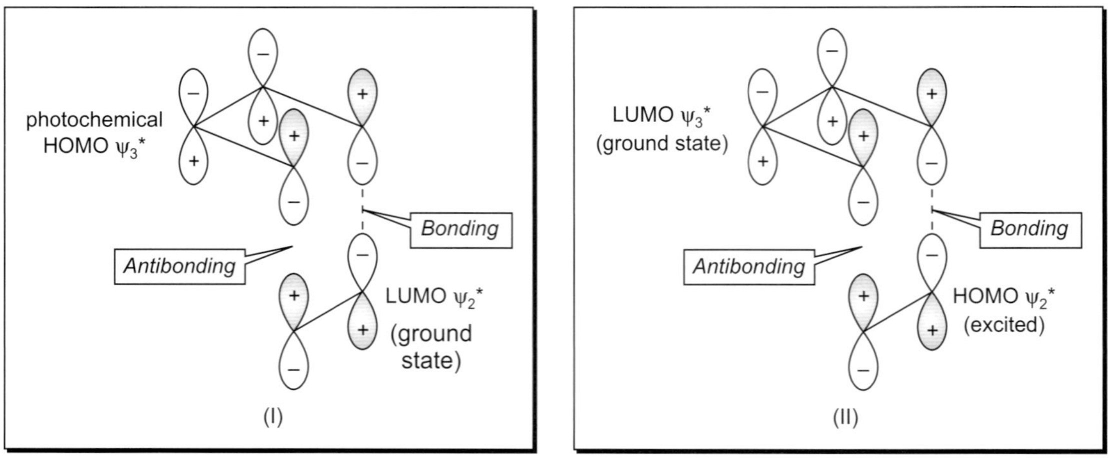

SCHEME 17.43

When in the dienophile there is conjugation to a group of – *M* type *e.g.,* carbonyl, nitro etc, the reaction occurs under milder conditions and gives good yields. The substituent lowers the energy of the LUMO of the dienophile so as to bring it closer in energy to the HOMO of the diene. Consequently the bonding interaction in the transition state increases. As expected, the reactivity is also increased by an electron releasing group in the diene. Conversely, when the diene contains an electron-withdrawing substituent the dienophile requires an electron-releasing substituent for ready reaction. In this situation the interaction is between diene's LUMO and the dienophile's HOMO. Thus, the bonding at the transition state is more effective when the HOMO of one reactant and the LUMO of other are more closely matched in energy. The following points may be noted:

- *s-cis Conformation of the Diene.* As correctly shown (Scheme 17.40) the diene reacts in the *s-cis* conformation, which allows the ends of the conjugated system to reach the doubly bonded carbons of the dienophile. That the *s-cis* geometry of the diene is essential is shown by the unreactive nature of the fixed *transoid* dienes (I and II, Scheme 17.44). Moreover, as expected the substituents in the diene may also effect the cycloaddition sterically. The substituents effect the equilibrium proportion of the diene in the required *cisoid* form (Scheme 17.44). Consequently *Z* alkyl or aryl substituents in the 1-position (III, Scheme 17.44 of the diene slow down the reaction by sterically hindering formation of the *cisoid* conformation, while bulky 2-substituents (IV) make it fast.

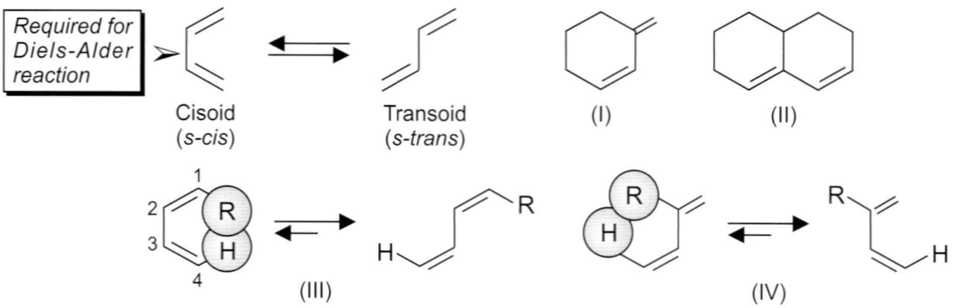

SCHEME 17.44

Cyclic dienes are among the useful dienes and particularly reactive in Diels-Alder reactions as the two double bonds are held in the *s-cis* conformation in five or six membered rings. Cyclopentadiene is highly reactive and forms a Diels-Alder adduct with itself. On heating, the commercially available dimer undergoes a *retro-Diels-Alder* reaction (The term *retro* means the reverse) to give cyclopentadiene (Scheme 17.45).

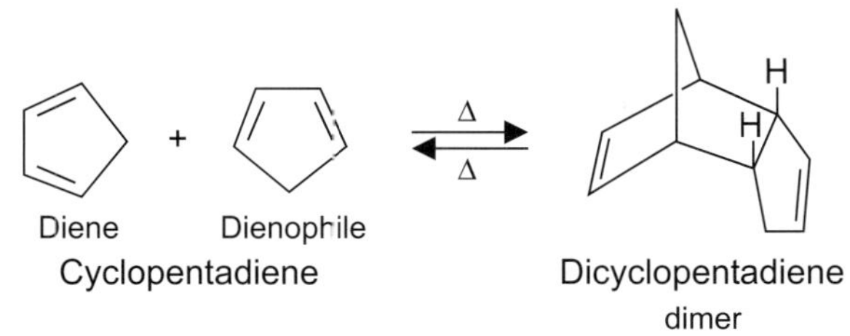

SCHEME 17.45

- *syn-Stereochemistry*. That the Diels-Alder reaction is concerted (both the new bonds are formed in the same transition state) is shown by the fact, that it proceeds with retention of configuration of both the diene and the dienophile (Schemes 17.46 and 17.47) *i.e.*, it proceeds stereoselectively *syn* with respect to both the diene and the dienophile as expected of a concerted (supra, supra) mode of addition. One may note that if in a diene both groups *e.g.*, methyls are "outside", these ends up *cis* in the product (Scheme 17.47) and if one methyl is "inside" and one "outside" these end up *trans* in the product.

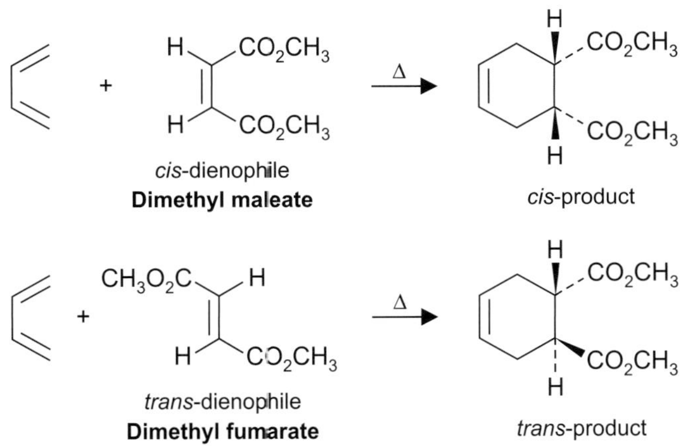

The stereochemistry of the dienophile is retained

SCHEME 17.46

- *The Endo Rule*. The Diels-Alder reaction takes place generally to give the less stable *endo* adduct as the major product. For the *endo* addition *e.g.*, with cyclopentadiene and methyl acrylate (Scheme 17.48), the transition state can be stabilized (speeding up the reaction) through secondary interactions. These interactions involve the lobes of HOMO and LUMO of the same phase which themselves are not involved directly in the formation of bonds. One sees that for the *endo* addition the π-system lies more completely over the other. These secondary interactions are not possible in the transition state for *exo* addition since the relevant set of centers in the diene and the

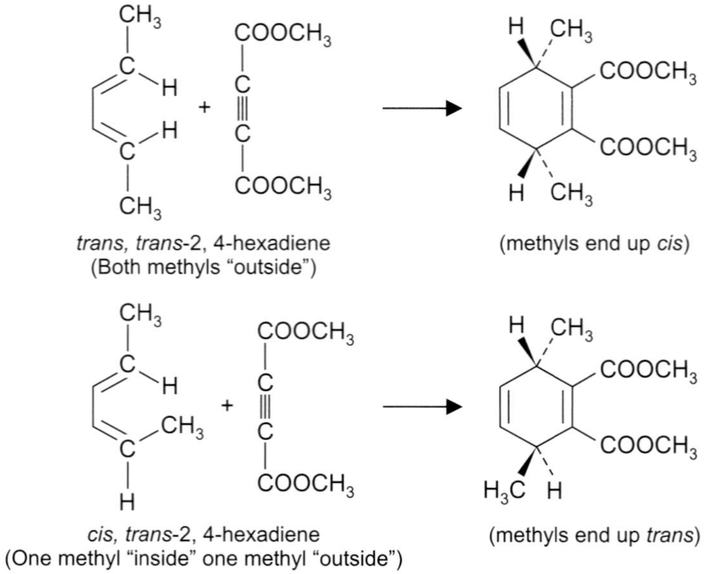

The stereochemistry of the diene is retained

SCHEME 17.47

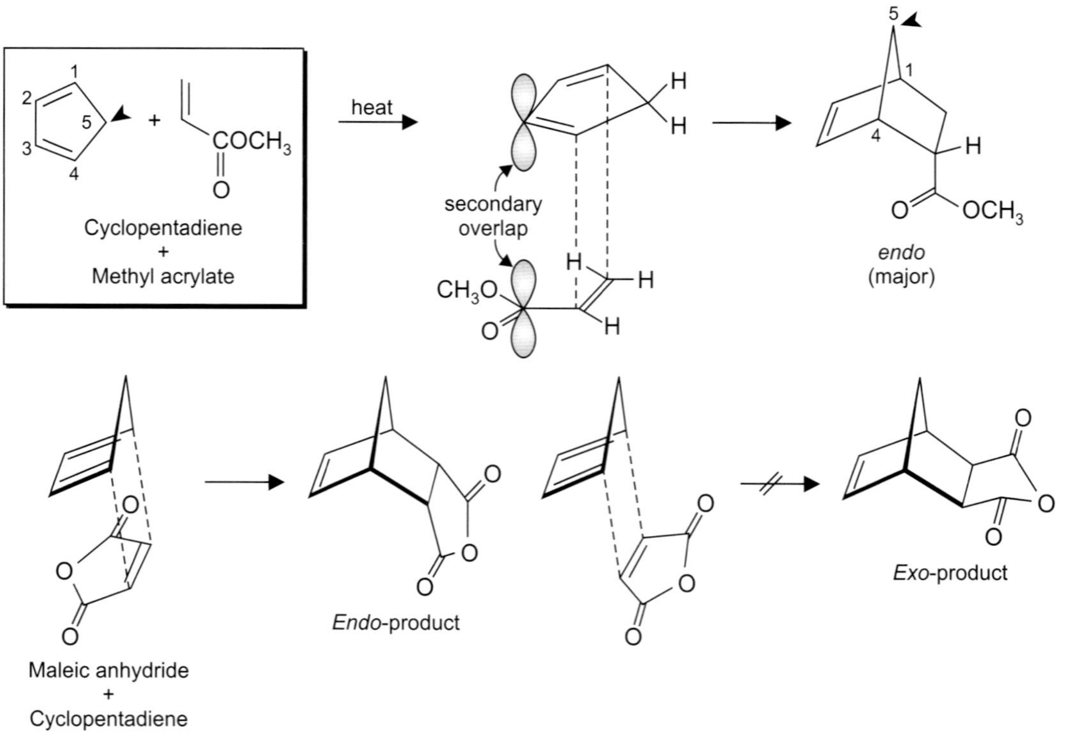

SCHEME 17.48

dienophile are now too far apart, from each other. Thus the preference for *endo* selectivity (which infact is due to several steric and electronic influences on the transition state—the *endo* transition state is lower in energy) is observed when the dienophile has a π bond in its electron withdrawing group *e.g.*, CN or C=O. The *p* orbitals of this group approach the central carbon atoms C2 and C3 of the diene and the resulting proximity leads to an overlap of the *p* orbitals and secondary overlap effects between the *p* orbitals of the diene and the dienophile (Scheme 17.48).

- *The endo Rule to Predict General Stereochemical Outcome.* One can use the *endo* rule to predict the stereochemical outcome of a reaction as detailed (Scheme 17.49). One can imagine the "inside" ligands of a diene to be the "CH_2" of cyclopentadiene and these will have a *cis*-relationship and shown on thick wedges. In keeping with the *endo* product formation when the product is a cyclohexene derivative the group on the dienophile (which is electron withdrawing) will be down (on a dotted wedge) due to it being inside the pocket of ring as a result of stability of the transition state (secondary overlap).

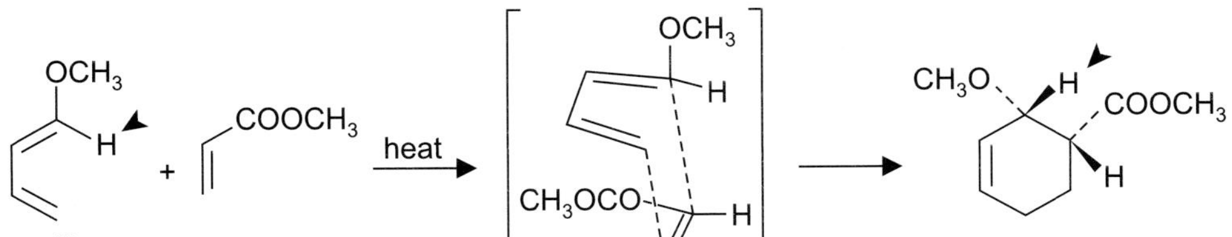

(I)

In (I) the H at C-1 ends "up" in the product (thick wedge) just like C5 methylene of cyclopentadiene (see, scheme 17.48) on its reaction with methyl acrylate. Based on endo rule $COOCH_3$ is below the diene in the transition state, thus it ends "down" (dotted wedge) in the product.

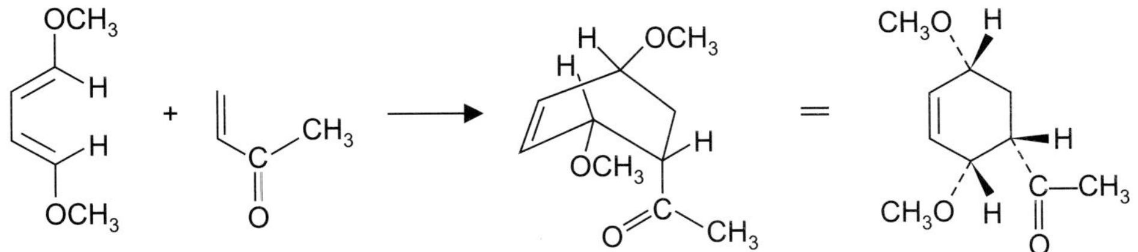

(II)

Relation with endo product formation

In (II) since both OCH_3 groups are outside these end up cis in the product. As in (I) both the inside H atoms end up on thick wedges. The $COCH_3$ group being under the pi system in the transition state is down (dotted wedge).

SCHEME 17.49

Stereochemistry Solving Hint

Considering schemes 17.47 and 17.49 one can evolve an useful stereochemistry solving hint of syn addition of concerted Diels-Alder reaction.

Two inner substituents always end up with a cis relationship in the product, and same is the case with two outer substituents. An inner substituent on one carbon ends up always trans to an outer substituent on the other.

When a ring junction is created then as expected the stereochemistry at the ring junction must be cis for a syn addition. Thus when maleic anhydride adds to trans, trans-2, 4-hexadine two diastereomeric syn addition products are possible (Scheme 17.49a).

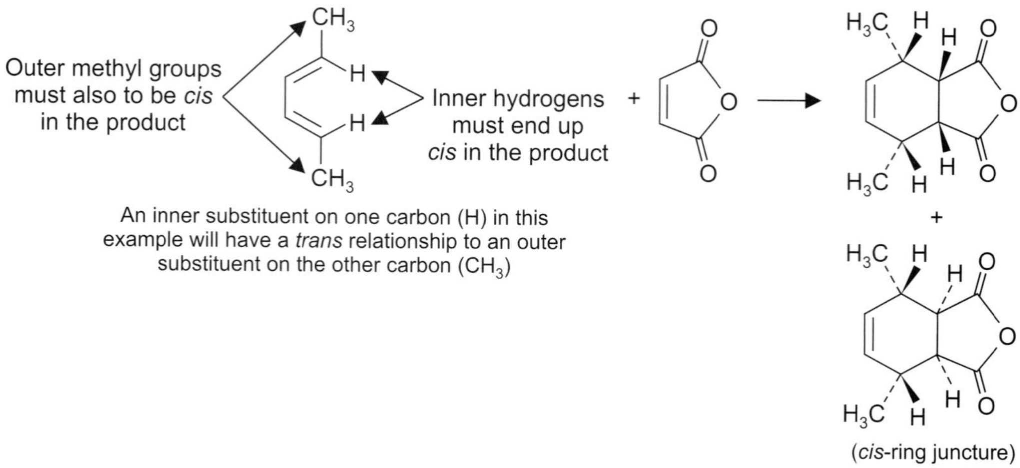

SCHEME 17.49a

- *Regioselectivity.* Cycloaddition of an unsymmetrically substituted diene and an unsymmetrically substituted dienophile can lead to regioisomers (Scheme 17.50).

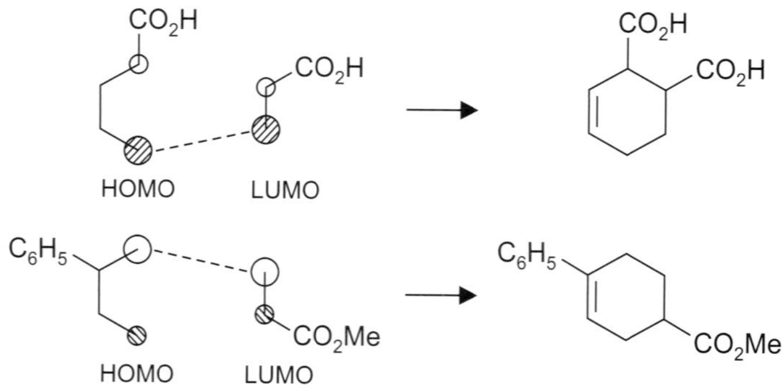

SCHEME 17.50

One has already seen that in a normal Diels-Alder reaction *i.e.,* between an electron rich diene and electron-deficient dienophile, the main interaction is between the HOMO of the diene and LUMO of the dienophile (In this situation these orbitals are more closely matched in energy, the better is the overlap and thus the reaction occurs more readily). However, the orientation of the product from an unsymmetrical diene and an unsymmetrical dienophile depends mostly on the atomic orbital coefficients at the reacting termini. The atoms with the larger terminal coefficients on each reactant, bond preferentially in the transition state, because of better orbital overlap. Consequently with 1-substituted butadienes the major product is 1,2 ("*ortho*") adduct while with 2-substituted butadienes, the major adduct is 1,4 ("*para*").

In the case of butadiene-1 carboxylic acid and acrylic acid the frontier orbitals are polarized as shown (Scheme 17.50). The size of the circles as shown is roughly proportional to the size of the coefficients and an allowed reaction leads to 1,2-adduct. Similarly, now with 2-phenyl-butadiene and methyl acrylate the major product formed would be 1,4.

- *Lewis Acid Catalysts.* Some Diels-Alder reactions are catalysed by Lewis acid catalysts. These catalysts form complexes with the polar groups on the dienophile which lower the energies of the frontier orbitals of the dienophile. Consequently, the energy difference between the HOMO of the diene and the LUMO of dienophile is reduced and the reaction becomes faster.

EXERCISE 17.10

Write the structure of the products from the reactions (Scheme 17.50a).

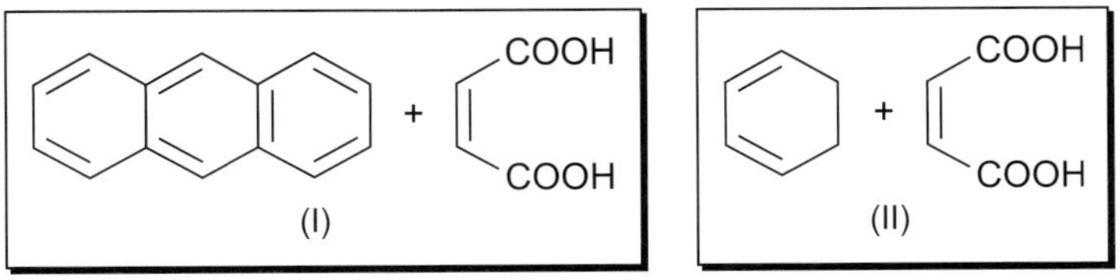

SCHEME 17.50a

ANSWER.

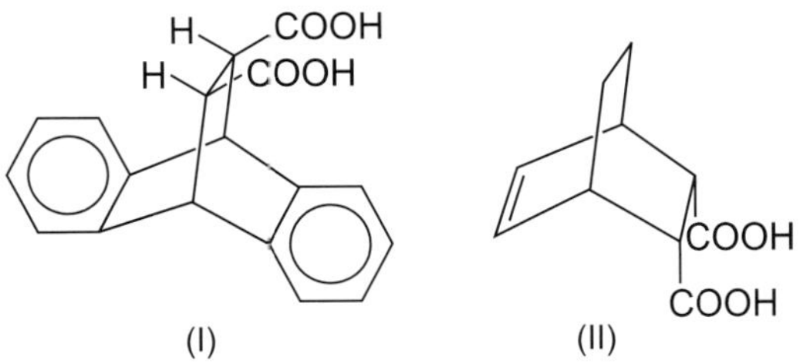

SCHEME 17.50b

B. Asymmetric Diels-Alder Reaction

In asymmetric Diels-Alder reaction generally camphor based chiral auxiliaries are employed and this is due to the reason that both geometrically possible stereoisomeric forms of camphor are readily available. The following points may be noted:

- In the case of a monosubstituted alkene there are two enantiotopic faces *Re* or *Si*.
- A diene *e.g.*, buta-1,3-diene thus can add either to *Re* or *Si* face of a dienophile like acrylic acid to give a cyclohexene in which the stereocenter generated via this Diels-Alder reaction could be either *S* (*Re* approach of the diene) or *R* (*Si* approach of the diene, Scheme 17.51).
- Diels-Alder reaction, therefore, creates stereocenters and when both reactants are achiral and no other chiral influence is there, racemic mixtures are obtained.

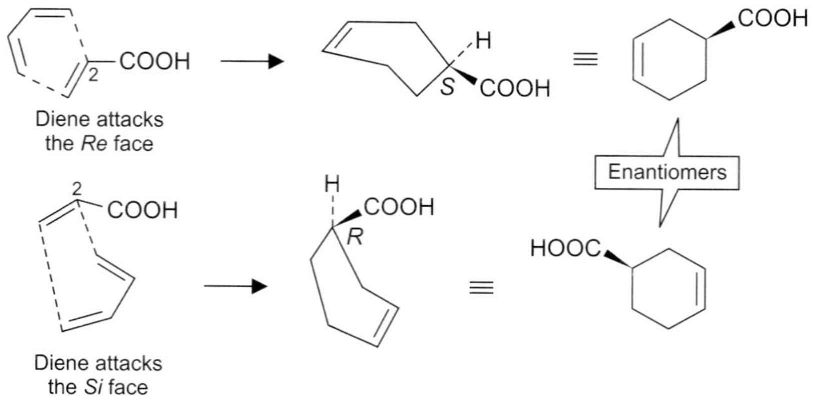

SCHEME 17.51

- When there is chiral influence, *e.g.*, in the dienophile as is so in chiral acrylate (optically active, Scheme 17.52) derived from acrylic acid and optically active alcohol (I, Scheme 17.52) derived from camphor, one of the faces of the double bond in the dienophile gets hindered.

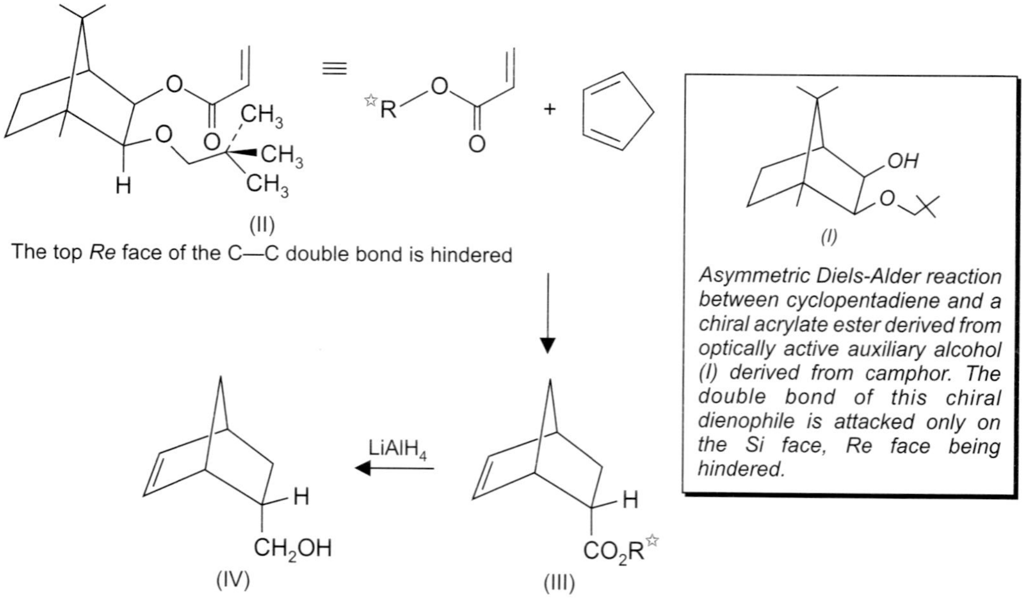

SCHEME 17.52

- The top *Re* face of the carbon-carbon double bond of the dienophile is hindered by the *t*-butyl group of neopentyl unit. This forces the addition to occur preferentially from the *Si*-face (back of the double bond to give (III) almost exclusively, to give only one of the enantiomers.
- One knows that Diels-Alder cycloadditions follow the *endo*-rule.
- Reduction of the ester with lithium aluminium hydride gives the product (IV) in an optically pure form and regenerates the camphor derived chiral auxiliary.

Another example is in (Scheme 17.53). The dienophile (*E*)-methyl crotonate becomes chiral when optically active alkyldichloroborane (I) complexes with it to yield (II). Now the approach of diene from the rear face of the dienophile is blocked (naphthyl group). Attack occurs from front to give (IV) in optically pure state (*trans*-geometry of dienophile is preserved in IV).

C. Hydrophobic Effects

It has been shown that some intermolecular Diels-Alder reactions are accelerated under hydrophobic effects in aqueous media. This was the case when cyclopentadiene reacted with methylvinyl ketone (Scheme 17.54) and it was observed that any additive which increased the hydrophobic effect also increased the rate *e.g.,* lithium chloride increases the hydrophobic effect by salting out nonpolar material.

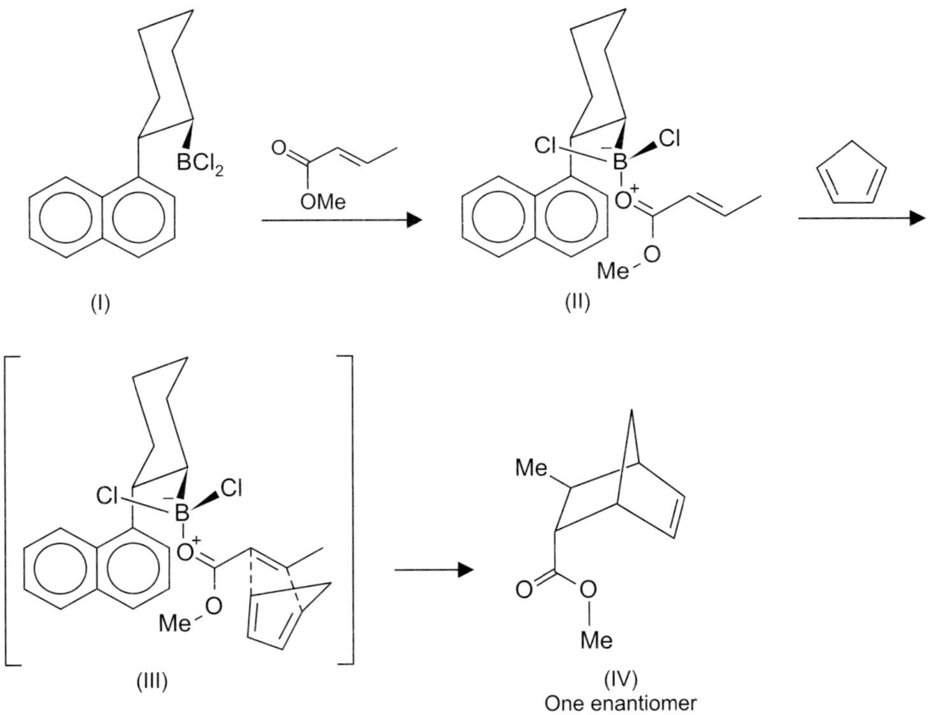

SCHEME 17.53

> *Hydrophilic and hydrophobic effects are water attractive and water repellant respectively. Soaps e.g., sodium oleate have a hydrophilic site (COO⁻) and a hydrophobic site (the hydrocarbon chain).*

β-Cyclodextrin has a hydrophobic cavity and if the system of a particular Diels-Alder combination can fit within the cavity a significant rate enhancement is observed. This is found in the case of methyl vinyl ketone and cyclopentadiene system in aqueous medium with β-cyclodextrin as additive. α-Cyclodextrin, however, has a smaller cavity which is not able to accommodate the reactive species and the rate is significantly diminished in its presence compared to that in β-cyclodextrin.

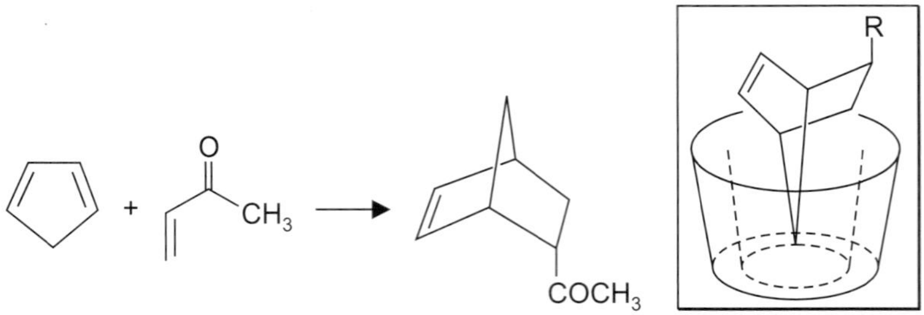

The hydrophobic effects of the cavity in β-cyclodextrin in aqueous medium enhance significantly this reaction rate. The cavity in β-cyclodextrin can accommodate well these reactive species.

SCHEME 17.54

D. [2+2] Cycloadditions

In the dimerization of ethylene, a thermal [2+2] cyclization would involve overlap of HOMO, of one molecule with the LUMO, of the other (see, Scheme 17.41). If in this concerted reaction both bonds to a component are formed on the same face *i.e.,* the process is suprafacial, the lobes of opposite phase would approach each other (Scheme 17.55). This interaction which is suprafacial with respect to both components [$\pi^{2s} + \pi^{2s}$] is therefore, antibonding and repulsive and the concerted reaction, does not take place (symmetry forbidden process).

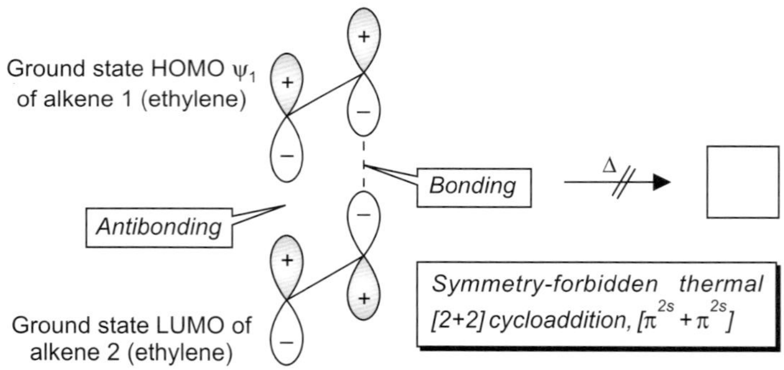

SCHEME 17.55

The photochemical [2+2] cycloadditions which are suprafacial with respect to both the components [$\pi^{2s} + \pi^{2s}$] will, however, permit a previously forbidden reaction to become a symmetry allowed process. During [2 + 2] cycloadditions, irradiation of an alkene with UV light excites an electron from ψ_1, the ground state HOMO to ψ_2* which now becomes the excited state HOMO. The interaction between the excited state HOMO of one alkene and LUMO of the second alkene is now a symmetry allowed process (Scheme 17.56).

The stereochemistry of the Diels-Alder reaction reveals that these are also $\pi^{4s} + \pi^{2s}$ processes. However, a thermal [2+2] cycloaddition could occur provided it is suprafacial with respect to one component and antarafacial with respect to the other *i.e.,* it is $\pi^{2s} + \pi^{2a}$ (Scheme 17.57). This process, through symmetry allowed is geometrically very difficult.

Thus the photochemical [2+2] cycloaddition reaction occurs smoothly and represents one of the best techniques to synthesize cyclobutane rings and cage compounds (Scheme 17.58). The CO double bond of an aldehyde or a ketone can act as one component in [2+2] cycloaddition with an alkene to form an oxetane (Scheme 17.58). The reaction can occur inter-or intramolecularly.

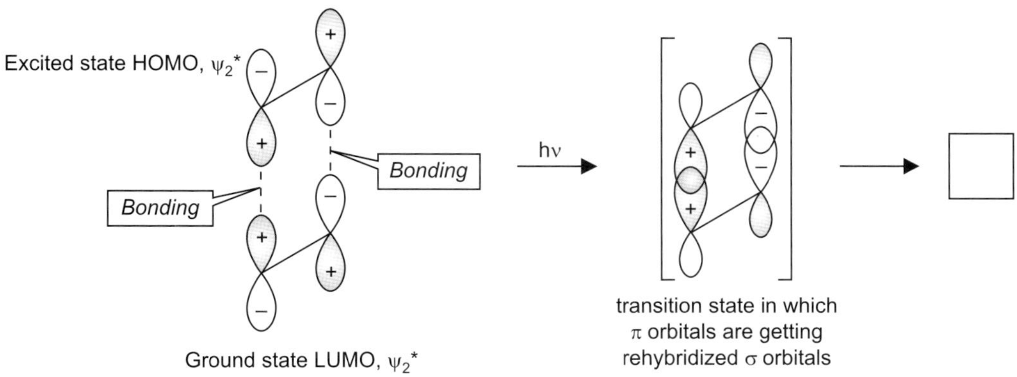

Symmetry-allowed photochemical [2+2] cycloaddition [$\pi^{2s} + \pi^{2s}$]

SCHEME 17.56

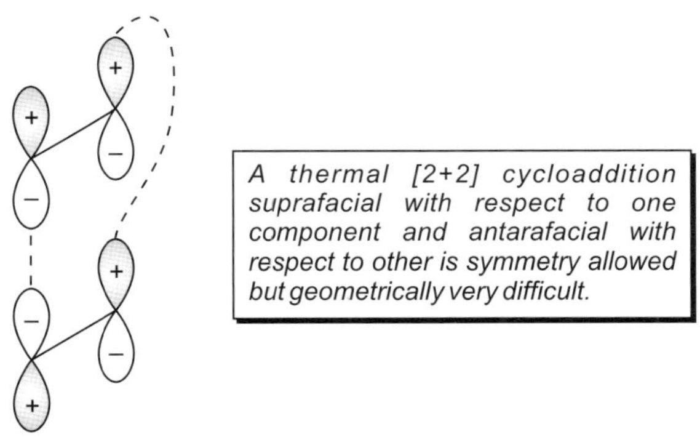

[$\pi^{2s} + \pi^{2a}$] cycloaddition

SCHEME 17.57

Both thermal as well as photochemical cycloaddition reactions take place by opposite stereochemical pathways. As with electrocyclic reactions one can categorize cycloadditions according to the total number of electron pairs (double bonds) taking part in the rearrangement. Thus, a Diels-Alder [4+2] reaction between a diene and a dienophile involves an odd number (three) of electron pairs and takes place by a ground state (thermal) suprafacial pathway. A [2+2] thermal reaction between two alkenes involves an even number (two) of electron

Table 17.2 Cycloaddition Reactions Stereochemical Rules

Electron pairs (double bonds)	Thermal reaction	Photochemical reaction
Even number	Antarafacial	Suprafacial
Odd number	Suprafacial	Antarafacial

pairs and must takes place by an antarafacial pathway. However, it may be said that both suprafacial and antarafacial cycloaddition pathways are symmetry allowed. Only the geometric constraints inherent in twisting a conjugated *pi* electron system out of planarity make antarafacial reaction geometrically difficult in most of the cases. One may note that preferences for cycloadditions may be summarized further (Table 17.3) to quickly know the success of a particular cycloaddition and provides a rule of thumb.

Table 17.3 Cycloaddition Reactions

Number of Electron Pairs	Allowed Cycloaddition
Odd	Thermal
Even	Photochemical

As a last example of cycloaddition, cyclopentadiene reacts with cycloheptatriene system to give a product—which is [6+4] suprafacial process (Scheme 17.59, this would include an aromatic transition state with 10 electrons).

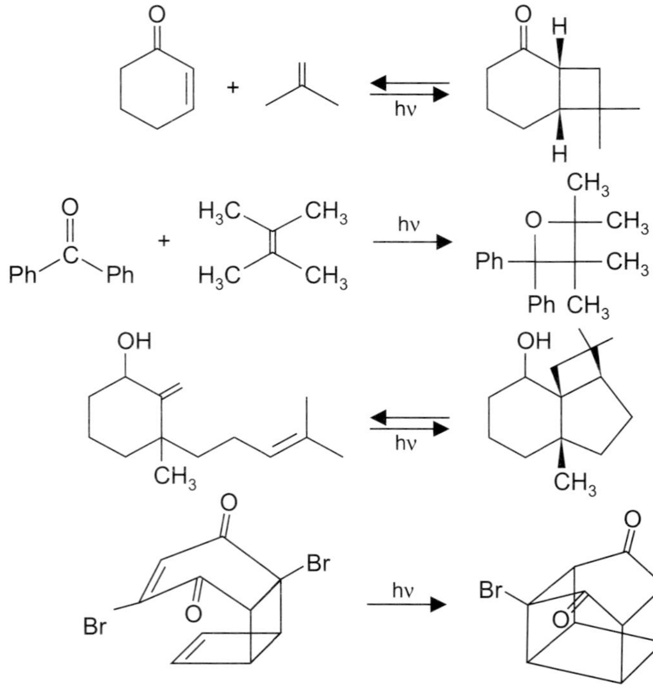

SCHEME 17.58

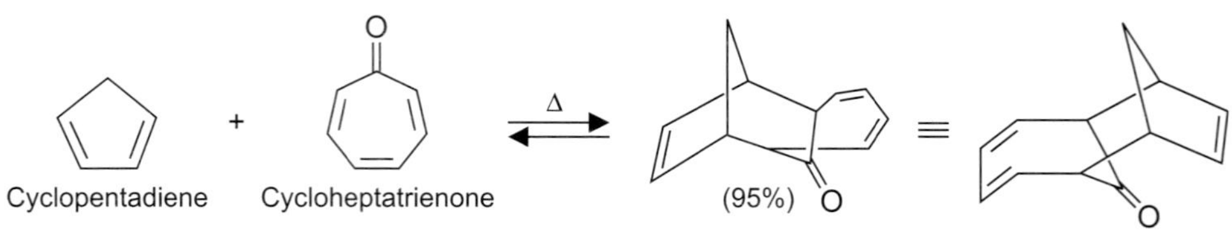

This is a [6+4] suprafacial cycloaddition reaction, which involves 10 electrons. The Diels-Alder reaction involves six electrons (4n+2 system), but the Hückel rule is also fulfilled with two, ten, etc. electrons, also see Schemes 17.98 and 17.99)

SCHEME 17.59

EXERCISE 17.11

Indicate if the following reactions (Scheme 17.59a) are allowed or forbidden.

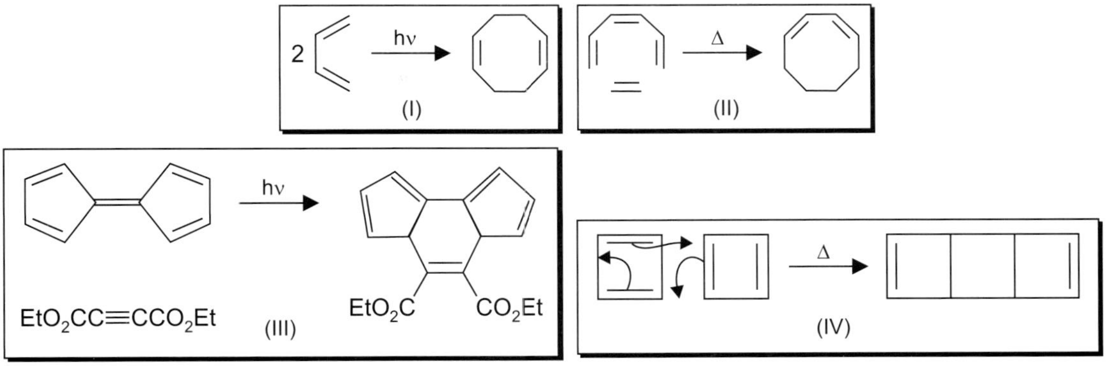

SCHEME 17.59a

Hint. *If in a reactant more π electrons are involved in the cyclization, the nonparticipating π electrons are not counted for the classification (see Table 17.3). A reaction involving odd number of electron pairs requires heat while a reaction involving even number of electron pairs requires light.*

ANSWER. *(I) A[4 + 4] cycloaddition, involves 4 electron pairs is photochemically allowed (see Table 17.3).*

(II) A[6 + 2] cycloaddition, involves 4 electron pairs is photochemically allowed and thermally forbidden.

(III) Since it is a [10 + 2] cycloaddition and involves 6 electron pairs, is photochemically allowed.

(IV) It is the allowed thermal dimerization [4 + 2].

EXERCISE 17.12

One of the reactions (Scheme 17.59b) requires heat and the other light. Which is which ? Explain.

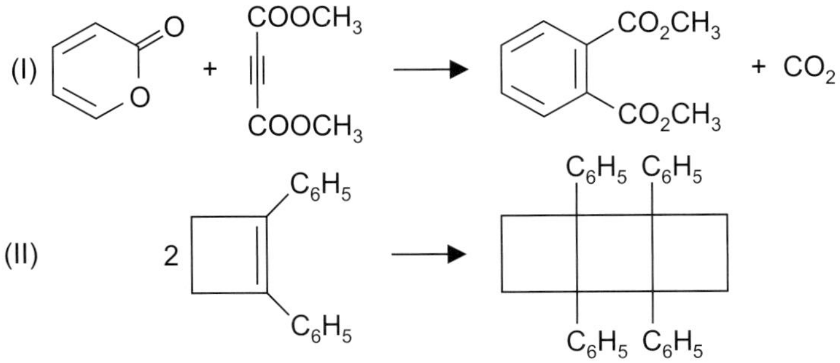

SCHEME 17.59b

ANSWER. *(I) This reaction is thermally allowed ; it being a [4 + 2] cycloaddition. The initially formed intermediate then undergoes a thermally allowed reverse Diels Alder reaction (Scheme 17.59c).*

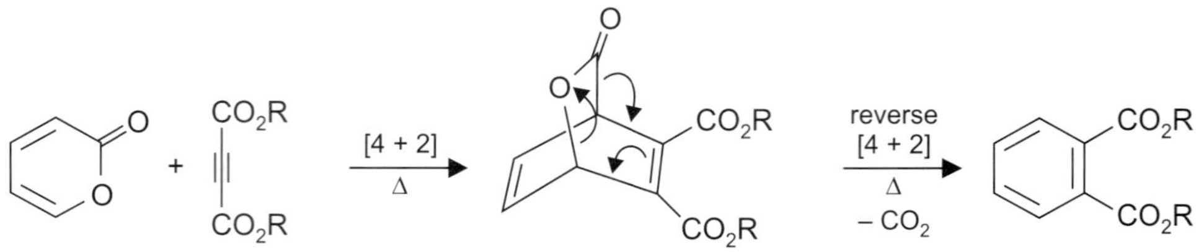

SCHEME 17.59c

(II) This is a photochemical [2 + 2] dimerization with the stereochemistry as dictated from end-to-end overlap of p-orbital components (Scheme 17.59d).

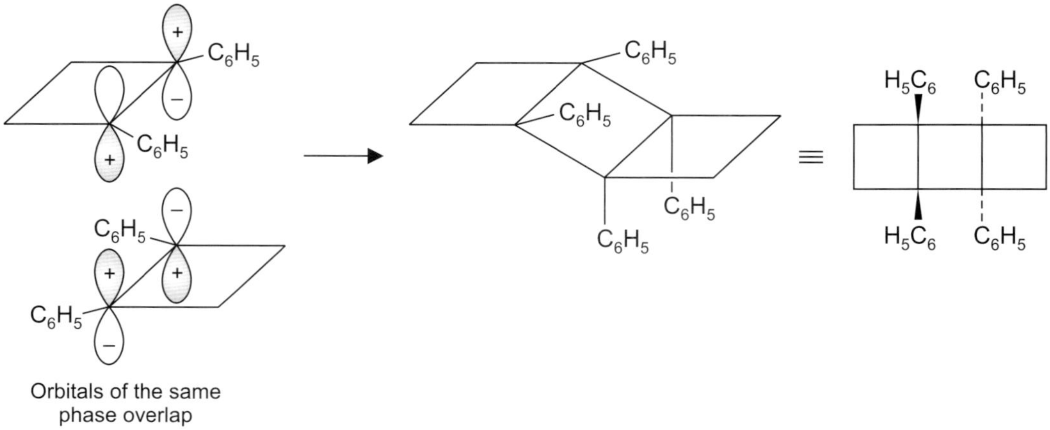

SCHEME 17.59d

17.5 1, 3-DIPOLAR CYCLOADDITIONS

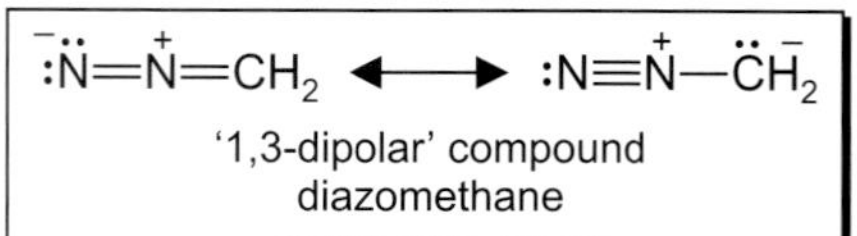

SCHEME 17.60

These cycloadditions are analogous to the Diels-Alder reaction in that they are concerted [π^{4s} + π^{2s}] reactions. The 1,3-dipolar components are compounds whose representation requires ionic structures which include ones with charges on atoms bearing 1,3-relationship, as in diazomethane (Scheme 17.60). These type of molecules which are called, 1,3-dipoles are isoelectronic with allyl anion. These have four π electron and each has at least one charge separated resonance structure with opposite charges in a 1,3 relationship. The other reactant (dipolarophile) in a dipolar cycloaddition has unsaturated bonds like, C=C, C≡C, C=O and C≡N. The 1,3-dipolar cycloadditions form useful reactions for the synthesis of five membered heterocyclic rings.

Mechanistically the transition state for 1, 3-dipolar cycloaddition is not very polar and the reaction rate is not strongly sensitive to solvent polarity. The loss of charge separation which is implied, is more apparent rather than real, since most 1, 3-dipolar compounds are not highly polar. The polarity associated with a single structure is balanced by other contributing structures.

A 1, 3-dipole represents a structural variant of the diene component in the Diels-Alder reaction; in the dipolar compound, four π-electrons are distributed over three atoms instead of the four in a diene. Moreover, the HOMO and LUMO of a 1, 3-dipole are similar in symmetry to that in a diene with respect to the two-fold axis and to the mirror plane which bisects the molecule (Scheme 17.61), a concerted cycloaddition *e.g.,* to an alkene is a symmetry allowed process. The reaction of an alkene with diazomethane to give a pyrazoline (Scheme 17.62 pyrazole derivative) belongs to this class.

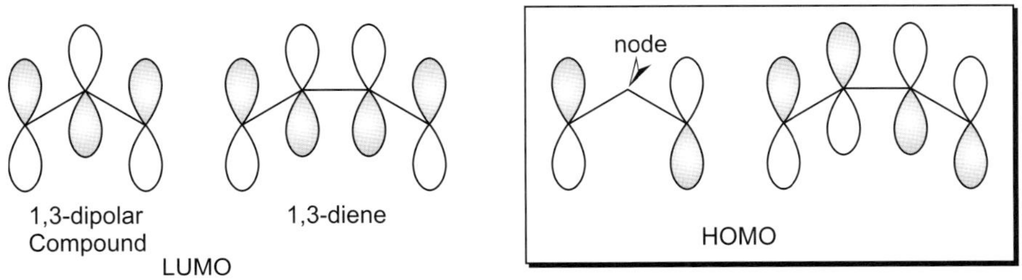

SCHEME 17.61

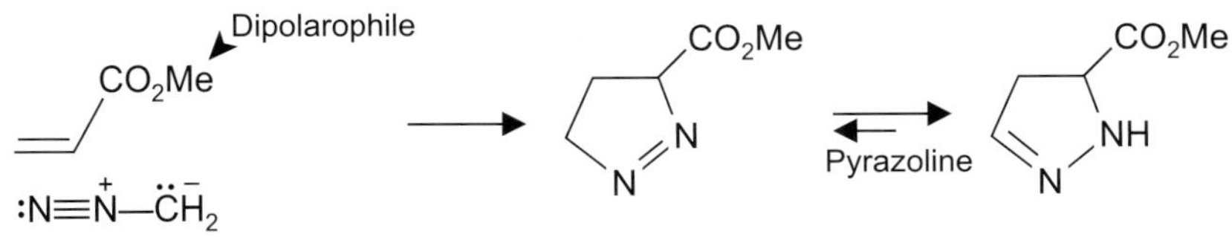

SCHEME 17.62

17.6 CHELETROPIC REACTIONS

In a cheletropic reaction two σ bonds that terminate at a single atom are made or broken during a concerted reaction (Scheme 17.63). In the case of molecules, sulphur dioxide or carbon monoxide the HOMO is that which has a lone pair of electrons in the plane having the atoms, while the LUMO represents the *p* orbital perpendicular to this plane (Scheme 17.64).

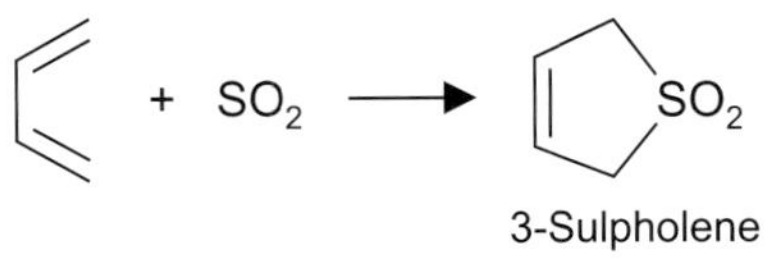

(a cheletropic reaction)

SCHEME 17.63

For a symmetry allowed cycloaddition of *e.g.,* SO_2 to a diene, the molecule of SO_2 must lie in a plane which bisects the *s-cis* conformation of the diene (Scheme 17.64). The interaction is suprafacial for diene and SO_2. In the transition state, the terminal carbon atoms of the diene must move in the disrotatory manner in order that the HOMO of SO_2 can interact with the LUMO of the diene, or the LUMO of SO_2 with the HOMO of the diene.

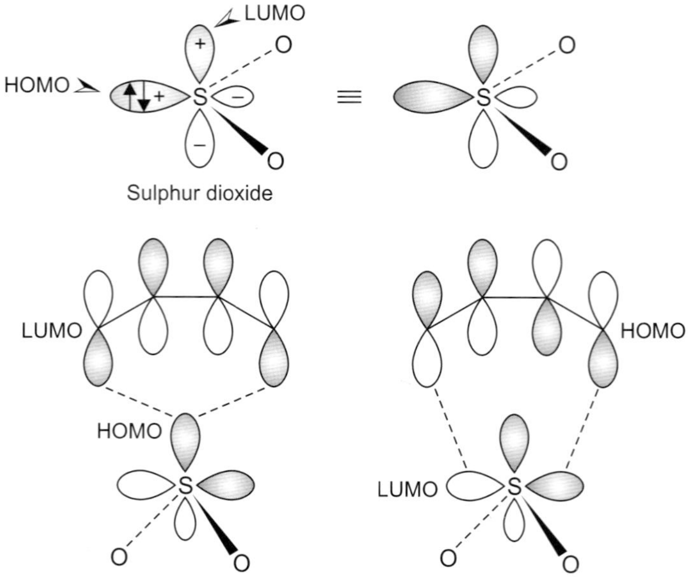

SCHEME 17.64

In keeping with these arguments the *trans, trans*-1,4-disubstituted dienes give specifically more crowded *cis*-substituted 3-sulphones (Scheme 17.65). By similar arguments *cis, trans*-disubstituted dienes, on the other hand afford *trans*-substituted-3-sulphones. As with electrocyclic reactions, the opposite stereochemistry is observed when the reaction is photochemical rather than thermochemical (Scheme 17.65).

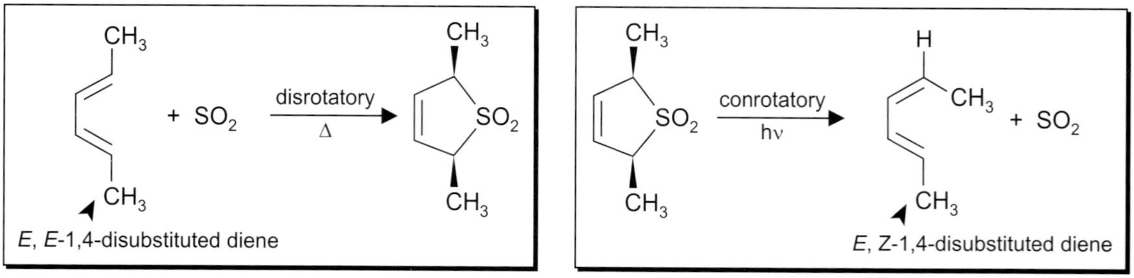

SCHEME 17.65

3-Sulpholene, a solid, is a convenient substitute for geseous butadiene. Butadiene is generated at high temperatures from 3-sulpholene in a reverse reaction (Scheme 17.66) and when a dienophile is present it is trapped in a Diels-Alder reaction. The Diels-Alder reaction in itself is usually reversible and has been used to protect double bonds (see Exercise 17.11).

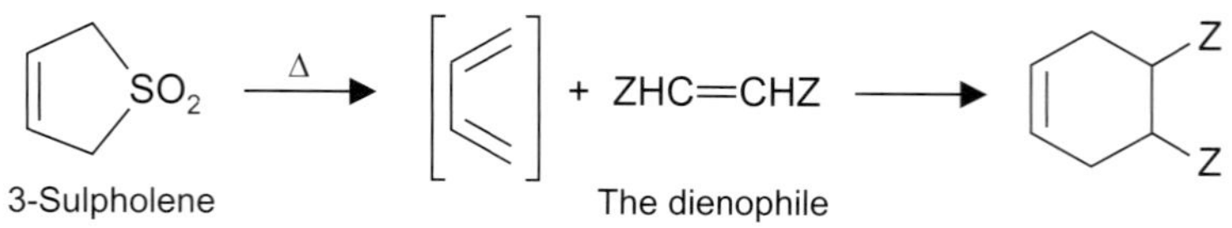

SCHEME 17.66

17.7 SIGMATROPIC REARRANGEMENTS

A sigmatropic rearrangement is a concerted intramolecular shift of an atom or a group of atoms. During this arrangement a σ bond is broken in the reactant and a new σ bond is formed in the product and the π bonds rearrange. The following points may be noted:

- The number of π bonds does not change, both the reactant and the product contain the same number of π bonds.
- The σ bond that cleaves can be in the middle of the π system or at end of the π system (Scheme 17.67).

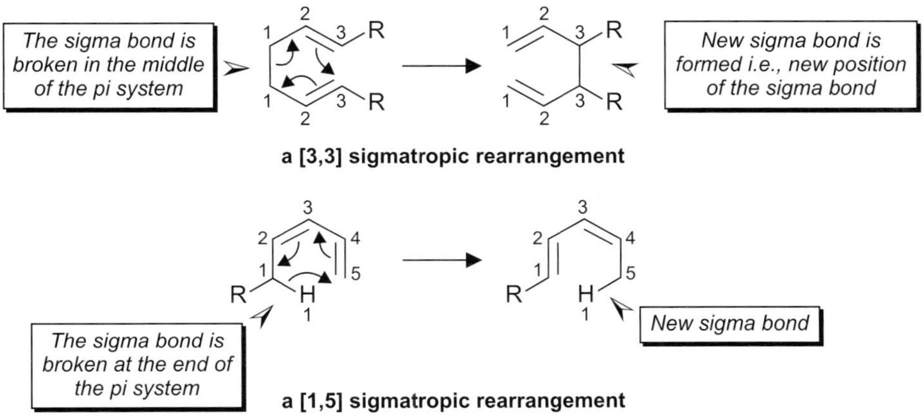

SCHEME 17.67

- The σ bond that breaks is bonded to an allylic carbon.
- To identify the order of a sigmatropic shift $[i, j]$ first identify the σ bond which is broken in the reaction. Then assign number 1 to both the atoms involved in this bond, then the atoms in each direction from the bond being cleaved, upto and including the atoms which form the new σ bond in the product are numbered as atoms 2,3 and so on. The numbers assigned to the atoms that form the new bond, separated by commas are put within the brackets to show the reaction order (Scheme 17.67). Similarly the migration of hydrogen (Scheme 17.68) is another example of [1,5] sigmatropic shift. The order [1,5] is not due to the fact that hydrogen migrates from C1 to C5 but since the hydrogen (one of the two atoms given the number 1) forms part of the new σ bond and had also formed part of the old σ bond. Only all the atoms taking part in the reaction have to be counted. Thus the rearrangement of cyclohexadiene (Scheme 17.68) cannot be labelled as [1,3] shift since the methylene group linking 1 and 5 is not involved in the reaction.

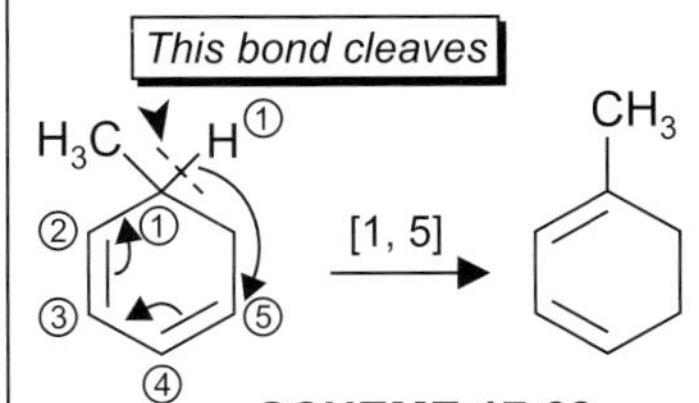

SCHEME 17.68

- Since in these reactions a change in the position of one σ bond takes place, Woodward and Hofmann coined the term "sigmatropic shifts".
- A [3,3] sigmatropic rearrangement of a 1,5-diene (when the six atoms involved are all carbons) is known as the Cope rearrangement (I, Scheme 17.69).

- The oxygen analog of the Cope rearrangement is called the Claisen rearrangement. Often one of the π bonds is part of an aromatic ring (II, Scheme 17.69). Allyl vinyl ethers also undergo Claisen rearrangement.

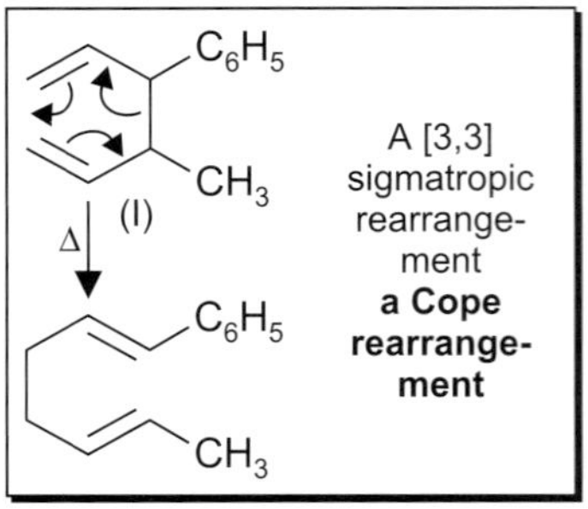

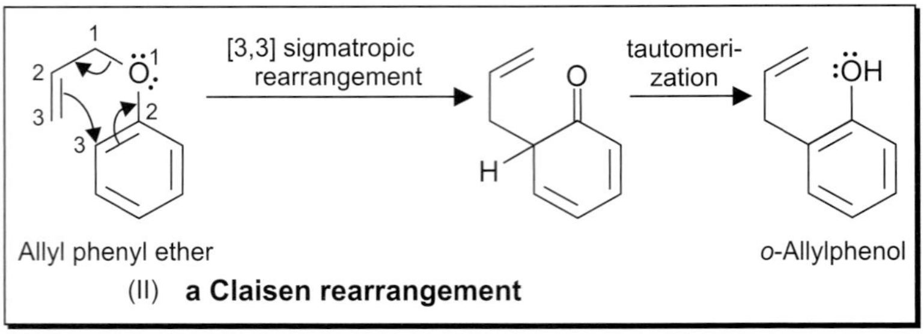

SCHEME 17.69

- More examples of sigmatropic rearrangements are in (Scheme 17.69*a*).

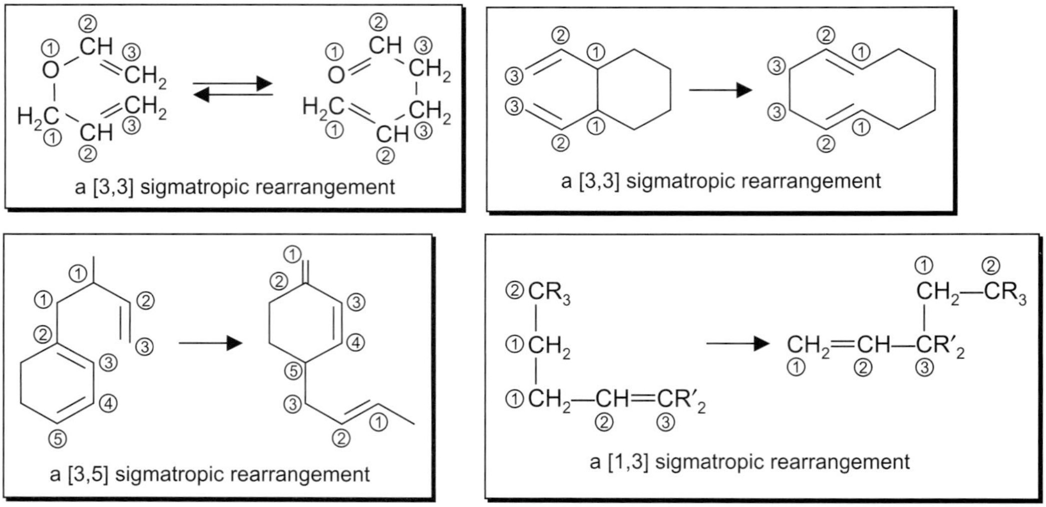

SCHEME 17.69a

A. Sigmatropic Migration of Hydrogen

(i) Introduction.

A hydrogen atom is reported to migrate from one end of a system of π bonds to the other, under thermal or photochemical rearrangements. In the transition state the hydrogen must be in contact with both ends of the chain at the same time. There are two distinct processes by which a sigmatropic migration can occur. If the hydrogen moves along the top or bottom face of the π-system *i.e.,* migrating group remains associated with same face of the conjugated system throughout the process, the migration is termed *suprafacial.* When the hydrogen moves across the π system either from top to bottom or vice versa *i.e.,* the migrating group moves to the opposite face of the π-system during the course of migration then it is called *antarafacial.*

In a given sigmatropic rearrangement, the migrating group is bonded to both the migration source and the migration termini in the transition state. It is imagined that the migrating H atom breaks away from the rest of the system which is treated as a free radical. Thus in a simplest case involving a [1,3] shift of hydrogen (Scheme 17.70), the frontier orbital analysis treats this system as a hydrogen atom interacting with an allyl radical. The electron of the hydrogen atom is in a 1*s* orbital which has only one lobe. The HOMO of an allylic free radical depends on the number of carbons in the π-framework (Scheme 17.71).

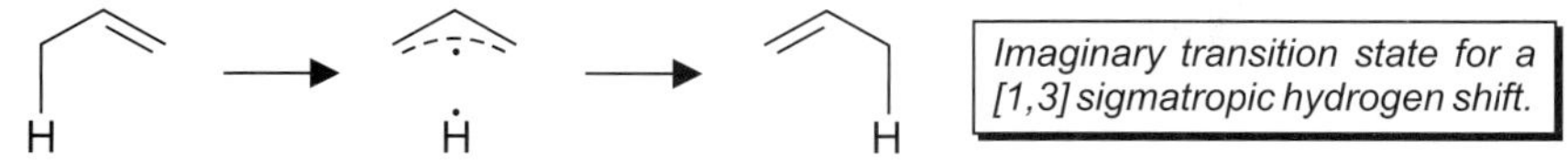

SCHEME 17.70

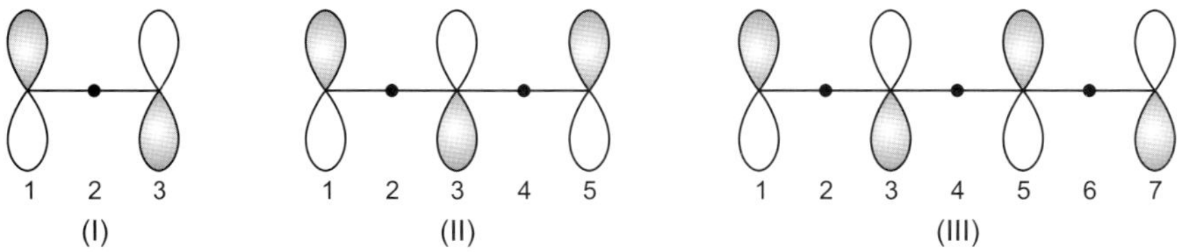

The HOMO of the *allylic* radicals

SCHEME 17.71

(ii) [1,3] Sigmatropic Rearrangement (Hydrogen Shift).

In the migration of hydrogen the H must move from a plus to plus or from minus to a minus lobe of the HOMO, it cannot move to a lobe of opposite sign.

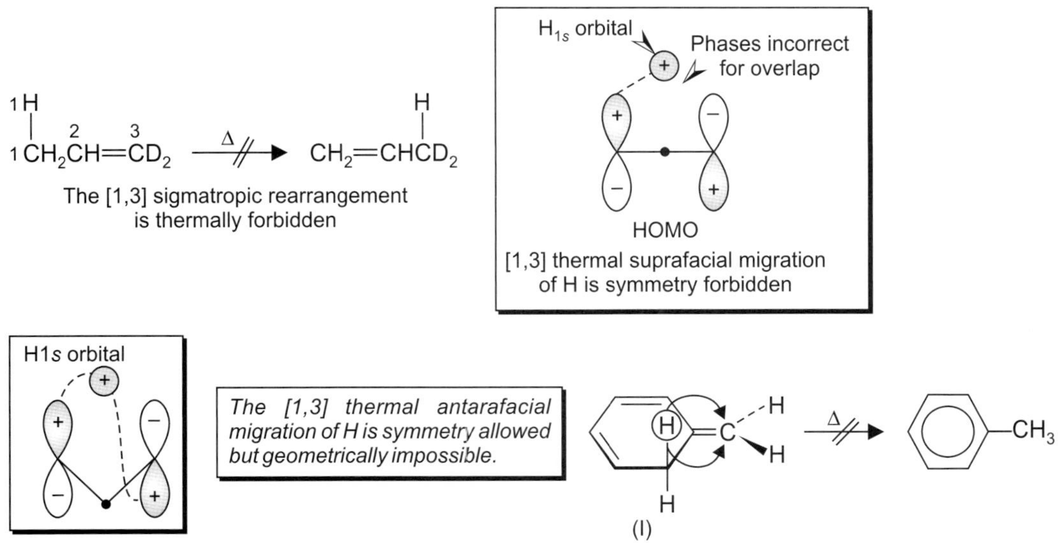

SCHEME 17.72

During a thermal [1,3] sigmatropic migration of a hydrogen, the overlap of the hydrogen 1*s* orbital with the HOMO of the allyl radical (I, Scheme 17.71, asymmetric) is bonding at one end and antibonding at the other end for the suprafacial migration (Scheme 17.72). Thus [1,3] sigmatropic suprafacial migration of hydrogen (under thermal conditions) is symmetry-forbidden (Scheme 17.72). However, in the antarafacial process (Scheme 17.72) the hydrogen atom shall have to cross over the *pi* system to the other face to form a four membered ring transition state, a geometrically very difficult situation. Thus over all thermal [1,3] sigmatropic rearrangements are rare. The stability of the triene (I, Scheme 17.72) which is not thermally isomerized to toluene, which is thermodynamically more stable is due to a symmetry-forbidden process (suprafacial H migration is symmetry forbidden, antarafacial H migration though symmetry allowed but sterically forbidden).

In a photochemical reaction promotion of an electron means that now (I, Scheme 17.73) becomes the HOMO. Suprafacial pathway for [1,3] shift now becomes an allowed process and antarafacial pathway forbidden. Thus, the compound (II, Scheme 17.73) displays a [1,3] hydrogen shift under photochemical conditions.

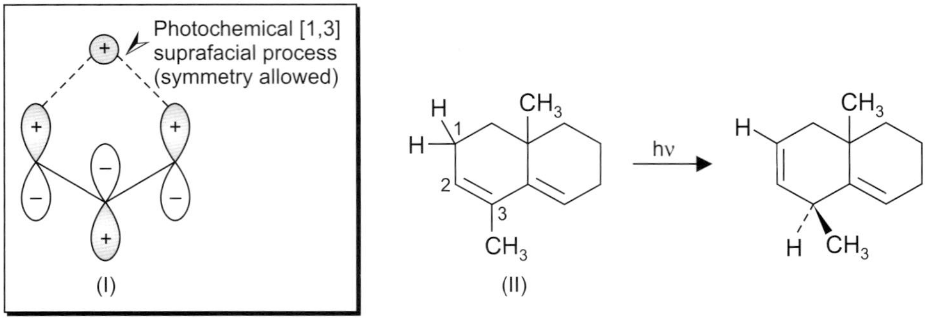

The [1,3] sigmatropic rearrangement is photochemically allowed

SCHEME 17.73

A [1,3] sigmatropic rearrangement involves a π bond and a pair of σ electrons so in all two pairs electrons are involved similarly a [1,5] sigmatropic rearrangement involves three pairs of electrons. Woodward Hoffmann rules for sigmatropic rearrangements are given in Table 17.4.

Table 17.4 Woodward-Hoffmann Rules for Sigmatropic Rearrangements

Number of pairs of electrons in the reacting system	Reaction condition	Allowed mode
Even number	Thermal	Antarafacial
	Photochemical	Suprafacial
Odd number	Thermal	Suprafacial
	Photochemical	Antarafacial

Thus, since a [1,3] sigmatropic migration involves two pairs of electrons, an antrafacial rearrangement for a 1,3-shift under thermal conditions does not take place due to geometrical constraints. 1,3-Shifts do take place photochemically [Table 17.4, moreover, since under photochemical conditions HOMO becomes symmetric (see, Scheme 17.73) hydrogen can migrate by suprafacial pathway].

(iii) [1,5] Sigmatropic Hydrogen Shift.

The [1,5] sigmatropic shift of hydrogen or deuterium atoms is well known. These involve three pairs of electrons, thus these occurs via a suprafacial pathway under thermal conditions (see Table 17.4). These shifts can be analyzed by examining a hydrogen atom and a pentadienyl radical whose HOMO (see III, Scheme 17.71) is bonding at both the migration origin and the migration terminus. Thus the migration maintains orbital symmetry when the migrating group remains on the same side of the conjugated system (suprafacial process, Scheme 17.74).

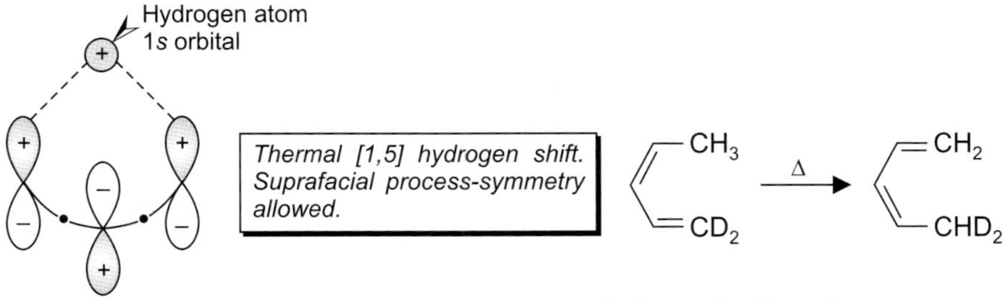

The [1,5] sigmatropic rearrangement is thermally allowed

SCHEME 17.74

Another remarkable example of suprafacial [1,5] hydrogen shift thermally, is in the 1,3-diene (I, Scheme 17.75) of known stereochemistry both at the double bond and at the stereocenter. This 1,3-diene gave a two component mixture compatible with only suprafacial migration. These results are explained as under:

- One has to consider two rotational isomers (I and I*a*, Scheme 17.75) for the reaction.
- Recall, a compound can have an infinite number of conformations but only one configuration.
- In (I) the methyl group is directed toward C4-C5 double bond while in (I*a*) it is now ethyl group that is directed toward C4-C5 bond.

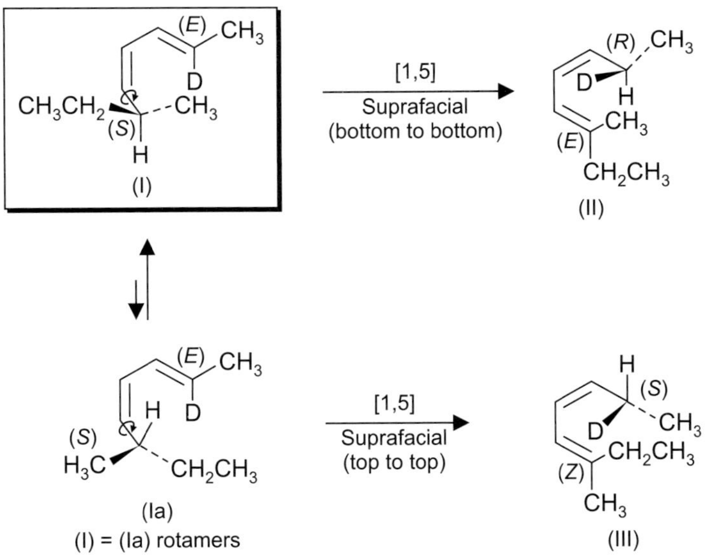

SCHEME 17.75

- There are two suprafacial [1,5] pathways for the hydrogen in these two conformations (I and I*a* Scheme 17.75) "top to top" as in (I) or "bottom to bottom" as in (Ia).
- Each of these suprafacial pathways gives a product with specific stereochemistry and both are formed.

- Considering the two stereogenic units in (I, Scheme 17.75) 4 stereoisomers could be considered. Two (II and III) are formed during suprafacial migration by the symmetry allowed pathway.
- If one considers, the antarafacial pathway the remaining two stereoisomers (as a diastereomeric pair) would have been formed (Scheme 17.76) which however, is not the case.

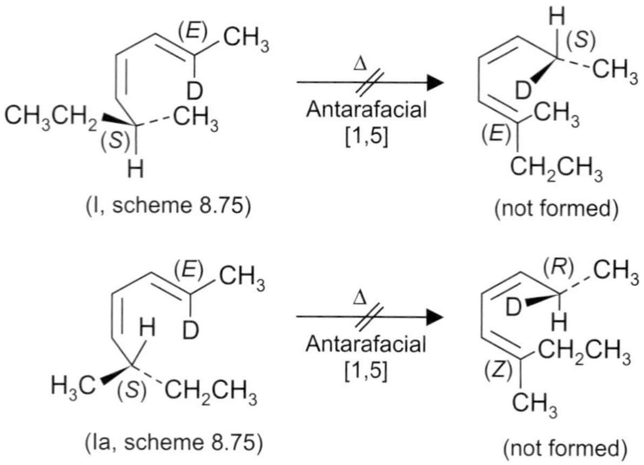

SCHEME 17.76

Heating of indene (Scheme 17.77) causes the scrambling of the label to all the three non-aromatic positions. It is only via [1,5] shift of H or D (by including the p-orbitals of the benzene ring) that one can account for the results.

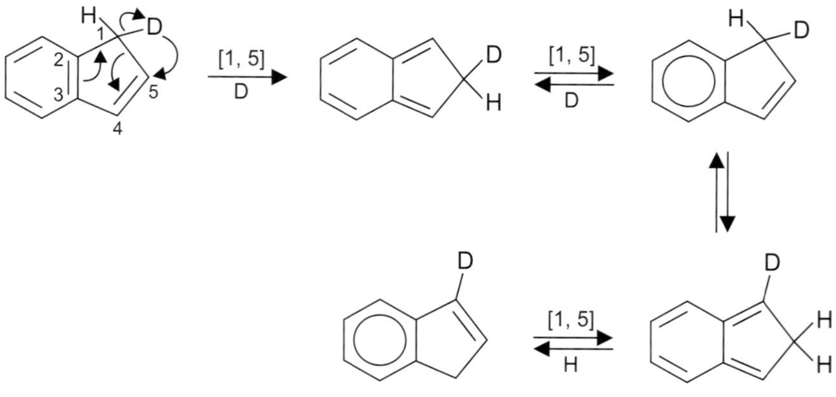

SCHEME 17.77

(iv) [1,7] Sigmatropic Hydrogen Shift

In the case of [1,7] hydrogen shifts, in a triene, the orbital symmetry rules (see III, Scheme 17.71) predict that the transfer of hydrogen must be antarafacial compared to [1,3] shift, the transition state is not much strained and the shift is sterically feasible. This is seen in the thermal interconversion of vitamin D series (Scheme 17.78).

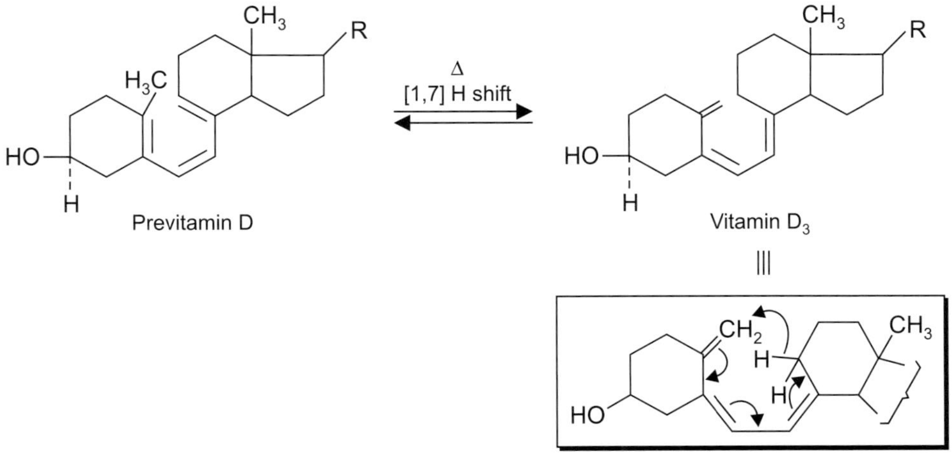

SCHEME 17.78

More on sigmatropic rearrangements (Degenerate rearrangements)

1, 3–Pentadiene on heating regenerates itself via a [1, 5] shift of hydrogen, this kind of process in which a reactant rearranges to itself is termed degenerate rearrangement (Scheme 17.78a). Significantly as expected the [1, 3] shift is symmetry forbidden.

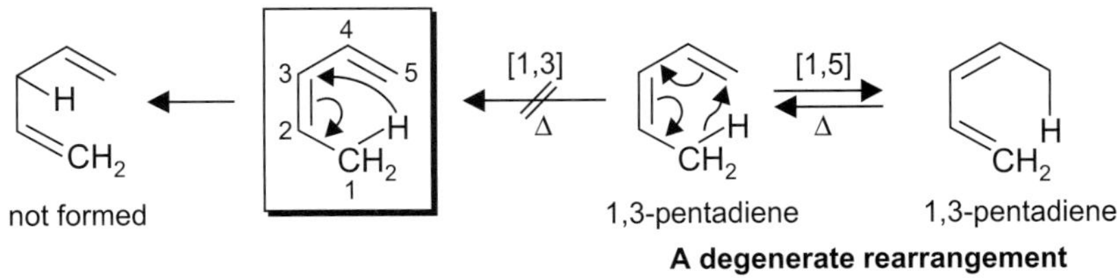

SCHEME 17.78a

The degenerate rearrangement can be established either by isotopically labeled molecules or suitably substituted molecules (Scheme 17.78b).

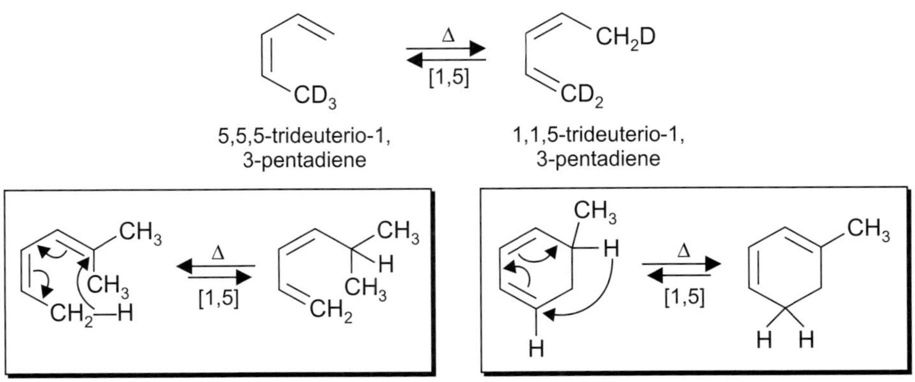

SCHEME 17.78b

Three membered rings can often play the role of a double bond and a [1, 5] H shift can take place just like in 1, 3-pentadiene and involves the opening of the cyclopropane (Scheme 17.78c).

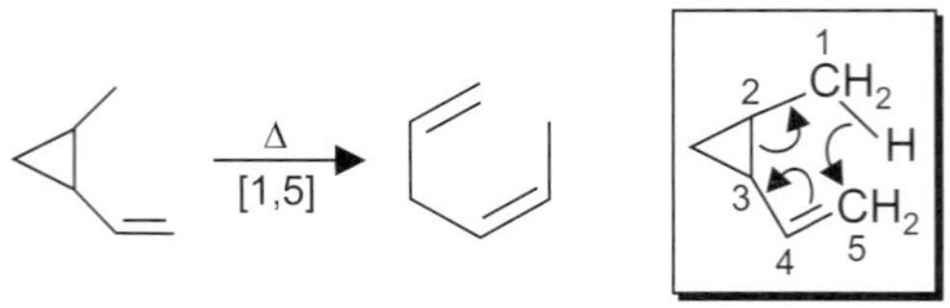

SCHEME 17.78c

B. Sigmatropic Migrations of Carbon

As compared to a hydrogen atom which has its electrons in a 1*s* orbital that has only one lobe, a carbon free radical (free imaginary transition state) has its odd electron in a *p* orbital which has two lobes of opposite sign. Recall that a [1,3] sigmatropic suprafacial migration of hydrogen (thermally) is symmetry forbidden while an antarafacial reaction though allowed is geometrically improbable (see Schemes 17.72–17.74). Interestingly an additional possibility would exist if an alkyl group (carbon) rather than a hydrogen was potential migrator. A [1,3] shift can now be suprafacial migration (I, Scheme 17.79) if the migrating group does so antarafacially *i.e.,* it would result in inversion of configuration of the migrating group. Thus carbon can simultaneously interact with the migration source and the migration terminus using either one of its lobes or both of its lobes (Scheme 17.79). Considering suprafacial rearrangement, carbon will migrate using one of its lobes if the HOMO is symmetric (II, Scheme 17.79). This happens during a thermal suprafacial [1,5] process. When carbon migrates with only one of its lobes interacting with migration source and migration terminus, the migrating group retains its configuration since bonding is always to the same lobe. When the carbon migrates using both of its lobes. (asymmetric HOMO, I Scheme 17.79), a [1,3] thermal suprafacial migration would involve opposite lobes. Thus, if the migrating carbon was originally bonded *via* its positive lobe, it must now use its negative lobe to form the new C—C bond. The stereochemical outcome of such a process is the inversion of configuration in the migrating group.

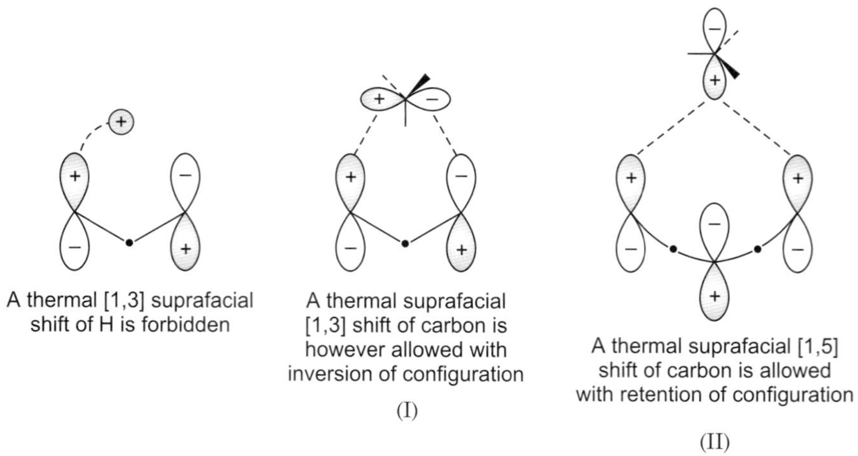

SCHEME 17.79

In summary, a suprafacial [1,5] thermal rearrangement proceeds with retention of configuration at the migrating carbon, while the related [1,3] suprafacial process proceeds with inversion. In the thermal conversion of (I, Scheme 17.80) to (II) a carbon atom migrates across an allyl system to leave C-1 and ending up at C-3 via a [1,3] shift. The inversion of configuration is observed using suitable substrates (Scheme 17.81). The [1,3] shifts of carbon

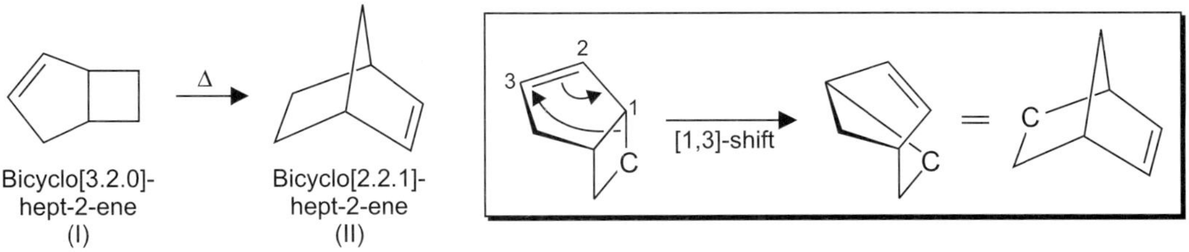

SCHEME 17.80

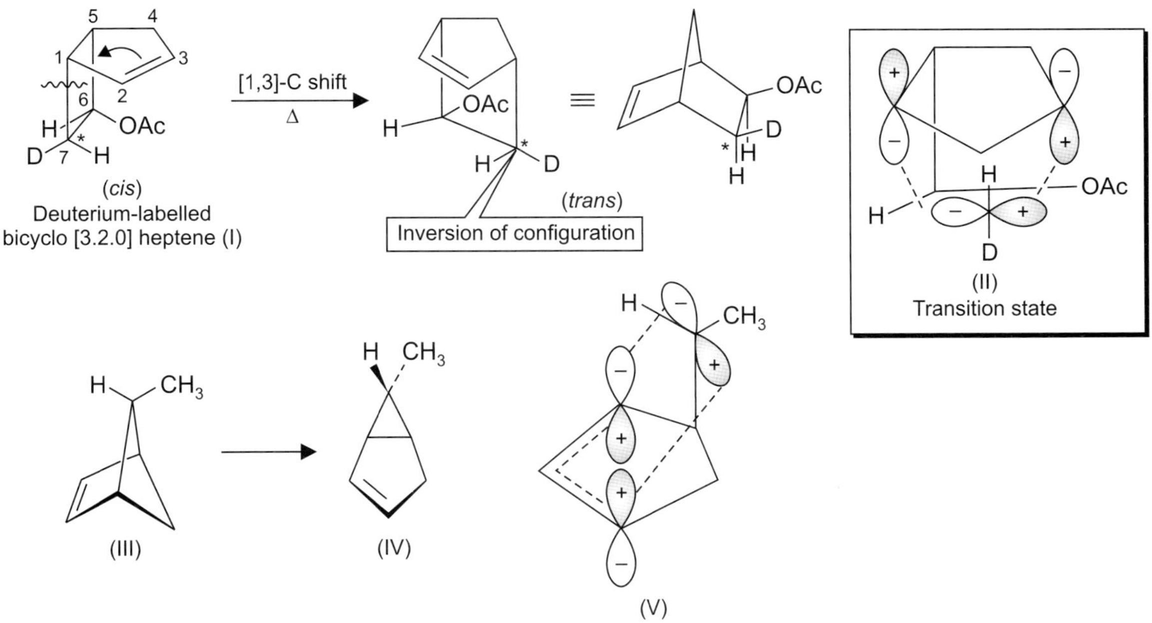

SCHEME 17.81

(*i.e.*, alkyl groups) in such reactions involve expansions of strained three-or four-membered rings. As predicted by orbital symmetry conservation rules these reactions proceed almost entirely with inversion of configurations in the migrating group as in (I, Scheme 17.81). In this case, a label deuterium was placed at C7 which was *trans* to the acetoxy group. After the reaction, it was found to be exclusively *cis* due to inversion of configuration at C7. The transition state (II, Scheme 17.81) shows that it is a [1,3] sigmatropic shift of carbon. Similarly (III, Scheme 17.81) gives (IV) via the transition state (V) by a suprafacial [1,3] shift with inversion at the migrating carbon under thermal conditions.

C. The Cope Rearrangement

A 1,5-diene on heating is rearranged to another 1,5-diene by concerted formation of a 1, 6-bond, breaking of the 3,4-bond and migration of both double bonds in a [3,3] sigmatropic rearrangement known as Cope rearrangement (see, Scheme 17.69). The compound rearranges by a [3,3] sigmatropic pathway and is also hypothetically pictured as split into two allyl radicals (Scheme 17.82). Interaction of the HOMO's of these allyl radicals is bonding at both ends, so the reaction is thermally allowed. The stereochemical outcome of this rearrangement is in keeping with their occurrence generally through the chair-shaped transition states (Scheme 17.83). *Meso* 3,4-dimethyl-1, 5-hexadiene gives *cis*, *trans*-2, 6-octadiene (in the starting compound the two methyl groups are having *cis*-relationship, in the chair form of a cyclohexane only 1, 2-axial, equatorial relationship is *cis*) while a boat shaped transition state would give *cis*, *cis*-product or *trans*, trans-product (Scheme 17.84).

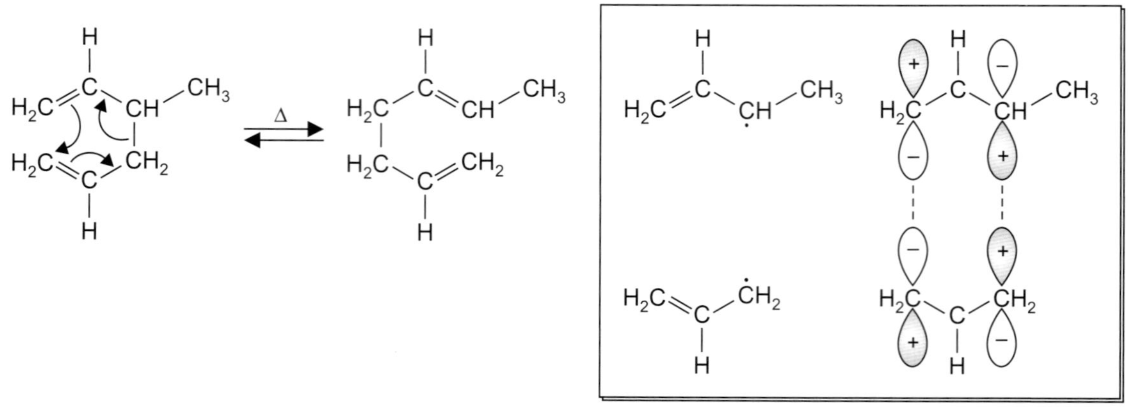

The [3,3] sigmatropic rearrangement is thermally allowed

SCHEME 17.82

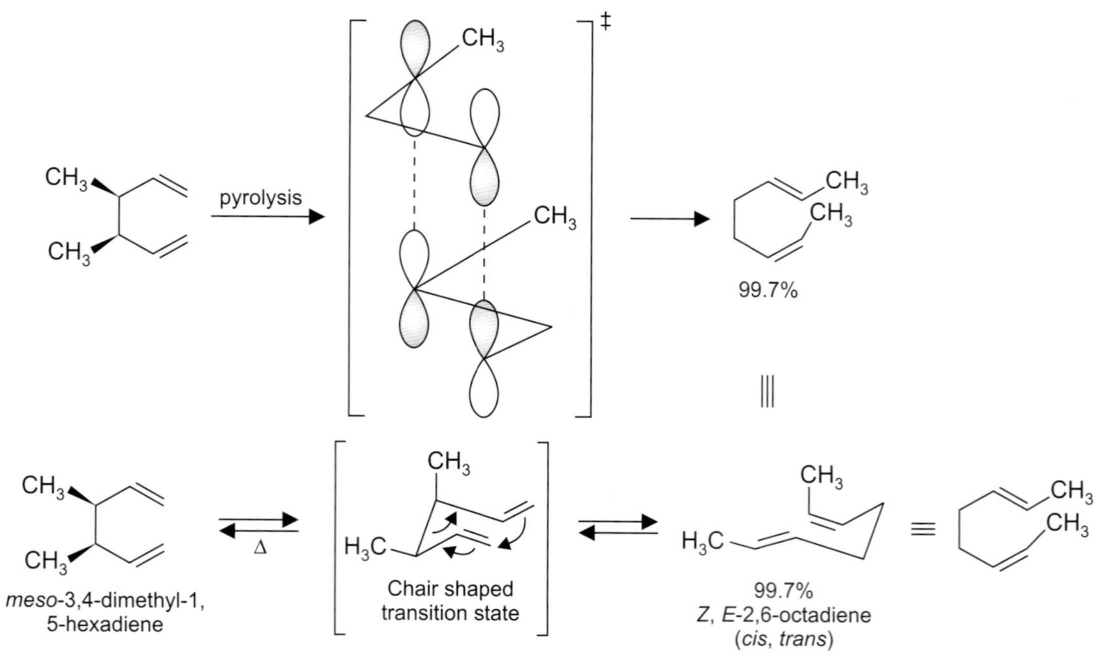

Cope rearrangement occurs via a chair shaped transition state

SCHEME 17.83

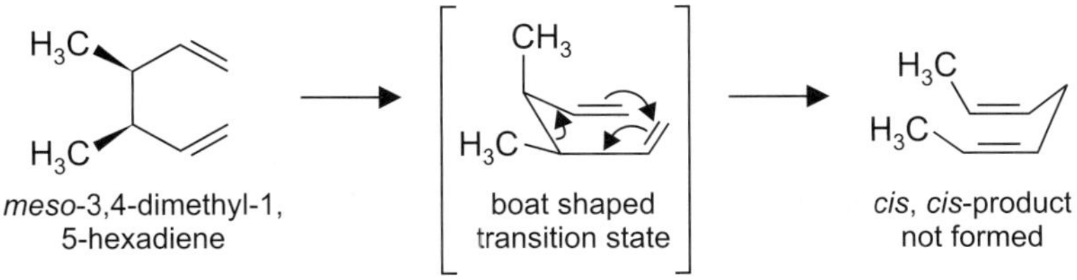

SCHEME 17.84

EXERCISE 17.13

Which diene you expect on pyrolysis of trans-3, 4-dimethylcyclohexadiene ?

ANSWER. *A Cope rearrangement occurs through a chair shaped transition state and the diene expected is E, E isomer of 2, 6-octadiene (Scheme 17.85).*

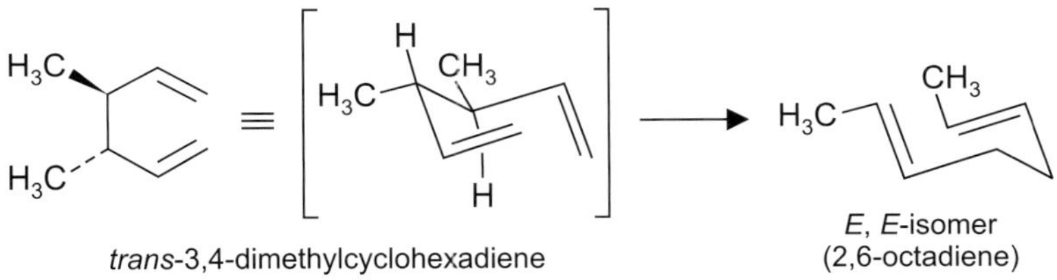

trans-3,4-dimethylcyclohexadiene

SCHEME 17.85

EXERCISE 17.14

Why Z, Z-2, 6-octadiene is not the product of pyrolysis of trans-3,4-dimethylcyclo-hexadiene ?

ANSWER. *The transition state (I, Scheme 17.86) with two pseudoequatorial groups is far more stable than (II) with two pseudoaxial groups. The Z, Z-isomer would arise from the less stable chair shaped transition state (II, Scheme 17.86).*

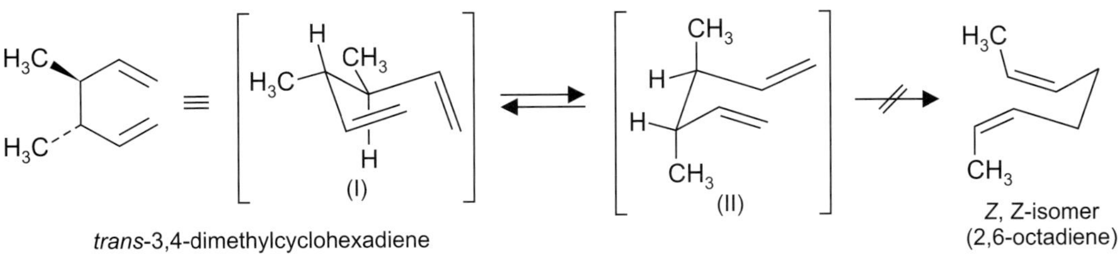

trans-3,4-dimethylcyclohexadiene

SCHEME 17.86

On introducing strain into the reactant, rate accelerations are observed and *cis*-divinylcyclopropane rapidly undergoes Cope rearrangement (Scheme 17.87). Similar reaction is however, not observed with *trans*-isomer where the reacting ends of the double bonds are too far apart. Thus the Cope rearrangement occurs at low temperatures in *cis*-1, 2-divinyl cyclopropane compared to Cope rearrangement of 1, 5-hexadiene itself which requires temperatures in the range of 200–300°C.

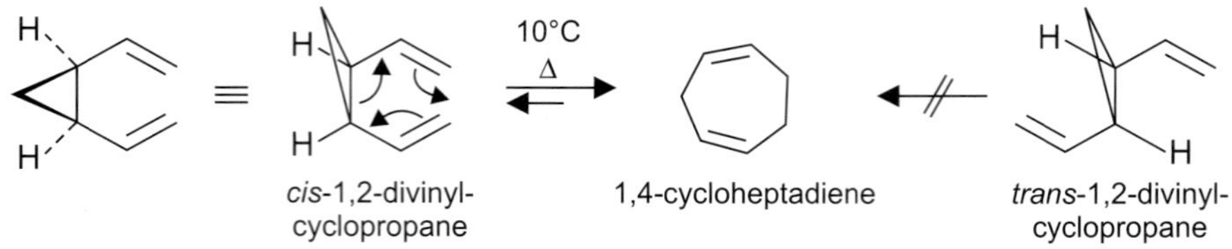

SCHEME 17.87

EXERCISE 17.15

Which two conformations of cis-1, 2-divinylcyclopropane can be considered for a possible Cope rearrangement ?

Which of these conformations is capable of undergoing this rearrangement ?

ANSWER. *See Scheme 17.87a. Two conformations can be adopted (I and II, Scheme 17.87a), (I) is less stable due to steric strain between double bond and a H atom and only this conformation can undergo Cope rearrangement. The conformation (II) will be unreactive since then the product will have two trans double bonds in a seven membered ring (an impossible geometrical situation).*

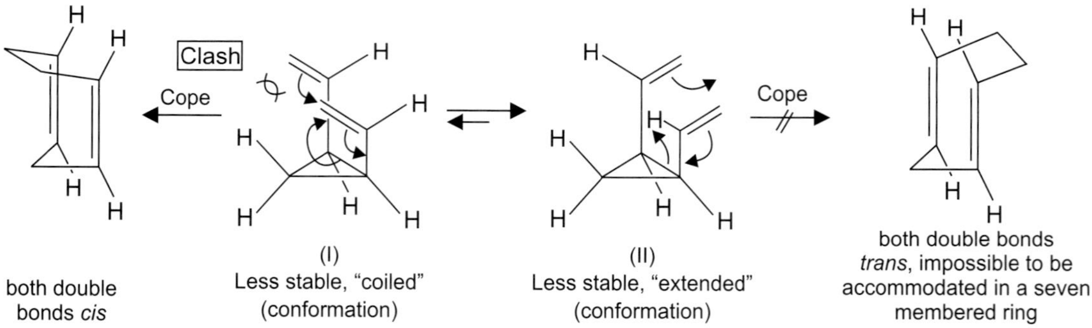

SCHEME 17.87a

D. Fluxional Molecules—A Degenerate Rearrangement

Divinylcyclopropane rearrangements can proceed even with more ease in case the entropy of activation is made still negative by incorporating both vinyl groups into a ring. The system then becomes homotropilidene (Scheme 17.88) which undergoes a degenerate Cope rearrangement

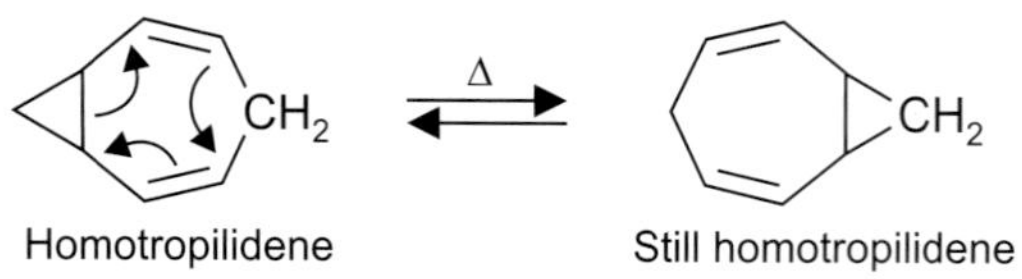

The *degenerate* Cope rearrangement of homotropilidene

SCHEME 17.88

A degenerate rearrangement leads to a product which is indistinguishable from the reactant. By bridging the two methylene groups in homotropilidene one gets a molecule of bullvalene (Scheme 17.89). This is converted into itself at 25°C. At 100°C the ^{1}HNMR spectrum of bullvalene shows a single peak at 4.22 ppm. Bullvalene has a three fold rotational axis; thus all the three double bonds are equivalent. The Cope rearrangement can occur in each of the three faces of the molecule and is degenerate in every case (Bullvalene is a fluxional molecule—a molecule which undergoes rapid degenerate rearrangement).

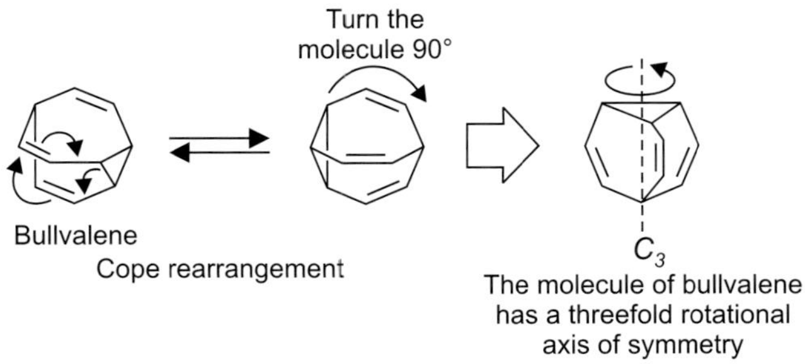

SCHEME 17.89

E. Oxy-Cope Rearrangement

As seen in other pericyclic reactions Cope rearrangement is also reversible and the position of equilibrium depends on the relative stability of the isomers. This problem can be checked and the forward reaction can be made to predominate provided the product reacts further. This is so in oxy-Cope rearrangement when the reactant contains an oxygen group on C3 or C4 position. The alcohol variant of Cope rearrangement (Scheme 17.90) is called the oxy-Cope rearrangement and when the alkoxide derivative is used it is referred to as the "anionic oxy-Cope rearrangement. The rates of sigmatropic rearrangements are enhanced and the temperatures for the reactions are drammatically decreased compared to parent alcohols.

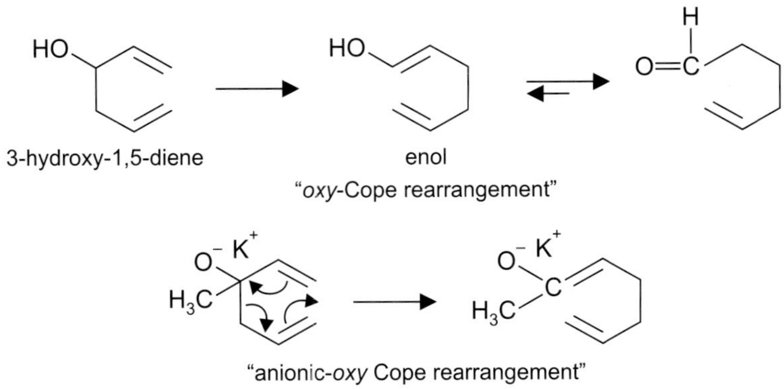

SCHEME 17.90

F. Aza-Cope Rearrangement

It is well known that the presence of oxygen atom adjacent to the π bond accelerates the Cope rearrangement. Similarly a nitrogen usually as an iminium salt fragment in the diene also induces an aza-Cope rearrangement. Thus the reactant (II, Scheme 17.91) derived from (I) underwent a fast aza-Cope rearrangement at low temperature.

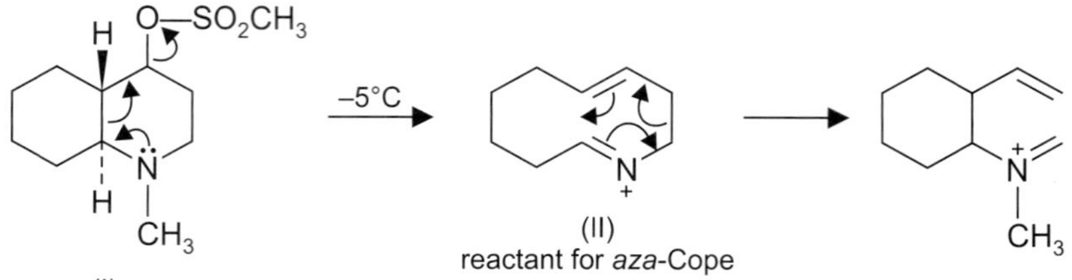

(I)

(II)
reactant for *aza*-Cope
rearrangement

***Aza*-Cope rearrangement**

SCHEME 17.91

EXERCISE 17.16

Write the structure of products from the reactions (Scheme 17.92).

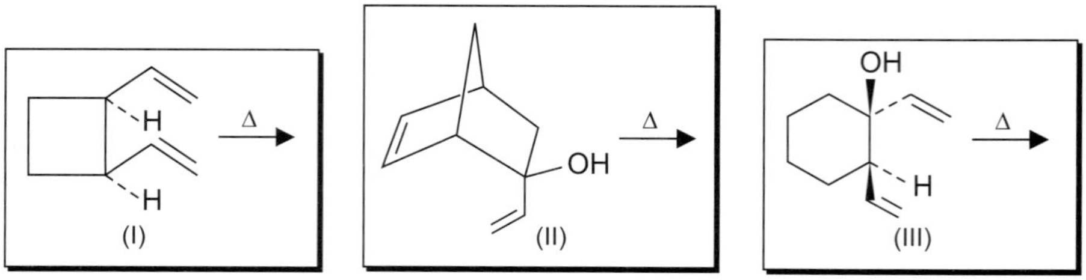

SCHEME 17.92

ANSWER. *Always look for the presence of a 1, 5-diene unit which will hint towards a Cope rearrangement (draw the arrow to form a bond between C1 and C6 and breaking a bond between C3 and C4, Scheme 17.93).*

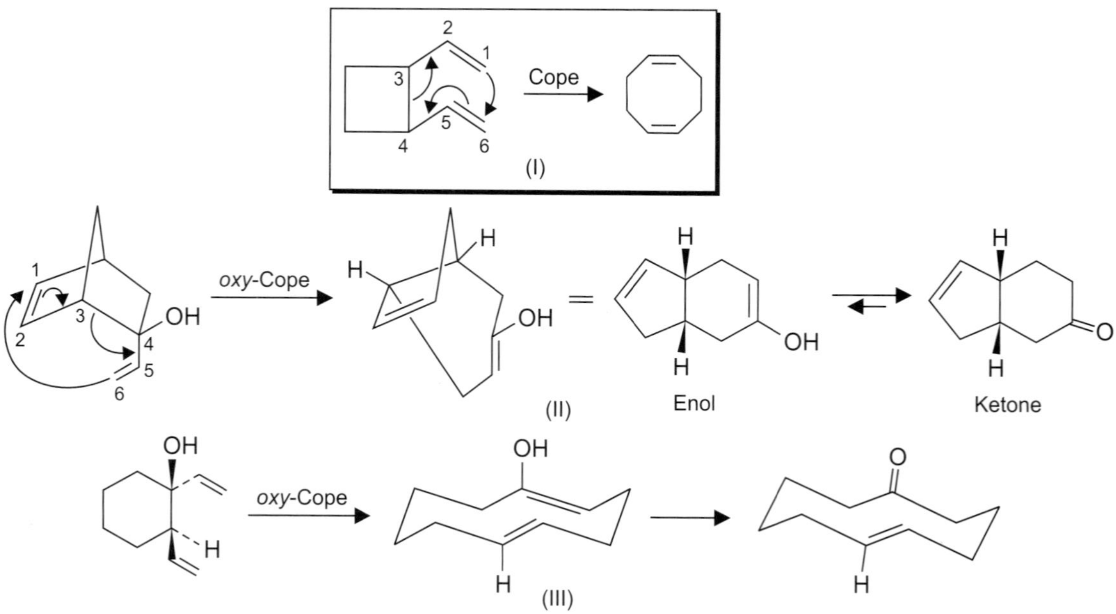

SCHEME 17.93

G. The Claisen Rearrangement

Claisen rearrangement also involves a [3,3] sigmatropic pathway like Cope rearrangement, however, in Claisen rearrangement the substrates incorporate one or more heteroatoms in place of carbon in the 1,5-hexadiene system. The simplest example of Claisen rearrangement is the thermal conversion of allyl vinyl ether to 4-pentenal (Scheme 17.94). The transition state involves a cycle of six orbitals and six electrons. With six electrons the transition state has aromatic character. Similarly allyl aryl ethers on heating rearrange to *o*-allyl phenols.

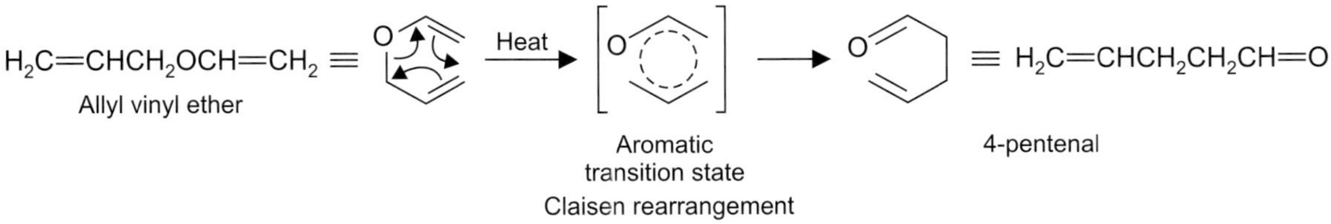

SCHEME 17.94

Studies using migrating groups labelled with ^{14}C or with substituents show that the allylic group is end-interchanged during the *ortho* rearrangement (Scheme 17.95). These and other results which show that the Claisen rearrangement is intramolecular provide strong support for a concerted mechanism. When both *o*-positions are occupied the allyl group migrates to the *p*-position (Scheme 17.96).

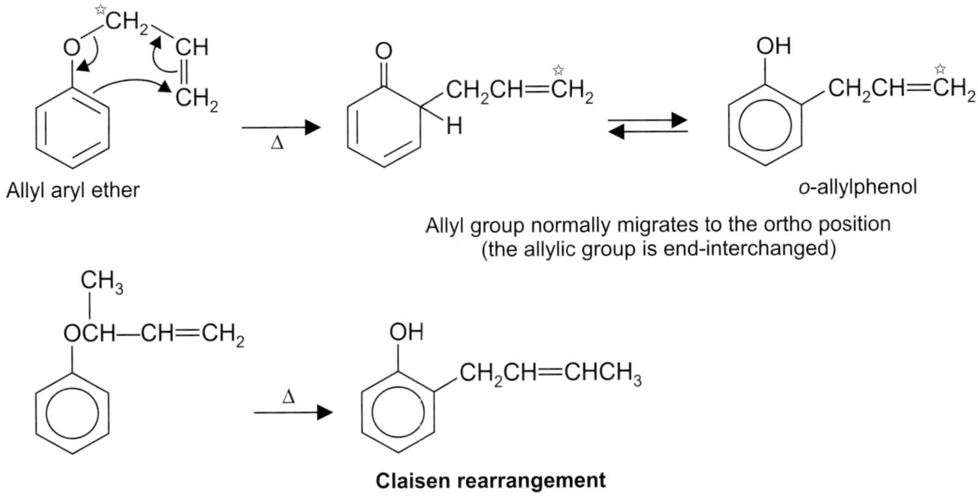

SCHEME 17.95

Like Cope rearrangement reliable stereochemical predictions can be made from a chair-like transition state (Scheme 17.96*a*). In (I) the methyl groups occupy pseudoequatorial positions in the transition state. Similarly in (II) the major product will have *E* configuration of the newly created double bond due to placement of the bulkier substituent in the pseudo equatorial position.

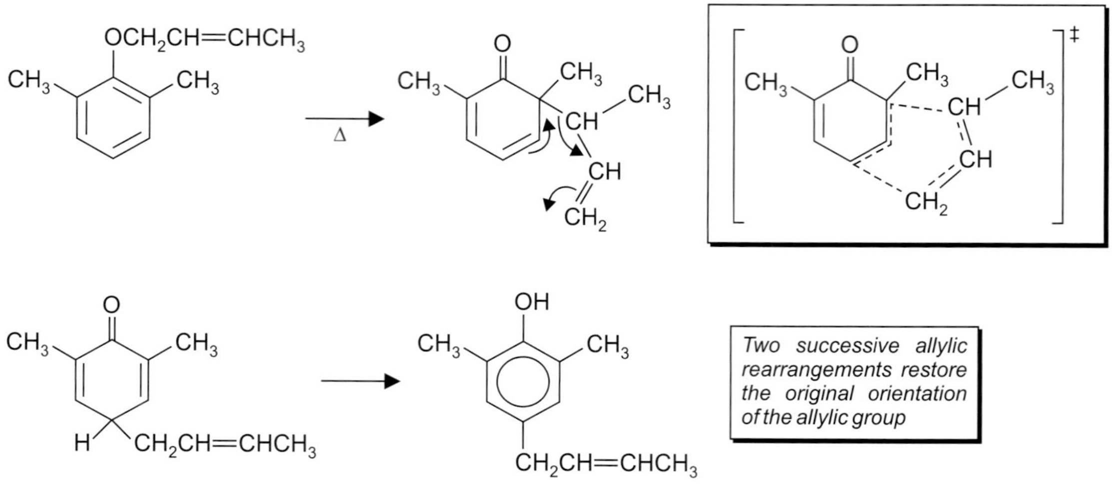

SCHEME 17.96

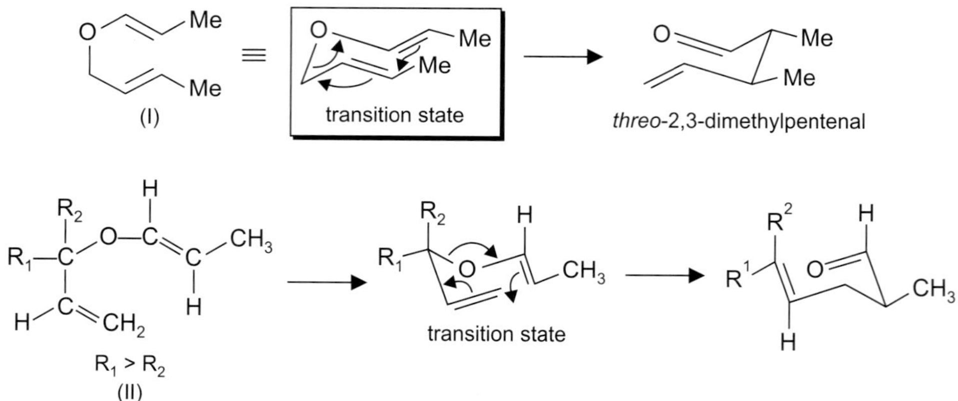

Claisen rearrangement

SCHEME 17.96a

H. [5, 5] Sigmatropic Shifts

Thermal [5, 5] shifts are facile, however the compounds undergoing such rearrangement are not common. One type of substrates are pentadienyl ethers which give 4-pentadienylphenols as the major products along with minor products arising from *ortho*-Claisen rearrangement (Scheme 17.96*b*). With the help of deuterium labeling it has been shown that major products arise from direct [5, 5] sigmatropic shifts and not by two consecutive [3, 3] shifts.

It is proved that [5, 5] shifts occur very fast in negatively charged compounds. An interesting reaction is oxy-Cope rearrangement of (I, Scheme 17.96*c*) the arrangement does not proceed by a sequence of consecutive [3, 3] shifts, however it is indeed a result of [5, 5] shifts, (Scheme 17.96*c*).

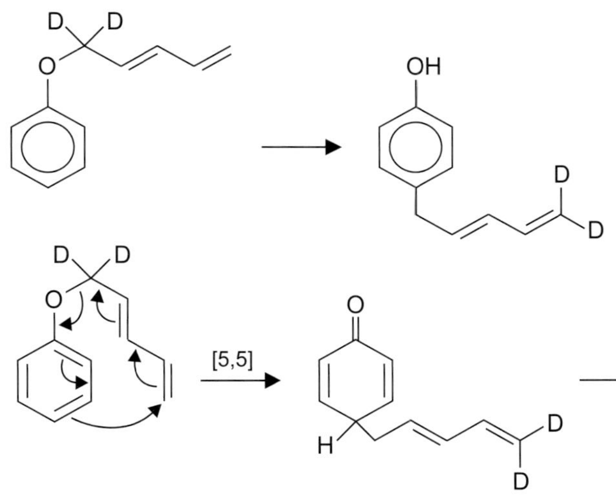

[5,5] Sigmatropic shifts

SCHEME 17.96b

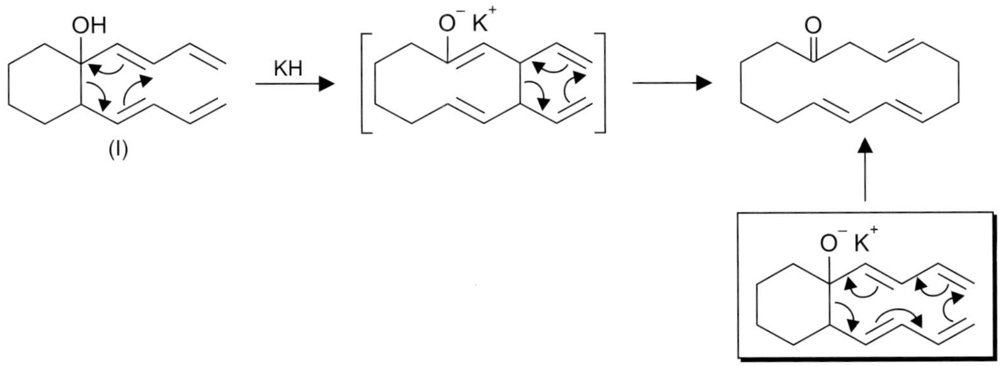

SCHEME 17.96c

17.8 THE ENE REACTION

In this reaction an alkene having an allylic hydrogen atom reacts thermally with a dienophile (C=C, C=O, N=N etc., called enophile) with the formation of a new σ bond to the terminal carbon of the allyl group. This is followed by the 1,5-migration of the allylic hydrogen and subsequent change in the position of allylic double bond. The reaction thus resembles both cycloaddition and a [1,5]-sigmatropic shift of hydrogen.

Mechanistically, the reaction is a concerted process, there being little charge development in the transition state. It shows a primary kinetic isotope effect to show C—H bond breaks in the rate determining step (the reverse process occurs in the pyrolysis of esters). The interaction of a hydrogen atom with the HOMO of the allyl radical and the LUMO of the enophile (Scheme 17.97) is a symmetry allowed process. A good example of ene reaction is found during allylic oxidation of alkenes with selenium dioxide.

The reaction shows a primary kinetic isotope effect of C—H bond breaking in the rate determining step. The ene reaction of β-pinene with maleic anhydride (Scheme 17.97) gives the product without skeletal rearrangement of the strained four membered ring in the β-pinene, to show the concerted nature of the reaction (rather than the formation of a cationic intermediate). The ene reaction, however requires higher temperature than in Diels-Alder reaction, but occurs faster with conjugated enones with Lewis acid catalysts. Coordination of the Lewis acids with the enophile lowers the energy of LUMO.

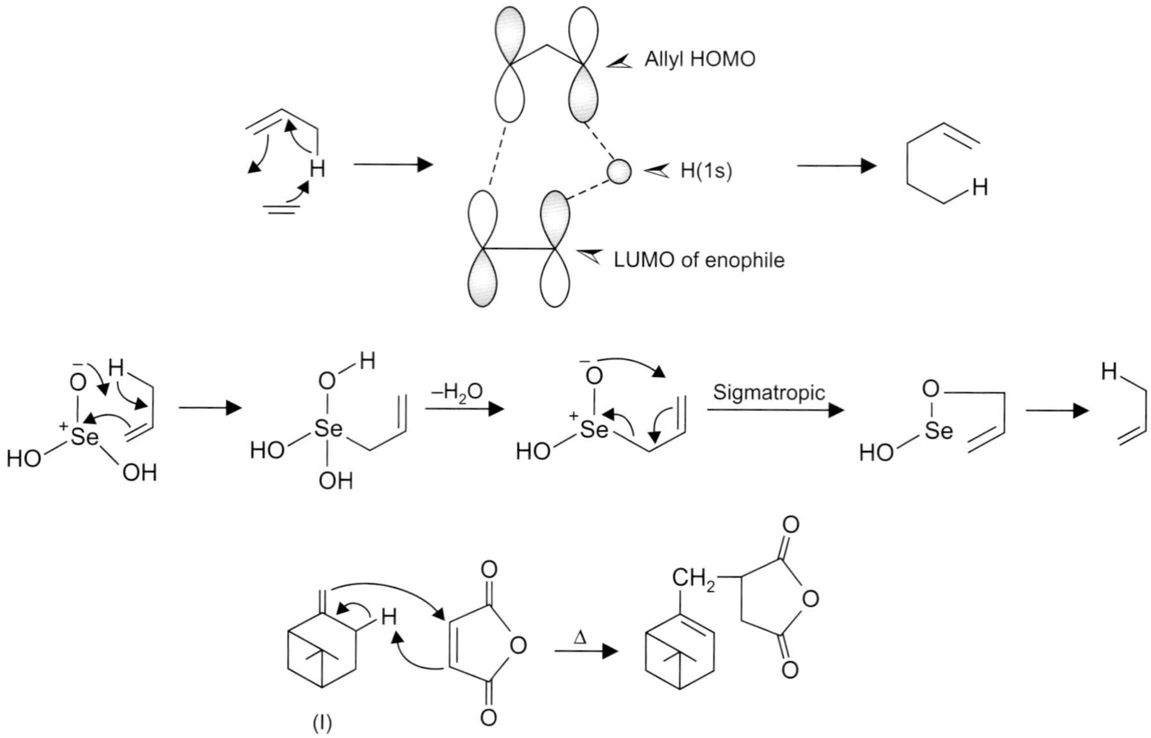

SCHEME 17.97

17.9 AROMATIC TRANSITION STATES

The reactions with aromatic transition states (2, 6, 10, 14 ... $4n + 2$) delocalized electrons are permitted, while the *anti*-aromatic systems (4, 8, 12 ... $4n$) delocalized electrons are the forbidden ones. It is clear that the Diels-Alder reaction has a 6π aromatic transition state isoelectronic with benzene while the forbidden cyclobutane formation has the unfavourable 4π transition state, isoelectronic with cyclobutadiene. This is the reason that although [4+2] reactions are common and general, the analogous concerted [2+2] and [4+4] thermal cycloaddition reactions generally do not occur, since the corresponding transition states involve 4 and 8 electrons respectively (Scheme 17.98 also see Table 17.3). However, the other hand, cycloaddition reactions which involve 6 or 10 electrons occur readily (Scheme 17.99). It may be mentioned that *e.g.,* a [4+2] cycloaddition involves four π electrons of one system and two on another.

Most pericyclic reactions involve six electrons, but the Huckel rule is also obeyed with two, ten etc. electrons. An important pericyclic reaction with two electrons is the rearrangement of carbocations. The pericylic nature of such a transition state is shown (Scheme 17.100). The cation with a total of two electrons has a filled shell and relative stability. The corresponding transition state for a carbanion involves four electrons and an unfilled shell. Accordingly, carbocation rearrangements are common, while carbanion rearrangements are not. Moreover, Wagner-Meerwien and related rearrangement order [1,2], occur in carbocation because of the allowed *s, s* pathway, but not in carbanions which would requires *s,a*. The migrating group retains its chirality. By contrast, the [1,3] shift (see, Scheme 17.79) is a accompanied with inversion of configuration.

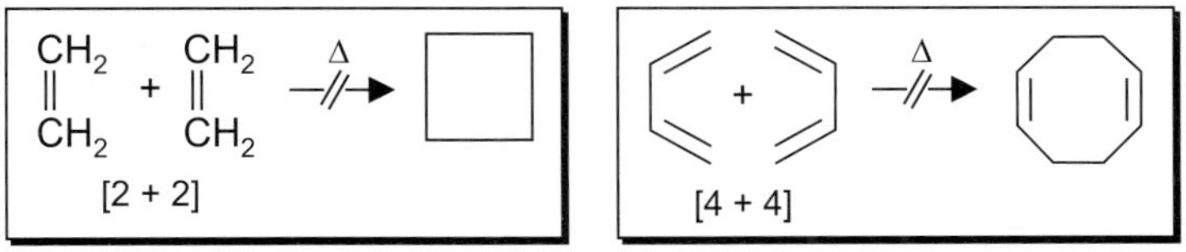

SCHEME 17.98

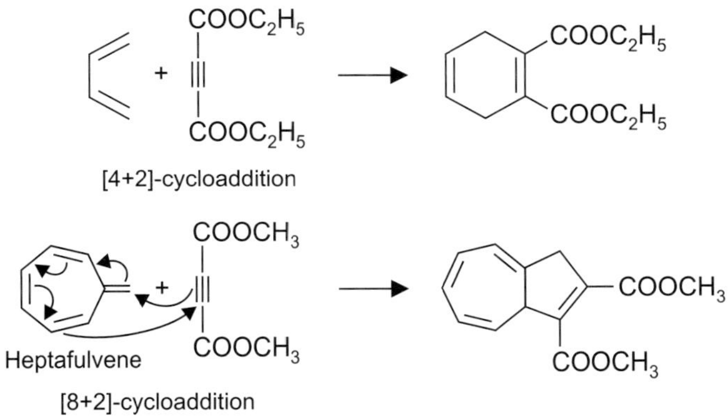

SCHEME 17.99

17.10 MÖBIUS-HÜCKEL ANALYSIS (PMO) APPROACH

The concerted reactions are analyzed by the classification of transition states as aromatic or antiaromatic. These predictions yield the same results as by other methods. In the previous section a mention was made of aromatic transition states. Hückel's rule of aromaticity states that a monocyclic planar conjugated system with $(4n + 2)\pi$ electrons is *aromatic* and therefore, stable in the ground state. On the other hand a system with $(4n)$ π electrons is unstable and is called antiaromatic. Further it has been shown that these rules are reversed by the presence of a node (a phase dislocation) in the array of atomic orbitals. The Möbius-Hückel concept is used to analyze the pericyclic reactions without using the actual molecular orbitals. The following points may be noted:

- Each atom of the interacting system is assigned a *p* orbital with one lobe black and one white (or some other designation). A hydrogen atom is represented by a circle of one color representing an *s* orbital.
- One draws each reactant with the black lobes on one side and the white on the other. Then one considers the transition state of a particular reaction, counts the number of

electrons and the nodes in the array to reach the conclusion if that reaction is symmetry allowed or forbidden.

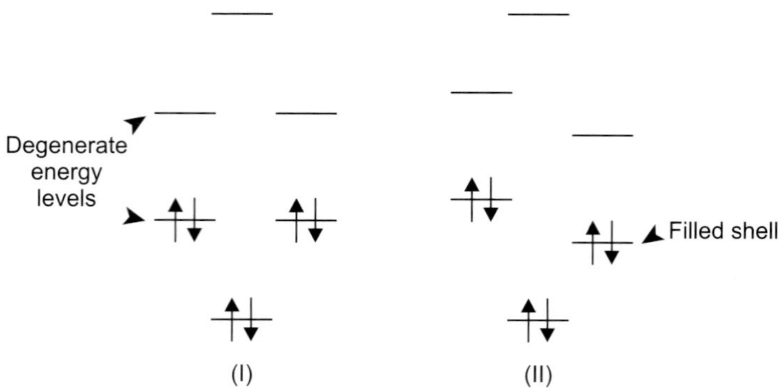

MO energy levels for (I) benzene and (II) a six-electron transition state

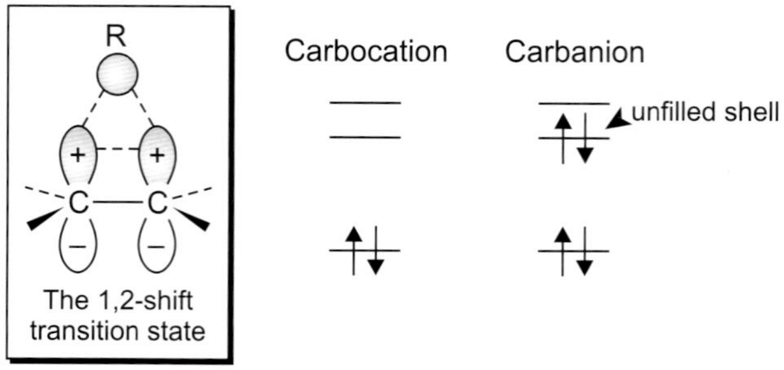

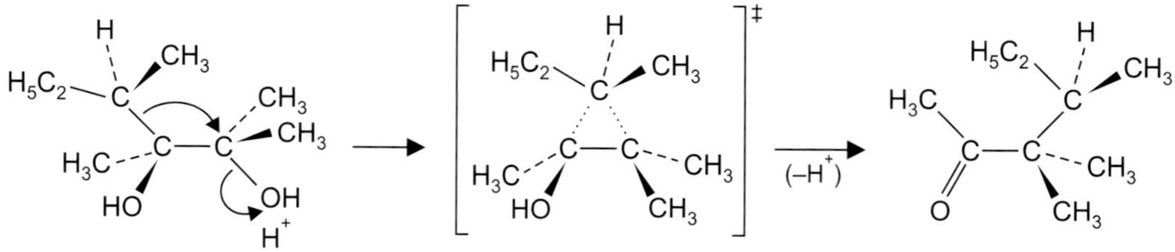

1,2-Rearrangement involves a three-center pericyclic transition state

SCHEME 17.100

- A Hückel system has zero (or any even number) of nodes (phase changes) around the orbital array. A Hückel system with $4n + 2$ electrons is aromatic and with $4n$ electrons is antiaromatic. An array with an odd number of phase dislocations is called an *anti* Hückel system (Mobius system). An *anti*-Hückel system with $4n$ electrons is aromatic and with $4n + 2$ electrons is antiaromatic.
- The condition for aromaticity in *anti*-Hückel system is opposite to that for Hückel system.

1. Electrocyclic Reactions

(a) Thermal Ring Opening of Cyclobutenes—4n Systems

On the basis of FMO approach under thermal conditions the observed stereochemistry of the products indicates a conrotatory motion. Consider the thermal ring opening in *cis* 3,4-dimethyl-cyclobutene which on the basis of FMO method gives 2*E*, 4*Z*-hexadiene (Schemes 17.6 and 17.10).

The Hückel-Möbius approach also predicts conrotatory motion under thermal conditions and predicts the formation of same diene. Consider the basis set for the starting butadiene (Scheme 17.101), the tilt at C-1 and C-4 as the butadiene system rotates toward the transition state is different for conrotatory and disrotatory modes. The transition state for conrotatory ring opening has one sign inversion (phase dislocation) and with four electrons it is aromatic. The conrotatory transition state for cyclobutene ring opening is therefore, aromatic. The transition state for disrotatory cyclobutene ring opening however, is anti-aromatic (no phase dislocation with 4 electrons). Thus the PMO approach like FMO method also predicts that for butadiene-cyclobutene interconversion the conrotatory transition state is the favoured aromatic transition state and thus thermal conrotatory ring opening in cyclobutenes is allowed and disrotatory opening is forbidden.

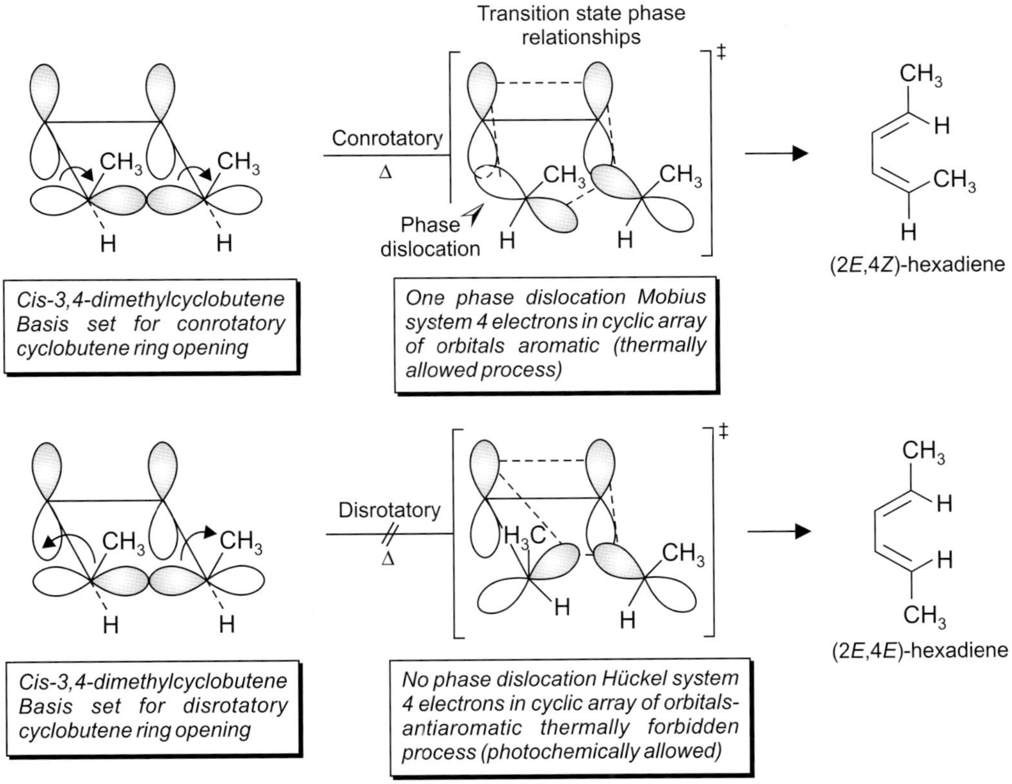

SCHEME 17.101

(b) Thermal Ring Closure of an Octatetraene—4n Systems

In this case also a conrotatory ring closure (Scheme 17.102) is allowed since in the transition state basis set there is one phase dislocation at the forming σ bond. The transition state is therefore, aromatic (8 electrons Möbius system). Disrotatory ring closure for the reaction (Scheme 17.102) will be however, antiaromatic eight electron Hückel system.

As with FMO approach the general statement is that thermal electrocyclic reactions in the conrotatory mode are allowed for $4n$ electron transition states.

The stereochemistry and other aspects are just the same as already discussed during FMO approach. Moreover, since Woodward-Hoffmann rules for photochemical reactions are always the reverse for thermal reactions, one can always predict the outcome of a reaction under photochemical conditions (Schemes 17.101 and 17.102).

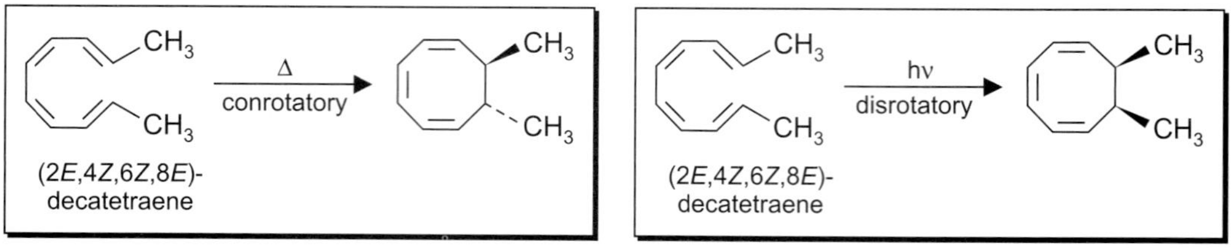

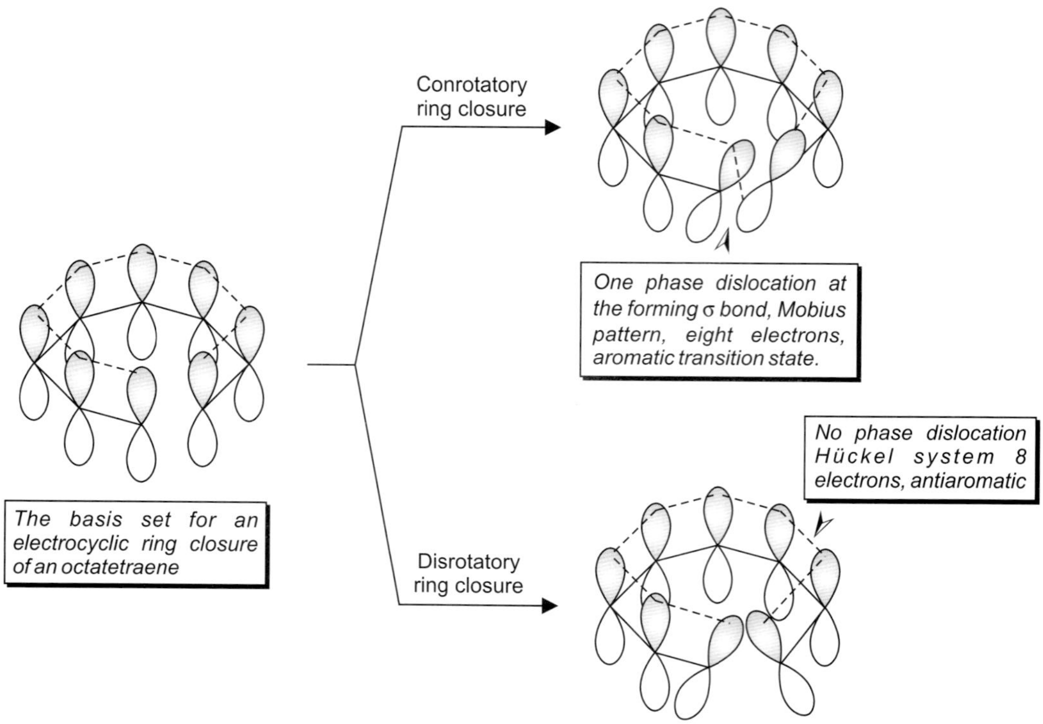

SCHEME 17.102

(c) Thermal Hexatriene Ring Closure—4n + 2 Systems

In FMO approach for $4n + 2$ electron transition states, in electrocyclic reactions, the disrotatory mode is allowed for thermal reactions and the conrotatory mode for photochemical reactions. In PMO method as well, this is found to be true. Starting from hexatriene basis set (I, Scheme 17.103), the transition state for disrotatory ring closure is a six-electron Hückel system and thus aromatic (II). The conrotatory ring closure proceeds through an antiaromatic anti-Hückel transition state. It is, therefore, correctly predicted that thermal ring closure of substituted hexatrienes should be disrotatory and the photochemical reaction should proceed via the opposite conrotatory path.

2. Diels-Alder Reaction

The selection rules may also be derived by the consideration of the aromaticity of the transition state of a Diels-Alder reaction. Same conclusions (as in the case of FMO approach) are again reached and these are summarized in Scheme 17.104).

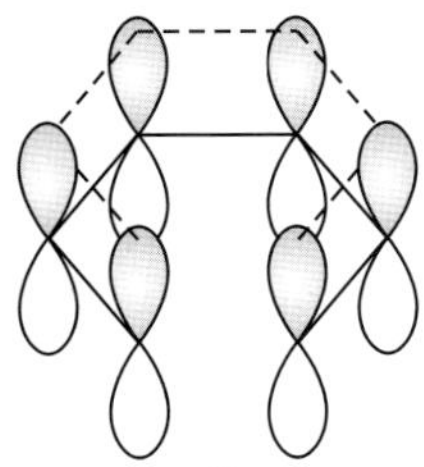

(I)

The hexatriene basis set

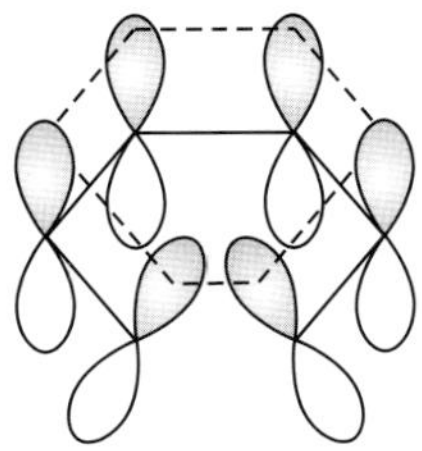

(II)

Transition state for hexatriene disrotatory ring closure, 6 electrons Hückel system (no phase dislocation) aromatic (thermally allowed)

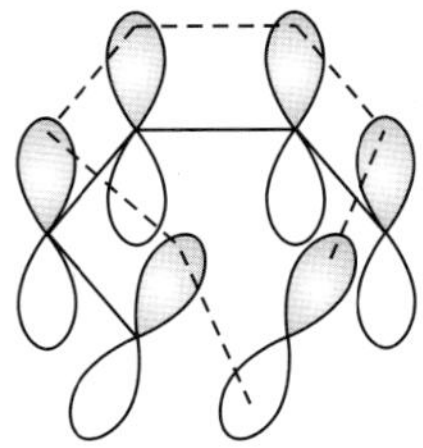

(III)

Transition state for hexatriene conrotatory ring closure (one phase dislocation) 6 electrons antiaromatic. (Photochemically allowed)

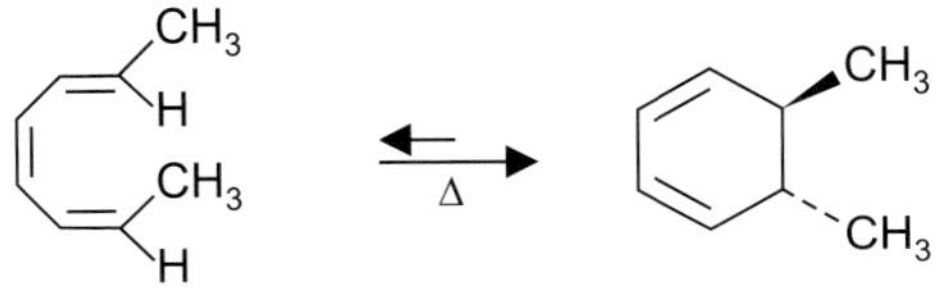

(2*E*,4*Z*,6*Z*)-octatriene → *trans*-5,6-dimethyl-1, 3-cyclohexadiene

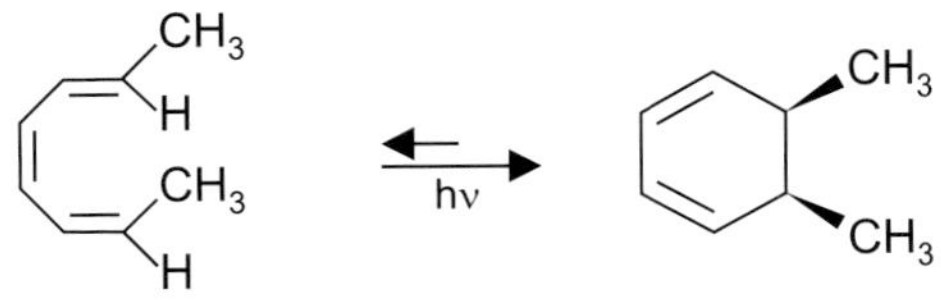

(2*E*,4*Z*,6*Z*)-octatriene → *cis*-5,6-dimethyl-1, 3-cyclohexadiene

SCHEME 17.103

Supra/Supra Hückel system (no phase dislocation) 6-electrons aromatic allowed.

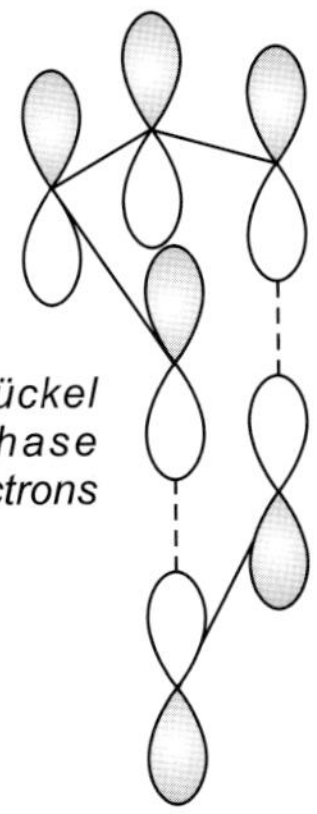

Supra/Supra Hückel system (no phase dislocation) Hückel system 4-electrons antiaromatic forbidden

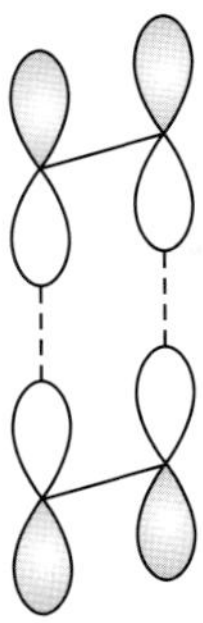

Supra/antara Mobius system (one phase dislocation) 4-electrons aromatic allowed.

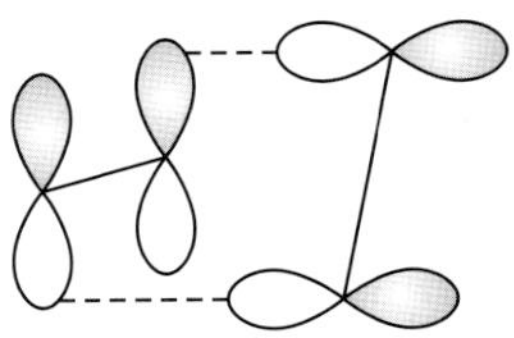

SCHEME 17.104

Antarafacial-Suprafacial cycloaddition is highly sterically hindered and is, therefore, less common. In case two ethylene molecules are brought together in such a way that a Mobius activated complex can be realized (see Scheme 17.104) the process become a suprafacial/ antarafacial addition. This process should be allowed since it is a Mobius system with four electron and a node. For this process to be realized the ethylene molecules have to approach each other in a perpendicular geometry. The completion of this addition involves distortion of the carbon framework. The process, therefore, is difficult although allowed. For this reason simple alkenes do not display this addition. The highly strained triene (Scheme 17.105) however, spontaneously dimerizes thermally and represents $[\pi^2a + \pi^2s]$ transition state. Reaction of heptafulvalene with tetracynoethylene is a remarkable example of a $[\pi^{14}a + \pi^2s]$ thermal cycloaddition leading to a product of *anti* addition. The transition state involves a negative overlap which corresponds to a Mobius cyclic electronic system, a favourable transition state for a 16-electron (4*n*) cyclic system (Scheme 17.105).

As already discussed many known cycloadditions $[p + q]$ involve pericyclic electrons equal to 6, 10, 14 etc., and involve Hückel aromatic transition states. The [14 + 2] cycloaddition, however, does not fit the Hückel rule.

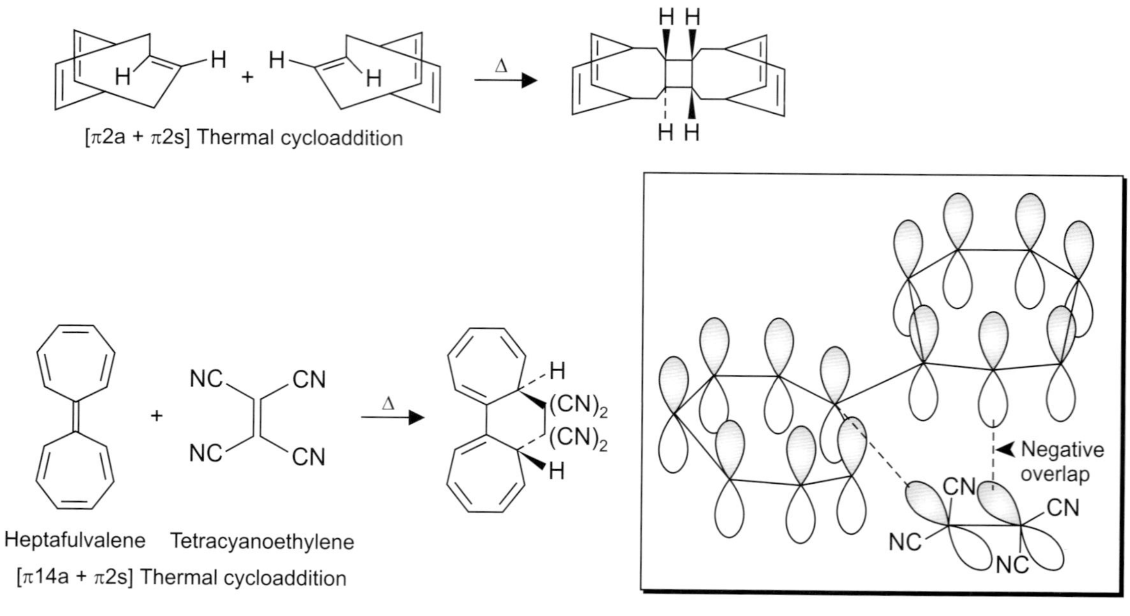

SCHEME 17.105

3. Sigmatropic Shifts

Consider the simplest case of 1,3-sigmatropic shift of a hydrogen. In the FMO approach the hydrogen 1*s* orbital interacts with an allyl radicals HOMO. A thermal [1,3] suprafacial shift is symmetry forbidden, the antarafacial is symmetry allowed, but energetically very unfavourable (see, Scheme 17.72). A consideration of basis set atomic orbitals and their classification as aromatic or antiaromatic reaches the same conclusions (Scheme 17.106). The 1,3-suprafacial shift of hydrogen is forbidden, but the suprafacial 1,5-shift is allowed. The 1,7-shifts should be antarafacial, when an alkyl group (carbon) migrates, an additional stereochemical feature has to be considered. Again in agreement with FMO approach, the allowed processes include, the suprafacial 1,3-shift with inversion and the suprafacial 1,5-shift with retention (Scheme 17.107).

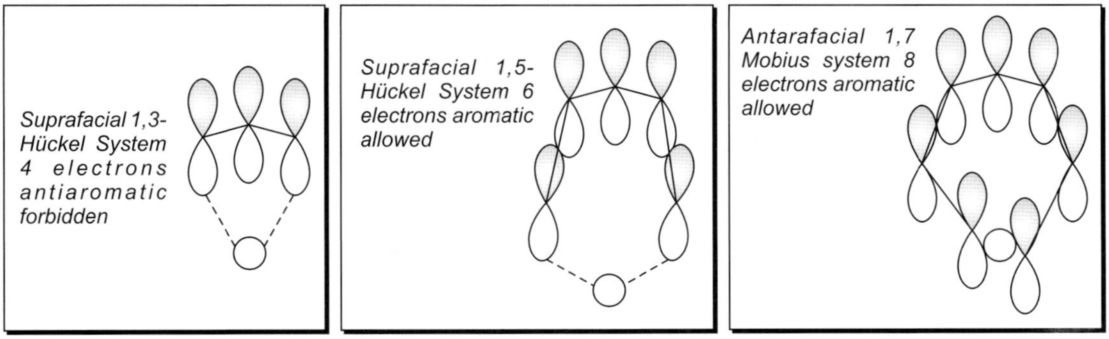

Classification of sigmatropic shifts of hydrogen atom with respect to basis set orbitals

SCHEME 17.106

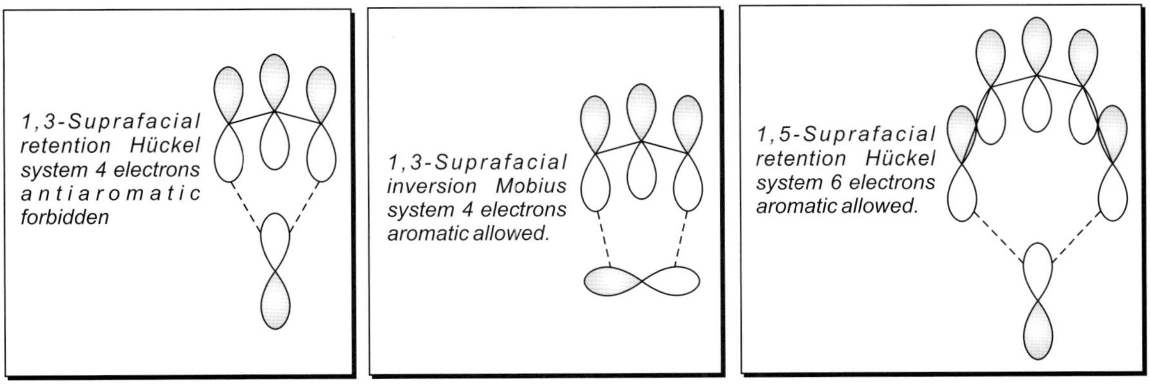

Classification of sigmatropic shifts of alkyl groups (Carbon) with respect to basis set orbitals

SCHEME 17.107

17.11 CORRELATION DIAGRAM METHOD

In this method a correlation of the geometrical symmetry of the orbitals between reactants and the products is involved. Based on this, a correlation diagram is developed which compares the symmetry characteristics from this comparison. A reaction can be easily predicted to be symmetry allowed or symmetry forbidden. The following points may be noted:

- Molecular orbitals (of the reactant and the product) are either symmetric or antisymmetric around the mirror plane or around a two fold axis of rotation (C_2). Thus π orbital of ethylene in the ground state is symmetric (S) with respect to mirror plane (I, Scheme 17.108) while it is antisymmetric (A) with respect to rotational axis C_2 (II, Scheme 17.108). A two fold axis of rotation may be regarded as a pin at right angles passing in the middle. If a molecular orbital is spun 180° (*i.e.*, C_2 axis 360°/2), around this axis it would either yield an identical orbital *i.e.*, symmetric or an orbital with all signs the opposite of what these were originally *i.e.*, antisymmetric (A). Thus when the π orbital of ethylene is rotated it is found that the symmetry around the axis of rotation is not maintained. One would see that after the rotation the shaded lobes would be instead on the bottom rather than at the top (as was so in the original). Thus π orbital of ethylene is antisymmetric (A) in relation to its axis of rotation (II Scheme 17.108).

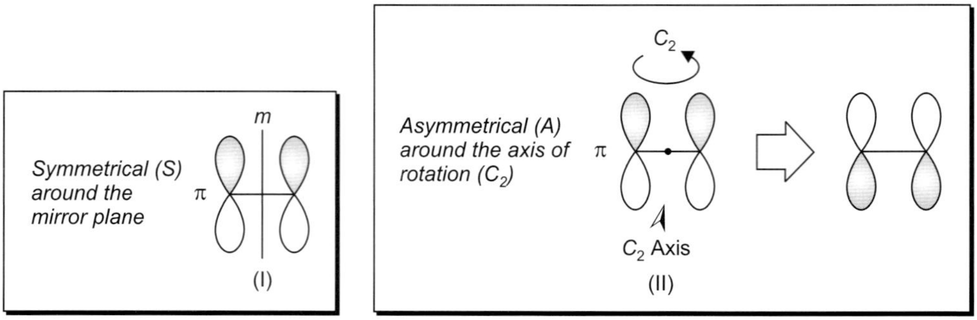

SCHEME 17.108

- On the other hand the same operations show that the antibonding π^* orbital of ethylene is antisymmetric around the mirror plane but symmetric around the C_2 axis of rotation (Scheme 17.109).

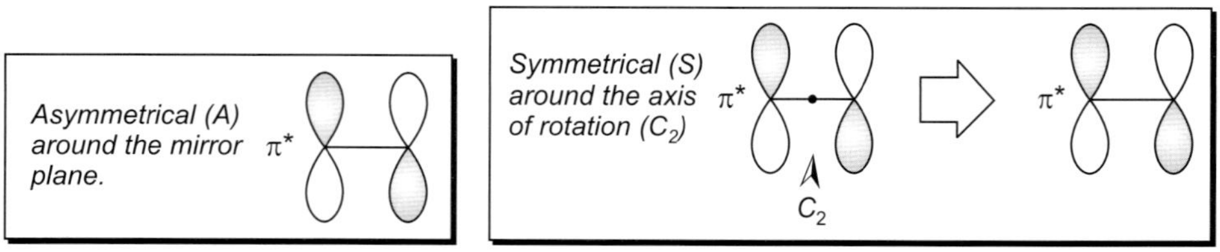

SCHEME 17.109

- Similarly the orbitals of the reactant and the product can be labelled. Thus the sigma orbital of a C-C covalent bond has a mirror plane of symmetry as well as C_2 symmetry, a σ^* orbital is antisymmetric both with respect to plane of symmetry as well as rotational axis of symmetry (Scheme 17.110).

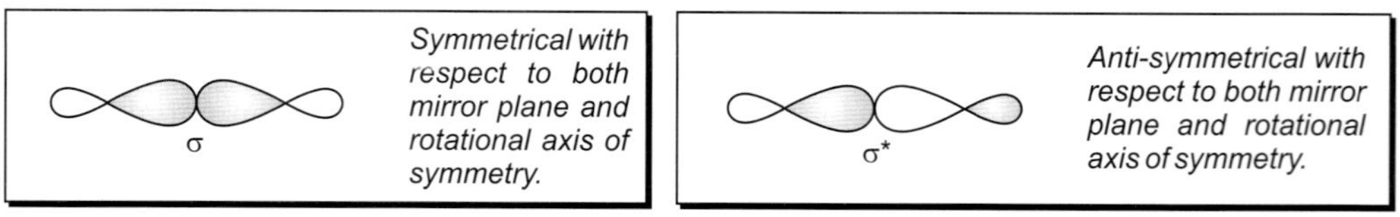

SCHEME 17.110

- During a disrotatory electrocyclic conversion, a plane of symmetry is maintained throughout (Scheme 17.111). If the reaction proceeds by a conrotatory motion a two fold axis of Symmetry (C_2) is preserved throughout (Scheme 17.112).

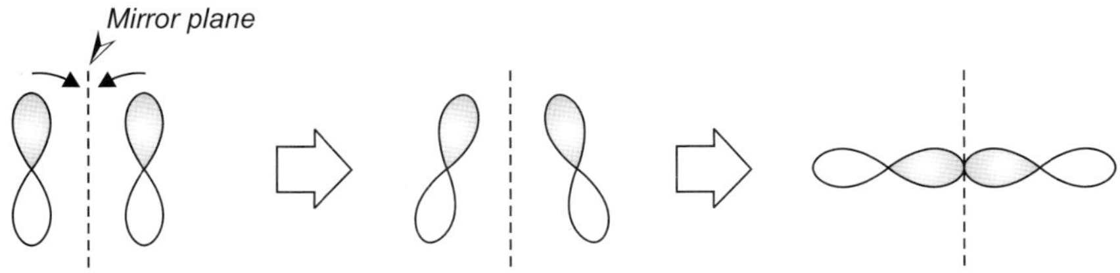

Disrotatory motion (terminal orbitals) a mirror plane of symmetry is maintained throughout

SCHEME 17.111

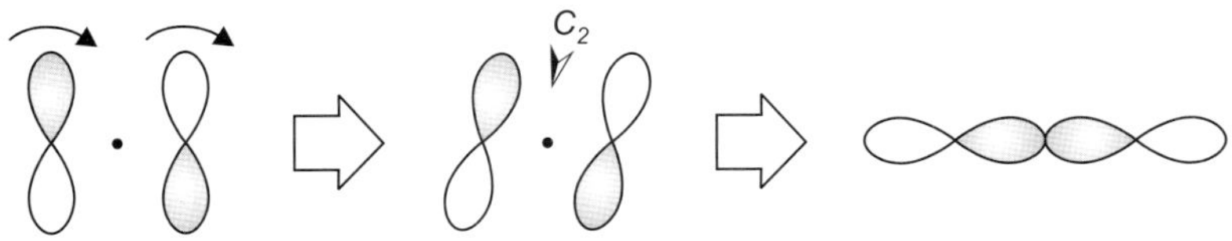

Conrotatory motion (terminal orbitals) a C_2 axis of symmetry is preserved throughout

SCHEME 17.112

- The most stable transition state is the one that conserves the symmetry of the reactant orbitals in passing to product orbitals—a symmetric (S) orbital in the reactant must transform to a symmetric orbital in the product, and an antisymmetric (A) orbital must transform to an antisymmetric orbital.

Example 1: *Correlation Diagrams For Electrocyclic Interconversion of 1,3-Butadiene and Cyclobutene.*

These correlation diagrams are now developed involving a plane of symmetry as well as an axis of symmetry. The four molecular orbitals of butadiene and cyclobutane are inspected for the two symmetry elements. [*In Scheme 17.113, the symmetric properties (plane of symmetry) of molecular orbitals of butadiene and cyclobutene along with the correlation diagram are depicted together. However, these symmetry properties can be translated onto a correlation diagram (as shown on the bottom of Scheme 17.113) for its study.*] Firstly one considers the correlation diagram for the disrotatory ring closure of 1,3-butadiene to cyclobutene during which the mirror plane of symmetry is preserved (Scheme 17.113). The following points may be considered:

- ψ_1 can be converted to σ, however, ψ_2 cannot be converted to π which is the second-lowest orbital of cyclobutene.
- In order to conserve symmetry around the mirror plane, ψ_2 must instead be converted to π^*, while it is ψ_3 which is converted to π.
- This symmetry correlation requires crossover between bonding and antibonding orbitals (Scheme 17.113). This is thermally an unfavourable energetic process (*i.e.,* symmetries of the molecular orbitals with respect to mirror plane do not show ground state correlation) and thus the disrotatory process is forbidden.
- When one considers the correlation diagram for the thermal conrotatory ring closure of 1,3-butadiene (Scheme 17.114) considering the symmetries of the orbitals in relation to the axis of rotation the following results arise.

- ψ_1 is now antisymmetric in relation to its axis of rotation (if the ψ_1 orbital is rotated 180° around the axis, the shaded lobes would come on the bottom instead of on top) similar is the case with π (Scheme 17.114).

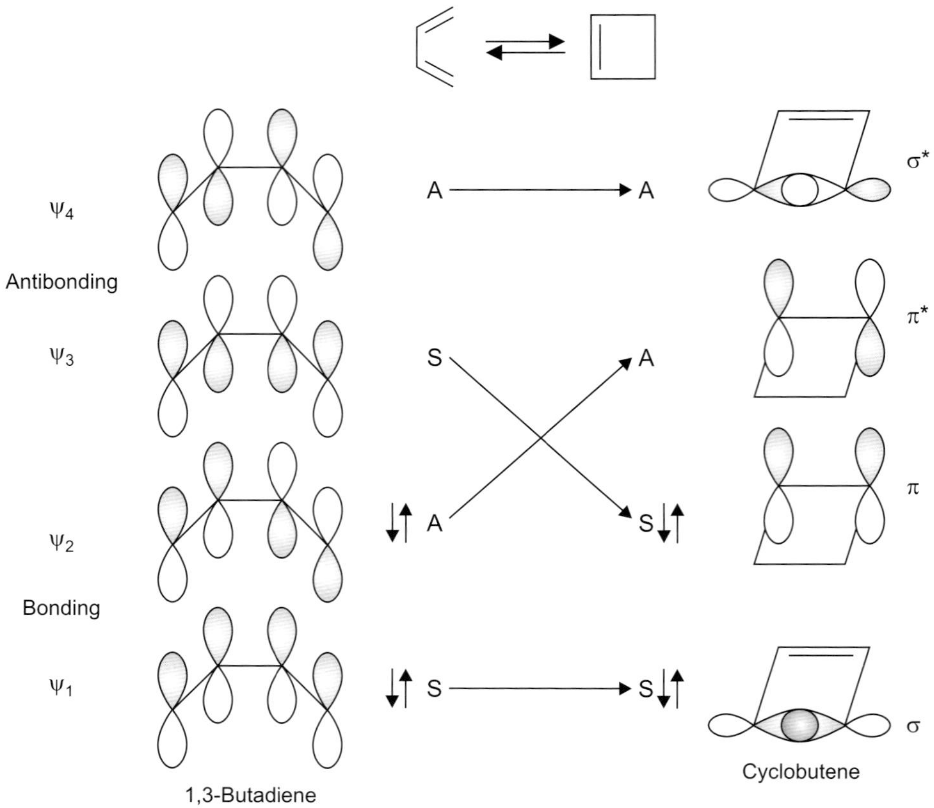

Symmetry properties of molecular orbitals and correlation diagram for disrotatory interconversion of butadiene-cyclobutene

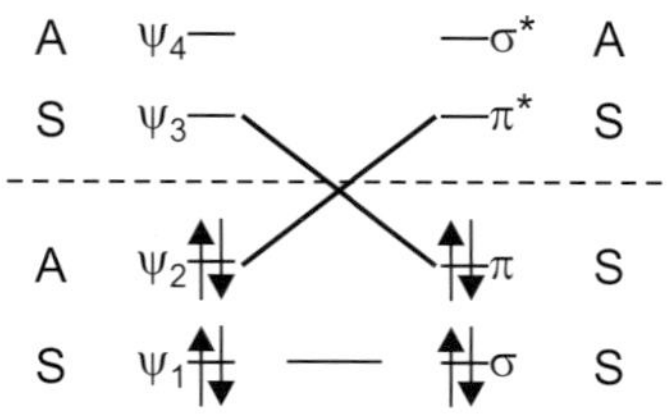

Correlation diagram for disrotatory interconversion of butadiene-cyclobutene (Mirror plane of symmetry is preserved)

Thermal disrotatory ring closure-1,3-butadiene (mirror symmetry maintained) symmetry forbidden

SCHEME 17.113

- However, ψ_2 and π^* are symmetric as in both the cases a 180° rotation would bring shaded lobes back to the top left and bottom right of the orbital.
- Similar operations reveal that σ is symmetric while σ^* is antisymmetric around the axis of rotation.
- These data show that correlation exists between the ground state bonding orbitals, therefore, a thermal conrotatory motion is symmetry allowed process.

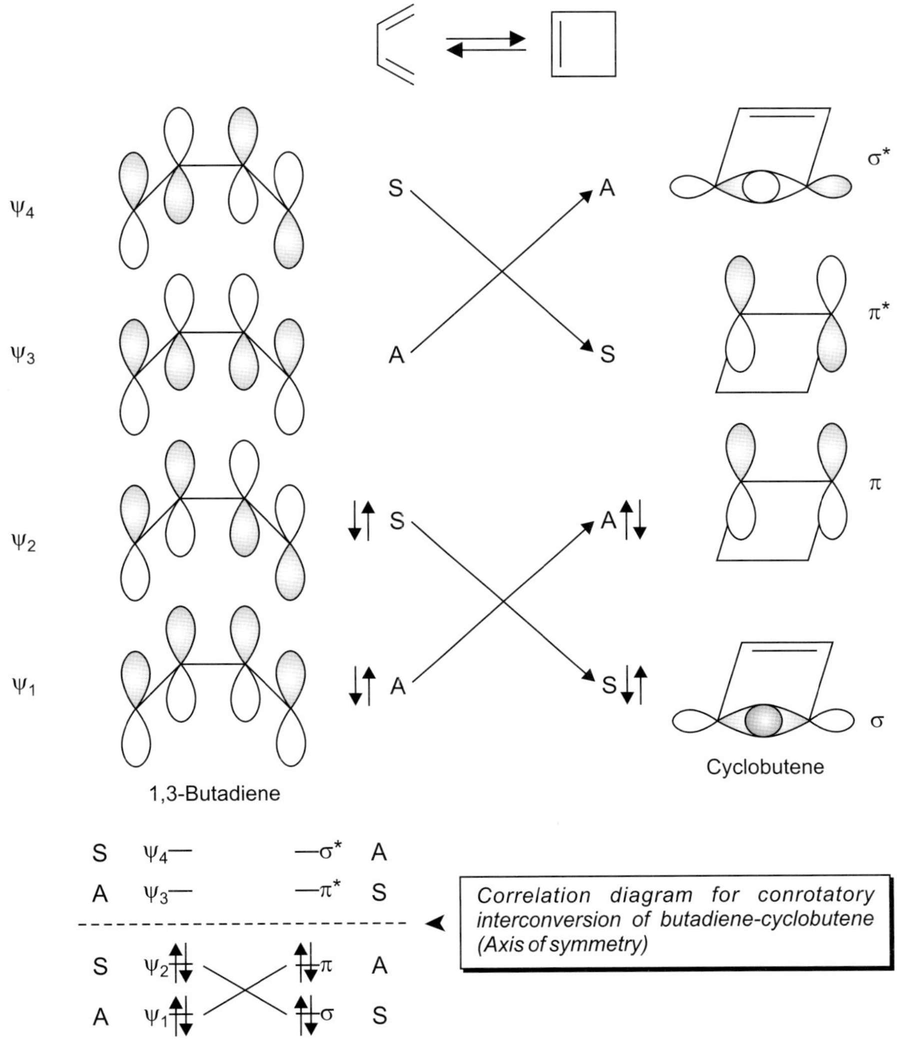

Thermal conrotatory ring closure-1, 3-butadiene (C_2 axis of symmetry maintained) symmetry allowed

SCHEME 17.114

Example 2: *[4 + 2] Cycloaddition of Ethylene to Butadiene to Give Cyclohexene Suprafacial Suprafacial Thermal Cycloaddition.*

The orbital symmetry relationships are given (Scheme 17.115) with respect to the mirror plane of symmetry of the whole reacting system. The two σ-bonds of the product are considered as a symmetric and antisymmetric combination. After the classification of the orbitals with respect to symmetry these are arranged according to energy and the correlation lines can be drawn. (Scheme 17.115). It is found that all bonding levels of the reactants correlate with product ground state orbitals (orbital symmetry is conserved within the bonding orbitals and also within the antibonding set and no cross over between the two sets occurs). This therefore, is an allowed reaction.

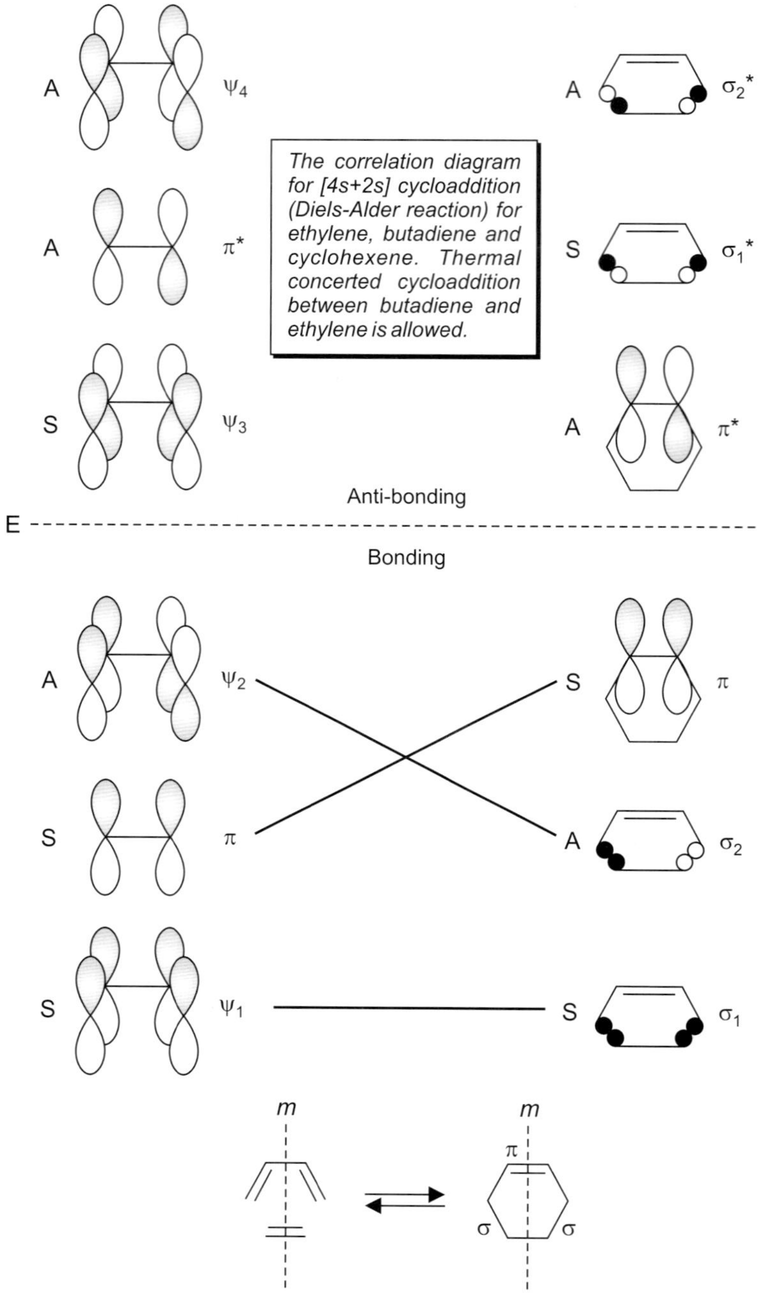

SCHEME 17.115

PROBLEMS

17.1. Write the stereostructure of the compound obtained by the Diels-Alder reaction of dimethyl maleate with butadiene.

17.2. Which diene and dienophile one would employ to synthesize the following compounds? Give alternative route for one of these.

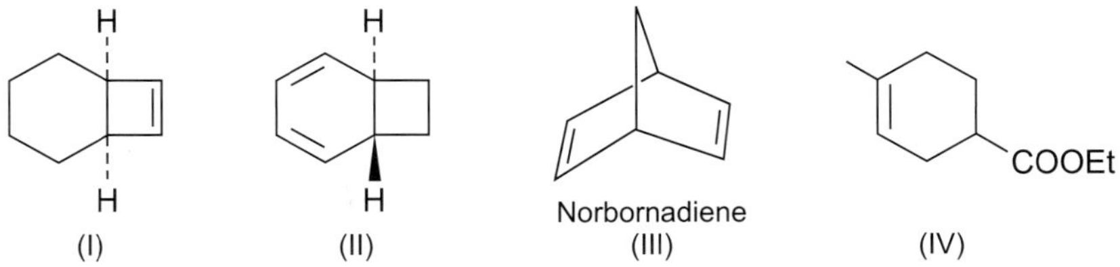

17.3. Furan and maleimide undergo a Diels-Alder reactions at 25°C to give *endo* adduct as the major product. When the reaction is carried out at 90°C, however, the major product is the *exo* isomer. The *endo* adduct isomerizes to the *exo* adduct when it is heated to 90°C. Propose an explanation.

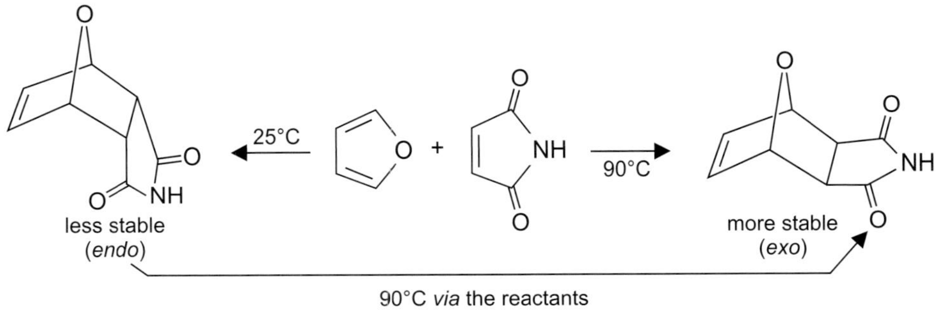

17.4. (*Z*)-1,3-pentadiene reacts with maleic anhydride at 100°C to give the adduct in 4% yield, while (*E*)-isomer gives the adduct in quantitative yield at 0°C. Explain.

17.5. What are the preferences for cycloaddition reactions ?

17.6. Give a classification of pericyclic reactions.

17.7. The transition state of the Diels-Alder pericyclic reaction is aromatic and compares with Cope rearrangement. Explain.

17.8. Predict the structure of photochemical electrocyclic cyclization product of (2*E*, 4*Z*)-hexadiene and compare the results with the thermal cyclization of the same compound.

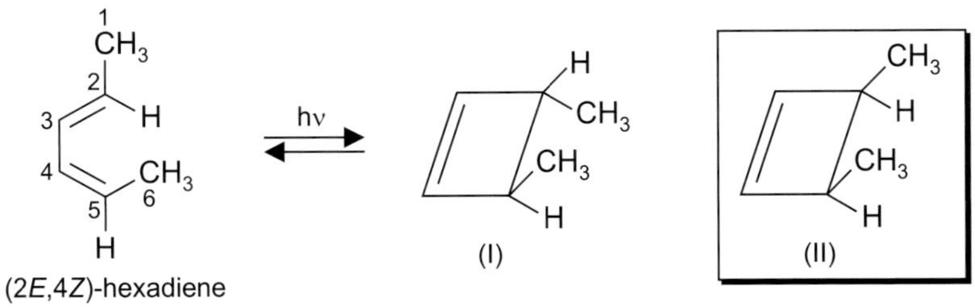

17.9. Give the stereostructure of the products from the following electrocyclic reactions of (I and II) carried out under photochemical reactions. Discuss if each reaction takes place in a conrotatory or disrotatory fashion.

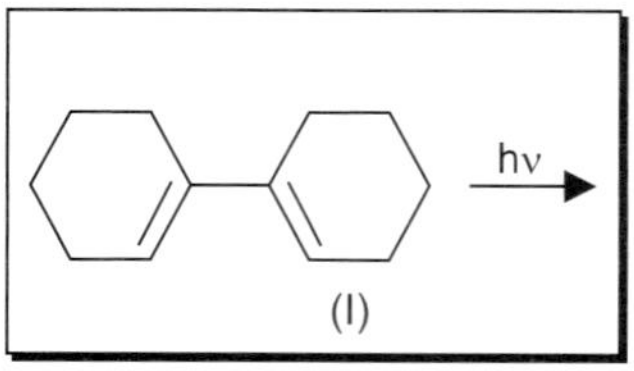

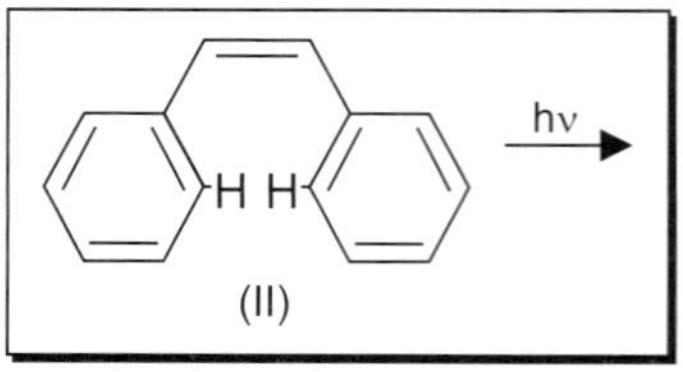

17.10. On thermal ring opening *cis* 3,4-dimethylcyclobutene gives two dienes (I and II). One of these is formed almost exclusively which is this diene and how it is formed?

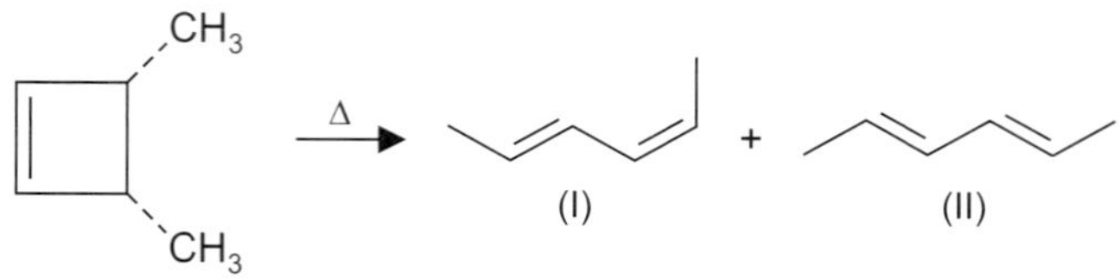

cis-3,4-dimethyl-cyclobutene

17.11. Predict whether conrotatory or disrotatory motion will take place under the conditions mentioned against each compound. Write the structure of the product with stereochemistry in each case.

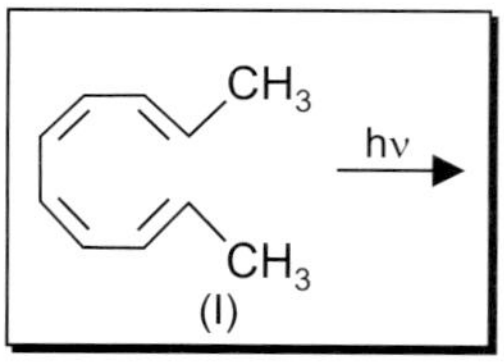

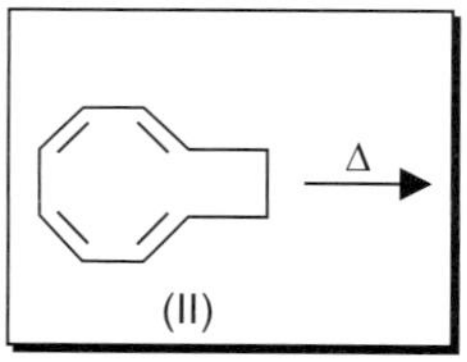

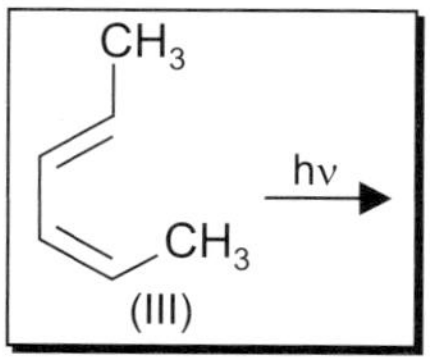

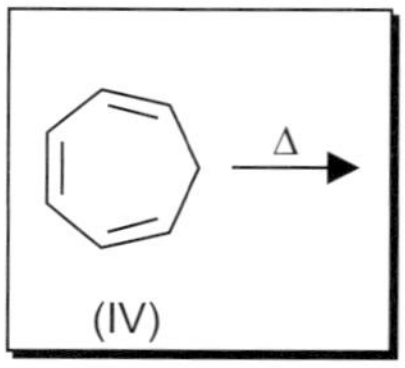

17.12. Explain briefly, taking one common example as to how FMO (frontier molecular orbital), method, PMO method and correlation diagram can be used for analyzing a pericyclic reaction.

17.13. Draw a correlation diagram for disrotatory conversion of butadiene to cyclobutene. Is the process allowed or forbidden? Explain.

17.14. A [3,3] sigmatropic rearrangement is thermally allowed via hypothetically formed allyl radicals. Explain by drawing appropriate bonding interactions.

17.15 (*a*) Explain briefly as to how a conjugated diene under photochemical conditions undergoes cyclization via a disrotatory path?

(*b*) Under which conditions, thermal or photochemical, the following ring closure will take place? Explain the stereochemistry at the ring fusion.

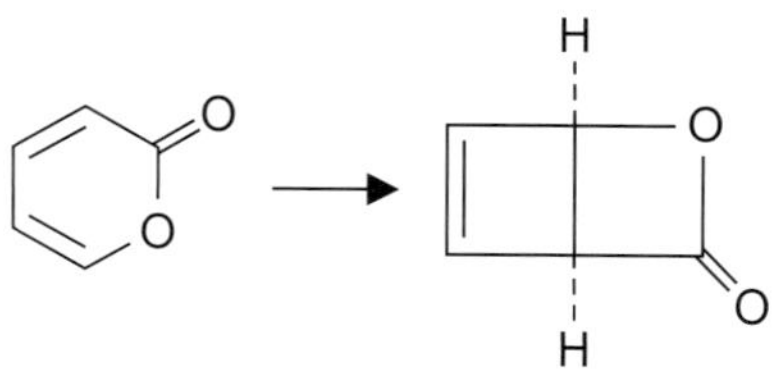

17.16. Which of the following statements are true or false.

(*i*) A conjugated diene with an even number of double bonds undergoes conrotatory ring closure under thermal conditions.

(*ii*) A conjugated diene with asymmetric HOMO undergoes conrotatory ring closure under thermal conditions.

(*iii*) A concerted antarafacial [1,3]-sigmatropic shift of hydrogen is thermally allowed.

(*iv*) The HOMO of a conjugated diene with an odd number of double bonds is symmetric.

(*v*) A [1,3] sigmatropic shift of carbon can occur under thermal conditions.

17.17. Fill in the blanks:

(*i*) A 1, 3-migration of carbon can take place thermally with ...... of configuration.

(*ii*) Pericyclic reactions are concerted, unaffected by catalysts or solvents and have ...... transition states.

(*iii*) [1,5] Sigmatropic shift of hydrogen involves three pairs of electrons and occurs by ...... pathway thermally.

(*iv*) A [1,7] sigmatropic shift of hydrogen occurs thermally by an ...... pathway.

(*v*) Frontier orbital analysis of a [4 + 2] cycloaddition shows that overlap of in phase orbitals to form new sigma bonds requires a ...... orbital overlap.

17.18. (*a*) Draw the transition states for suprafacial and antarafacial 1, 3 hydrogen shift by drawing the phase interactions in the basis sets. Show which process is thermally forbidden and which thermally allowed?

(*b*) Explain the results of the following photochemical reaction.

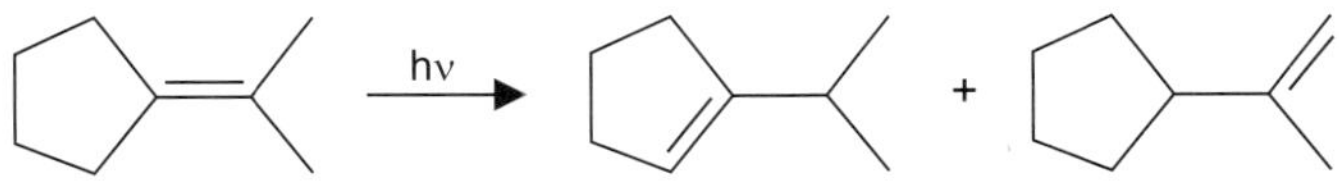

17.19. (*a*) Benzocyclobutene on heating with dimethyl *trans*-2-butene dioate (I) gives a bicyclic product of shown stereochemistry. Explain the reaction.

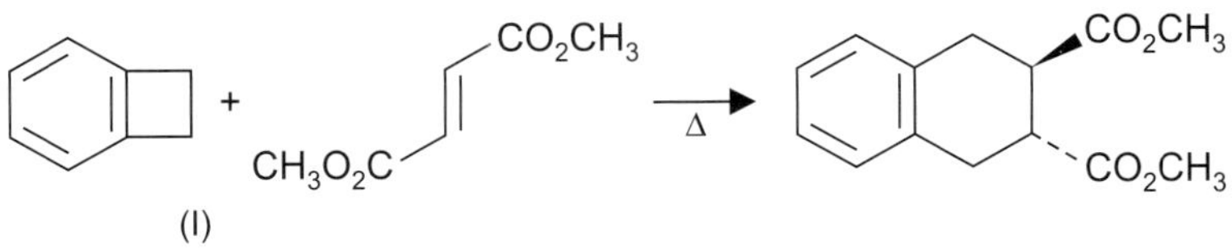

(*b*) Write the product with stereochemistry of the Diels-Alder reactions (II and III).

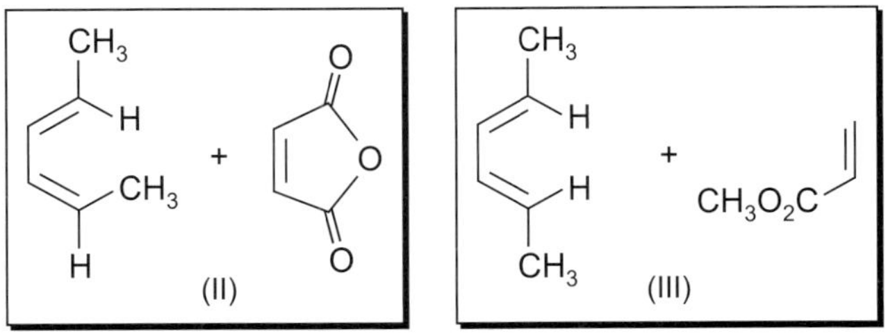

ANSWERS TO THE PROBLEMS

17.1. The reaction follows a stereospecifically *syn* pathway. The product is *cis*-dimethyl cyclohexene-4, 5-dicarboxylate. Therefore, the groups which are *cis* in the olefin also occupy *cis*-positions in the cyclohexene ring.

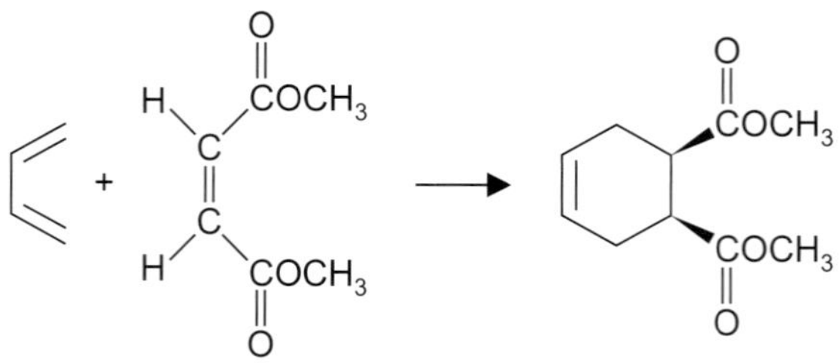

17.2. Cyclopentadiene and acetylene.

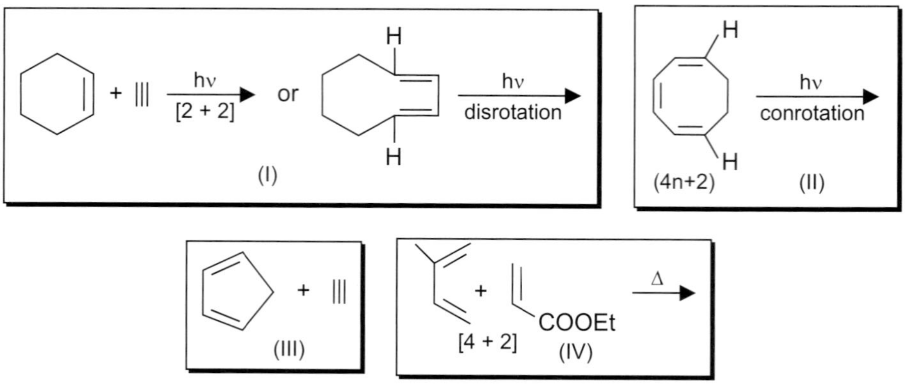

17.3. The *exo* product is thermodynamically more stable. The less stable *endo* isomer (kinetically favoured adduct) is formed faster and predominates at 25°C, the reaction is effectively irreversible. At 90°C this product is in rapid equilibrium with the reactants, consequently, the less rapidly formed but more stable *exo* isomer accumulates with time.

17.4. The bulky 1-*cis* (*Z*-) methyl substituent, sterically hinders formation of the *cisoid* conformation with a hydrogen at C-4. In the *E*-isomer the *cisoid* conformation is attained easily due to only tiny H, H interaction.

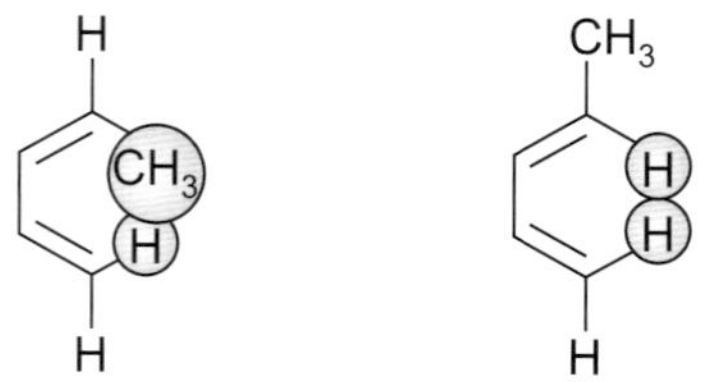

(*Z*)-1,3-pentadiene (*E*)-1,3-pentadiene

17.5. The rule of thumb is that when the reactants involve odd number of electron pairs, the cycloaddition is allowed thermally, while with even number of electron pairs, the cycloaddition is allowed photochemically.

17.6. Electrocyclic reactions are stereochemically classified as conrotatory and disrotatory, cycloadditions and sigmatropic rearrangements are classified as suprafacial or antarafacial.

17.7. In both the cases, the transition states involve six orbitals and six electrons.

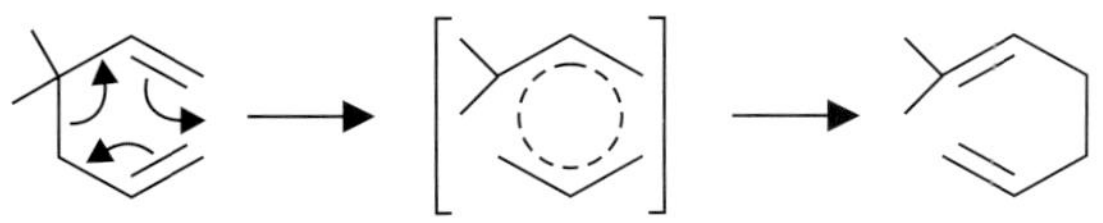

Aromatic transition state

the **Cope rearrangement**

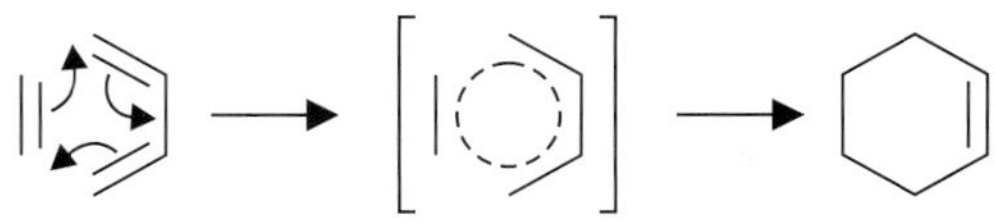

Aromatic transition state

the Diels-Alder reaction

17.8. In the photochemical cyclization disrotatory motion is required for bond formation, one methyl rotates up while the other down to give trans-3, 4-dimethylecyclohexene (I). The reverse would occur in thermal reaction (see Scheme 17.6) to give (II).

17.9. (I)

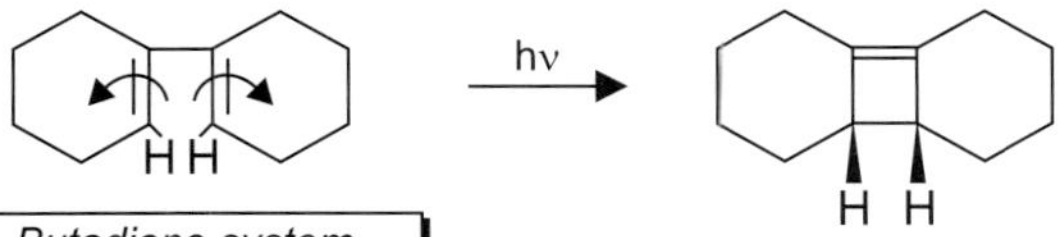

Butadiene system, disrotatory ring closure.

Hydrogen atoms on the same side of the ring.

(II) It is a $4n + 2$ electrocyclic reaction; $n = 1$, therefore, conrotatory in the excited state, the hydrogen atoms at the ring junctions in dihydrophenanthrene will be *trans* to one another.

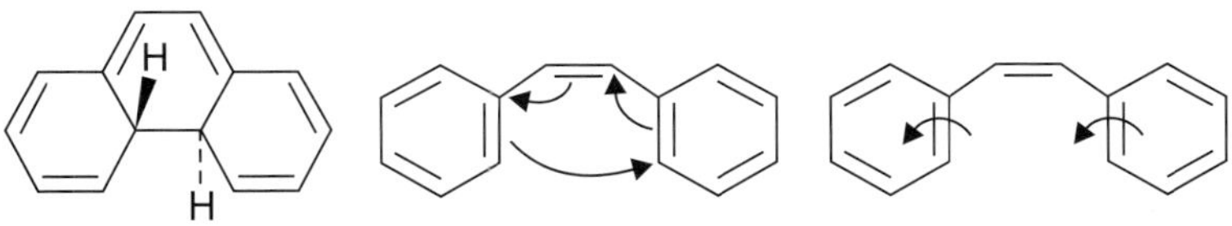

17.10. It is diene (I). Thermally a compound with two π bonds undergoes conrotatory ring closure. Since in the product the two methyls are *cis* placed, these point in the same direction in the reactant. In diene (I) the methyls point in the same direction.

17.11. The stereostructures of product in each case is presented (refer to Table 17.1).

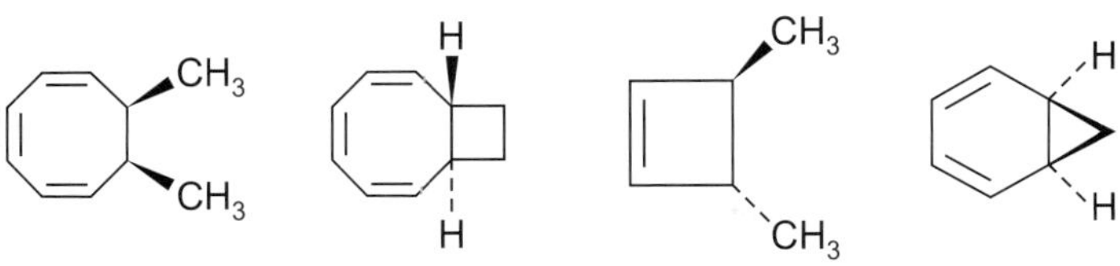

(I) (II) (III) (IV)

(I) The reactant being a ($4n$) π electron system will undergo disrotatory ring closure under photochemical conditions. Since the two methyl groups point in opposite directions these will be *cis* in the product.

(III) The reactant is ($2E$, $4Z$) hexadiene ($4n$) π electron system. Photochemically it will undergo ring closure by disrotation. Since the methyl substituents point in the same direction, the product will have these in *trans* relationship.

(IV) The reactant 1, 3, 5-cycloheptatriene is a ($4n + 2$) π electron system. It will undergo disrotatory ring closure under thermal conditions, since in the triene, the hydrogen substituents (not shown) point in opposite directions these will be *cis* in the product norcaradiene. However, due to the strain, norcaradiene cannot be isolated and it reverts back to the starting material. In case the ring size of the starting triene is large as in 1, 3, 5-cyclononatriene (Scheme 17.30) the product is stable and is isolated as the exclusive product.

17.15.(*b*) The electrocyclization will be under photochemical conditions (see table 17.1). Since in the reactant α-pyrone, the hydrogens on the diene system are pointing in opposite directions these will be *cis* in the product on disrotatory motion.

17.16.(*i*) True; (*ii*) true; (*iii*) true; (*iv*) true; (*v*) true.

17.17.(*i*) Inversion; (*ii*) cyclic; (*iii*) suprafacial; (*iv*) antarafacial; (*v*) suprafacial.

17.18. (*b*) This is a photochemical 1, 3-hydrogen shift (see, Scheme 17.34). Two products are expected since two different allylic hydrogens can undergo this shift.

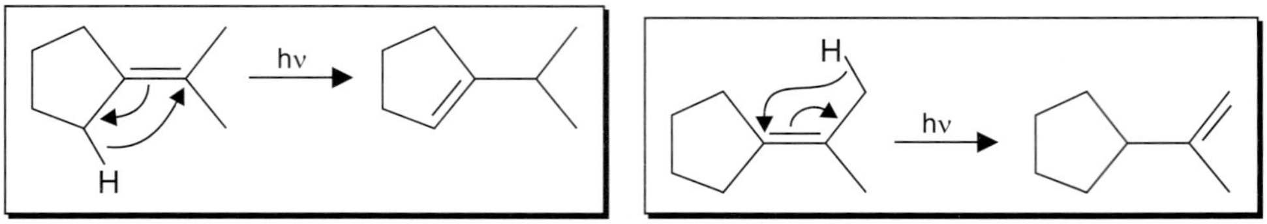

17.19 (*a*) It is a combination of electrocyclic reaction (to give IV) followed by Diels-Alder reaction.

(*b*) The reaction (II) gives the product (V) with the stereochemistry at centres other than ring junction as shown. This is deriveable from the discussion presented (Scheme 17.49*a*). Maleic anhydride is a *cis*-alkene, since the Diels-Alder reaction is a *syn* addition, the stereochemistry at the ring junction must be *cis*. The product from reactin (III) is (VI, campare with Scheme 17.49).

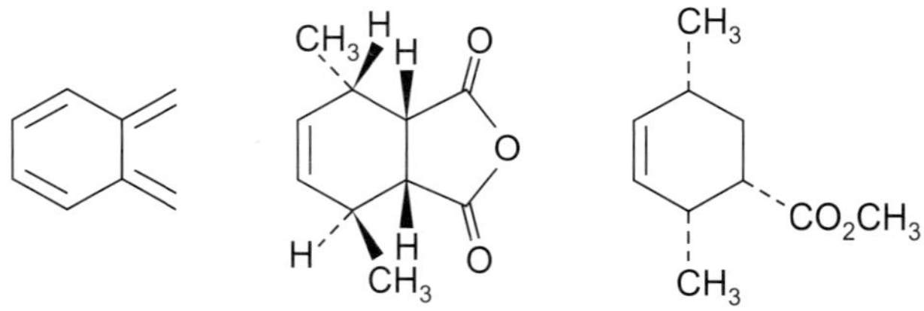

(IV) (V) (VI)

References and Further Reading

1. T.A. Albright, J.K. Burdett, and M.H. Whangbo, *Orbital Interactions in Chemistry*, John Wiley and Sons, New York, 1985.
2. P.J. Garratt, *Aromaticity,* Wiley, New York, 1986.
3. M. Sainsbury, *Aromatic Chemistry*, Oxford University Press, Oxford, 1992.
4. B.K. Carpenter, *Determination of Organic Reaction Mechanisms*, Wiley-Interscience, New York, 1984.
5. K.A. Connors, *Chemical Kinetics,* VCH Publishers, New York, 1990.
6. C. Reichardt, *Solvents and Solvent Effects in Organic Chemistry*, Wiley-VCH, Weinhem, 2003.
7. E. Erdik, *Organozinc Reagents in Organic Synthesis,* CRC Press, Boca Raton, FL, 1996.
8. P. Knochel and P. Jones, Editors, *Organozinc Reagents,* Oxford University Press, Oxford, 1999.
9. H.G. Richey, Jr., ed., *Grignard Reagents; New Developments,* Wiley, New York, 2000.
10. M. Schlosser, ed., *Organometallic in Synthesis; A Manual,* Wiley, New York, 1994.
11. B.J. Wakefield, *Organomagnesium Methods in Organic Synthesis,* Academic Press, London, 1995.
12. R.M. Crabtree, *The Organometallic Chemistry of the Transition Metals,* Wiley-Interscience, New York, 2005.
13. J.K. Kochi, *Organometallic Mechanisms and Catalysis,* Academic Press, New York, 1979.
14. E. Negishi and A. de Mejeire, eds., *Handbook of Organopalladium Chemistry for Organic Synthesis*, Vol. 1 and 2, Wiley-Interscience, New York, 2002.
15. H.C. Brown, *Organic Synthesis via Boranes,* Wiley, New York, 1975.
16. A. Pelter, K. Smith, and H.C. Brown, *Borane Reagents*, Academic Press, New York, 1988.
17. D.R. Arnold, N.C. Baird, J.R. Bolton, J.C.D. Brand, P.W.M. Jacobs, P. de Mayo, and W.R. Ware, *Photochemistry; An Introduction,* Academic Press, New York, 1974.
18. A Gilbert and J. Baggott, *Essentials of Molecular Photochemistry*, CRC Press, Boca Raton, FL, 1991.
19. W. Horspool and D. Armester, *Organic Photochemistry:* A *Comprehensive Treatment,* Ellis Horwood/Prentice-Hall, Chichester, 1992.
20. W. H. Horspool and P.-S. Song, eds., *Organic Photochemistry and Photobiology,* CRC Press, Boca Raton, FL, 1995.
21. W. Horspool and F. Lenci, eds., *CRC Handbook of Organic Photochemistry and Photobiology*, 2nd Edition, CRC Press, Boca Raton, FL, 2004.
22. P.B. de la Mare and R. Bolton, *Electrophilic Additions to Unsaturated Systems,* 2nd ed., Elsevier, New York, 1982.
23. P.V. Ramachandran and H.C. Brown, *Organoboranes for Synthesis,* American Chemical Society, Washington, 2001.
24. A.L.J. Beckwith and K.U. Ingold, in *Rearrangements in Ground and Excited States,* P. de Mayo, ed., Academic Press, New York, 1980, Chap. 4.

25. M. Birkhofer, H.-D. Beckhaus, and C. Rüchardt, *Substituent Effects in Radical Chemistry,* Reidel, Boston, 1986.
26. J. Fossey, D. Lefort, and J. Sorba, *Free Radicals in Organic Chemistry,* Wiley, Chichester, 1995.
27. B. Giese, *Radicals in Organic Synthesis: Formation of Carbon-Carbon Bonds,* Pergamon Press, Oxford, 1986.
28. W.B. Motherwell and D. Crich, *Free Radical Chain Reactions in Organic Synthesis,* Academic Press, London, 1992.
29. P. Renaud and M.P. Sibi, *Radicals in Organic Synthesis, Vol. 1, Basic Principles*, P. Renaud and M.P. Sibi, editors, Wiley-VCH, Weinheim, 2001.
30. M.B. Smith and J. March, *March's Advanced Organic Chemistry,* Wiley, New York, 2001.
31. Fleming, *Frontier Orbitals and Organic Chemical Reactions*, Wiley, London, 1976.
32. R.B. Woodward and R. Hoffmann, *The Conservation of Orbital Symmetry,* Verlag Chemie, Weinheim, 1971.
33. F.A. Carey and R.M. Sundberg, *Advanced Organic Chemistry*; 5th ed; Springer, 2007.

Index

A

B

E

F

G

H

I

J

K

L

M

N

O

P

Q

R

S

T

U

V

W

X

Y

Z